Basic Analysis V

Basic Analysis V: Functional Analysis and Topology

The cephalopods are now fully trained and are now valued colleagues in Jim's work.

James K. Peterson
Department of Mathematical Sciences
Clemson University

CRC Press
Taylor & Francis Group
Boca Raton London New York

CRC Press is an imprint of the
Taylor & Francis Group, an **informa** business
A CHAPMAN & HALL BOOK

First edition published 2022
by CRC Press
6000 Broken Sound Parkway NW, Suite 300, Boca Raton, FL 33487-2742

and by CRC Press
2 Park Square, Milton Park, Abingdon, Oxon, OX14 4RN

CRC Press is an imprint of Taylor & Francis Group, LLC

Library of Congress Cataloging-in-Publication Data

Names: Peterson, James K. (James Kent), author.
Title: Basic analysis I : functions of a real variable / James K. Peterson.

Other titles: Functions of a real variable
Description: First edition. | Boca Raton : CRC Press, 2020. | Includes bibliographical references and index. | Contents: Proving propositions -- Sequences of real numbers -- Bolzanoweierstrass results -- Topological compactness -- Function limits -- Continuity -- Consequences of continuity of intervals -- Lower semicontinuous and convex functions -- Basic differentiability -- The properties of derivatives -- Consequences of derivatives -- Exponential and logarithm functions -- Extremal theory for one variable -- Differentiation in R^2 and R^3 -- Multivariable extremal theory -- Uniform continuity -- Cauchy sequences of real numbers -- Series of real numbers -- Series in general -- Integration theory -- Existence of Reimann integral theories -- The fundamental theorem of calculus (FTOC) -- Convergence of sequences of functions -- Series of functions and power series -- Riemann integration : discontinuities and compositions -- Fourier series -- Application -- Summary.
Identifiers: LCCN 2019059882 | ISBN 9781138055025 (hardback) | ISBN 9781315166254 (ebook)
Subjects: LCSH: Analytic functions--Textbooks. | Functions of real variables--Textbooks. | Mathematical analysis--Textbooks.
Classification: LCC QA331 .P42 2020 | DDC 515/.823--dc23
LC record available at https://lccn.loc.gov/2019059882

ISBN: 9781138055131 (hbk)
ISBN: 9780367768539 (pbk)
ISBN: 9781315166155 (ebk)
DOI: 10.1201/9781315166155

Publisher's note: This book has been prepared from camera-ready copy provided by the author.

Dedication We dedicate this work to all of our students who have been learning these ideas of analysis through our courses. We have learned as much from them as we hope they have from us. We are firm believers that all our students are capable of excellence and that the only path to excellence is through discipline and study. We have always been proud of our students for doing so well on this journey. We hope these notes in turn make you proud of our efforts.

Abstract This book introduces graduate students in mathematics to concepts from topology and functional analysis, both linear and nonlinear. We illustrate these ideas with a variety of real world applications, which can profitably use these concepts such as differential geometry and degree theory. We also show you how to try to find a proper abstract framework for interesting and difficult problems in the study of the immune system and models of cognition. We feel it is very important to realize that the hardest part of applying mathematical reasoning to a new problem domain is how you choose the underlying mathematical framework to use on the problem. Once that choice is made, we have many tools we can bring to bear which may be helpful. However, a different choice would let us do the analysis from a different perspective. We will discuss in detail these sorts of choices in our chosen applications. We feel this is a skill to have when your life's work will involve quantitative modeling to gain insight into the real world. As usual, this book is designed for self-study.

Acknowledgments I want to acknowledge the great debt I have to my wife, Pauli, for her patience in dealing with those vacant stares and my long hours spent in typing and thinking. You are the love of my life.

The cover for this book is an original painting by me which was done the summer of 2017 to serve as the cover for this book. It shows that the cephalopods have become fully trained in mathematics and have joined me as colleagues, as you can see from the problem we are working out on the blackboard in my office. I am fascinated by cephalopods and I look forward to having them as colleagues in the future!

Table of Contents

Part I

Introduction

Chapter 1

Introduction

This text is based on quite a few years of teaching graduate courses in analysis and also trying to model complicated nonlinear interactions in economics, biology and physics. We have chosen to focus on biology modeling in the last part of the text because it is our current focus, but, in general, the last part of the text is designed to give some guidelines on how to come up with abstractions of reality that turn into computable models. The goal is always to obtain insight into the nonlinear interaction of interest. Many students and colleagues have been part of this journey and we appreciate all their advice and support. Most of what we do begins with handwritten notes and in this particular book, some of the handwritten notes stretch all the way back to notes we made for ourselves in 1977. They were color coded, of course, because that is just the way we roll.

This text is not focused on computational modeling, but occasionally we mention code fragments in our discussions. Here we use Octave (Eaton et al. (27) 2020), which is an open source GPL licensed (Free Software Foundation (33) 2020) clone of MATLAB® (MATLAB (85) 2018-2020), as a computational engine and we are not afraid to use it as an adjunct to all the theory we go over. Of course, theory, computation and science go hand in hand!

1.1 Table of Contents

In this text, we go over the following blocks of material.

Part One: These are our beginning remarks you are reading now, which are in Chapter 1.

Part Two: Some Algebraic Topology Here we want to provide a self-contained discussion of the Jordan Curve Theorem: the one that says a nice bounded closed curve in the plane divides the plane into three parts; the inside and the outside of the curve and the curve itself. This is surprisingly difficult to prove, so it is a good way to make you rethink how to use calculus in the plane and differentiable mappings to prove topological ideas. It is still set in $\Re^2$ though, so it is still not a push into ideas about $\Re^n$ topology. However, to set the stage for more complicated ideas in topology, we begin this part with a long treatment of metric spaces in general. This treatment is not tied to $\Re^2$ at all. Then, we move into the ideas about mappings in the plane which we need to prove the Jordan Curve Theorem. This part has these chapters:

- In Chapter 2, we provide a thorough discussion of many ideas, both basic and advanced, which are metric space based.
- In Chapter 3, we study differential forms, winding numbers and some homotopy ideas. This is primarily set in $\Re^n$. Winding numbers are a way to assign integer values to certain integrals around the boundary of nice bounded sets. It is quite similar to the

idea of Brouwer degree theory which we explore in Chapter 10, but Brouwer degree theory is much more general as it is set in $\Re^n$ and we can use it to prove the existence of solutions to $F(\boldsymbol{x}) = \boldsymbol{0}$ problems. Brouwer degree assigns an integer value to a function defined on a bounded set in $\Re^n$ whereas our winding number assigns an integer value for a construction set in $\Re^2$. We can extend the ideas about winding numbers to $\Re^n$ but we chose not to do that as we wanted to introduce degree theory instead. Further, we also note, we can use Brouwer degree to assign an integer value to a compact perturbation of the identity set in normed linear spaces acting on a bounded set. We do that in Chapter 11 and that degree is called Leray - Schauder degree. We can then extend Leray - Schauder degree to mappings which have the form $F = L + G$ where L is linear Fredholm of index zero and G has some compactness properties. This gives Coincidence degree and that is discussed in Chapter 12.

- In Chapter 4, we carefully prove the Jordan Curve Theorem which is a result set in $\Re^2$. We get a chance to introduce nice connections between analysis and topology.

Part Three: Deeper Topological Ideas In this part, we study topological spaces and topological vector spaces and learn how to construct topologies in various ways. The topologies we use in metric spaces and normed linear spaces are so similar to the ones we routinely use in $\Re^n$ that it is easy to lose sight of the fact that there are other ways to define open sets besides using a metric. Also, the space of objects we work in does not need to have linearity or algebraic properties, so it is useful to see what kind of properties a topology can have in general.

- Chapter 5 discusses topological spaces and topological vector spaces in reasonable detail. There is much to learn here as we are approaching the idea of open sets in much more generality than before. We learn a variety of tools that allows us to construct what are called topologies for a given set of objects. We prove a standard result which allows the construction of a metric on most topological spaces.
- In Chapter 6, we learn how to construct topologies using families of specialized functions on the set of objects. We also revisit the problem of building a metric now that we have a new way of metric construction. We also explore the idea of test functions which are C^∞ functions with compact support and show how to find an appropriate topology for this set of functions. We first construct a topology which is not complete. We then add some additional sets to get convexity which generates another topology which is not based on a metric. However, it is complete. This is a very interesting example and a good one to have in your toolkit.
- In Chapter 7, we return to the idea of linear functionals on a space of objects. Whether or not these linear functionals are continuous depends on the topology used on the space. We discuss locally convex topologies and weak convergence.
- In Chapter 8, we add a discussion of weak* convergence and prove the important Banach - Alaoglu result that polars of neighborhoods of zero are weak* compact. We finish a discussion of extreme points in a vector space and eventually prove the Krein - Milman Theorem. This is a very abstract version of the idea in linear programming that if we want to minimize a function over a convex bounded set in $\Re^n$, it is enough to look at the extreme points of the function. If the set is the intersection of planes, then the extreme points are the vertices of the set and the closure of the convex hull of the extreme points is another way to describe the convex bound set.
- Chapter 9 discusses Stone - Weierstrass results about the denseness of sets of functions in the set of all continuous functions. This uses more algebraic ideas.

Part Four: Topological Degree Theory We discuss Brouwer, Leray - Schauder and Coincidence Degree Theory and along the way look at a variety of interesting ideas from functional analysis.

- In Chapter 10, we develop Brouwer degree in a lot of detail. This is a way to associate an integer to the triple $(f, \boldsymbol{D}, \boldsymbol{0})$ where $\boldsymbol{D}$ is an open bounded subset of $\Re^n$ and f is continuous on $\overline{\boldsymbol{D}}$.
- In Chapter 11, we develop Leray - Schauder degree. This is a way to assign integers to triples $(I_T, \boldsymbol{D}, \boldsymbol{0})$ where $\boldsymbol{D}$ is an open bounded subset of a normed linear space and T is a compact perturbation of the identity on $\overline{\boldsymbol{D}}$. This degree is constructed using approximating sequences of Brouwer degrees for approximating maps.
- In Chapter 12, we develop Coincidence degree. This degree assigns an integer of mapping of the form $L + G$ on $\overline{\boldsymbol{D}}$ where $\boldsymbol{D}$ is an open bounded subset of a normed linear space and G is a mapping that is L - compact and L is linear Fredholm of index 0. We briefly describe some applications of boundary value problems in ordinary differential equations.

Part Five: Manifolds In this part, we provide a reasonable introduction to the idea of manifolds which are essentially sets of objects that are locally $\Re^n$ in nature. There are a lot of details and we build heavily on previous ideas from topology and differentiability in $\Re^n$.

- In Chapter 13, we discuss manifolds which are special topological spaces where open sets are homeomorphic to open sets of some $\Re^m$. So in a sense, manifolds are locally Euclidean with compatibility conditions on these homeomorphisms that must be imposed when two open sets overlap.
- In Chapter 14, we discuss mappings between manifolds. We have a lot of structure here and we go over very carefully the tangent and cotangent spaces associated with a given point on the manifold. We can define mappings between tangent spaces that act a lot like the usual linear differential mappings we have in $\Re^n$.
- In Chapter 15, we talk about global structure on manifolds. We now have the tools to discuss the tangent bundle and tensor analysis on manifolds in general. We finish with the metric tensor.

Part Six: Emerging Topologies We finish with a long discussion of how to model immunology and consciousness. This topic was chosen because it is not clear what objects we should use for this study. There are many choices of mathematical structure we could impose in our modeling process, none of which are obvious. Hence, this is a fitting way to end our discussions by giving you a glimpse into how we approach research-level questions.

- Chapter 16 explains how we can develop a model of asynchronously interacting objects using a general graph formulation.
- In Chapter 17, we develop models of cytokine messenging and immune and cognition networks and show how we learned to use a blend of mathematics, science and computation in our work.
- Chapter 18 looks at a model of consciousness.

Part Seven: Summing It All Up

In Chapter 19, we talk about the things you have learned here and where you should go next to continue learning more about this blend of ideas.

There is much more we could have done, but these topics are a nice introduction into the further use of abstraction in your way of thinking about models and should prepare you well for the next step, which is self-study and independent research.

1.2 Acknowledgments

The work in Part Two comes from our attempts to learn more about algebraic topology. We taught a graduate course in this and we used our handwritten notes for this part. Part Three is based on courses we taught in functional analysis and our handwritten notes and the discussion on test functions is partially based on notes from Steve Crawford's master's work as well our own teaching of this material. Part Four comes from handwritten notes we made for ourselves when we were first learning how to use degree theory in our research. Part Five is partly based on notes of my student Jay Wilkins who used them as part of his master's work with us, as well as our reading in differential geometry texts and our own lecture notes. Part Six is from our current research supported by grant W911NF-17-1-0455 from the Army Research Office in the area of complex models. Finally, Part Seven is a summary discussion of our journey into modeling, as we thought it might be both interesting and useful to some of you.

On a personal note, and moving away from the royal we viewpoint, I have just retired and I am looking forward to new projects. The one in mind at the moment is a book on abstract models of immunology. I'll see how that goes.

Jim Peterson
School of Mathematical and Statistical Sciences
Clemson University

Part II

Some Algebraic Topology

Chapter 2

Basic Metric Space Topology

Let's start by reviewing ideas from metric spaces. First, let's look at a nice result about open sets in $\Re$.

2.1 Open Sets of Real Numbers

Let's start by looking at open sets of the real line and prove a fundamental characterization. This will give you some practice at these kinds of arguments. Of course, this argument is set in $\Re$ so it does not translate to all metric spaces as written.

Theorem 2.1.1 Characterizing Open Sets in $\Re$

If G is a nonempty and open subset of $\Re$, it is a countable union of disjoint open intervals.

Proof 2.1.1
Let $p \in G$. Let

$$S_p = \{y|(p,y)|(p,y) \subset G\}, \quad T_p = \{x|(x,p) \subset G\}$$

where $x < p < y$. Since x is an interior point, we know S_p and T_p are not empty. Let $b_p = \sup S_p$ and $a_p = \inf T_p$. Note b_p could be $+\infty$ and a_p could be $-\infty$. Let $u \in (a_p, b_p)$. By the definition of the supremum, there is a y^* in S_p so that $u < y^* \leq b_p \leq \infty$ with $(p, y^*) \subset G$. Also, by definition of the infimum, there is an x^* in T_p so that $-\infty \leq a_p \leq x^* < u$ with $(x^*, p) \subset G$. Thus, we can see $u \in (x^*, y^*) \subset G$. This shows $u \in G$. Since the choice of u in (a_p, b_p) was arbitrary, we see $(a_p, b_p) \subset G$.

Now, if $a_p > -\infty$, we see a_p can't be in U as if it was it would be an interior point. But in that case, it couldn't be the infimum. Similarly, if $b_p < \infty$, b_p cannot be in U as it would no longer be the supremum if that was the case.

Also, if a_p or b_p were not finite, they would not be in $\Re$. So these two cases imply $(a_p, b_p) \subset G$ with a_p and b_p not in G. Now let $F = \{(a_p, b_p)|p \in G\}$. Note

$$U = \cup_{p \in G} \{(a_b, b_p)\} = \cup_F \{(a_b, b_p)\}$$

Now if $(a_p, b_p) \cap (a_q, b_q) \neq \emptyset$, there is a point v in both. Assume $p < q$. Then we may also assume we have the chain of inequalities $a_p < a_q < v < b_q < a_p$. You should draw a picture! This means $(p, v) \cup [p, w) \subset G$ for all w past q in (a_q, b_q). But this says $(p, w) \subset G$ for all such w. The definition *of b_p then tells us $w \leq b_p$ always. But then*

$$b_p \geq \sup w, (q, w) \subset G = b_q$$

We conclude $b_q \leq b_p$. But we already knew $b_p \leq b_q$ by assumption. Hence $b_q = b_p$.

A similar argument holds for the other case. We have for any z with $(z, p) \subset G$ that $(z, q) \subset G$ also. Hence, by the definition of the infimum, $a_q \leq z$. This implies

$$a_q \leq \inf z, (z, p) \subset G = a_p.$$

We conclude $a_q \leq a_p$. But we already knew $a_p \leq a_q$ by assumption. Hence $a_q = a_p$.

Thus, if we throw away copies, the disjoint intervals (a_p, b_p) form a disjoint collection of open intervals whose union is G. Now each of these disjoint intervals corresponds to some $p \in G$ and each contains a rational number r_p. Since two different intervals here are disjoint, the rationals r_p and r_q corresponding to two disjoint intervals must be different. We can enumerate these rational numbers as a sequence (r_i) for either a finite or a countably infinite index set Λ. For each index i, let a_i and b_i be the endpoints of the disjoint interval containing r_i. Then

$$G = \cup_{p \in G}(a_p, b_p) = \cup_{p \in G,\, \textit{disjoint}}\, (a_p, b_p) = \cup_{i \in \Lambda}(a_i, b_i)$$

■

We are interested in open sets in both $\Re^n$ and more general metric spaces. Let's review the basic ideas.

2.2 Metric Space Theory

There are many places you can go to learn more about metric spaces, of course, We talk about them extensively from the point of completion and convergence and not so much about their topological properties in (Peterson (100) 2020) but much more can be done. See, for example, (M. Gemignani (77) 1972), (N. Haaser and J. Sullivan (89) 1991) and (E. Kreyszig (26) 1989). You have already seen the definition of a metric space but we will repeat it so it is easy to refer to it in this volume.

Definition 2.2.1 Metric Spaces: Defining the Metric d on a set X

*A **metric space** (X, d) is any set X combined with any function $d : X \times X \rightarrow \Re$ such that for all choices of $x, y, z \in X$, the following conditions hold:*

M1: $d(x, y) \geq 0$

M2: $d(x, y) = 0$ *if and only if* $x = y$.

M3: $d(x, y) = d(y, x)$

M4: $d(x, y) \leq d(x, z) + d(z, y)$

The function d is simply referred to as a **metric.**

So, technically, a metric space is an ordered pair (X, d), but often we simply refer to a metric space X when its associated metric d is clear from the context. The classic example of a metric space is the set of real numbers $\Re$ operated on by the metric $d(x, y) = |x - y|$. This metric is nothing more than the distance between two points on the line. Indeed, Definition 2.2.1 is just a generalization of the usual concept of length or distance.

Homework

Exercise 2.2.1 *Let X be the set of vectors on the surface of the sphere $x^2 + y^2 + z^2 = 1$ in $\Re^3$. Let d be the usual shortest arc length distance between two points p and q on the surface. You can draw this easily in a picture. Is (X, d) a metric space?*

Exercise 2.2.2 *Let X be the set of functions which are continuous on $\Re$. Let $d(f, g) = \int_{\Re} e^{\alpha} t\phi(|f(t) - g(t)|)dt$ where $\phi(t) = \frac{t}{1+t}$ for all $t \geq 0$. Is (X, d) a metric space?*

Exercise 2.2.3 *Recall what an interior point of a set of real numbers means. We can define an interior point of a metric space in a similar way. We know all norms are equivalent on a finite dimensional normed linear space and we know all norms induce a metric. Let $X = \Re^2$ and consider the metric spaces (X, d_p) where $d_p(x, y) = \|x = y\|_p$ for $1 \leq p \leq \infty$. What do the open balls of radius $r > 0$ look like in these metric spaces?*

2.2.1 Open and Closed Sets

Let's start by reviewing the concepts of open and closed sets generated by a metric. The following set is the analogue of a circle about a point on the real line.

Definition 2.2.2 Open Balls in a Metric Space

*A **ball** of radius $r > 0$ centered at a point $x \in X$ is the set $B(x; r) = \{y \in X : d(x, y) < r\}$.*

We can use the idea of balls in a metric space to define interior points.

Definition 2.2.3 Interior Points

*A point p in a subset $\mathcal{O}$ of a metric space X is an **interior point** if there is a positive r, so the set $B(p; r) = \{y | d(y, x) < r\}$ is contained in $\mathcal{O}$. The set $B(p; r)$ as usual is called the **ball** about p of radius r.*

If every point in a set is an interior point, we say the set is an **open** set in the metric space.

Definition 2.2.4 Open Subsets in a Metric Space

*A subset $\mathcal{O}$ of a metric space S is **open** if for every $y \in \mathcal{O}$ there is an $r > 0$ so that if $x \in S$ satisfies $d(x, y) < r$, then $x \in \mathcal{O}$.*

It is easy to see the empty set $\emptyset$ is an open set of any metric space X and that the entire space X is itself an open set. It is also easy to see the sets $B(x; r)$ are open sets for any x and positive r.

It is also easy to see that in a metric space, finite or countable unions of open sets are also open and finite intersections of open sets are also open.

Homework

Exercise 2.2.4 *Prove finite or countable unions of open sets in a metric space are also open.*

Exercise 2.2.5 *Prove finite intersections of open sets in a metric space are also open.*

There is also a fundamental relationship between the set of all open balls and the set of all open sets of a metric space.

Theorem 2.2.1 Characterization of Open Sets in a Metric Space

A subset G of a metric space X is open $\Longleftrightarrow$ G if and only if it is the union of open balls.

Proof 2.2.1

$\Longrightarrow$: *If G is open, then $G \cup_{x \in X} B(x; r_x)$ where, since each x is an interior point, there is a positive number r_x so that $B(x; r_x) \subset G$.*

$\Longleftarrow$: *Conversely, it is clear a union of open balls is an open set.* ■

We observe that this result gives us a means of *reducing* all the open sets of a metric space to the set of its open balls. Any open set whatsoever can be constructed by taking a large enough union of open balls. The union can be finite, countably infinite, or even uncountably infinite if necessary. It is in this sense that we say the set of open balls forms a **basis** for all the open sets in a given metric space. This concept will be formally defined when we study topological spaces in general.

Next, we can state many analysis concepts in terms of open or closed sets.

2.3 Analysis Concepts in Metric Spaces

What does it mean for a sequence to converge?

Definition 2.3.1 Convergence in a Metric Space

A sequence $\{x_n\}$ of elements in a metric space X **converges** *if there is an $L \in X$ such that for all $\epsilon > 0$, there exists an $N > 0$ so that $d(x_n, L) < \epsilon$ whenever $n \geq N$. The object denoted by L is called the* **limit** *of the sequence $\{x_n\}$.*

Next, to discuss completeness, we need to define a Cauchy Sequence.

Definition 2.3.2 Cauchy Sequence in a Metric Space

A sequence $\{x_n\}$ of elements in a metric space X is **Cauchy** *(or a* **Cauchy sequence***) if for all $\epsilon > 0$, there exists an $N > 0$ so that $d(x_m, x_n) < \epsilon$ whenever $m, n \geq N$.*

We can then define the familiar idea of a complete metric space. There is a lot of abstract nuance here though, so don't underestimate the complexity. Look at (Peterson (100) 2020) to see how to complete $(\mathbb{Q}, \|dot\|)$ to form $\Re \equiv \widetilde{\mathbb{Q}}$ using the metric $\widetilde{\| \cdot \|}$ and the process by which we complete the metric space (X, d) to $(\widetilde{X}, \widetilde{d})$.

Definition 2.3.3 Complete Metric Space

A metric space X is **complete** *if every Cauchy sequence converges in X.*

We often want to look at mapping between metric spaces so this is how we define that.

Definition 2.3.4 Continuous Mappings between Metric Spaces

Let (X_1, d_1) and (X_2, d_2) be metric spaces. A function $f : X_1 \to X_2$ is **continuous** *at $x \in X$ if for all $\epsilon > 0$, there is a $\delta > 0$ so that $d_2(f(x), f(y)) < \epsilon$ whenever $d_1(x, y) < \delta$. Note here δ depends on x and ϵ.*

The notion of uniform continuity is also important.

Definition 2.3.5 Uniformly Continuous Mappings between Metric Spaces

Let (X_1, d_1) and (X_2, d_2) be metric spaces. A function $f : X_1 \to X_2$ is **uniformly continuous** *if for all $\epsilon > 0$, there is a $\delta > 0$ so that $d_2(f(x), f(y)) < \epsilon$ whenever $d_1(x, y) < \delta$. Note here δ only depends on ϵ.*

The ideas of dense subsets and separability are very important. Recall the embedding $T : X = \widetilde{X}$ we used in the metric space completion construction to see an example that is nicely abstract. In general we define dense subsets like this:

Definition 2.3.6 Dense Subsets in a Metric Space

A subset G of a metric space X is **dense** *within $H \subseteq X$ if for each $h \in H$ and $\epsilon > 0$ there exists $g \in G$ so that $d(h, g) < \epsilon$. If G is dense in the whole metric space X, then we say that G is* **everywhere dense**.

Hence, if the dense subset is countable, this leads to the idea of a separable metric space.

Definition 2.3.7 Separable Metric Space

A metric space is **separable** *if it has a dense subset that is also countable.*

Homework

Exercise 2.3.1 *Recall the proof that the metric space $C([0, 1], \|\cdot\|_\infty)$ is complete.*

Exercise 2.3.2 *Recall the proof that the metric space $C([0, 1], \|\cdot\|_1)$ is not complete.*

Exercise 2.3.3 *Recall the proof that the metric space $C([0, 1], \|\cdot\|_2)$ is not complete.*

Exercise 2.3.4 *Recall how the metric space $C([0, 1], \|\cdot\|_1)$ is completed to $\mathbb{L}_1([0, 1])$.*

Exercise 2.3.5 *Recall how the metric space $C([0, 1], \|\cdot\|_2)$ is completed to $\mathbb{L}_2([0, 1])$.*

Exercise 2.3.6 *Review the arguments that show $\mathscr{L}_1$ on the measure space $(\Re, \mathscr{M}, \mu)$ where μ is Lebesgue measure is complete.*

Exercise 2.3.7 *Review the arguments that show $\mathscr{L}_2$ on the measure space $(\Re, \mathscr{M}, \mu)$ where μ is Lebesgue measure is complete.*

Next, we can state many analysis concepts in terms of open or closed sets.

A good way to look at convergence of sequences here is to look at the cardinality of the inverse images of open sets.

Theorem 2.3.1 Characterization of a Limit in a Metric Space

A member L of a metric space X is the limit of a convergent sequence $\{x_n\} \Longleftrightarrow$ every open set $\mathscr{O}$ containing L also contains all but a finite number of the elements x_n.

Proof 2.3.1
This one is for you. ■

This leads naturally to a characterization of continuity using the inverse images of open sets.

Theorem 2.3.2 Characterization of Continuous Mapping between Metric Spaces

Let (S_1, d_1) and (S_2, d_2) be metric spaces. A function $f : S_1 \to S_2$ is continuous $\Longleftrightarrow$ $f^{-1}(\mathcal{O})$ is an open set in S_1 whenever $\mathcal{O}$ is an open set in S_2.

Proof 2.3.2
This one is for you too. ■

Since these theorems state both necessary and sufficient conditions for convergence and continuity, we can therefore use them to **define** such notions in terms of the open sets of a metric space. We also need other ideas such as limit points, closures and so forth which you are probably familiar with from analysis on $\Re$ of course.

Definition 2.3.8 Limit Points of a Set in a Metric Space

An element l in a metric space S is a **limit point** *(also known as a* **cluster point** *or* **accumulation point***) of $G \subseteq S$ if for all $\epsilon > 0$, there exists a $g \in G$ such that $d(l, g) < \epsilon$.*

A limit point of a given set can still be in the set, but this is not always true. Some sets contain only a few of their limit points while others contain all their limit points. This brings us to the following definitions:

Definition 2.3.9 Closed Sets in a Metric Space

A subset of a metric space containing all its limit points is a **closed set.**

and the idea of the closure of a set.

Definition 2.3.10 Closure of a Set in a Metric Space

The union of a subset G with all its limit points is the **closure** *of G and is denoted by G^*.*

It is easily proven that the closure of any set is itself a closed set. It is also true a subset G is closed if and only if $G^* = G$. We now have the following characterization of dense sets in terms of closed sets:

Theorem 2.3.3 Characterization of Everywhere Dense Subsets in a Metric Space

A subset G of a metric space X is everywhere dense (i.e. dense in S) $\Longleftrightarrow G^* = X$.

Proof 2.3.3
Nice exercise! ■

We end this section by listing the basic theorems and definitions concerning the open and closed sets of a metric space. This prepares the way for the natural generalization of these concepts within the larger context of topological spaces. In what follows, we represent the complement of an arbitrary set A by A^C.

Theorem 2.3.4 Subsets in Metric Spaces Closed if and only if their Complement is Open

A subset G in a metric space is closed $\Longleftrightarrow G^C$ is an open set.

Proof 2.3.4
This one is for you. ■

Theorem 2.3.5 Subsets in Metric Spaces Open if and only if their Complement is Closed

A subset $\mathcal{O}$ in a metric space is open $\Longleftrightarrow \mathcal{O}^C$ is a closed set.

Proof 2.3.5
This one is for you. ■

Comment 2.3.1 *An interesting result of these theorems is that the empty set $\emptyset$ and the whole metric space S are both open and closed sets.*

Theorem 2.3.6 Closed Ball in a Metric Space is a Closed Set

The **closed ball** $B_c(x,r) = \{y \in S : d(x,y) \leq r\}$ *is a closed set. Indeed, $B(x,r)^* = B_c(x,r)$ for all $x \in S$ and $r > 0$.*

Proof 2.3.6
This one is for you. ■

If a function is defined on a dense subset of a metric space, sometimes we can extend it to the whole space and preserve smoothness properties such as continuity. Here is a first theorem of that type.

Theorem 2.3.7 Uniformly Continuous Extensions from a Dense Subset

Let (X,d_1) and (Y,d_2) be metric spaces and assume also that Y is complete. Let $f : A \subset X \rightarrow Y$ be uniformly continuous on A with A dense in X. Then, there is an extension $\hat{f} : X \rightarrow Y$ with $\hat{f}|_A = f$ and $\hat{f}$uniformly continuous on X.

Proof 2.3.7
If A is the same as X, the result is easy. If A is a proper subset of X, the A^C is not empty. Let $p \in A^C$. Since A is dense, there is a sequence $(p_n) \subset A$ with $p_n \rightarrow p$. Thus, (p_n) is a Cauchy sequence in X. Let $\epsilon > 0$ be given. Since f is uniformly continuous on A, there is a $\delta > 0$, so that $d_2(f(x), f(y)) < \epsilon$ if $d_1(x,y) < \delta$ for $x, y \in A$. Hence, there is an N_1 so that $d_1(p_n, p_m) < \delta$ if $n, m > N$, Combining, we have $d_1(p_n, p_m) < \delta$ implying $d_2(f(p_n), f(p_m)) < \epsilon$ when $n, n > N_1$. This shows $(f(p_n))$ is a Cauchy sequence in Y which is complete. Hence, there is $y_p \in Y$ so that $f(p_n) \rightarrow y_p$. Define the extension $\hat{f}$ by

$$\hat{f} = \begin{cases} f(p), & if p \in A \\ y_p, & if p \in A^C \end{cases}$$

Now if (q_n) was another sequence in A which converged to p, we know $d_1(p_n, q_n) \rightarrow 0$ as $n \rightarrow \infty$. Also, given $\epsilon > 0$, there is an N_2 so that $d_1(q_n, q_m) < \delta$ if $n, m > N_2$, where δ is the same one we had before. Combining, we have $d_1(q_n, q_m) < \delta$ implying $d_2(f(q_n), f(q_m)) < \epsilon$ when $n, n > N_2$. This shows $(f(q_n))$ is a Cauchy sequence in Y which is complete. Hence, there is $z_p \in Y$ so that $f(q_n) \rightarrow z_p$.

Note also, we can find N_3, N_4 and N_5 so that

$$\begin{aligned} n > N_3 &\implies d_1(p_n, q_n) < \delta \Longrightarrow d_2(f(p_n), f(q_n)) < \frac{\epsilon}{3} \\ n > N_4 &\implies d_2(f(p_n), y_p) < \frac{\epsilon}{3} \\ n > N_5 &\implies d_2(f(q_n), z_p) < \frac{\epsilon}{3} \end{aligned}$$

So, choosing any $n > \max\{N_3, N_4, N_5\}$, by the triangle inequality, $d_2(y_p, z_p) < \epsilon$. Since $\epsilon > 0$ is arbitrary, we see $y_p = z_p$. Hence, $\hat{f}$ is well-defined.

It remains to show $\hat{f}$ is uniformly continuous. Choose $\epsilon > 0$. Since f is uniformly continuous on A, there is a $\delta > 0$, so that

$$d_1(x, y) < \delta \implies d_2(f(x), f(y)) < \frac{\epsilon}{3}$$

Choose any two points $x, y \in X$ with $d_1(x, y) < \frac{\delta}{6}$. Since A is dense in X, there are sequences with specific properties:

$$\begin{aligned} (x_n) \subset A &\ni x_n \to x \Longrightarrow f(x_n) \to \hat{f}(x) \\ (y_n) \subset A &\ni y_n \to y \Longrightarrow f(y_n) \to \hat{f}(y) \end{aligned}$$

So there is N_1 and N_2 so that if $n > \max\{N_1, N_2\}$, we have

$$\begin{aligned} d_1(x_n, x) < \frac{\delta}{6} &\longrightarrow d_2(f(x_n), \hat{f}(x)) < \frac{\epsilon}{3} \\ d_1(y_n, y) < \frac{\delta}{6} &\longrightarrow d_2(f(y_n), \hat{f}(y)) < \frac{\epsilon}{3} \end{aligned}$$

Thus, for $n > \max\{N_1, N_2\}$, if $d_1(x, y) < \frac{\delta}{6}$,

$$d_1(x_n, y_n) \leq d_1(x_n, x) + d_1(x, y) + d_1(y_n, y) < \frac{\delta}{6} + \frac{\delta}{6} + \frac{\delta}{6} = \frac{\delta}{2} < \delta$$

and so $d_2(f(x_n), f(y_n)) < \frac{\epsilon}{3}$. This, in turn, says,

$$d_2(\hat{f}(x), \hat{f}(y)) \leq d_2(\hat{f}(x), f(x_n)) + d_2(f(x_n), f(y_n)) + d_2(\hat{f}(y), f(y_n)) < \frac{\epsilon}{3} + \frac{\epsilon}{3} + \frac{\epsilon}{3} < \epsilon$$

This shows $\hat{f}$ is uniformly continuous on X. ■

Let's look more carefully at the structure of an open set.

Definition 2.3.11 Interior of a Set in a Metric Space

Recall an **interior point** *of a subset G is any $x \in G$ such that there exists an $r > 0$ with $B(x, r) \subseteq G$. The* **interior** *of a subset G is then the set of all interior points of G and is denoted by $int(G)$.*

This allows us to describe sets that are not dense at all!

Theorem 2.3.8 Subset of a Metric Space is Nowhere Dense if the Interior of its Closure is Empty

A subset of S is **nowhere dense** *if the interior of its closure is empty.*

Proof 2.3.8
This one is for you. ■

This idea is a bit nonstandard really, but it allows us to focus on points outside of a target set.

Definition 2.3.12 Exterior Point of a Subset in a Metric Space

An **exterior point** *of a subset G is any member $x \in G^C$ such that there exists an $\epsilon > 0$ with $B(x, \epsilon) \subseteq G^C$. The* **exterior** *of a subset G is the set of all exterior points of G and is denoted by $ext(G)$.*

With that said, it is straightforward to define boundary points of sets.

Definition 2.3.13 Boundary of a Subset in a Metric Space

The **boundary** *of a subset G is the set of all elements x such that for every $\epsilon > 0$, $B(x, \epsilon) \cap int(G) \neq \emptyset$ and $B(x, \epsilon) \cap ext(G) \neq \emptyset$. The boundary of G is denoted by $bnd(G)$ or ∂G.*

We can now characterize open and exterior sets better. First, boundary points:

Theorem 2.3.9 Characterization of an Open Set in a Metric Space

$$int(G) = \bigcup \{\mathcal{O} : \mathcal{O} \text{ is open and } \mathcal{O} \subseteq G\}$$

Proof 2.3.9
This one is for you. ■

Then the closure:

Theorem 2.3.10 Characterization of the Closure of Set in a Metric Space

$$G^* = \bigcup \{F : F \text{ is closed and } G \subseteq F\}$$

Proof 2.3.10
This one is for you. ■

And now the exterior:

Theorem 2.3.11 Characterization of the Exterior Points of a Set in a Metric Space

$$ext(G) = (G^*)^C$$

Proof 2.3.11
This one is for you. ■

Finally, the boundary.

Theorem 2.3.12 Characterization of the Boundary of a Set in a Metric Space

$$bnd(G) = G^* \setminus int(G)$$

Proof 2.3.12
This one is for you. ■

Homework

Exercise 2.3.8 *Prove Theorem 2.3.1*

Exercise 2.3.9 *Prove Theorem 2.3.2*

Exercise 2.3.10 *Prove Theorem 2.3.3*

Exercise 2.3.11 *Prove Theorem 2.3.4*

Exercise 2.3.12 *Prove Theorem 2.3.5*

Exercise 2.3.13 *Prove Theorem 2.3.6*

Exercise 2.3.14 *Prove Theorem 2.3.8*

Exercise 2.3.15 *Prove Theorem 2.3.9*

Exercise 2.3.16 *Prove Theorem 2.3.10*

Exercise 2.3.17 *Prove Theorem 2.3.11*

Exercise 2.3.18 *Prove Theorem 2.3.12*

2.4 Some Deeper Metric Space Results

We know a subset of a metric space or a normed linear space is nowhere dense if its closure has an empty interior. A standard project in analysis is to construct Cantor sets and we can prove a Cantor set cannot contain any interval. Hence it is a nowhere dense subset of $[0, 1]$. We also prove that compact subsets of infinite dimensional normed linear spaces are nowhere dense. You can review these results in (Peterson (100) 2020). In (Peterson (100) 2020), we also discussed first and second category sets and for completeness and ease of discussion, let's restate these discussions.

Definition 2.4.1 First Category Sets of a Metric Space

> *Let (X, d) be a metric space. We say X is of* **first category** *if there is a sequence (A_n) of subsets of X which are all nowhere dense so that $X = \cup_{n=1}^{\infty} A_n$. If X is not of first category, we say X is of* **second category**.

A most important result is then:

Theorem 2.4.1 Baire Category Theorem

> *Let (X, d) be a nonempty complete metric space. Then X is of* **second category**.

Proof 2.4.1
We assume X is of first category and derive a contradiction. This is a somewhat long proof so be patient! Since X is of first category, we can write $X = \cup_{n=1}^{\infty} M_n$, where each M_n is of first category.

(Step 1*:)*
Since M_1 is nowhere dense, $\overline{M_1}$ has no interior points; i.e. $\overline{M_1}$ contains no nonempty set. But X does contain nonempty open sets, trivially X itself, so we cannot have $\overline{M_1} = X$. Hence $(\overline{M_1})^C = X \setminus M_1$

is nonempty and open as it is the complement of a closed set.

Choose $p_1 \in (\overline{M_1})^C$. *Since this set is open, there is a radius* $\epsilon_1 > 0$ *so that* $B(p_1; \epsilon_1) \subset (\overline{M_1})^C$. *We can assume* $\epsilon_1 < \frac{1}{2}$.

*(***Step 2***:)*
By an argument similar to that in Step 1, we know $(\overline{M_2})^C$ *is both open and nonempty. Also* $\overline{M_2}$ *cannot contain an open set so* $B(p_1; \frac{\epsilon_1}{2}) \subset (\overline{M_2})^C$. *This means* $(\overline{M_2})^C \cap B(p_1, \frac{\epsilon}{2})$ *is a nonempty open set. Therefore there is* p_2 *and associated radius* ϵ_2 *which we can choose so that* $\epsilon_2 < \frac{\epsilon_1}{2} < \frac{1}{2^2}$ *so that*

$$p_2 \in (\overline{M_2})^C \cap B(p_1; \frac{\epsilon}{2}), \quad B(p_2; \epsilon_2) \subset (\overline{M_2})^C \cap B(p_1; \frac{\epsilon}{2})$$

*(***Step 3***:)*
Now continue by induction. We obtain a sequence (p_n) *and a sequence* (ϵ_n) *satisfying*

$$\begin{aligned} p_n &\in (\overline{M_n})^C \cap B(p_{n-1}; \frac{\epsilon_{n-1}}{2}), \quad \epsilon_n < \frac{\epsilon_{n-1}}{2} < \frac{1}{2^n} \\ B(p_n; \epsilon_n) &\subset (\overline{M_n})^C \cap B(p_{n-1}; \frac{\epsilon_{n-1}}{2}) \end{aligned}$$

Thus, for $n = 2, 3, \ldots$

$$B(p_n; \epsilon_n) \subset B(p_{n-1}; \frac{\epsilon_{n-1}}{2}) \subset B(p_{n-1}, \epsilon_{n-1})$$

Pick a tolerance $\eta > 0$. *There is* N *so that* $\frac{1}{2^N} < \eta$. *We see then that if* $n > N$, $p_n \in B(p_N; \epsilon_N) \subset B(p_N; \frac{1}{2^N})$. *It follows that if* $m, n > N$,

$$d(p_n, p_m) \leq d(p_n, p_N) + d(p_m, p_N) < \frac{1}{2^N} + \frac{1}{2^N} < \eta$$

This shows (p_n) *is a Cauchy Sequence in* X *which is complete. Therefore there is a* $p \in X$ *so that* $p_n \to d$. *Hence, given the tolerance* $\xi > 0$, *there is* M *so* $d(p_n, p) < \xi$ *if* $n > M$. *However, we also know* $\epsilon_n < \frac{1}{2^n}$. *This implies* $p \in B(p_n; \epsilon_n)$ *if* $\frac{1}{2^n} < \xi$. *Since* ξ *is arbitrary, we see* $p \in B(p_n; \epsilon_n) \subset (\overline{M_n})^C$ *for all* n. *But that means*

$$p \in \cap_{n=1}^{\infty} (\overline{M_n})^C = \left(\cup_{n-1}^{\infty} \overline{M_n}\right)^C$$

by DeMorgan's Laws. Thus, $p \in X^C$ *where we assume* $X = \cup_{n-1}^{\infty} \overline{M_n}$. *But this is not possible. Thus our assumption* X *is of first category is wrong and we conclude* X *is of second category.* ■

Comment 2.4.1 *If* (X, d) *is a nonempty complete space, if we can write* $X = \cup_{n=1}^{\infty} M_n$, *then at least one set* M_p *cannot be nowhere dense. So we know* $(\overline{M_p})^C$ *has a nonempty interior. Hence there is* $P \in \overline{M_p}$ *and a radius* $r > 0$ *so that* $B(p; r) \subset \overline{M_p}$.

Also recall the important notion of compactness is expressible in terms of open sets. The two notions of compactness are sequential compactness and topological compactness which are defined as follows:

Definition 2.4.2 Sequentially Compact Subsets of a Metric Space

A subset G *of a metric space is* **sequentially compact** *if every sequence in* G *has a convergent subsequence in* G.

and

Definition 2.4.3 Topologically Compact Subsets of a Metric Space

A subset G of a metric space is **topologically compact** *if every collection of open sets whose union contains G (this is called an* open cover *of G) has a finite subcollection whose union likewise contains G (i.e. a* **finite subcover** *of G).*

It then follows we can show

Theorem 2.4.2 Sequentially Compact in a Normed Linear or Metric Space Implies Closed and Bounded

Whether X is a normed linear space with norm $\|\cdot\|$ or a metric space with metric d, if $S \subset X$ is sequentially compact, then it is closed and bounded.

Proof 2.4.2
S **is closed**: *Let $(\boldsymbol{x_n})$ be a convergent sequence in S which is a subset of the normed linear space $(X, \|\cdot\|)$. Then, there is $\boldsymbol{x} \in X$ so that $\boldsymbol{x_n} \to \boldsymbol{x}$ in norm. Because S is sequentially compact, there is a subsequence $(\boldsymbol{x_n^1})$ which converges to $\boldsymbol{y} \in S$. But the subsequential limit must be the same as the limit of a convergent sequence. Hence, $\boldsymbol{y} = \boldsymbol{x}$ and we see S contains its limit points. So S is closed.*

If $(\boldsymbol{x_n})$ is a convergent sequence in S which is a subset of the metric space (X, d). Then, there is $\boldsymbol{x} \in X$ so that $d(\boldsymbol{x_n}, \boldsymbol{x}) \to 0$. Because S is sequentially compact, there is a subsequence $(\boldsymbol{x_n^1})$ which converges to $\boldsymbol{y} \in S$. But the subsequential limit must be the same as the limit of a convergent sequence. Hence, $\boldsymbol{y} = \boldsymbol{x}$ and we see S contains its limit points. So S is closed. Note the argument for the norm and the metric are essentially the same.

S **is bounded**: *Let $(\boldsymbol{x_n})$ be a convergent sequence in S which is a subset of the normed linear space $(X, \|\cdot\|)$. Assume S is not bounded. Then for all n, there is $\boldsymbol{x_n} \in S$ so that $\|\boldsymbol{x_n}\| > n$. But $(\boldsymbol{x_n})$ is a sequence in S and so there is a subsequence $(\boldsymbol{x_{n_k}})$ which converges to $\boldsymbol{y} \in S$. Then for $\epsilon = 1$ there is N so $n_k > N$ implies $\|\boldsymbol{x_{n_k}} - \boldsymbol{x}\| < 1$. The backwards triangle inequality for the norm gives $n_K > N$ implies*

$$\|\boldsymbol{x_{n_k}}\| \quad \leq \quad \|\boldsymbol{x}\| + 1$$

and so

$$\|\boldsymbol{x_{n_k}}\| \quad \leq \quad \max\{\max_{1 \leq n_k \leq N} \|\boldsymbol{x_{n_k}}\|, \ \|\boldsymbol{x}\| + 1\}$$

which says the subsequence is bounded in norm. But by construction, it is not. Hence our assumption the set is not bounded in norm is wrong.

In the metric space setting, we have to decide what it means for a set to be bounded. A reasonable definition is that S is bounded if there is $R > 0$ so that $d(\boldsymbol{x}, \boldsymbol{y}) < R$ for all $\boldsymbol{x}$ and $\boldsymbol{y}$ in S. Another way to look at this is to fix $\boldsymbol{x_0}$ in S and then boundedness of S means $d(\boldsymbol{x}, \boldsymbol{x_0}) < R$ for all $\boldsymbol{x} \in S$. Let's assume S is not bounded. Then for all n, there is $\boldsymbol{x_n}$ so that $d(\boldsymbol{x_n}, \boldsymbol{x_0}) > n$. But $(\boldsymbol{x_n}) \subset S$ and so it has a convergence subsequence $(\boldsymbol{x_{n_k}})$ which converges to $\boldsymbol{x} \in S$. Thus for $\epsilon = 1$, there is N so that $n_K > N$ implies $d(\boldsymbol{x_{n_k}}, \boldsymbol{x}) < 1$. Now apply the backwards triangle inequality to find

$$d(\boldsymbol{x_{n_k}}, \boldsymbol{x_0}) \quad < \quad d(\boldsymbol{x}, \boldsymbol{x_0}) + 1$$

We conclude

$$d(\boldsymbol{x}_{n_k}, \boldsymbol{x}_0) \quad \le \quad \max\{\max_{1 \le n_k \le N} d(\boldsymbol{x}_{n_k}, \boldsymbol{x}),\ d(\boldsymbol{x}, \boldsymbol{x}_0) + 1\}$$

which says the subsequence $d(\boldsymbol{x}_{n_k}, \boldsymbol{x}_0)$ is bounded. However, we know it is not. This contradiction means our assumption S is not bounded is wrong. ■

Now, we can also show if a set is topologically compact, the set is both closed and bounded.

Theorem 2.4.3 Topologically Compact Implies Closed and Bounded

A is topologically compact $\Rightarrow$ A is closed and bounded.

Proof 2.4.3
Let's assume A is topologically compact. If the space is a normed space, it induces a metric and this determines our open set structure. If we take the union of all $B_r(x)$ for a fixed radius r for all x in A, this is an open cover. Since A is topologically compact, this cover has a fsc which contains A. We can express the fsc as $\mathscr{V} = \{B_r(x_1), \ldots, B_r(x_N)\}$ for some N. Now we know $d(x_i, x_1) = d_i$. Let $D = \max_{1 \le i \le N} d_i$. Now let $x \in A$. Then $x \in B_r(x_j)$ for some $1 \le j \le N$. Thus, $|d(x, x_1) \le d(x, x_j) + d(x_j, x_1) < r + d_j < r + D$. Hence A is a bounded set.

To show A is closed, we will show A^C is open. Let $x \in A^C$. Given any $y \in A$, since $x \neq y$, the two points are separated by a distance d^{xy}. Thus, $B_{d^{xy}/2}(y)$ and $B_{d^{xy}/2}(x)$ have no points in common. The collection of all the circles $B_{d^{xy}/2}(y)$ for $y \in A$ gives an open cover $\mathscr{U}$ of A which must have a fsc, $\mathscr{V} = \{B_{d_1}(y_1), \ldots, B_{d_N}(y_N)\}$, so that $A \subseteq \cup_{n=1}^{N} B_{d_n}(y_n)$. where each $d_i = d^{xy_i}/2$. Let $\mathscr{W} = \cap_{n=1}^{N} B_{d_n}(x)$. The circles $B_{d_n}(x)$ are all disjoint from $B_{d_n}(y_i)$. Hence, points in the intersection are outside the union of the fsc $\mathscr{V}$. Then $\mathscr{W}$ is open as it is a finite intersection of open sets and it is outside of the union over the fsc $\mathscr{V}$ which contains A. We see $\mathscr{W} \subset A^C$ implying x is an interior point of A^C. Since x was arbitrary, we see A^C is open. Hence A is closed. ■

We already know the converse here is not true in infinite dimensional normed linear spaces due to Riesz's Lemma. For example, $\overline{B(\boldsymbol{0}, 1)}$ is not compact in $C([0,1], \|\cdot\|)$ because $C([0,1])$ is an infinite dimensional vector space. But this ball is closed and bounded!

We can't do the usual thing here and show **Sequentially Compact $\Longleftrightarrow$ Closed and bounded $\Longleftrightarrow$ Topologically Compact**. So we will introduce some new ideas. The size of a set in a metric space is a bit more difficult to define since we are not in a normed space and have no notion of the magnitude of elements in the space. So we do this:

Definition 2.4.4 Diameter of a Set

The diameter of a set A in the metric space (X, d) is $diam(A) = \sup_{x,y \in A} d(x, y)$. If the set is not bounded, we have $diam(A) = \infty$.

We also need some tools to help us characterize our two seemingly distinct notions of compactness. This is a good one:

Definition 2.4.5 Totally Bounded Sets

The set A in the metric space (X, d) is totally bounded if and only if for every $\epsilon > 0$, A can be covered by a finite number of subsets of X whose diameters are all less than epsilon. That is, given $\epsilon > 0$, there is a finite collection $\{S_1, \ldots, S_p^{\epsilon}\}$ so that $A \subset \cup_{i=1}^{p^{\epsilon}} S_i$ with $diam(S_i) < \epsilon$. This finite collection of points is also called an ϵ-net.

We now prove a useful result called Lebesgue's Lemma which we use in our development of the Jacobian Curve Theorem.

Lemma 2.4.4 Lebesgue's Lemma in a Metric Space

Let G_α be an open cover of the compact set $\boldsymbol{S}$ in the metric space (X, d). Then there is an $\epsilon > 0$ so that for all $\boldsymbol{x} \in \boldsymbol{S}$, $B(\boldsymbol{x}, \epsilon) \subset G_\alpha$ for some α.

Proof 2.4.4

Let G_α be an open cover of $\boldsymbol{S}$. Assume this is not true. Then for all n, there is a point $\boldsymbol{x_n} \in \boldsymbol{S}$ with $B(\boldsymbol{x_n}, \frac{1}{n}) \not\subset G_\alpha$ for all indices α. Since $\boldsymbol{S}$ is sequentially compact and $(\boldsymbol{x_n}) \subset \boldsymbol{S}$, there is a subsequence $(\boldsymbol{x_n^1})$ and a point $\boldsymbol{x} \in \boldsymbol{S}$ with $\boldsymbol{x_n^1} \rightarrow \boldsymbol{x}$. Since (G_α) is an open cover of $\boldsymbol{S}$, there is an index β so that $\boldsymbol{x} \in G_\beta$. Hence, $\boldsymbol{x}$ is an interior point of G_β. So, there is a $\delta > 0$ so that $B(\boldsymbol{x}, \delta) \subset G_\beta$. But also, there is an N so that $d(\boldsymbol{x_n^1}, \boldsymbol{x}) < \frac{\delta}{2}$ for all $n^1 > N$. Thus, if $\boldsymbol{y} \in B(\boldsymbol{x_n^1}, \frac{1}{n^1})$,

$$d(\boldsymbol{y}, \boldsymbol{x}) \leq d(y, \boldsymbol{x_n^1}) + d(\boldsymbol{x_n^1}, \boldsymbol{x}) < \frac{1}{n^1} + \frac{\delta}{2}$$

We see if we choose M so that $\frac{1}{n^1} < \frac{\delta}{2}$, if $n^1 > \max(N, M)$, we have

$$d(\boldsymbol{y}, \boldsymbol{x}) < \delta \Longrightarrow B(\boldsymbol{x_n^1}, \frac{1}{n^1}) \subset G_\beta$$

But, this is not possible. Hence, we conclude, for all $\boldsymbol{x} \in \boldsymbol{S}$, there is an $\epsilon > 0$ so that $B(\boldsymbol{x}, \epsilon) \subset G_\alpha$ for some α. ■

We can then prove a very nice equivalence theorem for these concepts which is not tied to $\Re^n$ and the Bolzano - Weierstrass Theorem.

Theorem 2.4.5 Subset in a Metric Space is Sequentially Compact iff Topologically Compact iff Complete and Totally Bounded

The following are equivalent for a set $A \subset X$ where (X, d) is a metric space.

A is topologically compact.

A is sequentially compact.

A is complete and totally bounded.

Proof 2.4.5

Topologically Compact Implies Sequentially Compact

We assume A is topologically compact. Consider $(x_n) \subset A$. Assume (x_n) has no convergent subsequence. Then for each x in X, there is an open set U_x so that $\{n | x_n \in U_x\}$ has finite cardinality. The collection $\{U_x\}$ is an open cover of A so there is a fsc, $\{U_{y_1}, \ldots, U_{y_p}\}$ for A. But then

$$\mathbb{N} = \{n | x_n \in A\} = \cup_{i=1}^{p} \{n | x_n \in U_{y_i}\}$$

But the set on the right-hand side is finite! This is a contradiction and so (x_n) must possess a convergent subsequence. Since A is topologically compact we also know it is closed, so x is in A. Thus, A is sequentially compact.

Sequentially Compact Implies Complete and Totally Bounded
If A is sequentially compact, Cauchy Sequences must converge and so we also know A is complete. Now if A was not totally bounded, there would be a positive ϵ so that no finite collection of open balls of radius ϵ covers A. Pick some x_1 and let $B_1 = B(x_1, \epsilon)$. Choose $x_2 \in A \setminus B_1$ and let $B_2 = B(x_2, \epsilon)$. These two sets cannot cover A. Choose $x_3 \in A \setminus (B_1 \cup B_2)$ and let $B_3 = B(x_3, \epsilon)$. Again, these sets cannot cover A so we can continue the process. We construct a sequence (x_n) with $x_{n+1} \in A \setminus (B_1 \cup \ldots \cup B_n)$. Then $d(x_n, x_m) \geq \epsilon$ always when $n \neq m$. This says the sequence (x_n) cannot have a convergent subsequence which contradicts that A is sequentially compact. Hence, A is complete and totally bounded.

Complete and Totally Bounded Implies Topologically Compact
Let's assume A is not topologically compact so there is an open cover (O_α) with $\alpha \in \Lambda$ where Λ is finite, countable or uncountable as an index set. We can cover A with a finite number of sets $\{C_1^1, \ldots . C_{p_1}^1\}$ if diameter ≤ 1 because A is totally bounded. One of these subsets, label this as $C^{n_1} \equiv C_{k_1}^1$, cannot be covered by a finite number of the sets O_α as if it could, we could construct a fsc for A. Now C^{n_1} can be covered by a finite number of sets of diameter $\leq 1/2$, $\{C_1^2, \ldots . C_{p_2}^2\}$. Again, one set, label this as $C^{n_2} \equiv C_{k_2}^2$, cannot be covered by a finite number of the sets O_α. Note $C^{n_2} \subset C^{n_1}$. Continue this process and find nonempty sets with

$$C^{n_p} \subset \ldots \subset C^{n_2} \subset C^{n_1}$$

where C^{n_p} has diameter $\leq 1/p$ and C^{n_p} cannot be covered by a finite number of sets O_α. Choose $x_p \in C^{n_p}$. Then (x_p) is a Cauchy Sequence by construction and so it must converge to some $x_0 \in A$. That means x_0 is an interior point of O_β and so there is a $\delta > 0$ so that $B(x_0, \delta) \subset O_\beta$. Then there is N with $1/N < \delta/2$ so that $d(x_0, x_n) < \delta/2$ if $n > N$. This tells us

$$C^{n_N} \subset B(x_N, 1/N) \subset B(x_N, \delta/2) \subset B(x, \delta) \subset O_\beta$$

But this set cannot be covered by a finite number of O_α so this is a contradiction. Hence, A must be topologically compact. ■

The following theorems state the fundamental properties of open and closed sets within a given metric space X: first, the open sets:

Theorem 2.4.6 Properties of Open Sets in a Metric Space

For any metric space X:

1. *The empty set $\emptyset$ is an open set.*
2. *The space S is an open set.*
3. *Any union of open sets is itself an open set.*
4. *Any finite intersection of open sets is itself an open set.*

Proof 2.4.6
We leave the proof of these results to you. ■

And now the closed sets:

Theorem 2.4.7 Properties of Closed Sets in a Metric Space

For any metric space X:

1. *The empty set $\emptyset$ is a closed set.*
2. *The space S is a closed set.*
3. *Any intersection of closed sets is itself a closed set.*
4. *Any finite union of closed sets is itself a closed set.*

Proof 2.4.7
We leave these results to you. ■

It is apparent that Theorem 2.4.7 is obtained by directly applying De Morgan's Laws to Theorem 2.4.6 and vice versa. Also, we note that when we say **any** union of open sets and **any** intersection of closed sets we really do mean *any* whether the union or intersection is finite, countably infinite, or even uncountably infinite. Likewise, the restriction of open sets to **finite** intersections and closed sets to **finite** unions is truly necessary. It is not correct in general that an intersection of open sets is open or that a union of closed sets is closed.
Homework

Exercise 2.4.1 *Prove $[a, b]$ is totally bounded in $\Re$ by finding the ϵ-net for a given choice of ϵ.*

Exercise 2.4.2 *Prove $[a, b] \times [c, d]$ is totally bounded in $\Re^2$ by finding the ϵ-net for a given choice of ϵ.*

Exercise 2.4.3 *Prove Theorem 2.4.6.*

Exercise 2.4.4 *Prove Theorem 2.4.7.*

We can characterize open subsets in $\Re^n$ too.

Theorem 2.4.8 Characterization of Open Sets in $\Re^n$

If U is a nonempty and open subset of $\Re^n$, there is a sequence of sets (K_i) with

$$K_1 \subset int(K_2) \subset K_2 \subset int(K_3) \subset \ldots \subset K_n \subset (K_n) \subset \ldots$$

with $U = \cup_{i=1}^{\infty} K_i$.

Proof 2.4.8
Let U_i be the balls of centers with rational coordinates and rational radii where each center is in U and $U_i \subset U$. If $x \in U$, then since it is an interior point, there is a radius r so that $B(x, r) \subset U$. We can choose r to be rational and since $\mathbb{Q}^n$ is dense in $\Re^n$, there is a rational vector t so that $x \in B(t, r)$ and $\overline{B(t, r)} \subset U$. Hence, $U = \cup_i U_i$ and each $\overline{U}_i \subset U$.
Let $K_1 = \overline{U}_1 \subset U$. The set K_1 is compact and so it is covered by $\cup_i U_i$. Hence, there is a finite subcover $\{U_{j_1}, \ldots, U_{j_{q_1}}\}$ with

$$K_1 \subset \cup_{i=1}^{q_1} U_{j_i} \subset \cup_{i=1}^{j_{q_1}} U_i$$

Now define K_2 by

$$K_2 = \overline{U}_1 \cup \ldots \cup \overline{U}_{j_{q_1}}$$

Then $K_1 \subset int(K_2) \subset U$.

Now proceed by induction. Assume $K_{m-1} \subset int(K_m)$, *Using the same finite subcover argument as we used in the basis step, there is a finite subcover* $\{U_{j_1}, \ldots, U_{j_{q_m}}\}$ *with*

$$K_m \subset \cup_{i=1}^{q_m} U_{j_i} \subset \cup_{i=1}^{j_{q_m}} U_i$$

Now define K_{m+1} *by*

$$K_{m+1} = \overline{U}_1 \cup \ldots \cup \overline{U}_{j_{q_m}}$$

Then $K_m \subset int(K_{m+1}) \subset U$.

Now if $x \in \cup_{i=1}^{\infty} K_i$, $x \in K_p \subset U$ *for some* p. *Hence* $\cup_{i=1}^{\infty} K_i \subset U$. *The converse is clear as each* $K_i \subset U$. ■

We can also prove a very useful result called the Cantor Intersection Theorem.

Theorem 2.4.9 Cantor Intersection Theorem

Let (X, d) *be a complete metric space and let* (F_n) *be a decreasing sequence of sets, i.e.* $F_{n+1} \subset F_n$ *for all* n *such that the diameter of* F_n *goes to zero. Then there is* $x \in X$ *so that* $\cap F_n = \{x\}$.

Proof 2.4.9
To show $F = \cap_n F_n$ *contains a point, let* $x_n \in F_n$ *be chosen. Since the diameter of* F_n *goes to zero, it is straightforward to show* (x_n) *is a Cauchy sequence and hence converges to* $x \in X$. *Further, it is easy to show* $x \in \cap_n F_n$. *Finally, to show* x *is the only point in the intersection, note if* x *and* y *were in* $\cap_n F_n$, *then given* $\epsilon > 0$, *there is an* N *so that the diameter of* F_n *is less than* ϵ *if* $n > N$. *This tells us* $d(x, y) < \epsilon$ *since* $x, y \in F_n$ *for* $n > N$. *But* $\epsilon > 0$ *is arbitrary, so* $x = y$. ■

We can also prove an important theorem about what are called **Partitions of Unity**. First, let's talk about open sets that satisfy a finite union condition.

Definition 2.4.6 Finite Union Condition for an Open Set

Assume an open set $U \subset \Re^n$ *can be written as* $U = \cup_{i=1}^{\infty} U_i$ *where each* U_i *is open. Also assume each* $x \in U$ *belongs to only a finite number of the sets* U_i. *Then we say* U *satisfies the finite union condition.*

We now show that partitions of unity exist in certain circumstances.

Theorem 2.4.10 Existence of Partitions of Unity for an Open Set Satisfying the Finite Union Condition

Let $U \subset \Re^n$ *be an open set satisfying the finite union condition. Then there are* C^∞ *functions* (ϕ_i) *so that the support of* ϕ_i *satisfies,* $supp(\phi_i) = \overline{\{x \in U | \phi(x) > 0\}} \subset U_i$ *and given any* $x \in U$, *there are a finite number of functions* $\phi > 0$, $\{\phi_{j_1}, \ldots, \phi_{j_n}\}$ *with* $\sum_{i=1}^{n} \phi_{j_i} = 1$. *This collection of functions is called a partition of unity.*

Proof 2.4.10
We will consider C^∞ *functions like the bump functions we discussed in (Peterson (99) 2020). Look*

at

$$f_0(x) = \begin{cases} e^{-\frac{1}{x}}, & x > 0 \\ 0, & x \le 0 \end{cases}$$

This is C^∞ *with* $\lim_{x \to 0^+} f_0(x) = 0$ *and* $\lim_{x \to \infty} f_0(x) = 1$. *A messy induction proof then shows* f_0 *is* C^∞ *with all the derivative zero at* $x = 0$. *This is easily generalized to*

$$f_a(x) = \begin{cases} e^{-\frac{1}{x-a}}, & x > a \\ 0, & x \le a \end{cases}$$

which is C^∞ *with all derivatives and the original function zero at* $x = a$. *In fact, if we define*

$$g(x) = f_0(x) f_0(1-x) = \begin{cases} e^{-\frac{1}{x}} e^{-\frac{1}{1-x}} = e^{-\frac{1}{x(1-x)}}, & 0 < x < 1 \\ 0, & x < 0 \text{ or } x > 1 \end{cases}$$

We see g *is* C^∞ *with all derivatives and the function itself zero at* $x = 0$ *and* $x = 1$. *We can adapt this to the interval* $[a, b]$ *to get*

$$\begin{aligned} g(x) &= f_0\left(\frac{x-a}{b-a}\right) f_0\left(1 - \frac{x-a}{b-a}\right) \\ &= f_0\left(\frac{x-a}{b-a}\right) f_0\left(\frac{b-x}{b-a}\right) \\ &= \begin{cases} e^{-\frac{(b-a)^2}{(x-a)(b-x)}}, & a < x < b \\ 0, & x < a \text{ or } x > b \end{cases} \end{aligned}$$

which has support $[a, b]$ *and all derivatives and the original function are zero at* $x = a$ *and* $x = b$. *We can then use this on the rectangle* $\mathscr{R} = (a_1, b_1) \times (a_2, b_2) \times \ldots \times (a_n, b_n)$ *to define the function* h *on* $\Re^n$ *as follows:*

$$h(x) = \prod_{i=1}^{n} g\left(\frac{x_i - a_i}{b_i - a_i}\right) = \begin{cases} e^{-\frac{(b_i - a_i)^2}{(x-a_i)(b_i - x)}}, & a_i < x_i < b_i \\ 0, & \text{else} \end{cases}$$

which has support $\overline{\mathscr{R}}$ *and all partials and the original function are zero on* $\partial\mathscr{R}$.

By Lebesgue's Lemma for the cover (U_i) *there is an* $\epsilon > 0$ *so that if the diameter of a set* W *in* U *is less than epsilon, there is a set* U_i *with* $W \subset U_i$. *Let* $A \subset U$ *be compact. Then* A *is contained in a finite union of open sets* U_i. *Letting* N *be the largest such index, we see* $A \subset \cup_{i=1}^{N} U_i$. *Now any rectangle* $\mathscr{R} \subset A$ *with diameter less than* ϵ *is contained in some* U_i. *Subdivide the compact set* A *using a partition* Π *on each axis so that* $\|\Pi\| < \frac{\epsilon}{2}$. *The partition then gives a collection of rectangles* $(\mathscr{R}_\alpha)$ *with* $diam(\mathscr{R}_\alpha) < \epsilon$. *Thus, each* $\overline{\mathscr{R}_\alpha} \subset U_{i_\alpha}$ *for some index* i_α. *Each rectangle will then have a* C^∞ *bump function* h_α *with* $supp(h_\alpha) = \overline{\mathscr{R}_\alpha}$ *and* $h_\alpha = \mathbf{0}$ *on* $(\mathscr{R}_\alpha)^C$. *Define these functions:*

$$\begin{aligned} \sigma_1 &= \sum_{\alpha : \overline{\mathscr{R}_\alpha} \subset U_1} h_\alpha \Longrightarrow supp(\sigma_1) \subset U_1 \\ \sigma_2 &= \sum_{\alpha : \overline{\mathscr{R}_\alpha} \subset U_2} h_\alpha \Longrightarrow supp(\sigma_2) \subset U_2 \\ &\vdots \\ \sigma_i &= \sum_{\alpha : \overline{\mathscr{R}_\alpha} \subset U_i} h_\alpha \Longrightarrow supp(\sigma_i) \subset U_i \end{aligned}$$

Since we only have to consider up to U_N, this process terminates at σ_N. We set $\sigma_j = 0$ when $j > N$.

By Theorem 2.4.8, we know we can write $U = \cup_{i=1}^{\infty} K_i$ where each K_i is compact and $K_i \subset int(K_{i+1})$ for all $i \geq 1$. Set $K_0 = \emptyset$ and $K_{-1} = \emptyset$. Define

$$\begin{aligned} A_1 = K_1 \setminus int(K_0) = K_1, \quad & W_1 = int(K_2) \setminus K_{-1} = int(K_2) \\ A_2 = K_2 \setminus int(K_1), \quad & W_2 = int(K_3) \setminus K_0 = int(K_3) \\ A_3 = K_3 \setminus int(K_2), \quad & W_3 = int(K_4) \setminus K_1 = int(K_3) \\ & \vdots \\ A_j = K_j \setminus int(K_{j-1}), \quad & W_j = int(K_{j+1}) \setminus K_{j-2} = int(K_3) \end{aligned}$$

For example, look at $x \in A_3$. Then $x \in K_3 \subset int(K_4) \subset K_4$ and $x \in (int(K_2))^C$. Our chain conditions then tell us $(int(K_2))^C \subset K_1^C$. Hence, $x \in K_4 \setminus K_1 = W_3$. Similar arguments hold for all the cases: so we know $A_i \subset W_i$ for all i. It is straightforward to see $U = \cup_i K_i = \cup_i A_i$. Thus, we have

$$\begin{aligned} A_1 &\subset W_1, \quad A_1 \text{ compact} \\ A_2 &\subset W_2, \quad A_2 \text{ compact} \\ A_3 &\subset W_3, \quad A_3 \text{ compact} \\ &\vdots \\ A_j &\subset W_j, \quad A_j \text{ compact} \end{aligned}$$

So all W_i are open and all A_i are compact. Also, from the conditions, we see $A_j \cap A_{j-1} = \emptyset$ so this is a sequence of disjoint sets. Since $A_j \subset W_j$, we see $A_j \subset \cup_i U_i \cap W_j$ and so $(U_i \cap W_j)$ is an open cover of A_j.

Now apply the argument we used for the compact set A earlier to each A_j. We obtain a set (σ_{ij}) with $supp(\sigma_{ij}) \subset U_i \cap W_j \subset A_j$. For each index j, only a finite number of $\sigma_{ij}(x) > 0$ for any $x \in A_j$ as we showed in our earlier construction. Let this number be n_j. Define $\psi_j = \sum_{i=1}^{n_j} \sigma_{ij}$. It follows $supp(\psi_j) = A_j$.

Define $\psi(x) = \sum_{j=1}^{\infty} \psi_j(x)$. Given $x \in U$, $x \in A_j$ for only one index j_0. By the finite union condition, there are at most a finite number of U_i with $x \in U_i$. Now if there were an infinite number of indices such that $\sigma_{i_k, j_k}(x) > 0$, then $x \in U_{i_k} \cap W_{j_k}$. This says x is in more than a finite number of sets U_i. This is not possible because of the finite union condition. Hence, at most a finite number of σ_{ij}'s can be positive at any x. Hence, we can write

$$\psi(x) = \sum_{k=1}^{n_k} \sigma_{i_k, j_k}(x) = \sum_{k=1}^{n_k} \psi_{j_k}(x)$$

Hence, $\psi(x)$ is finite, the infinite sum is well-defined and $\psi > 0$ on U. Finally, define $f_i = \frac{\psi_i}{\psi}$ which is positive on U and at any $x \in U$, we have

$$\sum_{i=1}^{\infty} f_i(x) = \frac{1}{\psi(x)} \sum_{j=1}^{\infty} \psi_j(x) = \frac{1}{\sum_{k=1}^{n_k} \psi_{j_k}(x)} \sum_{k=1}^{n_k} \psi_{j_k}(x) = 1$$

■

Homework

Exercise 2.4.5 *In the proof of the Cantor Intersection Theorem, prove that since the Cauchy sequence* (x_n) *obtained in the argument converges to* $x \in X$*,* $x \in \cap_n F_n$.

Exercise 2.4.6 *Go back and revisit the original proofs of the Bolzano - Weierstrass Theorem for a bounded infinite sequence and this time use the Cantor Intersection Theorem to find show the subsequential limit is in the set.*

Exercise 2.4.7 *Pick any open bounded set in* $\Re^2$ *you like, but make sure it is not convex so that is has folds. Find the sequence of sets* (K_i) *from Theorem 2.4.8 drawing the first four or five very carefully.*

Exercise 2.4.8 *Let* $A = \cup_{n=1}^{\infty}(1/(2n+1), 1/(2n-1))$*. Does A satisfy the finite union condition?*

Exercise 2.4.9 *Let* $A = \cup_{n=1}^{\infty}(-1/n, 1/n)$*. Does A satisfy the finite union condition?*

2.5 Deeper Vector Space and Set Results

As you know, we can prove every vector space has a basis using Zorn's Lemma. Again, for completeness of our discussions, we will include this basic material for easy reference.

Definition 2.5.1 Partial Ordering on a set

Let M *be a nonempty set. A partial ordering on* M *is a binary relation* $R \subset M \times M$ *satisfying:*

PO1: $(a, a) \in R$*. This is read as* aRa *or* $a \preceq a$ *and is called reflexivity.*

PO2: $(a, b) \in R$ *and* $(b, a) \in R$ *implies* $a = b$*. This is read as* aRb *and* bRa *implies* $a = b$ *or* $a \preceq b$ *and* $b \preceq a$ *implies* $a = b$*. This is called antisymmetry.*

PO3: $(a, b) \in R, (b, c) \in R$ *implies* $(a, c) \in R$*. This is read as* aRb *and* bRc *implies* aRc *or* $a \preceq b$ *and* $b \preceq c$ *implies* $a \preceq c$*. This is called transitivity.*

The pair $(M, \preceq)$ *is called a partially ordered set or* **poset**. *We often just say* M *is a poset and leave out mention of the ordering* $\preceq$.

Comment 2.5.1 *If you add the property that any two objects in* M *can be compared: i.e.* aRb *or* bRa*, the partial ordering becomes a total ordering. Recall, we added a total ordering when we were completing* $\mathbb{Q}$*. We say* $a, b \in M$ *are comparable, if either* $a \preceq b$ *or* $b \preceq a$ *or both.*

Comment 2.5.2 *The set of all subsets of a nonempty set* M *can be partially ordered by set containment* $\subset$*. Since two subsets can be disjoint, not all subsets can be compared. Hence, this is not a total ordering.*

Now let's define **chains**.

Definition 2.5.2 Totally Ordered Sets or Chains

Let $(M, \preceq)$ *be a collection of subsets of a nonempty poset. Then if any two elements in* M *are comparable, we say* M *is a totally ordered set or chain.*

We also need to know about the largest and smallest object in a partially ordered set, if possible.

Definition 2.5.3 Upper Bounds and Extremal Objects of Posets

An upper bound of a subset W in a poset P is an object $U \in P$ so that $x \preceq U$ for all $x \in W$. The least upper bound (lub) of W is an upper bound $M \in P$ so that for all upper bounds U, $M \preceq U$.

A lower bound of a subset W in a poset P is an object $L \in P$ so that $L \preceq x$ for all $x \in W$. The greatest lower bound (glb) of W is a lower bound $m \in M$ so that for all lower bounds L, $L \preceq m$.

Note the lub and glb need not be in W. If they are, we call the lub, the maximal element of W, and the glb, the minimal element of W.

Comment 2.5.3 *Note U need not be in W, of course, but it is in M. And such U need not exist.*

The first axiom of set theory is one we have used already: Zorn's Lemma.

Axiom 1 Zorn's Lemma

Let M be a nonempty poset. Assume every chain $C \subset M$ has an upper bound. Then M has at least one maximal object.

Zorn's Lemma is equivalent to other axioms. You should look the proof of their equivalence which is in (Hewitt and Stromberg (50) 1965).

Axiom 2 Axiom of Choice

Let $\mathbf{A}_\alpha$ be any nonempty family of nonempty sets where $\alpha \in I$ for some index set I and from some space X. The Cartesian product of this family is then $\Pi_\alpha\, \mathbf{A}_\alpha$ which is the set of all functions f whose value at α satisfies $f(\alpha) \in \mathbf{A}_\alpha$. We call such a function a choice function: it means we know we can make a choice of some element from $\mathbf{A}_\alpha$. This is clear if I is finite but problematic if I is countably infinite or uncountable. Hence, the existence of such a choice function is not obvious. The Axiom of Choice says such a choice function always exists.

Axiom 3 Tukey's Lemma

If $\mathscr{F}$ is a family of sets, we say $\mathscr{F}$ is a family of finite character if for each set $\mathbf{F} \in \mathscr{F}$, we have all finite subsets of $\mathbf{F}$ are also in $\mathscr{F}$. Tukey's Lemma says any nonempty family of finite character has a maximal member, where the partial ordering here is set inclusion.

Axiom 4 Hausdorff Maximality Principle

Every nonempty partially ordered set contains a maximal chain.

A well ordering of a set is total order $\preceq$ so that every nonempty subset of it has a least element in this ordering. The fact that all sets can be well ordered is an axiom equivalent to the Axiom of Choice. This is the next axiom.

Axiom 5 Well Ordering Principle

Every set $\mathscr{F}$ can be well ordered by a relationship $\preceq$; i.e. if $\mathbf{A} \subset \mathscr{F}$ there is an element of a of $\mathbf{A}$ so that $a \preceq x$ for all $x \in \mathbf{A}$. The element a is called the minimal element of $\mathbf{A}$.

All of these statements are equivalent to the Axiom of Choice.

The Hahn - Banach Theorem is a fundamental result in analysis. A typical but important application of the Hahn - Banach Theorem is its use in the proof that linear functionals on certain linear spaces can be represented in the useful inner product form, or as a pairing between the space and its dual space. There are many special cases of the Hahn - Banach Theorem, some of which we will prove as corollaries. We have done these proofs before in (Peterson (100) 2020) so we will just state them here.

Theorem 2.5.1 Hahn - Banach Theorem

Let V be a linear space over $\Re$, and let $p : V \rightarrow \Re$ satisfy

1. $p(u+v) \leq p(u) + p(v) \qquad \forall\, u, v \in V$
2. $p(\alpha v) = \alpha p(v) \qquad \forall\, v \in V,\ \alpha \geq 0.$

Suppose V_0 is a linear subspace of V and λ_0 is a linear functional on V_0 such that $\lambda_0(v) \leq p(v)\ \forall\, v \in V_0$. Then there exists a linear functional, λ, on V such that $\lambda|_{V_0} = \lambda_0$ and $\lambda(v) \leq p(v)\ \forall\, v \in V$.

Proof 2.5.1
You can find this proof in many places, including the one we use in (Peterson (100) 2020). ∎

Homework

Exercise 2.5.1 *Let Ω be the set of all vectors in $\Re^2$ with nonnegative components. If we say $\boldsymbol{x} \preceq \boldsymbol{y}$ if the components of $\boldsymbol{x}$ are less than or equal to the components of $\boldsymbol{y}$, is this a partial ordering? Is it a total ordering? Is it well ordered?*

Exercise 2.5.2 *Let Ω be the set of all complex numbers with nonnegative components. If we say $\boldsymbol{x} \preceq \boldsymbol{y}$ if the components of $\boldsymbol{x}$ are less than or equal to the components of $\boldsymbol{y}$, is this a partial ordering? Is it a total ordering? Is it well ordered?*

Exercise 2.5.3 *Let Ω be the set of all vectors in $\Re^2$ with nonnegative components that are in the sector $S = \{(x,y) | \theta_1 \leq \tan^{-1}(y/x) \leq \tan^{-1}(y/x)\}$ for $0 < \theta_1 < \theta_2 < \pi/2$. If we say $\boldsymbol{x} \preceq \boldsymbol{y}$ if the components of $\boldsymbol{x}$ are less than or equal to the components of $\boldsymbol{y}$, is this a partial ordering? Is it a total ordering? Is it well ordered?*

Exercise 2.5.4 *Let Ω be the set of all continuous functions on $[0,1]$ that are nonnegative that are in the sector $S = \{(x,y) | \theta_1 \leq \tan^{-1}(y/x) \leq \tan^{-1}(y/x)\}$ for $0 < \theta_1 < \theta_2 < \pi/2$. If we say $f \preceq g$ if $f(x) \leq g(x)$ for all $0 \leq x \leq 1$, is this a partial ordering? Is it a total ordering? Is it well ordered?*

There are many useful consequences of the Hahn - Banach Theorem. First, there is this generalization.

Theorem 2.5.2 Generalized Hahn - Banach Theorem

Let V be a real linear space, and suppose $p : V \rightarrow \Re$ satisfies $p(\alpha v) = |\alpha| p(v)$ for all $\alpha \in \Re$, $v \in V$, and $p(v+w) \leq p(v) + p(w)$ for all $v, w \in V$. Suppose V_0 is a subspace of V, and $\lambda_0 : V_0 \rightarrow \Re$ is a linear functional on V_0 satisfying $|\lambda_0(v)| \leq p(v)$ for all $v \in V_0$. Then there is a linear functional, λ, on V such that $\lambda|_{V_0} = \lambda_0$ and $|\lambda(v)| \leq p(v)$ for all $v \in V$.

Proof 2.5.2
This is an easy proof, so we will show it here. Since $\lambda_0(v) \leq |\lambda_0(v) \leq p(v)$ for all $v \in V_0$, the Hahn-Banach Theorem implies that there is a linear functional, λ, on V such that $\lambda|_{V_0} = \lambda_0$ and

$\lambda(v) \leq p(v)$ for all $v \in V$. So, for any $v \in V$, $-\lambda(v) = \lambda(-v) \leq p(-v) = p(v)$, which implies that $\lambda(v) \geq -p(v)$. Hence, we have $-p(v) \leq \lambda(v) \leq p(v)$ for all $v \in V$. Thus, $|\lambda(v)| \leq p(v)$ for all $v \in V$, and λ, therefore, is the desired linear functional. ■

Theorem 2.5.2 is referred to as the *Generalized Hahn-Banach Theorem*, because it is typically more applicable than the Hahn-Banach Theorem, since the dominating function p is usually a seminorm, meaning that it will satisfy the conditions in the corollary. The next result specializes the Hahn-Banach Theorem to normed linear spaces.

Theorem 2.5.3 Hahn-Banach Theorem for Normed Linear Spaces

Let $(X, \|\cdot\|)$ be a normed linear space, and let $Z \subset X$ be a subspace. Suppose there is a linear functional, $f : Z \to \Re$, that is bounded. Then, there is a linear functional $\tilde{f} : X \to \Re$ such that $\tilde{f}|_Z = f$ and $\|\tilde{f}\|_X = \|f\|_Z$. That is, $\tilde{f}$ agrees with f on Z and the operator norm of $\tilde{f}$ is the same as that of f on Z.

Proof 2.5.3

If $Z = \{0\}$, then the result is trivial by letting $\tilde{f} = 0$. So, suppose $Z \neq \{0\}$. By definition,

$$\|f\|_Z = \sup_{x \in Z, x \neq 0} \frac{|f(x)|}{\|x\|} \Rightarrow \frac{|f(x)|}{\|x\|} \leq \|f\|_Z$$

for all $x \in Z$, $x \neq 0$. Define $p : X \to \Re$ by $p(x) = \|f\|_Z\|x\|$. Then $p(x+y) = \|f\|_Z\|x+y\| \leq \|f\|_Z(\|x\| + \|y\|)$. So, p is subadditive. Moreover, $p(\alpha x) = \|f\|_Z\|\alpha x\| = |\alpha|\|f\|_Z\|x\| = |\alpha|p(x)$. Also, $|f(x)| \leq \|f\|_Z\|x\|$ for all $x \in Z$, so $|f(x)| \leq p(x)$ for all $x \in Z$.

By Theorem 2.5.2, there exists $\tilde{f} : X \to \Re$ such that $\tilde{f}|_Z = f$ and $|\tilde{f}(x)| \leq p(x)$ for all $x \in X$. It follows that $|\tilde{f}(x)| \leq \|f\|_Z\|x\|$ for all $x \in X$. That is,

$$\frac{|\tilde{f}(x)|}{\|x\|} \leq \|f\|_Z \qquad \forall x \in X,\ x \neq 0.$$

So, it follows that

$$\sup_{x \in X, x \neq 0} \frac{|\tilde{f}(x)|}{\|x\|} \leq \|f\|_Z \Rightarrow \|\tilde{f}\|_X \leq \|f\|_Z.$$

But, since $\tilde{f}|_Z = f$, we have

$$\sup_{x \in X, x \neq 0} \frac{|\tilde{f}(x)|}{\|x\|} \geq \sup_{x \in Z, x \neq 0} \frac{|\tilde{f}(x)|}{\|x\|} = \sup_{x \in Z, x \neq 0} \frac{|f(x)|}{\|x\|} = \|f\|_Z.$$

Thus, $\|\tilde{f}\|_X \geq \|f\|_Z$, and $\|\tilde{f}\|_X = \|f\|_Z$. ■

The following theorem uses the previous results to show that there always exists a bounded linear functional having norm 1 on any normed linear space. For that reason, it is usually referred to as the *Bounded Linear Functional Theorem.*

Theorem 2.5.4 Bounded Linear Functional Theorem

Suppose $(X, \|\cdot\|)$ is a normed linear space. Let $x_0 \in X$ be nonzero. Then there is a bounded linear functional $\tilde{f} : X \to \Re$ such that $\|\tilde{f}\|_X = 1$ and $\tilde{f}(x_0) = \|x_0\|$.

Proof 2.5.4
Let $Z = span\{x_0\}$. Define $f : Z \to \Re$ by $f(\alpha x_0) = \alpha \|x_0\|$. Then f is a bounded linear functional on Z, and, since

$$\frac{|f(\alpha x_0)|}{\|\alpha x_0\|} = \frac{|\alpha|\|x_0\|}{|\alpha|\|x_0\|} = 1 \quad for\, \alpha \neq 0$$

it follows that f is bounded. By Theorem 2.5.3, there is $\tilde{f} : X \to \Re$ such that $\tilde{f}$ is a bounded linear functional, $\tilde{f}|_Z = f$, and $\|\tilde{f}\|_X = \|f\|_Z = 1$. Moreover, $\tilde{f}(x_0) = f(x_0) = \|x_0\|$. ■

We can also characterize a norm in terms of linear functionals as the next result shows.

Theorem 2.5.5 Characterizing Norm in Terms of Linear Functionals

Suppose $(X, \|\cdot\|)$ is a normed linear space. Then

$$\|x\| = \sup_{x' \in x', x' \neq 0} \frac{|x'(x)|}{\|x'\|_X}.$$

Proof 2.5.5
If $x_0 \in X$, then Theorem 2.5.4 implies that there exists a bounded linear functional $f_{x_0} : X \to \Re$ such that $\|f_{x_0}\| = 1$ and $f_{x_0}(x_0) = \|x_0\|$. So, $f_{x_0} \in x'$ and we have

$$\sup_{x' \in x', x' \neq 0} \frac{|x'(x_0)|}{\|x'\|_X} \geq \frac{|f_{x_0}(x_0)|}{\|f_{x_0}\|_X} \Rightarrow \|x_0\| \leq \sup_{x' \in x', x' \neq 0} \frac{|x'(x_0)|}{\|x'\|_X}.$$

We also know that $|x'(x_0)| \leq \|x'\|_X \|x_0\|$, so

$$\frac{|x'(x_0)|}{\|x'\|_X} \leq \|x_0\| \Rightarrow \sup_{x' \in x', x' \neq 0} \frac{|x'(x_0)|}{\|x'\|_X} \leq \|x_0\|$$

It follows, then, that

$$\sup_{x' \in x', x' \neq 0} \frac{|x'(x_0)|}{\|x'\|_X} = \|x_0\|.$$

■

Homework

There are many exercises on this material in (Peterson (100) 2020), so we will just give you a few more which will require you to do some review.

Exercise 2.5.5 *If f is a continuous function on $[0,1]$, how do you characterize the set of continuous linear functionals on this set of functions?*

Exercise 2.5.6 *In (Peterson (100) 2020), we prove $B(\mathbb{N}, 2^{\mathbb{N}})$ is congruent to ℓ^∞ and the dual space of ℓ^∞ is $ba(\mathbb{N}, 2^{\mathbb{N}})$, the set of all bounded charges on $2^{\mathbb{N}}$. So if (x_i) is a sequence that converges, why do we know there is a bounded charge μ with $\|\mu\| = 1$ and $\mu((x_i)) = \|(x_i)\|$?*

Why do we know $\|(x_i)\| \leq |\mu((x_i))|/\|\mu\|$ for all μ?

Exercise 2.5.7 *If $\boldsymbol{x} \in \Re^4$, why do we know there is a vector $\boldsymbol{y}$ so that $\|\boldsymbol{y}\| = 1$ and $< \boldsymbol{y}, \boldsymbol{x} >= \|\boldsymbol{x}\|$? Find this vector $\boldsymbol{y}$.*

Chapter 3

Forms and Curves

A set $\boldsymbol{U}$ in $\Re^n$ is *not connected* if it can be written as $\boldsymbol{U} = \boldsymbol{A} \cup \boldsymbol{B}$ with $\boldsymbol{A}$ and $\boldsymbol{B}$ open and $\boldsymbol{A} \cap \boldsymbol{B} = \emptyset$. At this point, our notion of open is the usual one: a subset of $\Re^n$ is open if is contains no interior points. We discussed 1-forms and curves in $\Re^2$ in a preliminary way in (Peterson (101) 2020) and we will refer you to those discussions for some background now.

Another way to look at connectedness is to use paths. We say the set $\boldsymbol{U}$ is pathwise connected if every pair of points (x, y) and (x_0, y_0) can be connected by a continuous curve γ defined on $[0, 1]$ with $\gamma(0) = (x_0, y_0)$ and $\gamma(1) = (x, y)$. We usually want γ to be more than just continuous so the path γ is normally C^1. It is easy to see that if $\boldsymbol{U}$ is an open rectangle in $\Re^n$, $\boldsymbol{U}$ is path connected if and only if it is connected. For more general sets, this may not be true, but our interests are rectangles at this point.

Theorem 3.0.1 Open Subsets of $\Re^n$ are Pathwise Connected

Let $\mathscr{U} \subset \Re^n$ be open and $\boldsymbol{x}$ and $\boldsymbol{y}$ be two points in $\mathscr{U}$. Then there is a smooth path γ connecting $\boldsymbol{x}$ to $\boldsymbol{y}$ in $\mathscr{U}$.

Proof 3.0.1
Let $\mathscr{U} \subset \Re^n$ be open and $\boldsymbol{x}$ and $\boldsymbol{y}$ be two points in $\mathscr{U}$. Let

$$\mathscr{B} = \{B(\boldsymbol{x}, r) | \boldsymbol{x} \in \mathscr{U}, B(\boldsymbol{x}, r) \subset \mathscr{U}, \text{ any } r > 0\}$$

Since $\mathscr{U}$ is open and $\boldsymbol{x} \in \mathscr{U}$, there is an $r_{\boldsymbol{x}}$ so that $B(\boldsymbol{x}, r_{\boldsymbol{x}}) \in \mathscr{B}$. Label this ball as B_1.

Let

$$B_2 = \cup\{B(\boldsymbol{x}, r) \in \mathscr{B} | B(\boldsymbol{x}, r) \cap B_1 \neq \emptyset\}$$

Now continue this process. Given B_i, set

$$B_{i+1} = \cup\{B(\boldsymbol{x}, r) \in \mathscr{B} | B(\boldsymbol{x}, r) \cap B_i \neq \emptyset\}$$

Assume you could find some $B(\boldsymbol{x}, r) \in \mathscr{B}$ which is not in any B_i. Let $V = \cup_{i=1}^{\infty} B_i$. Note, it is easy to see that $B_i \cap B_{i+1} \neq \emptyset$ for all i. Define

$$\mathscr{W} = \{B(\boldsymbol{x}, r) \in \mathscr{B} | B(\boldsymbol{x}, r) \not\subset B_i, \forall\, i\} = (\mathscr{V})^C$$

Then $\mathscr{W}$ is open also. If $\mathscr{W}$ were nonempty, then the pair $\{\mathscr{V}, \mathscr{W}\}$would satisfy $\mathscr{U} = \mathscr{V} \cup \mathscr{W}$, both open and disjoint. This contradicts the fact that $\mathscr{U}$ is connected. Hence $\mathscr{W}$ is empty and $\mathscr{U} = \cup_{i=1}^{\infty} B_i$. Hence, there is an index n so that $\boldsymbol{y} \in B_n$.

Let's assume $n = 3$ for concreteness. Then start at $\boldsymbol{y} \in B_3$. Thus, $\boldsymbol{y} \in B(\boldsymbol{w_3}, r_3)$ which intersects $B(\boldsymbol{w_2}, r_2) \in B_2$. Then there is $\boldsymbol{z_3} \in B(\boldsymbol{w_3}, r_3), B(\boldsymbol{w_2}, r_2)$. The line segment $[\boldsymbol{y}, \boldsymbol{z_3}] \subset B(\boldsymbol{w_3}, r_3)$ and the line segment $[\boldsymbol{z_3}, \boldsymbol{w_2}]$ is contained in $B(\boldsymbol{w_2}, r_2)$.

Then, we know $B(\boldsymbol{w_2}, r_2) \cap B_1 \neq \emptyset$ and so there is a point $\boldsymbol{z_2} \in B(\boldsymbol{w_2}, r_2) \cap B_1$. Then the line segment $[\boldsymbol{w_2}, \boldsymbol{z_2}] \subset B(\boldsymbol{w_2}, r_2)$ and the line segment $[\boldsymbol{z_2}, \boldsymbol{x}]$ is in B_1. So we have a polygonal path from $\boldsymbol{x}$ to $\boldsymbol{y}$ inside $\mathscr{U}$.

We can do this argument in general, of course, but it is messy! Once we have a polygonal path, we can easily create a smooth one which does the job too. ∎

In $\Re^2$, we assume a 1-form ω has the form $pdx + qdy$ where p and q are C^∞ functions. We can also think of the 1-form as a vector field $\begin{bmatrix} p & q \end{bmatrix}$ or $\begin{bmatrix} p \\ q \end{bmatrix}$ depending on whether we want to use a row or column form for the field. A 2-form H can be written as $H = hdxdy$ or $H = hdydx$ where h is also a C^∞ function but this time of two variables. These are just formal definitions, of course, and we can make them more precise, but they suffice for now. If $\omega = pdx + qdy$, we define the 2-form obtained from the 1-form ω by $d\omega = (q_x - p_y)dxdy$. Again, this is discussed more carefully in (Peterson (101) 2020).

3.1 When Is a 1-Form Exact?

The big question is when does function f satisfy $\omega = df$? If this is true, we say ω is exact. We also say the 1-form ω is **closed** if $d\omega = 0$. Note if $\omega = df = f_x dx + f_y dy$, it is clear $d\omega = (\partial_x f_y - \partial_y f_x)dxdy = 0dxdy$ as long as the mixed partials are the same. We do know that as long as f has locally continuous partials up to second order this is, indeed, true. Thus, $\omega = df \Longrightarrow d\omega = 0$ or $d(df) = 0$.

There is no doubt this notation is very confusing. We have df is the 1-form $df = f_x dx + f_y dy$ and $d(df)$ is the 2-form $d(df) = (\partial_x f_y - \partial_y f_x)dxdy$ which is the zero 2-form as long as f is sufficiently smooth. Now, let $\mathscr{R} = [a, b] \times [c, d]$. Then $\partial\mathscr{R}$ consists of four line segments for which we will define four separate paths.

$$
\begin{aligned}
\gamma_1(t) &= \begin{bmatrix} t \\ c \end{bmatrix}, a \le t \le b, \quad \gamma_2(t) = \begin{bmatrix} b \\ t \end{bmatrix}, c \le t \le d \\
\gamma_3(t) &= \begin{bmatrix} t \\ d \end{bmatrix}, a \le t \le b, \quad \gamma_4(t) = \begin{bmatrix} a \\ t \end{bmatrix}, c \le t \le d
\end{aligned}
$$

You should draw a picture of this boundary and label the edges as follows: γ_1 traverses the boundary from the starting point (a, c) ending at (b, c). The curve γ_2 then moves from the point (b, c) to (b, d) along the boundary. Note, mathematically,

$$
-\gamma_3(t) = -\begin{bmatrix} t \\ d \end{bmatrix}, a \le t \le b
$$

which moves from the point $(-a, d)$ to the point $(-b, d)$ and these points are not on $\partial\mathscr{R}$. If we wanted to move backwards along the path given by γ_3, we could denote this by

$$\gamma_3^b(t) = \begin{bmatrix} (b+a) - t \\ d \end{bmatrix}, a \leq t \leq b$$

and this curve would move us backwards along the edge: (b, d) to (a, d) along the boundary edge. We will not use γ_3^b to denote this backwards movement. Instead, the curve going backwards will be denoted by $-\gamma_3$. Clearly, we do not mean the definition of the function $-\gamma_3$ here!

Hence, the simple and closed path around $\mathscr{R}$ that moves counterclockwise along $\partial\mathscr{R}$ will be denoted $\gamma_1 + \gamma_2 - \gamma_3 - \gamma_4$. It is also easy to see we could define the curves γ_i differently and still get the job done.

Now recall Green's Theorem in this context. If p and q are C^1 on $\mathscr{R}$, then

$$\begin{aligned} \iint\limits_{\mathscr{R}} (q_x - p_y)dxdy &= \int_{\gamma_1+\gamma_2-\gamma_3-\gamma_4} \begin{bmatrix} p \\ q \end{bmatrix} \begin{bmatrix} dx \\ dy \end{bmatrix} = \int_{\gamma_1} pdx + \int_{\gamma_2} qdy - \int_{\gamma_3} pdx - \int_{\gamma_4} qdy \\ &= \int_a^b p(x,c)dx + \int_c^d q(b,y)dy - \int_a^b p(x,d)dx - \int_c^d q(a,y)dy \end{aligned}$$

We usually let $\partial\mathscr{R} = \gamma_1 + \gamma_2 - \gamma_3 - \gamma_4$, so we state all of the above succinctly using line integrals as

$$\iint\limits_{\mathscr{R}} (q_x - p_y)dxdy = \int\limits_{\partial\mathscr{R}} (pdx + qdy)$$

Finally, use the 1-form $\omega = pdx + qdy$ with its associated 2-form $d\omega = (q_x - p_y)dxdy$, we can write

$$\iint\limits_{\mathscr{R}} d\omega = \int\limits_{\partial\mathscr{R}} \omega$$

Let $\mathscr{U} = int(\mathscr{R})$, the interior of a rectangle. We now prove if $d\omega = 0$ on $\mathscr{U}$, then $\int\limits_{\partial\mathscr{U}} \omega = 0$. This is the same as saying, if ω is closed, then $\int\limits_{\partial\mathscr{U}} \omega = 0$.

The argument is straightforward and uses Green's Theorem. We have $\mathscr{U} = (a, b) \times (c, d)$ but this time, we will allow the intervals (a, b) and (c, d) to be infinite. Let $\omega = pdx + qdy$. Pick $(x_0, y_0) \in \mathscr{U}$. Note this is an interior point of an open set so we can find (x, y) in $\mathscr{U}$, so that the rectangle $[x_0, x] \times [y_0, y] \subset \mathscr{U}$ with $x > x_0$ and $y + y_0$. Call this rectangle D. Choose the curves γ_i like before adjusting their endpoints to the new rectangle. Let $\gamma = \gamma_1 + \gamma_2$ which moves from the point (x_0, y_0) to (x, y) along the boundary of the rectangle ccw. Thus, $\gamma_1(t) = (x_0 + t, y_0)$ for $0 \leq t \leq x - x_0$ and $\gamma_2(t) = (x, y_0 + t)$ for $0 \leq t \leq y - y_0$. Also define $\hat{\gamma} = \gamma_4 + \gamma_3$ on the remaining two edges of the boundary of the rectangle. Then $\gamma - \hat{\gamma}$ is a ccw simple closed curve about the boundary of the rectangle and so by Green's Theorem, $0 = \iint\limits_{D} d\omega = \int\limits_{\partial D} \omega$.

Thus

$$0 = \int\limits_{\gamma-\hat{\gamma}} \omega \Longrightarrow \int\limits_{\gamma} \omega + \int\limits_{-\hat{\gamma}} \omega \Longrightarrow \int\limits_{\gamma} \omega = \int\limits_{\hat{\gamma}} \omega$$

Now define the function f by $f(x,y) = \int_\gamma \omega$. Then,

$$f(x,y) = \int_\gamma \omega = \int_0^{x-x_0} p(x_0+t, y_0)dt + \int_0^{y-y_0} q(x, y_0+t)dt$$

We have for any $x > x_0$

$$\begin{aligned} \lim_{x \to x_0} \frac{f(x,y) - f(x_0,y)}{x - x_0} &= \lim_{x \to x_0} \frac{1}{x - x_0}\Big(\int_0^{x-x_0} p(x_0+t, y_0)dt + \int_0^{y-y_0} q(x, y_0+t)dt \\ &\quad - \int_0^{y-y_0} q(x, y_0+t)dt \Big) \\ &= \lim_{x \to x_0} \frac{1}{x - x_0} \int_0^{x-x_0} p(x_0+t, y_0)dt = p(x_0, y_0) \end{aligned}$$

Now define the function g by $g(x,y) = \int_{\hat{\gamma}} \omega$. Then,

$$g(x,y) = \int_{\hat{\gamma}} \omega = \int_0^{y-y_0} q(x_0, y_0+t)dt + \int_0^{x-x_0} p(x_0+t, y)dt$$

Thus, we have for any $y > y_0$

$$\begin{aligned} \lim_{y \to y_0} \frac{f(x,y) - f(x,y_0)}{y - y_0} &= \lim_{y \to y_0} \frac{1}{y - y_0}\Big(\int_0^{y-y_0} q(x_0, y_0+t)dt + \int_0^{x-x_0} p(x_0+t, y)dt \\ &\quad - \int_0^{x-x_0} p(x_0+t, y)dt \Big) \\ &= \lim_{y \to y_0} \frac{1}{y - y_0} \int_0^{y-y_0} q(x_0, y_0+t)dt = q(x_0, y_0) \end{aligned}$$

Thus, we have $(f_x)^+(x_0,y_0) = p(x_0,y_0)$ and $(g_y)^+(x_0,y_0) = q(x_0,y_0)$. Of course, it is easy to repeat the argument for (x,y) position with $x < x_0$ and $y < y_0$. Hence, we see $f_x = p$ and $g_y = q$ at any (x_0,y_0). But we also know from Green's Theorem that $f(x,y) = g(x,y)$. We conclude $p = f_x$ and $q = f_y$ on $\mathscr{U}$ and the choice of the interior point (x_0,y_0) is arbitrary. We see that $df = \omega$.

Note the argument does not need this rectangle to be finite! So we know that on an open rectangle $\mathscr{U} \subset \Re^2$, if ω is a 1-form on $\mathscr{U}$ with $d\omega = 0$, then $\omega = df$. This also shows immediately, if $\omega = df$ on $\mathscr{U}$, then $0 = d\omega = d(df)$. Note the function we obtain is unique up to a constant.

Let's state this as a formal result.

Lemma 3.1.1 Closed 1-Form on an Open Rectangle is Exact

If ω is a closed 1-form (i.e. $d\omega = 0$) and $\mathscr{U}$ is an open rectangle, then ω is exact.

Proof 3.1.1
We have just gone through this argument. ■

We can also look at some equivalent conditions for exactness. First, let's define what we mean by a segmented path γ in $\mathscr{U}$ which is an open subset of $\Re^2$.

Definition 3.1.1 Segmented Paths

Let $\mathscr{U}$ be an open subset of $\Re^2$. A segmented smooth path is a finite set of smooth paths $\{\gamma_1, \ldots, \gamma_n\}$ for some positive integer n satisfying $\gamma_i : [a_i, b_i] \to \mathscr{U}$ and

$$\begin{aligned} \gamma_1(b_1) &= \gamma_2(a_2), \quad \gamma_2(b_2) = \gamma_3(a_3) \\ &\vdots \\ \gamma_i(b_i) &= \gamma_{i+1}(a_{i+1}), \\ &\vdots \end{aligned}$$

We write $\gamma = \gamma - 1 + \ldots + \gamma_n$. The initial point of γ is $\boldsymbol{P} = \gamma_1(a_1)$ and the final point of γ is $\gamma_n(b_n)$. If $\boldsymbol{P} = \boldsymbol{Q}$, we say γ is a closed path.

Now we can look at some equivalent conditions for exactness.

Theorem 3.1.2 Equivalent Conditions for Exactness

Let ω be 1-form on $\mathscr{U} \subset \Re^2$ where $\mathscr{U}$ is open. The following conditions are equivalent.

(i): $\int_\gamma \omega = \int_\delta \omega$ for all segmented paths γ and δ in $\mathscr{U}$ with the same start and end point.

(ii): $\int_\gamma \omega = 0$ for all segmented closed paths in $\mathscr{U}$.

(iii): ω is exact; i.e. there is a function f on $\mathscr{U}$ with $\omega = df$.

Proof 3.1.2
(iii) Implies (i): *If $\omega = df$ in $\mathscr{U}$, it is easy to see*

$$\int_\gamma \omega = \int_{\gamma_1} \omega + \ldots + \int_{\gamma_n} \omega = f(\gamma_1(b_1)) - (\, f(\gamma_1(a_1)) = f(\boldsymbol{P})\,)$$
$$+ f(\gamma_2(b_2)) - f(\gamma_2(a_2)) + \ldots + f(\gamma_n(b_n)) - f(\gamma_n(a_n)) = f(\boldsymbol{Q}) - f(\boldsymbol{P})$$

(ii) Implies (i): *If γ goes from $\boldsymbol{P}$ to $\boldsymbol{Q}$ and δ has the same endpoints, then $\gamma - \delta$ is a closed segmented path and hence $\int_{\gamma-\delta} \omega = 0$. This shows (ii) is satisfied.*

(i) Implies (ii):
If σ is a closed segmented path from $\boldsymbol{P}$ back to $\boldsymbol{P}$. Let η be the constant path $\eta(t) = \boldsymbol{P}$ for $0 \le t \le 1$. By assumption (i), we have $\int_\sigma \omega = \int_\eta \omega = 0$. This shows (ii) is satisfied.

(i) Implies (iii):
We can argue this for each connected component of $\mathscr{U}$, so it is enough to show our reasoning for a simple connected component. Equivalently, we may assume $\mathscr{U}$ is connected. By Theorem 3.0.1 we know $\mathscr{U}$ is pathwise connected too. Pick some point $\boldsymbol{P_0}$ in $\mathscr{U}$. Define the function $f : \mathscr{U} \to \Re$ by $f(\boldsymbol{P}) = \int_\gamma \omega$ where γ is any segmented path from $\boldsymbol{P_0}$ to $\boldsymbol{P}$. Here $\omega = Pdx + Vdy$ as usual.

Let $\boldsymbol{P} = (x, y)$ and consider $\frac{f(x+\Delta x, y) - f(x,y)}{\Delta x}$. Look at the path $\gamma + \sigma$ where we use γ to go from $\boldsymbol{P}_0$ to $\boldsymbol{P}$ and then use σ to go from $\boldsymbol{P}$ to $\boldsymbol{P} + \boldsymbol{\Delta P} = (x_\Delta x, y)$ defined by

$$\sigma(t) \quad = \quad (x + t, y), \; 0 \leq t \leq \Delta x$$

Since $\mathscr{U}$ is open we can choose Δx sufficiently small so the path σ stays in $\mathscr{U}$. Then

$$\frac{f(x + \Delta x, y) - f(x, y)}{\Delta x} \quad = \quad \frac{1}{\Delta x}\left(\int_{\gamma+\sigma} \omega - \int_{\gamma} \omega \right) = \frac{1}{\Delta x} \int_{\sigma} \omega = \frac{1}{\Delta x} \int_0^{\Delta x} P(x + t, y) dt$$

By the Mean Value Theorem, we have $P(x + t, y) = P(x^*, y)$ for some x^* in $(0, \Delta x)$. Hence,

$$\frac{1}{\Delta x} \int_0^{\Delta x} P(x + t, y) dt \quad = \quad P(x^*, y)$$

Thus, since P is smooth,

$$\lim_{\Delta x \to 0} \frac{1}{\Delta x} \int_0^{\Delta x} P(x + t, y) dt \quad = \quad \lim_{\Delta x \to 0} P(x^*, y) = P(x, y)$$

This shows $(f_x)^+(x, y) = P(x, y)$. A similar argument works for the case of $-\Delta x$ using $\sigma(t) = (x - t, y)$ for $0 \leq t \leq \Delta x$. We conclude $f_x = P$. We then argue in essentially the same way to show $f_y = Q$. This shows $\omega = df$. ■

Recall, we say the 1-form ω is **closed** if $d\omega = 0$ and the 1-form ω is **exact** if $\omega = df$. So we know by definition that if ω is exact on $\mathscr{U}$, then ω is **closed**. The discussion above gives a converse: if ω is closed on $\mathscr{U}$, then ω must be exact.

Homework

Exercise 3.1.1 *Given $\omega = ydx + xdy$, find $f(x, y)$ so that $\omega = df$. Note, this is the same as finding the general solution to $dy/dx = -y/x$.*

Exercise 3.1.2 *Given $\omega = ydx + xdy$, from the previous exercise, we know ω is exact. Let the segmented curve γ correspond to the rectangle centered at $(0, 0)$ traversed counterclockwise. Find by explicit integration, the value of line integral around γ.*

Exercise 3.1.3 *Given $\omega = (2x + 4x^3y^4)dx + 4x^4y^3dy$, find $f(x, y)$ so that $\omega = df$. Note, this is the same as finding the general solution to $dy/dx = -(1 + 2x^2y^4)/(2x^3y^3)$.*

Exercise 3.1.4 *Given $\omega = (2x+4x^3y^4)dx+4x^4y^3dy$, from the previous exercise, we know ω is exact. Let the segmented curve γ correspond to the rectangle centered at $(0, 0)$ traversed counterclockwise. Find by explicit integration, the value of line integral around γ.*

Exercise 3.1.5 *Given $\omega = (8xy^2 + 6y)dx + (8x^2y + 6x + 18y)dy$, find $f(x, y)$ so that $\omega = df$. Note, this is the same as finding the general solution to $dy/dx = -(8xy^2 + 6y)/(8x^2y + 6x + 18y)$.*

Exercise 3.1.6 *Given $\omega = (8xy^2 + 6y)dx + (8x^2y + 6x + 18y)dy$, from the previous exercise, we know ω is exact. Let the segmented curve γ correspond to the rectangle centered at $(0, 0)$ traversed counterclockwise. Find by explicit integration, the value of line integral around γ.*

3.2 Forms on More Complicated Sets

Now that we have done our warm-up exercises, let's do some new stuff.

Lemma 3.2.1 If $df = 0$ on a Connected Open Set, then f is Constant

If $\mathscr{U}$ is a connected open set and $df = 0$, then f is constant on $\mathscr{U}$.

Proof 3.2.1
If $\mathscr{U}$ is a connected open set and $df = 0$, then $0 = \underset{\gamma}{df}$ for any curve γ connecting (x_0, y_0) to (x, y). Since $\mathscr{U}$ is open, it is easy to find a rectangle in $\mathscr{U}$ doing this. For the simple γ_2 path we have used before, we find

$$0 = \int_0^1 f_y(x, (1-t)y_0 + ty)dt = f(x, y) - f(x, y_0)$$

Also, using the path γ_1, we have

$$0 = \int_0^1 f_x((1-t)x_0 + x, y_0)dt = f(x, y_0) - f(x_0, y_0)$$

Combining, $f(x, y) = f(x, y) = f(x_0, y_0)$. Since our choice of (x, y) was arbitrary, f is constant on $\mathscr{U}$. ■

Let's look at another closed implies exact result that is a bit more complicated.

Lemma 3.2.2 Closed on the Union of Open Sets with Connected Intersection Implies Exact

Let $\mathscr{U} = \mathscr{U}_1 \cup \mathscr{U}_2$ with $\mathscr{U}_1 \cap \mathscr{U}_2$ connected where $\mathscr{U}_i$ is an open rectangle in $\Re^2$. Then if the 1-form ω is closed in $\mathscr{U}$, ω is exact.

Proof 3.2.2
By our earlier discussions, we know there are two functions, f_1 and f_2 with f_i defined on $\mathscr{U}_i$ and $df_i = \omega$ on $\mathscr{U}_i$. Hence, $d(f_2 - f_1) = \omega - \omega = 0$ on $\mathscr{U}_1 \cap \mathscr{U}_2$. Hence, $f_2 = f_1$ on $\mathscr{U}_1 \cap \mathscr{U}_2$. Since $\mathscr{U}_1 \cap \mathscr{U}_2$ is connected, this tells us $f_2 - f_1 = c$ on $\mathscr{U}_1 \cap \mathscr{U}_2$ where c is some constant. Let $\hat{f}_2 = f_2 - c$. Then $\hat{f}_2 - f_1 = f_2 - f_1 - c = 0$ and so $\hat{f}_2 = f_1$ on $\mathscr{U}_1 \cap \mathscr{U}_2$. Define f by

$$f(x) = \begin{cases} f_1(x), & x \in \mathscr{U}_1 \\ \hat{f}_2(x), & x \in \mathscr{U}_2 \end{cases}$$

We see f is defined on $\mathscr{U}_1 \cup \mathscr{U}_2$ and $df = \omega$ on $\mathscr{U}_1 \cup \mathscr{U}_2$. Hence, ω is exact. ■

Another way of saying this is if $\mathscr{U}_1$ and $\mathscr{U}_2$ are open and $\mathscr{U}_1 \cap \mathscr{U}_2$ is connected, then if $\mathscr{U} = \mathscr{U}_1 \cup \mathscr{U}_2$ and ω is a 1-form on $\mathscr{U}$, then if $\omega|_{\mathscr{U}_1}$ and $\omega|_{\mathscr{U}_2}$ are exact, then ω is exact on $\mathscr{U}$.

Even if a 1-form is not exact on the open set $\mathscr{U}$, we can use local exactness to calculate integrals over paths.

Theorem 3.2.3 Integration of a Closed 1-Form over a Path

If ω is a closed 1-form on the open set $\mathscr{U}$ in $\Re^2$ and $\gamma : [a,b] \to \mathscr{U}$ is a smooth path, then there is a partition Π of $[a,b]$, $\{a = t_0, t_1, \ldots, t_{n-1}, t_n = b\}$ and a collection of open subsets $\{\mathscr{U}_1, \ldots, \mathscr{U}_n\}$ so that γ maps $[t_{i-1}, t_i] \to \mathscr{U}_i$ and $\omega_{\mathscr{U}_i} = df_i$ for functions f_i. Let $\boldsymbol{P_i} = \gamma(t_i)$. Then

$$\int_\gamma \omega = (f_1(\boldsymbol{P_1}) - f_1(\boldsymbol{P_0})) + \ldots + (f_n(\boldsymbol{P_n}) - f_n(\boldsymbol{P_{n-1}}))$$

Proof 3.2.3

For each point $\boldsymbol{P}$ in $\gamma([a,b])$, choose an open set $\mathscr{U}_{\boldsymbol{P}}$ on which the restriction of ω is exact. We know we can do this by our previous constructions. The collection of sets $(\mathscr{U}_{\boldsymbol{P}})$ is an open cover of the compact set $\gamma([a,b])$. By Lebesgue's Lemma, there is an $\epsilon > 0$ so that any subset W of diameter less than ϵ is contained in some $\mathscr{U}_{\boldsymbol{P}}$. For this ϵ, by the uniform continuity of γ, there is a $\delta > 0$ such that

$$|t - s| < \delta \implies |\gamma(t) - \gamma(s)| < \frac{\epsilon}{2}$$

Let Π be any partition of $[a,b]$ with $||\Pi|| < \frac{\delta}{2}$. Then

$$|t_{i+1} - t_i| < \frac{\delta}{2} < \delta \implies \sup_{t,s \in [t_i, t_{i+1}]} |\gamma(t) - \gamma(s)| \leq \frac{\epsilon}{2} < \epsilon$$

That is, the diameter of $\gamma([t_i, t_{i+1}]) < \epsilon$ and so there is a $\mathscr{U}_{\boldsymbol{P_i}}$ with $\gamma([t_i, t_{i+1}]) \subset \mathscr{U}_{\boldsymbol{P_i}}$. Call these sets $\mathscr{U}_i$ for convenience. There are a finite number of such subintervals. Call this number n.

On each open set $\mathscr{U}_i$, choose a function f_i so the restriction of ω to $\mathscr{U}_i$ is exact and $\omega_{\mathscr{U}_i} = df_i$. Let $\gamma_i = \gamma|_{[t_{i-1}, t_i]}$. Then,

$$\begin{aligned}
\int_\gamma \omega &= \int_{\gamma_1} \omega + \ldots + \int_{\gamma_n} \omega = \int_{\gamma_1} \omega_1 + \ldots + \int_{\gamma_n} \omega_n \\
&= (f_1(\gamma(t_1)) - f_1(\gamma(t_0))) + \ldots + (f_n(\gamma(t_n)) - f_n(\gamma(t_{n-1}))) \\
&= (f_1(\boldsymbol{P_1}) - f_1(\boldsymbol{P_0})) + \ldots + (f_n(\boldsymbol{P_n}) - f_n(\boldsymbol{P_{n-1}}))
\end{aligned}$$

■

This has some immediate consequences.

Theorem 3.2.4 Closed 1-Form whose Integral around a Circle is Zero Must be Exact

Let $\mathscr{U} = \Re^2 \setminus \boldsymbol{P}$ and let $r > 0$. Let $\gamma_{\boldsymbol{P},r}$ be a ccw circle of radius r about $\boldsymbol{P}$; i.e. $\gamma_{\boldsymbol{P},r} = \boldsymbol{P} + r(\cos(2\pi t), \sin(2\pi t))$, $0 \leq t \leq 1$. If ω is a closed 1-form on $\mathscr{U}$ with $\int_{\gamma_{\boldsymbol{P},r}} \omega = 0$, then ω is exact.

Proof 3.2.4

The point $\boldsymbol{P} = (x_0, y_0)$ can be used to divide the plane $\Re^2$ four ways.

$$\begin{aligned}
\mathscr{U}_R &= \{\boldsymbol{Q} = (u,v) | u > x_0\} = \textbf{Right Half Plane} \\
\mathscr{U}_L &= \{\boldsymbol{Q} = (u,v) | u < x_0\} = \textbf{Left Half Plane} \\
\mathscr{U}_T &= \{\boldsymbol{Q} = (u,v) | v > y_0\} = \textbf{Upper Half Plane}
\end{aligned}$$

$$\mathscr{U}_B = \{Q = (u,v)|v < y_0\} = \textbf{Lower Half Plane}$$

Let the partition we choose divide $[0,1]$ *as follows:* $\{[1/8,3/8],[3/8,5/8],[5/8,7/8],[7/8,1]\cup[1,1/8]\equiv[7/8,9/8]\}$ *where we use the periodicity of the* sin *and* cos *functions to identify the last interval. Note we have four curves:*

$$\begin{aligned}
\gamma_T &= \gamma_{P,r}|_{[1/8,3/8]}, \quad \gamma_T([1/8,3/8]) \subset \mathscr{U}_T, \quad P_0 = \gamma_{P,r}(1/8) \\
\gamma_L &= \gamma_{P,r}|_{[3/8,5/8]}, \quad \gamma_L([3/8,5/8]) \subset \mathscr{U}_L, \quad P_1 = \gamma_{P,r}(3/8) \\
\gamma_B &= \gamma_{P,r}|_{[5/8,7/8]}, \quad \gamma_B([5/8,7/8]) \subset \mathscr{U}_B, \quad P_2 = \gamma_{P,r}(5/8) \\
\gamma_R &= \gamma_{P,r}|_{[7/8,9/8]}, \quad \gamma_R([7/8,9/8]) \subset \mathscr{U}_R, \quad P_3 = \gamma_{P,r}(7/8)
\end{aligned}$$

By Lemma 3.1.1, there are functions f_R *for* $\mathscr{U}_R$*,* f_L *for* $\mathscr{U}_L$*,* f_T *for* $\mathscr{U}_T$ *and* f_B *for* $\mathscr{U}_B$ *so that* $df_R = \omega$ *on* $\mathscr{U}_R$*,* $df_L = \omega$ *on* $\mathscr{U}_L$*,* $df_T = \omega$ *on* $\mathscr{U}_T$ *and* $df_B = \omega$ *on* $\mathscr{U}_B$*. Further, these functions are unique up to a constant. We want all these functions to agree on the overlaps. Then*

$$\begin{aligned}
\text{On } \mathscr{U}_R \cap \mathscr{U}_T, \quad f_R(x) &= f_T(x) + C_{RT} \Longrightarrow C_{RT} = f_R(P_0) - f_T(P_0) \\
\text{On } \mathscr{U}_L \cap \mathscr{U}_T, \quad f_L(x) &= f_T(x) + C_{LT} \Longrightarrow C_{LT} = f_L(P_1) - f_T(P_1) \\
\text{On } \mathscr{U}_L \cap \mathscr{U}_B, \quad f_L(x) &= f_B(x) + C_{LB} \Longrightarrow C_{LB} = f_L(P_2) - f_B(P_2) \\
\text{On } \mathscr{U}_R \cap \mathscr{U}_B, f_R(x) &= f_B(x) + C_{RB} \Longrightarrow Longrightarrow C_{RB} = f_R(P_3) - f_B(P_3)
\end{aligned}$$

Now apply Theorem 3.2.3 to see

$$\begin{aligned}
0 &= \int_\gamma \omega = (f_R(P_0) - f_R(P_3)) + (f_T(P_1) - f_T(P_0)) \\
&\quad + (f_L(P_2) - f_L(P_1)) + (f_B(P_3) - f_B(P_2))
\end{aligned}$$

From our matching equations, we find

$$\begin{aligned}
& (f_T(P_0) + C_{RT}) - f_R(P_3) + (f_L(P_1) - C_{LT}) - f_T(P_0) \\
& + (f_B(P_2) + C_{LB}) - f_L(P_1) + (f_R(P_3) - C_{RB}) - f_B(P_2) \\
= & (C_{RT} - C_{LT}) + (C_{LB} - C_{RB})
\end{aligned}$$

So if we choose $C_{RT} = C_{LT}$ *and* $C_{LB} = C_{RB}$*, we satisfy the condition that the integral is zero. This gives us a choice of constants so that functions match on all overlaps. Thus, we can define* f *on* $\mathscr{U}$ *by*

$$f(x) = \begin{cases} f_R(x) + C_R, & x \in \mathscr{U}_R \\ f_L(x) + C_L, & x \in \mathscr{U}_L \\ f_B(x) + C_B, & x \in \mathscr{U}_B \\ f_T(x) + C_T, & x \in \mathscr{U}_T \end{cases}$$

and because of our choice of the constants, there is matching on all overlapping regions. This immediately implies $\omega = df$ *on* $\mathscr{U}$*.* ∎

Another useful result is:

Theorem 3.2.5 Closed 1-Form whose Integral around Two Disjoint Circles is Zero Must be Exact

Let $\mathscr{U} = \Re^2 \setminus \{\boldsymbol{P}, \boldsymbol{Q}\}$ and let $0 < r < \|\boldsymbol{P} - \boldsymbol{Q}\|$. Let $\gamma_{\boldsymbol{P},r}$ be a ccw circle of radius r about $\boldsymbol{P}$; i.e. $\gamma_{\boldsymbol{P},r} = \boldsymbol{P} + r(\cos(2\pi t), \sin(2\pi t))$, $0 \leq t \leq 1$ and let $\gamma_{\boldsymbol{Q},r}$ be a ccw circle of radius r about $\boldsymbol{Q}$; i.e. $\gamma_{\boldsymbol{Q},r} = \boldsymbol{Q} + r(\cos(2\pi t), \sin(2\pi t))$, $0 \leq t \leq 1$. If ω is a closed 1-form on $\mathscr{U}$ with $\int_{\gamma_{\boldsymbol{P},r}} \omega = 0$ and $\int_{\gamma_{\boldsymbol{Q},r}} \omega = 0$, then ω is exact.

Proof 3.2.5

To make our argument concrete, assume the point $\boldsymbol{Q} = (x_1, y_1)$ is below and to the left of $\boldsymbol{P} = (x_0, y_0)$ in the plane. The point $\boldsymbol{P} = (x_0, y_0)$ can be used to divide the plane $\Re^2$ as follows:

$$\begin{aligned} \mathscr{U}_R &= \{\boldsymbol{U} = (u,v) | u > x_0\} = \textbf{Right Half Plane for } \boldsymbol{P}, \textit{ choose } f_R \textit{ with } \omega = df_R \textit{ on } \mathscr{U}_R \\ \mathscr{U}_T &= \{\boldsymbol{U} = (u,v) | v > y_0\} = \textbf{Upper Half Plane} \textit{ for } \boldsymbol{P}, \textit{ choose } f_T \textit{ with } \omega = df_T \textit{ on } \mathscr{U}_T \end{aligned}$$

These functions are unique up to constants, so choose the constants so that $f_R = f_T$ on $\mathscr{U}_R \cap \mathscr{U}_T$. Then, let

$$\begin{aligned} \mathscr{U}_{LT} &= \{\boldsymbol{U} = (u,v) | u < x_0, v > y_1\} = \textbf{Left Half Plane for } \boldsymbol{P} \textbf{ Above } \boldsymbol{Q} \\ & \quad \textit{choose } f_{LT} \textit{ with } \omega = df_{LT} \textit{ on } \mathscr{U}_{LT} \end{aligned}$$

We want the constants chosen so that $f_{LT} = f_T$ on $\mathscr{U}_T \cap \mathscr{U}_{LT}$. Then let

$$\begin{aligned} \mathscr{U}_{BR} &= \{\boldsymbol{U} = (u,v) | u > x_1, v < y_0\} = \textbf{Lower Half Plane for } \boldsymbol{P} \textbf{ Right of } \boldsymbol{Q} \\ & \quad \textit{choose } f_{BR} \textit{ with } \omega = df_{BR} \textit{ on } \mathscr{U}_{BR} \end{aligned}$$

We must adjust the constants so $f_{BR} = f_{LT}$ on $\mathscr{U}_{LT} \cap \mathscr{U}_{BR}$. We also need $f_{BR} = f_R$ on $\mathscr{U}_R \cap \mathscr{U}_{BR}$. We use the assumption that $\int_{\gamma_{\boldsymbol{P},r}} \omega = 0$ in the same way we did in the previous argument so show we can find constants so that all the required functions match on the overlaps. Note all four open sets are involved because of our condition that $0 < r < \|\boldsymbol{P} - \boldsymbol{Q}\|$ which means the circle is in the interior of all four pieces for an appropriate partition of $[0, 1]$.

Then let

$$\begin{aligned} \mathscr{U}_L &= \{\boldsymbol{U} = (u,v) | u < x_1\} = \textbf{Left Half Plane for } \boldsymbol{Q} \\ & \quad \textit{choose } f_L \textit{ with } \omega = df_L \textit{ on } \mathscr{U}_L \end{aligned}$$

We must adjust constants so that $f_L = f_{LT}$ on $\mathscr{U}_{LT} \cap \mathscr{U}_L$. Finally, let

$$\begin{aligned} \mathscr{U}_B &= \{\boldsymbol{U} = (u,v) | v < y_1\} = \textbf{Lower Half Plane for } \boldsymbol{Q} \\ & \quad \textit{choose } f_B \textit{ with } \omega = df_B \textit{ on } \mathscr{U}_B \end{aligned}$$

We need $f_B = f_L$ on $\mathscr{U}_L \cap \mathscr{U}_B$ and $f_B = f_{BR}$ on $\mathscr{U}_{BR} \cap \mathscr{U}_B$. We use the assumption that $\int_{\gamma_{\boldsymbol{Q},r}} \omega = 0$ in the same way we did in the previous argument to show we can find constants so that all the required functions match on the overlaps. Note all four open sets are involved because of our condition that $0 < r < \|\boldsymbol{P} - \boldsymbol{Q}\|$ which means the circle is in the interior of all four pieces for an appropriate partition of $[0, 1]$.

We define the function f by

$$f(\boldsymbol{x} = \begin{cases} f_R(\boldsymbol{x}), & \boldsymbol{x} \in \mathscr{U}_R \\ f_T(\boldsymbol{x}), & \boldsymbol{x} \in \mathscr{U}_T \\ f_{LT}(\boldsymbol{x}), & \boldsymbol{x} \in \mathscr{U}_{LT} \\ f_{BR}(\boldsymbol{x}), & \boldsymbol{x} \in \mathscr{U}_{BR} \\ f_B(\boldsymbol{x}), & \boldsymbol{x} \in \mathscr{U}_B \\ f_L(\boldsymbol{x}), & \boldsymbol{x} \in \mathscr{U}_L \end{cases}$$

These functions agree on all overlaps and we see $\omega = df$ *on* $\Re^2 \setminus \{\boldsymbol{P}, \boldsymbol{Q}\}$. ■

Homework

Exercise 3.2.1 *In the proof of Theorem 3.2.4, we used the partition of* $[0,1]$ *given by*

$$\pi = \{[1/8, 3/8], [3/8, 5/8], [5/8, 7/8], [7/8, 1] \cup [1, 1/8] \equiv [7/8, 9/8]\}.$$

Redo the arguments we used using the partition

$$\pi = \{[1/9, 3/9], [3/9, 5/9], [5/9, 7/9], [7/9, 1] \cup [1, 1/9] \equiv [7/9, 10/9]\}.$$

Exercise 3.2.2 *Work out the details of the constants choices in the proof of Theorem 3.2.5.*

Exercise 3.2.3 *Prove Theorem 3.2.5 for the case that* $\boldsymbol{Q}$ *is below and to the right of* $\boldsymbol{P}$.

Exercise 3.2.4 *Prove Theorem 3.2.5 for the case that* $\boldsymbol{Q}$ *and* $\boldsymbol{P}$ *are on the same horizontal line.*

Exercise 3.2.5 *Prove Theorem 3.2.5 for the case that* $\boldsymbol{Q}$ *and* $\boldsymbol{P}$ *are on the same vertical line.*

Exercise 3.2.6 *Prove a version of Theorem 3.2.5 for three points* $\boldsymbol{P}$, $\boldsymbol{Q}$ *and* $\boldsymbol{R}$.

Exercise 3.2.7 *Prove a version of Theorem 3.2.5 for four points.*

Exercise 3.2.8 *Prove a version of Theorem 3.2.5 for* N *points.*

3.3 Angle Functions and Winding Numbers

Let γ be a smooth path in $\Re^2 \setminus \boldsymbol{0}$ where by smooth we mean the path is at least C^1. So $\gamma(t) = (x(t), y(t))$ for $t \in [a,b]$. Let's try to define this path in terms of polar coordinates; i.e. we want $\gamma(t) = (r(t)\cos(\theta(t)), r(t)\sin(\theta(t)))$ for $t \in [a,b]$. It is easy to define the function r: we set

$$r(t) = \sqrt{(x(t))^2 + (y(t))^2}$$

It is a bit harder to decide how to define the angle function. Start by choosing a starting angle $\theta(a) = \theta_a$ so that

$$\gamma(a) = (r(a)\cos(\theta_a), r(a)\sin(\theta_a))$$

We know that $f(x,y) = \tan^{-1}(\frac{y}{x})$ defines the 1-form

$$df = \frac{-y dx + x dy}{x^2 + y^2}$$

which suggests we define θ by

$$\begin{aligned}\theta'(t) &= \frac{-y(t)x'(t)+x(t)y'(t)}{(x(t))^2+(y(t))^2} = \frac{-y(t)x'(t)+x(t)y'(t)}{r^2(t)} \\ \theta(a) &= \theta_a\end{aligned}$$

This implies θ is the unique function

$$\theta(t) = \theta_a + \int_a^t \frac{-y(s)x'(s)+x(s)y'(s)}{r^2(s)}\,ds$$

It is straightforward to show θ and r are C^∞ on $\Re^2 \setminus \mathbf{0}$. Hence, we have determined γ in terms of θ and r.

Define the total signed change in the angle of the path γ to be $\theta(b) - \theta(a) = \theta(b) - \theta_a$. It is clear this is independent of the choice of θ_a. Thus,

$$\Delta\theta = \int_a^b \frac{-y(s)x'(s)+x(s)y'(s)}{r^2(s)}\,ds = \int_\gamma \frac{-ydx+xdy}{x^2+y^2} = \int_\gamma \omega_\delta$$

where ω_δ is the 1-form defined by df. Since γ is in $\Re^2 \setminus \mathbf{0}$, we can use this to count the number of times γ goes around the origin $\mathbf{0} = (0,0)$. We will count counterclockwise or ccw as positive and clockwise or cw as negative. We define the **winding number** to be $\frac{\Delta\theta}{2\pi}$ for closed paths γ. Call this $W(\gamma, \mathbf{0})$. If γ is a closed path, then γ starts at the point $\boldsymbol{P} = (x(a), y(a))$ and ends at the point $\boldsymbol{Q} = (x(b), y(b))$ with $\boldsymbol{P} = \boldsymbol{Q}$. Choose θ_P to be the start angle at $\boldsymbol{P}$. If the path γ is not smooth, we will just write γ as a finite sum of smooth paths. We will do these details later, of course. For a closed path $\gamma : [a,b] \to \Re^2 \setminus \mathbf{0}$, we suspect $\frac{\Delta\theta}{2\pi}$ is an integer. To see this, note

$$\theta_P + 2\pi W(\gamma, \mathbf{0}) = \theta_P + \int_\gamma \omega_\delta = \theta_P + \int_a^b \frac{-y(s)x'(s)+x(s)y'(s)}{(x(s))^2+(y(s))^2}\,ds = \theta(b)$$

So, if $\boldsymbol{P} = \boldsymbol{Q}$, $\theta_P = \theta(b)$ modulo 2π and we have $2\pi W(\gamma, \mathbf{0})$ is an integer multiple of 2π. Hence, the winding number is an integer.

Homework

Exercise 3.3.1 *Prove θ and r are infinitely differentiable on $\Re^2 \setminus \mathbf{0}$.*

Exercise 3.3.2 *How would you define the winding number for an arbitrary point $\boldsymbol{P}$ not $\mathbf{0}$?*

Exercise 3.3.3 *Is ω_δ closed? Is it exact?*

3.4 A More General Definition of Winding Number

Pick $\boldsymbol{P}$ in $\Re^2$. Consider any sector whose vertex is at $\boldsymbol{P}$. A sector is the region in the plane which is rooted at $\boldsymbol{P} = (x_P, y_P)$ defined by several parameters. This is a pie-shaped slice whose vertex is at (x_P, y_P) and it is the set of points bounded between the lines $y_1 = y_p + m_1(x - x_p)$ and $y_2 = y_p + m_2(x - x_p)$ for $x \geq x_p$. The slopes of these lines are the tangents of angles θ_1 and θ_2. We assume these angles are measured in $[0, 2\pi)$, $m_1 < m_2$ and we take the usual care in defining the angles depending of the quadrant the sector lives in. Hence, in Cartesian coordinates

$$S(\boldsymbol{P}, m_1, m_2) = \{(x,y) | y_p + m_1(x - x_p) \leq y \leq y_p + m_2(x - x_p),\ x \geq x_p\}$$

We can also describe the sector in polar coordinates where we need to remember to center our polar coordinate system at $\boldsymbol{P}$. Then

$$S(\boldsymbol{P}, \theta_1, \theta_2) = \{\boldsymbol{P} + (r, \theta) | \theta_1 \leq \theta \leq \theta_2,\ r \geq 0\}$$

where, again, remember $r(t) = \sqrt{(x - x_P)^2 + (y - y_P)^2}$. Hence, we can consider the map

$$p(r, \theta) = \boldsymbol{P} + (r\cos(\theta), r\sin(\theta))$$

Let $\gamma : [a, b] \to \Re^2 \setminus \boldsymbol{P}$ be a smooth path. Define the **winding number** here to be $W(\gamma, \boldsymbol{P})$ in the following way.

Since $\gamma : [a, b] \to \Re^2 \setminus \boldsymbol{P}$ is continuous, the image $\gamma([a, b])$ is compact. Hence, there is a rectangle $\mathscr{R} = [a_1, b_1] \times [a_2, b_2] \subset \Re^2 \setminus \boldsymbol{P}$ with $\gamma([a, b]) \subset \mathscr{R}$. Given any $(u, v) \in \gamma([a, b])$, pick a sector that contains (u, v). Call the interior of these sectors $\mathscr{U}(u, v)$. Then the collection $\{\mathscr{U}(u, v)\}$ is an open cover of $\gamma([a, b])$. By Lebesgue's Lemma, there is an $\epsilon > 0$ so that any subset W of diameter less than ϵ is contained in some $\mathscr{U}(u, v)$. For this ϵ, by the uniform continuity of γ, there is a $\delta > 0$ such that

$$|t - s| < \delta \implies |\gamma(t) - \gamma(s)| < \frac{\epsilon}{2}$$

Let $\mathbb{P}$ be any partition of $[a, b]$ with $||\mathbb{P}|| < \frac{\delta}{2}$. Then

$$|t_{i+1} - t_i| < \frac{\delta}{2} < \delta \implies \sup_{t,s \in [t_i, t_{i+1}]} |\gamma(t) - \gamma(s)| \leq \frac{\epsilon}{2} < \epsilon$$

That is, the diameter of $\gamma([t_i, t_{i+1}]) < \epsilon$ and so there is a $\mathscr{U}(u_i, v_i)$ with $\gamma([t_i, t_{i+1}]) \subset \mathscr{U}(u_i, v_i)$. Call the sets $\mathscr{U}(u_i, v_i)$, $\mathscr{U}_i$ for convenience. There are a finite number of such intervals. Call this number n.

Define an angle function θ_i on each $\mathscr{U}_i$ as usual. Let $\boldsymbol{P_i} = \gamma(t_i)$ for $0 \leq i \leq n$. Then we can define the winding number to be

$$\begin{aligned} W(\gamma, \boldsymbol{P}) &= \frac{1}{2\pi}\{\theta_1(\boldsymbol{P_1}) - \theta_1(\boldsymbol{P_0}) + \theta_2(\boldsymbol{P_2}) - \theta_2(\boldsymbol{P_1}) + \ldots + \theta_n(\boldsymbol{P_n}) - \theta_n(\boldsymbol{P_{n-1}})\} \\ &= \frac{1}{2\pi}\sum_{i=1}^{n}\{\theta_i(\boldsymbol{P_i}) - \theta_i(\boldsymbol{P_{i-1}})\} \end{aligned}$$

We can then prove the following result:

Theorem 3.4.1 Winding number $W(\gamma, \boldsymbol{P})$ is Well-defined

(i): The definition of $W(\gamma, \boldsymbol{P})$ is independent of the choice of the sectors $\mathscr{U}_i$ and the angle functions θ_i.

(ii): The value of $W(\gamma, \boldsymbol{P})$ is independent of the choice of partition $\mathbb{P}$.

(iii): If γ is a closed path, i.e. $\gamma(a) = \gamma(b)$, then $W(\gamma, \boldsymbol{P})$ is an integer.

Proof 3.4.1
Case (i): *What is an angle function on $\mathscr{U}_i$? If $\boldsymbol{Q} = (x_Q, y_Q)$ is in $\mathscr{U}_i$, we would define an angle function by*

$$\theta_i(\boldsymbol{Q}) = \theta_i^{\min} + \tan^{-1}(y_Q/x_Q)$$

On a different choice $\mathscr{U}_i'$ we would have

$$\theta_i'(\boldsymbol{Q}) = (\theta_i^{\min})' + \tan^{-1}(y_Q/x_Q)$$

Hence on $\mathscr{U}_i \cap \mathscr{U}_i'$, we find

$$\theta_i'(\boldsymbol{Q}) - \theta_i(\boldsymbol{Q}) = (\theta_i^{\min})' - \theta_i^{\min} \equiv C_i$$

Since $\gamma([t_i, t_{i+1}]) \subset \mathscr{U}_i \cap \mathscr{U}_i'$, we have for any point in the intersection

$$\begin{aligned} \frac{1}{2\pi}\sum_{i=1}^{n}\{\theta_i'(\boldsymbol{P_i}) - \theta_i'(\boldsymbol{P_{i-1}})\} &= \frac{1}{2\pi}\sum_{i=1}^{n}\{(\theta_i(\boldsymbol{P_i}) + C_i) - (\theta_i'(\boldsymbol{P_{i-1}}) + C_i)\} \\ &= \frac{1}{2\pi}\sum_{i=1}^{n}\{\theta_i(\boldsymbol{P_i}) - \theta_i(\boldsymbol{P_{i-1}})\} = W(\gamma, \boldsymbol{P}) \end{aligned}$$

So altering the angle function on the sectors does not change the result.

Case (ii): *What if the partition changed? We will show this does not change the winding number using the argument we used in proving results in Riemann integration theory.*

Add One Point to One Subinterval: *Suppose the new partition $\mathbb{P}'$ adds one point to one of the subintervals $[t_i, t_{i+1}]$. Call this point t^*. Then $[t_i, t_{i+1}]$ is a subinterval determined by $\mathbb{P}$ and the addition of the point leads to subintervals $[t_i, t^*]$ and $[t^*, t_{i+1}]$ determined by $\mathbb{P}'$. Having chosen $\mathscr{U}_i$ and θ_i for $[t_i, t_{i+1}]$, make the same choice on the two subintervals $[t_i, t^*]$ and $[t^*, t_{i+1}]$. Let $\boldsymbol{P}^* = \gamma(t^*)$. Then*

$$(\theta_i(\boldsymbol{P_i}) - \theta_i(\boldsymbol{P}^*)) + (\theta_i(\boldsymbol{P}^*) - \theta_i(\boldsymbol{P_{i-1}})) = \theta_i(\boldsymbol{P_i}) - \theta_i(\boldsymbol{P_{i-1}})$$

and so the sum does not change.

Add Multiple Points to One Subinterval: *This is an induction argument. The base case is the case above and the argument is similar when we add a new point to a subinterval which has already had k points added.*

Add Multiple Points to Multiple Subintervals: *This is another induction. The first two cases are the base case and adding another subinterval with multiple points is argued in the same way. It is also clear the value of the winding number does not change if we use the common refinement of two partitions for the computation.*

Case (iii): *It is clear $2\pi W(\gamma, \boldsymbol{P})$ is the change in angle from $\gamma(a)$ to $\gamma(b)$. So if $\gamma(a) = \gamma(b)$, this is the change in angle from $\gamma(a)$ back to itself. This is an integer multiple of 2π.* ■

Let ω be a 1-form for an open set in $\Re^2$ and let $\gamma : [a, b] \to \mathscr{U}$ be a smooth path. Assume $\phi : [a', b'] \to [a, b]$ is a C^∞ mapping. The path $\gamma \circ \phi$ is called a **reparameterization** of γ.

Theorem 3.4.2 Reparameterization does not change the integral of the 1-form ω over γ

If $\gamma \circ \phi$ is a reparameterization of γ, $\int_{\gamma \circ \phi} \omega = \int_{\gamma} \omega$

Proof 3.4.2
Let $\omega = pdx + qdy$ and $\gamma(t) = (x(t), y(t))$. Then

$$\begin{aligned}\int_{\gamma \circ \phi} \omega &= \int_{a'}^{b'} \{p(x(\phi(s)), y(\phi(s)))(x \circ \phi)'(s) + q(x(\phi(s)), y(\phi(s)))(y \circ \phi)'(s)\} ds \\ &= \int_{a'}^{b'} \{p(x(\phi(s)), y(\phi(s)))x'(\phi(s)) + q(x(\phi(s)), y(\phi(s)))y'(\phi(s))\} \phi'(s) ds\end{aligned}$$

Now let $u = \phi(s)$. This change of variables leads to

$$\int_{\gamma \circ \phi} \omega = \int_a^b \{p(x(u), y(u))x'(u) + q(x(u), y(u))y'(u)\} du = \int_{\gamma} \omega$$

■

Homework

Exercise 3.4.1 *In the proof of Theorem 3.4.1, go through all the details of the induction argument needed for* **Add Multiple Points to One Subinterval.**

Exercise 3.4.2 *In the proof of Theorem 3.4.1, go through all the details of the induction argument needed for* **Add Multiple Points to Multiple Subintervals..**

Exercise 3.4.3 *Given $\omega = ydx + x^2dy$, let the segmented curve γ_1 correspond to the rectangle centered at $(0, 0)$ traversed counterclockwise. Choose another parameterization of this rectangle and call the resulting segmented curve γ_2. Find, by explicit integration, the value of line integral around γ_1 and γ_2. Should they match?*

Exercise 3.4.4 *Given $\omega = y^2dx + x^3dy$, let the segmented curve γ_1 correspond to the rectangle centered at $(0, 0)$ traversed counterclockwise. Choose another parameterization of this rectangle and call the resulting segmented curve γ_2. Find, by explicit integration, the value of line integral around γ_1 and γ_2. Should they match?*

3.5 Homotopies

Consider a family of paths $\gamma_s : [a, b] \rightarrow \mathscr{U}$, where $\mathscr{U}$ is an open set in $\Re^2$, for $0 \leq s \leq 1$ and $a \leq t \leq b$. Assume the family $\{\gamma_s\}$ is smooth for all s and t. Another way to say this is that we have a smooth mapping $H : [a, b] \times [0, 1] \rightarrow \mathscr{U}$. Assume $H(a, s) = \boldsymbol{P}$ and $H(b, s) = \boldsymbol{Q}$ for all s. Then we have $\gamma_s = H(t, s)$ with each γ_s a smooth path from $\boldsymbol{P}$ to $\boldsymbol{Q}$. We call H a **smooth homotopy** for the path γ_0 to the path γ_1.

We say two paths from $[a, b]$ to $\mathscr{U}$ with the same start and end points are **smoothly homotopic** in $\mathscr{U}$ if there is such a homotopy.

Theorem 3.5.1 Integral of a Closed 1-form is the Same on Smoothly Connected Homotopic Paths

If γ and δ are smoothly homotopic paths from $\boldsymbol{P}$ to $\boldsymbol{Q}$ into $\mathscr{U}$ and ω is a closed 1-form on $\mathscr{U}$, then $\int_\gamma \omega = \int_\delta \omega$.

Proof 3.5.1

Let $\mathscr{V}$ be a neighborhood of $\mathscr{R} = [a,b] \times [0,1]$. $H : \mathscr{R} \to \mathscr{U}$ and so we can extend H to a given open set $\mathscr{V}$ in a smooth way like usual. The exercises let you explore this in more detail. To refresh your mind, to extend H from $[a,b] \times [0,1]$ to the enclosing open set $\mathscr{V}$ in a continuous way, we could do this:

- *(i): Since H is continuous, the range $H(\mathscr{R})$ is compact and so we can denote it by $H(\mathscr{R}) = [\alpha_1, \beta_1] \times [\alpha_2, \beta_2] \equiv \mathscr{W}$.*
- *(ii:) Since the range is contained in the open set $\mathscr{U}$, we can find, for some $\Delta > 0$, a bounding box of the form $\mathscr{W}_\Delta = [\alpha_1 - \Delta, \beta_1 + \Delta] \times [\alpha_2 - \Delta, \beta_2 + \Delta]$ which is also contained in $\mathscr{U}$. The interior of $\mathscr{W}_\Delta$ is then the open set $\mathscr{V}$ we will use. By the continuity of H, we can find a $\delta > 0$ giving $\mathscr{R}_\delta = [a-\delta, b+\delta] \times [-\delta, 1+\delta]$ with $H(\mathscr{R}_\delta)$ contained in the interior of $\mathscr{W}_\Delta$.*
- *(iii:) Consider one of the paths $\gamma_t(s)$. We have $\gamma_t(0) = \boldsymbol{P}$ and $\gamma_t(1) = \boldsymbol{Q}$. Let u_1 be a C^∞ bump function whose support is $[-\epsilon_1, 0]$ which is $\boldsymbol{P}$ at the endpoints with a maximum or minimum value $\boldsymbol{P} + \boldsymbol{r}$ where ϵ_1 and r are chosen so that the range of u_1 stays inside $\mathscr{W}_\Delta$. Further, let v_1 be a C^∞ bump function whose support is $[1, 1+\epsilon_1]$ which is $\boldsymbol{Q}$ at the endpoints with a maximum or minimum value $\boldsymbol{Q} + \boldsymbol{r}$ where ϵ_1 and r are chosen so that the range of v_1 stays inside $\mathscr{W}_\Delta$. Define the smooth extension $\tilde{\gamma}_t$ on $[-\epsilon_1, 1+\epsilon_1]$ by*

$$\tilde{\gamma}_t(s) = \begin{cases} u(s), & -\epsilon_1 \le s \le 0 \\ \gamma_t(s), & 0 \le s \le 1 \\ v(s), & 1 \le s \le 1+\epsilon_1 \end{cases}$$

We can do this for all $a \le t \le b$. Next, extend to the interval $[a-\epsilon_2, b+\epsilon_2]$ as follows:

$$\hat{\gamma}_t(s) = \begin{cases} u_2(t) + \tilde{\gamma}_t(s), & -\epsilon_1 \le s \le 1+\epsilon_1, a - \epsilon - 2 \le t \le a \\ \tilde{\gamma}_t(s), & 0 \le s \le 1, a \le t \le b \\ v_2(t) + \tilde{\gamma}_t(s), & -\epsilon_1 \le s \le 1+\epsilon_1, b \le t \le b+\epsilon_2 \end{cases}$$

where u_2 and v_2 are C^∞ bump functions defined just like u_1 and v_1 except we use the parameter ϵ_2 instead of ϵ_1. We see $\tilde{H}(t,s) = \hat{\gamma}_t(s)$ is a smooth extension of H with the properties we need.

So given $(x,y) \in \mathscr{V}$, there is a $(t,s) \in \mathscr{V}$ so that $x = f(t,s)$ and $y = g(t,s)$. We typically are a bit sloppy with the notation here and we call f and g the coordinate functions for $\mathscr{V}$ and denote them by $x(t,s)$ and $y(t,s)$. Now, if $(t,s) \in \mathscr{R}$, $\gamma_s(t) = (x(t,s), y(t,s))$ and the value of x and y off $\mathscr{R}$ comes from the C^∞ extension. Now look up the pullback of a 1-form in the context of line integrals in (Peterson (101) 2020). Let our 1-form ω be given by $pdx + qdy$ on (V). The pullback of the 1-form ω, ω^*, is a way to express the 1-form ω defined on $\mathscr{U}$ on the set $\mathscr{R}$ by pulling values from the image side $\mathscr{U}$ back to the domain side. Here, it is defined by

$$\begin{aligned} \omega^* &= \left(p(x(t,s), y(t,s))\frac{\partial x}{\partial s} + q(x(t,s), y(t,s))\frac{\partial y}{\partial s}\right) ds \\ &\quad + \left(p(x(t,s), y(t,s))\frac{\partial x}{\partial t} + q(x(t,s), y(t,s))\frac{\partial y}{\partial t}\right) dt \\ &= uds + vdt \end{aligned}$$

Recall if a 1-form is $F = uds + vdt$, then $dF = (v_s - u_t)dxdy$. Here

$$\begin{aligned} d\omega^* &= \frac{\partial}{\partial s}\left\{\left(p(x,y)\frac{\partial x}{\partial t} + q(x,y)\frac{\partial y}{\partial t}\right) - \frac{\partial}{\partial t}\left(p(x,y)\frac{\partial x}{\partial s} + q(x,y)\frac{\partial y}{\partial s}\right)\right\} \\ &= \left(p_x x_s x_t + p_y y_s x_t + p x_{ts} + q_x x_s y_t + q_y y_s y_t + q y_{st}\right) \\ &\quad -\left(p_x x_t x_s + p_y y_t x_s + p x_{ts} + q_x x_t y_s + q_y y_t y_s + q y_{ts}\right) \\ &= p_y(y_s x_t - y_t x_s) + q_x(x_s y_t - x_t y_s) = (p_y - q_x)(y_s x_t - y_t x_s) \end{aligned}$$

But we have assumed ω is closed and so $p_y = q_x$. We conclude $d\omega^* = 0$.

Consider the line integral of ω^* ccw around $\mathscr{R}$.

1. **Bottom** $[a,b] \times \{0\}$:

$$\int_{\textbf{Bottom}} \omega^* = \int_a^b (p(x(t,0),y(t,0))(x(t,0))' + q(x(t,0),y(t,0))(y(t,0))')dt$$

But $(x(t,0),y(t,0)) = \boldsymbol{P}$ always, so $(x(t,0))' = (y(t,0))' = 0$. Hence, we have $\int_{\textbf{Bottom}} \omega^* = 0$.

2. **Right side** $\{b\} \times [0,1]$:
Here, we have

$$\int_{\textbf{Right side}} \omega^* = \int_a^b (p(x(b,s),y(b,s))(x(b,s))'(s) + q(x(b,s),y(b,s))(y(b,s))'(s))ds$$

But $(x(b,s),y(b,s))$ is a parameterization of the right edge which is given by $\gamma_b(s)$. Hence, we have

$$\int_{\textbf{Right side}} \omega^* = \int_{\gamma_b} \omega$$

3. **Top** $[a,b] \times \{1\}$:
A calculation similar to the one we did for the bottom edge shows $\int_{\textbf{Top}} \omega^* = 0$.

4. **Left side** $\{b\} \times [0,1]$:
A calculation similar to the one we used on the right side shows

$$\int_{\textbf{Light side}} \omega^* = \int_{\gamma_a} \omega$$

Note, $\gamma_a = \gamma$ and $\gamma_b = \delta$. From Green's Theorem, we then have

$$0 = \int_{\gamma_b} \omega - \int_{\gamma_a} \omega \Longrightarrow \int_{\delta} \omega = \int_{\gamma} \omega$$

■

Here is an alternate proof using Lebesgue's Lemma. We use some of the notations and ideas of the proof above but now we use local exactness of the 1-form.

Proof

Each point $\boldsymbol{a}$ in $H(\mathscr{R})$ has an open rectangle $\mathscr{S}_{\boldsymbol{a}}$ on which ω is **exact** *since ω is exact on $\mathscr{U}$. Since $H(\mathscr{R})$ is compact, the collection $\{\mathscr{S}_{\boldsymbol{a}}\}$ of these open rectangles gives an open cover of $H(\mathscr{R})$. By Lebesgue's Lemma, there is an $\epsilon > 0$, so that any subset $\mathscr{W}$ with diameter less than ϵ is contained in one of these open rectangles. For our given ϵ, by the uniform continuity of H, there is a $\delta > 0$ with*

$$\|(t,s) - (t',s')\| < \delta \implies \|H(t,s) - H(t',s')\| < \frac{\epsilon}{2}$$

Choose a partition Π of $[a,b] \times [0,1]$ so that $\|\Pi\| < \frac{\delta}{2}$. Then for each rectangle $\mathbb{R}_{ij} = [t_i, t_{i+1}] \times [s_j, s_{j+1}]$ determined by Π, there is an open set $\mathscr{S}_{ij}$ so that $H(\mathbb{R}_{ij}) \subset \mathscr{S}_{ij}$. On $\mathscr{S}_{ij}$, ω is exact. Let $\omega_{ij} = \omega|_{\mathscr{S}_{ij}}$. Then there is a smooth function f_{ij} so that $\omega_{ij} = df_{ij} = (f_{ij})_x dx + (f_{ij})_y dy$. Then, using the same pullback idea as before, since $\omega = pdx + qdy$, we have

$$\begin{aligned}
\int_{\partial\mathscr{S}_{ij}} \omega^* &= \int_{\partial\mathscr{S}_{ij}} \Big\{ \Big(p(x(t,s),y(t,s))x_s + q(x(t,s),y(t,s))y_s \Big) ds \\
&\quad + \Big(p(x(t,s),y(t,s))x_t + q(x(t,s),y(t,s))y_t \Big) dt \Big\} \\
&= \int_{\partial\mathscr{S}_{ij}} \Big\{ \Big((f_{ij})_x(x(t,s),y(t,s))x_s + (f_{ij})_y(x(t,s),y(t,s))y_s \Big) ds \\
&\quad + \Big((f_{ij})_x(x(t,s),y(t,s))x_t + (f_{ij})_x(x(t,s),y(t,s))y_t \Big) dt \\
&= \int_{\partial\mathscr{S}_{ij}} (f_{ij})_t dt + (f_{ij})_s ds = \int_{\partial\mathscr{S}_{ij}} df_{ij} = 0
\end{aligned}$$

Also, using the same sort of annotations as in the first proof but for the rectangles $\mathscr{S}_{ij}$, we have

$$\int_{\partial\mathscr{S}_{ij}} = \int_{\textbf{Bottom}} + \int_{\textbf{Right side}} - \int_{\textbf{Top}} - \int_{\textbf{Left side}}$$

Let's look at a simple case. Suppose you had the partition Π give these subrectangles: $[a,b] = [a = t_0, t_1] \cup [t_1, t_2 = b]$ and $[0,1] = [0 = s_0, s_1] \cup [s_1, s_2] \cup [s_2, s_3 = 1]$. This gives a total of 6 rectangles to consider in our integrations. We have $\mathscr{S}_{11} = [t_0, t_1] \times [s_0, s_1]$, $\mathscr{S}_{12} = [t_0, t_1] \times [s_1, s_2]$, $\mathscr{S}_{13} = [t_0, t_1] \times [s_2, s_3]$, $\mathscr{S}_{21} = [t_0, t_1] \times [s_0, s_1]$, $\mathscr{S}_{22} = [t_0, t_1] \times [s_1, s_2]$ and $\mathscr{S}_{23} = [t_0, t_1] \times [s_2, s_3]$. Each of these rectangles has a bottom, right side, top and left side we will label B_{ij}, R_{ij}, T_{ij} and L_{ij}. But these rectangles share sides. So we have the rectangles

$$\begin{bmatrix}
\begin{bmatrix} -L_{13} & -T_{13} \\ B_{13} & R_{13} \end{bmatrix} & \begin{bmatrix} -L_{23} & -T_{23} \\ B_{23} & R_{23} \end{bmatrix} \\
\begin{bmatrix} -L_{12} & -T_{12} \\ B_{12} & R_{12} \end{bmatrix} & \begin{bmatrix} -L_{22} & -T_{22} \\ B_{22} & R_{22} \end{bmatrix} \\
\begin{bmatrix} -L_{11} & -T_{11} \\ B_{11} & R_{11} \end{bmatrix} & \begin{bmatrix} -L_{21} & -T_{21} \\ B_{21} & R_{21} \end{bmatrix}
\end{bmatrix}$$

We interpret integration over these rectangles as

$$\int_{\partial \mathscr{S}_{ij}} = \int_{\textbf{Bottom}} + \int_{\textbf{Right side}} - \int_{\textbf{Top}} - \int_{\textbf{Left side}} = \int_{B_{ij}} + \int_{R_{ij}} - \int_{T_{ij}} - \int_{L_{ij}}$$

But, we have paths going in both directions on the interior paths which cancel out. Hence, for this simple example,

$$\sum_{ij} \int_{\partial \mathscr{S}_{ij}} = \int_{B_{11}} + \int_{B_{21}} + \int_{R_{21}} + \int_{R_{22}} + \int_{R_{23}} - \int_{T_{11}} - \int_{B_{21}} - \int_{L_{21}} + \int_{L_{22}} + \int_{L_{23}}$$
$$\int_{\textbf{Bottom}} + \int_{\textbf{Right side}} - \int_{\textbf{Top}} - \int_{\textbf{Left side}}$$

where the last integrations are for the original rectangle $[a, b] \times [0, 1]$. *As before,* $\int_{\textbf{Bottom}} = \int_{\textbf{Top}} = 0$ *and* $\int_{\textbf{Right side}} = \int_{\gamma}$ *and* $\int_{\textbf{Left side}} = \int_{\delta}$. *We still know, using Green's Theorem,* $\int_{\mathscr{S}} d\omega^* = 0$, *and so we have* $\int_{\gamma} \omega = \int_{\delta} \omega$. *The argument for a general* Π *partition is quite similar and you should be able to work it out even though there are more subrectangles and more mess.* ■

Homework

Exercise 3.5.1 *Given* $\omega = ydx + xdy$, *we know* ω *is exact and therefore closed. Let* $P = (0, 0)$ *and* $Q = (1, 1)$ *and define* $\gamma_0(t) = (t, t)$ *and* $\gamma_1(t) = (t^2, t^2)$ *for* $0 \leq t \leq 1$. *Let*

$$H(t, s) = (1 - s)\gamma_0(t) + s\gamma_1(t),\ 0 \leq s \leq 1$$

and let $\gamma_s(t) = H(t, s) = ((1 - s)t + st^2, (1 - s)t + st^2)$. *Compute* $\int_{\gamma_s} \omega$ *for all s. Is* $\int_{\gamma_0} \omega = \int_{\gamma_1} \omega$?

Exercise 3.5.2 *Given* $\omega = (2x + 4x^3y^4)dx + 4x^4y^3dy$, *we know* ω *is exact and therefore closed. Let* $P = (0, 0)$ *and* $Q = (1, 1)$ *and define* $\gamma_0(t) = (t, t)$ *and* $\gamma_1(t) = (t^2, t^2)$ *for* $0 \leq t \leq 1$. *Let*

$$H(t, s) = (1 - s)\gamma_0(t) + s\gamma_1(t),\ 0 \leq s \leq 1$$

and let $\gamma_s(t) = H(t, s) = ((1 - s)t + st^2, (1 - s)t + st^2)$. *Compute* $\int_{\gamma_s} \omega$ *for all s. Is* $\int_{\gamma_0} \omega = \int_{\gamma_1} \omega$?

Exercise 3.5.3 *Given* $\omega = (2x + 4x^3y^4)dx + 4x^4y^3dy$, *we know* ω *is exact and therefore closed. Let* $P = (0, 0)$ *and* $Q = (1, 1)$ *and define* $\gamma_0(t) = (t, t)$ *and* $\gamma_1(t) = (t^2, t^2)$ *for* $0 \leq t \leq 1$. *Let*

$$H(t, s) = (1 - s)\gamma_0(t) + s\gamma_1(t),\ 0 \leq s \leq 1$$

and let $\gamma_s(t) = H(t, s) = ((1 - s)t + st^2, (1 - s)t + st^2)$. *Compute* $\int_{\gamma_s} \omega$ *for all s. Is* $\int_{\gamma_0} \omega = \int_{\gamma_1} \omega$?

Exercise 3.5.4 *If* $\mathscr{V}$ *is an open set containing* $[a, b] \times [0, 1]$, *prove there is an* $\epsilon > 0$ *so that* $[a, b] \times [0, 1] \subset [a - \epsilon, b + \epsilon] \times [-\epsilon, 1 + \epsilon] \subset \mathscr{V}$.

Exercise 3.5.5 *Look up the construction of* C^∞ *bump functions on an interval* $[a, b]$ *from (Peterson (99) 2020) and work out explicitly how this is done to give a function which is* 1 *on* $[a, b]$, 0 *on* $[a - \epsilon, b + \epsilon]^C$ *for a given* $\epsilon > 0$ *and* C^∞ *in between.*

Exercise 3.5.6 *Look up the construction of C^∞ bump functions on an interval $[a, b]$ from (Peterson (99) 2020) and now work out explicitly how this is done to give a function which is 1 on a rectangle and 0 on the complement of the rectangle padded by a strip of width $\epsilon > 0$.*

Exercise 3.5.7 *Look up the construction of C^∞ bump functions on an interval $[a, b]$ from (Peterson (99) 2020) and now work out explicitly how this is done to give a function which is 1 on a circle of radius r and 0 on the complement of the circle of radius $r + \epsilon$ for any $\epsilon > 0$.*

If the curve is closed, we can prove a similar theorem.

Theorem 3.5.2 Integral of a Closed 1-Form is the Same on Smoothly Connected Closed Homotopic Paths

If γ and δ are smoothly homotopic paths from $\boldsymbol{P}$ back to $\boldsymbol{P}$ into $\mathscr{U}$ and ω is a closed 1-form on $\mathscr{U}$, then $\int_\gamma \omega = \int_\delta \omega$.

Proof 3.5.2
The proof is the same except for the minor difference that $\boldsymbol{P} = \boldsymbol{Q}$. ■

There are nice consequences to the theorems we have just finished. Let $\mathscr{U} = \Re^2 \setminus \boldsymbol{P} = (x_0, y_0)$. Let $\omega = \frac{1}{2\pi}\omega_{\boldsymbol{P}\boldsymbol{\theta}}$ where

$$\frac{1}{2\pi}\omega_{\boldsymbol{P}\boldsymbol{\theta}} = \frac{-(y-y_0)dx + (x-x_0)dy}{(x-x_0)^2 + (y-y_0)^2}, \quad \text{on } \mathscr{U} = \Re^2 \setminus \boldsymbol{P}$$

The winding number for a path γ not containing the origin is $W(\gamma, \boldsymbol{0}) = \int_\gamma \omega_{\boldsymbol{0}\boldsymbol{\theta}}$ and, as we have discussed, it measures the change in the angle function as we move on the path γ. Hence, we can define the winding number for a path γ not containing $\boldsymbol{P}$ in the same way: $W(\gamma, \boldsymbol{P}) = \int_\gamma \omega_{\boldsymbol{P}\boldsymbol{\theta}}$. For, $\mathscr{R} = [a, b] \times [c, d]$ and $\Gamma : \mathscr{R} \to \Re^2$ we have this picture:

Top $\Gamma(t, d) = \gamma_3(t)$

left side $\Gamma(a, s) = \gamma_4(s)$ **Right side** $\Gamma(b, s) = \gamma_2(s)$

Bottom $\Gamma(t, c) = \gamma_1(t)$

and Γ defines a smooth mapping from the rectangle $\mathscr{R}$ into $\Re^2$. If $\boldsymbol{P} \not\in \Gamma(\mathscr{R})$, the arguments we gave in the proof of the previous theorem using Lebesgue's Lemma can be carried out with virtually no change to show

$$\int_{\textbf{Bottom}} + \int_{\textbf{Right side}} = \int_{\textbf{Left side}} + \int_{\textbf{Top}}$$

which becomes

$$\int_{\Gamma(t,c)} + \int_{\Gamma(b,s)} = \int_{\Gamma(a,s)} + \int_{\Gamma(t,d)}$$

which implies

$$\int_{\gamma_1} + \int_{\gamma_2} = \int_{\gamma_3} + \int_{\gamma_4}$$

This can be rewritten as

$$W(\gamma_1, \boldsymbol{P}) + W(\gamma_2, \boldsymbol{P}) \quad = \quad W(\gamma_3, \boldsymbol{P}) + W(\gamma_4, \boldsymbol{P})$$

If Γ is a smooth closed homotopy between the closed paths γ and δ around the point $\boldsymbol{P}$ which is not on the homotopy, the integrals over the bottom and top are zero like before and we have

$$\int_{\Gamma(b,s)=\delta} \quad = \quad \int_{\Gamma(a,s)=\gamma}$$

which says $W(\gamma, \boldsymbol{P}) = W(\delta, \boldsymbol{P})$. Hence, winding numbers are invariant under homotopy. Let's state this formally. First, recall what a homotopy is in detail.

Let $\mathscr{U} \subset \Re^2$ be open. Let $\gamma : [a, b] \rightarrow \mathscr{U}$ and $\delta : [a, b] \rightarrow \mathscr{U}$ be paths with the same initial and final points $\boldsymbol{Q}$ and $\boldsymbol{T}$. Let $H : [a, b] \times [0, 1] \rightarrow \mathscr{U}$ be a smooth map so that $H(t, 0) = \gamma$, $H(t, 1) = \delta$ on $[a, b]$ and $H(a, s) = \gamma(a) = \delta(a) = \boldsymbol{Q}$ $H(b, s) = \gamma(b) = \delta(b) = \boldsymbol{T}$ on $[0, 1]$. Thus, $\gamma_s(t) = H(t, s)$ is a path from $\boldsymbol{Q}$ to $\boldsymbol{T}$. The paths γ and δ are called homotopic with fixed endpoints and if $\boldsymbol{Q} = \boldsymbol{T}$, they are called closed.

Theorem 3.5.3 Winding Numbers are Invariant under Homotopy

If two paths γ and δ in $\Re^2 \setminus \boldsymbol{P}$ are homotopic either as paths with the same endpoints or as closed paths, then $W(\gamma, \boldsymbol{P}) = W(\delta, \boldsymbol{P})$.

Proof 3.5.3
As mentioned, the proof uses Lebesgue's Lemma in a manner similar to the second proof of Theorem 3.5.1. We leave it to you to work out the details. ■

What if we reparameterize the interval $[a, b]$ for a path? What happens to the winding number?

Theorem 3.5.4 Winding Number is Invariant under Domain Reparameterizations

Let $\gamma : [q, b] \rightarrow \Re^2 \setminus \boldsymbol{P}$ be a smooth path and $\phi : [c, d] \rightarrow [a, b]$ be a continuous function.

(i): If $\phi(c) = a$ and $\phi(d) = b$, then $W(\gamma \circ \phi, \boldsymbol{P}) = W(\gamma, \boldsymbol{P})$.

(ii): If $\phi(c) = b$ and $\phi(d) = a$, then $W(\gamma \circ \phi, \boldsymbol{P}) = -W(\gamma, \boldsymbol{P})$. For example, for $\gamma^{-1}(t) = \gamma(a + b - t)$, then $W(\gamma^{-1}, \boldsymbol{P}) = -W(\gamma, \boldsymbol{P})$.

Proof 3.5.4
(i): *Let $\Gamma : [a, b] \times [c, d] \rightarrow \mathscr{U}$ be defined by*

$$\Gamma(t, s) \quad = \quad \gamma(\min(t + \phi(s) - a, b))$$

For example, if $\gamma(t) = (t, t^2)$ on $[0, 1]$ and $\phi(t) = \frac{1}{2}t - 1$ on $[2, 4]$, then $(\gamma \circ \phi)(t) = (\frac{1}{2}t - 1, (\frac{1}{2}t - 1)^2)$. Then

$$\begin{aligned} t + \phi(s) - a &= t + \frac{1}{2}s - 1 = t - 1 + \frac{1}{2}s \\ \min(t + \phi(s) - a, b) &= \min(t - 1 + \frac{1}{2}s, 1) \\ \Gamma(t, s) &= \gamma(\min(t - 1 + \frac{1}{2}s, 1)) \end{aligned}$$

Hence,

$$\begin{aligned}
\Gamma(t,c) &= \Gamma(t,2) = \gamma(\min(t-1+\frac{1}{2}2,1)) = \gamma(\min(t,1)) = \gamma(t),\ \ 0 \leq t \leq 1 \\
\Gamma(t,d) &= \Gamma(t,4) = \gamma(\min(t-1+\frac{1}{2}4,1)) = \gamma(\min(t+1,1)) = \gamma(1),\ \ 0 \leq t \leq 1 \\
\Gamma(a,s) &= \Gamma(0,s) = \gamma(\min(0-1+\frac{1}{2}s,1)) = \gamma(\min(\frac{1}{2}s-1,1)) \\
&= \gamma(\frac{1}{2}s-1) = (\gamma \circ \phi)(s),\ \ 2 \leq s \leq 4 \\
\Gamma(b,s) &= \Gamma(1,s) = \gamma(\min(1-1+\frac{1}{2}s,1)) = \gamma(\min(\frac{1}{2}s,1)) \\
&= \gamma(\frac{1}{2}s) = \gamma(1),\ \ 2 \leq s \leq 4
\end{aligned}$$

Thus, in this example, Γ looks like:

Top $\Gamma(t,d) = \gamma(1) = \gamma_3(t)$

Left $\Gamma(a,s) = \gamma \circ \phi(s) = \gamma_4(t)$ **Right** $\Gamma(b,s) = \gamma(1) = \gamma_2(t)$

Bottom $\Gamma(t,c) = \gamma(t) = \gamma_1(t)$

In general, this will give a continuous mapping Γ. We can apply Theorem 3.5.3 to conclude

$$W(\gamma_1, \boldsymbol{P}) + W(\gamma_2, \boldsymbol{P}) = W(\gamma_3, \boldsymbol{P}) + W(\gamma_4, \boldsymbol{P})$$

or

$$W(\gamma, \boldsymbol{P}) + W(\textbf{ Constant map }, \boldsymbol{P}) = W(\textbf{ Constant map }, \boldsymbol{P}) + W(\gamma \circ \phi, \boldsymbol{P})$$

But $W(\textbf{ Constant map }, \boldsymbol{P}) = 0$, so we have $W(\gamma, \boldsymbol{P}) = W(\gamma \circ \phi, \boldsymbol{P})$.

(ii): *Here we use*

$$\Gamma(t,s) = \gamma(\max(t+\phi(s)-b,a))$$

Since $\phi(c) = b$ and $\phi(d) = a$, we find

$$\begin{aligned}
\gamma_1(t) &= \Gamma(t,c) = \gamma(\max(t+\phi(c)-b,a)) = \gamma(\max(t+b-b,a)) \\
&= \gamma(\max(t,a)) = \gamma(t),\ \ a \leq t \leq b \\
\gamma_3(t) &= \Gamma(t,d) = \gamma(\max(t+\phi(d)-b,a)) = \gamma(\max(t+a-b,a)) = \gamma(a),\ \ a \leq t \leq b \\
\gamma_2(s) &= \Gamma(b,s) = \gamma(\max(b+\phi(s)-b,a)) = \gamma(\max(\phi(s),a)) = (\gamma \circ \phi)(s) \\
\gamma_4(s) &= \Gamma(a,s) = \gamma(\max(a+\phi(s)-b,a)) = \gamma(a)
\end{aligned}$$

Thus,

$$W(\gamma_1, \boldsymbol{P}) + W(\gamma_2, \boldsymbol{P}) = W(\gamma_3, \boldsymbol{P}) + W(\gamma_4, \boldsymbol{P})$$

or

$$W(\gamma, \boldsymbol{P}) + W(\gamma \circ \phi, \boldsymbol{P}) = W(\textbf{ Constant map }, \boldsymbol{P}) + W(\textbf{ Constant map }, \boldsymbol{P})$$

But $W(\textbf{ Constant map }, \boldsymbol{P}) = 0$, so we have $W(\gamma \circ \phi, \boldsymbol{P}) = -W(\gamma, \boldsymbol{P})$. ∎

Homework

Exercise 3.5.8 *Let* $\Gamma : [a,b] \times [c,d] \to \mathscr{U}$ *be defined by*

$$\Gamma(t,s) \quad = \quad \gamma(\min(t + \phi(s) - a, b))$$

where $\gamma(t) = (2t, 3t^2)$ *on* $[0,1]$ *and* $\phi(t) = 2t - 3$ *on* $[3/2, 2]$, *compute* $(\gamma \circ \phi)(t)$ *and* $\Gamma(t,s)$ *as needed for the proof of Theorem 3.5.4.*

Exercise 3.5.9 *Let* $\Gamma : [a,b] \times [c,d] \to \mathscr{U}$ *be defined by*

$$\Gamma(t,s) \quad = \quad \gamma(\min(t + \phi(s) - a, b))$$

where $\gamma(t) = (5t, 4t^2)$ *on* $[0,1]$ *and* $\phi(t) = t + 2$ *on* $[-2, -1]$, *compute* $(\gamma \circ \phi)(t)$ *and* $\Gamma(t,s)$ *as needed for the proof of Theorem 3.5.4.*

Exercise 3.5.10 *Work out the proof of Theorem 3.5.3 using Lebesgue's Lemma is a manner similar to the second proof of Theorem 3.5.1.*

Chapter 4

The Jordan Curve Theorem

Now we are going to use more of the topology of $\Re^2$. If $\gamma : [a,b] \rightarrow \mathscr{U} \subset \Re^2$ is a smooth map, we know the support of γ is the compact set $\gamma([a,b])$. This is a closed and bounded subset and so its complement is an open set. The complement may have infinitely many open connected components. For example, consider a figure eight curve in $\Re^2$. The support of this path is the figure eight curve itself and the complement has three components; two bounded components *inside* the *curve* and one unbounded component which is *outside* the *curve*. An easy way to distinguish the inside and outside of a curve might be to find the normal and tangent vector to the curve $\boldsymbol{P}$ on the curve. Call these vectors $\boldsymbol{N}(\boldsymbol{P})$ and $\boldsymbol{T}(\boldsymbol{P})$. The vectors $\boldsymbol{N}(\boldsymbol{P})$ and $\boldsymbol{T}(\boldsymbol{P})$ determine a third orthonormal vector $\boldsymbol{B}(\boldsymbol{P})$ by the right-hand rule applied to the vector pair $\boldsymbol{T}(\boldsymbol{P})$ and $\boldsymbol{N}(\boldsymbol{P})$. The $\boldsymbol{B}(\boldsymbol{P})$ can point in two distinct directions that are perpendicular to the plane determined by $\boldsymbol{T}(\boldsymbol{P})$ and $\boldsymbol{N}(\boldsymbol{P})$. Based on what we find by doing this analysis with a circle or an ellipse, we think of the component of the complement corresponding to $\boldsymbol{B}(\boldsymbol{P})$ pointing **up** as an **inside** component and the other case as the **outside** component. It is not hard at all to think of paths γ giving really confusing notions of the vector $\boldsymbol{N}(\boldsymbol{P})$ where the *orientation* of the vector $\boldsymbol{B}(\boldsymbol{P})$ is confusing.

Thus, we are seeking a better way to describe how to define the inside and outside components of the complement of the support $\gamma([a,b])$. It seems reasonable, since paths γ can be very messy, to focus our attention on paths that are similar to circles. So we want to think about paths that are *homeomorphic* to a circle. As you will see, this is not so easy to do. In this chapter, we will go through all the details leading to the theorem that tells us all such curves have an inside and an outside. This curve is called the Jordan Curve Theorem and many people who study these things never see its proof. Since the proof uses a fair bit of $\Re^2$ topology and some ideas from a field called *algebraic topology*, doing this in a text devoted to ideas for functional analysis and topology seems very reasonable to do. Also, these ideas are very similar to more general ones about degree theory for continuous maps in $\Re^n$. So now we are specialized to $\Re^2$ but later we will work both in a general $\Re^n$ and general normed linear spaces when we study Leray - Schauder Topological degree.

4.1 Winding Numbers and Topology

In general, since the support of γ is bounded, at least one component of the complement of the support of γ must be unbounded.

Theorem 4.1.1 Winding Number is Constant on Each Connected Component of the Support of a Path

As a function of the point $\boldsymbol{P}$ not in the support of γ, $W(\gamma, \boldsymbol{P})$ is constant on each connected component of $\Re^2 \setminus \boldsymbol{P}$. It vanishes on unbounded components.

Proof 4.1.1

Given $\boldsymbol{P} \notin \gamma([a,b])$, choose a circle $B(\boldsymbol{P}, r) \subset \Re^2 \setminus \gamma([a,b])$. Let $\boldsymbol{V} = \boldsymbol{Q} - \boldsymbol{P}$ where $\boldsymbol{Q}$ is any other point of $B(\boldsymbol{P}, r)$. Let $\gamma + \delta$ be the path defined by $(\gamma+\delta)(t) = \gamma(t) + \boldsymbol{V}$ where $\delta(t) = \boldsymbol{V}$ is a constant mapping. Recall $W(\gamma, \boldsymbol{P})$ is defined using a partition Π of $[a,b]$ so that $\gamma([t_i, t_{i+1}]) \subset S_i$, where S_i is a sector rooted at $\boldsymbol{P}$. Remember, the existence of such a partition Π comes from an application of Lebesgue's Lemma as usual. Each sector S_i defines an angle function θ_i on S_i. Let $\boldsymbol{P_i} = \gamma(t_i)$ and define

$$W(\gamma, \boldsymbol{P}) = \frac{1}{2\pi} \sum_{i \in \Pi} (\, \theta_i(\gamma(t_{i+1})) - \theta_i(\gamma(t_i)) \,)$$

Then, for $t_i \leq s \leq t_{i+1}$,

$$\gamma(s) + \delta(s) = \gamma(t_i) + \boldsymbol{v} \subset S_i + \boldsymbol{V}$$

which is the translation of S_i by $\boldsymbol{V}$. We see $S_i + \boldsymbol{V}$ is simply the sector S_i rooted at the new point $\boldsymbol{P} + \boldsymbol{V}$. We can use the same angle function θ_i here and we find

$$W(\gamma+\delta, \boldsymbol{P}+\boldsymbol{V}) = \frac{1}{2\pi} \sum_{i \in \Pi} (\, \theta_i(\gamma(t_{i+1}) + \boldsymbol{V}) - \theta_i(\gamma(t_i) + \boldsymbol{V}) \,)$$

The constant additions do not alter the change in angle. So we have

$$W(\gamma+\delta, \boldsymbol{P}+\boldsymbol{V}) = \frac{1}{2\pi} \sum_{i \in \Pi} (\, \theta_i(\gamma(t_{i+1})) - \theta_i(\gamma(t_i)) \,) = W(\gamma, \boldsymbol{P})$$

But $\boldsymbol{V} = \boldsymbol{Q} - \boldsymbol{P}$. Thus, we have

$$W(\gamma, \boldsymbol{P}) = W(\gamma+\delta, \boldsymbol{P}+\boldsymbol{V}) = W(\gamma+\delta, \boldsymbol{P}+\boldsymbol{Q}-\boldsymbol{P}) = W(\gamma+\delta, \boldsymbol{Q})$$

Now define the homotopy

$$H(t,s) = \gamma(t) + s\delta,\ 0 \leq s \leq 1,\ a \leq t \leq b$$

We have $H(t,0) = \gamma$ and $H(t,1) = \gamma + \delta = \gamma + \boldsymbol{Q} - \boldsymbol{P}$. Note we never hit $\boldsymbol{Q}$ as $\gamma(t)$ never hits $\boldsymbol{P}$. Thus, $W(\gamma+\delta, \boldsymbol{Q}) = W(\gamma, \boldsymbol{Q})$. We already knew $W(\gamma, \boldsymbol{P}) = W(\gamma+\delta, \boldsymbol{Q})$. Thus, we have $W(\gamma, \boldsymbol{P}) = W(\gamma, \boldsymbol{Q})$. This shows the value of the winding number is the same in each component of the complement of the support of γ.

Finally, on an unbounded component, we know $W(\gamma, \boldsymbol{P})$ is a constant. Hence, there is a radius $R > 0$, so that the unbounded component contains all points outside of $(B(\boldsymbol{0}, R))^C$. Pick such a point $\boldsymbol{P}$ on the negative x-axis. Hence, the support of γ is contained to the right of $\boldsymbol{P}$. We can pick an angle function θ with $\Delta\theta = 0$. Hence, $W(\gamma, \boldsymbol{P}) = 0$ and since the winding number is constant on the component, the winding number must be zero on any unbounded component. ■

If $\mathbb{I}$ is the closed interval $[0,1]$ and $\mathbb{C}$ any circle, a closed continuous path $\gamma : \mathbb{I} \rightarrow \mathscr{U} \subset \Re^2$, an open subset, can be thought of as a map $F : \mathbb{C} \rightarrow \gamma(\mathbb{I}) \subset \mathscr{U}$ via $\gamma = F \circ \phi$ where $\phi(t) = \boldsymbol{P} + r(\cos(2\pi t), \sin(2\pi t))$ is a parameterization of the circle $\mathbb{C}$ centered at $\boldsymbol{P} = (x_0, y_0)$ or radius $r > 0$. For $0 \leq t \leq 1$, ϕ wraps $\mathbb{I}$ around $\mathbb{C}$ ccw and ϕ is 1-1 and onto from $\mathbb{I}$ to $\mathbb{C}$ except that

$\phi(0) = \phi(1)$. Hence, F satisfies $\gamma(0) = F(\phi(0)) = \gamma(1) = F(\phi(1))$.

Lemma 4.1.2 Path γ is Continuous if and only if F is Continuous

Let $\phi(t) = \boldsymbol{P} + r(\cos(2\pi t), \sin(2\pi t))$ be a parameterization of the circle $\mathbb{C}$ centered at $\boldsymbol{P} = (x_0, y_0)$ of radius $r > 0$. Recall $\mathbb{I}$ is the closed interval $[0, 1]$. Then, given $F = \mathbb{C} \rightarrow \gamma(\mathbb{I})$ and the path $\gamma : [0, 1] \rightarrow \gamma(\mathbb{I})$, then $\gamma = F \circ \phi$ and γ is continuous if and only if the map F is continuous.

Proof 4.1.2

The arguments we use here illustrate some topological ideas nicely. Since ϕ is continuous from $\mathbb{I}$ to $\mathbb{C}$, if W is open in $\mathbb{C}$, then $\phi^{-1}(W)$ is open in $\mathbb{I}$. But ϕ is 1-1 and onto so ϕ^{-1} is continuous and so if $\phi^{-1}(W)$ is open in $\mathbb{I}$, then $\phi(\phi^{-1}(W)) = W$ is open in $\mathbb{C}$. We conclude a subset W of $\mathbb{C}$ is open if and only if $\phi^{-1}(W)$ is open in $\mathbb{I}$.

Thus, $F^{-1}(V)$ is open in $\mathbb{C}$ if and only if $\phi^{-1}(F^{-1}(V))$ is open in $\mathbb{I}$. A simple argument will show you that $\phi^{-1}(F^{-1}(V)) = \gamma^{-1}(V)$. Hence, $\gamma^{-1}(V)$ is open. This shows γ is continuous when F is continuous and vice versa.

In case you have not seen the standard $\epsilon - \delta$ argument which shows that for a continuous map, V open in the range implies $F^{-1}(V)$ is open in the domain, you would reason like this. If V is an open set in $\phi(\mathbb{I})$, then if $\boldsymbol{Q} \in F^{-1}(V)$, $F(\boldsymbol{Q}) \in V$. Since V is open, this point is an interior point and so there is an $r > 0$ with $B(F(\boldsymbol{Q}), r) \subset V$. Also, F is continuous at $\boldsymbol{Q}$. Hence, for this r, there is a $\delta > 0$ so that $F(B(\boldsymbol{Q}, \delta)) \subset B(F(\boldsymbol{Q}), r) \subset V$. So, we have $B(\boldsymbol{Q}, \delta) \subset V$ which tells us $\boldsymbol{Q}$ is an interior point of $F^{-1}(\boldsymbol{Q})$. Since our choice of $\boldsymbol{Q}$ is arbitrary, we have shown $F^{-1}(V)$ is an open set. Thus, if V is open in the range, $F^{-1}(V)$ is open in the domain. We think the topological base reasoning is very clean and less messy, so it is a good idea to embrace this point of view. ■

Now for any continuous $F : \mathbb{C} \rightarrow \Re^2 \setminus \boldsymbol{P}$, the winding number of F around $\boldsymbol{P}$ is $W(F, \boldsymbol{P}) = W(F \circ \phi, \boldsymbol{P})$. Note this is a traditional winding number. Now assume there are two circles $\mathbb{C}$ centered at $\boldsymbol{P}$ and $\mathbb{C}'$ centered at $\boldsymbol{P'}$. Then, if $F : \mathbb{C} \rightarrow \mathbb{C}'$, we have $\gamma = F \circ \phi : \mathbb{I} \rightarrow \mathbb{C}'$. We use this to define the **degree** of the continuous map F from the circle $\mathbb{C}$ to another circle $\mathbb{C}'$ as the number of times the first circle $\mathbb{C}$ is wound around the second circle $\mathbb{C}'$ by the mapping F. Based on our discussions here, we define the **degree** by $deg(F) = W(F, \boldsymbol{P'})$.

Homework

Exercise 4.1.1 *Let $F(x, y) = (x^2, -y^3)$ be a map from $\Re^2$ to $\Re^2$. Let $\phi(t) = (1 + 2\cos(2\pi t), 1 + 2\sin(2\pi t))$ for $0 \leq t \leq 1$. Find $F \circ \phi$ and plot this curve by hand and also using MATLAB. Write down the integral which gives us the winding number of F relative to $P = (1, 1)$.*

Exercise 4.1.2 *Let $F(x, y) = (2x^2, -y^2)$ be a map from $\Re^2$ to $\Re^2$. Let $\phi(t) = (2+3\cos(2\pi t), -1+3\sin(2\pi t))$ for $0 \leq t \leq 1$. Find $F \circ \phi$ and plot this curve by hand and also using MATLAB. Write down the integral which gives us the winding number of F relative to $P = (2, -1)$.*

Exercise 4.1.3 *Define γ by*

$$\gamma(t) = \begin{cases} (2t, (2t)^2), & 0 \leq t \leq 1/2 \\ (2 - 2t, (2 - 2t)^{1/3}), & 1/2 \leq t \leq 1 \end{cases}$$

Then γ is a closed continuous curve. Let $\phi(t) = (2 + 3\cos(2\pi t), -1 + 3\sin(2\pi t))$ for $0 \leq t \leq 1$ which is a parameterization of the circle C of radius 3 centered at $(2, -1)$. Explain how you can find the function $G : C \rightarrow \gamma([0, 1])$ so that $\gamma = G \circ \phi$. This is best done with a graphical argument!

4.2 Some Fundamental Results

We will now explore some topological ideas for $\Re^2$.

Theorem 4.2.1 Extensions of a Function Defined on a Boundary have Winding Number Zero

Let $\mathbb{C} = \partial D$ where $D = \overline{B(\boldsymbol{P}, R)}$ for $R > 0$ and $\boldsymbol{P} = (x_0, y_0)$. Assume $F : \mathbb{C} \to \Re^2 \setminus \boldsymbol{P}$ extends to a continuous function $\tilde{F} : D \to \Re^2 \setminus \boldsymbol{P}$. Then $W(F, \boldsymbol{P}) = 0$.

Proof 4.2.1
Let ϕ be our usual circle mapping: $\phi(t) = \boldsymbol{P} + R(\cos(2\pi t), \sin(2\pi t))$ on $[0, 1]$. Define $\gamma = F \circ \phi$ as we have done before. Define

$$H(t, s) = \tilde{F}(\boldsymbol{P} + sR(\cos(2\pi t), \sin(2\pi t))),\ 0 \leq t \leq 1,\ 0 \leq s \leq 1$$

Then $H(t, 0) = \tilde{F}(\boldsymbol{P})$, a constant map and $H(t, 1) = \tilde{F}(\phi(t)) = F(\phi(t))$ as $\tilde{F} = F$ on ∂D. H is a homotopy from $\gamma = F \circ \phi$ to the constant map $\tilde{F}(\boldsymbol{P})$. Hence $W(F \circ \phi, \boldsymbol{P}) = W(\tilde{F}(\boldsymbol{P}), \boldsymbol{P}) = 0$.
■

We can use this result to prove there is no **retraction** from a closed disc onto its boundary circle. First, we need a definition.

Definition 4.2.1 Retractions

A retraction from a set X to a subset $Y \subset X$ is a continuous mapping $f : X \to Y$ so that $f(y) = y$ for all $y \in Y$.

Theorem 4.2.1 allows us to prove this.

Theorem 4.2.2 There is No Retraction from a Disc Onto its Boundary

Let $\mathbb{C} = \partial D$ where $D = \overline{B(\boldsymbol{P}, R)}$ for $R > 0$ and $\boldsymbol{P} = (x_0, y_0)$. Then there is no retraction $f : \partial D \to D$ with f the identity on ∂D.

Proof 4.2.2
The identity map, I, on ∂D is a map $I : \partial D \to \Re^2 \setminus \boldsymbol{P}$. We see $W(I, \boldsymbol{P}) = 1$. If there was a retraction, there would be a continuous mapping $f : D \to \partial D$ with $f|_{\partial D} = I$. The map f is then a continuous extension of I to D and so by Theorem 4.2.1, $W(I, \boldsymbol{P}) = 0$ which is a contradiction. Hence, no retraction exists. ■

Homework

Exercise 4.2.1 *Find a retraction from the unit square centered at the origin in $\Re^2$ to the unit ball centered at the origin in $\Re^2$.*

Exercise 4.2.2 *Find a retraction from the unit curve centered at the origin in $\Re^3$ to the unit ball centered at the origin in $\Re^3$*

Exercise 4.2.3 *Find a retraction from any rectangle in $\Re^2$ to a ball inside the rectangle.*

4.3 Some Applications

Now we consider some interesting applications of these results.

4.3.1 The Fundamental Theorem of Algebra

As a first example, let's prove the Fundamental Theorem of Algebra: all polynomials of degree n have n roots in the complex plane.

Theorem 4.3.1 Fundamental Theorem of Algebra

Every polynomial of degree $n > 0$ has n complex roots.

Proof 4.3.1
Identify the complex numbers $\mathbb{C}$ with $\Re^2$ via $z = x + \boldsymbol{i}y \equiv (x, y)$ as usual. It is straightforward to show the polynomial $g(z) = \sum_{i=0}^{n} a_i z^{n-i} = a_0 z^n + \ldots + a_{n-1} z + a_n$ is continuous for any choice of complex coefficients a_i. Assume $n > 0$ and $a_0 \neq 0$. Consider the problem of finding the root z so that $g(z) = 0$. Since $a_0 \neq 0$, we can divide by a_0 to get a new function $\hat{g}$ with leading coefficient 1. From now on, we will assume this has been done and we consider the root finding problem $g(z) = 0$ in this case; i.e. $a_0 = 1$. Restrict g to the curve $\mathscr{C}_r = \{z \in \mathbb{C} | |z| = r\}$ where we use the absolute value symbol to denote the complex modulus of z. Let $g|_{\mathscr{C}_r} = g_r$. Then $g_r : \mathscr{C}_r \rightarrow \mathbb{C} \setminus \{0\}$. It is easy to see g_4 extends from $\mathscr{C}_r$ to the interior of the disc $D_r = \{z \in \mathbb{C} | |z| \leq r\}$ in a continuous way. Call the extension $\tilde{g}_r$. If we assume we have no root, then $\tilde{g}_r$ maps into $\mathbb{C} \setminus \{0\}$ also. Applying Theorem 4.2.1, it follows, $W(g_r, 0) = 0$.

A path $\mathscr{C}_r$ around the origin of radius r is

$$\phi_r(t) \quad = \quad (r\cos(2\pi t), \sin(2\pi t)), \;\; 0 \leq t \leq 1$$

and on this path, $z(t) = r\cos(2\pi t) + \boldsymbol{i} r \sin(2\pi t)$ is a complex number of $\mathscr{C}_r$. Note, for the map $F(z) = z$ and $\phi(t) = (\cos(2\pi t), \sin(2\pi t))$, $(F \circ \phi)(t) = (\cos(2\pi t), \sin(2\pi t))$ which traverses the circle of radius 1 ccw one time. Thus $W(z, 0) = W(F \circ \phi, 0) = 1$.

We want to show $W(z^n, 0) = n$, which is an induction argument. Let's show one more step for the basis argument and look at the case $n = 2$ also. If $F(z) = z^2$, then

$$(F \circ \phi)(t) = z^2(t) = (\cos^2(2\pi t) - sin^2(2\pi t)) + 2\sin(2\pi t)\cos(2\pi t)\boldsymbol{i} = \cos(4\pi t) + \boldsymbol{i}\sin(4\pi t)$$

This curve wraps around the origin twice so $W(z^2, 0) = W(F \circ \phi, 0) = 2$.

Now assume for $F(z) = z^n$, $W(F \circ \phi, 0) = W(z^n, 0) = n$. Consider the map z^{n+1}. It is straightforward to show for $G(z) = z^{n+1}$ that

$$\begin{aligned} (F \circ \phi)(t) &= z^{n+1}(t) = (\cos(2\pi n t) - \sin(2\pi n t))\boldsymbol{i} \\ (G \circ \phi)(t) &= z^{n+1}(t) = (\cos(2\pi(n+1)t) - \sin(2\pi(n+1)t))\boldsymbol{i} \end{aligned}$$

which implies $W(G \circ \phi, 0) = W(z^{n+1}, 0) = n + 1$. Thus, by induction, $W(z^n, 0) = n$.

Now assume $\gamma : [a, b] \rightarrow \Re^2 \setminus \boldsymbol{P}$ and $\delta : [a, b] \rightarrow \Re^2 \setminus \boldsymbol{P}$ are two closed paths and the line segment $[\gamma(t), \delta t] = \{\boldsymbol{X} | \boldsymbol{X} = (1 - s)\gamma(t) + s\delta(t), \; 0 \leq s \leq 1, a \leq t \leq b\}$ never hits $\boldsymbol{P}$. Then the homotopy

$$H(t, s) \quad = \quad (1 - s)\gamma(t) + s\delta(t), \; 0 \leq s \leq 1, a \leq t \leq b$$

never hits $\boldsymbol{P}$ and so $W(\gamma, \boldsymbol{P}) = W(\delta, \boldsymbol{P})$. Let $f_r = z^n|_{\mathscr{C}_r}$. Then, on D_r, assuming $g_r \neq f_r$,

$$|f_r - g_r| \quad = \quad |a_1 z^{n-1} + \ldots + a_{n-1} z + a_n| \leq |a_1| r^{n-1} + \ldots + |a_{n-1}| r + |a_n|$$

For $r > 1$, choose

$$r > \max\{\frac{|a_1|}{n}, \left(\frac{|a_2|}{n}\right)^{1/2}, \left(\frac{|a_3|}{n}\right)^{1/3}, \ldots, \left(\frac{|a_{n-1}|}{n}\right)^{1/(n-1)}, \left(\frac{|a_n|}{n}\right)^{1/(n)}\}$$

Then,

$$\begin{aligned} r &> \frac{|a_1}{n} \Longrightarrow |a_1| < \frac{r}{n}, \quad r^2 > \frac{|a_2|}{n} \Longrightarrow |a_2| < \frac{r^2}{n}, \quad r^3 > \frac{|a_3|}{n} \Longrightarrow |a_3| < \frac{r^3}{n} \\ &\vdots \\ r^{n-1} &> \frac{|a_{n-1}|}{n} \Longrightarrow |a_{n-1}| < \frac{r^{n-1}}{n} \quad r^n > \frac{|a_n|}{n} \Longrightarrow |a_n| < \frac{r^n}{n} \end{aligned}$$

Then, for such an r,

$$\begin{aligned} |a_1|r^{n-1} + |a_2|r^{n-2} + \ldots + |a_{n-1}r + |a_n| &\leq \frac{r}{n}r^{n-1} + \frac{r^2}{n}r^{n-2} + \ldots + \frac{r^{n-1}}{n}r + \frac{r^n}{n} \\ &= \frac{r^n}{n} + \ldots + \frac{r^n}{n} \text{ (} n \text{ times)} = r^n \end{aligned}$$

So for r sufficiently large, on D_r,

$$f_r - g_r| < r^n = |f_r - 0|$$

Let $\gamma = f_r \circ \phi$ and $\delta = g_r \circ \phi$. Then, the line segment between $\gamma(t)$ and $\delta(t)$ was zero for some point on ϕ, then we have

$$(1 - s)f_r(t) + sg_r(t) = 0 \Longrightarrow |f_r(t)| = s|f_r(t) - g_r(t)|$$

But for such a time point, $|f_r(t)| = r^n$ and $|s(f_r(t) - g_r(t))| < sr^n$ for some s with $0 \leq s \leq 1$. But this is not possible. So the line segment never hits zero. Thus, we know $W(\gamma, 0) = W(\delta, 0)$ or $n = W(z^n, 0) = W(g_r, 0)$. However, we assumed g had no root and we already decided $W(g_r, 0) = 0$. This is a contradiction. Thus, our assumption is wrong and g must have a root.

Hence, there is a point z_1 so that $g(z_1) = 0$ which tells us $g(z) = (z - z_1)h(z)$ where h has degree $n - 1$. Now we just apply induction on the degree of g to show g has n roots. We have shown there are roots z_i so that $g(z) = \prod_{i=1}^n (z - z_i)$. ■

Homework

Exercise 4.3.1 *In the proof of the Fundamental Theorem of Algebra, show all the details of the argument that applies induction on the degree of g to show g has n roots.*

Exercise 4.3.2 *Go through all the details of the proof of the Fundamental Theorem of Algebra for polynomials of degree 2.*

Exercise 4.3.3 *Go through all the details of the proof of the Fundamental Theorem of Algebra for polynomials of degree 6.*

4.4 The Brouwer Fixed Point Theorem

Here is a simple fixed point theorem.

Theorem 4.4.1 Simple Fixed Point Theorem

Let $f : [a, b] \rightarrow [a, b]$ be continuous. Then there is a point $x \in [a, b]$ with $f(x) = x$.

Proof 4.4.1
Let $g(x) = f(x) = x$. Then g is continuous on $[a, b]$. Note $g(a) = f(a) - a \geq 0$ and $g(b) = f(b) - b \leq 0$. Since the image of $g([a, b])$ is connected, the interval $[g(a), g(b)]$ must contain 0. Hence, there is an $x \in [a, b]$ with $g(x) = 0$ or $f(x) = x$. ■

Comment 4.4.1 *Note this is an example of looking at a mapping $\boldsymbol{F} = \boldsymbol{I} + \boldsymbol{G}$ which is a perturbation of the identity. The space here is quite simple: $X = [a, b] \subset \Re$. We are interested in fixed point theorems for more general spaces X and more general mapping of the form $\boldsymbol{F} = \boldsymbol{I} + \boldsymbol{G}$. To do the more general fixed point theorems properly requires a lot more mathematical machinery. We begin with a result set in $\Re^2$ which we prove using ideas from winding numbers and retractions.*

We can now prove the Brouwer Fixed Point Theorem for $\Re^2$. We prove this result in more generality for $\Re^n$ and normed linear spaces in our work on topological degree theory which is in Chapter 10 ($\Re^n$) and Chapter 11 (normed linear spaces). But for now we are in $\Re^2$.

Theorem 4.4.2 Brouwer Fixed Point Theorem

$D = \overline{B(\boldsymbol{P}, R)}$ for $R > 0$ and $\boldsymbol{P} = (x_0, y_0)$. Any continuous mapping $f : D \rightarrow D$ must have a fixed point: i.e. there is an $\boldsymbol{P} \in D$ so that $f(\boldsymbol{P}) = \boldsymbol{P}$.

Proof 4.4.2
Assume f has no fixed point. Define $h : D \rightarrow \mathbb{C}$ where $\mathbb{C} = \partial D$ as follows: given $\boldsymbol{P} \in D$, look at the line $L(t) = \boldsymbol{P} + t\frac{f(\boldsymbol{P})-\boldsymbol{P}}{\|f(\boldsymbol{P})-\boldsymbol{P}\|}$. By assumption, the denominator $\|f(\boldsymbol{P}) - \boldsymbol{P}\|$ is never zero, so L is well-defined. When $\|L(t)\| = 1$, we have hit $\mathbb{C}$. We have

$$1 = <\boldsymbol{P}, \boldsymbol{P}> + 2t\left\langle \boldsymbol{P}, \frac{f(\boldsymbol{P}) - \boldsymbol{P}}{\|f(\boldsymbol{P}) - \boldsymbol{P}\|} \right\rangle + t^2\left\langle \frac{f(\boldsymbol{P}) - \boldsymbol{P}}{\|f(\boldsymbol{P}) - \boldsymbol{P}\|}, \frac{f(\boldsymbol{P}) - \boldsymbol{P}}{\|f(\boldsymbol{P}) - \boldsymbol{P}\|} \right\rangle$$

This simplifies to

$$t^2 + 2\left\langle \boldsymbol{P}, \frac{f(\boldsymbol{P}) - \boldsymbol{P}}{\|f(\boldsymbol{P}) - \boldsymbol{P}\|} \right\rangle t + <\boldsymbol{P}, \boldsymbol{P}> - 1 = 0$$

Solving, we find

$$t = -\left\langle \boldsymbol{P}, \frac{f(\boldsymbol{P}) - \boldsymbol{P}}{\|f(\boldsymbol{P}) - \boldsymbol{P}\|} \right\rangle \pm \sqrt{\left(\left\langle \boldsymbol{P}, \frac{f(\boldsymbol{P}) - \boldsymbol{P}}{\|f(\boldsymbol{P}) - \boldsymbol{P}\|} \right\rangle\right)^2 + 1 - <\boldsymbol{P}, \boldsymbol{P}>}$$

If $\boldsymbol{P} \notin \mathbb{C}$, then $\|\boldsymbol{P}\| < 1$ and the square root here is a positive number. Let $\epsilon_{\boldsymbol{P}} = 1 - <\boldsymbol{P}, \boldsymbol{P}> 0$. Then the positive root is

$$t^+ = -\left\langle \boldsymbol{P}, \frac{f(\boldsymbol{P}) - \boldsymbol{P}}{\|f(\boldsymbol{P}) - \boldsymbol{P}\|} \right\rangle + \sqrt{\left(\left\langle \boldsymbol{P}, \frac{f(\boldsymbol{P}) - \boldsymbol{P}}{\|f(\boldsymbol{P}) - \boldsymbol{P}\|} \right\rangle\right)^2 + \epsilon_{\boldsymbol{P}}} > 0$$

Call this positive root $t^+_{\boldsymbol{P}}$. *So we define for* $\|\boldsymbol{P}\| < 1$

$$h(\boldsymbol{P}) = \begin{cases} \boldsymbol{P} + t^+_{\boldsymbol{P}} \frac{f(\boldsymbol{P})-\boldsymbol{P}}{\|f(\boldsymbol{P})-\boldsymbol{P}\|}, & \|\boldsymbol{P}\| < 1, \left\langle \boldsymbol{P}, \frac{f(\boldsymbol{P})-\boldsymbol{P}}{\|f(\boldsymbol{P})-\boldsymbol{P}\|} \right\rangle \geq 0 \\ \boldsymbol{P} + t^-_{\boldsymbol{P}} \frac{f(\boldsymbol{P})-\boldsymbol{P}}{\|f(\boldsymbol{P})-\boldsymbol{P}\|}, & \|\boldsymbol{P}\| < 1, \left\langle \boldsymbol{P}, \frac{f(\boldsymbol{P})-\boldsymbol{P}}{\|f(\boldsymbol{P})-\boldsymbol{P}\|} \right\rangle < 0 \end{cases}$$

Now, let

$$u_{\boldsymbol{P}} = \left\langle \boldsymbol{P}, \frac{f(\boldsymbol{P})-\boldsymbol{P}}{\|f(\boldsymbol{P})-\boldsymbol{P}\|} \right\rangle$$

Hence, we have for any $\boldsymbol{Q}$ *with* $\|\boldsymbol{Q}\| < 1$,

$$h(\boldsymbol{Q}) = \begin{cases} \boldsymbol{Q} - u^2_{\boldsymbol{Q}} + \boldsymbol{u_Q}\sqrt{\boldsymbol{u_Q}^2 + 1 - \|\boldsymbol{Q}\|^2}, & \|\boldsymbol{Q}\| < 1, \ \boldsymbol{u}_Q \geq 0 \\ \boldsymbol{Q} - u^2_{\boldsymbol{Q}} - \boldsymbol{u_Q}\sqrt{\boldsymbol{u_Q}^2 + 1 - \|\boldsymbol{Q}\|^2}, & \|\boldsymbol{Q}\| < 1, \ \boldsymbol{u}_Q < 0 \end{cases}$$

It is clear $u_{\boldsymbol{Q}}$ *is a continuous function of* $\boldsymbol{Q}$ *on all of D. Hence, given* $\boldsymbol{P}$ *with* $\|\boldsymbol{P}\| < 1$, *if* $u_{\boldsymbol{P}} >= 0$, *there is a neighborhood about* $\boldsymbol{P}$ *so that* $u_{\boldsymbol{Q}} > 0$ *if* $\|\boldsymbol{Q} - \boldsymbol{P}\| < \delta$ *for some* $\delta > 0$. *Hence, if* $\boldsymbol{Q} \to \boldsymbol{P}$, *eventually* $\|\boldsymbol{Q} - \boldsymbol{P}\| < \delta$ *giving* $u_{\boldsymbol{Q}} > 0$. *In this case, it is straightforward to show* $h(\boldsymbol{Q}) \to h(\boldsymbol{P})$. *A similar argument shows that in the case that* $u_{\boldsymbol{P}} < 0$, *we have as* $\boldsymbol{Q} \to \boldsymbol{P}$, $h(\boldsymbol{Q}) \to h(\boldsymbol{P})$. *We conclude* h *is continuous on the interior of the disc.*

What happens on the boundary? If $\boldsymbol{Q} \to \boldsymbol{P}$ *with* $\|\boldsymbol{P}\| = 1$, *we know that either* $u_{\boldsymbol{P}}$ *is nonnegative or negative. In the nonnegative case, we have for* $\boldsymbol{Q}$ *sufficiently close to* $\boldsymbol{P}$ *that* $< \boldsymbol{Q}, u_{\boldsymbol{Q}} > \geq 0$ *and*

$$\begin{aligned} h(\boldsymbol{Q}) &= \boldsymbol{Q} - u^2_{\boldsymbol{Q}} + \boldsymbol{u_Q}\sqrt{\boldsymbol{u_Q}^2 + 1 - \|\boldsymbol{Q}\|^2} \\ &\to \boldsymbol{P} - u^2_{\boldsymbol{P}} + u_{\boldsymbol{P}}\sqrt{\boldsymbol{u_P}^2 + 1 - \|\boldsymbol{P}\|^2}\boldsymbol{P} - u^2_{\boldsymbol{P}} + u^2_{\boldsymbol{P}} = 0 \end{aligned}$$

On the other hand, if $u_{\boldsymbol{P}} < 0$, *for* $\boldsymbol{Q}$ *sufficiently closed to* $\boldsymbol{P}$ *we have*

$$\begin{aligned} h(\boldsymbol{Q}) &= \boldsymbol{Q} - u^2_{\boldsymbol{Q}} - \boldsymbol{u_Q}\sqrt{\boldsymbol{u_Q}^2 + 1 - \|\boldsymbol{Q}\|^2} \\ &\to \boldsymbol{P} - u^2_{\boldsymbol{P}} - u_{\boldsymbol{P}}\sqrt{\boldsymbol{u_P}^2 + 1 - \|\boldsymbol{P}\|^2}\boldsymbol{P} - u^2_{\boldsymbol{P}} - u_{\boldsymbol{P}}|u_{\boldsymbol{P}}| = 0 \end{aligned}$$

So in either case, the limiting value is $\boldsymbol{P}$. *We conclude we can define* $h|_{\partial D} = I$ *and that* h *is continuous on D. This says* h *is a retraction of D into* ∂D *which is not possible. So there must be a fixed point.* ■

Homework

Exercise 4.4.1 *Explain the proof of the Brouwer Fixed Point Theorem by drawing pictures.*

Exercise 4.4.2 *In the proof of the Brouwer Fixed Point Theorem, proof* $u_{\boldsymbol{Q}}$ *is a continuous function of* $\boldsymbol{Q}$ *on all of D.*

Exercise 4.4.3 *In the proof of the Brouwer Fixed Point Theorem, verify the roots to*

$$t = -\left\langle \boldsymbol{P}, \frac{f(\boldsymbol{P})-\boldsymbol{P}}{\|f(\boldsymbol{P})-\boldsymbol{P}\|} \right\rangle \pm \sqrt{\left(\left\langle \boldsymbol{P}, \frac{f(\boldsymbol{P})-\boldsymbol{P}}{\|f(\boldsymbol{P})-\boldsymbol{P}\|} \right\rangle\right)^2 + 1 - < \boldsymbol{P}, \boldsymbol{P} >}$$

are the ones we use.

Exercise 4.4.4 *In the proof of the Brouwer Fixed Point Theorem, provide details of the argument that in the case that $u_{\boldsymbol{P}} < 0$, we have as $\boldsymbol{Q} \to \boldsymbol{P}$, $h(\boldsymbol{Q}) \to h(\boldsymbol{P})$.*

4.5 De Rham Groups and 1-Forms

Let $\mathscr{U} \subset \Re^2$ be an open set. We need to define two important sets of 1-forms. First, we need to define locally constant functions.

Definition 4.5.1 Locally Constant Functions

Let $\mathscr{U} \subset \Re^2$ be open. We say $F : \mathscr{U} \to \Re$ is locally constant if it is constant on each component of $\mathscr{U}$.

Next, we need to define some special groups.

Definition 4.5.2 De Rham Groups

The sets $H^0(\mathscr{U})$ and the set of equivalence classes $H^1(\mathscr{U})$ are defined as follows:

(i): $H^0(\mathscr{U})$ is the set of all locally constant functions on $\mathscr{U}$.

(ii): Let $\mathbb{A}(\mathscr{U})$ be the set of all closed 1-forms on $\mathscr{U}$ and $\mathbb{B}(\mathscr{U})$ be the set of all exact 1-forms on $\mathscr{U}$. Then define the equivalence relation $\sim$ by $\omega_1 \sim \omega_2$ if $\omega_1 - \omega_2 = \delta$ for some $\delta \in \mathbb{B}(\mathscr{U})$. The set of all equivalence classes is denoted by $H^1(\mathscr{U}) \equiv \mathbb{A}(\mathscr{U}/\mathbb{B})(\mathscr{U})$ using the standard notation for cosets. We read this as $\mathbb{A}(\mathscr{U})$ mod $\mathbb{B}(\mathscr{U})$.

Of course, we do not know $\sim$ is an equivalence relation yet, but we can work that out. Let's get started. Recall if $\omega = Pdx + Qdy$ is closed means $d\omega = 0$ where $d\omega = (Q_x - P_y)dxdy$. Also, remember ω is exact if there is a function f so that $\omega = df = f_x dx + f_y dy$. This would then imply $d\omega = (f_{xy} - f(yx))dxdy = 0$. Hence, if ω is exact, ω is closed. To study $H^0(\mathscr{U})$ and $H^1(\mathscr{U})$ properly requires us to study another mapping called the **coboundary** map.

For any two open sets $\mathscr{U}$ and $\mathscr{V}$ in $\Re^2$, we want to define linear map $\delta : H^0(\mathscr{U} \cap \mathscr{V}) \to H^1(\mathscr{U} \cup \mathscr{V})$. We need some results to do this.

Lemma 4.5.1 C^∞ Difference Characterization

For any two open sets $\mathscr{U}$ and $\mathscr{V}$ in $\Re^2$, any C^∞ function F on $\mathscr{U} \cap \mathscr{V}$ can be written as the difference of two C^∞ functions $f_1 : \mathscr{U} \to \Re$ and $f_2 : \mathscr{V} \to \Re$ with $F = f_1 - f_2$ on $\mathscr{U} \cap \mathscr{V}$.

Proof 4.5.1
We use the idea of a partition of unity. The set is $\mathscr{U} \cup \mathscr{V}$ which is covered by the open cover $\{\mathscr{U}, \mathscr{V}\}$. Hence, the open set satisfies the finite union condition. There are thus two C^∞ functions ϕ and ψ on $\mathscr{U} \cup \mathscr{V}$ with $\overline{supp(\phi)} \subset \mathscr{U}$ and $\overline{supp(\psi)} \subset \mathscr{V}$ and $\phi + \psi = 1$ on $\mathscr{U} \cup \mathscr{V}$. Define $f_1 : \mathscr{U} \to \Re$ by

$$f_1(x) = \begin{cases} (\psi \circ F)(x), & x \in \mathscr{U} \cap \mathscr{V} \\ 0, & x \in \mathscr{U} \setminus \mathscr{U} \cap \mathscr{V} \end{cases}$$

Since, $\overline{supp(\psi)} \subset \mathscr{V}$, ψ *is zero on* $\mathscr{U} \setminus \mathscr{U} \cap \mathscr{V}$. *To see this, note*

$$\mathscr{U} \setminus \mathscr{U} \cap \mathscr{V} \quad = \quad \mathscr{U} \cap (\mathscr{U} \cap \mathscr{V})^C = \mathscr{U} \cap (\mathscr{V})^C$$

So $x \in \mathscr{U} \setminus \mathscr{U} \cap \mathscr{V}$, $x \in (\mathscr{V})^C$ *where* ψ *is zero. Also, such an* x *is in* $\mathscr{U}$ *and so is an interior point. Further,* $x \in \underline{(supp(\psi))}^C$ *which is open as well. Hence, there are radii* $\epsilon_1 > 0$ *and* $\epsilon_2 > 0$ *so that* $B(x, \epsilon_1) \cap \overline{supp(\psi)} = \emptyset$ *and* $B(x, \epsilon_2) \cap \mathscr{U} \subset \mathscr{U}$. *Thus,* $B(x, \min(\epsilon_1, \epsilon_2)) \subset \mathscr{U}$ *and* $\subset (\overline{supp(\psi)})^C$. *Hence, on* $B(x, \min(\epsilon_1, \epsilon_2))$, ψ *is zero. We see every point in* $\mathscr{U} \cap (\mathscr{V})^C$ *has a neighborhood disjoint from* $\overline{supp(\psi)}$. *Hence,* $f_1(B(x, \min(\epsilon_1, \epsilon_2))) = 0$. *This tells us* f_1 *is continuous on* $\mathscr{U} \setminus \mathscr{U} \cap \mathscr{V}$. *This argument can be repeated for any derivative* f_1^j. *Hence,* f_1 *is* C^∞ *on* $\mathscr{U} \setminus \mathscr{U} \cap \mathscr{V}$ *and so* f_1 *is* C^∞ *on* $\mathscr{U}$.

Similarly, define $f_2 : \mathscr{V} \to \Re$ *by*

$$f_2(x) \quad = \quad \begin{cases} -(\phi \circ F)(x), & x \in \mathscr{U} \cap \mathscr{V} \\ 0, & x \in \mathscr{V} \setminus \mathscr{U} \cap \mathscr{V} = \mathscr{V} \cap (\mathscr{U})^C \end{cases}$$

A similar argument shows f_2 *is* C^∞ *on* $\mathscr{U}$. *We see on* $\mathscr{U} \cap \mathscr{V}$

$$f_1(x) - f_2(x) \quad = \quad (\psi \circ F)(x) + (\phi \circ F)(x) = ((\psi + \phi) \circ F)(x) = F(x)$$

■

Homework

Exercise 4.5.1 *Let* $\mathscr{U} = B((0,0), 1)$ *and* $\mathscr{V} = B((1/2, 1/2), 1)$. *Let* $F(x, y) = x^2 + 2y^2$ *be defined on* $\mathscr{U} \cap \mathscr{V}$. *Find the functions* f_1 *and* f_2 *we use in Theorem 4.5.1. This means you will have to dig into the way the partition of unity functions are defined for an open cover!*

Exercise 4.5.2 *Let* $\mathscr{U} = B((1,1), 2)$ *and* $\mathscr{V}$ *be the open rectangle with vertices*

$$\{(1, -1), (4, -1), (4, 2), (1, 2)\}.$$

Let $F(x, y) = 2x^2 + 3y^2$ *be defined on* $\mathscr{U} \cap \mathscr{V}$. *Find the functions* f_1 *and* f_2 *we use in Theorem 4.5.1. This means you will have to dig into the way the partition of unity functions are defined for an open cover! This one will be hard!*

4.6 The Coboundary Map

Now we are ready to construct the coboundary map. The set $H^0(\mathscr{U})$ is the set of all locally constant functions on $\mathscr{U}$. If $\mathscr{U}$ has n connected components, $\mathscr{U} = \cup_{i=1}^n \mathscr{U}_i$ where the open sets $\mathscr{U}_i$ are pairwise disjoint. There is a constant c_i associated with each disjoint set $\mathscr{U}_i$. Define the functions e_i on $\mathscr{U}$ by $e_i(x) = 1$ if $x \in \mathscr{U}_i$ and 0 else. Then any locally constant f on $\mathscr{U}$ can be written as $f(x) = \left(\sum_{i=1}^n c_i e_i\right)(x)$ and so the set $\{e_1, \ldots, e_n\}$ spans $H^0(\mathscr{U})$. It is clear this set is also linearly independent. Thus, if $\mathscr{U}$ has n connected components, $H^0(\mathscr{U})$ is a vector space of dimension n. If $\mathscr{U}$ has infinitely many components, it is a vector space which is infinite dimensional.

The closed 1-forms on $\mathscr{U}$ satisfy the condition that if $\omega = Pdx + Qdy$, then $d\omega = (Q_x - P_y)dxdy = 0$. Clearly, linear combinations of closed 1-forms are closed and so this is a vector space also.

The exact 1-forms are a subset of the closed 1-forms and since linear combinations of exact 1-forms are also exact, they form a vector subspace of the vector space of closed 1-forms. Hence, the equivalence relation $\sim$ defined on the set of closed 1-forms by $\omega_1 \sim \omega_2$ if $\omega_1 - \omega_2 = \delta$ for some $\delta \in \mathbb{B}(\mathscr{U})$ is well-defined. The set of all equivalence classes denoted by $H^1(\mathscr{U}) \equiv \mathbb{A}(\mathscr{U}/\mathbb{B})(\mathscr{U})$ is therefore a well-understood mathematical construction. Recall, the usual notation for a member of this set is $[\cdot]$.

We have previously defined the 1-form

$$\omega_{\boldsymbol{P}} = \frac{1}{2\pi} \frac{-(y-y_0)dx + (x-x_0)dy}{(x-x_0)^2 + (y-y_0)^2}$$

for the point $\boldsymbol{P} = (x_0, y_0)$. It is straightforward to show, by direct calculation, that $\omega_{\boldsymbol{P}}$ is a closed 1-form.

Theorem 4.6.1 Characterizations of $H^1(\mathscr{U})$

> *(i): If $\mathscr{U}$ is an open rectangle, $H^1(\mathscr{U}) = [0]$.*
>
> *(ii): If $\mathscr{U} = \Re^2 \setminus \boldsymbol{P}$, $\{[\omega_{\boldsymbol{P}}]\}$ is a basis for $H^1(\mathscr{U})$.*
>
> *(iii): If $\mathscr{U} = \Re^2 \setminus \{\boldsymbol{P}, \boldsymbol{Q}\}$, $\{[\omega_{\boldsymbol{P}}], [\omega_{\boldsymbol{Q}}]\}$ is a basis for $H^1(\mathscr{U})$.*

Proof 4.6.1
(i): *If ω is a closed 1-form and $\mathscr{U}$ is an open rectangle, by Lemma 3.1.1, ω is exact, which implies it is closed. It follows immediately that $H^1(\mathscr{U}) = [0]$.*

(ii): *Fix $r > 0$. Let $\gamma_{\boldsymbol{P},r}$ be a ccw circle of radius r about $\boldsymbol{P}$; i.e.,*

$$\gamma_{\boldsymbol{P},r} = \boldsymbol{P} + r(\cos(2\pi t), \sin(2\pi t)), 0 \le t \le 1.$$

Look at $\omega_{\boldsymbol{P}}$ here. Note $[\omega_{\boldsymbol{P}}] \neq [0]$ as if $\omega_{\boldsymbol{P}}$ were exact, then $\omega_{\boldsymbol{P}} = df$ which implies $\int_{\gamma_{\boldsymbol{P},r}} \omega_{\boldsymbol{P}} = 0$. But we know this line integral is $+1$. Hence, $[\omega_{\boldsymbol{P}}] \neq [0]$.

Now let ω be any closed 1-form on $\mathscr{U}$. Let $c = \int_{\gamma_{\boldsymbol{P},r}} \omega$. Consider $\Omega = \omega - c\omega_{\boldsymbol{P}}$. This is a closed 1-form on $\mathscr{U}$ also. Thus,

$$\int_{\gamma_{\boldsymbol{P},r}} \Omega = \int_{\gamma_{\boldsymbol{P},r}} \omega - c \int_{\gamma_{\boldsymbol{P},r}} \omega_{\boldsymbol{P}} = c - c = 0$$

Then, by Theorem 3.2.4, $\omega - c\omega_{\boldsymbol{P}}$ is exact. Hence, $[\omega - c\omega_{\boldsymbol{P}}] = [0]$ which implies $[\omega] = c[\omega_{\boldsymbol{P}}]$. Thus, $H^1(\Re^2 \setminus \boldsymbol{P})$ has basis $[\omega_{\boldsymbol{P}}]$.

(iii): *Fix $0 < r < \|\boldsymbol{P} - \boldsymbol{Q}\|$. Consider $\omega = a\omega_{\boldsymbol{P}} + b\omega_{\boldsymbol{Q}}$. Let $\gamma_{\boldsymbol{P},r}$ and $\gamma_{\boldsymbol{Q},r}$ be our usual ccw paths. Then,*

$$\int_{\gamma_{\boldsymbol{P},r}} \omega_{\boldsymbol{P}} = 1, \quad \int_{\gamma_{\boldsymbol{Q},r}} \omega_{\boldsymbol{Q}} = 1$$

Now, if ω were exact, then

$$0 = \int_{\gamma_{P,r}} \omega_P = a \int_{\gamma_{P,r}} \omega_P + b \int_{\gamma_{P,r}} \omega_Q = a \times 1 + b \times 0 = a$$
$$0 = \int_{\gamma_{Q,r}} \omega_P = a \int_{\gamma_{Q,r}} \omega_P + b \int_{\gamma_{Q,r}} \omega_Q = a \times 0 + b \times 1 = b$$

which tells us $a = b = 0$. So if $a\omega_P + b\omega_Q$ is exact, then $a = b = 0$. Hence, if $a^2 + b^2 \neq 0$, $a\omega_P + b\omega_Q$ is not exact. Thus, $[a\omega_P + b\omega_Q] \neq [0]$ unless $a = b = 0$. Hence, $\{\omega_P, \omega_Q\}$ is a linearly independent set.

Next, if ω is any closed 2-form on $\Re^2 \setminus \{P, Q\}$, let

$$a = \int_{\gamma_{P,r}} \omega_P, \quad b = \int_{\gamma_{Q,r}} \omega_Q.$$

Consider $\Omega = \omega - a\omega_P - b\omega_Q$. This is a closed 1-form on $\Re^2 \setminus \{P, Q\}$. We have

$$\int_{\gamma_{P,r}} \Omega = \int_{\gamma_{P,r}} \omega - a \int_{\gamma_{P,r}} \omega_P - b \int_{\gamma_{P,r}} \omega_Q = \int_{\gamma_{P,r}} \omega - a \times 1 - b \times 0 = 0$$
$$\int_{\gamma_{Q,r}} \Omega = \int_{\gamma_{Q,r}} \omega - a \int_{\gamma_{Q,r}} \omega_P - b \int_{\gamma_{Q,r}} \omega_Q = \int_{\gamma_{Q,r}} \omega - a \times 0 - b \times 01 = 0$$

Then by Theorem 3.2.5, we see Ω is exact. Hence $[\Omega] = a[\omega_P] + b\omega_Q$ which shows spanning. ■

Homework

Exercise 4.6.1 *Generalize Theorem 4.6.1 to three points.*

Exercise 4.6.2 *Generalize Theorem 4.6.1 to three points.*

Exercise 4.6.3 *Generalize Theorem 4.6.1 to five points.*

Exercise 4.6.4 *Generalize Theorem 4.6.1 to n points. Pay close attention to how you generalize your argument!*

Now, let's generalize to connected closed subsets.

Theorem 4.6.2 If $A \subset \Re^2$ is Connected and Closed and P and Q are in A, then the Equivalence Class for $[\omega_P]$ and the Equivalence Class for $[\omega_Q]$ are Equal in $H^1(\Re^2 \setminus A)$

If $A \subset \Re^2$ is closed and connected and P and Q are in A, then $[\omega_P] = [\omega_Q]$ in $H^1(\Re^2 \setminus A)$.

Proof 4.6.2

Let $\omega = \omega_P - \omega_Q$. We need to show ω is exact on $\Re^2 \setminus A$. Recall from Theorem 3.1.2, it is enough to show $\int_\gamma \omega = 0$ for all segmented closed paths in $\mathscr{U} = \Re^2 \setminus A$. Consider for such a path γ

$$\int_\gamma \omega = \int_\gamma \omega_P - \int_\gamma \omega_Q$$

We have also shown in Theorem 4.1.1 that $W(\gamma, \boldsymbol{P})$ is constant on each connected component of $\Re^2 \setminus supp(\gamma) = \Re^2 \setminus \gamma([a,b])$ where $[a,b]$ is the domain of γ. We also know $W(\gamma, \boldsymbol{P}) = 0$ on the unbounded components. Now A is connected and closed. So if $\boldsymbol{P}$ and $\boldsymbol{Q}$ live in the same connected component, $W(\gamma, \boldsymbol{P}) = W(\gamma, \boldsymbol{Q})$. Hence, $\int_\gamma \gamma = 0$ and so by Theorem 3.1.2, ω is exact. Thus, $[\omega] = [\omega_{\boldsymbol{P}} - \omega_{\boldsymbol{Q}}] = [0]$. We conclude $[\omega_{\boldsymbol{P}}] = [\omega_{\boldsymbol{Q}}]$. ■

Now let's construct the coboundary map $\delta : H^0(\mathscr{U} \cap \mathscr{V}) \rightarrow H^1(\mathscr{U} \cup \mathscr{V})$ for the two open sets $\mathscr{U}$ and $\mathscr{V}$ in $\Re^2$. If f is locally constant on $\mathscr{U} \cap \mathscr{V}$, then by Lemma 4.5.1, $f = f_1 - f_2$ on $\mathscr{U} \cap \mathscr{V}$ and since f is locally constant, $df = df_1 - df_2 = 0$ on $\mathscr{U} \cap \mathscr{V}$. Define the 1-form ω by

$$\omega = \begin{cases} df_1 = df_2, & \text{on } \mathscr{U} \cap \mathscr{V} \\ df_1, & \text{on } \mathscr{U} \cap (\mathscr{V})^C \\ df_2, & \text{on } (\mathscr{U})^C \cap \mathscr{V} \end{cases}$$

Since ω is exact on $\mathscr{U}$, ω is closed on $\mathscr{U}$ and by the same reasoning, ω is closed on $\mathscr{V}$. Let $\delta(f) = [\omega] \in H^1(\mathscr{U} \cup \mathscr{V})$. We need to show δ is well-defined. Assume g_1 and g_2 are a different choice of representatives of f on $\mathscr{U} \cap \mathscr{V}$. Define Ω by

$$\Omega = \begin{cases} dg_1 = dg_2, & \text{on } \mathscr{U} \cap \mathscr{V} \\ dg_1, & \text{on } \mathscr{U} \cap (\mathscr{V})^C \\ dg_2, & \text{on } (\mathscr{U})^C \cap \mathscr{V} \end{cases}$$

Then $g_1 - g_2 = f_1 - f_2$ on $\mathscr{U} \cap \mathscr{V}$ or $g_1 - f_1 = g_2 - f_2$ on $\mathscr{U} \cap \mathscr{V}$. So we have a C^∞ function F such that

$$F = \begin{cases} g_1 - f_1 = g_2 - f_2, & \text{on } \mathscr{U} \cap \mathscr{V} \\ g_1 - f_1, & \text{on } \mathscr{U} \cap (\mathscr{V})^C \\ g_2 - f_2, & \text{on } (\mathscr{U})^C \cap \mathscr{V} \end{cases}$$

and so

$$dF = \begin{cases} dg_1 - df_1 = dg_2 - df_2 = dg_2, & \text{on } \mathscr{U} \cap \mathscr{V} \\ dg_2 - dg_1 = df_2 - df_1 = df = 0, & \text{on } \mathscr{U} \cap \mathscr{V} \\ dg_1 - df_1, & \text{on } \mathscr{U} \cap (\mathscr{V})^C \\ dg_2 = df_2, & \text{on } (\mathscr{U})^C \cap \mathscr{V} \end{cases}$$

Hence, $dF = \Omega - \omega$ is exact, which tells us $\Omega = \omega + dF$. Hence, $[\Omega] = [\omega]$ which shows the coboundary map δ is independent of the choice of representatives of f. Let's state all of this formally.

Definition 4.6.1 Coboundary Map

For any two open sets $\mathscr{U}$ and $\mathscr{V}$ in $\Re^2$, define the map $\delta : H^0(\mathscr{U} \cap \mathscr{V}) \rightarrow H^1(\mathscr{U} \cup \mathscr{V})$ as follows: for f is locally constant on $\mathscr{U} \cap \mathscr{V}$, f has a C^∞ representation $f = f_1 - f_2$ on $\mathscr{U} \cap \mathscr{V}$ with $df = df_1 - df_2 = 0$ on $\mathscr{U} \cap \mathscr{V}$. Define the 1-form ω by

$$\omega = \begin{cases} df_1 = df_2, & \text{on } \mathscr{U} \cap \mathscr{V} \\ df_1, & \text{on } \mathscr{U} \cap (\mathscr{V})^C \\ df_2, & \text{on } (\mathscr{U})^C \cap \mathscr{V} \end{cases}$$

Then, $[\omega] \in H^1(\mathscr{U} \cup \mathscr{V})$ and this value is independent of the choice of representatives of f. Define $\delta(f) = [\omega]$. We call δ the coboundary map.

We now prove δ is a linear map.

Lemma 4.6.3 Coboundary Map is Linear

Let $\mathscr{U}$ and $\mathscr{V}$ be open sets in $\Re^2$. The coboundary map δ is linear: i.e. $\delta(f+g) = \delta(f) + \delta(g)$ and $\delta(cf) = c\delta(f)$ for locally constant f and g and $c \in \Re$ a constant.

Proof 4.6.3

From the construction of δ, $\delta(f) = f_1 - f_2$ on $\mathscr{U} \cap \mathscr{V}$ and $\delta(g) = g_1 - g_2$ on $\mathscr{U} \cap \mathscr{V}$. We also know $f + g = (f_1 + g_1) - (f_2 + g - 2)$ on $\mathscr{U} \cap \mathscr{V}$. Let ω_f be the 1-form

$$\omega_f = \begin{cases} df_1 = df_2, & \text{on } \mathscr{U} \cap \mathscr{V} \\ df_1, & \text{on } \mathscr{U} \cap (\mathscr{V})^C \\ df_2, & \text{on } (\mathscr{U})^C \cap \mathscr{V} \end{cases}$$

and

$$\omega_g = \begin{cases} dg_1 = dg_2, & \text{on } \mathscr{U} \cap \mathscr{V} \\ dg_1, & \text{on } \mathscr{U} \cap (\mathscr{V})^C \\ dg_2, & \text{on } (\mathscr{U})^C \cap \mathscr{V} \end{cases}$$

Then, we have

$$\omega_f + \omega_g = \begin{cases} df_1 + dg_1 = df_2 + dg_2, & \text{on } \mathscr{U} \cap \mathscr{V} \\ d(f+g), & \text{on } \mathscr{U} \cap \mathscr{V} \\ df_1 + dg_1, & \text{on } \mathscr{U} \cap (\mathscr{V})^C \\ d((f+g)_1), & \text{on } \mathscr{U} \cap (\mathscr{V})^C \\ df_2 + dg_2, & \text{on } (\mathscr{U})^C \cap \mathscr{V} \\ d((f+g)_2), & \text{on } (\mathscr{U})^C \cap \mathscr{V} \end{cases}$$

Then, $\delta(f) + \delta(g) = \omega_f + \omega_g = d(f+g) = \delta(f+g)$.

Finally,

$$\omega_{cf} = \begin{cases} cdf_1 = cdf_2, & \text{on } \mathscr{U} \cap \mathscr{V} \\ cdf_1, & \text{on } \mathscr{U} \cap (\mathscr{V})^C \\ cdf_2, & \text{on } (\mathscr{U})^C \cap \mathscr{V} \end{cases} = c\omega_f$$

So $\delta(cf) = c\omega_f = c\delta(f)$. ■

What does the kernel of δ look like?

Lemma 4.6.4 Characterizing the Kernel of the Coboundary Map

Let $\mathscr{U}$ and $\mathscr{V}$ be open sets in $\Re^2$. A locally constant f on $\mathscr{U} \cap \mathscr{V}$ is in the kernel of δ if and only if there exist locally constant functions f_1 on $\mathscr{U}$ and f_2 on $\mathscr{V}$ with $f = f_1 - f_2$ on $\mathscr{U} \cap \mathscr{V}$. In particular, if $\mathscr{U}$ and $\mathscr{V}$ are connected, then $ker(\delta) = \{f | f \text{ is constant on } \mathscr{U} \cap \mathscr{V}\}$

Proof 4.6.4

If $f = f_1 - f_2$ with f_1 locally constant on $\mathscr{U}$ and f_2 locally constant on $\mathscr{V}$, then the functions f_1

and f_2 are a C^∞ representation of f and so there is a 1-form ω on $\mathscr{U} \cup \mathscr{V}$ given by

$$\omega = \begin{cases} df_1 = df_2 = 0, & \text{on } \mathscr{U} \cap \mathscr{V} \\ df_1 = 0, & \text{on } \mathscr{U} \cap (\mathscr{V})^C \\ df_2 = 0, & \text{on } (\mathscr{U})^C \cap \mathscr{V} \end{cases}$$

Hence, $\delta(f) = [0]$.

Conversely, if $\delta(f) = [0]$, then

$$\omega = \begin{cases} df_1 = df_2, & \text{on } \mathscr{U} \cap \mathscr{V} \\ df_1, & \text{on } \mathscr{U} \cap (\mathscr{V})^C \\ df_2, & \text{on } (\mathscr{U})^C \cap \mathscr{V} \end{cases}$$

and ω is exact. Hence, there is a function g so that $\omega = dg$. Thus

$$dg = \begin{cases} df_1 = df_2, & \text{on } \mathscr{U} \cap \mathscr{V} \\ df_1, & \text{on } \mathscr{U} \cap (\mathscr{V})^C \\ df_2, & \text{on } (\mathscr{U})^C \cap \mathscr{V} \end{cases}$$

We see $f_1 - g$ is locally constant on $\mathscr{U}$ and $f_2 - g$ is locally constant on $\mathscr{V}$. So, $f = f_1 - f_2 = (f_1 - g) - (f_2 - g)$ on $\mathscr{U} \cap \mathscr{V}$ and thus is the difference of two locally constant functions as desired.

Finally, if $\mathscr{U}$ and $\mathscr{V}$ are connected, the locally constant functions f_1 and f_2 are constant on all $\mathscr{U}$ and $\mathscr{V}$ respectively. This forces f to be constant on $\mathscr{U} \cap \mathscr{V}$. ■

Now let's characterize the image of δ.

Lemma 4.6.5 Image of the Coboundary Map

Let $\mathscr{U}$ and $\mathscr{V}$ be open sets in $\Re^2$. The equivalence class $[\omega]$ of a closed 1-form ω on $\mathscr{U} \cup \mathscr{V}$ is in the image of δ if and only if $\omega|_{\mathscr{U}}$ and $\omega|_{\mathscr{V}}$ are exact. In particular, if $H^1(\mathscr{U}) = [0]$ and $H^1(\mathscr{V}) = [0]$, then δ is onto.

Proof 4.6.5
If ω is in the image of δ, then there are functions, f_1 on $\mathscr{U}$ and f_2 on $\mathscr{V}$, so that $\omega = df_1$ on $\mathscr{U}$ and $\omega = df_2$ on $\mathscr{V}$, which implies the exactness condition.

Conversely, if $\omega|_{\mathscr{U}}$ and $\omega|_{\mathscr{V}}$ are exact, then there are functions f_1 so that $\omega|_{\mathscr{U}} = df_1$ and $\omega|_{\mathscr{V}} = df_2$. Then, on the intersection, the function $f = f_1 - f - 2$ satisfies $df_1 - df_2 = 0$, so f is locally constant with representatives f_1 and f_2. Then $\delta(f) = \Omega$ where

$$\Omega = \begin{cases} df_1 = \omega|_{\mathscr{U}} = df_2 = \omega|_{\mathscr{V}}, & \text{on } = \omega|_{\mathscr{V}} \\ df_1 = \omega|_{\mathscr{U}}, & \text{on } \mathscr{U} \cap (\mathscr{V})^C \quad = \omega \\ df_2 = \omega|_{\mathscr{V}}, & \text{on } (\mathscr{U})^C \cap \mathscr{V} \end{cases}$$

Finally, note if $H^1(\mathscr{U}) = [0]$ and $H^1(\mathscr{V}) = [0]$, we know ω closed on $\mathscr{U} \cup \mathscr{V}$ implies $\omega|_{\mathscr{U}}$ and $\omega|_{\mathscr{V}}$ are closed and thus, by assumption, are exact. Hence, there is a locally constant function f on $\mathscr{U} \cap \mathscr{V}$ so that $\delta(f) = \omega$. This shows δ is onto in this case. ■

Homework

Exercise 4.6.5 *Prove the three point variation of Theorem 4.6.2:* $A \subset \Re^2$ *is connected and closed and* $\boldsymbol{P}$*,* $\boldsymbol{Q}$ *and* $\boldsymbol{R}$ *are in A, then the equivalence class for* $[\omega_{\boldsymbol{P}}]$*, the equivalence class for* $[\omega_{\boldsymbol{Q}}]$ *and the equivalence class for* $[\omega_{\boldsymbol{R}}]$ *are equal in* $H^1(\Re^2 \setminus A)$*.*

Exercise 4.6.6 *Extend Lemma 4.6.3 to linear combinations: let* $\mathscr{U}$ *and* $\mathscr{V}$ *be open sets in* $\Re^2$*. The coboundary map* δ *is linear: i.e.* $\delta(\sum_{i=1}^n c_i f_i) = \sum_{i=1}^n c_i \delta(f_1)$ *for locally constant* f_i *and* $c_i \in \Re$ *constant.*

4.7 The Inside and Outside of a Curve

Let's begin by remembering what a homeomorphism means. Given two sets X and Y in $\Re^n$, if $f : X \to Y$ is 1-1 and onto and continuous with a continuous inverse, we say f is a homeomorphism. Our first result is to prove that subsets A of $\Re^2$ that are homeomorphic to a closed interval force the set $\Re^2 \setminus A$ to be connected. This sounds very abstract, but if you think of A as the image $\gamma([a,b])$ of some path, then we know $\gamma([a,b])$ is another interval $[c,d]$ and we also know $\Re^2 \setminus \gamma([a,b]) = \Re^2 \setminus [c,d]$ is connected. We are going to prove this fact below. As you might expect, since all we know is that the set Y is homeomorphic to a closed interval, the proof is going to be a bit intense. It is a nice use of a lot of the material we have been learning.

Theorem 4.7.1 Subset Y of $\Re^2$ Homeomorphic to a Closed Interval Implies $\Re^2 \setminus Y$ is Connected

If $f : [a,b] \to Y$ *is a homeomorphism, then* $\Re^2 \setminus Y$ *is connected.*

Proof 4.7.1
Since Y *is homeomorphic to* $[c,d]$ *by assumption, there is a 1-1, onto, continuous map* $\phi : Y \to [c,d]$ *which has a continuous inverse. Hence* $\phi^{-1} : [c,d] \to Y$ *is also a homeomorphism. It does not matter if the interval is* $[c,d]$*; so for convenience of exposition, let's assume it is the interval* $[0,1]$ *instead and let's rename* ϕ^{-1} *as* f*. Then* $f : [0,1] \to Y$ *is a homeomorphism.*

Let $A = f([0,1/2])$*,* $B = f([1/2,1])$*. Let* $\boldsymbol{Q} = f(1/2)$*. Then* $A \cap B = \boldsymbol{Q}$*. Let* $\mathscr{U} = \Re^2 \setminus A$ *and* $\mathscr{V} = \Re^2 \setminus B$ *which are both open sets. Note* $A \cup B = f([0,1]) = Y$*. Then,*

$$\begin{aligned} \mathscr{U} \cup \mathscr{V} &= \Re^2 \cap A^C \;\cup\; \Re^2 \cap B^C \\ &= A^C \cup B^C = (A \cap B)^C = \Re^2 \setminus \boldsymbol{Q} \\ \mathscr{U} \cap \mathscr{V} &= \Re^2 \cap A^C \;\cap\; \Re^2 \cap B^C \\ &= A^C \cap B^C = (A \cap B)^C = \Re^2 \setminus Y \end{aligned}$$

Assume Y *is* **not** *connected and let* $\boldsymbol{P}_0$ *and* $\boldsymbol{P}_1$ *be two points in different components. Look at the coboundary map*

$$\delta : H^0(\mathscr{U} \cap \mathscr{V}) = H^0(\Re^2 \setminus Y) \;\to\; H^1(\mathscr{U} \cap \mathscr{V}) = H^1(\Re^2 \setminus \boldsymbol{Q})$$

By the second part of Theorem 4.6.1, $H^1(\Re^2 \setminus \boldsymbol{Q})$ *is generated by* $[\omega_{\boldsymbol{Q}}]$*. By Lemma 4.6.5, the image of* δ *consists of the 1-forms* ω *is* $\omega|_{\mathscr{U}} = \omega|_{\Re^2 \setminus A}$ *and* $\omega|_{\mathscr{V}} = \omega|_{\Re^2 \setminus B}$ *are exact. We then know the image of* δ *is* $\{a[\omega_{\boldsymbol{Q}}] | a\omega_{\boldsymbol{Q}}\}$ *where* $a\omega_{\boldsymbol{Q}}$ *is exact on* $\Re^2 \setminus A$ *and* $\Re^2 \setminus A$*. If* $a\omega_{\boldsymbol{Q}}$ *is exact on* $\Re^2 \setminus A$*, then if* $\mathscr{C}_r$ *is a circle about* $\boldsymbol{Q}$ *of sufficiently large radius* r *so that* A *is in the interior of the circle* $\mathscr{C}_r$*, we have* $0 = \int_{\mathscr{C}_r} a\omega = a \int_{\mathscr{C}_r} \omega$*. But* $\int_{\mathscr{C}_r} \omega = 1$*. This is not possible unless* $a = 0$*. This implies the image of* δ *is* $[0]$*. Hence, every locally constant function* g *on* $\Re^2 \setminus Y$ *is sent to* $[0]$*. We conclude every locally constant function* g *on* $\Re^2 \setminus Y$ *is in the kernel of* δ*.*

From Lemma 4.6.4, g in the kernel of δ has the form $g = g_1 - g_2$ on $\mathscr{U} \cap \mathscr{V} = \Re^2 \setminus Y$ with g_1 and g_2 locally constant on $\mathscr{U} = \Re^2 \setminus A$ and $\mathscr{V} = \Re^2 \setminus B$, respectively. Let g be a locally constant function on $\mathscr{U} \cap \mathscr{V} = \Re^2 \setminus Y$ which takes on different values at the points $\boldsymbol{P_0}$ and $\boldsymbol{P_1}$ that live in different components. We claim $\boldsymbol{P_0}$ and $\boldsymbol{P_1}$ must actually be in either

(i): two different components of $\Re^2 \setminus A$,

(ii): two different components of $\Re^2 \setminus B$,

(iii): two different components of both sets.

If this claim is false, $\boldsymbol{P_0}$ and $\boldsymbol{P_1}$ are in the same component of $\Re^2 \setminus A$ and $\Re^2 \setminus B$. Then, since g_1 is locally constant on $\Re^2 \setminus A$, $g_1(\boldsymbol{P_0}) = g_1(\boldsymbol{P_1})$. The same is true for g_2. Thus, on $\mathscr{U} \cap \mathscr{V} = \Re^2 \setminus Y$, $g(\boldsymbol{P_0}) = g_1(\boldsymbol{P_0}) - g_2(\boldsymbol{P_0}) = g_1(\boldsymbol{P_1}) - g_2(\boldsymbol{P - 1}) = g(\boldsymbol{P_1})$. But we started by assuming $g(\boldsymbol{P_0}) \neq g(\boldsymbol{P_1})$. This is a contradiction. So our claim must be true.

To finish the argument, we use a chain of reasoning similar in spirit to the way we would prove the Bolzano - Weierstrass Theorem. Since Y is in two pieces, $Y = Y_1 \cup Y_2$. Let Y_1 be the piece, which is either A or B at this first step, which holds $\boldsymbol{P_0}$ and $\boldsymbol{P_1}$ in two components. Clearly $Y_1 \subset Y$. Then, $Y_1 = f([a_1, b_1])$ where $[a_1, b_1] = [0, 1/2]$ or $[a_1, b_1] = [1/2, 1]$. Note $b_1 - a_1 = 1/2$. Bisect Y_1 into two pieces. At least one of these new pieces contains $\boldsymbol{P_0}$ and $\boldsymbol{P_1}$ in separate components. Call this piece Y_2 and so $Y_2 = f([a_2, b_2])$ and $b_2 - a_2 = 1/2^2$ because we cut Y_1 in half. We have $Y_2 \subset Y_1 \subset Y$.

Now finish with an induction step. Given $Y_{n-1} = f([a_{n-1}, b_{n-1}])$ with $b_{n-1} - a_{n-1} = 1/2^{n-1}$ and $Y_{n-1} \subset Y_{n-2} \subset \ldots \subset Y_1 \subset Y$, bisect it also. One of these pieces contains $\boldsymbol{P_0}$ and $\boldsymbol{P_1}$ in two separate components. Call this piece $Y_n = f([a_n, b_n])$ with $b_n - a_n = 1/2^n$.

Hence, we have a sequence of compact sets $Y_n = f([a_n, b_n])$. Since (a_n) is monotonically increasing and (b_n) is monotonically decreasing, there are points α and β so that $a_n \rightarrow \alpha$ and $b_n \rightarrow \beta$. Thus, $|\beta - \alpha| < 1/2^n$ for all n implying $\alpha = \beta$. Let $\boldsymbol{P} = f(\alpha)$. Since $\Re^2 \setminus \boldsymbol{P}$ is connected, there is a path from $\boldsymbol{P_0}$ to $\boldsymbol{P_1}$ which does not hit $\boldsymbol{P}$. Hence, there is an $\epsilon > 0$ so that $B(\boldsymbol{P}, \epsilon)$ does not contain any points of this path. This tells us that for n sufficiently large, $Y_n \subset B(\boldsymbol{P}, \epsilon)$. So $\{\boldsymbol{P_0}, \boldsymbol{P_1}\} \in \Re^2 \setminus Y_n$ for some n and connected by a path. This means $\boldsymbol{P_0}$ and $\boldsymbol{P_1}$ are in the same component of this Y_n which violates the construction process for Y_n. Our assumption that $\boldsymbol{P_0}$ and $\boldsymbol{P_1}$ are in two different components is therefore wrong and we can conclude such points do not exist: i.e. $\Re^2 \setminus Y$ is connected.
■

Note how we used the coboundary map in the previous theorem! Now on to the next theorem about the inside and outside of a closed curve.

Theorem 4.7.2 Inside and Outside of a Closed Curve: the Jordan Curve Theorem

If $X \subset \Re^2$ is homeomorphic to a circle, then $\Re^2 \setminus X$ has two connected components: one bounded and one unbounded. Any neighborhood of a point in X has nonempty intersection with both components.

Proof 4.7.2
Let $\boldsymbol{P} \neq \boldsymbol{Q}$ be in X. Let f be a homeomorphism for the circle $\mathscr{C}_r$ which has radius $r > 0$ and is centered at a point $\boldsymbol{c} = (c_0, c_1)$. The points of $\mathscr{C}_r$ can be represented by $\phi(t) = \boldsymbol{c} + (r\cos(2\pi t), r\sin(2\pi t))$ and so $f(\phi(t)) \in X$ for all $0 \leq t \leq 1$. There is a point $t_{\boldsymbol{P}}$ and a point $t_{\boldsymbol{Q}}$ so that $\boldsymbol{P} = f(\phi(t_{\boldsymbol{P}}))$ and $\boldsymbol{Q} = f(\phi(t_{\boldsymbol{Q}}))$. We may assume without loss of generality that $t_{\boldsymbol{P}} < t_{\boldsymbol{Q}}$. Let $A = \{f(\phi(t)) : t_{\boldsymbol{P}} \leq t \leq t_{\boldsymbol{Q}}\}$ and $B = X \setminus A$. Now f is 1-1 and onto, so $A \cup B = X$

and $A \cap B = \{\boldsymbol{P}, \boldsymbol{Q}\}$. *Let* $\mathscr{U} = \Re^2 \setminus A$ *and* $\mathscr{V} = \Re^2 \setminus B$. *We have*

$$\begin{aligned} \mathscr{U} \cup \mathscr{V} &= (A \cap B)^C = \Re^2 \setminus \{\boldsymbol{P}, \boldsymbol{Q}\} \\ \mathscr{U} \cap \mathscr{V} &= (A \cup B)^C = \Re^2 \setminus X \end{aligned}$$

We will show $\Re^2 \setminus X$ *has two components by first showing the dimension of* $H^0(\mathscr{U} \cap \mathscr{V}) = H^0(\Re^2 \setminus X) = 2$.

From Theorem 4.6.1, we know $H^1(\Re^2 \setminus \{\boldsymbol{P}, \boldsymbol{Q}\})$ *has dimension two and has basis* $\{\boldsymbol{P}, \boldsymbol{Q}\}$. *Consider the coboundary map*

$$\delta : H^0(\mathscr{U} \cap \mathscr{V}) = H^0(\Re^2 \setminus X) \quad \rightarrow \quad H^1(\mathscr{U} \cup \mathscr{V}) = H^1(\Re^2 \setminus \{\boldsymbol{P}, \boldsymbol{Q}\})$$

by Theorem 4.7.1, A *and* B *are homeomorphic to closed intervals, so* $\mathscr{U} = \Re^2 \setminus A$ *and* $\mathscr{V} = \Re^2 \setminus B$ *are connected. Thus, by Lemma 4.6.4, the kernel of* δ *is the set of constant functions on* $\mathscr{U} \cap \mathscr{V} = \Re^2 \setminus X$. *Hence, the kernel of* δ *is one dimensional. Since* δ *is linear, the dimension of the kernel of* δ *plus the dimension of the image of* δ *must be the dimension of* $H^0(\mathscr{U} \cap \mathscr{V}) = H^0(\Re^2 \setminus X)$.

Now consider the image of δ. *We claim the image of* δ *is characterized by*

$$Im(\delta) \quad = \quad \{a[\omega_{bsP}] + b[\omega_{\boldsymbol{Q}}]\}$$

To see this, any ω *in* $H^1(\Re^2 \setminus \{\boldsymbol{P}, \boldsymbol{Q}\})$ *has a unique representation as*

$$\omega \quad = \quad a[\omega_{bsP}] + b[\omega_{\boldsymbol{Q}}]$$

From Lemma 4.6.5, we see δ *is onto, if* $a[\omega_{bsP}] + b[\omega_{\boldsymbol{Q}}]$ *restricted to* $\mathscr{U} = \Re^2 \setminus A$ *is exact and if when it is restricted to* $\mathscr{U} = \Re^2 \setminus B$, *it is also exact. Now* $\boldsymbol{P}$ *and* $\boldsymbol{Q}$ *are in* $A \cap B$ *and so are in the same connected component of both* $\Re^2 \setminus A$ *and* $\Re^2 \setminus B$. *Now we can apply Theorem 4.6.2 to infer* $[\omega_{bsP}] = [\omega_{bsQ}]$ *in* $\Re^2 \setminus A$. *So* $[\omega_{bsP} - \omega_{bsQ}] = [0]$ *in* $H^1(\Re^2 \setminus A)$. *A similar argument works for the set* B. *Hence,* $[\omega_{\boldsymbol{P}} - \omega_{bsQ}] = [0]$ *and so* $\omega_{\boldsymbol{P}} - \omega_{bsQ}$ *is exact in* $H^1(\Re^2 \setminus A)$ *and* $H^1(\Re^2 \setminus B)$, *respectively.*

Now, if $a + b = 0$, *then* $a\omega_{bsP} + b\omega_{\boldsymbol{Q}} = a(\omega_{bsP} - \omega_{\boldsymbol{Q}})$ *is also exact in both* $H^1(\Re^2 \setminus A)$ *and* $H^1(\Re^2 \setminus B)$. *Hence,* $a[\omega_{bsP}] + b[\omega_{\boldsymbol{Q}}]$ *with* $a + b = 0$ *is in* $Im(\delta)$. *Conversely, if* $\omega = a\omega_{\boldsymbol{P}} + b\omega_{\boldsymbol{Q}}$ *was exact on* $\Re^2 \setminus A$ *and* $\Re^2 \setminus B$, *let* γ *be a circle about the origin large enough to contain* X. *Since* ω *is exact on* $\Re^2 \setminus A$, $\int_\gamma \omega = 0$. *But since* $\boldsymbol{P}$ *and* $\boldsymbol{Q}$ *are in the circle, we have* $\int_\gamma \omega_{\boldsymbol{P}} = 1$ *and* $\int_\gamma \omega_{\boldsymbol{Q}} = 1$. *Hence,*

$$0 = \int_\gamma (a\omega_{\boldsymbol{P}} + b\omega_{\boldsymbol{Q}}) \quad = \quad a\int_\gamma \omega_{\boldsymbol{P}} + b\int_\gamma \omega_{\boldsymbol{Q}} = a + b$$

This tells us $a + b = 0$. *From these discussions, we conclude the dimension of* $Im(\delta)$ *is one. Combining our results, we have*

$$dim(ker(\delta)) + dim(Im(\delta)) \quad = \quad 1 + 1 = dim(H^0(\mathscr{U} \cap \mathscr{V}) = H^0(\Re^2 \setminus X))$$

Thus, the dimension of $H^0(\Re^2 \setminus X) = 2$. *This tells us* $H^0(\Re^2 \setminus X)$ *has two components. Since* X *is bounded, one component must be unbounded.*

Finally, we prove if $\boldsymbol{a} \in X$, *we must show* $B(\boldsymbol{a}, \epsilon)$ *contains points in both components. Divide* X *into two pieces. The first piece is labeled* C *and it is chosen so that* $C \subset B(\boldsymbol{a}, \epsilon)$. *The rest of* X *is* *labeled D. We know $\Re^2 \setminus D$ is connected as D is homeomorphic to a closed interval. Pick points $\boldsymbol{P_0}$ and $\boldsymbol{P_1}$ in the two components of $\Re^2 \setminus X$. Let $\boldsymbol{P_0}$ be in the bounded component and $\boldsymbol{P_1}$ be in the unbounded component. Since $\Re^2 \setminus D$ is connected, there is a path γ on $[0,1]$ with $\gamma(0) = \boldsymbol{P_0}$ to $\gamma(1) = \boldsymbol{P_1}$ in $\Re^2 \setminus D$. If this path did not intersect C, we would have a path in $\Re^2 \setminus C$ and in $\Re^2 \setminus D$, This would imply the path is in*

$$(\Re^2 \cap C^C) \cap (\Re^2 \cap D^C) \quad = \quad \Re^2 \cap (C^C \cap D^C) = \Re^2 \cap (C \cup D)^C = \Re^2 \setminus X$$

But this contradicts the fact that $\boldsymbol{P_0}$ and $\boldsymbol{P_1}$ are in different components of $\Re^2 \setminus X$. Hence, starting at $\boldsymbol{P_0}$ in the bounded component, γ hits a first point in C at t_{min} when leaving the bounded component. Before that, it hits $\partial B(\boldsymbol{a}, \epsilon)$ at a point t_0. The first time γ intersects C coming from the unbounded component is at s_{min} and from that point on, γ and at some $s < s_{min}$, γ it first hits $\partial B(\boldsymbol{a}, \epsilon)$. Call this value s_0. So $\gamma(s_0)$ is where the path first hits $\partial B(\boldsymbol{a}, \epsilon)$ coming from $\boldsymbol{P_1}$, $\gamma(s_{min})$ is when the path first hits C coming from $\boldsymbol{P_1}$, $\gamma(t_0)$ is where the path first hits $\partial B(\boldsymbol{a}, \epsilon)$ coming from $\boldsymbol{P_0}$ and $\gamma(t_{min})$ is when the path first hits C coming from $\boldsymbol{P_0}$.

So from (s_0, s_{min}), γ lives in the unbounded component and is inside $B(\boldsymbol{a}, \epsilon)$ and from (t_0, t_min), γ lives in the bounded component. Hence, $B(\boldsymbol{a}, \epsilon)$ contains points from both components. ■

Homework

Exercise 4.7.1 *In the proof of Theorem 4.7.1, provide the details of the induction step: given $Y_{n-1} = f([a_{n-1}, b_{n-1}])$ with $b_{n-1} - a_{n-1} = 1/2^{n-1}$ and $Y_{n-1} \subset Y_{n-2} \subset \ldots \subset Y_1 \subset Y$, bisect it also. One of these pieces contains $\boldsymbol{P_0}$ and $\boldsymbol{P_1}$ in two separate components. Call this piece $Y_n = f([a_n, b_n])$ with $b_n - a_n = 1/2^n$. Now finish the induction.*

Exercise 4.7.2 *Prove the boundary of a bounded rectangle in $\Re^2$ is homeomorphic to a circle centered at the origin.*

Exercise 4.7.3 *Prove the boundary of a bounded triangle in $\Re^2$ is homeomorphic to a circle centered at the origin.*

Exercise 4.7.4 *A simple curve in $\Re^2$ is one that does not self-intersect. A closed curve is one that ends at the same point it starts. A rectifiable curve is one that has finite arc length. Let $\gamma(t) = (x(t), y(t))$ be any parameterization of such a curve. The tangent vector to the curve is $T(t) = (x'(t), y'(t))$ and the normal vector to the curve is $N(t) = (-y'(t), x'(t))$. Prove γ is homeomorphic to a circle and so γ has an inside and outside. Use this to define an outward normal at all points. A curve that has a uniquely defined normal is called orientable. A curve that is simple, closed, rectifiable and orientable is called a* **SCROC** *and we use it quite a bit when we study functions of a complex variable.*

Part III

Deeper Topological Ideas

Chapter 5

Vector Spaces and Topology

The various areas of topology constitute a major branch of mathematical research, so much so that topology is a worthwhile study in its own right. However, it is quite common for many students to have never been introduced to fundamental topological concepts in a serious way. So we aim not only to build up ideas that will ultimately determine the analysis of distributions and test functions, topological degree and so forth, but we also want to take careful steps to fill in some holes that may be present in your mathematical background.

5.1 Topologies and Topological Spaces

We are now in a position to generalize the preceding discussion to broader classes of objects that are not necessarily metric spaces. In general, we always want to identify the essence of a particular theory and see if we can abstract it into a more expansive setting. We broaden our scope of metric spaces, normed linear spaces and inner product spaces by giving the following definition of a topological space:

Definition 5.1.1 Topological Spaces $(X, \mathscr{S})$

A **topological space** $(X, \mathscr{S})$ *is any set of elements* X *combined with a collection* $\mathscr{S}$ *of subsets from* X *such that:*

T1: $\emptyset \in \mathscr{S}$.

T2: $X \in \mathscr{S}$.

T3: Any union of sets from $\mathscr{S}$ *is again in* $\mathscr{S}$.

T4: Any finite intersection of sets from $\mathscr{S}$ *is again in* $\mathscr{S}$.

The collection $\mathscr{S}$ *of subsets from* X *is called a* **topology** *on* X.

We already have many examples of a topological space.

Example 5.1.1 (X, d) *is a metric space and* $\mathscr{S}$ *is the collection of all open sets as we have described them earlier: i.e.* A *is open if every point in* A *is an interior point. Note an interior point* p *means there is an* $r > 0$ *so that* $B(p, r) \subset A$ *where* $B(p, r) = \{x \in X | d(x, p) < r\}$. *So the metric defines what we mean by an open set. For these sets, we know*

T1: $\emptyset$ *is an open set.*

T2: X *are open sets.*

T3: Arbitrary unions of open sets are open.

T4: Finite intersections of open sets are open.

We say $\mathscr{S}$ is the topology generated by the metric d.

Example 5.1.2 *Since any normed linear space $(X, \|\cdot\|)$ induces a metric, all normed linear spaces have a topology generated by that induced metric, $d(x, y) = \|x - y\|$ for any $x, y \in X$.*

Example 5.1.3 *Since any inner product space $(X, <\cdot,\cdot>)$ induces a norm which induces a metric, an inner product space has a topology generated by that induced metric, $d(x, y) = \|x - y\| = \sqrt{< x, x >}$ for any $x, y \in X$.*

So, technically, a topological space is an ordered pair $(X, \mathscr{S})$ but we will tend to refer to a topological space as simply X when its topology is clear from the context. Let's be clear about our notation here.

Definition 5.1.2 Open Sets in a Topological Space

If $\mathscr{O} \in \mathscr{S}$, then $\mathscr{O}$ is called an **open set** *in the topological space. Open sets are also known as* **neighborhoods.**

Since any subset of a metric space is closed if and only if its complement is open, we use that idea as the **definition** of closed sets in a general topological space:

Definition 5.1.3 Closed Sets in a Topological Space

A subset G in a topological space is a **closed set** *if there is a set $\mathscr{O} \in \mathscr{S}$ such that $G = \mathscr{O}^C$ where, as usual, the complement of a set $A \subset X$ is the set of all points in X, not in A, which is denoted by A^C.*

It is not always true that the collection $\mathscr{S}$ of subsets can be constructed by some metric acting on the space X. There is a simple example demonstrating this:

Example 5.1.4 *Let $X = \{x, y\}$ and define $\mathscr{S} = \{\emptyset, \{x\}, T\}$. It is apparent that $\mathscr{S}$ is a topology on X. Now suppose d is any metric on X and let $r = d(x, y)$. Then we see that $B(y, r) = \{y\}$. Consequently, $\{y\}$ is an open set generated by the metric (since $\{y\}$ is just an open ball). But $\{y\}$ is not in $\mathscr{S}$ and so $\{y\}$ cannot be an open set in $\mathscr{S}$'s topology. Now since d is an arbitrary metric, then we see that the topology $\mathscr{S}$ cannot be generated by a metric acting on X.*

So, every metric space is a topological space but not every topological space is a metric space. This leads us to our first classification of topological spaces:

Definition 5.1.4 Metrizable Topological Spaces

If a topology $\mathscr{S}$ of a topological space X can be generated by a metric, then $\mathscr{S}$ is said to be **compatible** *with the metric (or simply a* **metric topology***) and the space X is said to be* **metrizable.**

Homework

Exercise 5.1.1 *If $\mathscr{S}_1$ and $\mathscr{S}_2$ are both topologies for a set X, prove $\mathscr{S}_1 \cap \mathscr{S}_2$ is also a topology.*

Exercise 5.1.2 *We know how we define open sets and so forth in a metric space. Show explicitly a metric space (X, d) is a topological space where $\mathscr{S}$ is the collection of open sets. Make sure you prove all the needed properties.*

Exercise 5.1.3 *Since a norm induces a metric, we know how to define open sets as usual. Show explicitly a normed linear space $(X, \|\cdot\|)$ is a topological space where $\mathscr{S}$ is the collection of open sets. Make sure you prove all the needed properties.*

Exercise 5.1.4 *Since an inner product induces a metric, again, we know how to define open sets as usual. Show explicitly an inner product space $(X, <\cdot,\cdot>)$ is a topological space where $\mathscr{S}$ is the collection of open sets. Make sure you prove all the needed properties.*

Exercise 5.1.5 *Let the finite interval $I = [a, b]$ for some $a < b$ and $\mathscr{S}$ be the collection of sets of the form $(c, d] \subset [a, b]$. Is this a topology for I?*

Exercise 5.1.6 *Let the finite interval $I = [a, b] \times [c, d]$ for some $a < b$ and $c < d$. Let $\mathscr{S}$ be the collection of sets of the form $(\alpha, \beta] \times (\gamma, \delta] \subset [a, b] \times [c, d]$. Is this a topology for I?*

5.1.1 Topological Generalizations of Analysis Concepts

We now turn our attention to the generalizations of analysis notions discussed earlier for metric spaces. We've already considered how open and closed sets are naturally defined within the framework of topological spaces. We can extend all these ideas, but we have to be careful not to use metric concepts. Thus we have the following:

Definition 5.1.5 Convergence in a Topological Space

A sequence $\{x_n\}$ of a topological space X **converges** *to a* **limit** *$L \in T$ if every open set containing L also contains all but a finite number of the elements x_n.*

We also need to define continuity.

Definition 5.1.6 Continuous Mappings between Topological Spaces

Let $(T_1, \mathscr{S}_1)$ and $(T_2, \mathscr{S}_2)$ be topological spaces. A function $f : T_1 \to T_2$ is **continuous** *if for all $\mathscr{O} \in \mathscr{S}_2$, $f^{-1}(\mathscr{O}) \in \mathscr{S}_1$. That is, f is* **continuous** *if every inverse image of an open set is itself open.*

As usual, we have the idea of topological compactness which is based on open covers.

Definition 5.1.7 Compactness in Topological Spaces

A subset G of a topological space X is **compact** *if every collection of open sets whose union contains G has a finite subcollection whose union contains G. That is, $G \subset T$ is* **compact** *if every open cover of G has a finite subcover.*

We have studied compactness extensively in the metric space setting and proved sequential compactness is equivalent to topological compactness in a metric space setting. So if our topological space is metrizable, we can use the notion of sequential compactness interchangeably with that of topological compactness. You can see how these definitions are just restatements of metric space theorems. There are, however, some subtleties here that ought to be mentioned, especially concerning the notion of convergence in a topological space. Since a topology is much more general than a metric, we need to be careful in making sure a given topology is *strong* enough. Some topologies are so *weak* that they can't tell the difference between distinct points in the topological space. We will define these notions of *strong* and *weak* better in the example below.

Example 5.1.5 *Consider the* **trivial topology** *$(X, \{\emptyset, X\})$. By the definition of topological space, the set $\{\emptyset, X\}$ is enough to form a topology on any space X. Consider the question of convergence in this topology. Let $\{x_n\}$ be any sequence of X and L be any element in X. Now, the only open set containing L is X itself, but X always contains all the elements x_n. So, according to our definition* *above, the sequence $\{x_n\}$ converges to L. In other words,* **any** *sequence converges in this topology to* **any** *point! Loosely stated, the trivial topology is so* weak *that it can't tell the difference between convergent and non-convergent sequences. It can't even tell what is or is not the limit of a convergent sequence.*

In order to rectify this issue of strong vs. weak topologies, topologists have introduced various notions of *separation* within a topological space. These separation properties will determine how much analysis can be done on the space. Basically the stronger the topology, the more we are able to tell the difference between distinct points and subsets. The weaker the topology, the less able we are to distinguish between points and subsets that we would like to be different from each other. Hence we have the following classification of topological spaces in terms of their separation properties:

Definition 5.1.8 T_0 **Separable or Just** T_0 **Topological Spaces**

*A topological space X is said to be T_0-***separable** *or just T_0 if given any two distinct points in X, there is an open set that contains one of the points but not the other.*

Definition 5.1.9 T_1 **Separable or Just** T_1 **Topological Spaces**

*A topological space X is said to be T_1-***separable** *or just T_1 if given any two distinct points $x, y \in X$, there is an open set U that contains x but not y* **and** *there is an open set V that contains y but not x.*

It is important to note that in T_1 spaces, the sets U and V may intersect each other. Also, we can show singleton sets are closed sets. Look at the complement of $\{x\}$. If y is in the complement, then y is distinct from x and so there is an open set U_y which contains y but does not contain x. The union of all these sets is still open and does not contain x. We see the complement of $\{x\}$ is open and so in a T_1 space, singleton sets are closed sets. We can't make this argument in a T_0 space as all we know is there is an open set U_y which contains y and not x or contains x and not y.

We get stronger separation when we require U and V to be disjoint.

Definition 5.1.10 T_2 **Separable or Just** T_2 **or Hausdorff Topological Spaces**

*A topological space is said to be T_2-***separable** *or just T_2 if given any two distinct points $x, y \in X$, there is an open set U containing x but not y and there is another open set V that contains y but not x, and $U \cap V = \emptyset$. T_2-separable spaces are often called* **Hausdorff spaces.**

So we see that T_2 is stronger than T_1 which in turn is stronger than T_0. The trivial topology $\{\emptyset, T\}$ is so weak that it's not even T_0. Even stronger separation properties are possible, and these are known as T_3**-separable** and T_4**-separable** spaces.

Definition 5.1.11 T_3 **Separable or Just** T_3 **Topological Spaces**

A topological space is said to be **regular** *if given any $x \in X$ and any closed set C which does not contain x, there are open sets U and V so that $x \in U$ and $C \subset V$ with $U \cap V = \emptyset$. If it is also a Hausdorff space (T_2), we say it is T_3-***separable** *or just T_3. Hence, a T_3 space is a regular Hausdorff space. Note T_2 implies T_1.*

Finally, another level of separation is found in the T_4 spaces.

Definition 5.1.12 T_4 **Separable or Just** T_4 **Topological Spaces**

A topological space is said to be **normal** *if given any two disjoint closed sets C and D in X, there are open sets U and V so that $C \subset U$, $D \subset V$ with $U \cap V = \emptyset$. If the space is also Hausdorff (T_2) it is called T_4 separable or a T_4 space.*

Homework

Exercise 5.1.7 *Prove a metric space (X, d) is T_0, T_1, T_2, T_3 and T_4.*

Exercise 5.1.8 *Prove a metric space (X, d) is normal and Hausdorff.*

Exercise 5.1.9 *Show explicitly $(C([a, b]), \|\cdot\|_\infty)$ is Hausdorff.*

However, the level of a T_2-separable or Hausdorff topology is sufficient for our purposes, because we have the following theorem regarding convergent sequences in a Hausdorff space:

Theorem 5.1.1 Limits are Unique in Hausdorff Spaces

In any Hausdorff space, the limit of a convergent sequence is unique.

Proof 5.1.1
Let $\{x_n\}$ be a convergent sequence in a Hausdorff space X and assume L_1 and L_2 are both limits of $\{x_n\}$ such that $L_1 \neq L_2$. Then, by definition, every open set containing L_1 has all but a finite number of elements x_n. The same would be also be true for any open set containing L_2. Now X is a Hausdorff space with distinct points L_1 and L_2, and so there exist disjoint open sets U and V such that $L_1 \in U$ and $L_2 \in V$. Consequently, U would contain all but a finite number of x_n, and this would imply V itself has just a finite number of x_n since V is outside of U. But this contradicts our assumption that V is an open set containing a limit L_2 of $\{x_n\}$. Therefore, there cannot be two distinct limits of a convergent sequence in a Hausdorff space. ■

Thus, the topology of a Hausdorff space is strong enough to guarantee that convergent sequences have one and only one limit. This is quite convenient for us since, as we will see in the next section, we focus primarily on topological vector spaces where we will show that topologies of such spaces are always Hausdorff in nature. Just as we defined the notions of convergence, continuity, and compactness using the open and closed sets of a topology, so also we can define all the other usual concepts in a straightforward manner. The topological definitions for the closure, interior, exterior, and boundary of a subset are given below: The closure is defined in terms of intersections over all closed sets that contain the set in question.

Definition 5.1.13 Closure of a Subset in a Topological Space

The **closure** *G^* of a subset G is the set: $\bigcap \{F : G \subset F \text{ and } F \text{ is closed}\}$.*

The closure is defined in terms of intersections over all open sets inside the set in question.

Definition 5.1.14 Interior of a Subset in a Topological Space

The **interior** *$int(G)$ of a subset G is the set: $\bigcup \{F : F \subset G \text{ and } F \text{ is open}\}$.*

The boundary is defined as usual.

Definition 5.1.15 Boundary of a Subset in a Topological Space

The **boundary** *∂G of a subset G is the set: $G^* \setminus int(G)$; that is, ∂G is the closure of G without the interior of G.*

The denseness of a set is now defined in terms of closures.

Definition 5.1.16 Dense Subsets in a Topological Space

A subset G is **dense** *in a subset $H \subset X$ if $G^* = H$. If H turns out to be the whole space X, then G is* **everywhere dense**. *If the $int(G^*) = \emptyset$, then G is* **nowhere dense**. *If G is not nowhere dense, then G is said to be* **somewhere dense**.

Homework

Exercise 5.1.10 *Let X be the normed linear space of 2×2 real matrices with the norm $\|A\|_{op} = \sup_{\|x\|_\infty = 1} \|A(\boldsymbol{x})\|$ for any A in X and $x \in \Re^2$. Let B be the subset whose matrices have determinant 1.*

- *Is B an open set?*
- *What is the boundary of B?*
- *What is the complement of B?*

Exercise 5.1.11 *Given an $n \times n$ matrix over the reals, A, let $[A]$ be the set of all matrices equivalent to it. Let X be the set of all $n \times n$ matrices over the reals with any norm you wish.*

- *Is $[A]$ an open set?*
- *What is the boundary of $[A]$?*
- *What is the complement of $[A]$?*

Exercise 5.1.12 *Find the general solution to*

$$\begin{aligned} x'(t) &= -4\,x(t) + y(t) \\ y'(t) &= -5\,x(t) + 2\,y(t) \end{aligned}$$

and show it forms a two dimensional vector space over $\Re$. Then show it is a finite dimensional normed linear space for a variety of norm choices. For each such norm, characterize $B(0, 1)$, the ball about the zero function of radius 1.

Exercise 5.1.13 *Find the general solution to*

$$\begin{aligned} x'(t) &= 5\,x(t) + y(t) \\ y'(t) &= -7\,x(t) - 3\,y(t) \end{aligned}$$

and show it forms a two dimensional vector space over $\Re$. Then show it is a finite dimensional normed linear space for a variety of norm choices. For each such norm, characterize $B(0, 1)$, the ball about the zero function of radius 1.

5.1.2 Urysohn's Lemma

Urysohn's Lemma is a necessary result in the construction of a metric on an arbitrary topological space. It is also used extensively in algebraic topology and differential geometry to obtain functions with desirable properties. The proof of Urysohn's Lemma is an elaborate computation. This result is set in normal topological spaces. We begin with a preliminary lemma.

Lemma 5.1.2 Neighborhoods of Closed Sets in Normal Topological Spaces

Let X be a normal topological space, and let F be a closed subset of X. If Ω is a neighborhood of F, then there exists an open set, W, such that $F \subset W \subset \bar{W} \subset \Omega$.

Proof 5.1.2

Since Ω is open, Ω^c and F are disjoint sets. So, by the normality hypothesis, there are open sets W and U such that $F \subset W$, $\Omega^c \subset U$, and $W \cap U = \emptyset$. Now, since $W \cap U = \emptyset$, it follows that $W \subset U^c \Rightarrow \bar{W} \subset U^c$, because $\bar{W}$ is the smallest closed set containing W. Thus, we have $F \subset W \subset \bar{W} \subset U^c$. Finally, if $x \in U^c$, then $x \notin \Omega^c$, as $\Omega^c \subset U$. Thus, if $x \in U^c$, then $x \in \Omega$. So, we have $F \subset W \subset \bar{W} \subset \Omega$. ■

The following lemma is the key to proving Urysohn's Lemma.

Lemma 5.1.3 Separating Disjoint Closed Sets in a Normal Topological Space

Let X be a normal topological space, and let A and B be disjoint closed sets. Let D be the set of all rational numbers in $[0,1]$. Then there exists a countable collection of open sets, $\{U_r\}$, each disjoint from B, indexed by D, such that for $p, q \in D$ with $p < q$, we have $A \subset \bar{U}_p \subset U_q$.

Proof 5.1.3

We construct the collection U_r by induction. First, because D is a countable set, we can order it in some fashion. It will be helpful for our purposes, however, if we let the first two elements of D be 1 and 0, respectively. Thus, we assume $D = \{1, 0, r_1, r_2, ...\}$. (The ordering of the elements in D is actually arbitrary. It is merely a convenient mechanism for our purpose.)

Define $U_1 = X - B$. Then U_1 is an open set containing A. By Lemma 5.1.2, we can find an open set U_0 such that $A \subset U_0 \subset \bar{U}_0 \subset U_1$. Given our particular ordering of D, let D_n be the set consisting of the first n elements of D. So, $D_1 = \{1\}$, $D_2 = \{1, 0\}$, etc. Suppose U_p has been defined for all $p \in D_n$, for some $n \in \mathbb{N}$, so that

$$p, q \in D_n \quad \text{and} \quad p < q \Rightarrow A \subset \bar{U}_p \subset U_q \tag{5.1}$$

Note that, with respect to the inductive process, we have already done the base steps $n = 1, 2$. For $n = 1$, we have $A \subset U_1$. For $n = 2$, we have $0 < 1$ and $A \subset U_0 \subset \bar{U}_0 \subset U_1$. That is, Equation 5.1 holds for D_1 and D_2. We continue with the inductive step by supposing Equation 5.1 holds for D_n and showing that it holds for D_{n+1}.

The set D_n is the first n elements of D, so let r be the next element in the sequence. Then $D_{n+1} = D_n \cup \{r\}$. This is a finite, well ordered subset of $[0,1]$. Now, each element in a finite, well ordered set, except for the smallest and largest elements, must have both an immediate predecessor and an immediate successor. In other words, since $r \in D_{n+1}$ (and we assume $n > 2$ so $r \neq 0$ or 1) there are elements $p, q \in D_{n+1}$ such that $p < r < q$, and these are the closest elements to r in the sense that if $k,\ j \in D_{n+1}$ with $k < r$ and $j > r$, then $k < p$ and $j > q$.

By our inductive hypothesis, since p and q must be in D_n (they are in D_{n+1} but not equal to r), the sets U_p and U_q are already well-defined and satisfy $A \subset \bar{U}_p \subset U_q$. Applying Lemma 5.1.2 again, we obtain an open set, U_r, such that $\bar{U}_p \subset U_r \subset \bar{U}_r \subset U_q$. Thus we have

$$A \subset \bar{U}_p \subset U_r \subset \bar{U}_r \subset U_q \tag{5.2}$$

Now, we claim that (11) holds for D_{n+1}. Let s and t be any two elements in D_{n+1}. If neither s nor t equals r, then $s, t \in D_n$ and the result holds by our inductive hypothesis. If $s \neq r$ and $t = r$, then there are 2 cases.

1. *$s \leq p$: Note that $s \leq p \Rightarrow s < t$. If $s = p$, then we have, by our result above, that $A \subset \bar{U}_p \subset U_r \Rightarrow A \subset \bar{U}_s \subset U_t$. If $s < p$, then both s and p are in D_n, so our inductive hypothesis and Equation 5.2 imply that*

$$A \subset \bar{U}_s \subset U_p \subset \bar{U}_p \subset U_r \quad \Rightarrow \quad A \subset \bar{U}_s \subset U_t.$$

2. *$s \geq q$: Note that $s \geq q \Rightarrow s > t$. If $s = q$, then our result above implies*

$$A \subset \bar{U}_r \subset U_q \quad \Rightarrow \quad A \subset \bar{U}_t \subset U_s.$$

   If $s < q$, then both s and q are in D_n, so the inductive hypothesis and Equation 5.2 imply that $A \subset U_r \subset \bar{U}_r \subset U_q \subset \bar{U}_q \subset U_s$. But $t = r$, so we have $A \subset \bar{U}_t \subset U_s$.

So, if $s \neq r$ and $t = r$, (11) holds. Likewise, a parallel argument shows that if $s = r$ and $t \neq r$, then (11) holds. Thus, for any pair of elements, s, and t, in D_{n+1}, we have $s < t \Rightarrow A \subset \bar{U}_s \subset U_t$. This completes the inductive step. Since, for each $n \in \mathbb{N}$, we add another rational number to D_n to form D_{n+1}, it follows by induction that we have a collection of sets, $\{U_r\}$, indexed by D, such that for $p, q \in D$ with $p < q$, we have $A \subset \bar{U}_p \subset U_q$. To see this explicitly, just let p and q be any two rational numbers in $[0, 1]$. Suppose, in our ordering, that q is the element with the higher index, so that q is, say, the nth element of D and p is the kth element for some $k < n$. Then $p, q \in D_n$, so, by the induction proof, if $p < q$ we have $A \subset \bar{U}_p \subset U_q$, and if $q < p$, we have $A \subset \bar{U}_q \subset U_p$. Also, note that, as a consequence of our construction, we have, for any $p \in D$,

$$A \subset U_0 \subset \bar{U}_0 \subset U_p \subset \bar{U}_p \subset U_1 \quad \Rightarrow \quad A \subset U_0 \subset U_p \subset U_1.$$

■

Now that we have the collection $\{U_r\}_{r \in D}$, we can extend our definition to all of $\Re$. This will be necessary in the proof of Urysohn's Lemma. For any rational number $r \in \Re$, define the set U_r by

$$U_r = \begin{cases} \emptyset & r < 0 \\ X & r > 1 \\ U_r & 0 \leq r \leq 1 \end{cases}$$

That is, for $r \in [0, 1]$, we define U_r to be the set we have constructed in Lemma 5.1.3.

Lemma 5.1.4 Urysohn's Lemma

> *Let X be a normal topological space, and let A and B be disjoint closed subsets of X. Then there is a continuous function $f : X \to [0, 1]$ such that $f(X) \subset [0, 1]$, $f(A) = \{0\}$, and $f(B) = \{1\}$.*

Proof 5.1.4

By Lemma 5.1.3, there is a countable collection of open sets, $\{U_r\}_{r \in D}$, such that $p, q \in \mathbf{Q}$ and $p < q$ imply that $A \subset \bar{U}_p \subset U_q \subset U_1 = X - B$. That is, $p < q \Rightarrow A \subset \bar{U}_p \subset U_q \subset B^c$.

Now, for each $x \in X$, consider the set $R(x) = \{r \, : \, x \in U_r\}$. That is, $R(x)$ is the set of all rational numbers, r, such that U_r contains x. This set contains no negative rational, since $r < 0 \Rightarrow U_r = \emptyset$. Moreover, $R(x)$ contains every rational, r, such that $r > 1$, since $r > 1 \Rightarrow U_r = X$. Thus, for *each $x \in X$, the set $R(x)$ is nonempty and it is bounded below, since the smallest rational that can be in $R(x)$ is 0. By the completeness of $\mathbf{Q}$ in $\Re$, it follows that $R(x)$ has a greatest lower bound, or infimum, and this element must lie in $[0,1]$.*

Define $f : X \to \Re$ by $f(x) = \inf R(x) = \inf\{r : x \in U_r\}$. For $x \in X$, we have $\inf R(x) \in [0,1]$, so $f(x) \in [0,1]$. If $x \in A$, then, since $A \subset U_r$ for all $r \in D$, it follows that $R(x)$ contains D. That is, $\inf R(x) \geq 0$ and $R(x)$ contains all rational numbers in $[0,1]$. Hence, $\inf R(x) = 0$ and $f(x) = 0$.

Now, suppose $x \in B$. We know that $R(x)$ contains all rationals greater than 1. Moreover, for any $r \in D$, we have $A \subset U_r \subset B^c$. If $r \in [0,1]$, then X cannot be in U_r, for if $x \in U_r$, then $x \in B^c$, which contradicts the fact that $x \in B$. Thus, $R(x)$ consists exactly of the rationals greater than 1. Again, by the density of $\mathbf{Q}$ in $\Re$, it follows that $\inf R(x) = 1$. This implies that $f(x) = 1$ for $x \in B$.

It remains only to show that f is continuous on X. Let x_0 be a point in X, and let (a,b) be an open interval around $f(x_0)$. We want to find a neighborhood, U, of x_0, such that $f(U) \subset (a,b)$. Using the density of $\mathbf{Q}$ in $\Re$, choose rational numbers p and q such that $a < p < f(x_0) < q < b$. We claim that the set $U = U_q - \bar{U}_p$ satisfies the condition.

First, note that U is open. We also need x_0 to be in U. Note that, for any $r \in \mathbf{Q}$, if $x \notin U_r$ then $f(x) \geq r$, since, if $x \notin U_r$, then x cannot be in U_s for any $s < r$. Hence, $R(x)$ can only contain rationals greater than r, so $\inf R(x) \geq r \Rightarrow f(x) \geq r$. The contrapositive of this result states that $f(x) < r \Rightarrow x \in U_r$ for any $r \in \mathbf{Q}$. So, since $f(x_0) < q$, it follows that $x_0 \in U_q$.

Next, note that for any $r \in \mathbf{Q}$, if $x \in \bar{U}_r$ then $f(x) \leq r$, since, if $x \in \bar{U}_r$, then $x \in U_s$ for every $s > r$ ($r < s \Rightarrow A \subset \bar{U}_r \subset U_s$). Hence, $R(x)$ contains all rationals greater than or equal to r, implying that $\inf R(x) \leq r \Rightarrow f(x) \leq r$. So, since $f(x_0) > p$, it follows that $x_0 \notin \bar{U}_p$. Hence, $x_0 \in U = U_q - \bar{U}_p$.

Finally, we show that $f(U) \subset (a,b)$. Let x be in U. Then we have $x \in U_q \subset \bar{U}_q$, which implies that $f(x) \leq q$ (beginning of previous paragraph). Also, since $x \in U$, we have $x \notin \bar{U}_p$, which implies that $x \notin U_p \Rightarrow f(x) \geq p$. Thus, for $x \in U$, we have $a < p \leq f(x) \leq q < b \Rightarrow f(U) \subset (a,b)$. Therefore, since x_0 was arbitrary, it follows that f is continuous. ■

This is then used to prove the Tietze Extension Theorem.

Theorem 5.1.5 Tietze Extension Theorem

Let X be a normal topological space, and let A be a closed subset of X. Then for a continuous function $f : A \to [a,b]$, with $\sup_{x\in A} |f(x)| = 1$, there is an extension $F : X \to \Re$ so that $F|_A = f$ and $\sup_{x\in A} |f(a)| = \sup_{x\in X} |F(x)|$.

Proof 5.1.5

It is enough to argue for the case where the range is $[-1,1]$. Let's start by assuming the function f has range $[-\alpha,\alpha]$ with $\alpha < 1$. The sets $f^{-1}((-\infty,-(1/3)\alpha])$ and $f^{-1}([(1/3)\alpha,\infty))$ are disjoint and closed in A. Since A is closed, they are closed in X also. Let $\phi : [0,1] \to [-(1/3)\alpha,(1/3)\alpha]$. Then by Urysohn's Lemma, there is a continuous mapping $\phi \circ g : X \to [-(1/3)\alpha,(1/3)\alpha]$ so that $\phi \circ g(f^{-1}((-\infty,-(1/3)\alpha]) = \phi(0) = -(1/3)\alpha$ and $\phi \circ g(f^{-1}([(1/3)\alpha,\infty)) = \phi(1) = (1/3)\alpha$. Let $G = \phi \circ g$. Then $|G(x)| \leq (1/3)\alpha$.

Hence, for any $x \in A$, if $-\alpha \leq f(x) \leq -(1/3)\alpha$, then $G(x) = -(1/3)\alpha$ and $|f(x) - G(x)| \leq (2/3)\alpha$. If $(1/3)\alpha \leq f(x) \leq \alpha$, $G(x) = (1/3)\alpha$, then $|f(x) - G(x)| \leq (2/3)\alpha$. Finally, if

$-(1/3)\alpha \leq f(x) \leq (1/3)\alpha$, $|G(x)| \leq (1/3)\alpha$ and so $|f(x) - G(x)| \leq (2/3)\alpha$.

We now start with the original function f with range $[-1,1]$. From the above discussion, since $\alpha = 1$, there is a continuous function $g_0 : X \rightarrow \Re$ so that on A, $|g_0(x)| \leq (1/3)$ and $f(x) - g_0(x)| \leq (2/3)$. Now apply our discussion to the new function $f_1 = g - g_0$. This function has range $-(2/3), (2/3)]$ so $\alpha = (2/3)$ now and there is a function $g_1 : X \rightarrow \Re$ with $|g_1(x)| \leq (1/3)(2/3)$ and $|f(x) - g_0(x) - g_1(x)| \leq (2/3)(2/3)$ on A.

Now repeat this process to construct a sequence of continuous functions (g_n) for $n \geq 0$ with $g_n(x)| \leq (1/3)(2/3)^n$ and $|f(x) - \sum_{i=0}^n g_i(x)| \leq (2/3)^n$ on A. Define $F(x) = \lim_{n\rightarrow\infty} \sum_{i=0}^n g_i(x)$ with $|\sum_{i=0}^n g_i(x)| \leq \sum_{i=0}^n (1/3)(2/3)^i$. The dominating geometric series converges, so if $S_n(x) = \sum_{i=0}^n g_i(x)$, for a given $\epsilon > 0$, there is N so that $|f(x) - S_n(x)| < \epsilon$ if $n > N$. This means there is a neighborhood of $f(x)$ which contains $S_n(x)$ for all $n > N$. With a little thought, we see we have shown $(S_n(x))$ is a sequence which converges to $f(x)$. Since each S_n is continuous, this means f is continuous at each $x \in X$. Further, since $F(x) \leq \sum_{i=0}^\infty (1/3)(2/3)^i = 1$, we see $|F(x)| \leq 1$.

Next, if $x \in A$, for a given $\epsilon > 0$, there is N so that

$$\begin{aligned} |f(x) - F(x)| &\leq |f(x) - S_N(x)| + |S_N(x) - F(x)| \\ &\leq (2/3)^{N+1} + \sum_{i=N+1}^{\infty} (1/3)(2/3)^i < (\epsilon/2) + (\epsilon/2) = \epsilon \end{aligned}$$

Thus, since ϵ is arbitrary, $f(x) = F(x)$ if $x \in A$. So F is a continuous function on X with $F|_A = f$. Since $\sup_{x\in X} |F(x)| \geq \sup_{x\in A} |f(x)| = 1$ and $\sup_{x\in X} |F(x)| \leq 1$, we see $\sup_{x\in X} |F(x)| = \sup_{x\in A} |f(x)|$.

The extension to the case $[a,b]$ is messy but straightforward and we encourage you to do the argument. ■

Homework

Exercise 5.1.14 *In the proof of Theorem 5.1.5, how would you argue when the interval was $[0,2]$ instead of $[-1,1]$?*

Exercise 5.1.15 *In the proof of Theorem 5.1.5, how would you argue when the interval was $[-3,4]$ instead of $[-1,1]$?*

Exercise 5.1.16 *In the proof of Theorem 5.1.5, how would you argue when the interval was $[a,b]$ instead of $[-1,1]$?*

5.2 Constructing Topologies from Simpler Sets

We now discuss two concepts that are quite important for what we do in Section 5.4.

Definition 5.2.1 Open Basis for a Topology

Let $(X, \mathscr{S})$ be a topological space. A collection $\mathscr{B}$ of open subsets of X is a **basis** *for $\mathscr{S}$ if every $\mathscr{O} \in \mathscr{S}$ can be written as a union of sets from $\mathscr{B}$. Note, we can also state this as, if G is a nonempty set and $x \in G$, then there is a $B \in \mathscr{B}$ with $x \in B$. If the open base is countable, the space is called a* **second countable space**.

We've already seen an example of this concept when we looked at metric spaces. There we mentioned the theorem that any open set in a metric space can be expressed as the union of open balls. Thus, the set of open balls forms a **basis** for a metric topology. The definition above generalizes this notion to include other structures besides metrizable spaces.

The reason why we consider a basis of a topological space is because it is not always convenient—or even possible— to write out the sets of a given topology. So the notion of a basis gives us a means of *reducing* or *simplifying* topological structure. Given any basis $\mathscr{B}$, we can construct a topology $\mathscr{S}$ by letting $\mathscr{S}$ be the collection of all unions of elements from $\mathscr{B}$.

If the space is second countable, we can characterize open sets nicely.

Theorem 5.2.1 Linderlöf's Theorem

Let $(X, \mathscr{S})$ be a topological space with a countable open base $\mathscr{B}$. If G is a nonempty set which can be written as $G = \cup_{\alpha \in \Lambda} G_\alpha$, $G_\alpha \in \mathscr{S}$, then G can be written as a countable union $G = \cup_i G_i$ of sets $G_i \in \{G_\alpha | \alpha \in \Lambda\}$.

Proof 5.2.1
Let (B_n) be a countable labeling of the countable base $\mathscr{B}$. Let $x \in G$. Then $x \in G_\beta$ for some $\beta \in \Lambda$ where Λ is the index set. Hence, there is an open basis set B_i with $x \in B_i \subset G_\beta$. Do this for each $x \in G$. Then, we have that G is contained in the union of basic open sets of which there are a countable number. Hence, $G = \cup_i B_i$, for a finite or countable collection of indices. Now each $B_i \subset G_{\beta_i}$ for some β_i. We see we have found a countable collection of indices, β_i, so that $G \subset \cup_i B_i \subset \cup_{\beta_i} G_{\beta_i}$. This is the countable collection we seek. ■

We can do more. If a second countable space has another open base, we can extract from it, a countable collection of open sets which is another countable open base.

Theorem 5.2.2 If X is Second Countable with a Second Open Base, the Second Open Base has a Countable Subbase

Let $\mathscr{B} = (B_n)$ be a countable open base for X and let $\mathscr{G}$ be another open base, where $\mathscr{G} = \{G_\alpha : \alpha \in \Lambda\}$ where Λ is the index set. Then there is a countable collection $\mathscr{C} \subset \mathscr{G}$ which is an open base called an open subbase.

Proof 5.2.2
Let $\mathscr{B} = (B_n)$ be a countable open base for X and let $\mathscr{G} = \{G_\alpha : \alpha \in \Lambda\}$ where Λ is the index set. Then, each $B_n = \cup G_{\alpha_n}$ where $(G_{\alpha_n}) \subset \mathscr{G}$. By Linderlöf's Theorem, each of these unions can be written as a countable union $B_n = \cup_i G_{i_n}$ where $(G_{i_n}) \subset (G_{\alpha_n}) \subset \mathscr{G}$. Let $\mathscr{C} = \cup_{i=1}^{\infty} G_{i_n}$. This is a countable union of countable sets and so is countable. Every B_n in $\mathscr{B}$ can be represented by a countable union of sets from $\mathscr{C}$. Hence, $\mathscr{G}$ has a countable open base which we call a subbase. ■

If X is a metric space, we can say more.

Theorem 5.2.3 (X, d) Separable Metric Space implies X is Second Countable

Let (X, d) be a metric space which is separable. Then X is second countable.

Proof 5.2.3
Let (X, d) be a metric space which is separable. Then X has a countable dense subset $A = (a_n)$. Let $\mathscr{B} = \{B(a_n, r_m) | a_n \in A, r_m \in \mathbb{Q}\}$ where (r_m) is an enumeration of $\mathbb{Q}$. We see $\mathscr{B}$ is a countable collection. To show $\mathscr{B}$ is an open base, pick any open set g and any $x \in G$ and show

there is a $B(a_n, r_m) \in \mathscr{B}$ *with* $x \in B(a_n, r_m) \subset G$. *Since* x *is an interior point of* G, *there is a radius* $r > 0$ *so that* $B(x, r) \subset G$. *Since* A *is dense, there is* $a_n \in A$ *so that* $d(a_n, x) < r/3$. *Since* $\mathbb{Q}$ *is dense in* $\Re$, *there is* $r_m \in \mathbb{Q}$ *with* $r/3 < r_m < 2r/3$. *Hence, if* $y \in B(a_n, r_m)$, $d(y, x) \leq d(y, a_n) + d(a_n, x) < r_m + r/3 < r$. *Hence,* $B(a_n, r_m) \subset B(x, r) \subset G$. *Also,* $d(x, a_n) < r/3 < r_m$, *so* $x \in B(a_n, r_m)$ *also. This shows* $\mathscr{B}$ *is a countable open base and so is second countable.* ■

We have been finding collections of open bases that serve as an open base too. Let's define this formally.

Definition 5.2.2 Open Subbases

If $(X, \mathscr{S})$ *is a topological space, an open subbase is a collection of open subsets* Ω *such that the collection of all possible finite intersections of sets from* Ω *forms an open base* $\mathscr{B}$ *for* X. *Sets in an open subbase are called subbasis open sets. We also say the open base* $\mathscr{B}$ *is generated by* Ω.

Homework

Exercise 5.2.1 *An open rectangle in* $\Re^2$ *is a set of the form* $(a, b) \times (c, d)$ *for* $a < b$ *and* $c < d$. *Let* $\mathscr{R}$ *be the collection of all open rectangles. Prove* $\mathscr{R}$ *is an open base for the usual topology induced by* $\|\cdot\|_2$ *on* $\Re^2$.

Exercise 5.2.2 *An open rectangle in* $\Re^3$ *is a set of the form* $(a, b) \times (c, d) \times (e, f)$ *for* $a < b$, $c < d$ *and* $e < f$. *Let* $\mathscr{R}$ *be the collection of all open rectangles. Prove* $\mathscr{R}$ *is an open base for the usual topology induced by* $\|\cdot\|_2$ *on* $\Re^3$.

Exercise 5.2.3 *Let* X *be the positive integers and define* $d(x, y) = 0$ *if* $x = y$ *and* 1 *if* $x \neq y$ *for any* $x, y \in X$.

- *Prove* (X, d) *is a metric space.*
- *Let the set of all continuous functions from* X *to* $\Re$ *be denoted by* $C(X, \Re)$. *Prove* $C(X, \Re)$ *is not separable.*
  Hint: *Let* (f_n) *be a sequence in* $C(X, \Re)$ *and define the function* f *by*

  $$\begin{cases} f(n) = 0, & |f_n(n)| \geq 1 \\ f(n) = |f_n(n)| + 1, & |f_n(n)| < 1 \end{cases}$$

  Then, $d(f_n, f) \geq 1$ *for all* n. *Why does this imply* $C(X, \Re)$ *is not separable?*

Exercise 5.2.4 *Let* X *be any nonempty set and define* $d(x, y) = 0$ *if* $x = y$ *and* 1 *if* $x \neq y$ *for any* $s, y \in X$.

- *Prove* (X, d) *is a metric space.*
- *Let the set of all continuous functions from* X *to* $\Re$ *be denoted by* $C(X, \Re)$. *Prove* $C(X, \Re)$ *is separable if and only if* X *is a finite set.*

5.3 Urysohn's Metrization Theorem

While all metric spaces are topological spaces, not all topological spaces are metrizable. Since metric spaces have very desirable properties within the category of topological spaces, particularly in analysis, the question of when a topological space is metrizable troubled mathematicians for years. A famous result is the following one.

Theorem 5.3.1 Urysohn's Metrization Theorem

A second countable normal topological space is metrizable. Further, it is homeomorphic to a subset of ℓ^2.

Proof 5.3.1

If X has only a finite number of points, this is easy. So assume X has infinitely many points. Let (g_n) be a countably infinite open base. Assume each G_n satisfies $\emptyset \neq G_n \neq X$. Then, given $x \in X$, since $\{x\}$ is closed, there is a G_n so that $x \in G_n$. Call this index $n(x)$. Then, by Lemma 5.1.2, there is an index $k(n(x))$ so that $x \in G_{k(n(x))} \subset \overline{G_{k(n(x))}} \subset G_{n(x)}$. The collection of pairs $\{(G_{k(n(x))}, G_{n(x)}) : x \in X\}$ is contained in $\{(G_i, G_j) : i, j \in \mathbb{N}\}$ and so it is countable. Label these pairs as $P_j = \{(G_{k(j)}, G_j)\}$. Note a given $G_{n(x)}$ might be associated with a countable number of indices $k(n(x))$. Further, each pair P_j satisfies

$$G_{j(n)} \subset \overline{G_{j(n)}} \subset G_j$$

Apply Urysohn's Lemma to find functions $f_j : X \to [0,1]$ with $f_j(x) = 0$ on $\overline{G_{j(n)}}$ and $f_j = 1$ on $(G_j)^C$. Define $\Psi : X \to \ell^2$ by $\Psi(x) = f_n(x)/n$. Clearly, this is well-defined as

$$\sum_{n=1}^{\infty} |f_n(x)|^2/n^2 \leq \sum_{n=1}^{\infty} 1/n^2 < \infty.$$

Now assume $x \neq y$. Since $\{x\}$ and $\{y\}$ are closed sets, there are neighborhoods $G_{n(x)}$ and $G_{n(y)}$ with $x \in G_{n(x)}$, $y \in G_{n(y)}$ and $G_{n(x)} \cap G_{n(y)} = \emptyset$. Hence, by construction,

$$x \in G_{k(n(x))} \subset \overline{G_{k(n(x))}} \subset G_{n(x)}$$

and

$$y \in G_{k(n(y))} \subset \overline{G_{k(n(y))}} \subset G_{n(y)} \subset (G_{n(x)})^C.$$

The index pair $(k(n(x)), n(x))$ corresponds to a P_j pair for some j. Thus, we have $x \in G_{k(j)} \subset \overline{G_{k(j)}} \subset G_j$ with $y \in G_j^C$. Hence, $f_j(x) = 0$ and $f_j(y) = 1$ which tells us Ψ is 1-1. Also, Ψ is 1-1 and onto its range $\Psi(X) \subset \ell^2$.

Define $\hat{d} : X \times X \to \Re$ by

$$\hat{d}(x, y) = \sqrt{\sum_{n=1}^{\infty} \frac{|f_n(x) - f_n(y)|^2}{n^2}}$$

Note $\hat{d}$ is non-negative and it satisfies the triangle inequality because it is built from the standard ℓ^2 metric. Finally, if $\hat{d}(x, y) = 0$, then $|f_n(x) - f_n(y)| = 0$ for all n. Thus, $\Psi(x) = \Psi(y)$ and so $x = y$ as Ψ is 1-1. Hence, $\hat{d}$ is a metric.

Is Ψ continuous? Given $x_0 \in X$ and $\epsilon > 0$, we must show there is a neighborhood U of x_0 so that if $y \in U$, then $\|\Psi(y) - \Psi(x_0)\|_2 < \epsilon$. Now, $\sum_{n=1}^{\infty} |f_n(y) - f_n(x_0)|^2/n^2 \leq \sum_{n=1}^{\infty} 1/n^2 < \infty$. Let $S_N(y) = \sum_{n=1}^{N} |f_n(y) - f_n(x_0)|^2/n^2$, the Nth partial sum of the series. Hence, the sum of the series, $S(y)$, defined by $\lim_{N\to\infty} S_N(y)$ or $\sum_{n=1}^{\infty} |f_n(y) - f_n(x_0)|^2/n^2$, exists. So here, since $S_N(y) \leq \sum_{n=1}^{\infty} 1/n^2 < \infty$, for a given $\epsilon > 0$, there is a Q so that $\sum_{n=1}^{N} 1/n^2 < \epsilon^2/2$ if $N > Q$.

Hence, for any $N > Q$,

$$\begin{aligned} S(y) &= \sum_{n=1}^{\infty} |f_n(y) - f_n(x_0)|^2/n^2 = \sum_{n=1}^{N} |f_n(y) - f_n(x_0)|^2/n^2 + \sum_{n=N+1}^{\infty} |f_n(y) - f_n(x_0)|^2/n^2 \\ &\leq \sum_{n=1}^{N} |f_n(y) - f_n(x_0)|^2/n^2 + \sum_{n=N+1}^{\infty} \frac{1}{n^2} \\ &< \sum_{n=1}^{N} |f_n(y) - f_n(x_0)|^2/n^2 + \frac{\epsilon^2}{2} \end{aligned}$$

But $S_N(y)$ converges monotonically up to $S(y)$, so we have

$$0 \leq S(y) - \sum_{n=1}^{N} |f_n(y) - f_n(x_0)|^2/n^2 < \frac{\epsilon^2}{2}$$

$|S_N(y) - S(y)| < \epsilon/2$ *if $N > Q$ implying* $\sup_{y \in X} |S_N(y) - S(y)| < \epsilon$ *if $N > Q$ and $y \in X$.*

Now each f_n is continuous at x_0. So there are neighborhoods $\{U_1, \ldots, U_N\}$ of x_0 so that $|f_i(y) - f_i(x_0)|^2 < \epsilon^2 i^2/(2N)$ for $1 \leq i \leq N$, if $y \in U_i$. So if $y \in \cap_{i=1}^N U_i$,

$$S_N(y) = \sum_{n=1}^{N} |f_n(y) - f_n(x_0)|^2/n^2 < \sum_{n=1}^{N} \frac{\epsilon^2 n^2}{2N\, n^2} = \sum_{n=1}^{N} \frac{\epsilon^2}{2N} = \frac{\epsilon^2}{2}$$

Combining,

$$S(y) < \sum_{n=1}^{N} |f_n(y) - f_n(x_0)|^2/n^2 + \frac{\epsilon^2}{2} < \frac{\epsilon^2}{2} + \frac{\epsilon^2}{2} = \epsilon^2$$

We conclude

$$\sqrt{\sum_{n=1}^{\infty} |f_n(y) - f_n(x_0)|^2/n^2 < \epsilon}$$

This show S is continuous at y.

Is Ψ^{-1} continuous? To prove this, we must show that Ψ acting on an open set is open. Let $x_0 \in X$ be given. It suffices to see what happens on a basic open set G_n which contains x_0. For this open G_n, there is an index ℓ so that $x_0 \in G_{k(\ell)} \subset \overline{G_{k(\ell)}} \subset G_n = G_\ell$. There is then a mapping f_ℓ which is 0 on $G_{k(\ell)}$ and 1 on $G_a \ell^C$. Now, if $\|\Psi(y) - \Psi(x_0)\|_2 < 1/(2\ell)$, then $\sum_{n=1}^{\infty} |f_n(y) - f_n(x_0)|^2/n^2 < 1/(4\ell^2)$. This tells us $|f_\ell(y) - f_\ell(x_0)|^2/\ell^2 < 1/(4\ell^2)$ or $|f_\ell(y) - f_\ell(x_0)| < 1/2$.

Now $f_\ell(x_0) = 0$ as $x_0 \in G_{k(\ell)}$. Thus, $|f_\ell(y)| < 1/2$. But we know f_l is 1 on $G_e ll^C$. This implies $y \in G_\ell$.

Thus,

$$\|\Psi(y) - \Psi(x_0)\|_2 < 1/(2\ell) \implies y \in G_\ell \implies \Psi^{-1}\Psi(y) \in G_\ell$$

Hence, $\Psi^{-1}(B(\Psi(x_0), 1/(2/\ell))) \subset G_n$ *which shows* Ψ^{-1} *is continuous.*

We have shown that Ψ *is a homeomorphism from* X *into the subset* $\Psi(X)$ *of* ℓ^2. ∎

Homework

Exercise 5.3.1 *Prove Theorem 5.3.1 when* X *has* 7 *points.*

Exercise 5.3.2 *Prove Theorem 5.3.1 when* X *has* 507 *points.*

When can subsets of a space X serve as a basis for a topology? In other words, when can an arbitrary collection $\mathscr{B}$ of subsets be used to construct a topology by taking the set of unions from $\mathscr{B}$?

Theorem 5.3.2 Topology Construction from a Collection of Sets

Let X *be a nonempty set. Let* Ω *be an arbitrary collection of sets from* X. *Then* Ω *can serve as an open subbase for a topology* $\mathscr{S}$ *where* $\mathscr{S}$ *consists of all unions of finite intersections of sets in* Ω.

Proof 5.3.2
If $\Omega = \emptyset$, *the only finite intersection we can get is the one which has no intersections which we will interpret as* X *itself. Also, the union over no sets is just* $\emptyset$. *Hence, this gives the topology* $\mathscr{S} = \{\emptyset, X\}$.

If Ω *is nonempty, let* $\mathscr{B}$ *be the collection of all finite intersections of sets in* Ω *and* $\mathscr{S}$ *be the collection of arbitrary unions of sets from* $\mathscr{B}$. *Again, the intersection over no sets is interpreted as* X *and the union over no sets is* $\emptyset$. *So* $\emptyset$ *and* X *are in* $\mathscr{S}$.

By definition, $\mathscr{C}$ *is closed under arbitrary unions so all these sets are in* $\mathscr{S}$. *Finally, let* $\{G_1, \ldots, G_n\}$ *be a nonempty finite collection of sets in* $\mathscr{S}$. *Since* $\emptyset \in \mathscr{S}$, *if* $\cap_i G_i = \emptyset$, *in this case* $\cap_i G_i \in \mathscr{S}$. *So we can assume* $\cap_i G_i \neq \emptyset$. *Each* G_i *is the union of sets in* $\mathscr{B}$; *hence,* $g_i = \cup_{j\Lambda_i} B_{ij}$ *for an index set* Λ_i. *We also know each* $B_{ij} = \cap_{k=1}^{p_{ij}} E_{ij}^k$ *where* $E_{ij}^k \in \Omega$.

Thus,

$$\begin{aligned} \cap_{i=1}^n G_i &= \cap_{i=1}^n \left(\cup_{j\Lambda_i} B_{ij} \right) = \cap_{i=1}^n \left(\cup_{j\Lambda_i} \cap_{k=1}^{p_{ij}} E_{ij}^k \right) \\ &= \cup_{j\Lambda_i} \cap_{i=1}^n \cap_{k=1}^{p_{ij}} E_{ij}^k \end{aligned}$$

The sets $\cap_{i=1}^n \cap_{k=1}^{p_{ij}} E_{ij}^k$ *are finite intersections of set from* $\mathscr{B}$ *and so must be in* $\mathscr{B}$. *We see we have written* $\cap_{i=1}^n G_i$ *as a union of finite intersections of sets from* $\mathscr{B}$. *Hence, it is in* $\mathscr{S}$.

This shows $\mathscr{S}$ *is a topology.* ∎

Note the sets in Ω do not have to be open in the way we think of open in a metric space. For example, let $X = [0, 1]$ and let Ω be the collection of sets of the form $(a, b]$ from X. Then we can use Ω to construct a topology on $[0, 1]$ and each of the sets $(a, b]$ is called an open set.
This theorem points the way towards constructing a topology from **any** collection S of subsets. According to the result, what we need to do is construct a new set $\mathscr{B}$ from S such that:

1. X is the union of sets in $\mathscr{B}$.

2. Every intersection of two sets in $\mathscr{B}$ is again in $\mathscr{B}$.

Once we have such a set $\mathscr{B}$, then the theorem guarantees it is a basis for a topology. The exact procedure for constructing such a basis from any set is laid out next.

Theorem 5.3.3 Constructing a Basis for a Topological Space

Let X be a set and Ω be any collection of subsets of X. Define a set:

$$\mathscr{B} = \{F : F = X \text{ or } F \text{ a finite intersection of sets from } \Omega\}$$

Then $\mathscr{B}$ is a basis for a topology $\mathscr{S}$ where $\mathscr{S} = \{U : U \text{ is a union of members of } \mathscr{B}\}$.

Proof 5.3.3
This is a restatement of sorts of the previous theorem. We show that the set $\mathscr{B}$ satisfies the conditions of the previous theorem:

(i): Since X is in $\mathscr{B}$ by assumption, then it is clear that X is the union of members in $\mathscr{B}$.

(ii): Let X and Y be sets in $\mathscr{B}$. Then both X and Y are finite intersections of sets in Ω. Thus, $X \cap Y$ must likewise be a finite intersection of sets in Ω, and this implies $X \cap Y$ is in $\mathscr{B}$.

Therefore, the set $\mathscr{B}$ satisfies the conditions necessary to be an open basis. ■

So this theorem maps out a way for constructing a topology from any collection of sets in a space X. Given a set S, we just take the set of all finite intersections of S and join to it the whole space X. This will then serve as a basis that makes X into a topological space.
Homework

Exercise 5.3.3 *Let $X = [a, b]$ with $a < b$ and let $\mathscr{R}$ be the collection of sets of the form $[c, d) \subset [a, b]$. Go through all the details of the proof that shows the collection of all unions of finite intersections of sets from $\mathscr{R}$ generates a topology $\mathscr{S}$ for $[a, b]$. Is any open set of X as defined by the usual norm topology in $\mathscr{S}$?*

Exercise 5.3.4 *Let $X = [a, b]$ with $a < b$ and let $\mathscr{R}$ be the collection of sets of the form $(c, d] \subset [a, b]$. Go through all the details of the proof that shows the collection of all unions of finite intersections of sets from $\mathscr{R}$ generates a topology $\mathscr{S}$ for $[a, b]$. Is any open set of X as defined by the usual norm topology in $\mathscr{S}$?*

Exercise 5.3.5 *Let $X = [a, b] \times [c, d]$ with $a < b$ and $c < d]$ and let $\mathscr{R}$ be the collection of sets of the form $[c, d) \times (e, f] \subset X$. Go through all the details of the proof that shows the collection of all unions of finite intersections of sets from $\mathscr{R}$ generates a topology $\mathscr{S}$ for X. Is any open set of X as defined by the usual norm topology in $\mathscr{S}$?*

Exercise 5.3.6 *Let $X = (C([a, b]), \|\cdot\|_\infty)$ with $a < b$ and let $\mathscr{R}$ be the collection of sets of the form $Y = \{f : [a, b] \to \Re : c < \|f\|_\infty < d,\ c < d\}$. Go through all the details of the proof that shows the collection of all unions of finite intersections of sets from $\mathscr{R}$ generates a topology $\mathscr{S}$ for X. Is any open set of X as defined by the norm topology in $\mathscr{S}$?*

5.4 Topological Vector Spaces

In the preceding discussion, we were unable to give an adequate definition for Cauchy sequences in a topological space. Consequently, we were not able to define what it means for a topological space to be *complete*. This was because there just wasn't enough structure in a general topological space to define a notion such as Cauchy sequences. To fix this, we can add a vector space structure to our topologies. The resulting spaces are then called **topological vector spaces**. We will see in this section that overlaying vector space operations onto a topological space produces other benefits as well. For instance, it simplifies the task of finding sub-bases and bases for our topologies. In fact, we can slightly relax the conditions of Theorems 5.3.2 and 5.3.3 once we have a vector space structure in place. But before we see this, let's first define a vector space and the related notion of a topological vector space. You have seen the definition of a vector space many times already, but there is no harm in repeating it here. We need to be precise here: given a nonempty set V, suppose there are mappings $\oplus$, $\odot_R$ and $\odot_L$ defined as

$$\oplus : V \times V \rightarrow V, \text{ written } (x,y) \rightarrow x \oplus y$$
$$\odot_L : F \times V \rightarrow V, \text{ written } (\alpha, x) \rightarrow \alpha \odot_L x, \qquad \odot_R : V \times F \rightarrow V, \text{ written } (x, \alpha) \rightarrow x \odot_R \alpha$$

The mapping $\oplus$ is called the **addition** operator and the mappings $\odot_R$ and $\odot_L$ are the **right scalar multiplication** and the **left scalar multiplication** operators. Note, if you program at all, you will recall, we typically have to define left and right versions of the command `rA` and `Ar` for an object in a class. A vector space consists of the set V and the mappings $\oplus$ and $\odot_L$ and various compatibility conditions that must be satisfied for the mappings. For example, in $\Re^2$, with vectors written as column vectors, we would have

$$\begin{aligned} \boldsymbol{V} \oplus \boldsymbol{W} &= \begin{bmatrix} v_1 \\ v_2 \end{bmatrix} \oplus \begin{bmatrix} w_1 \\ w_2 \end{bmatrix} = \begin{bmatrix} v_1 + w_1 \\ v_2 + w_2 \end{bmatrix} \\ \alpha \odot_L \boldsymbol{V} &= \alpha \odot_L \begin{bmatrix} v_1 \\ v_2 \end{bmatrix} = \begin{bmatrix} \alpha v_1 \\ \alpha v_2 \end{bmatrix} \end{aligned}$$

and for continuous functions on an interval $[a, b]$, the new functions $f \oplus g$ and $\alpha \odot_L f$ are defined pointwise by

$$(f \oplus g)(x) = f(x) + g(x), \quad (\alpha \odot_L f)(x) = \alpha f(x)$$

Since the right and left scalar multiplication, are typically often the same, we usually simply write $\odot$ instead of $\odot_L$; i.e. we identify the actions of $\odot_L$ and $\odot_R$. However, to be clear, we define a vector space V over the field F as follows:

Definition 5.4.1 Vector Space Structure

A **vector space** *is any set V combined with functions of vector addition, $\oplus$, and scalar multiplication, $\odot_L$, where the scalars come from a field F, such that for every choice of $x, y, z \in V$ and $\alpha, \beta \in F$*

1. $x \oplus y = y \oplus x.$
2. $(x \oplus y) \oplus z = x \oplus (y \oplus z).$
3. *There exists an element* $0 \in V$ *such that* $x \oplus 0 = 0 \oplus x = x.$
4. *There is an element* $-x \in V$ *such that* $x \oplus -x = 0.$
5. $1 \odot_L x = x,$
6. $\alpha \odot_L (\beta \odot_L x) = (\alpha \cdot \beta) \odot_L x.$
7. $\alpha \odot_L (x \oplus y) = \alpha \odot_L x \oplus \alpha \odot_L y.$
8. $(\alpha + \beta) \odot_L x = \alpha \odot_L x + \beta \odot_L x.$

Note, the usual $+$ symbol is used for addition in the field F and multiplication in the field is denoted by $\cdot$. We generally equate $\alpha \odot_L x$ and $x \odot_R \alpha$ and simply write $\alpha \odot x$. But in principle, you should remember they are distinct ideas. So from now on, we will use $\odot$ as the scalar multiplication operator. Also, we typically use the real numbers $\Re$ as the scalar field for our vector spaces or, when it becomes necessary, we let the complex numbers $\mathscr{C}$ be the field. Vector spaces already have significant structure by virtue of their intrinsic addition and multiplication operations. Algebraic structure, however, is not enough. The great majority of analysis studies need vector spaces to have a topological structure as well. We now define a new idea: the topological vector space. The mappings $\oplus$ and $\odot$ (we no longer worry about the right and left versions) are defined on cross products of spaces. We are interested in understanding what it means for $\oplus$ and $\odot$ to be continuous. We need a definition.

Definition 5.4.2 Cartesian Product of Two Topological Spaces

> *Let $(X, \mathscr{S})$ and $(Y, \mathscr{T})$ be two topological spaces. The cross-product topology induced on $X \times Y$ consists of the open sets of the form $A \times B$ where $A \in \mathscr{S}$ and $B \in \mathscr{T}$.*

Thus, a mapping $f : (X, \mathscr{S}) \times (Y, \mathscr{T}) \to (Y, \mathscr{U})$ would be called continuous if $f^{-1}(B)$ is open in $X \times Y$ when B is open in Z. That is, if $B \in \mathscr{U}$, $f^{-1}(B) = C \times D$, where $C \in \mathscr{S}$ and $D \in \mathscr{T}$. In particular,

- $\oplus$ is continuous if given $B \in \mathscr{S}$, the topology of X, $(\oplus)^{-1}(B) = C \times D$ where C and D are in $\mathscr{S}$.
- $\odot$ is continuous if given $B \in \mathscr{S}$, the topology of X, $(\odot)^{-1}(B) = C \times D$ where C is open in the field F and D is in $\mathscr{S}$.

We can now define a topological vector space.

Definition 5.4.3 Topological Vector Space $(X, \mathscr{S})$

> *A* **topological vector space** *is a topological space $(X, \mathscr{S})$ which is also a vector space over a field F where the vector addition map $\oplus$ and the scalar multiplication map $\odot$ are continuous.*

Some comments are in order. Often, we assume X has a topology such that $\{x\}$ is a closed set for every $x \in X$ (i.e. X is at least T_1), but we will not do that here. Besides, a topology is nothing more than a collection of subsets satisfying certain *openness* properties. Once we have enough subsets in our topology to guarantee the continuity condition for $\oplus$ and $\odot$, it hurts nothing when we add more subsets of the form $V \setminus \{x\}$ for every $x \in X$, thereby guaranteeing that $\{x\}$ is a closed set in X.

Secondly, let us be clear as to what exactly is meant by continuity. Suppose $x_1, x_2 \in X$. Vector addition $\oplus$ is **continuous** when for every neighborhood V of $x_1 \oplus x_2$, there exist neighborhoods V_1 of x_1 and V_2 of x_2 such that $V_1 + V_2 \subset V$. (Recall the definition of set addition in a vector space: $V_1 + V_2 = \{\, x \oplus y | x \in V_1,\ y \in V_2 \,\}$).

Now suppose $x \in V$, $\alpha \in F$, and U is a neighborhood of $\alpha \oplus x$. Scalar multiplication is **continuous** when there exist open sets W containing x and D containing α such that $\beta \odot W \subset U$ whenever $\beta \in D$.

Comment 5.4.1 *It is tedious to keep referring to the additive and scalar multiplication mappings as $\oplus$ and $\odot$. So we are going to relax our notation from this point on and simply refer to the $+$ and $\cdot$ mappings in this way: $x \oplus y \equiv x + y$ and $\alpha \odot x \equiv \alpha x$. It is much more convenient and, if we need additional clarity, we can go back to the formal notation at any time.*

It is perhaps more intuitive to think of continuity in terms of convergent sequences. Suppose $\{x_n\}$ and $\{y_n\}$ are sequences in X such that $\{x_n\}$ converges to x and $\{y_n\}$ converges to y. Suppose also $\{\alpha_n\}$ is a sequence in F that converges to c. Then vector addition is continuous if $\{x_n \oplus y_n\}$ converges to $x \oplus y$. Likewise, scalar multiplication is continuous if $\{\alpha_n \odot x_n\}$ converges to $\alpha \odot x$. However, to do it this way, means that we need the limits of convergent sequences to be unique. This need not be true. We will say more about this later.

Example 5.4.1 *The usual space $(C([0,1]), \|\cdot\|_\infty)$ is a normed linear space and so also a metric space but it is not an inner product space as no inner product induces this norm. It is a topological vector space using the pointwise definitions of $\oplus$ and $\odot$. We know sums of continuous functions and scalar multiples of continuous functions are also continuous, so $\oplus$ and $\odot$ are well-defined. The topology here is determined by the open sets defined by the norm. For two functions x and y look at the new function $x \oplus y$. Let $B(x \oplus y, r)$ be an open ball around the new function $x \oplus y$. Let $u \in B(x, r/4)$ and $v \in B(y, r/4)$. Then $|u(t) - x(t)| \leq r/4$ and $|v(t) - y(t)| < r/4$ for $0 \leq t \leq 1$. Hence $|(u(t) + v(t)) - (x(t) + y(t))| \leq r/2$ on $[0,1]$ and so $u \oplus v \in B(x \oplus y, r)$. This shows the open set $B(x, r/4) \times B(y, r/4)$ in the product topology is mapped by $\oplus$ into $B(x \oplus y, r)$. A little thought shows this proves the inverse image of any open set in $(C([0,1]), \|\cdot\|_\infty)$ is open in $(C([0,1]), \|\cdot\|_\infty) \times (C([0,1]), \|\cdot\|_\infty)$. Note the notation is cumbersome. For example, the addition of two sets A and B in a vector space is defined to $A + B = \{a + b : a \in A, b \in B\}$. If we insist on being really careful, we could say $A + B = \{a \oplus b : a \in A, b \in B\}$ or even $A \oplus B = \{a \oplus b : a \in A, b \in B\}$. You can see, we should exercise some restraint. Most of us can parse this expression just fine: we have showed $B(x, r/4) + B(y, r/4) \subset B(x + y, r)$ without the careful use of $\oplus$ for $+$. Just remember, that detail is there if you ever get confused.*

A similar argument shows $\odot$ is continuous and we leave that argument to you. So $(C([0,1]), \|\cdot\|_\infty)$ with these mappings is a topological vector space.

Example 5.4.2 *Look at the space $X = (C([0,1]), \|\cdot\|_1)$. We know this is a normed space and a metric space but not an inner product space. We define $\oplus$ and $\odot$ as usual using pointwise definitions. You can show it is a topological vector space. Let $B = \{x \in X | \|x\|_\infty < 1\}$. Is B open in X? Pick $x_0 \in B$. We must find an $r > 0$ so that $\{x \in X | \|x - x_0\|_1 < r\} \subset B$. Pick any $r > 0$ you want. Then pick any $0 < t_0 < 1$ and any δ small enough so that $t_0 \pm \delta$ is still in $(0,1)$. Let ϕ be a C^∞ bump function whose support is $[t_0 - \delta, t_0 + \delta]$. Adjust ϕ so the maximum value of ϕ is $\phi_max = r/(4\delta)$ with δ chosen so that $r/(4\delta) > \max\{\|x_0\|_\infty, 1\}$. Then $\int_0^1 \phi(t)dt \leq (2r\delta)/(4\delta) < r$. So $\int_0^1 |x_0(t) + \phi(t) - x_0(t)|dt = \int_0^1 \phi(t)dt < r$. But $\|x_0 + \phi\|_\infty = r/(4\delta) > 1$ for sufficiently small δ.*

So B is not open in the topology determined by $X = (C([0,1]), \|\cdot\|_1)$. A similar argument shows B is not open in the topology determined by $X = (C([0,1]), \|\cdot\|_2)$.

Now let's talk about what it means to be open relative to another set.

Definition 5.4.4 Inherited or Relative Topology due to $E \subset X$

If $(X, \mathscr{S})$ is a topological space and $E \subset X$, let $\mathscr{S}_E = \{E \cap G | G \in \mathscr{S}\}$. Then S_E is a topology on E which is called the topology inherited from $\mathscr{S}$. This is also called the relative topology.

So if $X = \Re$ with the standard topology, for $E = (0,1]$, $\mathscr{S}_E = \{(0,1] \cap G\}$ for all open G in $\Re$. Further, if $X = \Re^2$ with the standard topology, for $E = (0,1) \times [1,2)$, $\mathscr{S}_E = \{(0,1) \times [1,2) \cap G\}$ for all open G in $\Re$.

Homework

Exercise 5.4.1 *Show how to interpret* $\Re^n$ *as a topological vector space using the appropriate* $\oplus$ *and* $\odot$.

Exercise 5.4.2 *Show how to interpret the set of real-valued* $n \times m$ *matrices as a topological vector space using the appropriate* $\oplus$ *and* $\odot$ *using the row norm, the column norm and the Frobenius norm.*

Exercise 5.4.3 *For the space* $(C([0,1]), \|\cdot\|_\infty)$, *show the inverse image of any open set under* $\oplus$ *is open in the cross topology. You just need to show some details we left out.*

Exercise 5.4.4 *For the space* $(C([0,1]), \|\cdot\|_\infty)$, *show* $\odot$ *is continuous.*

Exercise 5.4.5 *Show the space* $(C([0,1]), \|\cdot\|_1)$ *is a topological vector space using the usual pointwise definitions of* $\oplus$ *and* $\odot$.

Exercise 5.4.6 *Show the space* $(C([0,1]), \|\cdot\|_2)$ *is a topological vector space using the usual pointwise definitions of* $\oplus$ *and* $\odot$.

5.4.1 Separation Properties of Topological Vector Spaces

What do we mean by bounded sets in a topological vector space? We no longer have a norm to help us decide this. We use this as our definition.

Definition 5.4.5 Bounded Sets in a Topological Vector Space

A subset E *in a topological vector space* $(X, \mathscr{S})$ *is said to be bounded if given an open set containing* 0 *in* X *(this is called a neighborhood of* 0*), there is an* $r > 0$ *so that* $E \subset tV$ *for all* $t > r$.

There are then two fundamental maps on X to itself: the translation and the scaling map.

Definition 5.4.6 Scaling and Translation Map on a Topological Vector Space

Let $(X, \mathscr{S})$ *be a topological vector space. For any* $a \in X$, $T_a : X \to X$ *defined by* $T_a(x) = a + x$ *is the translation map of* x *by the fixed amount* a. *Also, for any* $c \neq 0 \in F$, *the map* $M_c : X \to X$ *defined by* $M_c(x) = cx$ *is the scale by the constant* c *mapping.*

It is easy to prove that T_a and M_c are homeomorphisms.

Theorem 5.4.1 Scale and Translation Maps are Homeomorphisms

Let $(X, \mathscr{S})$ *be a topological vector space. For any* $a \in X$, *it is easy to see* $(T_a)^{-1} = T_{-a}$ *and for any* $c \neq 0 \in F$, $(M_c)^{-1} = M_{1/c}$. *We leave it to you to prove these maps are homeomorphisms.*

We can use these ideas to prove some interesting facts. First, about translation.

Theorem 5.4.2 E **is Open if and only if** $a + E$ **is Open for all** a

Let $(X, \mathscr{S})$ *be a topological vector space. Then* $E \in \mathscr{S}$ *if and only if* $a + E \in \mathscr{S}$ *for all* $a \in X$.

Proof 5.4.1
If E *is open, for* $a \in X$, *then* $(T_a)(E)$ *is open as* T_a *is a homeomorphism.* E *is open implies* $a + E$

is open for all $a \in X$.

Conversely, if $a + E$ is open for all $a \in X$, then, in particular $0 + E = E$ is open. ■

Second, we can look at scaling.

Theorem 5.4.3 E is Open if and only if cE is Open for all $c \neq 0$

> *Let $(X, \mathscr{S})$ be a topological vector space. Then $E \in \mathscr{S}$ if and only if $cE \in \mathscr{S}$ for all $c \neq 0$.*

Proof 5.4.2
If E is open, for $c \neq 0 \in \Re$, then $(M_c)(E)$ is also open as M_c is a homeomorphism. We conclude E is open implies cE is open for all $c \neq 0$.

Conversely, if cE is open for all $c \neq 0$, then, in particular $1 \cdot E = E$ is open. ■

This result allows us to simplify topological vector spaces immensely. It states that, in a certain sense, what happens at a single point is the same as what happens at any other point. How a topology *behaves* at one point is precisely how it *behaves* everywhere else.

Definition 5.4.7 Local bases at a point

> *Let $(X, \mathscr{S})$ be a topological space and let $p \in X$. Let $\mathscr{V}$ be a collection of open sets that contain p. The collection $\mathscr{V}$ is called a local base at p if every open set containing p contains a member of $\mathscr{V}$. Hence, if $p \in G$ for open G, there is a $V \in \mathscr{V}$ so that $p \in V \subset G$.*

In a metric space this is easy to see. Given $p \in (X, d)$, $\mathscr{V}$ is the collection $\{B(p, r)\}$ of all balls about p. Then if $p \in G$ for G open, p is an interior point and so there is an $r > 0$ with $B(p, r) \subset G$. Hence, $\mathscr{V}$ is a local base at p.

So what Theorem 5.4.2 tells us is that any local base completely determines the topology in a topological vector space. Consequently, we are free to specify our topologies by establishing a local base anywhere in the space. From now on, the term **local base** will always refer to a **local base at 0** because of the natural simplicity of this point, unless we have some compelling reason to focus on another point p. Let's look at local bases at 0 more carefully.

Lemma 5.4.4 Each Neighborhood of Zero Contains a Symmetric Set

> *Let V be a topological vector space. If W is a neighborhood of 0, then there is an open set U containing 0 that is symmetric (that is, $U = -U$) and that satisfies $U + U \subset W$.*

Proof 5.4.3
Let W be a neighborhood of 0. We note that $0 + 0 = 0$ and so, because addition is continuous, there are open sets V_1, V_2 containing 0 such that $V_1 + V_2 \subset W$. Now define $U = V_1 \cap V_2 \cap (-V_1) \cap (-V_2)$. Then it is apparent that U is symmetric. Let $x, y \in U$. Then x and y are in both V_1 and V_2. So $x + y \in W$ since $V_1 + V_2 \subset W$. Thus, $U + U \subset W$. ■

Comment 5.4.2 *We can extend this idea. We know if W is a neighborhood of 0, then there is an open set U containing 0 that is symmetric (that is, $U = -U$) and that satisfies $U + U \subset W$. Apply this idea to U. Then there is a symmetric neighborhood U_1 so that $U_1 + U_1 \subset U$ which implies $(U_1 + U_1) + (U_1 + U_1) \subset U + U \subset W$. In general, we can state this as there is a symmetric neighborhood U so that $\sum_{i=1}^{n} U + \sum_{i=1}^{n} U \subset W$ for any n.*

Now, if we tried to characterize continuity in terms of convergent sequences in a topological vector space, we have some problems. We need to guarantee we always have unique limits of convergent sequences. We can have unique limits in a Hausdorff space.

Theorem 5.4.5 Every T_1 Topological Vector Space is a Hausdorff Space

Let $(X, \mathscr{S})$ be a topological vector space and assume every singleton set $\{x\}$ is closed in the topology. This is the same as assuming the topology is T_1. Then, the topological vector space is a Hausdorff space.

Proof 5.4.4
Let x and y be distinct points in a topological vector space V. We break up the proof into two cases:

(I) *We first look at when one of the points is 0. Without loss of generality, we let $x = 0$. From the definition of topological vector space, $\{y\}$ is a closed set. Hence, $\{y\}^C$ is an open set, and since $y \neq 0$, then $\{y\}^C$ is an open set containing 0. Therefore by Lemma 5.4.4, there exists a neighborhood U of 0 such that $U = -U$, $U \subset \{y\}^C$, and $U + U \subset \{y\}^C$.*

Now consider the sets $x + U$ and $y + U$. Since U is open, then $x + U$ and $y + U$ are themselves open because the topology is translation-invariant. U contains 0 and so $x \in x + U$ and $y \in y + U$. We now show that $(x + U) \cap (y + U) = \emptyset$. Assume the two sets are not disjoint. Then there is an element shared by both. This common member would have the forms $x + u_1$ and $y + u_2$ where $u_1, u_2 \in U$.

Consequently, $x + u_1 = y + u_2$ or $u_1 - u_2 = y - x$. But U is symmetrical and so there is a $u_3 \in U$ such that $u_3 = -u_2$. Hence, $u_1 + u_3 = y - x$. Now recall that $u_1 + u_3 \in \{y\}^C$ (because $U + U \subset \{y\}^C$) and this implies that $y - x \in \{y\}^C$. But $y - x = y - 0 = y$ and so we would have $y \in \{y\}^C$ which is a contradiction. Therefore, $(x + U) \cap (y + U) = \emptyset$. Thus, $x + U$ and $y + U$ are two disjoint neighborhoods of x and y respectively.

(II) *Now suppose x and y are both nonzero. We translate the vector space V using the homeomorphism $-x + V$. Then x is now 0 and y is some nonzero value. From case (I), we see there are two disjoint neighborhoods containing these new x and y values. Furthermore, because of translation invariance, these neighborhoods remain open and disjoint when we translate back to the original x and y values.*

Therefore, every pair of distinct points in a topological vector space, where singleton points are closed sets, are separated by disjoint open sets. This demonstrates that any topological vector space is a Hausdorff space, and so all convergent sequences in a T_1 topological vector space have unique limits. ■

Now that we have sufficient structure in place, we are finally able to give a general topological definition of Cauchy sequences. Note that even though it requires more structure than just a topology alone, the following definition still makes no use of metrics! We will call a topological vector space in which singleton points are closed, a T_1 topological vector space.

Definition 5.4.8 Cauchy Sequences in a T_1 Topological Vector Space

Let $\mathscr{L}$ be a local base of a T_1 topological vector space V. A sequence $\{x_n\}$ is **Cauchy** *(or a* **Cauchy Sequence***) if for every $\mathscr{O} \in \mathscr{L}$ there is an N such that $x_n - x_m \in \mathscr{O}$ whenever $n, m \geq N$.*

Once we know what a Cauchy sequence is, we can define a complete topological vector space.

Definition 5.4.9 Complete T_1 Topological Vector Space

A T_1 topological vector space is **complete** *if every one of its Cauchy sequences converges in the space.*

Exercise 5.4.7 *Prove $\oplus$ and $\odot$ in $\Re^n$ as a topological vector space are homeomorphisms.*

Exercise 5.4.8 *Prove $\oplus$ and $\odot$ in the set of real-valued $n \times m$ matrices are homeomorphisms using the row norm, the column norm and the Frobenius norm.*

Exercise 5.4.9 *Prove $\oplus$ and $\odot$ in the space $(C([0,1]), \|\cdot\|_\infty)$ are homeomorphisms.*

Exercise 5.4.10 *Prove $\oplus$ and $\odot$ in the space $(C([0,1]), \|\cdot\|_1)$ are homeomorphisms.*

Exercise 5.4.11 *Prove $\oplus$ and $\odot$ in the space $(C([0,1]), \|\cdot\|_2)$ are homeomorphisms.*

Exercise 5.4.12 *Prove Lemma 5.4.4 showing all the details of the constructions required for the space $(C([0,1]), \|\cdot\|_1)$.*

Exercise 5.4.13 *Prove Lemma 5.4.4 showing all the details of the constructions required for the space $(C([0,1]), \|\cdot\|_2)$.*

Exercise 5.4.14 *Prove Lemma 5.4.4 showing all the details of the constructions required for the space $\Re^n$.*

Chapter 6

Locally Convex Spaces and Seminorms

We can derive powerful results if we add additional properties to the open base of a topological vector space.

6.1 Additional Classifications of Topological Vector Spaces

We have already mentioned that some topological spaces are **metrizable** which means their topology can be generated by a metric. If you recall, the metric induced by a norm has special properties: the induced metric is **invariant** in the sense that $d(x+z, y+z) = d(x, y)$. Here are some additional properties a topological vector space can have.

Definition 6.1.1 Topological Vector Space Classifications

Let $(X, \mathscr{S})$ be a topological vector space. We say

- X *is* **locally convex** *if there is a local base* $\mathscr{B}$ *whose members are convex.*
- X *is* **locally bounded** *if* $0 \in X$ *has a neighborhood which is bounded.*
- X *is* **locally compact** *if* $0 \in X$ *has a neighborhood whose closure is topologically compact.*
- X *is an F-Space if* $\mathscr{S}$ *comes from a completely invariant metric d. We say d is complete if the metric space* (X, d) *is complete.*
- X *is a* **Frechet Space** *if* X *is a locally convex F - Space.*
- X *is* **normable** *if there is a norm* ρ *on* X *so that the metric induced by* ρ*, d, determines* $\mathscr{S}$*.*
- X *has the* **Heine - Borel property** *if all closed and bounded subsets of* X *are compact.*

Let's prove the existence of various kinds of neighborhoods in a topological vector space. We have already discussed symmetric neighborhoods. There are two more interesting types.

Definition 6.1.2 Balanced Neighborhoods and the Balanced Hull

Let $(X, \mathscr{S})$ be a topological vector space. An open set or neighborhood B is called **balanced** *if $tB \subset B$ for all $-1 \leq t \leq 1$. The* **balanced hull** *of a set $S \subset X$ is $BH(S) = \cap\{B | B \text{ is balanced}, S \subset B\}$.*

Comment 6.1.1 *The idea of a balanced set works fine even if X is only a vector space as vector addition and scalar multiplication are available.*

Comment 6.1.2 *If we start with a balanced set S, then by definition, $BH(S) \subset S$ and $S \subset CH(S)$ too. So, in this case $BH(S) = S$.*

We can also look at convex hulls.

Definition 6.1.3 Convex Hull of a Set

Let $(X, \mathscr{S})$ be a topological vector space. The **convex hull** *of a set $S \subset X$ is*

$$C_S = \{x = \sum_{i=1}^{n} \alpha_i x_i | n \text{ some positive integer}, \sum_{i=1}^{n} \alpha_i = 1, \alpha_i \geq 0, x_i \in S, 1 \leq i \leq n\}$$

This is equivalent to defining the convex hull of S to be

$$CH(S) = \cap\{C | C \text{ is a convex set}, S \subset C\}$$

Comment 6.1.3 *The idea of a convex set and the convex hull of a set works fine even if X is only a vector space as vector addition and scalar multiplication are available.*

Comment 6.1.4 *It is easy to see C_S is a convex set that contains S. Hence, $CH(S) \subset C_S$. However, we also know that if $S \subset C$ with C convex, $C_S \subset C$ also. Hence, $C_S \subset CH(S)$. Hence, these are two ways of defining the convex hull of S.*

Theorem 6.1.1 Neighborhoods of 0 have Open Balanced Hulls

Let $(X, \mathscr{S})$ be a topological vector space. If S is a neighborhood of 0, then $BH(S)$ is open.

Proof 6.1.1
If B is balanced with $S \subset B$, $-B \subset B$ implying B is symmetric. Also, if B is balanced, B contains 0. So $0 \in S \subset B$. If $-1 \leq t \leq 1$, $tS \subset tB \subset B$ as B is balanced. Hence, $\cup_{-1 \leq t \leq 1} tS \subset B$ for all balanced B containing S. Let $B_0 = \cup_{-1 \leq t \leq 1} tS$ which is a balanced set containing S. Thus, $B_0 \subset BH(S)$. On the other hand, $BH(S) \subset B_0$. Hence, they are the same. But B_0 is open which proves the claim. ■

We can now prove a nice theorem about bases that consist of balanced hulls.

Theorem 6.1.2 Given a Base, the Balanced Hull of the Base is a Base

Let $(X, \mathscr{S})$ be a topological vector space. If $\mathscr{U}$ is an open base at 0 of X and $\mathscr{V}$ is the collection of balanced hulls of members of $\mathscr{U}$, then $\mathscr{V}$ is also an open base at 0.

Proof 6.1.2
Let $U \in \mathscr{U}$ and let $\mathscr{V} = \{BH(U) | u \in \mathscr{U}\}$. Since scalar multiplication is continuous, $0 \cdot 0 = 0$ means if U is a neighborhood of 0 in X, there are neighborhoods of 0, $B(0, r) \subset F$ and W in $\mathscr{U}$ so that $tW \subset U$ for $-r < t < r$. This implies $s(r/2)W \subset U$ for $-1 \leq s \leq 1$.

Let $W_0 = \cup_{-1 \leq s \leq 1} s(r/2)W$ which is a balanced set. Then $W_0 \subset U$. Recall $BH(W_0) = \cap\{B | B$ is balanced, $W_0 \subset B\}$. Clearly, $BH(W_0) = W_0$. Since $\mathscr{U}$ is an open base at 0, W_0 is a neighborhood of 0 implies there is a $U_1 \subset \mathscr{U}$ with $U_1 \subset W_0$. We thus have $BH(U_1) \subset BH(W_0) = W_0 \subset U$. This shows there is a member of $\mathscr{V}$ contained in U. We can do this for all U and so $\mathscr{V}$ is an open base too. ∎

Homework

Exercise 6.1.1 *Let S be a convex set in the topological vector space X. Assume $x_0 \in int(S)$ and $y_0 \in S$. Prove if $y = ax_0 + (1-a)y_0$, for $0 < a < 1$, $y \in int(S)$.*

Exercise 6.1.2 *Prove a convex set in a real vector space is balanced if and only if it is a symmetric set.*

Exercise 6.1.3 *Prove if S is convex in a topological vector space X, then the closure of S, $\overline{S}$, is also convex.*

6.1.1 Local Convexity Results

Let's look deep at convexity results in topological vector spaces. We start with a fundamental series of results. First, is the convex hull of an open set also open?

Theorem 6.1.3 Convex Hull of an Open Set is Open

> *Let $(X, \mathscr{S})$ be a topological vector space. If S is open, $CH(S)$ is open.*

Proof 6.1.3
Let $x \in CH(S)$. Then $x = \sum_{i=1}^{n} \alpha_i x_i$ for some n, with the $x_i S$, the scalars $\alpha_i \geq 0 \in F$ and $\sum_{i=1}^{n} \alpha_i = 1$. Since S is open, there are neighborhoods V_i of x_i so that $V_i \subset S$. Let $T = \sum_{i=1}^{m} \alpha_i V_i$. Then T is open as scalar multiplication and addition are homeomorphisms no matter what the value of n may be (this is a simple induction argument, of course). The subset T is a neighborhood of the point x in $CH(S)$. So $CH(S)$ is open. ∎

We are now ready to prove the next result: convex neighborhoods of 0 contain balanced convex neighborhoods of 0.

Theorem 6.1.4 Convex Neighborhoods of 0 Contain Balanced Convex Neighborhoods of 0

> *Let $(X, \mathscr{S})$ be a topological vector space. Let U be a convex neighborhood of 0. Then U contains a balanced convex neighborhood of 0.*

Proof 6.1.4
Since U is a neighborhood of 0, by Theorem 6.1.2, there is a balanced neighborhood of 0, V, so that $V \subset U$. Let $W = CH(V)$. By Theorem 6.1.3, we know $CH(V)$ is open. Hence, W is a neighborhood of 0 too. Since V is balanced, we have $BH(V) = V$, so $W = CH(V) = CH(BH(V))$. This tells us W is convex. Also, since U is convex, from the definition of the convex hull, $CH(V) \subset U$.

If A is a balanced set, let $x \in CH(A)$. Then, $x = \sum_{i=1}^{n} \alpha_i x_i$, each $\alpha_i \geq 0$, $\sum_{i=1}^{n} = 1$ and each $x_i \in A$. But since A is balanced, for $-1 \leq t \leq 1$, $tx = \sum_{i=1}^{n} \alpha_i (tx_i)$ and $tx_i \in A$ as A is balanced. Thus, $tx \in CH(A)$ for $-1 \leq t \leq 1$. This says $CH(A)$ is balanced and so $BH(CH(A)) = CH(A)$.

Apply this to our V. This says $BH(CH(V)) = CH(V) = W$ and so W is balanced. But we already know $W = CH(BH(V))$. So these two things must be the same. Hence, W is convex and balanced neighborhood of 0 contained in U. ■

Homework

Exercise 6.1.4 *Prove the convex hull of a convex set is itself.*

Exercise 6.1.5 *Given three points $\boldsymbol{A}$, $\boldsymbol{B}$ and $\boldsymbol{C}$ in $\Re^2$, find their convex hull. Assume the three points are not of the same line. What happens if they were on the same line?*

Exercise 6.1.6 *Given three points $\boldsymbol{A}$, $\boldsymbol{B}$ and $\boldsymbol{C}$ in $\Re^3$, find their convex hull in all possible cases.*

Exercise 6.1.7 *For the $n \times n$ matrix A and a given data vector $\boldsymbol{b} \in \Re^n$, the set of linear inequalities given by $A\boldsymbol{x} \leq \boldsymbol{b}$ with $x_i \geq 0$ determines a subset of $\Re^n$. Determine if this subset is convex.*

6.2 Metrization in a Topological Vector Space

We are now ready to think about when a topological vector space can be endowed with a metric.

Theorem 6.2.1 Metrization in a Topological Vector Space

Let $(X, \mathscr{S})$ be a T_1 topological vector space which has a countable local base at 0. Then there is a metric d on X so that:

(i): d determines the topology of X; i.e. this topology is compatible with the topology already on X as a topological vector space.

(ii): Open neighborhoods centered at 0 are balanced.

(iii): d is translation invariant: i.e. $d(x+y, x+z) = d(y,z)$ for all x, y and z in X. We also say d is an invariant metric.

We also say the topology generated by an invariant metric makes X a metric linear space. If X has a local base which is convex, then d can be chosen so that all open balls in the topology determined by d are convex.

Comment 6.2.1 *We could also assume $(X, \mathscr{S})$ is a T_1 topological vector space which is second countable. Then there is a countable collection of sets $\mathscr{V} = (V_n)$ so that every open set can be written as a union of sets from $\mathscr{V}$. Then $\mathscr{B}$, the collection of all sets from $\mathscr{V}$ which contain the point p, is a countable local base at p. In particular, this assumption means we have a countable local base at 0. Note, any open set containing 0 can thus be written as a union of sets from $\mathscr{B}$. Hence, since X contains 0, we can write $X = \cup_{n=1}^{\infty} V_n$.*

Proof 6.2.1
Since X has a countable local base at 0, $\mathscr{V} = (V_n)$, we have $X = \cup_{n=1}^{\infty} U_n$. We also know by Theorem 6.1.2, there is a new countable local base, $\{V_n\}$ which is balanced. Let $W_1 = V_1$. Then, by continuity of addition, there are open sets so that $V_{n(2)} + V_{n(2)} \subset V_1 = W_1$. Let $W_2 = V_{n(2)} \cap V_2 \subset V_2$. Then we have $W_2 + W_2 \subset W_1$.

Again, by continuity of addition, there are open sets so that $V_{n(3)} + V_{n(3)} \subset W_2$. Let $W_3 = V_{n(3)} \cap V_2$. Then $W_3 + W_3 \subset W_2$ and $W_2 \subset V_3$.

We can continue, via induction, to find $V_{n(k+1)} + V_{n(k+1)} \subset W_k \subset V_k$ *with* $W_{k+1} = V_{n(k+1)} \cap V_k \subset V_k$ *leading to* $W_{k+1} + W_{k+1} \subset W_k \subset V_k$. *We can do this for all* k. *Note all* W_k *are balanced.*
Let D *be the collection of all rational numbers in* $[0, 1]$ *which have finite binary representations. Thus, if* $r \in D$, *we can write* $r = \sum_{i=1}^{n_r} x_i(r)\,(1/2^i)$, *for some finite* n_r, *where the coefficients* $x_i(r) \in \{0, 1\}$. *Define the set* $A(r)$ *by*

$$A(r) \quad = \quad \begin{cases} X, & r \geq 1 \\ x_1(r)V_1 + x_2(r)V_2 + \ldots + x_{n_r}(r)V_{n_r}, & r \in D \end{cases}$$

Then, define $f : X \to \Re$ *by* $f(x) = \inf\{r | x \in A(r)\}$. *Note if* $x \in A(r)$, *by definition,* $f(x) \leq r$. *We want to show* $d(x, y) = f(x - y)$ *gives our desired metric.*

Since $x \in X$ *implies there is a set* V_M *with* $x \in V_M$, $x \in 0 \cdot V_1 + \ldots + 0 \cdot V_{M-1} + 1 \cdot V_M$. *These coefficients correspond to the binary representation* $r = (1/2^M)$. *Hence, we have found one possible* r *value so that* $x \in A(r)$. *This tells us* $f(x) \leq 1/2^M \leq 1/2 < 1$.

Let $\mathscr{A} = \{A(r) | r \in D \text{ or } r \geq 1\}$. *Consider the following two claims:*

(C1): $A(r) + A(s) \subset A(r + s)$, *for all* $r, s \in D$. *Note if both* r *and* s *are bigger than or equal to* 1, $A(r) + A(s) = X + X \subset X = A(r + s)$. *If at least one of* r *or* s *is bigger than or equal to* 1, *say* $s \geq 1$ *and* $r \in D$, *we have* $A(r) + A(s) = A(r) + X \subset X = A(r + s)$. *Also, if* $r, s \in D$ *and* $r + s \geq 1$, *the containment is clear. So all that remains is to prove the result for* $r, s \in D$ *with* $r + s < 1$.

(C2:) $A(r) \subset A(t)$ *if* $r < t$. *If both* r *and* t *are bigger than or equal to* 1, *we have* $X \subset X$ *which is true. If say* $t \geq 1$ *and* $r \in D$, *we have* $A(r) \subset X = A(t)$. *The last case is* $r < t$ *with* $r, t \in D$. *Then using C1, we have*

$$A(r) \subset A(r) + A(t - r) \subset A(r + t - r) = A(t)$$

So C2 is true.

(C3:) $0 \in A(r)$ *for all* r. *This one is easy. If* $r \geq 1$, $A(r) = X$ *which contains* 0 *and otherwise* $A(r)$ *is a finite sum of neighborhoods of* 0 *which contains* 0.

So all we must do is prove C1 for $r, s \in D$. *Let* P_N *be the statement:*

If $r + s < 1$ *with* $x_n(r) = x_n(s) = 0$ *for all* $n > N$, *then* $A(r) + A(s) \subset A(r + s)$.

We will prove P_N *is true for all* N *by induction.*

Base case, $N = 1$**:** *Here* r *and* s *have only one term,* $r = x_1(r)/2$ *and* $s = x_1(s)/2$. *So*

$$A(r) \quad = \quad \begin{cases} \{0\}, & x_1(r) = 0; r = 0 \\ V_1, & x_1(r) = 1; r = 1/2 \end{cases}, \quad A(s) = \begin{cases} \{0\}, & x_1(s) = 0; s = 0 \\ V_1, & x_1(s) = 1; s = 1/2 \end{cases}$$

and

$$A(r + s) \quad = \quad \begin{cases} \{0\}, & x_1(r) + x_1(s) = 0; r + s = 0 \\ V_1, & x_1(r) + x_1(s) = 1; r + s = 1/2 \\ X, & x_1(r) + x_1(s) = 2; r + s = 1 \end{cases}$$

Thus, in all cases $A(r) + A(s) \subset A(r + s)$.

Inductive case, assume true for $k \leq N-1$: *Assume P_{N-1} is true. Choose $r, s \in D$ with $r+s < 1$, $x_n(r) = x_n(s) = 0$ for $n > N$. Let $r' = r - x_N(r)/2^N$ and $s' = s - x_N(s)/2^N$. Then, $r' + s' = r + s - (x_N(r) + x_N(s))/2^N$, $r = r' + x_N(r)/2^N$ and $s = s' + x_N(s)/2^N$. We have*

$$A(r) = A(r') + x_N(r)V_N, \quad A(s) = A(s') + x_N(s)V_N$$

Now $r + s < 1$ implies $r' + s' < 1$ too. Since P_{N-1} is true, we have $A(r') + A(s') \subset A(r' + s')$. Thus,

$$\begin{aligned} A(r) + A(s) &= A(r') + x_N(r)V_N + A(s') + x_N(s)V_N \\ &\subset A(r' + s') + (x_N(r) + x_N(s))V_N \end{aligned}$$

There are then several cases:

1. *$x_N(r) = x_N(s) = 0$ and we have $r' = r$, $s' = s$ and we have $A(r) + A(s) \subset A(r + s)$.*
2. *$x_N(r) = 1$ and $x_n(s) = 0$. Then $s' = s$ and we have*

$$\begin{aligned} A(r) + A(s) &= A(r') + x_N(r)V_N + A(s) \\ &\subset A(r' + s) + V_N = A(r' + s + 1/2^N) = A(r + s). \end{aligned}$$

   You handle the case $x_N(r) = 0$ and $x_n(s) = 1$ in a similar way.

3. *Now $x_N(r) = x_N(s) = 1$.*

$$\begin{aligned} r &= \sum_{i=1}^{N-1} x_i(r)/2^i + 1/2^N, \quad r' = \sum_{i=1}^{N-1} x_i(r)/2^i \\ s &= \sum_{i=1}^{N-1} x_i(s)/2^i + 1/2^N, \quad s' = \sum_{i=1}^{N-1} x_i(s)/2^i \end{aligned}$$

   By assumption, $r + s = r' + s' + 2/2^N < 1$. Thus, $r' + s' + 1/2^{N-1} < 1$. Let $u = r' + s'$ and $v = 1/2N - 1$. Then u, v satisfy P_{N-1} because $u + v < 1$ and both have expansions that stop after $N - 1$. Thus, $A(u) + A(v) \subset A(u + v)$. Using this we have

$$\begin{aligned} A(r) + A(s) &= \sum_{i=1}^{N-1} x_i(r)V_i + \sum_{i=1}^{N-1} x_i(s)V_i + 1 \cdot V_N \\ &= A(r' + s') + A(1/2^{N-1}) = A(u) + A(v) \\ &\subset A(r' + s' + 1/2^{N-1}) = A(r + s) \end{aligned}$$

   Thus P_N is true and we have shown C1 is true.

Now we can look at the properties of the metric d.

1. *Is $f(0) = 0$? This is easy to see as $f(0) = \inf\{r | 0 \in A(r)\} = 0$ as $0 \in A(r)$ for all r. Now consider $x \neq 0$. Since X is T_1, X is Hausdorff. So if $x \neq 0$ in X, there is a neighborhood of x, V, and a neighborhood W_{n_0} of 0 so that $0 \in W_{n_0}$, $x \notin W_{n_0}$ with $x \in V$. Now if $x \notin W_{n_0} = A(1/2^{n_0})$.*

   Now we know $W_{n_0+1} + W_{n_0+1} \subset W_{n_0}$, which also implies $x \notin W_{n_0}$ tells us $x \notin W_{n_0+1}$. It then follows because of the properties of the (W_n) collection, that $x \notin W_{n_0+j}$ for all j. Thus, $f(x) = \inf\{r | x \in A(r)\}$ must satisfy $f(x) \geq 1/2^{n_0} > 0$. We see if $x \neq 0$, $f(x) > 0$.

Hence, $f(x-y) > 0$ is $x \neq y$ and $f(x-y) = 0$ if $x = y$. We have shown $d(x,y) \geq 0$ and $d(x,y) = 0$ if and only if $x = y$.

2. *First, note if $x \in X$, then since $1 = \sum_{i=1}^{\infty} 1/2^i$, $x \in A(1)$ implying $f(x) \leq 1$. Hence, given any $x, y \in X$, $f(x+y) \leq 1$. Hence, if $f(x)+f(y) \geq 1$, then $f(x+y) \leq f(x)+f(y)$. So the only case we have to consider is when $f(x)+f(y) < 1$. Now by C1, $A(r)+A(s) \subset A(r+s)$ and so if $x+y \in A(r+s)$, $f(x+y) \leq r+s$. Also, by the Infimum Tolerance Lemma, for a given $\epsilon > 0$, there are r_ϵ and s_ϵ so that $x \in A(r_\epsilon)$, $y \in A(s_\epsilon)$ and $f(x) \leq r_\epsilon < f(x)+\epsilon/2$ and $f(y) \leq s_\epsilon < f(y)+\epsilon/2$. Thus,*

$$f(x+y) \leq r_\epsilon + s_\epsilon < f(x)+f(y)+\epsilon \implies f(x+y) \leq f(x)+f(y)+\epsilon$$

But $\epsilon > 0$ is arbitrary, so we have $f(x+y) \leq f(x)+f(y)$.

3. *It is easy to see $d(x+z, y+z) = d(x,y)$ so we have translation invariance.*

Next, given any V_n, let's show we can choose a $\delta > 0$ so that $B(0,\delta) = \{x | d(x,0) < \delta\} = \{x | f(x) < \delta\}$ is contained in V_n. Now if $\delta < 1/2^N$, $f(x) < 1/2^N$. Hence, by the Infimum Tolerance Lemma, for $\epsilon = 1/2^N - f(x) > 0$, there is a r_0 so that $f(x) \leq r_0 < 1/2^N$ with $x \in A(r_0)$.

Hence, $A(r_0) \subset A(1/2^N) = V_N$. This tells us $x \in B(0,\delta) \subset V_N$ if $\delta < 1/2^N$. We have shown that given any neighborhood V_N of 0, there is a ball around 0 given by the metric d which is contained in it. So the topology generated by the metric d is compatible with the original topology.

Finally, note V_n balanced, tells us $A(r)$ and $B(0,\delta)$ is balanced. If we knew V_n was convex, we will also have $B(0,\delta)$ is convex and translates of $B(0,\delta)$ are convex. ■

Homework

Assume X and Y are topological vector spaces and $A : X \to Y$ is linear. There are three interesting properties about A that are related. We say A is bounded if A maps bounded sets into bounded sets. The properties are **a**: A is continuous, **b**: A is bounded, **c**: if $x_n \to 0$, then $(A(x_n))$ is bounded and **d**: if $x_n \to 0$, then $A(x_n) \to 0$.

Exercise 6.2.1 *Prove **a** implies **b** implies **c**.*
Hint: ***a** implies **b**: If E is a bounded set in Y and W is a neighborhood of 0 in Y, show there is a neighborhood V of 0 in Y so that $A(V) \subset W$. Then, since V is bounded, $E \subset tV$ for large t telling us $A(E) \subset tW$ which means $A(V)$ is bounded. So **a** implies **b**.*

Exercise 6.2.2 *If d is a translation invariant metric on a topological vector space X, prove $d(nx,0) \leq nd(x,0)$ for all x and $n = 1,2,3\ldots$.*
Hint: *This follows from $d(nx,0) \leq sum_{i=1}^{n} d(ix,(i-1)x) = nd(x,0)$.*

Exercise 6.2.3 *If d is a translation invariant metric on a topological vector space X, prove if (x_n) is a sequence with $x_n \to 0$, then there are positive scalars (γ_n) with $\gamma_n \to \infty$ and $\gamma_n x_n \to 0$.*
Hint: *Since $d(x_n,0) \to 0$, for each k there is an N_k so that $d(x_n,0) < 1/k^2$ when $k > N_k$. Set $\gamma_n = 1$ when $n < N_1$, $\gamma_2 = 2$, if $n_1 \leq 2N_3$ and, in general, $\gamma_k = k$, when $n_k \leq k, n_{k+1}$. Then, for this γ_n, $d(\gamma_n x_n, 0) = d(kx_n,0) \leq kd(x_n,0) < 1/k$.*

Exercise 6.2.4 *Now assume X is metrizable. Then, it has a countable local base and d is translation invariant. **c** implies **d** implies **a**.*
Hint: *By the previous exercise, there are positive scalars (γ_n) with $\gamma_n \to \infty$ and $\gamma_n x_n \to 0$. Thus, $(A(\gamma_n x_n))$ is a bounded set and since $A(x_n) = \gamma_n^{-1} A(\gamma_n x_n) \to 0$, $(A(x_n))$ is bounded. Next, if **a** is not true, then there is a neighborhood W of 0 in Y so that $A^{-1}(W)$ does not contain any neighborhood of 0 in X. Since X has a countable local base, there must be a sequence (x_n) with $x_n \to 0$ but $A(x_n) \notin W$ for all n.*

6.3 Constructing Topologies

We want to find ways to construct topologies with desired properties for their local bases. Recall what a norm on a vector space means.

Definition 6.3.1 Norms on Vector Space

Let V be a vector space over a field F. A **norm** *on V is any function $\|\cdot\| : V \rightarrow \Re$ such that for every choice of $x, y \in V$ and $\alpha \in F$:*

1. $\|x\| \geq 0$,
2. $\|x\| = 0$ *if and only if* $x = 0$,
3. $\|\alpha x\| = |\alpha| \|x\|$,
4. $\|x + y\| \leq \|x\| + \|y\|$.

It is apparent that a normed space is a metric space where the metric is defined to be $d(x, y) = \|x - y\|$. As a result, a norm always produces a metric topology on a vector space V. We now show that this topology also makes V into a topological vector space.

Theorem 6.3.1 Normed Spaces are also Topological Vector Spaces

Let $(V, \|\cdot\|)$ be a normed linear space. Then V is also a topological vector space under the metric topology generated by the norm.

Proof 6.3.1
Let V be a vector space with norm $\|\cdot\|$.

(a) We show that $\{x\}$ is a closed set by demonstrating $\{x\}^C$ to be an open set. Suppose $y \in \{x\}^C$ and let $r = d(x, y) = \|x - y\|$. Then $B(y, r/2) = \{ z \in V : \|y - z\| < r/2\}$ is an open set containing y and $B(y, r/2) \subseteq \{x\}^C$. Therefore, any point in $\{x\}^C$ has a neighborhood that's still contained in $\{x\}^C$, implying that $\{x\}^C$ is an open set. This in turn means that $\{x\}$ is a closed set.

(b) We now show that addition and multiplication are continuous operations. Assume $\{x_n\}$ converges in V to x and $\{y_n\}$ converges in V to y. Suppose also that $\{c_n\}$ converges in F to c. Then the inequalities

$$\| (x_n + y_n) - (x + y)\| \leq \| x_n - x\| + \| y_n - y\|$$

$$\begin{aligned} \|c_n x_n - cx\| &= \|c_n x_n - c_n x + c_n x - cx\| \\ &= \|c_n(x_n - x) + (c_n - c)x\| \\ &\leq |c_n| \, \|x_n - x\| + |c_n - c| \, \|x\| \end{aligned}$$

show that $x_n + y_n \rightarrow x + y$ and $c_n x_n \rightarrow cx$.

■

So we see that specifying a norm on a vector space makes things quite convenient. Once we have a norm operating on a space, all the necessary structure needed to perform analysis is automatically put into place. But, for better or worse, there are important vector spaces in which it is very difficult to define a suitable norm. In fact, the space of test functions which we discuss later is not metrizable at all. Nevertheless, topologies for these spaces still need to be found so that proper analysis can be done. Without an appropriate topology, analysis becomes impossible. For vector spaces where it is difficult to find an appropriate norm, we construct a sub-basis that will in turn give rise to a basis. Only here, since we're dealing with vector spaces, we build a **local** sub-basis which in turn produces a **local** base for the space. We do this in such a way that the local base makes the space into a topological vector space; that is, addition and multiplication are continuous operations and every singleton $\{x\}$ is closed. One useful tool is to use seminorms.

Definition 6.3.2 Seminorms on a Vector space

Let X be a vector space. The map $\rho : X \to \Re$ is called a seminorm if

SN1: $\rho(x) \geq 0$,

SN2: $\rho(x+y) \leq \rho(x) + \rho(y)$, $x, y \in X$,

SN3: $\rho(\alpha x) = |\alpha| \rho(x)$.

There are many possible seminorms. For example, let the vector space be the set of all continuous functions on $[0,1]$, $C([0,1])$. Let $\rho_t(x) = |x(t)|$ be the evaluation map. For example, if $x(t) = t^2$, $\rho_{1/2}(x) = |(1/2)^2| = 1/4$. It is clear that SN1 is valid for an evaluation map. We then see $\rho_{1/2}(x+y) = |(x+y)(1/2)| \leq |x(1/2)| + |y(1/2)|$ and so SN2 is satisfied. Also, $\rho_{1/2}(\alpha x) = |(\alpha x)(1/2)| = |\alpha||x(1/2)|$ and so SN3 is satisfied. We also see ρ_t is not a norm. We need some additional types of neighborhoods.

Definition 6.3.3 Absorbing Sets

Let $(X, \mathscr{S})$ be a topological vector space. A set $S \subset X$ is said to be absorbing if for each $x \in X$, there is an $\epsilon > 0$ so the $\alpha x \in S$ for $0 < |\alpha| < \epsilon$. This is equivalent to saying if $x \in X$, there is an $r > 0$ so that $x \in \alpha S$ if $|\alpha| \geq r$.

Comment 6.3.1 *Note the concept of an absorbing set makes sense even if X is just a vector space as vector addition and scalar multiplication are available.*

Comment 6.3.2 *Note $0 \cdot x = 0$, so if V is a neighborhood of 0, the continuity of scalar multiplication implies there is an $r > 0$ and a neighborhood of x, U_x, so that $sU_x \subset V$ if $|s| < r$. This says $sx \in V$ for $0 < |s| \leq r/2$; hence, V is absorbing. We conclude neighborhoods of 0 are absorbing.*

We can now prove our first result in this area. Let a topological vector space have an open base at 0 where each member of the base is balanced and absorbing, the base is closed under nonzero scalar multiplication and every member contains a symmetric neighborhood. Call this a balanced, absorbing, scalar closed and symmetric base (BASCSB). We can prove that such an open base exists.

Theorem 6.3.2 Topological Vector Space has balanced, absorbing, scalar closed and symmetric base (BASCSB) Open Base at 0

Let $(X, \mathscr{S})$ be a topological vector space. Then there is an open base at 0, $\mathscr{B}$, satisfying:

(i): Each $U \in \mathscr{B}$ is balanced and absorbing.

(ii): $\alpha U \in \mathscr{B}$ for all $U \in \mathscr{B}$ and $\alpha \neq 0$.

(iii): If $U \in \mathscr{B}$, there is $V \in \mathscr{B}$ so that $V + V \subset U$.

Proof 6.3.2

Let S be a neighborhood of 0. We know $BH(S)$ is also a neighborhood of 0. Let

$$\mathscr{B} = \{BH(S) | S \text{ is a neighborhood of } 0\}$$

i: *If $U \in \mathscr{B}$, U is balanced and absorbing by our comment above.*

ii: *If $U \in \mathscr{B}$, U is open and αU is also open for all $\alpha \neq 0$. For $-1 \leq t \leq 1$, we have $t\alpha U = \alpha(tU) \subset \alpha U$ as U is balanced. So αU is also balanced. By definition of $\mathscr{B}$, $U = BH(S)$ for some neighborhood of 0, S. Also, $\alpha U = BH(\alpha U) = \alpha BH(S) = BH(\alpha S)$ as αU is balanced. This shows $\alpha U \in \mathscr{B}$.*

iii: *From $0 \cdot 0 = 0$, given a neighborhood U of 0, there are neighborhoods V_1 and V_2 of 0 with $V_1 + V_2 \subset U$. Let $W = V_1 \cap V_2 \cap (-V_1) \cap (-V_2)$. Then W is symmetric and $W + W \subset U$ with $W \subset V_1$ and $W \subset V_2$.*

Further, again since $0 \cdot 0 = 0$, since W is a neighborhood of 0, there is a neighborhood of 0, W_1 so that $BH(W_1) \subset W$. This follows from the arguments in the proof of Theorem 6.1.2. It follows that $BH(W_1)$ is absorbing as it is a neighborhood of 0. We have $BH(W_1) + BH(W_1) \subset W + W \subset U$. By definition, $BH(W_1) \in \mathscr{B}$ so we have shown condition (iii) holds. ■

The last theorem told us we can find local bases with specific properties if we are in a topological vector space. The next result tells us the kind of family of subsets we need to construct a unique topological vector space.

Theorem 6.3.3 Family of Subsets in a Vector Space that Generates a Topological Vector Space

Let X be a vector space. Let $\mathscr{B}$ be a nonempty family of subsets so that:

(i): Each member of $\mathscr{B}$ is balanced and absorbing.

(ii): If $U \in \mathscr{B}$, there is $V \in \mathscr{B}$ with $V + V \subset U$.

(iii): If U_1 and U_2 are in $\mathscr{B}$, there is $U_3 \in \mathscr{B}$ with $U_3 \subset \cap U_1 \cap U_2$.

(iv): If $U \in \mathscr{B}$ and $x \in U$, there is $V \in \mathscr{B}$ so that $x + V \subset U$.

Then there is a unique topology $\mathscr{S}$ for X so that $(X, \mathscr{S})$ is a topological vector space.

Proof 6.3.3

Any nonempty balanced set contains 0. Thus $x \in x + U$ for all $U \in \mathscr{B}$. We need to define what our open sets are. So define S to be an open set if to each $x \in S$, there is $U \in \mathscr{B}$ so that $x + U \subset S$. From (iv), it then follows that each $U \in \mathscr{B}$ is an open set.

Now, if S_1 and S_2 are open sets, let $x \in S_1 \cap S_2$. Then there is $U_1 \in \mathscr{B}$ and $U_2 \in \mathscr{B}$ with $x + U_1 \subset S_1$ and $x + U_2 \subset S_2$. By (iii), there is a $U_3 \in \mathscr{B}$ with $U_3 \subset U_1 \cap U_2$. So $x + U_3 \subset x + U_1 \cap U_2 \subset S_1 \cap S_2$. This shows $S_1 \cap S_2$ is an open set. We will leave it to you to extend this argument to finite intersections.

Is $\emptyset$ open? It is open because there are no points in it to apply the definition to, so it satisfies the definition by default. Is X open? Clearly, $x + U$ is a subset of X for any $U \in \mathscr{B}$. So X is open.

Let (U_α) be a collection of open sets where $\alpha \in \Lambda$ for the index set Λ. Let $x \in \cup_{\alpha \in \Lambda} U_\alpha$. Then, there is an index β so that $x \in U_\beta$, an open set. By our definition of open, there is $V \in \mathscr{B}$ with

$x + V \subset U_\beta \subset \cup_{\alpha \in \Lambda} U_\alpha$. Thus, the union $\cup_{\alpha\in\Lambda} U_\alpha$ is open also.

Let $\mathscr{S} = \{S \subset X | S \text{ is open }\}$. Then we have shown $\mathscr{S}$ gives a topology and $(X, \mathscr{S})$ is a topological space.

Now $+$ is continuous at $(0,0)$ means, since $0+0=0$, then given an open set W containing 0, there are open sets U_1 and U_2 containing 0 so that $U_1 + U_2 \subset W$. Since $0 \in W$, an open set, there is $W_2 \in \mathscr{B}$ so that $0 + W_2 = W_2 \subset W$. By (ii), there are V_1 and V_2 in $\mathscr{B}$ with $V_1 + V_2 \subset W_2 \subset W$. We know these sets are open, so we have shown $+$ is continuous at $(0,0)$.

Since the map $+$ is defined by $(x_1, x_2) \rightarrow x_1 + x_2$, start with a neighborhood of $x_1 + x_2$. Then there is $W_2 \in \mathscr{B}$ so that $(x_1 + x_2) + W_2 \subset W$. Using the neighborhoods above from our proof that $+$ is continuous at $(0,0)$, let $V_3 = x_1 + V_1$ and $V_4 = x_2 + V_2$. Then, $V_3 + V_4 = (x_1 + V_1) + (x_2 + V_2) \subset (x_1 + x_2) + W_2 \subset W$. Note, all of these sets are open sets. This proves $+$ is continuous at (x_1, x_2).

Next, we show scalar multiplication is continuous. Fix α_0 and x_0. We know $(\alpha_0, x_0) \rightarrow \alpha_0 x_0$.

Step 1: *First, let's show if $U \in \mathscr{B}$ and $\alpha \neq 0$, there is $V \in \mathscr{B}$ with $\alpha V \subset U$.*

If $U \in mathscrB$, by (ii), there is $V \in \mathscr{B}$ with $V + V \subset U$. We can see if $x \in V$, $x + x = 2x \in V + V \subset U$. Thus, if $y \in 2V$, $y = 2x$ for some $x \in V$. We see $2V \subset V + V \subset U$. Now apply (ii) to V. We see there is V_2 with $V_2 + V_2 \subset V$ and $2V_2 + 2V_2 \subset V + V \subset U$; hence, there is $V_2 \in \mathscr{B}$ with $2^2 V_2 \subset U$. Continuing by induction, we see there is $V_n \in \mathscr{B}$ with $2^n V_n \subset U$.

If $\alpha \neq 0$, choose any n so that $|\alpha| < 2^n$. Then, $|\alpha| V_n \subset 2^n V_n \subset U$. But each V_n is balanced, so we actually have $\alpha V_n \subset U$. This completes the proof of Step 1.

Step 2: *If $U \in \mathscr{B}$, there is $V \in \mathscr{B}$ and there exists $\epsilon > 0$ so that*

$$\alpha x \in \alpha_0 x_0 + U, \;\; \text{if } |\alpha - \alpha_0| < \epsilon, \; x \in x_0 + V$$

Another way of saying this is

$$\alpha x - \alpha_0 x_0 \in +U, \;\; \text{if } |\alpha - \alpha_0| < \epsilon, \; x - x_0 \in V$$

Now given $U \in \mathscr{B}$, note by (ii), there is $W_1 \in \mathscr{B}$ with $W_1 + W_1 \subset U$. Further, there is $W_2 \in \mathscr{B}$ with $W_2 + W_2 \subset W_1$. Combining, we have $W_2 + W_2 + W_2 + W_2 \subset U$. We know $0 \in W_2$, so if $x \in W_2$, $x = x + 0 \in W_2 + W_2$. This implies $W_2 + W_2 + W_2 \subset W_2 + W_2 + W_2 + W_2 \subset U$.

Given $U \in \mathscr{B}$, there is $W_1 \in \mathscr{B}$ with $W_1 + w_2 \subset U$. If $\alpha_0 = 0$, let $V = W_2$. If $\alpha_0 \neq 0$, using Step 1, there is a W_3 so that $\alpha_0 W_3 \subset W_2$. Choose $V = W_3 \cap W_2$. So

- $\alpha_0 = 0$, $\alpha_0 V = 0 \in W_2$.
- $\alpha_0 \neq 0$, $V W_3 \cap W_2 \subset W_2$, $\alpha_0 V \subset W_2$.

So in either case, $V \subset W_2$ and $\alpha_0 V \subset W_2$.

By (i), each member of $\mathscr{B}$ is balanced and absorbing, so given x_0, there is an $\epsilon > 0$ so that $\beta x_0 \in V$, if $|\beta| < \epsilon$. We may assume without loss of generality that $\epsilon \leq 1$. Now if $|\alpha - \alpha_0| < \epsilon$ and $x = x_0 + V$, $(\alpha - \alpha_0)x_0 \in V$ and $x - x_0 \in V$. Also, since $|\alpha - \alpha_0| < \epsilon < 1$, $(\alpha - \alpha_0)(x - x_0) \in V$ as V is balanced. We conclude

$$\alpha x - \alpha_0 x_0 \;\; = \;\; (\alpha - \alpha_0)x_0 + \alpha_0(x - x_0) + (\alpha - \alpha_0)(x - x_0)$$

This first term is in $V \subset W_2$, the second term is in $\alpha_0 V \subset W_2$ and the last term is in $V \subset W_2$. But we also know $W_2 + W_2 + W_2 \subset U$. So we see $\alpha x - \alpha_0 x_0 \in U$.

Hence, if $\alpha x - \alpha x_0 \in U$ if $|\alpha - \alpha_0| < \epsilon$ and $x \in x_0 + V$. So if $\alpha_0 x_0 + U$ is a neighborhood of $\alpha_0 x_0$, there is a neighborhood of x_0, $x_0 + V$, and a neighborhood of α_0, $B(\alpha_0, \epsilon)$ so that $(\alpha, y) \in B(\alpha_0, \epsilon) \times (x_0 + V) \to \alpha_0 x_0 + U$. Now if W is a neighborhood of $\alpha_0 x_0$, there is U with $\alpha_0 x_0 + U \subset W$. This argument therefore shows scalar multiplication is continuous in this topology.

This shows $(X, \mathscr{B})$ is a topological vector space. It is unique because if two topologies share a base at 0, they generate the same topology. ■

Now let's add convexity.

Theorem 6.3.4 Family of Subsets in a Vector Space that Generates a Locally Convex Topological Vector Space

Let X be a vector space. Let $\mathscr{B}$ be a nonempty family of subsets so that:

(i): Each member of $\mathscr{B}$ is balanced, convex and absorbing.

(ii): If $U \in \mathscr{B}$, there is α with $0 < \alpha \leq 1/2$ and $\alpha U \in \mathscr{B}$.

(iii): If U_1 and U_2 are in $\mathscr{B}$, there is $U_3 \in \mathscr{B}$ with $U_3 \subset \cap U_1 \cap U_2$.

(iv): If $U \in \mathscr{B}$ and $x \in U$, there is $V \in \mathscr{B}$ so that $x + V \subset U$.

Then there is a unique topology $\mathscr{S}$ for X so that $(X, \mathscr{S})$ is a locally convex topological vector space.

Proof 6.3.4
If we show the open sets we choose in the proof of Theorem 6.3.3 satisfy (ii) of that theorem, we are done. This condition says if $U \in \mathscr{B}$, there is a $V \in \mathscr{B}$ with $V + V \subset U$. From condition (ii) in this theorem, given $U \mathscr{B}$, let α satisfy $0 < \alpha \leq 1/2$. Then, we know $\alpha U \subset U$. Let $V = \alpha U$. Then, $V + V = \alpha U + \alpha U$.

We claim $\alpha U + \alpha U \subset 2\alpha U$. If $\alpha x + \alpha y \in \alpha U + \alpha U$, note U is convex. So $(1/2)x$ and $(1/2)y$ are in U. Thus, $(1/2)x + (1/2)y \in U$ by convexity. Now, $\alpha x + \alpha y = 2\alpha((1/2)x + (1/2)y) \in 2\alpha U$. But $2\alpha((1/2)x + (1/2)y) \in U$ because U is balanced. Thus, $2\alpha U \subset U$. Hence, $\alpha U + \alpha U \subset U$ which shows (ii) of Theorem 6.3.3 is satisfied. This shows $(X, \mathscr{S})$ with $\mathscr{S}$ defined as in Theorem 6.3.3 is a topological vector space. Since the sets in $\mathscr{B}$ are convex, this is a locally convex topological vector space. It is unique because if two topologies share a base at 0, they generate the same topology. ■

Homework

Exercise 6.3.1 *Complete our argument that finite intersections of open sets are open in the proof above.*

Exercise 6.3.2 *Prove if $U \in \mathscr{B}$, $x + U$ is an open set for all $x \in X$.*

Exercise 6.3.3 *If $(X, \mathscr{S}_1)$ and $(X, \mathscr{S}_2)$ are both topological vector spaces on X with different topologies, show these are the same topology if the two topologies share a base at 0.*

6.4 Families of Seminorms

We know every norm is a seminorm, but not every seminorm is a norm. The difference is that for a seminorm the equation $\rho(x) = 0$ does not necessarily guarantee $x = 0$ or, equivalently, $x \neq 0$ does not necessarily imply $\rho(x) \neq 0$. Consequently, a seminorm establishes some degree of topological structure on a vector space but not enough. This deficiency, however, is overcome by having not just one seminorm but a whole collection of them. We gather enough seminorms so that for each nonzero $x \in V$ there is at least one seminorm ρ so that $\rho(x) \neq 0$. In other words, the seminorm collection taken as a whole acts like a single norm. This leads us to the following definition:

Definition 6.4.1 Separating Families of Seminorms

Suppose $\mathscr{P}$ is a family of seminorms on a vector space V in such a way that for every nonzero $x \in V$ there is a $\rho \in \mathscr{P}$ so that $\rho(x) \neq 0$. Then $\mathscr{P}$ is called a **separating** *family of seminorms.*

Finally, an important function is the Minkowski functional of a balanced and convex set.

Let $(X, \mathscr{S})$ be a locally convex topological space. Let $f : M \to \Re$ be continuous and linear on the subspace $M \subset X$. Thus, f is continuous at 0 and since f is linear, $f(0) = f(0 + 0) = 2f(0)$ implying $f(0) = 0$. Given the ball $B(0, 1) \subset \Re$, there is a neighborhood of 0, U, with $f(U) \subset B(0, 1)$ which tells us $|f(x) - 0| < 1$ if $x \in U \cap M$.

Since X is locally convex, X has a base $\mathscr{B}$ at 0 which consists of convex neighborhoods. We also know $\mathscr{C} = \{BH(S) | S \in \mathscr{B}\}$ is a base at 0. Thus, $\mathscr{C}$ is a base of convex and balanced sets. Hence, there is a convex and balanced neighborhood of 0, V, $V \subset U$, so that $|f(x)| < 1$ if $x \in V \cap M$. Further, since any neighborhood of 0 is absorbing, we know V is actually convex, balanced and absorbing.

Now define $g : X \to \Re$ by $g(x) = \inf\{\alpha > 0 | x \in \alpha V\}$. We define the set

$$V_x = \{\alpha > 0 | x \in \alpha V\}$$

so that $g(x) = \inf V_x$. If $V_x\ \emptyset$, we set $g(x) = \infty$. If $x \in V$, then $x \in 1 \cdot V$ and V_x contains 1 and is not empty. Since this set is also bounded below by 0, $g(x)$ is a well-defined nonnegative number on V. This function g is called the Minkowski functional. We want to use Minkowski functionals defined on sets V so that V_x is never empty. Hence, we want sets V that can capture any $x \in X$ using a suitable scaling. Such sets are absorbing sets. Recall, V is absorbing means for all $x \in X$, there is an $\epsilon > 0$ so that $\beta x \in V$ for $0 < |\beta| \leq \epsilon$. Hence, $x \in (1/\beta)V$ for $0 < |\beta| \leq \epsilon$. Letting $\alpha = 1/\beta$ and $r = 1/\epsilon$, this says $x \in \alpha V$ when $|\alpha \geq r$. For such an x, then $\inf\{\alpha > 0 | x \in \alpha V\} \leq r$ and so is finite. Hence, V_x is never empty and $\rho : X \to [0, \infty)$.

Definition 6.4.2 Minkowski Functional

Let X be a vector space and V a convex and absorbing neighborhood of 0. Let $g : X \to [0, \infty)$ be defined by $g(x) = \inf\{\alpha > 0 | x \in \alpha V\}$. The function g is called the Minkowski Functional of V.

We need to study the properties of this functional carefully which will require some work. We will show a Minkowski functional is positive homogeneous, subaddtive and balanced and maps 0 to 0. We can use these kinds of functionals to generate topologies for interesting topological vector spaces.

Lemma 6.4.1 Properties of the Minkowski Functional

> *Let g be defined as above. Then*
>
> *(i): $g(0) = 0$,*
>
> *(ii): $g(\lambda x) = \lambda g(x)$ if $\lambda > 0$ (positive homogeneous),*
>
> *(iii): $g(x + y) \leq g(x) + g(y)$ (subadditive),*
>
> *(iv): if V is also balanced, $g(\lambda x) = |\lambda| g(x)$.*

Proof 6.4.1
(i): $g(0) = \inf\{\alpha > 0 | 0 \in \alpha \cdot V\} = 0$.

(ii): *Let $\lambda > 0$ and $x \in X$. Let $\alpha \in V_x$. Thus, $x \in \alpha V$, $\lambda x \in \lambda\alpha V$ which implies $g(\lambda x) \leq \lambda\alpha$ for all such α. This tells us $g(\lambda x) \leq \lambda g(x)$.*

Next, for $\lambda > 0$, let $\beta \in V_{\lambda x}$. Then, for $y = \lambda x$, $y \in \beta V$. So $(1/\lambda)y \in \beta/\lambda$. Thus, $g((1/\lambda)y) = g(x) \leq (1/\lambda)g(\lambda x)$. We conclude $\lambda g(x) \leq g(\lambda x)$.

Combining, we see $g(\lambda x) = \lambda g(x)$ if $\lambda > 0$.

(iii): *Let $V_x = \{\alpha > 0 | x \in \alpha V\}$. So $g(x) = \inf V_x$. Pick x and y in X and let α and β be in V_x. Then, $x + y \in \alpha V + \beta V \subset (\alpha + \beta)V$. On the other hand, since V is convex, let $z = \alpha y_1 + \beta y_2$ where y_1 and y_2 are in V. Then, $\frac{1}{\alpha+\beta} z = \frac{\alpha}{\alpha+\beta} y_1 + \frac{\beta}{\alpha+\beta} y_2$ which is in V as V is convex. Hence, $z \in (\alpha + \beta)V$ showing $\alpha V + \beta V = (\alpha + \beta)V$.*

We started with $x + y \in \alpha V + \beta V$ which we now know is the same as $(\alpha + \beta)V$. Since $x + y \in (\alpha + \beta)V$, by definition, $\alpha + \beta \in V_{x+y}$. Thus, $g(x + y) \leq \alpha + \beta$. But α and β are any positive numbers. Hence, $g(x, y) = \inf V_x + \inf V_y = g(x) + g(y)$.

(iv): *Let $\lambda \neq 0$ and $x \in X$. Let $\alpha \in V_x$. Then $x \in \alpha V$ implying $(1/\alpha)x \in V$. Now $\frac{\lambda}{|\lambda|} = 1$ and since V is balanced, $\frac{\lambda}{|\lambda|}((1/\alpha)x) \in V$ also. We see $\lambda x \in |\lambda|\alpha V$ for all such α. Thus, $g(\lambda x) \leq |\lambda| g(x)$.*

Let $y = \lambda x$ for $\lambda \neq 0$ and let $\alpha \in V_y$. Then, $y \in \alpha V$ by definition implying $\frac{1}{\alpha} y \in V$ and $\frac{1/\lambda}{|1/\lambda|} \frac{1}{\alpha} y \in V$ as V is balanced. So, $(1/\lambda)y \in |1/\lambda|\alpha V$. This tells us $g((1/\lambda)y) \leq |1/\lambda|\alpha$. But $\alpha \in V_y$ is arbitrary, so

$$g((1/\lambda)y) = g((1/\lambda)\,\lambda x) = g(x) \leq \inf_{\alpha \in V_y} |1/\lambda|\alpha = |1/\lambda| g(y)$$

We conclude $|\lambda| g(x) \leq g(\lambda x)$ which implies the result. ■

We can show the Minkowski Functional is bounded on its defining set.

Lemma 6.4.2 Minkowski Functional on a Convex, Absorbing Set Containing 0 is Bounded by 1

> *Let $g : X \to \Re$ be as discussed above where V is a convex and absorbing set containing 0. Then, $g(x) \leq 1$ and if $g(x) < 1$, $x \in V$.*

Proof 6.4.2
If $x \in V$, then $1 \in V_x$ and so $g(x) \leq 1$. If $g(x) < 1$, let $\epsilon = 1 - g(x) > 0$. By the Infimum Tolerance Lemma, there is $\beta \in V_x$ so that $g(x) \leq \beta < g(x) + 1 - g(x) = 1$. If $\beta = 0$, $x \in 0 \cdot V$ telling us $x = 0 \in V$. Otherwise, $0 < \beta < 1$. Then, $1/\beta x \in V$. But V is convex and so $(1-\beta)\cdot 0 + \beta(1/\beta)x = x \in V$. Thus, $g(x) < 1$ implies $x \in V$. This implies $0 \leq g(x) \leq \beta < 1$. ■

In fact, we can prove more things about Minkowski functionals.

Lemma 6.4.3 Further Properties for Minkowski Type Functionals

Let $(X, \mathscr{S})$ be a topological vector space and assume V is a convex and absorbing set containing 0. Let $g(x) = \inf\{\alpha > 0 | x \in \alpha V\}$ for any $x \in X$. Let $K_1 = \{x \in V | g(x) < 1\}$ and $K_2 = \{x \in V | g(x) \leq 1\}$. Then

(i): $int(V) \subset K_1 \subset V \subset K_2 \subset \overline{V}$.

(ii): $V = K_1$ if V is open.

(iii): $V = K_2$ if V is closed.

(iv): g continuous implies $K_1 = int(V)$ and $K_2 = \overline{V}$.

(v): g continuous if and only if $0 \in int(V)$.

(vi): V bounded and X a T_1 space implies $g(x) = 0$ if and only if $x = 0$.

Proof 6.4.3
(i): *Let $x \in int(V)$. There is then an open set B so that $x \in B \subset V$. Now by Theorem 6.3.2, there is a balanced, absorbing, scalar closed and symmetric open base (BASCSB) at 0, $\mathscr{B}$. Thus, there is a balanced, absorbing neighborhood of 0, U, so that $x + U \subset B \subset int(V) \subset V$. Now, U is absorbing, so there is an $\epsilon > 0$ so that $\alpha x \in U$ if $0 < |\alpha| \leq \epsilon$. Hence, $x + \alpha x = (1+\alpha)x \in B$ if $0 < |\alpha| \leq \epsilon$. Pick any such α_0. Then $(1+\alpha_0)x \in B \subset int(V) \subset V$ implying $x \in (1/(1+\alpha_0))V$. By definition, this says $(1/(1+\alpha_0)) \in V_x$ which means $g(x) \leq (1/(1+\alpha_0)) < 1$. Hence, $x \in K_1$. Clearly $K_1 \subset V$, of course.*

If $g(x) < 1$, there is an $\alpha > 0$ in V_x with $0 < \alpha < 1$. Since $0 \in V$ and $(1/\alpha)x \in V$, since V is convex, $\alpha((1/\alpha)x + (1-\alpha)\,0) = x \in V$. This shows $K_1 \subset V$.
If $x \in V$, $1 \in V_x$ implying $g(x) \leq 1$ and so $x \in K_2$. Thus, $V \subset K_2$.

Clearly, K_2 is a subset of V, so $K_2 \subset \overline{V}$.

(ii): *If V is open, (i) tells us $V = K_1$.*

(iii): *If V is closed, $V = \overline{V}$ and so $\overline{V} \subset K_2\overline{V}$; hence, $K_2 = \overline{V}$.*

(iv): *If g is continuous, then $K_1 = \{x | g(x) < 1\} = g^{-1}(-\infty, 1)$. Since $(\infty, 1)$ is open in $\Re$, K_1 must be open. Thus, $K_1 = int(K_1) \subset V$ and so $K_1 \subset int(V)$. Hence, by (i), $int(V) \subset K_1 \subset int(V)$; so $K_1 = int(V)$. Also, K_2 is closed, so by (i), $\overline{V} \subset \overline{K_2} = K_2 \subset \overline{V}$. Hence, $K_2 = \overline{V}$.*

(v): *If g is continuous, note $g(0) = \inf\{\alpha > 0 | 0 \in \alpha V\}$. We know $0 \in V$, so $0 \in \alpha V$ for all α. Thus, $g(0) = 0 < 1$. We conclude $0 \in K_1$. By (iv), if g is continuous, $K_1 = int(V)$. So $0 \in int(V)$.*

Conversely, if $0 \in int(V)$, let $U \in \mathscr{B}$ with $0 \in U \subset int(V)$. By (i), $int(V) \subset K_1$, so $x \in U$ tells us $g(x) < 1$. Let ϵ be chosen so that $|\epsilon| > 0$. Then, ϵU is also a neighborhood of 0. If $x \in |\epsilon|U$, then $(1/|\epsilon|)x \in U$ and so $g((1/|\epsilon|)x) < 1$. Since U is balanced, the argument for (iii) in Lemma 6.4.1 holds and we have $(1/|\epsilon|)g(x) = g((1/|\epsilon|)x) < 1$. We conclude $g(x) < |\epsilon$.

Now if g was continuous at 0, given $\epsilon| > 0$, there is a neighborhood W of 0 so that $g(W) \subset (-|\epsilon|, |\epsilon|)$. Let $W = \epsilon U$. Then, from our discussion above, $g(|\epsilon|U) \subset (-|\epsilon|, |\epsilon|)$. Thus, we have shown g is continuous at 0.

We have

$$\begin{aligned} g(x) = g(x-y+y) &\leq g(x-y)+g(y) \Longrightarrow g(x)-g(y) \leq g(x-y) \\ g(y) = g(y-x+x) &\leq g(y-x)+g(x) \Longrightarrow g(y)-g(x) \leq g(y-x) \\ &\Longrightarrow \\ -g(y-x) &\leq g(x)-g(y) \leq g(x-y) \end{aligned}$$

Pick $\epsilon > 0$. There is a neighborhood of 0, U, so that $y \in U$ implies $g(y) \in (-\epsilon, \epsilon)$ because g is continuous at 0. Then, there is a $\hat{U} \subset U$ which is a balanced neighborhood of 0. Hence, $x + \hat{U}$ is a balanced neighborhood of x. If $y \in x + \hat{U}$, $y - x \in \hat{U}$ which tells us $g(y - x) \in (-\epsilon, \epsilon)$. But $\hat{U}$ is balanced, so $-(y - x) = x - y \in \hat{U}$ also. Hence, $g(x - y) \in (-\epsilon, \epsilon)$. Thus,

$$-\epsilon < -g(y-x) \leq g(x) - g(y) \quad \leq \quad g(x-y) < \epsilon$$

So $|g(x) - g(y)| < \epsilon$. This shows g is continuous at x.

(vi): *We assume V is bounded and X is a T_1 space. We want to show $g(x) = 0$ implies $x = 0$. Since, X is T_1, given $x \neq 0$, there is an open set U with $0 \in U$ and $x \notin U$. There is a balanced set $\hat{U}$ containing 0 with $\hat{U} \subset U$ also. Since V is bounded, there is an $r > 0$ so that $V \subset r\hat{U}$. Let $\alpha \in V_x$. Then, $x \in \alpha V$ with $\alpha > 0$. Thus, $x/\alpha \in V \subset r\hat{U}$ telling us $x/(r\alpha) \in V \subset \hat{U}$. If $r\alpha \leq 1$ then $x = (r\alpha)/(r\alpha)x = (r\alpha)\left((1/\alpha)x\right)$. But $(1/\alpha)x \in \hat{U}$ and $\hat{U}$ is balanced, so if $r\alpha \leq 1$, we must have $x \in \hat{U}$. But $x \notin \hat{U}$. Thus, we must have $r\alpha > 1$; i.e. $\alpha > 1/r$. Hence, $g(x) \geq 1/r > 0$. So if $x \neq 0$, we must have $g(x) > 0$. We conclude $g(x) = 0$ implies $x = 0$.* ■

We are now ready to state and prove results on how to construct topologies using seminorms.

Theorem 6.4.4 $V(\rho)$ **Set for Seminorm** ρ

Let X be a vector space and ρ a seminorm on X. Let $V(\rho) = \{x \in X | \rho(x) < 1\}$. Let $V_x = \{\alpha > 0 | x \in \alpha V(\rho)\}$. Then $V(\rho)$ is convex, balanced and absorbing and ρ is the Minkowski functional f of V_x.

Proof 6.4.4
(i): *If x_1 and x_2 are in $V(\rho)$, consider $tx_1 + (1-t)x_2$ for $0 \leq t \leq 1$. Then, $\rho(tx_1 + (1-t)x_2) \leq \rho(tx_1) + \rho((1-t)x_2) \leq t\rho(x_1) + (1-t)\rho(x_2) < 1$. Hence, $V(\rho)$ is convex.*

(ii): *If $x_1 \in V(\rho)$, consider tx_1 for $-1 \leq t \leq 1$. Then, $\rho(tx_1) = |t|\rho(x_1) < |t| < 1$. Hence, $V(\rho)$ is balanced.*

(iii): *Let $z \in X$. Now $V(\rho)$ is absorbing if there is an $\epsilon > 0$, so that $sz \in V(\rho)$ when $0 < |s| \leq \epsilon$. Note $sz \in V(\rho)$ means $\rho(sz) = |s|\rho(z) < 1$. If $z = 0$, any $\epsilon > 0$ works fine. Also, if $\rho(z) = 0$, any $\epsilon > 0$ works. So assume $\rho(z) \neq 0$ and let $\epsilon = \frac{1}{2\rho(z)}$. Then, if $0 < |\alpha| \leq \frac{1}{2\rho(z)}$,*

$\rho(\alpha z) = |\alpha|\rho(z) \leq \frac{1}{2} < 1$. *Hence, our choice of* $\epsilon = \frac{1}{2\rho(z)}$ *and we see* $V(\rho)$ *is absorbing.*

(iv): *We have*

$$\begin{aligned} f(x) &= \inf\{\alpha > 0 | x \in \alpha V\} = \inf\{\alpha > 0 | \rho(x/\alpha) < 1\} \\ &= \inf\{\alpha > 0 | \rho(x) < \alpha\} = \rho(x) \end{aligned}$$

■

We can use this to construct a locally convex topology for X determined by a family of seminorms.

Theorem 6.4.5 Locally Convex Topology Determined by a Seminorm Family

Let X *be a vector space and* P *be a family of seminorms on* X. *For each* $\rho \in P$, *let* $V(\rho) = \{x \in X | \rho(x) < 1\}$, *and let*

$$\mathscr{B} = \{r_1 V(\rho_1) \cap \ldots \cap r_k V(\rho_k) | r_i > 0 \in F, k \text{ any positive integer }, \rho_i \in P\}$$

Then, $\mathscr{B}$ *satisfies the conditions of Theorem 6.3.4 and thus* $(X, \mathscr{S})$ *is a locally convex topological vector space.*

Proof 6.4.5
We already know, from Theorem 6.4.4, that $V(\rho)$ *is convex, balanced and absorbing. It is easy to prove each set in* $\mathscr{B}$ *is convex, balanced and absorbing (we leave the proofs to you in the exercises). Then, we must look at the other conditions of Theorem 6.3.4. We have just shown (i).*

(ii): *If* $U \in \mathscr{B}$, *we must show there is an* $\alpha > 0$ *with* $0 < \alpha \leq (1/2)$ *and* $\alpha U \in \mathscr{B}$. *First, since* $U \in \mathscr{B}$, *we can write* $U = \cap_{i=1}^{k} r_i V(\rho_i)$ *for a finite number of positive scalars* r_i *and seminorms* $\rho_i \in P$. *Pick any* α *so that* $0 < \alpha \leq (1/2)$ *and note* $(\alpha r_1)V(\rho_1) \cap \ldots \cap (\alpha r_k)V(p\rho_k) \in U$ *because each set is balanced. This verifies (ii).*

(iii): *If* U_1 *and* U_2 *are in* $\mathscr{B}$, *we must show there is a* U_3 *in* $\mathscr{B}$ *with* $U_3 \subset U_1 \cap U_2$. *To see this, note, we have finite representations*

$$\begin{aligned} U_1 &= r_1 V(p_1) \cap \ldots \cap r_n V(p_n) \\ U_2 &= s_1 V(q_1) \cap \ldots \cap s_m V(p_m) \\ &\Longrightarrow \\ U_3 = U_1 \cap U_2 &= \cap_{i=1}^{n} r_i V(p_i) \ \cap \ \cap_{j=1}^{m} s_j V(q_j) \end{aligned}$$

We see $U_3 \subset U_1 \cap U_2$ *and it has the proper representation to be in* $\mathscr{B}$. *This verifies (iii).*

(iv): *We need to show if* $U \in \mathscr{B}$ *and* $x \in U$, *there is a* $V \in \mathscr{B}$ *with* $x + V \subset U$. *To see this, note* $U = r_1 V(p_1) \cap \cdots \cap r_k V(p_k)$. *Let* $x \in U$ *and let* $\alpha_i = r_1 - \rho_i(x)$. *Let*

$$V = \alpha_1 V(p_1) \cap \cdots \cap \alpha_k V(p_k) = (r_1 - \rho_1(x))V(p_1) \cap \cdots \cap (r_k - \rho_k(x))V(p_k)$$

Now $x + V \in x + (r_i - \rho_i(x))V(p_i)$ *for each* i. *Hence, if* $x + z \in x + V$, $x + z = x + (r_i - \rho_i(x))y$ *with* $\rho_i(y) < 1$. *Then,* $\rho_i(x + z) = \rho_i(x + (r_i - \rho_i(x))y) \leq \rho_i(x) + |r_i - \rho_i(x)|\rho_i(y)$. *But this tells us* $\rho_i(x + z) < \rho_i(x) + |r_i - \rho_i(x)|$. *However, we also know* $x \in U$ *and so* $x \in r_i V(\rho_i)$ *implying* $\rho_i(x) < r_i$. *Thus,* $r_i - \rho_i(x) > 0$. *We conclude*

$$\rho_i(x + z) < \rho_i(x) + |r_i - \rho_i(x)| = \rho_i(x) + r_i - \rho_i(x) = r_i$$

thus, $x + z = x + (r_i - \rho_i(x))y \in r_i\rho_i$. *But the choice of* y *was arbitrary, so we have for each* i, $x + (r_i - \rho_i(x))V(\rho_i) \subset r_iV(\rho_i)$. *We conclude* $x + \cap_{i=1}^{k}((r_i - \rho_i(x))V(\rho)i) \subset \cap_{i=1}^{k} r_iV(\rho_i) = U$. *This shows* $x + V \subset U$ *and completes the proof of (iv).*

The conditions of Theorem 6.3.4 are satisfied and so $\mathscr{B}$ *generates a locally convex topological space.* ■

Homework

Exercise 6.4.1 *Prove the intersection of two convex sets is also convex.*

Exercise 6.4.2 *Prove the intersection of two balanced sets is balanced.*

Exercise 6.4.3 *Prove* rA *is balanced if* A *is balanced, for any* $r > 0$.

Exercise 6.4.4 *Prove* $\alpha A + \beta B$ *is absorbing for any* α *and* β *positive and* A *and* B *absorbing.*

6.5 Another Metrization Result

We can construct a topology using a separating family of seminorms which has a countable local base at 0.

Theorem 6.5.1 Constructing a Locally Convex Topology with a Countable Local Base at 0

Let V *be a vector space over the real field. Suppose* $\mathscr{P} = (p_n)$ *is a countable separating family of seminorms on a vector space* V. *To each* $p_i \in \mathscr{P}$ *and to each natural number* n, *define the set:*

$$U(p_i, n) = \{x \in V : p_i(x) < 1/n\}$$

Let

$$\mathscr{U} = \{r_1V(\rho_1) \cap \ldots \cap r_kV(\rho_k) | r_i > 0, k \text{ any positive integer }, p_i \in P\}$$

Then $\mathscr{U}$ *generates a locally convex topology for* X *which has a countable local base at* 0.

Comment 6.5.1 *Before we prove this result, we make a couple of remarks. First, the fraction* $1/n$ *in the sets* $U(\rho, n)$ *is arbitrary in that it can be replaced by any function that approaches* 0 *as* $n \to \infty$. *The important thing is that we generate subsets for the topology that are closer and closer to* 0.

Let us now consider the proof for the theorem:

Proof 6.5.1
This is similar to the proof of Theorem 6.4.5. It is easy to show the sets $U(p_i, n)$ *are convex, balanced and absorbing and each set in* $\mathscr{U}$ *is convex, balanced and absorbing (we leave the proofs to you in the exercises). Then, we must look at the other conditions of Theorem 6.3.4. We have just shown (i).*

(ii): *If* $V \in \mathscr{U}$, *we must show there is an* $\alpha > 0$ *with* $0 < \alpha \leq (1/2)$ *and* $\alpha V \in \mathscr{B}$. *First, since* $V \in \mathscr{U}$, *we can write* $V = \cap_{i=1}^{k} r_iU(p_i, 1/n_i)$ *for a finite number of positive integers* n_i *and seminorms* $p_i \in P$. *Pick any* α *so that* $0 < \alpha \leq (1/2)$ *and note* $(\alpha)r_1V(p_1, 1/n_1)\cap \ldots \cap(\alpha)r_nV(p_k, 1/n_k) \in U$ *because each set is balanced. This verifies (ii).*

(iii): *If V_1 and V_2 are in $\mathscr{U}$, we must show there is a V_3 in $\mathscr{U}$ with $U_3 \subset U_1 \cap U_2$. To see this, note, we have finite representations*

$$\begin{aligned} V_1 &= r_1 U(p_1, 1/n_1) \cap \ldots \cap r_n U(p_n, 1/n_k) \\ V_2 &= s_1 V(q_1, 1/m_1) \cap \ldots \cap s_m V(q_m, 1/m_p) \\ &\Longrightarrow \\ V_3 = V_1 \cap V_2 &= \cap_{i=1}^{k} r_k V(p_i, 1/n_i) \ \cap \ \cap_{j=1}^{p} s_j V(q_j, 1/m_j) \end{aligned}$$

We see $V_3 \subset V_1 \cap V_2$ and it has the proper representation to be in $\mathscr{U}$. This verifies (iii).

(iv): *We need to show if $U \in \mathscr{U}$ and $x \in U$, there is a $V \in \mathscr{U}$ with $x + V \subset U$. To see this, note $U = r_1 U(p_1, 1/n_1) \cap \cdots \cap r_p U(p_k, 1/n_p)$. Let $x \in U$ and let $p_i(x) < k_i/m_i < 1/n_i$ define the rational number k_i/m_i. Let*

$$V = r_1 k_1 V(p_1, 1/m_1) \cap \cdots \cap k_p r_p V(p_k, 1/m_p)$$

Now $y \in x + V$, we have $y = x + z$ with $z \in k_i r_i V(p_i, 1/m_i)$ for each i. Hence, $p_i(z) < k_i/m_i < 1/n_i$ for each i. Thus, $x + z \in U$. Since z is arbitrary, we see $x + V \subset U$ which completes the proof of (iv).

The conditions of Theorem 6.3.4 are satisfied and so $\mathscr{U}$ generates a locally convex topological space. It is also clear we have a countable local base at 0, $(U(p_i, 1/n))$. ■

Next, note the topology generated by Theorem 6.5.1 is T_1. If $x \neq 0$, there is a seminorm p_i with $p_i(x) \neq 0$ as the seminorm family is separating. Thus, $U(p_i, 1/m) = \{y | p_i(y) < 1/m\}$ cannot contain x if $1/m < p_i(x)$. Hence, $x \in (U(p_i, 1/m))^C$ for such an m. So if we take two points x_1 and x_2 with $x_1 \neq x_2$. There is then a seminorm j with $p_j(x_1 - x_2) \neq 0$. Hence, there is $U(p_j, 1/m)$ with $x_1 - x_2 \in (U(p_j, 1/m))^C$. Thus, $x_2 \in x_2 + U(p_j, 1/m)$ and x_1 is not in that set. Also, $x_1 \in x_1 + U(p_j, 1/m)$ but x_2 is not in that set. This shows the topology is T_1 and hence, we also know the topology is Hausdorff.

So by Theorem 6.2.1, we know this topology is metrizable. We can state this as a theorem.

Theorem 6.5.2 Locally Convex Topology which is Metrizable

Let V be a vector space over the real field. Suppose $\mathscr{P} = (p_n)$ is a countable separating family of seminorms on a vector space V. To each $p_i \in \mathscr{P}$ and to each natural number n, define the set:

$$U(p_i, n) = \{x \in V : p_i(x) < 1/n\}$$

Let

$$\mathscr{U} = \{r_1 V(\rho_1) \cap \ldots \cap r_k V(\rho_k) | r_i > 0, k \text{ any positive integer }, p_i \in P\}$$

Then $\mathscr{U}$ generates a locally convex topology for X which has a countable local base at 0 and hence is metrizable.

Proof 6.5.2
We just went over this argument so we will not go over it again. But, you should do this argument for yourself as it will help you get better at using these ideas. ■

Homework

Exercise 6.5.1 *Using the notation of Theorem 6.5.2, prove each $U(p_i, n)$ is convex, balanced and absorbing.*

Exercise 6.5.2 *Using the notation of Theorem 6.5.2, prove each set in $\mathscr{U}$ is convex, balanced and absorbing.*

Exercise 6.5.3 *Using the assumptions of Theorem 6.5.2, prove we can also define a metric on X as follows:*

$$d(x,y) = \sum_{i=1}^{\infty} \frac{1}{2^i} \frac{p_i(x-y)}{1+p_i(x-y)} = \sum_{i=1}^{\infty} \frac{1}{2^i} \phi(p_i(x-y))$$

where $\phi(t) = t/(1+t)$ for $t \geq 0$. Prove this is a metric. Note it is easier to define the metric because we have a countable family of seminorms here. As part of your argument, prove the series that defines d converges uniformly on $X \times X$ (appropriately extending the notion of uniform convergence, of course). Also show d is continuous on $X \times X$.

Exercise 6.5.4 *Using the assumptions of Theorem 6.5.2, prove we can also define a metric on X as follows:*

$$d(x,y) = \sum_{i=1}^{\infty} \frac{1}{3^i} \frac{p_i(x-y)}{1+p_i(x-y)} = \sum_{i=1}^{\infty} \frac{1}{3^i} \phi(p_i(x-y))$$

where $\phi(t) = t/(1+t)$ for $t \geq 0$. Prove this is a metric. Note it is easier to define the metric because we have a countable family of seminorms here. As part of your argument, prove the series that defines d converges uniformly on $X \times X$ (appropriately extending the notion of uniform convergence, of course). Also show d is continuous on $X \times X$. Would it make any real difference in this argument if you replace $1/3^i$ by $1/n_i$ with $n_i \to \infty$?

Exercise 6.5.5 *For the matrix d defined above, prove the topology this d generates is compatible with the original topology on X. Remember, this means each open set of X contains an open set generated by d. Also note, it is enough to do this argument with open sets at 0.*

Exercise 6.5.6 *Prove Theorem 6.5.1 using $1/n^2$ instead of $1/n$.*

Exercise 6.5.7 *Let $X = C(\Re)$, which is the set of all functions that are continuous on $\Re$. Let $p_n(f) = \sup_{-n \leq x \leq n} |f(x)|$. Prove each f_n is a seminorm on X and that this seminorm family induces the metric*

$$d(x,y) = \sum_{i=1}^{\infty} \frac{1}{2^i} \frac{p_i(x-y)}{1+p_i(x-y)} = \sum_{i=1}^{\infty} \frac{1}{2^i} \phi(p_i(x-y))$$

where $\phi(t) = t/(1+t)$ for $t \geq 0$. Let $f(x) = \max\{0, 1-|x|\}$, $g(x) = 100f(x-2)$ and $h(x) = (f(x)+g(x))/2$. Show $d(f,0) = 1/2$, $d(g,0) = 50/101$ and $d(h,0) = 1/6+50/102$. Now show this implies $B(0,1/2)$ is not convex. So d is compatible with the usual locally convex topology the seminorms generate on $C(\Re)$ by it is not locally convex itself.

Exercise 6.5.8 *Let $X = C(\Re)$, which is the set of all functions that are continuous on $\Re$. Let $p_n(f) = \sup_{-n^2 \leq x \leq n^2} |f(x)|$. Prove each f_n is a seminorm on X and that this seminorm family* *induces the metric*

$$d(x,y) = \sum_{i=1}^{\infty} \frac{1}{2^i} \frac{p_i(x-y)}{1+p_i(x-y)} = \sum_{i=1}^{\infty} \frac{1}{2^i} \phi(p_i(x-y))$$

where $\phi(t) = t/(1+t)$ *for* $t \geq 0$. *Let* $f(x) = \max\{0, 1-|x|\}$, $g(x) = 100f(x-2)$ *and* $h(x) = (f(x)+g(x))/2$. *Show* $d(f,0) = 1/2$, $d(g,0) = 50/101$ *and* $d(h,0) = 1/6+50/102$. *Now show this implies* $B(0,1/2)$ *is not convex. So d is compatible with the usual locally convex topology the seminorms generated on* $C(\Re)$, *but is not locally convex itself.*

Exercise 6.5.9 *Let the set of continuous functions on* $\Re$ *be* $C(\Re)$.

- *Prove the seminorms* $\rho_n(f) = \sup_{x\in[-n,n]} |f(x)|$ *defined on* $C(\Re)$, *induce a metric d given by*

$$d(f,g) = \sum_{n=1}^{\infty} 3^{-n} \frac{\rho_n(f-g)}{1+\rho_n(f-g)}$$

- *Let*

$$f(x) = \max\{0, 1-|x|\}, \quad g(x) = 100f(x-2), \quad h = (f+g)/2$$

  Compute $d(f,0)$, $d(g,0)$ *and* $d(h,0)$.

- *Show this tells us* $B(0,1/4)$ *is not convex.*

- *Is there any radius r so that* $B(0,r)$ *is convex?*

Exercise 6.5.10 *Let the set of continuous functions on* $\Re$ *be* $C(\Re)$.

- *Prove the seminorms* $\rho_n(f) = \sup_{x\in[-n,n]} |f(x)|$ *defined on* $C(\Re)$, *induce a metric d given by*

$$d(f,g) = \sum_{n=1}^{\infty} 2^{-n} \frac{\rho_n(f-g)}{1+\rho_n(f-g)}$$

- *Let*

$$f(x) = \max\{0, 1-|x|\}, \quad g(x) = 100f(x-2), \quad h = (f+g)/2$$

  Compute $d(f,0)$, $d(g,0)$ *and* $d(h,0)$.

- *Show this tells us* $B(0,1/2)$ *is not convex.*

- *Is there any radius r so that* $B(0,r)$ *is convex?*

Exercise 6.5.11 *Let X be the set of all continuous functions on* $[0,1]$. *Let d be the metric defined by* $d(f,g) = \int_0^1 \frac{|f(x)-g(x)|}{1+|f(x)-g(x)|}$ *and let* σ *be the topology induced by d. Call this space* (X,σ). *Let* τ *be the topology induced on X by the seminorms* $\rho_x(f) = |f(x)|$, $0 \leq x \leq 1$. *Call this space* (X,τ).

- *Prove every set that is* τ *bounded is also* σ *bounded.*

- *Prove the identity map* $I : C(X,\tau) \to C(X,\sigma)$ *is not continuous. Note, by Exercise 6.2.4, this means* (C,τ) *is not metrizable!*

6.6 A Topology for Test Functions

Here, we discuss material covered by the master's project which Steve Crawford wrote under our direction in 2003 (Crawford (19) 2003). Steve was a delight to work with and we had many interesting conversations. He was crazy about the remake of *The Thing* by John Carpenter and he even took a trip out to the plains of Canada where the movie was filmed to toast its memory! Unfortunately, he left us due to a congenital, yet completely unknown and undiagnosed, heart condition, just a few short years after his master's work.

We want to consider a new type of function: the test function. Consider a formula from elementary calculus for taking the average of a function over an interval $[a, b]$:

$$f_{avg}[a, b] = \frac{1}{b - a} \int_a^b f(x)\, dx$$

We can define a new function ϕ where

$$\phi(x) = \begin{cases} \frac{1}{b-a}, & a \leq x \leq b \\ 0, & otherwise \end{cases}$$

so that the averaging formula can now be expressed as:

$$f_{avg}[a, b] = \int_{-\infty}^{+\infty} f(x)\, \phi(x)\, dx$$

Assuming that f is well-behaved on finite intervals, we see this improper integral exists for all possible cases. By expressing the average in this way, we can describe the function ϕ as sampling or testing f within a localized region. ϕ **tests** f to see what f does over a small domain, and by taking enough ϕ's over the real line we can get an idea of how f behaves in general. It is in this sense that we call objects like ϕ *test functions*.

There is an alternate way of looking at the relationship between ϕ and f. Instead of thinking of ϕ getting applied to f, we choose to look at it as though f gets applied to ϕ. As we see above, by integrating the two functions together, f can be interpreted as assigning a value to a given ϕ. We can thus visualize a new real-valued function F whose domain is the set of all ϕ such that:

$$F(\phi) = \int_{-\infty}^{\infty} f(x)\, \phi(x)\, dx$$

In this way F is a functional that acts on our *test functions*, a functional that ultimately depends on the nature of the original function f. An object like F that assigns a unique number to every test function is called a *distribution*. You can see F is linear with real values so it is a linear functional on some space X. It is up to us to find the right space X to use.

One problem with the above discussion is that the test function ϕ was only piecewise continuous (or, more specifically, piecewise differentiable). For reasons of mathematical convenience, we like our test functions to be smooth (i.e. differentiable) and not just smooth but *very* smooth. We can rectify ϕ's drawbacks by interpolating an appropriate function across its two discontinuities. For sake of simplicity, let $[a, b] = [-1/2,\ 1/2]$ so that ϕ becomes:

$$\phi(x) = \begin{cases} 1, & -1/2 \leq x \leq 1/2 \\ 0, & otherwise \end{cases}$$

We approximate ϕ arbitrarily closely by smooth functions. We have talked about C^∞ bump functions before, but we will discuss such functions from a different point of view now. As usual, it is always good to see derivations that have been done in different ways. Let $\delta > 0$ be a small deviation so that we interpolate a version of the logistic function on the domains $[-1/2 - \delta, -1/2 + \delta]$ and $[1/2 - \delta, 1/2 + \delta]$. Using standard properties of logistic functions, we find the approximation

$$\phi_\delta(x) = \begin{cases} \left[1 + exp(-tan \frac{\pi}{2\delta}(x + 1/2))\right]^{-1}, & x \in (-1/2 - \delta, -1/2 + \delta) \\ \left[1 + exp(tan \frac{\pi}{2\delta}(x - 1/2))\right]^{-1}, & x \in (-1/2 - \delta, -1/2 + \delta) \\ 1, & x \in [-1/2 + \delta, 1/2 - \delta] \\ 0, & \text{otherwise} \end{cases}$$

We make special note that ϕ_δ is not only continuous but smooth in the sense that it's differentiable everywhere. And not only is it differentiable, we can show it is *infinitely* differentiable. In Figure 6.1(a), we see a plot of ϕ_δ with $\delta = 1/10$: This graph may not seem differentiable at certain places,

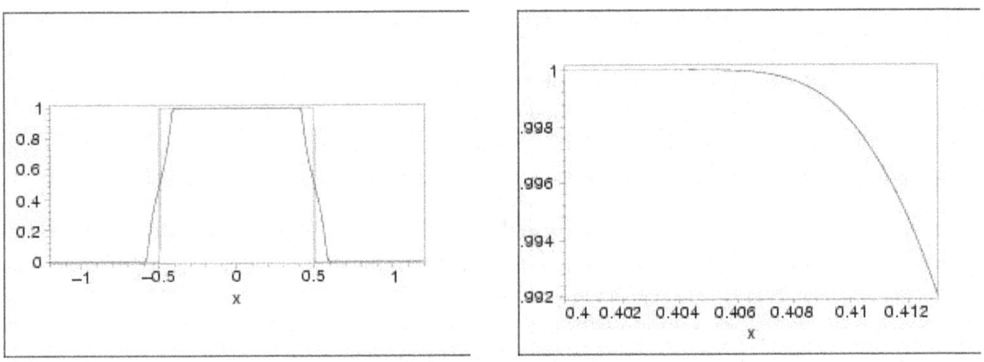

(a) Typical Φ test function. (b) Zoomed Φ test function.

Figure 6.1: The Φ test functions.

but if we take a closer look at one of the corners we see that it is indeed quite smooth. In Figure 6.1(b), for example, we can see the corner at the upper right-hand side: We can see that $\phi_\delta \longrightarrow \phi$ pointwise as $\delta \longrightarrow 0$, that is, all except when $|x| = 1/2$ where $\phi_\delta = 1/2$ for every δ. As a result, the average of the function f over $[-1/2, 1/2]$ can be approximated by

$$f_{avg}[-1/2, 1/2] \approx \int_{-\infty}^{+\infty} f(x)\, \phi_\delta(x)\, dx\,,$$

with the approximation getting better and better as $\delta \longrightarrow 0$. You should be able to see how this method of approximation is readily extended to any interval $[a, b]$.

This suggests the set of infinitely differentiable functions which equal 0 outside of some finite domain is rich enough to give us any localized average we desire. Again, think of the C^∞ bump functions discussed in (Peterson (99) 2020). Also, look at various partition of unity discussion we have in this text. The set of points where a function is not zero is important and this suggests how we can define the support of a function.

Definition 6.6.1 Support of a Function

The **support** *of a function $f : \Re \longrightarrow \Re$ is the closure of the set $\{x \in \Re : f(x) \neq 0\}$ and is denoted by supp(f).*

We can now define the space of test functions properly.

Definition 6.6.2 Test Functions $\mathcal{D}(\Re)$

The set $\mathcal{D}(\Re)$ of continuous functions $\phi : \Re \longrightarrow \Re$, where $\phi^{(n)}$ exists everywhere for all $n \geq 1$ and supp(ϕ) is compact, is called the set of **test functions**. *Sometimes this set is denoted by $C_c^\infty(\Re)$, where the subscript stands for compact support and the superscript refers to infinite differentiability.*

Thus, a test function is any function ϕ that has compact support and is infinitely smooth. Admittedly the test function ϕ_δ was a bit complicated, but the following shows that this need not be the case.

Example 6.6.1 *Let ψ be the function:*

$$\psi(x) = \begin{cases} k \, exp\left(\frac{a^2}{x^2 - a^2}\right), & |x| < a \\ 0, & |x| \geq a \end{cases}$$

We can show that ψ is infinitely differentiable, including at the points $|x| = a$, and it is apparent that supp$(\psi) = [-a, a]$ (see (Peterson (99) 2020)). Thus ψ is a test function and, by varying the values of k and a, we can make it as high and wide as we like. We can also horizontally shift ψ, by changing its argument to $x - \beta$, so that ψ becomes centered at an arbitrary $\beta \in \Re$.

After considering test functions such as ϕ_δ and ψ, we can make some general observations. Every test function must taper off smoothly to 0 the closer we get to the edges of its support. Unlike a horizontal asymptote, the test function must actually reach 0 at boundary points of its support. Except for the zero function, every test function must reach at least one absolute maximum or minimum somewhere inside of its support. We can easily show that $\mathcal{D}(\Re)$ is a vector space over $\Re$.

Theorem 6.6.1 Test Functions form a Vector Space

The set $\mathcal{D}(\Re)$ is a vector space over $\Re$ under the normal operations of function addition and scalar multiplication.

Proof 6.6.1
It is clear that by defining addition and scalar multiplication pointwise, linear combinations of test functions are new test functions. ■

Homework

Exercise 6.6.1 *Prove that the test functions form a vector space.*

Exercise 6.6.2 *For a finite interval $[a, b]$ with $a < b$, let*

$$X = \{\phi : [a, b] \rightarrow \Re | \phi \text{ is twice continuously differentiable } \}.$$

Prove $\|\phi\| = \max\{\|\phi\|_\infty, \|\phi'\|_\infty, \|\phi''\|_\infty\}$ is a complete normed linear space. The sup-norms are over the interval $[a, b]$, of course.

Exercise 6.6.3 *Let* $X = \{\phi : \Re \to \Re | \phi$ *is twice continuously differentiable and has compact support*$\}$. *Define* $\|f\|_\infty = \max_{x \in [-2,2]} |f(x)|$ *and define* $\|\rho_2\| = \max\{\|\phi\|_\infty, \|\phi'\|_\infty, \|\phi''\|_\infty\}$. *Prove* ρ_2 *is a semi-norm on* $\mathcal{D}(\Re)$.

6.6.1 The Test Functions as a Topological Vector Space

We need to embed vector spaces with a topological structure so that analysis can be performed on them. Let's find a suitable topology for the space $\mathcal{D}(\Re)$. After all is said and done, we want to have the following intuitive notions of convergence:

Definition 6.6.3 Convergence of Test Functions

A sequence $\{\phi_n\}$ *of test functions* **converges to 0** *if there exists an* $[a, b]$ *such that* $supp(\phi_n) \subseteq [a, b]$ *for every* n *and* $\{\phi_n^{(k)}\}$ *converges uniformly to* 0 *for every order* $k \geq 0$.

and

Definition 6.6.4 Sequence of Test Functions Converges if Difference Goes to Zero

A sequence $\{\phi_n\}$ *of test functions* **converges** *to* ϕ *if* $\{\phi_n - \phi\}$ *converges to* 0.

These definitions, however, are meaningless apart from some topology embedded in the space. Topologies for vector spaces can be normable but for spaces like $\mathcal{D}(\Re)$ it is exceedingly difficult to find a norm over the entire space that's powerful enough to give us the kind of rich topological structure we desire. As a result, we turn to the previous results of Theorem 6.4.5 and Theorem 6.5.1 to construct such rich topologies using a separating family of seminorms. The seminorms are used to construct a local subbasis, which led to a local basis which in turn produces the topology. This topology is then the natural one in the sense that every seminorm became continuous in the space. Let's look at this construction for $\mathcal{D}(\Re)$.

What family of seminorms should we use? The fundamental reason why $\mathcal{D}(\Re)$ cannot be easily normed is because the domain for the test functions is taken to be the entire real line. But suppose we restrict the domain to some compact set. Take, for instance, the set of test functions:

$$\mathcal{D}[a, b] = \{\phi \in \mathcal{D}(\Re) : supp(\phi) \subseteq [a, b]\}$$

Then it is not difficult to see that $\mathcal{D}([a, b])$ *can* be normed using $\|\cdot\|_\infty$. Indeed, it is nothing but a subspace of the normed linear space $C([a, b])$, the set of all continuous functions on $[a, b]$. Indeed, it is not hard to show $\mathcal{D}(K)$ is a normed linear space under $\|\cdot\|_\infty$ for any compact set K, not just for closed intervals. Recall, however, that we desire uniform convergence not just for ϕ_n but for all of ϕ_n's derivatives as well. The supremum norm $\|\cdot\|_\infty$ can guarantee uniformity of convergence for a given sequence but it doesn't guarantee such convergence for all of the sequence's derivatives. It is perfectly possible for a sequence to converge uniformly while its derivatives do not. In this sense, we need a norm on $\mathcal{D}[a, b]$ that is stronger than $\|\cdot\|_\infty$.
Note $\Re$ is expressible as the countable union of compact sets. For instance, $\Re = \bigcup_{n=1}^\infty [-n, n]$. So, if we can find the right topology for each $\mathcal{D}[-n, n]$, then perhaps we can construct a suitable topology for $\mathcal{D}(\Re)$ by piecing it together from these smaller topologies. As it turns out, this is precisely what happens.

In summary, we look for a family of seminorms that puts a cap on the size of a given test function ϕ as well as all of its derivatives, and does so for more and more test functions located along the real axis. After a little thought, we find that the family of seminorms $\{\rho_n\}$ does the trick, where each

$\rho_n : \mathcal{D}[-n, n] \longrightarrow \Re$ is defined to be

$$\rho_n(\phi) = \sup_{t \in [-n,n]} \{ |\phi(t)|, |\phi^{(1)}(t)|, \ldots, |\phi^{(n)}(t)| \}$$

We notice immediately that $\{\rho_n\}$ is not just a family of seminorms. It is actually a family of *norms*.

Lemma 6.6.2 ρ_n is a norm on $\mathcal{D}[-n, n]$

The function ρ_n is a norm on $\mathcal{D}[-n, n]$.

Proof 6.6.2
Let $\phi \in \mathcal{D}[-n, n]$ be a test function. We first note that every $\phi^{(k)}$ is bounded since if $\phi^{(k)}$ became unbounded at a point in supp(ϕ), then $\phi^{(k+1)}$ would not exist everywhere and would thus contradict the very definition of ϕ. Thus, $\rho_n(\phi)$ exists for every ϕ.

Now in order to simplify the calculations, we define a new function F_ϕ so that, as $k = 0 \ldots n$, F_ϕ on the interval $[(2k)2n,\ (2k + 1)2n]$ equals $|\phi^{(k)}|$ on the interval $[-n, n]$. Everywhere else, F_ϕ is defined to be 0. To get an idea of what F_ϕ is like, consider the case when $n = 2$:

$$F_\phi(x) = \begin{cases} |\phi(x - 2)|, & x \in [0, 4] \\ |\phi^{(1)}(x - 10)|, & x \in [8, 12] \\ |\phi^{(2)}(x - 18)|, & x \in [16, 20] \\ 0, & \textit{otherwise} \end{cases}$$

Thus, F_ϕ takes ϕ and its derivatives and pieces them all into one function so that $\rho_n(\phi) = \sup F_\phi$. With F_ϕ in hand, we can now check the properties of a norm. For every choice of $\phi, \psi \in \mathcal{D}[-n, n]$ and $\alpha \in \Re$ we have:

(a) $\rho_n(\phi) = \sup F_\phi$ and since $F_\phi \geq 0$ everywhere, then $\rho_n(\phi) \geq 0$.

(b) Suppose $\rho_n(\phi) = 0$. Then $\sup F_\phi = 0$ and since $F_\phi = |\phi(t)|$ on some interval, then ϕ is itself 0.

(c) Scalar multiplication:

$$\rho_n(\alpha\phi) = \sup F_{\alpha\phi} = \sup \{|\alpha| F_\phi\} = |\alpha| \cdot \sup F_\phi = |\alpha| \cdot \rho_n(\phi)$$

(d) Sub-additivity:

$$\rho_n(\phi + \psi) = \sup F_{\phi+\psi} \leq \sup \{ F_\phi + F_\psi \} \leq \sup F_\phi + \sup F_\psi = \rho_n(\phi) + \rho_n(\psi)$$

■

This immediately brings about the result.

Lemma 6.6.3 Family $\{\rho_n\}$ is separating on $\mathcal{D}(\Re)$

The family $\{\rho_n\}$ is separating on $\mathcal{D}(\Re)$.

Proof 6.6.3
Let $\phi \in \mathcal{D}(\Re)$ be a nonzero test function. Then supp(ϕ) is compact which implies supp(ϕ) is bounded. So there is a k such that supp(ϕ) $\subseteq [-k, k]$. Consequently, ϕ is in the domain of ρ_k and

since ρ_k is a norm, then $\rho_k(\phi) > 0$ because $\phi \neq 0$. Therefore, $\{\rho_n\}$ is a separating family of seminorms on $\mathcal{D}(\Re)$. ■

So the condition for Theorem 6.4.5 and Theorem 6.5.1 is satisfied by $\{\rho_n\}$. Following the procedure laid out for there, we form the collection $\{U_n\}$ of subsets from $\mathcal{D}(\Re)$:

$$U_n = \{ \phi \in \mathcal{D}[-n, n] : \rho_n(\phi) < 1/n \}$$

Then, according to the theorem, $\{U_n\}$ serves as a local sub-basis for the local base $\mathcal{B}$ defined to be the set of all finite intersections from $\{U_n\}$.

We take a closer look at these intersections. Consider two sets U_i, U_j and their intersection $U_i \cap U_j$. Without loss of generality, we can assume $i < j$. So we have:

$$U_i \cap U_j = \{ \phi : \phi \in \mathcal{D}[-i, i], \phi \in \mathcal{D}[-j, j], \rho_i(\phi) < 1/i, \rho_j(\phi) < 1/j \}$$

Two of the four conditions for this set are redundant. Because $\mathcal{D}[-i, i] \subseteq \mathcal{D}[-j, j]$, then we see that $U_i \cap U_j$ is a subset of $\mathcal{D}[-i, i]$ and so we can ignore the condition regarding $D[-j, j]$. Now compare ρ_i and ρ_j as follows

$$\begin{aligned} \rho_i(\phi) &= sup\{ |\phi(t)|, |\phi^{(1)}(t)|, \dots, |\phi^{(i)}(t)| \} \\ \rho_j(\phi) &= sup\{ |\phi(t)|, |\phi^{(1)}(t)|, \dots, |\phi^{(i)}(t)|, \dots, |\phi^{(j)}(t)| \} \end{aligned}$$

where the first supremum is taken over $[-i, i]$ and the second is taken over $[-j, j]$. It thus becomes apparent that $\rho_j < \epsilon$ implies $\rho_i < \epsilon$ but not vice-versa. So for every $\phi \in U_i \cap U_j$ we have $\rho_j(\phi) < 1/j$ and this means we can ignore the condition regarding ρ_i.

Therefore, we have the following characterization for the intersection:

$$\begin{aligned} U_i \cap U_j &= \{ \phi \in \mathcal{D}[-i, i] : \rho_j(\phi) < 1/j \} \\ &= \{ \phi \in \mathcal{D}[-min(i, j), min(i, j)] : \rho_{max(i,j)}(\phi) < \frac{1}{max(i, j)} \} \end{aligned}$$

Applying mathematical induction to this reasoning, we have come up with a characterization of open sets.

Lemma 6.6.4 Characterization of Open Sets

Suppose $\mathcal{O}$ is an open set in the local basis $\mathcal{B}$ generated by $\{U_n\}$. By definition $\mathcal{O} = \bigcap_{k=1}^{N} U_{n_k}$ for appropriate indices n_k. Let $m = min(n_k)$ and $M = max(n_k)$ as $k = 1 \dots N$. Then

$$\mathcal{O} = \{ \phi \in \mathcal{D}[-m, m] : \rho_M(\phi) < 1/M \}$$

Proof 6.6.4
We have just gone over this argument. ■

Now by appropriate choice of indices n_k, the values of m and M can both be made to vary over all natural numbers as long as $m \leq M$. Therefore, we have proven the following characterization of the local base:

Theorem 6.6.5 Local Base Characterization $\mathcal{B}$

The local base $\mathcal{B}$ generated by $\{U_n\}$ is the set $\mathcal{B} = \bigcup_{i \leq j} \{\mathcal{O}_{ij}\}$ where i and j are natural numbers and

$$\mathcal{O}_{ij} = \{ \phi \in \mathcal{D}[-i, i] : \rho_j(\phi) < 1/j \}$$

Proof 6.6.5
The argument has already been given. ∎

With this local base in place, Definition 6.6.3 and Definition 6.6.4 for the convergence of test functions allow us to look at test function convergence better. Being in the local base, every $\mathcal{O}_{ij}$ is thus a "neighborhood" of the zero function. So let us suppose $\{\phi_n\}$ converges to 0. Then, by definition, every $\mathcal{O}_{ij}$ contains all but a finite number of ϕ_n. It is clear the converging sequence $\{\phi_n\}$ has all $supp(\phi_n)$ in some interval $[a, b]$. So there is a large enough I such that $supp(\phi_n) \subseteq [-I, I]$. Consequently, $\phi_n \in \mathcal{D}[-I, I]$ for each n, and this implies every ϕ_n is in $\mathcal{O}_{ij}$ whenever $i \geq I$.

Now take a closer look at $\mathcal{O}_{ij}$ for a fixed $i \geq I$. Like we saw above, j can be any number greater than i. So let $j \longrightarrow \infty$. As we let j grow larger, the sizes of $|\phi_n|$, ... , $|\phi_n^{(j)}|$ each become smaller than $1/j$. And since all of our seminorms are supremums, we see that the convergence of $\{\phi_n\}$ and its derivatives to 0 is truly uniform.

Similar comments can be said for $\{\phi_n\}$ as it converges uniformly to a limit ϕ. In that case, every $\mathcal{O}_{ij}$ would contain all but a finite number of $\phi_n - \phi$, guaranteeing that $\{\phi_n\}$ and all its derivatives converge uniformly to ϕ and its derivatives $\phi^{(k)}$.

Example 6.6.2 *Using the function ψ from Example 6.6.1, we construct a sequence $\{\psi_n\}$ of test functions that converge to 0. We let ψ_n be the function:*

$$\psi_n(x) = \begin{cases} \frac{1}{n^2} \, exp\left(\frac{100}{x^2 - 100}\right), & |x| < 10 \\ 0, & |x| \geq 10 \end{cases}$$

Then is it apparent that $\{\psi_n\}$ converges to 0.

We are now ready to put the finishing touches on our topology for the space of test functions. A major problem with the topology generated by $\mathcal{B}$ is that it is not complete. For instance, choose a $\phi \in \mathcal{D}[0, 1]$ so that $\phi > 0$ on the open interval $(0, 1)$. We construct a sequence $\{\psi_m\}$ so that

$$\begin{aligned} \psi_m(x) &= \sum_{k=1}^{m} \frac{1}{k} \, \phi(x - k) \\ &= \phi(x - 1) + \frac{1}{2} \, \phi(x - 2) + \ldots + \frac{1}{m} \, \phi(x - m) \end{aligned}$$

Then ψ_m is a linear combination of test functions in $\mathcal{D}(\Re)$, implying that it is itself a test function. We show that $\{\psi_m\}$ is a Cauchy sequence. That is, for every $\mathcal{O}_{ij}$ in the local base $\mathcal{B}$ there exists an N so that $\psi_m - \psi_n \in \mathcal{O}_{ij}$ whenever $m, n > N$. Let $\mathcal{O}_{ij}$ be in the local base $\mathcal{B}$ and suppose indices m, n are greater than i. Consider $\psi_m - \psi_n$. Without loss of generality, we may assume $m > n$ so that we have

$$\psi_m - \psi_n = \sum_{k=1}^{m} \frac{1}{k} \, \phi(x - k) - \sum_{k=1}^{n} \frac{1}{k} \, \phi(x - k)$$

$$= \frac{1}{n+1}\phi[x - (n+1)] + ... + \frac{1}{m}\phi(x - m)$$

Since $n > i$ then we can see that $supp(\phi_m - \phi_n)$ is outside of $[-i, i]$ and this means that $\phi_m - \phi_n$ is 0 when restricted to $[-i, i]$. And because the zero function is in all $\mathcal{D}[-i, i]$, then $\phi_m - \phi_n \in \mathcal{D}[-i, i]$ and $\rho_i(\psi_m - \psi_n) = 0$. That is, $\psi_m - \psi_n \in \mathcal{O}_{ij}$ whenever $m, n > i$. Hence, $\{\psi_m\}$ is indeed a Cauchy sequence. But $\{\psi_m\}$ has no limit in $\mathcal{D}(\Re)$ since, for instance, any such limit would not have compact support. So we see that the topology generated by $\mathcal{B}$ is not complete in that not every Cauchy sequence converges in the space.

We can fix this situation by constructing a new local base that is very similar to $\mathcal{B}$. We turn our attention to all subsets in $\mathcal{D}(\Re)$ that are both convex and balanced.

Theorem 6.6.6 Convex and Balanced Local Base for Test Functions

Let τ_n denote the topology we generated on $\mathcal{D}[-n, n]$ by the norm ρ_n. Define $\mathcal{B}^*$ to be set of all convex, balanced subsets $\mathcal{O} \subseteq \mathcal{D}(\Re)$ such that $\mathcal{O} \cap \mathcal{D}[-n, n] \in \tau_n$ for every n. The set $\mathcal{B}^*$ is a local base for a topology on the space $\mathcal{D}(\Re)$.

Proof 6.6.6

All we need to show is that any finite intersection of elements in $\mathcal{B}^*$ is again in $\mathcal{B}^*$. Let b_1 and b_2 be members of $\mathcal{B}^*$. Then for every n we have

$$b_1 \cap \mathcal{D}[-n, n] \in \tau_n, \quad b_2 \cap \mathcal{D}[-n, n] \in \tau_n$$

and since τ_n is a topology, then for each n we see that

$$\left(b_1 \cap \mathcal{D}[-n, n]\right) \bigcap \left(b_2 \cap \mathcal{D}[-n, n]\right) \in \tau_n$$

and re-expressing this we get:

$$\left(b_1 \cap b_2\right) \bigcap \mathcal{D}[-n, n] \in \tau_n$$

Thus $b_1 \cap b_2 \in \tau_n$ for every n.

What remains to be shown is that $b_1 \cap b_2$ is convex and balanced. Let $x, y \in b_1 \cap b_2$. Then because both b_1 and b_2 are convex, we see that for every $\alpha \in (0, 1)$

$$\alpha x + (1 - \alpha) y \in b_1, \quad \alpha x + (1 - \alpha) y \in b_2$$

and this only implies that:

$$\alpha x + (1 - \alpha) y \in b_1 \cap b_2$$

Thus, $b_1 \cap b_2$ is convex. Now suppose $|\alpha| \leq 1$ and $x \in b_1 \cap b_2$. Then because b_1 and b_2 are both balanced we have $\alpha x \in b_1$ and $\alpha x \in b_2$. Thus, $\alpha x \in b_1 \cap b_2$ and this means that $b_1 \cap b_2$ is balanced.

Therefore $b_1 \cap b_2 \in \mathcal{B}^*$. Using mathematical induction, we readily extend this argument to any finite intersection of elements from $\mathcal{B}^*$. Therefore, $\mathcal{B}^*$ is a local base for a topology on $\mathcal{D}(\Re)$. ■

We can see how $\mathcal{B}^*$ works. It is nothing more than the collection of all *nice* sets of test functions (i.e. sets that are convex and balanced), each of which behaves like it should with respect to $\mathcal{B}$'s original topology. We also note that the definition for $\mathcal{B}^*$ is meaningful in that the set is nonempty. To see this, consider the whole set $\mathcal{D}(\Re)$ which is clearly convex and balanced. Now $\mathcal{D}(\Re) \cap \mathcal{D}[-n, n]$ is just $\mathcal{D}[-n, n]$ which is nothing but the whole space operated on by the norm ρ_n. And by the nature of a topology, τ_n must contain $\mathcal{D}[-n, n]$ as one of its elements (since all topological spaces have the whole space as part of their structure). Therefore, $\mathcal{D}(\Re) \cap \mathcal{D}[-n, n] \in \tau_n$ for every n and so $\mathcal{D}(\Re) \in \mathcal{B}^*$.

Homework

Exercise 6.6.4 *Let*

$$X = \{\phi : \Re \rightarrow \Re | \phi \text{ is twice continuously differentiable and has compact support in} [-2, 2]\}.$$

Define $\|f\|_\infty = \max_{x \in [-2,2]} |f(x)|$ *and define* $\|\rho_2\| = \max\{\|\phi\|_\infty, \|\phi'\|_\infty, \|\phi''\|_\infty\}$. *Prove* ρ_2 *is a norm on* $\mathcal{D}(\Re)$.

Exercise 6.6.5 *Prove* $\mathcal{D}(K)$ *is a normed linear space under* $\| \cdot \|_\infty$ *for any compact set* K.

Exercise 6.6.6 *In the proof of Theorem 6.6.6, provide the details of the mathematical induction proof that enables us to handle any finite intersection of elements from* $\mathcal{B}^*$.

6.6.2 Properties of the Topological Vector Space $\mathcal{D}(\Re)$

As already mentioned, the advantage here is that the topology generated by $\mathcal{B}^*$ has essential properties $\mathcal{B}$'s topology did not have. From now on, we denote the topology generated by $\mathcal{B}^*$ as simply τ and, whenever we refer to $\mathcal{D}(\Re)$ as a topological vector space, it is always understood that we have τ in mind. We first demonstrate that $\mathcal{D}(\Re)$ is indeed a topological vector space.

Theorem 6.6.7 Test Functions as a Topological Vector Space: $\mathcal{D}(\Re)$ with τ

The topology τ makes $\mathcal{D}(\Re)$ into a T_1 topological vector space.

Proof 6.6.7
We verify each of the essential properties:

(a) We demonstrate that every singleton is a closed set. Let ϕ_1 and ϕ_2 be distinct test functions in $\mathcal{D}(\Re)$. Define a set $\mathcal{O}$ so that:

$$\mathcal{O} = \{\psi \in \mathcal{D}(\Re) : \|\psi\|_\infty < \|\phi_1 - \phi_2\|_\infty\}$$

We first show that $\mathcal{O} \in \mathcal{B}^*$. Let $\psi_1, \psi_2 \in \mathcal{O}$ and $\alpha \in (0, 1)$ so that

$$\begin{aligned} \|\alpha\,\psi_1 + (1-\alpha)\,\psi_2\|_\infty &\leq \alpha\,\|\psi_1\|_\infty + (1-\alpha)\,\|\psi_2\|_\infty \\ &< \alpha\,\|\phi_1 - \phi_2\|_\infty + (1-\alpha)\,\|\phi_1 - \phi_2\|_\infty \\ &= \|\phi_1 - \phi_2\|_\infty \end{aligned}$$

which means that $\mathcal{O}$ is convex. Now consider $\psi \in \mathcal{O}$ with $|\alpha| \leq 1$ so that

$$\|\alpha\,\psi\|_\infty = |\alpha|\,\|\psi\|_\infty < |\alpha|\,\|\phi_1 - \phi_2\|_\infty \leq \|\phi_1 - \phi_2\|_\infty$$

which implies that $\mathcal{O}$ is balanced. Now we look at:

$$\mathcal{O} \cap \mathcal{D}[-n, n] = \{\psi \in \mathcal{D}[-n, n] : \|\psi\|_\infty < \|\phi_1 - \phi_2\|_\infty\}$$

$$= \{ \psi \in \mathcal{D}[-n,n] \,:\, \rho_0(\psi) < \|\phi_1 - \phi_2\|_\infty \}$$

We show that this set is open in τ_n. *Recall that* τ_n *is the collection of all open sets of* $\mathcal{D}[-n,n]$ *generated by the norm* ρ_n. *So let* $\psi \in \mathcal{O} \cap \mathcal{D}[-n,n]$ *and consider the neighborhood* $B(\psi, 1/N)$ *where* N *is large enough so that* $1/N < \|\phi_1 - \phi_2\|$. *Then it is apparent that* $B(\psi, 1/N) \subseteq \mathcal{O} \cap \mathcal{D}[-n,n]$ *indicating that* $\mathcal{O} \cap \mathcal{D}[-n,n]$ *is an open set in* τ_n.

Therefore, $\mathcal{O}$ *is an element of* $\mathcal{B}^*$ *and is thus an open set in the topology generated by* $\mathcal{B}^*$. *But* ϕ_1 *is not in* $\phi_2 + \mathcal{O}$ *or, equivalently,* ϕ_2 *is not in the closed set* $(\phi_2 + \mathcal{O})^c$ *containing* ϕ_1. *This means that* ϕ_2 *is not in the closure of* $\{\phi_1\}$. *Since* ϕ_2 *was arbitrary we thus see that* $\{\phi_1\}^* = \{\phi_1\}$ *and so every singleton is a closed set in the topology of* $\mathcal{B}^*$.

(b) *We demonstrate that addition is continuous. Let* $\phi_1, \phi_2 \in \mathcal{D}(\Re)$ *and* $\mathcal{O} \in \tau$. *Then* $(\phi_1+\phi_2)+\mathcal{O}$ *is a neighborhood of* $\phi_1 + \phi_2$. *Likewise,* $\phi_1 + \frac{1}{2}\mathcal{O}$ *is a neighborhood of* ϕ_1 *and* $\phi_2 + \frac{1}{2}\mathcal{O}$ *is a neighborhood of* ϕ_2. *And due to the convexity of each* $\mathcal{O}$, *we see that*

$$(\phi_1 + \frac{1}{2}\mathcal{O}) + (\phi_2 + \frac{1}{2}\mathcal{O}) = (\phi_1 + \phi_2) + \mathcal{O}$$

and this satisfies the condition for continuity of addition.

(c) *We now demonstrate that scalar multiplication is continuous. Fix a scalar* $\alpha_0 \in \Re$ *and a test function* $\phi_0 \in \mathcal{D}(\Re)$. *Then:*

$$\begin{aligned} \alpha\,\phi - \alpha_0\,\phi_0 &= \alpha\,\phi - \alpha\,\phi_0 + \alpha\,\phi_0 - \alpha_0\,\phi_0 \\ &= \alpha\,(\phi - \phi_0) + (\alpha - \alpha_0)\,\phi_0 \end{aligned}$$

Now let $\mathcal{O} \in \mathcal{B}^*$. *Then there exists a* $\delta > 0$ *such that* $\delta\,\phi_0 \in \frac{1}{2}\,\mathcal{O}$ *and define a number* β *so that:*

$$\beta = \frac{1}{2\,(\delta + |\alpha_0|)}$$

So suppose that $|\alpha - \alpha_0| < \delta$ *and* $\phi - \phi_0 \in \beta\,\mathcal{O}$. *Then*

$$\alpha\,(\phi - \phi_0) \;\in\; \alpha\,\beta\,\mathcal{O} \in \frac{1}{2}\,\frac{\alpha}{|\alpha_0| + \delta}\,\mathcal{O} \in \frac{1}{2}\,\mathcal{O}$$

and

$$(\alpha - \alpha_0)\,\phi_0 \;<\; \delta\,\phi_0 \in \frac{1}{2}\,\mathcal{O}$$

implying that

$$(\alpha - \alpha_0)\,\phi_0 \in \frac{1}{2}\,\mathcal{O}$$

and so we conclude:

$$\alpha\,\phi - \alpha_0\,\phi_0 \in \mathcal{O}$$

■

We can see that in one sense τ is stronger than the topology generated by $\|\cdot\|_\infty$. This is because each ρ_n is a supremum occurring not just for $|\phi(t)|$ but also for all $|\phi^{(k)}(t)|$. One disadvantage, though, is that $\mathcal{D}(\Re)$'s topology is non-metrizable.

Theorem 6.6.8 Test Functions Topological Vector Space Structure is not Metric

τ is not a metric topology.

Proof 6.6.8

Suppose $d : \mathcal{D}(\Re) \times \mathcal{D}(\Re) \longrightarrow \Re$ is a metric that generates τ. We let $\{\phi_n\}$ be any sequence of test functions such that supp$(\phi_n) \nsubseteq [-n, n]$.

Now $\mathcal{D}(\Re)$ is a topological vector space and that means scalar multiplication is continuous. So fix a $\phi_k \in \{\phi_n\}$ and define a function $f : \Re \longrightarrow \mathcal{D}(\Re)$ such that $f(\alpha) = \alpha\,\phi_k$. Since f is continuous in a presumably metric topology, then for any $\epsilon = 1/k$ there is a δ_k such that

$$d(f(\alpha), f(0)) = d(\alpha\,\phi_k, 0) < \epsilon = 1/k$$

whenever $|\alpha| < \delta_k$. So choose a number less than δ_k and call this value α_k.

Repeating this argument for each ϕ_n, we see that in every case there is an α_n such that $d(\alpha_n\phi_n, 0) < 1/n$, implying that $\{\alpha_n\psi_n\}$ converges to 0 in the presumably metric topology. This, however, contradicts the very definition of convergence in $\mathcal{D}(\Re)$ for there cannot be any $[a, b]$ that contains each supp(ϕ_n). Thus, τ is non-metrizable. ■

Though the space $\mathcal{D}(\Re)$ does have its drawbacks, it nevertheless has some very nice advantages that were not true for the topology generated by $\mathcal{B}$.

Theorem 6.6.9 Test Functions Topological Vector Space, $\mathcal{D}(\Re)$ with τ, is complete

$\mathcal{D}(\Re)$ is a complete topological vector space.

Proof 6.6.9

Suppose $\{\phi_n\}$ is a Cauchy sequence in $\mathcal{D}(\Re)$. Since it is Cauchy, then it is bounded which in turn implies that there exists an interval $[-m, m]$ that contains every supp$\phi_n^{(k)}$. Thus $\{\phi_n\} \subseteq C^\infty[-m, m]$, the set of infinitely smooth functions over $[-m, m]$. And since $C^\infty[-m, m] \subseteq C[-m, m]$, then each ϕ_n is also in $C[-m, m]$. Now the norms ρ_j all include the supremum norm $\|\cdot\|_\infty$. And, since $C[-m, m]$ is a complete space under $\|\cdot\|_\infty$, then there exists a ϕ such that $\phi_n \longrightarrow \phi$ uniformly in $\mathcal{D}(\Re)$'s topology.

We show that ϕ is a test function. It is apparent that ϕ is 0 outside of $[-m, m]$ so that ϕ has compact support. We just need to show that it is infinitely differentiable. Consider the differential operator $D^\alpha : C^\infty[-m, m] \longrightarrow C[-m, m]$. Then, by assumption, the sequence $\{D^\alpha(\phi_n)\}$ is Cauchy, and by the same reasoning above it converges to some $\phi_\alpha \in C[-m, m]$. And because the topology of $\mathcal{D}(\Re)$ guarantees the uniform convergence of derivatives of all orders (recall the general definition of ρ_j), D^α is thus a continuous operator. This implies that we can interchange the limiting process across the operator D^α:

$$D^\alpha(\phi) = D^\alpha(lim\,\phi_n) = lim(D^\alpha\phi_n) = lim(\phi_n^{(\alpha)}) = \phi_\alpha$$

So ϕ has derivatives of all orders, and thus it is a test function. Therefore, we conclude that $\{\phi_n\}$ converges in $\mathcal{D}[-m, m]$ and because $\mathcal{D}[-m, m] \subseteq \mathcal{D}(\Re)$, then we see that $\{\phi_n\}$ converges in $\mathcal{D}(\Re)$.

Hence, $\mathcal{D}(\Re)$ is a complete topological vector space. ■

Homework

Recall, $\mathcal{D}(\Re)$ is assumed to have the topology τ.

Exercise 6.6.7 *In the context of test functions, a* **distribution** *is any continuous linear functional on $\mathcal{D}(\Re)$; i.e. if f is a distribution, then $f \in (\mathcal{D}(\Re))'$, the dual of $\mathcal{D}(\Re)$. Let δ be the functional on $\mathcal{D}(\Re)$ defined by $\delta(\phi) = \phi(0)$. Prove that δ is a continuous linear functional.*

Exercise 6.6.8 *A real-valued function f defined on $\Re$ is called locally integrable if the Lebesgue integral $\int_S |f| d\mu$ exists and is finite for all compact $S \subset \Re$ where μ is standard Lebesgue measure. In the context of Riemann integration, we would say the Riemann integral $\int_a^b f(x)dx$ exists and is finite for all finite intervals $[a, b]$, with $a < b$ finite.*

- *Prove any continuous function on $\Re$ is locally integrable.*
- *Define $F : \mathcal{D}(\Re) \rightarrow \Re$ by $F(\phi) = \int_\Re f(x)\phi(x)dx$. Note this is always well-defined as ϕ has compact support. Prove F is a continuous linear functional.*

Exercise 6.6.9 *If F is a continuous linear functional on $\mathcal{D}(\Re)$ and there exists a locally integrable f so that $F(\phi) = \int_\Re f(x)\phi(x)dx$ for all $\phi \in \mathcal{D}(\Re)$, we say F is a regular distribution. If this is not true, we say F is a singular distribution. Prove δ is a singular distribution.*
Hint: *Assume δ is a regular distribution. Then $\delta(\phi) = \phi(0) = \int_\Re f(x)\phi(x)dx$. This is true for*

$$\phi_\epsilon(x) = \begin{cases} e^{\frac{-x^2}{\epsilon^2 - x^2}}, & |x| < \epsilon \\ 0, & |x| \geq \epsilon \end{cases}$$

Since $|\phi_\epsilon(x)| \leq \phi_\epsilon(0) = 1/\epsilon$, show $\frac{1}{\epsilon} \leq \frac{1}{\epsilon} \int_\Re |f(x)| dx$ for any $\epsilon > 0$. But, as $\epsilon \rightarrow 0$ show $\int_{-\epsilon}^{\epsilon} |f(x)| dx \rightarrow 0$. This gives a contradiction and so δ is a singular distribution.

Chapter 7

A New Look at Linear Functionals

Let X be a vector space over $\Re$ and X^* be the set of linear functionals on X. Recall $x' \in X^*$ which means $x' \rightarrow \Re$. We let X' be the set of continuous linear functionals on X. We cannot talk about the continuity of x' unless there is a topology on X. Hence, the linear functionals that are continuous depends on the topology we add to X. Note, we have discussed dual spaces in some detail in (Peterson (100) 2020) but now we have a much better understanding of how a topology on a space can be constructed. If $x' \in X'$, then $x' \in X^*$ but now x' is continuous. This means that if U is open in the topology we choose to use on $\Re$, then $(x')^{-1}(U)$ is open in the topology we choose to use on X. We have been building tools to make intelligent decisions about what topology we want to use on X and now we will use them in our discussions.

First, let's recall the easy case: everything is in a normed linear space.

Definition 7.0.1 Linear Operators on Normed Linear Spaces

Let $(X, \|\cdot\|_X)$ and $(Y, \|\cdot\|_Y)$ be two normed linear spaces. Let $T : dom(T) \subset X \rightarrow Y$ be a mapping and assume $dom(T)$ is a subspace of X and $T(\alpha \boldsymbol{x} + \beta \boldsymbol{y}) = \alpha T(\boldsymbol{x}) + \beta T(\boldsymbol{y})$ for all α, β in the scalar field F of the vector space and for all $\boldsymbol{x}, \boldsymbol{y}$ in X. We say T is a linear operator from X to Y.

If $dom(T) = X$ and $Y = \Re$, we say T is a real linear functional on X. If $Y = \mathbb{C}$, we say T is a complex linear functional on X.

So X^* is the set of all linear functionals on the normed linear space X. The continuity of such a linear functional is then defined by the usual $\epsilon - \delta$ mechanism. The linear functional x' is continuous at $\boldsymbol{x_0} \in X$, if for all $\epsilon > 0$, there is a $\delta > 0$ so that $\|\boldsymbol{x} - \boldsymbol{x_0}\| < \delta$ implies $|x'(x) - x'(x_0)\| < \epsilon$. This is a pointwise definition rooted at a particular point $\boldsymbol{x_0}$ in X and this point of view hides a lot of important structure. A much deeper look at continuity is to think of the range $(\Re, |\cdot|)$ as a normed linear space with its open sets determined by the norm $|\cdot|$. Note the balls at 0 in $\Re$ form a locally convex, balanced, absorbing base. The domain space $(X, \|\cdot\|)$ uses its own norm determined topology and the balls at 0 in X also form a locally convex, balanced, absorbing base. Then we can speak of continuity of x' in this way: if U is open in the topology of $(\Re, |\cdot|)$, then $(x')^{-1}(U)$ is open in the topology of $(X, \|\cdot\|)$. Note this definition is no longer pointwise and more importantly, we are no longer restricted to a topology on the domain and range space that is norm determined. We have been spending a lot of time lately finding ways to build locally convex T_1 topological vector spaces which we can now exploit. As usual, the space of continuous linear functionals on the space X is denoted X'. But remember the topology we use on X determines what we mean by continuous.

Also, remember weak convergence. We say a sequence $(\boldsymbol{x_n})$ in X converges weakly if $x'(x_n) \rightarrow x'(x)$ for all $x' \in X'$. The evaluation map $\phi_t(f) = f(t)$ is linear and since for a given $\boldsymbol{f_0} \in X$, $|\phi_t(f) - \phi_t(f_0)| = |f(t) - f(t_0)|$, with the right norm on X, the evaluation map can be continuous. If so, then weak convergence implies $\phi_t(x_n) = x_n(t) \rightarrow \phi_t(x) = x(t)$ which is what we call pointwise convergence. We can also define the set of all linear functionals on X^*, X^{**} and the set of continuous linear functionals on X' to be X''. Note weak convergence on X' means if (x'_n) is a sequence in X', then this sequence converges weakly in X' if $x''(x'_n) \rightarrow x''(x')$ for all $x'' \in X''$. Using the canonical embedding of X into X'', for any x'' which is the embedding of $\boldsymbol{x} \in X$ via the embedding, weak convergence in this setting would give $x'_n(x) \rightarrow x'(x)$ for all $x \in X$. This is what we call weak* convergence. Understanding this convergence means we must understand the topology we put on X'.

7.1 The Basics

Let's define a collection of sets in X^*. Start by choosing a subset $F \subset X^*$. This does not have to be all of X^*. Let

$$\mathscr{B} = \{A | A \in F \subset X^*, \ A \text{ is finite}\},$$

where A finite means the subset has only a finite number of elements in it. Hence, a typical A might be $A = \{x'_1, \ldots, x'_n\}$ where we label the finite number of elements in A from 1 to n for some n. We also know each $x'_i in F$. Then, define a local open base for our upcoming topology by defining for each $A \in \mathscr{B}$, the subsets of X given by:

$$\begin{aligned} U(A, \epsilon) &= \{x \in X|, |x'(x)| < \epsilon, \} \\ \mathscr{U} &= \{U(A, \epsilon) | A \in \mathscr{B}, \ \epsilon > 0\} \end{aligned}$$

Now, for any $V \subset X$ that is convex and absorbing, we know there is a Minkowski functional defined on V. We will show each $U(A, \epsilon)$ has an associated Minkowski functional and hence, this topology can be thought of as being generated by a family of seminorms.

(i): $0 \in U(A, \epsilon)$, for any $\epsilon > 0$ and $A \in \mathscr{B}$.

(ii): $U(A, \epsilon)$ is balanced, absorbing and convex. This is an easy argument:

- If $-1 \leq t \leq 1$ and $x \in U(A, \epsilon)$, then $|x'(tx)| = |t||x'(x)| \leq |x'(x)| < 1$, so $tx \in U(A, \epsilon)$. Hence, the set is balanced.
- If x_1 and x_2 are in $U(A, \epsilon)$, then for $0 < s < 1$, $|x'((1-s)x_1 + sx_2)| \leq (1-s)|x'(x_1) + s|x'(x_2)| < 1$ and so the set is convex.
- $U(A, \epsilon)$ is absorbing, if there is a $\xi > 0$ so that $\alpha x \in U(A, \epsilon)$, if $0 < |\alpha| \leq \xi$. If we choose $\xi = (1/2)$, we see $U(A, \epsilon)$ is absorbing.

(iii): If x is in $U(A_1 \cup A_2, \min\{\epsilon_1, \epsilon_2\})$, then $|x'(x)| < \min\{\epsilon_1, \epsilon_2 \leq \epsilon_1 \leq\}$ if $x' \in A_1$ and $|x'(x)| < \min\{\epsilon_1, \epsilon_2\} \leq \epsilon_2 \leq$ if $x' \in A_2$. Thus, $x \in U(A_1, \epsilon_1)$ and $x \in U(A_2, \epsilon_1)$. So we see $U(A_1 \cup A_2, \min\{\epsilon_1, \epsilon_2\}) \subset U(A_1, \epsilon_1) \cap U(A_12, \epsilon_2)$.

(iv): It is straightforward to prove $\alpha(U, \epsilon) = U(A, |\alpha|\epsilon)$ for any $\alpha \neq 0$. We leave this to you as an exercise.

(v): If $x \in U(A, \epsilon)$, let $\delta = \sup_{x' \in A} |x'(x)|$. We see $0 \leq \delta < \epsilon$. It is another straightforward exercise to show $x + U(A, \epsilon - \delta) \subset U(A, \epsilon)$.

(vi): It is also easy to show $U(A, \epsilon/2) + U(A, \epsilon/2) \subset U(A, \epsilon)$.

Then by Theorem 6.3.3, there is a unique topology $\mathscr{S}$ for X so that $(X, \mathscr{S})$ is a topological vector space. Theorem 6.3.3 does not use the convexity of the local base. However, from Theorem 6.3.4, if the local base also satisfies if $U \in \mathscr{B}$, there is α with $0 < \alpha \leq 1/2$ and $\alpha U \in \mathscr{B}$, then there is a unique topology $\mathscr{S}$ for X so that $(X, \mathscr{S})$ is a locally convex topological vector space. This condition is easy to see because of the way $U(A, \epsilon)$ is defined. Hence, we can assume there is a unique topology $\mathscr{S}$ for X so that $(X, \mathscr{S})$ is a locally convex topological vector space. We denote this topology by $\mathscr{S}(X, F)$ because the topology is dependent on the choice of $F \subset X^*$. Now we can look at the continuity of $x' \in X^*$ properly. We need a series of results to look at this carefully. We now show each of the $U(A, \epsilon)$ sets have an associated Minkowski functional.

Theorem 7.1.1 $U(A, \epsilon)$ **has an Associated Minkowski Functional**

$$\mathscr{B} = \{A | A \in F \subset X^*, \ A \textit{ is finite}\},$$

where A finite means the subset has only a finite number of elements in it. Then, define a local open base for by defining

$$U(A, \epsilon) = \{x \in X |, |x'(x)| < \epsilon, \}. \quad \mathscr{U} = \{U(A, \epsilon) | A \in \mathscr{B}, \ \epsilon > 0\}$$

The Minkowski functional for $U(A, \epsilon)$ *is*

$$\rho(x) = \frac{1}{\epsilon} \sup_{x' \in A} |x'(x)|$$

Proof 7.1.1

The standard Minkowski functional here is $\rho(x) = \inf\{\alpha > 0 | x \in \alpha\, U(A, \epsilon)\}$. *For convenience, let* $q(x) = \frac{1}{\epsilon} \sup_{x' \in A} |x'(x)|$. *Hence, if* $x \in \alpha U(A, \epsilon)$, $\rho(x) \leq \alpha$. *Also, by definition, we have*

$$x \in \{y || x'(y)| < \epsilon, \ \forall\, x' \in A\}$$

Hence, $x = \alpha y_0$ *with* $y_0 \in U(A, \epsilon)$ *with* $|x'(y_0)| < \epsilon$ *for all* $x' \in A$. *Since* A *is a finite set, we see* $\sup_{x' \in A} |x'(y_0)| < \epsilon$. *This tells us* $\frac{1}{\epsilon} \sup_{x' \in A} |x'(y_0)| < 1$. *Therefore*

$$\sup_{x' \in A} |x'(x)| = \sup_{x' \in A} |x'(\alpha y_0)| = \alpha \sup_{x' \in A} |x'(y_0)| < \alpha\, \epsilon$$

We conclude $\frac{1}{\epsilon} \sup_{x' \in A} |x'(x)| < \alpha$. *But, the choice of* α *is arbitrary, so* $\frac{1}{\epsilon} \sup_{x' \in A} |x'(x)|$ *is a lower bound for* $\{\alpha > 0 | x \in \alpha U(A, \epsilon)\}$. *Hence, by definition,*

$$q(x) = \frac{1}{\epsilon} \sup_{x' \in A} |x'(x)| \leq \rho(x)$$

To show the reverse, given x, *there is a* $\hat{x}' \in A$ *so that* $\sup_{x' \in A} |x'(x)| = \hat{x}'(x)$ *because* A *is a finite set. Now, if* $\hat{x}'(x) = 0$, *this would say* $x'(x) = 0$ *for all members of* A. *Thus,* $x'(\beta x) = 0$ *for all* $x' \in A$ *and* $\beta > 0$. *This tells us* $\beta x \in U(A, \epsilon)$ *for all* $\beta > 0$ *and so* $x \in \frac{1}{\beta} U(A, \epsilon)$ *for all* $\beta > 0$. *This tells us* $q(x) = 0$ *too.*

Then, by our definition of ρ, *we see* $\rho(x) \leq \frac{1}{\beta}$ *for all* $\beta > 0$ *implying* $\rho(x) = 0$. *If* $|\hat{x}'(x)| > 0$, *then for any* $\xi > 0$,

$$\left| x'\left(\frac{\epsilon x}{|\hat{x}'(x)| + \xi} \right) \right| = \frac{\epsilon}{|\hat{x}'(x)| + \xi} |x'(x)| = \epsilon \frac{|x'(x)|}{|\hat{x}'(x)| + \xi} < \epsilon$$

So $\left(\frac{\epsilon x}{|\hat{x}'(x)|+\xi} \right) \in U(A,\epsilon)$. *This tells us* $x \in \frac{2|\hat{x}'(x)|+\xi}{\epsilon} U(A,\epsilon)$. *We can thus say*

$$\rho(x) \leq \frac{|\hat{x}'(x)|+\xi}{\epsilon} = \frac{1}{\epsilon}(\sup_{x' \in A} |x'(x)|)\frac{1}{\epsilon} + \frac{\xi}{\epsilon}$$

But $\xi > 0$ *is arbitrary, so this says*

$$\rho(x) \leq q(x) + \inf_{\xi > 0} \frac{\xi}{\epsilon} = q(x)$$

This completes the argument for the reverse inequality. So $\rho(x) = q(x)$. ∎

The family of Minkowski functionals determined by the $U(A,\epsilon)$ sets are also seminorms as the sets $U(A,\epsilon)$ are balanced. So by Theorem 6.4.5 they determine a locally convex topology too which is the same as the one we constructed before because the topologies are unique. Call this family P. Recall the open base sets are constructed from finite subsets of P, F, given by

$$\mathscr{B} = \{r_1 V(\rho_1) \cap \ldots \cap r_k V(\rho_k) | r_i > 0, \rho_i \in F\}$$

where each $\rho \in P$, $V(\rho) = \{x \in X | \rho(x) < 1\}$. It is straightforward to show $rV_\rho = rU(\{x'\},\epsilon) = U\left(\{x'\}, \frac{\epsilon}{r}\right)$ (we leave that to you in the exercises) and so these two families are just different ways to describe the critical open sets in the base.

Homework

Exercise 7.1.1 *Prove* $0 \in U(A,\epsilon)$, *for any* $\epsilon > 0$ *and* $A \in \mathscr{B}$.

Exercise 7.1.2 *Prove* $\alpha(U,\epsilon) = U(A, |\alpha|\epsilon)$ *for any* $\alpha \neq 0$.

Exercise 7.1.3 *If* $x \in U(A,\epsilon)$, *let* $\delta = \sup_{x' \in A} |x'(x)|$. *We see* $0 \leq \delta < \epsilon$. *Prove* $x + U(A, \epsilon - \delta) \subset U(A,\epsilon)$.

Exercise 7.1.4 *Prove* $U(A,\epsilon/2) + U(A,\epsilon/2) \subset U(A,\epsilon)$.

Exercise 7.1.5 *Prove* $rV_\rho = rU(\{x'\},\epsilon) = U\left(\{x'\}, \frac{\epsilon}{r}\right)$.

Next we need to understand better how the set $F \subset X^*$ determines continuity of $x' \in X^*$.

Theorem 7.1.2 $y' \in span\{x'_1, \ldots, x'_n\}$ **if** y' **is Zero whenever** x'_i **is Zero**

> *Let* $\{x'_1, \ldots, x'_n\}$ *be linearly independent in* X^*. *Then if* $y' \in X^*$ *satisfies* $y'(0) = 0$ *when* $x'_i(0) = 0$, $y' \in span\{x'_1, \ldots, x'_n\}$.

Proof 7.1.2
This is an induction argument. If $n = 1$, *we only have* x'_1. *We assume* $y'(0) = x'_1(0)$. *If* $ker(x'_1) = X$, *then* $y' = 0 \in span\{x'_1\}$. *Otherwise, let* $x_0 \in X \setminus ker(x'_1)$. *Let* $z = x - \frac{x'_1(x)}{x'_1(x_0)} x_0$. *Then*

$$x'_1(z) = x'_1(x) - \frac{x'_1(x)}{x'_1(x_0)} x'_1(x_0) = 0$$

Hence $z \in ker(x'_1)$ and we see any $x \in X$ can be written as $x = z + \frac{x'_1(x)}{x'_1(x_0)}\, x_0$. Then,

$$y'(x) \quad = \quad y'(z) + \frac{x'_1(x)}{x'_1(x_0)}\, y'(x_0)$$

By assumption, $y'(z) = 0$, so $y'(x) = \frac{y'(x_0)}{x'_1(x_0)}\, x'_1(x)$, which shows y' is a multiple of x'_1.

Now assume the results hold for any subset of $n-1$ elements in X^*. Consider a set of n linearly independent elements in X^* given by $\{x'_1, \ldots, x'_{n-1}, x'_n\}$. We assume $y'(0) = 0$ when $x'_i(0) = 0$. If we look at a given x'_k, with $1 \leq k \leq n$, the remaining elements are a subset of $n-1$ linearly independent elements of X^*. If $x'_k(0) = 0$ when $x'_i(0)$ for $i \neq k$, we can apply the induction hypothesis to conclude $x'_k \in span\{x'_i, i \neq k\}$ which contradicts the fact that the full set was linearly independent. So this is not possible. Hence, there is an x_k with $x'_k(x_k) \neq 0$ when $x'_i(x_k) = 0$ for $i \neq k$. We can do this for each $1 \leq k \leq n$. Thus,

$$x'_k\left(\frac{x_k}{sign\; x'_k(x_k)}\right) = 1, \quad x'_i\left(\frac{x_k}{sign\; x'_k(x_k)}\right) = 0, \; i \neq k$$

Let $y_k = \frac{x_k}{sign\; x'_k(x_k)}$ which tells us $x'_i(y_k) = \delta^k_i$. So we have points $\{y_1, \ldots, y_n\}$ with $x'_k(x_k) = 1$. Let $x \in X$. Let

$$y \quad = \quad x - \sum_{k=1}^{n} x'_k(x)\, y_k \Longrightarrow x'_i(y) = x'_i(x) - \sum_{k=1}^{n} x'_k(x_k)\delta^k_i = x'_i(x) - x'_i(x) = 0$$

We see $x'_i(y) = 0$ for all i. Hence, by assumption, $y'(y) = 0$ too. Hence,

$$x = y + \sum_{k=1}^{n} x'_k(x)\, y_k \Longrightarrow y'(x) = y'(y) + \sum_{k=1}^{n} x'_k(x) y'(y_k) = \sum_{k=1}^{n} y'(y_k) x'_k(x)$$

Since x is arbitrary, this shows $y' \in span\{x'_1, \ldots, x'_n\}$. ■

Homework

Exercise 7.1.6 *Let $\{f_1, f_2\}$ be the linear functionals on $\Re^2$ given by $f_1 = \begin{bmatrix} 1 \\ 2 \end{bmatrix}$ and $f_2 = \begin{bmatrix} -1 \\ 3 \end{bmatrix}$. Prove directly Theorem 7.1.2 for this example showing all details.*

Exercise 7.1.7 *Let $\{f_1, f_2, f-3\}$ be the linear functionals on $\Re^3$ given by $f_1 = \begin{bmatrix} 1 \\ 2 \\ -2 \end{bmatrix}$, $f_2 = \begin{bmatrix} -1 \\ 3 \\ 2 \end{bmatrix}$ and $f_3 = \begin{bmatrix} 4 \\ -1 \\ 5 \end{bmatrix}$. Prove directly Theorem 7.1.2 for this example showing all details.*

Exercise 7.1.8 *Let $\{f_1, f_2\}$ be the linear functionals on $C([0,1], \|\cdot\|)$ given by $f_1(x) = \int_0^1 x dx$ and $f_2 = \int_0^1 f dg$ where g is a differentiable function on $[0,1]$ so that $\int_0^1 f dg = \int_0^1 f g' dx$. Prove directly Theorem 7.1.2 for this example showing all details.*

Here is a nice result which has lots of applications. For now, it helps us prove that $y' \in X^*$ is continuous with respect to the topology generated by the seminorms for the sets $U(A, \epsilon)$ based on a subset F of X^* if and only if it is in the $span(F)$. However, that result is non-trivial to prove so we have to work hard to get that result.

Theorem 7.1.3 Continuity Condition for a Linear Map between Two Locally Convex Linear Topological Spaces in Terms of Seminorms

Let $(X, \mathscr{S}(P))$ be a locally convex linear topological space generated by the seminorm family P and $(Y, \mathscr{S}(Q))$ be the generated by the seminorm family Q. Let $T : X \rightarrow Y$ be linear. Then T is continuous on X if and only if for all $q \in Q$, there is a continuous seminorm ρ on X so that $q(Tx) \leq \rho(x)$ for all $x \in X$.

Proof 7.1.3

$\Longrightarrow$: *Let T be continuous on X. Pick $q \in Q$. Let $w = \{y \in Y | q(y) < 1\}$. W is a member of the local base at 0 in Y. Since T is continuous, $T^{-1}(W)$ is open in X. We can assume without loss of generality, that $T^{-1}(W) \neq \emptyset$. For $y \in W$, let $x = T^{-1}(y)$. Then, there is an open set $v \subset X$ containing x having the form*

$$v = r_1 V(\rho_1) \cap \ldots \cap r_n V(\rho_n),\ r_i > 0,\ \rho_i \in P$$

with $T(x) \in W$. Thus, $q(T(x)) < 1$. Define $\rho : X \rightarrow \Re$ by $\rho(x) = \max_{1 \leq k \leq n} \frac{1}{r_k} \rho_k(x)$. Then, ρ is a continuous seminorm on X (we leave this for you to do in an exercise). Let $\hat{x} \in X$ and consider $\rho(\hat{x})$. It is clear this is finite and so there is an $\alpha > 0$ with $\rho(\hat{x}) < \alpha$ implying $\rho(\hat{x}/\alpha) < 1$. So,

$$\begin{aligned} \max_{1 \leq k \leq n} \frac{1}{r_k} \rho_k(\hat{x}/\alpha) &< 1 \Longrightarrow \frac{1}{r_k} \rho_k(\hat{x}/\alpha) < 1,\ 1 \leq k \leq n \\ &\Longrightarrow \rho_k(\hat{x}/\alpha) < r_k,\ 1 \leq k \leq n \Longrightarrow \frac{\hat{x}}{\alpha} \in r_k V(\rho_k),\ 1 \leq k \leq n \end{aligned}$$

Thus, $\frac{\hat{x}}{\alpha} \in V$. So $T(\hat{x}/\alpha) \in W$ and so $q(T(\hat{x}/\alpha)) < 1$. We conclude $q(T(\hat{x})) < \alpha$.

We can do this for all α with $\rho(\hat{x}) < \alpha$. This tells us $q(T(\hat{x})) \leq \rho(\hat{x})$ with $\hat{x}$ arbitrary. This proves the first claim.

$\Longleftarrow$: *Now, we assume for all $q \in Q$, there is a continuous seminorm $\rho : X \rightarrow \Re$ so that $q(T(x)) \leq \rho(x)$ for all $x \in X$. We want to show T is continuous on X. Let W be open in Y containing 0. Fix $q \in Q$ and consider $\{y | q(y) < 1\}$. For this q, there is a seminorm ρ with $q(T(x)) \leq \rho(x)$. Since ρ is continuous, $\rho^{-1}(-\infty 1)$ is open and if $\rho(x) < 1$, $q(T(x)) < 1$ also. This tells us $T(x) \in \{y | q(y) < 1\}$.*

Now if $x_0 \in \{x | \rho(x) < 1\}$, which is an open set containing x_0, then $T(x_0) \in \{y | q(y) < 1\}$. Thus, x_0 is an interior point of $T^{-1}(\{y | q(y) < 1\})$. So $T^{-1}(\{y | q(y) < 1\})$ is open. We see T is continuous on X. ■

Comment 7.1.1 *If $P = \{\| \cdot \|_X\}$ for the normed linear space $(X, \cdot \|_X)$ and $Q = \{\| \cdot \|_Y\}$ for the normed linear space $(Y, \cdot \|_Y)$, we can apply Theorem 7.1.3 to the continuous linear map $T : X \rightarrow Y$. Let $W = \{y | \|y\|_Y < 1\}$. $T^{-1}(W)$ is open as T is continuous. Hence $T^{-1}(W)$ contains $V = \{x | \|x\|_X < r\} = rV(\| \cdot \|_X)$ for a suitably small $r > 0$. We have $V = r\{x | \|x\|_X < 1\}$. Moreover $T(V) \subset W$ and from the proof of Theorem 7.1.3, we know the seminorm ρ from the theorem has the form $\rho(x) = \max_{1 \leq k \leq n}(1/r_k)\rho_k(x)$ for a finite subset of P and $r_k > 0$. But P is a single element, so we have $\rho(x) = (1/r)\|x\|_X$. The theorem then says $q(x) = \|T(x)\|_Y \leq (1/r)\|x\|_X$ for all $x \in X$. We can rewrite this as*

$$\frac{\|T(x)\|_Y}{\|x\|_X} \leq \frac{1}{r},\ x \neq 0 \in X$$

Hence, $\|T\|_{op} < \infty$ and T is bounded.

We already knew this but we now have new tools to derive the same result and the new tools are much more abstract and much more general.

Comment 7.1.2 *Let $(X, \mathscr{S}(P))$ be a locally convex linear topological space generated by the seminorm family P and $(\Re, \mathscr{S}(|\cdot|))$ be the generated by the usual topology in $\Re$ generated by the metric $|\cdot|$. If $x' \in X^*$, then Theorem 7.1.3 tells us x' is continuous with respect to $\mathscr{S}(P)$ if and only if there is a continuous seminorm ρ on X so that $|x'(x)| \leq \rho(x)$ for all $x \in X$. This is the form we wish to use when we study the continuity of linear functionals $x' \in X^*$ with respect to the topologies $\mathscr{S}$ generated by $F \subset X^*$.*

Homework

Exercise 7.1.9 *In the proof of Theorem 7.1.3, prove ρ is a continuous seminorm on X.*

Exercise 7.1.10 *If ρ is the Minkowski functional for $U(\{x'\}, \epsilon)$, prove $rV_\rho = rU(\{x'\}, \epsilon) = U\left(\{x'\}, \frac{\epsilon}{r}\right)$*

Now, at last, the result we have been working for: when is $y' \in X^*$ continuous in the topology generated by $F \subset X^*$.

Theorem 7.1.4 Linear Functionals on X are Continuous in $\mathscr{S}(X, F)$ if and only if x' is in the Span of F

$x' \in X^*$ is continuous in the topology $\mathscr{S}(X, F)$ if and only if $x' \in span(F)$.

Proof 7.1.4
$\Longleftarrow$: *It is clear if $x \in span(F)$, that x is continuous with respect to the topology $\mathscr{S}(X, F)$. Let's reflect on this a bit.*

If $x' \in F$, given an open set in $\Re$ containing 0, V, to show continuity of the sum x' means we can find an open neighborhood of 0, U, in $\mathscr{S}(X, F)$ so that $(x')(U) \subset V$. Since points in V are interior points in the usual sense of a metric on $\Re$, it is sufficient to consider neighborhoods V of the form $(-r, r)$ for any $r > 0$. Note for the open base set $U(\{x'\}, \delta)$, if $x \in U(\{x'\}, \delta)$, $|x'(x)| < \delta$. So if we choose $\delta < r$, we have shown x' is continuous with respect to this topology.

The argument that x' is continuous at any $x_0 \in X$, then uses translation. We use the neighborhood $x_0 + U(\{x'\}, \delta)$ whose image is inside $x'(x_0) + (-r, r)$.

If x' and y' are in F, given an open set in $\Re$ containing 0, V, to show continuity of the sum $x' + y'$ means we can find an open neighborhood of 0, U, in $\mathscr{S}(X, F)$ so that $(x' + y')(U) \subset V$. Note for the open base sets $U(\{x'_1\}, \delta)$, and $U(\{x'_2\}, \delta)$, then $x_1 + x_2$ in $U(\{x'_1\}, \delta) \cap U(\{x'_2\}, \delta)$ satisfies $x'_1(x_1) + x'_2(x_2) < 2r$. Hence, if we choose $\delta < r/2$, we have $(x'_1 + x'_2)(U(\{x'_1\}, \delta) \cap U(\{x'_2\}, \delta))$ is contained in the image $(-r, r)$. The extension to continuity at $x_0 \neq 0$ is then straightforward.

We can then look at an arbitrary element of $span(F)$ and prove continuity in a similar way.

$\Longrightarrow$: *Assume $y' \in X^*$ and y' is continuous with respect to the topology generated by F. We want to show $y' \in span(F)$. Apply Theorem 7.1.3 to y' remembering the comment after that theorem's proof. In the proof of the theorem, we found there were constants $r_i > 0$ and functionals $x'_i \in F$,*

$1 \leq i \leq n$ for some n so the continuous seminorm ρ from the theorem has the form

$$\rho(x) = \max_{1 \leq k \leq n} \frac{1}{r_k} |x'_k(x)|$$

and we know $|y'(x)| \leq \rho(x)$.

Let $A = \{x'_1, \ldots, x'_n\} \subset F$. The set $U(A, \epsilon)$ is open in the topology generated by F for any $\epsilon > 0$. Let q_ϵ be the Minkowski Functional for $U(A, \epsilon)$ which has the form

$$q_\epsilon(x) = \frac{1}{\epsilon} \sup_{x' \in A} |x'(x)|$$

Let $(1/\xi) = \max\{1/r_1, \ldots, 1/r_n\}$. Then

$$q_\xi(x) = \frac{1}{\xi} \max_{1 \leq k \leq n} |x'_k(x)|$$

Now, for each $1 \leq k \leq n$, we have

$$\frac{1}{r_k}|x'_k(x)| \leq \frac{1}{\xi}|x'_k(x)| \leq \frac{1}{\xi} \max_{1 \leq k \leq n} |x'_k(x)| = q_\xi(x)$$

So

$$\rho(x) = \max_{1 \leq k \leq n} \frac{1}{r_k} |x'_k(x)| \leq q_\xi(x) = \frac{1}{\xi} \max_{1 \leq k \leq n} |x'_k(x)|$$

Thus, $|y'(x)| \leq \rho(x) \leq q_\xi(x)$. Hence, $y'(x) = 0$ when $x'_k(x) = 0$ for $1 \leq k \leq n$. By Theorem 7.1.2, we then have $y' \in span\{x'_1, \ldots x'_n\} \subset span(F)$. ∎

Homework

These are a bit difficult but will help you understand these things much better.

Exercise 7.1.11 *Let $F = \{f_1, f_2\}$ be the linear functionals on $X = \Re^2$ given by $f_1 = \begin{bmatrix} 1 \\ 2 \end{bmatrix}$ and $f_2 = \begin{bmatrix} -1 \\ 3 \end{bmatrix}$. Determine the topology $\mathscr{S}(X, F)$ and show directly $f \in (\Re^2)'$ is continuous in this topology if and only if $f \in span(F)$.*

Exercise 7.1.12 *Let $F = \{f_1, f_2, f-3\}$ be the linear functionals on $X = \Re^3$ given by $f_1 = \begin{bmatrix} 1 \\ 2 \\ -2 \end{bmatrix}$, $f_2 = \begin{bmatrix} -1 \\ 3 \\ 2 \end{bmatrix}$ and $f_3 = \begin{bmatrix} 4 \\ -1 \\ 5 \end{bmatrix}$. Determine the topology $\mathscr{S}(X, F)$ and show directly $f \in (\Re^3)'$ is continuous in this topology if and only if $f \in span(F)$.*

Exercise 7.1.13 *Let $F = \{f_1, f_2\}$ be the linear functionals on $X = C([0,1], \|\cdot\|)$ given by $f_1(x) = \int_0^1 x dx$ and $f_2 = \int_0^1 f dg$ where g is a differentiable function on $[0,1]$ so that $\int_0^1 f dg = \int_0^1 f g' dx$. Determine the topology $\mathscr{S}(X, F)$ and show directly $f \in (C([0,1], \|\cdot\|))'$ is continuous in this topology if and only if $f \in span(F)$.*

Exercise 7.1.14 *Let $F = \{f_1, f_2\}$ be the linear functionals on $X = \ell^1$ given by $f_1 = (x_1) \in \ell^\infty$ and $f_2 = (x_2) \in \ell^\infty$. Determine the topology $\mathscr{S}(X, F)$ and show directly $f \in (\ell^1)' \cong \ell^q$ is continuous* *in this topology if and only if* $f \in span(F)$.

Exercise 7.1.15 *Repeat the previous exercise but with arbitrary conjugate indices* p *and* q.

Since these topologies arise frequently, it is helpful to set up some notation for them.

Definition 7.1.1 Topologies Generated by $F \subset X^*$: Locally convex $\mathscr{Y}(X, F)$ and $\mathscr{Y}(X^*, \phi(X))$

If X *is a vector space over* $\Re$ *and* $F \subset X^*$*, the locally convex topological vector space generated by* F *is called* $\mathscr{Y}(X, F)$.

If $F = X^*$*, we get the topology* $\mathscr{Y}(X, X^*)$ *in which every linear functional is continuous.*

Recall, the canonical embedding of X *into* X^{**} *defined by* $\phi : X \to X^{**}$ *gives* $\phi(x)$ *is the element in* X^{**} *defined by*

$$(\phi(x))(x') = x'(x)$$

The set $F = \phi(X) \subset X^{**}$ *and the topology generated by* F *for the vector space* X^* *is* $\mathscr{Y}(X^*, \phi(X))$.

As previously discussed, if X is a normed linear space with norm $\| \cdot \|$, we can talk about the continuity of a linear functional x' in the traditional way: x' is continuous at x_0 if given $\epsilon > 0$, there is a $\delta > 0$ so that $\|x - x_0\| < \delta$ implies $|x'(x) - x'(x_0)| < \epsilon$. We let X', as usual, be the set of continuous linear functionals on X. Let $\mathscr{S}(X, \| \cdot \|)$ be the locally convex topology generated by the norm $\| \cdot \|$ on X and let $\mathscr{S}(\Re, | \cdot |)$ be the locally convex topology generated by the norm $\| \cdot \|$ on $\Re$. So the $\epsilon - \delta$ norm-based notion of continuity is the same as looking at $x' : X \to \Re$ as a mapping from the locally convex linear topological space $(X, \mathscr{S}(X), \| \cdot \|)$ to the locally convex linear topological space $(\Re, \mathscr{S}(\Re), | \cdot |)$. Note we can generate a topology on X using linear functionals from $F \subset X^*$, from families of seminorms on X (the $F \subset X^*$ approach gives a family of seminorms) or from a single norm, in which case the family of seminorms on X reduces to the singleton family $\{\| \cdot \|\}$.

We also have two types of convergence in this setting. A sequence $(x_n) \subset (X, \| \cdot \|)$ converges weakly to $x \in X$ if $x'(x_n) \to x'(x)$ for all continuous linear functionals x'. We use the symbol X' to denote the continuous linear functionals on X and we now know the definition of X' depends on how we choose a topology for X. If a norm on X is present, we define such continuity using the norm on X and the absolute value function on $\Re$. However, we need not make that choice and for some vector spaces X this is not possible. We write this as $x_n \stackrel{weak}{\longrightarrow} x$.

The second type of convergence is weak*. If (x'_n) is a sequence in X', we say (x'_n) converges weak* if there is an element $x' \in X'$ so that $x'_n(x) \to x'(x)$ for all $x \in X$. To understand this, we need to know what the continuous linear functionals on X are, which means we have to choose a topology for X. We write this as $x'_n \stackrel{weak*}{\longrightarrow} x'$.

Once we impose a topology on X' using some mechanism which includes $G \subset X^{**}$, a family of seminorms defined on X' or a norm defined on X', we know what X'' looks like.

First, we can think of this in terms of the canonical embedding of X into X^{**}. The equation $x'_n(x) \to x'(x)$ is the same as $\phi(x)(x'_n) \to \phi(x)(x')$. Since $\phi(x) \in X^{**}$ we can denote $\phi(x) \equiv x'' \in X^{**}$ and rewrite again as $x''(x'_n) \to x''(x')$. The range of ϕ is not necessarily all of X^{**} so we can only say $x''(x'_n) \to x''(x')$ for all $x'' \in \phi(X)$. If X is reflexive, $\phi(X) = X^{**}$ and weak* convergence becomes weak convergence applied to X'.

If we have imposed a topology on X' so that we know X'', all the above still holds. We then still have weak* convergence is the same as weak convergence in X' when X is reflexive.

You can see how confusing this is. We can't discuss these ideas easily unless we know the topologies we will use on the domains X and X^*. Note also, we can think of $x' : X \rightarrow \Re$ as an operator and we know x' is continuous in the norm topology on X if and only if its operator norm $\|x'\|_{op}$ is finite. Hence, we can also think of X' in terms of operator norms and our comment about Theorem 7.1.3 shows how our locally convex topologies on X and $\Re$ lead naturally to the idea that x' has a bounded operator norm if x' is continuous in the locally convex topologies.

If a topological vector space is normable, as you can see, it lets us do analysis in a very familiar setting. So when is a topological vector space normable?

Theorem 7.1.5 Topological Vector Space is Normable if and only There is a Convex Bounded Neighborhood of Zero

A topological vector space $(X, \mathscr{S})$ is normable if and only there is a convex bounded neighborhood of zero in X.

Proof 7.1.5
$\Longrightarrow$: *If X is normable, then there is a norm ρ. The set $\{x|\rho(x) < 1\}$ is a bounded convex neighborhood of 0.*

$\Longleftarrow$: *Let U be a convex bounded neighborhood of 0. There is also a convex balanced neighborhood V of 0 with $V \subset U$. Also, all neighborhoods of 0 are absorbing. Let ρ be the Minkowski functional for V. By Lemma 6.4.1, we know $\rho(0) = 0$, $\rho(x + y) \leq \rho(x) + \rho(y)$ and $\rho(\lambda x) = |\lambda|\rho(x)$. Also, by definition, $\rho(x) \geq 0$. By property (vi) of Lemma 6.4.3, since V is bounded, we know $\rho(x) = 0$ if and only if $x = 0$. This shows ρ is a norm on X.*

Now V is open, so by (ii) of Lemma 6.4.3, we know $V = \{x \in X|\rho(x) < 1\}$. If $\alpha > 0$, $\alpha V = \{x|\rho(x) < \alpha\}$ (you can prove this as an exercise). We will show $\mathscr{V} = \{\alpha V|\alpha > 0\}$ is an open base at 0 for the topology of X. If U is any neighborhood of 0 in the original topology of X, there is a balanced neighborhood of 0, W, with $W \subset U$. Since V is bounded by assumption, there is an $r > 0$ so the $V \subset rW$. Thus, $(1/r)V \subset W \subset U$. This shows $\mathscr{V}$ is an open base at 0. So the topology generated by the norm is compatible with the original topology. ■

Homework

Exercise 7.1.16 *Prove $\alpha(U, \epsilon) = U(A, |\alpha|\epsilon)$ for any $\alpha \neq 0$.*

Exercise 7.1.17 *If $x \in U(A, \epsilon)$, let $\delta = \sup_{x' \in A} |x'(x)|$. We see $0 \leq \delta < \epsilon$. Show $x + U(A, \epsilon - \delta) \subset U(A, \epsilon)$.*

Exercise 7.1.18 *In the proof of Theorem 7.1.5, we need to show $\alpha V = \{x|\rho(x) < \alpha\}$. Prove this result.*

7.2 Locally Convex Topology Examples

Let's work through some examples of spaces with locally convex topologies that are created using a family of norms.

7.2.1 A Locally Convex Topology on Continuous Functions

Consider the vector space $C([a,b])$ of continuous functions on $[a,b]$. Let $P = \{p_t | p_t(f) = |f(t)|, f \in C([a,b]),\ a \le t \le b\}$. It is straightforward to see each $p_t \in P$ is a seminorm (you can prove this as an exercise). Let $f \neq 0$. Then there is a $t_0 \in [a,b]$ so that $f(t_0) \neq 0$. Hence, there is an $r > 0$, so that if $t \in (t_0 - r, t_0 + r) \cap [a,b]$, $f(t) \neq 0$. Hence, $p_t(f) = |f(t)| > 0$ if $t \in (t_0 - r, t_0 + r) \cap [a,b]$. We conclude there are many seminorms p_t satisfying $p_t(f) > 0$. This tells us the family is a separating family of seminorms. Theorem 6.4.5 tells us the family of seminorms P determines a locally convex topology and so turns the vector space $C([a,b])$ into a topological vector space. This topology is determined by the sets $V(p_t) = \{f \in C([a,b]) | p_t(f) < 1\} = \{f \in C([a,b]) || f(t)| < 1\}$ and the associated open base is

$$\mathscr{B} \quad = \quad \{r_1 V(p_{t_1}) \cap \ldots \cap r_k V(p_{t_k}) | r_i > 0,\ k \text{ any positive integer },\ p_{t_i} \in P\}$$

Since the family is separating, we can show the topology is T_1 . For convenience, let's assume $f \neq g$ with both nonzero from $C([a,b])$. Hence, there is a point t_c so that $|f(t) - g(t)| > 0$ on $(t_c - \delta, t_c + \delta)$ for some $\delta > 0$. Assume $g(t_c) > f(t_c)$ for concreteness. Then $f \in |g(t_c)| V(p_{t_c}) = \{h || h(t_c)| < |g(t_c)|\}$ but g is not in that set. Also, $g \in g + (1/2)(g(t_c) - f(t_c)) V(p_{t_c}) = g + \{h || h(t_c)| < (1/2)|g(t_c) - f(t_c)|\}$ but f is not. Hence, for this case, we have shown the T_1 requirement. It is straightforward to verify this is true for other cases of the relationship between $f(t_c)$ and $g(t_c)$ and we will leave that to you.

Since the topology is T_1, singleton sets are closed. By Theorem 5.4.5, we know the topology is Hausdorff and so limits are unique by Theorem 5.1.1. Now if $(x_n) \subset C([a,b])$ and (x_n) converges to $x \in C([a,b])$ is this topology, then given any neighborhood which contains x, all but finitely many of the elements x_n belong to the neighborhood. If W is a neighborhood containing 0, there is a base open set $U = r_1 V(p_{t_1}) \cap \ldots \cap r_k V(p_{t_k})$, for $r_i > 0$ and $p_{t_i} \in P$ for some k for which $x + U \subset W$. Hence, for this neighborhood, there is an N so that $x_n \in x + \cap_{i=1}^k r_i V(p_{t_i})$ if $n > N$. Now, $r_i V(p_{t_i}) = \{f \mid p_{t_i}(f) < r_i\} = \{f \mid |f(t_i)| < r_i\}$. Thus, for $n > N$, $x_n - x \in r_i V(p_{t_i})$ which tells us $|x_k(t_i) - x(t_i)| < r_i$ for $n > N$ and $1 \le i \le k$.

Now fix $\epsilon > 0$ and $t \in [a,b]$. From the above discussion, for any choice of neighborhood $W = x + \epsilon V(p_t)$, the role of U is played by $\epsilon V(p_t)$. Hence, for this t, there is an N_t so that for $n > N_t$, $x_n - x \in \epsilon V(p_t)$ which tells us $|x_n(t) - x(t)| < \epsilon$ for $n > N_t$. This says that $x_n \to x$ is this topology means $x_n(t) \to x(t)$ for each $t \in [a,b]$. In classical analysis, this is called the pointwise convergence of the sequence (x_n) to x in $C([a,b])$.

Now define $\phi : C([a,b]) \to \Re$ by $\phi(t) = f(t)$ which is the standard evaluation map of a function f. Is ϕ continuous in the locally convex topology we just built for $C([a,b])$?

Pick $f_0 \in C([a,b])$ and $\epsilon > 0$. Let $U = \{f \in C([a,b]) \mid p_t(f) < \epsilon\} = \epsilon V(p_t)$ which is therefore a basic open set at 0. We see $f_0 + U = f_0 + \epsilon V(p_t)$. Consider for $f \in C([a,b])$ with $|f(t) - f_0(t)| < \epsilon$. This is the same as saying $p_t(f - f_0) < \epsilon$ or $f - f_0 \in \epsilon V(p_t)$. We conclude $f \in f_0 + \epsilon V(p_t) = f_0 + U$.

Conversely, if $f \in f_0 + U$, we can easily see $|f(t) - f_0(t)| < \epsilon$. Thus, we conclude $\{f || f(t) - f_o(t)| < \epsilon\} = f_0 + U$. Hence, given $\epsilon > 0$, $|\phi(f) - \phi(f_0)| = |f(t) - f_0(t)| < \epsilon$ if $f - f_0 \in \epsilon V(p_t)$. This says, ϕ is continuous at f_0. Since f_0 is arbitrary, ϕ is continuous in this topology on $C([a,b])$. So this topology generated by the seminorm family P is a topology in which all mappings $\phi_t : C([a,b]) \to \Re$ given by $\phi_t(f) = f(t)$ are continuous.

If such a map ϕ_t is continuous with respect to another topology, $\mathscr{S}_2$ on $C([a,b])$, then consider $\{f \in C([a,b]) \mid |f(t)|, \epsilon\} = (\phi_t)^{-1}(-\epsilon, \epsilon)$. Note,

$$
\begin{aligned}
(\phi_t)^{-1}(-\epsilon,\epsilon) &= \{f \in C([a,b]) \mid \phi_t(f) \in (-\epsilon,\epsilon)\} = \{f \in C([a,b]) \mid -\epsilon < f(t) < \epsilon\} \\
&= \{f \in C([a,b]) \mid |f(t)| < \epsilon\} = \epsilon V(p_t)
\end{aligned}
$$

If we assume ϕ_t is continuous with respect to $\mathscr{S}_2$, then $(\phi_t)^{-1}(-\epsilon,\epsilon) = \epsilon V(p_t)$ must be open in $\mathscr{S}_2$. This implies $\epsilon V(p_t)$ is open in $\mathscr{S}_2$. Hence, the open sets generated by P must be in $\mathscr{S}_2$.

Comment 7.2.1 *Now do this all again but in the vector space of bounded functions, $B([a,b])$. Use the same seminorm family. Note, we build a locally convex topology for $B([a,b])$ as before and it is still true that the seminorm family is separating and so the topology is Hausdorff. Convergence in this topology is still what we used to call pointwise convergence.*

Comment 7.2.2 *This leads to an important observation. All topologies that make ϕ_t continuous must contain the topology generated by P. We therefore say the topology generated by P is the* **weakest** *topology on $C([a,b])$ which makes all ϕ_t (the evaluation maps) continuous. This topology is called the* **weak topology** *of the space $C([a,b])$.*

Let's go back and look at this classically. Here is a famous example of pointwise convergence of continuous functions where the limit function is not continuous.

Consider the sequence of functions (x_n) on $[0,1]$ by $x_n(t) = t^n$ for all integers $n \geq 0$. Note $(x_n) \subset C([0,1])$. For each fixed t, does the sequence of real numbers $(x_n(t))$ converge? If $0 \leq t < 1$, $x_n(t) = t^n \to 0$. However, at $t = 1$, $x_n(1) = 1^n = 1$ for all n. Hence, we have

$$
\lim_{n\to\infty} x_n(t) = \begin{cases} 0, & 0 \leq t < 1 \\ 1, & t = 1 \end{cases}
$$

Thus, this sequence of functions converges to the function

$$
f(t) = \begin{cases} 0, & 0 \leq t < 1 \\ 1, & t = 1 \end{cases}
$$

which is clearly not continuous at $t = 1$ even though each of the x_n is continuous on $[0,1]$. The locally convex topology we put on $C([0,1])$ using the seminorm family P does not apply to this case as the limit function is not continuous. Our interpretation of convergence of a sequence of functions in $C([0,1])$ with the topology generated by P does imply pointwise convergence but only if the limit is known to be continuous.

However, we can build this topology on $B([0,1])$ and still get that convergence with respect to the topology generated by P is pointwise convergence. Note, we can find examples of sequences of continuous functions that converge pointwise to a continuous function. Let (x_n) be the sequence of functions on $[0,1]$ defined by $x_n(t) = \frac{2nt}{e^{nt^2}}$. This sequence is in $C([0,1])$. First, it is easy to see the pointwise limit function is $x(t) = 0$ on $[0,1]$. The derivative is $x_n'(t) = \frac{2n(1-2nt^2)}{e^{nt^2}}$ which is zero at $t = 0$ and $t = \pm 1/\sqrt{2n}$. The critical point at $t = 0$ is uninteresting and the maximum occurs at $t_n = 1/\sqrt{2n}$ and has value $\sqrt{2n/e}$. We have

$$
||x_n - x||_\infty = \sup_{0 \leq t \leq 1} |2nt/e^{nt^2} - 0| = \sqrt{2n/e}
$$

For the convergence to be uniform, given $\epsilon > 0$, we would have to be able to find N_ϵ so that $n > N_\epsilon$ implies $\|x_n - x\|_\infty < \epsilon$. Here that means we want $\sqrt{2n/e} < \epsilon$ when $n > N_\epsilon$. But for $n > N_\epsilon$,

$\sqrt{2n/e} \to \infty$. So this cannot be satisfied and the convergence is not uniform. The convergence of this sequence in the topology generated by P on $B([0,1])$ is the same as the pointwise convergence we discuss here.

The point here is that we now have new tools to look at convergence of sequences of functions in new ways! Relish that as the ability to find new insight is a powerful gift.
Homework

Exercise 7.2.1 *Prove each p_t as defined above is a seminorm on $C([a,b])$.*

Exercise 7.2.2 *Discuss carefully the details of building this topology on $B([0,1])$.*

An obvious question to ask next is if we let

$$X = \{f : T \to \Re |\text{ each function } f \text{ satisfies some property }\},$$

can we repeat this construction process?

7.2.2 A Locally Convex Topology on All Sequences

Let $T = \mathbb{N}$, the counting numbers and let $X = \{f : \mathbb{N} \to \Re\}$, which is just the set of all real-valued sequences. The points $t \in [a,b]$ are now the integers $n \in \mathbb{N}$ and the functions are of the form $f = (x_i)_{1 \leq i < \infty}$. Define $p_n(f) = |f(n)| = |x_n|$ and let P be the set of all such mappings. These mappings are seminorms and form a separating family. Hence, the topology generated by P is a T_1 locally convex topology which is Hausdorff. Hence, convergent sequences have unique limits. The arguments above then show that if the sequence (f_n) converges in this topology to f, this means pointwise convergence. Here, that means convergence at each index n. Thus $f_n \to f$ in the topology generated by P implies $f_n(k) \to f(k)$ for each k. Using a more familiar labeling, this means $(x_{kn}) \to (x_n)$ if $x_{kn} \to x_k$ for each k.

Here are the details. The sequence (f_n) converges to the sequence f in the topology generated by P means that given any neighborhood which contains f, all but finitely many of the elements f_n belong to the neighborhood. If W is a neighborhood containing 0, there is a base open set $U = r_1 V(p_1) \cap \ldots \cap r_k V(p_k)$, for $r_i > 0$ and $p_i \in P$ for some k for which $x + U \subset W$. Hence, for this neighborhood, there is an N so that $f_n \in f + \cap_{i=1}^k r_i V(p_i)$ if $n > N$. Now, $r_i V(p_i) = \{g \mid p_i(g) < r_i\} = \{g \mid |g(n)| < r_i\}$. Now let $f_n = (x_{nj})$ and $f = (x_j)$. Thus, for $n > N$, $(x_{nj}) - (x_j) \in r_i V(p_i)$ which tells us $|x_{ni} - x_i| < r_i$ for $n > N$ and $1 \leq i \leq k$.

Now fix $\epsilon > 0$ and n_0. From the above discussion, for any choice of neighborhood $W = f + \epsilon V(p_{n_0})$, the role of U is played by $\epsilon V(p_{n_0})$. Hence, for this n_0, there is an N_0 so that for $n > N_0$, $x_{nj} - x_j \in \epsilon V(p_{n_0})$ which tells us $|x_{n,n_0} - x_{n_0}| < \epsilon$ for $n > N_0$. This says that $x_{n,k} \to x_k$ is this topology means $x_{n,n_0} \to x_{n_0}$ for each n_0. If we define $\phi_n : X \to \Re$ by $\phi_n((x_k)) = x_k$, the evaluation map of the sequence (x_k), we see, by our remarks above, that ϕ_n is continuous in the topology generated by P and P is the weakest topology in which these evaluation maps are continuous.

Homework

Exercise 7.2.3 *What do the base open sets $U = r_1 V(p_1) \cap \ldots \cap r_k V(p_k)$, for $r_i > 0$ look like here?*

Exercise 7.2.4 *Expand our discussion above on the construction of the topology as we were brief in our remarks. Make sure you thoroughly understand the details.*

7.3 Total Sets and Weak Convergence

There is another idea which is useful.

Definition 7.3.1 Total Sets of Linear Functionals

Let X be a vector space. A set $F \subset X^*$ is a total set if for all $x \neq 0$ in X, there is an $x' \in F$ so that $f(x) \neq 0$.

Comment 7.3.1 *Earlier, we used a similar idea for a family of seminorms, $\mathscr{P}$. The family $\mathscr{P}$ is called a* **separating** *family of seminorms if for every nonzero $x \in V$ there is a $\rho \in \mathscr{P}$ so that $\rho(x) \neq 0$. Now for a given $x' \in X^*$, the mapping ρ defined by $\rho(x) = |x'(x)|$ is a seminorm as $\rho(x) \geq 0$, $\rho(x+y) \leq \rho(x) + \rho(y)$ and $\rho(\alpha x) = |\alpha|\rho(x)$. Hence, we are extending the idea of a separating family of seminorms to a family of linear functionals.*

Comment 7.3.2 *Note, if we knew $x'(f) = 0$ for all $x' \in F$ implied $x = 0$, this is the same as saying if $x \neq 0$, there is at least one x' with $x'(x) \neq 0$.*

Now consider the topological vector space $(X, \mathscr{S}(X, F))$ which is generated by the family F.

Theorem 7.3.1 Topology $\mathscr{S}(X, F)$ is T_1 if and only if F is a Total Subset

The topology $\mathscr{S}(X, F)$ is T_1 if and only if F is a total subset.

Proof 7.3.1
The local open base for $\mathscr{S}(X, F)$ uses the sets

$$\begin{aligned} U(A, \epsilon) &= \{x \in X|, |x'(x)| < \epsilon, \} \\ \mathscr{U} &= \{U(A, \epsilon) | A \in \mathscr{B}, \ \epsilon > 0\} \end{aligned}$$

where

$$\mathscr{B} = \{A | A \in F \subset X^*, \ A \text{ is finite}\}.$$

Now if F is a total subset, then the family $(|x'|)$ for $x' \in F$ is a separating family of seminorms and so the topology is T_1.

On the other hand, if $\mathscr{S}(X, F)$, then for any $x \neq 0$, there is an open set containing 0 and not containing x. Hence, there is an $x' \in F$ so that $0 \in U(x', \epsilon)$ and $x \in (U(x', \epsilon))^C$ for some $\epsilon > 0$. Thus, $|x'(x)| \geq \epsilon > 0$. We see there is an $x' \in F$ with $x'(x) \neq 0$. We conclude F is a total subset of X^*. ■

If F is a total subset, the topology $\mathscr{S}(X, F)$ is T_1 which implies it is Hausdorff. Thus, if a sequence $(x_n) \subset X$ converges to $x \in X$ is this topology, the limit is unique. As before, if $F = X^*$, we would have the topology $\mathscr{S}(X, X^*)$ in which all linear functionals are continuous. Recall, sometimes we would refer to this topology generated by F as $\mathscr{Y}(X, F)$. Thus, when $F = X^*$, we get the topology $\mathscr{Y}(X, X^*)$ in which every linear functional is continuous. In general, for a given F, we say (x_n) is F-weakly convergent to $x \in X$ if $x_n \to x$ in the topology $\mathscr{S}(X, F) = \mathscr{Y}(X, F)$. This means any open set containing x contains an infinite number of x_n. Then, as we have said, if F is total, we know more: the limit is unique.

We have already talked about the canonical embedding of X into X^{**}. We can connect a few dots now.

Theorem 7.3.2 Image of the Canonical Embedding of X into X^{**} is a Total Set

Let $\phi : X \to X^{**}$ be the canonical embedding defined by $(\phi(x))(x') = x'(x)$. Let $F = \phi(X)$. Then F is a total subset of X^{**}.

Proof 7.3.2
Let $x' \in X^*$ be nonzero. Then, there is an $x_0 \in X$ so that $x'(x_0) \neq 0$. But this is the same as saying $(\phi(x_0))(x') \neq 0$. Thus, F is a total subset of X^{**}. ■

Comment 7.3.3 *Since $\phi(X)$ is a total subset in X^{**}, the topology $(X^*, \mathscr{S}(X^*), \phi(X))$ is T_1 and so Hausdorff. We also call this topology $\mathscr{Y}(X^*, \phi(X))$. Hence, sequences that converge in this topology on X^* have unique limits. We usually abuse notation and denote this topology as $\mathscr{S}(X^*, X)$ where it is understood that X in that notation refers to the image $\phi(X)$. So lots of room for confusion.*

Also, note the use of X^* and X' is sort of moot. Once we have established the topology $\mathscr{S}(X, F)$, all linear functionals in the span of F are continuous. Thus, if the span of F is X^*, all elements of X^* are continuous and so X^* and X' are the same. We can make similar comments about X^{**} versus X''.

Homework

Exercise 7.3.1 *If F is the set of linear functionals in $(C([0, 1], \|\cdot\|_\infty))'$ defined by $f \in F$ means $f(x) = \int_0^1 f dp$ where p is any polynomial on $[0, 1]$. Prove F is a total set.*

Exercise 7.3.2 *If F is the set of linear functionals in $(C([0, \pi], \|\cdot\|_\infty))'$ defined by $f \in F$ means $f(x) = \int_0^1 f dg$ where g is in the span of the functions $\{1, \sin(x), \sin(2x), \ldots\}$ on $[0, \pi]$. Prove F is a total set.*

Chapter 8

Deeper Results on Linear Functionals

We already know, from (Peterson (100) 2020), that $\|x\| = \sup_{x' \in X', \|x'\|=1} |x'(x)|$. This fact leads to some interesting observations. Maybe we do not need to use all of X' to get this result. Hence, to identify what subsets of X' might be interesting, we define some new quantities.

Definition 8.0.1 Norm Determining Sets of Linear Functionals

If $(X, \|\cdot\|)$ is a normed linear space and X' is the set of continuous linear functionals on X with respect to the norm topology, let $M \subset X'$. Define

$$\begin{aligned} p_M(x) &= \sup_{x' \in M,\ \|x'\|=1} |x'(x)| \\ V(M) &= \inf_{x \in X,\ \|x\|=1} p_M(x) \end{aligned}$$

The value $V(M)$ is called the characteristic of M and we say M is norm determining for X is $V(M) > 0$.

Why do we call M norm-determining here? It is clear

$$p_M(x) \leq \sup_{x' \in X',\ \|x'\|=1} |x'(x)| = \|x\| \Longrightarrow 0 \leq V(M) \leq 1$$

It is straightforward to check that if M is norm-determining, p_M is a norm on X.

- **N1:** $p_M(x) \geq 0$, for all $x \in X$.
- **N2:** If $p_M(x) = 0$ when $x \neq 0$, that would imply $V(M) = 0$ which is not possible as M is norm-determining.
- **N3:** $p_M(\alpha x) = |\alpha| p_M(x)$, for all scalars α and $x \in X$.
- **N4:** $p_M(x + y) \leq p_M(x) + p_M(y)$, for all x and y in X.

Since $p_M(x) \leq \|x\|$, we see the topology determined by p_M is the same as the one determined by $\|\cdot\|$. Hence, we say M is norm determining because it determines a norm compatible with the usual one.

Hence,

$$\|x\| = \sup_{x' \in X', \|x'\|=1} |x'(x)| = p_{X'}$$

$$\Longrightarrow \quad V(X') = \inf_{x \in X,\ \|x\|=1} p_M(X') = \inf_{x \in X,\ \|x\|=1} \|x\| = 1$$

so we see the Hahn - Banach Theorem has allowed us to see that X' is norm determining for X.

Homework

We already know the answers to these exercises, but they are a good way to understand the new definitions.

Exercise 8.0.1 *Let* $M = \{f \in (\Re^2)' : \|f\|_{op} = 1\}$.

- *Show* $M = \left\{ \begin{bmatrix} \cos(\theta) \\ \sin(\theta) \end{bmatrix} : 0 \leq \theta \leq 2\pi \right\}$.
- *Show* $p_M(x, y) = \sqrt{x^2 + y^2}$ *and* $V_M = 1$ *and so* M *is norm determining.*

Exercise 8.0.2 *Let* $M = \{f \in (\Re^3)' : \|f\|_{op} = 1\}$.

- *Show* $M = \left\{ \begin{bmatrix} u_1 \\ u_2 \\ u_3 \end{bmatrix} : u_1^2 + u_2^2 + u_3^2 = 1 \right\}$.
- *Find* $p_M(x, y)$ *and* V_M *and decide if* M *is norm determining.*

Exercise 8.0.3 *Let* $M = \{f \in (\ell^1)' : \|f\|_{op} = 1\}$.

- *Show* $M = \{y \in \ell^\infty : \|y\|_\infty = 1\}$.
- *Find* $p_M(x, y)$ *and* V_M *and decide if* M *is norm determining.*

8.1 Closed Operators and Normed Linear Spaces

In Chapter 2, we have discussed first and second category sets in a metric space setting. Recall

Definition 8.1.1 First Category Sets of a Metric Space

Let (X, d) *be a metric space. We say* X *is of* **first category** *if there is a sequence* (A_n) *of subsets of* X *which are all nowhere dense so that* $X = \cup_{n=1}^{\infty} A_n$. *If* X *is not of first category, we say* X *is of* **second category**.

A most important result is then:

Theorem The Baire Category Theorem in Complete Metric Spaces

Let (X, d) *be a nonempty complete metric space. Then* X *is of* **second category**.

We also need to look more closely at the graph of a mapping. There are several classical results that are relevant which we mention here.

Definition 8.1.2 Open Mappings in Metric Spaces

Let (X, d_X) *and* (Y, d_Y) *be metric spaces and* : $T : D(T) \subset X \rightarrow Y$ *be a mapping. We say* T *is an* **open mapping** *if* $T(U)$ *is an open set in* Y *for all open sets* U *in* X.

A standard result is the open ball lemma whose proof uses the Baire Category Theorem.

Lemma 8.1.1 Open Unit Ball Lemma in Complete Normed Linear Spaces

Let $T : X \rightarrow Y$ be a mapping between two complete normed linear spaces where $\|\cdot\|_X$ is the norm on X and $\|\cdot\|_Y$ on Y. Let T be a bounded, linear and onto operator. Let $B_0 = B(\mathbf{0}; 1) \subset X$. Then $T(B_0)$ contains an open ball $B(\mathbf{0}; r) \subset Y$ for some $r > 0$.

Proof 8.1.1
This is done in (Peterson (100) 2020). ∎

Here, an open set is defined like usual in a metric space: each point p in an open set must be an interior point. We want to prove that bounded linear operators from X to Y must be open mappings. Of course, the requirement that T is linear means we must let X and Y be normed linear spaces whose metrics are induced by a norm. Note if T is also $1-1$, the inverse T^{-1} exists and $T^{-1} : T(dom(T)) \subset Y \rightarrow X$ where $T(dom(T))$ is the range of T. Let $U \subset dom(T)$ be an open set. Then for $g = t^{-1}$ to be continuous, $g^{-1}(U)$ must be open in Y. But $g^{-1} = (T^{-1})^{-1} = T$. So if T is an open mapping, it will make T^{-1} a continuous mapping.

Using this lemma, we can prove the Open Mapping Theorem (OMT) in complete normed linear spaces.

Theorem 8.1.2 Open Mapping Theorem in Complete Normed Linear Spaces

Let $T : X \rightarrow Y$ be a mapping between two complete normed linear spaces where $\|\cdot\|_X$ is the norm on X and $\|\cdot\|_Y$ on Y. Let T be a bounded, linear and onto operator. Let $U \subset X$ be open. Then $T(U)$ is open in Y.

Proof 8.1.2
This is also done in (Peterson (100) 2020) and we encourage you to go back and review the details of the arguments. ∎

We then define what we mean by the graph of a mapping: this is the usual thing. If $y = f(x)$ was a real-valued mapping, the set of ordered pairs $(x, f(x))$ for all x in the domain of f is what we call the graph of f. We can start with the setting of normed linear spaces.

Definition 8.1.3 Graph of a Mapping between Normed Linear Spaces

Let $(X, \|\cdot\|_X)$ and $(Y, \|\cdot\|_Y)$ be normed linear spaces and $T : D \subset X \rightarrow T(D) \subset Y$ be a given mapping. The set $G_T = \{(x, T(x)) | x \in D\}$ is called the graph of T.

For such a mapping, it is important to know if its graph is closed in the product topology of $X \times Y$. Remember, in the simple case of a function like $f(x) = x^2$ on $[-1, 2]$, the graph here is the set of pairs $(x, f(x))$ where $x \in [-1, 2]$ and $f(x) \in \Re$. The product topology in this case is the product topology of $[-1.2] \times \Re$ where the topology on $[-1, 2]$ is the usual relative topology.

Definition 8.1.4 Closed Mappings between Normed Linear Spaces

Let $(X, \|\cdot\|_X)$ and $(Y, \|\cdot\|_Y)$ be normed linear spaces and $T : D \subset X \rightarrow T(D) \subset Y$ be a given mapping. The mapping T is called a closed mapping if its graph $G_T = \{(x, T(x)) | x \in D\}$ is closed in the product topology of $X \times Y$. Here closure is interpreted in terms of the norm $\|\cdots\|$ on $X \times Y$ defined by $\|(\boldsymbol{x}, \boldsymbol{y})\| = \|\boldsymbol{x}\|_X + \|\boldsymbol{y}\|_Y$. Of course, T could be linear or nonlinear.

We also need to understand convergence in the product topology here.

Theorem 8.1.3 Completeness of the Product Norm

Let $(X, \|\cdot\|_X)$ and $(Y, \|\cdot\|_Y)$ be complete normed linear spaces. The $X \times Y$ is complete with respect to the norm $\|(\cdot,\cdot)\| = \cdot\|_X + \|\cdot\|_Y$.

Proof 8.1.3
Although the proof is in (Peterson (100) 2020), let's repeat it here so we can gain some familiarity with the product topology.

The product topology is the topology induced by the product norm $\|\boldsymbol{x}\|_X + \|\boldsymbol{y}\|_Y$ on $(\boldsymbol{x}, \boldsymbol{y}) \in X \times Y$. It is straightforward to show this is indeed a norm and we let you do that in an exercise. Let $(\boldsymbol{x}_n, \boldsymbol{y}_n) \subset X \times Y$ be a Cauchy sequence. Then given $\epsilon > 0$, there is N so that

$$n, m > N \implies \|(\boldsymbol{x}_n, \boldsymbol{y}_n) - (\boldsymbol{x}_n, \boldsymbol{y}_n)\| < \epsilon$$

Hence,

$$n, m > N \implies \|\boldsymbol{x}_n - \boldsymbol{x}_m\|_X + \|\boldsymbol{y}_n - \boldsymbol{y}_m\|_Y < \epsilon$$

This immediately tells us $(\boldsymbol{x}_n)$ is a Cauchy sequence in X and $(\boldsymbol{y}_n)$ is a Cauchy sequence in Y. Thus, there is a pair $(\boldsymbol{x}, \boldsymbol{y})$ so that $\boldsymbol{x}_n \to \boldsymbol{x}$ in $\|\cdot\|_X$ and $\boldsymbol{y}_n \to \boldsymbol{y}$ in $\|\cdot\|_Y$; i.e. $(\boldsymbol{x}_n, \boldsymbol{y}_n) \to (\boldsymbol{x}, \boldsymbol{y})$ in $\|(\cdot,\cdot)\| = \cdot\|_X + \|\cdot\|_Y$ norm on $X \times Y$. This shows $X \times Y$ is complete with respect to this norm.
■

The standard result in this setting is then:

Theorem 8.1.4 Closed Graph Theorem in Complete Normed Linear Spaces

Let $(X, \|\cdot\|_X)$ and $(Y, \|\cdot\|_Y)$ be complete normed linear spaces. Let $T : dom(T) \subset X \to Y$ be a closed linear operator. Then if $dom(T)$ is closed, T is bounded; i.e. T is continuous.

Proof 8.1.4
This proof is in (Peterson (100) 2020). ■

Homework

Exercise 8.1.1 *Go through all the details of the proof of Theorem 8.1.3 in the case $X = Y = \Re^2$.*

Exercise 8.1.2 *Go through all the details of the proof of Theorem 8.1.3 in the case $X = Y = \Re^3$.*

Exercise 8.1.3 *Go through all the details of the proof of Theorem 8.1.3 in the case $X = \ell^1$ and $Y = \ell^\infty$.*

Exercise 8.1.4 *Go through all the details of the proof of Theorem 8.1.3 in the case $X = \Re^n$ and $Y = C([0,1], \|\cdot\|_\infty)$*

Exercise 8.1.5 *Go through all the details of the proof of Theorem 8.1.3 in the case $X = \Re^n$, $Y = \ell^2$ and $Z = \ell^2$.*

Exercise 8.1.6 *Go through all the details of the proof of Theorem 8.1.3 in the case $X = \Re^n$, $Y = \ell^1$ and $Z = \ell^\infty$.*

This leads to a nice characterization of closed linear operators.

Theorem 8.1.5 Characterization of Closed Linear Operators in Complete Normed Linear Spaces

Let X and Y be normed linear spaces with norms $|$ cdot$\|_X$ and $\| \cdot \|_Y$ respectively. Let $T : D(t) \subset X \to Y$ be linear. Then

$$T \text{ is closed} \iff \left\{ \begin{array}{l} (\boldsymbol{x_n}) \subset D(T),\ (T(\boldsymbol{x_n})) \subset Y \\ \text{with } \boldsymbol{x_n} \to \boldsymbol{x},\ T(\boldsymbol{x_n}) \to \boldsymbol{y} \\ \Longrightarrow \boldsymbol{x} \in D(T),\ T(\boldsymbol{x}) = \boldsymbol{y} \end{array} \right\}$$

Proof 8.1.5
This proof is in (Peterson (100) 2020) also. ■

There is also a strong connection between closed domains and closed operators.

Lemma 8.1.6 Connections between a Closed Domain and a Closed Operator in Complete Normed Linear Spaces

Let X and Y be normed linear spaces with norms$| \cdot \|_X$ and $\| \cdot \|_Y$ respectively. Let $T : D(T) \subset X \to Y$ be linear and bounded, i.e. continuous. Then:

1. *If $D(T)$ is closed, T is a closed operator.*
2. *If T is closed and Y complete, then $D(T)$ is closed.*

Proof 8.1.6
See (Peterson (100) 2020) for the argument. ■

Homework

Exercise 8.1.7 *Let $X = Y = C([0,1], \| \cdot \|_\infty)$. Let D be the set of all $x \in X$ with a continuous derivative on $[0,1]$. Define $T : D \to Y$ by $T(x) = x'$.*

- *Prove T is linear but not continuous.*
- *Prove T is a closed operator. This uses ideas about the interchange of differentiation and sequence limits we discussed in (Peterson (99) 2020).*

Exercise 8.1.8 *Let $X = Y = \mathscr{L}_2([0,1])$ be the Hilbert space of Lebesgue square integrable functions. Let D be the set of all $x \in X$ that are absolutely continuous on $[0,1]$ with $x(0) = x(1)$ and $x' \in X$. If $x \in D$, this means $x(t) = c + \int_0^t u(s)ds$ with $\int_0^1 u(s)ds = 0$ for some $u \in X$. Define $T : D \to Y$ by $T(x) = x'$.*

- *Prove T is linear but not continuous.*
- *Prove T is a closed operator.*
  - *Let $(x_n) \subset D$, $x, y \in X$ with $x_n \to x$ and $T(x_n) \to y$. Show $\int_0^t x_n'(s)ds \to \int_0^t y(t)dt$ uniformly.*
  - *Show $\int_0^1 y(t)dt = 0$.*
  - *Show $(x_n(0))$ is a Cauchy sequence with limit c for some scalar c.*
  - *Show (x_n) converges uniformly to a function z with $z(t) = c + \int_0^t y(s)ds$ and so $z \in D$.*

– *Show* $\int_0^1 |x_n(t) - z(t)|^2 dt \to 0$; *i.e.* $x_n \to z$ *in* X. *Thus,* $z = x$ *implying* $T(x)$ *and so* T *is closed.*

8.2 Closed Operators and Topological Spaces

We can extend these ideas of topological spaces.

Definition 8.2.1 Graph of a Mapping between Topological Spaces

Let $(X, \mathscr{S}_X)$ *and* $(Y, \mathscr{S}_Y)$ *be topological spaces and* $T : D \subset X \to T(D) \subset Y$ *be a given mapping. The set* $G_T = \{(x, T(x))|x \in D\}$ *is called the graph of* f.

For such a mapping, it is important to know if its graph is closed in the product topology of $X \times Y$.

Definition 8.2.2 Closed Mappings between Topological Spaces

Let $(X, \mathscr{S}_X)$ *and* $(Y, \mathscr{S}_Y)$ *be topological spaces and* $T : D \subset X \to T(D) \subset Y$ *be a given mapping. The mapping* T *is called a closed mapping if its graph* $G_T = \{(x, T(x))|x \in D\}$ *is closed in the product topology of* $X \times Y$.

We can then prove the following result.

Theorem 8.2.1 Characterization of a Closed Mapping on Topological Spaces

Let $(X, \mathscr{S}_X)$ *be a topological vector space and* $(Y, \mathscr{S}_Y)$ *be a Hausdorff space. Assume* $f : D \subset X \to f(D) \subset Y$. *Then, if* f *is continuous and* D *is closed in* X, *then* f *is a closed mapping.*

Proof 8.2.1

If f *is not closed, then* G_f^C *is not open in* $X \times Y$. *Thus, there is* $(x_0, y_0) \in G_f^C$, *so that every neighborhood of* (x_0, y_0) *contains points of* G_f. *Hence, every neighborhood of* x_0 *in* X *contains a point in* D. *This says* $x_0 \in \overline{D}$. *But, we assumed* D *was closed, so we now know* $x_0 \in D$. *However,* $(x_0, y_0) \in G_f^C$ *and so* $y_0 \neq f(x_0)$.

Now Y *is Hausdorff, so there are disjoint neighborhoods* V_1 *of* y_0 *and* V_2 *of* $f(x_0)$. *We also know* f *is continuous at* x_0 *so there is a neighborhood* U *of* x_0 *so that* $x \in D \cap U$ *implies* $f(x) \in V_2$. *Now* $U \times V_1$ *is a neighborhood of* (x_0, y_0) *and must contain a point* $(x_1, f(x_1)) \in G_f$; *i.e.* $x_1 \in U$ *and* $f(x_1) \in V_1$. *We see* $x_1 \in D \cap U$ *and so* $f(x_1) \in V_2$. *But* $f(x_1) \in V_1$ *also. However, we know* V_1 *and* V_2 *are disjoint, so this is a contradiction. Thus,* G_f^C *is open and so* G_f *is closed. We conclude* f *must be closed.* ■

This can then be applied to the inverse of a mapping.

Theorem 8.2.2 Characterization of when the Inverse of a Mapping between Topological Spaces is Closed

Let $(X, \mathscr{S}_X)$ *be a topological vector space and* $(Y, \mathscr{S}_Y)$ *be a Hausdorff space. Assume* $f : D \subset X \to f(D) \subset Y$. *Then, if* f *is closed and* f^{-1} *exists,* f^{-1} *is a closed mapping.*

Proof 8.2.2

Let $(y_n, f^{-1}(x_n)) \subset G_{f^{-1}}$ *with* $(y_n, f^{-1}(y_n)) \to (y, x) \in G_{f^{-1}}$. *This defines a sequence* $(x_n) \subset D$ *so that* $y_n = f(x_n)$ *and so we have* $f(x_n) \to y$ *in* Y *and* $x_n \to x$ *is* X. *Since* f *is closed, this tells us* $y = f(x)$ *or* $x = f^{-1}(y)$*, Hence,* $(y, x) = (y, f^{-1}(y)) \in G_{f^{-1}}$*. This shows* f^{-1} *is closed.* ■

We also need conditions that ensure the inverse mapping is closed.

Theorem 8.2.3 If $f : D \subset X \to Y$ **is Closed and Invertible, then** f^{-1} **is Closed**

> *Let* $(X, \mathscr{S}_X)$ *be a topological vector space and* $(Y, \mathscr{S}_Y)$ *be Hausdorff. Assume* $f : D \subset X \to f(D) \subset Y$ *is closed and* f^{-1} *exists. Then* f^{-1} *is closed.*

Proof 8.2.3
$G_{f^{-1}} = \{(f(x), x) | x \in D\}$ *is a subset of* $Y \times X$*. By assumption,* G_f *is closed in* $X \times Y$ *and it is easy to see this tells us* $G_{f^{-1}}$ *is closed in* $Y \times X$*. Hence,* f^{-1} *is closed.* ■

In the setting of normed linear spaces, we have this result.

Theorem 8.2.4 $T : D \subset X \to Y$ **Continuous and Closed with** Y **Complete implies** D **is Closed**

> *Let* X *and* Y *be normed linear spaces with norms* $|\cdot\|_X$ *and* $\|\cdot\|_Y$ *respectively. Assume* Y *is complete. Let* $T : D \subset X \to T(D) \subset Y$ *be continuous and closed. The* D *is closed.*

Proof 8.2.4
Let $x \in \overline{D}$*. Then there is a sequence* $(x_n) \subset D$ *with* $x_n \to x$ *in* $|\cdot\|_X$*. Consider* $(T(x_n)) \subset Y$*. Since* $x_n \to x$*,* (x_n) *is a Cauchy sequence. Note,* T *continuous means* $\|T\| < \infty$*. Thus,* $\|T(x_n) - T(x_m)\|_Y \le \|T\|\|x_n - x_m\|_X$ *which immediately tells us* $(T(x_n))$ *is a Cauchy sequence in* Y*. So there is* $y \in Y$ *with* $T(x_n) \to y$ *in* $|\cdot\|_Y$*. Since* T *is closed,* $x_n \to x$ *and* $T(x_n) \to y$ *implies* $x \in D$ *and* $T(x) = y$*. So* D *is closed.* ■

And also the following.

Theorem 8.2.5 If $T : D \subset X \to T(D) \subset Y$ **is Linear and Closed with a Continuous Inverse, then** $T(D)$ **is Closed**

> *Let* X *and* Y *be normed linear spaces with norms* $|\cdot\|_X$ *and* $\|\cdot\|_Y$ *respectively. Assume* X *is complete. Let* $T : D \subset X \to T(D) \subset Y$ *be linear and closed. If* T^{-1} *exists and* T^{-1} *is continuous, then* $T(D)$ *is closed in* Y*.*

Proof 8.2.5
By Theorem 8.2.2. T^{-1} *is closed. Since* T^{-1} *is continuous and closed, by Theorem 8.2.4, the domain of* T^{-1} *or* $T(D)$ *is closed.* ■

Homework

Exercise 8.2.1 *Let* $(X, \|\cdot\|)$ *be a complete normed linear space and* $A : X \to X$ *be linear with* $\|A\|_{op} < 1$*.*

- *Prove* $(I - A)^{-1}$ *exists.*
  Hint: *Since* $\|A\|_{op} < 1$*,* $\sum_{i=0}^{\infty} \|A\|_{op}^i$ *converges. Show this implies* $\sum_{i=0}^{\infty} A^i$ *converges in* $\|\cdot\|_{op}$*. Call this sum* B*. Then, show* $AB = BA = \sum_{i=0}^{\infty} A^{i+1}$ *and so* $(I - A)B = B(I - A) = I$ *and so* $(I - A)^{-1}$ *exists.*

- *Prove* $\|(I - A)^{-1}\| \le 1/(1 - \|A\|_{op})$*.*
  Hint: *This follows because* $\sum_{i=0}^{\infty} \|A\|_{op}^i = 1/(1 - \|A\|_{op})$*.*

Exercise 8.2.2 *Let $(X, \|\cdot\|)$ be a complete normed linear space and R is the set of operators $A : X \to X$ that are linear and onto with a continuous inverse. Prove if $A \in R$ and $B : X \to X$ is any linear continuous operator.*

- *Prove if $\|A - B\|_{op} < 1/\|A^{-1}\|_{op}$, then $B \in R$.*
  Hint: *$B = A - (A - B) = A(I - A^{-1}(A - B))$. Now show $\|A^{-1}(A - B)\| < 1$ and so by the previous exercise, $(I - A^{-1}(A - B))^{-1}$ exists, is continuous and $(I - A^{-1}(A - B))^{-1} = \sum_{i=0}(A^{-1}(A - B))^i$. From this, show B^{-1} exists and is continuous with $B^{-1} = (I - A^{-1}(A - B))^{-1}A^{-1}$.*

- *Prove*

$$\|B^{-1}\|\|_{op} \leq \frac{\|A^{-1}\|_{op}}{1 - \|A^{-1}\|_{op}\|A - B\|_{op}}, \quad \|B^{-1} - A^{-1}\|_{op} \leq \frac{\|A^{-1}\|_{op}^2\|A - B\|_{op}}{1 - \|A^{-1}\|_{op}\|A - B\|_{op}}$$

  Hint: *This follows from the first part.*

- *Prove this shows R is open in the $\|\cdot\|_{op}$ topology and A^{-1} is a continuous function of A.*

8.3 Extensions to Metric Linear Spaces

Recall, a metric linear space is one which has an invariant metric. To get started, let's state the definition of first and second category sets in a topological space.

Definition 8.3.1 First Category Sets of a T_1 Topological Space

We say a subset D of X is dense in X, if $\overline{D} = X$. Let $(X, \mathscr{S})$ be a topological space. We say X is of **first category** *if there is a sequence (A_n) of subsets of X which are all nowhere dense so that $X = \cup_{n=1}^{\infty} A_n$. If X is not of first category, we say X is of* **second category**.

Next, we need to establish some technical results starting with some local neighborhood behavior.

Theorem 8.3.1 If $T : X \to Y$ is a Linear Mapping between Topological Vector Spaces whose Range is Second Category, then there are Neighborhoods of 0 so that $V \subset \overline{T(U)}$

Let $(X, \mathscr{S}_X)$ and $(Y, \mathscr{S}_Y)$ be topological vector spaces and $T : X \to Y$ be a given linear mapping. Assume $T(X)$ is of second category. Then to each neighborhood of 0 in X, there is a neighborhood of 0, V, in Y so that $V \subset \overline{T(U)}$.

Proof 8.3.1
Let's start by assuming X is a normed linear space with norm $\|\cdot\|$. Let U be a neighborhood of 0 in X. Then, there is an $\epsilon > 0$ so that $B(0, 2\epsilon) = \{x | \|x\| < 2\epsilon\} \subset U$. Let $W = B(0, \epsilon)$. Then $nW = \{x | \|x\| < n\epsilon\}$ and $T(X) = \cup_{n=1}^{\infty} T(nW)$. Since $T(X)$ is of second category, there is an N so that $int(\overline{T(NW)})$ is nonempty. But, it is easy to see $\overline{T(NW)} = \overline{NT(W)} = N\overline{T(W)}$ as T is linear.

Now $\overline{NT(W)}$ and $\overline{T(W)}$ are homeomorphic to each other. Hence, since the interior of $\overline{T(NW)}$ is nonempty, the interior of $\overline{T(W)}$ is nonempty also. Thus, there is $y_0 \in \overline{T(W)}$ so that y_0 is an interior point of $\overline{T(W)}$ as well. We can therefore say 0 is an interior point of $-y_0 + \overline{T(W)}$.

Let V be any neighborhood of 0 on Y so that $V \subset -y_0 + \overline{T(W)}$. Since, $y_0 \in T(W)$, there is an $x_0 \in W$ so that $y_0 = T(x_0)$. Next, note

$$\begin{aligned} -y_0 + T(W) &= \{-y_0 + T(w),\ w \in W\} = \{T(x_0) + T(w),\ w \in W\} \\ &= \{T(w - x_0),\ w \in W\} \end{aligned}$$

Now $w \in W$ implies $\|w\| < \epsilon$ and $x_0 \in W$ implies $\|x_0\| < \epsilon$. We then see $\|w - x_0\| < 2\epsilon$ and so $w - x_0 \in U$. This tells us $-y_0 + T(W) \subset T(U)$ and so

$$\overline{-y_0 + T(W)} = -y_0 + \overline{T(W)} \subset \overline{T(U)}$$

We conclude $V \subset \overline{T(U)}$. This completes the proof when X is a normed linear space.

If X is a topological vector space, we only have to change how we choose W. Given a neighborhood U of 0 in Y, there is a balanced absorbing neighborhood W of 0 so that $W + W \subset U$. Since W is balanced and absorbing, given $x \in X$, there is an $\epsilon > 0$ so that $\alpha x \in W$ when $0 < |\alpha| < \epsilon$. This says $x \in (2/\epsilon)W$ and so $x \in \cup_{n=1}^{\infty} nW$.

Since W is balanced, if $x_0 \in W$, $-x_0 \in W$ also. Thus, we find $w - x_0 \in W + W \subset U$. Now repeat the entire normed linear space argument choosing the neighborhood V as usual. So $V \subset \overline{T(U)}$. This completes the argument in the case X is a topological vector space. ■

Using this technical result, we can prove additional results for second category ranges of a closed and linear mapping.

Theorem 8.3.2 Technical Second Category Range Results for a Closed and Linear Mapping between Metric Linear Spaces

Let (X, d_X) and (Y, d_Y) be metric linear spaces which means d_X and d_Y are invariant metrics. Assume X is complete. Let $T : D \subset X \rightarrow T(X) \subset Y$ be closed and linear with the range $T(X)$ second category. For all $\alpha > 0$, let $B_\alpha = \{x \in X | d_X(x, 0) \leq \alpha\}$ and $C_\alpha = \{x \in X | d_Y(y, 0) \leq \alpha\}$. Then

(i): For all $\alpha > 0$, there is a $\beta > 0$ so that $C_\beta \subset \overline{T(D \cap B_\alpha)}$.

(ii): $T(X) = Y$.

(iii): If S is relatively open in D, then $T(D)$ is open in Y.

(iv): If T^{-1} exists, T^{-1} is continuous.

(v): In particular, if X and Y are normed linear spaces with norms $\|\cdot\|_X$ and $\|\cdot\|_Y$ respectively, (i) implies there is an $m > 0$ so that each $y \in Y$ is of the form $y = T(x)$ with $x \in D$ having $\|x\|_X \leq m\|y\|_Y$.

Proof 8.3.2
For any $\alpha > 0$, let $U = \{x \in X | d_X(x, 0) \leq \alpha/2\}$. Thus, $U \subset B_{\alpha/2}$. By Theorem 8.3.1, there is a neighborhood $V = \{y \in Y | d_Y(y, 0) \leq 2\beta\}$, for some $\beta > 0$, of 0 in Y, so that $V \subset \overline{T(D \cap U)}$. Note we use D for X in our use of Theorem 8.3.1. Hence, $V \subset \overline{T(D \cap B_{\alpha/2})}$.

(i): *To prove (i), we show we can find an $r > 0$ so that $V \subset \overline{T(D \cap B_r)}$ when $\alpha/2 < r \leq \alpha$. For such an r, choose $c < r - \alpha/2$ which tells us $\sum_{n=1}^{\infty} c(1/2^n) = c < r - \alpha/2$. Let $U_n = \{x \in X | d_X(x, 0) \leq c/2^n\}$. Again, using Theorem 8.3.1, there is a positive number δ_n so that if*

$V_n = \{y \in Y | d_Y(y, 0) \leq \delta_n\}$, $V_n \subset \overline{T(D \cap U_n)}$. *Once we find* δ_1, *we can easily choose* $\delta_2 < \delta_1/2$ *and so on. So we can assume the sequence* $\delta_n \to 0$.

Let $y \in V$. *First, note* $U_n = \{x \in X | d_X(x, 0) \leq c/2^n\} \subset B_{\alpha/2}$ *as* $c/2^n < r - \alpha/2 < \alpha - \alpha/2 = \alpha/2$. *Here is the argument chain we need:*

- *Since* $y \in V$, *there is a sequence* $(z_n) \subset D \cap B_{\alpha/2}$ *with* $T(z_n) \to y$. *Hence, for sufficiently large* n, $y - T(z_n) \in V_1$. *For any such index* n_0, *let* $x_0 = z_{n_0}$ *and* $y_0 = T(x_0)$. *Then,* $y - y_0 \in V_1$.
- *Next, since* $y - y_0 \in V_1$, *there is a sequence* $(z_n) \subset D \cap U_1$ *with* $T(z_n) \to y - y_0$. *Hence, for sufficiently large* n, $y - y_0 - T(z_n) \in V_2$. *For any such index* n_1, *let* $x_1 = z_{n_1}$ *and* $y_1 = T(x_1)$. *Then,* $y - y_0 - y_1 \in V_2$.
- *We continue in this way to construct sequences* (x_n) *and* (y_n) *with* $y_n = T(x_n)$, $x_n \in D \cap U_n$, $y - y_0 - y_1 - \ldots - y_n \in V_{n+1}$ *for* $n \geq 1$.

Let $S_n = \sum_{i=0}^{n} y_i$. *Then* $y - S_n \in V_{n+1}$ *for* $n \geq 1$ *with* $V_{n+1} = \{y \in Y | d_Y(y, 0) \leq \delta_{n+1}\}$ *with* $\delta_n \to 0$. *This says* $S_n \to y$.

Further, let $K_n = \sum_{i=0}^{n} x_i$. *Since* $d_X(x_n, 0) < c/2^n$, *we then have* $\sum_{n=0}^{\infty} d_X(x_n, 0) \leq d_X(x_0, 0) + \sum_{n=1}^{\infty} c/2^n < r - \alpha/2$. *This shows the sequence* (K_n) *is a Cauchy sequence in* X. *Since* X *is complete, there is an* $x \in X$ *so that* $K_n \to x$. *Further, given any tolerance* ξ, *there is an* N *so that* $d_X(x - S_n, 0) < \xi$ *is* $n > N$. *Thus,*

$$\begin{aligned} d_X(x, 0) &\leq d_X(x_0, 0) + d_X(\sum_{i=1}^{n} x_i, 0) + \xi \\ &\Longrightarrow d_X(x, 0) \leq d_X(x_0, 0) + r - \alpha/2 + \xi \end{aligned}$$

Since this is true for any ξ, *we have* $d_X(x, 0) < d_X(x_0, 0) + r - \alpha/2$. *We also know* $x_0 \in B_{\alpha/2}$. *We conclude* $d_X(x, 0) < r$ *or* $x \in B_r$.

Since $K_n \to x$ *and* $s_n \to y$, *the fact that* T *is closed implies* $x \in D$ *and* $y = T(x)$. *Hence, for this* $y \in V$, *we have shown* $y \in T(D \cap B_r)$. *We chose* β *so that* $C_{2\beta} = V$. *Now choose* $r = \alpha$ *and we have* $C_\beta \subset T(D \cap B_\alpha)$.

(ii):
Given $y \neq 0$, $(1/n)y \to 0$, *so there is an* N *with* $d_Y((1/n)y, 0) < \beta$ *for* $n > N$. *From (i),* $(1/n)y \in C_\beta$ *implies* $(1/n)y \in T(D \cap B_\alpha)$. *So there is an* $x \in D \cap B_\alpha$ *with* $(1/n)y = T(x)$. *Since* T *is linear, this tells us* $y = T(nx) \in T(X)$. *We conclude* $T(X) = Y$.

(iii):
S *is relatively open means we can write* $S = D \cap W$ *where* W *is open in* X. *Let* $y_0 \in T(S)$. *Then, there is* $x_0 \in S$ *so that* $y_0 = t(x_0)$. *The set* $-x_0 + W$ *is a neighborhood of* 0 *and so there is an* $\alpha > 0$ *so that* $B_\alpha \subset -x_0 + W$. *Then,* $D \cap B_\alpha \subset D \cap (-x_0 + W) = -x_0 + D \cap W$. *By (i), there is a* $\beta > 0$ *so that*

$$\begin{aligned} C_\beta &\subset T(D \cap B_\alpha) \subset T(D \cap (-x_0 + W)) \\ &= -y_0 + T(D \cap W) = -y_0 + T(S) \end{aligned}$$

Thus, $y_0 + C_\beta \subset T(S)$ *and we have shown* y_0 *is an interior point of* $T(S)$. *So,* $T(S)$ *is open.*

(iv):
T^{-1} is continuous if $(T^{-1})^{-1}(U)$ is open in Y when U is relatively open in X. But $(T^{-1})^{-1} = T$, so we must show $T(U)$ is open in Y when U is relatively open in X. This follows immediately from (iii).

(v):
Now, X and Y are normed linear spaces with norms $\|\cdot\|_X$ and $\|\cdot\|_Y$ respectively. If $y \neq 0$ in Y, then $\|\beta y\|_Y/\|y\|_Y = |\beta|$. Now apply (i). For a given $\alpha > 0$, there is a $\beta > 0$ so that $C_\beta \subset T(D \cap B_\alpha)$. So there is $u \in D \cap B_\alpha$ with $T(u) = \beta y/\|y\|_Y$. Let $x = (\|y\|_Y/\beta)u$. Then, $T(x) = y$ with $\|x\|_X \leq \|y\|_Y(1/\beta)\|u\|_X \leq (\alpha/\beta)\|y\|_Y$. Choose $m = (\alpha/\beta)$ and we have shown (v). ■

With these technical results out of the way, we can prove a metric linear space version of the Closed Graph Theorem.

Theorem 8.3.3 If $T : X \to Y$, X and Y Complete Metric Linear Spaces, $T(X) = Y$ is Second Category implies if T^{-1} Exists, it is Continuous

> *Let (X, d_X) and (Y, d_Y) be complete metric linear spaces. Let $T : X \to Y$ be linear, continuous and onto, i.e. $T(X) = Y$. Assume T^{-1} exists. Then T^{-1} is continuous.*

Proof 8.3.3
By Theorem 8.2.1, T is closed. Since $T(X) = Y$, then by the Baire Category Theorem, $T(X)$ is second category. By Theorem 8.3.2, part (iv), T^{-1} is continuous. ■

Now, finally, the version of the Closed Graph Theorem for metric linear spaces.

Theorem 8.3.4 Closed Graph Theorem for Metric Linear Spaces

> *Let (X, d_X) and (Y, d_Y) be complete metric linear spaces. Let $T : D \subset X \to T(X) \subset Y$ be linear and closed. Then T is continuous.*

Proof 8.3.4
Consider $X \times Y$. X has the invariant metric d_X and Y has the invariant metric d_Y. Define the invariant metric d on $X \times Y$ by $d((x_1, y_1), (x_2, y_2)) = d_X(x_1, y_1) + d_Y(x_2, y_2)$. You can check this is true as an exercise. We know $G_F = \{(x, T(x)) | x \in X\}$. Since T is closed, G_T is a closed subspace of Y. Since $X \times Y$ is complete with respect to d (another exercise for you), we see we can think of G_T as a complete metric linear space itself. Define the mapping $A > G_T \subset X \times Y \to A(G_T) \subset X$ by $A(x, T(x)) - = x$.

Now $d_X(x, 0) < d_X(x, 0) + d_Y(T(x), 0) = d((x, T(x)), (0, 0))$. So $d_X(A(x, T(x)) - A(0, T(0))) = d_X(x - 0) \leq d((x, T(x)), (0, 0))$. So given $\epsilon > 0$, if $d((x, T(x)), (0, 0)) < \delta = \epsilon$, then $d_X(A(x, T(x)) - A(0, T(0))) < \epsilon$. This shows A is continuous at $(0, 0)$.

It is easy to see A is linear and so continuity at $(0, 0)$ implies continuity on G_T.

Now A is continuous on G_T onto X, A is linear and $A^{-1}(x) = (x, T(x))$ telling us A^{-1} exists. Hence, by Theorem 8.3.3, A^{-1} is continuous. Thus, for any x_0, given $\epsilon > -$, there is a $\delta > 0$, so that $d(A^{-1}(x) - A^{-1}(x_0)) < \epsilon$ if $d_X(x, x_0) = d_X(x - x_0, 0) < \delta$ since d_X is invariant. Thus, $d_X(x, x_0) < \delta$ implies

$$
\begin{aligned}
d((x, T(x)) - (x_0, T(x_0))) &= d_X(x - x_0) + d_Y(T(x), T(x_0)) < \epsilon \\
&\Longrightarrow d_Y(T(x), T(x_0)) < \epsilon
\end{aligned}
$$

This tells us T is continuous at x_0 and since x_0 is arbitrary, we have shown T is continuous. ■

Homework

Exercise 8.3.1 *Let $[a, b]$ be a finite interval with a, b and let $X = C^2([a, b], \|\cdot\|) \cap \{x(a) = 0, x(b) = 0\}$ where if $x \in X$, $\|\cdot\| = \max\{\|x\|_\infty, \|x'\|_\infty, \|x''\|_\infty\}$. Let $Y = C([a, b], \|\cdot\|)$. Define $T : X \rightarrow Y$ by $T(x) = a_0 x'' + a_1 x' + a_2 x$ for given $a_i \in Y$.*

- *Prove $\|\cdot\|$ is a norm.*
- *Prove T is a bounded linear operator.*
- *If we assume the differential equation $T(x) = y$ has a unique and continuous solution $x \in X$ for each $y \in Y$, we know T^{-1} exists and $\boldsymbol{R}(T) = Y$. Explain how Theorem 8.3.3 shows that the solution y depends continuously on x. Note this is a strong conclusion as we get this result without any assumptions about the nature of a_i other than that they are continuous.*

Exercise 8.3.2 *Apply the exercise above to the differential equation $x'' + 3x' + 2x = y$ and estimate how the solution changes when it is solved for $y + \delta$ with $\|\delta\|_\infty < r$ for some $r > 0$.*

Exercise 8.3.3 *Apply Theorem 8.3.3 to answer whether or not the solutions to $x'' + (3 + \delta_1)x' + (2 + \delta_2)x = y$ for a fixed y and scalars δ_1 and δ_2 depend continuously on the pair (δ_1, δ_2).*

Exercise 8.3.4 *Explain how the results of the first exercise can be applied to the Stürm - Liouville problems discussed in (Peterson (100) 2020) at any λ not an eigenvalue. You can adapt the way we solved that exercise to other boundary conditions, of course.*

8.4 Linear Functional Results

Let $(X, \|\cdot\|)$ be a normed linear space and X' the set of continuous linear functionals on X like usual. In Theorem 2.5.5, we showed that $\|x\| = \sup_{x',\, \|x'\|=1} |x'(x)|$. This showed X' was norm-determining for X. We want to extend these results now.

Theorem 8.4.1 Characterization of Bounded Sets in a Normed Linear Space

Let $(X, \|\cdot\|)$ be a normed linear space and assume $F \subset X$. Let $M \subset X'$ be a closed subspace that is norm determining for X. Assume $\sup_{x \in F} |x'(x)| < \infty$ for all $x' \in M$. Then F is bounded; i.e. there is an $R > 0$ so that $\|x\| < R$ is $x \in F$.

Proof 8.4.1

The set $M \subset X'$ is itself a complete normed linear space using $\|x'\|_{op,X} = \sup_{x \in X, \|x\|=1} |x'(x)|$. Note, for any $x \in X$,

$$\begin{aligned} p_M(x) &= \sup_{x' \in M, \|x'\|_{op,X}=1} |x'(x)| \\ V(M) &= \inf_{x \in X,\, \|x\|=1} p_M(x) \end{aligned}$$

and by assumption $V(M) > 0$. Now,

$$p_M(x) = \sup_{x' \in M, \|x'\|_{op,X}=1} |x'(x)| \leq \sup_{x' \in X', \|x'\|_{op,X}=1} |x'(x)| = \|x\|$$

and so $0 < V(M) \leq 1$.

Let $B_F = \{f : F \rightarrow \Re |\ f$ is bounded$\}$. Further, define something like the operator norm of $f \in B_F$ by $\|f\|_{op,F} = \sup_{x \in F,\, x \neq 0} |f(x)|/\|x\|$. Then, B_F is a complete normed linear space with norm $\|\cdot\|_{op,F}$. Now define the mapping $T : M \rightarrow B_F$ by $T(f)$ is the function whose value at $x \in F$ is $(T(f))(x) = f(x)$. It is easy to see T is linear and $\|T(f)\|_{op,F} = \sup_{x \in F,\, x \neq 0} |f(x)|/\|x\| < \infty$ by assumption. So $T(f) \in B_F$.

Assume $(f_n) \subset M$ and $f_n \rightarrow f$ in $\|\cdot\|_{op,X}$. Since M is closed, $f \in M$ also. Consider

$$\begin{aligned} \|T(f_n) - T(f)\|_{op,F} &= \sup_{x \in F,\, x \neq 0} |T(f_n)(x) - T(f)(x)|/\|x\| = \sup_{x \in F,\, x \neq 0} |f_n((x) - f)(x)|/\|x\| \\ &= \|f_n - f\|_{op,F} \end{aligned}$$

Thus, given $\epsilon > 0$, there is an N so that $n > N$ implies

$$\|T(f_n) - T(f)\|_{op,F} = \|f_n - f\|_{op,X} \quad < \quad < \epsilon/2$$

Thus, $f_n \rightarrow f$ in M implies $T(f_n) \rightarrow T(f)$ in B_F. This implies T is a linear closed operator between M and B_F. By the Closed Graph Theorem, T must be continuous and so $\|T\|_{op} < \infty$.

Note, the operator norm of T is

$$\|T\|_{op} \quad = \quad \sup_{f \in M,\, \|f\|_{op,X} = 1} \|T(f)\|_{op,F}|$$

which implies for $x \in F$ not zero, $\sup_{f \in M,\, \|f\|_{op,X}=1} |(T(f))(x)|/\|x| \ \leq \|T\|_{op}$. This says

$$p_M(x) = \sup_{f \in M,\, \|f\|_{op,X}=1} |f(x)| \leq \|T\|_{op}$$

and so

$$V(M) = \inf_{x \in F, \|x\|=1} p_M(x)$$

We can rewrite this as

$$V(M) = \inf_{x \in F, x \neq 0} p_M(x/\|x\|) \leq \|T\|_{op}$$

Thus, $V(M)\|x\| \leq p_M(x) \leq \|T\|_{op}$. This tells us $\|x| \ \leq \|T\_op/V(M)\infty$. We conclude F is bounded. ■

We can look at this kind of result another way.

Theorem 8.4.2 Characterization of Bounded Sets in a Normed Linear Space Two

Let $(X, \|\cdot\|)$ be a normed linear space and assume $F \subset X$. Assume $\sup_{x \in F} |x'(x)| < \infty$ for all $x' \in X'$. Then F is bounded; i.e. there is an $R > 0$ so that $\|x\| < R$ is $x \in F$.

We can specialize to the case where the closed subspace M is all of X'.

Proof 8.4.2
Apply Theorem 8.4.1 using $M = X'$. ■

Now consider X' and X''. Let ϕ be the canonical embedding map of X into X''. Consider the range $\phi(X)$. We see for any $x' \in X'$,

$$\begin{aligned} p_{\phi(X)}(x') &= \sup_{\phi(x) \in \phi(X),\, \|\phi(x)\|=1} |(\phi(x))(x')| \\ &= \sup_{x \in X,\, \|\phi(x)\|=1} |(x')(x)| \end{aligned}$$

But we know ϕ is a norm isometry, so $\|\phi(x)\| = 1$ implies $\|x\| = 1$. So we have

$$p_{\phi(X)}(x') = \sup_{x \in X,\, \|x\|=1} |(x')(x)| = \|x'\|_{op}$$

where $\|x'\|_{op}$ is the usual operator norm. Thus,

$$V(\phi(X)) = \inf_{x' \in X',\, \|x'\|_{op}=1} p_M(x) = \inf_{x' \in X',\, \|x'\|=1} \|x'\|_{op} = 1$$

Since $V(\phi(X)) > 0$, we know $\phi(X)$ is norm determining for X' and so $p_{\phi(X)}$ determines a norm compatible with the usual norm topology on X'.

If $(x_n) \subset X$ satisfies $\phi(x_n) \to x''$, then the sequence $(\phi(x_n))$ is a Cauchy sequence in X''. Thus, given $\epsilon > 0$, there is N so that $\|\phi(x_n) - \phi(x_m)\|_{op} < \epsilon$ when n and m are larger than N. Thus,

$$\sup_{\|x'\|_{op}=1} |\phi(x_n)(x') - \phi(x_m)(x')| < \epsilon \Longrightarrow \sup_{\|x'\|_{op}=1} |x'(x_n) - x'(x_m)| < \epsilon$$

But $\sup_{\|x'\|_{op}=1} |x'(x)| = \|x\|$. Therefore, we have $\|x_n - x_m\| < \epsilon$ if $n, m > N$. This says (x_n) is a Cauchy sequence in X which is complete. So there is an x so that $x_n \to x$. Next, note

$$\begin{aligned} \|\phi(x_n) - \phi(x)\|_{op} &= \sup_{\|x'\|_{op}=1} |\phi(x_n)(x') - \phi(x)(x')| \\ &= \sup_{\|x'\|_{op}=1} |x'(x_n) - x'(x)| = \|x_n - x\| \end{aligned}$$

We see this tells us $\phi(x_n) \to \phi(x)$. But limits are unique and so $x'' = \phi(x)$. This shows $\phi(X)$ is a closed subspace of X''.

Homework

Exercise 8.4.1 *Explain all the details used to apply Theorem 8.4.1 using $M = X'$.*

Exercise 8.4.2 *Work through all the details of the proof that show $\phi(\ell^1)$ is a closed subspace of $(\ell^1)''$.*

Exercise 8.4.3 *Work through all the details of the proof that show $\phi(\mathbf{c})$ is a closed subspace of $(\mathbf{c})''$ where $\mathbf{c}$ is the space of sequences that converge.*

Exercise 8.4.4 *Work through all the details of the proof that show $\phi(\mathbf{c_0})$ is a closed subspace of $(\mathbf{c_0})''$ where $\mathbf{c_0}$ is the space of sequences that converge to 0.*

The work above leads to the next result.

Theorem 8.4.3 Bounded Sets of Linear Functionals

Let $(X, \|\cdot\|)$ be a complete normed linear space and let $F \subset X'$. Assume $\sup_{x' \in F} |x'(x)| < \infty$ for all $x \in X$. Then $\sup_{x' \in F} \|x'\|_{op} < \infty$.

Proof 8.4.3
In Theorem 8.4.1, replace X by X' and M by $\phi(X)$. Thus, we can say we start with $(X, \|\cdot\|)$, a normed linear space. Then $(X', \|\cdot\|_{op})$ is a normed linear space. Assume $F \subset X'$. Let $M \subset X''$ be a closed subspace that is norm determining for X'. Assume $\sup_{x' \in F} |x''(x')| < \infty$ for all $x'' \in M$. Then F is bounded; i.e. there is an $R > 0$ so that $\|x'\|_{op} < R$ is $x' \in F$.

Since we assume X is complete, our discussion above shows $\phi(X)$ is a closed subspace of X'' that is norm-determining for X'. The condition $\sup_{x' \in F} |\phi(x)(x')| < \infty$ for each $x \in X$ is the same as $\sup_{x' \in F} |x'(x)| < \infty$ at each x which we assume. So we can apply the theorem to conclude $\sup_{x' \in F} \|x'\|_{op} < \infty$. ■

Now that we have established some important results, let's look at results involving sequences.

Theorem 8.4.4 Bounded Sequence Results

Let $(X, \|\cdot\|)$ be a normed linear space.

(i): Assume $(x_n) \subset X$ satisfies $(x'(x_n))$ is a convergent sequence for each $x' \in X'$. Then $\sup_{n \geq 1} \|x_n\| < \infty$.

(ii): Now let X be complete and assume $(x'_n) \subset X'$ satisfies $(x'_n(x))$ is a convergent sequence for each $x \in X$. Then $\sup_{n \geq 1} \|x'_n\|_{op} < \infty$.

Proof 8.4.4
(i): *Let $F = (x_n)$. Since we assume $(x'(x_n))$ is a convergent sequence for each $x' \in X'$, we know $\sup_{n \geq 1} |x'(x_n)| < \infty$ for each x'. Apply Theorem 8.4.1 to conclude $\sup_{x \in F} \|x\| < \infty$ or $\sup_{n \geq 1} \|x_n\| < \infty$.*

(ii): *Let $F = (x'_n)$. We assume$(x'_n(x))$ is a convergent sequence for each $x \in X$. Thus, for all x, we have $\sup_{n \geq 1} |x'_n(x)| < \infty$. Apply Theorem 8.4.3 to conclude $\sup_{n \geq 1} \|x'_n\|_{op} < \infty$.* ■

Comment 8.4.1 *We can think of (i) in Theorem 8.4.4 in a different way. We can endow X with the topology $\mathscr{Y}(X, X')$. Convergence in this topology is what we call weak convergence. Recall, we showed $x_n \overset{w}{\to} x$ implies $x'(x_n) \to x'(x)$ for all $x' \in X'$. So we could rewrite (i) as if (x_n) converges weakly implies (x_n) is bounded in norm.*

Comment 8.4.2 *We can think of (ii) in Theorem 8.4.4 in a different way also. We can endow X' with the topology $\mathscr{Y}(X', \phi(X))$. Convergence in this topology is what we call weak* convergence. Recall, we showed $x'_n \overset{w*}{\to} x'$ implies $x'_n(x) \to x'(x)$ for all x. The assumption $(x'_n(x))$ is a convergent sequence is thus equivalent to assuming (x'_n) is weak* convergent. So we can restate (ii) as (x'_n) weak* convergent implies (x'_n) is bounded.*

Homework

Exercise 8.4.5 *In Theorem 8.4.3, it is necessary that X be complete. To see this work through this example: $X = \{x \in \ell^2 |$ there is an integer N_x so that $x_k = 0,\ k > N_x\}$. Define the linear functionals f_n by $f_n(x) = n x_n$. Let $F = \{f_n : n = 1, 2, \ldots\}$.*

- *Show X is not complete.*
- *Show $\sup_n |f_n(x)| = 0$ for all $x \in X$ and $\sup_n \|f_n\|_{op} = \infty$.*

Exercise 8.4.6 *Repeat the exercise above in the space ℓ^1. Why does this argument fail in the space ℓ^∞?*

Exercise 8.4.7 *Repeat our arguments that show $x_n \stackrel{w}{\to} x$ implies $x'(x_n) \to x'(x)$ for all $x' \in X'$.*

Exercise 8.4.8 *Repeat our arguments that show $x'_n \stackrel{w*}{\to} x'$ implies $x'_n(x) \to x'(x)$ for all x.*

8.5 Early Banach - Alaoglu Results

Now the next theorem is quite important. It is our first taste of the Banach - Alaoglu Theorem.

Theorem 8.5.1 Existence of Weak* Subsequential Limits in Bounded Sets

> *Let $(X, \|\cdot\|)$ be a normed linear space that is separable. Then, every bounded sequence in X' contains a subsequence which converges in the weak* topology on X'.*

Comment 8.5.1 *So, if X is separable, the bounded and closed set $B = \{x' \in X' \mid \|x'\|_{op} \leq 1\}$ is sequentially weak* compact!*

Proof 8.5.1
Let (x'_n) be bounded by $B > 0$. Let (x_k) be a countable dense sequence in X. Since $(x'_n(x_1))$ is a bounded sequence of scalars, it contains a convergent subsequence $(x'_{n,1}(x_1))$. Next, the same reasoning shows there is a convergent subsequence of $(x'_{n,1}(x_2))$ we can label as $(x'_{n,2}(x_2))$. Note, since subsequential limits of a convergent sequence are the same as the original limit, we have $(x'_{n,2}(x_1))$ has the same limit as $(x'_{n,1}(x_1))$.

Continuing, we can find subsequences that satisfy:

$$(x'_{n,k}) \quad \subset \quad (x'_{n,k-1}) \subset \ldots \subset (x'_{n,1}) \subset (x'_n)$$

and $(x'_{n,k}(x_j))$ converges for $1 \leq j \leq k$.

The construction we now follow is called the Cauchy Diagonalization Process. Consider the subsequence $(x'_{n,n})$. It helps to look at this in a big table.

$$\begin{array}{cccccccc}
\boxed{x'_{1,1}(x_1)} & x'_{2,1}(x_1) & \ldots & \ldots & \ldots & \ldots & x'_{n,1}(x_1) & \ldots \\
x'_{1,2}(x_2) & \boxed{x'_{2,2}(x_2)} & \ldots & \ldots & \ldots & \ldots & x'_{n,2}(x_2) & \ldots \\
\vdots & \vdots & \ddots & \vdots & \vdots & \vdots & \vdots & \vdots \\
x'_{1,j}(x_j) & x'_{2,j}(x_j) & \ldots & \boxed{x'_{j,j}(x_j)} & \ldots & \ldots & x'_{n,j}(x_j) & \ldots \\
\vdots & \vdots & \ddots & \vdots & \vdots & \ddots & \vdots & \vdots \\
x'_{1,n}(x_j) & x'_{2,n}(x_n) & \ldots & \ldots & \ldots & \ldots & \boxed{x'_{n,n}(x_n)} & \ldots \\
\vdots & \vdots & \ddots & \vdots & \vdots & \vdots & \vdots & \vdots
\end{array}$$

Because of the way we have constructed these subsequences, we know the subsequence in row 1 converges at x_1 to a number α_1, the subsequence in row 2 converges at x_1 and x_2 to α_1 and α_2 and the subsequence in row n converges at x_1 through x_n to the values α_1 to α_n. Consider for fixed k, the sequence $(x'_{n,n}(x_k))$. We know $(x'_{n,k+j}) \subset (x'_{n,k})$ for $j \geq 1$. So $x'_{k+j,k+j} \in (x'_{n,k})$ and this sequence converges to α_k. So, we see $x'_{n,n}(x_k) \to \alpha_k$.

We see the sequence $(x'_{n,n}(x_k))$ converges for all k to α_k. Now consider the sequence $(x'_{n,n}(x))$ for any $x \in X$. We have for any x_k,

$$\begin{aligned}
|x'_{n,n}(x) - x'_{m,m}(x)| &\leq |x'_{n,n}(x) - x'_{n,n}(x_k)| + |x'_{n,n}(x_k) - x'_{m,m}(x_k)| \\
&\quad + |x'_{m,m}(x_k) - x'_{m,m}(x)| \\
&\leq \|x'_{n,n}\|_{op}\|x - x_k\| + |x'_{n,n}(x_k) - x'_{m,m}(x_k)| + \|x'_{m,m}\|_{op}\|x_k - x\| \\
&\leq 2B\|x - x_k\| + |x'_{n,n}(x_k) - x'_{m,m}(x_k)|
\end{aligned}$$

Then, given $\epsilon > 0$, since (x_k) is dense in X, there is an index k so that $\|x - x_k\| < \epsilon/(3B)$. Using this, we see

$$|x'_{n,n}(x) - x'_{m,m}(x)| \quad < \quad 2\epsilon/3 + |x'_{n,n}(x_k) - x'_{m,m}(x_k)|$$

The final piece is to recognize that the sequence $(x'_{n,n}(x_k))$ converges so it is a Cauchy sequence. Thus, there is N so that $|x'_{n,n}(x_k) - x'_{m,m}(x_k)| < \epsilon/3$ if $n, m > N$. Hence, $|x'_{n,n}(x) - x'_{m,m}(x)| < \epsilon$ if $n, m > N$ and so $(x'_{n,n}(x))$ is a Cauchy sequence for all x and hence converges to a real number β_x. Define $x' : X \rightarrow \Re$ by $x'(x) = \beta_x$.

It is clear x' is linear. Is it bounded? Let $\epsilon = 1$. Then, there is N so that $|x'_{n,n}(x) - x'(x)| < 1$ if $n > N$. Choose $\hat{n} > N$. Then,

$$\begin{aligned}
|x'(x)| &\leq |x'(x) - x'_{\hat{n},\hat{n}}(x)| + |x'_{\hat{n},\hat{n}}(x)| \\
&\leq 1 + |x'_{\hat{n},\hat{n}}(x)| \leq 1 + \|x'_{\hat{n},\hat{n}}\|_{op}\|x\|
\end{aligned}$$

Thus, $|x'(x)|/\|x\| \leq 1 + \|x'_{\hat{n},\hat{n}}\|_{op} < \infty$. We conclude x' is bounded. We have shown $x'_{n,n}(x) \rightarrow x'(x)$ which is the same as saying $x'_n \stackrel{w*}{\rightarrow} x'$. ■

We can prove a few more useful results in normed linear spaces.

Theorem 8.5.2 Bounded Weakly Convergent Sequences Have Limits with the Same Bound

> *Let $(X, \|\cdot\|)$ be a normed linear space and $(x_n) \subset X$ and $x \in X$ so that $x_n \stackrel{w}{\rightarrow} x$. Then, if (x_n) is bounded by C, so is x.*

Proof 8.5.2
By assumption, $x'(x_n) \rightarrow x'(x)$ for all $x' \in X'$. Then, $|x'(x_n)| \leq \|x'\|_{op}\|x_n\|$. It follows

$$\underline{\lim}_{n\to\infty} |x'(x_n)| \quad \leq \quad \|x'\|_{op} \underline{\lim}_{n\to\infty} \leq \|x'\|_{op}\, C$$

But,

$$\underline{\lim}_{n\to\infty} |x'(x_n)| = lim_{n\to\infty}|x'(x_n)| = |x'(x)|$$

Thus, $|x'(x)| \leq C \leq \|x'\|_{op}$ implying $|x'(x)|/\|x'\|_{op} \leq C$. This says, $\sup_{\|y'\|_{op}=1} |x'(x)| \leq C$. We conclude $\|x\| \leq C$. ■

Here are some facts:

(i): If X is a vector space and X_0 is a subspace of X, then we can define the quotient space X/X_0 with equivalence classes $[x] = x + X_0 = \{y = x + x_0 | x_0 \in X_0\}$. This is a vector space as well where it is easy to see the operations defined by $\alpha[x] = \alpha[x]$ and $[x + y] = [x] + [y]$ are well-defined.

(ii): If $(X, \|\cdot\|)$ is a normed linear space and X_0 is a closed subspace of X, then X/X_0 is also a normed linear space. By (i), we know it is a vector space. We define the norm by $\|[x]\| = \inf_{x_0 \in X_0} \|x - x_0\|$. The properties of a norm can be established with little work except for the property that $\|[x]\| = 0$ if and only if $[x] = [0]$. To see this, it is enough to show $\|[x]\| = 0$ implies $[x] = [0]$. We have $\inf_{x_0 \in X_0} \|x - x_0\| = 0$. Hence, there is a minimizing sequence $(x_{0,n})$ so that $0 \leq \|x - x_{0,n}\| < 1/n$. This tells us $x_{0,n} \to x$. Since X_0 is closed, we thus have $x \in X_0$ implying $[x] = [0]$.

(iii): Now, we have $(X, \|\cdot\|)$ is a normed linear space and X_0 is a subspace of X. We let $X^0 = \{x' \in X' | x'(x) = 0, \ x \in X_0\}$. We see X^0 is a closed subspace of X' and so X/X^0 is a normed linear space with the usual norm $\|[x']\|$ as discussed above.

Now define $z'_x \in X'_0$ by choosing $y' \in [x']$ and setting $z'_x(x) = y'(x)$ for all $x \in X_0$. We can easily show z'_x is independent of the choice of representative y' and so z'_x is a well-defined element of X'_0.

Also, note $\|z'_x(x)\|_{op} \leq \|z'_x\|_{op}\|x\| = \|y'\|\|x\|$ for all $x \in X_0$ and any $y' \in [x']$. This implies

$$\sup_{x \neq 0} \frac{\|z'_x(x)\|_{op}}{\|x\|} \leq \|y'\|$$

for all $y' \in [x']$. Thus, $\|z'_x\|_{op} \leq \|y'\|$ for all $y' \in [x']$. This then tells us $\|z'_x\|_{op} \leq \|[x']\|$.

Now by the Hahn - Banach Theorem, there is $y'_x \in [x']$ with $\|y'_x\| = \|z'_x\|$ and $y'_x(x) = z'_x(x)$ for all $x \in X_0$. This gives the other side of the inequality so we can conclude $\|z'_x\|_{op} = \|[x']\|$.

Define the mapping $T : X'/X^0 \to X'_0$ by $T([x']) = z'_x$. We have shown T is an isometric (i.e., it preserves norms) mapping. It is straightforward to show T is also linear. Finally, given $x' \in X'_0$, the Hahn - Banach Theorem tells us there is a $y' \in X'$ with the same norm which matches x' on X_0. Hence, from the way we defined z'_x, we see $z'_x = y'$ on X_0. Then $x' = T([y'])$ and T is onto. This shows we can identify X'_0 and X/X^0.

(iv): Now we assume X is norm reflexive and X_0 is a closed subspace of X. We know the mapping $T : X'/X^0 \to X'_0$ is 1-1, onto, linear and a norm isometry. Choose $x'' \in X''_0$. Define $y'' \in X''$ by $y''(x') = x''(T([x']))$. Since X is reflexive, there is an $y \in X$, so that $\phi(y) = y''$ with $y''(x') = (\phi(y))(x') = x'(y)$.

We claim $y \in X_0$. It is enough to assume X_0 is a proper subspace as if $X_0 = X$, we already have $\phi(X_0) = X''_0$. Now if X_0 is a closed and proper subspace of X, if $y \notin X_0$, then $\delta = \inf_{x \in X_0} \|y - x_0\| > 0$ and there is $z' \in X'$ with $\|z'\|_{op} = 1$, $z' = 0$ on X_0 and $z'(y) = \delta > 0$. Thus, $z' \in X^0$ and $[z'] = [0]$ in $X'/X^0 \cong X'_0$. This means $T([z']) = 0 \in X'_0$. This says $y''(z') = x''(T([z'])) = x''(0) = 0$. But this also says $\phi(y)(z') = z'(y) = 0$ which is not possible. Hence, we must have $y \in X_0$.

So now we can say $y''(x') = (\phi(x_0))(x') = x'(x_0)$ with $x_0 \in X_0$. Now if $y' \in X'_0$, then $y' = T([x'])$ for some x' and $y'(x_0) = x'(x_0)$ by definition for any $x' \in [x']$, for all $x_0 \in X_0$. Thus, $x''(y') = y''(x') = x'(x_0) = y'(x_0)$. We conclude $x''(y') = y'(x_0)$.

Hence, the canonical map $\phi(x_0) = x''$ gives $\phi(x_0)(y') = y'(x_0)$. Therefore, X''_0 is reflexive as $\phi(X_0) = X''_0$.

Homework

Exercise 8.5.1 *In the discussion above, we define $z'_x \in X'_0$ by choosing $y' \in [x']$ and setting $z'_x(x) = y'(x)$ for all $x \in X_0$. Provide the arguments that show z'_x is independent of the choice of representative y' and so z'_x is a well-defined element of X'_0.*

Exercise 8.5.2 *In the discussion above, we define the mapping $T : X'/X^0 \rightarrow X'_0$ by $T([x']) = z'_x$. We show T is an isometric (i.e., it preserves norms) mapping. Provide the arguments that show T is linear.*

Exercise 8.5.3 *Prove if $X = \ell^1$, the bounded and closed set $B = \{x' \in X' \mid \|x'\|_{op} \leq 1\}$ is sequentially weak* compact.*

Exercise 8.5.4 *Prove if $X = \mathscr{L}_2([0,1])$, the bounded and closed set $B = \{x' \in X' \mid \|x'\|_{op} \leq 1\}$ is sequentially weak* compact. How can you characterize the set B?*

If the normed space is reflexive, we can say more.

Theorem 8.5.3 Bounded Sequences in a Reflexive Normed Linear Space have Weakly Convergent Subsequences

Let $(X, \|\cdot\|)$ be a reflexive normed linear space and $(x_n) \subset X$ be bounded. Then, there is a subsequence $(x_{n,1})$ which converges weakly.

Proof 8.5.3
By Theorem 8.5.2, if (x_n) is a weakly convergent sequence with limit x, then if there is a C so that $\|x_n\| \leq C$ for all n, we also know $\|x\| \leq C$. Here, we start with a bounded sequence (x_n) satisfying $\|x_n\| \leq C$ for all n. Let $X_0 = \overline{span\{(x_n)\}}$. Then X_0 is a separable closed subspace of X. We know the canonical embedding ϕ applied to X_0 gives $\phi(X_0) \subset X''_0 \subset X''$. From our remarks above, we know X_0 is reflexive, so $\phi(X_0) = X''_0$.

Now

$$\begin{aligned} \|\phi(x_n)\|_{op} &= \sup_{x' \in X'} \frac{(\phi(x_n))(x')}{\|x'\|_{op}} = \sup_{x' \in X'} \frac{x'(x_n)}{\|x'\|_{op}} \\ &= \|x_n\| \leq C \end{aligned}$$

Thus, $(\phi(x_n))$ is a sequence in X''_0 bounded by C. Apply Theorem 8.5.1 to the separable normed linear space X'_0. We have $(\phi(x_n))$ is a bounded sequence in X''_0 and so it contains a subsequence which converges in the weak topology for X'_0.*

Let's discuss this topology a bit. Look at an element $\phi(x')$ from $\phi(X'_0)$. Then, $(\phi(x'))(x'') = x''(x')$. But $\phi(X_0) = X''_0$ so we can say $(\phi(x'))(x'') = x'(x)$. This shows $\phi(X'_0) \cong X_0'''$. So the weak topology on X'_0 is $(\mathscr{Y}(X''_0), \phi(X'_0))$.*

Let $(\phi(x_{n,1}))$ be the subsequence that converges in $\mathscr{Y}(X''_0, \phi(X'_0))$ to an element in $y \in X''_0$. Thus, we can write $y = \phi(x)$ for some $x \in X_0$. Recall convergence here means $\phi(x')(\phi(x_{n,1})) \rightarrow \phi(x')(\phi(x))$ for all $x' \in X'_0$. We can rewrite this as

$$\begin{aligned} \phi(x_{n,1})(x') &\rightarrow \phi(x)(x') \text{ for all } x' \in X'_0 \Longrightarrow \\ x'(x_{n,1}) &\rightarrow x'(x) \end{aligned}$$

Hence, this convergence is our usual weak convergence; i.e. $x_{n,1} \xrightarrow{w} x$. ■

Comment 8.5.2 *Hence, if $(x_n) \subset B(0,1)$, there is a subsequence $(x_{n,1})$ which converges weakly so some $x \in X$. Since (x_n) is bounded by 1, we know $\|x\| \leq 1$. This, of course, tells us that $\overline{B(0,1)}$ is weakly compact in a reflexive normed linear space.*

Homework

Exercise 8.5.5 *Prove if $X = \ell^2$, the bounded and closed set $B = \{x \in \ | \ \|x\|_2 \leq 1\}$ is sequentially weakly compact.*

Exercise 8.5.6 *Prove if $X = \mathscr{L}_2([0,1])$, the bounded and closed set $B = \{x \in X \mid \|x\|_2 \leq 1\}$ is sequentially weakly compact. How can you characterize the set B?*

Exercise 8.5.7 *Prove if $X = \Re^n$, the bounded and closed set $B = \{x \in X | \|x\|_2 \leq 1\}$ is sequentially weakly compact. How can you characterize the set B?*

The next result is set in a Hilbert space which, of course, is reflexive. Hence, the result above holds and we know $\overline{B(0,1)}$ is weakly compact. We used this result in the work we did on self-adjoint operators in (Peterson (100) 2020). We will prove this directly without using the preceding result just for fun!

Theorem 8.5.4 Closed Unit Ball in Hilbert Space is Weakly Compact

Let H be a Hilbert Space. Then $\overline{B(\mathbf{0},1)}$ is weakly sequentially compact.

Proof 8.5.4

The proof of this result uses the Cauchy Diagonalization Process like we used in the proof of Theorem 8.5.1 and in the proof of the Arzela - Ascoli Theorem in (Peterson (100) 2020). Theorem 8.5.1 has a separability requirement, but there is a clever way to get around that. So pay attention to the details of the proof as they are instructive.

Let $(\boldsymbol{u_n})$ be a sequence in $\overline{B(\mathbf{0},1)}$. We want to show there is a subsequence $(\boldsymbol{u_n^1})$ and an element $\boldsymbol{u}$ in $\overline{B(\mathbf{0},1)}$ so that $\boldsymbol{u_n^1} \stackrel{weak}{\rightarrow} \boldsymbol{u}$. This means $\boldsymbol{f}(\boldsymbol{u_n^1}) \rightarrow \boldsymbol{f}(\boldsymbol{u})$ for all $\boldsymbol{f} \in H'$.

We do not know if H is separable, but we do not need that here. Let $U = \overline{span(\boldsymbol{u_n})}$ be the closure of the span of sequence in H. Then $H = U \oplus U^{\perp}$. If $\boldsymbol{y} \in U$, then there is a sequence $(\boldsymbol{s_n})$ from U so that $\boldsymbol{s_n} \rightarrow \boldsymbol{y}$. Each $\boldsymbol{s_n}$ has a representation: $\boldsymbol{s_n} = \sum_{i=1}^{p_n} c_{n,i}\boldsymbol{u_{n,i}}$. Hence, given $\epsilon > 0$, there is an element s_p with $\|\boldsymbol{s_p} - \boldsymbol{y}\| < \epsilon$. If we restrict our attention to rational coefficients, it is straightforward to then show there is an element $\widetilde{\boldsymbol{s_p}}$ with rational coefficients in the representation which is within ϵ of $\boldsymbol{y}$ in norm. The set of all such finite representations with rational coefficients is countable and dense in U. Label the elements of this set as $(\boldsymbol{y_k})$ as it is countable. Use these elements to create an orthogonal sequence $\boldsymbol{x_k} = \frac{\boldsymbol{y_k}}{\|\boldsymbol{x_k}\|}$. Hence, $\|\boldsymbol{x_k}\| = 1$ for all indices. We now know U is separable.

The sequence $(< \boldsymbol{u_n}, \boldsymbol{x_1} >)$ is bounded and so there is a convergent subsequence $(\boldsymbol{u_n^1})$ and a scalar c_1 so that $< \boldsymbol{u_n^1}, \boldsymbol{x_1} > \rightarrow c_1$. Thus $|c_1| \leq 2$ using a standard argument. Next, since the sequence $(< \boldsymbol{u_n^1}, \boldsymbol{x_2} >)$ is bounded, there is a convergent subsequence $(\boldsymbol{u_n^2}) \subset (\boldsymbol{u_n^1})$ and a scalar c_2 so that $< \boldsymbol{u_n^2}, \boldsymbol{x_2} > \rightarrow c_2$. Again, it is easy to see $|c_2| \leq 2$. In fact, note we also know $< \boldsymbol{u_n^2}, \boldsymbol{x_1} > \rightarrow c_1$.

It is now clear how the process goes. After P steps, we have extracted a subsequence $(\boldsymbol{u_n^P}) \subset \ldots \subset (\boldsymbol{u_n^1})$ and a scalar c_P so that $< \boldsymbol{u_n^P}, \boldsymbol{x_i} > \rightarrow c_i$ for $1 \leq i \leq P$ with $|c_P| \leq 2$. Since (c_n) is a bounded sequence by the Bolzano - Weierstrass theorem there is a subsequence (c_n^1) and a number c so that $c_n^1 \rightarrow c$. Now keep only $\boldsymbol{u_n}$ elements whose indices correspond to the indices in the subsequence

(c_n^1).

We now use the Cauchy diagonalization process on these elements. We look only at the **diagonal** *elements* $\boldsymbol{u_p^p}$ *in this process. For convenience of exposition, let's let* $\boldsymbol{u'_p} = \boldsymbol{u_p^p}$. *We then know* $< \boldsymbol{u'_p}, \boldsymbol{x_k} > \to c_k$ *for all p and k. Given any* $\boldsymbol{y}$ *in* U, *there is a sequence* $(\boldsymbol{x_k^1})$ *so that* $\boldsymbol{x_k^1} \to \boldsymbol{y}$. *Note by the continuity of the inner product*

$$< \boldsymbol{u'_n}, \boldsymbol{y} > \quad = \quad \lim_{k \to \infty} < \boldsymbol{u'_n}, \boldsymbol{x_k^1} >$$

Let $\epsilon > 0$ *be given. Then there is a* K_1 *so that* $\|\boldsymbol{y} - \boldsymbol{x_k^1}\| < \frac{\epsilon}{3}$ *if* $k > K_1$. *Also, there is a* K_2 *so that if* $k > K_2$, *then* $|c - c_k| < \frac{\epsilon}{3}$. *Hence, if* $k > max\{K_1, K_2\}$, *we have*

$$\begin{aligned} | < \boldsymbol{u'_n}, \boldsymbol{y} > -c| &= | < \boldsymbol{u'_n}, \boldsymbol{y} - \boldsymbol{x_k^1} + \boldsymbol{x_k^1} > -c_k + c_k - c| \\ &\leq | < \boldsymbol{u'_n}, \boldsymbol{y} - \boldsymbol{x_k^1} > | + | < \boldsymbol{u'_n}, \boldsymbol{x_k^1} > -c_k| + |c_k - c| \\ &\leq \|\boldsymbol{y} - \boldsymbol{x_k^1}\| + | < \boldsymbol{u'_n}, \boldsymbol{x_k^1} > -c_k| + \frac{\epsilon}{3} \\ &< \frac{\epsilon}{3} + | < \boldsymbol{u'_n}, \boldsymbol{x_k^1} > -c_k| + \frac{\epsilon}{3} = 2\frac{\epsilon}{3} + | < \boldsymbol{u'_n}, \boldsymbol{x_k^1} > -c_k| \end{aligned}$$

Finally, there is N *so that if* $n > N$, *we have* $| < \boldsymbol{u'_n}, \boldsymbol{x_k^1} > -c_k| < \frac{\epsilon}{3}$. *Thus, if* $n > N$, $| < \boldsymbol{u'_n}, \boldsymbol{y} > -c| < \epsilon$. *This shows the* $\lim_{n \to \infty} < \boldsymbol{u'_n}, \boldsymbol{y} >$ *always exists. Hence, the map* f *defined on* U *by*

$$f(\boldsymbol{y}) \quad = \quad \lim_{n \to \infty} < \boldsymbol{u'_n}, \boldsymbol{y} >$$

with $f = 0$ *on* $U^\perp$ *is well-defined. This is clearly a linear map. Note for all* $\boldsymbol{y}$ *not zero*

$$\frac{f(\boldsymbol{y})}{\|\boldsymbol{y}\|} \quad \leq \quad \frac{\lim_{n \to \infty} \|\boldsymbol{u'_n} \ \|\boldsymbol{y}\|\|}{\|\boldsymbol{y}\|} = 1$$

which implies f *is a bounded linear functional in* H'.

Now use the Riesz Representation theorem: there is a unique $\boldsymbol{z} \in H$ *so that* $f(\boldsymbol{y}) = < \boldsymbol{y}, \boldsymbol{z} >$ *for all* $\boldsymbol{y} \in H$ *and* $\|\boldsymbol{z}\| = \|f\| \leq 1$. *This shows* $\lim_{n \to \infty} < \boldsymbol{u'_n}, \boldsymbol{y} > = < \boldsymbol{z}, \boldsymbol{y} >$ *on* H *with* $\boldsymbol{z} \in \overline{B(\boldsymbol{0}, 1)}$ *and establishes the weak sequential compactness.* ■

Homework

Exercise 8.5.8 *In the proof of Theorem 8.5.4, we let* $U = \overline{span(\boldsymbol{u_n})}$ *be the closure of the span of sequence in* H. *Then* $H = U \oplus U^\perp$. *If* $\boldsymbol{y} \in U$, *then there is a sequence* $(\boldsymbol{s_n})$ *from* U *so that* $\boldsymbol{s_n} \to \boldsymbol{y}$. *Each* $\boldsymbol{s_n}$ *has a representation:* $\boldsymbol{s_n} = \sum_{i=1}^{p_n} c_{n,i} \boldsymbol{u_{n,i}}$. *Hence, given* $\epsilon > 0$, *there is an element* s_p *with* $\|\boldsymbol{s_p} - \boldsymbol{y}\| < \epsilon$. *If we restrict our attention to rational coefficients, provide the details of the argument that shows there is an element* $\widetilde{\boldsymbol{s_p}}$ *with rational coefficients in the representation which is within* ϵ *of* $\boldsymbol{y}$ *in norm. Prove the set of all such finite representations with rational coefficients is countable and dense in* U.

Exercise 8.5.9 *In the proof of Theorem 8.5.4, we define the map* f *on* U *by*

$$f(\boldsymbol{y}) \quad = \quad \lim_{n \to \infty} < \boldsymbol{u'_n}, \boldsymbol{y} >$$

with $f = 0$ *on* $U^\perp$. *Prove this map is well-defined. and it is a linear map.*

Exercise 8.5.10 *Again, note we can prove if $X = \mathscr{L}_2([0,1])$, the bounded and closed set $B = \{x \in X \mid \|x\|_2 \leq 1\}$ is sequentially weakly compact using Theorem 8.5.4. Make sure you understand that we can prove this two ways.*

8.6 The Full Banach - Alaoglu Result

As we mentioned in our comment to Theorem 8.5.4, we can cast this result as a weak* compactness theorem in a topological vector space. However, to do this we need to know more about compactness for product topologies. First, we need some background. You are all familiar with finite, countable and uncountable index sets. The Well Ordering Principle tells us every index set Λ can be well ordered by a relationship $\preceq$; i.e. if $\boldsymbol{A} \subset \Lambda$ there is an element of a of $\boldsymbol{A}$ so that $a \preceq x$ for all $x \in \boldsymbol{A}$. The element a is called the minimal element of $\boldsymbol{A}$.

We are also familiar with induction. If you want to do an induction argument over an index set Λ which is not countable, you invoke the Well Ordering Principle and then apply what is called transfinite induction. Note since the Well Ordering Principle is equivalent to the Axiom of Choice, if the index set Λ requires an invocation of the Well Ordering Principle to apply transfinite induction, we are essentially using a principle of proof that is based on the Axiom of Choice.

Definition 8.6.1 Transfinite Induction

Let $P(\alpha)$ be a property defined for all $\alpha \in \Lambda$ where Λ is the index set. If $P(\beta)$ is true for all $\beta \preceq \alpha$ implies $P(\alpha)$ is true, then the property $P(\lambda)$ is true for all $\lambda \in \Lambda$.

The way transfinite induction is applied is similar to the way our usual induction arguments are done. We can call the minimal element of Λ, 0, for convenience.

- Prove $P(0)$ is true. This is the traditional base case.
- Prove $P(\beta)$ true for all $\beta \preceq \alpha$ implies $P(\alpha)$ is true. This is the usual induction step.

Next, let's talk about products of compact sets. Recall what the product topology for a finite number of topological spaces means. Earlier, we defined this for two spaces. Let's do it for a finite number of spaces now.

Definition 8.6.2 Cartesian Product of a Finite Number of Topological Spaces

Let $(X_i, \mathscr{S}_i)$ topological spaces for $1 \leq i \leq n$ for some positive integer n. The cross-product topology induced on $X_1 \times X_2 \times \ldots \times X_n$ consists of the open sets of the form $A_1 \times A_2 \times \ldots \times A_n$ where $A_i \in \mathscr{S}_i$ for $1 \leq i \leq n$.

Now let's consider the proof that if C_1 is compact in X_1 and C_2 is compact in X_2, the set $C_1 \times C_2$ is compact in $X_1 \times X_2$. Normally, we would say C_1 is compact in X_1 means every open cover of C_1 has a finite subcover. However, another way to say it is that if there is a family of open set (O_α) for $\alpha \in \Lambda$ for an index set Λ, so that if no finite collection from (O_α) covers C_1, then (O_α) is not an open cover of C_1. So assume we have such a collection of open sets $(U_\alpha \times V_\alpha)$ and no finite collection from it covers $C_1 \times C_2$.

Now assume for each $x \in C_1$, there is a neighborhood W_x of x and $U_x \times C_2$ is finitely covered by $(U_\alpha \times V_\alpha)$. The collection (W_x) covers C_1 and so there is a finite subcover of (W_x) which covers C_1. Let $\{W_{x_1}, \ldots, W_{x_p}\}$ be this finite subcover. Then there is a finite collection of $(U_\alpha \times V_\alpha)$ which covers $C_1 \times C_2$ which we assume is not possible.

So there is some $x_0 \in C_1$ so that for all neighborhoods W_{x_0}, the set $W_{x_0} \times C_2$ cannot be finitely covered by $(U_\alpha \times V_\alpha)$.

Now assume for each $y \in C_2$, there is a neighborhood Z_y of y and $W_{x_0} \times Z_y$ is finitely covered by $(U_\alpha \times V_\alpha)$. The collection (Z_y) covers C_2 and so there is a finite subcover of (Z_y) which covers C_2. Let $\{Z_{y_1}, \ldots, Z_{y_q}\}$ be this finite subcover. Then there is a finite collection of $(U_\alpha \times V_\alpha)$ which covers $W_{x_0} \times C_2$ which is not possible.

So there is some $y_0 \in C_2$ so that for all neighborhoods V_{y_0}, the set $W_{x_0} \times V_{y_0}$ cannot be finitely covered by $(U_\alpha \times V_\alpha)$. This shows (x_0, y_0) is not in the union of the sets in the collection $(U_\alpha \times V_\alpha)$. This shows $C_1 \times C_2$ is compact.

Now assume we have a finite product of compact sets. $C = C_1 \times \ldots \times C_n$ for some n. How would the argument go now? We assume $W = (U^1_\alpha \times U^n_\alpha)$ is a family of open sets so that if no finite collection from W covers $C_1 \times \ldots \times C_n$, then W is not an open cover of C. We construct a point $x = (x_1, \ldots, x_n) \in C$ so that no neighborhood of x is finitely covered. By the same argument, we used above, there is a point $x_1 \in C_1$ so that for all neighborhoods $W^1_{x_1}$, the set $W^1_{x_1} \times C_2 \times \ldots \times C_n$ cannot be finitely covered by W.

By the same argument, we used above, we again prove there is a point $x_2 \in C_2$ so that for all neighborhoods $W^2_{x_2}$, the set $W^1_{x_1} \times W^2_{x_1} \times C_3 \times \ldots \times C_n$ cannot be finitely covered by W.

We can continue to do this until we find the last point x_n. We then have a point $x = (x_1, \ldots, x_n)$ so that the set $W^1_{x_1} \times \ldots \times W^n_{x_n}$ cannot be finitely covered by W. This shows x is not in the union of the sets from W. This shows C is compact.

Now let's go for the gold. The next case shows that a countable product of compact sets is compact, and the last case, the uncountable product of compact sets is compact. These results are called the Tychonoff Product Theorems. First, we extend the idea of the product topology to arbitrary index sets Λ.

Definition 8.6.3 Cartesian Product over any Index set of Topological Spaces

Let $(X_\alpha, \mathscr{S}_\alpha)$ topological spaces for $\alpha \in \Lambda$. The cross-product topology induced on $\prod_{\alpha \in \Lambda} X_\alpha$ consists of the open sets of the form $\prod_{\alpha \in \Lambda} U_\alpha$ where U_α is open in X_α for $\alpha \in \Lambda$.

Now if the index set λ is countable or uncountable, we can prove the product theorems as follows:

Theorem 8.6.1 Countable Tychonoff Product Theorem for Topological Spaces

Let $(X_i, \mathscr{S}_i)$ be a topological space with $C_i \subset X_i$ compact for all positive integers i. Then $\prod_{i=1}^\infty C_i$ is compact.

Proof 8.6.1

The argument here is a typical induction argument which is a straightforward use of the argument we used for the case of a finite product as the induction argument. ∎

Theorem 8.6.2 Uncountable Tychonoff Product Theorem for Topological Spaces

Let $(X_\alpha, \mathscr{S}_\alpha)$ be a topological space with $C_\alpha \subset X_\alpha$ compact for all $\alpha \in \Lambda$ where Λ is an uncountable index set. Then $\prod_{\alpha \in \Lambda} C_\alpha$ is compact.

Proof 8.6.2
Use the Well Ordering Theorem to well order Λ and then use transfinite induction similar to the way we use regular induction. ■

Now we have the tools to prove the complete Banach-Alaoglu Theorem.

Theorem 8.6.3 Banach-Alaoglu Theorem: Polar of a Neighborhood of 0 in a T_1 Topological Vector Space is Compact in the weak* Topology

Let $(X, \mathscr{S})$ be a T_1 topological vector space and let V be a neighborhood of 0. Let $K = \{x' \in X' || x'(x)| \leq 1, \forall\, x \in V\}$. Then K is weak compact.*

Comment 8.6.1 *If $(X, \|\cdot\|)$ is the normed linear space $\Re^2$ with $V = B(0,1)$. Identify $\Re^2$ as the two component column vectors. Then X' is identified with the two component row vectors. For $x = \begin{bmatrix} x_1 \\ x_2 \end{bmatrix}$, note*

$$
\begin{aligned}
V &= \left\{ x = \begin{bmatrix} x_1 \\ x_2 \end{bmatrix} | x_1^2 + x_2^2 < 1 \right\} \\
K &= \left\{ y = \begin{bmatrix} y_1 \\ y_2 \end{bmatrix} | |x_1 y_1 + x_2 y_2| \leq 1,\ \|x\| < 1 \right\} = \{ y \in \Re^2 || < x, y > | \leq 1,\ \|x\| < 1\}
\end{aligned}
$$

This K is called the polar of the unit ball in $\Re^2$. Note V is a neighborhood of 0 in $\Re^2$ and $K = \{x' \in X' || x'(x)| \leq 1, \forall\, x \in V\}$. Hence, by this theorem, K is weak compact.*

Comment 8.6.2 *If (X, d) is a metric space, then $A \subset X$ compact implies A is sequentially compact. On the other hand, we could have $(X, \mathscr{S})$ is a T_1 topological vector space which is separable. In this case, we know X is metrizable as a subset of X'. This means there is a metric such that the topology induced by the metric is compatible with the existing topology on X' which is weak*. Hence, in this case, weak* compactness implies sequential weak* compactness; i.e. if $(x'_n) \subset K$, there is a subsequence $(x'_{n,1})$ and $x' \in K$ so that $x'_n \stackrel{w*}{\to} x'$ or $x'_n(x) \to x'(x)$ for all $x \in X$.*

We are now ready to prove our result in this general case.

Proof 8.6.3
K is called the polar of V. It is straightforward to see K is convex and balanced. Since a neighborhood of 0 is absorbing, for any $x \in X$, there is $r(x) > 0$, so that $x \in r(x)V$. Then, if $x' \in K$, $|x'((1/r(x))x)| \leq 1$ as $(1/r(x))x \in V$. This is true even if $r(x) = 0$, so we can say $x'(x)| \leq r(x)$ for any $x \in X$ and $x' \in K$.

Let $D_x = \{\alpha : |\alpha| \leq r(x)\} = \overline{B(0, r(x))}$. Let $P = \prod_{x \in X} \overline{B(0, r(x))}$ and use the product topology on this space. Each $\overline{B(0, r(x))}$ is compact in $\Re$ so by Tychonoff's Theorem, P is compact. Also, each element of P can be thought of as a sequence $(\alpha(x))$ which assigns a scalar from $\overline{B(0, r(x))}$. This, of course, uses the Axiom of Choice. Another way to think of this is a function $f : X \to \Re$ with $f(x) = \alpha(x)$ and $|\alpha(x)| \leq r(x)$. It is odd to think of sequences or functions in this context, but it helps with the visualization. Also, we don't know such a function is linear. We see that we can think of a given $x' \in K$ as a function since $x'(x) = \alpha(x)$ with $|\alpha(x)| \leq r(x)$ for all x. So, we can think of $K \subset P$ and $K \subset X'$ or $K \subset P \cap X'$. Hence, K inherits two topologies: (1) the weak topology and (2) the product topology from P. For convenience, call the topology of case (2), τ. We will show*

(i): The τ topology and the weak topology coincide.*

(ii): K is a closed subset of P.

If (ii) is true, K is closed with respect to the τ topology and so K^C is open. Since P is compact, any open cover of K, (O_β), is made into an open cover of P by adding K^C. Hence, there is a finite subcover $(O_n) \subset (O_\beta)$ so that $(O_n) \cup K^C$ cover P. We see (O_n) covers K so K must be compact with respect to the τ topology. If we establish (i), that the weak topology is the same as the τ topology, this will tell us K is weak* compact.*

Proof of (i): *Fix $x'_0 \in K$. Choose $x_i \in X$, $1 \leq i \leq n$ for some n and choose $\delta > 0$. Let*

$$\begin{aligned} W_1 &= \{x' \in X' | |x'(x_i) - x'_0(x_i)| < \delta, \ 1 \leq i \leq n\} \\ W_2 &= \{f \in P | f(x_i) - x'_0(x)| < \delta, \ 1 \leq i \leq n\} \end{aligned}$$

Then, define the following collections.

$$\begin{aligned} \Omega_1 &= \{W_1 | n \in \mathbb{N}, \ \{x_1, \ldots, x_n\} \subset X, \ \delta > 0\} \\ \Omega_2 &= \{W_2 | n \in \mathbb{N}, \ \{x_1, \ldots, x_n\} \subset X, \ \delta > 0\} \end{aligned}$$

Then Ω_1 is a local base for the weak topology of X' at x'_0. Further Ω_2 is a local base for the product topology τ of P at x'_0. Since $K \subset P \cap X'$, if $x' \in \Omega_1 \cap K$, $x' \in P$ as well. So given any set W_1, $|x'(x_i) - x'_0(x_i)| < \delta$, $1 \leq i \leq n$. But $x' \in P$ too, so x' is in a W_2 set. This shows $\Omega_1 \cap K \subset \Omega_2 \cap K$. A similar argument shows $\Omega_2 \cap K \subset \Omega_1 \cap K$.*

Hence, we conclude $\Omega_2 \cap K = \Omega_1 \cap K$ and hence, τ and weak are the same on K. This proves (i).*

Proof of (ii): *We want to show K is closed in τ. Let the closure of K in τ be denoted by $\overline{K_\tau}$. Let $f_0 \in \overline{K_\tau}$ and let $x, y \in X$, $\alpha, \beta \in \Re$ and choose $\epsilon > 0$. Consider this set*

$$B = \{f \in P | \ |f(x) - f_0(x)| < \epsilon, \ |f(y) - f_0(y)| < \epsilon, \ |f(\alpha x + \beta y) - f_0(\alpha x + \beta y)| < \epsilon\}$$

Note B_1 is a W_2 type neighborhood of f_0 in P. Since $K \subset P \cap X'$, every neighborhood of f_0 in τ must contain elements of K. For B_1, for such a $g \in K$, we know g is linear as it is in X' and

$$f_0(\alpha x + \beta y) - \alpha f_0(x) - \beta f_0(y) = (f_0 - g)(\alpha x + \beta y) + \alpha(g - f_0)(x) + \beta(g - f_0)(y)$$

So,

$$\begin{aligned} & |f_0(\alpha x + \beta y) - \alpha f_0(x) - \beta f_0(y)| \\ \leq \ & |(f_0 - g)(x)(\alpha x + \beta y)| + |\alpha||(g - f_0)(x)| + |\beta||(g - f_0)(y)| \\ < \ & \epsilon + |\alpha|\epsilon + |\beta|\epsilon = (1 + \alpha + \beta)\epsilon \end{aligned}$$

But the choice of ϵ was arbitrary. Hence, we conclude f_0 is actually linear.

Now, if $x \in V$ and $\delta > 0$, $\{f \in P| \ |f(x) - f_0(x)| < \epsilon\}$ is a τ neighborhood of f_0. This implies there is a $g \in K$ so that $|g(x) - f_0(x)| < \epsilon$. Hence, $|f_0(x)| \leq |g(x)| + \epsilon < 1 + \epsilon$ as $g \in K$. Again, ϵ is arbitrary, so we have $|f_0(x)| \leq 1$ for all $x \in V$ showing $f_0 \in K$. This shows K is closed in τ and establishes (ii). ■

Comment 8.6.3 *This theorem tells us that if $(X, \mathscr{S})$ is a T_1 topological vector space and $V = B(0,1)$, then $K = \{x' \in X' | \ |x'(x)| \leq 1, \ \forall x \in B(0,1)\}$ is weak* compact.*

Comment 8.6.4 *If $(X, \|\cdot\|)$ is a normed linear space and $V = B(0,1)$, then $K = \{x' \in X' | |x'(x)| \leq 1, \ \|x\| < 1\}$ is weak* compact. Also, note $\|x\| < 1$ implies $\sup_{\|x'\|=1} |x'(x)| < 1$.*

Homework

Exercise 8.6.1 *Provide the details for the proof of Theorem 8.6.2.*

Exercise 8.6.2 *If* $(X, \|\cdot\|)$ *is the normed linear space* $\Re^2$ *with* $V = B(0,1)$*. Identify* $\Re^2$ *as the two component column vectors. Then* X' *is identified with the two component row vectors. For* $x = \begin{bmatrix} x_1 \\ x_2 \end{bmatrix}$*, let*

$$\begin{aligned} V &= \left\{ x = \begin{bmatrix} x_1 \\ x_2 \end{bmatrix} |x_1^2 + x_2^2 < 1 \right\} \\ K &= \left\{ y = \begin{bmatrix} y_1 \\ y_2 \end{bmatrix} ||x_1 y_1 + x_2 y_2| \leq 1,\ \|x\| < 1 \right\} = \{ y \in \Re^2 || < x, y > | \leq 1,\ \|x\| < 1 \} \end{aligned}$$

This K *is called the polar of the unit ball in* $\Re^2$*. Note* V *is a neighborhood of* 0 *in* $\Re^2$ *and* $K = \{x' \in X' | |x'(x)| \leq 1,\ \forall\, x \in V\}$*. Hence, by this theorem,* K *is weak* compact. Explain* K *visually in* $\Re^2$*. What does weak* compact mean in* $\Re^2$*?*

Exercise 8.6.3 *If* $(X, \|\cdot\|)$ *is the normed linear space* $\Re^3$ *with* $V = B(0,1)$*. For* $x = \begin{bmatrix} x_1 \\ x_2 \\ x-3 \end{bmatrix}$*, note*

$$\begin{aligned} V &= \left\{ x = \begin{bmatrix} x_1 \\ x_2 \\ x_3 \end{bmatrix} |x_1^2 + x_2^2 + x_3^2 < 1 \right\} \\ K &= \left\{ y = \begin{bmatrix} y_1 \\ y_2 \\ y_3 \end{bmatrix} ||x_1 y_1 + x_2 y_2 + x_3 y_3| \leq 1,\ \|x\| < 1 \right\} \\ &= \{ y \in \Re^3 || < x, y > | \leq 1,\ \|x\| < 1 \} \end{aligned}$$

This K *is called the polar of the unit ball in* $\Re^3$*. Note* V *is a neighborhood of* 0 *in* $\Re^2$ *and* $K = \{x' \in X' | |x'(x)| \leq 1,\ \forall\, x \in V\}$*. Hence, by this theorem,* K *is weak* compact. Explain* K *visually in* $\Re^2$*.*

What does weak compact mean in* $\Re^2$*?*

Exercise 8.6.4 *In* $\Re^2$*, let* $V = (-1,1) \times (-1,1)$ *and define* K *as usual. Explain* K *visually in* $\Re^2$ *and explain why* K *is weak* compact.*

What does weak compact mean here?*

Exercise 8.6.5 *In* $\Re^2$*, let* V *be the interior of a simplex formed by the intersection of three lines that contains* 0 *and define* K *as usual. Explain* K *visually in* $\Re^2$ *and explain why* K *is weak* compact.*

What does weak compact mean here?*

Recall, we have proved two useful results: the first is in a normed linear space setting

Theorem The Existence of Weak* Subsequential Limits in Bounded Sets

> *Let* $(X, \|\cdot\|)$ *be a normed linear space that is separable. Then, every bounded sequence in* X' *contains a subsequence which converges in the weak* topology on* X'*.*

and the second is in a topological vector space setting.

Theorem The Banach-Alaoglu Theorem: Polar of a Neighborhood of 0 in a T_1 Topological Vector Space is Compact in the weak* Topology

Let $(X, \mathscr{S})$ be a T_1 topological vector space and let V be a neighborhood of 0. Let $K = \{x' \in X' | |x'(x)| \leq 1, \forall\, x \in V\}$. Then K is weak compact.*

We can do a bit more if we add separability to the topological vector space.

Theorem 8.6.4 Banach-Alaoglu Theorem: Polar of a Neighborhood of 0 in a T_1 Separable Topological Vector Space is Metrizable and so Weak* Compact

Let $(X, \mathscr{S})$ be a T_1 topological vector space that is separable. Let V be a neighborhood of 0. Let $K = \{x' \in X' | |x'(x)| \leq 1, \forall\, x \in V\}$. Then K is metrizable and hence, weak compact.*

Proof 8.6.4

Let $(X, \mathscr{S})$ be a T_1 topological vector space which is separable. Let $(x_n) \subset V$ be dense. Define the functions f_n by $f_n : X' \rightarrow \Re$ by $f_n(x') = x'(x_n)$. Hence, $f_n \in X''$ and f_n is defined via the usual canonical embedding; i.e. $f_n = \phi(x_n)$.

As an aside, we note if $f' \neq g'$, $f_n(f') - f_n(g') = f'(x_n) - g'(x_n)$. We know there is a $y \in V$ so that $f'(y) \neq g'(y)$. Assume, $f'(y) \neq g'(y)$. Then there are disjoint neighborhoods $C_{f'(y)}$ and $C_{g'(y)}$ of $f'(y)$ and $g'(y)$, respectively. The continuity of f' and g', then implies there are neighborhoods U_y and V_y for y so that $f'(U_y) \subset C_{f'(y)}$ and $g'(V_y) \subset C_{g'(y)}$. Let $D_y = U_y \cap V_y$. But the denseness of (x_n) in V means D_y contains x_p for some element from (x_n). Thus, $f'(x_p) \in f'(D_y) \subset C_{f'(y)}$ and $g'(x_p) \in g'(D_y) \subset C_{g'(y)}$. But $C_{f'(y)}$ is disjoint from $C_{g'(y)}$. So $f'(x_p) \neq g'(x_p)$ or $f_p(f') \neq f_p(g')$. Hence, (f_n) separates points on X'.

This implies we can define a metric on K like this: for any $f', g' \in K$,

$$d(f', g') = \sum_{n=1}^{\infty} \frac{1}{2^n} |f_n(f') - g_n(g')| = \sum_{n=1}^{\infty} \frac{1}{2^n} |f'(x_n) - g'(x_n)| \leq \sum_{n=1}^{\infty} \frac{2}{2^n} M < \infty$$

We note the series always converges and so K is metrizable. This also shows K is compact in the weak topology so our argument is another way of proving Theorem 8.6.3.* ■

Homework

Exercise 8.6.6 *In the proof of Theorem 8.6.4, provide the details for why for any $f', g' \in K$,*

$$d(f', g') = \sum_{n=1}^{\infty} \frac{1}{2^n} |f_n(f') - g_n(g')| = \sum_{n=1}^{\infty} \frac{1}{2^n} |f'(x_n) - g'(x_n)| \leq \sum_{n=1}^{\infty} \frac{2}{2^n} M < \infty$$

the series always converges. Explain why this means K is metrizable.

Exercise 8.6.7 *Explain why this theorem tells us that if $(X, \mathscr{S})$ is a separable T_1 topological vector space, then $V = B(0,1)$, then $K = \{x' \in X' |\, |x'(x)| \leq 1, \ \forall x \in B(0,1)\}$ is weak* compact.*

Exercise 8.6.8 *Explain why this theorem tells us that if $(X, \mathscr{S})$ is a separable normed linear space and $V = B(0,1)$, then $K = \{x' \in X' |\, |x'(x)| \leq 1, \ \|x\| < 1\}$ is weak* compact.*

8.7 Separation Ideas

Now let's consider a brief introduction to separation ideas. We start with a fundamental lemma.

Lemma 8.7.1 Compactness and Neighborhoods in a Topological Vector Space

Let $(X, \mathscr{S})$ be a T_1 topological vector space and K and C be subsets with K compact and C closed. Assume $K \cap C = \emptyset$. Then there is a neighborhood V of 0 so that $(K+V) \cap (C+V) = \emptyset$.

Proof 8.7.1
If $K = \emptyset$, we interpret $K + V = \cup_{x \in K}(x+V) = \emptyset$ and so any V works.

Hence, we can assume $K \neq \emptyset$. This tells us $K+V \neq \emptyset$. Let $x \in K$, since $x \in C^C$, there is a neighborhood U of 0 so that $x + U \subset C^C$. Note, we use the translation invariance here. We also know there is a neighborhood of 0, W_0 which is symmetric, W_0 which is symmetric, so $W_0 = -W_0$ with $W_0 + W_0 \subset U$. This, of course, also means $W_0 \subset U$. We also know there is a symmetric neighborhood W_1 of W_0 with $W_1 + W_1 \subset W_0$ and $W_1 \subset W_0$. Thus, $W_1 + (W_1 + W_1) \subset W_1 + W_0 \subset W_0 + W_0 \subset U$. We conclude there is a symmetric neighborhood of 0, W_1, so that $x + W_1 + W_1 + W_1 \subset C^C$.

Now, if $(x + W_1 + W_1) \cap (C + W_1)$ is not empty, there is a $y = x + w_1 + w_2 = c_1 + w_3$ with $w_i \in W_1$ and $c_1 \in C$. Thus, $x + w_1 + w_2 - w_3 \in x + W_1 + W_1 + W_1$ as $W_1 = -W_1$ and $x + w_1 + w_2 - w_3 = c_1 \in C$. But this cannot be true, as $x + W_1 + W_1 + W_1 \subset C^C$. We conclude $(x + W_1 + W_1) \cap (C + W_1) = \emptyset$. We did all this for the point $x \in K$. Let's call the neighborhood W_1 that we have found, W_x, to make this clear.

We can do this for each $x \in K$. Then, for all $x \in K$, there is a symmetric neighborhood of 0, W_x, so that

$$x + W_x + W_x + W_x \subset C^C, \quad (x + W_x + W_x) \cap (C + W_x) = \emptyset$$

The neighborhoods $(x+W_x)$ are an open cover of K. So there is a finite subcover $\{x_1+W_{x_1}, \ldots, x_n+W_{x_n}\}$ for some n with $K \subset \cup_{i=1}^n (x_i + W_{x_i})$. Let $V = \cup_{i=1}^n W_{x_i}$. Then,

$$K + V \quad \subset \quad \cup_{i=1}^n (x_i + W_{x_i} + V) \subset \cup_{i=1}^n (x_i + W_{x_i} + W_{x_i})$$

We know, $(x + W_x + W_x) \cap (C + W_x) = \emptyset$. Hence, $\cup_{i=1}^n (x_i + W_{x_i} + W_{x_i}) \cap (C + W_{x_i}) = \emptyset$. So, we have found a neighborhood of 0, V, so that $(K+V) \cap (C+V)\ \emptyset$. ■

This leads to our first separation theorem.

Theorem 8.7.2 Separation Theorem One: Separating Sets with Linear Functionals

Let $(X, \mathscr{S})$ be a T_1 topological vector space. Let A and B be disjoint, nonempty convex subsets of X.

(i): If A is open, there is $x' \in X'$ and $r \in \Re$ so that $x'(x) \leq x'(y)$ for all $x \in A$ and $y \in B$.

(ii): If A is compact, B is closed, and X is also locally convex, there is $x' \in X'$ and $r_1, r_2 \Re$ so that $x'(x) < r_1 < r_2 < x'(y)$ for all $x \in A$ and $y \in B$.

Proof 8.7.2
Fix $a_0 \in A$ and $b_0 \in B$. Let $x_0 = b_0 - a_0$ and $C = A - B + x_0$ which is a set containing 0 by the way we defined x_0. It is straightforward to prove C is convex which implies it is absorbing.

Let p be the Minkowski functional of C. We know $p(x) \leq 1$ if $x \in C$ and $p(x) < 1$ implies $x \in C$. So if $x \notin C$, $p \geq 1$. Also, if

$$\begin{aligned} x_0 = b_0 - a_0 \in C &\implies \exists\, a \in A,\ b \in B \\ &\implies b_0 - a_0 = a - b + b_0 - a_0 \implies 0 = a - b \implies A \cap B \neq \emptyset \end{aligned}$$

This is not possible, so we know $x_0 = b_0 - a_0 \in C^C$ and so $p(x_0) \geq 1$. We also know $p(x+y) \leq p(x) + p(y)$ and $p(tx) = tp(x)$ if $t > 0$.

Let $M = span\{x_0\}$. Define $f : M \to \Re$ by $f(tx_0) = t$. We see if $t > 0$, $f(tx_0) = t = t \times 1 \leq tp(x_0) \leq p(tx_0)$. When $t = 0$, $f(tx_0) = 0 = p(0)$. Finally, if $t < 0$, $f(tx_0) = t < 0 \leq p(tx_0)$. We conclude $f \leq p$ on M. Apply the Hahn - Banach Theorem to conclude there is a linear functional x' so that $x'|_M = f$ and $x' \leq p$ on X.

Then, note for any $x \in X$,

$$-p(-x) \;\leq\; -x'(x) = x'(x) \leq p(x) \implies -p(-x) \leq x'(x) \leq p(x)$$

Now if $x \in C$, $p(x) < 1$. So, $x'(x) \leq p(x) < 1$ on C.

To show x' is continuous, it is enough to show it is continuous at 0 as x' is linear. Since $-p(-x) \leq x'(x) \leq p(x)$ for all $x \in X$,

$$\begin{aligned} x \in C &\implies x'(x) \leq p(x) < 1 \\ x \in -C &\implies x'(x) \geq -p(-x) \end{aligned}$$

But if $x \in -C$, $-x \in C$ and so $p(-x) < 1$ or $-p(-x) > -1$. We conclude if $x \in C \cap (-C)$, a neighborhood of 0, we have $-1 < x^(x) < 1$ or $|x'(x)| < 1$. Now if $\hat{x} \in C \cap (-C)$, $\alpha\hat{x} \in C$ if $0 \leq |\alpha| < \epsilon_0$ because C is absorbing. We then see $-1 < x'((1/\alpha)\hat{x}) < 1$ for all $0 < |\alpha| < \epsilon_0$. This tells us $|x'(x)| < |\alpha|$ if $x = \alpha\hat{x}$. In particular, for any $\epsilon < \epsilon_0$, we have $|x'(x)| < \epsilon$ if $x = \beta\hat{x}$ for $0 \leq |\beta| < \epsilon$. Thus, $x = \beta\hat{x} \in (x')^{-1}(-\epsilon, \epsilon)$ for $0 \leq |\beta| < \epsilon$. There is a neighborhood of 0, U, so $\hat{x} + U \subset (x')^{-1}(-\epsilon, \epsilon)$. This shows x' is continuous at 0. Since x' is linear, x' is continuous on X. We see $x' \in X'$.

Finally, if $a \in A$ and $b \in B$, we have, since x' matches f and x_0,

$$x'(a) - x'(b) + 1 \;=\; x'(a-b) + x'(x_0) = x'(a - b + x_0) \leq p(a - b + x_0)$$

But, $a - b + x_0 \in A - B + x_0$ and so $p(a - b + x_0) < 1$. So $x'(a) - x'(b) + 1 < 1$ implying $x'(a) < x'(b)$. This result is true for any $a \in A$ and $b \in B$ and so $x'(A) \cap x'(B) = \emptyset$.

Since A is open and x' is a homeomorphism, $x'(A)$ is an open set. We can show easily $x'(A)$ and $x'(B)$ are convex sets because A and B are convex. Now $x'(A)$ open in $\Re$ means we can write $x'(A) = \cup_{n=1}^{\infty}(a_n, b_n)$. Since $x'(B)$ is disjoint from $x'(A)$, this means $x'(b) \geq b_n$ for all endpoints b_n. Let $r = \sup_{n=1}^{\infty} b_n \leq x'(b)$ for any $b \in B$, so we know r is finite. We thus have $x'(a) < r \leq x'(b)$ which is what we wanted to show.

(ii):
Here, A is compact, B is closed and X is a locally convex T_1 topological vector space. Hence, by Lemma 8.7.1, there is a neighborhood of 0, V so that $(A+V) \cap (B+V) = \emptyset$. In particular, $0 \in V$, so $(A+V) \cap B = \emptyset$. Now since we also know X is locally convex, we may assume V is convex. We have $A+V$ is open and if $a_1 + v_1$ and $a_2 + v_2$ are in $A+V$, then for $0 \leq t \leq 1$,

$$t(a_1 + v_1) + (1-t)(a_2 + v_2) = (ta_1 + (1-t)a_2) + (tv_1 + (1-t)v_2) \in A + V$$

as A and V are convex. Hence, $A+V$ is convex. Now apply (i) to the sets $A+V$ and B. There is $x' \in X'$ with $x'(A+V)$ and $x'(B)$ disjoint, $x'(A+V)$ is open and lying to the left of $x'(B)$.

Since x' is continuous, $x'(A)$ is compact. Therefore, there are α and β so that $\alpha \leq x'(A) \leq \beta$. Choose $u \in x'(A+V) \setminus x'(A)$ which is the set difference of an open set and a compact set inside it. Since u is an interior point, we see there is $r > 0$ so that $(u-r, u+r) \cap x'(A) = \emptyset$ and $(u-r, u+r) \subset x'(A+V)$. Hence, if $a \in A$ and $b \in B$, $x'(a) < u - r < u + r < x'(b)$. Let $r_1 = u - r$ and $r_2 = u + r$. This proves the assertion. ■

Homework

Exercise 8.7.1 *Provide the details of the proof of Theorem 8.7.2 in the case A and B are disjoint, nonempty convex sets of $\Re^2$. Draw pictures for several illustrative cases: the sets are bounded and the sets are lines.*

Exercise 8.7.2 *Provide the details of the proof of Theorem 8.7.2 in the case A and B are disjoint, nonempty convex sets of $\Re^3$. Draw pictures for several illustrative cases: the sets are bounded and the sets are lines or planes.*

Exercise 8.7.3 *Provide the details of the proof of Theorem 8.7.2 in the case A and B are disjoint, nonempty convex sets of $\Re^n$. Of course, you can no longer draw pictures!*

We are now in a position to use the separation theorem to obtain some useful results.

Theorem 8.7.3 If X is a Locally Convex T_1 Topological Vector Space, then X' Separates Points

Let $(X, \mathscr{S})$ be a locally convex T_1 topological vector space. Then X' separates points.

Proof 8.7.3
Let $x_1 \neq x_2$ be in X. Let $A = \{x_1\}$ and $B = \{x_2\}$. Then A is closed, nonempty, convex and compact. The set B is closed, nonempty and convex. Now apply (ii) of Theorem 8.7.2 to find $x' \in X'$ with real numbers r_1 and r_2 so that $x'(x_1) < r_1 < r_2 < x'(x_2)$. Thus X' separates points. ■

We can also find continuous linear functionals which are zero on a given subspace and have the value 1 at a given point outside the closure of the subspace.

Theorem 8.7.4 Subspace Separation Theorem in a Locally Convex T_1 Topological Vector Space

Let $(X, \mathscr{S})$ be a locally convex T_1 topological vector space. Let $M \subset X$ be a subspace and $x_0 \notin \overline{M}$. Then there is $x' \in X'$ so that $x'|_M = 0$ and $x'(x_0) = 1$.

Proof 8.7.4
Let $A = \{x_0\}$ and $b = \overline{M}$. We can then apply (ii) of Theorem 8.7.2 to find $x' \in X'$ with real numbers r_1 and r_2 so that $x'(x_0) < r_1 < r_2 < x'(x)$ for all $x \in \overline{M}$. This shows $x'(x_0)$ and $x'(\overline{M})$

are disjoint. However, $x'(M)$ *is a subspace of* $\Re$ *which means it is* $\Re$*. Since it does not contain* $x'(x_0)$*, we must have* $x'|_M = 0$ *and* $x'(x_0) \neq 0$*. Let* $y' = (1/x'(x_0))x'$*. Then* y' *is the continuous linear functional we seek.* ■

Comment 8.7.1 *As an application, if you want to show that a given point* x_0 *in a locally convex* T_1 *topological vector space lies in* $\overline{M}$ *for a subspace* M *of* X*, Theorem 8.7.4 tells us if* $x_0 \notin \overline{M}$*, there is* x' *with* $x'(x_0) = 1$ *and* $x'|_M = 0$*. So if you can show* $x'(x_0) = 0$ *for all* x' *satisfying* $x'|_M = 0$ *this would imply* $x_0 \in M$*.*

Next, let's look at weak closures.

Theorem 8.7.5 Weak Closure Equals the Closure in a Locally Convex T_1 **Topological Vector Space**

Let $(X, \mathscr{S})$ *be a locally convex* T_1 *topological vector space and* $E \subset X$*. Then the weak closure of E, denoted by* $\overline{E_w}$ *is the same as the closure of E,* $\overline{E}$*.*

Proof 8.7.5

If $(x_n) \subset X$ *converges weakly to* $x \in X$*, then given any neighborhood of* 0*,* V*, there is* N *so* $x_n - x \in V$ *if* $n > N$*. Pick any* $x' \in X'$*. Then* $U = \{x||x'(x)| < \epsilon\}$ *is open as* x' *is continuous. We note it contains* 0 *and if* $n > N$*, then* $x_n - x \in V$ *and* $|x'(x_n) - x'(x)| < \epsilon$*. This shows* $x_n \stackrel{w}{\rightarrow} x$*.*

So if $x_n \rightarrow x$ *in* X*, with* $(x_n) \subset E$*,* $x \in \overline{E}$*. However, we also know* $x_n \stackrel{w}{\rightarrow} x$ *which shows* $x \in \overline{E_w}$*. Hence* $\overline{E} \subset \overline{E_w}$*.*

To show the reverse, choose $x_0 \in \overline{E}^C$*. If* $E = X$*, our conclusion is easy, so we can assume* $\overline{E}^C$ *is not empty. Let* $A = \{x_0\}$ *and* $B = \overline{E}$ *and apply (ii) from Theorem 8.7.2 to find* $x' \in X'$ *with real numbers* r_1 *and* r_2 *so that* $x'(x_0) < r_1 < r_2 < x'(x)$ *for all* $x \in \overline{E}$*. Now look at* $W = \{x||x'(x)| < r_1\}$*. This is a neighborhood of* 0 *which is disjoint from* E *(some call this a weak neighborhood which is a bit confusing we think). However, it is true this tells us* x_0 *cannot be a weak limit of a sequence in* E*. Thus,* $x_0 \in \overline{E_w}^C$*. We have shown* $x_0 \in \overline{E}^C$ *implies* $x_0 \in \overline{E_w}^C$*. Hence,* $\overline{E_w} \subset \overline{E}$*.*

Combining these results, we see $\overline{E_w} = \overline{E}$*.* ■

Homework

Exercise 8.7.4 *Let* $(X, \mathscr{S})$ *be a locally convex* T_1 *topological vector space and* $M \subset X$ *a subspace. Then* M *is closed if and only if* M *it is weakly closed.*

Exercise 8.7.5 *Let* $(X, \mathscr{S})$ *be a locally convex* T_1 *topological vector space and* $M \subset X$ *a convex set. Then* M *is dense in* X *if and only if* M *it is weakly dense, i.e. dense with respect to weak convergence.*

8.8 Krein - Milman Results

We need to define what we mean by an extreme point in more general spaces. In linear programming, you are used to thinking of the vertices of the set $\{x \in \Re^n : Ax \leq b\}$ for some $b \in \Re^n$ as extreme points, but we want to look at this idea more carefully. The set above is convex and it turns out, that helps us think about these points in the correct way. Let's start by defining what we mean by extreme sets in a vector space.

Definition 8.8.1 Extreme Sets in a Vector Space

Let K be a subset of the vector space X. A nonempty subset S of K is an extreme set of K if no point in S can be written as a proper convex combination of two points in K, not in S. We say y is a proper convex combination means of x and y from K but not in S, if $y = tx + (1-t)y$, for some $0 < t < 1$. If S has only one point in it, S is called an extreme point.

Comment 8.8.1 *If K is convex, then you cannot find proper convex combinations of points in K that are not in K, so K is an extreme set.*

Comment 8.8.2 *Let K be a nonempty triangle in $\Re^2$ where K consists of the union of the interior of the triangle and its perimeter. The perimeter of the triangle then consists of the vertices and edges that define the boundary of the triangle. Draw such a triangle. Then if you pick an edge, no point on the edge is a proper combination of two points not on that edge. So an edge is an extreme set of K. If you pick a vertex, then it is not a proper convex combination of two points from $K \setminus \{v\}$ where v is the vertex.*

Comment 8.8.3 *Another way of saying this is that $S \subset K$ is an extreme set if*

(i): Given x and y in K and $0 < t < 1$, then $tx + (1-t)y \in S$ implies x and y are in S.

(ii): No point of an extreme set S is an internal point of a line segment whose endpoints are not in S.

We are now ready for the main event.

Theorem 8.8.1 Krein - Milman Theorem

Let $(X, \mathscr{S})$ be a T_1 topological vector space and assume X' separates points. If $K \subset X$ is compact and convex, then if $H = \{p \,|\, p \text{ is an extreme point of } K\}$ and $co(H)$ is the convex hull of H, then $\overline{co(H)} = K$.

Comment 8.8.4 *Note if X is locally convex, X' separates points.*

Comment 8.8.5 *You can see in our triangle in $\Re^2$ example, if K is the boundary of the triangle, then K is not convex as if you take two points on different sides, the line between them contains points in the interior of the triangle which are not on the boundary. If you take convex combinations of points on two different edges, you will get points interior to the triangle. So the three edges, namely the boundary of the triangle, cannot be an extreme set.*

If K is the boundary of the triangle and the interior, then K is compact and convex. Each vertex of the triangle is an extreme point. The convex hull of the three vertices is the boundary of the triangle and the interior of the triangle. Here H is the set of the three vertices and $\overline{co(H)}$ is the same as $co(H)$ and equals K.

Proof 8.8.1
Let $\mathbb{P} = \{A \subset K | A \text{ is compact and extreme }\}$. We know since K is convex and compact, that $\mathbb{P}$ is not empty. We will partially order this with set inclusion. Let's show $\mathbb{P}$ has two properties.

(i): The intersection S of any nonempty subcollection of $\mathbb{P}$ is a member of $\mathbb{P}$ unless $S = \emptyset$.

(ii): If $S \in \mathbb{P}$, $x' \in X'$ and $\mu = \sup_{x \in S} x'(x) = \max_{x \in S} x'(x)$ since S is compact. Let $S_{x'} = \{x \in S | x'(x) = \mu\}$. Then $S_{x'} \in \mathbb{P}$.

(i): *Let (A_α) be a collection of sets from $\mathbb{P}$ and let $S = \cap_\alpha A_\alpha$. Then, each A_α is compact and extreme. If x and y are both in S, then x and y are both in each A_α. Let u and v be in K. Then* *for $0 < t < 1$, $tu + (1-t)v \in A_\alpha$ implies u and v must be in A_α. So, if $tu + (1-t)v \in S$, $tu + (1-t)v \in A_\alpha$ for all α and hence, u and v must be in each A_α. We conclude u and v are in S and so S is an extreme set.*

(ii): *Let x and y be in K and $0 < t < 1$ If $z = tx + (1-t)y \in S_{x'}$. Then, by definition, $z \in S$ and $x' = \mu$. Since S is extreme, it follows that x and y must be in S also. Hence, $x'(x) \leq \mu$ and $x'(y) \leq \mu$. Then, by the linearity of x',*

$$\mu \quad = \quad x'(z) = x'(tx + (1-t)y) = tx'(x) + (1-t)x'(y) \leq \mu$$

So $x'(z) = x'(tx + (1-t)y) = \mu$ for all $0 < t < 1$. Now x' is continuous and therefore, $x'(x) = x'(y) = 1$. We conclude x and y are in $S_{x'}$. Therefore $S_{x'}$ is extreme. We leave showing $S_{x'}$ is compact to you as an exercise. So $S_{x'} \in \mathbb{P}$.

Next, pick any S in $\mathbb{P}$. Let $\mathbb{P}' = \{B \in \mathbb{P} | B \subset S\}$. This is the same as

$$\mathbb{P}' = \{B \subset S | B \text{ is compact and convex }\}.$$

So $\mathbb{P}'$ is a collection like our original collection $\mathbb{P}$. Hence, properties (i) and (ii) hold for $\mathbb{P}'$. Since $S \in \mathbb{P}'$, $\mathbb{P}'$ is not empty. Partially order $\mathbb{P}'$ by set inclusion. Now a collection C from $\mathbb{P}'$ is totally ordered if given any pair B_1 and B_2 from C, either $B_1 \subset B_2$ or $B_2 \subset B_1$. Now recall the Hausdorff Maximality Principle: every nonempty partially ordered set contains a maximal chain or totally order collection. Apply this to $\mathbb{P}'$. Call this maximal chain Ω. Then Ω is maximal with respect to the property of being totally ordered; i.e., if Ψ is another chain of subsets of $\mathbb{P}'$, then $\Omega \subset \Psi$ implies $\Omega = \Psi$. Let $\Omega = (O_\alpha)$ and let $M = \cap_\alpha O_\alpha$. We can assume each O_α is nonempty, of course. Another way to characterize compactness in a Hausdorff topological vector space (which we have here as the space is T_1) is that a subset $D \subset X$ is compact if and only if every family of closed sets in D has a nonempty intersection whenever each finite subfamily has a nonempty intersection. This is called the finite intersection property. Here, each O_α is a compact subset of K which is compact. So the collection Ω has nonempty intersection as long as each finite subcollection has nonempty intersection.

Choose $\{B_1, \ldots, B_n\}$ from (O_α). Then since the family is totally ordered we may assume without loss of generality that $B_1 \subset B_2 \subset \ldots \subset B_n$ and so $\cap_{i=1}^n B_i = B_1$ which is nonempty (and compact). Since each finite subcollection has nonempty intersection, we then know $\cap_\alpha O_\alpha$ is nonempty. From (i), we therefore know $M \in \mathbb{P}'$. Now assume Y is a proper subset of M and $Y \subset\in \mathbb{P}$. Then Y is a proper subset of O_α for all α and so a subset of S. So $Y \in \mathbb{P}'$ and so $Y \in \mathbb{P}$.

Consider the collection $\{(O_\alpha), Y\}$ of $\mathbb{P}'$. Since $Y \subset O_\alpha$ for all α, this collection of totally ordered. But it contains Ω properly which contradicts the fact that Ω is maximal. We conclude no proper subset Y of M is in $\mathbb{P}$. Now apply (ii) to M. For any x', $M_{x'} = \{x \in M | x'(x) = \max_{x \in M} x'(x)\}$ is in $\mathbb{P}$. Thus, $M_{x'}$ is an extreme compact set contained in K. In addition, since no proper subset of M is in $\mathbb{P}$, we must have $M_{x'} = M$. We conclude x' is constant on M.

Now X' separates points on X. So if x and y are in M with $x \neq y$, there must be x' with $x'(x) \neq x'(y)$. But x' is constant on M so this tells us M must be a single point. We see M is therefore an extreme point of K with $M \subset S$. Thus, each $S \in \mathbb{P}$ contains an extreme point of K.

Let H be the convex hull of the extreme points of K. Then, by the discussion above, since each $S \subset K$ contains a point of H, $H \cap S$ is not empty for all $S \in \mathbb{P}$. Now, K is compact and convex and all extreme points of K are in K. By definition, H is the intersection of all convex sets which contain *the extreme points of K. So $H \subset K$. Since K is closed, we have $\overline{H} \subset K$.*

Now assume $x_0 \in K$ but not in $\overline{H}$. Apply the separation theorem, Theorem 8.7.2 to $A = \overline{H}$ and $B = \{x_0\}$. Then there is x' and r_1 and r_2 so that $x'(x) < r_1 < r_2 < x'(x_0)$, for all $x \in \overline{H}$. The set $B = \{x_0\}$ is in $\mathbb{P}$ because $x_0 \in K$. Now apply (ii), to find for $S = \overline{H}$, that $\overline{H}_{x'} = \{x \in \overline{H} | x'(x) = \max_{x \in \overline{H}} x'(x)\}$ is in $\mathbb{P}$. But $x'(x_0) > \max_{x \in \overline{H}} x'(x)$ and so $x_0 \notin \overline{H}$; i.e. $\overline{H} \cap \{x_0\} = \emptyset$. But earlier, we found $H \cap S$ is not empty for all $S \in \mathbb{P}$. This is a contradiction, so such x_0 cannot exist. We have shown $\overline{H} = K$. ■

Homework

Exercise 8.8.1 *Prove the set $S_{x'}$ discussed in the proof above is a compact set.*

Exercise 8.8.2 *$H \subset \Re^n$ is called a hyperplane if there is a nonzero $d\Re^n$ and a scalar c so that $< d, x >= c$ for all $x \in H$. Let $E \subset \Re^n$ be a convex set with nonempty interior and let $y \in \partial E$. Prove there is a $d \neq 0 \in \Re^n$ so that $< d, x >= c$ so that $y \in H$ and all points of $u \in E$ satisfy $< d, u >< c$ or $< d, u >> c$. H is called a separating hyperplane for E.*
Hint: *Assume 0 is an interior point of E and apply the Hahn - Banach theorem.*

Exercise 8.8.3 *Let K be a compact and convex set in $\Re^n$. Prove every point $x \in K$ is a convex combination of at most $n + 1$ extreme points of K.*
Hint: *Use induction on n. Draw a line from a given extreme point of K through the point x to where it leaves K. We use the previous exercise here.*

Exercise 8.8.4 *Another way to characterize compactness in a Hausdorff topological vector space is that a subset $D \subset X$ is compact if and only if every family of closed sets in D has a nonempty intersection whenever each finite subfamily has a nonempty intersection. This is called the finite intersection property. Prove this assertion. Feel free to do some reading here. We already used an alternate characterization of compactness when we talked about Tychonoff's Theorem for compactness in the product topology. Look this up and make sure you can understand the reasoning.*

Chapter 9

Stone - Weierstrass Results

The Stone - Weierstrass Theorem is another theorem that is used repeatedly in modern analysis. It is not mentioned as frequently as it is used, but it is lurking in the background of many results and applications. Roughly speaking, it describes conditions for a subalgebra of functions in $C(X, \Re)$ to be dense. We will present a series of preliminary lemmas and definitions before finally proving the Stone - Weierstrass Theorem. We start with a classical result which we proved carefully in (Peterson (99) 2020).

9.1 Weierstrass Approximation Theorem

This result is indispensable in modern analysis. Fundamentally, it states that a continuous real-valued function defined on a compact set can be uniformly approximated by a smooth function. This is used throughout analysis to prove results about various functions. We can often verify a property of a continuous function, f, by proving an analogous property of a smooth function that is uniformly close to f. We will only prove the result for a closed finite interval in $\Re$. After this result, we will prove a more general result for a compact subset of topological vector spaces. We follow the development of (G. Simmons (38) 1963) for this proof.

Theorem 9.1.1 Weierstrass Approximation Theorem for $C([a, b])$

Let f be a continuous real-valued function defined on $[0, 1]$. For any $\epsilon > 0$, there is a polynomial, p, such that $|f(t) - p(t)| < \epsilon$ for all $t \in [0, 1]$, that is $\| p - f \|_\infty < \epsilon$. Note this says the polynomials on $[0, 1]$ are dense in $C([0, 1], \|\cdot\|_\infty)$, i.e. $\overline{P([0, 1])} = C([a, b], \|\cdot\|_\infty)$ where $P([a, b])$ is the set of all polynomials on $[0, 1]$. The result extends easily to $[a, b]$ by a change of variable.

Proof 9.1.1
The proof we give here is a classical analysis proof and it is not the way to go to prove an analogue of this result in a topological vector space. For future use, note the space $[0, 1]$ is an example of a compact Hausdorff space and the collection of powers of t, $\mathscr{F} = \{1, t, t^2, \ldots\}$ has structure which underlies why this argument works. Note, the span() $= \mathscr{P}$ where $\mathscr{P}$ is the set of all polynomials on $[0, 1]$.

We first derive some equalities. We will denote the interval $[0, 1]$ by I. By the binomial theorem, for any $x \in I$, we have

$$\sum_{k=0}^{n} \binom{n}{k} x^k (1-x)^{n-k} = (x + 1 - x)^n = 1. \qquad (\boldsymbol{\alpha})$$

Differentiating both sides of Equation $\boldsymbol{\alpha}$, we get

$$\begin{aligned} 0 &= \sum_{k=0}^{n} \binom{n}{k} \left(k x^{k-1} (1-x)^{n-k} - x^k (n-k)(1-x)^{n-k-1} \right) \\ &= \sum_{k=0}^{n} \binom{n}{k} x^{k-1} (1-x)^{n-k-1} \left(k(1-x) - x(n-k) \right) \\ &= \sum_{k=0}^{n} \binom{n}{k} x^{k-1} (1-x)^{n-k-1} \left(k - nx \right) \end{aligned}$$

Now, multiply through by $x(1-x)$, to find

$$0 = \sum_{k=0}^{n} \binom{n}{k} x^k (1-x)^{n-k} (k - nx).$$

Differentiating again, we obtain

$$0 = \sum_{k=0}^{n} \binom{n}{k} \frac{d}{dx} \left(x^k (1-x)^{n-k} (k - nx) \right).$$

This leads to a series of simplifications.

$$\begin{aligned} 0 &= \sum_{k=0}^{n} \binom{n}{k} \left[-n x^k (1-x)^{n-k} + (k - nx) \left((k-n) x^k (1-x)^{n-k-1} + k x^{k-1} (1-x)^{n-k} \right) \right] \\ &= \sum_{k=0}^{n} \binom{n}{k} \left[-n x^k (1-x)^{n-k} + (k - nx)(1-x)^{n-k-1} x^{k-1} \left((k-n)x + k(1-x) \right) \right] \\ &= \sum_{k=0}^{n} \binom{n}{k} \left(-n x^k (1-x)^{n-k} + (k - nx)^2 (1-x)^{n-k-1} x^{k-1} \right) \\ &= -n \sum_{k=0}^{n} \binom{n}{k} x^k (1-x)^{n-k} + \sum_{k=0}^{n} \binom{n}{k} (k - nx)^2 x^{k-1} (1-x)^{n-k-1} \end{aligned}$$

Thus, since the first sum is 1, we have $n = \sum_{k=0}^{n} \binom{n}{k} (k - nx)^2 x^{k-1} (1-x)^{n-k-1}$ and multiplying through by $x(1-x)$, we have $nx(1-x) = \sum_{k=0}^{n} \binom{n}{k} (k - nx)^2 x^k (1-x)^{n-k}$ or

$$\frac{x(1-x)}{n} = \sum_{k=0}^{n} \binom{n}{k} \left(\frac{k - nx}{n} \right)^2 x^k (1-x)^{n-k}$$

This last equality then leads to the

$$\sum_{k=0}^{n} \binom{n}{k} \left(x - \frac{k}{n} \right)^2 x^k (1-x)^{n-k} = \frac{x(1-x)}{n} \qquad (\boldsymbol{\beta})$$

We now define the nth *order* Bernstein Polynomial *associated with* f *by*

$$B_n(x) = \sum_{k=0}^{n} \binom{n}{k} x^k (1-x)^{n-k} f\Big(\frac{k}{n}\Big).$$

Note that

$$f(x) - B_n(x) = \sum_{k=0}^{n} \binom{n}{k} x^k (1-x)^{n-k} \Big[f(x) - f\Big(\frac{k}{n}\Big)\Big].$$

Also note that $f(0) - B_n(0) = f(1) - B_n(1) = 0$, *so* f *and* B_n *match at the endpoints. It follows that*

$$\mid f(x) - B_n(x) \mid \;\; \leq \;\; \sum_{k=0}^{n} \binom{n}{k} x^k (1-x)^{n-k} \Big| f(x) - f\Big(\frac{k}{n}\Big)\Big|. \qquad (\gamma)$$

Now, f *is uniformly continuous on* I *since it is continuous. So, given* $\epsilon > 0$, *there is a* $\delta > 0$ *such that* $|x - \frac{k}{n}| < \delta \Rightarrow |f(x) - f(\frac{k}{n})| < \frac{\epsilon}{2}$. *Consider* x *to be fixed in* $[0,1]$. *The sum in Equation* γ *has only* $n+1$ *terms, so we can split this sum up as follows. Let* $\{K_1, K_2\}$ *be a partition of the index set* $\{0, 1, ..., n\}$ *such that* $k \in K_1 \Rightarrow |x - \frac{k}{n}| < \delta$ *and* $k \in K_2 \Rightarrow |x - \frac{k}{n}| \geq \delta$. *Then*

$$\mid f(x) - B_n(x) \mid \leq \sum_{k \in K_1} \binom{n}{k} x^k (1-x)^{n-k} \Big| f(x) - f\Big(\frac{k}{n}\Big)\Big| + \sum_{k \in K_2} \binom{n}{k} x^k (1-x)^{n-k} \Big| f(x) - f\Big(\frac{k}{n}\Big)\Big|.$$

which implies

$$\begin{aligned} |f(x) - B_n(x)| \;\; &\leq \;\; \frac{\epsilon}{2} \sum_{k \in K_1} \binom{n}{k} x^k (1-x)^{n-k} + \sum_{k \in K_2} \binom{n}{k} x^k (1-x)^{n-k} \Big| f(x) - f\Big(\frac{k}{n}\Big)\Big| \\ &= \;\; \frac{\epsilon}{2} + \sum_{k \in K_2} \binom{n}{k} x^k (1-x)^{n-k} \Big| f(x) - f\Big(\frac{k}{n}\Big)\Big|. \end{aligned}$$

Now, f *is bounded on* I, *so there is a real number* $M > 0$ *such that* $|f(x)| \leq M$ *for all* $x \in I$. *Hence*

$$\sum_{k \in K_2} \binom{n}{k} x^k (1-x)^{n-k} \Big| f(x) - f\Big(\frac{k}{n}\Big)\Big| \leq 2M \sum_{k \in K_2} \binom{n}{k} x^k (1-x)^{n-k}.$$

Since $k \in K_2 \Rightarrow |x - \frac{k}{n}| \geq \delta$, *using Equation* $\boldsymbol{\beta}$, *we have*

$$\delta^2 \sum_{k \in K_2} \binom{n}{k} x^k (1-x)^{n-k} \leq \sum_{k \in K_2} \binom{n}{k} \Big(x - \frac{k}{n}\Big)^2 x^k (1-x)^{n-k} \leq \frac{x(1-x)}{n}.$$

This implies that

$$\sum_{k \in K_2} \binom{n}{k} x^k (1-x)^{n-k} \leq \frac{x(1-x)}{\delta^2 n}.$$

and so combining inequalities

$$2M \sum_{k \in K_2} \binom{n}{k} x^k (1-x)^{n-k} \leq \frac{2Mx(1-x)}{\delta^2 n}$$

We conclude then that

$$\sum_{k \in K_2} \binom{n}{k} x^k (1-x)^{n-k} \Big| f(x) - f\Big(\frac{k}{n}\Big) \Big| \leq \frac{2Mx(1-x)}{\delta^2 n}.$$

Now, the maximum value of $x(1-x)$ *on* I *is* $\frac{1}{4}$, *so*

$$\sum_{k \in K_2} \binom{n}{k} x^k (1-x)^{n-k} \Big| f(x) - f\Big(\frac{k}{n}\Big) \Big| \leq \frac{M}{2\delta^2 n}.$$

Finally, choose n *so that* $n > \frac{M}{\delta^2 \epsilon}$. *Then* $\frac{M}{n\delta^2} < \epsilon$ *implies* $\frac{M}{2n\delta^2} < \frac{\epsilon}{2}$. *So, Equation* γ *becomes*

$$| f(x) - B_n(x) | \leq \frac{\epsilon}{2} + \frac{\epsilon}{2} = \epsilon.$$

Note that the polynomial B_n *does not depend on* $x \in I$, *since* n *only depends on* M, δ, *and* ϵ, *all of which, in turn, are independent of* $x \in I$. *So,* B_n *is the desired polynomial, as it is uniformly within* ϵ *of* f. ■

Comment 9.1.1 *A change of variable translates this result to any closed interval* $[a, b]$.

Homework

Exercise 9.1.1 *Prove the two variable version of Theorem 9.1.1. We have* $f(x, y)$ *is a real-valued function defined and continuous on* $X = [a, b] \times [c, d]$ *where* $a < b$ *and* $c < d$ *are finite intervals. Prove* f *can be uniformly approximated on* X *by polynomials in* x *and* y *with real coefficients.*

Exercise 9.1.2 *Prove* $C([a, b], \Re)$, *for* $a < b$, *a finite interval, is separable.*

Exercise 9.1.3 *Let* f *be a continuous real-valued function on* $[0, 1]$. *The moments of* f *are the numbers* $\int_0^1 f(x)x^n dx$ *for* $n = 0, 1, 2, \ldots$. *Prove that if* f *and* g *are continuous with the same moments, they must be the same function.*

9.2 Partial Orderings

We have already discussed partial orderings in Section 2.5 as part of our treatment of the standard Hahn - Banach material. So we will refer you to that material for review if necessary. Now, in (Peterson (100) 2020), we worked with lattices. We defined them using ordered vector spaces.

Definition 9.2.1 Ordered Vector Spaces

A vector space V *over* $\Re$ *is said to be ordered if there is a partial ordering* $\preceq$ *which is compatible with the vector space operations* $\oplus$ *and* $\odot$ *(the scalar multiplication operation). This means:*

1. $x \preceq y \longrightarrow x \oplus z \preceq y \oplus z$ *for all* $x, y, z \in V$.
2. $x \preceq y \longrightarrow c \odot x \preceq c \odot y$ *for all* $c \neq 0$ V *and all* $x, yinV$.

We denote this ordered vector space by $(V, \preceq)$.

In an ordered vector space we can find the infimum and supremum of two elements of the vector space.

Definition 9.2.2 Infimum and Supremum of Elements in an Ordered Vector Space

- *If $x, y \in (V, \preceq)$ and if there is $z \in (V, \preceq)$ so that $x \preceq z$ and $y \preceq z$ and if $x \preceq z'$ and $y \preceq z'$ implies $z \preceq z'$, we call z the supremum of x and y and denote it by $\sup\{x, y\}$ or $x \vee y$. Of course, z may not exist!*

- *If $x, y \in (V, \preceq)$ and if there is $w \in (V, \preceq)$ so that $w \preceq x$ and $w \preceq y$ and if $w' \preceq x$ and $w' \preceq y$ implies $w' \preceq w$, we call w the infimum of x and y and denote it by $\inf\{x, y\}$ or $x \wedge y$. Such an element w need not exist of course.*

If Γ is an index set and if $\{x_\alpha \,|\, \alpha \in \Gamma\}$ is a subset of $(V, \preceq)$, then the definition above easily extends to $\sup(x_\alpha) = \vee_{\alpha \in \Gamma} x_\alpha$ and $\inf(x_\alpha) = \wedge_{\alpha \in \Gamma} x_\alpha$ as follows:

- *If there is $z \in (V, \preceq)$ so that $x_\alpha \preceq z$ for all α and if $x_\alpha \preceq z'$ for all α implies $z \preceq z'$, we call z the supremum of (x_α).*

- *If there is $w \in (V, \preceq)$ so that $w \preceq x_\alpha$ and if $w' \preceq x_\alpha$ for all α implies $w' \preceq w$, we call w the infimum of (x_α).*

We can now define a vector lattice.

Definition 9.2.3 Vector Lattices in ordered vector spaces

We say an ordered vector space $(V, \preceq)$ is a Vector Lattice if $x \vee y$ and $x \wedge y$ exist for all $x, y \in (V, \preceq)$.

Further, in a vector lattice, we define the positive part of x to be $x^+ = x \vee 0$ and the negative part of x to be $x^- = x \wedge 0$. The modulus or size of x is then $|x| = x^+ + x^-$. Finally, we can say x and y are orthogonal if $|x| \wedge |y| = 0$ and then we write $x \perp y$.

We will relax these definitions now and drop the vector space setting and instead place everything in the world of posets.

Definition 9.2.4 Lattices

A lattice is a nonempty poset L in which each $(x, y) \in L$ has a lub and a glb. These are, of course, maximal and minimal elements of L. We use $x \vee y = lub\{x, y\}$ to denote the maximal element and $x \wedge y = glb\{x, y\}$ to denote the minimal element. The symbol, $x \vee y$ is read as x up y and the symbol $x \wedge y$ is read as x down y.

A sublattice L_1 of L is a subset satisfying $x, y \in L_1$ implies $x \vee y$ and $x \wedge y$ are also in L_1.

If every nonempty subset of L has a glb and lub, we say L is a complete lattice.

Example 9.2.1 *Let $(X, \mathscr{S})$ be a nonempty topological vector space and let $P = \{f : X \to \Re\}$. The partial ordering relationship $\preceq$ is defined by $f \preceq y$ means $f(x) \leq g(x)$ for all $x \in X$. Then, $f \vee g$ is defined pointwise by $(f \vee g)(x) = \max\{f(x), g(x)\}$ and $f \wedge g$ is defined pointwise by $(f \wedge g)(x) = \min\{f(x), g(x)\}$. Then, $f \vee g$ and $f \wedge g$ are in P. Note $(f \vee g)(x) \leq f(x)$ and $(f \vee g)(x) \leq g(x)$ for all x. Also, $(f \wedge g)(x) \geq f(x)$ and $(f \wedge g)(x) \geq g(x)$ for all x. We can see this set is a lattice.*

Now often we have a vector space with additional structure. A typical vector space has two operations which we can denote by $+$ and $\cdot$ with $(x, y) \rightarrow x + y$ and $(\alpha, x) \rightarrow \alpha \cdot x$. Of course, we rarely actually write $\alpha \cdot x$ and instead simply write αx. The $+$ is the addition operator on the vector space and $\cdot$ is the scalar multiplication operator. Many times the vector space has an additional operator, $*$, which we think of as the multiplication operator, $(x, y) \rightarrow x * y$. Typically, we don't write it this way and just write xy instead of $x * y$. We want the usual compatibility things: here are a few of them:

$$\begin{aligned} x * (y + z) &= x * y + a * z \\ \alpha \cdot (x * y) &= (\alpha \cdot x) * y = x * (\alpha \cdot y) \\ x * (y * z) &= (x * y) * z \end{aligned}$$

Of course, it is fairly tedious to write down all the compatibility rules for the three operations, so we will leave that to you. When we add the operator $*$ to vector space, the vector space is called an algebra.

Example 9.2.2 *Let $(X, \mathscr{S})$ be a nonempty topological vector space and $P = \{f : X \rightarrow \Re\}$. Define the usual operators pointwise:*

$$\begin{aligned} (f + g)(x) &= f(x) + g(x) \\ (\alpha f)(x) &= \alpha f(x) \\ (fg)(x) &= f(x)g(x) \end{aligned}$$

These hold for all $f, g \in P$ and $\alpha \in \Re$. With these operators, P is an algebra.

A standard space to look at is then $C(X, \mathscr{S}, \Re) = \{f : X \rightarrow \Re |\ f$ is continuous on $X\}$. If the topological space is $\Re$, we can look at polynomials on $\Re$, $P(\Re, \Re) = \{f : \Re \rightarrow \Re | f(x) = \sum_{i=0}^{n} a - ix^i,\ a_i \in \Re,$ some $n\}$. The topology on $\Re$ is the usual one generated by the absolute value function $|\cdot|$. If we look at polynomials whose domain in $[a, b] \subset \Re$ for a finite interval, we usually just write $P([a, b])$ for this space.

Let's use the standard $\leq$ as a partial ordering on $P([a, b])$. We define the partial ordering pointwise. We see $f(x) \leq f(x)$, $f(x) \leq g(x)$ and $g(x) \leq f(x)$ implies $f(x) = g(x)$ and $f(x) \leq g(x)$, $g(x) \leq h(x)$ does imply $f(x) \leq h(x)$. So $P([a, b])$ is a lattice with $f \vee g$ and $f \wedge g$ defined as usual. Note also, we can think of $P([a, b])$ in a new way: every element of $P([a, b])$ can be generated from the two functions: $f_0(x) = 1$ for all x and $f_1(x) = x$ for all x. We typically abuse notation and simply write these as $\{1, x\}$. Any polynomial can be generated from a finite number of applications of $*$, $+$ and $\cdot$. So we can say $P([a, b])$ is generated by $\{1, x\}$.

Homework

Exercise 9.2.1 *If L is a lattice, prove the following assertions:*

- *For $x, y \in L$, $x \wedge x = x$ and $x \vee x = x$.*
- *For $x, y \in L$, $x \wedge y = y \wedge x$ and $x \vee y = y \vee x$.*
- *For $x, y, z \in L$, $x \wedge (y \vee z) = (x \wedge y) \vee z$.*

Exercise 9.2.2 *Prove the set of $n \times n$ real-valued nonnegative matrices partially ordered by $A \preceq B$ means $A_{ij} \leq B_{ij}$ is a lattice.*

Exercise 9.2.3 *Prove the set of nonnegative functions f on $[0, 1]$ is a lattice.*

Let's look at objects and operations on objects in some detail. We will be a bit casual in our definitions so we can focus on what is important for us. All of these ideas can be treated at book length and you should feel free to find books on these topics and learn more. We now define groups, rings, fields and algebras along with some simple examples.

Definition 9.2.5 Groups

Let X be a nonempty set of objects and let $\circledcirc$ denote the mapping $\circledcirc : X \times X \to X$ which is written as $x \circledcirc y$ for any x and y in X. The fact that $x \circledcirc y$ is always an element of X is a property called closure. We assume $\circledcirc$ satisfies:

G1: Associativity: $(x \circledcirc y) \circledcirc z = x \circledcirc (y \circledcirc z)$ for all x, y and z in X.

G2: Identity: there is a unique element, $x_{\circledcirc}$ so that $x_{\circledcirc} \circledcirc x = x$ for all $x \in X$.

G3: Inverses: To each $x \in X$, there is a unique element $y_x \in X$ so that $x \circledcirc y_x = x_{\circledcirc}$.

The object with the operation is called a group. So to be clear, the pair $(X, \circledcirc)$ is a group with group operation $\circledcirc$.

If the group operation does not depend on order, we say the group is abelian or commutative with respect to the operator $\circledcirc$. This would mean $x \circledcirc y = y \circledcirc x$ for all x and y in X.

Comment 9.2.1 *The standard operators are addition and multiplication, which are different. For example, we know how to add and multiply matrices. For addition, all we need is that the matrices are the same size. For multiplication of two matrices, we know to multiply the $m \times n$ matrix A and the $r \times p$ matrix B, we must have $n = r$, so multiplication requires some conditions. Addition is denoted $\oplus$ and multiplication $\otimes$ and we would rephrase the group definitions for these operations depending on the operator.*

If the operator is addition, we would call $(X, \oplus)$ an additive group. The definitions then become:

G1: Associativity: $(x \oplus y) \oplus z = x \oplus (y \oplus z)$ for all x, y and z in X.

G2: Identity: there is a unique element, 0 so that $0 \oplus x = x$.

G3: Inverses: To each $x \in X$, there is a unique element $y_x \in X$ so that $x \oplus y_x = 0$. This element is denoted by $-x$ so that $x \oplus (-x) = 0$. Also, we define the subtraction operator by $x \ominus y = x \oplus (-y)$ so that $x \oplus (-x) = 0$ becomes $x - x = 0$.

If the operator is multiplication, we would call $(X, \otimes)$ a multiplicative group. The definitions then become:

G1: Associativity: $(x \otimes y) \otimes z = x \otimes (y \otimes z)$ for all x, y and z in X.

G2: Identity: there is a unique element, 1 so that $1 \otimes x = x$.

G3: Inverses: To each $x \in X$, there is a unique element $y_x \in X$ so that $x \otimes y_x = 1$. This element is denoted by x^{-1} or sometimes by $1/x$ so that $x \times x^{-1} = 1$.

So the set of all $n \times m$ matrices over $\Re$ is an additive group and the set of all square matrices of size $n \times n$ is an additive group but not a multiplicative group because not all $n \times n$ matrices have inverses. The set of all invertible $n \times n$ matrices is a multiplicative group. You can easily come up with many other examples of groups and group operations.

A nonempty set of objects X can have two operators associated with it: $\circledcirc_1$ and $\circledcirc_2$. This leads to the idea of a ring.

Definition 9.2.6 Rings

Let X be a nonempty set of objects and let $\circledcirc_1$ and $\circledcirc_2$ denote the mappings $\circledcirc_i : X \times X \rightarrow X$ which is written as $x \circledcirc_i y$, for $i = 1$ or $i = 2$, for any x and y in X. The fact that $x \circledcirc_i y$ is always an element of X is a property called closure. We assume $\circledcirc_1$ satisfies:

G11: Associativity: $(x \circledcirc_1 y) \circledcirc_1 z = x \circledcirc_1 (y \circledcirc_1 z)$ for all x, y and z in X.

G21: Identity: there is a unique element, $x_{\circledcirc_1}$ so that $x_{\circledcirc_1} \circledcirc_1 x = x$ for all $x \in X$.

G31: Inverses: To each $x \in X$, there is a unique element $y_x \in X$ so that $x \circledcirc_1 y_x = x_{\circledcirc_1}$.

We assume $\circledcirc_2$ satisfies

G21: Associativity: $(x \circledcirc_2 y) \circledcirc_2 z = x \circledcirc_2 (y \circledcirc_2 z)$ for all x, y and z in X.

G22: Identity: there is a unique element, $x_{\circledcirc_2}$ so that $x_{\circledcirc_2} \circledcirc_2 x = x$ for all $x \in X$.

We also assume compatibility of $\circledcirc_1$ and $\circledcirc_2$ in the sense that $\circledcirc_2$ is distributive over $\circledcirc_1$: $x \circledcirc_2 (y \circledcirc_1 z) = (x \circledcirc_2 y) \circledcirc_1 (x \circledcirc_2 z)$ for all x, y and z in X. The object with these operations is called a ring. So to be clear, the pair $(X, \circledcirc_1, \circledcirc_2)$ is a ring with ring operations $\circledcirc_1$ and $\circledcirc_2$. Note X is a group with respect to operation $\circledcirc_1$ but is not a group with respect to $\circledcirc_1$ because we don't require inverses for operation $\circledcirc_2$.

If the group operation $\circledcirc_2$ does not depend on order, we say the group is abelian or commutative with respect to the operator $\circledcirc_2$. In a ring, typically, the operation $\circledcirc_1$ is abelian but the operation $\circledcirc_2$ need not be abelian.

Comment 9.2.2 *The set of $n \times n$ invertible matrices over $\Re$ is not a group with respect to addition but is a group with respect to multiplication.*

The set of all polynomials on $[a, b]$ is a group with respect to addition but it is not a group with respect to multiplication as inverses of polynomials need not exist as polynomials. However, it is a ring.

Comment 9.2.3 *Again, we usually let $\circledcirc_1 = +$ and $\circledcirc_2 = \times$. In this case, X is a nonempty set of objects and $+$ and $\times$ denote the mappings defined by $x + y$ and $x \times y$ for any x and y in X. We generally just write xy instead of $x \times y$. For the ring definitions, we assume $+$ satisfies*

G11: Associativity: $(x + y) + z = x + (y + z)$ for all x, y and z in X.

G12: Identity: there is a unique element, 0 so that $0 + x = x$ for all $x \in X$.

G13: Inverses: To each $x \in X$, there is a unique element $-x \in X$ so that $x + (-x) = x - x = 0$.

We assume $\times$ satisfies

G21: Associativity: $(x \times y) \times z = x \times (y \times z)$ for all x, y and z in X. We usually just write $(xy)z = x(yz)$.

G22: Identity: there is a unique element, 1 so that $1 \times x = x$ for all $x \in X$. We usually just write $1x = x$.

We also assume compatibility of $+$ and $\times$ in the sense that $\times$ is distributive over $+$: i.e. $x \times (y + z) = (x \times y) + (x \times z)$ or simply $x(y + z) = xy + yz$ for all x, y and z in X. The object with these operations is called a ring. So to be clear the pair $(X, +, \times)$ is a ring with ring operations $+$ and $\times$. Note X is a group with respect to operation $+$ but is not a group with respect to $\times$.

Homework

Exercise 9.2.4 *Prove the set of 2×2 rotation matrices is a group with matrix multiplication. Is this an abelian group?*

Exercise 9.2.5 *Prove the set of 3×3 invertible real-valued matrices is a group under matrix multiplication but they do not form a group under matrix addition.*

Exercise 9.2.6 *Prove the set of all continuous real-valued functions on $[a, b]$ with $a < b$ a finite interval, form a group with respect to pointwise addition of functions.*

Exercise 9.2.7 *Prove the set of all continuous real-valued functions on $[a, b]$ with $a < b$ a finite interval, form a ring with respect to pointwise addition of functions and pointwise multiplication of functions.*

We can add more structure by assuming inverses for nonzero elements exist in the ring. A ring with inverses for all elements not equal to the identity, $x_{\circledcirc_1}$, are called division rings. If both operations are commutative, we have a field.

Definition 9.2.7 Fields

A set of objects X is called a field if it is a commutative ring with respect to the two operations $\circledcirc_1$ and $\circledcirc_2$ and in addition, for each $x \neq x_{\circledcirc_1}$, where $x_{\circledcirc_1}$ is the identity element for operation $\circledcirc_1$, there is a unique element $x^{-1} \in X$ so that $x \circledcirc_2 x^{-1|} = x_{\circledcirc_2}$.

The operators are usually denoted by $+$ and $\times$ and examples of a field are $\mathbb{Q}$ (the rational numbers), $\Re$ (the real numbers) and $\mathbb{C}$ (the complex numbers).

We do not always need inverses to do useful things. It is enough to have an algebra over a field F or simply an algebra.

Definition 9.2.8 Algebras

A set of nonempty objects X is an algebra over the field F if X is a vector space with the two operations $\oplus$ and $\odot$ which also has mapping $\otimes$ so that

A1: $x \otimes (y \oplus z) = (x \otimes y) \oplus (x \otimes z)$ for all x, y and z in X.

A2: $x \otimes (\alpha \odot y) = \alpha \odot (x \otimes y)$ for all x and y in X and $\alpha \in F$.

Note the mapping $\otimes$ need not be associative: i.e. $x \otimes (y \otimes z)$ need not be the same as $(x \otimes y) \otimes z$. Also, there does not have to be an identity $x_\otimes$ associated with $\otimes$. If there is, the algebra is called unital.

We also would like to think about when a subset generates an algebra.

Definition 9.2.9 Algebras Generated by a Set F

First, a subalgebra of X is any set $B \subset X$ which is itself an algebra. Let X be a set and let $F \subset X$. The algebra generated by F is the algebra $A(F)$ defined by $A(F) = \cap\{B | B, B$ is an algebra containing $F\}$. Of course, X be an algebra already.

Comment 9.2.4 *Given a set F of objects, to generate an algebra, we need three operations which we will call $+$ (object addition), $\cdot$ (scalar multiplication by field elements) and $\times$ (object multiplication). Let $C = \{y | y = \sum_{i=1}^{n} a_i x^i\}$ for arbitrary scalars a_i, any nonnegative integer n and $x^i = x \times x \ldots \times x$ (i terms). By definition, C is closed under these operations. If there is an object in F*

that can function as a multiplicative identity, this algebra is unital. There may be many algebras containing this one, so a precise way to say this is what is given in the definition above.

Example 9.2.3 *If $F = \{1, x\}$, the algebra generated by F is $P([a, b])$ for any $[a, b]$. This algebra is unital.*

Comment 9.2.5 *The polynomials with real coefficients on $[a, b]$ is an algebra over $\Re$ using the usual notions of $+$, $\times$ and scalar multiplication $\cdot$. It is a unital algebra as 1 is a polynomial on $[a, b]$.*

Homework

Exercise 9.2.8 *If A is an algebra, prove that $C = \{x \ : \ xy = yx\}$, where xy is the usual notation for the multiplication operation, is a subalgebra. This is called the center of A.*

Exercise 9.2.9 *Let A be an algebra of linear transformations on a vector space V. If A contains the identity map I, prove the center of A (defined in the previous exercise) contains all scalar transformations.*

Exercise 9.2.10 *Let X be a set and let F be the collection of subsets of X that satisfy A and its complement A^C are in F. Let G be the collection of sets that satisfy $A, B \in F$ implies $A \cap B^C \in G$ and $A \cup B \in G$. Prove G is a ring using the operations of set union and set intersection.*

9.3 Continuous Functions on a Topological Space

Let's look at continuous functions on a topological space.

Theorem 9.3.1 Bounded Continuous Functions on a T_1 Topological Space Are a Complete Normed Linear Space

Let $C_b(X, \Re)$ denote the set of bounded continuous functions from X to $\Re$, where $(X, \mathscr{S})$ is a topological space. Define $\|f\|_\infty = \sup_{x \in X}\{|f(x)|\}$. Then $C_b(X, \Re)$ is a complete normed linear space.

Proof 9.3.1

We first show completeness under $\|\cdot\|_\infty$. First, suppose $\{f_n\} \subset C(X,\Re)$ converges to some function, f, with respect to the norm $\|\cdot\|_\infty$. Then, for $\epsilon > 0$, there is N such that $\|f_n - f\|_\infty < \epsilon/3$ for $n \geq N$. Fix $n_0 > N$. Then $|f_{n_0}(x) - f(x)| < \epsilon/3$ for every $x \in X$. Since f_{n_0} is continuous, we can pick any $x_0 \in X$, and there is some neighborhood, W, of x_0 such that $x \in W \Longrightarrow |f_{n_0}(x) - f_{n_0}(x_0)| < \epsilon/3$. For any $x \in W$, consider $|f(x) - f(x_0)|$. We have

$$\begin{aligned} |f(x) - f(x_0)| &= |f(x) - f(x_0) + f_{n_0}(x) - f_{n_0}(x) + f_{n_0}(x_0) - f(x_0)| \\ &\leq |f(x) - f(x_0)| + |f_{n_0}(x) - f_{n_0}(x_0)| + |f_{n_0}(x_0) - f(x_0)| \\ &< \epsilon/3 + \epsilon/3 + \epsilon/3 < \epsilon \end{aligned}$$

That is, $|f(x) - f(x_0)| < \epsilon$ for $x \in W$. This implies the limit function, f, is continuous.

Now, still assuming that $f_n \to f$ with respect to the $\|\cdot\|$, choose $N \in \mathbf{N}$ such that $\|f_n - f\| < \frac{1}{2}$ for $n \geq N$. Then, for $x \in X$ and some $n_0 \geq N$, we have $|f_{n_0}(x) - f(x)| < \frac{1}{2}$, and this is true for all $x \in X$. Thus, for any $x \in X$, we have

$$\frac{1}{2} + f_{n_0}(x) < f(x) < \frac{1}{2} + f_{n_0}(x) \quad \Longrightarrow \quad -\frac{1}{2} - M_{n_0} < f(x) < \frac{1}{2} + M_{n_0},$$

where M_{n_0} is the bound on f_{n_0}, which exists by hypothesis. Hence, the limit function, f, is bounded. Combining results, we have shown if $\{f_n\} \subset C(X, \Re)$ converges to some function, f, with respect to the norm $\| \cdot \|_\infty$, then f is bounded and continuous.

Now, let f_n be a Cauchy sequence in $C(X, \Re)$. Then, given $\epsilon > 0$, there is N such that $\|f_n - f_m\|_\infty < \epsilon/2$ for $n, m \geq N$. So, given any $x \in X$, $\{f_n(x)\}$ is a Cauchy sequence in $\Re$, which must converge.

Define the function f by $f(x) = \lim_{n\to\infty} f_n(x)$ for each $x \in X$. Next, let $m \to \infty$ in the expression $|f_n(x) - f_m(x)| < \epsilon/2$, where x is any fixed element of X and $n \geq N$. Using the continuity of the absolute value function, it follows that

$$|f_n(x) - \lim_{m\to\infty} f_m(x)| \quad \leq \quad \epsilon/2$$

and so for $n > N$, any $x \in X$, we have

$$|f_n(x) - f(x)| \quad leq \quad \epsilon/2$$

Thus, since $\|f_n - f\|_\infty = \sup_{x\in X} |f_n(x) - f(x)|$, the previous inequality implies that $\|f_n - f\|_\infty < \epsilon$ for $n \geq N$. That is, $\|f_n - f\| \to 0 \implies f_n \to f$ with respect to the norm $\| \cdot \|_\infty$. Moreover, our previous results tell us that f is continuous and bounded, so $f \in C(X, \Re)$. Therefore, $C(X, \Re)$ is complete. ■

Homework

Let $(X, \mathscr{S})$ be a topological space. The support of a function $f : X \to \Re$ is the closure of the set of points where $f(x) \neq 0$. The support of f is denoted by $supp(f)$.

- Let $C_{00}(X, \Re)$ denote the set of bounded continuous functions f from X to $\Re$ with $supp(f)$ compact. These functions are said to vanish in a neighborhood of infinity.
- Let $C_0(X, \Re)$ denote the set of bounded continuous functions f from X to $\Re$ so that for all $\epsilon > 0$, there is a compact subset $F_{f,\epsilon}$, with $|f(x)| < \epsilon$ on $F^C_{f,\epsilon}$. These functions are said to vanish at infinity.

Exercise 9.3.1 *Let X be a locally compact Hausdorff space. This means at each point x in X there is a neighborhood of x whose closure is compact. Consider the set of continuous real-valued functions on X, $C(X, \Re)$ and let $B(X, \Re)$ be the set of bounded real-valued functions on X with the usual infinity norm. Prove*

- $C_{00}(X, \Re) \subset C_0(X, \Re) \subset C(X, \Re)$.
- *If X is compact, prove $C_{00}(X, \Re) = C_0(X, \Re) = C(X, \Re)$.*
- *Prove $C_0(X, \Re)$ is closed in $B(X, \Re)$.*

Exercise 9.3.2 *Let X be a locally compact Hausdorff space. This means at each point x in X there is a neighborhood of x whose closure is compact. Consider the set of continuous real-valued functions on X, $C(X, \Re)$ and $B(X, \Re)$ be the set of bounded real-valued functions on X with the usual infinity norm. Prove*

- *$C_0(X, \Re)$ is closed in $B(X, \Re)$.*
- *C_0 is a complete metric space.*

Next, we can show the set of bounded continuous functions on a topological space is a commutative algebra.

Theorem 9.3.2 Bounded Continuous Functions on a T_1 Topological Space are a Commutative Algebra

Let $C_b(X, \Re)$ denote the set of bounded continuous functions from X to $\Re$, where $(X, \mathscr{S})$ is a topological space. Using multiplication $\times$ defined pointwise, it is a commutative real algebra with identity and $\|f \times g\|_\infty \leq \|f\|_\infty \|g\|_\infty$. Further, if 1 denotes the constant function $f(x) = 1$ for all $x \in X$, then $\|1\|_\infty = 1$.

Proof 9.3.2
The linear structure of $C(X, \Re)$ is straightforward, so we will only prove that it is closed under products. Suppose $f, g \in C(X, \Re)$. Fix $x_0 \in X$ and consider $|(fg)(x) - (fg)(x_0)| = |f(x)g(x) - f(x_0)g(x_0)|$. We have

$$
\begin{aligned}
|f(x)g(x) - f(x_0)g(x_0)| &= |f(x)g(x) - f(x_0)g(x) + f(x_0)g(x) - f(x_0)g(x_0)| \\
&\leq |g(x)||f(x) - f(x_0)| + |f(x_0)|g(x) - g(x_0)|
\end{aligned}
$$

Since f is continuous at x_0, for $\delta_1 > 0$ there is a neighborhood, U_1 of x_0 such that $x \in U_1 \Longrightarrow |f(x) - f(x_0)| < \delta_1$. Thus, there is a neighborhood, U_2, of x_0 such that $x \in U_2 \Longrightarrow |f(x) - f(x_0)| < 1$. Hence, $x \in U_2 \Longrightarrow |f(x)| \leq 1 + |f(x_0)| \leq 1 + M_f$, where M_f is the bound on f. Let $B_f = M_f + 1$, so $x \in U_2 \Longrightarrow |f(x)| \leq B_f$. Now, $U_1 \cap U_2$ is a neighborhood of x_0, so for $x \in U_1 \cap U_2$ we have $|f(x) - f(x_0)| < \delta_1$ and $|f(x)| \leq B_f$.

Since g is continuous, we can use the same reasoning to obtain neighborhoods V_1 and V_2 of x_0 such that $x \in V_1 \cap V_2 \Longrightarrow |g(x) - g(x_0)| < \delta_2$ and $|g(x)| \leq B_g$, where $\delta_2 > 0$ is arbitrary and $B_g = 1 + M_g$.

It follows, then, that $|f(x)g(x) - f(x_0)g(x_0)| < B_g\delta_1 + B_f\delta_2$ for $x \in U_1 \cap U_2 \cap V_1 \cap V_2$. So, if $\epsilon > 0$ is given, we can let $\delta_1 = \frac{\epsilon}{2B_g}$ and $\delta_2 = \frac{\epsilon}{2B_f}$. Then, for $x \in U_1 \cap U_2 \cap V_1 \cap V_2$, we have $|(fg)(x) - (fg)(x_0)| < \epsilon$. Thus, fg is continuous, and it is bounded by $M_f M_g$. So $fg \in C(X, \Re)$, and $C(X, \Re)$ is an algebra.

It is clear the function $f(x) = 1$ for all x is the identity, so this is a unital algebra. ■

Now we show the set of bounded continuous functions on a topological space is a lattice.

Theorem 9.3.3 Bounded Continuous Functions on a T_1 Topological Space are a Lattice

Let $C_b(X, \Re)$ denote the set of bounded continuous functions from X to $\Re$, where $(X, \mathscr{S})$ is a topological space. Define $\|f\|_\infty = \sup_{x \in X}\{|f(x)|\}$. Then, it is also a lattice in which $\sup\{f, g\} = f \vee g$ and $\inf\{f, g\} = f \wedge g$ are defined pointwise.

Proof 9.3.3
To show that $C(X, \Re)$ is a lattice, we will show that $f, g \in C(X, \Re) \Longrightarrow f \vee g \in C(X, \Re)$. A similar proof shows that $f \wedge g \in C(X, \Re)$. We first note that $f \vee g$ must be bounded, as f and g are. Thus, we need only show continuity. Moreover, an open set in $\Re$ is a countable collection of open intervals, all of which must be of one of the forms (a, ∞), $(-\infty, a)$, or (a, b), where a and b are real numbers. So, we need only show that $(f \vee g)^{-1}(\Omega)$ is open when Ω is one of these sets. For the first case, we have

$$
\begin{aligned}
(f \vee g)^{-1}(a, \infty) &= \{x \in X : (f \vee g)(x) > a\} = \{x \in X : f(x) > a\} \cup \{x \in X : g(x) > a\} \\
&= f^{-1}(a, \infty) \cup g^{-1}(a, \infty)
\end{aligned}
$$

Since $f^{-1}(a,\infty)$ and $g^{-1}(a,\infty)$ are open, so is $(f \vee g)^{-1}(a,\infty)$. Likewise, for the other cases we have

$$\begin{aligned}(f \vee g)^{-1}(-\infty,a) &= \{x \in X : (f \vee g)(x) < a\} = \{x \in X : f(x) < a\} \cup \{x \in X : g(x) < a\} \\ &= f^{-1}(a,\infty) \cap g^{-1}(a,\infty)\end{aligned}$$

and

$$\begin{aligned}(f \vee g)^{-1}(a,b) &= \{x \in X : a < (f \vee g)(x) < b\} \\ &= [f^{-1}(a,\infty) \cup g^{-1}(a,\infty)] \cap [f^{-1}(-\infty,b) \cap g^{-1}(-\infty,b)]\end{aligned}$$

So, for any open set, $\Omega \subset \Re$, $(f \vee g)^{-1}(\Omega)$ is open. Hence, $f \vee g \in C(X,\Re)$ and $C(X,\Re)$ is a lattice. ■

Here is an obvious question. If A is an algebra in $C_b(X,\mathscr{S})$, is the closure of the algebra, $\overline{A}$, with respect to $\|\cdot\|_\infty$ also an algebra? Recall, the closure of A contains the limits of all sequences in A which converge.

Theorem 9.3.4 Closure of an Algebra in $C_b(X,\mathscr{S})$ is an Algebra

Let A be an algebra in the normed linear space $C_b(X,\mathscr{S},\|\cdot\|_\infty)$. Then $\overline{A}$ is an algebra also.

Proof 9.3.4
Let A be an algebra in the normed linear space $C_b(X,\mathscr{S},\|\cdot\|_\infty)$. We need to show $\overline{A}$ is closed under products. Let $f,g \in \overline{A}$. **Case 1:** *Both $f,g \in A$. Since A is an algebra, $fg \in A \subset \overline{A}$.*

Case 2: *$f \in A$ and $g = \lim_{n\to\infty} g_n$ in $\|\cdot\|_\infty$ with each $g_n \in A$. It is easy to see $fg_n \to fg$ in norm. Since, $g_n \in A$, $fg_n \in A$ too. Thus, $fg \in \overline{A}$ by definition. A similar argument shows if $g \in A$ and $f = \lim_{n\to\infty} f_n$ in $\|\cdot\|_\infty$ with each $f_n \in A$, then $fg \in \overline{A}$.*

Case 3: *If $f = \lim_{n\to\infty} f_n$ and $g = \lim_{n\to\infty} g_n$ with $f_n, g_n \in A$, then $f_n g_n \in A$ as A is an algebra and so the limit $fg \in \overline{A}$.*

We conclude $\overline{A}$ is an algebra also. ■

Example 9.3.1 *Hence, Theorem 9.3.4 tells us $\overline{A(\{1,x\})}$ is an algebra where $A(\{1,x\})$ is the algebra generated by $\{1,x\}$. Note $A(\{1,x\})$ is an algebra in $C_b(X,\mathscr{S},\|\cdot\|_\infty)$ and so $\overline{A(\{1,x\})}$ is also an algebra in $C_b(X,\mathscr{S},\|\cdot\|_\infty)$. Our first Weierstrass Approximation Theorem leads us to speculate that we can say more: since $[a,b]$ is a Hausdorff space which is compact, it seems likely $\overline{A(\{1,x\})} = C_b(X,\mathscr{S})$ under the right conditions on X!*

We can also look at this in the context of lattices.

Theorem 9.3.5 Closure of a lattice in $C_b(X,\mathscr{S})$ with the Infinity Norm is a Lattice

Let L be an algebra in the normed linear space $C_b(X,\mathscr{S},\|\cdot\|_\infty)$. Then $\overline{L}$ is a lattice also.

Proof 9.3.5
Recall $\overline{L}$ is a lattice if $\overline{L}$ is a nonempty poset in which each $(x,y) \in L$ has a lub and a glb. Let L be a lattice in the normed linear space $C_b(X,\mathscr{S},\|\cdot\|_\infty)$. We need to show any f,g in $\overline{L}$ have an infimum

and supremum. **Case 1:** *Both $f, g \in L$. Since L is a lattice, $f \vee g \in L$ and $f \wedge g \in L$ and so also in $\overline{L}$.*

Case 2: *$f \in L$ and $g = \lim_{n\to\infty} g_n$ in $\|\cdot\|_\infty$ with each $g_n \in L$. It is easy to see $fg_n \to fg$ in norm. Since, $g_n \in L$, $f \vee g_n \in L$ too. Thus, $fg \in \overline{L}$ by definition. A similar argument shows if $g \in L$ and $f = \lim_{n\to\infty} f_n$ in $\|\cdot\|_\infty$ with each $f_n \in L$, then $f \vee g \in \overline{L}$.*

Case 3: *If $f = \lim_{n\to\infty} f_n$ and $g = \lim_{n\to\infty} g_n$ with $f_n, g_n \in L$, then $f_n \vee g_n \in L$ as L is a lattice and so the limit $fg \in \overline{L}$.*

A similar set of arguments works for the $\wedge$ case. We conclude $\overline{L}$ is an lattice also. ■

Homework

Exercise 9.3.3 *What is the algebra generated by $\{1, \cos(x)\}$?*

Exercise 9.3.4 *What is the algebra generated by $\{1, e^x\}$?*

Exercise 9.3.5 *In the proof of Theorem 9.3.3, prove $f, g \in C(X, \Re) \Longrightarrow fwedgeg \in C(X, \Re)$.*

Exercise 9.3.6 *In the proof of Theorem 9.3.5, provide the arguments the prove the $\wedge$ case.*

We note one last result before we prove the more general Stone - Weierstrass Theorem. The absolute value function $t \to |t|$ is continuous on $\Re$. So, using the Weierstrass Approximation Theorem, there is, for each $\epsilon > 0$, a polynomial p such that $||t| - p(t)| < \epsilon$ for all $t \in [-1, 1]$. First, we define what we mean by a separating family of functions.

Definition 9.3.1 Separating Families of Functions

A collection, $\mathscr{F}$, of functions on a space X is said to separate points if for any $x, y \in X$ with $x \neq y$, there is a function $f \in \mathscr{F}$ such that $f(x) \neq f(y)$.

If $(X, \mathscr{S})$ is a topological vector space, if we assume $C_b(X, \mathscr{S})$ separates points, then given $x \neq y$ in X, there is a function $f \in C_b(X, \mathscr{S})$ with $f(x) \neq f(y)$. Thus, there a neighborhoods W_x and W_y in $\Re$ so that $f^{-1}(W_x)$ and $f^{-1}(W_y)$ are both open. In particular, if we assume $f(x) < f(y)$, we can find an r so that $f(x) < r < f(y)$ and so $f^{-1}(r, \infty)$ and $f^{-1}(-\infty, r)$ are open and disjoint. This tells us X must be Hausdorff.
If $(X, \mathscr{S})$ is a topological vector space and L is a lattice, when is $\overline{L}$ the same as $C_b(X, \mathscr{S})$ under the norm $\|\cdot\|_\infty$?

Theorem 9.3.6 $\overline{L} = C_b(X, \mathscr{S})$, X compact

Let L be a lattice in $C_b(X, \mathscr{S})$ under the norm $\|\cdot\|_\infty$. We assume X is topologically compact and T_1 and X has more than one point. Further, we assume L satisfies $x, y \in X$, $x \neq y$, $a, b \in \Re$ implies there is $\phi \in L$ with $\phi(x) = a$ and $\phi(y) = b$. Then $\overline{L} = C_b(X, \mathscr{S})$.

Proof 9.3.6
Let $f \in C_b(X, \mathscr{S})$ where X is topologically compact with respect to the topology of $\mathscr{S}$. Since $\overline{L}$ is closed, it contains all its limit points under the norm. We will construct a sequence $(g_n) \subset L$ so that $g_n \to f$ in norm. Note, if we could find $g \in L$ for a given $\epsilon > 0$ with $\|f - g\|_\infty < \epsilon$, we could use a sequence of scalars $(\epsilon_n = 1/n)$ to construct (g_n) so that $g_n \to f$ in norm. So all we have to do is to fix an $\epsilon > 0$ and find such a function $g \in L$.

Fix $x_0 \in X$. If $y \neq x_0$, then by assumption, there is a $\phi_y \in L$ with $\phi_y(x_0) = f(x_0)$ and $\phi_y(y) = f(y)$; i.e. we let $a = f(x)$ and $b = f(y)$ when we use the property we assume L has.

Let $G_y = \{z | \phi_y(z) < f(z) + \epsilon\} \subset X$. *We know* G_y *is open as* f *is continuous.*

Then, $\phi_y(x_0) = f(x_0) < f(x_0) + \epsilon$ *and so* $x_0 \in G_y$. *Hence,* G_y *is nonempty. Also,* $\phi_y(y) = f(y) < f(y) + \epsilon$ *and so* $y \in G_y$. *We conclude the collection* $\{G_y | y \neq x_0\}$ *is an open cover of* X. *But* X *is compact and so there is a finite subcover* $\{G_{y_1}, \ldots, G_{y_n}\}$ *for some* n *and points* $y_i \in X$. *For each of these points* y_i, *there is a* ϕ_{y_i} *so that* $\phi_{y_i}(x_0) = f(x_0)$ *and* $\phi_(y_i) = f(y_i)$ *with* $G_{y_i} = \{z | \phi_(y_i)(z) < f(z) + \epsilon\}$.

Let

$$g_{x_0} \quad = \quad \phi_{y_1} \wedge \phi_{y_2} \wedge \ldots \wedge \phi_{y_n} \in L$$

Then $g_{x_0}(x) = \phi_{y_1}(x) \wedge \phi_{y_2}(x) \wedge \ldots \wedge \phi_{y_n}(x)$. *Since* $X = \cup_{i=1}^n G_{y_i}$, *if* $z \in X$, $z \in G_{y_j}$ *for some* y_j. *This tells us* $\phi_{y_j}(z) < f(z) + \epsilon$ *and so* $g_{x_0} < f(z) + \epsilon$. *This is true for all* $z \in X$.

Next, let $H_{x_0} = \{z | g_{x_0} > f(z) - \epsilon\}$. *Then* H_{x_0} *is an open set. We can do this argument for all* $x \in X$ *and obtain a collection* $\{H_x\}$ *which is then an open cover of* X *which is compact. Thus, there is a finite subcover* $\{H_{x_1}, \ldots, H_{x_m}\}$ *with* $X = \cup_{j=1}^m H_{x_j}$. *Each* H_{x_j} *is associated with a lattice function* g_{x_j}.

Let $g = g_{x_1} \vee \ldots \vee g_{x_m} \in L$. *If* $z \in X$, $z \in H_{x_p}$ *for some* p. *This implies* $g_{x_j}(z) > f(z) - \epsilon$. *Thus,* $g(z) \geq f_{x_j}(z) > f(z) - \epsilon$. *We also know* $g_{x_j}(z) < f(z) + \epsilon$ *for all* j. *Hence,* $g(z) = g_{x_1}(z) \vee \ldots \vee g_{x_m}(z) < f(z) + \epsilon$ *for all* $z \in X$.

Combining, $|g(z) - f(z)| < \epsilon$ *for all* $z \in X$ *or* $\|f - g\|_\infty \leq \epsilon$. *This is the* g *we seek; just redo the argument using* $\epsilon = \epsilon/2$. ■

Homework

Exercise 9.3.7 *Prove if* $f, g \in C_b(X, \mathscr{S})$, $f \vee g = (f+g+|f-g|)/2$ *and* $f \wedge g = (f+g-|f-g|)/2$.

Exercise 9.3.8 *If* V *is a subspace of* $C_b(X, \mathscr{S})$ *in which* $f \in V$ *implies* $|f| \in V$, *then the previous exercise can be used to compute* $f \vee g$ *and* $f \wedge g$. *Prove this shows* $f \vee g$ *and* $f \wedge g$ *in in* V *and so* V *is a lattice.*

Exercise 9.3.9 *Prove the functions* e^x *form a separating family on* $\Re$.

Exercise 9.3.10 *Prove the family consisting of the cosines* $\cos(nx)$ *and the constant functions* 1 *are separating on* $\Re$.

Exercise 9.3.11 *Prove the family consisting of the sines* $\sin(nx)$ *on* $\Re$ *is not separating.*

Exercise 9.3.12 *Let* F *be a family of functions on* $[0, 1]$ *containing the constant function* 1 *and the identity map* $I(x) = x$ *that also satisfies:*

- F *is a separating family of continuous functions on* $[0, 1]$.
- $f \in F$ *and* $a \in \Re$ *implies* $af \in F$.
- $f, g \in F$ *implies* $f + g \in F$.
- $f, g \in F$ *implies* $f \vee g \in F$

Show F *is a lattice consisting of all piecewise linear continuous real-valued functions on* $[0, 1]$. *This means for each* $f \in F$, *there is a partition* $P = \{0, x_1, x_2, \ldots, x_p\}$ *so that* f *is linear on* $[x_i, x_{i+1}]$ *for* $0 \leq i \leq (p - 1)$.

There is a nice connection between closed lattices and algebras.

Theorem 9.3.7 Closed Algebra is Also a Closed Lattice

Let $(X, \mathscr{S})$ be a T_1 topological space and $\overline{A}$ a closed algebra of $C_b(X, \mathscr{S})$. Then $\overline{A}$ is also a closed lattice.

Proof 9.3.7

We will prove this by showing if $f \in \overline{A}$, then $|f| \in \overline{A}$. For the continuous function $\phi(t) = |t|$ on $[a, b]$, for any $\epsilon > 0$, there is a polynomial q_ϵ so that $|\,|t| - q_\epsilon(t)\,| < \epsilon/2$ for all $t \in [a, b]$ by the Weierstrass Approximation Theorem. In particular, since our f is bounded, we can find this approximating polynomial for the interval $[-\|f\|_\infty, \|f\|_\infty]$. The polynomial q_ϵ has representation $q_\epsilon(t) = a_0 + \sum_{i=1}^n a_i t^i$ for scalars a_i (which depend on ϵ also, but we will not label them for that) and some n. At $t = 0$, we find $|\,|0| - q_\epsilon(t)\,| = |a_0| < \epsilon/2$. For convenience, let $p(t) = \sum_{i=1}^n a_i t^i$. Then, for any $t \in [-\|f\|_\infty, \|f\|_\infty]$, we have

$$\begin{aligned} |\,|t| - q_\epsilon(t)\,| &= |\,|t| - q_\epsilon(t)\,| = |\,|t| - (a_0 + p(t)) + a_0\,| \\ &\leq |\,|t| - (a_0 + p(t))\,| + |a_0| < \epsilon/2 + \epsilon/2 = \epsilon \end{aligned}$$

In particular, this holds for $t = f(x)$. Thus,

$$|\,|f(x)| - p(f(x))\,| \quad < \quad \epsilon$$

Since $f \in \overline{A}$, $p(f) = \sum_{i=1}^n a_i f^i \in \overline{A}$ also. Hence, letting $w = p(f)$, we see $|\,|f(x)| - w(x)\,| < \epsilon$ for all $x \in X$. This implies $\|f - w\|_\infty < \epsilon$.

Letting $(\epsilon_n - 1/n)$, this shows there is a sequence $(w_n) \in \overline{A}$ with $w_n \rightarrow |f|$. Hence $|f| \in \overline{A}$ also. By an exercise above, this shows $\overline{A}$ is also a lattice which is closed since $\overline{A}$ is closed. ■

9.4 The Stone - Weierstrass Theorem

We can now prove an extension to our original Weierstrass Approximation Theorem.

Theorem 9.4.1 Real Stone - Weierstrass Theorem for compact Hausdorff spaces

Let X be a compact Hausdorff space. Suppose $A \subset C(X, \Re)$ satisfies

1. *A is an algebra.*
2. *A separates points of X.*
3. *The constant function $f(x) = 1$ is in A.*

Then $\overline{A} = C_b((X, \mathscr{S}))$. That is, A is dense in $C_b((X, \mathscr{S}))$.

Proof 9.4.1

If X contained only one point, then $C_b((X, \mathscr{S}))$ would contain only constant functions. In this case, we have $\overline{A} = C_b((X, \mathscr{S}))$ because we assume A contains the function 1.

Hence, we can assume X contains more than one point. If we know M is a closed lattice such that $x \neq y \in X$, then $a, b \in \Re$ implies there is $f \in L$ so that $f(x) = a$, $f(y) = b$, then by Theorem 9.3.6, *we have $M = C_b((X, \mathscr{S}))$. From Theorem 9.3.7, we know $\overline{A}$ is a closed lattice.*

If $x \neq y$, there is a $g \in A$ so that $g(x) \neq g(y)$. Choose $a, b \in \Re$ and define f by

$$f(z) = a\frac{g(z) - g(y)}{g(x) - g(y)} + b\frac{g(z) - g(x)}{g(y) - g(x)}$$

Then,

$$f(x) = a\frac{g(x) - g(y)}{g(x) - g(y)} = a, \quad f(y) = b\frac{g(y) - g(x)}{g(y) - g(x)} = b$$

Thus, we can apply Theorem 9.3.7 to infer $\overline{A}$ is a closed lattice. ■

Example 9.4.1 *The collection of polynomials on any closed interval $[a, b]$ is easily seen to be dense in $C(X, \Re)$. Less obviously, the collection of trigonometric polynomials, or functions of the form*

$$f(x) = \sum_{k=0}^{n} a_n cos(nx) + b_n sin(nx) \qquad n \in N$$

defined on any closed interval is also dense.

Homework

Exercise 9.4.1 *Prove the functions e^x form a separating family on $\Re$ and thus, polynomials formed from e^x and 1 are dense in $C([a, b], \Re)$ for all finite intervals $[a, b]$ with $a < b$.*

Exercise 9.4.2 *Prove the family consisting of the cosines $\cos(nx)$ and the constant functions 1 are separating on $[0, \pi]$. Thus, polynomials formed from $\cos(x)$ and 1 are dense in $C([a, b], \Re)$ for all finite intervals $[a, b]$ with $a < b$. To show this, use an induction argument to show $\cos^n(x)$ can be written as a linear combination of terms of the form $\cos(kx)$.*

Exercise 9.4.3 *Prove the family consisting of the sines $\sin(nx)$ and the constant function 1 is separating on $[0, \pi]$. Thus, polynomials formed from $\sin(x)$ and 1 are dense in $C([a, b], \Re)$ for all finite intervals $[a, b]$ with $a < b$. To show this, use an induction argument to show $\cos^n(x)$ can be written as a linear combination of terms of the form $\cos(kx)$.*

Exercise 9.4.4 *Let F be a family of functions on $[0, 1]$ containing the constant function 1 and the identity map $I(x) = x$ that also satisfies:*

- *F is a separating family of continuous functions on $[0, 1]$.*
- *$f \in F$ and $a \in \Re$ implies $af \in F$.*
- *$f, g \in F$ implies $f + g \in F$.*
- *$f, g \in F$ implies $f \vee g \in F$*

Earlier we showed F is a lattice consisting of all piecewise linear continuous real-valued functions on $[0, 1]$. Prove F is dense in $C([0, 1], \Re)$.

Exercise 9.4.5 *Let F be a family of functions on $[0, 1]$ containing the constant function 1 and the identity map $I(x) = x$ that also satisfies:*

- *F is a separating family of continuous functions on $[0, 1]$.*
- *$f \in F$ and $a \in \Re$ implies $af \in F$.*

- *$f, g \in F$ implies $f + g \in F$.*
- *$f, g \in F$ implies $f \vee g \in F$*

Earlier we showed F is a lattice consisting of all piecewise linear continuous real-valued functions on $[0, 1]$. Show F is not a subalgebra by showing if $f, f^2 \in F$, then f must be a constant function.

Part IV

Topological Degree Theory

Chapter 10

Brouwer Degree Theory

We are now going to construct a mapping which assigns an integer to various kinds of functions which is similar to the idea of a winding number. This has applications to fixed point theorems and it is originally defined on $\Re^n$, whereas our winding number extensions were defined on $\Re^2$. We will use the usual norm based topology of $\Re^n$ here in our initial discussions. However, what is really interesting and useful is that these ideas can be ported to infinite dimensional spaces so that they can be applied in the theory of ordinary and partial differential equations among other things.

10.1 Construction of n - Dimensional Degree

Let $\boldsymbol{D} \subset \Re^n$ be an open, bounded and nonempty set. With every continuous map $f : \overline{\boldsymbol{D}} \rightarrow \Re^n$ and every $\boldsymbol{p} \notin f(\partial \boldsymbol{D})$, it is possible to associate an integer $d(f, \boldsymbol{D}, \boldsymbol{p})$ with the following properties:

Theorem The Invariance under Homotopy Property

If $f(\boldsymbol{x}, t) : \overline{\boldsymbol{D}} \rightarrow \Re^n$ is continuous with $\boldsymbol{p} \neq f(\partial \boldsymbol{D}, [0, 1])$, then $d(f(\cdot, 0), \boldsymbol{D}, \boldsymbol{p}) = d(f(\cdot, 1), \boldsymbol{D}, \boldsymbol{p})$.

Theorem The Existence of a Solution to $f(x) = \boldsymbol{p}$

If $d(f, \boldsymbol{D}, \boldsymbol{p}) \neq 0$, then there is an $\boldsymbol{x}_0 \in \boldsymbol{D}$ so that $f(\boldsymbol{x}_0) = \boldsymbol{p}$.

Theorem The Decomposition of Domain

If $\boldsymbol{D} = \cup_{i=1}^m \boldsymbol{D}_i$ where each $\boldsymbol{D}_i$ is open and pairwise disjoint with $\partial \boldsymbol{D}_i \subset \partial \boldsymbol{D}$, then for all $\boldsymbol{p} \notin f(\partial \boldsymbol{D})$, $\sum_{i=1}^m d(f, \boldsymbol{D}_i, \boldsymbol{p}) = d(f, \boldsymbol{D}, \boldsymbol{p})$.

Theorem The Borsuk Theorem: When is the Degree an Odd Integer?

Let $\boldsymbol{D}$ be a symmetric, bounded open set in $\Re^N$ containing the origin, $\boldsymbol{0}$. If $f : \overline{\boldsymbol{D}} \rightarrow \Re^N$ is an odd mapping; i.e. $f(\boldsymbol{x}) = -f(-\boldsymbol{x})$ for all $\boldsymbol{x} \in \boldsymbol{D}$ with $\boldsymbol{0} \notin \partial \boldsymbol{D}$, then $d(f, \boldsymbol{D}, \boldsymbol{0})$ is an odd number. Of course, this means the value is not zero and so the existence theorem can be applied.

This type of integer valued mapping is called a **topological degree** and so our discussion here is about how to establish all these properties of such a degree. Hence, this subject is called *degree theory*. We first wrote up our treatment of this material when we were working on our PhD on the existence of solutions to optimal control problems where we used an extension of these ideas to function spaces which we will cover in later chapters. A primary reference we used when we were learning about this was (Fučik et al. (34) 1973). Of course, this was a long time ago! We found this book quite useful although we read widely among many others. At the time, there were few such books around (this was around 1977 or so) and so we decided to write a nice set of notes to ourselves. We are pleased these notes, originally hand written in pencil, in an old Colorado State bookstore notebook, can find a use today. We have never had a chance to teach this material in all the 30 years we have taught analysis at the graduate level and we would like to correct that oversight in this volume. It is a nice part of seeing how topology and analysis fit together!

10.1.1 Defining the Degree of a Mapping

If $f : \boldsymbol{D} \subset \Re^N \rightarrow \Re^N$ where $\boldsymbol{D}$ is an open, bounded and nonempty set, for any positive integer k, we say f is of class $C^k(\boldsymbol{D})$ if the component functions $f_1, f_2, \ldots, f_N$ of f all have continuous partial derivatives up to order k. The derivative, $f' = Df$ is then

$$Df = \begin{bmatrix} \frac{\partial f_1}{\partial x^1} & \ldots & \frac{\partial f_1}{\partial x^N} \\ \vdots & & \vdots \\ \frac{\partial f_N}{\partial x^1} & \ldots & \frac{\partial f_N}{\partial x^N} \end{bmatrix} = \begin{bmatrix} D_{11}(f), & \ldots & D_{1N}(f) \\ \vdots & & \vdots \\ D_{N1}(f), & \ldots & D_{NN}(f) \end{bmatrix} = (D_{ij}f)$$

In particular, for now, focusing on $f \in C^1(\boldsymbol{D})$, recall the Jacobian, $Jf(\boldsymbol{x})$ is defined by $Jf = detDf$ or

$$Jf(\boldsymbol{x}) = \det \begin{bmatrix} \frac{\partial f_1}{\partial x^1} & \ldots & \frac{\partial f_1}{\partial x^N} \\ \vdots & & \vdots \\ \frac{\partial f_N}{\partial x^1} & \ldots & \frac{\partial f_N}{\partial x^N} \end{bmatrix} (\boldsymbol{x}) = \det \begin{bmatrix} D_{11}(f), & \ldots & D_{1N}(f) \\ \vdots & & \vdots \\ D_{N1}(f), & \ldots & D_{NN}(f) \end{bmatrix} (\boldsymbol{x}) = det(D_{ij}f(\boldsymbol{x}))$$

Now let's extend f to include $\partial \boldsymbol{D}$ and consider $f : \overline{\boldsymbol{D}} \subset \Re^N \rightarrow \Re^N$ with $f \in C^1(\boldsymbol{D}) \cap C(\overline{\boldsymbol{D}})$ and let $\boldsymbol{p} \in \Re^N$. Look at $\boldsymbol{x} \in f^{-1}(\boldsymbol{p})$. We classify these points as follows:

Definition 10.1.1 Regular and Critical Points

If $Jf(\boldsymbol{x}) \neq 0$, we say $\boldsymbol{x}$ is a **regular point**. *Otherwise, $\boldsymbol{x}$ is a* **critical point**.

For now, assume $\boldsymbol{p} \in f(\overline{\boldsymbol{D}}) \setminus f(\partial \boldsymbol{D})$, which means $\boldsymbol{p} \in f(\boldsymbol{D})$ and for all $\boldsymbol{x} \in f^{-1}(\boldsymbol{p})$, the point $\boldsymbol{x}$ is a regular point. Then by the implicit function theorem (you can see a careful development of these kinds of results in (Peterson (101) 2020) as well as other texts), for each $\boldsymbol{x} \in f^{-1}(\boldsymbol{p})$, we can find neighborhoods $N_1(\boldsymbol{x})$ and $N_2(\boldsymbol{p})$ so that $f : N_1(\boldsymbol{x}) \rightarrow N_2(\boldsymbol{p})$ is **1-1** and **onto**. An immediate consequence is that these neighborhoods are disjoint from each other. If we assumed $f^{-1}(\boldsymbol{p})$ was countably or uncountably infinite, since $\overline{\boldsymbol{D}}$ is compact and $\boldsymbol{x} \in \overline{\boldsymbol{D}}$, then there would be a convergent subsequence in $f^{-1}(\boldsymbol{p})$. Thus, there would be $(\boldsymbol{x_k})$ with $f(\boldsymbol{x_k}) = \boldsymbol{p}$ and $\boldsymbol{x_k} \rightarrow \boldsymbol{x} \in \overline{\boldsymbol{D}}$. The continuity of f would then tell us $f(\boldsymbol{x}) = \boldsymbol{p}$. But that would imply the neighborhoods are not disjoint. Hence, $f^{-1}(\boldsymbol{p})$ is a finite set and so it is bounded in norm. This leads to the following definition.

Definition 10.1.2 Degree of a Mapping at a Regular Point

> *Let $f \in C^1(\boldsymbol{D}) \cap C(\overline{\boldsymbol{D}})$, $\boldsymbol{p} \in \Re^N$ with $\boldsymbol{p} \notin f(\partial \boldsymbol{D})$. Assume each $\boldsymbol{x} \in f^{-1}(\boldsymbol{p})$ is regular. Set*
>
> $$d(f, \boldsymbol{D}, \boldsymbol{p}) = \sum_{\boldsymbol{x} \in f^{-1}(\boldsymbol{p})} sign\ Jf(\boldsymbol{x})$$
>
> *where by the remarks above, we know this is a finite sum. We call $d(f, \boldsymbol{D}, \boldsymbol{p})$ the* **degree** *of the mapping f with respect to $\boldsymbol{D}$ and $\boldsymbol{p}$. If $f^{-1}(\boldsymbol{p}) = \emptyset$, we set $d(f, \boldsymbol{D}, \boldsymbol{p}) = 0$. Clearly, $d(f, \boldsymbol{D}, \boldsymbol{p})$ is an integer.*

Homework

Exercise 10.1.1 *Let*

$$f(x, y) = \begin{bmatrix} x^2 + y^2 \\ 2x^2 - y^2 \end{bmatrix}$$

- *Find Df and $Jf((x, y))$ for all $(x, y) \in \Re^2$.*
- *Find all regular points and all critical points.*
- *Find $f(\overline{D})$, for D, the ball about $(0, 0)$ of radius $1/2$. Compute*

  $$d(f, \boldsymbol{D}, (1, 1)) = \sum_{(x,y) \in f^{-1}((1,1))} sign\ Jf((x, y))$$

  and as part of your calculation, find $f(\partial D)$.
- *Find $f(\overline{D})$, for D, the ball about $(0, 0)$ of radius 2. Compute*

  $$d(f, \boldsymbol{D}, (2, -1)) = \sum_{(x,y) \in f^{-1}((2,-1))} sign\ Jf((x, y))$$

  and as part of your calculation, find $f(\partial D)$.

Exercise 10.1.2 *Let*

$$f(x, y) = \begin{bmatrix} x^2 + y^2 \\ 2x^2 - 3y^2 \end{bmatrix}$$

- *Find Df and $Jf((x, y))$ for all $(x, y) \in \Re^2$.*
- *Find all regular points and all critical points.*
- *Find $f(\overline{D})$, for D, the ball about $(0, 0)$ of radius $1/2$. Compute*

  $$d(f, \boldsymbol{D}, (1, 1)) = \sum_{(x,y) \in f^{-1}((1,1))} sign\ Jf((x, y))$$

  and as part of your calculation, find $f(\partial D)$.
- *Find $f(\overline{D})$, for D, the ball about $(0, 0)$ of radius 2. Compute*

  $$d(f, \boldsymbol{D}, (3, -2)) = \sum_{(x,y) \in f^{-1}((3,-2))} sign\ Jf((x, y))$$

and as part of your calculation, find $f(\partial D)$.

Exercise 10.1.3 *Let*

$$f(x,y) \quad = \quad \begin{bmatrix} 2xy \\ x^2 - y^2 \end{bmatrix}$$

- *Find* Df *and* $Jf((x,y))$ *for all* $(x,y) \in \Re^2$.
- *Find all regular points and all critical points.*
- *Find* $f(\overline{D})$, *for D, the ball about* $(0,0)$ *of radius* $1/2$. *Compute*

$$d(f, \boldsymbol{D}, (1,1)) = \sum_{(x,y) \in f^{-1}((1,1))} sign\ Jf((x,y))$$

and as part of your calculation, find $f(\partial D)$.

- *Find* $f(\overline{D})$, *for D, the ball about* $(0,0)$ *of radius* 2. *Compute*

$$d(f, \boldsymbol{D}, (3,-2)) = \sum_{(x,y) \in f^{-1}((3,-2))} sign\ Jf((x,y))$$

and as part of your calculation, find $f(\partial D)$.

10.1.2 Sard's Theorem

The next important result in Sard's Theorem. Before we state and prove this result, let's recall some notation. For $f : \boldsymbol{D} \subset \Re^n \rightarrow \Re^m$ defined locally at $\boldsymbol{x_0}$, we say f is differentiable at $\boldsymbol{x_0}$ if there is a linear map $\boldsymbol{L}(\boldsymbol{x_0}) : \Re^n \rightarrow \Re^m$ and an error vector $\mathscr{E}(\boldsymbol{x_0}, \boldsymbol{\Delta})$ so that for given $\boldsymbol{\Delta}$,

$$f(\boldsymbol{x_0} + \boldsymbol{\Delta}) \quad = \quad f(\boldsymbol{x_0}) + \boldsymbol{L}(\boldsymbol{x_0})\, \boldsymbol{\Delta} + \mathscr{E}(\boldsymbol{x_0}, \boldsymbol{\Delta})$$

where

$$\lim_{\boldsymbol{\Delta} \rightarrow \boldsymbol{0}} \mathscr{E}(\boldsymbol{x_0}, \boldsymbol{\Delta}) \quad = \quad 0, \quad \lim_{\boldsymbol{\Delta} \rightarrow \boldsymbol{0}} \frac{\mathscr{E}(\boldsymbol{x_0}, \boldsymbol{\Delta})}{\|\boldsymbol{\Delta}\|} = 0$$

where we use norm convergence with respect to the norm $\|\cdot\|$ used in $\Re^n$ and $\Re^m$. The derivative of f at $\boldsymbol{x_0}$ is denoted by $f'(\boldsymbol{x_0}) = \boldsymbol{D}\boldsymbol{f}(\boldsymbol{x_0})$. If we know f is differentiable at $\boldsymbol{x_0}$, then

$$(\boldsymbol{L}(\boldsymbol{x_0}))_{ij} \quad = \quad f_{i,x_j}(\boldsymbol{x_0}) = \frac{\partial f_i}{\partial x_j} = Df_{ij}(\boldsymbol{x_0})$$

i.e. $\boldsymbol{L}(\boldsymbol{x_0}) = \boldsymbol{J_f}(\boldsymbol{x_0})$.

In particular, if $f : \boldsymbol{D} \subset \Re^n \rightarrow \Re^n$ is differentiable on $\boldsymbol{D}$, then given two points $\boldsymbol{q}$ and $\boldsymbol{p}$ in $\boldsymbol{D}$, we have

$$f(\boldsymbol{q}) \quad = \quad f(\boldsymbol{p}) + f'(\boldsymbol{p})\,(\boldsymbol{q} - \boldsymbol{p}) + \mathscr{E}(\boldsymbol{p}, \boldsymbol{q} - \boldsymbol{p})$$

Theorem 10.1.1 Sard's Theorem

Let $f : \boldsymbol{D} \rightarrow \Re^n$, *with* $\boldsymbol{D} \subset \Re^n$ *an open and bounded set. Assume* $f \in C^1(\boldsymbol{D})$. *Choose* $\boldsymbol{G}$ *so that* $\overline{\boldsymbol{G}} \subset \boldsymbol{D}$. *Then, if* $\boldsymbol{B} = \{\boldsymbol{x} \in \boldsymbol{G} | Jf(\boldsymbol{x}) = 0\}$, *we have* $\mu(f(\boldsymbol{B})) = 0$ *where* μ *is* n*-dimensional Lebesgue measure.*

Proof 10.1.1

Since $\boldsymbol{D}$ is bounded, there is a rectangular parellelopiped $\mathscr{R}$ with sides parallel to the coordinate axes so that $\boldsymbol{D} \subset \mathscr{R}$. Let $\rho = dist(\overline{\boldsymbol{G}}, \Re^n \setminus \boldsymbol{D})$. Since $\overline{\boldsymbol{G}} \subset \boldsymbol{D}$ and $\boldsymbol{D}$ is open, we must have $\rho > 0$. Subdivide $\mathscr{R}$ by passing planes parallel to the coordinate axes into cubes whose diameter is less than $\rho/2$. Hence, for some positive integer M, $\overline{\boldsymbol{G}} \subset \cup_{i=1}^{M} \boldsymbol{C_i}$ where $\boldsymbol{C_i}$ is a cube of diameter less than $\rho/2$ with $\boldsymbol{C_i} \subset \boldsymbol{D}$.
Let's prove

$$\mu\Bigg(f\bigg(\{\boldsymbol{x} \in \cup_{i=1}^{M} \boldsymbol{C_i}\} | Jf(\boldsymbol{x}) = 0\bigg)\Bigg) = 0$$

Let ℓ be the length of the side of $\boldsymbol{C_i}$ and let $\boldsymbol{p_1}$ and $\boldsymbol{p_2}$ be in $\boldsymbol{C_i}$. Now

$$f(\boldsymbol{p_1}) - f(\boldsymbol{p_2}) = f'(\boldsymbol{p_2})(\boldsymbol{p_1} - \boldsymbol{p_2}) + \mathscr{E}(\boldsymbol{p_2}, \boldsymbol{p_1} - \boldsymbol{p_2})$$

where

$$\lim_{\boldsymbol{p_1} \to \boldsymbol{p_2}} \mathscr{E}(\boldsymbol{p_2}, \boldsymbol{p_1} - \boldsymbol{p_2}) = 0, \quad \lim_{\boldsymbol{p_1} \to \boldsymbol{p_2}} \frac{\mathscr{E}(\boldsymbol{p_2}, \boldsymbol{p_1} - \boldsymbol{p_2})}{\|\boldsymbol{p_1} - \boldsymbol{p_2}\|} = 0$$

Hence, given $\epsilon > 0$, there is a δ_ϵ which we can choose less than $\frac{\rho}{2}$, so that if $\|\boldsymbol{p_1} - \boldsymbol{p_2}\| < \delta_\epsilon$, then $\frac{\mathscr{E}(\boldsymbol{p_2}, \boldsymbol{p_1} - \boldsymbol{p_2})}{\|\boldsymbol{p_1} - \boldsymbol{p_2}\|} < \epsilon$. Now in three dimensions, the diameter of a cube is $\sqrt{3} \times \ell$ where ℓ is the length of the side of the cube. Hence, in $\Re^n$, this diameter is $\sqrt{n}\ell$. Since we have chosen this cube diameter less than $\rho/2$, we must have $\ell < \frac{\rho}{2\sqrt{n}}$.

So for any $\boldsymbol{p_1}$ and $\boldsymbol{p_2}$ in $\boldsymbol{C_i}$, we have $\|\boldsymbol{p_1} - \boldsymbol{p_2}\| \leq \ell\sqrt{n}$. Choose any m so that $\frac{\ell\sqrt{n}}{m} < \delta_\epsilon$. Then, for such an m, if $\|\boldsymbol{p_1} - \boldsymbol{p_2}\| < \frac{\ell\sqrt{n}}{m} < \delta_\epsilon < \frac{\rho}{2}$, we have

$$\frac{\mathscr{E}(\boldsymbol{p_2}, \boldsymbol{p_1} - \boldsymbol{p_2})}{\|\boldsymbol{p_1} - \boldsymbol{p_2}\|} < \epsilon$$

Now divide $\boldsymbol{C_i}$ into cubes with side length $\frac{\ell}{m}$. This implies we form m^n cubes. Label these cubes $\boldsymbol{C_{ij}}$. For each $\boldsymbol{p_1}$ and $\boldsymbol{p_2}$ in $\boldsymbol{C_{ij}}$, we must have

$$\|\boldsymbol{p_1} - \boldsymbol{p_2}\| \leq \frac{\ell\sqrt{n}}{m} \Longrightarrow \mathscr{E}(\boldsymbol{p_2}, \boldsymbol{p_1} - \boldsymbol{p_2}) < \epsilon\|\boldsymbol{p_1} - \boldsymbol{p_2}\| = \epsilon\frac{\ell\sqrt{n}}{m}$$

Now fix $\boldsymbol{p_2} \in \boldsymbol{C_{ij}}$. Define the mapping $\boldsymbol{T}$ by $\boldsymbol{T}(\boldsymbol{p}) = f(\boldsymbol{p_2}) + f'(\boldsymbol{p_2})(\boldsymbol{p} - \boldsymbol{p_2})$ which is the usual linearization of f at the point $\boldsymbol{p_2}$. Then look at what $\boldsymbol{T}$ does to the cube $\boldsymbol{C_{ij}}$. Let

$$L = \max_{\boldsymbol{p} \in \boldsymbol{C_i}} \left(\sum_{j,k=1}^{n} \left| \frac{\partial f_j}{\partial x_k}(\boldsymbol{p}) \right| \right)$$

Since $\boldsymbol{C_i}$ is compact, L is finite. Recall the diameter of $\boldsymbol{C_{ij}}$ is $\frac{\ell\sqrt{n}}{m}$. Then, if $\boldsymbol{p}$ and $\boldsymbol{q}$ are both in $\boldsymbol{C_{ij}}$, we have

$$\boldsymbol{T}(\boldsymbol{p}) - \boldsymbol{T}(\boldsymbol{q}) = f'(\boldsymbol{p_2})(\boldsymbol{p} - \boldsymbol{p_2}) - f'(\boldsymbol{p_2})(\boldsymbol{q} - \boldsymbol{p_2}) = f'(\boldsymbol{p_2})(\boldsymbol{p} - \boldsymbol{q})$$

and so on $\boldsymbol{C_i}$,

$$\|\boldsymbol{T}(\boldsymbol{p}) - \boldsymbol{T}(\boldsymbol{q})\| \leq \|f'(\boldsymbol{p_2})(\boldsymbol{p} - \boldsymbol{q})\| \leq \|f'(\boldsymbol{p_2})\| \, \|\boldsymbol{p} - \boldsymbol{q}\| \leq L\frac{\ell\sqrt{n}}{m}$$

Now assume $\boldsymbol{C_{ij}}$ contains a critical point $\boldsymbol{p_2}$. Then $Jf(\boldsymbol{p_2}) = 0$ which implies $f'(\boldsymbol{p_2})$ has a nontrivial nullspace which is at least one dimensional. So there is a nonzero $\boldsymbol{v}$ in this nullspace. Then

$$T(\alpha \boldsymbol{v} + \boldsymbol{p_2}) = f(\boldsymbol{p_2}) + f'(\boldsymbol{p_2})(\alpha \boldsymbol{v}) = f(\boldsymbol{p_2})$$

i.e. $T(<\boldsymbol{v}> + \boldsymbol{p_2}) = f(\boldsymbol{p_2})$ *where* $<\boldsymbol{v}>$ *denotes the span of* $\boldsymbol{v}$.

Thus, the dimension of $T(\boldsymbol{C_{ij}})$ is at most $n-1$. If $\boldsymbol{p} \in \boldsymbol{C_{ij}}$, then

$$f(\boldsymbol{p}) = f(\boldsymbol{p_2}) + f'(\boldsymbol{p_2})(\boldsymbol{p} - \boldsymbol{p_2}) + \mathscr{E}(\boldsymbol{p_2}, \boldsymbol{p} - \boldsymbol{p_2}) = T(\boldsymbol{p}) + \mathscr{E}(\boldsymbol{p_2}, \boldsymbol{p} - \boldsymbol{p_2})$$

Since $\|\boldsymbol{p} - \boldsymbol{p_2}\| < \frac{\ell\sqrt{n}}{m}$ on $\boldsymbol{C_{ij}}$, we see $\mathscr{E}(\boldsymbol{p_2}, \boldsymbol{p} - \boldsymbol{p_2}) < \epsilon \|\boldsymbol{p} - \boldsymbol{p_2}\| < \epsilon \frac{\ell\sqrt{n}}{m}$. This is the size estimate for the part of $f(\boldsymbol{p})$ which is not in the range of $T(\boldsymbol{p})$. So $f(\boldsymbol{p})$ has the part that is in the box determined by $T(\boldsymbol{p})$ and a one dimensional part determined by the error term. Now,

$$\|f(\boldsymbol{p}) - T(\boldsymbol{p})\| < \epsilon \frac{\ell\sqrt{n}}{m} \Longrightarrow \|f(\boldsymbol{p})\| < \|T(\boldsymbol{p})\| + \epsilon \frac{\ell\sqrt{n}}{m}$$

We then have

$$\begin{aligned} \|f(\boldsymbol{p}) - f(\boldsymbol{q})\| &\leq \|f(\boldsymbol{p}) - T(\boldsymbol{p}) - f(\boldsymbol{q}) + T(\boldsymbol{q}) + T(\boldsymbol{p}) - T(\boldsymbol{q})\| \\ &\leq \|f(\boldsymbol{p}) - T(\boldsymbol{p})\| + \|f(\boldsymbol{q}) - T(\boldsymbol{q})\| + \|T(\boldsymbol{p}) - T(\boldsymbol{q})\| \\ &< 2\epsilon \frac{\ell\sqrt{n}}{m} + L \frac{\ell\sqrt{n}}{m} \end{aligned}$$

We conclude the diameter of $f(\boldsymbol{C_{ij}})$ is overestimated by a box with $n-1$ dimensions of size $2\epsilon \frac{\ell\sqrt{n}}{m} + L \frac{\ell\sqrt{n}}{m}$ and one dimension of the size given by the diameter $2\epsilon \frac{\ell\sqrt{n}}{m}$. Since there are m^n cubes in $\boldsymbol{C_i}$, we have

$$\begin{aligned} \mu(f(\boldsymbol{C_i})) &\leq m^n \left(2\epsilon \frac{\ell\sqrt{n}}{m} + L \frac{\ell\sqrt{n}}{m}\right)^{n-1} 2\epsilon \frac{\ell\sqrt{n}}{m} \\ &\leq m^n \left(\frac{\ell\sqrt{n}}{m}\left(L + 2\epsilon\right)\right)^{n-1} 2\epsilon \frac{\ell\sqrt{n}}{m} \\ &\leq m^n \left(\frac{\ell\sqrt{n}}{m}\right)^n \left(L + 2\epsilon\right)^{n-1} 2\epsilon = \ell^n n^{n/2} \left(L + 2\epsilon\right)^{n-1} 2\epsilon \end{aligned}$$

If we assume $\epsilon < L/2$, this simplifies nicely to

$$\mu(f(\boldsymbol{C_{ij}})) < \ell^n n^{n/2}\, 2(2L)^{n-1}\, \epsilon$$

which has the form $C\epsilon$ for a constant C. This shows $\mu(f(\boldsymbol{C_i}))$ is arbitrarily small. We conclude $\mu(f(\boldsymbol{C_i})) = 0$. This implies $\mu\left(\cup \boldsymbol{C_i}\right) = 0$ too. ∎

Now we use Sard's Theorem to extend our definition of degree to the situation that there is a point $\boldsymbol{p}$ with an $\boldsymbol{x} \in f^{-1}(\boldsymbol{p})$ with $Jf(\boldsymbol{x}) = 0$; i.e. $\boldsymbol{x}$ is not regular. Let

$$\boldsymbol{B} = \{\boldsymbol{x} \in \boldsymbol{D} | Jf(\boldsymbol{x}) = 0\}$$

Suppose $\boldsymbol{p} \neq f(\boldsymbol{x})$ for all $\boldsymbol{x} \in \partial \boldsymbol{D}$. From Sard's Theorem, we can infer the Lebesgue measure of $f(\boldsymbol{B})$ is zero. Thus, $f(\boldsymbol{B})$ has no interior point. This means no matter what size neighborhoods we put about $\boldsymbol{p}$, there are always points $\boldsymbol{q}$ in the neighborhood so that $\boldsymbol{q} \notin f(\boldsymbol{B})$. Thus, we can choose a sequence $(\boldsymbol{p_n})$ with $\boldsymbol{p_n} \notin f(\boldsymbol{B})$ and $\boldsymbol{p_n} \to \boldsymbol{p}$. Since $\partial \boldsymbol{D}$ is closed and bounded, it is compact. If a subsequence of $(\boldsymbol{p}^1_{\boldsymbol{n}})$ was contained in $f(\partial \boldsymbol{D})$, then there is a sequence $(\boldsymbol{x}^1_{\boldsymbol{n}})$ in $\partial \boldsymbol{D}$ so that $\boldsymbol{p}^1_{\boldsymbol{n}} = f(\boldsymbol{x}^1_{\boldsymbol{n}})$. Since $\partial \boldsymbol{D}$ is compact, there is another subsequence $(\boldsymbol{x}^2_{\boldsymbol{n}})$ in $\partial \boldsymbol{D}$ which converges to a point $\boldsymbol{x_0}$ in $\partial \boldsymbol{D}$. Hence, letting n go to infinity in these subsequences, we have

$$\boldsymbol{p}^2_{\boldsymbol{n}} = f(\boldsymbol{x}^2_{\boldsymbol{n}}) \implies \boldsymbol{p} = \lim_{n \to \infty} f(\boldsymbol{x}^2_{\boldsymbol{n}}) = f(\lim_{n \to \infty} \boldsymbol{x}^2_{\boldsymbol{n}}) = f(\boldsymbol{x_0})$$

But this would say $\boldsymbol{p} \in f(\partial \boldsymbol{D})$ which is a contradiction to our assumption. Hence, we can conclude we can choose our sequence $(\boldsymbol{p_n})$ so that $\boldsymbol{p_n} \notin f(\partial \boldsymbol{D})$ for all n and $\boldsymbol{p_n} \to \boldsymbol{p}$. Further, note we have since $f^{-1}(\boldsymbol{p_n}) \notin \boldsymbol{B}$, we have $Jf(f^{-1}(\boldsymbol{p_n})) \neq 0$.

Our plan is to define the degree properly in a sequence of steps.

A1: In this case, we assume the existence of a sequence $\boldsymbol{p_n} \to \boldsymbol{p}$ with $f^{-1}(\boldsymbol{p_n}) \notin \boldsymbol{B}$ and $\boldsymbol{p} \notin f(\partial \boldsymbol{D})$. We then have $Jf(f^{-1}(\boldsymbol{p_n})) \neq 0$. We define the degree as

$$d(f, \boldsymbol{D}, \boldsymbol{p}) = \lim_{n \to \infty} d(f, \boldsymbol{D}, \boldsymbol{p_n})$$

Each of the terms $d(f, \boldsymbol{D}, \boldsymbol{p_n})$ looks like

$$d(f, \boldsymbol{D}, \boldsymbol{p_n}) = \sum_{\boldsymbol{x} \in f^{-1}(\boldsymbol{p_n})} sign\ Jf(\boldsymbol{x})$$

We must show this value is independent of the choice of sequence $(\boldsymbol{p_n})$. Once this is done, we have defined a degree for all $f \in C^1(\boldsymbol{D}) \cap C(\overline{\boldsymbol{D}})$ and $\boldsymbol{p} \in \Re^n \setminus f(\partial \boldsymbol{D})$ by $d(f, \boldsymbol{D}, \boldsymbol{p}) = \lim_{n \to \infty} d(f, \boldsymbol{D}, \boldsymbol{p_n})$.

A2: The next step is to define the degree of a map $f \in C(\overline{\boldsymbol{D}})$. By means of the Stone - Weierstrass Approximation Theorem, we can find functions (f_n) so that

- $f_n \in C^1(\boldsymbol{D}) \cap C(\overline{\boldsymbol{D}})$.
- $\lim_{n \to \infty} f_n = f$ uniformly on $\overline{\boldsymbol{D}}$. Note $\overline{\boldsymbol{D}}$ is compact.
- $\boldsymbol{p} \notin f_n(\partial \boldsymbol{D})$.

We define the degree as follows:

$$d(f, \boldsymbol{D}, \boldsymbol{p}) = \lim_{n \to \infty} d(f_n, \boldsymbol{D}, \boldsymbol{p})$$

once we show the limiting value is independent of the choice of sequence (f_n).

To this end, let $f \in C^1(\boldsymbol{D}) \cap C(\overline{\boldsymbol{D}})$ and $\boldsymbol{p} \in \Re^n \setminus f(\partial \boldsymbol{D})$. Let $f^{-1}(\boldsymbol{p})$ contain only regular points. Then, since $f^{-1}(\boldsymbol{p})$ is a finite set, let

$$f^{-1}(\boldsymbol{p}) = \{\boldsymbol{x_1}, \ldots, \boldsymbol{x_k}\}$$

for some positive integer k. By the Implicit Function Theorem, for each $\boldsymbol{x_i} \in f^{-1}(\boldsymbol{p})$, there is an open neighborhood $N(\boldsymbol{x_i})$ so that

(i): $N(\boldsymbol{x_i}) \subset \boldsymbol{D}$.

(ii): $N(\boldsymbol{x_i}) \cap N(\boldsymbol{x_j}) = \emptyset$ if $i \neq j$.

(iii): $Jf(\boldsymbol{y}) \neq 0$ for all $\boldsymbol{y} \in \cup_{i=1}^k N(\boldsymbol{x_i})$. This condition is clear as f is bijective on $N(\boldsymbol{x_i}) \to f(N(\boldsymbol{x_i}))$.

(iv): f is $1-1$ on $N(\boldsymbol{x_i})$.

(v): Set $g = f - \boldsymbol{p}$. Then $g(N(\boldsymbol{x_i}))$ is a neighborhood of $\boldsymbol{0}$. Hence, $\cap_{i=1}^k g(N(\boldsymbol{x_i}))$ is an open set and a neighborhood of $\boldsymbol{0}$ also. This implies there is a $\eta > 0$ so that $B(\boldsymbol{0}, \eta) \subset \cap_{i=1}^k g(N(\boldsymbol{x_i}))$. Now $g(\boldsymbol{x}) = \boldsymbol{0}$ implies $f(\boldsymbol{x}) = \boldsymbol{p}$. Thus, $\boldsymbol{x} \in f^{-1}(\boldsymbol{p})$.

(vi): We can then show that given conditions (i) through (v), there is a $\delta > 0$ so that

$$\|g(\boldsymbol{x})\| \geq \delta, \ \forall\, \boldsymbol{x} \in \overline{\boldsymbol{D}} \setminus \cup_{i=1}^k N(\boldsymbol{x_i})$$

To see this, assume it is not true. Then, there is a sequence $(\boldsymbol{y_m})$ with $\|f(\boldsymbol{y_m}) - \boldsymbol{p}\| < \frac{1}{m}$ for some $\boldsymbol{y_m} \in \cap_{i=1}^k (N(\boldsymbol{x_i}))^C \cap \overline{\boldsymbol{D}}$. But the set $\boldsymbol{E} = \cap_{i=1}^k (N(\boldsymbol{x_i}))^C \cap \overline{\boldsymbol{D}}$ is compact and there is a subsequence $(\boldsymbol{y_m^1})$ with $\boldsymbol{y_m^1} \to \boldsymbol{y_0} \in \boldsymbol{E}$. Thus, $\|f(\boldsymbol{y_m^1}) - \boldsymbol{p}\| \to 0$ and by continuity of f, this tells us $f(\boldsymbol{y_0}) = \boldsymbol{p}$. Hence, $\boldsymbol{y_0} = \boldsymbol{x_p}$ for some p which is not possible. Thus, the claim is true.

Homework

Exercise 10.1.4 *Let*

$$f(x,y) = \begin{bmatrix} 2xy \\ x^2 - y^2 \end{bmatrix}, \quad f_n(x,y) = \begin{bmatrix} 2xy + 1/n \\ x^2 - y^2 - 1/n \end{bmatrix}$$

- *Find Df and $Jf((x,y))$ for all $(x,y) \in \Re^2$.*
- *Find Df_n and $Jf_n((x,y))$ for all $(x,y) \in \Re^2$.*
- *Show $f_n \to f$ uniformly on $\Re^2$.*
- *Show $(0,0)$ is not a regular point and find $f_n^{-1}(0,0)$.*
- *Find $f(\overline{D})$, for D, the ball about $(0,0)$ of radius $1/2$. Compute $d(f, \boldsymbol{D}, (0,0))$ and as part of your calculation, find $f(\partial D)$.*

Exercise 10.1.5 *Let*

$$f(x,y) = \begin{bmatrix} x^2 + y^2 \\ 2x^2 - y^2 \end{bmatrix}, \quad f_n(x,y) = \begin{bmatrix} x^2 + y^2 - 1/n \\ 2x^2 - y^2 - 1/n \end{bmatrix}$$

- *Find Df and $Jf((x,y))$ for all $(x,y) \in \Re^2$.*
- *Find Df_n and $Jf_n((x,y))$ for all $(x,y) \in \Re^2$.*
- *Show $f_n \to f$ uniformly on $\Re^2$.*
- *Show $(0,0)$ is not a regular point and find $f_n^{-1}(0,0)$.*
- *Find $f(\overline{D})$, for D, the ball about $(0,0)$ of radius $1/2$. Compute $d(f, \boldsymbol{D}, (0,0))$ and as part of your calculation, find $f(\partial D)$.*

Exercise 10.1.6 *Let*

$$f(x,y) = \begin{bmatrix} 3xy \\ x^2 - y^2 \end{bmatrix}, \quad f_n(x,y) = \begin{bmatrix} 3xy + 1/n \\ x^2 - y^2 - 1/n \end{bmatrix}$$

- *Find Df and $Jf((x,y))$ for all $(x,y) \in \Re^2$.*
- *Find Df_n and $Jf_n((x,y))$ for all $(x,y) \in \Re^2$.*

- *Show* $f_n \to f$ *uniformly on* $\Re^2$.
- *Show* $(0,0)$ *is not a regular point and find* $f_n^{-1}(0,0)$.
- *Find* $f(\overline{D})$, *for* D, *the ball about* $(0,0)$ *of radius* $1/2$. *Compute* $d(f, \boldsymbol{D}, (0,0))$ *and as part of your calculation, find* $f(\partial D)$.

Exercise 10.1.7 *Let*

$$f(x,y) = \begin{bmatrix} x^2 + y^2 \\ x^2 - 2y^2 \end{bmatrix}, \quad f_n(x,y) = \begin{bmatrix} x^2 + y^2 - 1/n \\ x^2 - 2y^2 - 1/n \end{bmatrix}$$

- *Find* Df *and* $Jf((x,y))$ *for all* $(x,y) \in \Re^2$.
- *Find* Df_n *and* $Jf_n((x,y))$ *for all* $(x,y) \in \Re^2$.
- *Show* $f_n \to f$ *uniformly on* $\Re^2$.
- *Show* $(0,0)$ *is not a regular point and find* $f_n^{-1}(0,0)$.
- *Find* $f(\overline{D})$, *for* D, *the ball about* $(0,0)$ *of radius* $1/2$. *Compute* $d(f, \boldsymbol{D}, (0,0))$ *and as part of your calculation, find* $f(\partial D)$.

Lemma 10.1.2 Expressing the Degree as an Integration with Respect to a C^∞ Function of Compact Support

Let f *be a function with all the properties stated above. Let* $\Phi \in C([0,\infty)) \cap C^\infty((0,\infty))$ *with* $\int_{\Re^n} \Phi(\|\boldsymbol{x}\|) d\boldsymbol{x} = 1$ *with the support of* Φ *contained in the set* $(0, \min(\delta, \eta))$. *Then*

$$\int_{\boldsymbol{D}} \Phi(\|f(\boldsymbol{x}) - \boldsymbol{p}\|) Jf(\boldsymbol{x}) d\boldsymbol{x} = \sum_{i=1}^{k} sign\ Jf(\boldsymbol{x_i}) = d(f, \boldsymbol{D}, \boldsymbol{p})$$

Proof 10.1.2
We have for $g(\boldsymbol{x}) = f(\boldsymbol{x}) - \boldsymbol{p}$,

$$\int_{\boldsymbol{D}} \Phi(\|f(\boldsymbol{x}) - \boldsymbol{p}\|) Jf(\boldsymbol{x}) d\boldsymbol{x} = \int_{\boldsymbol{D}} \Phi(\|g(\boldsymbol{x})\|) Jg(\boldsymbol{x}) d\boldsymbol{x}$$
$$= \sum_{i=1}^{k} \int_{N(\boldsymbol{x_i})} \Phi(\|g(\boldsymbol{x})\|) Jg(\boldsymbol{x}) d\boldsymbol{x} + \sum_{i=1}^{k} \int_{\overline{\boldsymbol{D}} \setminus \cup_{i=1}^{k} N(\boldsymbol{x_i})} \Phi(\|g(\boldsymbol{x})\|) Jg(\boldsymbol{x}) d\boldsymbol{x}$$

But $\|g(\boldsymbol{x})\| \geq \delta$ *on* $\overline{\boldsymbol{D}} \setminus \cup_{i=1}^{k} N(\boldsymbol{x_i})$ *and* $\Phi(\|\boldsymbol{w}\|) = 0$ *if* $\|\boldsymbol{w}\| \geq \min(\delta, \eta)$. *Hence, we conclude*

$$\int_{\boldsymbol{D}} \Phi(\|f(\boldsymbol{x}) - \boldsymbol{p}\|) Jf(\boldsymbol{x}) d\boldsymbol{x} = \sum_{i=1}^{k} \int_{N(\boldsymbol{x_i})} \Phi(\|g(\boldsymbol{x})\|) Jg(\boldsymbol{x}) d\boldsymbol{x}$$

Now on each $N(\boldsymbol{x_i})$, *the sign of* $Jg(\boldsymbol{x})$ *must be a constant. If not, we would have a point in* $N(\boldsymbol{x_i})$ *with sign* $Jg(\boldsymbol{x}) = 0$ *which would contradict condition (iii). Hence, we can write*

$$\int_{\boldsymbol{D}} \Phi(\|f(\boldsymbol{x}) - \boldsymbol{p}\|) Jf(\boldsymbol{x}) d\boldsymbol{x} = \sum_{i=1}^{k} \int_{N(\boldsymbol{x_i})} \Phi(\|g(\boldsymbol{x})\|) sign\ Jg(\boldsymbol{x})\ |Jg(\boldsymbol{x})| d\boldsymbol{x}$$

$$= \sum_{i=1}^{k} sign\ Jg(\boldsymbol{x}) \int_{N(\boldsymbol{x_i})} \Phi(\|g(\boldsymbol{x})\|)\ |Jg(\boldsymbol{x})| d\boldsymbol{x}$$

Let $\boldsymbol{u} = g(\boldsymbol{x})$. Then, using this change of variable, we have

$$\int_{\boldsymbol{D}} \Phi(\|f(\boldsymbol{x}) - \mathbf{p}\|) Jf(\boldsymbol{x}) d\boldsymbol{x} = \sum_{i=1}^{k} sign\ Jg(\boldsymbol{x}) \int_{g(N(\boldsymbol{x_i}))} \Phi(\|\boldsymbol{u}\|)\ d\boldsymbol{u}$$

But by (v), there is an η so that $B(\mathbf{0}, \eta) \subset \cap_{i=1}^{k} g(N(\boldsymbol{x_i}))$. We also know $\Phi(\|\boldsymbol{u}\|) = 0$ if $\|\boldsymbol{u}\| \geq \min(\delta, \eta)$. Thus,

$$1 = \int_{\Re^n} \Phi(\|\boldsymbol{u}\|) d\boldsymbol{u} = \int_{B(\mathbf{0},\eta)} \Phi(\|\boldsymbol{u}\|) d\boldsymbol{u}$$

Now $B(\mathbf{0}, \eta) \subset g(N(\boldsymbol{x_i}))$. Thus, we can say

$$\int_{g(N(\boldsymbol{x_i}))} \Phi(\|\boldsymbol{u}\|) d\boldsymbol{u} = 1$$

Since $Jf(\boldsymbol{x_i}) = Jg(\boldsymbol{x_i})$, we see

$$\int_{\boldsymbol{D}} \Phi(\|f(\boldsymbol{x}) - \mathbf{p}\|) Jf(\boldsymbol{x}) d\boldsymbol{x} = \sum_{i=1}^{k} sign\ Jg(\boldsymbol{x}) = d(f, \boldsymbol{D}, \mathbf{p})$$

■

Lemma 10.1.2 says $\int_{\boldsymbol{D}} \Phi(\|f(\boldsymbol{x}) - \mathbf{p}\|) Jf(\boldsymbol{x}) d\boldsymbol{x} = d(f, \boldsymbol{D}, \mathbf{p})$. Since $d(f, \boldsymbol{D}, \mathbf{p})$ is independent of the choice of Φ, so is $\int_{\boldsymbol{D}} \Phi(\|f(\boldsymbol{x}) - \mathbf{p}\|) Jf(\boldsymbol{x}) d\boldsymbol{x}$. Hence, any Φ with these properties will work.

Lemma 10.1.3 7ϵ **Lemma**

Let $\mathbf{p} \in \Re^n$, $f_i : \overline{\boldsymbol{D}} \to \Re^n$ for $i = 1, 2$ with $f_i \in C^1(\boldsymbol{D}) \cap C(\overline{\boldsymbol{D}})$. Let $\epsilon > 0$ be given so that

(i): $\|f_i(\boldsymbol{x}) - \mathbf{p}\| \geq 7\epsilon$ for $i = 1, 2$ for $\boldsymbol{x} \in \partial \boldsymbol{D}$.

(ii): $\|f_1(\boldsymbol{x}) - f_2(\boldsymbol{x})\| < \epsilon$ for $\boldsymbol{x} \in \overline{\boldsymbol{D}}$.

(iii): $f_i^{-1}(\mathbf{p}) \cap \boldsymbol{B} = \emptyset$ for $i = 1, 2$.

Then $d(f_1, \boldsymbol{D}, \mathbf{p}) = d(f_2, \boldsymbol{D}, \mathbf{p})$.

Proof 10.1.3
We may assume $\mathbf{p} = 0$, because otherwise we look at the translates $g_i(\boldsymbol{x}) = f_i(\boldsymbol{x}) - \mathbf{p}$. Let $h : [0, \infty) \to \Re$ be in $C^1([0, \infty))$ satisfy $0 \leq h(r) \leq 1$ with

$$h(r) = \begin{cases} 1, & 0 \leq r \leq 2\epsilon \\ 0, & 3\epsilon \leq r \end{cases}$$

It is straightforward to construct such a C^∞ function of compact support as detailed in (Peterson (99) 2020). Define f_3 by

$$f_3(\boldsymbol{x}) = (1 - h(\|f_1(\boldsymbol{x})\|))\ f_1(\boldsymbol{x}) + h(\|f_1(\boldsymbol{x})\|)\ f_2(\boldsymbol{x})$$

Thus, $f_3(\boldsymbol{x}) = f_2(\boldsymbol{x})$ *when* $0 \le \|f_1(\boldsymbol{x})\| \le 2\epsilon$ *and* $f_3(\boldsymbol{x}) = f_1(\boldsymbol{x})$ *when* $3\epsilon \le \|f_1(\boldsymbol{x})\|$. *Note* f_3 *is essentially a convex combination of* f_1 *and* f_2 *when* $2\epsilon \le \|f_1(\boldsymbol{x})\| \le 3\epsilon$. *It is easy to see* $f_3 \in C(\overline{\boldsymbol{D}}) \cap C^1(\boldsymbol{D})$. *Moreover,*

$$\|f_1(\boldsymbol{x}) - f_3(\boldsymbol{x})\| = \begin{cases} 0, & 3\epsilon \le \|f_1(\boldsymbol{x})\| \\ \|f_1(\boldsymbol{x}) - f_2(\boldsymbol{x})\| < \epsilon, & 0 \le \|f_1(\boldsymbol{x})\| \le 2\epsilon \\ h(\|f_1(\boldsymbol{x}))\|f_1(\boldsymbol{x}) - f_2(\boldsymbol{x})\|, & 2\epsilon \le \|f_1(\boldsymbol{x})\| \le 3\epsilon \end{cases}$$

But $h(\|f_1(\boldsymbol{x})) \in [0,1]$ *so* $h(\|f_1(\boldsymbol{x}))\|f_1(\boldsymbol{x}) - f_2(\boldsymbol{x})\| < \epsilon$ *and so*

$$\|f_1(\boldsymbol{x}) - f_3(\boldsymbol{x})\| = \begin{cases} < \epsilon, & 0 \le \|f_1(\boldsymbol{x})\| \le 2\epsilon \\ < \epsilon, & 2\epsilon \le \|f_1(\boldsymbol{x})\| \le 3\epsilon \\ 0, & 3\epsilon \le \|f_1(\boldsymbol{x})\| \end{cases}$$

That is, $\|f_1(\boldsymbol{x}) - f_3(\boldsymbol{x})\| < \epsilon$ *on* $\overline{\boldsymbol{D}}$. *A similar argument shows*

$$\|f_2(\boldsymbol{x}) - f_3(\boldsymbol{x})\| = \begin{cases} 0, & 0 \le \|f_1(\boldsymbol{x})\| \le 2\epsilon \\ < \epsilon, & 2\epsilon \le \|f_1(\boldsymbol{x})\| \le 3\epsilon \\ < \epsilon, & 3\epsilon \le \|f_1(\boldsymbol{x})\| \end{cases}$$

So we also know, $\|f_2(\boldsymbol{x}) - f_3(\boldsymbol{x})\| < \epsilon$ *on* $\overline{\boldsymbol{D}}$. *Finally,*

$$\begin{aligned} \|f_3(\boldsymbol{x})\| &= \|(1 - h(\|f_1(\boldsymbol{x})\|))\, f_1(\boldsymbol{x}) + h(\|f_1(\boldsymbol{x})\|)\, f_2(\boldsymbol{x})\| \\ &= \|f_1(\boldsymbol{x}) - \boldsymbol{p} + \boldsymbol{p} + h(\|f_1(\boldsymbol{x})\|)\,(f_2(\boldsymbol{x}) - f_1(\boldsymbol{x}))\| \\ &\ge \|f_1(\boldsymbol{x}) + h(\|f_1(\boldsymbol{x})\|)\,(f_2(\boldsymbol{x}) - f_1(\boldsymbol{x}))\| \\ &\ge \|f_1(\boldsymbol{x})\| - h(\|f_1(\boldsymbol{x})\|)\, \|f_2(\boldsymbol{x}) - f_1(\boldsymbol{x})\| \\ &\ge \|f_1(\boldsymbol{x}) - \boldsymbol{p}\| - \epsilon \ge 7\epsilon - \epsilon = 6\epsilon \end{aligned}$$

Hence, $\|f_3(\boldsymbol{x})\| \ge 6\epsilon$ *on* $\partial\boldsymbol{D}$.

Let Φ_1 *and* Φ_2 *be two functions with the properties required in Lemma 10.1.2 such that*

$$\Phi_1(r) = 0, \ r \in [0, 4\epsilon] \cup [5\epsilon, \infty), \quad \Phi_2(r) = 0, \ r \in [\epsilon, \infty)$$

Then,

$$\Phi_1(\|f_3(\boldsymbol{x})\|) J f_3(\boldsymbol{x}) = 0, \ , \boldsymbol{x} \in \partial\boldsymbol{D}, \quad \Phi_1(\|f_1(\boldsymbol{x})\|) J f_1(\boldsymbol{x}) = 0, \ , \boldsymbol{x} \in \partial\boldsymbol{D}$$

Further,

$$\begin{aligned} 0 \le \|f_1(\boldsymbol{x})\| \le 2\epsilon &\implies f_3(\boldsymbol{x}) = f_2(\boldsymbol{x}), \ \Phi_1(\boldsymbol{x}) = 0 \\ 3\epsilon \le f_1(\boldsymbol{x}) &\implies f_3(\boldsymbol{x}) = f_1(\boldsymbol{x}) \implies \Phi_1(\|f_3(\boldsymbol{x})\|) J f_3(\boldsymbol{x}) = \Phi_1(\|f_1(\boldsymbol{x})\|) J f_1(\boldsymbol{x}) \end{aligned}$$

For $2\epsilon < \|f_1(\boldsymbol{x})\| < 3\epsilon$, *we see* $\Phi_1(\|f_1(\boldsymbol{x})\|) = 0$ *and*

$$\begin{aligned} \|f_3(\boldsymbol{x})\| &\ge \|f_1(\boldsymbol{x})\| - h(\|f_1(\boldsymbol{x})\|)\|f_2(\boldsymbol{x}) - f_1(\boldsymbol{x})\| \ge 2\epsilon - \epsilon = \epsilon \\ \|f_3(\boldsymbol{x})\| &\le \|f_1(\boldsymbol{x})\| + h(\|f_1(\boldsymbol{x})\|)\|f_2(\boldsymbol{x}) - f_1(\boldsymbol{x})\| \le 2\epsilon + \epsilon = 4\epsilon \end{aligned}$$

This implies that when $2\epsilon < \|f_1(\boldsymbol{x})\| < 3\epsilon$, $\Phi_1(\|f_3(\boldsymbol{x})\|) = 0$. *Hence,*

$$\Phi_1(\|f_3(\boldsymbol{x})\|) J f_3(\boldsymbol{x}) = \Phi_1(\|f_1(\boldsymbol{x})\|) J f_1(\boldsymbol{x}), \ \boldsymbol{x} \in \overline{\boldsymbol{D}}$$

Finally,

$$\Phi_2(\|f_3(\boldsymbol{x})\|)Jf_3(\boldsymbol{x}) = \Phi_2(\|f_2(\boldsymbol{x})\|)Jf_2(\boldsymbol{x}) = 0, \ \boldsymbol{x} \in \partial \boldsymbol{D}$$

Also,

$$0 \leq \|f_2(\boldsymbol{x})\| \leq 2\epsilon \implies f_3 = f_2 \Longrightarrow \Phi_2(\|f_3(\boldsymbol{x})\|)Jf_3(\boldsymbol{x}) = \Phi_2(\|f_2(\boldsymbol{x})\|)Jf_2(\boldsymbol{x})$$

If $\|f_2(\boldsymbol{x})\| > 2\epsilon$*, the* $\Phi_2 = 0$*, and so again we have* $\Phi_2(\|f_3(\boldsymbol{x})\|)Jf_3(\boldsymbol{x}) = \Phi_2(\|f_2(\boldsymbol{x})\|)Jf_2(\boldsymbol{x})$*. We conclude*

$$\Phi_2(\|f_3(\boldsymbol{x})\|)Jf_3(\boldsymbol{x}) = \Phi_2(\|f_2(\boldsymbol{x})\|)Jf_2(\boldsymbol{x}), \ \boldsymbol{x} \in \overline{\boldsymbol{D}}$$

Now apply Lemma 10.1.2 to find

$$\begin{aligned} d(f_1, \boldsymbol{D}, \boldsymbol{p}) &= \int_{\boldsymbol{D}} \Phi_1(\|f_1(\boldsymbol{x})\|)Jf_1(\boldsymbol{x})d\boldsymbol{x} = \int_{\boldsymbol{D}} \Phi_1(\|f_3(\boldsymbol{x})\|)Jf_3(\boldsymbol{x})d\boldsymbol{x} \\ &= \int_{\boldsymbol{D}} \Phi_2(\|f_3(\boldsymbol{x})\|)Jf_3(\boldsymbol{x})d\boldsymbol{x} = \int_{\boldsymbol{D}} \Phi_2(\|f_2(\boldsymbol{x})\|)Jf_2(\boldsymbol{x})d\boldsymbol{x} = d(f_2, \boldsymbol{D}, \boldsymbol{p}) \end{aligned}$$

■

Homework

Exercise 10.1.8 *Let* $h : [0, \infty) \to \Re$ *be in* $C^1([0, \infty))$ *and satisfy* $0 \leq h(r) \leq 1$ *with*

$$h(r) = \begin{cases} 1, & 0 \leq r \leq 2\epsilon \\ 0, & 3\epsilon \leq r \end{cases}$$

Show how to construct this C^∞ *function of compact support.*

Exercise 10.1.9 *Let* $h : [0, \infty) \to \Re$ *be in* $C^1([0, \infty))$ *and satisfy* $0 \leq h(r) \leq 1$ *with*

$$h(r) = \begin{cases} 1, & 0 \leq r \leq 4\epsilon \\ 0, & 4\epsilon \leq r \end{cases}$$

Show how to construct this C^∞ *function of compact support.*

This leads to the next result.

Lemma 10.1.4 7ϵ **Corollary**

Let $f_i \in C^1(\boldsymbol{D}) \cap C(\overline{\boldsymbol{D}})$ *and* $\boldsymbol{z_i}$ *be in* $\Re^n$ *for* $i = 1, 2$*. Let* $\epsilon > 0$ *be given so that*

(i): $\|f_i(\boldsymbol{x}) - z_i\| \geq 7\epsilon$*,* $i, j = 1, 2$*, for* $\boldsymbol{x} \in \partial \boldsymbol{D}$*.*

(ii): $\|f_1(\boldsymbol{x}) - f_2(\boldsymbol{x})\| < \epsilon$ *for* $\boldsymbol{x} \in \overline{\boldsymbol{D}}$*.*

(iii): $\|\boldsymbol{z_1} - \boldsymbol{z_2}\| < \epsilon$

Let each point from the sets $f_i^{-1}(\boldsymbol{z_j})$ *for* $i, j = 1, 2$ *be regular. Then* $d(f_1, \boldsymbol{D}, \boldsymbol{z_1}) = d(f_2, \boldsymbol{D}, \boldsymbol{z_2})$*.*

Proof 10.1.4
By Lemma 10.1.3, $d(f_1, \boldsymbol{D}, \boldsymbol{z_1}) = d(f_2, \boldsymbol{D}, \boldsymbol{z_1})$*. Consider the mappings* $g_1 = f_2$ *and* $g_2 = f_2 +$

$(\boldsymbol{z_1} - \boldsymbol{z_2})$. We see

$$\begin{aligned} \|g_1 - g_2\| &= \|\boldsymbol{z_1} - \boldsymbol{z_2}\| < \epsilon, \ \boldsymbol{x} \in \overline{\boldsymbol{D}} \\ \|g_1 - \boldsymbol{z_1}\| &= \|f_2 - \boldsymbol{z_1}\| \geq 7\epsilon, \ \boldsymbol{x} \in \partial \boldsymbol{D} \\ \|g_2 - \boldsymbol{z_1}\| &= \|f_2 - \boldsymbol{z_2}\| \geq 7\epsilon, \boldsymbol{x} \in \partial \boldsymbol{D} \end{aligned}$$

Now apply Lemma 10.1.3 to these mappings, to conclude

$$\begin{aligned} d(f_2, \boldsymbol{D}, \boldsymbol{z_1}) &= d(g_1, \boldsymbol{D}, \boldsymbol{z_1}) = d(g_2, \boldsymbol{D}, \boldsymbol{z_1}) \\ &= d(f_2 + \boldsymbol{z_1} - \boldsymbol{z_2}, \boldsymbol{D}, \boldsymbol{z_1}) \\ &= \sum_{\boldsymbol{x} \in g_2^{-1}(\boldsymbol{z_1})} sign\ Jg_2(\boldsymbol{x}) \end{aligned}$$

Now $\boldsymbol{x} \in g_2^{-1}(\boldsymbol{z_1})$ implies $f_2(\boldsymbol{x}) + \boldsymbol{z_1} - \boldsymbol{z_2} = \boldsymbol{z_1}$. This says $f_2(\boldsymbol{x}) = \boldsymbol{z_2}$ or $\boldsymbol{x} \in f_2^{-1}(\boldsymbol{z_2})$. Using this, we have

$$d(f_2, \boldsymbol{D}, \boldsymbol{z_1}) = \sum_{\boldsymbol{x} \in f_2^{-1}(\boldsymbol{z_2})} sign\ Jg_2(\boldsymbol{x}) = \sum_{\boldsymbol{x} \in f_2^{-1}(\boldsymbol{z_2})} sign\ Jf_2(\boldsymbol{x}) = d(f_2, \boldsymbol{D}, \boldsymbol{z_2})$$

■

We can now examine our definition of the degree when $\boldsymbol{p}$ is not a regular point. Remember how we were going to handle case A1. We know if $\boldsymbol{p} \in f(\boldsymbol{B})$, there is a sequence $(\boldsymbol{p_n})$ with $\boldsymbol{p_n} \to \boldsymbol{p}$ with $f^{-1}(\boldsymbol{p_n}) \cap \boldsymbol{B} = \emptyset$ and $\boldsymbol{p_n} \notin f(\partial \boldsymbol{D})$ for all n. Since $\boldsymbol{p} \notin f(\partial \boldsymbol{D})$, there is a $\delta > 0$ so that $\|f(\partial \boldsymbol{D}) - \boldsymbol{p}\| \geq \delta$. Define $\epsilon > 0$ by $\delta = 8\epsilon$. Then $\|f(\partial \boldsymbol{D}) - \boldsymbol{p}\| \geq 8\epsilon$. Further, since $(\boldsymbol{p_n})$ is a Cauchy sequence, there is n_0 so that $n, m \geq n_0$ implies $\|\boldsymbol{p_n} - \boldsymbol{p_m}\| < \epsilon$ and $\|\boldsymbol{p} - \boldsymbol{p_n}\| < \epsilon$. Thus, for $\boldsymbol{x} \in \partial \boldsymbol{D}$ and $n > n_0$,

$$\|f(\boldsymbol{x}) - \boldsymbol{p_n}\| \geq \|f(\boldsymbol{x}) - \boldsymbol{p}\| - \|\boldsymbol{p} - \boldsymbol{p_n}\| \geq 8\epsilon - \epsilon = 7\epsilon$$

The hypotheses of Lemma 10.1.4 are then satisfied for $f_1 = f_2 = f$ and so $d(f, \boldsymbol{D}, \boldsymbol{p_n}) = d(f, \boldsymbol{D}, \boldsymbol{p_{n_0}})$ for all $n \geq n_0$.

Now let $(\boldsymbol{q_n})$ be another sequence converging to $\boldsymbol{p}$ satisfying the same assumptions. Then $f^{-1}(\boldsymbol{q_n}) \cap \boldsymbol{B} = \emptyset$. Again, we find a n_1 so that by Lemma 10.1.4, we have $d(f, \boldsymbol{D}, \boldsymbol{q_n}) = d(f, \boldsymbol{D}, \boldsymbol{p_{n_0}})$ for all $n \geq n_1$. We conclude the value of $\lim_{n \to \infty} d(f, \boldsymbol{D}, \boldsymbol{p_n})$ must be independent of the choice of sequence $(\boldsymbol{p_n})$ where $\boldsymbol{p_n} \to \boldsymbol{p}$, $f^{-1}(\boldsymbol{p_n}) \cap \boldsymbol{B} = \emptyset$ and $\boldsymbol{p_n} \notin f(\partial \boldsymbol{D})$ for all n.

This argument lets us prove the following result.

Lemma 10.1.5 When are the Degrees of f_1 and f_2 the same?

Let $f_i \in C^1(\boldsymbol{D}) \cap C(\overline{\boldsymbol{D}})$ for $i = 1, 2$, $\boldsymbol{p} \in \Re^n$ and $\epsilon > 0$. Assume

(i): $\|f_i(\boldsymbol{x}) - \boldsymbol{p}\| \geq 8\epsilon$, for $\boldsymbol{x} \in \partial \boldsymbol{D}$.

(ii): $\|f_1(\boldsymbol{x}) - f_2(\boldsymbol{x})\| < \epsilon$ for all $\boldsymbol{x} \in \overline{\boldsymbol{D}}$.

Then $d(f_1, \boldsymbol{D}, \boldsymbol{p}) = d(f_2, \boldsymbol{D}, \boldsymbol{p})$.

Proof 10.1.5

Let $(\boldsymbol{p_n})$ be chosen so that $\boldsymbol{p_n} \to \boldsymbol{p}$ with $f^{-1}(\boldsymbol{p_n}) \cap \boldsymbol{B} = \emptyset$ and $\boldsymbol{p_n} \notin f(\partial \boldsymbol{D})$ for all n. Now argue *just like we did above for f_1 and f_2. We can find an index n_0 so that the assumptions of Lemma 10.1.4 are satisfied. We conclude $d(f_1, \boldsymbol{D}, \boldsymbol{p_n}) = d(f_2, \boldsymbol{D}, \boldsymbol{p_n})$ for all $n \geq n_0$. Thus, letting $n \to \infty$, we have*

$$\lim_{n\to\infty} d(f_1, \boldsymbol{D}, \boldsymbol{p_n}) = \lim_{n\to\infty} d(f_2, \boldsymbol{D}, \boldsymbol{p_n}) \Longrightarrow d(f_1, \boldsymbol{D}, \boldsymbol{p}) = d(f_2, \boldsymbol{D}, \boldsymbol{p})$$

■

We can now completely define our notion of degree when $f \in C(\overline{\boldsymbol{D}})$. This is the context of case A2. Let's recall the setup here: By means of the Stone - Weierstrass Approximation Theorem, we can find functions (f_n) so that $f_n \in C^1(\boldsymbol{D}) \cap C(\overline{\boldsymbol{D}})$ and $\lim_{n\to\infty} f_n = f$ uniformly on $\overline{\boldsymbol{D}}$. We assume also $\boldsymbol{p} \notin f_n(\partial \boldsymbol{D})$.

We define the degree as follows:

$$d(f, \boldsymbol{D}, \boldsymbol{p}) = \lim_{n\to\infty} d(f_n, \boldsymbol{D}, \boldsymbol{p})$$

once we show the limiting value is independent of the choice of sequence (f_n). As usual, let $(\boldsymbol{p_n})$ be a sequence with $\boldsymbol{p_n} \to \boldsymbol{p}$, $f^{-1}(\boldsymbol{p_n}) \cap \boldsymbol{B} = \emptyset$ with $\boldsymbol{p_n} \notin f(\partial \boldsymbol{D})$ for all n. Since $\boldsymbol{p} \notin f(\boldsymbol{p_n})$, there is a $\delta > 0$ so that $\|f(\boldsymbol{x}) - \boldsymbol{p}\| \geq \delta$ for $\boldsymbol{x} \in \partial \boldsymbol{D}$. Define ϵ by $\delta = 10\epsilon$. Since $f_n \to f$ uniformly on $\overline{\boldsymbol{D}}$, there is an n_0 so that $\|f_n - f_m\| < \epsilon$ and $\|f - f_n\| < \epsilon$ for $m, n \geq n_0$.

Thus, for $\boldsymbol{x} \in \partial \boldsymbol{D}$,

$$\begin{aligned} \|f_n(\boldsymbol{x}) - \boldsymbol{p_n}\| &= \|f_n(\boldsymbol{x}) - \boldsymbol{p_n} + \boldsymbol{p} - \boldsymbol{p} + f(\boldsymbol{x}) - f(\boldsymbol{x})\| \\ &\geq \|f(\boldsymbol{x}) - \boldsymbol{p}\| - \|\boldsymbol{p} - \boldsymbol{p_n}\| - \|f_n(\boldsymbol{x}) - f(\boldsymbol{x})\| \\ &\geq 10\epsilon - \epsilon - \epsilon = 8\epsilon \end{aligned}$$

Thus, by Lemma 10.1.5, $d(f_n, \boldsymbol{D}, \boldsymbol{p_m}) = d(f_{n_0}, \boldsymbol{D}, \boldsymbol{p_m})$ when $n, m > n_0$. We already know that

$$\begin{aligned} \lim_{m\to\infty} d(f_n, \boldsymbol{D}, \boldsymbol{p_m}) &= d(f_n, \boldsymbol{D}, \boldsymbol{p}) \\ \lim_{m\to\infty} d(f_{n_0}, \boldsymbol{D}, \boldsymbol{p_m}) &= d(f_{n_0}, \boldsymbol{D}, \boldsymbol{p}) \end{aligned}$$

Thus, $d(f_n, \boldsymbol{D}, \boldsymbol{p}) = d(f_{n_0}, \boldsymbol{D}, \boldsymbol{p})$ and so

$$\lim_{n\to\infty} d(f_n, \boldsymbol{D}, \boldsymbol{p}) = d(f_{n_0}, \boldsymbol{D}, \boldsymbol{p})$$

We see the limiting value of $d(f_n, \boldsymbol{D}, \boldsymbol{p})$ exists. The argument that this limit is independent of the choice of sequence $(\boldsymbol{p_n})$ is similar to the argument we used to show the limit in case A1 was independent of the choice of sequence. We leave that to you as an exercise.

We have now gone the full route to define the degree for $f \in C(\overline{\boldsymbol{D}})$. We have proven there is an integer-valued function on all $f \in C(\overline{\boldsymbol{D}})$ and $\boldsymbol{p} \notin f(\partial \boldsymbol{D})$ which we will denote as $d_B(f, \boldsymbol{D}, \boldsymbol{p})$ which is called the **Brouwer Degree** defined by

$$\begin{aligned} d_B(f, \boldsymbol{D}, \boldsymbol{p}) &= \sum_{\boldsymbol{x} \in f^{-1}(\boldsymbol{p})} sign\, Jf(\boldsymbol{x}), \;\; f \in C^1(\boldsymbol{D}) \cap C(\overline{\boldsymbol{D}}) \\ &= \lim_{n\to\infty} d_B(f, \boldsymbol{D}, \boldsymbol{p_n}), \;\; f \in C^1(\boldsymbol{D}) \cap C(\overline{\boldsymbol{D}}) \\ & \quad\; \boldsymbol{p} \in f(\boldsymbol{B}), \;\; \boldsymbol{p_n} \to \boldsymbol{p}, \; \boldsymbol{p_n} \notin f(\partial \boldsymbol{D}), \boldsymbol{p_n} \notin f(\boldsymbol{B}) \; \forall n \\ &= \lim_{n\to\infty} d_B(f_n, \boldsymbol{D}, \boldsymbol{p}), \;\; f_n \in C^1(\boldsymbol{D}) \cap C(\overline{\boldsymbol{D}}) \end{aligned}$$

$$f_n \to f \text{ uniformly on } \overline{\boldsymbol{D}},\ \boldsymbol{p} \notin f_n(\partial \boldsymbol{D})\ \forall n$$

This is a very important integer-value function which has many applications.

Homework

Exercise 10.1.10 *Prove in case A2, that the limit is independent of the choice of sequence* (f_n).

Exercise 10.1.11 *Let*

$$f(x,y) = \begin{bmatrix} x^2 + y^2 \\ 4x^2 - y^2 \end{bmatrix}$$

- *Find* Df *and* $Jf(x,y)$ *for all* $(x,y) \in \Re^2$.
- *Find all regular points and all critical points.*
- *Find* $f(\overline{D})$ *and* $f(\partial D)$*, for* D*, the ball about* $(0,0)$ *of radius* 2*. Compute* $d(f, \boldsymbol{D}, (4,0))$ *using the limiting process we have discussed here.*

Exercise 10.1.12 *Let*

$$f(x,y) = \begin{bmatrix} x^2 + y^2 \\ 2x^2 - 6y^2 \end{bmatrix}$$

- *Find* Df *and* $Jf((x,y)$ *for all* $(x,y) \in \Re^2$.
- *Find all regular points and all critical points.*
- *Find* $f(\overline{D})$ *and* $f(\partial D)$*, for* D*, the ball about* $(0,0)$ *of radius* 1*. Compute* $d(f, \boldsymbol{D}, (3,0))$ *using the discussed limiting process.*

Exercise 10.1.13 *Let*

$$f(x,y) = \begin{bmatrix} 7xy \\ x^2 - 3y^2 \end{bmatrix}$$

- *Find* Df *and* $Jf(x,y)$ *for all* $(x,y) \in \Re^2$.
- *Find all regular points and all critical points.*
- *Find* $f(\overline{D})$ *and* $f(\partial D)$ *for* D*, the ball about* $(0,0)$ *of radius* $1/2$*. Compute* $d(f, \boldsymbol{D}, (0,0))$ *using the discussed limiting process.*

10.2 The Properties of the Degree

It is time to work out some of the properties of Brouwer Degree.

Theorem 10.2.1 Degree on Disjoint Subsets

Let $\boldsymbol{D_1}$ *and* $\boldsymbol{D_2}$ *be disjoint, bounded open subsets of* $\Re^n$*. Furthermore, let* $f : \overline{\boldsymbol{D_1}} \cup \overline{\boldsymbol{D_2}} \to \Re^n$ *be continuous and* $\boldsymbol{p} \notin f(\partial \boldsymbol{D_1} \cup \partial \boldsymbol{D_2})$*. Then,* $d_B(f, \boldsymbol{D_1} \cup \boldsymbol{D_2}, \boldsymbol{p}) = d_B(f, \boldsymbol{D_1}, \boldsymbol{p}) + d_B(f, \boldsymbol{D_2}, \boldsymbol{p})$.

Proof 10.2.1

$$\begin{aligned}
d_B(f, \boldsymbol{D_1} \cup \boldsymbol{D_2}, \boldsymbol{p}) &= \lim_{n\to\infty} d_B(f_n, \boldsymbol{D_1} \cup \boldsymbol{D_2}, \boldsymbol{p}), \;\; f_n \in C^1(\boldsymbol{D_1} \cup \boldsymbol{D_2}) \cap C(\overline{\boldsymbol{D_1} \cup \boldsymbol{D_2}}) \\
&= \lim_{n\to\infty} \sum_{\boldsymbol{x} \in f_n^{-1}(\boldsymbol{p})} sign\; Jf_n(\boldsymbol{x}) \\
&= \lim_{n\to\infty} \left(\sum_{\boldsymbol{x} \in f_n^{-1}(\boldsymbol{p}) \cap \boldsymbol{D_1}} sign\; Jf_n(\boldsymbol{x}) + \sum_{\boldsymbol{x} \in f_n^{-1}(\boldsymbol{p}) \cap \boldsymbol{D_2}} sign\; Jf_n(\boldsymbol{x}) \right) \\
&= \lim_{n\to\infty} \sum_{\boldsymbol{x} \in f_n^{-1}(\boldsymbol{p}) \cap \boldsymbol{D_1}} sign\; Jf_n(\boldsymbol{x}) + \lim_{n\to\infty} \sum_{\boldsymbol{x} \in f_n^{-1}(\boldsymbol{p}) \cap \boldsymbol{D_2}} sign\; Jf_n(\boldsymbol{x}) \\
&= \lim_{n\to\infty} d_B(f_n, \boldsymbol{D_1}, \boldsymbol{p}) + \lim_{n\to\infty} d_B(f_n, \boldsymbol{D_2}, \boldsymbol{p}) \\
&= d_B(f, \boldsymbol{D_1}, \boldsymbol{p}) + d_B(f, \boldsymbol{D_2}, \boldsymbol{p})
\end{aligned}$$

■

Now we are ready to show what we can do with a degree theory. Let's prove an existence result for the equation $f(\boldsymbol{x}) = \boldsymbol{p}$; that is, can we find $\boldsymbol{x}$ so that $f(\boldsymbol{x}) = \boldsymbol{p}$?

Theorem 10.2.2 Existence of a Solution to $f(\boldsymbol{x}) = \boldsymbol{p}$

Let $f : \overline{\boldsymbol{D}} \to \Re^n$ be continuous and $\boldsymbol{p} \notin f(\partial \boldsymbol{D})$. Let $d_B(f, \boldsymbol{D}, \boldsymbol{p}) \neq 0$. Then there is at least one $\boldsymbol{x_0} \in \boldsymbol{D}$ so that $f(\boldsymbol{x_0}) = \boldsymbol{p}$.

Proof 10.2.2

Assume $f(\boldsymbol{x}) \neq \boldsymbol{p}$ for all $\boldsymbol{x} \in \boldsymbol{D}$. Then, since $f(\boldsymbol{x}) \neq \boldsymbol{p}$ on $\partial \boldsymbol{D}$ by assumption, there is a $\delta > 0$, which we use to define ϵ as $2\epsilon = \delta$, so that $\|f(\boldsymbol{x}) - \boldsymbol{p}\| > 2\epsilon$ for all $\boldsymbol{x} \in \overline{\boldsymbol{D}}$. Let $f_n \to f$ uniformly on $\overline{\boldsymbol{D}}$, $\boldsymbol{p} \notin f_n(\partial \boldsymbol{D})$ and $f_n \in C^1(\boldsymbol{D}) \cap C(\overline{\boldsymbol{D}})$ for all n. Then, there is an n_0 so that

$$\|f_n(\boldsymbol{x}) - \boldsymbol{p}\| \;\; \geq \;\; \|f(\boldsymbol{x}) - \boldsymbol{p}\| - \|f(\boldsymbol{x}) - f_n(\boldsymbol{x})\| \geq 2\epsilon - \epsilon = \epsilon, \;\; \forall \boldsymbol{x} \in \overline{\boldsymbol{D}}, \;\; n \geq n_0$$

using a standard argument. For each such n, choose a function Φ_n as required by Lemma 10.1.2. Use Lemma 10.1.2 with the support of $\Phi_n \subset (0, \min(\delta, \eta))$ replaced by the support of $\Phi_n \subset (0, \min(\delta, \eta, \epsilon))$. Then we have

$$d_B(f_n, \boldsymbol{D}, \boldsymbol{p}) \;\; = \;\; \int_{\boldsymbol{D}} \Phi_n(\|f_n(\boldsymbol{x}) - \boldsymbol{p})\| Jf_n(\boldsymbol{x}) d\boldsymbol{x}$$

But, for $n \geq n_0$, the support of Φ_n implies $\Phi_n(\|f_n(\boldsymbol{x}) - \boldsymbol{p})\| = 0$. Thus, $d_B(f_n, \boldsymbol{D}, \boldsymbol{p}) = 0$ for all $n \geq n_0$. This says $d_B(f, \boldsymbol{D}, \boldsymbol{p}) = 0$ which is a contradiction. Hence, there must be such an $\boldsymbol{x_0} \in \boldsymbol{D}$ so that $f(\boldsymbol{x_0}) = \boldsymbol{p}$. ■

Another important property is what is called **invariance under homotopy**.

Theorem 10.2.3 Invariance under Homotopy

Let $\boldsymbol{D} \subset \Re^n$ be bounded and let $F : \overline{\boldsymbol{D}} \times [a, b] \to \Re^n$ be continuous with $\boldsymbol{p} \notin F(\partial \boldsymbol{D} \times [a, b])$. Then $d_B(F(\cdot, t), \boldsymbol{D}, \boldsymbol{p})$ is a constant for all $t \in [a, b]$.

Proof 10.2.3
By assumption, $\boldsymbol{p} \notin F(\partial \boldsymbol{D} \times [a,b])$. *Hence, there is a* $\delta > 0$, *and* ϵ *defined by* $9\epsilon = \delta$ *so that*

$$\|f(\boldsymbol{x},t) - \boldsymbol{p}\| \quad > \quad \delta = 9\epsilon, \ \forall\, \boldsymbol{x} \in \partial \boldsymbol{D}, \ \forall\, t \in [a,b]$$

By the uniform continuity of F *on* $\overline{\boldsymbol{D}}$, *there is a* $\eta(\epsilon)$ *so that*

$$|t_1 - t_2| \leq \eta(\epsilon) \text{ and } \boldsymbol{x} \in \overline{\boldsymbol{D}} \implies \|F(\boldsymbol{x},t_1) - F(\boldsymbol{x},t_2)\| < \frac{\epsilon}{3}$$

Use the Stone - Weierstrass Approximation Theorem to approximate $F(\boldsymbol{x},t_1)$ *by a sequence of functions* (f_{n1}) *and* $F(\boldsymbol{x},t_2)$ *by a sequence of functions* (f_{n2}) *so that*

$$\begin{aligned} f_{n1} &\to F(\cdot,t_1) \text{ uniformly on } \overline{\boldsymbol{D}} \\ f_{n2} &\to F(\cdot,t_2) \text{ uniformly on } \overline{\boldsymbol{D}} \end{aligned}$$

with $\boldsymbol{p} \notin f_{ni}(\partial \boldsymbol{D})$ *and* $f_{ni} \in C^1(\boldsymbol{D}) \cap C(\overline{\boldsymbol{D}})$ *for* $i = 1,2$. *Hence, there is an* n_0 *so that for* $i = 1,2$, *we have*

$$\begin{aligned} \|f_{ni}(\boldsymbol{x}) - f_{mi}(\boldsymbol{x})\| &< \frac{\epsilon}{3} \\ \|f_{ni}(\boldsymbol{x}) - F(\boldsymbol{x},t_i)\| &< \frac{\epsilon}{3} \end{aligned}$$

for all $n, m \geq n_0$. *Thus,*

$$\begin{aligned} \|f_{n1}(\boldsymbol{x}) - f_{n2}(\boldsymbol{x})\| &\leq \|f_{n1}(\boldsymbol{x}) - F(\boldsymbol{x},t_1) + F(\boldsymbol{x},t_1) - F(\boldsymbol{x},t_2) + F(\boldsymbol{x},t_2) - f_{n2}(\boldsymbol{x})\| \\ &\leq \|f_{n1}(\boldsymbol{x}) - F(\boldsymbol{x},t_1)\| + \|F(\boldsymbol{x},t_1) - F(\boldsymbol{x},t_2)\| + \|F(\boldsymbol{x},t_2) - f_{n2}(\boldsymbol{x})\| \end{aligned}$$

So, for all $n \geq n_0$ *since* $|t_1 - t_2| < \eta(\epsilon)$, *we have*

$$\|f_{n1}(\boldsymbol{x}) - f_{n2}(\boldsymbol{x})\| \quad < \quad \frac{\epsilon}{3} + \frac{\epsilon}{3} + \frac{\epsilon}{3} = \epsilon$$

Further,

$$\|f_{ni}(\boldsymbol{x}) - \boldsymbol{p}\| \quad \geq \quad \|F(\boldsymbol{x},t_i) - \boldsymbol{p}\| - \|f_{ni}(\boldsymbol{x}) - F(\boldsymbol{x},t_i)\| \geq 9\epsilon - \frac{\epsilon}{3} > 8\epsilon$$

on $\overline{\boldsymbol{D}}$. *We conclude* $\|f_{n1}(\boldsymbol{x}) - f_{n2}(\boldsymbol{x})\| < \epsilon$ *and* $\|f_{ni}(\boldsymbol{x}) - \boldsymbol{p}\| > 8\epsilon$ *on* $\overline{\boldsymbol{D}}$. *We can then apply Lemma 10.1.5 to conclude* $d_B(f_{n1}, \boldsymbol{D}, \boldsymbol{p}) = d_B(f_{n2}, \boldsymbol{D}, \boldsymbol{p})$ *if* $n \geq n_0$. *Thus, as* $n \to \infty$, *we obtain* $d_B(F(\cdot,t_1), \boldsymbol{D}, \boldsymbol{p}) = d_B(F(\cdot,t_2), \boldsymbol{D}, \boldsymbol{p})$. *This holds for all* $|t_1 - t_2| < \eta(\epsilon)$. *But we can cover* $[a,b]$ *with overlapping neighborhoods of width* $0.5\eta(\epsilon)$. *Hence,* $d_B(F(\cdot,t), \boldsymbol{D}, \boldsymbol{p})$ *is constant on* $[a,b]$. ■

Homework

Exercise 10.2.1 *Let*

$$f(x,y) \quad = \quad \begin{bmatrix} x^2 + y^2 \\ 4x^2 - y^2 \end{bmatrix}$$

- *Write down the homotopy on* $[0,1]$ *connecting the identity map to* f.
- *For* D, *the ball about* $(1,1)$ *of radius* 2, *show* $(1,1)$ *is not in* $f(\partial D) \times [0,1])$.
- *Use the homotopy to compute* $d(f, \boldsymbol{D}, (1,1))$.

Exercise 10.2.2 *Let*

$$f(x,y) = \begin{bmatrix} x^2+y^2 \\ 2x^2-6y^2 \end{bmatrix}$$

- *Write down the homotopy on* $[0,1]$ *connecting the identity map to* f.
- *For* D*, the ball about* $(1,1)$ *of radius* 2*, show* $(1,1)$ *is not in* $f(\partial D)\times[0,1]$.
- *Use the homotopy to compute* $d(f,\boldsymbol{D},(1,1))$.

Exercise 10.2.3 *Let*

$$f(x,y) = \begin{bmatrix} 7xy \\ x^2-3y^2 \end{bmatrix}$$

- *Write down the homotopy on* $[0,1]$ *connecting the identity map to* f.
- *For* D*, the ball about* $(1,1)$ *of radius* 2*, show* $(1,1)$ *is not in* $f(\partial D)\times[0,1]$.
- *Use the homotopy to compute* $d(f,\boldsymbol{D},(1,1))$.

The degree of two mappings can coincide under a variety of conditions. Here is another one called **Rouché's** Theorem.

Theorem 10.2.4 Rouché's Theorem

Let $\boldsymbol{D}$ *be a bounded subset of* $\Re^n$ *and let the mappings* $f_i : \overline{\boldsymbol{D}} \to \Re^n$ *be continuous, for* $i=1,2$*. Further, assume* $\|f_1(\boldsymbol{x})-f_2(\boldsymbol{x})\| < \|f_1(\boldsymbol{x})-\boldsymbol{p}\|$ *for all* $\boldsymbol{x}\in\partial\boldsymbol{D}$*. Then* $d_B(f_1,\boldsymbol{D},\boldsymbol{p}) = d_B(f_2,\boldsymbol{D},\boldsymbol{p})$.

Proof 10.2.4
Let $F(\boldsymbol{x},t) = f_1(\boldsymbol{x}) + t(f_2(\boldsymbol{x})-f_1(\boldsymbol{x}))$ *for* $0\le t\le 1$ *be a convex combination of* f_1 *and* f_2*. Then*

$$\begin{aligned} \|F(\boldsymbol{x},t)-\boldsymbol{p}\| &= \|f_1(\boldsymbol{x})+t(f_2(\boldsymbol{x})-f_1(\boldsymbol{x}))-\boldsymbol{p}\| \\ &\ge \|f_1(\boldsymbol{x})-\boldsymbol{p}\| - t\|f_2(\boldsymbol{x})-f_1(\boldsymbol{x})\| \\ &\ge \|f_1(\boldsymbol{x})-\boldsymbol{p}\| - \|f_2(\boldsymbol{x})-f_1(\boldsymbol{x})\| > 0 \end{aligned}$$

by our assumption. This is true for all $\boldsymbol{x}\in\partial\boldsymbol{D}$ *and* $0\le t\le 1$*. Since the degree in invariant under homotopy,* $d_B(F(\cdot,0),\boldsymbol{D},\boldsymbol{p}) = d_B(F(\cdot,1),\boldsymbol{D},\boldsymbol{p})$*. But* $F(\cdot,0)=f_1$ *and* $F(\cdot,1)=f_2$*. Hence,* $d_B(f_1,\boldsymbol{D},\boldsymbol{p}) = d_B(f_2,\boldsymbol{D},\boldsymbol{p})$. ∎

Recall the Tietze Extension Theorem which says if $\boldsymbol{E}$ is a closed subspace of a normed linear space X, then every map $g:\boldsymbol{E}\to[0,1]$ has an extension G to all of X. This means $G|_{\boldsymbol{E}} = g$. Further, $\sup_{\boldsymbol{x}\in\boldsymbol{E}}|g(\boldsymbol{x})| = \sup_{\boldsymbol{x}\in X}|G(\boldsymbol{x})|$. Hence, if g is bounded so is the extension G with the same bound and the extension is continuous.

Also, recall a subset $\Omega\subset X$ where X has a topology τ is a subspace when Ω is *topologized* by the topology induced by $i^{-1}(\tau)$ where $i:\omega\to X$ is an inclusion map. This topology is called the *relative topology*. Remember also, a space is a *normal* space if and only if every disjoint pair of closed sets in X have disjoint neighborhoods.

Here $\partial\boldsymbol{D}$ plays the role of ω and the role of X is $\overline{\boldsymbol{D}}$. If we have a continuous function $f:\partial\boldsymbol{D}\to\Re^n$, then each component $f_i:\partial\boldsymbol{D}\to\Re$. Since f is continuous, there is a finite interval $[a,b]$ so that $f_i(\partial\boldsymbol{D})\subset[a,b]$ for all $1\le i\le n$. Now apply the Tietze Extension Theorem to each component f_i

to obtain continuous extensions $\tilde{f}_i : \overline{\boldsymbol{D}} \rightarrow [a, b]$. Define $\tilde{f}$ by $\tilde{f}_i = \tilde{f}_i$; i.e. each component of $\tilde{f}$ is an extension of the appropriate component of f. Note $\tilde{f} : \overline{\boldsymbol{D}} \rightarrow \Re^n$. If $\boldsymbol{p} \notin \partial\boldsymbol{D}$, then for any two extensions $\tilde{f}$ and $\tilde{g}$, we have

$$\boldsymbol{x} \in \partial\boldsymbol{D},\ \tilde{f}(\boldsymbol{x}) = \tilde{g}(\boldsymbol{x}) \implies \|\tilde{f}(\boldsymbol{x}) - \tilde{g}(\boldsymbol{x})\| = 0 < \|\tilde{f}(\boldsymbol{x}) - \boldsymbol{p}\|$$

with $\|\tilde{f}(\boldsymbol{x}) - \boldsymbol{p}\| = \|f(\boldsymbol{x}) - \boldsymbol{p}\| > 0$. Thus, by Rouché's Theorem, $d_B(\tilde{f}, \boldsymbol{D}, \boldsymbol{p}) = d_B(\tilde{g}, \boldsymbol{D}, \boldsymbol{p})$. In essence, this argument says we can *define* the Brouwer degree of $f : \partial\boldsymbol{D} \rightarrow \Re^n$ as the Brouwer degree of any extension of f to $\overline{\boldsymbol{D}}$; i.e. $d_B(f, \partial\boldsymbol{D}, \boldsymbol{p}) \equiv d_B(\tilde{f}, \boldsymbol{D}, \boldsymbol{p})$ and this value is independent of the choice of extension.

This argument also shows that if f and g agree on $\partial\boldsymbol{D}$ with $\boldsymbol{p} \notin \partial\boldsymbol{D}$, then $d_B(f, \partial\boldsymbol{D}, \boldsymbol{p}) = d_B(g, \partial\boldsymbol{D}, \boldsymbol{p})$. Hence, the Brouwer degree of f depends only on the values of f on the boundary $\partial\boldsymbol{D}$.

Homework

Exercise 10.2.4 *Let*

$$f(x, y) = \begin{bmatrix} 2xy \\ x^2 - y^2 \end{bmatrix}, \quad f_n(x, y) = \begin{bmatrix} 2xy + 1/n \\ x^2 - y^2 - 1/n \end{bmatrix}$$

For D, the ball about $(0,0)$ *of radius* $1/2$, *use Rouche's Theorem to show* $d_B(f_n, D, (0,0)) = d_B(f, D, (0,0))$ *for* n *sufficiently large.*

Exercise 10.2.5 *Let*

$$f(x, y) = \begin{bmatrix} x^2 + y^2 \\ 2x^2 - y^2 \end{bmatrix}, \quad f_n(x, y) = \begin{bmatrix} x^2 + y^2 - 1/n \\ 2x^2 - y^2 - 1/n \end{bmatrix}$$

For D, the ball about $(0,0)$ *of radius* $1/2$, *use Rouche's Theorem to show* $d_B(f_n, D, (0,0)) = d_B(f, D, (0,0))$ *for* n *sufficiently large.*

Exercise 10.2.6 *Let*

$$f(x, y) = \begin{bmatrix} 3xy \\ x^2 - y^2 \end{bmatrix}, \quad f_n(x, y) = \begin{bmatrix} 3xy + 1/n \\ x^2 - y^2 - 1/n \end{bmatrix}$$

For D, the ball about $(0,0)$ *of radius* 1, *use Rouche's Theorem to show* $d_B(f_n, D, (0,0)) = d_B(f, D, (0,0))$ *for* n *sufficiently large.*

Exercise 10.2.7 *Let*

$$f(x, y) = \begin{bmatrix} x^2 + y^2 \\ x^2 - 2y^2 \end{bmatrix}, \quad f_n(x, y) = \begin{bmatrix} x^2 + y^2 - 1/n \\ x^2 - 2y^2 - 1/n \end{bmatrix}$$

For D, the ball about $(0,0)$ *of radius* $1/4$, *use Rouche's Theorem to show* $d_B(f_n, D, (0,0)) = d_B(f, D, (0,0))$ *for* n *sufficiently large.*

10.3 Fixed Point Results

Now let's look at a *fixed point result*. There are many places you can read about this result and we encourage you to widen your scope and look into this result in more detail. You can look at (Heinz (47) 1959), which discusses this theorem from the point of view of Lemma 10.1.2. You can also do the development using the theory of analytic functions (we have spent little time on this theory although it is very elegant and clear) and this approach can be found in (Birkoff and Kellog (12) 1922) and (Seki (115) 1957). Don't be put off by the age of these papers; they are still useful reading.

Theorem 10.3.1 Brouwer Fixed Point Theorem

Let f be a continuous mapping of $B(\mathbf{0},1) = \{\boldsymbol{x} | \|\boldsymbol{x}\| \leq 1\} \subset \Re^n$ into itself. Then there is an $\boldsymbol{x_0} \in B(\mathbf{0},1)$ with $f(\boldsymbol{x_0}) = \boldsymbol{x_0}$.

Proof 10.3.1

We will prove this with a homotopy argument. Let $\boldsymbol{D} = B(\boldsymbol{x},1)$. If there is a $\boldsymbol{x_0} \in \partial\boldsymbol{D}$ with $f(\boldsymbol{x_0}) = \boldsymbol{x_0}$ we are done. If not, we have $f(\boldsymbol{x}) \neq \boldsymbol{x}$ on $\partial\boldsymbol{D}$. Define the homotopy F by $F(\boldsymbol{x},t) = \boldsymbol{x} - tf(\boldsymbol{x})$ for $0 \leq t \leq 1$. Note this says $F = I - tf$ where I is the identity mapping. We see F is continuous on $\overline{\boldsymbol{D}} \times [0,1]$. Now if $\boldsymbol{x} \in \partial\boldsymbol{D}$, $\|\boldsymbol{x}\| = 1$.

- *At $t = 0$, $F(\boldsymbol{x},0) = \boldsymbol{x}$. This is not zero on $\partial\boldsymbol{D}$ as $\|\boldsymbol{x}\| = 1$.*
- *If $0 < t < 1$, $\|F(\boldsymbol{x},t)\| = \|\boldsymbol{x} - tf(\boldsymbol{x})\| \geq \|\boldsymbol{x}\| - t\|f(\boldsymbol{x})\|$. Since $f(\boldsymbol{x}) \in \boldsymbol{D}$, $\|f(\boldsymbol{x})\| \leq 1$. We conclude $\|F(\boldsymbol{x},t)\| \geq \|\boldsymbol{x}\| - t \geq 1 - t > 0$ as $0 < t < 1$.*
- *If $t = 1$, we have $\|F(\boldsymbol{x},1)\| = ||\boldsymbol{x} - f(\boldsymbol{x})\| \neq 0$ on $\partial\boldsymbol{D}$.*

So $F(\boldsymbol{x},t) \neq 0$ on $\partial\boldsymbol{D}$ for $0 \leq t \leq 1$. By the Invariance Under Homotopy Theorem, we have $d_B(F(\cdot,0), B(\mathbf{0},1), \mathbf{0}) = d_B(F(\cdot,1), B(\mathbf{0},1), \mathbf{0})$. Thus, $d_B(I, B(\mathbf{0},1), \mathbf{0}) = d_B(I - f, B(\mathbf{0},1), \mathbf{0})$. But

$$d_B(I, B(\mathbf{0},1), \mathbf{0}) = \sum_{\boldsymbol{x} \in I^{-1}(\mathbf{0})} sign\ JI(\boldsymbol{x}) = sign\ JI(\mathbf{0}) = 1$$

Thus, we see $d_B(I - f, B(\mathbf{0},1), \mathbf{0}) = 1 \neq 0$. Since the degree is not zero there must be a solution $\boldsymbol{x_0}$ to $(I - f)(\boldsymbol{x_0}) = 0$ or $f(\boldsymbol{x_0}) = \boldsymbol{x_0}$. ■

By the way, note if $\mathbf{0} \notin \boldsymbol{D}$, $d_B(I, \boldsymbol{D}, \mathbf{0}) = 0$.

Homework

Exercise 10.3.1 *Prove Theorem 10.3.1 for the case $B(\mathbf{0},r)$ instead of the special case $r = 1$.*

Exercise 10.3.2 *Let f be continuous on $\Re^n$ and satisfy $\frac{<f(\boldsymbol{x}),\boldsymbol{x}>}{\boldsymbol{x}} \rightarrow \infty$ as $\|\boldsymbol{x}\| \rightarrow \infty$. Then f is surjective.*
Hint: *Define the homotopy $h(t,\boldsymbol{x}) = t\boldsymbol{x} + (1-t)f(\boldsymbol{x})$ and use that to show the equation $f(\boldsymbol{x}) = \boldsymbol{y}$ always has a solution for any $\boldsymbol{y} \in \Re^n$.*

Exercise 10.3.3 *Let $f : D = B(\mathbf{0},r) \subset \Re^n \rightarrow \Re^n$ be continuous on $\overline{D}$. Assume $\mathbf{0} \notin f(\partial D)$ and assume n is odd. Then there is a point $\boldsymbol{x_0} in \partial D$ and $\lambda \neq 0$ with $f(\boldsymbol{x_0}) = \lambda\boldsymbol{x_0}$.*
Hint: *We know $d_B(-I, D, \mathbf{0}) = -1$. If $d_B(f, D, \mathbf{0}) \neq -1$ use the homotopy $(1-t)f(\boldsymbol{x}) - t(\boldsymbol{x})$. On the other hand, if $d_B(f, D, \mathbf{0}) = -1$, use the homotopy $(1-t)f(\boldsymbol{x}) + t(\boldsymbol{x})$.*

Exercise 10.3.4 *Extend the Brouwer Fixed Point theorem to the case where $D \subset \Re^n$ is a compact convex set.*
Hint: *Extend f to its continuous extension $\tilde{f}$ on $\Re^n$. For any set $U \subset \Re^n$, the convex hull of V is*

$$conv\ V = \{\sum_{i=1}^{n} \lambda_i x_i : x_i \in V, 0 \leq \lambda_i \leq 1, \sum_{i=1}^{n} \lambda_i = 1\}$$

Show $f(\Re^n) \subset \overline{conv\ f(D)} \subset D$. Choose a ball $B(\mathbf{0}, r)$ which contains D and then apply our usual theorem to find a fixed point.

Exercise 10.3.5 *Let A be an $n \times n$ matrix with nonnegative entries. Then, there is a nonnegative eigenvalue λ with a corresponding eigenvector $\boldsymbol{v}$ with nonnegative entries; i.e. $A(\boldsymbol{v}) = \lambda \boldsymbol{v}$.*
Hint: *Let D be the set of all vectors $\boldsymbol{x}$ in $\Re^n$ with nonnegative entries and $\|\boldsymbol{x}\|_1 = 1$. If $A(\boldsymbol{x}) = \mathbf{0}$ for some $\boldsymbol{x} \in D$, we have shown the result for the choice $\lambda = 0$. Otherwise, we have $A(\boldsymbol{x}) \neq 0$ on D which implies $\|A(\boldsymbol{x})\|_1 \geq \alpha > 0$ for some α. Let $f(\boldsymbol{x}) = \frac{A(\boldsymbol{x})}{\|A(\boldsymbol{x})\|_1}$. Then $f(D) \subset D$ and so there is a fixed point by the previous exercise which is the eigenvalue and eigenvector we seek.*

10.4 Borsuk's Theorem

We will use these ideas in the proof of Borsuk's Theorem. Useful references here are (Fučik et al. (34) 1973) and (Nirenberg (93) 2001). You should be able to find these in either used or updated editions. Up to now, we know

$$d_B(I, \boldsymbol{D}, \boldsymbol{p}) = \begin{cases} 1, & \boldsymbol{p} \in \boldsymbol{D} \\ 0, & \boldsymbol{p} \notin \boldsymbol{D} \end{cases}$$

so that Brouwer degree acts like the characteristic function of a set. We can prove the following theorem which is important in applications.

Theorem The Borsuk Theorem

Let $\boldsymbol{D}$ be a symmetric bounded open set in $\Re^n$ that contains the origin and $f : \overline{\boldsymbol{D}} \rightarrow \Re^n$ be continuous and odd on $\partial \boldsymbol{D}$; i.e. $-f(\boldsymbol{x}) = f(-\boldsymbol{x})$ for all $\boldsymbol{x} \in \partial \boldsymbol{D}$. Assume $\mathbf{0} \notin f(\partial \boldsymbol{D})$. Then $d_B(f, \boldsymbol{D}, \mathbf{0})$ is an odd integer. Note, this means the Brouwer degree is nonzero and hence, there must be a $\boldsymbol{x_0}$ so that $f(\boldsymbol{x_0}) = \mathbf{0}$; i.e. f must have a zero.

We will prove Borsuk's Theorem via a sequence of lemmas.

Lemma 10.4.1 Extension Lemma One

Let $bsF \subset \Re^n$ be closed and bounded. Let $g : \boldsymbol{F} \rightarrow \Re^k$ be continuous. Then there is a continuous mapping $h : \Re^n \rightarrow \Re^k$ so that $h|_{\boldsymbol{F}} = g$. Moreover, if $\|g(\boldsymbol{x})\| \leq K$ for all $\boldsymbol{x} \in F$, $h(\boldsymbol{y}) \leq \sqrt{k} K$ for all $\boldsymbol{y} \in \Re^n$.

Proof 10.4.1
First, let's suppose $k = 1$. Consider

$$\rho(\boldsymbol{x}, \boldsymbol{a}) = \max \left\{ 2 - \frac{\|\boldsymbol{x} - \boldsymbol{a}\|}{\min_{\boldsymbol{y} \in \boldsymbol{F}} \|\boldsymbol{x} - \boldsymbol{y}\|}, 0 \right\}, \ \forall \boldsymbol{x} \notin \boldsymbol{F}, \ \boldsymbol{a} \in \Re^n$$

Note, since $\boldsymbol{x} \notin \boldsymbol{F}$,

$$\min_{\boldsymbol{y} \in \boldsymbol{F}} \|\boldsymbol{x} - \boldsymbol{y}\| = \inf_{\boldsymbol{y} \in \boldsymbol{F}} \|\boldsymbol{x} - \boldsymbol{y}\| = dist(\boldsymbol{x}, \boldsymbol{F}) > 0$$

We know $\rho(\boldsymbol{x}, \boldsymbol{a})$ is a continuous function. As soon as $\frac{\|\boldsymbol{x} - \boldsymbol{a}\|}{\min_{\boldsymbol{y} \in \boldsymbol{F}} \|\boldsymbol{x} - \boldsymbol{y}\|} > 2$, $\rho(\boldsymbol{x}, \boldsymbol{a}) = 0$. Also, note $\rho(\boldsymbol{x}, \boldsymbol{x}) = 2$. Hence, $0 \leq \rho(\boldsymbol{x}, \boldsymbol{a}) \leq 2$.

For any k, since $\boldsymbol{F}$ is compact, let $(\boldsymbol{a_n}) \subset \boldsymbol{F}$ be a sequence chosen as follows: since $\boldsymbol{F}$ is compact, for each k, there is a $\frac{1}{k}$-net, $\boldsymbol{F_k} = \{b_{k1}, \ldots, b_{k,n_k}\}$. Now define the collection $(\boldsymbol{a_n})$ by $(\boldsymbol{a_n}) =$

$\cup_{k=1}^{\infty} \boldsymbol{F_k}$. The $\cup_{k=1}^{\infty} \boldsymbol{F_k}$ is countable and $(\boldsymbol{a_n})$ is an enumeration of it. Define $U(\boldsymbol{x}) = \sum_{n=1}^{\infty} \frac{\rho(\boldsymbol{x}, \boldsymbol{a_n})}{2^n}$. Then,

$$U(\boldsymbol{x}) \leq \sum_{n=1}^{\infty} \frac{1}{2^{n-1}} = \sum_{k=0}^{\infty} \frac{1}{2^k} = 2 < \infty$$

and since g is continuous on $\boldsymbol{F}$, g is bounded on $\boldsymbol{F}$ by a constant K. Thus,

$$V(\boldsymbol{x}) = \frac{\rho(\boldsymbol{x}, \boldsymbol{a_n})}{2^n} g(\boldsymbol{a_n}) \leq K \sum_{n=1}^{\infty} \frac{1}{2^{n-1}} = 2K < \infty$$

Then, the function h defined by

$$h(\boldsymbol{x}) = \begin{cases} g(\boldsymbol{x}), & \boldsymbol{x} \in \boldsymbol{F} \\ \frac{V(\boldsymbol{x})}{U(\boldsymbol{x})}, & \boldsymbol{x} \notin \boldsymbol{F} \end{cases}$$

is well-defined. We conclude

$$\|h(\boldsymbol{x})\| \leq \frac{\|V(\boldsymbol{x})\|}{\|U(\boldsymbol{x})\|} \leq K = \sqrt{1}K$$

where we specifically note that we can phrase this in terms of the square root of the dimension of $\Re^k$.

Next, we show h is continuous. Let S_M be the Mth partial sum:

$$S_M(\boldsymbol{x}) = \sum_{n=1}^{M} \frac{\rho(\boldsymbol{x}, \boldsymbol{a_n})}{2^n}$$

Note, S_M is continuous on $\boldsymbol{F}^C$ and for $N > M$,

$$|S_M(\boldsymbol{x}) - S_N(\boldsymbol{x})| \leq \left| \sum_{n=M+1}^{N} \frac{\rho(\boldsymbol{x}, \boldsymbol{a_n})}{2^n} \right| \leq \sum_{n=M+1}^{N} \frac{1}{2^{n-1}} \leq \frac{1}{2^{M+1}}$$

So, by the Weierstrass Uniform Convergence test, (S_M) converges uniformly to a continuous function. Thus, $U(\boldsymbol{x})$ is continuous on $\boldsymbol{F}^C$. In a similar way, we can show $V(\boldsymbol{x})$ is continuous on $\boldsymbol{F}^C$. The only way, $\frac{V(\boldsymbol{x})}{U(\boldsymbol{x})}$ can fail to be continuous is if $U(\boldsymbol{x}) = 0$ at some $\boldsymbol{x} \notin \boldsymbol{F}$. If this was true, $\rho(\boldsymbol{x}, \boldsymbol{a_n}) = 0$ for all n. This would imply

$$\max\left\{2 - \frac{\|\boldsymbol{x} - \boldsymbol{a}\|}{\min_{\boldsymbol{y} \in \boldsymbol{F}} \|\boldsymbol{x} - \boldsymbol{y}\|}, 0\right\} = 0, \ \forall n \Longrightarrow \frac{\|\boldsymbol{x} - \boldsymbol{a}\|}{\min_{\boldsymbol{y} \in \boldsymbol{F}} \|\boldsymbol{x} - \boldsymbol{y}\|} \geq 2, \ \forall n$$

Note, as long as $\boldsymbol{x} \notin \boldsymbol{F}$, $dist(\boldsymbol{x}, \boldsymbol{F}) > 0$ as $\boldsymbol{F}$ is compact. Thus, $\|\boldsymbol{x} - \boldsymbol{a_n}\| \geq 2dist(\boldsymbol{x}, \boldsymbol{F})$. But, the sequence $(\boldsymbol{a_n})$ was chosen using ϵ-nets to be dense in $\boldsymbol{F}$. Hence, given $\boldsymbol{y} \in \boldsymbol{F}$, there is a sequence $\boldsymbol{a}^1_{\boldsymbol{n}}$ so that $\boldsymbol{a}^1_{\boldsymbol{n}} \to \boldsymbol{y}$. Thus,

$$| \boldsymbol{x} - \boldsymbol{a}^1_{\boldsymbol{n}}\| \geq 2dist(\boldsymbol{x}, \boldsymbol{F}) \Longrightarrow \lim_{n^1 \to \infty} | \boldsymbol{x} - \boldsymbol{a}^1_{\boldsymbol{n}}\| = \|\boldsymbol{x} - \boldsymbol{y}\| \geq 2dist(\boldsymbol{x}, \boldsymbol{F})$$

Since this is true for all $\boldsymbol{y} \in \boldsymbol{F}$, we have $dist(\boldsymbol{x}, \boldsymbol{F}) \geq 2dist(\boldsymbol{x}, \boldsymbol{F})$ which is not possible as $dist(\boldsymbol{x}, \boldsymbol{F}) > 0$. We conclude, $U(\boldsymbol{x}) \neq 0$ on $\boldsymbol{F}^C$. Hence, $h(\boldsymbol{x}) = \frac{V(\boldsymbol{x})}{U(\boldsymbol{x})}$ is continuous on $\boldsymbol{F}^C$.

Now, if $(\boldsymbol{x_m}) \subset \boldsymbol{F}^C$ *and* $\boldsymbol{x_m} \rightarrow \boldsymbol{x} \in \boldsymbol{F}$ *(i.e.* $\boldsymbol{x} \in \partial \boldsymbol{F}$*), we must show* $\frac{V(\boldsymbol{x_m})}{U(\boldsymbol{x_m})} \rightarrow g(\boldsymbol{x})$*. Let* $\epsilon > 0$ *be given. Then, there is a* δ *so that* $\|\boldsymbol{x} - \boldsymbol{a_n}\| < \delta$ *implies* $\|g(\boldsymbol{x}) - g(\boldsymbol{a_n})\| < \epsilon$*. Then,*

$$|g(\boldsymbol{x}) - h(\boldsymbol{x_m})| = \left|g(\boldsymbol{x}) - \frac{V(\boldsymbol{x})}{U(\boldsymbol{x})}\right| = \left|\frac{\sum_{n=1}^{\infty} \frac{\rho(\boldsymbol{x},\boldsymbol{a_n})}{2^n}(g(\boldsymbol{x}) - g(\boldsymbol{a_n}))}{\sum_{n=1}^{\infty} \frac{\rho(\boldsymbol{x},\boldsymbol{a_n})}{2^n}}\right|$$

We also know there is a P *so that* $\|\boldsymbol{x} - \boldsymbol{x_m}\| < \frac{\delta}{4}$ *if* $m > P$*. This tells us* $dist(\boldsymbol{x_m}, \boldsymbol{F}) < \frac{\delta}{4}$ *if* $m > P$*. Consider the term* $\rho(\boldsymbol{x_m}, \boldsymbol{a_n})$*. This term is nonzero when*

$$2 - \frac{\|\boldsymbol{x_m} - \boldsymbol{a_n}\|}{dist(\boldsymbol{x_m}, \boldsymbol{F})} > 0 \Longrightarrow \|\boldsymbol{x_m} - \boldsymbol{a_n}\| < 2dist(\boldsymbol{x_m}, \boldsymbol{F})$$

Hence, if $m > P$*, we have* $\|\boldsymbol{x_m} - \boldsymbol{a_n}\| < 2\frac{\delta}{4} = \frac{\delta}{2}$*. Also, if* $m > P$*,*

$$\|\boldsymbol{x} - \boldsymbol{a_n}\| \leq \|\boldsymbol{x} - \boldsymbol{x_m}\| + \|\boldsymbol{x_m} - \boldsymbol{a_n}\| \leq \frac{\delta}{4} + \frac{\delta}{2} < \delta$$

Hence, if $J_m = \{i \| \|\boldsymbol{x_m} - \boldsymbol{a_i}\| < 2dist(\boldsymbol{x_m}, \boldsymbol{F})\}$*, we have when* $m > P$*,* $\|\boldsymbol{x} - \boldsymbol{a_n}\| < \delta$ *and so*

$$|g(\boldsymbol{x}) - h(\boldsymbol{x_m})| \leq \frac{\sum_{i \in J_m} \frac{\rho(\boldsymbol{x},\boldsymbol{a_n})}{2^n}|g(\boldsymbol{x}) - g(\boldsymbol{a_n})|}{\sum_{i \in J_m} \frac{\rho(\boldsymbol{x},\boldsymbol{a_n})}{2^n}} \leq \frac{\sum_{i \in J_m} \frac{\rho(\boldsymbol{x},\boldsymbol{a_n})}{2^n}\epsilon}{\sum_{i \in J_m} \frac{\rho(\boldsymbol{x},\boldsymbol{a_n})}{2^n}} = \epsilon$$

This shows $|g(\boldsymbol{x}) - h(\boldsymbol{x_m})\| \rightarrow 0$ *and* $m \rightarrow \infty$*. Since the choice of sequence* $(\boldsymbol{x_m})$ *is arbitrary, we have for* $\boldsymbol{x} \in \boldsymbol{F}$ *and* $\boldsymbol{y} \in \boldsymbol{F}^C$*, that* $\lim_{\boldsymbol{y} \rightarrow \boldsymbol{x}} h(\boldsymbol{y}) = h(\boldsymbol{x})$*. We conclude* h *is continuous on* $\Re^n$ *and we have established the case for* $k = 1$*.*

If $k > 1$*, we let* $g = [g_1, \ldots, g_k]$ *and by the first part of our argument, there are functions* h_1 *to* h_k *which are continuous extensions of the functions* g_1 *to* g_k *to all of* $\Re^n$*. Note, if* $\|g(\boldsymbol{x})\| \leq K$*, then* $\|h(\boldsymbol{x})\| = \sqrt{(h_1(\boldsymbol{x}))^2 + \ldots (h_k(\boldsymbol{x}))^2} \leq \sqrt{k}K$*.*

This concludes the proof. ■

Can we extend a function defined on a set to a closed cube enclosing the set? The next lemma answers this question.

Lemma 10.4.2 Extension Lemma Two

Let $\boldsymbol{G} \subset \Re^n$ *be a closed and bounded subset. Let* $f : \boldsymbol{G} \rightarrow \Re^{n+j}$*, with* $j \geq 1$*, be a continuous mapping satisfying* $\boldsymbol{0} \notin f(\boldsymbol{G})$*. Then* f *can be extended to a continuous nowhere vanishing mapping defined on any closed cube* $\mathscr{C}$ *with* $\boldsymbol{G} \subset \mathscr{C}$*.*

Proof 10.4.2

Extend $f : \boldsymbol{G} \rightarrow \Re^{n+1}$ *to all of* $\Re^n$ *by extending the* $\partial \boldsymbol{G}$ *values down to zero in such a way that* f *has compact support that contains* $\boldsymbol{G}$*. Call this new function* $\tilde{f}$*. Now how would you go about finding this extension? Let's sketch how you do this as even if you have seen the argument, there are so many different ideas here, it is easy to get lost.*

Let $\boldsymbol{A} \subset \Re^n$ *be a nonempty set. Define* $\rho : \Re^n \rightarrow \Re$ *by* $\rho(\boldsymbol{x}) = dist(\boldsymbol{x}, \boldsymbol{A})$*. Note* $\rho|_{\boldsymbol{A}} = 0$*. Now, if* $\boldsymbol{y} \in \boldsymbol{A}$*, for the pair* $\boldsymbol{x}$ *and* $\boldsymbol{x_0}$*, we have*

$$\rho(\boldsymbol{x}) \leq \|\boldsymbol{x} - \boldsymbol{y}\| \leq \|\boldsymbol{x} - \boldsymbol{x_0}\| + \|\boldsymbol{x_0} - \boldsymbol{y}\|$$

Since this is true for all $\boldsymbol{y} \in \boldsymbol{A}$, this implies

$$\rho(\boldsymbol{x}) \leq \|\boldsymbol{x} - \boldsymbol{x_0}\| + \rho(\boldsymbol{x_0}) \implies \rho(\boldsymbol{x}) - \rho(\boldsymbol{x_0}) \leq \|\boldsymbol{x} - \boldsymbol{x_0}\|$$

Reversing the order of the argument shows

$$\rho(\boldsymbol{x_0}) - \rho(\boldsymbol{x}) \leq \|\boldsymbol{x} - \boldsymbol{x_0}\|$$

Combining, we have $|\rho(\boldsymbol{x}) - \rho(\boldsymbol{x_0})| \leq \|\boldsymbol{x} - \boldsymbol{x_0}\|$. This tells us ρ is a continuous function. Now, there is an $R > 0$ so that $\boldsymbol{G} \subset B(\boldsymbol{0}, R) \equiv \boldsymbol{B}$. Hence, it follows that the function

$$\phi(\boldsymbol{x}) = \frac{dist(\boldsymbol{x}, \boldsymbol{B}^C)}{dist(\boldsymbol{x}, \boldsymbol{B}^C) + dist(\boldsymbol{x}, \boldsymbol{F})} = \begin{cases} 1, & \boldsymbol{x} \in \boldsymbol{F} \\ \in (0,1), & \boldsymbol{x} \in \boldsymbol{F}^C \cap \boldsymbol{B} \\ 0, & \boldsymbol{B}^C \end{cases}$$

is continuous as well. Hence, $\tilde{f} = f\phi$ gives

$$\tilde{f}(\boldsymbol{x}) = \begin{cases} f(\boldsymbol{x}), & \boldsymbol{x} \in \boldsymbol{F} \\ \in (0, f(\boldsymbol{x})), \boldsymbol{x} \in \boldsymbol{F}^C \cap \boldsymbol{B} & \\ 0, & \boldsymbol{B}^C \end{cases}$$

is a continuous function whose support is compact. This is the extension we seek.

Let $c = min_{\boldsymbol{x} \in \boldsymbol{G}} \|f(\boldsymbol{x})\|$. Since we assume $f(\boldsymbol{G}) > 0$, we know $c > 0$. Let $\epsilon > 0$ with $\epsilon < \frac{c}{8\sqrt{n+1}}$. By the Stone - Weierstrass Approximation Theorem, there is a C^1 function $\Psi : \Re^n \to \Re^{n+j}$ such that $\|\tilde{f} - \Psi\|_\infty < \epsilon$. Note $|\Psi|_{B^C} = \boldsymbol{0}$. Assume the dimension of $\Psi(\Re^n)$ was bigger than or equal to $n+1$. Then at points $\boldsymbol{x}$ where $J\Psi(\boldsymbol{x}) \neq \boldsymbol{0}$, the implicit function theorem tells us, Ψ can be restricted to a bijection from a neighborhood $\mathscr{O} \subset \Re^n$ to a neighborhood $\mathscr{O}' \subset \Re^{n+j}$. But this would say an open subset on $\Re^n$ was homeomorphic to an open subset of $\Re^{n+j}$. This is not possible. Sard's Theorem says $\{\boldsymbol{x} \in \boldsymbol{G} | Jf(\boldsymbol{x}) = \boldsymbol{0}\}$ must have measure zero. Hence, there are many points in $\boldsymbol{G}$ where such a contradiction could occur. We conclude the $n+j$ dimensional measure of $\Psi(\Re^n)$ must be zero. Hence, $\Psi(\Re^n)$ cannot contain any interior points in the $\Re^{n+j}$ topology.

Suppose for all $\boldsymbol{y}$ with $\|\boldsymbol{y}\| < \epsilon$ and $\boldsymbol{y} \in \Re^{n+j}$, there is an $\boldsymbol{x} \in \Re^n$ so that $\Psi(\boldsymbol{x}) = \boldsymbol{y}$. This would say the image of Ψ, $\boldsymbol{\Psi}(\Re^n)$, contains an open $\Re^{n+j}$ ball which cannot happen. Hence there must be a $\boldsymbol{y_0}$ with $\|\boldsymbol{y_0}\| < \epsilon$ and $\boldsymbol{y_0} \in \Re n + j$ and $\Psi(\boldsymbol{x}) - \boldsymbol{y_0} \neq \boldsymbol{0}$ for all $\boldsymbol{x} \in \Re^n$. Hence, in $\boldsymbol{x} \in \boldsymbol{G}$

$$\begin{aligned} \|\tilde{f}(\boldsymbol{x}) - \boldsymbol{y_0}\| \geq \|f(\boldsymbol{x})\| - \|\boldsymbol{y_0}\| &\geq c - \epsilon \\ \|(\tilde{f}(\boldsymbol{x}) - \boldsymbol{y_0}) - (\Psi(\boldsymbol{x}) - \boldsymbol{y_0})\| &< \epsilon \end{aligned}$$

Now choose a continuous function $\eta : (0, \infty) \to \Re$ so that

$$\eta(t) = \begin{cases} 1, & t \geq \frac{c}{2} \\ \frac{2t}{c}, & t < \frac{c}{2} \end{cases}$$

Define F by

$$F(\boldsymbol{x}) = \frac{\Psi(\boldsymbol{x}) - \boldsymbol{y_0}}{\eta(\|\Psi(\boldsymbol{x}) - \boldsymbol{y_0}\|)}, \ \forall \boldsymbol{x} \in \Re^n$$

Note,

$$\|\boldsymbol{\Psi}(\boldsymbol{x})\| + \|\boldsymbol{y_0}\| \geq \|\boldsymbol{\Psi}(\boldsymbol{x}) - \boldsymbol{y_0}\| = \|(\boldsymbol{\Psi}(\boldsymbol{x}) - \tilde{f}(\boldsymbol{x})) - (\tilde{f}(\boldsymbol{x}) - \boldsymbol{y_0})\|$$

$$\geq \quad \|\tilde{f}(\boldsymbol{x}) - \boldsymbol{y_0}\| - \|\boldsymbol{\Psi}(\boldsymbol{x}) - \tilde{f}(\boldsymbol{x})\| \geq c - \epsilon - \epsilon$$

Thus, for our choice of ϵ*, we see*

$$\|\boldsymbol{\Psi}(\boldsymbol{x})\| \quad \geq \quad c - 2\epsilon - \|\boldsymbol{y_0}\| = c - 3\epsilon > c - \frac{3c}{8\sqrt{n+1}} > \frac{c}{2} > 0$$

We see $\Psi(\boldsymbol{x}) > \boldsymbol{0}$ *on* $\Re^n$.

We have

$$\|F(\boldsymbol{x})\| \quad = \quad \frac{\|\Psi(\boldsymbol{x}) - \boldsymbol{y_0}\|}{\eta(\|\Psi(\boldsymbol{x}) - \boldsymbol{y_0}\|)} = \begin{cases} \|\Psi(\boldsymbol{x}) - \boldsymbol{y_0}\|, & \|\Psi(\boldsymbol{x}) - \boldsymbol{y_0}\| \geq \frac{c}{2} \\ \frac{c}{2}, & \|\Psi(\boldsymbol{x}) - \boldsymbol{y_0}\| < \frac{c}{2} \end{cases}$$

Thus, $\|F(\boldsymbol{x})\| \geq \frac{c}{2}$. *Then, on* $\boldsymbol{G}$*, we have*

$$\begin{aligned} \frac{c}{8\sqrt{n+1}} \quad &> \quad \epsilon > \|f(\boldsymbol{x}) - \boldsymbol{y_0} + \boldsymbol{y_0} - \Psi(\boldsymbol{x})\| \geq \|f(\boldsymbol{x})\| - \|\Psi(\boldsymbol{x}) - \boldsymbol{y_0}\| - \|\boldsymbol{y_0}\| \\ &\geq \quad c - \epsilon - \|\Psi(\boldsymbol{x}) - \boldsymbol{y_0}\| \end{aligned}$$

Thus, $\|\Psi(\boldsymbol{x}) - \boldsymbol{y_0}\| > c - 2\epsilon < \frac{c}{2}$. *Hence, on* $\boldsymbol{G}$, $\eta(\|\Psi(\boldsymbol{x}) - \boldsymbol{y_0}\|) = 1$. *This tells us that on* $\boldsymbol{G}$*,*

$$\|F(\boldsymbol{x}) - f(\boldsymbol{x}) \quad = \quad \|Y(\boldsymbol{x}) - \boldsymbol{y_0} - f(\boldsymbol{x})\| < \epsilon + \|\boldsymbol{y_0}\| < 2\epsilon$$

Now, let $\boldsymbol{C}$ *be any closed cube that contains* $\boldsymbol{G}$*. Apply Lemma 10.4.1 to the mapping* $F - f$ *on* $\boldsymbol{G}$*. Then there is a continuous mapping* $\delta : \Re^n \rightarrow \Re^{n+j}$ *with* $\delta = F - f$ *on* $\boldsymbol{G}$*. Since,* $\|F(\boldsymbol{x}) - f(\boldsymbol{x})\| < 2\epsilon$ *on* $\boldsymbol{G}$*,* $\|F(\boldsymbol{x}) - f(\boldsymbol{x})\| < 2\epsilon\sqrt{n+j}$ *on the closed cube* $\boldsymbol{C}$*. Define* $F_1(\boldsymbol{x}) = F(\boldsymbol{x}) - \delta(\boldsymbol{x})$ *on* $\boldsymbol{C}$*. Then* $F_1 = f$ *on* $\boldsymbol{G}$ *and on* $\boldsymbol{C}$*,*

$$\|F_1(\boldsymbol{x})\| \quad = \quad \|F(\boldsymbol{x}) - \delta(\boldsymbol{x})\| \geq \|F(\boldsymbol{x})\| - \|\delta(\boldsymbol{x})\| \geq \frac{c}{2} - 2\epsilon\sqrt{n+1} > \frac{c}{4} > 0$$

The function F_1 *is the desired extension.* ∎

We now want to show we can find odd extensions.

Lemma 10.4.3 Extension Lemma Three

Let $\boldsymbol{D} \subset \Re^n$ *be a bounded open subset with* $\boldsymbol{0} \notin \overline{\boldsymbol{D}}$ *and assume* $\boldsymbol{D}$ *is symmetric about the origin. Let* $\Psi : \partial\boldsymbol{D} \rightarrow \Re^{n+j} \setminus \boldsymbol{0}$ *with* $j \geq 1$ *and* Ψ *odd. Then there is an extension of* Ψ *to all of* $\overline{\boldsymbol{D}}$ *as an odd mapping into* $\Re^{n+j} \setminus \boldsymbol{0}$.

Proof 10.4.3

The proof is by induction on n*. When* $n = 1m$ *since* $\boldsymbol{0} \notin \boldsymbol{D}$*,* $\boldsymbol{D} \subset [-\epsilon, \epsilon]^C$ *for* $\epsilon > 0$ *sufficiently small. Then,* $\partial\boldsymbol{D}$ *is contained in a bounded subset* $\boldsymbol{D}^- \cup \boldsymbol{D}^+$ *of* $(-\infty, -\epsilon] \cup [\epsilon, \infty)$*, where* $\boldsymbol{D}^- \subset (\infty, -\epsilon]$ *and* $\boldsymbol{D}^+ \subset [\epsilon, \infty)$*. By Lemma 10.4.2,* Ψ *on* $\partial\boldsymbol{D} \cap [\epsilon, \infty) = \partial\boldsymbol{D}^+$ *can be extended to any closed cube* $\boldsymbol{C_k}$ *to the function* $\tilde{\Psi}$ *on* $[\epsilon, k] = \boldsymbol{C_k}$ *into* $\Re^{1+j} \setminus \boldsymbol{0}$ *where* k *is large enough so that* $\boldsymbol{D}^+ \subset \boldsymbol{C_k}$*. For all* $\boldsymbol{x} \in \partial D^-$ *define* $\tilde{\Psi}(\boldsymbol{x}) = -\tilde{\Psi}(-\boldsymbol{x})$*. This clearly extends* $\tilde{\Psi}$ *as a nonvanishing odd map on all of* $\overline{\boldsymbol{D}}$*.*

Now if the result is true for $n - 1$*, let* $\boldsymbol{D_0} = \boldsymbol{D} \cap \{\boldsymbol{x} | x_n = 0\}$ *where* x_n *is the* n^{th} *component of* $\boldsymbol{x}$*. Then,* $\Psi|_{\partial\boldsymbol{D_0}}$ *is an odd nonvanishing map on* $\overline{\boldsymbol{D}}$*. By the induction hypothesis, there is an extension of* $\Psi|_{\partial\boldsymbol{D_0}}$ *to* $\tilde{\Psi}_0$ *defined on* $\overline{\boldsymbol{D_0}}$ *mapping into* $\Re^{n+j} \setminus \boldsymbol{0}$*. Let* $\boldsymbol{D}^+ = \boldsymbol{D} \cap \{\boldsymbol{x} | x_n > 0\}$ *and*

$D^- = D \cap \{x | x_n < 0\}$. Define $\Psi_1 : \partial D^+ \rightarrow \Re^{n+j} \setminus \mathbf{0}$ by

$$\Psi_1 \quad = \quad \begin{cases} \Psi, & x \in \partial D\{x | x_n \geq 0\} \\ \Psi_0, & x \in D_0 = D \cap \{x | x_n = 0\} \end{cases}$$

Apply Lemma 10.4.2 to Ψ_1. Then there is an extension $\tilde{\Psi_1}$ to any closed cube containing $\overline{D}$. In particular, the extension is defined on ∂D^+. Label this closed cube as C^+. Then $\tilde{\Psi_1} : C^+ \rightarrow \Re^{n+j} \setminus \mathbf{0}$. Define $\tilde{\Psi_1}(x) = -\tilde{\Psi_1}(-x)$ when $x \in \overline{D^-}$. Then $\tilde{\Psi_1}$ is the continuous, nonvanishing odd extension we seek. ■

Now let's focus on mappings from $\Re^n$ into $\Re^n$.

Lemma 10.4.4 Extension Lemma Four

Let $D \subset \Re^n$ be a bounded, symmetric open subset with $\mathbf{0} \notin \overline{D}$. Let $\Psi : \partial D \rightarrow \Re^n \setminus \mathbf{0}$ be continuous and odd. Then, there is an extension $\tilde{\Psi} : \overline{D} \rightarrow \Re^n$ which is continuous and odd so that $\tilde{\Psi} \neq \mathbf{0}$ on $D \cap \{x \in \Re^n | x_n = 0\}$ where x_n is the nth component of x.

Proof 10.4.4
By Lemma 10.4.3 applied to Ψ restricted to $\partial(D \cap \{x | x_n = 0\})$ which is a map from $\Re^{n-1} \rightarrow \Re^n$. For concreteness, call this map Θ. There is a continuous, nonvanishing odd extension $\tilde{\Theta}$ to $\overline{D \cap \{x | x_n = 0\}} = \overline{D} \cap \{x | x_n = 0\}$. Thus, $\tilde{\Theta} : \overline{D} \cap \{x | x_n = 0\} \rightarrow \Re^n \setminus \mathbf{0}$. Apply Lemma 10.4.1 to the closed set $\overline{D} \cap \{x | x_n = 0\} \subset \Re^n$ to get a continuous extension of $\tilde{\Theta}$ to $\hat{\Theta}$ to $\overline{D} \cap \{x | x_n \geq 0\}$. Extend to $\overline{D} \cap \{x | x_n \leq 0\}$ in the usual manner since D is symmetric. ■

Next, look at even mappings.

Lemma 10.4.5 Degree of a Continuous, Odd Mapping Can Be Even

Let $D \subset \Re^n$ be an open, bounded and symmetric subset with $\mathbf{0} \notin \overline{D}$. Then, if $\Psi : \partial D \rightarrow \Re^n \setminus \mathbf{0}$ is continuous and odd, $d_B(\Psi, D, \mathbf{0})$ is even.

Proof 10.4.5
We have already discussed why the degree of a mapping Ψ is independent of the extension $\tilde{\Psi}$ in our comments after the proof of Theorem 10.2.4. By Lemma 10.4.4, there is an extension $\tilde{\Psi} : \overline{D} \rightarrow \Re^n$ so that $\tilde{\Psi} : \overline{D} \cap \{x | x_n = 0\} \rightarrow \Re^n \setminus \mathbf{0}$. For a given ϵ, approximate $\tilde{\Psi}$ by a function Θ on $\overline{D}$ which is in $C^1(\overline{D})$; i.e. $\| \tilde{\Psi} - \Theta\|_\infty < \epsilon/2$. Replace Θ by $\hat{\Theta}(x) = \frac{1}{2}(\Theta(x) - \Theta(-x))$ which is an odd function. Then $\| \tilde{\Psi} - \hat{\Theta}\|_\infty < \epsilon/$. Using standard arguments, we can choose $\epsilon > 0$ small enough so that $\hat{\Theta}(x) \neq \mathbf{0}$ on ∂D and $\hat{\Theta}(\overline{D} \cap \{x | x_n = 0\}) \neq \mathbf{0}$. We see then that

$$deg_B[\Psi, \overline{D}, \mathbf{0}] \quad = \quad deg_B[\tilde{\Psi}, \overline{D}, \mathbf{0}] = deg_B[\hat{\Theta}, \overline{D}, \mathbf{0}]$$

■

We are finally ready to tackle the proof of Borsuk's Theorem.

Theorem 10.4.6 Borsuk's Theorem Proof

Let D be a symmetric bounded open set in $\Re^n$ that contains the origin and $f : \overline{D} \rightarrow \Re^n$ be continuous and odd on ∂D; i.e. $-f(x) = f(-x)$ for all $x \in \partial D$. Assume $\mathbf{0} \notin f(\partial D)$. Then $d_B(f, D, \mathbf{0})$ is an odd integer. Note, this means the Brouwer degree is nonzero and hence, there must be a x_0 so that $f(x_0) = \mathbf{0}$; i.e. f must have a zero.

Proof 10.4.6

The set $\partial \boldsymbol{D}$ is a bounded closed subset of $\Re^n$. Since $\boldsymbol{D}$ is open and $\boldsymbol{0} \in \boldsymbol{D}$, there is a neighborhood U of $\boldsymbol{0}$ with $U \cup \boldsymbol{D}$ with $U \cap \partial \boldsymbol{D} = \emptyset$. Apply Lemma 10.4.1 to $Q : \partial \boldsymbol{D} \cup \overline{U} \rightarrow \Re^n$ where

$$Q(\boldsymbol{x}) \quad = \quad \begin{cases} f(\boldsymbol{x}), & \boldsymbol{x} \in \partial \boldsymbol{D} \\ I(\boldsymbol{x}), & \boldsymbol{x} \in \overline{U} \end{cases}$$

Then there is an extension $\phi : \overline{\boldsymbol{D}} \rightarrow \Re^n$ so that $\phi = f$ on $\partial \boldsymbol{D}$ and $\phi = I$ on $\overline{U}$. Let $g(\boldsymbol{x}) = (1/2)(\phi(\boldsymbol{x}) - \phi(-\boldsymbol{x}))$. Then, g is a mapping defined on $\overline{D}$ which is odd, matches f on $\partial \boldsymbol{D}$ and is the identity on $\overline{U}$.

$$g(\boldsymbol{x}) \quad = \quad \begin{cases} f(\boldsymbol{x}), & \boldsymbol{x} \in \partial \boldsymbol{D} \\ I(\boldsymbol{x}), & \boldsymbol{x} \in \overline{U} \end{cases}$$

Now apply Rouchës Theorem, to note $d_B(g, \boldsymbol{D}, \boldsymbol{0}) = d_B(f, \boldsymbol{D}, \boldsymbol{0})$. Also, $d_B(g, \boldsymbol{D}, \boldsymbol{0}) = d_B(g, \partial U \cup (\boldsymbol{D} \setminus \overline{U}), \boldsymbol{0})$. Now $\boldsymbol{D} \setminus \overline{U} = D \cap \overline{U}^C$ and so $\partial(\boldsymbol{D} \setminus \overline{U}) \subset \partial \boldsymbol{D}$. We know $\boldsymbol{0} \notin f(\partial \boldsymbol{D}) = g(\partial \boldsymbol{D})$ and on ∂U, $g = I$. Hence, $\boldsymbol{0} \notin g(\partial U \cup \partial \boldsymbol{D} \setminus \overline{U})$. By Theorem 10.2.1, we therefore know $d_B(g, \boldsymbol{D}, \boldsymbol{0}) = deg_B(g, U, \boldsymbol{0}) + deg_b(g, \boldsymbol{D} \setminus \overline{U}, \boldsymbol{0})$. But $g = I$ on $\overline{U}$ implying $deg_B(g, U, \boldsymbol{0}) = 1$. We also know g is odd on $\boldsymbol{D} \setminus \overline{U}$. Also, $\boldsymbol{0} \notin \boldsymbol{D} \setminus \overline{U}$ and $\boldsymbol{D} \setminus \overline{U}$ is open, bounded and symmetric. Hence, by Lemma 10.4.5, $deg_b(g, \boldsymbol{D} \setminus \overline{U}, \boldsymbol{0})$ is even. We conclude $d_B(g, \boldsymbol{D}, \boldsymbol{0})$ is odd and so $d_B(f, \boldsymbol{D}, \boldsymbol{0})$ is odd also. ■

There is an immediate corollary.

Theorem 10.4.7 Corollary to Borsuk's Theorem for Brouwer Degree

Let $\boldsymbol{D}$ be a symmetric bounded open set in $\Re^n$ that contains the origin and $f : \overline{\boldsymbol{D}} \rightarrow \Re^n$ be continuous with $\boldsymbol{0} \notin f(\partial \boldsymbol{D})$. Assume

$$\frac{f(\boldsymbol{x})}{\|f(\boldsymbol{x})\|} \quad \neq \quad \frac{f(-\boldsymbol{x})}{\|f(-\boldsymbol{x})\|}, \quad \forall \boldsymbol{x} \in \partial \boldsymbol{D}$$

Then $d_B(f, \boldsymbol{D}, \boldsymbol{0})$ is an odd integer.

Proof 10.4.7

Set $g(\boldsymbol{x}) = f(\boldsymbol{x}) - f(-\boldsymbol{x})$ and $h(\boldsymbol{x}, t) = f(\boldsymbol{x}) - tf(-\boldsymbol{x})$ for all $\boldsymbol{x} \in \overline{\boldsymbol{D}}$ and $0 \leq t \leq 1$. Then, for all $\boldsymbol{x} \in \partial \boldsymbol{D}$, $\|h(\boldsymbol{x}, t)\| \geq \|f(\boldsymbol{x})\| - t\|f(-\boldsymbol{x})\|$. Now, for all $\boldsymbol{x} \in \partial \boldsymbol{D}$:

(i): If $\|f(\boldsymbol{x})\| = \|f(-\boldsymbol{x})\|$, we have $\|h(\boldsymbol{x}, t)\| \geq \|f(\boldsymbol{x})\|(1 - t) > 0$ when $0 \leq t < 1$. If $t = 1$, $\|h(\boldsymbol{x}, 1)\| = \|f(\boldsymbol{x}) - f(-\boldsymbol{x})\|$. In this case, if $\|f(\boldsymbol{x})\| = \|f(-\boldsymbol{x})\|$, our boundary assumptions imply we have $f(\boldsymbol{x}) \neq f(-\boldsymbol{x})$. Thus, $h(\boldsymbol{x}, 1) > 0$.

(ii): If $\|f(\boldsymbol{x})\| \neq \|f(-\boldsymbol{x})\|$, let's assume $h(\boldsymbol{x}, t_0) = 0$ for $0 \leq t_0 \leq 1$. We know $t_0 > 0$ as $f(\boldsymbol{x}) \neq 0$ and $t_0 < 1$ as $\|f(\boldsymbol{x})\| \neq \|f(-\boldsymbol{x})\|$. Now, if this were true, we would have $f(\boldsymbol{x}) = t_0 f(-\boldsymbol{x})$ implying

$$\frac{f(\boldsymbol{x})}{\|f(\boldsymbol{x})\|} \quad = \quad \frac{t_0 f(-\boldsymbol{x})}{t_0 \|f(-\boldsymbol{x})} = \frac{f(-\boldsymbol{x})}{\|f(-\boldsymbol{x})}$$

This violates our assumption and so $h(\boldsymbol{x}, t) \neq 0$ in this case.

Combining, we see $0 \neq h(\boldsymbol{x}, t)$ for $\boldsymbol{x} \in \partial \boldsymbol{D}$ and $0 \leq t \leq 1$. By Theorem 10.2.3,

$$deg_B(h(\cdot, 0), \boldsymbol{D}, \boldsymbol{0}) \quad = \quad deg_B(h(\cdot, 0), \boldsymbol{D}, \boldsymbol{1}) \Longrightarrow deg_B(f, \boldsymbol{D}, \boldsymbol{0}) = deg_B(g, \boldsymbol{D}, \boldsymbol{0})$$

But $g(-\boldsymbol{x}) = f(-\boldsymbol{x}) - f(- - \boldsymbol{x}) = -g(\boldsymbol{x})$, *so* g *is an odd mapping. Also,* $f(\boldsymbol{x}) \neq f(-\boldsymbol{x})$ *on* $\partial \boldsymbol{D}$. *Hence, by Borsuk's Theorem,* $deg_B(g, \boldsymbol{D}, \boldsymbol{0})$ *is odd implying* $deg_B(f, \boldsymbol{D}, \boldsymbol{0})$ *is odd.* ■

Homework

Let P be a polynomial in three real variables, z_1, z_2, z_3, with real coefficients and let $|\alpha| = \alpha_1 + \alpha_2$ be the multi-index with $\alpha_i \geq 0$. Hence, $|\alpha| = 1$ implies $\alpha_i = 1$ for only one index with all others zero. Thus, each value of $|\alpha|$ corresponds to a set of triples consisting of nonnegative integers $\Lambda_{|\alpha|}$; i.e.

$$\begin{aligned}
|\alpha| &= 0 \Longrightarrow \Lambda_0 = \{(0,0,0)\} \\
|\alpha| &= 1 \Longrightarrow \Lambda_1 = \{(1,0,0),(0,1,0),(0,0,1)\} \\
|\alpha| &= 2 \Longrightarrow \Lambda_2 = \{(2,0,0),(0,2,0),(0,0,2),(1,1,0),(0,1,1),(1,0,1)\} \\
|\alpha| &= 3 \Longrightarrow \Lambda_3 = \{(3,0,0),(0,3,0),(0,0,3),(2,1,0),(2,0,1),(0,2,1), \\
&\qquad (1,2,0),(0,1,2),(1,0,2),(1,1,1)\}
\end{aligned}$$

Then P has the form

$$P(z_1, z_2, z_3) = \sum_{i=0}^{3} \sum_{\beta \in \Lambda_i} a_\beta z_1^{\beta_1} z_3^{\beta_2} z_3^{\beta_3}$$

where β_i are the components of the triple β. Written out in detail, a typical polynomial would be

$$\begin{aligned}
P(z_1, z_2, z_3) &= a_{(0,0,0)} + a_{(1,0,0)} z_1 + a_{(0,1,0)} z_2 + a_{(0,0,1)} z_3 \\
&+ a_{(2,0,0)} z_1^2 + a_{(0,2,0)} z_2^2 + a_{(0,0,2)} z_3^2 + a_{(1,1,0)} z_1 z_2 + a_{(0,1,1)} z_2 z_3 + a_{(1,0,1)} z_1 z_3 \\
&+ a_{(3,0,0)} z_1^3 + a_{(0,3,0)} z_2^3 + a_{(0,0,3)} z_3^3 + a_{(2,1,0)} z_1^2 z_2 \\
&+ a_{(2,0,1)} z_1^2 z_3 + a_{(1,2,0)} z_1 z_2^2 + a_{(0,2,1)} z_2^2 z_3 + a_{(1,0,2)} z_1 z_3^2 + a_{(0,1,2)} z_2 z_3^2 \\
&+ a_{(1,1,1)} z_1 z_2 z_3
\end{aligned}$$

A homogeneous polynomial of degree m, Q_m, is one collection of these sums:

$$\begin{aligned}
Q_0(z_1, z_2, z_3) &= a_{(0,0,0)} \\
Q_1(z_1, z_2, z_3) &= a_{(1,0,0)} z_1 + a_{(0,1,0)} z_2 + a_{(0,0,1)} z_3 \\
Q_2(z_1, z_2, z_3) &= a_{(2,0,0)} z_1^2 + a_{(0,2,0)} z_2^2 + a_{(0,0,2)} z_3^2 + a_{(1,1,0)} z_1 z_2 + a_{(0,1,1)} z_2 z_3 + a_{(1,0,1)} z_1 z_3 \\
Q_3(z_1, z_2, z_3) &= a_{(3,0,0)} z_1^3 + a_{(0,3,0)} z_2^3 + a_{(0,0,3)} z_3^3 + a_{(2,1,0)} z_1^2 z_2 \\
&+ a_{(2,0,1)} z_1^2 z_3 + a_{(1,2,0)} z_1 z_2^2 + a_{(0,2,1)} z_2^2 z_3 + a_{(1,0,2)} z_1 z_3^2 + a_{(0,1,2)} z_2 z_3^2 \\
&+ a_{(1,1,1)} z_1 z_2 z_3
\end{aligned}$$

Thus, $Q_m(z_1, z_2, z_3) = \sum_{\beta \in \Lambda_m} a_\beta \, z_1^{\beta_1} z_2^{\beta_2} z_3^{\beta_3}$ for $m = 0, 1, 2, 3$.

We say Q_m is elliptic if $Q_m(z_1, z_2, z_3) \neq 0$ on $\Re^3 \setminus (0,0,0)$.

Exercise 10.4.1 *If* Q_m *is elliptic, show this forces* m *to be even.*
Hint*: Assume* Q_m *is elliptic with* m *odd and let*

$$f(x_1, x_2) = (Q_m(x_1, x_2, 0), Q_m(x_1, x_2, 0))$$

$$\tilde{f}(x_1, x_2) = Q_m(x_1, x_2, \xi)$$

- *Apply Borsuk's Theorem to show* $d_B(f, D, (0,0))$ *is odd for suitable* D *implying* $f(x_1, x_2) = (0,0)$ *for some* (x_1^*, x_2^*).
- *Let's focus on* $m = 3$. *Note* $Q_3(x_1, x_2, \xi) = Q_3(x_1, x_2, 0) + \xi Q_2(x_1, x_2, \xi)$ *where* Q_2 *is homogeneous of degree 2 using coefficients from the* Q_3 *polynomial. If* $(x_1^*, x_2^*) \neq (0,0)$, *we have* $Q_3(x_1^*, x_2^*, 0) = 0$ *violating the fact that* Q_3 *is elliptic. On the other hand, if* $(x_1^*, x_2^*) \neq (0,0)$,

$$\tilde{f}(x_1^*, x_2^*) = Q_3(x_1^*, x_2^*, 0) + \xi\, Q_2(x_1^*, x_2^*, \xi) = \xi\, Q_2(x_1^*, x_2^*, \xi)$$

If $Q_2(x_1^*, x_2^*) = 0$, *we have* $Q_3(x_1^*, x_2^*, \xi) = 0$ *for all* $\xi \neq 0$, *showing* Q_3 *is not elliptic. If* $Q_2(x_1^*, x_2^*) \neq 0$, *we can conclude the same.* m *cannot be* 3.

Exercise 10.4.2 *Explain why* Q_1 *cannot be elliptic.*

Exercise 10.4.3 *Show* $P(x,y) = x + iy$ *is elliptic. Here, we allow the coefficients to be complex.*

10.5 Further Properties of Brouwer Degree

There are many more interesting properties of Brouwer degree. We will sample a few here.

Theorem 10.5.1 Excision Property for Brouwer Degree

Let $\boldsymbol{D} \subset \Re^n$ *be an open and bounded subset. Let* $f : \overline{\boldsymbol{D}} \to \Re^n$ *be continuous. Further, let* $\boldsymbol{p} \notin f(\partial \boldsymbol{D})$, $K \subset \overline{\boldsymbol{D}}$ *be closed and* $\boldsymbol{p} \notin f(\partial \boldsymbol{K})$. *Then,* $d_B(f, \boldsymbol{D}, \boldsymbol{p}) = d_B(f, \boldsymbol{D} \setminus \boldsymbol{K}, \boldsymbol{p})$.

Proof 10.5.1
Let $f_n \to f$ *uniformly on* $\overline{\boldsymbol{D}}$, $f_n \in C^1(\boldsymbol{D}) \cap C(\overline{\boldsymbol{D}})$ *with* $\boldsymbol{p} \notin f_n(\partial \boldsymbol{D})$, $\boldsymbol{p} \notin f_n(\boldsymbol{K})$ *for all* n. *Let* $\boldsymbol{B_n} = \{\boldsymbol{x} | J f_n(\boldsymbol{x}) = 0\}$. *For convenience, let's assume* $\boldsymbol{p} \notin f_n(\boldsymbol{B_n})$ *too.*

Then, for sufficiently large n,

$$d_B(f, \boldsymbol{D}, \boldsymbol{p}) = deg_B(f_n, \boldsymbol{D}, \boldsymbol{p}) = \sum_{\boldsymbol{x} \in f_n^{-1}(\boldsymbol{p})} sign\ J f_n(\boldsymbol{x})$$

But $f_n^{-1}(\boldsymbol{p})$ *not in* $\boldsymbol{K}$ *implies*

$$\sum_{\boldsymbol{x} \in f_n^{-1}(\boldsymbol{p})} sign\ J f_n(\boldsymbol{x}) = = \sum_{\boldsymbol{x} \in f_n^{-1}(\boldsymbol{p}),\ \boldsymbol{x} \notin \boldsymbol{D} \setminus \boldsymbol{K}} sign\ J f_n(\boldsymbol{x}) = d_B(f_n, \boldsymbol{D} \setminus \boldsymbol{K}, \boldsymbol{p}) = d_B(f, \boldsymbol{D} \setminus \boldsymbol{K}, \boldsymbol{p})$$

Now let $\boldsymbol{B} = \{\boldsymbol{x} | J f(\boldsymbol{x}) = 0\}$. *We then let* $(\boldsymbol{p_n})$ *be a sequence with* $\boldsymbol{p_n} \to \boldsymbol{p}$ *with* $f^{-1}(\boldsymbol{p_n}) \cap \boldsymbol{B} = \emptyset$ *for all* n. *Then, we know there is* N_0, *so that if* $n > N_0$,

$$d_B(f_n, \boldsymbol{D}, \boldsymbol{p_k}) = d_B(f_{N_0}, \boldsymbol{D}, \boldsymbol{p_k})$$

and as $k \to \infty$, *we find*

$$d_B(f_n, \boldsymbol{D}, \boldsymbol{p}) = d_B(f_{N_0}, \boldsymbol{D}, \boldsymbol{p})$$

and then as $n \to \infty$, *we find*

$$d_B(f, \boldsymbol{D}, \boldsymbol{p}) = d_B(f_{N_0}, \boldsymbol{D}, \boldsymbol{p})$$

A little thought shows you that, with a suitable change in notation, the assumption we made for convenience ($\boldsymbol{p} \notin f_n(\boldsymbol{B_n})$) is simply a reflection of the above limiting argument. ■

What about the degree of a product?

Theorem 10.5.2 Brouwer Degree of a Product

Let $\boldsymbol{D_1}$ be open and bounded in $\Re^n$ and $\boldsymbol{D_2}$ be open and bounded subsets in $\Re^m$. Let $f : \overline{\boldsymbol{D_1}} \rightarrow \Re^n$ and $g : \overline{\boldsymbol{D_2}} \rightarrow \Re^m$ be continuous. Then, whenever the Brouwer degrees are defined,

$$d_B(f \times g, \boldsymbol{D_1} \times \boldsymbol{D_2}, (\boldsymbol{p}, \boldsymbol{q})) \quad = \quad d_B(f, \boldsymbol{D_1}, \boldsymbol{p})\, d_B(g, \boldsymbol{D_2}, \boldsymbol{q})$$

Proof 10.5.2
The remarks we made in the proof of the previous theorem, show it is enough to prove this result when $\boldsymbol{p} \notin f(\boldsymbol{B_1})$, $\boldsymbol{q} \notin g(\boldsymbol{B_2})$ for f is in $C^1(\boldsymbol{D_1}) \cap C(\overline{\boldsymbol{D_1}})$ and g is in $C^1(\boldsymbol{D_2}) \cap C(\overline{\boldsymbol{D_2}})$. Here, $\boldsymbol{B_1} = \{\boldsymbol{x} | Jf(\boldsymbol{x}) = 0\}$ and $\boldsymbol{B_2} = \{\boldsymbol{x} | Jg(\boldsymbol{x}) = 0\}$ and $(\boldsymbol{p}, \boldsymbol{q}) \notin (f \times g)(\partial \boldsymbol{D_1} \times \boldsymbol{D_2})$. Also, remember, $\partial \boldsymbol{D_1} \times \partial \boldsymbol{D_2} \neq \partial \boldsymbol{D_1} \times \partial \boldsymbol{D_2}$, in general. Just draw a few pictures in $\Re^2$ to see this.

Now

$$d_B(f \times g, \boldsymbol{D_1} \times \boldsymbol{D_2}, (\boldsymbol{p}, \boldsymbol{q})) \quad = \sum_{(\boldsymbol{x},\boldsymbol{y}) \in (f \times g)^{-1}(\boldsymbol{p},\boldsymbol{q}),\ (\boldsymbol{x},\boldsymbol{y}) \in \Re^n \times \Re^m} sign\ J(f \times g)(\boldsymbol{x}, \boldsymbol{y})$$

But, organizing the components of $\Re^n \times \Re^m$ as x_1 to x_{n+m}, we have

$$J(f \times g)(\boldsymbol{x}, \boldsymbol{y}) \quad = \quad det \begin{bmatrix} \frac{\partial f_1}{\partial x_1} & \cdots & \frac{\partial f_1}{\partial x_n} & 0 & \cdots & 0 \\ \vdots & \vdots & \vdots & \vdots & \vdots & \vdots \\ \frac{\partial f_n}{\partial x_1} & \cdots & \frac{\partial f_n}{\partial x_n} & 0 & \cdots & 0 \\ 0 \ \cdots & 0 & \frac{\partial g_1}{\partial x_{n+1}} & \cdots & & \frac{\partial g_1}{\partial x_{n+m}} \\ \vdots & \vdots & \vdots & \vdots & & \vdots \\ 0 \ \cdots & 0 & \frac{\partial g_m}{\partial x_{n+1}} & \cdots & & \frac{\partial g_m}{\partial x_{n+m}} \end{bmatrix} = Jf(\boldsymbol{x})\, Jg(\boldsymbol{y})$$

where we identify $\boldsymbol{x}$ with the components x_1 to x_n and $\boldsymbol{y}$, with components x_{n+1} to x_{n+m}. Hence, it follows that

$$\begin{aligned} d_B(f \times g, \boldsymbol{D_1} \times \boldsymbol{D_2}, (\boldsymbol{p}, \boldsymbol{q})) &= \sum_{(\boldsymbol{x},\boldsymbol{y}) \in (f \times g)^{-1}(\boldsymbol{p},\boldsymbol{q}),\ (\boldsymbol{x},\boldsymbol{y}) \in \Re^n \times \Re^m} sign(\ Jf(\boldsymbol{x})\, Jg(\boldsymbol{y})) \\ &= \sum_{(\boldsymbol{x},\boldsymbol{y}) \in (f \times g)^{-1}(\boldsymbol{p},\boldsymbol{q}),\ (\boldsymbol{x},\boldsymbol{y}) \in \Re^n \times \Re^m} sign(\ Jf(\boldsymbol{x}))\, sign(Jg(\boldsymbol{y})) \end{aligned}$$

But,

$$(\boldsymbol{x}, \boldsymbol{y}) \in (f \times g)^{-1}(\boldsymbol{p}, \boldsymbol{q}) \quad \Longrightarrow \quad (f \times g)(\boldsymbol{x}, \boldsymbol{y}) = (\boldsymbol{p}, \boldsymbol{q}) \Longrightarrow f(\boldsymbol{x}) = \boldsymbol{p},\ g(\boldsymbol{y}) = \boldsymbol{q}$$

Hence, $\boldsymbol{x} \in f^{-1}(\boldsymbol{p})$ and $\boldsymbol{y} \in g^{-1}(\boldsymbol{q})$. Conversely, if $\boldsymbol{x} \in f^{-1}(\boldsymbol{p})$ and $\boldsymbol{y} \in g^{-1}(\boldsymbol{q})$, then $(\boldsymbol{x}, \boldsymbol{y}) \in (f \times g)^{-1}(\boldsymbol{p}, \boldsymbol{q})$. Thus,

$$\{(\boldsymbol{x}, \boldsymbol{y}) \in (f \times g)^{-1}(\boldsymbol{p}, \boldsymbol{q})\} \quad = \quad \{\boldsymbol{x} \in f^{-1}(\boldsymbol{p}),\ \boldsymbol{y} \in g^{-1}(\boldsymbol{q})\}$$

We can then say

$$\begin{aligned} d_B(f \times g, \boldsymbol{D_1} \times \boldsymbol{D_2}, (\boldsymbol{p}, \boldsymbol{q})) &= \sum_{\boldsymbol{x} \in f^{-1}(\boldsymbol{p}),\, \boldsymbol{y} \in g^{-1}(\boldsymbol{q})} sign(\, Jf(\boldsymbol{x}))\, sign(Jg(\boldsymbol{y})) \\ &= \sum_{\boldsymbol{x} \in f^{-1}(\boldsymbol{p})} sign(\, Jf(\boldsymbol{x})) \times \sum_{\boldsymbol{y} \in g^{-1}(\boldsymbol{q})} sign(Jg(\boldsymbol{y})) \\ &= d_B(f, \boldsymbol{D_1}, \boldsymbol{p})\, d_B(g, \boldsymbol{D_2}, \boldsymbol{q}) \end{aligned}$$

The more complicated cases where f and g are just in $C(\overline{\boldsymbol{D_1}})$ and $C(\overline{\boldsymbol{D_2}})$ are then handled using the limiting arguments we mentioned in the last two proofs. ■

Now that we have proven the standard product property for Brouwer degree on a product space, let's look at the degree on compositions of maps.

Theorem 10.5.3 Multiplicative Property for Brouwer Degree

Let $\boldsymbol{D}$ be open and bounded in $\Re^n$. Let $f : \overline{\boldsymbol{D}} \rightarrow \Re^n$ and $g : \Re^n \rightarrow \Re^n$ be continuous. Assume $\mathbf{p} \notin (g \circ f)(\partial \boldsymbol{D})$. Then

$$d_B(g \circ f, \boldsymbol{D}, \mathbf{p}) = \sum_{y_j \in g^{-1}(\mathbf{p})} sign\ Jg(\boldsymbol{y_j}) d_B(f, \boldsymbol{D}, \boldsymbol{y_j})$$

Proof 10.5.3
As usual, it is enough to consider the case where $f \in C^1(\boldsymbol{D}) \cap C(\overline{\boldsymbol{D}})$ and $g \in C^1(\Re^n) \cap C(\Re^n)$ and $\mathbf{p} \notin (g \circ f)(\boldsymbol{B})$ where

$$\boldsymbol{B} = \{\boldsymbol{x} \in \Re^n | Jg \circ f(\boldsymbol{x}) = \mathbf{0}\}$$

Since we assume $\mathbf{p} \notin (g \circ f)(\partial \boldsymbol{D})$, we know

$$d_B(g \circ f, \boldsymbol{D}, \mathbf{p}) = \sum_{\boldsymbol{x} \in (g \circ f)^{-1}(\mathbf{p})} sign\ J(g \circ f)(\boldsymbol{x}) = \sum_{\boldsymbol{x} \in f^{-1}(g^{-1}(\mathbf{p}))} sign\ J(g \circ f)(\boldsymbol{x})$$

Now

$$J(g \circ f)(\boldsymbol{x}) = det \begin{bmatrix} \sum_{i=1}^n \frac{\partial g_1}{\partial f_i} \frac{\partial f_i}{\partial x_i} & \cdots & \sum_{i=1}^n \frac{\partial g_1}{\partial f_i} \frac{\partial f_i}{\partial x_n} \\ \vdots & & \vdots \\ \sum_{i=1}^n \frac{\partial g_j}{\partial f_i} \frac{\partial f_i}{\partial x_1} & \cdots & \sum_{i=1}^n \frac{\partial g_j}{\partial f_i} \frac{\partial f_i}{\partial x_n} \\ \vdots & & \vdots \\ \sum_{i=1}^n \frac{\partial g_n}{\partial f_i} \frac{\partial f_i}{\partial x_1} & \cdots & \sum_{i=1}^n \frac{\partial g_n}{\partial f_i} \frac{\partial f_i}{\partial x_n} \end{bmatrix}$$

Then, denoting the matrix above by B with entries $B_{ij} = (g_{jk} f_{kj})$ where $g_{jk} f_{kj} = \sum_{i=1}^n \frac{\partial g_j}{\partial f_k} \frac{\partial f_k}{\partial x_j}$, we have

$$det B = det(g_{jk} f_{kj}) = \sum_{\sigma \in S_n} sign\, \sigma\, \sigma_{B_{1,\sigma(1)}, \ldots, B_{n,\sigma(n)}}$$

This representation of the determinant of a matrix is covered in many texts; our take on this discussion is in (Peterson (101) 2020). The term S_n refers to the permutation of n symbols group. Feel free to leave our argument here to refresh your understanding of this idea. It is very important to write *down an abstract template for the computation of the determinant here. Now, we will go through a forbiddingly long series of manipulations.*

$$\begin{aligned} det B &= \sum_{\sigma \in S_n} sign\, \sigma\, g_{1i} f_{i,\sigma(1)} \cdots g_{ni} f_{i,\sigma(n)} \\ &= \sum_{\sigma \in S_n} \sum_{i_1} \cdots \sum_{i_n} sign\, \sigma \left(g_{1,i_1} \cdots g_{n,i_n} \right) \left(f_{i_1,\sigma(1)} \cdots f_{i_n,\sigma(n)} \right) \\ &= \sum_{\sigma \in S_n} \sum_{(i_1,\ldots,i_n)=\tau \in S_n} sign\, \sigma \left(g_{1,\tau(1)} \cdots g_{n,\tau(n)} \right) \left(f_{\tau(1),\sigma(1)} \cdots f_{\tau(n),\sigma(n)} \right) \end{aligned}$$

We can continue this manipulation using permutations to find

$$\begin{aligned} det B &= \sum_{\sigma \in S_n} \sum_{\tau \in S_n} sign\, \sigma \left(sign\, \tau\, g_{1,\tau(1)} \cdots g_{n,\tau(n)} \right) \left(sign\, \tau\, f_{\tau(1),\sigma(1)} \cdots f_{\tau(n),\sigma(n)} \right) \\ &= \sum_{\tau \in S_n} \left(sign\, \tau\, g_{1,\tau(1)} \cdots g_{n,\tau(n)} \right) \sum_{\sigma \in S_n} \left(sign\, \tau\, sign\, \sigma\, f_{\tau(1),\sigma(1)} \cdots f_{\tau(n),\sigma(n)} \right) \\ &= \sum_{\tau \in S_n} \left(sign\, \tau\, g_{1,\tau(1)} \cdots g_{n,\tau(n)} \right) Jf(\boldsymbol{x}) \\ &= Jg(f(\boldsymbol{x}))\, Jf(\boldsymbol{x}) \end{aligned}$$

Hence,

$$\begin{aligned} d_B(gf, \boldsymbol{D}, \boldsymbol{p}) &= \sum_{\boldsymbol{x} \in (g \circ f)^{-1}(\boldsymbol{p})} sign\, Jg(f(\boldsymbol{x}))\, sign\, Jf(\boldsymbol{x}) \\ &= \sum_{\boldsymbol{x} \in f^{-1} g^{-1}(\boldsymbol{p})} sign\, Jg(f(\boldsymbol{x}))\, sign\, Jf(\boldsymbol{x}) \\ &= \sum_{f(\boldsymbol{x}) \in g^{-1}(\boldsymbol{p})} sign\, Jg(f(\boldsymbol{x}))\, sign\, Jf(\boldsymbol{x}) \\ &= \sum_{\boldsymbol{y} \in g^{-1}(\boldsymbol{p}), \boldsymbol{x} \in f^{-1}(\boldsymbol{y})} sign\, Jg(\boldsymbol{y})\, sign\, Jf(\boldsymbol{x}) \end{aligned}$$

Now, $\boldsymbol{x} \in (g \circ f)^{-1}(\boldsymbol{p}) = f^{-1}g^{-1}(\boldsymbol{p})$ implies $\boldsymbol{x} \in f^{-1}(\boldsymbol{y})$ where $\boldsymbol{y} \in g^{-1}(\boldsymbol{p})$. Since $\mathbf{p} \in \overline{\boldsymbol{D}}$, we have $\boldsymbol{y} = g^{-1}(\boldsymbol{p})$ implies $g(\boldsymbol{y}) \in \overline{\boldsymbol{D}}$ which is compact. So we know the set $g^{-1}(\boldsymbol{p})$ must be finite by our comments in the original construction of Brouwer degree. So we can rewrite our conclusion as

$$\begin{aligned} d_B(g \circ f, \boldsymbol{D}, \boldsymbol{p}) &= \sum_{\boldsymbol{y_j} \in g^{-1}(\boldsymbol{p})} sign\, Jg(\boldsymbol{y_j}) \sum_{\boldsymbol{x} \in f^{-1}(\boldsymbol{y_j})} sign\, Jf(\boldsymbol{x}) \\ &= \sum_{\boldsymbol{y_j} \in g^{-1}(\boldsymbol{p})} sign\, Jg(\boldsymbol{y_j})\, d_B(f, \boldsymbol{D}, \boldsymbol{y_j}) \end{aligned}$$

To extend this to the case where f and g are merely continuous, is the same argument we have used before. ■

The range of a continuous function on a subset of $\Re^n$ divides $\Re^n$ into components. What can we say about the degree of such a function on a component?

Theorem 10.5.4 Brouwer Degree is Constant on Components

Let $\boldsymbol{D}$ be a bounded open subset of $\Re^n$. Let $f : \overline{\boldsymbol{D}} \rightarrow \Re^n$ be continuous. Assume $\mathbf{p}$ and $\boldsymbol{q}$ are in the same component of $\Re^n \setminus f(\partial \boldsymbol{D})$. Then $d_B(f\boldsymbol{D}, \mathbf{p}) = d_B(f, \boldsymbol{D}, \boldsymbol{q})$; i.e. Brouwer degree is constant on a component of $\Re^n \setminus f(\partial \boldsymbol{D})$.

Proof 10.5.4

The set $\Re^n \setminus f(\partial \boldsymbol{D})$ can be split into connected open subsets which are the components of $\Re^n \setminus f(\partial \boldsymbol{D})$. Since $\partial \boldsymbol{D}$ is compact, $f(\partial \boldsymbol{D})$ is compact and so closed. Hence, $\Re^n \setminus f(\partial \boldsymbol{D})$ is an open set. Let Δ be a component of $\Re^n \setminus f(\partial \boldsymbol{D})$. If $\mathbf{p}$ and $\boldsymbol{q}$ are both in Δ, then there is a continuous curve ϕ with $\phi(t) \in \Delta$ for all $0 \leq t \leq 1$ with $\phi(0) = \mathbf{p}$ and $\phi(1) = \boldsymbol{q}$. Define the homotopy H by $H(\boldsymbol{x}, t) = f(\boldsymbol{x}) - \phi(t)$. Since $\mathbf{p}$ and $\boldsymbol{q}$ are not in $\partial \boldsymbol{D}$, there is a $\delta > 0$ so that $\|f(\boldsymbol{x}) - \mathbf{p}\| \geq \delta$ and $\|f(\boldsymbol{x}) - \boldsymbol{q}\| \geq \delta$ for all $\boldsymbol{x} \in \partial \boldsymbol{D}$. Hence, $\|H(\boldsymbol{x}, t)\| \geq \inf_{0 \leq t \leq 1} dist(\phi(t), f(\partial \boldsymbol{D}))$. Since $[0, 1]$ is a compact set, $\phi([0, 1])$ is a compact set and so there is a $\nu > 0$ so that $dist(\phi([0, 1]), f(\partial \boldsymbol{D})) = \nu > 0$. We see $\|H(\boldsymbol{x}, t)\| \geq \nu > 0$ on $\partial \boldsymbol{D} \times [01,]$. Note $\mathbf{0} \notin H(\partial \boldsymbol{D} \times [0, 1])$ as if so, $f(\boldsymbol{x}) = \phi(t)$ for an $\boldsymbol{x} \in \partial \boldsymbol{D}$ and a $t \in [0, 1]$ which is not possible. Thus, invariance under homotopy tells us

$$
\begin{aligned}
d_B(H(\cdot, 0), \boldsymbol{D}, \mathbf{0}) &= d_B(H(\cdot, 0), \boldsymbol{D}, \mathbf{0}) \Longrightarrow d_B(f - \phi(0), \boldsymbol{D}, \mathbf{0}) = d_B(f - \phi(1), \boldsymbol{D}, \mathbf{0}) \\
&\Longrightarrow d_B(f - \mathbf{p}, \boldsymbol{D}, \mathbf{0}) = d_B(f - \boldsymbol{q}, \boldsymbol{D}, \mathbf{0}) \Longrightarrow d_B(f, \boldsymbol{D}, \mathbf{p}) = d_B(f, \boldsymbol{D}, \boldsymbol{q})
\end{aligned}
$$

■

Let's look at Theorem 10.5.3 again. Recall this says:

Theorem The Multiplicative Property for Brouwer Degree

Let $\boldsymbol{D}$ be open and bounded in $\Re^n$. Let $f : \overline{\boldsymbol{D_1}} \rightarrow \Re^n$ and $g : \Re^n \rightarrow \Re^n$ be continuous. Assume $\mathbf{p} \notin (g \circ f)(\partial \boldsymbol{D})$. Then

$$
d_B(g \circ f, \boldsymbol{D}, \mathbf{p}) = \sum_{\boldsymbol{y_j} \in g^{-1}(\mathbf{p})} sign\ Jg(\boldsymbol{y_j}) d_B(f, \boldsymbol{D}, \boldsymbol{y_j})
$$

Let $\{C_\alpha\}$ be the collection of components of $\Re^n \setminus f(\partial \boldsymbol{D})$. Let's think about components more carefully. Since $\partial \boldsymbol{D}$ is bounded and closed, it is compact. Hence, there is an $R > 0$ so that $f(\partial \boldsymbol{D}) \subset \overline{B(\mathbf{0}, R)}$. Then $\overline{B(\mathbf{0}, R)}^C$ is an unbounded open set that is contained in $(f(\partial \boldsymbol{D}))^C$. This must be contained in a component C_∞. This tells us the component C_∞ of $(f(\partial \boldsymbol{D}))^C$ must be unbounded. The other components must be inside $(f(\partial \boldsymbol{D}))^C \cap \overline{B(\mathbf{0}, R)}$.

Now if C_α is a bounded component, if $C_\alpha \cap \overline{B(\mathbf{0}, R)}^C \neq \emptyset$, then $C\alpha \cap C_\infty \neq \emptyset$. Since they are both components, this would force them to be the same component. Hence $C_\alpha = C_\infty$ would be unbounded which is a contradiction. Hence, if C_α is a bounded component, it is a subset of $\overline{B(\mathbf{0}, R)}$. Then,

$$
\overline{B(\mathbf{0}, R)} \subset \cup_\alpha C_\alpha \cup B(\mathbf{0}, R+1) \setminus B(\mathbf{0}, R+1)
$$

where each C_α is a bounded component. But since $\overline{B(\mathbf{0}, R)}$ is compact, we must have $(f(\partial \boldsymbol{D}))^C \subset \cup_{i=1}^n C_i \cup C_\infty$ where (C_i) for $1 \leq i \leq p$ is the finite subcover guaranteed to exist by the compactness of $\overline{B(\mathbf{0}, R)}$. We conclude $(f(\partial \boldsymbol{D}))^C$ has a finite number of bounded components and one unbounded component.

We know Brouwer degree is constant on components. For the unbounded component of $(f(\partial \boldsymbol{D}))^C$, we can choose $\boldsymbol{p}$ outside of $\overline{B(\mathbf{0}, R)}^C$ and then $f^{-1}(\boldsymbol{p}) = \emptyset$, which tells us the Brouwer degree is zero. Thus, the Brouwer degree on the unbounded component is zero.

We have for the components (C_j) for $1 \leq j \leq p$

$$\begin{aligned} d_B(g \circ f, \boldsymbol{D}, \boldsymbol{p}) &= \sum_{y_j \in g^{-1}(\boldsymbol{p})} sign\ Jg(\boldsymbol{y_j}) d_B(f, \boldsymbol{D}, \boldsymbol{y_j}) \\ &= \sum_{y_j \in g^{-1}(\boldsymbol{p}) \cap C_j} sign\ Jg(\boldsymbol{y_j}) d_B(f, \boldsymbol{D}, \boldsymbol{y_j}) \end{aligned}$$

Since $\boldsymbol{y_j} \in C_j$, we know the Brouwer degree is constant on C_j and so we have

$$\begin{aligned} d_B(g \circ f, \boldsymbol{D}, \boldsymbol{p}) &= \sum_{y_j \in g^{-1}(\boldsymbol{p}) \cap C_j} sign\ Jg(\boldsymbol{y_j}) d_B(f, \boldsymbol{D}, C_j) \\ &= d_B(f, \boldsymbol{D}, C_j) \sum_{y_j \in g^{-1}(\boldsymbol{p}) \cap C_j} sign\ Jg(\boldsymbol{y_j}) \\ &= sum_{j=1}^{p}\, d_B(g, C_j, \boldsymbol{p})\, d_B(f, \boldsymbol{D}, C_j) \end{aligned}$$

We can state this as a theorem.

Theorem 10.5.5 Component Multiplicative Property for Brouwer Degree

Let $\boldsymbol{D}$ be open and bounded in $\Re^n$. Let $f : \overline{\boldsymbol{D_1}} \rightarrow \Re^n$ and $g : \Re^n \rightarrow \Re^n$ be continuous. Assume $\boldsymbol{p} \notin (g \circ f)(\partial \boldsymbol{D})$. Then, if C_j, $1 \leq j \leq p$ are the bounded components of $(f(\partial \boldsymbol{D}))^C$, we have

$$d_B(g \circ f, \boldsymbol{D}, \boldsymbol{p}) = \sum_{j=1}^{p} d_B(g, C_j, \boldsymbol{p})\, d_B(f, \boldsymbol{D}, C_j)$$

Proof 10.5.5
We have worked out the details of this argument in the discussion above. ■

You can read more about this point of view in (Schwartz (114) 1968). Another result is a sort of projection result, which is called a reduction theorem.

Theorem 10.5.6 Brouwer Degree Reduction

Let $\boldsymbol{D} \subset \Re^n$ be open and bounded. Let $\Re^m \subset \Re^n$ in the standard way: i.e. $\boldsymbol{x} = (x_1, \ldots, x_m) \in \Re^m$ is identified with $\boldsymbol{x} = (x_1, \ldots, x_m, 0, \ldots, 0) \in \Re^n$. Let $f : \overline{\boldsymbol{D}} \rightarrow \Re^m$ be continuous and $g : \overline{\boldsymbol{D}} \rightarrow \Re^n$ be the map given by $g(\boldsymbol{x}) = \boldsymbol{x} + f(\boldsymbol{x})$. Then, for all $\boldsymbol{p} \in \Re^m$, $\boldsymbol{p} \notin g(\partial \boldsymbol{D})$,

$$d_B(g, \boldsymbol{D}, \boldsymbol{p}) = d_B(g|_{\Re^m \cap \overline{\boldsymbol{D}}}, \boldsymbol{D} \cap \Re^m, \boldsymbol{p})$$

Proof 10.5.6
Since $f(\boldsymbol{x}) \in \Re^m$, for all $\boldsymbol{x} \in \Re^m \cap \overline{\boldsymbol{D}}$, $g(\boldsymbol{x}) = \boldsymbol{x} = f(\boldsymbol{x}) \in \Re^m$. So $g|_{\Re^m \cap \overline{\boldsymbol{D}}}$ is well-defined.

It suffices to assume $f \in C^1(D) \cap C(\overline{\boldsymbol{D}})$ and $\boldsymbol{p} \notin f(\boldsymbol{B})$ where $\boldsymbol{B} = \{\boldsymbol{x} | Jf(\boldsymbol{x}) = \boldsymbol{0}\}$. Consider $\boldsymbol{x} \in g^{-1}(\boldsymbol{p})$. Then, $g(\boldsymbol{x}) = \boldsymbol{p}$ or $\boldsymbol{x} + f(\boldsymbol{x}) = \boldsymbol{p}$. Hence, if $\boldsymbol{p} \in \Re^m$, $\boldsymbol{x} = \boldsymbol{p} - f(\boldsymbol{x}) \in \Re^m$. Since $g^{-1}(\boldsymbol{p}) \subset \overline{\boldsymbol{D}}$ and $\boldsymbol{p} \notin g(\partial \boldsymbol{D})$ by assumption, we have $\boldsymbol{x}$ is not in $\partial \boldsymbol{D}$. We conclude $\boldsymbol{x} \in \boldsymbol{D}$ and so

$\boldsymbol{x} \in \boldsymbol{D} \cap \Re^m$. Now, $d_B(g, \boldsymbol{D}, \boldsymbol{p}) = \sum_{\boldsymbol{x} \in g^{-1}(\boldsymbol{p})} sign\ Jg(\boldsymbol{x})$. We want to show

$$d_B(g, \boldsymbol{D}, \boldsymbol{p}) = \sum_{\boldsymbol{x} \in (g|_{\Re^m \cap \boldsymbol{D}})^{-1}(\boldsymbol{p})} sign\ Jg|_{\Re^m \cap \boldsymbol{D}}(\boldsymbol{x})$$

It is easy to see

$$(g|_{\Re^m \cap \boldsymbol{D}})^{-1}(\boldsymbol{p}) = g^{-1}(\boldsymbol{p})$$

by our earlier remarks. So the only difference that might arise would be in the sign of the Jacobians. But

$$\begin{aligned} sign\ Jg(\boldsymbol{x}) &= \frac{Jg(\boldsymbol{x})}{|Jg(\boldsymbol{x})|} \\ sign\ Jg|_{\Re^m \cap \boldsymbol{D}}(\boldsymbol{x}) &= \frac{Jg|_{\Re^m \cap \boldsymbol{D}}(\boldsymbol{x})}{|Jg|_{\Re^m \cap \boldsymbol{D}}(\boldsymbol{x})|} \end{aligned}$$

However, for all $\boldsymbol{x} \in \Re^m \cap \boldsymbol{D}$,

$$Jg|_{\Re^m \cap \boldsymbol{D}}(\boldsymbol{x}) = Jg(\boldsymbol{x})$$

Hence, the signs are equal and

$$d_B(g, \boldsymbol{D}, \boldsymbol{p}) = d_B(g|_{\Re^m \cap \boldsymbol{D}}, \boldsymbol{D} \cap \Re^m, \boldsymbol{p})$$

The more general results for f continuous and $\boldsymbol{p} \in f(\boldsymbol{B})$ are then handled in the usual way. ■

We should mention what we already know formally: the Brouwer degree for the identity map.

Theorem 10.5.7 Brouwer Degree of the Identity Map

Let $\boldsymbol{D} \subset \Re^n$ be open and bounded. Let $I : \overline{\boldsymbol{D}} \to \Re^n$ be the identity map. Then, if $\boldsymbol{p} \notin \partial \boldsymbol{D}$,

$$d_B(I, \boldsymbol{D}, \boldsymbol{p}) = \begin{cases} 0, & \boldsymbol{p} \notin \boldsymbol{D} \\ 1, & \boldsymbol{p} \in \boldsymbol{D} \end{cases}$$

Proof 10.5.7
Immediate. ■

Homework

Exercise 10.5.1 *In the proof of Theorem 10.5.2, provide the details for the more complicated cases where f and g are just in $C(\overline{\boldsymbol{D_1}})$ and $C(\overline{\boldsymbol{D_2}})$ using standard limiting arguments.*

Exercise 10.5.2 *In the proof of Theorem 10.5.3, provide the details to extend the argument to the case where f and g are merely continuous.*

Exercise 10.5.3 *In the proof of Theorem 10.5.3, provide the details to extend the argument for f continuous and $\boldsymbol{p} \in f(\boldsymbol{B})$.*

10.6 Extending Brouwer Degree to Finite Dimensional Normed Linear Spaces

Let $(X, \|\cdot\|)$ be a normed linear space with $dim(X) = n < \infty$. Let $\{\boldsymbol{a_1}, \ldots, \boldsymbol{a_n}\}$ be a basis for X and $\{\boldsymbol{b_1}, \ldots, \boldsymbol{b_n}\}$ be a basis for $\Re^n$. Define a mapping $\phi : X \to \Re^n$ by $\phi(\boldsymbol{a_i}) = b_i$ for $1 \leq i \leq n$. Then, given the representation $\boldsymbol{x} = \sum_{i=1}^n \alpha_i a_i \in X$, extend ϕ linearly by defining $\phi(\boldsymbol{x}) = \sum_{i=1}^n \alpha_i \phi(\boldsymbol{a_i}) = \sum_{i=1}^n \alpha_i b_i$. Then, it is straightforward to see ϕ is a bijection from X to $\Re^n$. Let $f : \overline{\boldsymbol{D}} \subset X \to X$, where $\boldsymbol{D}$ is an open and bounded subset of X, be continuous. Let $\boldsymbol{p} \notin f(\partial \boldsymbol{D})$. Then, we know

$$\Re^n \xrightarrow{\phi^{-1}} \overline{\boldsymbol{D}} \subset X \xrightarrow{f} X \xrightarrow{\phi} \Re^n$$

Note $\boldsymbol{p} \notin f(\partial \boldsymbol{D})$ implies $\phi(\boldsymbol{p}) \notin \phi \circ \phi^{-1}(\partial(\phi(\boldsymbol{D})))$ as if it was, $\boldsymbol{p} \in f \circ \phi^{-1}(\partial(\phi(\boldsymbol{D})))$. But then, $\boldsymbol{p} = f(\boldsymbol{y})$ with $\boldsymbol{y} \in \phi^{-1}(\partial(\phi(\boldsymbol{D})))$. You can then show that $\partial(\phi(\boldsymbol{D})) = \phi(\partial \boldsymbol{D})$. Hence, $\boldsymbol{p} = f(\boldsymbol{y})$ with $\boldsymbol{y} \in \partial \boldsymbol{D}$ which is not possible by assumption.

Next, note $\phi \circ f \circ \phi^{-1} : \phi(\overline{\boldsymbol{D}}) \subset \Re^n \to \Re^n$. Since ϕ is a bijection, $\phi(\boldsymbol{D})$ is an open and bounded subset of $\Re^n$. We also know $\phi(\boldsymbol{p}) \notin \phi \circ \phi^{-1}(\partial(\phi(\boldsymbol{D})))$. Thus, $d_B(\phi \circ f \circ \phi^{-1}, \phi(\boldsymbol{D}), \phi(\boldsymbol{p}))$ is well-defined.

In the simplest case,

$$\begin{aligned} d_B(\phi \circ f \circ \phi^{-1}, \phi(\boldsymbol{D}), \phi(\boldsymbol{p})) &= \sum_{\boldsymbol{x} \in (\phi \circ f \circ \phi^{-1})(\phi(\boldsymbol{p}))} sign\ J(\phi \circ f \circ \phi^{-1})(\boldsymbol{x}) \\ &= \sum_{\boldsymbol{x} \in (\phi \circ f^{-1})(\boldsymbol{p})} sign\ J(\phi \circ f \circ \phi^{-1})(\boldsymbol{x}) \end{aligned}$$

The matrix representation of ϕ with respect to the basis $\boldsymbol{A} = \{\boldsymbol{a_1}, \ldots, \boldsymbol{a_n}\}$ for X and $\boldsymbol{B} = \{\boldsymbol{b_1}, \ldots, \boldsymbol{b_n}\}$ for $\Re^n$ is the usual identity matrix:

$$[\phi]_{\boldsymbol{A},\boldsymbol{B}} = \begin{bmatrix} 1 & 0 & \ldots & 0 \\ 0 & 1 & \ldots & 0 \\ \vdots & \vdots & \ddots & \vdots \\ 0 & \ldots & \ldots & 1 \end{bmatrix}$$

We see we can write for any $\boldsymbol{x} \in \Re^n$, $J(\phi \circ f \circ \phi^{-1})(\boldsymbol{x}) = J[\phi](f[\phi^{-1}](\boldsymbol{x}))$.

What happens if we use a different set of bases for X and $\Re^n$; say $\hat{\boldsymbol{A}} = \{\hat{\boldsymbol{a}}_1, \ldots, \hat{\boldsymbol{a}}_n\}$ for X and $\hat{\boldsymbol{B}} = \{\hat{\boldsymbol{b}}_1, \ldots, \hat{\boldsymbol{b}}_n\}$ for $\Re^n$ to define a new mapping $\hat{\phi}$ in the same way? With this change of basis, we find coefficients, q_{ij}, p_{ij} and s_{ij} so that

$$\hat{\boldsymbol{a}}_i = \sum_{j=1}^n q_{ji} \boldsymbol{a_j}, \quad \hat{\boldsymbol{b}}_i = \sum_{j=1}^n p_{ji} \boldsymbol{b_j}, \quad \boldsymbol{b_i} = \sum_{j=1}^n s_{ji} \hat{\boldsymbol{b}}_j$$

which implies

$$\hat{\phi}(\hat{\boldsymbol{a}}_i) = \hat{\boldsymbol{b}}_i = \sum_{j=1}^n p_{ji} \boldsymbol{b_j} = \sum_{j=1}^n p_{ji} \phi(\boldsymbol{a_j})$$

This tells us how to find the matrix representation of $\hat{\phi}$.

$$[\hat{\phi}]_{\hat{\boldsymbol{A}},\hat{\boldsymbol{B}}} = \begin{bmatrix} p_{11} & \dots & p_{1n} \\ \vdots & \vdots & \vdots \\ p_{n1} & \dots & p_{nn} \end{bmatrix} [\phi]_{\boldsymbol{A},\boldsymbol{B}}$$

We can also write

$$\begin{aligned} \boldsymbol{a_i} &= \sum_{j=1}^{n} r_{ji}\hat{\boldsymbol{a}}_{\boldsymbol{j}} \Longrightarrow \\ \hat{\phi}(\boldsymbol{a_i}) &= \sum_{j=1}^{n} r_{ji}\hat{\phi}(\hat{\boldsymbol{a}}_{\boldsymbol{j}}) = \sum_{j=1}^{n} r_{ji}\hat{\boldsymbol{b}}_{\boldsymbol{j}} \\ &= \sum_{j=1}^{n} r_{ji}\sum_{\ell=1}^{n} p_{\ell j}\boldsymbol{b_\ell} = \sum_{\ell=1}^{n}\sum_{j=1}^{n} p_{\ell j}r_{ji}\boldsymbol{b_\ell} \end{aligned}$$

$$[\hat{\phi}]_{\hat{\boldsymbol{A}},\hat{\boldsymbol{B}}} = \begin{bmatrix} p_{11} & \dots & p_{1n} \\ \vdots & \vdots & \vdots \\ p_{n1} & \dots & p_{nn} \end{bmatrix} [\phi]_{\boldsymbol{A},\boldsymbol{B}} \begin{bmatrix} r_{11} & \dots & r_{1n} \\ \vdots & \vdots & \vdots \\ r_{n1} & \dots & r_{nn} \end{bmatrix}$$

We know $det[\phi]_{\boldsymbol{A},\boldsymbol{B}} = 1$, so

$$det\,[\hat{\phi}]_{\hat{\boldsymbol{A}},\hat{\boldsymbol{B}}} = det \begin{bmatrix} p_{11} & \dots & p_{1n} \\ \vdots & \vdots & \vdots \\ p_{n1} & \dots & p_{nn} \end{bmatrix} det \begin{bmatrix} r_{11} & \dots & r_{1n} \\ \vdots & \vdots & \vdots \\ r_{n1} & \dots & r_{nn} \end{bmatrix}$$

Now we can say more.

$$\boldsymbol{a_i} = \sum_{j=1}^{n} r_{ji}\hat{\boldsymbol{a}}_{\boldsymbol{j}} = \sum_{j=1}^{n}\sum_{\ell=1}^{n} r_{ji}q_{\ell j}\boldsymbol{a_\ell} = \sum_{\ell=1}^{n}\left(\sum_{j=1}^{n} q_{\ell j}r_{ji}\right)\boldsymbol{a_\ell}$$

Therefore, we can say $\sum_{\ell=1}^{n}\sum_{j=1}^{n} q_{\ell j}r_{ji} = \delta_{i\ell}$. We see

$$\begin{bmatrix} q_{11} & \dots & q_{1n} \\ \vdots & \vdots & \vdots \\ q_{n1} & \dots & q_{nn} \end{bmatrix} \begin{bmatrix} r_{11} & \dots & r_{1n} \\ \vdots & \vdots & \vdots \\ r_{n1} & \dots & r_{nn} \end{bmatrix} = I$$

where I is the identity matrix. We conclude

$$det \begin{bmatrix} q_{11} & \dots & q_{1n} \\ \vdots & \vdots & \vdots \\ q_{n1} & \dots & q_{nn} \end{bmatrix} det \begin{bmatrix} r_{11} & \dots & r_{1n} \\ \vdots & \vdots & \vdots \\ r_{n1} & \dots & r_{nn} \end{bmatrix} = 1$$

Hence,

$$det \begin{bmatrix} r_{11} & \dots & r_{1n} \\ \vdots & \vdots & \vdots \\ r_{n1} & \dots & r_{nn} \end{bmatrix} = \alpha \neq 0$$

In a similar way, we can show

$$det \begin{bmatrix} p_{11} & \cdots & p_{1n} \\ \vdots & \vdots & \vdots \\ p_{n1} & \cdots & p_{nn} \end{bmatrix} = \beta \neq 0$$

From these remarks, we see we have shown

$$det\,[\hat{\phi}]_{\hat{\boldsymbol{A}},\hat{\boldsymbol{B}}} = \alpha\,\beta \neq 0$$

Thus, dropping some of the parenthesis, as they simply clutter the argument,

$$\begin{aligned} J(\hat{\phi} \circ f \circ \hat{\phi}^{-1})(\boldsymbol{x}) &= J([\hat{\phi}]_{\hat{\boldsymbol{A}},\hat{\boldsymbol{B}}} f [\hat{\phi}]^{-1}_{\hat{\boldsymbol{A}},\hat{\boldsymbol{B}}})(\boldsymbol{x}) \\ &= J([p_{ij}][\phi]_{\boldsymbol{A},\boldsymbol{B}} f [p_{ij}][\phi]^{-1}_{\boldsymbol{A},\boldsymbol{B}})(\boldsymbol{x}) \\ &= J([p_{ij}][\phi]_{\boldsymbol{A},\boldsymbol{B}} f [\phi]^{-1}_{\boldsymbol{A},\boldsymbol{B}}[p_{ij}]^{-1})(\boldsymbol{x}) \\ &= det[p_{ij}] J(f[\phi]^{-1}_{\boldsymbol{A},\boldsymbol{B}}[p_{ij}]^{-1})(\boldsymbol{x}) = \beta J(f[\phi]^{-1}_{\boldsymbol{A},\boldsymbol{B}}[p_{ij}]^{-1})(\boldsymbol{x}) \end{aligned}$$

But here (since we are looking at the $\hat{\phi}$ case),

$$\boldsymbol{x} \in [\hat{\phi}]_{\hat{\boldsymbol{A}},\hat{\boldsymbol{B}}} f^{-1}(\boldsymbol{p}) = [p_{ij}][\phi]_{\boldsymbol{A},\boldsymbol{B}} f^{-1}(\boldsymbol{p})$$

and so $[\hat{\phi}]_{\hat{\boldsymbol{A}},\hat{\boldsymbol{B}}} f^{-1}(\boldsymbol{p}) \subset [p_{ij}]^{-1}\boldsymbol{x} \in [\phi]_{\boldsymbol{A},\boldsymbol{B}} f^{-1}(\boldsymbol{p})$. On the other hand, if $z = \phi f^{-1}(\boldsymbol{p})$, then $[p_{ij}]z = [p_{ij}]\phi f^{-1}(\boldsymbol{p}) = \hat{\phi} f^{-1}(\boldsymbol{p})$. But, $\hat{\phi} f^{-1}(\boldsymbol{p}) \subset [p_{ij}][\phi]_{\boldsymbol{A},\boldsymbol{B}} f^{-1}(\boldsymbol{p})$. Combining, we conclude $\hat{\phi} f^{-1}(\boldsymbol{p}) = [p_{ij}][\phi]_{\boldsymbol{A},\boldsymbol{B}} f^{-1}(\boldsymbol{p})$. Thus, letting $\boldsymbol{y} = [p_{ij}]^{-1}(\boldsymbol{x})$,

$$\begin{aligned} d_B(\hat{\phi} f \hat{\phi}^{-1}, \hat{\phi}(\boldsymbol{D}), \hat{\phi}(\boldsymbol{p})) &= \sum_{\boldsymbol{x} \in \hat{\phi} f^{-1}(\boldsymbol{p})} sign\ J(\hat{\phi} f \hat{\phi}^{-1})(\boldsymbol{x}) \\ &= \sum_{\boldsymbol{x} \in [p_{ij}]\phi f^{-1}(\boldsymbol{p})} sign\ \beta J(\phi f \phi^{-1}[p_{ij}]^{-1})(\boldsymbol{x}) \\ &= sign\beta \sum_{\boldsymbol{y} \in \phi f^{-1}(\boldsymbol{p})} sign\ J(\phi f \phi^{-1})(\boldsymbol{y}) \\ &= (\pm)\, d_B(\phi f \phi^{-1}, \phi(\boldsymbol{D}), \phi(\boldsymbol{p})) \end{aligned}$$

We conclude $d_B(\phi f \phi^{-1}, \phi(\boldsymbol{D}), \phi(\boldsymbol{p}))$ is unique up to sign when we apply the bijections ϕ. So we define $d_B, (f, \boldsymbol{D}, \boldsymbol{p}) = d_B(\phi f \phi^{-1}, \phi(\boldsymbol{D}), \phi(\boldsymbol{p}))$ where $\phi : X \to \Re$ is a bijection chosen so that $det[\phi]_{\boldsymbol{A},\boldsymbol{B}} = 1$. For all other $\hat{\phi}$, if $det[\hat{\phi}]_{\hat{\boldsymbol{A}},\hat{\boldsymbol{B}}} > 0$, we say $\hat{\phi}$ is orientation preserving. Otherwise, it is orientation reversing. We now have a unique, up to ± 1, Brouwer degree defined on finite dimensional normed linear spaces. We can thus easily prove the following theorem.

Theorem 10.6.1 Brouwer Degree of the Identity Map for a Finite Dimensional Normed Linear Space

Let $\boldsymbol{D} \subset (X, \|\cdot\|)$ be open and bounded in the finite dimensional normed linear space X. Let $I : \overline{\boldsymbol{D}} \to \Re^n$ be the identity map. Then, if $\boldsymbol{p} \notin \partial \boldsymbol{D}$,

$$d_B(I, \boldsymbol{D}, \boldsymbol{p}) = \begin{cases} 0, & \boldsymbol{p} \notin \boldsymbol{D} \\ \pm 1, & \boldsymbol{p} \in \boldsymbol{D} \end{cases}$$

Proof 10.6.1
Immediate. ■

Finally, from our discussions above, it is easy to see all of our results for Brouwer degree in $\Re^n$ still hold in a finite dimensional normed linear space.

Homework

Just to remind you a bit about finite dimensional normed linear spaces, here are some examples.

Exercise 10.6.1 *Find the general solution to*

$$\begin{aligned} x'(t) &= -4\,x(t) + y(t) \\ y'(t) &= -5\,x(t) + 2\,y(t) \end{aligned}$$

and show it forms a two dimensional vector space over $\Re$. Then show it is a finite dimensional normed linear space for a variety of norm choices.

Exercise 10.6.2 *Find the general solution to*

$$\begin{aligned} x'(t) &= 5\,x(t) + y(t) \\ y'(t) &= -7\,x(t) - 3\,y(t) \end{aligned}$$

and show it forms a two dimensional vector space over $\Re$. Then show it is a finite dimensional normed linear space for a variety of norm choices.

Chapter 11

Leray - Schauder Degree

The Brouwer degree uniquely assigns an integer to the triple $(f, \boldsymbol{D}, \boldsymbol{p})$ where $f : \boldsymbol{D} \subset \Re^n \to \Re^n$ is a continuous function on the open and bounded set $\boldsymbol{D}$ and $\boldsymbol{p} \in \Re^n \setminus \partial \boldsymbol{D}$. It is easily extended to continuous functions on a bounded open subset of finite dimensional normed linear spaces into the space. The extension, however, does not quite give the same unique result. Now the degree is arbitrary up to a $\pm 1 \cdot d_B(f, \boldsymbol{D}, \boldsymbol{p})$ with the arbitrariness due to the whether or not we identify the finite dimensional normed linear space using an orientation preserving or reversing homeomorphism.

Can a degree be defined for arbitrary continuous mappings on an infinite dimensional normed linear space $(X, \|\cdot\|)$? The answer is **no** and the standard example that shows this comes from Leray's original paper (Leray (73) 1936).

Let $\boldsymbol{x_0} \in (C([0,1]), \|\cdot\|_\infty)$ be the constant function $x_0(s) = 1/2$ for $0 \leq s \leq 1$. Let $\boldsymbol{D} = \{\boldsymbol{x} \in X | \|\boldsymbol{x} - \boldsymbol{x_0}\|_\infty < (1/2)\}$. Then $\boldsymbol{D}$ is an open and bounded subset of X. Further, let $\boldsymbol{E} = \{\boldsymbol{x} \in X | 0 < x(t) < 1,\ 0 \leq t \leq 1\}$. Define $F : \overline{\boldsymbol{D}} \to X$ by choosing $\phi \in X$ with $\phi(0) = 0$, $\phi(1) = 1$ and $0 \leq \phi(t) \leq 1$ for $0 \leq t \leq 1$ and defining $(F(\boldsymbol{x}))(s) = \phi(x(s))$. We see F is continuous.

Let's assume there is a degree defined on X with all the familiar properties. Define a homotopy $H(\boldsymbol{x}, t) = tF(\boldsymbol{x}) + (1-t)\boldsymbol{x}$. Note $H(\boldsymbol{x}, t) : X \to X$ and $H(\boldsymbol{x}, 0) = \boldsymbol{x}$; i.e. $H(\boldsymbol{x}, 0) = I$, the identity mapping on X. Also, $H(\boldsymbol{x}, 1) = F(\boldsymbol{x})$.

Let $\boldsymbol{y} \in \partial \boldsymbol{D}$. Then $\|\boldsymbol{y} - \boldsymbol{x_0}\|_\infty = (1/2)$. So, for $0 \leq s \leq 1$, $0 \leq y(s) \leq 1$ and either there is a point s_0 with $y(s_0) = 1$ or there is a point s_0 with $y(s_0) = 0$ or both points exist. Thus, $\phi(y(s_0)) = \phi(0) = 0$ or $\phi(y(s_0)) = \phi(1) = 1$. This tells us $\|\boldsymbol{y} - \boldsymbol{x_0}\|_\infty = (1/2)$ and thus, $\phi(\boldsymbol{y}) \in \partial \boldsymbol{D}$. So if $\boldsymbol{y} \notin \phi(\partial \boldsymbol{D})$, $\phi(\boldsymbol{y}) \notin \partial \boldsymbol{D}$. This would mean $\boldsymbol{y} \notin \partial \boldsymbol{D}$.
If $\boldsymbol{y_0} \in \boldsymbol{D}$, then $-(1/2) < y_0(s) - (1/2) < (1/2)$ for $0 \leq s \leq 1$. Thus, $0 < y_0(s) < 1$. Now consider $H(\boldsymbol{y_0}, t)$. We have

$$
\begin{aligned}
0 < t\phi(y_0(s)) + (1-t)y_0(s) < t + (1-t) = 1 \implies \\
-(1/2) < t\phi(y_0(s)) + (1-t)y_0(s) - (1/2) < (1/2)
\end{aligned}
$$

This says $|H(\boldsymbol{y_0}, t) - (1/2)| < (1/2)$ for all s. Since the maximum of $H(\boldsymbol{y_0}, t) - (1/2)$ occurs at some s' in $[0,1]$, this implies $\|H(\boldsymbol{y_0}, t) - \boldsymbol{x_0}\|_\infty < (1/2)$. Hence, $H(\boldsymbol{y_0}, t) \notin \partial \boldsymbol{D}$ for $0 \leq t \leq 1$. Hence, $\boldsymbol{y_0} \in \boldsymbol{D}$, implies $\boldsymbol{y_0} \notin F(\partial \boldsymbol{D} \times [0,1])$.
We have assumed we have a degree theory defined on X and so invariance under homotopy holds. We conclude as long as $\boldsymbol{y_0} \in \boldsymbol{D}$, $\boldsymbol{y_0} \notin F(\partial \boldsymbol{D} \times [0,1])$ and $d(F(\cdot, 0), \boldsymbol{D}, \boldsymbol{y_0}) = d(F(\cdot, 1), \boldsymbol{D}, \boldsymbol{y_0})$.

Hence, $1 = d(I, \boldsymbol{D}, \boldsymbol{y_0}) = d(\phi, \boldsymbol{D}, \boldsymbol{y_0})$. Note, $d(\phi, \boldsymbol{D}, \boldsymbol{y_0})$ is defined as long as $\boldsymbol{y_0} \notin \phi(\partial \boldsymbol{D})$. By our remarks above, this means $\boldsymbol{y_0} \notin \partial \boldsymbol{D}$ too. Hence, this degree is well-defined for this $\boldsymbol{y_0}$.

Since $d(\phi, \boldsymbol{D}, \boldsymbol{y_0}) = 1$, there must be a function $\boldsymbol{w}$ with $\phi(\boldsymbol{w}) = \boldsymbol{y_0}$. However, there is a problem. Let $\boldsymbol{y_0}$ be the function $y_0(s) = (1/4) + (1/2)s$ and ϕ be defined by

$$\phi(t) = \begin{cases} t, & 0 \leq t \leq (1/2) \\ 1 - t, & (1/2) < t \leq (5/8) \\ (5/3)\Big(t - (5/8)\Big) + (3/8), & (5/8) < t \leq 1 \end{cases}$$

We assume there is a solution $\boldsymbol{x}$ so that $\phi(x(s)) = y_0(s)$. This means $\phi(x(0)) = y_0(0) = (1/4)$. Now, $\phi(x(s))$ must increase strictly monotonically on $[0, 1]$ from $(1/4)$ to $(3/4)$. Since ϕ starts at $(1/4)$, it can only increase until $(1/2)$ and then it must decrease. We see we cannot find a function $\boldsymbol{x}$ that solves $\phi(\boldsymbol{w}) = y_0$.

Therefore, a degree for arbitrary continuous functions on an infinite dimensional normed linear space (complete or not) is not always possible.

We need to find the right kind of mappings for which a degree will work.

11.1 Zeroing in on an Infinite Dimensional Degree

Let's refresh your mind about compactness as that will be an essential part of our new degree.

Definition 11.1.1 Compactness and Relative Compactness in a Normed Linear Space

Let $(X, \|\cdot\|)$ be a normed linear space. Then $\boldsymbol{K} \subset X$ is compact if every infinite subset of $\boldsymbol{K}$ has a limit point in $\boldsymbol{K}$. We say K is relatively compact if $\overline{\boldsymbol{K}}$ is compact.

Comment 11.1.1 *We are being less careful than usual here. The norm on X induces a metric and there are still the two notions of compactness called topological and sequential compactness which by Theorem 2.4.5 we know are equivalent to one another. Hence, we will just use the term compactness instead of the more correct term sequential compactness. We also know compactness is equivalent to being totally bounded: for every $\epsilon > 0$, a compact set in a metric space can be covered by a finite number of subsets whose diameters are all less than epsilon.*

We want to consider a special class of mappings on subsets of the normed linear space X. For historical reasons, mappings on a normed linear space are often called operators too, so we will be using that label more now. An important type of operator is one that is called compact.

Definition 11.1.2 Compact Mappings on a Normed Linear Space

Let $(X, \|\cdot\|)$ be a normed linear space. Let $T : \boldsymbol{E} \subset X \to X$.

(i): We say T is compact if for every bounded subset $\boldsymbol{M} \subset \boldsymbol{E}$, $T(\boldsymbol{M})$ is relatively compact. Note T need not be a linear mapping and T need not be continuous.

(ii): We say T is completely continuous if it is both compact and continuous. Again, T need not be linear.

(iii): If T is linear and compact, T must be bounded and therefore continuous. Hence, compact and linear implies completely continuous.

Homework

Exercise 11.1.1 *Prove that if T is compact and linear, it must be bounded and is therefore continuous. Hence $\|T\|_{op}$ is finite.*

Exercise 11.1.2 *Prove the Stürm - Liouville models give rise to integral operators that are linear and compact. The background for this is in (Peterson (100) 2020).*

We are now ready to prove our first result.

Theorem 11.1.1 Approximation of Compact Operators

Let $(X, \|\cdot\|)$ be a normed linear space. Let $T : \overline{\boldsymbol{M}} \subset X \to X$ be compact. Let $\boldsymbol{K} = \overline{T(\boldsymbol{M})} = T(\overline{\boldsymbol{M}})$. For a given $\epsilon > 0$, let $\boldsymbol{V_\epsilon} = \{\boldsymbol{v_1}, \ldots, \boldsymbol{v_{p(\epsilon)}}\}$ denote a corresponding ϵ net for $\boldsymbol{K}$. Define the functions $m_i : \boldsymbol{K} \to X$ by

$$m_i(\boldsymbol{y}) = \begin{cases} \epsilon - \|\boldsymbol{y} - \boldsymbol{v_i}\|, & \|\boldsymbol{y} - \boldsymbol{v_i}\| \leq \epsilon \\ 0, & \|\boldsymbol{y} - \boldsymbol{v_i}\| > \epsilon \end{cases}$$

for $1 \leq i \leq p(\epsilon)$. Then define the mappings $F_\epsilon : \boldsymbol{M} \to X$ by

$$F_\epsilon(\boldsymbol{x}) = \frac{\sum_{i=1}^{p(\epsilon)} m_i(T(\boldsymbol{x}))\, \boldsymbol{v_i}}{\sum_{i=1}^{p(\epsilon)} m_i(T(\boldsymbol{x}))}$$

Then, $\|T(\boldsymbol{x}) - F_\epsilon(T(\boldsymbol{x}))\| < \epsilon$.

Proof 11.1.1

$$\begin{aligned}\|T(\boldsymbol{x}) - F_\epsilon(T(\boldsymbol{x}))\| &= \left\| T(\boldsymbol{x}) - \frac{\sum_{i=1}^{p(\epsilon)} m_i(T(\boldsymbol{x}))\, \boldsymbol{v_i}}{\sum_{i=1}^{p(\epsilon)} m_i(T(\boldsymbol{x}))} \right\| \\ &= \frac{\left\| \sum_{i=1}^{p(\epsilon)} m_i(T(\boldsymbol{x}))\, T(\boldsymbol{x}) - \sum_{i=1}^{p(\epsilon)} m_i(T(\boldsymbol{x}))\, \boldsymbol{v_i} \right\|}{\left| \sum_{i=1}^{p(\epsilon)} m_i(T(\boldsymbol{x})) \right|}\end{aligned}$$

In the denominator, the absolute values are not necessary as all $m_i \geq 0$. Also, since V_ϵ is an ϵ net for $\boldsymbol{K}$, there is at least one $m_i(T(\boldsymbol{x})) > 0$. Let $J = \{i | m_i(T(\boldsymbol{x})) > 0\}$. Then,

$$\begin{aligned}\|T(\boldsymbol{x}) - F_\epsilon(T(\boldsymbol{x}))\| &= \frac{\left\| \sum_{i \in J} m_i(T(\boldsymbol{x}))\, T(\boldsymbol{x}) - \sum_{i \in J} m_i(T(\boldsymbol{x}))\, \boldsymbol{v_i} \right\|}{\sum_{i \in J} m_i(T(\boldsymbol{x}))} \\ &= \frac{\left\| \sum_{i \in J}(\epsilon - \|T(\boldsymbol{x}) - \boldsymbol{v_i}\|)\, T(\boldsymbol{x}) - \sum_{i \in J}(\epsilon - \|T(\boldsymbol{x}) - \boldsymbol{v_i}\|)\, \boldsymbol{v_i} \right\|}{\sum_{i \in J}(\epsilon - \|T(\boldsymbol{x}) - \boldsymbol{v_i}\|)} \\ &= \frac{\left\| \sum_{i \in J}(\epsilon - \|T(\boldsymbol{x}) - \boldsymbol{v_i}\|)\, (T(\boldsymbol{x}) - \boldsymbol{v_i}) \right\|}{\sum_{i \in J}(\epsilon - \|T(\boldsymbol{x}) - \boldsymbol{v_i}\|)}\end{aligned}$$

But the terms $\epsilon - \|T(\boldsymbol{x}) - \boldsymbol{v_i}\| > 0$, *so we have*

$$\begin{aligned}\|T(\boldsymbol{x}) - F_\epsilon(T(\boldsymbol{x}))\| &= \frac{\sum_{i \in J}(\epsilon - \|T(\boldsymbol{x}) - \boldsymbol{v_i}\|)\,\|T(\boldsymbol{x}) - \boldsymbol{v_i}\|}{\sum_{i \in J}(\epsilon - \|T(\boldsymbol{x}) - \boldsymbol{v_i}\|)} \\ &< \frac{\sum_{i \in J}(\epsilon - \|T(\boldsymbol{x}) - \boldsymbol{v_i}\|)}{\sum_{i \in J}(\epsilon - \|T(\boldsymbol{x}) - \boldsymbol{v_i}\|)}\,\epsilon = \epsilon\end{aligned}$$

■

To define a suitable degree, we use the operators F_{ϵ_n} for any sequence (ϵ_n) with $\epsilon \to 0$ monotonically. Let $T_n = F_{\epsilon_n} \circ T$ which means $T_n(\boldsymbol{x}) = F_{\epsilon_n}(T(\boldsymbol{x}))$ for the compact operator $T : \overline{\boldsymbol{D}} \subset X \to X$ where $\boldsymbol{D} \subset X$ is an open and bounded subset of X. Assume $\boldsymbol{p} \notin (I - T)(\partial \boldsymbol{D})$ where $I : X \to X$ is the identity mapping. We will prove a succession of important facts:

(i): There is $r > 0$ so that $\inf_{\boldsymbol{y} \in \partial \boldsymbol{D}} \|(I - T)(\boldsymbol{y}) - \boldsymbol{p}\| \geq r > 0$.

Proof

Assume no such $r > 0$ *can be found. Then, we can find a sequence* $(\boldsymbol{y_n}) \subset \partial \boldsymbol{D}$ *with* $\lim_{n \to \infty}(I - T_n)(\boldsymbol{y_n}) = \boldsymbol{p}$. *Since* T *is compact,* $\overline{T(\overline{\boldsymbol{D}})}$ *is compact. Thus,* $(T(\boldsymbol{y_n})) \subset \overline{T(\overline{\boldsymbol{D}})}$ *and hence, there is a subsequence* $(T(\boldsymbol{y_{n,1}}))$ *and* $\boldsymbol{y_0}$ *so that* $T(\boldsymbol{y_{n,1}}) \to \boldsymbol{y_0} \in \overline{T(\overline{\boldsymbol{D}})} = T(\overline{\boldsymbol{D}})$. *Therefore, since* $(I - T)(\boldsymbol{y_{n,1}}) \to \boldsymbol{p}$, $\lim_{n \to \infty} \boldsymbol{y_{n,1}} - \lim_{n \to \infty} T(\boldsymbol{y_{n,1}}) = \boldsymbol{p}$. *We conclude* $\lim_{n \to \infty} \boldsymbol{y_{n,1}} = \boldsymbol{p} + \boldsymbol{y_0}$.

However, $\partial \boldsymbol{D}$ *is closed and so the limit* $\boldsymbol{y_0} + \boldsymbol{p} = \boldsymbol{z_0} \in \partial \boldsymbol{D}$. *But then* $\lim_{n \to \infty}(\boldsymbol{y_{n,1}} - T(\boldsymbol{y_{n,1}})) = \boldsymbol{p}$ *implies by the continuity of* T *that* $\lim_{n \to \infty}(I - T)(\boldsymbol{y_{n,1}}) = (I_T)(\boldsymbol{z_0}) = \boldsymbol{p}$ *or* $\boldsymbol{p} \in (I - T)(\partial \boldsymbol{D})$ *which is not possible. So our assumption is wrong and so a positive* r *must exist.* ■

(ii): There is an index n_0 so that $\|(I - T_n)(\boldsymbol{x}) - \boldsymbol{p}\| \geq (r/2)$ when $n \geq n_0$.

Proof

We have $T_n = F_{\epsilon_n} \circ T$. *Since* ϵ *decreases monotonically to* 0, *there is* n_0 *so that* $\epsilon < (r/2)$ *if* $n \geq n_0$. *By Theorem 11.1.1,* $\|T - T_n\| < \epsilon_n$ *for all* $\boldsymbol{x} \in \overline{\boldsymbol{D}}$. *So, for* $\boldsymbol{x} \in \partial \boldsymbol{D}$,

$$\begin{aligned}\|(I - T_n)(\boldsymbol{x}) - \boldsymbol{p}\| &= \|(I - T)(\boldsymbol{x}) - \boldsymbol{p} + (T - T_n)(\boldsymbol{x})\| \\ &\geq \|(I - T)(\boldsymbol{x}) - \boldsymbol{p}\| - \|(T - T_n)(\boldsymbol{x})\| \geq r - \epsilon\end{aligned}$$

Hence, if $n \geq n_0$, $\|(I - T_n)(\boldsymbol{x}) - \boldsymbol{p}\| \geq (r/2)$. ■

(iii): We now find an appropriate sequence of Brouwer degrees that we can use to define the new degree we want.

Proof

Let $\boldsymbol{V_n} = \{\boldsymbol{v_{n,1}}, \ldots, \boldsymbol{v_{n,p_n}}\}$ *be the* ϵ *net for* $T(\overline{\boldsymbol{D}})$ *used to define* F_{ϵ_n}. *Let* X_n *be the finite dimensional subspace generated by* $\boldsymbol{V_n}$. *In general, we don't know* $\boldsymbol{D} \cap T(\overline{\boldsymbol{D}})$ *is nonempty. So, let's adjoin a point* $\boldsymbol{q} \in \boldsymbol{D}$ *to* X_n. *We then define* $Y_n = span\{\boldsymbol{V_\epsilon}, \boldsymbol{p}, \boldsymbol{q}\}$. *This tells us* $\boldsymbol{p} \in Y_n$ *for all* n. *Now* Y_n *is a normed linear space using the norm induced by the norm on* X. *We have* $Y_n \cap \boldsymbol{D}$ *is nonempty always and* $\boldsymbol{D_n} = Y_n \cap \boldsymbol{D}$ *is a bounded and open subset of* Y_n *which is a finite dimensional normed linear space. We see* $\partial \boldsymbol{D_n} \subset \partial \boldsymbol{D}$. *Now,* $(I - T_n)(\overline{\boldsymbol{D_n}}) \subset (I - T_n)(\overline{Y_n}) = (I - T_n)(Y_n)$. *Also,* $\inf_{\boldsymbol{x} \in \partial \boldsymbol{D_n}} \|(I - T_n)(\boldsymbol{x}) - \boldsymbol{p}\| \geq$

$\inf_{\boldsymbol{x} \in \partial \boldsymbol{D}} \|(I - T_n)(\boldsymbol{x}) - \boldsymbol{p}\| \geq (r/2) > 0$. *Thus,* $\boldsymbol{p} \notin (I - T_n)(\partial \boldsymbol{D_n})$.

Once we show $(I - T_n)(\overline{\boldsymbol{D_n}}) \subset Y_n$, *we can define a degree on the finite dimensional normed linear space* Y_n *like usual.*

Let $\boldsymbol{x} \in \overline{\boldsymbol{D_n}} = \overline{\{\boldsymbol{V_\epsilon}, \boldsymbol{p}, \boldsymbol{q}\}} \cap \boldsymbol{D}$. *Then, if* $J = \{i | m_i(T(\boldsymbol{x})) > 0\}$.

$$(I - T_n)(\boldsymbol{x}) = \boldsymbol{x} - F_{\epsilon_n}(T(\boldsymbol{x})) = \frac{\sum_{i \in J} m_i(T(\boldsymbol{x}))\, \boldsymbol{v_{n,i}}}{\sum_{i \in J} m_i(T(\boldsymbol{x}))}$$

Let

$$\alpha_i = \frac{\epsilon_n - \|T(\boldsymbol{x}) - \boldsymbol{v_{n,i}}\|}{\sum_{i \in J} m_i(T(\boldsymbol{x}))}, \ \forall\, i \in J$$

Then, $(I - T_n)(\boldsymbol{x}) = \boldsymbol{x} - \sum_{i \in J} \alpha_i \boldsymbol{v_{n,i}}$.

Now, $\overline{\boldsymbol{D_n}} = \overline{\boldsymbol{Y_n} \cap \boldsymbol{D}} = \boldsymbol{Y_n} \cap \overline{\boldsymbol{D}}$. *Since* $\boldsymbol{x} \in \overline{\boldsymbol{D_n}}$, $\boldsymbol{x} \in \boldsymbol{Y_n} \cap \overline{\boldsymbol{D}}$. *Hence,*

$$\begin{aligned} \boldsymbol{x} &= \sum_{i=1}^{p_n} \beta_i \boldsymbol{v_{n,i}} + \gamma \boldsymbol{q} + \delta \boldsymbol{q} \\ (I - T_n)(\boldsymbol{x}) &= \sum_{i=1}^{p_n} \beta_i \boldsymbol{v_{n,i}} + \gamma \boldsymbol{q} + \delta \boldsymbol{q} - \sum_{i \in J} \alpha_i \boldsymbol{v_{n,i}} \in Y_n \end{aligned}$$

We conclude $(I - T_n)(\overline{\boldsymbol{D_n}}) \subset Y_n$. *Since the dimension of* Y_n *is finite and* $\mathbf{p} \notin (I - T_n)(\partial \boldsymbol{D_n})$, $(I - T_n) : \overline{\boldsymbol{D_n}} \to Y_n$, $\boldsymbol{D_n}$ *is open and bounded in* Y_n, *Brouwer degree on this finite dimensional normed linear space is defined. Thus,* $d_B(I - T_n, \boldsymbol{D_n}, \boldsymbol{p})$ *exists for all* n. ■

(iv): Our new degree, called Leray - Schauder degree, is defined as this: $d_{LS}(I - T, \boldsymbol{D}, \boldsymbol{p}) = \lim_{n \to \infty} d_B(I - T_n, \boldsymbol{D_n}, \boldsymbol{p})$. We need to show this definition is independent of the choice of T_n.

Proof

Let T_{n_1} *and* T_{n_2} *be two sets of mappings of the type we have been discussing. Hence, there is an* $r > 0$ *so that* $\|T - T_{n_1}\| < (r/2)$ *and* $\|T - T_{n_2}\| < (r/2)$ *with corresponding subspaces* Y_{n_1} *and* Y_{n_2} *of finite dimension. Let* $Y_m = span\{Y_{n_1}, Y_{n_2}\}$ *and define* $\boldsymbol{D_m}$ *in the usual way. Now all of our results from finite dimensional Brouwer degree on* $\Re^n$ *transfer to the finite dimensional normed linear space case.*

Let's rewrite the reduction theorem in this context.

Theorem The Finite Dimensional Normed Linear Space Degree Reduction Theorem

Let $(X, \|\cdot\|)$ *be a normed linear space of dimension* n *and let* $\boldsymbol{D} \subset X$ *be open and bounded. Let* Y *be a normed linear space of dimension* $m < n$, *identified with the corresponding subspace of* X *in the standard way. Let* $f : \overline{\boldsymbol{D}} \to Y$ *be continuous and* $g : \overline{\boldsymbol{D}} \to X$ *be the map given by* $g(\boldsymbol{x}) = \boldsymbol{x} + f(\boldsymbol{x})$. *Then, for all* $\mathbf{p} \in Y$ *with* $\mathbf{p} \notin g(\partial \boldsymbol{D})$, $d_B(g, \boldsymbol{D}, \boldsymbol{p}) = d_B(g|_{Y \cap \overline{\boldsymbol{D}}}, \boldsymbol{D} \cap Y, \boldsymbol{p})$.

Apply the reduction theorem separately to

(i): $f = -T_{n_1}$ using $\boldsymbol{D} = \boldsymbol{D_m}$, $X = Y_m$ and $Y = Y_{n_1}$. Then

$$\begin{aligned} d_B(I - T_{n_1}, \boldsymbol{D_m}, \boldsymbol{p}) &= d_B(I - T_{n_1}|_{Y_{n_1} \cap \overline{\boldsymbol{D}}}, \boldsymbol{D_m} \cap Y_{n_1}, \boldsymbol{p}) \\ &= d_B(I - T_{n_1}|_{\overline{\boldsymbol{D_{n_1}}}}, \boldsymbol{D_{n_1}}, \boldsymbol{p}) \end{aligned}$$

(ii): $f = -T_{n_2}$ using $\boldsymbol{D} = \boldsymbol{D_m}$, $X = Y_m$ and $Y = Y_{n_2}$

$$\begin{aligned} d_B(I - T_{n_2}, \boldsymbol{D_m}, \boldsymbol{p}) &= d_B(I - T_{n_2}|_{Y_{n_2} \cap \overline{\boldsymbol{D}}}, \boldsymbol{D_m} \cap Y_{n_2}, \boldsymbol{p}) \\ &= d_B(I - T_{n_2}|_{\overline{\boldsymbol{D_{n_2}}}}, \boldsymbol{D_{n_2}}, \boldsymbol{p}) \end{aligned}$$

Next, consider the homotopy H defined on $\boldsymbol{D_m} \times [0,1]$ by $H(\boldsymbol{x}, t) = t(I - T_{n_1})(\boldsymbol{x}) + (1 - t)(I - T_{n_2})(\boldsymbol{x})$. If $\boldsymbol{x} \in \partial \boldsymbol{D}$, we have

$$\begin{aligned} \|H(\boldsymbol{x}, t) - (I - T)(\boldsymbol{x})\| &= \|t(I - T_{n_1})(\boldsymbol{x}) + (1 - t)(I - T_{n_2})(\boldsymbol{x}) - (I - T)(\boldsymbol{x})\| \\ &\leq \|t(I - T_{n_1})(\boldsymbol{x}) - t(I - T)(\boldsymbol{x})\| \\ &\quad + \|(1 - t)(I - T_{n_2})(\boldsymbol{x}) - (1 - t)(I - T)(\boldsymbol{x})\| \\ &\leq t\|(T - T_{n_1})(\boldsymbol{x})\| + (1 - t)\|(T - T_{n_2})(\boldsymbol{x})\| \\ &< t \cdot \epsilon_{n_1} + (1 - t) \cdot \epsilon_{n_1} \leq t(r/2) + (1 - t)(r/2) = (r/2) \end{aligned}$$

Thus, for $\boldsymbol{x} \in \partial \boldsymbol{D_m}$,

$$\begin{aligned} \|H(\boldsymbol{x}, t) - \boldsymbol{p}\| &= \|H(\boldsymbol{x}, t) - (I_T)(\boldsymbol{x}) + (I_T)(\boldsymbol{x}) - \boldsymbol{p}\| \\ &\geq \|(I - T)(\boldsymbol{x}) - \boldsymbol{p}\| - \|H(\boldsymbol{x}, t) - (I_T)(\boldsymbol{x})\| \\ &> \|(I - T)(\boldsymbol{x}) - \boldsymbol{p}\| - (r/2) > r - (r/2) = (r/2) \end{aligned}$$

Hence, $\|H(\boldsymbol{x}, t) - \boldsymbol{p}\| > 0$ for all $\boldsymbol{x} \in \partial \boldsymbol{D_m}$; i.e., $\boldsymbol{p} \notin H(\partial \boldsymbol{D_m} \times [0,1])$.

By invariance under homotopy, we therefore have

$$d_B(H(\cdot, 0), \boldsymbol{D_m}, \boldsymbol{p}) = d_B(H(\cdot, 0), \boldsymbol{D_m}, \boldsymbol{p}) \Longrightarrow d_B(I - T_{n_1}, \boldsymbol{D_m}, \boldsymbol{p}) = d_B(I - T_{n_2}, \boldsymbol{D_m}, \boldsymbol{p})$$

From the reduction theorem result, we then have

$$d_B(I - T_{n_1}|_{\overline{\boldsymbol{D_{n_1}}}}, \boldsymbol{D_{n_1}}, \boldsymbol{p}) = d_B(I - T_{n_2}|_{\overline{\boldsymbol{D_{n_2}}}}, \boldsymbol{D_{n_2}}, \boldsymbol{p})$$

We can certainly choose n large enough so that $\epsilon < (r/2)$, and so for any mappings with $\|T - T_n\| < (r/2)$, $d_B(I - T_n, \boldsymbol{D_n}, \boldsymbol{p})$ is a constant. Hence, $\lim_{n \to \infty} d_B(I - T_n, \boldsymbol{D_n}, \boldsymbol{p})$ exists and is independent of the choice of approximation mappings T_n and subspaces Y_n. We can conclude our notion of a degree $d_{LS}(I - T, \boldsymbol{D}, \boldsymbol{p})$ is well-defined and exists when $T : \overline{\boldsymbol{D}} \to X$ is compact and $\boldsymbol{p} \notin (I_T)(\partial \boldsymbol{D})$ when $\boldsymbol{D}$ is an open and bounded subset of X. ■

With all this done, let's formally define Leray - Schauder degree for maps of the form $I - T$. Since the operator T is compact, we often call $I - T$ a compact perturbation of the identity. The inverses of some differential operators are often maps of this type and hence, we can study the existence of solutions to ODE problems using these tools. We simply have to show the degree is nonzero and we will have existence! Of course, it is not that simple as we have to define the right mappings and spaces for the problem under consideration. If you go back and look at previous discussions of Stürm - Liouville mappings you can see how such a compact pertubation of the identity arises in an ODE problem.

Definition 11.1.3 Leray - Schauder Degree for Compact Perturbations of the Identity

Let $(X, \|\cdot\|)$ be a normed linear space and $\boldsymbol{D} \subset X$ an open and bounded subset. Assume $T : \overline{\boldsymbol{D}} \subset X \to X$ is a compact operator and $\boldsymbol{p} \notin (I - T)(\partial \boldsymbol{D})$. Let $\boldsymbol{K} = T(\overline{\boldsymbol{D}})$. For a given $\epsilon > 0$, let $\boldsymbol{V_\epsilon} = \{\boldsymbol{v_1}, \ldots, \boldsymbol{v_{p(\epsilon)}}\}$ denote a corresponding ϵ net for $\boldsymbol{K}$. Define the functions $m_i : \boldsymbol{K} \to X$ by

$$m_i(\boldsymbol{y}) = \begin{cases} \epsilon - \|\boldsymbol{y} - \boldsymbol{v_i}\|, & \|\boldsymbol{y} - \boldsymbol{v_i}\| \leq \epsilon \\ 0, & \|\boldsymbol{y} - \boldsymbol{v_i}\| > \epsilon \end{cases}$$

for $1 \leq i \leq p(\epsilon)$. Define the mappings $T_\epsilon = F_\epsilon \circ T : \boldsymbol{D} \to X$ by

$$T_\epsilon(\boldsymbol{x}) = \frac{\sum_{i=1}^{p(\epsilon)} m_i(T(\boldsymbol{x}))\, \boldsymbol{v_i}}{\sum_{i=1}^{p(\epsilon)} m_i(T(\boldsymbol{x}))}$$

Let X_ϵ be the finite dimensional subspace generated by $\boldsymbol{V_\epsilon}$. Adjoin a point $\boldsymbol{q} \in \boldsymbol{D}$ to X_ϵ and define $Y_\epsilon = span\{\boldsymbol{V_\epsilon}, \boldsymbol{p}, \boldsymbol{q}\}$. Let $\boldsymbol{D_n} = Y_n \cap \boldsymbol{D}$. Then, for any monotonically decreasing sequence (ϵ_n) with $\epsilon_n \to 0$, we let $T_n = F_{\epsilon_n} \circ T$ and define the Leray - Schauder degree to be $d_{LS}(I - T, \boldsymbol{D}, \boldsymbol{p}) = \lim_{n \to \infty} d_B(I - T_n, \boldsymbol{D_n}, \boldsymbol{p})$. This value well-defined as it is independent of the choice of approximations T_n.

It is also clear from our discussion that Leray - Schauder and Brouwer degrees for the mapping I_T coincide on a finite dimensional normed linear space. To make this concrete, we will state it as a theorem.

Theorem 11.1.2 Leray - Schauder and Brouwer Degrees Coincide on Finite Dimensional Normed Linear Spaces

Let $(X, \|\cdot\|)$ be a finite dimensional normed linear space and $\boldsymbol{D} \subset X$ an open and bounded subset. Assume $T : \overline{\boldsymbol{D}} \subset X \to X$ is a compact operator and $\boldsymbol{p} \notin (I - T)(\partial \boldsymbol{D})$. Then $d_{LS}(I_T, \boldsymbol{D}, \boldsymbol{p}) = d_B(I_T, \boldsymbol{D}, \boldsymbol{p})$.

Proof 11.1.2
This follows from our discussions. ■

Homework

Exercise 11.1.3 *Let A be an $n \times n$ real-valued matrix. Let $\boldsymbol{D} \subset \Re^n$ be bounded. Prove $A(\overline{\boldsymbol{D}})$ and $\overline{A(\overline{\boldsymbol{D}})}$ are compact.*

Exercise 11.1.4 *Let $\boldsymbol{D}$ be the finite rectangle $\Pi_{i=1}^n [a_i, b_i]$ where $a_i < b_i$. Let $\epsilon = 1/2^m$ and choose an ϵ net for $\boldsymbol{D}$. Construct the functions m_i. Do this first for $\Re^2$ so you can see what these functions look like. Feel free to use MATLAB to help you with the visualization. Then, do the case $\Re^3$ and move up to the general case $\Re^n$.*

Exercise 11.1.5 *Let $\boldsymbol{D}$ be the finite rectangle $\Pi_{i=1}^n [a_i, b_i]$ where $a_i < b_i$. Let $\epsilon = 1/2^m$ and choose an ϵ net for $\boldsymbol{D}$. Construct the function F_ϵ. Do this first for $\Re^2$ so you can see what the function looks like. Feel free to use MATLAB to help you with the visualization. Then, do the case $\Re^3$ and move up to the general case $\Re^n$.*

Exercise 11.1.6 *Let A be an $n \times n$ real-valued matrix. Let $\boldsymbol{D} \subset \Re^n$ be bounded. Prove $A(\overline{\boldsymbol{D}})$ and $\overline{A(\overline{\boldsymbol{D}})}$ are compact. Let $\epsilon = 1/2^m$ and choose an ϵ net for $\boldsymbol{D}$. Construct the functions $T_m =$*

$F_{1/2^m} \circ A$. Do this first for $\Re^2$ so you can see what the function looks like. Feel free to use MATLAB to help you with the visualization. Then, do the case $\Re^3$ and move up to the general case $\Re^n$.

Exercise 11.1.7 *Let A be an $n \times n$ real-valued matrix. Find $I - A$ and $I - T_{1/2^m}$. Let $\boldsymbol{D}$ be the finite rectangle $\Pi_{i=1}^n [a_i, b_i]$ where $a_i < b_i$, and discuss $d_B(I - T_{1/2^m}, \boldsymbol{D}, \boldsymbol{p})$.*

Exercise 11.1.8 *Do all of the above for various $\boldsymbol{D}$ with the matrix A given by*

$$A = \begin{bmatrix} 1 & 2 \\ -3 & 4 \end{bmatrix}$$

Exercise 11.1.9 *Do all of the above for various $\boldsymbol{D}$ with the matrix A given by*

$$A = \begin{bmatrix} 6 & 1 & 3 \\ 4 & -1 & 2 \\ -1 & 3 & 6 \end{bmatrix}$$

Exercise 11.1.10 *Do all of the above for various $\boldsymbol{D}$ with the matrix A given by*

$$A = \begin{bmatrix} 4 & 2 & 1 & 7 \\ 6 & -1 & 3 & 5 \\ 2 & -3 & -5 & 1 \\ 1 & 1 & 1 & 4 \end{bmatrix}$$

11.2 Properties of Leray - Schauder Degree

We now turn to proving useful properties of Leray - Schauder degree. We will continue to use the notation from the previous section. To set the stage, we repeat the basics. Let $(X, \|\cdot\|)$ be a normed linear space and $\boldsymbol{D} \subset X$ an open and bounded subset. Assume $T : \overline{\boldsymbol{D}} \subset X \rightarrow X$ is a compact operator and $\boldsymbol{p} \notin (I - T)(\partial \boldsymbol{D})$. Let $\boldsymbol{K} = T(\overline{\boldsymbol{D}})$. For a given $\epsilon > 0$, let $\boldsymbol{V_\epsilon} = \{\boldsymbol{v}_1, \ldots, \boldsymbol{v}_{\boldsymbol{p}(\epsilon)}\}$ denote a corresponding ϵ net for $\boldsymbol{K}$. Define the functions $m_i : \boldsymbol{K} \rightarrow X$ by

$$m_i(\boldsymbol{y}) = \begin{cases} \epsilon - \|\boldsymbol{y} - \boldsymbol{v_i}\|, & \|\boldsymbol{y} - \boldsymbol{v_i}\| \leq \epsilon \\ 0, & \|\boldsymbol{y} - \boldsymbol{v_i}\| > \epsilon \end{cases}$$

for $1 \leq i \leq p(\epsilon)$. Define the mappings $T_\epsilon = F_\epsilon \circ T : \boldsymbol{D} \rightarrow X$ by

$$T_\epsilon(\boldsymbol{x}) = \frac{\sum_{i=1}^{p(\epsilon)} m_i(T(\boldsymbol{x}))\, \boldsymbol{v_i}}{\sum_{i=1}^{p(\epsilon)} m_i(T(\boldsymbol{x}))}$$

Let X_ϵ be the finite dimensional subspace generated by $\boldsymbol{V_\epsilon}$. Adjoin a point $\boldsymbol{q} \in \boldsymbol{D}$ to X_ϵ and define $Y_\epsilon = span\{\boldsymbol{V_\epsilon}, \boldsymbol{p}, \boldsymbol{q}\}$. Let $\boldsymbol{D_n} = Y_n \cap \boldsymbol{D}$. Then, for any monotonically decreasing sequence (ϵ_n) with $\epsilon_n \rightarrow 0$, we let $T_n = F_{\epsilon_n} \circ T$ and define the Leray - Schauder degree to be $d_{LS}(I - T, \boldsymbol{D}, \boldsymbol{p}) = \lim_{n\rightarrow\infty} d_B(I - T_n, \boldsymbol{D_n}, \boldsymbol{p})$. This value is well-defined as it is independent of the choice of approximations T_n.

Theorem 11.2.1 Existence of a Solution to $(I - T)(\boldsymbol{x}) = \boldsymbol{0}$

Let $(X, \|\cdot\|)$ be a normed linear space and $\boldsymbol{D} \subset X$ an open and bounded subset. Assume $T : \overline{\boldsymbol{D}} \subset X \rightarrow X$ is a compact operator and $\boldsymbol{0} \notin (I - T)(\partial \boldsymbol{D})$. Then $d_{LS}(I - T, \boldsymbol{D}, \boldsymbol{0}) \neq 0$ implies there is a solution $\boldsymbol{x} \in \boldsymbol{D}$ to $(I - T)(\boldsymbol{x}) = \boldsymbol{0}$. In other words, there is a fixed point: $T(\boldsymbol{x}) = \boldsymbol{x}$ for some $\boldsymbol{x} \in \boldsymbol{D}$.

Proof 11.2.1
We know there is n_0 so that if $n \geq n_0$, $d_{LS}(I - T, \boldsymbol{D}, \boldsymbol{0}) = d_B(I - T_n, \boldsymbol{D_n}, \boldsymbol{0})$. Then, by the existence theorem for Brower degree, Theorem 10.2.2, there is $\boldsymbol{x_n} \in \boldsymbol{D_n}$ so that $(I - T_n)(\boldsymbol{x_n}) = \boldsymbol{0}$. Since $\boldsymbol{D_n} \subset \boldsymbol{D}$,

$$\|(I - T)(\boldsymbol{x_n}) - \boldsymbol{0}\| = \|(I - T_n)(\boldsymbol{x_n}) + (T_n - T)(\boldsymbol{x_n}) - \boldsymbol{0}\| = \|(T_n - T)(\boldsymbol{x_n}) - \boldsymbol{0}\| < \epsilon_n$$

Let $\boldsymbol{y_n} = (I - T)(\boldsymbol{x_n})$. Then we know $\boldsymbol{y_n} \to \boldsymbol{0}$.

We can do this for each $n \geq n_0$ giving us a sequence $(\boldsymbol{x_n}) \subset \boldsymbol{D}$. Since T is compact, $(T(\boldsymbol{x_n}))$ contains a subsequence $(T(\boldsymbol{x_{n,1}}))$ so that $T(\boldsymbol{x_{n,1}}) \to \boldsymbol{y} \in \overline{(T(\boldsymbol{x_n}))}$. We see

$$\lim_{n\to\infty} (I - T)(\boldsymbol{x_{n,1}}) = \boldsymbol{0} \Longrightarrow \lim_{n\to\infty} \boldsymbol{x_{n,1}} - \lim_{n\to\infty} \boldsymbol{y_{n,1}} \Longrightarrow \lim_{n\to\infty} \boldsymbol{x_{n,1}} = \boldsymbol{y}$$

So $\boldsymbol{y} \in \overline{\boldsymbol{D}}$. By the continuity of $(I - T)$, we also have

$$\begin{aligned} \boldsymbol{0} &= \lim_{n\to\infty} (I - T)(\boldsymbol{x_{n,1}}) \\ &= \lim_{n\to\infty} (\boldsymbol{x_{n,1}}) - T(\lim_{n\to\infty} (\boldsymbol{x_{n,1}})) = \boldsymbol{y} - T(\boldsymbol{y}) \end{aligned}$$

We assumed $\boldsymbol{0} \notin (I - T)(\partial \boldsymbol{D})$, so $\boldsymbol{y} \notin \partial \boldsymbol{D}$. We conclude $T(\boldsymbol{y}) = \boldsymbol{y}$ for a $\boldsymbol{y} \in \boldsymbol{D}$. ■

Now let's look at homotopies of compact mappings.

Definition 11.2.1 Homotopies of Compact Mappings

Let $(X, \|\cdot\|)$ be a normed linear space and $\boldsymbol{E} \subset X$. Let $K(\boldsymbol{E}) = \{T : \boldsymbol{E} \subset X \to X, T \text{ compact}\}$; the set of compact mappings on $\boldsymbol{E}$. Let $H : [0,1] \to K(\boldsymbol{E})$; i.e., $H(t) : \boldsymbol{E} \to X$ is compact for all $0 \leq t \leq 1$. We say H is a homotopy of compact mappings if for all $\epsilon > 0$ and for all bounded $\boldsymbol{M} \subset \boldsymbol{E}$, there is $\delta_{\epsilon,M} > 0$ so that $|t_1 - t_2| < \delta_{\epsilon,M}$ implies $\|H(t_1)(\boldsymbol{x}) - H(t_2)(\boldsymbol{x})\| < \epsilon$.

We can now prove an invariance under homotopy for Leray - Schauder degree.

Theorem 11.2.2 Invariance under Compact Homotopy for Leray - Schauder Degree

Let $(X, \|\cdot\|)$ be a normed linear space and $\boldsymbol{D} \subset X$ be open and bounded. Let $H : [0,1] \to K(\overline{\boldsymbol{D}})$ be a homotopy of compact operators on $\overline{\boldsymbol{D}}$. Assume $(I - H(t))(\partial \boldsymbol{D}) \neq \boldsymbol{0}$ for $0 \leq t \leq 1$. Then $d_{LS}(I - H(t), \boldsymbol{D}, \boldsymbol{0})$ is constant on $[0,1]$.

Proof 11.2.2
We prove this in a series of steps.

(i): We show there is $r > 0$ so that for all $\boldsymbol{x} \in \partial \boldsymbol{D}$ and $0 \leq t \leq 1$, $\|(I - H(t))(\boldsymbol{x}) - \boldsymbol{0}\| \geq r > 0$.

Assume there does not exist such a positive r. Then, there is a sequence $(t_n, \boldsymbol{x_n})$ with $t_n \in [0,1]$ and $\boldsymbol{x_n} \in \partial \boldsymbol{D}$ so that $(I - H(t_n))(\boldsymbol{x_n}) = \boldsymbol{y_n}$ and $\|(I - H(t_n))(\boldsymbol{x_n})\| = \|\boldsymbol{y_n}\| < (1/n)$. This implies $\boldsymbol{y_n} \to \boldsymbol{0}$. Thus, $\boldsymbol{x_n} = H(t_n)(\boldsymbol{x_n}) + \boldsymbol{y_n}$.

Since $(t_n) \subset [0,1]$, which is compact, there is a subsequence $(t_{n,1})$ which converges to a point t_0 in $[0,1]$. Since $(\boldsymbol{x_n}) \subset \partial \boldsymbol{D}$, which is closed and bounded, we know it is a bounded sequence. Consider the subsequence $(\boldsymbol{x_{n,1}})$. Since H is a compact homotopy, $H(t_0)$ is compact. Hence, there is a subsequence $(\boldsymbol{x_{n,2}}) \subset (\boldsymbol{x_{n,1}})$ so that $H(t_0)(\boldsymbol{x_{n,2}})$ converges to a *point* $\boldsymbol{y_0}$.

Note the subsequence $(t_{n,2})$ *also converges to* t_0. *We thus have* $\lim_{n\to\infty} H(t_0)(\boldsymbol{x_{n,2}}) = \boldsymbol{y_0}$, $\lim_{n\to\infty}(t_{n,2}) = t_0$ *and by the continuity of* H *with respect to* t, $\lim_{n\to\infty} H(t_{n,2}) = H(t_0)$. *Therefore, given* $\epsilon > 0$, *there is* N *so that* $n > N$ *implies*

$$\|H(t_0)(\boldsymbol{x_{n,2}}) - \boldsymbol{y_0}\| \;<\; \frac{\epsilon}{2}, \quad \|H(t_0) - H(t_{n,2})\| < \frac{\epsilon}{2(1+\sup_n \|\boldsymbol{x_{n,2}})\|}$$

as we know $sup_n\|\boldsymbol{x_{n,2}}\| < \infty$ *as the sequence is bounded. Hence,*

$$\begin{aligned}
\|H(t_0)(\boldsymbol{x_{n,2}}) - \boldsymbol{y_0}\| &\leq \|H(t_0)(\boldsymbol{x_{n,2}}) - H(t_{n,2})(\boldsymbol{x_{n,2}})\| + \|H(t_0)(\boldsymbol{x_{n,2}}) - \boldsymbol{y_0}\| \\
&\leq \|H(t_0)(\boldsymbol{x_{n,2}}) - H(t_{n,2})\|\;\|\boldsymbol{x_{n,2}}\| + \frac{\epsilon}{2} \\
&< \frac{\epsilon}{2(1+\sup_n \|\boldsymbol{x_{n,2}})\|}\|\boldsymbol{x_{n,2}}\| + \frac{\epsilon}{2} < \epsilon
\end{aligned}$$

We conclude $\lim_{n\to\infty} H(t_{n,2})(\boldsymbol{x_{n,2}}) = \boldsymbol{y_0}$. *Therefore*

$$\begin{aligned}
\lim_{n\to\infty} \boldsymbol{x_{n,2}} &= \lim_{n\to\infty} H(t_{n,2})(\boldsymbol{x_{n,2}}) + \lim_{n\to\infty} \boldsymbol{y_{n,2}} \\
&\Longrightarrow \lim_{n\to\infty} \boldsymbol{x_{n,2}} = \boldsymbol{y_0} + \boldsymbol{0}
\end{aligned}$$

Since $\boldsymbol{x_{n,2}} \in \partial\boldsymbol{D}$ *and* $\partial\boldsymbol{D}$ *is closed, we have* $\boldsymbol{y_0} \in \partial\boldsymbol{D}$. *Then*

$$\lim_{n\to\infty}(\boldsymbol{x_{n,2}} - H(t_0)(\boldsymbol{x_{n,2}})) \;=\; \boldsymbol{y_0} - H(t_0)(\boldsymbol{y_0}) = (I - H(t_0))(\boldsymbol{y_0})$$

Hence, we have shown there is $\boldsymbol{y_0} \in \partial\boldsymbol{D}$ *with* $(I - H(t_0))(\boldsymbol{y_0}) = \boldsymbol{0}$ *which is not possible. So there must exist* $r > 0$ *so that for all* $\boldsymbol{x} \in \partial\boldsymbol{D}$ *and* $0 \leq t \leq 1$, $\|(I - H(t))(\boldsymbol{x}) - \boldsymbol{0}\| \geq r > 0$.

(ii): *Given* $t_1 \in [0,1]$, *there is a* $\delta > 0$ *and* n_0 *so that for all* $\boldsymbol{x} \in \overline{\boldsymbol{D}}$, *for all* $|t - t_1| < \delta$ *and for all* $n \geq n_0$, $\|H(t)(\boldsymbol{x}) - H_n(t_1)(\boldsymbol{x})\| < (r/2)$.

Let $t_1 \in [0,1]$. *Let* $H_n(t_1) = F_{\epsilon_n}(H(t_1))$ *where* $F_{\epsilon_n} : H(t_1)(\overline{\boldsymbol{D}}) \to X$ *is defined as we discussed earlier. Furthermore, since* ϵ_n *decreases monotonically to zero, there is* n_0 *so that if* $n \geq n_0$, *for all* $\boldsymbol{x} \in \overline{\boldsymbol{D}}$, $\|H_n(t_1)(\boldsymbol{x}) - H(t_1)(\boldsymbol{x})\| < \epsilon_n < (r/4)$. *Since* H *is a compact homotopy, there is a* $\delta > 0$ *so that for all* $\boldsymbol{x} \in \overline{\boldsymbol{D}}$, $|t - t_1| < \delta$, $\|H(t)(\boldsymbol{x}) - H(t_1)(\boldsymbol{x})\| < (r/4)$. *Hence, for all* $\boldsymbol{x} \in \overline{\boldsymbol{D}}$, *for all* $|t - t_1| < \delta$ *and for all* $n \geq n_0$,

$$\begin{aligned}
\|H(t)(\boldsymbol{x}) - H_n(t_1)(\boldsymbol{x})\| &\leq \|H(t)(\boldsymbol{x}) - H(t_1)(\boldsymbol{x})\| + \|H(t_1)(\boldsymbol{x}) - H_n(t_1)(\boldsymbol{x})\| \\
&< (r/4) + (r/4) < (r/2)
\end{aligned}$$

(iii): $d_{LS}(I - H(t), \boldsymbol{D}, \boldsymbol{0})$ *is constant on* $[0,1]$.

We know that Leray - Schauder degree is independent of the choice of approximations T_{ϵ_n} *which are constructed from a choice of* ϵ *net* $\boldsymbol{V_{\epsilon_n}} = \{\boldsymbol{v_1}, \ldots, \boldsymbol{v_{p(\epsilon_n)}}\}$ *for* $\boldsymbol{K}$ *where* $\boldsymbol{K} = T(\overline{\boldsymbol{D}})$. *Recall,* X_{ϵ_n} *is the finite dimensional subspace generated by* $\boldsymbol{V_{\epsilon_n}}$. *We adjoin a point* $\boldsymbol{q} \in \boldsymbol{D}$ *to* X_{ϵ_n} *and define* $Y_{\epsilon_n} = span\{\boldsymbol{V_{\epsilon_n}}, \boldsymbol{p}, \boldsymbol{q}\}$. *Let* $\boldsymbol{D_n} = Y_n \cap \boldsymbol{D}$ *and* $T_{\epsilon_n} = F_{\epsilon_n} \circ T : \overline{\boldsymbol{D_n}} \to Y_{\epsilon_n}$. *Then,* $d_{LS}(I - T, \boldsymbol{D}, \boldsymbol{p}) = \lim_{n\to\infty} d_B(I - T_n, \boldsymbol{D_n}, \boldsymbol{p})$. *This value well-defined as it is independent of the choice of approximations* T_n.

Further, from Step (iv) in our discussion of the existence of Leray - Schauder degree, any approximate mappings Φ of this type that satisfy $\|H(t) - \Phi\| < (r/2)$, we have Leray - Schauder degree is a constant. Thus, we can say for all $|t - t_1| < \delta$, for all $n \geq n_0$, $d_{LS}(I - H(t), \boldsymbol{D}, \boldsymbol{0}) = d_B(I - H_n(t_1), \boldsymbol{D_n}, \boldsymbol{0})$. But for $n \geq n_0$, $d_B(I - H_n(t_1), \boldsymbol{D_n}, \boldsymbol{0}) = d_{LS}(I - H(t_1), \boldsymbol{D}, \boldsymbol{0})$. We conclude $d_{LS}(I - H(t), \boldsymbol{D}, \boldsymbol{0}) = d_{LS}(I - H(t_1), \boldsymbol{D}, \boldsymbol{0})$, for $|t - t_1| < \delta$.

The interval $[0, 1]$ can be covered by a finite number of intervals of length $\delta/2$. From the above results, we see $d_{LS}(I - H(t_2), \boldsymbol{D}, \boldsymbol{0}) = d_{LS}(I - H(t_1), \boldsymbol{D}, \boldsymbol{0})$ for all t_1 and t_2 in $[0, 1]$. This says Leray - Schauder degree is constant.

■

Now we can prove a powerful fixed point theorem.

Theorem 11.2.3 Schauder Fixed Point Theorem

Let $(X, \|\cdot\|)$ be a normed linear space and $\boldsymbol{K} \subset X$ be an open, bounded and convex set. Let $T : \overline{\boldsymbol{K}} \to X$ be compact with $T(\overline{\boldsymbol{K}}) \subset \overline{\boldsymbol{K}}$. Then T has a fixed point $\boldsymbol{x} \in \overline{\boldsymbol{K}}$; i.e., $T(\boldsymbol{x}) = \boldsymbol{x}$.

Proof 11.2.3

For simplicity, first let $K = \{\boldsymbol{x} \in X | \|\boldsymbol{x}\| < \rho\}$ for some $\rho > 0$. Let $H(\boldsymbol{x}, t) = tT(\boldsymbol{x})$. Then, H is compact for all $t \in [0, 1]$. If there is $\boldsymbol{x} \in \partial \boldsymbol{K}$ with $\boldsymbol{x} = T(\boldsymbol{x})$, we have found the $\boldsymbol{x}$ we seek. If not, this $\boldsymbol{0} \notin (I - T)(\partial \boldsymbol{K})$. Thus, $(I - H(\cdot, 1))(\boldsymbol{x}) \neq \boldsymbol{0}$ on $\partial \boldsymbol{K}$. We know $\boldsymbol{0} \notin \partial \boldsymbol{K}$. So what is left is the case where there is a point $t_0 \in (0, 1)$ with t_0 in $[0, 1]$ with $\boldsymbol{x} = t_0 T(\boldsymbol{x})$. Then $\|T(\boldsymbol{x})\| = (1/t_0)\|\boldsymbol{x_0}\| > \|\boldsymbol{x_0}\|$. However, by assumption, $T(\overline{\boldsymbol{K}}) \subset \overline{\boldsymbol{K}}$, which implies $\|T(\boldsymbol{x})\| \leq \rho$. This is a contradiction. We conclude $\boldsymbol{0} \notin (I - tT)(\partial \boldsymbol{K})$ for $0 \leq t \leq 1$. Hence, $\boldsymbol{0} \notin (I - H(\cdot, t))(\partial \boldsymbol{K})$ for $0 \leq t \leq 1$. We conclude $d_{LS}(I - H(\cdot, 0), \boldsymbol{K}, \boldsymbol{0}) = d_{LS}(I - H(\cdot, 1), \boldsymbol{K}, \boldsymbol{0})$ or $d_{LS}(I, \boldsymbol{K}, \boldsymbol{0}) = d_{LS}(I - T, \boldsymbol{K}, \boldsymbol{0})$.

Now the zero map, $\boldsymbol{0}$, is compact, so the approximations to $\boldsymbol{0}$ follow this scheme. We have $\overline{\boldsymbol{K}} = \{\boldsymbol{0}\}$. For a given $\epsilon > 0$, the ball $B(\boldsymbol{0}, \epsilon)$ is an ϵ net for the closure of the range of $\boldsymbol{0}$ which is $\{\boldsymbol{0}\}$. So all of our subspaces are very simple. The function $m : \{\boldsymbol{0}\} \to X$ is defined by $m(\boldsymbol{0}) = \epsilon$. We set X_ϵ to be the finite dimensional subspace generated by $\boldsymbol{V_\epsilon} = \{\boldsymbol{0}\}$. We do not need to adjoin a point $\boldsymbol{q} \in \boldsymbol{K}$ to X_ϵ and so here $Y_\epsilon = span\{\boldsymbol{0}, \boldsymbol{p}, \} = span\{\boldsymbol{p}\}$. Let $\boldsymbol{D_\epsilon} = Y_\epsilon \cap \boldsymbol{D} = span\{\boldsymbol{p}\} \cap \boldsymbol{D}$. Then define the mappings $\boldsymbol{0}_\epsilon : \boldsymbol{D} \to X$ by

$$\boldsymbol{0}_\epsilon(\boldsymbol{x}) = \boldsymbol{0}$$

Then, as we expect $\|0(\boldsymbol{x}) - 0_\epsilon(\boldsymbol{x})\| < \epsilon$.

We know $d_{LS}(I - \boldsymbol{0}, \boldsymbol{K}, \boldsymbol{0}) = d_B(I - \boldsymbol{0}\epsilon, \boldsymbol{D_\epsilon}, \boldsymbol{0})$ for $\epsilon > 0$ sufficiently small. This is Brouwer degree on a finite dimensional normed linear space and by choosing orientation preserving maps, we can make this $+1$ always. Thus, $d_{LS}(I - T, \boldsymbol{K}, \boldsymbol{0}) \neq 0$ and so by Theorem 11.2.1, we can then say there is a solution to $(I - T)(\boldsymbol{x}) = \boldsymbol{0}$ in K.

Now if $\boldsymbol{K}$ was open, bounded and convex with $\boldsymbol{0} \in \boldsymbol{K}$, what do we do? All we have to do is look at the case where there is a point $t_0 \in (0, 1)$ with t_0 in $[0, 1]$ with $\boldsymbol{x} = t_0 T(\boldsymbol{x})$ and $\boldsymbol{x} \in \partial \boldsymbol{K}$. This would mean $\boldsymbol{x}/t_0 \in \overline{\boldsymbol{K}}$. But since $\boldsymbol{x} \in \partial \boldsymbol{K}$, we must have $\boldsymbol{x}/t_0 \in \boldsymbol{K}^C$ implying $T(\boldsymbol{x}) \in \boldsymbol{K}^C$ which is impossible. Hence, we conclude $\boldsymbol{0} \notin (I - tT)(\partial \boldsymbol{K})$ for $0 \leq t \leq 1$. The rest of the argument is the same. ■

A common term we use when we search for solutions to the equation $G(\boldsymbol{x}) = \boldsymbol{0}$ in some normed linear space is to first find an upper bound on the norm of the possible solutions. This does not mean a solution exists, of course, but it tells us if one did, there is a constant R so that $\|\boldsymbol{x}\| < R$. The number R is called an apriori bound. Next, we use this idea to examine the solutions to $(I - H(t))(\boldsymbol{x}) = \boldsymbol{0}$ for a given homotopy of compact operators.

Theorem 11.2.4 Apriori Bounded Solutions and Constant Leray - Schauder Degree

Let $(X, \|\cdot\|)$ be a normed linear space and $\boldsymbol{K} \subset X$ be an open and bounded set. Let $H : \overline{\boldsymbol{K}} \times [0,1] \to X$ be a compact homotopy. Assume there is an apriori bound A on the solutions to $(I - H(t))(\boldsymbol{x}) = \boldsymbol{0}$ for all $0 \le t \le 1$. This means if $(I - H(t))(\boldsymbol{x}) = \boldsymbol{0}$, then $\|\boldsymbol{x}\| < A$. Then, for any $r \ge 1$, $\boldsymbol{\Omega_r} = \{\boldsymbol{x} \in \boldsymbol{K} | \|\boldsymbol{x}\| < rA\}$, $d_{LS}(I - H(t), \boldsymbol{\Omega_r}, \boldsymbol{0})$ exists and is constant.

Proof 11.2.4
Since $\boldsymbol{0} \notin (I - H(t))\partial\boldsymbol{\Omega_r}$, Theorem 11.2.2 applies and we see $d_{LS}(I - H(t), \boldsymbol{\Omega_r}, \boldsymbol{0})$ exists and is constant. ∎

Homework

Exercise 11.2.1 *Let $X = (C([0,1]), \|\cdot\|_\infty)$ and let $f : [0,T] \times X \to X$ be compact and continuous and satisfy the growth condition $|f(t,x)| \le c(1 + \|x\|_\infty)$ for some $c > 0$. Then we will show there is $x \in Y = C([0,T], \|\cdot\|_\infty)$ so that $x(t) = x_0 + \int_0^t f(s, x(s))ds$ on $[0,T]$. Note, this is equivalent to saying there is a differentiable function $x \in Y$ which solves the initial value problem $x'(t) = f(t, x(t))$, $x(0) = x_0$ on $[0,T]$.*

- *Prove $F : Y \to Y$ defined by $F(y) = x_0 + \int_0^t f(s, y(s))ds$ is continuous and compact.*
- *Consider the solutions y to $y(t) = \lambda(x_0 + \int_0^t f(s, x(s))ds)$ or $y = \lambda F(y)$. Prove $|y(t)| \le (\|x_0\| + cT) + c\int_0^t |y(s)|ds = \phi(t)$.*
- *Prove $y'(t) \le c\phi(t)$ implying $(\phi(t)e^{-ct})' \le 0$. Hence, $|\phi(t)| \le (\|x_0\| + cT)e^{cT} = c_2$ on $[0,T]$. Hence, all solutions to $y = \lambda F(y)$ are apriori bounded by $\|y\|_\infty \le c_2$.*
- *Choose $r > c_2$. Show $d_{LS}(I - F, B(\boldsymbol{0}, r), \boldsymbol{0}) = d_{LS}(I, B(\boldsymbol{0}, r), \boldsymbol{0}) = 1$ telling us the initial value problem has a solution.*

Exercise 11.2.2 *Can you repeat the previous exercise with the growth condition $|f(t,x)| \le c(1 + \|x\|_\infty + \|x\|_\infty^2)$ for some $c > 0$?*

Exercise 11.2.3 *Can you repeat the previous exercise with the growth condition $|f(t,x)| \le ce^{d\|x\|_\infty}$ for some $c > 0$ and $d > 0$?*

Exercise 11.2.4 *Can you repeat the previous exercise with the growth condition $|f(t,x)| \le ce^{-d\|x\|_\infty}$ for some $c > 0$ and $d > 0$?*

Exercise 11.2.5 *Let $(X, \|\cdot\|)$ be a complete normed linear space and define for any $r > 0$ and $x_0 \in X$, $R : X \to B(x_0, r) \subset X$ by*

$$R(x) = \begin{cases} x, & \|x - x_0\| < r \\ x_0 + r\frac{x - x_0}{\|x - x_0\|}, & \|x - x_0\| \ge r \end{cases}$$

- *Prove $\|R(x) - R(y)\| \le 2\|x - y\|$ and if X is a Hilbert space, we can say $\|R(x) - R(y)\| \le \|x - y\|$.*

- *Prove R is continuous and so it is a retraction as defined by Definition 4.2.1.*

Exercise 11.2.6 *Let X be a complete normed linear space and $F : X \to X$ be compact and continuous. Prove either $x - tF(x) = 0$ has a solution for each $0 \le t \le 1$ or $S = \{x : x = tF(x)\}$ is an unbounded set for some $0 < t < 1$.*
Hint: *If $x - t_0F(x)$ has no solution for some $t_0 \in (0, 1]$, let $F_0 = t_0F$.*

- *For any $r > 0$, let R be the map defined in the previous exercise. Prove RF_0 has a fixed point x^* in $B(0, r)$. Show this means $\|F_0(x^*)\| > r$.*
- *Then, show for $\mu = rt_0\|F_0(x^*)\|^{-1} < 1$, $x^* = \mu F(x^*)$. This shows S is unbounded.*

Now we are ready to tackle the Leray - Schauder version of Borsuk's Theorem.

Theorem 11.2.5 Borsuk's Theorem for Leray - Schauder Degree

Let $(X, \|\cdot\|)$ be a normed linear space and $\boldsymbol{D} \subset X$ be an open, bounded and symmetric set with $\boldsymbol{0} \in \boldsymbol{D}$. Let $T : \overline{\boldsymbol{D}} \to X$ be odd and compact. Assume $\boldsymbol{0} \notin (I - T)(\partial \boldsymbol{D})$. Then, $d_{LS}(I - T, \boldsymbol{D}, \boldsymbol{0})$ is an odd integer.

Proof 11.2.5
Let $\boldsymbol{K} = T(\overline{\boldsymbol{D}})$. For any $\epsilon > 0$, let $\boldsymbol{V_\epsilon} = \{\boldsymbol{v_1}, \ldots, \boldsymbol{v_{p(\epsilon)}}\}$ denote a corresponding ϵ net for $\boldsymbol{K}$. Extend $\boldsymbol{V_\epsilon}$ as follows: for convenience of notation, we will now let p_ϵ be denoted by just p:

$$\boldsymbol{W_\epsilon} = \{\boldsymbol{v_1}, \ldots, \boldsymbol{v_p}, \boldsymbol{v_{p+1}} = -\boldsymbol{v_1}, \ldots, \boldsymbol{v_{2p}} = -\boldsymbol{v_p}\}$$

and for all $\boldsymbol{x} \in T(\overline{\boldsymbol{D}})$, define, using our standard m_i functions $m_i : \boldsymbol{K} \to X$ for $1 \le i \le p$ by

$$m_i(\boldsymbol{y}) = \begin{cases} \epsilon - \|\boldsymbol{y} - \boldsymbol{v_i}\|, & \|\boldsymbol{y} - \boldsymbol{v_i}\| \le \epsilon \\ 0, & \|\boldsymbol{y} - \boldsymbol{v_i}\| > \epsilon \end{cases}$$

and for $1 \le i \le p$ by

$$m_{p+i}(\boldsymbol{y}) = \begin{cases} \epsilon - \|\boldsymbol{y} - -\boldsymbol{v_i}\|, & \|\boldsymbol{y} - -\boldsymbol{v_i}\| \le \epsilon \\ 0, & \|\boldsymbol{y} - -\boldsymbol{v_i}\| > \epsilon \end{cases} = \begin{cases} \epsilon - \|\boldsymbol{y} + \boldsymbol{v_i}\|, & \|\boldsymbol{y} + \boldsymbol{v_i}\| \le \epsilon \\ 0, & \|\boldsymbol{y} + \boldsymbol{v_i}\| > \epsilon \end{cases}$$

Then, define the functions $F_\epsilon : \boldsymbol{K} \to X$ by

$$F_\epsilon(\boldsymbol{x}) = \frac{\sum_{i=1}^{2p} m_i(\boldsymbol{x})\boldsymbol{v_i}}{\sum_{i=1}^{2p} m_i(\boldsymbol{x})}$$

Define finite dimensional subspaces as follows: $Y_\epsilon = \{\boldsymbol{W_\epsilon}, \boldsymbol{0}\}$. If $\|-\boldsymbol{x} - \boldsymbol{v_i}\| \le \epsilon$, this says $\|-\boldsymbol{x} + \boldsymbol{v_{p+i}}\| \le \epsilon$. Thus,

$$\begin{aligned} 1 \le i \le p,\ \|\boldsymbol{v_{i+p}} - \boldsymbol{x}\| \le \epsilon &\implies m_{i+p}(\boldsymbol{x}) > 0 \\ p < i \le 2p,\ \|\boldsymbol{v_{i-p}} - \boldsymbol{x}\| \le \epsilon &\implies m_{i-p}(\boldsymbol{x}) > 0 \end{aligned}$$

Let $J = \{i | m_i(-\boldsymbol{x}) > 0\}$. Then,

$$\begin{aligned} F_\epsilon(T(-\boldsymbol{x})) &= F_\epsilon(-T(\boldsymbol{x})) = \frac{\sum_{i \in J} m_i(-T(\boldsymbol{x}))\boldsymbol{v_i}}{\sum_{i \in J} m_i(-T(\boldsymbol{x}))} \\ &= \frac{\sum_{i \in J \cap \{1,\ldots,p\}} m_i(-T(\boldsymbol{x}))\boldsymbol{v_i} + \sum_{i \in J \cap \{p+1,\ldots,2p\}} m_i(-T(\boldsymbol{x}))(-\boldsymbol{v_i})}{\sum_{i \in J \cap \{1,\ldots,p\}} m_i(-T(\boldsymbol{x})) + \sum_{i \in J \cap \{p+1,\ldots,2p\}} m_i(-T(\boldsymbol{x}))} \end{aligned}$$

Now,

$$m_{i+p}(T(\boldsymbol{x})) = \begin{cases} \epsilon - \|T(\boldsymbol{x}) + \boldsymbol{v_i}\|, & \|T(\boldsymbol{x}) + \boldsymbol{v_i}\| \leq \epsilon \\ 0, & \|T(\boldsymbol{x}) + \boldsymbol{v_i}\| > \epsilon \end{cases} = m_i(-T(\boldsymbol{x}))$$

$$m_{i+p}(-T(\boldsymbol{x})) = \begin{cases} \epsilon - \|-T(\boldsymbol{x}) + \boldsymbol{v_i}\|, & \|-T(\boldsymbol{x}) + \boldsymbol{v_i}\| \leq \epsilon \\ 0, & \|-T(\boldsymbol{x}) + \boldsymbol{v_i}\| > \epsilon \end{cases} = m_i(T(\boldsymbol{x}))$$

So we have

$$\begin{aligned} F_\epsilon(-T(\boldsymbol{x})) &= \frac{\sum_{i \in J \cap \{1,\ldots,p\}} m_i(-T(\boldsymbol{x}))\boldsymbol{v_i} + \sum_{i \in J \cap \{p+1,\ldots,2p\}} m_i(-T(\boldsymbol{x}))(-\boldsymbol{v_i})}{\sum_{i \in J \cap \{1,\ldots,p\}} m_i(-T(\boldsymbol{x})) + \sum_{i \in J \cap \{p+1,\ldots,2p\}} m_i(-T(\boldsymbol{x}))} \\ &= \frac{\sum_{i \in J \cap \{1,\ldots,p\}} m_{i+p}(T(\boldsymbol{x}))\boldsymbol{v_i} + \sum_{i \in J \cap \{p+1,\ldots,2p\}} m_{i-p}(T(\boldsymbol{x}))(-\boldsymbol{v_i})}{\sum_{i \in J \cap \{1,\ldots,p\}} m_{i+p}(T(\boldsymbol{x})) + \sum_{i \in J \cap \{p+1,\ldots,2p\}} m_{i-p}(T(\boldsymbol{x}))} \\ &= \frac{-\sum_{i \in J \cap \{1,\ldots,p\}} m_{i+p}(T(\boldsymbol{x}))\boldsymbol{v_{i+p}} - \sum_{i \in J \cap \{p+1,\ldots,2p\}} m_{i-p}(T(\boldsymbol{x}))(\boldsymbol{v_{i-p}})}{\sum_{i \in J \cap \{1,\ldots,p\}} m_{i+p}(T(\boldsymbol{x})) + \sum_{i \in J \cap \{p+1,\ldots,2p\}} m_{i-p}(T(\boldsymbol{x}))} \end{aligned}$$

We see the right-hand side is $-F_\epsilon(T(\boldsymbol{x}))$. Since $F_\epsilon \circ T$ is continuous for all ϵ, we see $F_\epsilon \circ T$ is an odd continuous map. By the proof of the existence of Leray - Schauder Degree, we know $d_{LS}(I - T, \boldsymbol{D}, \boldsymbol{0}) = d_B(I - F_\epsilon \circ T, \boldsymbol{D_\epsilon}, \boldsymbol{0})$ for sufficiently small ϵ. Now apply Borsuk's Theorem for Brouwer degree on a finite dimensional normed linear space to conclude $d_B(I - F_\epsilon \circ T, \boldsymbol{D_\epsilon}, \boldsymbol{0})$ is odd. Thus, $d_{LS}(I - T, \boldsymbol{D}, \boldsymbol{0})$ is odd. ■

Next is the standard corollary to Borsuk's Theorem which gives a condition we can use to determine if a degree is odd.

Theorem 11.2.6 Corollary to Borsuk's Theorem for Leray - Schauder Degree

Let $(X, \|\cdot\|)$ be a normed linear space and $\boldsymbol{D} \subset X$ be an open, bounded and symmetric set with $\boldsymbol{0} \in \boldsymbol{D}$. Let $T : \overline{\boldsymbol{D}} \to X$ be compact. Assume $\boldsymbol{0} \notin (I - T)(\partial \boldsymbol{D})$. Further, assume

$$\frac{\boldsymbol{x} - T(\boldsymbol{x})}{\|\boldsymbol{x} - T(\boldsymbol{x})\|} \neq \frac{-\boldsymbol{x} - T(-\boldsymbol{x})}{\|-\boldsymbol{x} - T(-\boldsymbol{x})\|}, \quad \forall \boldsymbol{x} \in \partial \boldsymbol{D}$$

Then, $d_{LS}(I - T, \boldsymbol{D}, \boldsymbol{0})$ is an odd integer.

Proof 11.2.6
Let $g(\boldsymbol{x}) = T(\boldsymbol{x}) - T(-\boldsymbol{x})$. We see g is odd and compact. Let $H(\boldsymbol{x}, t) = T(\boldsymbol{x}) - tT(-\boldsymbol{x})$ for all $0 \leq t \leq 1$. Then H is a compact homotopy. Let $\boldsymbol{x} \in \partial \boldsymbol{D}$. Consider $\|(I - H(\cdot, t))(\boldsymbol{x})\| = \|\boldsymbol{x} - T(\boldsymbol{x}) + tT(-\boldsymbol{x})\|$.

(i): Assume $\|\boldsymbol{x} - T(\boldsymbol{x})\| = \|-\boldsymbol{x} - T(-\boldsymbol{x})\|$. Then,

$$\begin{aligned} \|\boldsymbol{x} - H(\boldsymbol{x}, t)\| &= \|\boldsymbol{x} - T(\boldsymbol{x}) + tT(-\boldsymbol{x})\| = \|\boldsymbol{x} - T(\boldsymbol{x}) - (-t\boldsymbol{x} - tT(-\boldsymbol{x})) - t\boldsymbol{x}\| \\ &\leq \|\boldsymbol{x} - T(\boldsymbol{x})\| + \|-t\boldsymbol{x}\| - t\|-\boldsymbol{x} - T(-\boldsymbol{x})\| \geq (1 - t)\|\boldsymbol{x} - T(\boldsymbol{x})\| \end{aligned}$$

Now if $t < 1$, we see $\|\boldsymbol{x} - H(\boldsymbol{x}, t)\| > 0$. If $t = 1$,

$$\begin{aligned} \|\boldsymbol{x} - H(\boldsymbol{x}, 1)\| &= \|\boldsymbol{x} - T(\boldsymbol{x}) + T(-\boldsymbol{x})\| = \|\boldsymbol{x} - T(\boldsymbol{x}) - (-\boldsymbol{x} - T(-\boldsymbol{x})) - \boldsymbol{x}\| \\ &\geq \|\boldsymbol{x}\| + \|\boldsymbol{x} - T(\boldsymbol{x})\| - \|-\boldsymbol{x} - T(-\boldsymbol{x})\| = \boldsymbol{x} > 0 \end{aligned}$$

as $\boldsymbol{0} \notin \partial \boldsymbol{D}$. So in this case, $\|\boldsymbol{x} - H(\boldsymbol{x}, t)\| > 0$ on $\partial \boldsymbol{D} \times [0, 1]$.

(ii): *Assume* $\|\boldsymbol{x} - T(\boldsymbol{x})\| \neq \| -\boldsymbol{x} - T(-\boldsymbol{x})\|$. *We will do the case where* $\|\boldsymbol{x} - T(\boldsymbol{x})\| > \| -\boldsymbol{x} - T(-\boldsymbol{x})\|$ *and leave the other case to you. We have*

$$\begin{aligned}\|\boldsymbol{x} - T(\boldsymbol{x}) + tT(-\boldsymbol{x})\| &= \|\boldsymbol{x} - T(\boldsymbol{x}) - (-t\boldsymbol{x} - tT(-\boldsymbol{x})) - t\boldsymbol{x}\| \\ &\geq \|\boldsymbol{x} - T(\boldsymbol{x})\| + t\|\boldsymbol{x}\| - t\| -\boldsymbol{x} - T(-\boldsymbol{x})\| > t\boldsymbol{x} > 0\end{aligned}$$

when $t > 0$. *When* $t = 0$, *we have* $\|\boldsymbol{x} - T(\boldsymbol{x})\| > 0$ *because by assumption* $\mathbf{0} \notin \partial\boldsymbol{D}$. *Hence, in this case,* $\|\boldsymbol{x} - H(\boldsymbol{x}, t)\| > 0$ *on* $\partial\boldsymbol{D} \times [0, 1]$.

We conclude $(I - H(\cdot, t))(\boldsymbol{x}) \neq \mathbf{0}$ *on* $\partial\boldsymbol{D} \times [0, 1]$. *By invariance under compact homotopy,* $d_{LS}(I - H(\cdot, 0), \boldsymbol{D}, \mathbf{0}) = d_{LS}(I - H(\cdot, 1), \boldsymbol{D}, \mathbf{0})$ *or* $d_{LS}(I - T, \boldsymbol{D}, \mathbf{0}) = d_{LS}(I - g, \boldsymbol{D}, \mathbf{0})$. *But* g *is odd and compact. So by Borsuk's Theorem,* $d_{LS}(I - g, \boldsymbol{D}, \mathbf{0})$ *is odd. Thus,* $d_{LS}(I - T, \boldsymbol{D}, \mathbf{0})$ *is odd.* ■

Homework

Exercise 11.2.7 *In the proof of Theorem 11.2.5, we show*

$$F_\epsilon(-T(\boldsymbol{x})) = \frac{-\sum_{i \in J \cap \{1,\ldots,p\}} m_{i+p}(T(\boldsymbol{x}))\boldsymbol{v_{i+p}} - \sum_{i \in J \cap \{p+1,\ldots,2p\}} m_{i-p}(T(\boldsymbol{x}))(\boldsymbol{v_{i-p}})}{\sum_{i \in J \cap \{1,\ldots,p\}} m_{i+p}(T(\boldsymbol{x})) + \sum_{i \in J \cap \{p+1,\ldots,2p\}} m_{i-p}(T(\boldsymbol{x}))}$$

verify the right-hand side is $-F_\epsilon(T(\boldsymbol{x}))$. *Since* $F_\epsilon \circ T$ *is continuous for all* ϵ, *we thus see* $F_\epsilon \circ T$ *is an odd continuous map.*

Exercise 11.2.8 *If* $A = \begin{bmatrix} 1 & 2 \\ 3 & 4 \end{bmatrix}$, *is* A *an odd continuous and compact operator? Does Borsuk's Theorem apply?*

Exercise 11.2.9 *Let* A *be any* $n \times n$ *real matrix with* $I - T$ *invertible. Is* A *an odd continuous and compact operator? Does Borsuk's Theorem apply?*

Exercise 11.2.10 *In the proof of Theorem 11.2.6, do the case* $\|\boldsymbol{x} - T(\boldsymbol{x})\| < \| -\boldsymbol{x} - T(-\boldsymbol{x})\|$.

11.3 Further Properties of Leray - Schauder Degree

There are some additional properties of Leray - Schauder that are important. The first one involves what happens if the range of the operator lies in a finite dimensional subspace.

Theorem 11.3.1 Finite Dimensional Subspace Restriction for Leray - Schauder Degree

Let $(X, \|\cdot\|)$ *be a normed linear space and* $\boldsymbol{D} \subset X$ *be an open and bounded set. Let* $T : \overline{\boldsymbol{D}} \to X$ *be compact and* $T(\overline{\boldsymbol{D}})$ *be bounded with* $T(\overline{\boldsymbol{D}}) \subset Y$, *where* Y *is a finite dimensional subspace of* X. *If* $\mathbf{0} \notin (I - T)(\partial\boldsymbol{D})$, *then* $d_{LS}(I - T, \boldsymbol{D}, \mathbf{0}) = d_{LS}((I - T)|_Y, \boldsymbol{D} \cap Y, \mathbf{0})$.

Proof 11.3.1
Recall the limit process that defined Leray - Schauder degree. We let $\boldsymbol{K} = \overline{T(\overline{\boldsymbol{D}})} = T(\overline{\boldsymbol{D}}) \subset Y$ *by assumption. Since* $\boldsymbol{K}$ *is compact, for each* ϵ *net of* $\boldsymbol{K}$, *we define the continuous map* $F_\epsilon : K \to Y$ *as usual so that* $\|T(\boldsymbol{x}) - F_\epsilon(T(\boldsymbol{x}))\| < \epsilon$ *for all* $\boldsymbol{x} \in \overline{\boldsymbol{D}}$. *By Step (i) in that argument, we know there is* $r > 0$ *so that* $\inf_{\boldsymbol{y} \in \partial\boldsymbol{D}} \|(I - T)(\boldsymbol{y})\| \geq r > 0$. *Then by Step (iv), for all mappings* $\Phi : \overline{\boldsymbol{D}} \to Z$ *where* Z *is a finite dimensional subspace of* X, *with* $\|T(\boldsymbol{x}) - \Phi(\boldsymbol{x})\| < (r/2)$ *for all* $\boldsymbol{x} \in \overline{\boldsymbol{D}}$, *we have* $d_{LS}(I - T, \boldsymbol{D}, \mathbf{0}) = d_B(I - \Phi, \boldsymbol{D} \cap Z, \mathbf{0})$. *We can then apply these results to* $\Phi = T$, *itself. Then, we have* $d_{LS}(I - T, \boldsymbol{D}, \mathbf{0}) = d_B(I - T, \boldsymbol{D} \cap Y, \mathbf{0})$. *But,* $d_B(I - T, \boldsymbol{D} \cap Y, \mathbf{0}) = d_B((I - T)|_Y, \boldsymbol{D}, \mathbf{0})$. *This establishes the result.* ■

The second one concerns what happens if the range of the operator lies in a closed subspace.

Theorem 11.3.2 Closed Subspace Restriction for Leray - Schauder Degree

Let $(X, \|\cdot\|)$ be a normed linear space and $\boldsymbol{D} \subset X$ be an open and bounded set. Let $T : \overline{\boldsymbol{D}} \to X$ be compact and $T(\overline{\boldsymbol{D}})$ with $T(\overline{\boldsymbol{D}}) \subset Y$, where Y is a closed subspace of X. If $\boldsymbol{0} \notin (I - T)(\partial \boldsymbol{D})$, then $d_{LS}(I - T, \boldsymbol{D}, \boldsymbol{0}) = d_{LS}((I - T)|_Y, \boldsymbol{D} \cap Y, \boldsymbol{0})$.

Proof 11.3.2
For a monotonically decreasing sequence (ϵ_n) with $\epsilon_n \to 0$, for sufficiently large n, $d_{LS}(I - T, \boldsymbol{D}, \boldsymbol{0}) = d_B(I - F_{\epsilon_n} \circ T, \boldsymbol{D} \cap Y_n, \boldsymbol{0})$ where all terms are defined as usual. In this case, the ϵ_n nets for $T(\overline{\boldsymbol{D}})$ all come from Y as $T(\overline{\boldsymbol{D}}) \subset \overline{Y} = Y$. Hence, $\boldsymbol{D} \cap Y_n \subset \boldsymbol{D} \cap Y$. Thus, for large enough n,

$$\begin{aligned} d_{LS}((I_T)|Y, \boldsymbol{D} \cap Y, \boldsymbol{0}) &= d_B((I - F_{\epsilon_n} \circ T)_Y, \boldsymbol{D} \cap Y \cap Y_n, \boldsymbol{0}) \\ &= d_B((I - F_{\epsilon_n} \circ T), \boldsymbol{D} \cap Y_n, \boldsymbol{0}) = d_{LS}(I - T, \boldsymbol{D}, \boldsymbol{0}) \end{aligned}$$

■

We now note that Leray - Schauder degree is translation invariant, a useful property!

Theorem 11.3.3 Invariance under Translation of Leray - Schauder Degree

Let $(X, \|\cdot\|)$ be a normed linear space and $\boldsymbol{D} \subset X$ be an open and bounded set. Let $T : \overline{\boldsymbol{D}} \to X$ be compact. Then, $d_{LS}(I - T \pm \{\boldsymbol{y}\}, \boldsymbol{D}, \boldsymbol{p}) = d_{LS}(I - T, \boldsymbol{D} \pm \{\boldsymbol{y}\}, \boldsymbol{p} \pm \{\boldsymbol{y}\})$.

Proof 11.3.3
This is pretty easy to see. First, the operator $I - T + \{\boldsymbol{y}\}$ applied to $\boldsymbol{x}$ gives $(I - T)(\boldsymbol{x}) + \boldsymbol{y}$. So $\boldsymbol{p} + \boldsymbol{y} \in (I - T + \{\boldsymbol{y}\})(\partial \boldsymbol{D})$ means there is $\boldsymbol{d} \in \partial \boldsymbol{D}$ so that $\boldsymbol{p} + \boldsymbol{y} = (I - T)(\boldsymbol{d}) + \boldsymbol{y}$ or $\boldsymbol{p} \in (I - T)(\partial \boldsymbol{D})$. So $\boldsymbol{p} + \boldsymbol{y} \notin (I - T + \{\boldsymbol{y}\})(\partial \boldsymbol{D})$ is the same as $\boldsymbol{p} \notin (I - T)(\partial \boldsymbol{D})$. So the translation idea is transparent here. ■

Homework

Exercise 11.3.1 *Our proof of Theorem 11.3.1 leaves out many details, so to check your understanding fill in these gaps in the proof carefully.*

Exercise 11.3.2 *Our proof of Theorem 11.3.2 does not show all the details and instead twice uses the phrase* **for sufficiently large** n*. Fill in those details.*

The complement of a bounded set is a union of components. First, we show Leray - Schauder degree is constant on a component.

Theorem 11.3.4 Leray - Schauder Degree is Constant on Components

Let $(X, \|\cdot\|)$ be a normed linear space and $\boldsymbol{D} \subset X$ be an open and bounded set. Let $T : \overline{\boldsymbol{D}} \to X$ be compact with $\boldsymbol{0} \notin (I - T)(\partial \boldsymbol{D})$. Let $X \setminus (I - T)(\partial \boldsymbol{D}) = \cup_{i=1}^N \boldsymbol{\Delta}_i \cup \boldsymbol{\Delta}$ where each $\boldsymbol{\Delta}_i$ is a bounded simply connected open subset of $X \setminus (I - T)(\partial \boldsymbol{D})$ and $\boldsymbol{\Delta}$ is the unbounded open simply connected subset of $X \setminus (I - T)(\partial \boldsymbol{D})$. Let $\boldsymbol{p}$ and $\boldsymbol{q}$ be in the same component. Then $d_{LS}(I - T, \boldsymbol{D}, \boldsymbol{p}) = d_{LS}((I - T)|_Y, \boldsymbol{D} \cap Y, \boldsymbol{q})$.

Proof 11.3.4
Since $\boldsymbol{p}$ and $\boldsymbol{q}$ are in the same component, call this component Γ. Since the component is sim-

ply connected, there is a continuous function $\phi : [0,1] \to \Gamma$ *with* $\phi(0) = \boldsymbol{p}$ *and* $\phi(1) = \boldsymbol{q}$. *Let* $H(\boldsymbol{x},t) = T(\boldsymbol{x}) - \phi(t)$ *on* $\overline{\boldsymbol{D}} \times [0,1]$. *Then* H *is a compact homotopy and* $H(\boldsymbol{x},0) = T(\boldsymbol{x}) - \boldsymbol{p}$ *and* $H(\boldsymbol{x},1) = T(\boldsymbol{x}) - \boldsymbol{q}$. *Consider any* $\boldsymbol{x} \in \partial \boldsymbol{D}$. *Then,* $\|(I - H(\cdot,t))(\boldsymbol{x})\| = \|\boldsymbol{x} - T(\boldsymbol{x}) - \phi(t)\|$. *Now* $\inf_{0 \le t \le 1} \|\phi(t) - (\boldsymbol{x} - T(\boldsymbol{x}))\|$ *is positive. To see this, suppose the infimum is zero. Then there is a sequence* $(t_n) \subset [0,1]$ *with* $\|\phi(t_n) - (\boldsymbol{x} - T(\boldsymbol{x}))\| \to 0$. *But* $[0,1]$ *is compact, so there is a subsequence* $(t_{n,1})$ *and a point* t_0 *so that* $t_{n,1} \to t_0$ *and we still have* $\|\phi(t_{n,1}) - (\boldsymbol{x} - T(\boldsymbol{x}))\| \to 0$. *But by continuity of the norm and* ϕ, *this tells us* $\|\phi(t_0) - (\boldsymbol{x} - T(\boldsymbol{x}))\| = 0$. *Hence,* $\phi(t) \in (I - T)(\partial \boldsymbol{D})$ *which is not possible as* $\phi(t_0)$ *is in a component of the complement of* $(I - T)(\partial \boldsymbol{D})$. *Hence, this infimum is positive. Let the value of the infimum be* $\mu > 0$. *Since the range of* ϕ *is in this complement and* $\inf_{0 \le t \le 1} \|(I - H(\cdot,t))(\boldsymbol{x})\| \ge \mu > 0$, *we have shown* $\boldsymbol{0} \notin (I - H(\cdot,t))(\partial \boldsymbol{D} \times [0,1])$.

By invariance under compact homotopy, we have $d_{LS}(I - H(\cdot,0), \boldsymbol{D}, \boldsymbol{0}) = d_{LS}(I - H(\cdot,1), \boldsymbol{D}, \boldsymbol{0})$ *or* $d_{LS}(I - T - \boldsymbol{p}, \boldsymbol{D}, \boldsymbol{0}) = d_{LS}(I - T - \boldsymbol{q}, \boldsymbol{D}, \boldsymbol{0})$. *But since Leray - Schauder degree is translation invariant, this says* $d_{LS}(I - T, \boldsymbol{D}, \boldsymbol{p}) = d_{LS}(I - T, \boldsymbol{D}, \boldsymbol{q})$. *This shows the value of the degree is constant on the component.* ■

Comment 11.3.1 *Since Brouwer degree is zero on the unbounded component, it follows the Leray - Schauder degree is also zero on the unbounded component.*

We can now prove a nice decomposition result involving the components.

Theorem 11.3.5 Decomposition of Leray - Schauder Degree on Components

Let $(X, \|\cdot\|)$ *be a normed linear space and* $\boldsymbol{D} \subset X$ *be an open and bounded set. Let* $G : \overline{\boldsymbol{D}} \to X$ *be compact. Let* $\boldsymbol{\Delta}$ *be an open and bounded subset of* X *with* $(I - G)(\overline{\partial \boldsymbol{D}}) \subset \boldsymbol{\Delta}$. *Let* $K : \overline{\boldsymbol{\Delta}} \to X$ *be compact. Then, if* $\boldsymbol{0} \notin (I - K)(I - G)(\partial \boldsymbol{D})$,

$$d_{LS}((I - K)(I - G), \boldsymbol{D}, \boldsymbol{0}) = \sum_{i=1}^{N} d_{LS}((I - K), \boldsymbol{\Delta_i}, \boldsymbol{0})\, d_{LS}((I - G), \boldsymbol{D}, \boldsymbol{\Delta_i})$$

where $\boldsymbol{\Delta_i}$ *are the bounded connected components of* $\boldsymbol{\Delta} \setminus (I - G)(\partial \boldsymbol{D})$.

Proof 11.3.5
Let

$$\begin{aligned} H &= (I - K)(I - G) = I - G - K + K \circ G = I - (G + K - K \circ G) \\ &= I - (G + K \circ (I - G)) \end{aligned}$$

Since K *and* T *are compact,* $T = G + K \circ (I - G)$ *is also compact. Thus, there is an* $r > 0$, *with* $\|I(\boldsymbol{x}) - T(\boldsymbol{x})\| \ge r > 0$ *for all* $\boldsymbol{x} \in \partial \boldsymbol{D}$. *Now pick an* $\epsilon > 0$. *Let* $\boldsymbol{V_\epsilon}$ *be an* ϵ *net for* $\overline{\boldsymbol{D}}$. *As usual, pick* $\boldsymbol{q} \in \boldsymbol{D}$ *and set* $Y_\epsilon = span\{\boldsymbol{V_\epsilon}, \boldsymbol{p}, \boldsymbol{q}\}$ *and* $\boldsymbol{D_\epsilon} = Y_\epsilon \cap \boldsymbol{D}$. *Then, define the approximate map* $G_\epsilon = F_{K,\epsilon} \circ G$ *as before.*

Now, let $\boldsymbol{W_{1,\epsilon}}$ *be an* ϵ *net for* $\overline{\boldsymbol{\Delta}}$ *and let* $\boldsymbol{W_\epsilon}$ *be the* ϵ *net we get by adjoining* $K(\boldsymbol{V_\epsilon})$ *to* $\boldsymbol{W_{1,\epsilon}}$. *Hence,* $\boldsymbol{W_\epsilon} = \{\boldsymbol{W_{1,\epsilon}}, K(\boldsymbol{V_\epsilon})\}$. *Pick a point* $\boldsymbol{z} \in \boldsymbol{\Delta}$ *and let* $\boldsymbol{V_\epsilon} = span\{\boldsymbol{W_\epsilon}, \boldsymbol{z}, G(\boldsymbol{p})\}$. *Then construct the approximate operators* K_ϵ *as usual.*

We know

$$(I - G_\epsilon)(\overline{\boldsymbol{D_\epsilon}}) \subset Y_\epsilon, \quad K_\epsilon((I - G_\epsilon)(\overline{\boldsymbol{D_\epsilon}})) \subset K_\epsilon(Y_\epsilon) \subset W_\epsilon$$

Now for $\boldsymbol{x} \in \overline{\boldsymbol{D}_\epsilon}$, we have

$$
\begin{aligned}
&\|(G + K \circ (I - G))(\boldsymbol{x}) - (G_\epsilon + K_\epsilon \circ (I - G_\epsilon))(\boldsymbol{x})\| = \\
&\|(G - G_\epsilon)(\boldsymbol{x}) + (K - K_\epsilon)(\boldsymbol{x}) + K_\epsilon \circ G_\epsilon(\boldsymbol{x}) - K \circ G(\boldsymbol{x})\| = \\
&\|(G - G_\epsilon)(\boldsymbol{x}) + (K - K_\epsilon)(\boldsymbol{x}) + K_\epsilon \circ G_\epsilon(\boldsymbol{x}) - K_\epsilon \circ G(\boldsymbol{x}) + K_\epsilon \circ G(\boldsymbol{x}) - K \circ G(\boldsymbol{x})\| \leq \\
&\|(G - G_\epsilon)(\boldsymbol{x})\| + \|(K - K_\epsilon)(\boldsymbol{x})\| + \|K_\epsilon \circ (G_\epsilon - G)(\boldsymbol{x}) + (K_\epsilon - K) \circ G(\boldsymbol{x})\| \\
&\|(G - G_\epsilon)(\boldsymbol{x})\| + \|(K - K_\epsilon)(\boldsymbol{x})\| + \|K_\epsilon \circ (G_\epsilon - G)(\boldsymbol{x})\| + \|(K_\epsilon - K) \circ G(\boldsymbol{x})\|
\end{aligned}
$$

We can do this for a monotonically decreasing sequence (ϵ_n) with $\epsilon_n \to 0$. We know $\|G - G_{\epsilon_n}\| \to 0$ and $\|K - K_{\epsilon_n}\| \to 0$ and since $\partial \boldsymbol{D}$ is compact, we know there is an R so that $\|\boldsymbol{x}\| \leq R$. We also know since $\|K - K_{\epsilon_n}\| \to 0$, there is $A > 0$ so that $\|K_{\epsilon_n}\| \leq A$. Finally, since G is compact, there is $B > 0$, so that $\|G(\boldsymbol{x})\| \leq B$. This implies we can choose n large enough so that given the r above

$$
\|(G - G_{\epsilon_n})\| \;\leq\; \min\left\{\frac{r}{8R}, \frac{r}{8R\,A}\right\}, \quad \|(K - K_{\epsilon_n})\| \leq \min\left\{\frac{r}{8R}, \frac{r}{8R\,B}\right\}
$$

Hence, we see for such n, $\|(T - T_{\epsilon_n})(\boldsymbol{x})\| < \frac{r}{2}$. By Step (iv) in the proof of the existence of Leray - Schauder degree, we then know

$$
\begin{aligned}
d_{LS}((I - K)(I - G), \boldsymbol{D}, \boldsymbol{0}) &= d_B(I - (G_{\epsilon_n} + K_{\epsilon_n} \circ (I - G_{\epsilon_n})), \boldsymbol{D}_{\epsilon_n}, \boldsymbol{0}) \\
&= d_B((I - K_{\epsilon_n})(I - G_{\epsilon_n}), \boldsymbol{D}_{\epsilon_n}, \boldsymbol{0})
\end{aligned}
$$

Now, since span$\{\boldsymbol{V}_\epsilon, \boldsymbol{p}, \boldsymbol{q}\}$ is a subspace of span$\{\boldsymbol{W}_\epsilon, \boldsymbol{p}, \boldsymbol{q}\}$ in our construction process above, we can apply our comments for the chain property of Brouwer degree, Theorem 10.5.3, which says

$$
d_B(g \circ f, \boldsymbol{D}, \boldsymbol{0}) \;=\; \sum_{\alpha, g^{-1}(\boldsymbol{0}) \cap \boldsymbol{\Delta}_\alpha \neq \emptyset} d_B(g, \boldsymbol{\Delta}_\alpha, \boldsymbol{0}) d_B(f, \boldsymbol{D}, \boldsymbol{\Delta}_\alpha)
$$

where $\{\boldsymbol{\Delta}_\alpha\}$ is the collection of bounded components of $\Re^n - f(\partial \boldsymbol{D})$. Apply this result here to our maps to find

$$
d_B((I - K_{\epsilon_n})(I - G_{\epsilon_n}), \boldsymbol{D}_{\epsilon_n}, \boldsymbol{0}) = \sum_{\alpha, g^{-1}(\boldsymbol{p}) \cap \boldsymbol{\Delta}_\alpha \neq \emptyset} d_B(I - K_{\epsilon_n}, \boldsymbol{\Delta}_\alpha, \boldsymbol{0})\, d_B(I - G_{\epsilon_n}, \boldsymbol{D}_{\epsilon_n}, \boldsymbol{\Delta}_\alpha)
$$

where $d_B(I - G_{\epsilon_n}, \boldsymbol{D}_{\epsilon_n}, \boldsymbol{\Delta}_\alpha)$ is the constant value of Brouwer degree of $I - G_{\epsilon_n}$ on the component $\boldsymbol{\Delta}_\alpha$ of $\boldsymbol{V}_{\epsilon_n} \setminus (I - G)(\partial \boldsymbol{D})$. But since n is sufficiently large,

$$
d_{LS}(I - G, \boldsymbol{D}, \boldsymbol{\Delta}_\alpha) = d_B(I - G_{\epsilon_n}, \boldsymbol{D}_{\epsilon_n}, \boldsymbol{\Delta}_\alpha)
$$

Also, $d_{LS}(I_K, \boldsymbol{\Delta}_\alpha, \boldsymbol{0}) = d_B(I - K_{\epsilon_n}, \boldsymbol{\Delta}_\alpha, \boldsymbol{0})$. We conclude,

$$
d_{LS}((I - K)(I - G), \boldsymbol{D}, \boldsymbol{0}) \;=\; \sum_\alpha d_{LS}(I_K, \boldsymbol{\Delta}_\alpha, \boldsymbol{0})\, d_{LS}(I - G, \boldsymbol{D}\boldsymbol{\Delta}_\alpha)
$$

■

Under some circumstances we can remove part of the domain $\boldsymbol{D}$.

Theorem 11.3.6 Excision Property for Leray - Schauder Degree

Let $(X, \|\cdot\|)$ be a normed linear space and $\boldsymbol{D} \subset X$ be an open and bounded set. Let $T : \overline{\boldsymbol{D}} \to X$ be compact. Let $\boldsymbol{p} \notin (I - T)(\partial \boldsymbol{D})$ and $\boldsymbol{K} \subset \overline{\boldsymbol{D}}$ with $\boldsymbol{K}$ closed and $\boldsymbol{p} \notin (I - T)(\partial \boldsymbol{K})$. Then, $d_{LS}(I - T, \boldsymbol{D}, \boldsymbol{p}) = d_{LS}(I - T, \boldsymbol{D} \setminus \boldsymbol{K}, \boldsymbol{p})$.

Proof 11.3.6
By the definition of Leray - Schauder degree, for sufficiently large n, $d_{LS}(I - T, \boldsymbol{D}, \boldsymbol{p}) = d_B(I - T_n, \boldsymbol{D_n}, \boldsymbol{p})$ where T_n and $\boldsymbol{D_n}$ are defined as usual. Apply the excision property for Brouwer degree, Theorem 10.5.1, to find

$$d_B(I - T_n, \boldsymbol{D_n}, \boldsymbol{p}) = d_B(I - T, \boldsymbol{D_n} \setminus \boldsymbol{K}, \boldsymbol{p}) = d_{LS}(I - T, \boldsymbol{D} \setminus \boldsymbol{K}, \boldsymbol{p})$$

The result then follows. ■

Under the right circumstances, Leray - Schauder degree is additive.

Theorem 11.3.7 Additivity Theorem for Leray - Schauder Degree

Let $(X, \|\cdot\|)$ be a normed linear space and $\boldsymbol{D} \subset X$ be an open and bounded set with $\boldsymbol{D} = \cup_{i=1}^{N} \boldsymbol{D_i}$ where each $\boldsymbol{D_i}$ is open also and the collection $\{\boldsymbol{D_1}, \ldots, \boldsymbol{D_N}\}$ is pairwise disjoint. Let $T : \overline{\boldsymbol{D}} \to X$ be compact. Then if $\boldsymbol{p} \notin (I - T)(\partial \boldsymbol{D})$, $d_{LS}(I - T, \boldsymbol{D}, \boldsymbol{p}) = \sum_{i=1}^{N} d_{LS}(I - T, \boldsymbol{D_i}, \boldsymbol{p})$.

Proof 11.3.7
This proof is very similar to one we used for Theorem 11.3.6. We will leave the details to you. ■

The final result of this section is what is called a continuation theorem.

Theorem 11.3.8 Continuation Theorem for Leray - Schauder Degree

Let $(X, \|\cdot\|)$ be a normed linear space and $\boldsymbol{D} \subset X$ be an open and bounded set. Let $T : \overline{\boldsymbol{D}} \to X$ be compact. Let $\boldsymbol{x} \neq \lambda T(\boldsymbol{x})$ for all $\boldsymbol{x} \in \partial \boldsymbol{D}$ and for all $0 < \lambda \leq 1$. Let $\boldsymbol{0} \in \boldsymbol{D}$. Then, there is a solution to $\boldsymbol{x} = T(\boldsymbol{x})$ for some $\boldsymbol{x} \in \overline{\boldsymbol{D}}$.

Proof 11.3.8
Let $H(\boldsymbol{x}, \lambda) = \lambda T(\boldsymbol{x})$ for $\boldsymbol{x} \in \overline{\boldsymbol{D}}$ and $0 \leq \lambda \leq 1$. Then H is a compact homotopy and by assumption $(I - H(\boldsymbol{s}, \lambda)(\boldsymbol{x}) = \boldsymbol{x} - \lambda T(\boldsymbol{x}) \neq \boldsymbol{0}$ on $\partial \boldsymbol{D} \times (0, 1]$. At $\lambda = 0$, we have $I - H(\boldsymbol{x}, 0)(\boldsymbol{x}) = \boldsymbol{x} \neq \boldsymbol{0}$ also by assumption. Then, by invariance of compact homotopy, we see

$$\pm 1 = d_{LS}(I, \boldsymbol{D}, \boldsymbol{0}) = d_{LS}(I - H(\cdot, 0), \boldsymbol{D}, \boldsymbol{0}) = d_{LS}(I - H(\cdot, 1), \boldsymbol{D}, \boldsymbol{0}) = d_{LS}(I - T, \boldsymbol{D}, \boldsymbol{0})$$

Since the degree is nonzero, by the existence theorem, we know the desired solution exists. ■

You can read more about Leray - Schauder degree in the original paper (Leray (72) 1934) and a nice survey of topological methods in nonlinear analysis in (Mawhin (87) 1999). Of course, we have only touched on some of the wonderful tools we can use to approach nonlinear problems. You should also look at (Deimling (24) 2010) to learn more about these things.

Homework

Exercise 11.3.3 *Explain in detail why if Brouwer degree is zero on the unbounded component, it follows that the Leray - Schauder degree is also zero on the unbounded component.*

Exercise 11.3.4 *In the proof of Theorem 11.3.5, prove that K and T are compact implies $T = G + K \circ (I - G)$ is also compact.*

Exercise 11.3.5 *In the proof of Theorem 11.3.6, provide the details for the construction of T_n and $\boldsymbol{D}_n$ which we say are defined as usual.*

Exercise 11.3.6 *Prove Theorem 11.3.7.*

11.4 Linear Compact Operators

Our aim here is to establish an important result that will prove useful in our discussion of coincidence degree. Note here our operators are linear and compact which means they are continuous and so bounded.

Theorem The Leray - Schauder Degree of $I - T$ when T is Linear and Compact and $I - T$ is Invertible

Let $(X, \|\cdot\|)$ be a normed linear space. Let $T : X \rightarrow X$ be linear and compact and assume $ker(I - T) = \{\boldsymbol{0}\}$. Then for all $\boldsymbol{D} \subset X$ that are open and bounded, with $\boldsymbol{0} \notin \partial\boldsymbol{D}$, we have

$$d_{LS}(I - T, \boldsymbol{D}, \boldsymbol{0}) = \begin{cases} 0, & \boldsymbol{0} \notin \boldsymbol{D} \\ (-1)^m, & \boldsymbol{0} \in \boldsymbol{D} \end{cases}$$

where m is the sum of the multiplicities of the eigenvalues of T larger than 1.

To establish this result we need to study the Riesz - Schauder theory of linear operators.

11.4.1 The Resolvent Operator

Let $(X, \|\cdot\|)$ be a normed linear space and $T : \boldsymbol{D}(T) \subset X \rightarrow \boldsymbol{R}(T) \subset X$ be a linear operator. We are not assuming T is compact here nor do we assume T is continuous. Consider the eigenvalue problem $(\lambda I - T)(\boldsymbol{x}) = \boldsymbol{0}$. Here λ is a scalar parameter (and from your study of matrices, you know we want the scalars here to be complex) and I is the usual identity operator on X. The eigenvalues of T are the scalars λ and associated nonzero $\boldsymbol{x}$ for which this equation is true. The associated nonzero vectors $\boldsymbol{x}$ for a given λ are called the eigenvectors associated with λ and they will form at least a one dimensional subspace of X. If the eigenvalue $\lambda = 0$ was a possibility, this is the same as saying T has a nonempty kernel. In this discussion, we assume this is not possible as our operator T is invertible. We use $\boldsymbol{R}(T)$ to denote the range of T and $\boldsymbol{D}(T)$ to indicate its domain. Note the domain and range are subspaces.

Definition 11.4.1 Resolvent Set and Spectrum of a Linear Operator

Let $(X, \|\cdot\|)$ be a normed linear space which is not just $\{\boldsymbol{0}\}$ and $T : \boldsymbol{D}(T) \subset X \rightarrow \boldsymbol{R}(T) \subset X$ be linear. If $\boldsymbol{R}(\lambda I - T)$ is dense in X and if $(\lambda I - T)^{-1}$ exists and is continuous, we say λ is in the resolvent set of T, which we denote by $\rho(T)$. We denote the set of all values of $\lambda \notin \rho(T)$ comprise the set $\sigma(T)$, which is called the spectrum of T.

We know there are n eigenvalues of an $n \times n$ matrix and they can have algebraic multiplicities larger than one. Each eigenvalue λ has a corresponding subspace we denote by $\boldsymbol{E}_\lambda$. The dimension of the eigenspace need not match the algebraic multiplicity. We have studied the Jordan Canonical form in (Peterson (101) 2020), so we have a pretty good understanding of the spectrum in the case of such a matrix. To understand the spectrum of a linear operator on an infinite dimensional normed linear space is a much more complicated task. To get started, we let $\boldsymbol{R}_\lambda = \boldsymbol{R}(T_\lambda)$, where we now use

$T_\lambda = \lambda I - T$ to denote the range of $\lambda I - T$. The operator norm of T_λ^{-1}, if it exists and is continuous, is a useful tool here. We let $M(\lambda) = \|T_\lambda^{-1}\|_{op}$ when T_λ^{-1} exists and is continuous. Our first result helps us understand this inverse does not exist in an isolated manner. We will usually just write $\|T_\lambda^{-1}\|$ for the operator norm, so make sure you remember the norm here is not the norm of X.

Theorem 11.4.1 T_λ^{-1} Exists and is Continuous Locally

Assume T_μ exists and is continuous. Then T_λ has a continuous inverse if $|\lambda-\mu|M(\mu) < 1$. Further, $\overline{\boldsymbol{R_\lambda}}$ is not a proper subset of $\overline{\boldsymbol{R_\mu}}$.

Proof 11.4.1

Assume $\boldsymbol{x} \in \boldsymbol{D}(T)$. Then, $T_\lambda(\boldsymbol{x}) = \lambda\boldsymbol{x} - T(\boldsymbol{x}) = (\lambda - \mu)\boldsymbol{x} + T_\mu(\boldsymbol{x})$. Hence, $\|T_\lambda\| \geq \|T_\mu\| - |\lambda - \mu|\|\boldsymbol{x}\|$. Just a reminder here: $\|\boldsymbol{x}\|$ uses the norm on X and $\|T_\mu^{-1}\|$ is the operator norm if the operator is bounded. We assume T_μ^{-1} exists and is continuous, which tells us it has a bounded operator norm. Thus, $\|\boldsymbol{x}\| = \|T_\mu^{-1}\,T_\mu\| \leq \|T_\mu^{-1}\|\;\|T_\mu\|$. Thus,

$$\begin{aligned}\|T_\lambda(\boldsymbol{x})\| &\geq \|T_\mu(\boldsymbol{x})\| - |\lambda - \mu|M(\mu)\;\|T_\mu(\boldsymbol{x})\| \\ &= \|T_\mu(\boldsymbol{x})\|(1 - |\lambda - \mu|M(\mu))\end{aligned}$$

If we choose λ so that $0 < 1 - |\lambda - \mu|M(\mu) < 1$, then, letting $\alpha = 1 - |\lambda - \mu|M(\mu)$, we see $T_\lambda(\boldsymbol{x}) = \boldsymbol{0}$ implies $\boldsymbol{0} \geq \alpha\;\|T_\mu(\boldsymbol{x})\|$. This forces $\|T_\mu(\boldsymbol{x})\| = \boldsymbol{0}$. Since T_μ is invertible, this tells us $\boldsymbol{x} = \boldsymbol{0}$. So we have shown T_λ has a zero kernel and hence, T_λ^{-1} exists.

We now show T_λ^{-1} is continuous. Since $\|T_\lambda\| \geq \|T_\mu\| - |\lambda - \mu|\|\boldsymbol{x}\|$, we have

$$M(\mu)\|T_\lambda\| \;\geq\; M(\mu)\;\|T_\mu\| - |\lambda - \mu|M(\mu)\;\|\boldsymbol{x}\|$$

But, $\|\boldsymbol{x}\| \leq M(\mu)\|T_\mu(\boldsymbol{x})\|$ and so

$$M(\mu)\|T_\lambda\| \;\geq\; (1 - |\lambda - \mu|M(\mu))\|\boldsymbol{x}\|$$

If $0 < 1 - |\lambda - \mu|M(\mu) < 1$, we have already shown T_λ^{-1} exists. In this case, we know there is $\boldsymbol{y}$ so that $\boldsymbol{x} = T_\lambda^{-1}(\boldsymbol{y})$. So

$$M(\mu)\|T_\lambda\| \;\geq\; (1 - |\lambda - \mu|M(\mu))\|\boldsymbol{x}\| \Longrightarrow \|T_\lambda\| \geq \frac{(1 - |\lambda - \mu|M(\mu))}{M(\mu)}\|\boldsymbol{x}\|$$

Thus,

$$\|\boldsymbol{x}\| \;=\; \|T_\lambda^{-1}(\boldsymbol{y})\| \leq \frac{M(\mu)}{(1 - |\lambda - \mu|M(\mu))}\|\boldsymbol{y}\|$$

This implies $\|T_\lambda^{-1}\| \leq \frac{M(\mu)}{1-|\lambda-\mu|M(\mu)}$ which shows T_λ^{-1} is a bounded linear operator and so continuous.

Finally, let's show $\overline{\boldsymbol{R_\lambda}}$ is not a proper subset of $\overline{\boldsymbol{R_\mu}}$. Assume this is not true. Then $\overline{\boldsymbol{R_\mu}} \subsetneq \overline{\boldsymbol{R_\mu}}$. Choose θ so that $|\lambda - \mu|M(\mu) < \theta < 1$. By Riesz's Lemma, there is a $\boldsymbol{y_0} \in \overline{\boldsymbol{R_\mu}}$ with $\|\boldsymbol{y_0}\| = 1$ and $\|\boldsymbol{y} - \boldsymbol{y_0}\| \geq \theta$ for all $\boldsymbol{y} \in \overline{\boldsymbol{R_\lambda}}$. Let $(\boldsymbol{y_n}) \subset \boldsymbol{R_\lambda}$ with $\boldsymbol{y_n} \rightarrow \boldsymbol{y_0}$. Let $\boldsymbol{x_n} = T_\lambda^{-1}(\boldsymbol{y_n})$. Then, $T_\lambda(x_n) = (\lambda - \mu) + T_\mu(\boldsymbol{x_n})$ which implies

$$\|T_\lambda(x_n) - T_\mu(\boldsymbol{x_n})\| \;=\; |\lambda - \mu|\|\boldsymbol{x_n}\| \leq |\lambda - \mu|M(\mu)\|T_\mu(\boldsymbol{x_n})\| = |\lambda - \mu|M(\mu)\|\boldsymbol{y_n}\|$$

But $T_\lambda(\boldsymbol{x_n}) \in \boldsymbol{R_\lambda}$. *Hence, we have*

$$\begin{aligned} \theta &\leq \|\boldsymbol{y_0} - T_\lambda(\boldsymbol{x_n})\| \leq \|\boldsymbol{y_0} - T_\mu(\boldsymbol{x_n})\| + \|T_\mu(\boldsymbol{x_n}) - T_\lambda(\boldsymbol{x_n})\| \\ &\leq \|\boldsymbol{y_0} - \boldsymbol{y_n}\| + |\lambda - \mu| M(\mu) \|\boldsymbol{y_n}\| \end{aligned}$$

Let $n \to \infty$ *and we obtain*

$$\theta \leq |\lambda - \mu| M(\mu) \|\boldsymbol{y_0}\| = |\lambda - \mu| M(\mu)$$

which is not possible. Hence, our assumption is false and $\overline{\boldsymbol{R_\lambda}}$ *is not a proper subset of* $\overline{\boldsymbol{R_\mu}}$. ■

We can now prove the resolvent set is open and hence, the spectrum is closed. By the way, you will note we do not assume T is continuous. Recall, the Stürm - Liouville differential operators are linear but not bounded, although in many cases these inverses exist and are continuous!

Theorem 11.4.2 Resolvent is open and the Spectrum is closed

The resolvent set $\rho(T)$ *is open and the spectrum,* $\sigma(T)$ *is closed.*

Proof 11.4.2
If $\mu \in \rho(T)$, *then* T_μ^{-1} *exists and is continuous and* $\overline{\boldsymbol{R_\mu}} = X$. *By Theorem 11.4.1, if* $|\lambda - \mu| < 1/M(\mu)$, *we know* T_λ^{-1} *exists, is continuous and* $\overline{\boldsymbol{R_\lambda}}$ *is not a proper subset of* X. *Hence,* $\overline{\boldsymbol{R_\lambda}} = X$. *This says* $\lambda \in \rho(T)$ *too. Thus,* $\rho(T)$ *is open and therefore* $\sigma(T)$ *is closed.* ■

Homework

Exercise 11.4.1 *Let* $A = \begin{bmatrix} 2 & -1 \\ 3 & 2 \end{bmatrix}$.

- *Find the set of* λ *for which* $A_\lambda = \lambda I - A$ *is invertible.*
- *Find the set of* λ *for which* $\lambda I - A$ *is not invertible.*
- *Find* $M(\lambda) = \|A_\lambda^{-1}\|_{op}$ *for all* λ *for which* A_λ *is invertible for a variety of matrix norms.*

Exercise 11.4.2 *Let* $A = PJP^{-1}$ *where* J *is the Jordan canonical form for some invertible* P. *Let's focus on simple Jordan forms for a* 3×3 *matrix.*

- *Let*

$$J = \begin{bmatrix} 2 & 0 & 0 \\ 0 & 2 & 0 \\ 0 & 0 & 3 \end{bmatrix}$$

  - *Find the set of* λ *for which* $A_\lambda = \lambda I - A$ *is invertible.*
  - *Find the set of* λ *for which* $\lambda I - A$ *is not invertible.*
  - *Find* $M(\lambda) = \|A_\lambda^{-1}\|_{op}$ *for all* λ *for which* A_λ *is invertible for a variety of matrix norms.*

- *Let*

$$J = \begin{bmatrix} 2 & 1 & 0 \\ 0 & 2 & 0 \\ 0 & 0 & 3 \end{bmatrix}$$

  - *Find the set of* λ *for which* $A_\lambda = \lambda I - A$ *is invertible.*

- *Find the set of λ for which $\lambda I - A$ is not invertible.*
- *Find $M(\lambda) = \|A_\lambda^{-1}\|_{op}$ for all λ for which A_λ is invertible for a variety of matrix norms.*

- *Let*

$$J = \begin{bmatrix} 2 & 1 & 0 \\ 0 & 2 & 1 \\ 0 & 0 & 2 \end{bmatrix}$$

- *Find the set of λ for which $A_\lambda = \lambda I - A$ is invertible.*
- *Find the set of λ for which $\lambda I - A$ is not invertible.*
- *Find $M(\lambda) = \|A_\lambda^{-1}\|_{op}$ for all λ for which A_λ is invertible for a variety of matrix norms.*

Exercise 11.4.3 *Prove that $M(\lambda)$ is continuous on the open set of values of λ for which $\lambda I - A$ has a continuous inverse.*
Hint: *Prove if λ, μ are in this open set and $|\lambda - \mu|$ is sufficiently small, then*

$$\frac{M(\mu)}{1 + |\lambda - \mu| M(\mu)} \leq M(\lambda) < \frac{M(\mu)}{1 - |\lambda - \mu| M(\mu)}$$

The operator T_λ^{-1} is called the resolvent operator of T. We denote this operator by $\mathbb{R}_\lambda$. We can now prove some properties of the resolvent operator.

Theorem 11.4.3 Properties of the Resolvent Operator

Let $(X, \|\cdot\|)$ be a normed linear space which is not just $\{\mathbf{0}\}$ and $T : \boldsymbol{D}(T) \subset X \rightarrow \boldsymbol{R}(T) \subset X$ be linear. Assume $\boldsymbol{R}_\mu = X$ if $\lambda \in \rho(T)$. Then, for all λ and μ in $\rho(T)$,

(i): $\mathbb{R}_\lambda - \mathbb{R}_\mu = (\mu - \lambda)\mathbb{R}_\lambda \mathbb{R}_\mu$.

(ii): $\mathbb{R}_\lambda \mathbb{R}_\mu = \mathbb{R}_\mu \mathbb{R}_\lambda$.

(iii): If $\mu \in \rho(T)$ and $|\mu - \lambda| \|\mathbb{R}_\mu\| < 1$, then $\lambda \in \rho(T)$ and $\mathbb{R}_\lambda = \sum_{n=0}^{\infty} (\lambda - \mu)^n \mathbb{R}_\mu^{n+1}$ where the series converges in the metric given by the usual operator norm.

(iv): As a function on $\rho(T)$, $\mathbb{R}_\lambda$ has derivatives of all orders with $\frac{d^n}{d\lambda^n}\mathbb{R}_\lambda = (-1)^n n! \mathbb{R}_\lambda^{n+1}$.

Proof 11.4.3
(i): *Consider $(\mathbb{R}_\lambda - \mathbb{R}_\mu)(\boldsymbol{y})$ for $\boldsymbol{y} \in X$. Let $\boldsymbol{x} = \mathbb{R}_\mu(\boldsymbol{y})$ so that $\boldsymbol{y} = T_\mu(\boldsymbol{x})$. Then,*

$$T_\mu(\boldsymbol{x}) - T_\lambda(\boldsymbol{x}) = (\mu I - T - \lambda I + T)(\boldsymbol{x}) = (\mu - \lambda)(\boldsymbol{x})$$

Hence, $\boldsymbol{y} = T_\lambda \mathbb{R}_\mu(\boldsymbol{y}) = (\mu - \lambda)\mathbb{R}_\mu(\boldsymbol{y})$. Hence, $\mathbb{R}_\lambda(\boldsymbol{y}) - \mathbb{R}_\mu(\boldsymbol{y}) = (\mu - \lambda)\mathbb{R}_\lambda \mathbb{R}_\mu(\boldsymbol{y})$. This establishes (i).

(ii): *By symmetry, we also easily show $\mathbb{R}_\mu(\boldsymbol{y}) - \mathbb{R}_\lambda(\boldsymbol{y}) = (\lambda - \mu)\mathbb{R}_\mu \mathbb{R}_\lambda(\boldsymbol{y})$. Multiplying this last equation by -1 and comparing, we see $(\mu - \lambda)\mathbb{R}_\lambda \mathbb{R}_\mu(\boldsymbol{y}) = (\mu - \lambda)\mathbb{R}_\mu \mathbb{R}_\lambda(\boldsymbol{y})$. This shows (ii) is true.*

(iii):
By Theorem 11.4.1, $\mu \in \rho(T)$ and $|\lambda - \mu| \|\mathbb{R}_\mu\| < 1$ implies $\lambda \in \rho(T)$. Consider the partial sum

$\sum_{i=0}^{n}(\mu-\lambda)^i\mathbb{R}_\mu^{i+1}+(\mu-\lambda)^{n+1}\mathbb{R}_\mu^{n+1}\mathbb{R}_\lambda$. *For* $n=0$, *we have* $\mathbb{R}_\mu+(\mu-\lambda)\mathbb{R}_\mu\mathbb{R}_\lambda$. *This equals* $\mathbb{R}_\lambda$ *by our proof for (i). Now assume for* $n\leq N$, *that*

$$\mathbb{R}_\lambda = \sum_{i=0}^{n}(\mu-\lambda)^i\mathbb{R}_\mu^{i+1}+(\mu-\lambda)^{n+1}\mathbb{R}_\mu^{n+1}\mathbb{R}_\lambda$$

Now consider

$$\sum_{i=0}^{N+1}(\mu-\lambda)^i\mathbb{R}_\mu^{i+1}+(\mu-\lambda)^{N+2}\mathbb{R}_\mu^{N+2}\mathbb{R}_\lambda =$$
$$\sum_{i=0}^{N}(\mu-\lambda)^i\mathbb{R}_\mu^{i+1}+(\mu-\lambda)^{N+1}\mathbb{R}_\mu^{N+1}\mathbb{R}_\lambda+(\mu-\lambda)^{N+1}\mathbb{R}_\mu^{N+2}+(\mu-\lambda)^{N+2}\mathbb{R}_\mu^{N+2}\mathbb{R}_\lambda$$

By the induction hypothesis, we have

$$\mathbb{R}_\lambda-(\mu-\lambda)^{N+1}\mathbb{R}_\mu^{N+1}\mathbb{R}_\lambda = \sum_{i=0}^{N}(\mu-\lambda)^i\mathbb{R}_\mu^{i+1}$$

We can then say

$$\sum_{i=0}^{N+1}(\mu-\lambda)^i\mathbb{R}_\mu^{i+1}+(\mu-\lambda)^{N+2}\mathbb{R}_\mu^{N+2}\mathbb{R}_\lambda =$$
$$\mathbb{R}_\lambda-(\mu-\lambda)^{N+1}\mathbb{R}_\mu^{N+1}\mathbb{R}_\lambda+(\mu-\lambda)^{N+1}\mathbb{R}_\mu^{N+2}+(\mu-\lambda)^{N+2}\mathbb{R}_\mu^{N+2}\mathbb{R}_\lambda =$$
$$\mathbb{R}_\lambda-(\mu-\lambda)^{N+1}\mathbb{R}_\mu^{N+1}\Big(\mathbb{R}_\lambda-\mathbb{R}_\mu-(\mu-\lambda)\mathbb{R}_\mu\mathbb{R}_\lambda\Big) =$$
$$\mathbb{R}_\lambda-(\mu-\lambda)^{N+1}\mathbb{R}_\mu^{N+1}\Big((\mu-\lambda)\mathbb{R}_\mu\mathbb{R}_\lambda-(\mu-\lambda)\mathbb{R}_\mu\mathbb{R}_\lambda\Big) = \mathbb{R}_\lambda$$

Hence, by induction, for all n, *we know*

$$\mathbb{R}_\lambda = \sum_{i=0}^{n}(\mu-\lambda)^i\mathbb{R}_\mu^{i+1}+(\mu-\lambda)^{n+1}\mathbb{R}_\mu^{n+1}\mathbb{R}_\lambda$$

Now,

$$\lim_{n\to\infty}\|(\mu-\lambda)^{n+1}\mathbb{R}_\mu^{n+1}\mathbb{R}_\lambda\| \leq \lim_{n\to\infty}\|(\mu-\lambda)\mathbb{R}_\mu\|^{n+1}\,\|\mathbb{R}_\lambda\|=0$$

as we assumed $|(\mu-\lambda)\|\mathbb{R}_\mu\|<1$. *Hence, we have* $\mathbb{R}_\lambda=\sum_{n=0}^{\infty}(\lambda-\mu)^n\mathbb{R}_\mu^{n+1}$ *where the series converges in the metric given by* $d(A,B)=\sup_{\|\boldsymbol{x}\|\neq 0}\frac{\|A(\boldsymbol{x})-B(\boldsymbol{x})\|}{\|\boldsymbol{x}\|}$.

(iv): *From the proof of Theorem 11.4.1, for* $|\lambda-\mu|<1/\|\mathbb{R}_\mu$, $\|T_\lambda^{-1}\|=\|\mathbb{R}_\lambda\|\leq\frac{\|\mathbb{R}_\mu\|}{1-|\lambda-\mu|\|\mathbb{R}_\mu}$. *Thus,*

$$\left\|\frac{\mathbb{R}_\lambda-\mathbb{R}_\mu}{\lambda-\mu}+\mathbb{R}_\mu^2\right\| \leq \left\|\frac{\mathbb{R}_\lambda-\mathbb{R}_\mu+(\lambda-\mu)\mathbb{R}_\mu^2}{\lambda-\mu}\right\|$$
$$\leq \left\|\frac{(\mu-\lambda)\mathbb{R}_\lambda\mathbb{R}_\mu+(\lambda-\mu)\mathbb{R}_\mu^2}{\lambda-\mu}\right\|$$
$$\leq \frac{|\lambda-\mu|}{|\lambda-\mu|}\|\mathbb{R}_\lambda\mathbb{R}_\mu-\mathbb{R}_\mu^2\| = \|\mathbb{R}_\mu\|\|\mathbb{R}_\lambda-\mathbb{R}_\mu\|$$

$$= \quad ||\mathbb{R}_\mu|| ||(\mu - \lambda)\mathbb{R}_\lambda \mathbb{R}_\mu|| \leq |\lambda - \mu| ||\mathbb{R}_\mu|| |^2 ||\mathbb{R}_\lambda||$$

Hence, as $\lambda \rightarrow \mu$, $||\frac{\mathbb{R}_\lambda - \mathbb{R}_\mu}{\lambda - \mu} + \mathbb{R}_\mu^2|| \rightarrow 0$. *This shows* $\frac{d}{d\lambda}\mathbb{R}_\lambda = -\mathbb{R}_\lambda^2$. *We leave it to you to verify the full result by induction.* ■

Homework

Exercise 11.4.4 *In Theorem 11.4.3, establish the differentiability result by completing the induction argument.*

Exercise 11.4.5 *Define* $T : \ell^1 \rightarrow \ell^1$ *by the infinite matrix*

$$\begin{bmatrix} 0 & 1 & 0 & 0 & 0 & \ldots \\ 0 & 0 & 1 & 0 & 0 & \ldots \\ 0 & 0 & 0 & 1 & 0\ldots & \\ \vdots & \vdots & \vdots & \vdots & \vdots & \end{bmatrix}$$

If $x = (x_i)_{i=1}^\infty$, *then this matrix representation is a nice short hand way of saying*

$$T(x_1, x_2, x_3, x_4, \ldots) = (x_2, x_3, x_4, \ldots)$$

which is just the left shift by one operator.

- *Prove* $||T||_{op} = 1$.
- *Consider* $(\lambda I - T)(x) = y$. *This means* $\lambda x_k - x_{k+1} = y_k$ *for* $k \geq 1$. *Prove* $|\lambda| > 1$ *implies* $\lambda \in \rho(T)$.
- *When* $\lambda| < 1$, *prove* λ *is an eigenvalue with eigenvectors generated by* $(1, \lambda, \lambda^2, \lambda^3, \ldots)$.
- *When* $\lambda = 1$, *prove* $(I - T)(x) = y$ *has an* ℓ^1 *solution* $x_1 = S$ *and* $x_{i+1} = -\sum_{i=1}^n y_i$, *where* S *is the sum* $\sum_{i=1}^\infty y_i$. *Thus,* $(I - T)^{-1}$ *exists. You can do a similar argument for* $(-I - T)$ *too.*
- *Although* $(I - T)$ *inverse exists,* 1 *is not an eigenvalue as the eigenvector we want is* $(1, 1, 1, \ldots)$ *which is not in* ℓ^1.
- *Prove* $(I - T)^{-1}$ *is not continuous and use a similar argument to show* $(-I - T)^{-1}$ *is not continuous.*

Exercise 11.4.6 *Do the previous exercise for the case* ℓ^2.

Exercise 11.4.7 *Do the previous exercise for the case* ℓ^3.

11.4.2 The Spectrum of a Bounded Linear Operator

We are now ready to discuss the spectrum of a continuous linear operator T. This means $||T||_{op}$ is finite now.

Theorem 11.4.4 Resolvent Results when T is continuous and X is Complete

Let $(X, \|\cdot\|)$ be a normed linear space which is not just $\{\mathbf{0}\}$ and $T : X \to \boldsymbol{R}(T) \subset X$ be linear and continuous.

(i): If $|\lambda| > \|T\|$, then $(\lambda I - T)^{-1}$ exists and $(\lambda I - T)^{-1}(\boldsymbol{y}) = \sum_{n=1}^{\infty} \lambda^{-n} T^{n-1}(\boldsymbol{y})$ for $\boldsymbol{y} \in \boldsymbol{R}(T_\lambda)$.

(ii): If X is complete, if $|\lambda > \|T\|$, then $\lambda \in \rho(T)$ and $\mathbb{R}_\lambda = \sum_{n=1}^{\infty} \lambda^{-n} T^{n-1}$.

The series above are assumed to converge in the usual norm.

Proof 11.4.4

(i): *If $|\lambda| > \|T\|$, then we have $\|\lambda \boldsymbol{x} - T(\boldsymbol{x})\| \geq |\lambda| \|\boldsymbol{x}\| - \|T(\boldsymbol{x})\| \geq (|\lambda - \|T\|) \|\boldsymbol{x}\|$. Thus, if $\boldsymbol{x} \neq \mathbf{0}$, $\frac{\|\lambda \boldsymbol{x} - T(\boldsymbol{x})\|}{\|\boldsymbol{x}\|} \geq |\lambda| - \|T\|$. This tells us $\|\lambda I - T\| \geq |\lambda| - \|T\| > 0$.*

In case you have not seen this argument yet, the fact that $\|\lambda I - T\| > 0$ implies $(\lambda I - T)^{-1}$ exists and is continuous. Since $\|\lambda I - T\| > 0$, there is a $c > 0$ so that $\|\lambda \boldsymbol{x} - T(\boldsymbol{x})\| \geq c\|\boldsymbol{x}\|$ for all $\boldsymbol{x} \in \boldsymbol{D}(T)$. Since T is linear, $\boldsymbol{D}(T)$ is a subspace and contains $\mathbf{0}$. If $(\lambda I - T)(\boldsymbol{x}) = \mathbf{0}$, this lower bound would imply $c\|\boldsymbol{x}\| \leq 0$ telling us $\boldsymbol{x} = 0$. Hence, $\lambda I - T$ is 1-1 and so $(\lambda I - T)^{-1}$ exists. For any $\boldsymbol{y}$ in $\boldsymbol{R}(\lambda I - T)$, we can say $\boldsymbol{x} = (\lambda I - T)^{-1}(\boldsymbol{y})$ and so we can rewrite the earlier bounds as $\boldsymbol{y}\| \geq c\|(\lambda I - T)^{-1}(\boldsymbol{y})\|$. This implies $\|(\lambda I - T)^{-1}\| \leq 1/c$. Hence, $(\lambda I - T)^{-1}$ is also continuous.

For $n = 1$, if $\boldsymbol{y} = (\lambda I - T)(\boldsymbol{x})$, then $T(\boldsymbol{x}) = \lambda \boldsymbol{x} - \boldsymbol{y}$ or $\boldsymbol{x} = \lambda^{-1}\boldsymbol{y} + \lambda^{-1}T(\boldsymbol{x})$. Now assume for all $n \leq N$, we have $\boldsymbol{x} = \sum_{n=1}^{n} \lambda^{-n} T^{n-1}(\boldsymbol{y}) + \lambda^{-n} T^n(\boldsymbol{x})$. Look at the case of $n = N + 1$. We have

$$
\begin{aligned}
& \sum_{n=1}^{N+1} \lambda^{-n} T^{n-1}(\boldsymbol{y}) + \lambda^{-N-1} T^{N+1}(\boldsymbol{x}) = \\
& \sum_{n=1}^{N} \lambda^{-n} T^{n-1}(\boldsymbol{y}) + \lambda^{-N-1} T^{N}(\boldsymbol{y}) + \lambda^{-N-1} T^{N+1}(\boldsymbol{x}) = \\
& \boldsymbol{x} - \lambda^{-N} T^{N}(\boldsymbol{x}) + \lambda^{-N-1} T^{N}(\boldsymbol{y}) + \lambda^{-N-1} T^{N+1}(\boldsymbol{x}) = \\
& \boldsymbol{x} - \lambda^{-N-1} T^{N}(\lambda \boldsymbol{x} - \boldsymbol{y} - T(\boldsymbol{x})) = \\
& \boldsymbol{x} - \lambda^{-N-1} T^{N}(T(\boldsymbol{x}) - T(\boldsymbol{x})) = \boldsymbol{x}
\end{aligned}
$$

by the induction hypothesis. Hence, for all n, $\boldsymbol{x} = \sum_{n=1}^{n} \lambda^{-n} T^{n-1}(\boldsymbol{y}) + \lambda^{-n} T^n(\boldsymbol{x})$ and so

$$
\|\boldsymbol{x} - \sum_{n=1}^{n} \lambda^{-n} T^{n-1}(\boldsymbol{y})\| \quad = \quad |\lambda|^{-n} \|T^n(\boldsymbol{x})\| \leq \frac{\|T^n\| \|\boldsymbol{x}\|}{|\lambda|^n} \leq \left(\frac{\|T\|}{|\lambda|}\right)^n \|\boldsymbol{x}\|
$$

But $\frac{\|T\|}{|\lambda|} < 1$, so for a given $\boldsymbol{x}$, as $n \to \infty$, we have $\boldsymbol{x} = \sum_{n=1}^{\infty} \lambda^{-n} T^{n-1}(\boldsymbol{y})$.

(ii): *If X is complete, let $A = \lambda^{-1}T$. Then since $\|A\| = \frac{\|T\|}{|\lambda} < 1$, we see the series $\sum_{n=0}^{\infty} \|A\|^n = \sum_{n=0}^{\infty} |\lambda|^{-n} \|T\|^n$ converges in operator norm. The space of continuous linear operators on a complete normed linear space is complete (see (Peterson (100) 2020)) so since $\|A\|^n \leq \|A\|^n$, the sequence of partial sums $\sum_{i=0}^{n} \|A\|^i$ is a Cauchy sequence and so the series $\sum_{n=0}^{\infty} \lambda^{-n} T^n$ converges to a bounded linear operator B.*

Now

$$
AB \quad = \quad \lambda^{-1} T \sum_{n=0}^{\infty} |\lambda|^{-n}\, T^n = \sum_{n=0}^{\infty} |\lambda|^{-n-1} T^{n+1} = \sum_{n=0}^{\infty} |\lambda|^{-n}\, T^n\, \lambda^{-1} T = BA
$$

Therefore,

$$\begin{aligned}(I-\lambda^{-1}T)B &= B-AB = B-BA = B(I-A) = B(I-\lambda^{-1}T)\\ &= \left(\sum_{n=0}^{\infty}\lambda^{-n}T^{n}\right)(I-\lambda^{-1}T)\end{aligned}$$

The partial sum here is $(\sum_{n=0}^{N}\lambda^{-n}T^{n})(I-\lambda^{-1}T)$ *which is telescoping and gives* $I-\lambda^{-N-1}T^{N+1}$*. By the same argument as before, this partial sum converges to* I*. Thus,* $(I-\lambda^{-1}T)B = B(I-\lambda^{-1}T) = I$*. We conclude* $(I-\lambda^{-1}T)^{-1} = B$*. This tells us* $(\lambda I - T)^{-1} = \lambda\sum_{n=0}^{\infty}\lambda^{-n}T^{n}$*. Thus,* $(\lambda I-T)^{-1} = \sum_{n=1}^{\infty}\lambda^{-n}T^{n-1}$*. Since the domain of* $(\lambda I-T)^{-1}$ *is all of* X *when* X *is complete, we see for all* $\boldsymbol{x}\in X$*,* $\boldsymbol{y} = (\lambda I-T)^{-1}(\boldsymbol{x})$ *implies* $(\lambda I-T)(\boldsymbol{y}) = \boldsymbol{x}$*. Thus, the range of* $(\lambda I-T)$ *is all* X*. This says* $\lambda\in\rho(T)$*.* ■

If X is complete, we can say more about the spectrum.

Theorem 11.4.5 Spectrum of a Linear Continuous Operator on the Complete Normed Linear Space X whose Domain is all X is Compact

Let $(X, \|\cdot\|)$ *be a normed linear space which is not just* $\{\boldsymbol{0}\}$ *and* $T : X \rightarrow \boldsymbol{R}(T)\subset X$ *be linear and continuous. Then* $\sigma(T)$ *is compact.*

Proof 11.4.5

If X *is complete, Theorem 11.4.4 says if* $\lambda > \|T|$*, then* $|\lambda|\in\rho(T)$*. All other* λ *are in the spectrum. Thus* $|\lambda|\leq\|T\|$ *implies* $\lambda\in\sigma(T)$*. This says* $\sigma(T)$ *is closed and bounded and therefore compact.* ■

Let's introduce another important concept, the spectral radius of an operator. The spectrum of the operator is also divided into subsets called the continuous, the residual and the point spectrum. It takes a bit of work to define these, so lay out these definitions carefully. Note the eigenvalues, λ, of an $n\times n$ matrix T are actually quite simple: there are n of them, λ_n, possibly repeated. The absolute values of these eigenvalues clearly have a maximum value given by $\lambda_{\max}$. Each eigenvalue has an eigenspace which is at least one dimensional and so $(\lambda_n I - T)^{-1}$ does not exist and $\boldsymbol{R}(\lambda_n I - T)\subsetneq\mathbb{C}^n$ assuming some of the eigenvalues are complex-valued. Keep these facts in mind while you read through the following definitions. First, let's define the spectral radius. Recall if $\boldsymbol{R}(\lambda I - T)$ is dense in X and if $(\lambda I - T)^{-1}$ exists and is continuous, we say λ is in the resolvent set of T which we denote by $\rho(T)$.

Definition 11.4.2 Spectral Radius of a linear operator T

Let $(X, \|\cdot\|)$ *be a normed linear space which is not just* $\{\boldsymbol{0}\}$ *and* $T : \boldsymbol{D}(T)\subset X\rightarrow\boldsymbol{R}(T)\subset X$ *be linear. The set of all values of* $\lambda\notin\rho(T)$ *is the set* $\sigma(T)$ *which is called the spectrum of* T*. If* $\sigma(T)$ *is nonempty and bounded, we define the spectral radius of* T *to be* $r_\sigma(T) = \sup_{\lambda\in\sigma(T)}|\lambda|$*.*

Then, we can talk about the various parts of $\sigma(T)$. The eigenvalues of an $n\times n$ matrix A are particularly simple. There are n complex eigenvalues λ, possibly repeated, and $(\lambda I - A)^{-1}$ exists and is continuous for all λ that are not eigenvalues. Also, the range of $(\lambda I - A)$ is all of $\Re^n$ for any λ that are not eigenvalues. The situation on a general normed linear space which is infinite dimensional is much more complicated and we will explore that to some extent. We will only look at some of what is possible and we encourage you to read more widely. We note the spectrum $\sigma(T)$ has several important parts.

Definition 11.4.3 Subdivisions of the Spectrum of a Linear Operator: Continuous, Residual and Point

Let $(X, \|\cdot\|)$ be a normed linear space which is not just $\{\mathbf{0}\}$ and $T : \boldsymbol{D}(T) \subset X \rightarrow \boldsymbol{R}(T) \subset X$ be linear. We divide the spectrum into subdivisions.

(i): The continuous spectrum, $c_\sigma(T)$, is defined by

$$\begin{aligned} &c_\sigma(T) = \\ &\{\lambda \in \sigma(T) | \boldsymbol{R}(\lambda I - T) = X,\ (\lambda I - T)^{-1} \exists \text{ but not continuous } \} \\ &\cup \\ &\{\lambda \in \sigma(T) | \overline{\boldsymbol{R}(\lambda I - T)} = X \text{ but} \\ &\qquad \boldsymbol{R}(\lambda I - T) \neq X,\ (\lambda I - T)^{-1} \exists \text{ but not continuous } \} \end{aligned}$$

(ii): The residual spectrum, $r_\sigma(T)$, is defined by

$$\begin{aligned} &r_\sigma(T) = \{\lambda \in \sigma(T) | \overline{\boldsymbol{R}(\lambda I - T)} \subsetneq X,\ (\lambda I - T)^{-1} \exists \text{ but not continuous } \} \\ &\cup \\ &\{\lambda \in \sigma(T) | \overline{\boldsymbol{R}(\lambda I - T)} \subsetneq X \text{ but } \boldsymbol{R}(\lambda I - T) \neq X \text{ and} \\ &\qquad (\lambda I - T)^{-1} \exists \text{ and is continuous } \} \end{aligned}$$

(iii): The point spectrum, $p_\sigma(T)$, is defined by

$$\begin{aligned} &p_\sigma(T) = \{\lambda \in \sigma(T) | \boldsymbol{R}(\lambda I - T) = X,\ (\lambda I - T)^{-1} \ \nexists\} \\ &\cup \\ &\{\lambda \in \sigma(T) | \overline{\boldsymbol{R}(\lambda I - T)} = X \text{ but } \boldsymbol{R}(\lambda I - T) \neq X \text{ and } (\lambda I - T)^{-1} \ \nexists\} \\ &\cup \\ &\{\lambda \in \sigma(T) | \overline{\boldsymbol{R}(\lambda I - T)} \subsetneq X \text{ and } (\lambda I - T)^{-1} \ \nexists\} \end{aligned}$$

The values in the point spectrum are the ones we typically think of as eigenvalues of the operator.

Homework

Exercise 11.4.8 *Explain how Exercise 11.4.4 gives us an example of an operator T on the normed linear space ℓ^1 with ± 1 in the residual spectrum $r_\sigma(T)$.*

Exercise 11.4.9 *Explain how Exercise 11.4.6 gives us an example of an operator T on the normed linear space ℓ^2 with ± 1 in the residual spectrum $r_\sigma(T)$.*

Exercise 11.4.10 *Explain how Exercise 11.4.7 gives us an example of an operator T on the normed linear space ℓ^3 with ± 1 in the residual spectrum $r_\sigma(T)$.*

11.4.3 The Ascent and Descent of an Operator

To understand the spectrum better, we will look at what are called the ascent and descent of an operator. Let's set up some notation we will use a lot, and although we have used this notation already, it is useful to have a reminder. As usual, X is a normed linear space with norm $\|\cdot\|$.

1. For an operator T, its domain is $\boldsymbol{D}(T)$ and its range is $\boldsymbol{R}(T)$.

2. The operator $\lambda I - T$ is denoted by T_λ and it will have domain and range, $\boldsymbol{D}(T_\lambda)$ and $\boldsymbol{R}(T_\lambda)$, respectively.

3. The inverse T_λ^{-1} is denoted by $\mathbb{R}_\lambda$ and it will have domain $\boldsymbol{D}(\mathbb{R}_\lambda)$ and range $\boldsymbol{R}(\mathbb{R}_\lambda)$.

4. The nullspace or kernel of an operator is $\boldsymbol{N}(T) = \{\boldsymbol{x} \in X | T(\boldsymbol{x}) = \boldsymbol{0}\}$.

5. We can take powers of an operator to give new operators T^n for $n = 0, 1, \ldots$ where we identify T^0 with the identity I. We note the domain of T^n is

$$\boldsymbol{D}(T^n) = \{\boldsymbol{x} \in X | \boldsymbol{x}, T(\boldsymbol{x}), \ldots, T^{n-1}(\boldsymbol{x}) \in \boldsymbol{D}(T)\}.$$

We let $\boldsymbol{D_n}(T) = \boldsymbol{D}(T^n)$. We see $\boldsymbol{D_n}(T) \subset \boldsymbol{D_{n-1}}(T)$.

The nullspace of T^n is $\boldsymbol{N}(T^n) = \{\boldsymbol{x} \in \boldsymbol{D_n}(T) | T^n(\boldsymbol{x}) = \boldsymbol{0}\}$. Note $\boldsymbol{N}(T^0) = \boldsymbol{N}(I) = \boldsymbol{0}$.

Our first result is straightforward.

Theorem 11.4.6 Ascending Chain of Nullspaces of T^n

Let $(X, \|\cdot\|)$ be a normed linear space and $T : \boldsymbol{D}(T) \subset X \rightarrow \boldsymbol{R}(T) \subset X$ be a linear operator. Then

(i): $\boldsymbol{N}(T^n) \subset \boldsymbol{N}(T^{n+1})$ for all $n = 0, 1, \ldots$.

(ii): If $\boldsymbol{N}(T^k) = \boldsymbol{N}(T^{k+1})$ for some k, then $\boldsymbol{N}(T^n) = \boldsymbol{N}(T^k)$ for all $n \geq k$.

Proof 11.4.6
(i): *This statement is easy to see so we will leave that to you.*

(ii): *Assume there is k so that $\boldsymbol{N}(T^k) = \boldsymbol{N}(T^{k+1})$. Let $\boldsymbol{x} \in \boldsymbol{N}(T^{k+2})$. Then $\boldsymbol{x} \in \boldsymbol{D}(T^{k+2})$ and $T^{k+2}(\boldsymbol{x}) = \boldsymbol{0}$. Since $T^i(\boldsymbol{x}) \in \boldsymbol{D}(T)$ for $0 \leq i \leq k+1$, because $\boldsymbol{x} \in \boldsymbol{N}(T^{k+2})$, it follows that $T^i(T(\boldsymbol{x})) \in \boldsymbol{D}(T)$ for $0 \leq i \leq k+1$. This tells us $T(\boldsymbol{x}) \in \boldsymbol{D}(T^{k+1})$. Since $T^{k+1}(T(\boldsymbol{x})) = \boldsymbol{0}$ as $T^{k+2}(\boldsymbol{x}) = \boldsymbol{0}$. So $T(\boldsymbol{x}) \in \boldsymbol{N}(T^{k+1}) = \boldsymbol{N}(T^k)$ by assumption. Thus, $T^k(T(\boldsymbol{x})) = \boldsymbol{0}$ which implies $\boldsymbol{x} \in \boldsymbol{N}(T^{k+1})$. Thus, $\boldsymbol{N}(T^{k+2}) \subset \boldsymbol{N}(T^{k+1})$. We already knew the reverse containment, so we see $\boldsymbol{N}(T^{k+2}) = \boldsymbol{N}(T^{k+1})$. The general result follows by an induction argument we leave to you also.*
■

This leads to a definition of the ascent of a linear operator.

Definition 11.4.4 Ascent of a Linear Operator

Let $(X, \|\cdot\|)$ be a normed linear space and $T : \boldsymbol{D}(T) \subset X \rightarrow \boldsymbol{R}(T) \subset X$ be a linear operator. If there is $n \geq 0$ so that $\boldsymbol{N}(T^n) = \boldsymbol{N}(T^{n+1})$, the smallest such integer n is the ascent of T and is denoted by $\alpha(T)$. If no such integer exists, we say $\alpha(T) = \infty$.

Now, let's look at at similar ideas concerning descending chains of ranges of T^n.

Theorem 11.4.7 Descending Chain of ranges of T^n

Let $(X, \|\cdot\|)$ be a normed linear space and $T : \boldsymbol{D}(T) \subset X \rightarrow \boldsymbol{R}(T) \subset X$ be a linear operator. Then

(i): $\boldsymbol{R}(T^{n+1}) \subset \boldsymbol{N}(T^n)$ for all $n = 0, 1, \ldots$.

(ii): If $\boldsymbol{R}(T^{k+1}) = \boldsymbol{N}(T^k)$ for some k, then $\boldsymbol{R}(T^n) = \boldsymbol{R}(T^k)$ for all $n \geq k$.

Proof 11.4.7
(i): *If $\boldsymbol{y} \in \boldsymbol{R}(T^{n+1})$, then there is $\boldsymbol{x} \in \boldsymbol{D_{n+1}}(T)$ with $\boldsymbol{y} = T^{n+1}(\boldsymbol{x})$. This tells us $T(\boldsymbol{x}) \in \boldsymbol{D_n}(T)$. Also, $\boldsymbol{y} = T^n(T(\boldsymbol{x}))$ tells us $\boldsymbol{y} \in \boldsymbol{R}(T^n)$. This proves the result.*

(ii): *First note, $\boldsymbol{R}(T^{n+1}) = T(\boldsymbol{R}(T^n) \cap \boldsymbol{D}(T))$. We leave this to you in an exercise. Now assume there is k so that $\boldsymbol{R}(T^{k+1}) = \boldsymbol{R}(T^k)$. Let $\boldsymbol{y} \in \boldsymbol{R}(T^{k+1})$. Then if $\boldsymbol{y} = T(\boldsymbol{x})$, $\boldsymbol{x} \in \boldsymbol{R}(T^k) \cap \boldsymbol{D}(T)$. Thus, $\boldsymbol{y} \in \boldsymbol{R}(T^{k+2})$. This shows $\boldsymbol{R}(T^{k+1}) \subset \boldsymbol{R}(T^{k+2})$. We already know the reverse containment and so $\boldsymbol{R}(T^{k+1}) = \boldsymbol{R}(T^{k+2})$. The general result follows by an induction argument we leave to you also.* ■

We can now define the descent of a linear operator.

Definition 11.4.5 Descent of a Linear Operator

Let $(X, \|\cdot\|)$ be a normed linear space and $T : \boldsymbol{D}(T) \subset X \rightarrow \boldsymbol{R}(T) \subset X$ be a linear operator. If there is $n \geq 0$ so that $\boldsymbol{R}(T^{n+1}) = \boldsymbol{R}(T^n)$, the smallest such integer n is the descent of T and is denoted by $\delta(T)$. If no such integer exists, we say $\delta(T) = \infty$.

We can now show that if the ascent is finite and the descent is zero, the ascent must be zero also.

Theorem 11.4.8 Ascent Finite with the Descent Zero Implies the Ascent is Zero

Let $(X, \|\cdot\|)$ be a normed linear space and $T : \boldsymbol{D}(T) \subset X \rightarrow \boldsymbol{R}(T) \subset X$ be a linear operator. If $\alpha(T)$ is finite and $\delta(T) = 0$, then $\alpha(T) = 0$ also.

Proof 11.4.8
Assume $0 < \alpha(T) < \infty$ and $\delta(T) = 0$. Then $\boldsymbol{R}(T^n) = \boldsymbol{R}(I)$ for all n. This says $\boldsymbol{R}(T^n) = X$ always. Now if $\boldsymbol{N}(T) = \boldsymbol{0}$, then $\boldsymbol{N}(T^0) \subset \boldsymbol{N}(T)$ tells us $\alpha(T) = 0$. By assumption, $\alpha(T)$ finite and positive implies $\boldsymbol{N}(T) \neq \boldsymbol{N}(T^0) = \boldsymbol{0}$. Thus, T^{-1} does not exist. Choose $\boldsymbol{x_1} \neq \boldsymbol{0} \in \boldsymbol{D}(T)$ with $T(\boldsymbol{x_1}) = \boldsymbol{0}$. Since $\boldsymbol{R}(T^2) = X$, there is $\boldsymbol{x_2} \in \boldsymbol{D}(T)$ with $T(\boldsymbol{x_2}) = \boldsymbol{x_1}$. Hence, by induction, we can define a sequence $(\boldsymbol{x_n})$ with $T(\boldsymbol{x_{n+1}}) \in \boldsymbol{D}(T)$ so that $T(\boldsymbol{x_{n+1}}) = \boldsymbol{x_n}$. Note

$$\{\boldsymbol{x_{n+1}}, T(\boldsymbol{x_{n+1}}), T^2(\boldsymbol{x_{n+1}}), \ldots, T^n(\boldsymbol{x_{n+1}})\} \quad = \quad \{\boldsymbol{x_{n+1}}, \boldsymbol{x_n}, \boldsymbol{x_{n-1}}, \ldots, \boldsymbol{x_1}\} \subset \boldsymbol{D}(T)$$

Hence, $\boldsymbol{x_{n+1}} \in \boldsymbol{D_{n+1}}(T)$ and $T^{n+1}(\boldsymbol{x_{n+1}}) = T(\boldsymbol{x_1}) \neq \boldsymbol{0}$. This means $\boldsymbol{x_{n+1}} \in \boldsymbol{N}(T^{n+1})$. This argument shows that $\boldsymbol{x_{n+1}} \in \boldsymbol{N}(T^{n+1})$ but $\boldsymbol{x_{n+1}} \notin \boldsymbol{N}(T^n)$. Since this is true for all n, this says $\alpha(T) = \infty$. But we assumed the ascent was finite. So our assumption that the ascent is positive leads to a contradiction. This means the ascent must be zero. ■

Next, we show that if the ascent and descent are both finite, then the ascent is bounded above by the descent.

Theorem 11.4.9 $\alpha(T) \leq \delta(T)$ if Both are Finite

Let $(X, \|\cdot\|)$ be a normed linear space and $T : \boldsymbol{D}(T) \subset X \to \boldsymbol{R}(T) \subset X$ be a linear operator. If $\alpha(T)$ and $\delta(T)$ are finite, then $\alpha(T) \leq \delta(T)$.

Proof 11.4.9
Let $r = \delta(T)$. Note, $\boldsymbol{R}(T^n) = \boldsymbol{R}(T^r)$ if $n \geq r$. $\boldsymbol{R}(T^r) = \boldsymbol{R}(T^{r+1}) = T(\boldsymbol{R}(T^r) \cap \boldsymbol{D}(T))$. Let $X_1 = \boldsymbol{R}(T^r)$, $\boldsymbol{E_1} = \boldsymbol{R}(T^r) \cap \boldsymbol{D}(T)$ and $S_1 = T|_{\boldsymbol{E_1}} = T^r$. Then $\boldsymbol{R}(S_1) = X_1$ and so, $\boldsymbol{R}(S_1) = \boldsymbol{R}(S_1^0) = \boldsymbol{R}(I|_{\boldsymbol{E_1}})$. Thus, $\delta(S_1) = 0$.

Now $\boldsymbol{D}(S_1) = \boldsymbol{R}(T^r) \cap \boldsymbol{D}(T)$. Assume $\boldsymbol{D}(S_1^k) = \boldsymbol{R}(T^r) \cap \boldsymbol{D_k}(T)$ for all $k \leq n$. This is our induction hypothesis. We want to show $\boldsymbol{D}(S_1^{n+1}) = \boldsymbol{R}(T^r) \cap \boldsymbol{D_{n+1}}(T)$.

Consider $\boldsymbol{D}(S_1^{n+1}) = \{\boldsymbol{x} | S_1^i(\boldsymbol{x}) \in \boldsymbol{D}(S_1), \, 0 \leq i \leq n\}$. Pick $\boldsymbol{x} \in \boldsymbol{D}(S_1^{n+1})$. Since $S_1^i(\boldsymbol{x}) \in \boldsymbol{D}(S_1)$, for $0 \leq i \leq n-1$, $\boldsymbol{x} \in \boldsymbol{D}(S_1^n)$. We know $\boldsymbol{D}(S_1^n) = \boldsymbol{R}(T^r) \cap \boldsymbol{D_n}(T)$ by the induction assumption. But $\boldsymbol{R}(T^r) \cap \boldsymbol{D_n}(T) \subset \boldsymbol{R}(T^r) \cap \boldsymbol{D}(T) = E_1 = \boldsymbol{D}(S_1)$. Since $S_1 = T|_{\boldsymbol{E_1}}$, $S_1^n(\boldsymbol{x}) = T^n(\boldsymbol{x})$ for all $\boldsymbol{x} \in \boldsymbol{D}(S_1^n)$. Thus, $T^n(\boldsymbol{x}) \in \boldsymbol{D}(S_1) = \boldsymbol{R}(T^r) \cap \boldsymbol{D}(T)$ implying $T^n(\boldsymbol{x}) \in \boldsymbol{D}(T)$. Hence, $\boldsymbol{x} \in \boldsymbol{D}(T^{n+1})$. But $\boldsymbol{x} \in \boldsymbol{R}(T^r)$ also. We conclude $\boldsymbol{x} \in \boldsymbol{R}(T^r) \cap \boldsymbol{D}_{n+1}(T)$. This shows $\boldsymbol{D}(S_1^{n+1}) \subset \boldsymbol{R}(T^r) \cap \boldsymbol{D}_{n+1}(T)$.

Now if $\boldsymbol{x} \in \boldsymbol{R}(T^r) \cap \boldsymbol{D}_{n+1}(T)$, $T^i(\boldsymbol{x}) \in \boldsymbol{D}(T)$ for $0 \leq i \leq n$. But $\boldsymbol{x} \in \boldsymbol{R}(T^r) \cap \boldsymbol{D}(T)$ implies $T(\boldsymbol{x}) \in T(\boldsymbol{R}(T^r) \cap \boldsymbol{D}(T)) = \boldsymbol{R}(T^{r+1}) = \boldsymbol{R}(T^r)$. Using a similar argument, we see $T^i(\boldsymbol{x}) \in \boldsymbol{R}(T^r)$ for $0 \leq i \leq n$. This says $T^i(\boldsymbol{x}) \in \boldsymbol{R}(T^r) \cap \boldsymbol{D}(T) = \boldsymbol{D}(S_1)$ for $0 \leq i \leq n$. We conclude $\boldsymbol{x} \in \boldsymbol{D}(S_1^{n+1})$. This shows $\boldsymbol{R}(T^r) \cap \boldsymbol{D}_{n+1}(T) \subset \boldsymbol{D}(S_1^{n+1})$. This completes the induction argument and so we know $\boldsymbol{D}(S_1^{n+1}) = \boldsymbol{R}(T^r) \cap \boldsymbol{D_{n+1}}(T)$ for all n.

Next, assume $S_1^{n+1}(\boldsymbol{x}) = \boldsymbol{0}$ for an $\boldsymbol{x} \in \boldsymbol{D}(S_1^{n+1}) = \boldsymbol{R}(T^r) \cap \boldsymbol{D_{n+1}}$. Since $S_1 = T|_{\boldsymbol{D}_{\boldsymbol{S_1}}}$, this tells us $T^{n+1}(\boldsymbol{x}) = \boldsymbol{0}$ also. Thus, $\boldsymbol{N}(S_1^{n+1}) \subset \boldsymbol{N}(T^{n+1})$. Now, we also know $\boldsymbol{D}(S_1^{n+1}) \subset \boldsymbol{D}(S_1^n)$. Therefore,

$$\boldsymbol{N}(T^n) \cap \boldsymbol{D}(S_1^{n+1}) \quad \subset \quad \boldsymbol{N}(T^n) \cap \boldsymbol{D}(S_1^n) \subset \boldsymbol{N}(S_1^n)$$

We claim

$$\boldsymbol{N}(S_1^{n+1}) \setminus \boldsymbol{N}(S_1^n) \quad \subset \quad \boldsymbol{N}(T^{n+1}) \setminus \boldsymbol{N}(T^n)$$

To see this, assume there is $\boldsymbol{x} \neq \boldsymbol{0}$ in $\boldsymbol{N}(S_1^{n+1}) \setminus \boldsymbol{N}(S_1^n)$. Then, $S_1^{n+1}(\boldsymbol{x}) = \boldsymbol{0}$ but $S_1^n(\boldsymbol{x}) \neq \boldsymbol{0}$. Hence, $T^{n+1}(\boldsymbol{x}) = \boldsymbol{0}$ too. Now, if $T^n(\boldsymbol{x}) = \boldsymbol{0}$, then $\boldsymbol{x} \in \boldsymbol{N}(T^n) \cap \boldsymbol{D}(S_1^n) \subset \boldsymbol{N}(S_1^n)$. So, $S_1^n(\boldsymbol{x}) = \boldsymbol{0}$. But, we assumed $S_1^n(\boldsymbol{x}) \neq \boldsymbol{0}$. This is a contradiction, so such an $\boldsymbol{x}$ cannot exist and we see the claim is true.

Consequently, if $\alpha(T) = m$, $\boldsymbol{N}(T^{m+1}) \setminus \boldsymbol{N}(T^m) = \emptyset$. This implies $\boldsymbol{N}(S_1^{n+1}) \setminus \boldsymbol{N}(S_1^n) = \emptyset$ too. Thus, we have shown $\alpha(S_1) \leq m$ and so $\alpha(S_1)$ is finite with $\delta(S_1) = 0$. Applying Theorem 11.4.8, we find $\alpha(S_1) = 0$ and so S_1^{-1} exists.

Finally, suppose $\boldsymbol{x} \in \boldsymbol{N}(T^{r+1})$. Let $\boldsymbol{y} = T^r(\boldsymbol{x})$. Then, $\boldsymbol{y} \in \boldsymbol{R}(T^r) \cap \boldsymbol{D}(T) = \boldsymbol{D}(S_1)$. Since $S_1 = T|_{\boldsymbol{D}(S_1)}$, $S_1(\boldsymbol{y}) = T(\boldsymbol{y}) = T^{r+1}(\boldsymbol{x}) = \boldsymbol{0}$. Since S_1^{-1} exists, this implies $\boldsymbol{y} = \boldsymbol{0}$. This tells us $\boldsymbol{N}(T^{r+1}) \subset \boldsymbol{N}(T^r)$. We already know the reverse inclusion, so we have shown $\boldsymbol{N}(T^{r+1}) = \boldsymbol{N}(T^r)$. Hence, $\alpha(T) \leq r = \delta(T)$. ■

Homework

Exercise 11.4.11 *In the proof of Theorem 11.4.6, give the details of the full induction argument.*

Exercise 11.4.12 *Prove* $\boldsymbol{R}(T^{n+1}) = T(\boldsymbol{R}(T^n) \cap \boldsymbol{D}(T))$.

Exercise 11.4.13 *In the proof of Theorem 11.4.7, give the details of the full induction argument.*

Exercise 11.4.14 *Find* $\alpha(A)$ *and* $\delta(A)$ *for* $A = \begin{bmatrix} 0 & 1 & 3 \\ 0 & 0 & 4 \\ 0 & 0 & 0 \end{bmatrix}$.

Exercise 11.4.15 *Find* $\alpha(A)$ *and* $\delta(A)$ *for* $A = \begin{bmatrix} 1 & 1 & 3 \\ 0 & 2 & 4 \\ 0 & 0 & 3 \end{bmatrix}$.

Exercise 11.4.16 *For the operator* T *in Exercise 11.4.4, find* $\alpha(A)$ *and* $\delta(A)$.

Exercise 11.4.17 *For the operator* T *in Exercise 11.4.6, find* $\alpha(A)$ *and* $\delta(A)$.

Exercise 11.4.18 *Find* $\alpha(A)$ *and* $\delta(A)$ *for the derivative operator* x' *on* $C([0,1])$ *for any choice of norm.*

We can say even more if the domain of T is all X.

Theorem 11.4.10 Ascent and Descent Match if $\boldsymbol{D}(T) = X$

Let $(X, \|\cdot\|)$ *be a normed linear space and* $T : \boldsymbol{D}(T) \subset X \rightarrow \boldsymbol{R}(T) \subset X$ *be a linear operator. If* $\alpha(T)$ *and* $\delta(T)$ *are finite and* $\boldsymbol{D}(T) = X$, *then* $\alpha(T) = \delta(T)$.

Proof 11.4.10
By Theorem 11.4.9, we know $\alpha(T) \leq \delta(T)$. *So we need to prove the reverse inequality. If* $\delta(T) = 0$, *by Theorem 11.4.8,* $\alpha(T) = 0$ *as well and so they are equal. So assume* $\delta(T) = r > 0$. *Then there is* $\boldsymbol{y} \in \boldsymbol{R}(T^{r-1}) \setminus \boldsymbol{R}(T^r)$ *and* $\boldsymbol{x}$ *with* $\boldsymbol{y} = T^{r-1}(\boldsymbol{x})$. *Let* $\boldsymbol{z} = T(\boldsymbol{y}) = T^r(\boldsymbol{x})$. *So* $\boldsymbol{z} \in \boldsymbol{R}(T^r)$. *We know* $T^r(\boldsymbol{R}(T^r) \cap \boldsymbol{D}(T)) = \boldsymbol{R}(T^{2r})$.

But $\boldsymbol{D}(T) = X$, *so* $\boldsymbol{R}(T^{2r}) = T^r(\boldsymbol{R}(T^r))$. *Since* $\delta(T) = r$, *so* $\boldsymbol{R}(T^r) = T^r(\boldsymbol{R}(T^r))$. *It follows there is* $\boldsymbol{u} \in \boldsymbol{R}(T^r)$ *with* $\boldsymbol{z} = T^r(\boldsymbol{u})$. *Let* $\boldsymbol{v} = \boldsymbol{x} - \boldsymbol{u}$. *Then,* $T^r(\boldsymbol{v}) = T^r(\boldsymbol{x}) - T^r(\boldsymbol{u}) = \boldsymbol{z} - \boldsymbol{z} = \boldsymbol{0}$. *Also,* $T^{r-1}(\boldsymbol{v}) = T^{r-1}(\boldsymbol{x}) - T^{r-1}(\boldsymbol{u}) = \boldsymbol{y} - T^{r-1}(\boldsymbol{u})$.

Now, $T^{r-1}(\boldsymbol{u}) \in \boldsymbol{R}(T^{2r-1}) = \boldsymbol{R}(T^r)$. *Since* $\boldsymbol{y} \notin \boldsymbol{R}(T^{2r-1})$, *this says* $\boldsymbol{y} - T^{r-1}(\boldsymbol{u}) \neq \boldsymbol{0}$. *So* $T^{r-1}(\boldsymbol{u}) \neq \boldsymbol{0}$. *We have shown* $\boldsymbol{v} \in \boldsymbol{N}(T^r) \setminus \boldsymbol{N}(T^{r-1})$. *Hence,* $\alpha(T) \geq r = \delta(T)$. *This is the reversed inequality we seek.* ■

We can now explore the structure of the subspaces $\boldsymbol{R}(T^r)$ where $r = \delta(T)$. First, a word about notation. We would say two subspaces V and W, not zero dimensional, are linearly independent in the vector space X if $V \cap W = \boldsymbol{0}$.

Theorem 11.4.11 Decomposition of $\boldsymbol{D}(T^r)$ when $r = \delta(T)$

Let $(X, \|\cdot\|)$ *be a normed linear space and* $T : \boldsymbol{D}(T) \subset X \rightarrow \boldsymbol{R}(T) \subset X$ *be a linear operator. If* $\alpha(T)$ *and* $\delta(T)$ *are finite with* $\delta(T) = r$, *then* $\boldsymbol{R}(T^r)$ *and* $\boldsymbol{N}(T^r)$ *are linearly independent subspaces and* $\boldsymbol{D}(T^r) = \{\boldsymbol{R}(T^r) \cap \boldsymbol{D}(T^r)\} \oplus \boldsymbol{N}(T^r)$.

Proof 11.4.11
Let $\boldsymbol{y} \in \boldsymbol{R}(T^r) \cap \boldsymbol{N}(T^r)$. *Then, there is* $\boldsymbol{x} \in \boldsymbol{D}(T^r)$ *with* $\boldsymbol{y} = T^r(\boldsymbol{x})$ *and* $T^r(\boldsymbol{y}) = \boldsymbol{0}$. *Thus,* $T^{2r}(\boldsymbol{x}) = \boldsymbol{0}$ *and* $\boldsymbol{x} \in \boldsymbol{N}(T^{2r})$. *However, since both the ascent and descent are finite, we must have* $\alpha(T) \leq r$. *Hence,* $\boldsymbol{N}(T^{2r}) = \boldsymbol{N}(T^r)$ *and we see* $\boldsymbol{x} \in \boldsymbol{N}(T^r)$ *implying* $T(\boldsymbol{x}) = \boldsymbol{0}$. *We con-*

clude $\boldsymbol{y} = \boldsymbol{0}$. Thus, $\boldsymbol{R}(T^r) \cap \boldsymbol{N}(T^r) = \boldsymbol{0}$ and so these subspaces are linearly independent subspaces.

As in the proof of Theorem 11.4.9, let $X_1 = \boldsymbol{R}(T^r)$, $\boldsymbol{E_1} = \boldsymbol{R}(T^r) \cap \boldsymbol{D}(T)$ and $S_1 = T|_{\boldsymbol{E_1}} = T^r$. Then, $\boldsymbol{R}(S_1) = X_1$ and $\boldsymbol{D}(S_1) = \boldsymbol{R}(T^r) \cap \boldsymbol{D}(T)$. Then $\boldsymbol{R}(S_1) = X_1$ and so, $\boldsymbol{R}(S_1) = \boldsymbol{R}(S_1^0) = \boldsymbol{R}(I|_{\boldsymbol{E_1}})$. Thus, $\delta(S_1) = 0$.

Let $\boldsymbol{x} \in \boldsymbol{D}(T^r)$. Now $\boldsymbol{D}(S_1^r) = \boldsymbol{R}(T^r) \cap \boldsymbol{D}(T^r)$ by the proof of Theorem 11.4.9. Since $\boldsymbol{R}(T^r) = X_1 = \boldsymbol{R}(S_1^r)$, there is $\boldsymbol{x_1} \in \boldsymbol{D}(S_1^r)$ so that $S_1^r(\boldsymbol{x_1}) = T^r(\boldsymbol{x})$. But, $\boldsymbol{x_1} \in \boldsymbol{D}(S_1^r)$ implies $\boldsymbol{x_1} \in \boldsymbol{D}(S_1)$ and on that set T and S_1 match. So $T^r(\boldsymbol{x_1}) - T^r(\boldsymbol{x}) = \boldsymbol{0}$. This tells us $\boldsymbol{x_1} - \boldsymbol{x} \in \boldsymbol{N}(T^r)$.

Let $\boldsymbol{x_2} = \boldsymbol{x} - \boldsymbol{x_1}$ or $\boldsymbol{x} = \boldsymbol{x_1} + \boldsymbol{x_2}$. Then, we have written $\boldsymbol{x} \in \boldsymbol{D}(T^r)$ as the sum of an element $\boldsymbol{x_1} \in \boldsymbol{R}(T^r) \cap \boldsymbol{D}(T^r)$ and $\boldsymbol{x_2} \in \boldsymbol{N}(T^r)$. Thus, $\boldsymbol{D}(T^r) \subset \{\boldsymbol{R}(T^r) \cap \boldsymbol{D}(T^r)\} \oplus \boldsymbol{N}(T^r)$. The reverse containment is obvious. This proves the equality. ■

We can now show that with the proper restricted domain, the operator T can become a bijection.

Theorem 11.4.12 Linear Operator is a Bijection on $\boldsymbol{R}(T^{\text{ascent}})$

> *Let $(X, \|\cdot\|)$ be a normed linear space and $T : X \to \boldsymbol{R}(T) \subset X$ be a linear operator whose domain is X. If $\alpha(T)$ and $\delta(T)$ are finite with $\delta(T) = r$, then $T : \boldsymbol{R}(T^r) \to \boldsymbol{R}(T^r)$ is 1-1 and onto.*

Proof 11.4.12
By Theorem 11.4.11, $\boldsymbol{D}(T^r) = \{\boldsymbol{R}(T^r) \cap \boldsymbol{D}(T^r)\} \oplus \boldsymbol{N}(T^r)$ and we know $\boldsymbol{R}(T^r) \cap \boldsymbol{N}(T^r) = \boldsymbol{0}$. Since $\boldsymbol{D}(T) = X$, $\boldsymbol{D}(T^r) = X$ and so $X = \boldsymbol{R}(T^r) \oplus \boldsymbol{N}(T^r)$. From the proof of Theorem 11.4.9, we have $S_1 = T^r$ and $\boldsymbol{D}(T^k) = \boldsymbol{R}(T^r) \cap \boldsymbol{D}(T^k)$ for all k. Hence, $\boldsymbol{R}(T^r) = T(\boldsymbol{R}(T^r) \cap \boldsymbol{D}(T^k)) = T(\boldsymbol{R}(T^r) \cap \boldsymbol{D}(T))$. But $\boldsymbol{D}(T) = X$, so $\boldsymbol{R}(T^r) = T(\boldsymbol{R}(T^r))$.

We also know $T : \boldsymbol{N}(T^r) \to \boldsymbol{N}(T^r)$. In Theorem 11.4.9, we have $S_1 = T^r$ and the domain of S_1 is the domain of T^r. So, $S_1 : \boldsymbol{D}(S_1) \to \boldsymbol{R}(T^r) \cap \boldsymbol{D}(T)$ with $\delta(S_1) = 0 = \alpha(S_1) = 0$. Hence, S_1^{-1} exists. Since $T|_{\boldsymbol{R}(T^r)} = T^r$, we have $T^r : \boldsymbol{R}(T^r) \to \boldsymbol{R}(T^r)$ 1-1 and onto. ■

Homework

Exercise 11.4.19 *Verify the conclusion of Theorems 11.4.11 and 11.4.12 for* $A = \begin{bmatrix} 0 & 0 & 1 & 0 \\ 0 & 0 & 0 & 1 \\ 0 & 0 & 0 & 0 \\ 0 & 0 & 0 & 0 \end{bmatrix}$.

Exercise 11.4.20 *Verify the conclusion of Theorems 11.4.11 and 11.4.12 for* $A = \begin{bmatrix} 0 & 0 & 0 & 1 \\ 0 & 0 & 0 & 0 \\ 0 & 0 & 0 & 0 \\ 0 & 0 & 0 & 0 \end{bmatrix}$.

Exercise 11.4.21 *Verify the conclusion of Theorems 11.4.11 and 11.4.12 for* $A = \begin{bmatrix} 0 & 1 & 0 \\ 0 & 0 & 0 \\ 0 & 0 & 2 \end{bmatrix}$.

11.4.4 More on Compact Operators

Let X be a normed linear space with norm $\|\cdot\|$. Let $T : \boldsymbol{D}(T) \subset X \to \boldsymbol{R}(T) \subset X$ be a linear compact operator. Let $\boldsymbol{S_n} = \{\boldsymbol{x} | \boldsymbol{x}\| \leq n\}$. Then $\boldsymbol{R}(T) = \cup_{n=1}^{\infty} T(\boldsymbol{S_n})$. The compactness of T implies given k, the sets $B(\boldsymbol{y}, 1/k)$, where $\boldsymbol{y} \in T(\boldsymbol{S_n})$ is arbitrary, are an open cover of $\overline{T(\boldsymbol{S_n})}$ as any accumulation point of $T(\boldsymbol{S_n})$ is contained in some $B(\boldsymbol{y}, 1/k)$. Therefore there is a $(1/k)$ net for $T(\boldsymbol{S_n})$ consisting of a finite number of points $\boldsymbol{V_k} = \{\boldsymbol{v_{k,1}}, \ldots, \boldsymbol{v_{k,p(k)}}\}$. The union $\boldsymbol{V} = \cup_{k=1}^{\infty} \boldsymbol{V_k}$ is then a countable set which is dense in $T(\boldsymbol{S_n})$. The union of all these sets then gives a countable dense subset of the range of T. We have proven

Theorem 11.4.13 Range of a Linear Compact Operator is Separable

Let $(X, \|\cdot\|)$ be a normed linear space. Let $T : \boldsymbol{D}(T) \subset X \to \boldsymbol{R}(T) \subset X$ be linear and compact operator. Then $\boldsymbol{R}(T)$ is separable.

Proof 11.4.13
The argument is the one we presented above. ■

Now recall $T_\lambda = \lambda I - T$ and the resolvent operator T_λ^{-1} is denoted by $\mathbb{R}_\lambda$. We can now show the dimension of the $\boldsymbol{N}(T_\lambda^n)$ is finite and the range $\boldsymbol{R}(T_\lambda^n)$ is closed for all n when $\lambda \neq 0$.

Theorem 11.4.14 Dimension of the Nullspace of T_λ^n is Finite if $\lambda \neq 0$ when T is linear compact

Let $(X, \|\cdot\|)$ be a normed linear space. Let $T : X \to \boldsymbol{R}(T) \subset X$ be a linear and compact operator with $\boldsymbol{D}(T) = X$. Then $\boldsymbol{N}(T_\lambda^n)$ has finite dimension and $\boldsymbol{R}(T_\lambda^n)$ is closed for $n = 0, 1, \ldots$ when $\lambda \neq 0$.

Proof 11.4.14
The dimension of the $\boldsymbol{N}(T_\lambda^n)$ is finite: *Let $n = 1$. It suffices to show $\boldsymbol{S} = \{\boldsymbol{x} | \boldsymbol{x} \in \boldsymbol{N}(T_\lambda),\ \|\boldsymbol{x}\| = 1\}$ is compact this will imply this ball must live in a finite dimensional normed linear space; namely $\boldsymbol{N}(T_\lambda^n)$. So it is enough to show every sequence form $\boldsymbol{S}$ has a convergent subsequence. Let $(\boldsymbol{x_n}) \subset \boldsymbol{S}$. Then, $\boldsymbol{x_n} \in \boldsymbol{N}(T_\lambda^n)$ and so $\boldsymbol{x_n} = (1/\lambda)T(\boldsymbol{x_n})$. Since T is compact, there is a subsequence $(\boldsymbol{x_{n,1}})$ so that $T(\boldsymbol{x_{n,1}}) \to \boldsymbol{z} \in \overline{T_\lambda(\boldsymbol{S})}$. Thus, $\boldsymbol{x_{n,1}} \to (1/\lambda)\boldsymbol{z}$. This implies $\boldsymbol{S}$ is compact. This handles the case $n = 1$.*

If $n > 1$, we have

$$T_\lambda^n = \lambda^n - n\lambda^{n-1}T + \ldots + (-1)^n T^n = \lambda^n - TA$$

where $A = n\lambda^{n-1} + \ldots + (-1)^n T^{n-1}$. Since T is linear compact, so is TA. Hence, if $\boldsymbol{S_n} = \{\boldsymbol{x} | \boldsymbol{x} \in \boldsymbol{N}(T_\lambda^n),\ \|\boldsymbol{x}\| = 1\}$, this is the same as $\boldsymbol{S_n} = \{\boldsymbol{x} | \boldsymbol{x} \in \boldsymbol{N}(\lambda^n - TA),\ \|\boldsymbol{x}\| = 1\}$. The earlier argument we used then works here by using $\mu = \lambda^n$ and $S = TA$.

$\boldsymbol{R}(T_\lambda^n)$ is closed: *Let $n = 1$. Assume $\boldsymbol{R}(T_\lambda)$ is not closed. Then there is a sequence $(T_\lambda(\boldsymbol{x_n}))$ with $T_\lambda(\boldsymbol{x_n}) \to \boldsymbol{y} \notin \boldsymbol{R}(T_\lambda)$. Since $\boldsymbol{0} \in \boldsymbol{R}(T_\lambda)$, we know $\boldsymbol{y} \neq \boldsymbol{0}$. Hence, there is N so that if $n \geq N$, $T_\lambda(\boldsymbol{x_n}) \neq \boldsymbol{0}$. Since $\boldsymbol{N}(T_\lambda)$ is finite dimensional, it is closed. Thus, $d_n = dist(\boldsymbol{x_n}, \boldsymbol{N}(T_\lambda)) > 0$ for all n.*

Let $\boldsymbol{u_n} \in \boldsymbol{N}(T_\lambda)$ be chosen so that $\theta_n = \|\boldsymbol{x_n} - \boldsymbol{u_n}\| < 2d_n$ for all n. We claim $\theta_n \to \infty$. If not, $\{\boldsymbol{x_n} - \boldsymbol{u_n}\}$ is a bounded set. Hence, $(T(\boldsymbol{x_n}) - \boldsymbol{u_n})$ would contain a convergent subsequence by the compactness of T. But

$$\boldsymbol{x_n} - \boldsymbol{u_n} = \frac{1}{\lambda}\Big(T_\lambda(\boldsymbol{x_n} - \boldsymbol{u_n}) + T(\boldsymbol{x_n} - \boldsymbol{u_n})\Big) = \frac{1}{\lambda}$$

We know $T_\lambda(\boldsymbol{x_n} - \boldsymbol{u_n}) = T_\lambda(\boldsymbol{x_n}) \rightarrow \boldsymbol{y}$ as $\boldsymbol{u_n} \in \boldsymbol{N}(T_\lambda)$. Let $(\boldsymbol{x_{n,1}} - \boldsymbol{u_{n,1}})$ be the subsequence so that $(T(\boldsymbol{x_{n,1}}) - \boldsymbol{u_{n,1}})$ converges. Then,

$$\lim_{n\rightarrow\infty} (\boldsymbol{x_{n,1}} - \boldsymbol{u_{n,1}}) = \frac{1}{\lambda}\boldsymbol{y} + \lim_{n\rightarrow\infty} T(\boldsymbol{x_{n,1}} - \boldsymbol{u_{n,1}})$$

The right-hand side has a limit, so we know there is $\boldsymbol{x}$ so that $\lim_{n\rightarrow\infty}(\boldsymbol{x_{n,1}} - \boldsymbol{u_{n,1}}) = \boldsymbol{x}$. Since T is linear and compact, T is continuous, so it follows that $\lim_{n\rightarrow\infty} T(\boldsymbol{x_{n,1}} - \boldsymbol{u_{n,1}}) = T(\boldsymbol{x})$. But we have therefore shown $\boldsymbol{x} = (1/\lambda)\boldsymbol{y} + T(\boldsymbol{x})$ or $T_\lambda(\boldsymbol{x}) = \boldsymbol{y}$. This says $\boldsymbol{y} \in \boldsymbol{R}(T_\lambda)$ which is a contradiction. We conclude $\theta_n \rightarrow \infty$.

Now, let $\boldsymbol{v_n} = (1/\theta_n)(\boldsymbol{x_n} - \boldsymbol{u_n})$. Then,

$$T_\lambda(\boldsymbol{v_n}) = (1/\theta_n)T_\lambda(\boldsymbol{x_n}), \Longrightarrow \|T_\lambda(\boldsymbol{v_n})\| = (1/\theta_n)\|T_\lambda(\boldsymbol{x_n})\|$$

But since $T_\lambda(\boldsymbol{x_n}) \rightarrow \boldsymbol{y}$, it is bounded. Since $\theta_n \rightarrow \infty$,we see $T_\lambda(\boldsymbol{v_n}) \rightarrow \boldsymbol{0}$. But $\boldsymbol{v_n} = (1/\lambda)(T_\lambda(\boldsymbol{v_n}) + T(\boldsymbol{v_n}))$. By the compactness of T, there is a subsequence $(\boldsymbol{v_{n,1}})$ so that $T(\boldsymbol{v_{n,1}}) \rightarrow \boldsymbol{w}$. Hence, $\boldsymbol{v_{n,1}} \rightarrow (1/\lambda)(\boldsymbol{0}) + (1/\lambda)\boldsymbol{w}$. Then, $T_\lambda(\boldsymbol{v_{n,1}}) \rightarrow \boldsymbol{w} = \boldsymbol{w} = \boldsymbol{0} = T_\lambda(\boldsymbol{w})$. Therefore, $T_\lambda(\boldsymbol{w}) = \boldsymbol{0}$ telling us $\boldsymbol{w} \in \boldsymbol{N}(T_\lambda)$.

Let $\boldsymbol{w_n} = \boldsymbol{u_n} + \theta_n\boldsymbol{w}$. Then, $T_\lambda(\boldsymbol{w_n}) = T_\lambda(\boldsymbol{u_n}) + \theta_n T_\lambda(\boldsymbol{w}) = \boldsymbol{0} + \boldsymbol{0} = \boldsymbol{0}$. Sp $\boldsymbol{w_n} \in \boldsymbol{N}(T_\lambda)$. Hence, $d_n \leq \|\boldsymbol{x_n} - \boldsymbol{w_n}\|$. Now,

$$\boldsymbol{x_n} - \boldsymbol{w_n} = (\theta_n\boldsymbol{v_n} + \boldsymbol{u_n}) - (\boldsymbol{u_n} + \theta_n\boldsymbol{w}) = \theta_n(\boldsymbol{v_n} - \boldsymbol{w})$$

This tells us

$$d_n \leq \|\boldsymbol{x_n} - \boldsymbol{w_n}\| = \theta_n\|\boldsymbol{v_n} - \boldsymbol{w}\| < 2d_n\|\boldsymbol{v_n} - \boldsymbol{w}\|$$

Hence, $(1/2) < \|\boldsymbol{v_n} - \boldsymbol{w}\|$ for all n. So no subsequence of $(\boldsymbol{v_n})$ can converge. But, we already have shown $\boldsymbol{v_{n,1}} \rightarrow \boldsymbol{w}$. This is a contradiction. So our assumption that $\boldsymbol{R}(T_\lambda)$ is not closed is wrong. We conclude $\boldsymbol{R}(T_\lambda)$ is closed.

For $n > 1$, as we did in the first part of the proof, $T_\lambda^n = \lambda^n - TA$. Since TA is compact, to show $\boldsymbol{R}(T_\lambda^n)$ is closed, we show $\boldsymbol{R}(\lambda^n - TA)$ is closed. We apply our arguments above to $\mu = \lambda^n$ and $S = TA$. ■

Next, we prove the ascent and descent of T_λ is finite when $\lambda \neq 0$.

Theorem 11.4.15 Ascent and Descent of T_λ is Finite when $\lambda \neq 0$ when T is linear compact

Let $(X, \|\cdot\|)$ be a normed linear space. Let $T : X \rightarrow \boldsymbol{R}(T) \subset X$ be linear and compact operator with $\boldsymbol{D}(T) = X$. Then $\alpha(T_\lambda) = \delta(T_\lambda) < \infty$ when $\lambda \neq 0$.

Proof 11.4.15
Assume $\alpha(\lambda I - T) = \infty$. Then $\boldsymbol{N}(T_\lambda^{n-1}) \subsetneq \boldsymbol{N}(T_\lambda^n)$ for all $n \geq 1$. By Riesz's Lemma, there is $\boldsymbol{x_n} \in \boldsymbol{N}(T_\lambda^n)$ with $\|\boldsymbol{x_n}\| = 1$ and $\|\boldsymbol{x} - \boldsymbol{x_n}\| \geq (1/2)$ for all $\boldsymbol{x} \in \boldsymbol{N}(T_\lambda^n)$. Assume $1 \leq m < n$. Let

$$\boldsymbol{z} = \boldsymbol{x_m} + \frac{1}{\lambda}T_\lambda(\boldsymbol{x_n}) - \frac{1}{\lambda}T_\lambda(\boldsymbol{x_m})$$

Then,

$$T_\lambda^{n-1}(\boldsymbol{z}) = T_\lambda^{n-1}(\boldsymbol{x_m}) + \frac{1}{\lambda}T_\lambda^n(\boldsymbol{x_n}) - \frac{1}{\lambda}T_\lambda^n(\boldsymbol{x_m}) = T_\lambda^{n-1}(\boldsymbol{x_m}) - \frac{1}{\lambda}T_\lambda^n(\boldsymbol{x_m})$$

$$= T_\lambda^{n-1}(\boldsymbol{x_m} - T_\lambda(\boldsymbol{x_m}))$$

Thus,

$$T_\lambda^{n-1}(\boldsymbol{z}) = T_\lambda^{n-1}(I - \lambda^{-1}T)(\boldsymbol{x_m}) = \lambda^{-1}T_\lambda^{n-1}T_\lambda(\boldsymbol{x_m}) = \lambda^{-1}T_\lambda^{n}(\boldsymbol{x_m})$$

But $\boldsymbol{N}(T_\lambda^m) \subsetneq \boldsymbol{N}(T_\lambda^n)$ for all $n \geq m$. Hence, $T_\lambda^n(\boldsymbol{x_m}) = \boldsymbol{0}$ and so $T_\lambda^{n-1}(\boldsymbol{z}) = \boldsymbol{0}$. Therefore, $\boldsymbol{z} \in \boldsymbol{N}(T_\lambda^{n-1})$. But then $\|\boldsymbol{x_n} - \boldsymbol{z}\| \geq (1/2)$. Now,

$$\begin{aligned} T(\boldsymbol{x_n}) - T(\boldsymbol{x_m}) &= -\lambda \boldsymbol{x_n} T(\boldsymbol{x_n}) + \lambda \boldsymbol{x_n} + \lambda \boldsymbol{x_m} - T(\boldsymbol{x_m}) - \lambda \boldsymbol{x_m} \\ &= -T_\lambda(\boldsymbol{x_n}) + T_\lambda(\boldsymbol{x_m}) + \lambda(\boldsymbol{x_n} - \boldsymbol{x_m}) \\ &= \lambda\Big(-\frac{1}{\lambda}T_\lambda(\boldsymbol{x_n}) + \frac{1}{\lambda}T_\lambda(\boldsymbol{x_m}) + (\boldsymbol{x_n} - \boldsymbol{x_m})\Big) \\ &= \lambda(\boldsymbol{x_n} - \boldsymbol{z}) \end{aligned}$$

So $\|T(\boldsymbol{x_n}) - T(\boldsymbol{x_m})\| \geq \lambda/2 > 0$. Hence, the sequence $(T(\boldsymbol{x_n}))$ can have no convergent subsequence contracting the fact that T is compact. Hence, it is not possible that $\alpha(\lambda I - T) = \infty$.

In a similar way, we prove $\delta(\lambda I - T)$ is also finite. If you assume $\delta(T) = \infty$, then $\boldsymbol{R}(T_\lambda^n) \subsetneq \boldsymbol{R}(T_\lambda^{n-1})$ for all $n \geq 1$. By Theorem 11.4.14, each $\boldsymbol{R}(T_\lambda^n)$ is closed. By Riesz's Lemma, there is $\boldsymbol{y_n} \in \boldsymbol{R}(T_\lambda^n)$ with $\|\boldsymbol{y_n}\| = 1$ and $\|\boldsymbol{y} - \boldsymbol{y_n}\| \geq (1/2)$ for all $\boldsymbol{y} \in \boldsymbol{R}(T_\lambda^{n+1})$. For $1 \leq m < n$, define $\boldsymbol{w} = \boldsymbol{y_n} - \lambda^{-1}T_\lambda(\boldsymbol{y_n}) + \lambda^{-1}T_\lambda(\boldsymbol{y_m})$. Now, if $\boldsymbol{y_k} \in \boldsymbol{R}(T_\lambda^k)$, there is $\boldsymbol{x_k}$ so that $\boldsymbol{y_k} = T_\lambda^k(\boldsymbol{x_k})$. So,

$$\boldsymbol{w} = T_\lambda^n(\boldsymbol{x_n}) - \lambda^{-1}T_\lambda^{n+1}(\boldsymbol{x_n}) + \lambda^{-1}T_\lambda^{m+1}(\boldsymbol{x_m})$$

Since $\boldsymbol{R}(T_\lambda^n) \subsetneq \boldsymbol{R}(T_\lambda^{m+1})$ for all $n \geq m+1$, we see $\boldsymbol{w} \in \boldsymbol{R}(T_\lambda^{m+1})$. Thus, $\|\boldsymbol{w} - \boldsymbol{y_m}\| \geq (1/2)$. But $T(\boldsymbol{y_m}) - T(\boldsymbol{y_n}) = \lambda(\boldsymbol{y_m} - \boldsymbol{w})$ using the same sort of argument we used before. Thus, $\|T(\boldsymbol{y_m}) - T(\boldsymbol{y_n})\| \geq (\lambda)/2 > 0$ which contradicts the compactness of T. So $\delta(T)$ is finite. Thus, since $\boldsymbol{D}(T) = X$, by Theorem 11.4.10, $\alpha(T) = \delta(T)$. ■

Homework

Let A be an $n \times n$ real-valued matrix and let $A = PJP^{-1}$, with P invertible, be the Jordan canonical form representation for A.

Exercise 11.4.22 *For the 3×3 case, if λ is an eigenvalue of the 3×3 matrix $A = PJP^{-1}$, prove $\alpha(\lambda I - A) = \delta(\lambda I - A) = k$ where the Jordan block for the eigenvalue is $k \times k$.*

Exercise 11.4.23 *For the 4×4 case, if λ is an eigenvalue of the 4×4 matrix $A = PJP^{-1}$, prove $\alpha(\lambda I - A) = \delta(\lambda I - A) = k$ where the Jordan block for the eigenvalue is $k \times k$.*

Exercise 11.4.24 *For the $n \times n$ case, if λ is an eigenvalue of the $n \times n$ matrix $A = PJP^{-1}$, prove $\alpha(\lambda I - A) = \delta(\lambda I - A) = k$ where the Jordan block for the eigenvalue is $k \times k$.*

Now, we can discuss the eigenvalues of T in detail.

11.4.5 The Eigenvalues of a Compact Operator

Theorem 11.4.16 $\lambda \neq 0$ is either an Eigenvalue or is in the Resolvent Set of T

Let $(X, \|\cdot\|)$ be a normed linear space. Let $T : X \to \boldsymbol{R}(T) \subset X$ be a linear and compact operator with $\boldsymbol{D}(T) = X$. Let $\lambda \neq 0$. Then, $\lambda \in p_\sigma(T)$, i.e. is a traditional eigenvalue of T, or $\lambda \in \rho(T)$, the resolvent set of T.

Proof 11.4.16
Step I: *We show there is $M > 0$ so that $dist(\boldsymbol{x}, \boldsymbol{N}(T_\lambda)) \leq M\|T_\lambda(\boldsymbol{x})\|$.*

For convenience, let $d(\boldsymbol{x}) = dist(\boldsymbol{x}, \boldsymbol{N}(T_\lambda))$. Now assume this conjecture is false. Then, there is a sequence $(\boldsymbol{x_n})$ not in $\boldsymbol{N}(T_\lambda)$ with $d(\boldsymbol{x_n})/\|T_\lambda(\boldsymbol{x_n})\| \to \infty$. Since $\boldsymbol{N}(T_\lambda)$ is finite dimensional, there is $\boldsymbol{u_n} \in \boldsymbol{N}(T_\lambda)$ with $d(\boldsymbol{x_n}) = \|\boldsymbol{u_n} - \boldsymbol{x_n}\| = d(\boldsymbol{x_n})$. Let $\boldsymbol{y_n} = (\boldsymbol{x_n} - \boldsymbol{u_n})/d(\boldsymbol{x_n})$. Then, $\|\boldsymbol{y_n}\| = 1$ and

$$T_\lambda(\boldsymbol{y_n}) = \frac{T_\lambda(\boldsymbol{x_n}) - T_\lambda(\boldsymbol{u_n})}{d(\boldsymbol{x_n})} = \frac{T_\lambda(\boldsymbol{x_n})}{d(\boldsymbol{x_n})} \to \boldsymbol{0}$$

as we assume $d(\boldsymbol{x_n})/\|T_\lambda(\boldsymbol{x_n})\| \to \infty$. Now $\boldsymbol{y_n} = (1/\lambda)(T_\lambda(\boldsymbol{y_n}) + T(\boldsymbol{y_n}))$. By the compactness of T and the fact that $\lim_{n\to\infty} T_\lambda(\boldsymbol{y_n}) = \boldsymbol{0}$, we know there is a subsequence $(\boldsymbol{y_{n,1}})$ and a point $\boldsymbol{y}$ with $\boldsymbol{y_{n,1}} \to \boldsymbol{y}$ and $\boldsymbol{0} = \lim_{n\to\infty} T_\lambda(\boldsymbol{y_{n,1}}) \to T\lambda(\boldsymbol{y})$. Thus, $\boldsymbol{y} \in \boldsymbol{N}(T_\lambda)$ and so $\boldsymbol{u_n} + d(\boldsymbol{x_n})\boldsymbol{y} \in \boldsymbol{N}(T_\lambda)$. Hence,

$$\begin{aligned} \|\boldsymbol{y_n} - \boldsymbol{y}\| &= \left\|\frac{\boldsymbol{x_n} - \boldsymbol{u_n}}{d(\boldsymbol{x_n})} - \boldsymbol{y}\right\| = \left\|\frac{\boldsymbol{x_n} - \boldsymbol{u_n} - d(\boldsymbol{x_n})\boldsymbol{y}}{d(\boldsymbol{x_n})} - \boldsymbol{y}\right\| \\ &= \left\|\frac{\boldsymbol{x_n} - (\boldsymbol{u_n} + d(\boldsymbol{x_n})\boldsymbol{y})}{d(\boldsymbol{x_n})} - \boldsymbol{y}\right\| \end{aligned}$$

But since $\boldsymbol{u_n} + d(\boldsymbol{x_n})\boldsymbol{y} \in \boldsymbol{N}(T_\lambda)$, we must have $\|\boldsymbol{x_n} - (\boldsymbol{u_n} + d(\boldsymbol{x_n})\boldsymbol{y})\| \geq d(\boldsymbol{x_n})$. Hence, $\|\boldsymbol{y_n} - \boldsymbol{y}\| \geq d(\boldsymbol{x_n})/d(\boldsymbol{x_n}) = 1$. This contradicts the fact that $(\boldsymbol{y_n})$ has a convergent subsequence. Hence, our assumption that no finite M exists is wrong and there is $M > 0$ and finite so that $d(\boldsymbol{x}) \leq M\|T_\lambda(\boldsymbol{x})\|$.
Step II: *Let M be the constant from Step I. Then for all $\boldsymbol{y} \in \boldsymbol{R}(T_\lambda)$, there is $\boldsymbol{x}$ so that $T_\lambda(\boldsymbol{x}) = \boldsymbol{y}$ and $\|\boldsymbol{x}\| \leq M\|\boldsymbol{y}\|$. Hence, $(\lambda I - T)^{-1}$ is continuous if it exists. The point λ is in $\rho(T)$, in which case $\boldsymbol{R}(T_\lambda) = X$ or the point λ is in $p_\sigma(T)$, in which case $\boldsymbol{R}(T_\lambda) \subsetneq X$.*

Let $\boldsymbol{y} \in \boldsymbol{R}(T_\lambda)$. If $\boldsymbol{y} = \boldsymbol{0}$ there are lots of solutions: i.e. $T_\lambda(\boldsymbol{N}(T_\lambda)) = \boldsymbol{0}$. So we can assume $\boldsymbol{y} \neq \boldsymbol{0}$. Then, there is $\boldsymbol{x_1}$ with $\boldsymbol{y} = T_\lambda(\boldsymbol{x_1})$. Choose $\boldsymbol{u} \in \boldsymbol{N}(T_\lambda)$ so that $\|\boldsymbol{x_1} - \boldsymbol{u}\| = d(\boldsymbol{x_1})$. Let $\boldsymbol{x} = \boldsymbol{x_1} - \boldsymbol{u}$. Then, $T_\lambda(\boldsymbol{x}) = T_\lambda(\boldsymbol{x_1} - \boldsymbol{u}) = T_\lambda(\boldsymbol{x_1}) = \boldsymbol{y}$. Further, $\|\boldsymbol{x}\| = \|\boldsymbol{x_1} - \boldsymbol{u}\| = d(\boldsymbol{x_1}) \leq M\|T_\lambda(\boldsymbol{x_1})\| = M\boldsymbol{y}$. This proves the first part of Step II.

By Theorem 11.4.15, $\alpha(T) = \delta(T) = 0$ or $0 < \alpha(T) = \delta(T) < \infty$. If the first case holds, $(\lambda I - T)^{-1}$ exists and $\boldsymbol{R}(T_\lambda) = X$. From our earlier discussion, since $T_\lambda(\boldsymbol{x}) = \boldsymbol{y}$, we have $\boldsymbol{x} = T_\lambda^{-1}(\boldsymbol{y})$ and so $\|T_\lambda^{-1}(\boldsymbol{y})\| \leq M\|\boldsymbol{y}\|$. Thus, $\|T_\lambda^{-1}\|$ is bounded and so it is continuous. This shows $\lambda \in \rho(T)$. If the second case holds, T_λ^{-1} does not exist and by the definition of $\delta(\lambda I - T)$, $\boldsymbol{R}(T_\lambda) \subsetneq X$. We know $\boldsymbol{R}(T_\lambda)$ is closed and hence, $\boldsymbol{R}(T_\lambda)$ is a closed proper subspace of X. Thus, $\lambda \in p_\sigma(T)$.
We conclude therefore that $\lambda \in p_\sigma(T)$ or $\lambda \in \rho(t)$ if T is linear and compact with domain X. ■

How many points can be in the point spectrum?

Theorem 11.4.17 Cardinality of the Point Spectrum

Let $(X, \|\cdot\|)$ be a normed linear space. Let $T : X \to \boldsymbol{R}(T) \subset X$ be a linear and compact operator with $\boldsymbol{D}(T) = X$. Then $p_\sigma(T)$ contains at most a countable number of points and the only possible accumulation point in $\lambda = 0$. In fact, $\overline{p_\sigma}$ is compact.

Proof 11.4.17
Observe if $T(\boldsymbol{x_k}) = \lambda_k \boldsymbol{x_k}$, $\boldsymbol{x_k} \neq \boldsymbol{0}$, for $k = 1, 2, \ldots n$ with the values λ_k distinct, then the set

$\{\boldsymbol{x_1}, \ldots, \boldsymbol{x_n}\}$ is linearly independent. For, it not, let $\boldsymbol{x_m}$ be the first $\boldsymbol{x_k}$ which is a linear combination of its predecessors; i.e. $\boldsymbol{x_m} = \alpha_1 \boldsymbol{x_1} + \ldots + \alpha_{m-1}\boldsymbol{x_{m-1}}$. But $T(\boldsymbol{x_m}) = \lambda_m \boldsymbol{x_m}$ and so

$$\begin{aligned} T(\boldsymbol{x_m}) &= \lambda_m \boldsymbol{x_m} = \alpha_1 T(\boldsymbol{x_1}) + \ldots + \alpha_{m-1} T(\boldsymbol{x_{m-1}}) \\ &= \lambda_1 \alpha_1 \boldsymbol{x_1} + \ldots + \lambda_{m-1}\alpha_{m-1}\boldsymbol{x_{m-1}} \end{aligned}$$

Then,

$$\boldsymbol{0} = T(\boldsymbol{x_m}) - \lambda_m \boldsymbol{x_m} = (\lambda_1 - \lambda_m)\alpha_1 \boldsymbol{x_1} + \ldots + (\lambda_{m-1} - \lambda_m)\alpha_{m-1}\boldsymbol{x_{m-1}}$$

The values λ_k are distinct and $\{\boldsymbol{x_1}, \ldots, \boldsymbol{x_{m-1}}\}$ is a linearly independent set. Thus, the constants $\alpha_1 = \ldots = \alpha_{m-1} = 0$. This implies $\boldsymbol{x_m} = \boldsymbol{0}$ which is not possible. We can conclude the set $\{\boldsymbol{x_1}, \ldots, \boldsymbol{x_n}\}$ is linearly independent.

Now to prove the result, it suffices to show the points $\lambda \in p_\sigma(T)$ for which $|\lambda| \geq \epsilon$ is a finite set for any $\epsilon > 0$. Assume this is not true and there is an $\epsilon > 0$ and a sequence $(\lambda_n) \subset p_\sigma(T)$ and $|\lambda_n| \geq \epsilon$ for all n. Then, there is a sequence $(\boldsymbol{x_n})$ with $T(\boldsymbol{x_n}) = \lambda_n \boldsymbol{x_n}$ for all n. From our discussion above, the set $\{\boldsymbol{x_1}, \ldots\}$ is a linearly independent set. Let $M_n = span\{\boldsymbol{x_1}, \ldots, \boldsymbol{x_n}\}$ is a proper closed subspace of M_{n+1}. By Riesz's Lemma, there is $\boldsymbol{u_{n+1}} \in M_{n+1}$ so that $\|\boldsymbol{u_{n+1}}\| = 1$ and $\|\boldsymbol{u_{n+1}} - \boldsymbol{x}\| \geq (1/2)$ for all $\boldsymbol{x} \in M_n$. Since $\boldsymbol{u_{n+1}} \in M_{n+1}$, there are scalars $\{\alpha_1, \ldots, \alpha_{n+1}\}$ with $\boldsymbol{u_{n+1}} = \alpha_1 \boldsymbol{x_1} + \ldots + \alpha_{n+1}\boldsymbol{x_{n+1}}$. Thus,

$$T(\boldsymbol{u_{n+1}}) = \sum_{i=1}^{n+1} \alpha_i T(\boldsymbol{x_i}) = \sum_{i=1}^{n+1} \alpha_i \lambda_i \boldsymbol{x_i} \in M_{n+1}$$

But if $\boldsymbol{y} \in M_{n+1}$, $\boldsymbol{y} = \sum_{i=1}^{n+1} \beta_i \boldsymbol{x_i}$ for scalars $\{\beta_1, \ldots, \beta_{n+1}\}$,

$$(\lambda_{n+1} I - T)(\boldsymbol{y}) = \sum_{i=1}^{n+1} \lambda_{n+1}\beta_i \boldsymbol{x_i} - T(\boldsymbol{y}) = \sum_{i=1}^{n+1} \beta_i(\lambda_{n+1} - \lambda_i)\boldsymbol{x_i} = \sum_{i=1}^{n} \beta_i(\lambda_{n+1} - \lambda_i)\boldsymbol{x_i} \in M_n$$

This shows $(\lambda_{n+1} I - T) M_{n+1} \subset M_n$. So, in particular, $(\lambda_{n+1} I - T)(\boldsymbol{u_{n+1}}) \in M_n$. We conclude $\boldsymbol{z} = (\lambda_{n+1} I - T)(\boldsymbol{u_{n+1}}) + T(\boldsymbol{u_m}) \in M_n$ for all $1 \leq m < n+1$ as $M_m \subset M_{n+1}$. This tells us $(1/\lambda_{n+1})\boldsymbol{z} \in M_n$.

Now,

$$\begin{aligned} T(\boldsymbol{u_{n+1}}) - T(\boldsymbol{u_m}) &= \lambda_{n+1}\boldsymbol{u_{n+1}} - \lambda_{n+1}\boldsymbol{u_{n+1}} + T(\boldsymbol{u_{n+1}}) - T(\boldsymbol{u_m}) \\ &= \lambda_{n+1}\boldsymbol{u_{n+1}} - (\lambda_{n+1}\boldsymbol{u_{n+1}} - T(\boldsymbol{u_{n+1}}) + T(\boldsymbol{u_m})) \\ &= \lambda_{n+1}(\boldsymbol{u_{n+1}} - \lambda_{n+1}^{-1}\boldsymbol{z}) \end{aligned}$$

Since $\lambda_{n+1}^{-1}\boldsymbol{z} \in M_n$, we know $\|\boldsymbol{u_{n+1}} - \lambda_{n+1}^{-1}\boldsymbol{z}\| \geq (1/2)$. It follows that

$$\|T(\boldsymbol{u_{n+1}}) - T(\boldsymbol{u_m})\| \geq \frac{\lambda_{n+1}}{2} \geq \frac{\epsilon}{2} > 0$$

Thus, $(T(\boldsymbol{u_n}))$ can have no convergent subsequence contradicting the fact that T is compact. Our claim is therefore true and the set of $\lambda \in p_\sigma(T)$ with $|\lambda| \geq \epsilon > 0$ is finite for all choices of $\epsilon > 0$. Since $p_\sigma(T) = \cup_n \{\lambda \mid |\lambda| \geq (1/n)\}$ it is a countable union of finite sets and so it is countable.

Finally, we already know from Theorem 11.4.2, that $\sigma(T)$ is closed. We know from Theorem 11.4.16, that if $\lambda \neq 0$, then $\lambda \in p_\sigma(T)$ or $\lambda \in \rho(T)$. Hence, $\sigma(T) = p_\sigma(T) \cup \{0\}$ if $0 \in c_\sigma(T)$ or $0 \in r_\sigma(T)$.

If $0 \notin \sigma(T)$, then $\sigma(T) = p_\sigma(T)$. We have $\overline{p_\sigma(T)} = p_\sigma(T)$ or $\overline{p_\sigma(T)} = p_\sigma(T) \cup \{0\}$. Since the set $\{\lambda \in p_\sigma(T) | |\lambda| \geq 1\}$ is finite, the set of all $\lambda \in p_\sigma(T)$ is bounded. Thus, $\overline{p_\sigma(T)}$ is closed and bounded and therefore compact. If λ_0 was an accumulation of $\overline{p_\sigma(T)}$, then every neighborhood of λ_0 would contain an infinite number of values from $p_\sigma(T)$ which would violate the fact that the set of $\lambda \in p_\sigma(T)$ with $|\lambda| \geq \epsilon > 0$ is finite for all choices of $\epsilon > 0$. So the only possible accumulation point is 0. ■

Given all we have discussed, we can now look at the solutions of a certain class of integral equations: the ones called Fredholm equations of the second kind. Consider the following equation

$$y(s) \quad = \quad x(s) - \mu \int_a^b k(s,t)x(t)dt \tag{11.1}$$

where the function k, called the kernel, has the properties that guarantee the operator T defined by $T(\boldsymbol{x}) = \int_a^b k(s,t)x(t)dt$ is linear and compact. We studied operators of this form when we looked carefully at Stürm - Liouville models in (Peterson (100) 2020), so you might want to refresh your mind about that now. Using the identity operator, 11.1 can be written as $\boldsymbol{y} = (I - \mu T)(\boldsymbol{x}) = \mu(\mu^{-1}I - T)(\boldsymbol{x})$. We have been studying compact perturbations of the identity so trying to understand the equation involves the compact perturbation of the identity $\mu I - T$. Of course, we have to choose the normed linear space to work in. Typically, the domain here is a subset of $(C([a,b]), \|\cdot\|)$ for some choice of norm $\|\cdot\|$. Under some conditions, Equation 11.1 is equivalent to a differential equation with boundary conditions. Since $(C([a,b]), \|\cdot\|)$ is dense in $\mathcal{L}_2([a,b], \mathcal{S}, \mu)$ for some choice of measure space $\mathcal{S}$ and measure μ. We often choose Lebesgue measure and sometimes Borel measure here. In this case, the norm in $\|\cdot\|_2$ and, of course, we are really working with equivalence classes of functions. We have rewritten Equation 11.1 in operator form as

$$\boldsymbol{y} \quad = \quad \mu(\mu^{-1}I - T)(\boldsymbol{x}) \tag{11.2}$$

Our theory of compact operators says that for a given $\mu \neq 0$, there are two possibilities:

(i): $(\mu^{-1}I - T)^{-1}$ exists and is continuous, i.e. this value of μ is in $\rho(T)$, and so we can solve to find the unique solution

$$\boldsymbol{x} \quad = \quad \frac{1}{\mu}(\mu^{-1}I - T)^{-1}(\boldsymbol{y})$$

(ii): The homogeneous equation $bs0 = (\mu^{-1}I - T)(\boldsymbol{x})$ has a finite number of linearly independent solutions because $\boldsymbol{N}(\mu^{-1}I - T)$ is finite dimensional. In this case, as is usual, the value μ^{-1} is called an eigenvalue of T and if the dimension of the kernel here is $p(\mu^{-1})$, we can label the finite number of linearly independent solutions as $\boldsymbol{E}_{\mu^{-1}} = \{\boldsymbol{x}_{\mu^{-1},1}, \ldots, \boldsymbol{x}_{\mu^{-1},p(\mu^{-1})}\}$. This is called the eigenspace corresponding to the eigenvalue μ^{-1}. This is called the Fredholm Alternative for Compact Linear Operators.

Our theory says the number of eigenvalues is at most countably infinite and hence we can label them as $\{\lambda_1, \ldots, \lambda_n\}$ for some n or as $(\lambda_k)_{k=1}^{\infty}$. In either case, the corresponding eigenspace for eigenvalue λ_k has dimension $p(k)$ and we know $\boldsymbol{E}_{\boldsymbol{k}} = \{\boldsymbol{x}_{\boldsymbol{k},1}, \ldots, \boldsymbol{x}_{\boldsymbol{k},p(\boldsymbol{k})}\}$. We usually call each $\boldsymbol{x}_i$ an eigenvector corresponding to eigenvalue λ_k. We know also the number of eigenvalues with $|\mu^{-1}| > 1$ is finite, so the number of eigenvalues with $|\lambda_k| < 1$ is finite. If the sequence of eigenvalues is infinite, then 0 is a limit point and hence $\mu_k^{-1} \to 0$ or $|\lambda_k| \to \infty$.

In general, if T is linear and compact with domain X where X is a normed linear space with norm $\|\cdot\|$, for any linear bounded operator, $\rho(T)$ is open and $\sigma(T)$ is closed. If T is also compact, then if $\lambda \neq 0$, λ is in $p_\sigma(T)$ or $\lambda \in \rho(T)$ by Theorem 11.4.16. Since the eigenvalues of T are a bounded set, if there are countably many, by the Bolzano - Weierstrass Theorem, there must be an accumulation point. However, we know the set of eigenvalues with $|\lambda| > 1$ is finite, so the only accumulation point is 0. In this case, $0 \in \overline{\sigma(T)} = \sigma(T)$ as it is a closed set. However, $\lambda \in \sigma(T)$ must be in $p_\sigma(T)$. Hence, in this case, $\sigma(T) = p_\sigma(T) \cup \{\mathbf{0}\}$. If $\lambda \neq 0$, by Theorem 11.4.15, the ascent and descent of T are both finite with $\alpha(\lambda I - T) = \delta(\lambda I - T)$. Let $\alpha_\lambda = \alpha(\lambda I - T)$ and $\delta_\lambda = \delta(\lambda I - T)$ and as usual $t_\lambda = \lambda I - T$. By Theorem 11.4.14, the dimension of $\boldsymbol{N}(T_\lambda^{\alpha_\lambda}) = m(\lambda) < \infty$ and $\boldsymbol{R}(T_\lambda^{\alpha_\lambda})$ is closed. By Theorem 11.4.11, since the domain of T is X, $X = \boldsymbol{N}(T_\lambda^{\alpha_\lambda}) \oplus \boldsymbol{R}(T_\lambda^{\alpha_\lambda})$.

We can also discuss how to decompose T into direct sums of projections. Define $E_\lambda : X \rightarrow X$ by

$$E_\lambda(\boldsymbol{x}) = \begin{cases} \boldsymbol{x}, & \boldsymbol{x} \in \boldsymbol{N}(T_\lambda^{\alpha_\lambda}) \\ 0, & ,\boldsymbol{x} \in \boldsymbol{R}(T_\lambda^{\alpha_\lambda}) \end{cases}$$

Then E_λ is a projection (we discuss this in (Peterson (100) 2020) if you wish to review). Since the eigenvectors corresponding to λ are distinct and linearly independent from the eigenvectors corresponding to $\mu \neq \lambda$ when $\mu \in p_\sigma(T)$, if $\{\lambda_1, \ldots, \lambda_n\} \in p_\sigma(T)$, we have

$$\begin{aligned} X &= E_{\lambda_1}(X) \oplus E_{\lambda_2}(X) \oplus \ldots \oplus E_{\lambda_n}(X) \oplus (I - E_{\lambda_1} - \ldots - E_{\lambda_n})(X) \\ &= \boldsymbol{N}(T_{\lambda_1}^{\alpha_{\lambda_1}}) \oplus \ldots \oplus \boldsymbol{N}(T_{\lambda_n}^{\alpha_{\lambda_n}}) \oplus M_n \end{aligned}$$

where $M_n = Q_n(X)$ where $Q_n = I - E_{\lambda_1} - \ldots - E_{\lambda_n}$. We know $T_{\lambda_i}(\boldsymbol{N}(T_\lambda^{\alpha_\lambda})) \subset \boldsymbol{N}(T_\lambda^{\alpha_\lambda})$. Hence, $\boldsymbol{N}(T_\lambda^{\alpha_\lambda})$ is called an invariant subspace under the operator T_{λ_i}. It is easy to see $\boldsymbol{N}(T_\lambda^{\alpha_\lambda})$ is also invariant under T itself. From the way Q_n is defined, $Q_n(X) = M_n$ is also T invariant. Hence, letting $T_i = T|_{\boldsymbol{N}(T_{\lambda_1}^{\alpha_{\lambda_1}})}$ and $T_M = T|_{M_n}$, we see

$$\begin{aligned} T(X) &= T(\boldsymbol{N}(T_{\lambda_1}^{\alpha_{\lambda_1}}) \oplus \ldots \oplus \boldsymbol{N}(T_{\lambda_n}^{\alpha_{\lambda_n}}) \oplus M_n) \\ &= T_1(X) \oplus \ldots \oplus T_n(X) \oplus T_M(X) \Longrightarrow \\ T &= T_1 \oplus \ldots \oplus T_n \oplus T_M \end{aligned}$$

The final result concerns the Leray - Schauder Degree of $I - T$ when it is invertible. This is the same as saying $\lambda = 0$ is not an eigenvalue of T.

Theorem 11.4.18 Leray - Schauder Degree of $I - T$ when T is Linear and Compact and $I - T$ is Invertible

Let $(X, \|\cdot\|)$ be a normed linear space. Let $T : X \rightarrow X$ be linear and compact and assume $ker(I - T) = \mathbf{0}$. Then for all $\boldsymbol{D} \subset X$ that are open and bounded, with $\mathbf{0} \notin \partial\boldsymbol{D}$, we have

$$d_{LS}(I - T, \boldsymbol{D}, \mathbf{0}) = \begin{cases} 0, & \mathbf{0} \notin \boldsymbol{D} \\ (-1)^m, & \mathbf{0} \in \boldsymbol{D} \end{cases}$$

where $m = \sum_{\lambda \in S} m(\lambda)$ where $S = \{\lambda | \lambda \in p_\sigma(T),\ \lambda \in \Re,\ \lambda > 1\}$.

Proof 11.4.18

From our discussions, we know the number of eigenvalues $|\lambda| > 1$ is finite (possibly empty, of course). Assume, for the moment, it is nonempty. Let $\{\lambda_1, \ldots, \lambda_n\}$ be the real eigenvalues larger than one. Then,

$$\begin{aligned} X &= \boldsymbol{N}(T_{\lambda_1}^{\alpha_{\lambda_1}}) \oplus \ldots \oplus \boldsymbol{N}(T_{\lambda_n}^{\alpha_{\lambda_n}}) \oplus M_n \\ T &= T_1 \oplus \ldots \oplus T_n \oplus T_M \end{aligned}$$

where $M_n = Q_n(X)$ *where* $Q_n = I - E_{\lambda_1} - \ldots - E_{\lambda_n}$, *and* T_i *and* T_M *is defined as before. Let*

$$\begin{aligned} N &= \boldsymbol{N}(T_{\lambda_1}^{\alpha_{\lambda_1}}) \oplus \ldots \oplus \boldsymbol{N}(T_{\lambda_n}^{\alpha_{\lambda_n}}) \\ M &= M_n \end{aligned}$$

Then $X = N \oplus M$ *and letting* $T_N = T|_N$, *we also have* $T = T_N \oplus T_M$. *Let's assume* $\mathbf{0} \notin \partial \boldsymbol{D}$. *Consider the compact homotopy defined on* $\overline{\boldsymbol{D}} \times [0,1]$ *by*

$$H(\cdot, t) = (1-t)(I-T) + t(-I_N + I_M)$$

where $I_N = I|_N$ *and* $I_M = I|_M$. *Then, for all* $\boldsymbol{x} \neq \mathbf{0}$,

$$\begin{aligned} & (1-t)(I-T)(\boldsymbol{x}) + t(-I_N + I_M)(\boldsymbol{x}) \\ = \; & (1-t)(I_N + I_M - T_N - T_M)(\boldsymbol{x}) - t(I_N - I_M)(\boldsymbol{x}) \\ = \; & (1-t)(I_N - T_N)(\boldsymbol{x}) + (1-t)(I_M - T_M)(\boldsymbol{x}) - t(I_N - I_M)(\boldsymbol{x}) \end{aligned}$$

Now, if there was $\boldsymbol{x} \neq \mathbf{0}$ *and* $t \in (0,1)$ *with*

$$(1-t)(I_N - T_N)(\boldsymbol{x}) + (1-t)(I_M - T_M)(\boldsymbol{x}) - t(I_N - I_M)(\boldsymbol{x}) = 0$$

this would imply

$$(1-t)(I_N - T_N)(\boldsymbol{x}) + (1-t)(I_M - T_M)(\boldsymbol{x}) = t(I_N - I_M)(\boldsymbol{x})$$

or

$$(1-t)(I_N - T_N)(\boldsymbol{x}) = tI_N(\boldsymbol{x}), \quad (1-t)(I_M - T_M)(\boldsymbol{x}) = -tI_M(\boldsymbol{x})$$

Thus,

$$T_N(\boldsymbol{x}) = \frac{1-2t}{1-t}(\boldsymbol{x}), \quad T_M(\boldsymbol{x}) = \frac{1}{1-t}\boldsymbol{x}$$

Therefore, $\boldsymbol{x} \in \boldsymbol{N}(\mu I - T_N)$ *and* $\boldsymbol{x} \in \boldsymbol{N}(\nu I - T_M)$. *Letting* $\mu = (1-2t)/(1-t)$ *and* $\nu = 1/(1-t)$, *we have shown* $(\mu I - T_N)(\boldsymbol{x}) = \mathbf{0}$ *and* $(\nu I - T_M)(\boldsymbol{x}) = \mathbf{0}$. *Now look at* T_N. *It cannot have eigenvalues other than* λ_1 *to* λ_n *which are all bigger than one. But* $\mu < 1$. *This is a contradiction. So* $\boldsymbol{x}$ *must be zero. Also, by looking at how* M *is defined, we can see that on* M, T *can only have eigenvalues less than* 1. *This is also a contradiction, so* $\boldsymbol{x}$ *is zero again. We conclude* $(1-t)(I_T)\boldsymbol{x} - t(I_N - I_M)(\boldsymbol{x}) \neq \mathbf{0}$ *for all* $\boldsymbol{x} \neq \mathbf{0}$ *and* $t \in (0,1)$. *If* $t = 0$, $(I-T)(\boldsymbol{x}) \neq \mathbf{0}$ *for all* $\boldsymbol{x} \neq 0$ *as* $\boldsymbol{N}(I-T) = \mathbf{0}$. *Finally, if* $t = 1$, $(I_M - I_N)(\boldsymbol{x}) = \mathbf{0}$ *is not possible as* $X = N \oplus M$. *Thus, we can say*

$$\begin{aligned} H(\boldsymbol{x}, t) &= (1-t)(I_N - T_N)(\boldsymbol{x}) + (1-t)(I_M - T_M)(\boldsymbol{x}) - t(I_N - I_M)(\boldsymbol{x}) \\ &= (1-2t)I_N(\boldsymbol{x}) + I_M(\boldsymbol{x}) - (1-t)I_T(\boldsymbol{x}) = (I_M - I_N + (2-2t)I_N - (1-t)T)(\boldsymbol{x}) \end{aligned}$$

Since, N *is finite dimensional and* T *is compact, this is a compact perturbation of the identity. Since* $\mathbf{0} \notin \partial \boldsymbol{D}$, *by invariance under compact homotopy, we have* $d_{LS}(I-T, \boldsymbol{D}, \mathbf{0}) = d_{LS}(I_M - I_N, \boldsymbol{D}, \mathbf{0})$. *Since* I_N *has a finite dimensional range,* $I_M - I_N$ *is a compact perturbation of the identity* I_M. *Hence, by Theorem 11.3.2,*

$$d_{LS}(I_M - I_N, \boldsymbol{D}, \mathbf{0}) = d_{LS}((I_M - I_N)|_N, \boldsymbol{D} \cap N, \mathbf{0}) = d_{LS}(-I_N|_N, \boldsymbol{D} \cap N, \mathbf{0})$$

But the dimension of N is $m = m(\lambda_1)+\ldots+m(\lambda_n)$. Since $\mathbf{0} \in \boldsymbol{D}$, $d_B(-I_N|_N, \boldsymbol{D} \cap N, \mathbf{0}) = (-1)^m$ where we make the usual assumptions about the orientation of the finite dimensional normed linear space N.

If $\mathbf{0} \notin \boldsymbol{D}$, the Leray - Schauder existence theorem tells us $d_{LS}(I_T, \boldsymbol{D}, \mathbf{0}) = 0$.

The case where there are no real eigenvalues $\lambda > 1$ is similar to the argument we just made. Since there are no eigenvalues which are real and larger than 1, we note $X = \{\mathbf{0}\} \oplus \boldsymbol{R}(I - T)$ as $I - T$ is invertible. We no longer have to use the projections. We let the homotopy be $H(\cdot, t) = (1-t)(I - T) + tI = I - (1-t)T$. If $\boldsymbol{x} \neq 0$, if there was $0 < t < 1$ with $H(\boldsymbol{x}, t) = 0$, we would have $(1-t)(1/(1-t)I - T)(\boldsymbol{x}) = \mathbf{0}$ which would tell us T has a real eigenvalue $\mu = 1/(1-t) > 1$ which we assume is not true. Hence, we can use compact homotopy invariance to say

$$d_{LS}(I - T, \boldsymbol{D}, \mathbf{0}) \quad = \quad d_{LS}(I, \boldsymbol{D}, \mathbf{0}) = 1$$

In the case where S is empty, we interpret the sum $m = \sum_{\lambda \in S} m(\lambda) = 0$. The formula then still stands. ■

Note, the important thing here is T linear compact and $I - T$ invertible implies $|d_{LS}(I - T, \boldsymbol{D}, \mathbf{0}) = 1 \neq 0$.

Homework

Exercise 11.4.25 *Prove this formula from the discussion above:*

$$X \quad = \quad E_{\lambda_1}(X) \oplus E_{\lambda_2}(X) \oplus \ldots \oplus E_{\lambda_n}(X) \oplus (I - E_{\lambda_1} - \ldots - E_{\lambda_n})(X)$$

Exercise 11.4.26 *Prove $Q_n(X) = M_n$ is also T invariant.*

Exercise 11.4.27 *If X is a normed linear space and $A, B : X \to X$ are linear, continuous and compact, prove AB and BA are also linear, continuous and compact.*

Exercise 11.4.28 *Let X be the set of all linear and continuous operators from the normed linear space X to the normed linear space Y. Let C be the set of all compact operators in X. Prove B is a subspace of X and if Y is complete, C is a closed subspace.*
Hint*: If (A_k) is a sequence of compact operators and $T \in X$ with $\|A_k - T\|_{op} \to 0$, let (x_k) be a bounded sequence in X. Use a standard diagonalization procedure to obtain a subsequence (x_k^1) so that $A_p(x_k^1)$ is a convergent sequence for each p.*

Chapter 12

Coincidence Degree

Using Leray - Schauder degree as a foundation, we can also develop another degree for infinite dimensional settings which is called coincidence degree which works with linear Fredholm maps of index zero. The book on coincidence degree, (Gaines and Mawhin (40) 1977), came out just as we were starting our dissertation work and most of the development in that book was not fleshed out in much detail. So we tried to write complete arguments for ourselves which helped teach us more functional analysis and helped us to see it is absolutely essential to understand a tool we want to use, rather than just applying it in one of our arguments. We use coincidence degree ideas in our dissertation and in our early papers. The discussion below is based on our notes from that time.

12.1 Functional Analysis Background

So let's start by defining our new setting. Let X and Z be normed linear spaces with their own norms which we will stop specifically mentioning. Assume $L : \boldsymbol{D}(L) \subset X \rightarrow Z$ is linear. A linear map that is Fredholm of index j is defined as follows:

Definition 12.1.1 Linear Fredholm of Index j Mappings

Let X and Z be normed linear spaces and assume $L : \boldsymbol{D}(L) \subset X \rightarrow Z$ is linear. L is Fredholm of index j when:

(i): The dimension of $\boldsymbol{N}(L)$ is finite.

(ii): The dimension of the quotient space $Z/\boldsymbol{R}(L)$ is finite. This dimension is called the codimension of $\boldsymbol{R}(L)$ and is denoted by $codim(\boldsymbol{R}(L))$.

(iii): The subspace $\boldsymbol{R}(L)$ is closed in Z.

Here, $j = dim(\boldsymbol{N}(L)) - codim(\boldsymbol{R}(L))$.

For coincidence degree, we will work with linear Fredholm maps of index 0: so $dim(\boldsymbol{N}(L)) = codim(\boldsymbol{R}(L))$. Here are some simple examples:

1. $0 : X \rightarrow Z$ where $dimX = dimZ$ are both finite. Let n be this common dimension. Then, $dim\boldsymbol{N}(0) = n$ and $\boldsymbol{R}(0) = \{\boldsymbol{0}\}$. Thus, $Z/\boldsymbol{R}(0) = Z/\{\boldsymbol{0}\}$ and $codim(\boldsymbol{R}(0)) = n$. Hence, the zero map here is linear Fredholm of index 0.

2. $I : X \rightarrow X$. Then, $dim(\boldsymbol{N}(I)) = dim(\{\boldsymbol{0}\}) = 0$ and $\boldsymbol{R}(I) = X$ so the $codim(X/\boldsymbol{R}(I)) = codim(X/X) = 0$. Therefore, the identity map is linear Fredholm of index 0.

3. Let $A : \Re^5 \rightarrow \Re^3$ be a 5×3 matrix. Then, $\Re^5 = \boldsymbol{N}(A) \oplus U$ and $\Re^3 = \boldsymbol{R}(A) \oplus V$ for subspaces U and V. Clearly, U is the orthogonal complement of $\boldsymbol{N}(A)$ and V is the orthogonal complement of $\boldsymbol{R}(A)$. So $5 = dim(\boldsymbol{N}(A)) + dim(U)$ and $3 = dim(\boldsymbol{R}(A)) + dim(V)$. We get this table:

$$\begin{aligned} dim(\boldsymbol{N}(A)) = 0 &\implies dim(\boldsymbol{R}(A)) = 3 \implies codim(\boldsymbol{R}(A)) = 0 \implies \text{LF0} \\ dim(\boldsymbol{N}(A)) = 1 &\implies dim(\boldsymbol{R}(A)) = 2 \implies codim(\boldsymbol{R}(A)) = 1 \implies \text{LF0} \\ dim(\boldsymbol{N}(A)) = 2 &\implies dim(\boldsymbol{R}(A)) = 1 \implies codim(\boldsymbol{R}(A)) = 2 \implies \text{LF0} \\ dim(\boldsymbol{N}(A)) = 3 &\implies dim(\boldsymbol{R}(A)) = 0 \implies codim(\boldsymbol{R}(A)) = 3 \implies \text{LF0} \end{aligned}$$

   So this mapping is linear Fredholm of index 0.

4. Let $A : \Re^n \rightarrow \Re^m$ be a $n \times m$ matrix with $n > m$. The same analysis as in the previous case shows this is also a linear Fredholm map of index 0.

5. The same is true if $L : X \rightarrow Z$ is a linear mapping between two finite dimensional normed linear spaces.

Homework

Exercise 12.1.1 *For the operator T in Exercise 11.4.4, find the index of T^k.*

Exercise 12.1.2 *For the operator T in Exercise 11.4.6, find the index of T^k.*

Exercise 12.1.3 *The operator T in Exercise 11.4.4 is the left shift by one operator. We can also define the right shift by one operator on ℓ^1 by $T(x_1, x_2, x_3, \ldots) = (0, x_1, x_2, \ldots)$. Find $\alpha(T)$, $\delta(T)$ and the index of T^k.*

Exercise 12.1.4 *For the derivative operator x' on $C([0, 1])$, find the index.*

To understand these mappings better, we need to look at annihilators.

Definition 12.1.2 Annihilators of Subsets of normed linear spaces X and X'

Let X be a normed linear space and X' its continuous dual. Let $\boldsymbol{A} \subset X$ and $\boldsymbol{B} \subset X'$. Then:

1. *The annihilator of $\boldsymbol{A}$ is*

$$\boldsymbol{A}^0 = \{\boldsymbol{x}' \in X' | \boldsymbol{x}'(\boldsymbol{x}) = 0, \forall \boldsymbol{x} \in \boldsymbol{A}\}$$

2. *The annihilator of $\boldsymbol{B}$ is*

$$^0\boldsymbol{B} = \{\boldsymbol{x} \in X | \boldsymbol{x}'(\boldsymbol{x}) = 0, \forall \boldsymbol{x}' \in \boldsymbol{B}\}$$

We need to establish some properties about annihilators. First the annihilators are closed subspaces.

Lemma 12.1.1 Annihilators are Closed Subspaces

X is a normed linear space with $\boldsymbol{A} \subset X$ and $\boldsymbol{B} \subset X'$. Then $\boldsymbol{A}^0$ and $^0\boldsymbol{B}$ are closed subspaces.

Proof 12.1.1
If $\boldsymbol{x}'$ and $\boldsymbol{y}'$ are in $\boldsymbol{A}^0$, then for all scalars α and β, $\alpha\boldsymbol{x}'(\boldsymbol{x}) + \beta\boldsymbol{y}'(\boldsymbol{x}) = \boldsymbol{0}$ for all $\boldsymbol{x} \in \boldsymbol{A}$. This tells

us $\alpha \boldsymbol{x}' + \beta \boldsymbol{y}' \in \boldsymbol{A}^0$. *Hence,* $\boldsymbol{A}^0$ *is a subspace. If* $(\boldsymbol{x}'_n)$ *was a sequence in* $\boldsymbol{A}^0$ *which converged to* $\boldsymbol{x}'$, *then since* $|\boldsymbol{x}'_n(\boldsymbol{x}) - \boldsymbol{x}'(\boldsymbol{x})| \leq \|\boldsymbol{x}'_n - \boldsymbol{x}'\|\|\boldsymbol{x}\|$, *we see* $\boldsymbol{x}'_n(\boldsymbol{x}) - \boldsymbol{x}'(\boldsymbol{x}) \rightarrow \boldsymbol{0}$. *Since* $\boldsymbol{x}'_n(\boldsymbol{x}) = 0$ *for all* n, *this tells us* $\boldsymbol{x}'(\boldsymbol{x}) = 0$ *too. Hence,* $\boldsymbol{A}^0$ *is closed.*

We leave the proof that ${}^0\boldsymbol{B}$ *is a closed subspace to you.* ■

Next, the annihilator of the annihilator of a closed subspace $\boldsymbol{A}$ of X is $\boldsymbol{A}$.

Lemma 12.1.2 ${}^0A^0 = A$ if A is Closed Subspace

> *X is a normed linear space with $\boldsymbol{A} \subset X$ a closed subspace. Then, ${}^0\boldsymbol{A}^0 = \boldsymbol{A}$.*

Proof 12.1.2
If $\boldsymbol{x} \in {}^0\boldsymbol{A}^0$, *then* $\boldsymbol{x}'(\boldsymbol{x}) = 0$ *for all* $\boldsymbol{x}' \in \boldsymbol{A}^0$. *This is satisfied for all* $\boldsymbol{x} \in \boldsymbol{A}$. *So* $\boldsymbol{A} \subset {}^0\boldsymbol{A}^0$.

To show the reverse, let $\boldsymbol{x}_1 \in \boldsymbol{A}^C$. *Since* $\boldsymbol{A}$ *is a closed subspace,* $d(\boldsymbol{x}_1, \boldsymbol{A}) = \inf_{\boldsymbol{x} \in \boldsymbol{A}} \|\boldsymbol{x}_1 - \boldsymbol{x}\| = d > 0$. *By a standard corollary to the Hahn - Banach Theorem, there is a continuous linear functional* ϕ *with* $\|\phi\| = 1$, $\phi(\boldsymbol{x}_1) = d$ *and* $\phi = 0$ *on* $\boldsymbol{A}$. *This tells us* $\phi \in \boldsymbol{A}^0$. *But* $\phi(\boldsymbol{x}_1) \neq 0$, *so* $\boldsymbol{x}_1 \notin {}^0\boldsymbol{A}^0$. *Thus,* ${}^0\boldsymbol{A}^0 \subset \boldsymbol{A}$.

Combining these results, we have ${}^0\boldsymbol{A}^0 = \boldsymbol{A}$. ■

If we start with a subset $\boldsymbol{S} \subset X$ and we let $\boldsymbol{A} = \overline{span(\boldsymbol{S})}$, we can connect the annihilator of A to the annihilator of $\boldsymbol{S}$.

Lemma 12.1.3 If A is the Closure of the Span of a Set S, then $A^0 = S^0$ and $A = {}^0S^0$

> *X is a normed linear space with $\boldsymbol{S} \subset X$. Let $\boldsymbol{A} = \overline{span(\boldsymbol{S})}$. Then $\boldsymbol{A}^0 = \boldsymbol{S}^0$ and $\boldsymbol{A} = {}^0\boldsymbol{S}^0$.*

Proof 12.1.3
Since $\boldsymbol{S} \subset \boldsymbol{A}$, *we see* $\boldsymbol{A}^0 \subset \boldsymbol{S}^0$.

Now, if $\{\boldsymbol{x}_j, j = 1, n\} \in S$ *for some* n *and* $\{\alpha_j, j = 1, n\}$ *are scalars, then if* $\boldsymbol{x}' \in \boldsymbol{S}^0$, $\boldsymbol{x}'(\sum_{j=1}^n \alpha_j \boldsymbol{x}'(\boldsymbol{x}_j)) = 0$ *also. So* $\boldsymbol{x}'(spanS) = 0$. *If* $(\boldsymbol{x}_n)$ *is a sequence in* S *which converges to a point* $\boldsymbol{x}$, *we note if* $\underline{\boldsymbol{x}' \in \boldsymbol{S}^0}$, $\boldsymbol{x}'(\boldsymbol{x}_n) = 0$ *for all* n. *By the continuity of* $\boldsymbol{x}'$ *we then see* $\boldsymbol{x}'(\boldsymbol{x}) = 0$ *also. Thus,* $\boldsymbol{x}'(\overline{span(\boldsymbol{S})}) = \boldsymbol{x}'(\boldsymbol{A}) = 0$. *Therefore,* $\boldsymbol{x}' \in \boldsymbol{A}^0$ *and we see* $\boldsymbol{S}^0 \subset \boldsymbol{A}^0$.

Combining, $\boldsymbol{S}^0 = \boldsymbol{A}^0$. *By Lemma 12.1.1,* $\boldsymbol{A}$ *is a closed subspace and thus, by Lemma 12.1.2,* $\boldsymbol{A} = {}^0\boldsymbol{A}^0 = {}^0\boldsymbol{S}^0$. ■

Homework

Exercise 12.1.5 *In the proof of Lemma 12.1.1, prove* ${}^0\boldsymbol{B}$ *is a closed subspace.*

Let x be a normed linear space and X' its continuous dual. First, recall, the annihilator of $\boldsymbol{A} \subset X$ is $\boldsymbol{A}^0 = \{\boldsymbol{x}' \in X' | \boldsymbol{x}'(\boldsymbol{x}) = 0, \ \forall \boldsymbol{x} \in \boldsymbol{A}\}$, and the annihilator of $\boldsymbol{B} \subset X'$ is ${}^0\boldsymbol{B} = \{\boldsymbol{x} \in X | \boldsymbol{x}'(\boldsymbol{x}) = 0, \ \forall \boldsymbol{x}' \in \boldsymbol{B}\}$.

Exercise 12.1.6 *Let* $X = \Re^2$. *Then* X' *is also* $\Re^2$ *and if you want you can identify* X *with column vectors and* X' *with row vectors.*

- *Let* $A = [0, 1] \times [0, 1]$ *and* $B = \{[1, 2]\}$. *Find* $\boldsymbol{A}^0$ *and* ${}^0\boldsymbol{B}$.

- *Let A be the set of vectors with nonnegative components and B be the set of linear functionals given by the boundary of the triangle in $\Re^2$ formed by the intersection of the lines $y = 0$, $x = 0$ and $x + y = 1$. Find $\boldsymbol{A^0}$ and $\boldsymbol{{}^0B}$.*

- *Let A be the intersection of three lines in $\Re^2$ so that the set of points inside and intersection and the boundary of this region form a bounded convex set. Let B be the set of dual vectors with nonnegative components. Find $\boldsymbol{A^0}$ and $\boldsymbol{{}^0B}$.*

Exercise 12.1.7 *Let $X = C([0,1], \|\cdot\|_\infty)$ and X' be its continuous dual, which is discussed in (Peterson (100) 2020). Let P_n be the set of polynomials of degree n on $[0,1]$, and let $B \subset X'$ be the set of functionals ϕ that have the form $\phi(f) = \int_0^1 f(x)dx$; i.e. these functionals can be written as Riemann integrals.*

- *Let $A = P_0$. Characterize $\boldsymbol{A^0}$ and $\boldsymbol{{}^0B}$.*
- *Let $A = P_1$. Characterize $\boldsymbol{A^0}$ and $\boldsymbol{{}^0B}$.*
- *Let $A = P_2$. Characterize $\boldsymbol{A^0}$ and $\boldsymbol{{}^0B}$.*
- *Let $A = P_n$. Characterize $\boldsymbol{A^0}$ and $\boldsymbol{{}^0B}$.*

Exercise 12.1.8 *Let $X = C([0,1], \|\cdot\|_\infty)$ and X' be its continuous dual, which is discussed in (Peterson (100) 2020). Let P_n be the set of polynomials of degree n on $[0,1]$ in the functions e^x and let $B \subset X'$ be the set of functionals ϕ that have the form $\phi(f) = \int_0^1 f(x)dx$; i.e. these functionals can be written as Riemann integrals.*

- *Let $A = P_0$. Characterize $\boldsymbol{A^0}$ and $\boldsymbol{{}^0B}$.*
- *Let $A = P_1$. Characterize $\boldsymbol{A^0}$ and $\boldsymbol{{}^0B}$.*
- *Let $A = P_2$. Characterize $\boldsymbol{A^0}$ and $\boldsymbol{{}^0B}$.*
- *Let $A = P_n$. Characterize $\boldsymbol{A^0}$ and $\boldsymbol{{}^0B}$.*

Exercise 12.1.9 *Let $X = C([0,1], \|\cdot\|_\infty)$ and X' be its continuous dual, which is discussed in (Peterson (100) 2020). Let P_n be the set of polynomials of degree n on $[0,1]$ in the functions $\cos(x)$ and let $B \subset X'$ be the set of functionals ϕ that have the form $\phi(f) = \int_0^1 f(x)dx$; i.e. these functionals can be written as Riemann integrals.*

- *Let $A = P_0$. Characterize $\boldsymbol{A^0}$ and $\boldsymbol{{}^0B}$.*
- *Let $A = P_1$. Characterize $\boldsymbol{A^0}$ and $\boldsymbol{{}^0B}$.*
- *Let $A = P_2$. Characterize $\boldsymbol{A^0}$ and $\boldsymbol{{}^0B}$.*
- *Let $A = P_n$. Characterize $\boldsymbol{A^0}$ and $\boldsymbol{{}^0B}$.*

Exercise 12.1.10 *Let $X = \ell^1$ and $X' = \ell^\infty$. Let A be the set of sequences in X with nonnegative components and B be the set of sequences y in X' with $\|y\|_\infty = 1$. Characterize $\boldsymbol{A^0}$ and $\boldsymbol{{}^0B}$.*

Next, let's review the idea of the adjoint of an operator. If X and Z are normed linear spaces and $L : X \to Z$ is a linear operator, define $L' : Z' \to X'$ by $(L'(\boldsymbol{z}'))(\boldsymbol{x}) = \boldsymbol{z}'(L(\boldsymbol{x}))$. We call L' the adjoint of L. Clearly L' is linear. The element $\boldsymbol{x}' \in X'$ defined by $L'(\boldsymbol{z}')$ is given by $\boldsymbol{x}'(\boldsymbol{x}) = \boldsymbol{z}'(L(\boldsymbol{x}))$. Assume, there were two elements in X', say $\boldsymbol{x}_1'$ and $\boldsymbol{x}_2'$ so that $\boldsymbol{x}_1'(\boldsymbol{x}) = \boldsymbol{z}'(L(\boldsymbol{x})) = \boldsymbol{x}_2'(\boldsymbol{x})$. Then, $(\boldsymbol{x}_1' - \boldsymbol{x}_2')(\boldsymbol{x}) = 0$ for all $\boldsymbol{x} \in X$. But this tells us $\boldsymbol{x}_1' = \boldsymbol{x}_2'$. Thus, the adjoint mapping is well-defined.

Also, note $\boldsymbol{N}(L')$ is closed in Z'. If $\boldsymbol{y} = L(\boldsymbol{x})$ and $\boldsymbol{z}' \in \boldsymbol{N}(L')$, then $L'(\boldsymbol{z}') = \boldsymbol{0}$. This implies $(L'(\boldsymbol{z}'))(\boldsymbol{x}) = \boldsymbol{z}'(L(\boldsymbol{x})) = \boldsymbol{z}'(\boldsymbol{y}) = 0$. This is true for all $\boldsymbol{x} \in X$. Thus, $L(\boldsymbol{x}) \in {}^0\boldsymbol{N}(L')$. Hence, $\boldsymbol{R}(L) \subset {}^0\boldsymbol{N}(L')$.

Next, if $\boldsymbol{z}' \in \boldsymbol{R}(L)^0$, then $\boldsymbol{z}'(L(\boldsymbol{x})) = 0$ for all $\boldsymbol{x} \in X$. This tells us $\boldsymbol{z}' \in \boldsymbol{N}(L')$. If $\boldsymbol{z}' \in \boldsymbol{N}(L')$, then $\boldsymbol{z}'(L(\boldsymbol{x})) = 0$ for all $\boldsymbol{x} \in X$ and so $\boldsymbol{z}' \in \boldsymbol{R}(L)^0$. Thus, $\boldsymbol{R}(L)^0 = \boldsymbol{N}(L')$. By Lemma 12.1.1, we see $\boldsymbol{N}(L')$ is closed.

By Lemma 12.1.3, $\overline{\boldsymbol{R}(L)} = {}^0(\boldsymbol{R}(L))^0 = {}^0\boldsymbol{N}(L')$. This leads to another result.

Lemma 12.1.4 $R(L) = {}^0N(L')$ if and only if $R(L)$ is Closed

Let X and Z be normed linear spaces and $L : X \to Z$ be a linear operator and L' its adjoint. Then, $\boldsymbol{R}(L) = {}^0\boldsymbol{N}(L')$ if and only if $\boldsymbol{R}(L)$ is closed.

Proof 12.1.4

By the remarks above, $\overline{\boldsymbol{R}(L)} = {}^0\boldsymbol{N}(L')$. So, if $\boldsymbol{R}(L)$ is closed, we have $\boldsymbol{R}(L) = {}^0\boldsymbol{N}(L')$. On the other hand, since $\overline{\boldsymbol{R}(L)} = {}^0\boldsymbol{N}(L')$, if $\boldsymbol{R}(L) = {}^0\boldsymbol{N}(L')$, then $\overline{\boldsymbol{R}(L)} = \boldsymbol{R}(L)$ and $\boldsymbol{R}(L)$ is closed.
■

Now, if we were given a finite dimensional subspace $\boldsymbol{U}$ of $\Re^n$, we know that $\Re^n = \boldsymbol{U} \oplus \boldsymbol{U}^\perp$ because we can use the inner product of $\Re^n$ and our knowledge of the dual $(\Re^n)'$. What about an arbitrary normed linear space X? Here is the result.

Theorem 12.1.5 Existence of Direct Sum Decompositions for a Given Finite Dimensional Subspace of a Normed Linear Space

Let X be a normed linear space and $\boldsymbol{U}$ a finite dimensional subspace of X. Then, there is a closed subspace $\boldsymbol{V}$ so that $X = \boldsymbol{U} \oplus \boldsymbol{V}$.

Proof 12.1.5

Let $\{\boldsymbol{x}_1, \ldots, \boldsymbol{x}_n\}$ be a basis for $\boldsymbol{U}$, where n is the dimension of $\boldsymbol{U}$. Define linear functionals $\{\boldsymbol{x}'_1, \ldots, \boldsymbol{x}'_n\}$ by $\boldsymbol{x}'_i(\boldsymbol{x}_j) = \delta^i_j$. Let $\boldsymbol{V} = \{\boldsymbol{x} \in X | \boldsymbol{x}'_j(\boldsymbol{x}) = 0, 1 \leq j \leq n\}$. Then $\boldsymbol{V} = {}^0(\{\boldsymbol{x}'_1, \ldots, \boldsymbol{x}'_n\})$. It is easy to see $\boldsymbol{V}$ is a subspace of X. Let $(\boldsymbol{z_k})$ be a sequence in $\boldsymbol{V}$ so that $\boldsymbol{z_k} \to \boldsymbol{z}$. Then, $\boldsymbol{x}'_j(\boldsymbol{z_k}) = 0$ for all j and all k. We know $|\boldsymbol{x}'_j(\boldsymbol{z_k}) - \boldsymbol{x}'_j(\boldsymbol{z})| \leq \|\boldsymbol{x}'_j\| \|\boldsymbol{z_k} - \boldsymbol{z}\|$, so it follows, $\boldsymbol{x}'_j(\boldsymbol{z_k}) \to \boldsymbol{x}'_j(\boldsymbol{z})$. Since $\boldsymbol{x}'_j(\boldsymbol{z_k}) = 0$, we have $\boldsymbol{x}'_j(\boldsymbol{z}) = 0$ and so $\boldsymbol{z} \in \boldsymbol{V}$. Thus, $\boldsymbol{V}$ is closed.

Let $\boldsymbol{w} \in \boldsymbol{U} \cap \boldsymbol{V}$. Then, $\boldsymbol{x}'_j(\boldsymbol{w}) = 0$ for all j. Also, we can write $\boldsymbol{w} = \sum_{p=1}^n \alpha_o \boldsymbol{x_p}$ for some scalars α_p as $\{\boldsymbol{x}_1, \ldots, \boldsymbol{x}_n\}$ is a basis for $\boldsymbol{U}$. Hence, $\boldsymbol{x}'_j(\sum_{p=1}^n \alpha_p \boldsymbol{x_p}) = 0$ implies $\sum_{j=1}^n \alpha_p \delta^j_p = \alpha_j = 0$ for each j. Thus, $\boldsymbol{w} = \boldsymbol{0}$ and so $\boldsymbol{U} \cap \boldsymbol{V} = \boldsymbol{0}$.

Finally, given $\boldsymbol{x} \in X$, let $\boldsymbol{u} = \sum_{p=1}^n \boldsymbol{x}'_p(\boldsymbol{x})\boldsymbol{x_p}$. Then $\boldsymbol{u} \in \boldsymbol{U}$ and $\boldsymbol{x}'_j(\boldsymbol{x} - \boldsymbol{u}) = \boldsymbol{x}'_j(\boldsymbol{x}) - \boldsymbol{x}'_j(\boldsymbol{x}) = 0$. So $\boldsymbol{x} - \boldsymbol{u} \in \boldsymbol{V}$ which tells us $\boldsymbol{x} = \boldsymbol{u} + (\boldsymbol{x} - \boldsymbol{u}) \in \boldsymbol{U} \oplus \boldsymbol{V}$. Therefore $X = \boldsymbol{U} \oplus \boldsymbol{V}$. ■

Homework

Exercise 12.1.11 *Let $X = C([0, 1], \| \cdot \|_\infty)$ and P_n be the polynomials of degree n on $[0, 1]$. Prove P_n is finite dimensional and so there is a closed subspace Y_n so that $X = P_n \oplus Y_n$.*

Exercise 12.1.12 *Let $X = C([0,1], \|\cdot\|_\infty)$ and U be the solution space to the linear differential operator $L : C^n([0,1], \|\cdot\|_\infty) \rightarrow C([0,1], \|\cdot\|_\infty)$. Prove U is a finite dimensional subspace and so there is a closed subspace V so that $X = U \oplus V$.*

We might also be given a closed subspace of a normed linear space which is not necessarily finite dimensional and want to find another subspace so we can write the normed linear space as a direct sum decomposition.

Theorem 12.1.6 Normed Linear Space Direct Sum Decompositions for a Closed Subspace with a Finite Dimensional Annihilator

Let X be a normed linear space and $\boldsymbol{U}$ a closed subspace such that $\boldsymbol{U}^{\mathbf{0}} = \{\boldsymbol{x}' \in X' | \boldsymbol{x}'(\boldsymbol{u}) = 0, \forall \boldsymbol{u} \in \boldsymbol{U}\}$ has finite dimension n. Then, there is a finite dimensional subspace $\boldsymbol{V}$ so that $X = \boldsymbol{U} \oplus \boldsymbol{V}$.

Proof 12.1.6
Let $\{\boldsymbol{x}'_1, \ldots, \boldsymbol{x}'_n\}$ be a basis for $\boldsymbol{U}^{\mathbf{0}}$. By Lemma 12.1.3, $\boldsymbol{U} = {}^{\mathbf{0}}\boldsymbol{U}^{\mathbf{0}}$. Hence, we have $\boldsymbol{u} \in \boldsymbol{U}$ if and only if $\boldsymbol{x}'_j(\boldsymbol{x}) = 0$ for all j. We now show there is a set $\{\boldsymbol{x}_1, \ldots, \boldsymbol{x}_n\}$ so that $\boldsymbol{x}'_j(\boldsymbol{x}_i) = \delta^j_i$. The proof is by induction on n.

If $n = 1$, we seek $\boldsymbol{x}_1$ so that $\boldsymbol{x}'_1(\boldsymbol{x}_1) = 1$. Since $\boldsymbol{x}'_1 \neq \boldsymbol{0}$, there is $\boldsymbol{x}_0$ so that $\boldsymbol{x}'_1(\boldsymbol{x}_0) = \alpha \neq 0$. So, $\boldsymbol{x}'_1(\boldsymbol{x}_0/\alpha) = 1$. Hence, choose $\boldsymbol{x}_1 = \boldsymbol{x}_0/\alpha$.

Now assume the proposition is true for all $m \leq n-1$. Then, there are sets $\{\boldsymbol{x}_1, \ldots, \boldsymbol{x}_{n-1}\}$ so that $\boldsymbol{x}'_j(\boldsymbol{x}_i) = \delta^j_i$. Therefore, for all $\boldsymbol{x} \in X$ and for $1 \leq j \leq n-1$,

$$\boldsymbol{x}'_p(\boldsymbol{x} - \sum_{j=1}^{n-1} \boldsymbol{x}'_j(\boldsymbol{x})\, \boldsymbol{x}_j) = \boldsymbol{x}'_p(\boldsymbol{x}) - \boldsymbol{x}'_p(\boldsymbol{x}) = 0$$

Now assume $\boldsymbol{x}'_n(\boldsymbol{x}) = 0$ for all $\boldsymbol{x}$ in ${}^0\{\boldsymbol{x}'_1, \ldots, \boldsymbol{x}'_{n-1}\}$. Then, for all $\boldsymbol{x} \in X$, since $\boldsymbol{x}'_p(\boldsymbol{x} - \sum_{j=1}^{n-1} \boldsymbol{x}'_j(\boldsymbol{x})\boldsymbol{x}_j) = 0$ for all $1 \leq p \leq n-1$, we have $\boldsymbol{x} - \sum_{j=1}^{n-1} \boldsymbol{x}'_j(\boldsymbol{x})\boldsymbol{x}_j \in {}^0\{\boldsymbol{x}'_1, \ldots, \boldsymbol{x}'_{n-1}\}$. It follows that for all $\boldsymbol{x} \in X$,

$$\boldsymbol{x}'_n\left(\boldsymbol{x} - \sum_{j=1}^{n-1} \boldsymbol{x}'_j(\boldsymbol{x})\boldsymbol{x}_j\right) = 0$$

This implies

$$\boldsymbol{x}'_n(\boldsymbol{x}) = \sum_{j=1}^{n-1} \boldsymbol{x}'_j(\boldsymbol{x})\boldsymbol{x}'_n(\boldsymbol{x}_j)$$

This tells us $\boldsymbol{x}'_n = \sum_{j=1}^{n-1} \boldsymbol{x}'_n(\boldsymbol{x}_j)\boldsymbol{x}'_j$. But we are told the set $\{\boldsymbol{x}'_1 \ldots, \boldsymbol{x}'_n\}$ is linearly independent. So the assumption that $\boldsymbol{x}'_n(\boldsymbol{x}) = 0$ for all $\boldsymbol{x}$ in ${}^0\{\boldsymbol{x}'_1, \ldots, \boldsymbol{x}'_{n-1}\}$ is wrong. Hence, there must be $\boldsymbol{y} \in {}^0\{\boldsymbol{x}'_1, \ldots, \boldsymbol{x}'_{n-1}\}$ so that $\boldsymbol{x}'_n(\boldsymbol{y}) \neq 0$. Let $\boldsymbol{x}_n = \boldsymbol{y}/\boldsymbol{x}'_n(\boldsymbol{y})$. Then $\boldsymbol{x}'_n(\boldsymbol{x}_n) = 1$ and $\boldsymbol{x}'_j(\boldsymbol{x}_n) = 0$, for $1 \leq j \leq n-1$.

We claim $\{\boldsymbol{x}_1, \ldots, \boldsymbol{x}_{n-1}\}$ is a linearly independent set. Consider $\sum_{j=1}^n \beta_j \boldsymbol{x}_j = 0$. Then, it is straightforward to see $\boldsymbol{x}'_p(\sum_{j=1}^n \beta_j \boldsymbol{x}_j) = \beta_p = 0$ for $1 \leq p \leq n$. Thus, all the scalars $\beta_j = 0$ and we can conclude the set if linearly independent.

Now let $\boldsymbol{V} = span\{\boldsymbol{x}_1, \ldots, \boldsymbol{x}_n\}$*. Then, if* $\boldsymbol{v} \in \boldsymbol{V}$*,* $\boldsymbol{v} = \sum_{j=1}^{n} \beta_j \boldsymbol{x}_j$ *and* $\boldsymbol{x}'_j(\boldsymbol{v}) = \beta_j$ *for all* j*. But if* $\boldsymbol{w} \in \boldsymbol{U}$*, also, then* $\boldsymbol{x}'_i(\boldsymbol{v}) = 0$ *for all* j*. Hence, if* $\boldsymbol{w} \in \boldsymbol{U} \cap V$*,* $\boldsymbol{w} = \boldsymbol{0}$*.*

Finally, if $\boldsymbol{x} \in X$*, let* $\boldsymbol{v} = \sum_{j=1}^{n} \boldsymbol{x}'_j(\boldsymbol{x})\boldsymbol{x}_j$*. Then,* $\boldsymbol{v} \in \boldsymbol{V}$ *and* $\boldsymbol{x}'_{\boldsymbol{p}}(\boldsymbol{x} - \boldsymbol{v}) = \boldsymbol{x}'_{\boldsymbol{p}}(\boldsymbol{x}) - \boldsymbol{x}'_{\boldsymbol{p}}(\boldsymbol{x}) = 0$*. We see* $\boldsymbol{u} = \boldsymbol{x} - \boldsymbol{v} \in \boldsymbol{U}$*. We conclude we can write* $\boldsymbol{x} = \boldsymbol{u} + \boldsymbol{v}$*,* $\boldsymbol{u} \in \boldsymbol{U}$*,* $\boldsymbol{v} \in \boldsymbol{V}$ *uniquely. Hence,* $X = \boldsymbol{U} \oplus \boldsymbol{V}$ *where the dimension of* $\boldsymbol{V} = n$*.*

This completes the induction proof and shows the result holds for any n*.* ■

We can also start with a closed subspace and a finite dimensional subspace which only intersects it at the point $\boldsymbol{0}$ and construct a direct sum decomposition that lives inside the normed linear space.

Theorem 12.1.7 Constructing Direct Sum Decompositions inside a Normed Linear Space

Let X *be a normed linear space and* $\boldsymbol{U}$ *a closed subspace of* X *and* $\boldsymbol{V}$ *a finite dimensional subspace with* $\boldsymbol{U} \cap \boldsymbol{V} = \boldsymbol{0}$*. Then* $\boldsymbol{W} = \boldsymbol{U} \oplus \boldsymbol{V}$ *is a closed subspace and the linear operator* $P : \boldsymbol{W} \to \boldsymbol{W}$ *defined by*

$$P(\boldsymbol{x}) = \begin{cases} \boldsymbol{x}, & \boldsymbol{x} \in \boldsymbol{V} \\ \boldsymbol{0}, & \boldsymbol{x} \in \boldsymbol{U} \end{cases}$$

satisfies $P^2 = P$ *and* P *is continuous on* $\boldsymbol{W}$*; i.e.* P *is a continuous projection onto* $\boldsymbol{W}$*.*

Proof 12.1.7

It is easy to see P *is linear and* $P^2 = P$ *for all* $\boldsymbol{x} \in \boldsymbol{W}$*. We will show* P *is continuous by showing it is bounded. Assume it is not; then, there is no* C *so that* $\|P(\boldsymbol{x})\| \leq C\|\boldsymbol{x}\|$*, for all* $\boldsymbol{x} \in \boldsymbol{W}$*. Then, there is a sequence* $(\boldsymbol{x}_n) \subset \boldsymbol{W}$ *where* $\boldsymbol{x}_n = \boldsymbol{u}_n + \boldsymbol{v}_n$*, with* $\boldsymbol{u}_n \in \boldsymbol{U}$ *and* $\boldsymbol{v}_n \in \boldsymbol{V}$*, with* $\|P(\boldsymbol{x}_n)\|/\|\boldsymbol{x}_n\| \to \infty$*. Thus, we have*

$$\frac{\|P(\boldsymbol{x}_n)\|}{\|\boldsymbol{x}_n\|} = \frac{\|\boldsymbol{v}_n\|}{\|\boldsymbol{u}_n + \boldsymbol{v}_n\|} \to \infty$$

Thus, $\|\boldsymbol{x}_n\|/\|\boldsymbol{v}_n\| \to 0$ *implying* $\boldsymbol{x}_n/\|\boldsymbol{v}_n\| \to \boldsymbol{0}$*. Also,*

$$\left\| P\left(\frac{\boldsymbol{x}_n}{\|\boldsymbol{v}_n\|}\right) \right\| = \frac{\|\boldsymbol{v}_n\|}{\|\boldsymbol{v}_n\|} = 1 \tag{12.1}$$

The sequence $(\frac{\boldsymbol{v}_n}{\|\boldsymbol{v}_n\|})$ *is a subset of* $B(\boldsymbol{0}, 1) \subset \boldsymbol{V}$*. Since* $\boldsymbol{V}$ *is finite dimensional, this is a compact set and so there is a subsequence* $(\boldsymbol{x}_{n,1})$ *and* $\boldsymbol{z}$ *with* $\boldsymbol{z} \in \boldsymbol{V}$ *as* $\boldsymbol{V}$ *finite dimensional implies it is closed.*

$$\begin{aligned} P\left(\frac{\boldsymbol{x}_{n,1}}{\|\boldsymbol{v}_{n,1}\|}\right) &= \frac{\boldsymbol{v}_{n,1}}{\|\boldsymbol{v}_{n,1}\|} \\ \left(I - P\right)\left(\frac{\boldsymbol{x}_{n,1}}{\|\boldsymbol{v}_{n,1}\|}\right) &= \frac{\boldsymbol{x}_{n,1}}{\|\boldsymbol{v}_{n,1}\|} - \frac{\boldsymbol{v}_{n,1}}{\|\boldsymbol{v}_{n,1}\|} \end{aligned}$$

Letting $n \to \infty$*, we find*

$$\begin{aligned} \lim_{n\to\infty} P\left(\frac{\boldsymbol{x}_{n,1}}{\|\boldsymbol{v}_{n,1}\|}\right) &= \lim_{n\to\infty} \frac{\boldsymbol{v}_{n,1}}{\|\boldsymbol{v}_{n,1}\|} = \boldsymbol{z} \\ \lim_{n\to\infty} \left(I - P\right)\left(\frac{\boldsymbol{x}_{n,1}}{\|\boldsymbol{v}_{n,1}\|}\right) &= \lim_{n\to\infty} \frac{\boldsymbol{x}_{n,1}}{\|\boldsymbol{v}_{n,1}\|} - \lim_{n\to\infty} \frac{\boldsymbol{v}_{n,1}}{\|\boldsymbol{v}_{n,1}\|} = \boldsymbol{0} - \boldsymbol{z} \end{aligned}$$

By assumption, $\boldsymbol{U}$ is also closed. Now

$$\left(I - P\right)\left(\frac{\boldsymbol{x_{n,1}}}{\|\boldsymbol{v_{n,1}}\|}\right) \quad \in \quad \boldsymbol{U}$$

It follows that $-\boldsymbol{z} \in \boldsymbol{U}$ telling us $\boldsymbol{z} \in \boldsymbol{U}$. Thus, $\boldsymbol{z} \in \boldsymbol{U} \cap \boldsymbol{V} = \boldsymbol{0}$. But

$$\|\boldsymbol{z}\| \quad = \quad \left\| \lim_{n,1 \to \infty} P\left(\frac{\boldsymbol{x_{n,1}}}{\|\boldsymbol{v_{n,1}}\|}\right)\right\| = 1$$

which is a contradiction. Thus, our assumption that P is not bounded is false. So P is bounded and therefore continuous.

To show $\boldsymbol{W}$ is closed, assume $(\boldsymbol{x_n}) \subset \boldsymbol{W}$ and $\boldsymbol{x_n} \to \boldsymbol{x}$. We know $\boldsymbol{x_n} = \boldsymbol{u_n} + \boldsymbol{v_n}$ with $\boldsymbol{u_n} \in \boldsymbol{U}$ and $\boldsymbol{v_n} \in \boldsymbol{V}$. We know $P(\boldsymbol{x_n}) \to P(\boldsymbol{x})$ as P is continuous. Thus, $\boldsymbol{v_n} \to P(\boldsymbol{x})$. Since $\boldsymbol{V}$ is closed, $P(\boldsymbol{x}) \in \boldsymbol{V}$. Also, $(I - P)(\boldsymbol{x_n}) \to (I - P)(\boldsymbol{x})$ and so $\boldsymbol{u_n} \to (I - P)(\boldsymbol{x})$. Since $\boldsymbol{U}$ is closed, we see $(I - P)(\boldsymbol{x}) \in \boldsymbol{U}$. Thus, $\boldsymbol{x} = P(\boldsymbol{x}) + (I - P)(\boldsymbol{x}) \in \boldsymbol{U} \oplus \boldsymbol{V}$. Thus, $\boldsymbol{W}$ is closed. ■

Homework

Exercise 12.1.13 *Let $f(x) = |x+3| + |x-1| + |x-5| + |x-7|$ be a convex function on $\Re$. Let's recall our discussions about the subdifferential $\partial f(x)$ that we had in (Peterson (99) 2020).*

- *Show*

$$f(x) \quad = \quad \begin{cases} -4x - 10, & \infty < x \leq -3 \\ -2x + 16, & -3 \leq x < 1 \\ 14, & 1 \leq x < 5 \\ 2x + 4, & 5 \leq x < 7 \\ 4x - 10, & 7 \leq x \end{cases}$$

- *Show*

$$\partial f(x) \quad = \quad \begin{cases} -4, & \infty < x < -3 \\ [-4, -2], & x = -3 \\ -2, & -3 < x < 1 \\ [-2, 0], & x = 1 \\ 0, & 1 < x < 5 \\ [0, 2], & x = 5 \\ 2, & 5 < x < 7 \\ [2, 4], & x = 7 \\ 4, & 7 < x \end{cases}$$

- *The subgradients of $f \in C([-10, 10], \|\cdot\|_\infty)$ at the point x_0 are the elements of the dual that satisfy $\phi(x - x_0) \leq f(x) - f(x_0)$ for all $x \in [-10, 10]$. In (Peterson (99) 2020), we did not define this in this way but now we have more mathematical machinery at our disposal. Let the function of bounded variation g be defined by*

$$g(x) \quad = \quad \begin{cases} -4, & -10 \leq x < -3 \\ -2, & -3 \leq x < 1 \\ 0, & 1 \leq x < 5 \\ 2, & 5 \leq x < 7 \\ 4, & 7 \leq x \leq 10 \end{cases}$$

Define $\phi(h) = \int_{-10}^{10} h dg$, the usual Riemann - Stieljes integral for this function of bounded variation. From our discussions in (Peterson (100) 2020), we characterized the dual space of $(C([-10,10], \|\cdot\|_\infty))'$. Hence, ϕ is a bounded linear functional. Show $\phi(x - x_0) = 136 - 20x_0$.

- *It turns out $\partial f(x_0) = \phi$. This is pretty messy in general, so fix $x_0 = -1$ and show $\phi(x - (-1)) \leq f(x) - f(-1)$ by direct calculation, for all $x \in [-10, 10]$.*

- *Note our usual subgradient written as $\partial f(x)$ as described earlier is the same as the function whose value at x is $[g(x^-), g(x^+)]$.*

Exercise 12.1.14 *Repeat the exercise above for $f(x) = |x+2| + |x-7|$.*

Exercise 12.1.15 *Repeat the exercise above for $f(x) = |x-1|$.*

Exercise 12.1.16 *Let $X = \mathbf{c}$, the set of all sequences that converge with the* sup-*norm. We know $c' \cong \ell^1$ but we need to remember the Schauder basis for $\mathbf{c}$ is the usual $\boldsymbol{e_i}$ for $1 \leq i$ and $\boldsymbol{e_0} = (1,1,1,\ldots)$ and $x \in \mathbf{c}$ has representation $x = x_\infty \boldsymbol{e_0} + \sum_{i=1}^{\infty}(x_i - x_\infty)\boldsymbol{e_i}$, where x_∞ is the limit of x. The element in ℓ^1 identified with a given $x' \in X'$ is then $(y_0, y_1, \ldots, y_i, \ldots)$ where $y_0 = x'(\boldsymbol{e_0})$. Let*

$$U = \{y \in X' | Y(x) = \sum_{i=1}^{\infty} x_i y_i, \sum_{i=1}^{\infty} |y_i| < \infty\}$$

For $(x_i) \in X$, define $g \in X'$ by $g(x) = x_\infty$.

- *Prove $g \in X'$.*

- *Prove $g \in {}^0U^0 \setminus U$.*
  Hint: *First, prove ${}^0U^0 = X'$. Then, if $g \in {}^0U^0$, $g(x) = \sum_{i=1}^{\infty} x_i y_i = 0$ for all x. Show this tells us $y_i = 0$ for $1 \leq i$. But if $x = (1,1,1,\ldots)$, $x_\infty = 1$ and we must have $\sum_{i=1} |y_i| = 1$. This is not possible, so $g \notin {}^0U^0$.*

We now want to characterize the set of continuous linear functionals on a closed linear subspace to an appropriate quotient space.

Theorem 12.1.8 Characterizing the Continuous Linear Functionals on a Closed Linear Subspace

> *Let X be a normed linear space and $\boldsymbol{U}$ a closed linear subspace. Let $\boldsymbol{F} = X/\boldsymbol{U}$ be the quotient space. Then $\boldsymbol{F'} \equiv \boldsymbol{U}^0$.*

Proof 12.1.8
We know $X/\boldsymbol{U}$ is a normed linear space itself with the norm defined on equivalence classes given by $\|[\boldsymbol{x}]\| = \inf_{\boldsymbol{y}\in[\boldsymbol{x}]} \|\boldsymbol{y}\|$. You can refresh your mind about this by looking at the material in Section 8.5. Define $\phi : \boldsymbol{F'} \to \boldsymbol{U}^0$ by $\phi(f)(\boldsymbol{x}) = f([\boldsymbol{x}])$ for any $f \in \boldsymbol{F'}$ and $\boldsymbol{x} \in \boldsymbol{U}$. Since f is linear, it is clear $\phi(f)$ is also linear. Then, by definition, $\|[\boldsymbol{x}]\| \leq \|\boldsymbol{x} + \boldsymbol{u}\|$ for all $\boldsymbol{u} \in \boldsymbol{U}$, we see $\|[\boldsymbol{x}]\| \leq \|\boldsymbol{x}\|$. Thus, if $[\boldsymbol{x}] \neq [\boldsymbol{0}]$, we see $\boldsymbol{x} \neq \boldsymbol{0}$ and

$$\frac{|f([\boldsymbol{x}])|}{\|[\boldsymbol{x}]\|} \geq \frac{|f([\boldsymbol{x}])|}{\|\boldsymbol{x}\|} = \frac{|\phi(f)(\boldsymbol{x})|}{\|\boldsymbol{x}\|}$$

Thus, for all $\boldsymbol{x} \neq \boldsymbol{0}$,

$$\sup_{[\boldsymbol{x}]\neq[\boldsymbol{0}]} \frac{|f([\boldsymbol{x}])|}{\|[\boldsymbol{x}]\|} \geq \frac{|\phi(f)(\boldsymbol{x})|}{\|\boldsymbol{x}\|} \implies \|f\| \geq \frac{|\phi(f)(\boldsymbol{x})|}{\|\boldsymbol{x}\|}$$

This tells us $\|f\| \geq \|\phi(f)\|$. *So* $\phi(f)$ *is a bounded linear functional.*

Next, note there is a sequence $(\boldsymbol{u_n}) \subset \boldsymbol{U}$ *so that* $\|[\boldsymbol{x_0}]\| = \lim_{n\to\infty} \|\boldsymbol{x} + \boldsymbol{u_n}\|$. *Hence,*

$$\begin{aligned}
\frac{|f([\boldsymbol{x}])|}{\|[\boldsymbol{x}]\|} &= \frac{|\phi(f)(\boldsymbol{x})|}{\|[\boldsymbol{x}]\|} = \lim_{n\to\infty} \frac{|\phi(f)(\boldsymbol{x})|}{\|\boldsymbol{x}+\boldsymbol{u_n}\|} \\
&= \lim_{n\to\infty} \frac{|\phi(f)(\boldsymbol{x}+\boldsymbol{u_n})|}{\|\boldsymbol{x}+\boldsymbol{u_n}\|} \leq \lim_{n\to\infty} \sup_{\boldsymbol{x}\neq \boldsymbol{0}} \frac{|\phi(f)(\boldsymbol{x})|}{\|\boldsymbol{x}\|} \\
&= \lim_{n\to\infty} \|\phi(f)\| = \|\phi(f)\|
\end{aligned}$$

We conclude $\|f\| \leq \|\phi(f)\|$. *Combining, we see* $\|f\| = \|\phi(f)\|$ *and the mapping* ϕ *is a norm isometry.*

Define $\boldsymbol{x}'_{\boldsymbol{f}} \in X'$ *by* $\boldsymbol{x}'_{\boldsymbol{f}}(\boldsymbol{x}) = f([\boldsymbol{x}])$ *for any* $\boldsymbol{x} \in X$. *Thus,* $\phi(f) = \boldsymbol{x}'_{\boldsymbol{f}}$. *Given a scalar* α, *the functional* αf *is mapped to* $\boldsymbol{x}'_{\boldsymbol{\alpha f}}$ *defined by* $\boldsymbol{x}'_{\boldsymbol{\alpha f}}(\boldsymbol{x}) = (\alpha f)([\boldsymbol{x}]) = \alpha f([\boldsymbol{x}])$. *Hence* $\phi(\alpha f) = \boldsymbol{x}'_{\boldsymbol{\alpha f}} = \alpha \boldsymbol{x}'_{\boldsymbol{f}} = \alpha\phi(f)$. *Now if* $\boldsymbol{u} \in \boldsymbol{U}$, $[\boldsymbol{u}] = [\boldsymbol{0}]$ *and* $\boldsymbol{x}'_{\boldsymbol{f}}(\boldsymbol{u}) = f([\boldsymbol{u}]) = f([\boldsymbol{0}]) = 0$. *This tells us* $\boldsymbol{x}'_{\boldsymbol{f}} \in \boldsymbol{U}^0$. *If* $f \in \boldsymbol{F'}$, *then for all scalars,* α *and* β,

$$\phi(\alpha_1 f_1 + \alpha_2 f_2)(\boldsymbol{x}) = (\alpha_1 f_1 + \alpha_2 f_2)([\boldsymbol{x}]) = (\alpha_1 \phi(f_1) + \alpha_2 \phi(f_2))([\boldsymbol{x}])$$

This shows ϕ *is linear.*

If $\phi(f) = \boldsymbol{0}$, *then* $\phi(f)(\boldsymbol{x}) = f([\boldsymbol{x}]) = 0$ *for all* $\boldsymbol{x} \in X$. *Hence,* $f = \boldsymbol{0}$. *This tells us* ϕ *is 1-1. Now if* $\boldsymbol{x}' \in \boldsymbol{U}^0$, *define* $f \in \boldsymbol{F'}$ *by* $f([\boldsymbol{x}]) = f(\boldsymbol{x} + \boldsymbol{U}) = \boldsymbol{x}'(\boldsymbol{x})$. *Then,* $\phi(f)(\boldsymbol{x}) = f([\boldsymbol{x}]) = \boldsymbol{x}'(\boldsymbol{x})$. *Thus,* $\phi(f) = \boldsymbol{x}'$. *We see* ϕ *is also onto and so* ϕ *is a bijection.*

So ϕ *is a bijection preserving norms, which shows* $\boldsymbol{F'} \equiv \boldsymbol{U}^0$. ∎

Homework

Exercise 12.1.17 *Let* $X = \Re^n$ *with basic* $\boldsymbol{E} = \{\boldsymbol{E_1}, \ldots, \boldsymbol{E_n}\}$. *Let* $U = span\{\boldsymbol{E_1}\}$. *Let* $F = \Re^n \setminus U$. *Find* F *explicitly and show all the steps that show* $F' = U^0$.

Exercise 12.1.18 *Let* $X = \Re^n$ *with basic* $\boldsymbol{E} = \{\boldsymbol{E_1}, \ldots, \boldsymbol{E_n}\}$. *Let* $U = span\{\boldsymbol{E_1}, \boldsymbol{E_2}\}$. *Let* $F = \Re^n \setminus U$. *Find* F *explicitly and show all the steps that show* $F' = U^0$.

Exercise 12.1.19 *Let* $X = \Re^n$ *with basic* $\boldsymbol{E} = \{\boldsymbol{E_1}, \ldots, \boldsymbol{E_n}\}$. *Let* $U = span\{\boldsymbol{E_1}, \ldots, \boldsymbol{E_j}\}$ *for* $1 < j < n$. *Let* $F = \Re^n \setminus U$. *Find* F *explicitly and show all the steps that show* $F' = U^0$.

Exercise 12.1.20 *Let* $X = C([0,1], \|\cdot\|_\infty)$ *and* P_n *be the set of polynomials of degree* n. *Let* $F = X \setminus P_n$. *Find* F *explicitly and show all the steps that show* $F' = (P_n)^0$.

12.2 The Development of Coincidence Degree

Let's apply the results we have just discussed to a linear Fredholm map F of index 0. Recall $L : dom(L) \subset X \to Z$, where X and Z are normed linear spaces. We can do this in steps.

12.2.1 The Generalized Inverse of a Linear Fredholm Operator of Index Zero

The special properties of these operators and the results we have established so far allow us to rewrite X and Z into useful direct sum decompositions and to define a generalized inverse for L.

(i): By Theorem 12.1.6, there is a closed subspace $\boldsymbol{X_0}$ so that we can write $X = \boldsymbol{N}(L) \oplus \boldsymbol{X_0}$.

(ii): By Theorem 12.1.8, we know $(Z/\boldsymbol{R}(L))' \equiv (\boldsymbol{R}(L))^0$. By Lemma 12.1.4, we also know $\boldsymbol{R}(L) = {}^0\boldsymbol{N}(L')$. Thus, $\boldsymbol{R}(L)^0 \equiv {}^0\boldsymbol{N}(L')^0$. Then, by Lemma 12.1.3, we have $\boldsymbol{N}(L') = {}^0\boldsymbol{N}(L')^0$. We conclude $(Z/\boldsymbol{R}(L))' \equiv \boldsymbol{R}(L)^0 = \boldsymbol{N}(L')$.

(iii): Since $Z/\boldsymbol{R}(L)$ has finite dimension, say n, we know there is a basis $\{\boldsymbol{w_1}, \ldots, \boldsymbol{w_n}\} \subset Z/\boldsymbol{R}(L)$ so that any element in $Z/\boldsymbol{R}(L)$ can be written as a unique linear combination of this basis. Define $\boldsymbol{w_i'} \in (Z/\boldsymbol{R}(L))'$ by $\boldsymbol{w_i'}(\boldsymbol{w_j}) = \delta^i_j$. Then, if $\boldsymbol{w'} \in (Z/\boldsymbol{R}(L))'$, and $\boldsymbol{w} = \sum_{i=1}^n \beta_i \boldsymbol{w_i} \in Z/\boldsymbol{R}(L)$. Then, we see $\beta_i = \boldsymbol{w_i'}(\boldsymbol{w})$ and we can write

$$\boldsymbol{w} = \sum_{i=1}^{n} \boldsymbol{w_i'}(\boldsymbol{w})\boldsymbol{w_i}, \quad \boldsymbol{w'}(\boldsymbol{w}) = \sum_{i=1}^{n} \boldsymbol{w_i'}(\boldsymbol{w})\boldsymbol{w_i'}(\boldsymbol{w_i})$$

This is true for all $\boldsymbol{w} \in Z/\boldsymbol{R}(L)$ and so it follows $\boldsymbol{w'} = \sum_{i=1}^n \boldsymbol{w_i'}$. Therefore, $(Z/\boldsymbol{R}(L))'$ is the span of $\{\boldsymbol{w_1'}, \ldots, \boldsymbol{w_n'}\}$.

If $\sum_{i=1}^n \alpha_i \boldsymbol{w_i'} = \boldsymbol{0}$, then for all $\boldsymbol{w} \in Z/\boldsymbol{R}(L)$, we have $\left(\sum_{i=1}^n \alpha_i \boldsymbol{w_i'}\right)(\boldsymbol{w}) = \boldsymbol{0}$. Thus, $\alpha_i = 0$ for all i and so the set $\{\boldsymbol{w_1'}, \ldots, \boldsymbol{w_n'}\}$ is a linearly independent spanning set; i.e. a basis for $(Z/\boldsymbol{R}(L))'$.

From the discussion above, we see the dimension of $\boldsymbol{R}(L)^0$ equals the dimension of $(Z/\boldsymbol{R}(L))'$. Thus, $dim(\boldsymbol{N}(L')) = n = dim(Z/\boldsymbol{R}(L))$.

(iv): Since $\boldsymbol{R}(L)^0$ has finite dimension n, by Theorem 12.1.6 there is an n dimensional subspace $\boldsymbol{M_0}$ so that $Z = \boldsymbol{R}(L) \oplus \boldsymbol{M_0}$. By Theorem 12.1.7, there a continuous projection operators P and Q so that

$$P(\boldsymbol{x}) = \begin{cases} \boldsymbol{x}, & \boldsymbol{x} \in \boldsymbol{N}(L) \\ \boldsymbol{0}, & \boldsymbol{x} \in \boldsymbol{X_0} \end{cases}, \quad Q(\boldsymbol{x}) = \begin{cases} \boldsymbol{x}, & \boldsymbol{x} \in \boldsymbol{M_0} \\ \boldsymbol{0}, & \boldsymbol{x} \in \boldsymbol{R}(L) \end{cases}$$

We can show the projections graphically as a mnemonic using Figure 12.1 and Figure 12.2. The projection P operates on X and the projection Q operates on Z.

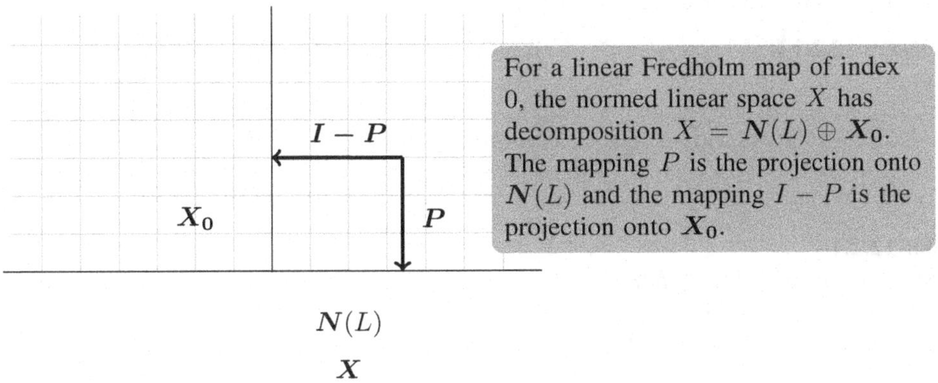

Figure 12.1: Linear Fredholm maps of index zero determine decompositions of the domain normed linear space.

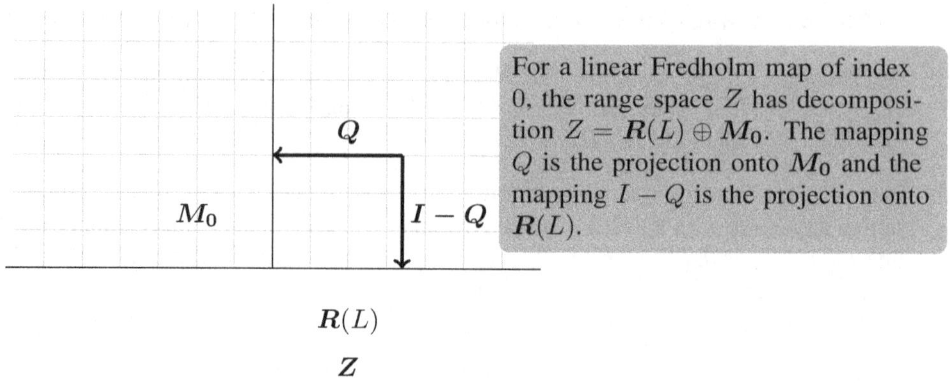

Figure 12.2: Linear Fredholm maps of index zero determine decompositions of the range normed linear space.

Now, let's do the next step. We have identified appropriate decompositions and now we will define new operators using these subspaces.

(i): Define $L_P = L|_{dom(L) \cap \boldsymbol{N}(P)} = L|_{dom(L) \cap \boldsymbol{X_0}}$. Then,

$$L_P(\boldsymbol{x}) \;=\; \boldsymbol{0} \Longrightarrow \boldsymbol{x} \in dom(L) \cap \boldsymbol{X_0} \cap \boldsymbol{N}(L)$$

But, $\boldsymbol{X_0} \cap \boldsymbol{N}(L) = \boldsymbol{0}$. Hence, L_P is 1-1 and so L_P^{-1} exists.

(ii): Let $K_p = L_p^{-1}$. Then $K_p : \boldsymbol{R}(L) \rightarrow dom(L) \cap \boldsymbol{N}(P) = dom(L) \cap \boldsymbol{X_0}$.

(iii): Define $K_{PQ} : Z \rightarrow dom(L) \cap \boldsymbol{N}(P) = dom(L) \cap \boldsymbol{X_0}$ by $K_{PQ} = K_P(I - Q)$. We call K_{PQ} the generalized inverse of L.

Now using all the notations we have established, we can prove some new results. In the next theorem, we are interested in isomorphisms between the vector subspaces $\boldsymbol{R}(Q)$ and $\boldsymbol{N}(L)$. This is not just a bijection; it is also linear and preserves the operations of the individual vector spaces. It then follows that if $J : \boldsymbol{R}(Q) \rightarrow \boldsymbol{N}(L)$ is such an isomorphism, $J(\boldsymbol{0}) = \boldsymbol{0}$. We will call these mappings vector isomorphisms to make this clear.

Theorem 12.2.1 If $J : \boldsymbol{R}(Q) \rightarrow \boldsymbol{N}(L)$ is a Vector Isomorphism, then $(L+J^{-1}P)^{-1} = JQ+K_{PQ}$

If $J : \boldsymbol{R}(Q) \rightarrow \boldsymbol{N}(L)$ is a vector isomorphism, then $L + J^{-1}P : dom(L) \subset X \rightarrow Z$ is invertible and its inverse is $JQ + K_{PQ}$.

Proof 12.2.1
First, note, since $\boldsymbol{R}(Q) = \boldsymbol{M_0}$ and the $dim(\boldsymbol{M_0}) = n = dim(Z/\boldsymbol{R}(L))$, it is straightforward to construct such isomorphisms.

First, consider for any $\boldsymbol{z} \in Z$,

$$\begin{aligned}(L + J^{-1}P)\,(JQ + K_{PQ})(\boldsymbol{z}) &= LJQ(\boldsymbol{z}) + LK_{PQ}(\boldsymbol{z}) + J^{-1}PJQ(\boldsymbol{z}) + J^{-1}PK_{PQ}(\boldsymbol{z})\\ &= LJQ(\boldsymbol{z}) + LK_P(I - Q)(\boldsymbol{z}) + J^{-1}PJQ(\boldsymbol{z}) + J^{-1}PK_P(I - Q)(\boldsymbol{z})\end{aligned}$$

(i): $Q(\boldsymbol{z}) \in \boldsymbol{M_0}$, so $JQ(\boldsymbol{z}) \in \boldsymbol{N}(L)$ implying $LJQ(\boldsymbol{z}) = \boldsymbol{0}$.

(ii): $LK_P(I-Q)(\boldsymbol{z}) = LK_P(\boldsymbol{w})$ *where* $\boldsymbol{w} = (I-Q)(\boldsymbol{z}) \in \boldsymbol{R}(L)$*. But on* $\boldsymbol{R}(L)$*,* $K_P = L_P^{-1}$ *and* $L = L_P$*. Thus,* $LK_P(I-Q)(\boldsymbol{z}) = (I-Q)(\boldsymbol{z})$*.*

(iii): $J^{-1}PJQ(\boldsymbol{z}) = J^{-1}JQ(\boldsymbol{z}) = Q(\boldsymbol{z})$ *as* $JQ(\boldsymbol{z}) \in \boldsymbol{N}(L)$ *and so* $PJQ(\boldsymbol{z}) = JQ(\boldsymbol{z})$*.*

(iv): $J^{-1}PK_P(I-Q)(\boldsymbol{z}) = J^{-1}(\boldsymbol{0})$ *as* $K_P(I-Q)(\boldsymbol{z}) \in dom(L) \cap \boldsymbol{N}(P)$*. Since* J *is a vector isomorphism,* $J^{-1}(\boldsymbol{0}) = \boldsymbol{0}$*.*

Combining these results, we find

$$(L + J^{-1}P)\,(JQ + K_{PQ})(\boldsymbol{z}) \quad = \quad \boldsymbol{0} + (I-Q)(\boldsymbol{z}) + Q(\boldsymbol{z}) + \boldsymbol{0} = \boldsymbol{z}$$

Thus, $(L + J^{-1}P)\,(JQ + K_{PQ})$ *is the identity on* Z*.*

Next, if $\boldsymbol{x} \in dom(L)$*, then*

$$(JQ + K_{PQ})(L + J^{-1}P)\,(\boldsymbol{x}) \quad = \quad JQL(\boldsymbol{x}) + JQJ^{-1}P(\boldsymbol{x}) + K_{PQ}L(\boldsymbol{x}) + K_{PQ}J^{-1}P\,(\boldsymbol{x})$$

(i): $JQL(\boldsymbol{x}) = J(\boldsymbol{0}) = \boldsymbol{0}$*.*

(ii): $JQJ^{-1}P(\boldsymbol{x}) = JQJ^{-1}(\boldsymbol{w})$ *where* $\boldsymbol{w} \in \boldsymbol{N}(L)$*. Now* $J^{-1}(\boldsymbol{w}) \in \boldsymbol{R}(Q) = \boldsymbol{M}_0$ *and so* $QJ^{-1}P(\boldsymbol{x}) = J^{-1}P(\boldsymbol{x})$*. We conclude* $JQJ^{-1}P(\boldsymbol{x}) = JJ^{-1}P(\boldsymbol{x}) = P(\boldsymbol{x})$*.*

(iii): $K_{PQ}L(\boldsymbol{x}) = K_P(I-Q)(L(\boldsymbol{x})) = K_PL(\boldsymbol{x})$*. So,* $K_PL(P(\boldsymbol{x})+(I-P)(\boldsymbol{x})) = K_PLP(\boldsymbol{x}) + K_PL(I-P)(\boldsymbol{x})$*. Since* $P(\boldsymbol{x}) \in \boldsymbol{N}(L)$*, this gives* $K_PL(\boldsymbol{x}) = K_PL(I-P)(\boldsymbol{x}) = (I-P)(\boldsymbol{x})$*.*

(iv): $K_{PQ}J^{-1}P\,(\boldsymbol{x}) = K_P(I-Q)(\boldsymbol{w})$ *where* $\boldsymbol{w} \in \boldsymbol{R}(Q) = \boldsymbol{M}_0$*. Thus,* $K_P(\boldsymbol{0}) = \boldsymbol{0}$*.*

Combining,

$$(JQ + K_{PQ})(L + J^{-1}P)\,(\boldsymbol{x}) \quad = \quad \boldsymbol{0} + P(\boldsymbol{x}) + (I-P)(\boldsymbol{x}) + \boldsymbol{0} = \boldsymbol{x}$$

Thus, $(JQ + K_{PQ})(L + J^{-1}P)$ *is the identity on* $dom(L)$*.* ■

Homework

Here are two problems of interest.

- The Picard Problem: the interval $[a, b]$ is finite with $a < b$ and

$$x'' \quad = \quad f(t, x, x'), \quad x(a) = x(b) = 0$$

  where $f : [a, b] \times \Re^n \times \Re^n \rightarrow \Re^n$ is sufficiently smooth.

- The Periodic Problem: the interval $[0, T]$ is finite with $0 < T$ and

$$x'' = f(t, x, x'). \quad x(0) = x(T),\ x'(0) = x'(T)$$

  where $f : [0, T] \times \Re^n \times \Re^n \rightarrow \Re^n$ is sufficiently smooth.

Exercise 12.2.1 *For the Picard problem, define* $X = C^2([a, b], \|\cdot\|) \cap \{x | x(a) = x(b) = 0\}$ *and* $Z = C([a, b], \|\cdot\|)$*. Then* $dom(L) = X \cap C^2([a, b], \|\cdot\|)$*,* $L : dom(L) \rightarrow Z$ *is defined by* $L(x) = x''$*. Further,* $N : X \rightarrow Z$ *is defined by* $N(x) = f(\cdot, x(\cdot), x'(\cdot))$*.*

- *Show* L *is linear Fredholm of index* 0*.*
- *Show the Picard problem can be written as* $L(x) = N(x)$*.*

- *Show* $\boldsymbol{N}(L) = \{0\}$ *so* L^{-1} *exists.*
- *Show this means* $P = I$ *and* $Q = 0$ *and so* $K_P(I - Q)N(x) = L^{-1}N(x)$.
- *Show*

$$\begin{aligned} x(t) &= K_P(I-Q)N(x) \\ &= \frac{b-t}{b-a}\int_a^t \int_a^s f(u, x(u), x'(u))\, du\, ds + \frac{a-t}{b-a}\int_t^b \int_a^s f(u, x(u), x'(u))\, du\, ds \end{aligned}$$

- *Prove, using the Arzela - Ascoli Theorem, that* $L^{-1}N(\overline{\boldsymbol{D}})$ *is compact for any bounded set* $\boldsymbol{D}$.

Exercise 12.2.2 *For the periodic problem, define* $X = C^1([a,b], \|\cdot\|) \cap \{x | x(0) = x(T),\ x'(0) = x'(T)\}$ *and* $Z = C([0,T], \|\cdot\|)$. *Then* $dom(L) = X \cap C^2([a,b], \|\cdot\|)$, $L : dom(L) \rightarrow Z$ *is defined by* $L(x) = x''$. *Further,* $N : X \rightarrow Z$ *is defined by* $N(x) = f(\cdot, x(\cdot), x'(\cdot))$.

- *Show* $\boldsymbol{N}(L) \neq \{0\}$ *and so L is not invertible.*
- *Show* $\boldsymbol{N}(L) = \Re^n$ *and*

$$\boldsymbol{R}(L) = \{z \in C^2([a,b], \|\cdot\|) : x'' = z, x(0) = x(T), x'(0) = x'(T)\}$$

  which is a closed subspace.
- *Show there are periodic solutions as long as* f *satisfies* $\int_0^T f(s, x(s), x'(s))ds = 0$ *and hence*

$$\boldsymbol{R}(L) = \{z \in C^2([a,b], \|\cdot\|) : \int_0^T z(s)ds = 0\} \cap Z$$

- *Show the solutions have the form*

$$\begin{aligned} x(t) &= \beta + (1 - t/T)\int_0^t \int_0^s f(u, x(u), x'(u))\, duds \\ &\quad -(t/T)\int_t^T \int_0^s f(u, x(u), x'(u))\, duds \end{aligned}$$

- *Show we can choose* $Q(z) = (1/T)\int_0^T z(u)du$ *and the dimension of* $Z/\boldsymbol{R}(L)$ *is the dimension of* $\boldsymbol{R}(Q) = n$ *which is the dimension of* $\boldsymbol{N}(L)$ *and so L is linear Fredholm of index* 0.
- *Show we can choose* $P(x) = x(0)$.
- *Show the periodic problem can thus be written as* $L(x) = N(x)$.
- *Show*

$$\begin{aligned} x(t) = K_P(I-Q)N(x) &= \frac{T-t}{T}\int_0^t \int_0^s f(u, x(u), x'(u))\, du\, ds \\ &\quad + \frac{-t}{T}\int_t^T \int_0^s f(u, x(u), x'(u))duds \end{aligned}$$

- *Prove, using the Arzela - Ascoli Theorem, that* $K_P(I - Q)N(\overline{\boldsymbol{D}})$ *is compact for any bounded set* $\boldsymbol{D}$.

12.2.2 Applying Leray - Schauder Degree Tools

The real problem we want to solve is to determine if the equation $L(\boldsymbol{x}) + G(\boldsymbol{x}) = \boldsymbol{0}$ has solutions. This is not directly a compact perturbation of the identity, so the use of Leray - Schauder degree is not available yet. Consider this blueprint towards a tool to establish the existence of a solution. We want $L(\boldsymbol{x}) + G(\boldsymbol{x}) = \boldsymbol{0}$, where $\boldsymbol{x} \in dom(L) \cap \overline{\boldsymbol{D}}$ where $\boldsymbol{D}$ is an open bounded subset of X. Let $J : \boldsymbol{R}(Q) \to \boldsymbol{N}(L)$ be any vector isomorphism. We will use all the notations we have already developed so if you need to you can go back and refresh your mind about the terms. We have the following chain of operations:

$$
\begin{aligned}
L(\boldsymbol{x}) + G(\boldsymbol{x}) = \boldsymbol{0} \iff & (L + J^{-1}P)(\boldsymbol{x}) + (G - J^{-1}P)(\boldsymbol{x}) = \boldsymbol{0} \\
\iff & \boldsymbol{x} + (L + J^{-1}P)^{-1}(G - J^{-1}P)(\boldsymbol{x}) = \boldsymbol{0} \\
\iff & \boldsymbol{x} + (JQ + K_{PQ})(G - J^{-1}P)(\boldsymbol{x}) = \boldsymbol{0} \\
\iff & \boldsymbol{x} + (JQ + K_{PQ})G(\boldsymbol{x}) - (JQ + K_{PQ})J^{-1}P(\boldsymbol{x}) = \boldsymbol{0}
\end{aligned}
$$

Now, we know $-JQJ^{-1}P(\boldsymbol{x}) = -JJ^{-1}P(\boldsymbol{x}) = -P(\boldsymbol{x})$ as Q is the identity on $\boldsymbol{R}(Q)$. Also, $-K_{PQ}J^{-1}P(\boldsymbol{x}) = -K_P(I - Q)J^{-1}P(\boldsymbol{x}) = \boldsymbol{0}$ as $I - Q$ is zero on $\boldsymbol{R}(Q)$. Hence, we can simplify a bit to find

$$
\begin{aligned}
L(\boldsymbol{x}) + G(\boldsymbol{x}) = \boldsymbol{0} \iff & \boldsymbol{x} + (JQ + K_{PQ})G(\boldsymbol{x}) - P(\boldsymbol{x}) = \boldsymbol{0} \\
\iff & (I - P)(\boldsymbol{x}) + (JQ + K_{PQ})G(\boldsymbol{x}) = \boldsymbol{0} \\
\iff & (I - (P - (JQ + K_{PQ})G))(\boldsymbol{x}) = \boldsymbol{0}
\end{aligned}
$$

We have a degree theory for compact perturbations of the identity, Leray - Schauder degree. Let $M = P - (JQ + K_{PQ})G$. Then, we have

$$
L(\boldsymbol{x}) + G(\boldsymbol{x}) = \boldsymbol{0} \iff (I - M)(\boldsymbol{x}) = \boldsymbol{0}
$$

Thus, we can apply Leray - Schauder degree technique if M is a compact mapping. Since $\boldsymbol{R}(P) = \boldsymbol{N}(L)$ is finite dimensional, we see P is compact. Also, since linear combinations of compact operators are compact, it is clear that M is compact when $(JQ + K_{PQ})G$ is compact. Now J has finite range, so JQG is compact if QG is continuous and $QG(\overline{\boldsymbol{D}})$ is bounded. Thus, $(JQ + K_{PQ})G$ is compact if

(i): QG is continuous and $QG(\overline{\boldsymbol{D}})$ is bounded.

(ii): $K_{PQ}G$ is compact.

To summarize, the equation $L(\boldsymbol{x}) - G(\boldsymbol{x}) = \boldsymbol{0}$ where L is linear Fredholm of index zero and $G : \overline{\boldsymbol{D}} \subset X \to Z$ is equivalent to the equation $(I - M)(\boldsymbol{x}) = \boldsymbol{0}$ where $M : \overline{\boldsymbol{D}} \subset X \to Z$ is given by $M = P - (JQ + K_{PQ})G$. Then, if M is compact and $\boldsymbol{0} \notin (I - M)(\partial \boldsymbol{D})$, Leray - Schauder degree is defined and we can study the existence of solutions of $L(\boldsymbol{x}) - G(\boldsymbol{x}) = \boldsymbol{0}$ using topological degree methods. Note $\partial \boldsymbol{D} = dom(L) \cap \partial \boldsymbol{D} \cup (dom(L))^C \cap \partial \boldsymbol{D}$. Consider a point where $\boldsymbol{x} = M(\boldsymbol{x})$. Since, for all $\boldsymbol{x} \in \overline{\boldsymbol{D}}$, $M(\boldsymbol{x}) \in dom(L)$, such a fixed point must be in $dom(L)$. Thus, it suffices to require $\boldsymbol{0} \notin (I - M)(dom(L) \cap \partial \boldsymbol{D})$.

Homework

Exercise 12.2.3 *For the Picard problem, discussed in Exercise 12.2.1,*

- *Go through the arguments discussed in this section carefully showing all details.*
- *Prove $(JQ + K_{PQ})N$ is compact by showing*

(i): *QN is continuous and $QN(\overline{\boldsymbol{D}})$ is bounded.*

(ii): *$K_{PQ}N$ is compact (we already did this in Exercise 12.2.1).*

Exercise 12.2.4 *For the periodic problem, discussed in Exercise 12.2.2,*

- *Go through the arguments discussed in this section carefully showing all details.*
- *Prove $(JQ + K_{PQ})N$ is compact by showing*

(i): *QN is continuous and $QN(\overline{\boldsymbol{D}})$ is bounded.*

(ii): *$K_{PQ}N$ is compact (we already did this in Exercise 12.2.2).*

Although, we can use Leray - Schauder degree theory on $I - M$, to what extent will our results depend on the choice of P, Q and J?

12.2.3 The Leray - Schauder Tools Dependence on J, P and Q

Let's start by looking at the vector isomorphisms J. We need to look at all the possibilities. Let $\mathcal{L}_L = \{J : \boldsymbol{R}(Q) \rightarrow \boldsymbol{N}(L) | J$ is a vector isomorphism $\}$.

Definition 12.2.1 Homotopic Vector Isomorphisms $\boldsymbol{R}(Q)$ to $\boldsymbol{N}(L)$

J_1 and J_2 in $\mathcal{L}_L$ are homotopic in $\mathcal{L}_L$ if there is a continuous mapping $\phi : \boldsymbol{R}(Q) \times [0,1] \rightarrow \boldsymbol{N}(L)$ so that $\phi(\cdot, 0) = J_1$ and $\phi(\cdot, 1) = J_2$ and $\phi(\cdot, t) \in \mathcal{L}_L$ for all $0 \leq t \leq 1$.

This definition allows us to partition $\mathcal{L}_L$ into two equivalence classes.

Lemma 12.2.2 J_1 and J_2 in $\mathcal{L}_L$ are Homotopic in $\mathcal{L}_L$ if and only if $det(J_2 J_1^{-1}) > 0$

Let J_1 and J_2 in $\mathcal{L}_L$.

(i): *Then J_1 and J_2 are homotopic in $\mathcal{L}_L$ if and only if $det(J_2 J_1^{-1}) > 0$.*

(ii): *$\mathcal{L}_L$ is partitioned into two homotopy classes.*

Proof 12.2.2
Since J_1 and J_2 are homotopic, there is a mapping ϕ as described in Definition 12.2.1. Let n be the dimension of $\boldsymbol{R}(Q)$ and $\boldsymbol{N}(L)$.

(i): *Let $\boldsymbol{A} = \{\boldsymbol{a_1}, \ldots \boldsymbol{a_n}\}$ be a basis for $\boldsymbol{R}(Q)$ and $\boldsymbol{B} = \{\boldsymbol{b_1}, \ldots \boldsymbol{b_n}\}$ be a basis for $\boldsymbol{N}(L)$. Given any $J \in \mathcal{L}_L$, let $[J]_{\boldsymbol{A},\boldsymbol{B}} = (r_{ij})$ be the usual matrix representation of J with respect to these bases. The change of basis are always confusing, so it won't hurt to do a review. For example $[J]_{\boldsymbol{A},\boldsymbol{B}}$ acts on $\boldsymbol{a_p}$ to give the new vector in $\boldsymbol{N}(L)$ given by $< [J]^p_{\boldsymbol{A},\boldsymbol{B}}, \boldsymbol{a_p} >$ which is the dot product of column p of $[J]_{\boldsymbol{A},\boldsymbol{B}}$ with $\boldsymbol{a_p}$. In terms of these bases, we note*

$$
\begin{aligned}
[J]_{\boldsymbol{A},\boldsymbol{B}} = [R] &= \begin{bmatrix} r_{11} & \ldots & r_{1n} \\ \vdots & \vdots & \vdots \\ r_{n1} & \ldots & r_{nn} \end{bmatrix} \Longrightarrow \\
< [R]^p, \boldsymbol{a_p} > &= \begin{bmatrix} r_{1p} & \ldots & r_{np} \end{bmatrix} \begin{bmatrix} 0 \\ \vdots \\ 1(\text{ column } p\text{ }) \\ \vdots \\ 0 \end{bmatrix} = \begin{bmatrix} r_{1p} \\ \vdots \\ r_{np} \end{bmatrix} = \sum_{j=1}^{n} r_{jp} \boldsymbol{b_j}
\end{aligned}
$$

So we can either use the basis notation or the matrix notation and we have to know how to move back and forth between them. Also, we are not making explicit transformations between $\Re^n$ and $\boldsymbol{R}(Q)$ and $\boldsymbol{N}(L)$ as we did in our discussions of how to define Brouwer degree for a finite dimensional normed linear space. We will let that go here.

So, using the notations for the bases, we have $J(\boldsymbol{a_i}) = \sum_{j=1}^n r_{ji}\boldsymbol{b_j}$ and $[J]_{\boldsymbol{A},\boldsymbol{B}} = [R]$. It is easy to see that $det([R]) \neq 0$ as $\boldsymbol{A}$ and $\boldsymbol{B}$ are linearly independent. Since, $\phi(\cdot, t) \in \mathcal{L}_L$ for all $0 \leq t \leq 1$, the same argument shows $det([\phi(\cdot, t)]) \neq 0$ on $[0, 1]$. Since ϕ is continuous and the determinants are all nonzero, it is not possible for $J_1 = \phi(\cdot, 0)$ and $\phi(\cdot, 1) = J_2$ to have a different sign for the determinant. Hence,

$$det([J_2][J_1]^{-1}) \quad = \quad \frac{det([J_2])}{det([J_1])} = \frac{\phi(\cdot,1)}{\phi(\cdot,0)} > 0$$

since the algebraic signs are the same. This completes the first argument.

(ii): *We now know if J_1 and J_2 are homotopic, then $det([J_2][J_1]^{-1}) > 0$ implying sign $det(J_2) =$ sign $det(J_1)$ for any choice of bases $\boldsymbol{A}$ and $\boldsymbol{B}$. Hence, if we fix J_1, the set of all J so that $det([J][J_1]^{-1}) > 0$ is one equivalence class and the set of all J so that $det([J][J_1]^{-1}) < 0$ is the other one. That is, we define an equivalence class on $\mathcal{L}_L$ by $J_1 \sim J_2$ if $det([J_2])$ and $det[J_1]$ have the same sign. Then, if J_2 has sign $det([J_2]) = -$sign $det([J_1])$, we have $-J_1 \sim J_2$.* ■

Let $\boldsymbol{C} = \{\boldsymbol{c_1}, \dots \boldsymbol{c_n}\}$ be another basis for $\boldsymbol{R}(Q)$ and $\boldsymbol{D} = \{\boldsymbol{d_1}, \dots \boldsymbol{d_n}\}$ be another basis for $\boldsymbol{N}(L)$. Then, there are change of basis matrices P and Q (not the same P and Q we have used before!) defined by $P(\boldsymbol{a_i}) = \boldsymbol{c_i} = \sum_{j=1}^n s_{ji}\boldsymbol{a_j}$ and $Q(\boldsymbol{d_i}) = \boldsymbol{b_i} = \sum_{j=1}^n t_{ji}\boldsymbol{d_j}$. It then follows that

$$\begin{aligned} J(\boldsymbol{c_i}) &= J\Big(\sum_{j=1}^n s_{ji}\boldsymbol{a_j}\Big) = \sum_{j=1}^n s_{ji}J(\boldsymbol{a_j}) = \sum_{j=1}^n s_{ji}\Big(\sum_{\ell=1}^n r_{\ell j}\boldsymbol{b_\ell}\Big) \\ &= \sum_{j=1}^n\sum_{\ell=1}^n s_{ji}r_{\ell j}\boldsymbol{b_\ell} = \sum_{j=1}^n\sum_{\ell=1}^n s_{ji}r_{\ell j}\Big(\sum_{k=1}^n t_{k\ell}\boldsymbol{d_k}\Big) \\ &= \sum_{\ell=1}^n\Big(\sum_{j=1}^n r_{\ell j}s_{ji}\Big)\Big(\sum_{k=1}^n t_{k\ell}\boldsymbol{d_k}\Big) = \sum_{k=1}^n\Big\{\sum_{\ell=1}^n t_{k\ell}\Big(\sum_{j=1}^n r_{\ell j}s_{ji}\Big)\Big\}\boldsymbol{d_k} \end{aligned}$$

Thus, $J(\boldsymbol{c_i}) = \sum_{k=1}^n \gamma_{ki}\boldsymbol{d_k}$ where $\gamma_{ki} = \sum_{\ell=1}^n t_{k\ell}(\sum_{j=1}^n r_{\ell j}s_{ji})$. If you let the matrix $[S]$ be defined by

$$[S] = \begin{bmatrix} s_{11} & \dots & s_{1n} \\ \vdots & \vdots & \vdots \\ s_{n1} & \dots & s_{nn} \end{bmatrix},$$

then, $\sum_{j=1}^n r_{\ell j}s_{ji} =< [R]_\ell, [S]^i >$, where $[R]_\ell$ is the ℓth row of $[R]$ and $[S]^i$ is the ith column of $[S]$. If you let the matrix $[T]$ be defined by

$$[T] \quad = \quad \begin{bmatrix} t_{11} & \dots & t_{1n} \\ \vdots & \vdots & \vdots \\ t_{n1} & \dots & t_{nn} \end{bmatrix},$$

then,

$$\begin{aligned}\sum_{\ell=1}^{n} t_{k\ell}\left(\sum_{j=1}^{n} r_{\ell j}s_{ji}\right) &= \sum_{\ell=1}^{n} t_{k\ell} < [R]_\ell, [S]^i >= \sum_{\ell=1}^{n} t_{k\ell}([R][S]_{\ell,i}) \\ &= ([T][R][S])_{ki} = ([T][J]_{\boldsymbol{A},\boldsymbol{B}}[S])_{ki}\end{aligned}$$

This shows us $[J]_{\boldsymbol{C},\boldsymbol{D}} = [T][J]_{\boldsymbol{A},\boldsymbol{B}}[S]$. We know the change of bases matrices P and Q are invertible; hence, $det([P]) = det(s_{ij}) \neq 0$ and $det([Q]) = det(t_{ij}) \neq 0$. We conclude $det([J]_{\boldsymbol{C},\boldsymbol{D}}) = \pm\alpha det([J]_{\boldsymbol{A},\boldsymbol{B}})$.

If $det(s_{ij}) = det(t_{ij}) = 1$, then $det([J]_{\boldsymbol{C},\boldsymbol{D}}) = det([J]_{\boldsymbol{A},\boldsymbol{B}})$. Define an equivalence relation on $S = \{\boldsymbol{A}|\boldsymbol{A}$ is a basis for $\boldsymbol{R}(Q)\}$ by $\boldsymbol{A} \sim \boldsymbol{B}$ if and only if $det(t_{ij}) > 0$. This splits all the bases in S into two classes. Arbitrarily choose one to be the positive orientation, S^+, and the other to be the negative orientation, S^-.

Now choose the positive orientation for $\boldsymbol{R}(Q)$ and $\boldsymbol{N}(L)$. Let $\boldsymbol{A}$ be a basis for $\boldsymbol{R}(Q)$ which is of positive orientation and $\boldsymbol{B}$ be a basis for $\boldsymbol{N}(L)$ which is of that orientation too. Let S_1^+ denote the positive orientation bases for $\boldsymbol{R}(Q)$ and S_2^+ denote the positive orientation bases for $\boldsymbol{N}(L)$.

Definition 12.2.2 Orientation Preserving Vector Isomorphisms $\boldsymbol{R}(Q)$ to $\boldsymbol{N}(L)$

Let $\boldsymbol{A}$ be a basis for $\boldsymbol{R}(Q)$. We say $J : \boldsymbol{R}(Q) \rightarrow \boldsymbol{N}(L)$ is an orientation vector isomorphism if $J(\boldsymbol{A}) \in S_2^+$. If this is not true, we say J is orientation reversing.

We can now start to classify vector isomorphisms that are homotopic in a better way.

Lemma 12.2.3 Vector Isomorphisms are Homotopic if and only if They are both Orientation Preserving or Orientation Reversing

If an orientation is chosen for both $\boldsymbol{R}(Q)$ and $\boldsymbol{N}(L)$, then J_1 and J_2 are homotopic in $\mathcal{L}_L$ if and only if they both preserve or reverse orientation.

Proof 12.2.3
Let $\boldsymbol{A}$ and $\boldsymbol{B}$ be bases for S_1^+ and S_2^+, respectively. By our earlier remarks, the basis $J_1(\boldsymbol{A}) \in S_2^+$ if and only if $\det([J_1]) = 1$ where $[J_1] = (t_{ij})$ is the matrix representation of J_1 defined by $J_1(\boldsymbol{a_i}) = \sum_{j=1}^n t_{ji}\boldsymbol{b_j}$. Consider the corresponding matrix representation for $J_2(\boldsymbol{A})$. Let $[J_2] = (\overline{t_{ij}})$. Then,

$$J_2(\boldsymbol{a_i}) = J_2J_1^{-1}J_1(\boldsymbol{a_i}) = J_2J_1^{-1}\left(\sum_{j=1}^{n} t_{ji}\boldsymbol{b_j}\right)$$

or

$$J_2(\boldsymbol{a_i}) = \sum_{j=1}^{n} \overline{t_{ij}}\boldsymbol{b_j} = \sum_{j=1}^{n} t_{ji}(J_2J_1^{-1})(\boldsymbol{b_j})$$

Hence, if $(\beta_{ij}) = [J_2J_1^{-1}]_{\boldsymbol{B},\boldsymbol{B}}$, then

$$\sum_{j=1}^{n} \overline{t_{ij}}\boldsymbol{b_j} = \sum_{j=1}^{n} t_{ji}\sum_{\ell=1}^{n} \beta_{\ell j}\boldsymbol{b_\ell} = \sum_{\ell=1}^{n}\left(\sum_{j=1}^{n} \beta_{\ell j}t_{ji}\right)\boldsymbol{b_\ell}$$

We conclude the matrix $(\overline{t_{ij}})$ can be written as the matrix product $(\overline{t_{ij}}) = (\beta_{ij})(t_{ij})$. Thus,

$$\frac{det(\overline{t_{ij}})}{det(t_{ij})} = det(\beta_{ij}) = det([J_2 J_1^{-1}]_{\boldsymbol{B},\boldsymbol{B}})$$

If J_1 and J_2 are homotopic in $\mathcal{L}_L$, then $det([J_2 J_1^{-1}]_{\boldsymbol{B},\boldsymbol{B}}) > 0$. Thus sign $det(\overline{t_{ij}}) =$ sign $det(t_{ij})$. So J_1 and J_2 are either both orientation preserving or orientation reversing.

The converse is left to you. ■

Homework

Exercise 12.2.5 *Define a relation on $S = \{\boldsymbol{A}|\boldsymbol{A}$ is a basis for $\boldsymbol{R}(Q)\}$ by $\boldsymbol{A} \sim \boldsymbol{B}$ if and only if $det(t_{ij}) > 0$. Prove this is an equivalence relation.*

Exercise 12.2.6 *In Lemma 12.2.3, prove the converse of the statement.*

Next, let's look at the projections P.

Lemma 12.2.4 Combinations of Projections with the Same Range

Let V be a vector space and $P, P' : V \to V$ be two projections with $\boldsymbol{R}(P) = \boldsymbol{R}(P') \neq \boldsymbol{0}$. Then, $P'' = aP + bP'$ for any a and b in $\Re$ is also a projection with $\boldsymbol{R}(P'') = \boldsymbol{R}(P) = \boldsymbol{R}(P')$ if and only if $a + b = 1$.

Proof 12.2.4
$\Longrightarrow$: *If P'' has the same range as P and P', then*

$$\begin{aligned} P'' &= aP + bP' \Longrightarrow (P'')^2 = a^2P^2 + baP'P + abPP' + b^2(P')^2 \\ &= a^2P + baP'P + abPP' + b^2P' \end{aligned}$$

as P and P' are projections. Since $\boldsymbol{R}(P) = \boldsymbol{R}(P')$, we have $P'P = PP' = P = P'$. So

$$(P'')^2 = a^2P + abP' + baP + +b^2P' = a(aP + bP') + b(aP + bP') = (a + b)P''$$

If P'' is a projection, $(P'')^2 = P''$. We conclude $a + b = 1$.

$\Longleftarrow$: *If $a + b = 1$, the previous computation shows $(P'')^2 = P''$. Further, if $\boldsymbol{x} \in \boldsymbol{R}(P) = \boldsymbol{R}(P')$, then*

$$P''(\boldsymbol{x}) = (aP + (1 - a)P')(\boldsymbol{x}) = a\boldsymbol{x} + (1 - a)\boldsymbol{x} = \boldsymbol{x}$$

Hence, P'' is a projection. ■

Now we can apply these results to K_P calculations.

Lemma 12.2.5 Constructing the Inverse K_P for Combinations of Projections

Let P and P' be continuous projections onto $\boldsymbol{N}(L)$. Then, if $P'' = aP + bP'$ with $a + b = 1$, $K_{P''} = aK_P + bK_{P'}$.

Proof 12.2.5
Recall $K_P = L_P^{-1} : \boldsymbol{R}(L) \to dom(L) \cap \boldsymbol{N}(P)$. On $\boldsymbol{R}(L)$, by definition $LK_P = I$. Thus, $L(K_P - K_{P'}) = I - I = \boldsymbol{0}$. This implies $K_P - K_{P'} : \boldsymbol{R}(L) \to \boldsymbol{N}(L)$. Thus, $P(K_P - K_{P'}) =$

$P'(K_P - K_{P'}) = K_P - K_{P'}$. We can rewrite this as $PK_P - PK_{P'} = P'K_P - P'K_{P'}$. But $PK_P = P'K_{P'} = \mathbf{0}$, so we are left with

$$PK_{P'} + P'K_P = \mathbf{0} \tag{12.2}$$

and $P'K_P - P'K_{P'} = K_P - K_{P'}$. Thus, $P'K_P = K_P - K_{P'}$ or

$$K_{P'} = (I - P')K_P \tag{12.3}$$

Apply Equation 12.2 and Equation 12.3 to P'' and P. We find

$$\begin{aligned} K_{P''} &= (I - P'')K_P = (I - aP - bP')K_P = K_P - bP'K_P \\ &= (a+b)K_P - bP'K_P = aK_P + b(I - P')K_P = aK_P + bK_{P'} \end{aligned}$$

■

We are now ready to look more closely at the details of the K_{PQ} construction's possible dependence of P and Q.

Lemma 12.2.6 Properties of the K_{PQ} Construction are Retained under Projection Change

If P and P' are continuous projections onto $\boldsymbol{N}(L)$ and if $I - Q$ and $I - Q'$ are continuous projections onto $\boldsymbol{R}(L)$ with $G : \overline{\boldsymbol{D}} \subset X \rightarrow Z$ satisfying

(i): QG is continuous on $\overline{\boldsymbol{D}}$,

(ii): $QG(\overline{\boldsymbol{D}})$ is bounded,

(iii): $K_{PQ}G$ is compact,

then $K_{P'Q'}$ is compact, $Q'G$ is continuous on $\overline{\boldsymbol{D}}$ and $Q'G(\overline{\boldsymbol{D}})$ is bounded.

Proof 12.2.6

$$K_{P'Q'}G = K_{P'}(I - Q') = K_{P'}(I - Q)G + K_{P'}(Q - Q')G$$

By Equation 12.3 in the proof of Lemma 12.2.5, $K_{P'} = (I - P')K_P$, so

$$\begin{aligned} K_{P'Q'}G &= (I - P')K_P(I - Q) + (I - P')K_P(Q - Q')G \\ &= (I - P')K_{PQ}G + (I - P')K_P(Q - Q')G \end{aligned}$$

By assumption, $K_{PQ}G$ is compact and P' is continuous. Hence, $(I - P')K_{PQ}G$ is compact. Since $\boldsymbol{R}(Q - Q')$ is finite dimensional and K_P is linear, $\tilde{K}_P = K_P|_{\boldsymbol{R}(Q-Q')}$ is continuous. So to show $K_{P'Q'}G$ is continuous, it is enough to show $(Q - Q')G$ is continuous.

The projection Q determines the decomposition $Z = \boldsymbol{R}(L) \oplus \boldsymbol{M}_0$ and the projection Q', the decomposition, $Z = \boldsymbol{R}(L) \oplus \boldsymbol{M}_1$. Let $(\boldsymbol{x}_n) \subset \overline{\boldsymbol{D}}$ satisfy $\boldsymbol{x}_n \rightarrow \boldsymbol{x}$. Then, by our continuity assumptions, $QG(\boldsymbol{x}_n) \rightarrow QG(\boldsymbol{x})$. Let $\boldsymbol{y}_n = G(\boldsymbol{x}_n)$. Then, $\boldsymbol{y}_n = \boldsymbol{y}_{0n} + \boldsymbol{y}_{1n}$ where $\boldsymbol{y}_{0n} \in \boldsymbol{M}_0$ and $\boldsymbol{y}_{1n} \in \boldsymbol{R}(L)$. But we also know $\boldsymbol{y}_n = \hat{\boldsymbol{y}}_{0n} + \hat{\boldsymbol{y}}_{1n}$ where $\hat{\boldsymbol{y}}_{0n} \in \boldsymbol{M}_1$ and $\hat{\boldsymbol{y}}_{1n} \in \boldsymbol{R}(L)$. Further, $G(\boldsymbol{x}) = \boldsymbol{y} = \boldsymbol{y}_0 + \boldsymbol{y}_1 = \hat{\boldsymbol{y}}_0 + \hat{\boldsymbol{y}}_1$ using the same decompositions. We also have $Q(\boldsymbol{y}_n) \rightarrow Q(\boldsymbol{y})$ or $\boldsymbol{y}_{0n} \rightarrow \boldsymbol{y}_0$.

Consider,

$$\lim_{n\to\infty} Q'(\boldsymbol{y_{0n}}) = \lim_{n\to\infty} Q'(\boldsymbol{\hat{y}_{0n}} + \boldsymbol{\hat{y}_{1n}} - \boldsymbol{y_{1n}}) = \lim_{n\to\infty} \boldsymbol{\hat{y}_{0n}}$$

But Q' is continuous, so $\lim_{n\to\infty} Q'(\boldsymbol{y_{0n}}) = Q'(\boldsymbol{y_0})$. Now, $Q'(\boldsymbol{y_0}) = Q'(\boldsymbol{\hat{y}_0} + \boldsymbol{\hat{y}_1} - \boldsymbol{y_1}) = Q'(\boldsymbol{\hat{y}_0})$. We conclude $\lim_{n\to\infty} \boldsymbol{\hat{y}_{0n}} = \boldsymbol{\hat{y}_0}$. or $Q'(\boldsymbol{y_n}) = Q'(\boldsymbol{y_{0n}}) \to Q'(\boldsymbol{y}) = \boldsymbol{y_0}$. Thus, $Q'G$ is continuous and so $K_{P'Q'}G$ is continuous.

To show $K_{P'Q'}G$ is compact, it suffices to show $Q'G(\overline{\boldsymbol{D}})$ is compact. Assume not. Then there is a sequence $(\boldsymbol{x_n}) \subset \overline{\boldsymbol{D}}$ so that $Q'G(\boldsymbol{x_n})$ does not converge for any subsequence. Letting $G(\boldsymbol{x_n}) = \boldsymbol{y_n}$, our assumption implies no subsequence of $(\boldsymbol{y_n})$ converges. By the compactness of $\overline{QG(\overline{\boldsymbol{D}})}$, we know there is a subsequence $(\boldsymbol{x_{n,1}})$ with corresponding subsequence $G(\boldsymbol{x_{n,1}}) = \boldsymbol{y_{n,1}}$ and $\boldsymbol{x} \in \overline{\boldsymbol{D}}$ so that $G(\boldsymbol{x_{n,1}}) = \boldsymbol{y_{n,1}} \to G(\boldsymbol{x}) = \boldsymbol{y}$. From the first part of the proof, we know $Q'(\boldsymbol{y_{n,1}}) \to \boldsymbol{y}$. Thus, $Q'G(\boldsymbol{x_{n,1}}) \to Q'G(\boldsymbol{x})$. This is a contradiction and so our assumption that $Q'G(\overline{\boldsymbol{D}})$ is not compact is wrong.

We have all the pieces to conclude $K_{P'Q'}G$ is compact, $Q'G$ is continuous on $\overline{\boldsymbol{D}}$ and $Q'G(\overline{\boldsymbol{D}})$ is bounded. ■

Homework

These are very technical results, so we will focus on giving you some exercises to illuminate one of them: Lemma 12.2.4.

Exercise 12.2.7 *Let f and g be two linear independent functions in $C([0,1], \|\cdot\|_2)$.*

- *Prove $\{f+g, f-g\}$ is also a linear independent set and the span$\{f,g\}$ = span$\{f+g, f-g\}$.*
- *Prove the orthonormal sets $\{E_1, E_2\}$ and $\{F_1, F_2\}$ obtained from Graham Schmidt Orthogonalization (GSO) of these two sets are not the same.*
- *Define projections P_1 and P_2 by $P_1(x) =< x, E_1 > E_1 + < x, E_2 > E_2$ and $P_2(x) =< x, F_1 > F_1 + < x, F_2 > F_2$. Prove $\boldsymbol{R}(P_1) = \boldsymbol{R}(P_2)$ even though $P_1 \neq P_2$.*

Exercise 12.2.8 *Do the exercise above for $f(t) = 1$ and $g(t) = t$ on $[0,1]$. Use MATLAB to compute the orthonormal bases using GSO.*

Exercise 12.2.9 *Let A and B be two linear independent vectors in $\Re^3$. Then a basis for $\Re^3$ is $\{A, B, A \times B\}$.*

- *Prove $\{A+B, A-B\}$ is also a linear independent set and the span $\{A,B\}$ = span $\{A+B, A-B\}$.*
- *Prove the orthonormal sets $\{E_1, E_2\}$ and $\{F_1, F_2\}$ obtained from Graham Schmidt orthogonalization of these two sets are not the same.*
- *Define projections P_1 and P_2 by $P_1(x) =< x, E_1 > E_1 + < x, E_2 > E_2$ and $P_2(x) =< x, F_1 > F_1 + < x, F_2 > F_2$. Prove $\boldsymbol{R}(P_1) = \boldsymbol{R}(P_2)$ even though $P_1 \neq P_2$.*

Exercise 12.2.10 *Do the exercise above for $A = \begin{bmatrix} 1 \\ 1 \\ 1 \end{bmatrix}$ and $B = \begin{bmatrix} 1 \\ 2 \\ -2 \end{bmatrix}$. Use MATLAB to compute the orthonormal bases using GSO.*

We now have enough results to define L - compact mappings or operators.

Definition 12.2.3 L **Compact Operators**

Let $L : dom(L) \subset X \rightarrow Z$ be linear Fredholm of index zero. Let $G : \overline{\boldsymbol{D}} \subset X \rightarrow Z$ where $\boldsymbol{D}$ is open and bounded satisfy for some continuous projections P and Q so that $\boldsymbol{R}(P) = \boldsymbol{N}(L)$ and $\boldsymbol{N}(Q) = \boldsymbol{R}(L)$ with QG compact and $K_{PQ}G$ is compact. Then, we say G is L - compact.

From our work so far, it is easy to see G is $L - compact$ independent of the choice of the continuous projections P and Q, as long as the other assumptions are satisfied.

12.2.4 The Leray - Schauder Degree for $L + G = 0$ is Independent of P and Q

Now, with all this done, we can show $d_{LS}(I - M, \boldsymbol{D}, \underline{0})$ only depends on L, G, $\boldsymbol{D}$ and the homotopy class of J in $\mathcal{L}_L$.

Theorem 12.2.7 Leray - Schauder Degree Associated with $L + G = 0$ **Depends only on** L**,** G**,** $\boldsymbol{D}$ **and** J

Let X and Z be normed linear spaces. Assume $L : dom(L) \subset X \rightarrow Z$ is linear Fredholm of index 0, $G : \overline{\boldsymbol{D}} \subset X \rightarrow Z$ is L - compact and J in $\mathcal{L}_L$. Let $M = P - (JQ + K_{PQ})G$ and $\boldsymbol{0} \notin (L + G)(dom(L) \cap \partial \boldsymbol{D})$. Then $d_{LS}(I - M, \boldsymbol{D}, \boldsymbol{0})$ depends only on L, G, D and the homotopy class of J.

Proof 12.2.7
Let P, P', Q and Q' be continuous projections with $\boldsymbol{R}(P) = \underline{R}(P') = \boldsymbol{N}(L)$ and $\boldsymbol{N}(Q) = \boldsymbol{N}(Q') = \boldsymbol{R}(L)$. Let J_1 and J_2 be two vector isomorphisms in $\mathcal{L}_L$ that belong to the same equivalence class. Let $\phi : \boldsymbol{R}(Q) \times [0, 1] \rightarrow \boldsymbol{N}(L)$ be the homotopy connecting J_1 and J_2. By Lemma 12.2.4, for $0 \leq t \leq 1$, $P(t) = (1-t)P + tP'$ and $Q(t) = (1-t)Q + tQ'$ are continuous projections with $\boldsymbol{R}(Q(t)) = \boldsymbol{N}(L)$ and $\boldsymbol{N}(Q(t)) = \boldsymbol{R}(L)$. Note, for the Q result, apply the Lemma to $I - Q$ and $I - Q'$. Also, $P(0) = P$, $P(1) = P'$, $Q(0) = Q$ and $Q(1) = Q'$.

By Lemma 12.2.5, $K_{P(t)} = (1 - t)K_P + tK_{P'}$. We know $L(\boldsymbol{x}) + G(\boldsymbol{x}) = \boldsymbol{0}$ is equivalent to $(I - M(\cdot, t))(\boldsymbol{x}) = \boldsymbol{0}$ where $M(\cdot, t) : \overline{\boldsymbol{D}} \times [0, 1] \rightarrow dom(L)$ where $M(\cdot, t) = P(t) - (\phi(\cdot, t)Q(t) + K_{P(t)Q(t)})G$.

By assumption, $\boldsymbol{0} \notin (L + G)(dom(L) \cap \partial \boldsymbol{D})$. Thus, $\boldsymbol{0} \neq (I - M(\cdot, t))(\boldsymbol{x})$ for any $\boldsymbol{x} \in dom(L) \cap \partial \boldsymbol{D}$ and $0 \leq t \leq 1$. In fact, as our remarks from right before Definition 12.2.1 show, any fixed point must lie in $dom(L)$, so we know not only is $\boldsymbol{0} \neq (I - M(\cdot, t))(\boldsymbol{x})$ for any $\boldsymbol{x} \in dom(L) \cap \partial \boldsymbol{D}$, but indeed $\boldsymbol{0} \neq (I - M(\cdot, t))(\boldsymbol{x})$ for any $\boldsymbol{x} \in \partial \boldsymbol{D}$.

Consider $M(\cdot, t)$. We show $M(\cdot, t)$ is compact on $\overline{\boldsymbol{D}} \times [0, 1]$. We have

$$
\begin{aligned}
M(\boldsymbol{x}, t) &= P(t) - (\phi(\cdot, t)Q(t) + K_{P(t)Q(t)})G(\boldsymbol{x}) \\
&= (1-t)P(\boldsymbol{x}) + tP'(\boldsymbol{x}) - \Big(\phi(\cdot, t)((1-t)Q + tQ') + K_{P(t)}(I - Q(t))\Big)G(\boldsymbol{x}) \\
&= (1-t)P(\boldsymbol{x}) + tP'(\boldsymbol{x}) \\
&\quad - \Big(\phi(\cdot, t)((1-t)Q + tQ') + ((1-t)K_P + tK_{P'})(I - Q(t))\Big)G(\boldsymbol{x}) \\
&= (1-t)P(\boldsymbol{x}) + tP'(\boldsymbol{x})
\end{aligned}
$$

$$-\Big(\phi(\cdot,t)((1-t)Q+tQ')+((1-t)K_P+tK_{P'})(I-(1-t)Q-tQ')\Big)G(\boldsymbol{x})$$

By Lemma 12.2.6, all the pieces are continuous; thus, $M(\cdot,t)$ is continuous. The finite dimensionality of the range of P and P' implies it suffices for us to check the compactness of

$$\begin{aligned}
\mathbb{M}(t) &= ((1-t)K_P+tK_{P'})(I-(1-t)Q-tQ')G \\
&= (1-t)K_P((1-t)(I-Q)+t(I-Q'))+tK_{P'}((1-t)(I-Q)+t(I-Q'))G \\
&= (1-t)^2K_P(I-Q)+t(1-t)K_P(I-Q')+t(1-t)K_{P'}(I-Q)+t^2K_{P'}(I-Q')G \\
&= (1-t)^2K_{PQ}G+t(1-t)K_{PQ'}G+t(1-t)KP'QG+t^2K_{P'Q'}G
\end{aligned}$$

By Lemma 12.2.6, all pieces are compact. Hence, $\mathbb{M}(t)$ is compact for $0\le t\le 1$. We conclude $M(\cdot,t)$ is compact also for $\boldsymbol{x}\in\overline{\boldsymbol{D}}$ and $0\le t\le 1$.

Then, by the invariance of Leray - Schauder degree with respect to a compact homotopy,

$$d_{LS}(I-M(\cdot,0),\boldsymbol{D},\boldsymbol{0}) = d_{LS}(I-M(\cdot,1),\boldsymbol{D},\boldsymbol{0})$$

Thus,

$$d_{LS}(I-P+(\phi(\cdot,0)Q+K_{PQ})G,\boldsymbol{D},\boldsymbol{0}) = d_{LS}(I-P+(\phi(\cdot,1)Q+K_{PQ})G,\boldsymbol{D},\boldsymbol{0})$$

or

$$d_{LS}(I-P+(J_1Q+K_{PQ})G,\boldsymbol{D},\boldsymbol{0}) = d_{LS}(I-P'+(J_2Q'+K_{P'Q'})G,\boldsymbol{D},\boldsymbol{0})$$

Hence, for any M obtained in this manner, $d_{LS}(I-M,\boldsymbol{D},\boldsymbol{0})$ is independent of the choice of P, Q and J from the same equivalence class. ■

There is one last thing to do: examine more carefully how $d_{LS}(I-M,\boldsymbol{D},\boldsymbol{0})$ depends on the homotopy class of J in $\mathcal{L}_L$.

Lemma 12.2.8 Dependency of $d_{LS}(I-M,\boldsymbol{D},\boldsymbol{0})$ on Automorphisms of $\boldsymbol{N}(L)$

If $\xi:\boldsymbol{N}(L)\to\boldsymbol{N}(L)$ is a vector isomorphism (i.e. a vector autoisomorphism) and if $M'=P-(\xi JQ+K_{PQ})G$, then $I-M'=(I-P+\xi P)(I-M)$.

Proof 12.2.8

$$\begin{aligned}
(I-P+\xi P)(I-M) &= (I-P+\xi P)(I-P+(JQ+K_{PQ})G) \\
&= (I-P)^2+(I-P)(JQ+K_{PQ})G \\
&\quad +\xi P(I-P)+\xi PJQG+\xi PK_{PQ}G \\
&= (I-P)+(I-P)JQG+(I-P)K_{PQ}G+\xi PJQG \\
&\quad +\xi PK_{PQ}G
\end{aligned}$$

Now, $J:\boldsymbol{R}(Q)\to\boldsymbol{N}(L)$ and so $JQG(\boldsymbol{x})\in\boldsymbol{N}(L)$ implying $(I-P)JQG=0$. Also, $K_P:\boldsymbol{R}(L)\to dom(L)\cap\boldsymbol{N}(P)$. Thus, $\xi PK_{PQ}G=0$. We thus have

$$\begin{aligned}
(I-P+\xi P)(I-M) &= (I-P)+(I-P)K_{PQ}G+\xi PJQG \\
&= (I-P)+\xi PJQG+(I-P)K_{PQ}G
\end{aligned}$$

We also know $(I-P)K_{PQ}G = K_{PQ}G$ *as* $I-P$ *is the identity on* $\boldsymbol{N}(P)$. *So,* $(I-P+\xi P)(I-M) = (I-P)+(\xi PJQ+K_{PQ})G = I-M'$. ■

Using this result, we can prove how $d_{LS}(I-M,\boldsymbol{D},\boldsymbol{0})$ depends on J in $\mathcal{L}_L$. First, let's go over matrix representations for a vector automorphism ϕ again. If $\boldsymbol{A}$ is a basis for $\boldsymbol{N}(L)$ and $\boldsymbol{B}$ is a another basis for $\boldsymbol{N}(L)$, we have $\phi(\boldsymbol{a_j}) = \sum_{i=1}^n c_{ij}\boldsymbol{a_i}$, $\boldsymbol{b_j} = \sum_{i=1}^n d_{ij}\boldsymbol{a_i}$ and

$$\boldsymbol{a_\ell} = \sum_{p=1}^n f_{p\ell}\boldsymbol{b_p} = \sum_{p=1}^n f_{p\ell}\sum_{i=1}^n d_{ip}\boldsymbol{a_i} = \sum_{i=1}^n\left(\sum_{p=1}^n d_{ip}f_{p\ell}\right)\boldsymbol{a_i}$$

This implies $\sum_{p=1}^n d_{ip}f_{p\ell} = \delta_\ell^i$. Hence, the matrix representation for the change of basis from $\boldsymbol{A}$ to $\boldsymbol{B}$, (d_{ij}) and the matrix representation for the change of basis from $\boldsymbol{B}$ to $\boldsymbol{A}$, (f_{ij}), must satisfy $(d_{ij})(f_{ij}) = (f_{ij})(d_{ij}) = I$.

Then, note

$$\begin{aligned}\phi(\boldsymbol{b_j}) &= \sum_{i=1}^n d_{ij}\phi(\boldsymbol{a_i}) = \sum_{i=1}^n d_{ij}\sum_{\ell=1}^n c_{\ell i}\boldsymbol{a_\ell} = \sum_{\ell=1}^n\sum_{i=1}^n c_{\ell i}d_{ij}\boldsymbol{a_\ell}\\ &= \sum_{\ell=1}^n\sum_{i=1}^n c_{\ell i}d_{ij}\sum_{p=1}^n f_{p\ell}\boldsymbol{b_p} = \sum_{p=1}^n f_{p\ell}\left(\sum_{\ell=1}^n\left\{\sum_{i=1}^n c_{\ell i}d_{ij}\right\}\right)\boldsymbol{b_p}\end{aligned}$$

This tells us $[\phi]_{\boldsymbol{B}}$, the matrix representation of ϕ with respect to the basis $\boldsymbol{B}$, is related to $[\phi]_{\boldsymbol{A}}$, the matrix representation of ϕ with respect to the basis $\boldsymbol{A}$, via the equation $[\phi]_{\boldsymbol{B}} = (f_{ij})[\phi]_{\boldsymbol{A}}(d_{ij})$. Thus,

$$\begin{aligned}det([\phi]_{\boldsymbol{B}}) &= det((f_{ij}))det([\phi]_{\boldsymbol{A}})det((d_{ij}))\\ &= det((f_{ij})(d_{ij}))det([\phi]_{\boldsymbol{A}}) = det(I)det([\phi]_{\boldsymbol{A}}) = det([\phi]_{\boldsymbol{A}})\end{aligned}$$

Hence, the determinant of the matrix representation of the vector automorphism ϕ is independent of the choice of basis of $\boldsymbol{N}(L)$. So, we will denote this simply saying $det(\phi)$ without referring to the matrix representation at all. With that said, we can tackle the proof of our next result.

Lemma 12.2.9 Dependency of $d_{LS}(I-M,\boldsymbol{D},\boldsymbol{0})$ on J

If J *and* J' *are in* $\mathcal{L}_L$ *and if* $M' = P-(J'Q+K_{PQ})G$, *then* $d_{LS}(I-M',\boldsymbol{D},\boldsymbol{0}) = sign\ det(J'J^{-1})\ d_{LS}(I-M,\boldsymbol{D},\boldsymbol{0})$.

Proof 12.2.9

By Lemma 12.2.8, $J'J^{-1}$ *is a vector isomorphism from* $\boldsymbol{N}(L)$ *onto* $\boldsymbol{N}(L)$. *Thus,* $I-M' = (I-P+(J'J^{-1})P)(I-M)$. *We now show* $I-P+(J'J^{-1})P$ *is a vector automorphism on* X. *First, it is clearly linear. Next, if* $(I-P+(J'J^{-1})P)(\boldsymbol{x}) = \boldsymbol{0}$, *we want to show* $\boldsymbol{x} = \boldsymbol{0}$ *so that it is a 1-1 mapping. We know* $\boldsymbol{x} = \boldsymbol{x_0}+\boldsymbol{x_1}$ *where* $\boldsymbol{x_0}\in\boldsymbol{N}(L)$ *and* $\boldsymbol{x_1}\in X_0$ *as* $X = \boldsymbol{N}(L)\oplus X_0$. *We see*

$$\begin{aligned}&(I-P+(J'J^{-1})P)(\boldsymbol{x}) = (I-P+(J'J^{-1})P)(\boldsymbol{x_0}+\boldsymbol{x_1})\\ &= (I-P)(\boldsymbol{x_0}+\boldsymbol{x_1})+(J'J^{-1})P(\boldsymbol{x_0}+\boldsymbol{x_1}) = \boldsymbol{x_1}+(J'J^{-1})(\boldsymbol{x_0})\end{aligned}$$

Hence, we want $\boldsymbol{x_1}+(J'J^{-1})(\boldsymbol{x_0}) = \boldsymbol{0}$ *and so* $\boldsymbol{x_1} = \boldsymbol{0}$ *and* $(J'J^{-1})(\boldsymbol{x_0}) = \boldsymbol{x_0}$. *We conclude* $\boldsymbol{x_0} = \boldsymbol{x_1} = \boldsymbol{0}$. *Thus, the mapping is 1-1.*

To show it is onto, if $\boldsymbol{x} = \boldsymbol{x_0}+\boldsymbol{x_1}\in X$, *then* $\boldsymbol{z} = (J(J')-1)(\boldsymbol{x_0})+\boldsymbol{x_1}$ *satisfies*

$$(I-P+(J'J^{-1})P)(\boldsymbol{z}) = (I-P+(J'J^{-1})P)((J(J')-1)(\boldsymbol{x_0})+\boldsymbol{x_1})$$

$$= \boldsymbol{x_1} + (J'J^{-1})P(J'J^{-1})(\boldsymbol{x_0}) = \boldsymbol{x_1} + (J'J^{-1})(J'J^{-1})(\boldsymbol{x_0})$$
$$= \boldsymbol{x_1} + \boldsymbol{x_0} = \boldsymbol{x}$$

Thus, the mapping is onto also. We conclude $I - P + (J'J^{-1})P$ is a vector automorphism on X.

It is straightforward to see $P - (J'J^{-1})P$ is compact on $\overline{\boldsymbol{D}}$. The usual conditions on $I - M$ are assumed to hold; i.e. $\boldsymbol{0} \notin (I - M)(\partial \boldsymbol{D})$. Thus, $(I - P + (J'J^{-1})P)(I - M)(\partial \boldsymbol{D}) \neq \boldsymbol{0}$ as $I - P + (J'J^{-1})P$ is an automorphism of X. Thus, by Theorem 11.3.5, letting $K = P - (J'J^{-1})P$, that

$$d_{LS}((I-K)(I-M), \boldsymbol{D}, \boldsymbol{0}) = \sum_{i=1}^{N} d_{LS}((I-K), \boldsymbol{\Delta_i}, \boldsymbol{0})\, d_{LS}((I-M), \boldsymbol{D}, \boldsymbol{\Delta_i})$$

where $\boldsymbol{\Delta_i}$ are the bounded connected components of $\boldsymbol{\Delta} \setminus (I - M)(\partial \boldsymbol{D})$. Here $\boldsymbol{\Delta}$ is a bounded open subset with $(I - M)(\overline{\boldsymbol{D}}) \subset \boldsymbol{\Delta}$. We know $\boldsymbol{\Delta}$ exists because $\overline{\boldsymbol{D}}$ is bounded and M is compact. We can rewrite this as

$$d_{LS}((I-K)(I-M), \boldsymbol{D}, \boldsymbol{0}) = \sum_{i=1}^{N} d_{LS}((I-K), \boldsymbol{\Delta_i}, \boldsymbol{0})\, d_{LS}((I-M), \boldsymbol{D}, \boldsymbol{y_i})$$

for arbitrary $\boldsymbol{y_i} \in \boldsymbol{\Delta_i}$ as Leray - Schauder degree is constant on components.

Now consider $d_{LS}(I - P + (J'J^{-1})P, \boldsymbol{\Delta_i}, \boldsymbol{0})$. The only solution to $(I - P + (J'J^{-1})P)(\boldsymbol{x}) = \boldsymbol{0}$ is $\boldsymbol{x} = \boldsymbol{0}$ as it is 1-1. Hence, by the existence theorem for Leray - Schauder degree, $d_{LS}(I - P + (J'J^{-1})P, \boldsymbol{\Delta_i}, \boldsymbol{0}) = 0$ unless $\boldsymbol{0} \in \boldsymbol{\Delta_i}$. Let the bounded component that contains $\boldsymbol{0}$ be called $\boldsymbol{\Delta_0}$. We know one exists because $\boldsymbol{0} \notin (I - M)(\partial \boldsymbol{D})$. Choose the corresponding $\boldsymbol{y}$ value from this component to be $\boldsymbol{y_0} = \boldsymbol{0}$. Then, we have

$$d_{LS}((I-K)(I-M), \boldsymbol{D}, \boldsymbol{0}) = d_{LS}((I-K), \boldsymbol{\Delta_0}, \boldsymbol{0})\, d_{LS}((I-M), \boldsymbol{D}, \boldsymbol{0})$$

But $(I - M') = (I - K)(I - M)$ so

$$d_{LS}((I-M'), \boldsymbol{D}, \boldsymbol{0}) = d_{LS}((I-K), \boldsymbol{\Delta_0}, \boldsymbol{0})\, d_{LS}((I-M), \boldsymbol{D}, \boldsymbol{0})$$

By the excision property of Leray - Schauder degree, since $\boldsymbol{0}$ is the only fixed point of $P - (J'J^{-1})P$, if $B(\boldsymbol{0}, 1)$ is the ball of radius 1 in X about $\boldsymbol{0}$,

$$d_{LS}((I-M'), \boldsymbol{D}, \boldsymbol{0}) = d_{LS}((I-K), B(\boldsymbol{0}, 1), \boldsymbol{0})\, d_{LS}((I-M), \boldsymbol{D}, \boldsymbol{0})$$

By Theorem 11.3.2, since $(I - P + (J'J^{-1})P)(B(\boldsymbol{0}, 1)) \subset \boldsymbol{N}(L)$ and $\boldsymbol{N}(L)$ is closed, $d_{LS}((I - K), B(\boldsymbol{0}, 1), \boldsymbol{0}) = d_B((I - K)|_{\boldsymbol{N}(L)}, B(\boldsymbol{0}, 1) \cap \boldsymbol{N}(L), \boldsymbol{0})$. But $P|_{\boldsymbol{N}(L)} = I$. So, $d_{LS}((I - K), B(\boldsymbol{0}, 1), \boldsymbol{0}) = d_B((J'J^{-1}), B(\boldsymbol{0}, 1) \cap \boldsymbol{N}(L), \boldsymbol{0})$. But $d_B((J'J^{-1}), B(\boldsymbol{0}, 1) \cap \boldsymbol{N}(L), \boldsymbol{0}) = sign\ det(J'J^{-1})$. We conclude

$$d_{LS}((I-M'), \boldsymbol{D}, \boldsymbol{0}) = sign\ det(J'J^{-1})\, d_{LS}((I-M), \boldsymbol{D}, \boldsymbol{0})$$

If J' and J are in the same homotopy class, $det(J'J^{-1}) > 0$ and we have $d_{LS}((I - M'), \boldsymbol{D}, \boldsymbol{0}) = d_{LS}((I - M), \boldsymbol{D}, \boldsymbol{0})$. ■

Homework

Exercise 12.2.11 *In Lemma 12.2.9, prove $P - (J'J^{-1})P$ is compact on $\overline{\boldsymbol{D}}$.*

Exercise 12.2.12 *Let* $A = \begin{bmatrix} 1 \\ 1 \\ 1 \end{bmatrix}$ *and* $B = \begin{bmatrix} 1 \\ 2 \\ -2 \end{bmatrix}$. *Then a basis for* $\Re^3$ *is* $\{A, B, A \times B\}$. *We know* $\{A+B, A-B\}$ *is also a linear independent set and the span* $\{A, B\} =$ *span* $\{A+B, A-B\}$. *Let* $\{E_1, E_2\}$ *and* $\{F_1, F_2\}$ *be the orthonormal bases obtained from Graham Schmidt orthogonalization of these two sets. Let* J_1 : *span* $\{E_1, E_2\} \rightarrow$ *span* $\{E_1, E_2\}$ *be the vector isomorphism defined by* $J(x_1 E_1 + x_2 E_2) = 3x_1 E_1 + 4x_2 E_2$. *Let* J_2 : *span* $\{E_1, E_2\} \rightarrow$ *span* $\{E_1, E_2\}$ *be the vector isomorphism defined by* $J(x_1 E_1 + x_2 E_2) = 5x_1 E_1 + 3x_2 E_2$. *Work out all the details of the arguments presented in Lemma 12.2.2 for these vector isomorphisms.*

Exercise 12.2.13 *Let* $A = \begin{bmatrix} 1 \\ 2 \\ 1 \end{bmatrix}$ *and* $B = \begin{bmatrix} 3 \\ 1 \\ -1 \end{bmatrix}$. *Then a basis for* $\Re^3$ *is* $\{A, B, A \times B\}$. *Let* $\{E_1, E_2\}$ *be the orthonormal basis obtained from Graham Schmidt orthogonalization of this set. Let* J_1 : *span* $\{E_1, E_2\} \rightarrow$ *span* $\{E_1, E_2\}$ *be the vector isomorphism defined by* $J(x_1 E_1 + x_2 E_2) = 3x_1 E_1 + 4x_2 E_2$. *Work out all the details of the arguments presented in Lemma 12.2.8 for this vector isomorphism.*

12.2.5 The Definition of Coincidence Degree

If we choose orientations for $\boldsymbol{R}(Q)$ and $\boldsymbol{N}(L)$, then by Lemma 12.2.3, J_1 and J_2 are homotopic if and only if they both preserve or reverse orientation. We now have all the pieces to define what we will call coincidence degree.

Definition 12.2.4 Coincidence Degree: $d_C((L, G), \boldsymbol{D}) = d_{LS}(I - M, \boldsymbol{D}, \boldsymbol{0})$

Let X *and* Z *be normed linear spaces. Let* $L : dom(L) \subset X \rightarrow Z$ *be linear Fredholm of index* 0. *Let* $\boldsymbol{D}$ *be an open bounded subset of* X. *Let* $G : \overline{\boldsymbol{D}} \subset X \rightarrow Z$ *satisfy for some continuous projections* P *and* Q *with* $\boldsymbol{R}(P) = \boldsymbol{N}(L)$ *and* $\boldsymbol{N}(Q) = \boldsymbol{R}(L)$, *the following conditions:*

(i): QG *is compact.*

(ii): $K_{PQ}G$ *is compact.*

Let orientations be chosen for $\boldsymbol{R}(Q)$ *and* $\boldsymbol{N}(L)$. *Let* J *be an orientation preserving vector isomorphism from* $\boldsymbol{R}(Q)$ *to* $\boldsymbol{N}(L)$. *Then, the set of solutions to* $(L + G)(\boldsymbol{x}) = \boldsymbol{0}$ *is identical to the set of fixed points of* $M = P - (JQ + K_{PQ})G$ *and by Theorem 12.2.7 and Lemma 12.2.9,* $d_{LS}(I - M, \boldsymbol{D}, \boldsymbol{0})$ *is independent of the choice of* P, Q *and* J *if* J *preserves orientations and depends only on* L, G *and* D. *We define this degree to be the coincidence degree of the pair* (L, G) *on* $\boldsymbol{D}$ *and denote it by* $d_C((L, G), \boldsymbol{D}) = d_{LS}(I - M, \boldsymbol{D}, \boldsymbol{0})$.

Comment 12.2.1 *If* $X = Z$ *and* $L - I$, *we see* L *is linear Fredholm of index zero with* $\boldsymbol{N}(L) = \boldsymbol{0}$, $\boldsymbol{R}(L) = X$. *Thus,* $\boldsymbol{R}(P) = \boldsymbol{0}$ *implying* $P = 0$ *and* $\boldsymbol{N}(Q) = X$ *implying* $Q = 0$. *The* L *- compactness of* G *reduces to the compactness of* G. *We see* $L_P = L^{-1}$, $K_{PQ} = L^{-1} = I$ *and* $I - M = I - P + (JQ + K_{PQ})G = I - G$. *So if* $(I_M)(\partial \boldsymbol{D}) \neq \boldsymbol{0}$, $d_C((I, G), \boldsymbol{D}) = d_{LS}(I + G, \boldsymbol{D}, \boldsymbol{0}) = d_{LS}(I - (-G), \boldsymbol{D}, \boldsymbol{0})$. *In this case, coincidence degree reduces to the usual Leray - Schauder degree of* $I + G$.

Comment 12.2.2 *Applying the theory to* $L - G = L + (-G)$ *with* $L = I$ *leads to a similar conclusion:* $d_C((I, G), \boldsymbol{D}) = d_{LS}(I - G, \boldsymbol{D}, \boldsymbol{0})$. *In this case, coincidence degree reduces to the usual Leray - Schauder degree of* $I - G$.

Comment 12.2.3 *If the mapping F is split as $F = L_1 + G_1$ and $F = L_2 + G_2$ where both L_1 and L_2 are linear Fredholm of index zero, and G_1 is L_1 - compact and G_2 is L_2 - compact, we can show $d_C((L_1, G_1), \boldsymbol{D}) = d_C((L_2, G_2), \boldsymbol{D})$. Hence, if we find two such decompositions of F, we can be confident the degree value we find does not depend on the choice of decomposition. We will prove this in Section 12.5.*

Homework

Exercise 12.2.14 *For the Picard problem in Exercise 12.2.1, explain in detail how $d_C((L, G), \boldsymbol{D})$ is defined.*

Exercise 12.2.15 *For the periodic problem defined in Exercise 12.2.2, explain in detail how $d_C((L, G), \boldsymbol{D})$ is defined.*

12.3 Properties of Coincidence Degree

Coincidence degree inherits many properties from Leray - Schauder degree. We can state some of them quickly in a catch-all theorem. First, to make the theorem more compact, let's collect the standard setup into a definition.

Definition 12.3.1 Coincidence Degree Standard Setup

X and Z are normed linear spaces. We assume:

- *$L : dom(L) \subset X \to Z$ is linear Fredholm of index 0 and $\boldsymbol{D}$ be an open bounded subset of X.*
- *$G : \overline{\boldsymbol{D}} \subset X \to Z$ is L - compact with associated continuous projections P and Q with $\boldsymbol{R}(P) = \boldsymbol{N}(L)$ and $\boldsymbol{N}(Q) = \boldsymbol{R}(L)$.*
- *Orientations are chosen for $\boldsymbol{R}(Q)$ and $\boldsymbol{N}(L)$ and J is an orientation preserving vector isomorphism from $\boldsymbol{R}(Q)$ to $\boldsymbol{N}(L)$.*
- *$M = P - (JQ + K_{PQ})G$, in which case the set of solutions to $(L + G)(\boldsymbol{x}) = \boldsymbol{0}$ is identical to the set of fixed points of M, or let $M = P + (JQ + K_{PQ})G$, in which case the set of solutions to $(L - G)(\boldsymbol{x}) = \boldsymbol{0}$ is identical to the set of fixed points of M.*
- *In either case, $(I - M)(dom(L)) \cap \partial\boldsymbol{D} \neq \boldsymbol{0}$.*

Then, by Theorem 12.2.7 and Lemma 12.2.9, $d_{LS}(I - M, \boldsymbol{D}, \boldsymbol{0})$ is independent of the choice of P, Q and J if J preserves orientations and depends only on L, G and D.

Thus, coincidence degree of the pair (L, G) on $\boldsymbol{D}$ is well-defined and denoted by $d_C((L, G), \boldsymbol{D}) = d_{LS}(I - M, \boldsymbol{D}, \boldsymbol{0})$.

We can then state our first properties of coincidence degree theorem nicely. These properties follow from the properties of Leray - Schauder degree we have proved in our earlier discussions as coincidence degree is an appropriate Leray - Schauder degree. However, as usual, there are details to consider, and we think it is not helpful to your education to simply say it is an easy set of exercises.

Hence, we will go over the arguments carefully so you can see how to make the appropriate modifications.

Theorem 12.3.1 Properties of Coincidence Degree

Assume the setup of Definition 12.3.1. Then, we have:

1. **Existence Theorem**: *if $d_C((L,G),\boldsymbol{D}) \neq 0$, then there is an $\boldsymbol{x} \in dom(L) \cap \boldsymbol{D}$ so that $(L+G)(\boldsymbol{x}) = \boldsymbol{0}$.*
2. **Excision Property**: *If $\boldsymbol{D_0} \subset \boldsymbol{D}$ is open so that $(L+G)^{-1}(\boldsymbol{0}) \subset \boldsymbol{D_0}$, then $d_C((L,G),\boldsymbol{D}) = d_C((L,G),\boldsymbol{D_0})$.*
3. **Additivity Property** *If $\boldsymbol{D} = \cup_{i=1}^{N} \boldsymbol{D_i}$, $\boldsymbol{D_i} \cap \boldsymbol{D_j} = \delta_j^i$ with each $\boldsymbol{D_i}$ open, then $d_C((L,G),\boldsymbol{D}) = \sum_{i=1}^{n} d_C((L,G),\boldsymbol{D_i})$.*
4. **Generalized Borsuk Theorem** *If $\boldsymbol{D}$ is symmetric with respect to $\boldsymbol{0}$ and if $G(-\boldsymbol{x}) = -G(\boldsymbol{x})$ in $\boldsymbol{D}$, then $d_C((L,G),\boldsymbol{D})$ is odd.*

We can say similar things about the case $L-G$, although we do not state this case explicitly.

Proof 12.3.1
Since $d_C((L,G),\boldsymbol{D}) = d_{LS}(I-M,\boldsymbol{D},\boldsymbol{0})$, all these properties follow from the corresponding properties of the Leray - Schauder degree. The case of $L-G$ uses the same argument. ■

Let's look at some other properties. It is very useful to have an invariance under homotopy result. In general, for these results, we assume the setup of Definition 12.3.1.

Theorem 12.3.2 Invariance with respect to Homotopy for Coincidence Degree

Let L is linear Fredholm of index 0 and let $H(\cdot,t) : \overline{\boldsymbol{D}} \times [0,1] \rightarrow Z$ be L - compact in $\overline{\boldsymbol{D}} \times [0,1]$. Assume $(L-H(\cdot,t))(dom(L) \cap \partial\boldsymbol{D}) \neq \boldsymbol{0}$ for $0 \leq t \leq 1$. Then $d_C((L,H(\cdot,t)),\boldsymbol{D})$ is independent of t in $[0,1]$.

Proof 12.3.2
$d_C((L,H(\cdot,t)),\boldsymbol{D}) = d_{LS}(I-M(\cdot,t),\boldsymbol{D},\boldsymbol{0})$ where $M(\cdot,t) = P-(JQ+K_{PQ})H(\cdot,t)$. $M(\cdot,t)$ is compact and $(I-M(\cdot,t))(dom(L) \cap \partial\boldsymbol{D}) \neq \boldsymbol{0}$ for $0 \leq t \leq 1$. By invariance under homotopy for Leray - Schauder degree, the result follows. ■

We need to show the value of the degree only depends on what happens at the boundary.

Theorem 12.3.3 Dependence Only on Boundary Values for Coincidence Degree

$d_C((L,N),\boldsymbol{D})$ depends only on L, $\boldsymbol{D}$ and N on $\partial\boldsymbol{D}$.

Proof 12.3.3
If $N = N'$ on $\partial\boldsymbol{D}$, let $H(\cdot,t) = (1-t)N + tN'$. Then, $H(\cdot,t)$ is L - compact on $\overline{\boldsymbol{D}} \times [0,1]$ and for all $\boldsymbol{x} \in \partial\boldsymbol{D}$, $(I-H(\cdot,t))(\boldsymbol{x}) = \boldsymbol{0}$ implies $(I-(1-t)N-tN')(\boldsymbol{x}) = \boldsymbol{0}$. Since $N|_{\partial\boldsymbol{D}} = N'|_{\partial\boldsymbol{D}}$, this would force $(I-N)(\boldsymbol{x}) = \boldsymbol{0}$ which is not possible. Hence, $(I-H(\cdot,t))(\boldsymbol{x}) \neq \boldsymbol{0}$ on $dom(L) \cap \partial\boldsymbol{D}$ and $0 \leq t \leq 1$. By invariance under homotopy, $d_C((L,N),\boldsymbol{D}) = d_C((L,N'),\boldsymbol{D})$. ■

Homework

Exercise 12.3.1 *In Theorem 12.3.1, show all the details of the proof of the* **Existence Theorem**: *If $d_C((L,G),\boldsymbol{D}) \neq 0$, then there is an $\boldsymbol{x} \in dom(L) \cap \boldsymbol{D}$ so that $(L+G)(\boldsymbol{x}) = \boldsymbol{0}$.*

Exercise 12.3.2 *In Theorem 12.3.1, show all the details of the proof of the* **Excision Property***: If* $\boldsymbol{D_0} \subset \boldsymbol{D}$ *is open so that* $(L+G)^{-1}(\boldsymbol{0}) \subset \boldsymbol{D_0}$*, then* $d_C((L,G),\boldsymbol{D}) = d_C((L,G),\boldsymbol{D_0})$.

Exercise 12.3.3 *In Theorem 12.3.1, show all the details of the proof of the* **Additivity Property***: If* $\boldsymbol{D} = \cup_{i=1}^N \boldsymbol{D_i}$*,* $\boldsymbol{D_i} \cap \boldsymbol{D_j} = \delta_j^i$ *with each* $\boldsymbol{D_i}$ *open, then* $d_C((L,G),\boldsymbol{D}) = \sum_{i=1}^n d_C((L,G),\boldsymbol{D_i})$.

Exercise 12.3.4 *In Theorem 12.3.1, show all the details of the proof of the* **Generalized Borsuk Theorem***: If* $\boldsymbol{D}$ *is symmetric with respect to* $\boldsymbol{0}$ *and if* $G(-\boldsymbol{x}) = -G(\boldsymbol{x})$ *in* $\boldsymbol{D}$*, then* $d_C((L,G),\boldsymbol{D})$ *is odd.*

In a little bit, we will need a Brouwer degree on 0 dimensional spaces; i.e. $\{\boldsymbol{0}\}$ which we usually just designate at $\boldsymbol{0}$. For such a space, we only have the identity mapping $I : \boldsymbol{0} \to \boldsymbol{0}$ and so we define $d_B(I, \{\boldsymbol{0}\}, \boldsymbol{0}) = 1$. Note, the notation is strange as we have the set $\boldsymbol{D}$ being $\{\boldsymbol{0}\}$ and the point $\boldsymbol{0}$. This clearly agrees with our usual meaning of Brouwer degree. Further, set $d_B(I, \emptyset, \boldsymbol{0}) = 0$. Note, by additivity, for any mapping F, we need to have $d_B(F, \boldsymbol{D}, \boldsymbol{0}) = d_B(F, \emptyset, \boldsymbol{0}) + d_B(F, \boldsymbol{D}, \boldsymbol{0})$ which implies we should set $d_B(F, \emptyset, \boldsymbol{0}) = 0$. So, for any finite dimensional space, we set $d_B(I, \emptyset, \boldsymbol{0}) = 0$.

Now, let's look at solutions to homotopy equations.

Lemma 12.3.4 Equivalencies for Sets of Solutions to Homotopies for Coincidence Degree

Let $H(\cdot,t) : \overline{\boldsymbol{D}} \times [0,1] \to Z$ *be* L *- compact in* $\overline{\boldsymbol{D}} \times [0,1]$*. Let* $H(\cdot,1) = G$ *and let* $\boldsymbol{y} \in \boldsymbol{R}(L)$*. Consider the family of solutions to* $L(\boldsymbol{x}) = tH(\boldsymbol{x},t) + \boldsymbol{y}$ *for* $0 < t \le 1$*. Then every solution to this family is a solution to* $L(\boldsymbol{x}) = QH(\boldsymbol{x},t) + t(I-Q)H(\boldsymbol{x},t) + \boldsymbol{y}$ *for* $0 < t \le 1$*. Further, if* $t = 0$*, the solutions to both families match. Indeed,* $QH(\boldsymbol{x},0) = \boldsymbol{0}$ *and we have* $L(\boldsymbol{x}) = \boldsymbol{y}$*.*

Proof 12.3.4
Let $t \in [0,1]$*. Then, since* $\boldsymbol{N}(Q) = \boldsymbol{R}(L)$ *and* $\boldsymbol{y} \in \boldsymbol{R}(L)$*,* $0 = QL(\boldsymbol{x}) = tH(\boldsymbol{x},t) + Q(\boldsymbol{y}) = tH(\boldsymbol{x},t)$*. Thus,* $QH(\boldsymbol{x},t) = \boldsymbol{0}$*. Then,* $L(\boldsymbol{x}) = (I-Q)L(\boldsymbol{x}) = t(I-Q)H(\boldsymbol{x},t) + (I-Q)(\boldsymbol{y})$ *or* $L(\boldsymbol{x}) = t(I-Q)H(\boldsymbol{x},t) + \boldsymbol{y}$*. But since* $QH(\boldsymbol{x},t) = \boldsymbol{0}$*, this is the same as* $L(\boldsymbol{x}) = QH(\boldsymbol{x},t) + t(I-Q)H(\boldsymbol{x},t) + \boldsymbol{y}$*. This shows a solution to the first family is a solution to the second family.*

If $t = 0$*, the first family is the set of solutions to* $L(\boldsymbol{x}) = \boldsymbol{y}$ *and the second family is the set of solutions to* $L(\boldsymbol{x}) = QH(\boldsymbol{x},0)$*. If* $L(\boldsymbol{x}) = QH(\boldsymbol{x},0) + \boldsymbol{y}$*, this says* $QH(\boldsymbol{x},0) \in \boldsymbol{R}(L)$*. Hence,* $QH(\boldsymbol{x},0) = \boldsymbol{0}$ *and we have* $L(\boldsymbol{x}) = \boldsymbol{y}$*. So at* $t = 0$*, a solution to the second family is also a solution to the first family.* ■

We now prove the continuation theorem for coincidence degree.

Theorem 12.3.5 Continuation Theorem for Coincidence Degree

Let L *be linear Fredholm of index* 0*. Let* $H(\cdot,t)$ *be* L *- compact on* $\overline{\boldsymbol{D}} \times [0,1]$*. Assume the setup of Definition 12.3.1 holds for* $H(\cdot,1) = G$*. Further, assume:*

(i): $L(\boldsymbol{x}) \neq tH(\boldsymbol{x},t) + \boldsymbol{y}$ *for all* $\boldsymbol{x} \in dom(L) \cap \partial\boldsymbol{D}$ *and* $0 < t < 1$*.*

(ii): $QH(\boldsymbol{x},0) \neq \boldsymbol{0}$ *for all* $\boldsymbol{x} \in L^{-1}(\boldsymbol{y}) \cap \partial\boldsymbol{D}$*.*

(iii): $d_B(JQH(\boldsymbol{x},0)|_{L^{-1}(\boldsymbol{y})}, \boldsymbol{D} \cap L^{-1}(\boldsymbol{y}), \boldsymbol{0}) \neq 0$

Then, for all $0 \le t \le 1$*, if* $\boldsymbol{y} \in \boldsymbol{R}(L)$*,* $L(\boldsymbol{x}) = tH(\boldsymbol{x},t) + \boldsymbol{y}$ *has at least one solution in* $\boldsymbol{D}$ *and* $L(\boldsymbol{x}) = G(\boldsymbol{x}) + \boldsymbol{y}$ *has at least one solution in* $\overline{\boldsymbol{D}}$*.*

Proof 12.3.5

Recall, $X = \boldsymbol{N}(L) \oplus \boldsymbol{X_0}$*. Note*

$$L^{-1}(\boldsymbol{y}) = \{\boldsymbol{x} \in X | L(\boldsymbol{x}) = \boldsymbol{y}\} = \{\boldsymbol{x_0} + \boldsymbol{x_1} | \boldsymbol{x_0} \in \boldsymbol{N}(L),\ \boldsymbol{x_1} \in \boldsymbol{X_0},\ L(\boldsymbol{x_1}) = \boldsymbol{y}\}$$

Since $H(\cdot, t)$ *is* L *- compact for all* $0 \le t \le 1$, $QH(\cdot, 0)$ *is compact and continuous. Further,* L *is 1-1 on* $\boldsymbol{X_0}$*, and so there is a unique* $\boldsymbol{x_1} \in \boldsymbol{X_0}$ *so that* $L(\boldsymbol{x_1}) = \boldsymbol{y}$*. Hence,* $L^{-1}(\boldsymbol{y}) = \boldsymbol{x_1} + \boldsymbol{N}(L)$ *telling us* $L^{-1}(\boldsymbol{y})$ *is a finite dimensional translate of a subspace in* X*; i.e. an affine subspace of* X*. The Brouwer degree* $d_B(QH(\cdot, 0)|_{L^{-1}(\boldsymbol{y})}, \boldsymbol{D} \cap L^{-1}(\boldsymbol{y}), \boldsymbol{0})$ *is defined in the usual manner.*

Let's apply invariance under homotopy using the L *- compact homotopy* $M(\boldsymbol{x}, t) = QH(\boldsymbol{x}, t) + t(I - Q)H(\boldsymbol{x}, t) + \boldsymbol{y}$*. Now,* $QM(\boldsymbol{x}, t) = QH(\boldsymbol{x}, t)$ *as* $\boldsymbol{y} \in \boldsymbol{R}(L)$ *and*

$$\begin{aligned} K_{PQ}M(\boldsymbol{x}, t) &= K_P(I - Q)QH(\boldsymbol{x}, t) + tK_P(I - Q)(I - Q)H(\boldsymbol{x}, t) + K_P(I - Q)(\boldsymbol{y}) \\ &= tK_{PQ}H(\boldsymbol{x}, t) + K_P(\boldsymbol{y}) \end{aligned}$$

Since $\underline{y}$ *is fixed and* $H(\cdot, t)$ *is* L *- compact for* $0 \le t \le 1$ *and* $\boldsymbol{x} \in \overline{\boldsymbol{D}}$*, we have* $QM(\cdot, t)$ *is compact and* $K_{PQ}M(\cdot, t)$ *is also compact. By Lemma 12.3.4, for all* $t \in (0, 1]$*, the set of solutions to* $L(\boldsymbol{x}) = tH(\boldsymbol{x}, t) + \boldsymbol{y}$ *is equivalent to the set of solutions to* $L(\boldsymbol{x}) = QH(\boldsymbol{x}, t) + t(I - Q)H(\boldsymbol{x}\boldsymbol{t}, t) + \boldsymbol{y} = M(\boldsymbol{x}, t)$*. By assumption (i),* $L(\boldsymbol{x}) \ne tH(\boldsymbol{x}, t) + \boldsymbol{y}$ *for all* $\boldsymbol{x} \in dom(L) \cap \partial\boldsymbol{D}$ *and* $t \in (0, 1)$*. Hence,* $M(\boldsymbol{x}, t) \ne L(\boldsymbol{x})$ *for all* $\boldsymbol{x} \in dom(L) \cap \partial\boldsymbol{D}$ *and* $t \in (0, 1)$*. If* $t = 0$, $M(\boldsymbol{x}, 0) = QH(\boldsymbol{x}, 0) + \boldsymbol{y}$ *and again by Lemma 12.3.4,* $L(\boldsymbol{x}) = M(\boldsymbol{x}, 0)$ *is equivalent to* $L(\boldsymbol{x}) = \boldsymbol{y}$ *and* $QH(\boldsymbol{x}, 0) = 0$*. This says* $QH(\boldsymbol{x}, 0) = 0$ *and* $\boldsymbol{x} \in L^{-1}(\boldsymbol{y})$*. By assumption (ii),* $QH(\boldsymbol{x}, 0) \ne 0$ *for all* $\boldsymbol{x} \in L^{-1}(\boldsymbol{y})$*. We conclude* $M(\boldsymbol{x}, 0) \ne L(\boldsymbol{x})$ *for all* $\boldsymbol{x} \in L^{-1}(\boldsymbol{y})$*.*

From the above, we can say $L(\boldsymbol{x}) \ne M(\boldsymbol{x}, t)$ *for all* $\boldsymbol{x} \in dom(L) \cap \partial\boldsymbol{D}$ *and* $t \in [0, 1)$*. Now, if there was* $\boldsymbol{x} \in dom(L) \cap \partial\boldsymbol{D}$ *with* $L(\boldsymbol{x}) = H(\boldsymbol{x}, 1) + \boldsymbol{y} = G(\boldsymbol{x}) + \boldsymbol{y}$*, then we have a solution in* $\overline{\boldsymbol{D}}$ *and the last part of this theorem is true. However, if this is not true, we have* $L(\boldsymbol{x}) \ne M(\boldsymbol{x}, t)$ *for all* $\boldsymbol{x} \in dom(L) \cap \partial\boldsymbol{D}$ *and* $t \in [0, 1]$*. Now, we can apply Invariance under homotopy for coincidence degree to give* $d_C((L, M(\cdot, t)), \boldsymbol{D})$ *is independent of* t *in* $[0, 1]$*. Hence,* $d_C((L, M(\cdot, 0)), \boldsymbol{D}) = d_C((L, M(\cdot, 1)), \boldsymbol{D})$*. Thus,*

$$d_C((L, QH(\cdot, 0) + \boldsymbol{y}), \boldsymbol{D}) = d_C((L, QH(\cdot, t) + (I - Q)H(\cdot, t) + \boldsymbol{y}), \boldsymbol{D})$$

But

$$\begin{aligned} d_C((L, QH(\cdot, 0) + \boldsymbol{y}), \boldsymbol{D}) &= d_{LS}((I - P - (JQ + K_{PQ})(QH(\cdot, 0) + \boldsymbol{y})), \boldsymbol{D}, \boldsymbol{0}) \\ &= d_{LS}((I - P - JQ^2H(\cdot, 0) \\ &\quad -JQ(\boldsymbol{y}) - K_{PQ}QH(\cdot, 0) - K_{PQ}(\boldsymbol{y})), \boldsymbol{D}, \boldsymbol{0}) \\ &= d_{LS}(I - P - JQH(\cdot, 0) - K_P(\boldsymbol{y}), \boldsymbol{D}, \boldsymbol{0}) \end{aligned}$$

and so

$$d_C((L, M(\cdot, t)), \boldsymbol{D}) = d_{LS}((I - P - JQH(\cdot, 0) - K_P(\boldsymbol{y})), \boldsymbol{D}, \boldsymbol{0})$$

Case 1*: If* $\boldsymbol{N}(L) = \boldsymbol{0}$*, in this case,* $P = Q = 0$, $K_{PQ} = L^{-1}$ *and so*

$$d_{LS}((I - P - JQH(\cdot, 0) - K_P(\boldsymbol{y})), \boldsymbol{D}, \boldsymbol{0}) = d_{LS}((I - L^{-1}(\boldsymbol{y})), \boldsymbol{D}, \boldsymbol{0})$$

Assumption (ii) wants $QH(\boldsymbol{x}, 0) \ne \boldsymbol{0}$ *for all* $\boldsymbol{x} \in L^{-1}(\boldsymbol{y}) \cap \partial\boldsymbol{D}$*. In this case, this implies* $L^{-1}(\boldsymbol{y}) \cap \partial\boldsymbol{D} = \emptyset$*. Now assumption (iii) wants* $d_B(JQH(\boldsymbol{x}, 0)|_{L^{-1}(\boldsymbol{y})}, \boldsymbol{D} \cap L^{-1}(\boldsymbol{y}), \boldsymbol{0}) \ne 0$*. Then, by the remarks we made on how we define the Brower degree on a zero dimensional space, if* $\boldsymbol{D} \cap$

$\cap L^{-1}(\boldsymbol{y}) = \emptyset$, this becomes $d_B(0, \emptyset, \mathbf{0}) \neq 0$ which is not possible. So we must have $\boldsymbol{D} \cap \cap L^{-1}(\boldsymbol{y}) =/ \emptyset$. Since L is invertible here, $L^{-1}(\boldsymbol{y}) = \boldsymbol{x} \in domL \cap \boldsymbol{D}$. We conclude

$$d_{LS}(I - L^{-1}(\boldsymbol{y}), \boldsymbol{D}, \mathbf{0}) \quad = \quad d_{LS}(I, \boldsymbol{D} + L^{-1}(\boldsymbol{y}), \mathbf{0}) = d_{LS}(I, \boldsymbol{D}, L^{-1}(\boldsymbol{y}))$$

Since $L^{-1}(\boldsymbol{y}) \in \boldsymbol{D}$, we see $d_{LS}(I, \boldsymbol{D}, L^{-1}(\boldsymbol{y})) = \pm 1 \neq 0$.

By the existence theorem for coincidence degree, since

$$d_C((L, M(\cdot, t)), \boldsymbol{D}) \quad = \quad \pm 1$$

for all $t \in [0,1]$, there is $\boldsymbol{x} \in domL \cap \boldsymbol{D}$ so $L(\boldsymbol{x}) = M(\boldsymbol{x}, t)$ for all $t \in [0,1]$. Then, by Lemma 12.3.4, $L(\boldsymbol{x}) = tH(\boldsymbol{x}, t) + \boldsymbol{y}$ for all $t \in [0,1]$, has a solution in particular at $t = 1$ so that $L(\boldsymbol{x}) = G(\boldsymbol{x}) + \boldsymbol{y}$ has a solution in $\boldsymbol{D}$.

Case 2*: We still assume there is no solution to $L(\boldsymbol{x}) = H(\boldsymbol{x}, 1) + \boldsymbol{y} = G(\boldsymbol{x}) + \boldsymbol{y}$ in $domL \cap \boldsymbol{D}$. So all the homotopy invariance arguments still hold. Here $\boldsymbol{N}(L) \neq \mathbf{0}$. By invariance under translation,*

$$\begin{aligned}
&d_{LS}(I - P - JQH(\cdot, 0) - K_P(\boldsymbol{y}), \boldsymbol{D}, \mathbf{0}) \\
&= d_{LS}(I - P - JQH(\cdot + K_P(\boldsymbol{y}), 0), \boldsymbol{D} - K_P(\boldsymbol{y}), \mathbf{0})
\end{aligned}$$

Note, for any $\boldsymbol{x} \in \boldsymbol{D} + K_P(\boldsymbol{y})$, $P + JQH(\cdot + K_P(\boldsymbol{y}), 0)(\boldsymbol{x}) \in \boldsymbol{N}(L)$. Now apply Theorem 11.3.1, to conclude

$$\begin{aligned}
&d_{LS}(I - P - JQH(\cdot + K_P(\boldsymbol{y}), 0), \boldsymbol{D} - K_P(\boldsymbol{y}), \mathbf{0}) = \\
&\quad d_B(I - P - JQH(\cdot + K_P(\boldsymbol{y}), 0)|_{\boldsymbol{N}(L)}, \boldsymbol{D} - K_P(\boldsymbol{y}) \cap \boldsymbol{N}(L), \mathbf{0})
\end{aligned}$$

But $P = I$ on $\boldsymbol{N}(L)$ and so

$$\begin{aligned}
&d_B(I - P - JQH(\cdot + K_P(\boldsymbol{y}), 0)|_{\boldsymbol{N}(L)}, \boldsymbol{D} - K_P(\boldsymbol{y}) \cap \boldsymbol{N}(L), \mathbf{0}) = \\
&d_B(-JQH(\cdot + K_P(\boldsymbol{y}), 0)|_{\boldsymbol{N}(L)}, \boldsymbol{D} - K_P(\boldsymbol{y}) \cap \boldsymbol{N}(L), \mathbf{0}) = \\
&d_B(-JQH(\cdot + K_P(\boldsymbol{y}), 0)|_{\boldsymbol{N}(L)}, \boldsymbol{D} \cap \boldsymbol{N}(L), \mathbf{0})
\end{aligned}$$

But $-JQH(\cdot + K_P(\boldsymbol{y}), 0)|_{\boldsymbol{N}(L)} = -JQH(\cdot, 0)|_{L^{-1}(\boldsymbol{y})}$ and so

$$\begin{aligned}
&d_B(I - P - JQH(\cdot + K_P(\boldsymbol{y}), 0)|_{\boldsymbol{N}(L)}, \boldsymbol{D} - K_P(\boldsymbol{y}) \cap \boldsymbol{N}(L), \mathbf{0}) = \\
&d_B(-JQH(\cdot, 0)|_{L^{-1}(\boldsymbol{y})}, \boldsymbol{D} \cap \boldsymbol{N}(L), \mathbf{0}) \neq 0
\end{aligned}$$

by Assumption (iii). Thus, by the existence theorem, there is $\boldsymbol{x} \in domL \cap \boldsymbol{D}$ such that $L(\boldsymbol{x}) = M(\boldsymbol{x}, t)$ for all $t \in [0,1]$ which implies, $L(\boldsymbol{x}) = tH(\boldsymbol{x}, t) + \boldsymbol{y}$ for all $t \in [0,1]$.

Now, if there was an $\boldsymbol{x} \in domL \cap \boldsymbol{D}$ with $L(\boldsymbol{x}) = H(\boldsymbol{x}, 1) + \boldsymbol{y} = G(\boldsymbol{x}) + \boldsymbol{y}$, the invariance under homotopy would not be true, In this case, choose $\lambda \in (0,1)$ and let $s = t/\lambda$. Set $\lambda(\boldsymbol{x}, s) = QH(\boldsymbol{x}, t/\lambda) + (t/\lambda)(I - Q)H(\boldsymbol{x}, t/\lambda) + \boldsymbol{y}$. Then, for $0 \le t \le \lambda$, $0 \le s \le 1$ and we can apply our previous arguments to $\Lambda(\boldsymbol{x}, s)$. Thus, for each $\lambda \in (0,1)$, there is $\boldsymbol{x}_{\boldsymbol{\lambda}} \in domL \cap \boldsymbol{D}$ with $L(\boldsymbol{x}_{\boldsymbol{\lambda}}) = H(\boldsymbol{x}_{\boldsymbol{\lambda}}, \lambda) + \boldsymbol{y}$. Choose the subsequence $\lambda_n = 1 - 1/n$. By the compactness of H, there is a subsequence $(\boldsymbol{x}_{\boldsymbol{n},\mathbf{1}})$ which converges to $\boldsymbol{x} \in \overline{\boldsymbol{D}}$ with $\lambda_{n,1} \to 1$. Then by the continuity of H, we have $L(\boldsymbol{x}) = H(\boldsymbol{x}, 1) + \boldsymbol{y} = G(\boldsymbol{x}) + \boldsymbol{y}$. ∎

Comment 12.3.1 *If $\boldsymbol{y} = \mathbf{0}$, the continuation theorem states $L(\boldsymbol{x}) = G(\boldsymbol{x})$ has a solution which lies in $dom(L) \cap \overline{\boldsymbol{D}}$ when:*

(i): $L(\boldsymbol{x}) \neq tH(\boldsymbol{x}, t)$ for all $\boldsymbol{x} \in dom(L) \cap \partial \boldsymbol{D}$ and $0 < t < 1$.

(ii): $QH(\boldsymbol{x},0) \neq \mathbf{0}$ *for all* $\boldsymbol{x} \in L^{-1}(\mathbf{0}) \cap \partial\boldsymbol{D}$.

(iii): $d_B(JQH(\boldsymbol{x},0)|_{L^{-1}(\mathbf{0})}, \boldsymbol{D} \cap L^{-1}(\mathbf{0}), \mathbf{0}) \neq 0$.

A good reference for the results in this chapter is (Gaines and Mawhin (40) 1977) although many details are lacking in the proofs of the ideas we present here. Also, we have included requisite functional analysis background to make it easier to follow the discussions. You are just learning much of this material and we feel we need to move slowly and carefully through the arguments! Please note, in this reference, the map J is replaced by the composition $J = \Lambda\Pi$ where $\Pi : Z \rightarrow Z/\boldsymbol{R}(L)$ is the canonical projection such that $\Pi G(\overline{\boldsymbol{D}})$ is bounded, ΠG is continuous and $\Lambda : Z/\boldsymbol{R}(l) \rightarrow \boldsymbol{N}(L)$ is a vector isomorphism.

Homework

This material is complicated to apply. We refer to (Gaines and Mawhin (40) 1977) for some applications using the idea of bound sets, curvature bound sets and Nagumo-sets, pages 42-47 in (Gaines and Mawhin (40) 1977). The continuation theorem can then be used to prove existence theorems for the Picard and the Periodic boundary value problems if the function f satisfies certain growth conditions, pages 48-51 in (Gaines and Mawhin (40) 1977). There is quite a bit of reading and background to go through here and we encourage you to look into it.

Exercise 12.3.5 *The Picard problem is converted to the standard* (L,N) *in Exercise 12.2.1. Translate all of the statements for the continuation theorem using the Picard problem as the underlying model. Choose the homotopy that connects to the identity.*

Exercise 12.3.6 *The periodic problem is converted to the standard* (L,N) *in Exercise 12.2.2. Translate all of the statements for the continuation theorem using the Periodic problem as the underlying model. Choose the homotopy that connects to the identity.*

12.4 Further Properties of Coincidence Degree

Let's look at some additional properties of coincidence degree which will help us understand how it depends on the way we split the operator $F = L + G$ (or $F = L - G$) into a linear Fredholm map of index 0 plus another compact component. Our setting is the usual: X and Z are normed linear spaces, $L : dom(L) \subset X \rightarrow Z$ is linear Fredholm of index 0 and $\boldsymbol{D} \subset X$ is a bounded open subset.

Definition 12.4.1 Collection C_F: different splittings $F = L + G$

Consider $F : dom(F) \subset X \rightarrow Z$*. In what follows,* $L : dom(L) \subset X \rightarrow Z$ *is linear Fredholm of index* 0 *and* $G : \overline{\boldsymbol{D}} \rightarrow Z$ *is* L *- compact. We let*

$$C_F(\boldsymbol{D}) = \{(L,G)|F = L + G,\ \mathbf{0} \notin F(\partial\boldsymbol{D} \cap dom(L))\}$$

Hence, we can assume $dom(F) = dom(L)$*. So, this is a set of possible splittings of* F *into a linear Fredholm of index* 0 *part plus another part which is* L *- compact.*

Recall, $d_C((L,G),\boldsymbol{D}) = d_{LS}(I - M, \boldsymbol{D}, \mathbf{0})$ where $M = P - (JQ + K_{PQ})G$. Let

$$H_{JPQ} = JQ + K_{PQ}, \quad R_{JPQ} = I - P + H_{JPQ}$$

Note,

$$H_{JPQ}F = (JQ + K_{PQ})(L + G) = JQL + K_{PQ}L + JQG + K_{PQ}G$$

$$= \quad JQG + K_{PQ}G + K_P L$$

But $K_P L - I - P$ and so $H_{JPQ}F = I - P + (JQ + K_{PQ})G = R_{JPQ}$. By our discussion in the proof of Theorem 12.2.1, $J + K_{PQ}$ is a vector isomorphism with $(J + K_{PQ})^{-1} = L + J^{-1}P$. We define a new degree by

Definition 12.4.2 Degree of $F = L + G$, d_F

We define the degree of $F = L + G$ to be $d_F(\boldsymbol{D}) = d_C((L, G), \boldsymbol{D}) = d_{LS}(R_{JPQ}, \boldsymbol{D}, \boldsymbol{0})$.

This leads to our first result.

Theorem 12.4.1 Degree d_F when the Range of G is Finite Dimensional

Let $F = L + G$ with $G(\overline{\boldsymbol{D}}) \subset Y$ where Y is a finite dimensional subspace of Z with $Z = \boldsymbol{R}(L) \oplus Y$. Then $|d_F(\boldsymbol{D})| = |d_B(G|_{\boldsymbol{N}(L)}, \boldsymbol{D} \cap \boldsymbol{N}(L))|$.

Proof 12.4.1
Since Y is finite dimensional, Y is a closed subspace of Z. We have shown

$$H_{JPQ}F \quad = \quad I - P + (JQ + K_{PQ})G$$

We can choose Q to be the projection onto Y and P to be the projection on $\boldsymbol{N}(L)$. Then, since $G(\overline{\boldsymbol{D}}) \subset Y$, $QG = G$, $K_{PQ}G = K_P(I - Q)G = 0$. So

$$R_{JPQ} = H_{JPQ}F \quad = \quad I - P + JG$$

Now $(P - JG)(\overline{\boldsymbol{D}}) \subset \boldsymbol{N}(L)$. Thus, $d_F(\boldsymbol{D}) = d_{LS}(I - (P - JG), \boldsymbol{D}, \boldsymbol{0})$. Since $\boldsymbol{N}(L)$ is finite dimensional, we then have

$$\begin{aligned} d_{LS}(I - (P - JG), \boldsymbol{D}, \boldsymbol{0}) &= d_B((I - (P - JG))|_{\boldsymbol{N}(L)}, \boldsymbol{D} \cap \boldsymbol{N}(L), \boldsymbol{0}) \\ &= d_B(JG|_{\boldsymbol{N}(L)}, \boldsymbol{D} \cap \boldsymbol{N}(L), \boldsymbol{0}) \\ &= sign\ det(J)\, d_B(G|_{\boldsymbol{N}(L)}, \boldsymbol{D} \cap \boldsymbol{N}(L), \boldsymbol{0}) \end{aligned}$$

Thus, $|d_F(\boldsymbol{D})| = |d_B(G|_{\boldsymbol{N}(L)}, \boldsymbol{D} \cap \boldsymbol{N}(L))|$. ■

We can also relate $d_F(\boldsymbol{D})$ to Leray - Schauder degree of a larger collection of mappings containing the R_{JPQ} mappings.

Lemma 12.4.2 H_{APQ} Mapping

Let $A : \boldsymbol{R}(Q) \to dom(L)$ be linear so that $PA : \boldsymbol{R}(Q) \to \boldsymbol{N}(L)$ is a vector isomorphism. Then, the mapping $H_{APQ} = AQ + K_{PQ}$ is a vector isomorphism from Z onto $dom(L)$ and $H_{APQ}^{-1} = L - LA(PA)^{-1}P + (PA)^{-1}P$. Moreover, if $F = L + G$, we have $H_{APQ}F = I - P + H_{APQ}G$ for all $\boldsymbol{x} \in dom(L) \cap \partial\boldsymbol{D}$ with $H_{APQ}G : \overline{\boldsymbol{D}} \to X$ compact.

Proof 12.4.2
Let $\boldsymbol{x} \in dom(L)$. Then, for $\boldsymbol{z} \in Z$, we have $H_{APQ}(\boldsymbol{z}) = \boldsymbol{x}$ implies $AQ(\boldsymbol{z}) + K_{PQ}(\boldsymbol{z}) = \boldsymbol{x}$. Thus,

$$PAQ(\boldsymbol{z}) + PK_{PQ}(\boldsymbol{z}) \quad = \quad P(\boldsymbol{x}), \quad (I - P)AQ(\boldsymbol{z}) + (I - P)K_{PQ}(\boldsymbol{z}) = (I - P)(\boldsymbol{x})$$

But, $K_{PQ}(\boldsymbol{z}) \in dom(L) \cap \boldsymbol{N}(P)$ *and so* $PK_P(I-Q)(\boldsymbol{z}) = \boldsymbol{0}$. *Also,* $I - P$ *acts like the identity on* $\boldsymbol{N}(P)$*; hence,* $(I-P)K_{PQ}(\boldsymbol{z}) = K_{PQ}(\boldsymbol{z})$. *Rewriting, we have*

$$PAQ(\boldsymbol{z}) = P(\boldsymbol{x}), \quad (I-P)AQ(\boldsymbol{z}) + K_{PQ}(\boldsymbol{z}) = (I-P)(\boldsymbol{x})$$

Therefore, $Q(\boldsymbol{z}) = (PA)^{-1}P(\boldsymbol{x})$. *Now consider* $L(I-P)AQ(\boldsymbol{z}) + LK_{PQ}(\boldsymbol{z}) = L(I-P)(\boldsymbol{x})$.

(i): We note $\boldsymbol{z} \in Z$, *so* $\boldsymbol{z} = \boldsymbol{z}_0 + \boldsymbol{z}_1$ *with* $\boldsymbol{z}_0 \in \boldsymbol{M}_0$ *and* $\boldsymbol{z}_1 \in \boldsymbol{R}(L)$. *So,* $(I-Q)(\boldsymbol{z}) = \boldsymbol{z}_1$. *Since* $K_P : \boldsymbol{R}(L) \to dom(L)$, $K_P(I-Q)(\boldsymbol{z}) = K_P(\boldsymbol{z}_1) = \boldsymbol{x}_1 \in \boldsymbol{X}_0$ *and* $LK_P(I-Q)(\boldsymbol{z}) = L(\boldsymbol{x}_1) = \boldsymbol{z}_1$. *We see* $LK_P(I-Q)(\boldsymbol{z}) = (I-Q)(\boldsymbol{z})$.

(ii): $\boldsymbol{x} = \boldsymbol{x}_0 + \boldsymbol{x}_1$ *with* $\boldsymbol{x}_0 \in \boldsymbol{N}(L)$ *and* $\boldsymbol{x}_1 \in \boldsymbol{X}_0$. *So* $L(I-P)(\boldsymbol{x}) = L(\boldsymbol{x}_1) = L(\boldsymbol{x})$. *So* $L(I-P)(\boldsymbol{x}) = L(\boldsymbol{x})$.

Therefore, $L(I-P)AQ(\boldsymbol{z})(\boldsymbol{z}) + LK_{PQ}(\boldsymbol{z}) = L(I-P)(\boldsymbol{x})$ *becomes* $LAQ(\boldsymbol{z}) + (I-Q)(\boldsymbol{z}) = L(\boldsymbol{x})$. *We have shown* $H_{APQ}(\boldsymbol{z}) = \boldsymbol{x}$ *is equivalent to* $Q(\boldsymbol{z}) = (PA)^{-1}P(\boldsymbol{x})$ *and* $(I-Q)(\boldsymbol{z}) = L(\boldsymbol{x}) - LAQ(\boldsymbol{z})$. *We can rewrite this by adding these equations to find*

$$\begin{aligned} \boldsymbol{z} &= L(\boldsymbol{x}) - LAQ(\boldsymbol{z}) + (PA)^{-1}P(\boldsymbol{x}) \\ &= L(\boldsymbol{x}) - LA(PA)^{-1}P(\boldsymbol{x}) + (PA)^{-1}P(\boldsymbol{x}) \end{aligned}$$

This shows $H_{APQ}^{-1} = L - LA(PA)^{-1}P + (PA)^{-1}P$.

Finally, on $dom(L) \cap \overline{\boldsymbol{D}}$,

$$\begin{aligned} H_{APQ}F &= (AQ + K_{PQ})(L+G) = AQL + AQG + K_{PQ}L + K_{PQ}G \\ &= (I-P) + (AQ + K_{PQ})G = (I-P) + H_{APQ}G \end{aligned}$$

If $(L, G) \in C_F(\boldsymbol{D})$, *then* QG *and* K_{PQ} *are compact. Since* A *is linear with a finite dimensional domain,* A *is continuous. This tells us* AQG *is compact. Thus,* $H_{APQ}G$ *is compact.* ∎

Now define $R_{APQ} = I - P + H_{APQ}G = I - (P - H_{APQ}G)$. Lemma 12.4.2 tells us if $F = L + G \in C_{F,G}(\boldsymbol{D})$, then $R_{APQ} \in C_{I,G}(\boldsymbol{D})$ and hence $d_{LS}(R_{APQ}, \boldsymbol{D}, \boldsymbol{0})$ is defined.

Lemma 12.4.3 Relating Leray - Schauder Degrees for R_{BPQ} and R_{APQ}

Let $A, B : \boldsymbol{R}(Q) \to dom(L)$ *be linear so that* $PA, PB : \boldsymbol{R}(Q) \to \boldsymbol{N}(L)$ *are vector isomorphisms. Assume* $F = L + G$, *where* $(L, G) \in C_F(\boldsymbol{D})$. *Then, for all* $r > 0$,

$$d_{LS}(R_{BPQ}, \boldsymbol{D}, \boldsymbol{0}) = d_{LS}(I - (A-B)(PA)^{-1}P, B(\boldsymbol{0}, r), \boldsymbol{0})\, d_{LS}(R_{APQ}, \boldsymbol{D}, \boldsymbol{0})$$

Proof 12.4.3
Let $\boldsymbol{x} \in \overline{\boldsymbol{D}}$. *Then,*

$$\begin{aligned} (I - (A-B)(PA)^{-1}P)R_{APQ}(\boldsymbol{x}) &= (I - (A-B)(PA)^{-1}P)(I - P + H_{APQ}G)(\boldsymbol{x}) \\ &= (I - P + H_{APQ}G)(\boldsymbol{x}) - (A-B)(PA)^{-1}P(I-P)(\boldsymbol{x}) \\ &\quad -(A-B)(PA)^{-1}PH_{APQ}G(\boldsymbol{x}) \\ &= (I-P)(\boldsymbol{x}) + (AQ + K_{PQ})G(\boldsymbol{x}) \\ &\quad -(A-B)(PA)^{-1}P(AQ + K_{PQ})G(\boldsymbol{x}) \\ &= (I-P)(\boldsymbol{x}) + (AQ + K_{PQ})G(\boldsymbol{x}) \end{aligned}$$

$$-(A - B)(PA)^{-1}PAQ(\boldsymbol{x})$$
$$-(A - B)(PA)^{-1}PK_{PQ}G(\boldsymbol{x})$$

But $K_{PQ}G(\boldsymbol{x}) \in \boldsymbol{N}(P)$ *and so* $(A - B)(PA)^{-1}PK_{PQ}G(\boldsymbol{x}) = \mathbf{0}$. *Thus,*

$$\begin{aligned}
&(I - (A - B)(PA)^{-1}P)R_{APQ}(\boldsymbol{x}) \\
&= (I - P)(\boldsymbol{x}) + (AQ + K_{PQ})G(\boldsymbol{x}) \\
&\quad -(A - B)(PA)^{-1}PAQ(\boldsymbol{x}) \\
&(I - P)(\boldsymbol{x}) + (AQ + K_{PQ})G(\boldsymbol{x}) \\
&\quad -A(PA)^{-1}PAQ(\boldsymbol{x}) + B(PA)^{-1}PAQ(\boldsymbol{x}) \\
&= (I - P)(\boldsymbol{x}) + (AQ + K_{PQ})G(\boldsymbol{x}) - (A - B)Q(\boldsymbol{x}) \\
&= (I - P)(\boldsymbol{x}) + AQG(\boldsymbol{x}) + K_{PQ}G(\boldsymbol{x}) - AQG(\boldsymbol{x}) + BQG(\boldsymbol{x}) \\
&= (I - P)(\boldsymbol{x}) + K_{PQ}G(\boldsymbol{x}) + BQG(\boldsymbol{x}) \\
&= (I - P)(\boldsymbol{x}) + (BQ + K_{PQ})G(\boldsymbol{x}) = R_{BPQ}(\boldsymbol{x})
\end{aligned}$$

Now, $(PA)^{-1}P(X) \subset \boldsymbol{R}(Q)$ *which is finite dimensional. Since A and B are linear,* $(A-B)(PA)^{-1}$ $P(X)$ *is also finite dimensional. Hence,* $(A - B)(PA)^{-1}P(X)$ *is compact on* $\overline{\boldsymbol{D}}$.

Also, if $\boldsymbol{x} \in \boldsymbol{N}(I - (A - B)(PA)^{-1}P)$, *then* $\boldsymbol{x} = (A - B)(PA)^{-1}P(\boldsymbol{x})$. *Hence,* $P(\boldsymbol{x}) = (PA - PB)(PA)^{-1}P(\boldsymbol{x}) = P(\boldsymbol{x}) - PB(PA)^{-1}P(\boldsymbol{x})$. *Thus,* $PB(PA)^{-1}P(\boldsymbol{x}) = \mathbf{0}$ *and* $(PA)^{-1}P(\boldsymbol{x}) = (PB)^{-1}(\mathbf{0}) = \mathbf{0}$. *We conclude* $P(\boldsymbol{x}) = \mathbf{0}$ *and so* $\boldsymbol{x} \in \boldsymbol{N}(P) = \boldsymbol{X_0}$. *However, if* $P(\boldsymbol{x}) = \mathbf{0}$, *we have* $\boldsymbol{x} = (A - B)(PA)^{-1}P(\boldsymbol{x}) = (A - B)(PA)^{-1}(\mathbf{0}) = \mathbf{0}$. *We see* $I - (A - B)(PA)^{-1}P$ *is 1-1.*

By our discussion of the Fredholm alternative for linear compact operators in Section 11.4.5, since $I - (A - B)(PA)^{-1}P$ *is invertible, the equation* $\boldsymbol{y} = (I - (A - B)(PA)^{-1}P)(\boldsymbol{x})$ *always has a unique solution* $\boldsymbol{x} \in dom(L)$. *Hence,* $I - (A - B)(PA)^{-1}P$ *is onto. Therefore by the product theorem for Leray - Schauder degree,*

$$d_{LS}(R_{BPQ}, \boldsymbol{D}, \mathbf{0}) \;=\; \sum_{i=1}^{N} d_{LS}(I - (A - B)(PA)^{-1}P, \boldsymbol{\Delta_i}, \mathbf{0})\, d_{LS}(R_{APQ}, \boldsymbol{D}, \boldsymbol{y_i})$$

where the sum is over the bounded connected components of $X \setminus R_{APQ}(\partial \boldsymbol{D})$, $\boldsymbol{\Delta_i}$ *and the* $\boldsymbol{y_i}$ *are arbitrarily chosen from* $\boldsymbol{\Delta_i}$ *(recall, the degree is constant on components). Since we assume* $R_{APQ}(\partial \boldsymbol{D}) \neq \mathbf{0}$, *two things can happen.*

- *(i):* $\mathbf{0} \notin R_{APQ}(\boldsymbol{D})$ *which implies* $\mathbf{0} \notin R_{BPQ}(\boldsymbol{D})$ *as* R_{BPQ} *is a vector isomorphism of* X *onto* X.

- *(ii):* $\mathbf{0} \in R_{APQ}(\boldsymbol{D})$ *which implies* $\mathbf{0} \in \boldsymbol{\Delta_i}$ *for some* i. *Call this value of* i, i_0.

If (i) is the case, $\mathbf{0} \notin R_{APQ}(\boldsymbol{D})$ *and* $\mathbf{0} \notin R_{BPQ}(\boldsymbol{D})$ *which tells us* $d_{LS}(R_{BPQ}, \boldsymbol{D}, \mathbf{0}) = 0 = d_{LS}(R_{APQ}, \boldsymbol{D}, \boldsymbol{y_i})$ *and the equality we seek is trivially satisfied. On the other hand, if (ii) is the case, then*

$$d_{LS}(R_{BPQ}, \boldsymbol{D}, \mathbf{0}) \;=\; d_{LS}(I - (A - B)(PA)^{-1}P, \boldsymbol{\Delta_{i_0}}, \mathbf{0})\, d_{LS}(R_{APQ}, \boldsymbol{D}, \mathbf{0})$$

where we have chosen $\boldsymbol{y_{i_0}} = \mathbf{0}$.

Since $\mathbf{0} \in \boldsymbol{\Delta_{i_0}}$, *there is* $r > 0$ *so that* $B(\mathbf{0}, r) \subset \boldsymbol{\Delta_{i_0}}$. *By the excision property of Leray - Schauder degree, we have* $d_{LS}(I - (A - B)(PA)^{-1}P, \boldsymbol{\Delta_{i_0}}, \mathbf{0}) = d_{LS}(I - (A - B)(PA)^{-1}P, B(\mathbf{0}, r), \mathbf{0})$.

However, since $I - (A - B)(PA)^{-1}P$ is a vector isomorphism, $\mathbf{0} \in B(\mathbf{0}, r)$ for any value of $r > 0$. This proves the result. ∎

Now we can talk about the relationship of Leray - Schauder degrees for R_{JPQ} and R_{APQ}.

Lemma 12.4.4 Relating Leray - Schauder Degrees for R_{JPQ} and R_{APQ}

Let $A : \boldsymbol{R}(Q) \rightarrow dom(L)$ be linear so that $PA : \boldsymbol{R}(Q) \rightarrow \boldsymbol{N}(L)$ is a vector isomorphism. Let $J : \boldsymbol{R}(Q) \rightarrow \boldsymbol{N}(L)$ be a vector isomorphism in $\mathcal{L}_L$. Assume $F = L + G$, where $(L, G) \in C_F(\boldsymbol{D})$. Then, for all $r > 0$,

$$d_F(\boldsymbol{D}) = d_{LS}(R_{JPQ}, \boldsymbol{D}, \mathbf{0}) = d_{LS}(I - (A - J)(PA)^{-1}P, B(\mathbf{0}, r), \mathbf{0}) \times d_{LS}(R_{APQ}, \boldsymbol{D}, \mathbf{0})$$

Proof 12.4.4
First, note J satisfies $PJ : \boldsymbol{R}(Q) \rightarrow \boldsymbol{N}(L)$ is a vector isomorphism. Now use Lemma 12.4.3 with $B = J$. ∎

We finish this section with a result for the case where $\boldsymbol{N}(F) = \mathbf{0}$.

Theorem 12.4.5 $d_F(\boldsymbol{D})$ when $\boldsymbol{N}(F) = \mathbf{0}$ and G is Linear

Let $F = L + G$, $G : \overline{\boldsymbol{D}} \subset X \rightarrow Z$ be linear and L - compact for $\boldsymbol{D} \subset X$ is open and bounded. Assume $\boldsymbol{N}(F) = \mathbf{0}$ and $\mathbf{0} \notin F(\partial \boldsymbol{D})$. Then, $d_F(\boldsymbol{D}) = 0$ if $\mathbf{0} \notin \boldsymbol{D}$ and $|d_F(\boldsymbol{D})| = 1$ if $\mathbf{0} \in \boldsymbol{D}$.

Proof 12.4.5
As discussed earlier, $R_{JPQ} = I - P + H_{JPQ}G$ is a compact perturbation of I. Moreover, if $R_{JPQ}(\boldsymbol{x}) = \mathbf{0}$ it also satisfies $(L + G)(\boldsymbol{x}) = \mathbf{0}$. Thus, $\boldsymbol{N}(R_{JPQ}) = \mathbf{0}$. Since G is linear, R_{JPQ} is a linear compact perturbation of I with zero kernel. Thus, by Leray - Schauder existence theorem, $d_{LS}(R_{JPQ}, \boldsymbol{D}, \mathbf{0}) = 0$ if $\mathbf{0} \notin \boldsymbol{D}$. Also, by Theorem 11.4.18, $|d_{LS}(R_{JPQ}, \boldsymbol{D}, \mathbf{0})| = 1$ if $\mathbf{0} \in \boldsymbol{D}$. Hence, $|d_F(\boldsymbol{D})| = 1$ if $\mathbf{0} \in \boldsymbol{D}$. ∎

Homework

Exercise 12.4.1 *In the proof of Lemma 12.4.2, provide all the details that show $H_{APQ}G$ is compact.*

Exercise 12.4.2 *In the proof of Lemma 12.4.3, explain in detail why the Fredholm alternative for linear compact operators tells us the equation $\boldsymbol{y} = (I - (A - B)(PA)^{-1}P)(\boldsymbol{x})$ always has a unique solution $\boldsymbol{x} \in dom(L)$.*

Exercise 12.4.3 *In the proof of Theorem 12.4.5, explain in detail why R_{JPQ} is a linear compact perturbation of I with zero kernel.*

12.5 The Dependence of Coincidence Degree on Operator Splitting

We can now examine the dependence of $d_F(\boldsymbol{D})$ with respect to the splitting of F as $F = L + G$. As usual, X and Z are normed linear spaces, $L : dom(L) \subset X \rightarrow Z$ is linear Fredholm of index 0 and $G : \overline{\boldsymbol{D}} \subset X \rightarrow Z$ is L - compact where $\boldsymbol{D}$ is open and bounded. If $F = L + G = L' + G'$ where

(L, G) and $(L', G') \in C_F(\boldsymbol{D})$, we know $L - L' = G' - G$. Let $\delta L = L - L'$. Since L' and L are linear, this forces δL to be linear. Also, we know G and G' are both defined on $\overline{\boldsymbol{D}}$. Then, we have

$$\begin{aligned} F &= L + G = L' + (L - L^prime) + G = L' + \delta L + G \\ &= L' + G' \end{aligned}$$

We conclude $G' = G + \delta L$ and $G = G' - \delta L$. We can also do this the other way.

$$\begin{aligned} F &= L' + G' = L + (L^prime - L) + G' = L - \delta L + G' \\ &= L + G \end{aligned}$$

which tells the same thing: $G = G' - \delta L$. This implies

(i): $G + \delta L$ is L' - compact. So $Q'(G + \delta L)$ is compact and $K'_{P'Q'}(G + \delta L)$ is compact.

(ii): $Q(G' - \delta L)$ is compact and $K_{PQ}(G' - \delta L)$ is compact.

(iii): QG is compact and $K_{PQ}Q$ is compact and $Q'G'$ and $K'_{P'Q'}G'$ is compact.

Now, it seems reasonable that if F can be expressed in two ways, that some compatibility conditions apply. Hence, we assume G is L' - compact and G' is L - compact on $\overline{\boldsymbol{D}}$. Then, from (i), $Q'G' = Q'(G) + Q'(\delta L)$ implies $Q'(\delta L)$ is compact. Also, (ii) tells us $QG = Q(G') + Q(\delta L)$ and so $Q(\delta L)$ is compact. Similar arguments show $K_{PQ}(\delta L)$ and $K'_{P'Q'}(\delta L)$ are compact. Hence, δL is both L - compact and L' compact on $\overline{\boldsymbol{D}}$. However, just because δL matches $G' - G$ on $\overline{\boldsymbol{D}} \cap dom(L)$ doesn't mean δL is L - compact and L' - compact on only $\overline{\boldsymbol{D}}$. If δL is L - compact and L' - compact on all X this works too. So we will assume our perturbation of L, δL, is linear, defined on all X and δL is L - compact and L' compact on all X. We still also retain the compatibility conditions G is L' - compact and G' is L - compact on $\overline{\boldsymbol{D}}$.

Let P' and Q' be the usual projections, but this time the projections are associated with L'. Define

$$\begin{aligned} H &= H_{JPQ} = JQ + K_{PQ} \\ H' &= H_{J'P'Q'} = J'Q' + K'_{P'Q'} \end{aligned}$$

so that $H : Z \to dom(L)$ and $H' : Z \to dom(L') = dom(L)$. As discussed before, H and H' map Z onto $dom(L) = dom(L')$ as vector isomorphisms. Recall, $H^{-1} = L + J^{-1}P$ and $(H')^{-1} = L' + (J')^{-1}P'$. Define $K' = \delta L + J^{-1}P$.

We show $J^{-1}P$ is L' - compact on any bounded subset $\boldsymbol{B} \subset X$. The argument is as follows:

(a): Since $\boldsymbol{B}$ is bounded, it is clear that the closure of $J^{-1}P(\overline{\boldsymbol{B}})$ is compact because $\boldsymbol{R}(Q)$ is finite dimensional.

(b): It follows then by the linearity of Q', $I - Q'$ and $K'_{P'}$ that since $J^{-1}P(\overline{\boldsymbol{B}})$ is bounded, both $Q'J^{-1}Q(\overline{\boldsymbol{B}})$ and $K'_{P'}(I - Q')J^{-1}Q(\overline{\boldsymbol{B}})$ are bounded sets in a finite dimensional space. We can therefore conclude $J^{-1}P$ is L' - compact in $\boldsymbol{B}$.

Hence $J^{-1}P$ is L' - compact and since δL is L' - compact on $\overline{\boldsymbol{D}}$, we can say K' is L' - compact on $\overline{\boldsymbol{D}}$.

Next, note

$$L' + K' = L' + \delta L + J^{-1}P = L + J^{-1}P = H^{-1}$$

We can prove $I - P' + H'K'$ is a linear bijection on X.

Lemma 12.5.1 $I - P' + H'K'$ is a Vector Isomorphism on X

$I - P' + H'K'$ is a vector isomorphism from X onto X.

Proof 12.5.1
Note,

$$\begin{aligned} H'H^{-1} &= (J'Q' + K'_{P'Q'})(L' + K') \\ &= J'Q'L' + (J'Q' + K'_{P'Q'})K' + K'_{P'Q'}L' \\ &= (I - P' + H'K') \end{aligned}$$

Since K' is L' - compact, $H'K'$ is too. Hence, $I - P' + H'K'$ is a linear compact perturbation of the identity. Since $(I - P' + H'K')(\boldsymbol{x}) = \boldsymbol{0}$ implies $H'H^{-1}(\boldsymbol{x}) = \boldsymbol{0}$, we see $\boldsymbol{x} = \boldsymbol{0}$. Thus, $(I - P' + H'K')(\boldsymbol{x})$ has zero kernel. By the Fredholm alternative for linear compact perturbations of the identity, its range must be X. Hence, $I - P' + H'K'$ is a vector isomorphism from X onto X.
■

Now, for all $\boldsymbol{x} \in dom(L) \cap \overline{\boldsymbol{D}}$, from our discussions about H_{JPQ} maps, we know

$$HF = R_{JPQ} = I - P + HG, \quad H'F = I - P' + H'G'$$

Thus,

$$(I - P + HG)(\boldsymbol{x}) = HF(\boldsymbol{x}) \in dom(L), \quad (I - P' + H'G')(\boldsymbol{x}) = H'F(\boldsymbol{x}) \in dom(L)$$

But then,

$$\begin{aligned} (I - P' + H'G')(\boldsymbol{x}) &= H'H^{-1}(I - P + HG)(\boldsymbol{x}) = H'(L' + K')(I - P + HG)(\boldsymbol{x}) \\ &= (H'L' + H'K')(I - P + HG)(\boldsymbol{x}) \end{aligned}$$

Also,

$$H'L' = (J'Q' + K'_{P'Q'})L' = K'_{P'Q'}L' = I - P'$$

Thus,

$$(I - P' + H'G')(\boldsymbol{x}) = (I - P' + H'K')(I - P + HG)(\boldsymbol{x}) = H'H^{-1}(I - P + HG)(\boldsymbol{x})$$

Lemma 12.5.2 Connecting the Degree of F for Splittings

Under the assumptions we have been using, for all $r > 0$,

$$d_C((L', G'), \boldsymbol{D}) = d_C((L', K'), B(\boldsymbol{0}, r))\, d_C((L, G), \boldsymbol{D})$$

Proof 12.5.2
By the Leray - Schauder Product theorem

$$d_{LS}(I - P' + H'G', \boldsymbol{D}, \boldsymbol{0}) = \sum_{i=1}^{N} d_{LS}(I - P' + H'K', \boldsymbol{\Delta_i}, \boldsymbol{0})\, d_{LS}(I - P + HG, \boldsymbol{D}, \boldsymbol{y_i})$$

where the sum is over the bounded components of $X - (I - P + HG)(\partial \boldsymbol{D})$ *and the* $\boldsymbol{y_i}$ *are arbitrarily chosen from* $\boldsymbol{\Delta_i}$. *Since* $I - P' + H'K'$ *is a vector isomorphism onto* X, *if* $\boldsymbol{0} \notin (I - P' + H'K')(\boldsymbol{D})$, *then* $\boldsymbol{0} \notin (I - P + HG)(\boldsymbol{D})$. *We see* $d_C((L,G),\boldsymbol{D}) = 0$ *and* $0 = d_{LS}(I - P' + H'G', \boldsymbol{D}, \boldsymbol{0}) = d_C((L',G'),\boldsymbol{D})$. *Hence, the result is established.*

On the other hand, if $\boldsymbol{0} \in (I - P + HG)(\boldsymbol{D})$, *there is a bounded component,* $\boldsymbol{\Delta_{i_0}}$, *which contains* $\boldsymbol{0}$. *We choose* $\boldsymbol{y_{i_0}} = \boldsymbol{0}$ *and we have*

$$d_{LS}(I - P' + H'G', \boldsymbol{D}, \boldsymbol{0}) = d_{LS}(I - P' + H'K', \boldsymbol{\delta_{i_0}}, \boldsymbol{0})\, d_{LS}(I - P + HG, \boldsymbol{D}, \boldsymbol{0})$$

By the excision property, we see for all $r > 0$,

$$d_{LS}(I - P' + H'G', \boldsymbol{D}, \boldsymbol{0}) = d_{LS}(I - P' + H'K', B(\boldsymbol{0}, r), \boldsymbol{0})\, d_{LS}(I - P + HG, \boldsymbol{D}, \boldsymbol{0})$$

Thus,

$$d_C((L',G'),\boldsymbol{D}) = d_{LS}(I - P' + H'K', B(\boldsymbol{0}, r), \boldsymbol{0})\, d_C((L,G),\boldsymbol{D})$$

But $d_{LS}(I - P' + H'K', B(\boldsymbol{0}, r), \boldsymbol{0}) = d_C((L',K'), B(\boldsymbol{0}, r))$. *Thus,*

$$d_C((L',G'),\boldsymbol{D}) = d_C((L',K'), B(\boldsymbol{0}, r))\, d_C((L,G),\boldsymbol{D})$$

■

We are now ready to prove our final result. Let's make sure we state our assumptions.

Theorem 12.5.3 $|\, d_C \,|$ **is Independent of the Splitting**

We assume the usual requirements of the pairs (L,G) *and* (L',G') *and* $F = L + G = L' + G'$ *where* $L' + \delta L = L$. *The maps* L *and* L' *have the same domain,* $dom(L)$ *and* G *and* G' *both have the domain* $\overline{\boldsymbol{D}}$. *We require the perturbation* δL *to be linear and* L' *- and* L *- compact on* X. *We also require the compatibility conditions:* G *is* L'*- compact. and* G' *is* L *- compact on* $\overline{\boldsymbol{D}}$. *Then* $|d_C((L',G'),\boldsymbol{D})| = |d_C((L,G),\boldsymbol{D})|$.

Proof 12.5.3
Since $H^{-1} = L' + K'$ *is a vector isomorphism,* $|d_C((L',K'), B(\boldsymbol{0}, r))| = 1$ *by Theorem 12.4.5.* ■

Thus, the coincident degree of a map F is, to some degree, independent of the manner in which F can be written as a sum of a linear Fredholm map of index 0, L, and an L -compact map G. Of course, if F does not have such a split, we cannot speak of a coincidence degree. Further, for the problem $(L + G)(\boldsymbol{x}) = \boldsymbol{0}$, Theorem 12.5.3 assures us that once we know $d_C((L,G),\boldsymbol{D})$, we will not get different results for other reasonable factorizations $L' + G'$. Another reference for this is a set of notes from lectures given at Claremont University in 1977 (Mawhin (86) 1979). Theorem 12.5.3 shows us $d_F(\boldsymbol{D})$ is well-defined as it is independent of the choice of factorization of F, as long as the factorizations satisfy a few reasonable conditions.

Homework

Exercise 12.5.1 *In the proof of Lemma 12.5.1, prove* $H'K'$ *is* L' *- compact.*

Exercise 12.5.2 *In the proof of Lemma 12.5.2, explain in detail why the excision property tells us, for all $r > 0$,*

$$d_{LS}(I - P' + H'G', \boldsymbol{D}, \boldsymbol{0}) = d_{LS}(I - P' + H'K', B(\boldsymbol{0}, r), \boldsymbol{0})\, d_{LS}(I - P + HG, \boldsymbol{D}, \boldsymbol{0})$$

12.6 Applications of Topological Degree Methods to Boundary Value Problems

Let's look at some applications to ODE problems. We will show you how to convert them into the (L, G) coincidence degree formulation.

Consider this problem on $I = [0, 1]$. We seek a solution x with $x(t) \in \Re^n$ for $t \in I$ so

$$\begin{aligned} \boldsymbol{x}'(t) &= f(t), 0 \leq t \leq 1, \text{ Dynamics} \\ Mx(0) + Nx(1) &= \boldsymbol{c}, \text{ Boundary Conditions} \end{aligned}$$

where $\boldsymbol{c}$ is a constant and M and N are nonzero $n \times n$ matrices. In a $\Re^3$ setting the boundary conditions might be

$$\begin{bmatrix} 1 & 0 & 0 \\ 0 & 2 & 0 \\ 0 & 0 & 3 \end{bmatrix} \begin{bmatrix} x_1(0) \\ x_2(0) \\ x_3(0) \end{bmatrix} + \begin{bmatrix} 4 & 0 & 0 \\ 0 & 1 & 0 \\ 0 & 0 & 7 \end{bmatrix} \begin{bmatrix} x_1(1) \\ x_2(1) \\ x_3(1) \end{bmatrix} = \begin{bmatrix} c_1 \\ c_2 \\ c_3 \end{bmatrix}$$

or

$$\begin{aligned} x_1(0) + 4x_1(1) &= c_1 \\ 2x_2(0) + x_2(1) &= c_2 \\ 3x_3(0) + 7x_3(1) &= c_3 \end{aligned}$$

Here $f \in \mathcal{L}_1(I, \Re^n, \mathcal{S}, \mu)$, which is the standard space of summable Lebesgue measurable functions. By a solution to this problem, which is called a boundary value problem, we mean an absolutely continuous function $\boldsymbol{x} : I \to \Re^n$ which satisfies the dynamics a.e. on I, and $Mx(0) + Nx(1) = \boldsymbol{c}$. Let's formulate this problem within the framework of coincidence degree.

Let $X = C(I, \Re^n)$, which is the set of continuous functions on I with values in $\Re^n$. Let $Z = \mathcal{L}_1(\Re^n, \mathcal{S}, \mu) \times \Re^n$. Define the following maps:

1. $L : dom(L) \to \mathcal{L}_1(\Re^n, \mathcal{S}, \mu) \times \Re^n$ with $dom(L) = \{x \in X | x \text{ is absolutely continuous }\}$ and $L(\boldsymbol{x}) = (\boldsymbol{x}', Mx(0) + Nx(1))$.

2. $G : I \to Z$ by $G(t) = (f(t), \boldsymbol{c})$.

Then $L(\boldsymbol{x}) = 0$ if and only if $(\boldsymbol{x}', Mx(0) + Nx(1)) = (\boldsymbol{0}, \boldsymbol{0})$. Now $\boldsymbol{x}' = 0$ a.e. tells us $\boldsymbol{x} = \boldsymbol{0}$ as $\boldsymbol{x}$ is continuous. Thus, $x(0) = x(1) = 0$ and so $Mx(0) + Nx(1) = \boldsymbol{0}$. So,

$$\boldsymbol{N}(L) = \{\boldsymbol{x} \in X | \boldsymbol{x} \text{ is a constant }, (M + N)(x(0)) = \boldsymbol{0}\}$$

Thus, $\boldsymbol{N}(L) = \boldsymbol{N}(M + N)$.

Next, $\boldsymbol{x}' = f$ if and only if $x(t) = x(0) + \int_0^t f(s)ds$. Let $\boldsymbol{d} = Mx(0) + N(x(0) + \int_0^1 f(s)ds) = Mx(0) + Nx(0) + N\int_0^1 f(s)ds$. Thus, $(M+N)(x(0)) = \boldsymbol{d} - N\int_0^1 f(s)ds$. This tells us

$$\begin{aligned}\boldsymbol{R}(L) &= \{(f,\boldsymbol{d})|\boldsymbol{d} = (M+N)(x(0)) + N\int_0^1 f(s)ds\}\\ &= \{(f,\boldsymbol{d})|\boldsymbol{d} - N\int_0^1 f(s)ds \in \boldsymbol{R}(M+N)\}\end{aligned}$$

Now define $A : Z \to \Re^n$ by $A(f,\boldsymbol{d}) = \boldsymbol{d} - N\int_0^1 f(s)ds$ with the usual norms on $\mathcal{L}_1(\Re^n, \mathcal{S}, \mu)$ and $\Re^n$. We see A is continuous so $\boldsymbol{R}(A)$ is closed in $\Re^n$. Note, A is onto $\Re^n$ as given $\boldsymbol{d} \in \Re^n$, $A(0,\boldsymbol{d}) = \boldsymbol{d}$.

If $A(f,\boldsymbol{d}) = \boldsymbol{0}$, then $\boldsymbol{d} - N\int_0^1 f(s)ds = \boldsymbol{0}$. Thus, $\boldsymbol{N}(A) = \{(f,\boldsymbol{d})|\boldsymbol{d} = N\int_0^1 f(s)ds\}$. Given $(h,e) \in Z$, note $(h,e) = (h, e + N\int_0^1 h(s)ds - N\int_0^1 h(s)ds)$. We can rewrite this as $(h,e) = (h, e - N\int_0^1 h(s)ds) + (h, N\int_0^1 h(s)ds)$. Since $(h, N\int_0^1 h(s)ds) \in \boldsymbol{N}(A)$, we see $(h,e) \in (h, e - N\int_0^1 h(s)ds) + \boldsymbol{N}(A)$. So the collection of nonzero equivalence classes is $\{\boldsymbol{e} \mid \boldsymbol{e} \neq N\int_0^1 h(s)ds\}$. Clearly $\boldsymbol{0}$ is a representative for $\boldsymbol{0} + \boldsymbol{N}(A)$. We conclude $Z/\boldsymbol{N}(A) \equiv \Re^n$. This implies the codimension of $\boldsymbol{N}(A) = n$.

We also know $(Z/\boldsymbol{N}(A))' \equiv (\boldsymbol{N}(A))^0$. Hence, $dim(Z/\boldsymbol{N}(A)) = n$ implies $dim(Z/\boldsymbol{N}(A))' = n$. Thus, $dim(\boldsymbol{N}(A))^0 = n$ also. There is then an n dimensional subspace U of Z so that $Z = \boldsymbol{N}(A) \oplus U$.

Let $A_U = A|_U$ for convenience. Then

$$A^{-1}(\boldsymbol{R}(M+N)) = \boldsymbol{N}(A) \oplus A_U^{-1}(\boldsymbol{R}(M+N))$$

Let $k = dim(\boldsymbol{R}(M+N))$. Since A_U is bijective, $k = dim A_U^{-1}(\boldsymbol{R}(M+N))$. For convenience, let $V = A_U^{-1}(\boldsymbol{R}(M+N))$. Then, $A^{-1}(\boldsymbol{R}(M+N)) = A^{-1}(\boldsymbol{N}(A)) \oplus V$. Finally, note

$$\boldsymbol{R}(L) = \{(f,\boldsymbol{d}) \in Z|\boldsymbol{d} - N\int_0^1 f(s)ds \in \boldsymbol{R}(M+N)\}$$

We conclude $\boldsymbol{R}(L) = A^{-1}(\boldsymbol{R}(M+N))$ which implies $\boldsymbol{R}(L) = \boldsymbol{N}(A) \oplus V$. Thus,

$$\begin{aligned}dim(Z/\boldsymbol{R}(L)) &= dim(Z/(\boldsymbol{N}(A) \oplus V)) = dim(Z/\boldsymbol{N}(A)) - k \Longrightarrow\\ dim(Z/\boldsymbol{R}(L)) &= n - k = n - dim(\boldsymbol{R}(M+N)) = dim(\boldsymbol{N}(M+N)) = dim(\boldsymbol{N}(L))\end{aligned}$$

We have shown L is linear Fredholm of index 0.

Now let $S : \Re^n \to \Re^n$ be the projection so that $\boldsymbol{R}(S) = \boldsymbol{N}(M+N) = \boldsymbol{N}(L)$. Then, $I - S$ is the projection onto $\boldsymbol{R}(M+N)$. Let $(M+N)_S = (M+N)|_{\boldsymbol{N}(S)}$. Then $(M+N)_S$ is a bijection from $\boldsymbol{R}(M+N)$ onto $\boldsymbol{R}(M+N)$. Thus, for all $(f,\boldsymbol{d}) \in \boldsymbol{R}(L)$, we have:

$$\begin{aligned}&(M+N)(x(0)) = \boldsymbol{d} - N\int_0^1 f(s)ds \Longleftrightarrow\\ &(M+N)\Big(S(x(0)) + (I-S)(x(0))\Big) = \boldsymbol{d} - N\int_0^1 f(s)ds \Longleftrightarrow\\ &(M+N)(I-S)(x(0)) = \boldsymbol{d} - N\int_0^1 f(s)ds \Longleftrightarrow\end{aligned}$$

$$(I - S)(x(0)) = (M + N)_S^{-1}\left(\boldsymbol{d} - N \int_0^1 f(s)ds\right) \Longleftrightarrow$$
$$x(0) = S(x(0)) + (M + N)_S^{-1}\left(\boldsymbol{d} - N \int_0^1 f(s)ds\right)$$

Recall, $x(t) = x(0) + \int_0^t f(s)ds$. The solution to the boundary value problem occurs when $(M + N)(x(0)) = \boldsymbol{c} - N \int_0^1 f(s)ds$. The above manipulations tell us $x(0) = S(x(0)) + (M+N)_S^{-1}(\boldsymbol{c} - N \int_0^1 f(s)ds)$. Thus, the solution we seek is

$$x(t) \quad = \quad S(x(0)) + (M + N)_S^{-1}\left(\boldsymbol{c} - N \int_0^1 f(s)ds\right) + \int_0^t f(s)ds$$

Define the projection $P_S : X \to \boldsymbol{N}(L) \equiv \boldsymbol{N}(M + N)$ where the constant maps in $\boldsymbol{N}(L)$ are identified with their value $x(0) \in \Re^n$. Thus, $P_S(\boldsymbol{x}) = S(x(0))$.

Hence, since $X = \boldsymbol{N}(L) \oplus \boldsymbol{X_0}$, we have $K_{P_S} : \boldsymbol{R}(L) \to \boldsymbol{X_0}$ is the unique solution $\boldsymbol{x}$ to

$$\begin{aligned} x'(t) &= f(t), a.e. \\ M(x(0)) + N(x(1)) &= \boldsymbol{c} \\ P_S(\boldsymbol{x}) &= S(x(0)) = \boldsymbol{0} \end{aligned}$$

Thus,

$$\begin{aligned} K_{P_S}(f, \boldsymbol{c}) &= (M + N)_S^{-1}\left(\boldsymbol{c} - N \int_0^1 f(s)ds\right) + \int_0^t f(s)ds \\ &= \int_0^t f(s)ds + (M + N)_S^{-1} A(f, \boldsymbol{c}) \end{aligned}$$

Let $T : \Re^n \to \boldsymbol{R}(M + N)$ be a continuous projection where $\Re^n = \boldsymbol{N}(M + N) \oplus \boldsymbol{R}(M + N)$. Define the map Q_T by $Q_T(f, \boldsymbol{c}) = (0, (I - T)A(f, \boldsymbol{c}))$. Now

$$\begin{aligned} (f, \boldsymbol{c}) \in \boldsymbol{R}(L) &\Longleftrightarrow (f, \boldsymbol{c}) \in A^{-1}(\boldsymbol{R}(M + N)) \Longleftrightarrow A(f, \boldsymbol{c}) \in \boldsymbol{R}(M + N) \\ &\Longleftrightarrow TA(f, \boldsymbol{c}) = A(f, \boldsymbol{c}) \Longleftrightarrow (f, \boldsymbol{c}) \in \boldsymbol{N}(Q_T) \end{aligned}$$

Recall, $Z = \boldsymbol{R}(L) \oplus \boldsymbol{M_0}$. We see $Q_T : Z \to \boldsymbol{M_0}$ is a continuous projection.

Thus, letting $\boldsymbol{fc} = (f, \boldsymbol{c})$ for convenience,

$$\begin{aligned} K_{P_S Q_T}(f, \boldsymbol{c})(t) &= K_{P_S}((I - Q_T)\boldsymbol{fc})(t) = K_{P_S}(\boldsymbol{fc} - (0, (I - T)A(\boldsymbol{fc}))(t) \\ &= K_{P_S}(\boldsymbol{fc} - (0, A(\boldsymbol{fc}) - TA(\boldsymbol{fc})))(t) \\ &= K_{P_S}\left(\boldsymbol{fc} - (0, \boldsymbol{c} - N \int_0^1 f(s)ds - TA(\boldsymbol{fc}))\right)(t) \\ &= K_{P_S}\left((f, \boldsymbol{c}) + (0, -\boldsymbol{c} + N \int_0^1 f(s)ds + TA(\boldsymbol{fc}))\right)(t) \\ &= K_{P_S}\left((f, N \int_0^1 f(s)ds + TA(\boldsymbol{fc}))\right)(t) \\ &= \int_0^t f(s)ds + (M + N)_S^{-1}\left(N \int_0^1 f(s)ds + TA(\boldsymbol{fc}) - N \int_0^1 f(s)ds\right)(t) \end{aligned}$$

$$= \int_0^t f(s)ds + (M+N)_S^{-1}TA(f,\mathbf{c}))(t)$$

Homework

Exercise 12.6.1 *Do all of the analysis we have discussed for the case*

$$M = \begin{bmatrix} 1 & 0 & 0 \\ 0 & 1 & 0 \\ 0 & 0 & 1 \end{bmatrix}, \quad N = \begin{bmatrix} -1 & 0 & 0 \\ 0 & -1 & 0 \\ 0 & 0 & -1 \end{bmatrix}, \quad c = \begin{bmatrix} 0 \\ 1 \\ 3 \end{bmatrix}$$

Exercise 12.6.2 *Do all of the analysis we have discussed for the case*

$$M = \begin{bmatrix} 1 & 0 & 0 \\ 0 & 2 & 0 \\ 0 & 0 & 1 \end{bmatrix}, \quad N = \begin{bmatrix} 2 & 0 & 0 \\ 0 & 1 & 0 \\ 0 & 0 & 3 \end{bmatrix}, \quad c = \begin{bmatrix} 1 \\ 2 \\ -1 \end{bmatrix}$$

Exercise 12.6.3 *Do all of the analysis we have discussed for the case*

$$M = \begin{bmatrix} 1 & 1 & 0 \\ 1 & 1 & 0 \\ 0 & 0 & 2 \end{bmatrix}, \quad N = \begin{bmatrix} 0 & 2 & 3 \\ 1 & 0 & 1 \\ 2 & 1 & 0 \end{bmatrix}, \quad c = \begin{bmatrix} 1 \\ 2 \\ -1 \end{bmatrix}$$

Now consider the same problem but this time with $\mathbf{c} = 0$. Let's formulate the problem in another way. Let $X = \{x \in C(I, \Re^n) | M(x(0)) + N(x(1)) = \mathbf{0}\}$. If $\mathbf{c}$ is not zero, then X is not a subspace of $C(I, \Re^n)$, so having $\mathbf{c} = \mathbf{0}$ allows us to fold the boundary conditions into the definition of the space X itself. We let $Z = \mathcal{L}_1(\Re^n, \mathcal{S}, \mu)$. We define the maps

1. $L : dom(L) \rightarrow \mathcal{L}_1(\Re^n, \mathcal{S}, \mu)$ with $dom(L) = \{x \in X | x \text{ is absolutely continuous }\}$ by $L(\boldsymbol{x}) = \boldsymbol{x}'$.

2. $G : I \rightarrow Z$ by $G(t) = f(t)$.

Then, $\boldsymbol{N}(L) = \{\boldsymbol{x} \in dom(L) | \boldsymbol{x}' = 0, \ a.e.\}$. But $\boldsymbol{x}$ is continuous on I and so $\boldsymbol{x}$ must be a constant map. Also, we must have $M(x(0)) + N(x(0)) = 0$. Thus,

$$\boldsymbol{N}(L) = \{\boldsymbol{x} \text{ is a constant map of value } \boldsymbol{d} | \boldsymbol{d} \in \boldsymbol{N}(M+N)\}$$

The problem is $\boldsymbol{x}'(t) = f(t), \ a.e.$ with $M(x(0)) + N(x(1)) = \mathbf{0}$. Thus, $x(t) = x(0) + \int_0^t f(s)ds$ and $M(x(0)) + N(x(1)) = \mathbf{0}$. Now $x(1) = x(0) + \int_0^1 f(s)ds$, so to satisfy the boundary conditions, we must have

$$M(x(0)) + N\left(x(0) + \int_0^1 f(s)ds\right) = \mathbf{0} \Longrightarrow (M+N)(x(0)) = -N \int_0^1 f(s)ds$$

Now define $B : Z \rightarrow \Re^n$ by $B(f) = -N \int_0^1 f(s)ds$. Note,

$$\boldsymbol{R}(L) = \{f \in Z | -N \int_0^1 f(s)ds \in \boldsymbol{R}(M+N)\} = B^{-1}(\boldsymbol{R}(M+N))$$

We have $dim(\boldsymbol{N}(L)) = dim(\boldsymbol{N}(M+N)) < \infty$. Since B is continuous, $\boldsymbol{R}(L) = B^{-1}(\boldsymbol{R}(M+N))$ is closed. Further, $dim(Z/\boldsymbol{R}(L)) = dim(Z/(B^{-1}(\boldsymbol{R}(M+N))))$.

Now $B(f) = -N\int_0^1 f(s)ds$ and so $\boldsymbol{R}(B) = -\boldsymbol{R}(N)$. It is easy to see B maps Z onto $-\boldsymbol{R}(N)$. So $Z/\boldsymbol{N}(B) \equiv -\boldsymbol{R}(N)$ which implies $dim(Z/\boldsymbol{N}(B)) = dim(-\boldsymbol{R}(N)) < \infty$. Now,

$$\begin{aligned} \boldsymbol{N}(B) &= \left\{f \in Z| - N\int_0^1 f(s)ds = \boldsymbol{0}\right\} = \left\{f \in Z| \int_0^1 f(s)ds \in \boldsymbol{N}(-N)\right\} \\ &= \left\{f \in Z| \int_0^1 f(s)ds \in \boldsymbol{N}(N)\right\} \end{aligned}$$

There is, therefore, a subspace U so that $Z = \boldsymbol{N}(B) \oplus U$. Hence, if $B_U = B|_U$, we see

$$B^{-1}(\boldsymbol{R}(M+N)) = \boldsymbol{N}(B) \oplus B_U^{-1}(\boldsymbol{R}(M+N)) = \boldsymbol{N}(B) \oplus V$$

where $V = B_U^{-1}(\boldsymbol{R}(M+N))$. Since B_U is bijective,

$$dim(B_U^{-1}(\boldsymbol{R}(M+N))) = dim(\boldsymbol{R}(M+N))$$

Thus,

$$dim(Z/\boldsymbol{R}(L)) = dim(Z/(\boldsymbol{N}(B) \oplus V)) = dim(-\boldsymbol{R}(N)) - dim(\boldsymbol{R}(M+N))$$

Clearly, defining L this way, does not necessarily lead to L being Fredholm of index 0. Further assumptions would have to be made to allow us to conclude that. For example, if $\boldsymbol{N}(M+N)\boldsymbol{0}$, then $\boldsymbol{N}(L) = \boldsymbol{0}$ and $dim(\boldsymbol{R}(M+N)) = n$. Since $dim(V) = dim(\boldsymbol{R}(M+N))$ and $\boldsymbol{R}(L) \subset Z$, we would have $dim(V) \leq dim(Z/\boldsymbol{N}(B)) = dim(\boldsymbol{R}(N))$. Since $dim(V) = n$, we have $n \leq dim(\boldsymbol{R}(N))$. This tells us $dim(Z/\boldsymbol{R}(L)) = n - n = 0$. So, in this case L is linear Fredholm of index zero.

Homework

Exercise 12.6.4 *Do all of the analysis we have discussed for the case*

$$M = \begin{bmatrix} 1 & 0 & 0 \\ 0 & 1 & 0 \\ 0 & 0 & 1 \end{bmatrix}, \quad N = \begin{bmatrix} -1 & 0 & 0 \\ 0 & -1 & 0 \\ 0 & 0 & -1 \end{bmatrix}, \quad c = \begin{bmatrix} 0 \\ 0 \\ 0 \end{bmatrix}$$

Exercise 12.6.5 *Do all of the analysis we have discussed for the case*

$$M = \begin{bmatrix} 1 & 0 & 0 \\ 0 & 2 & 0 \\ 0 & 0 & 1 \end{bmatrix}, \quad N = \begin{bmatrix} 2 & 0 & 0 \\ 0 & 1 & 0 \\ 0 & 0 & 3 \end{bmatrix}, \quad c = \begin{bmatrix} 0 \\ 0 \\ 0 \end{bmatrix}$$

Exercise 12.6.6 *Do all of the analysis we have discussed for the case*

$$M = \begin{bmatrix} 1 & 1 & 0 \\ 1 & 1 & 0 \\ 0 & 0 & 2 \end{bmatrix}, \quad N = \begin{bmatrix} 0 & 2 & 3 \\ 1 & 0 & 1 \\ 2 & 1 & 0 \end{bmatrix}, \quad c = \begin{bmatrix} 0 \\ 0 \\ 0 \end{bmatrix}$$

Part V

Manifolds

Chapter 13

Manifolds

In our study of topological spaces and topological vector spaces, we have learned how to construct a wide variety of such mathematical objects. In this part of the text, we will study a new type of topological space: this one has its open sets homeomorphic to an open set in $\Re^n$ for some value of integer n. The value of this n can change as we change the open set in the topological space, but in many cases, there is one integer n that does the job. These topological spaces are called manifolds and will give us another choice of mathematical structure we can use in our modeling efforts. The next three chapters are partly based on notes of my student Jay Wilkins, who used them as part of his master's work with us, as well as our reading in differential geometry texts and our own lecture notes. Just remember, in building a model of the immune system or cognition, it is not clear at all what underlying mathematical structures are useful. Hence, in this part, we study another set of potentially useful abstractions.

Let's see if we can illustrate a simple way we might try to find levels of meaning in a signal which happens to be composed of cytokine messenging molecules. We discuss this more in Chapter 18, but here we will just assume we have been given some measurements. We have actually discussed this example in (Peterson (97) 2016) but, it is useful to go over it again as it gives perspective. We start with three specific cytokine signals, $\boldsymbol{V_1}$, $\boldsymbol{V_2}$ and $\boldsymbol{V_3}$, and we assume there is a set of 10 parameters that characterize any cytokine; hence, a given cytokine can be identified with a parameter vector in $\Re^{10}$. In a typical flow cytometry experiment, different cytokines in the raw data stream collected are associated with a list of parameters and for our illustration, we are assuming 10 parameters can be used to distinguish any cytokines of interest to us. The dimension 10 is not important here, of course, and is just used for illustration. Let's assume the cytokine signals are represented by the parameter vectors below and the possible signals sent to the brain are comprised of linear combinations of the signals $\boldsymbol{V_1}$, $\boldsymbol{V_2}$ and $\boldsymbol{V_3}$. Thus, for example we can look at a specific collection of such signals given as follows:

$$\boldsymbol{V_1} = \begin{bmatrix} 1 \\ 2 \\ 3 \\ -4 \\ 5 \\ -2 \\ 1 \\ 10 \\ 4 \\ -5 \end{bmatrix}, \quad \boldsymbol{V_2} = \begin{bmatrix} -1 \\ 12 \\ 7 \\ 8 \\ 8 \\ 3 \\ -2 \\ 14 \\ 1 & 6 \end{bmatrix}, \quad \boldsymbol{V_3} = \begin{bmatrix} 4 \\ 9 \\ -3 \\ 1 \\ 7 \\ 4 \\ 6 \\ 4 \\ 19 \\ 8 \end{bmatrix}, \quad \begin{aligned} \boldsymbol{V_4} &= 2\,\boldsymbol{V_1} \\ \boldsymbol{V_5} &= 7\,\boldsymbol{V_2} \\ \boldsymbol{V_6} &= 8\,\boldsymbol{V_3} \\ \boldsymbol{V_7} &= 2\,\boldsymbol{V_1} + 4\,\boldsymbol{V_2} - 8\,\boldsymbol{V_3} \\ \boldsymbol{V_8} &= 13\,\boldsymbol{V_2} \\ \boldsymbol{V_9} &= 22\,\boldsymbol{V_3} \\ \boldsymbol{V_{10}} &= -3\,\boldsymbol{V_1} - 4\,\boldsymbol{V_2} \end{aligned}$$

We can then construct the matrix $\boldsymbol{A}$ from these ten signals using each signal as a column of $\boldsymbol{A}$. We can then find the standard $\boldsymbol{LU}$ decomposition of $\boldsymbol{A}$. We see that $\boldsymbol{U}$ has 7 zero rows, which tells us that only 3 of the vectors $\boldsymbol{V_i}$ used to construct $\boldsymbol{A}$ are linearly independent, and so the dimension of the kernel $\boldsymbol{K}$ is 7. We can compute the basis for the kernel of $\boldsymbol{A}$ and find its basis is $\{\boldsymbol{K_1}, \boldsymbol{K_2}, \boldsymbol{K_3}, \boldsymbol{K_4}, \boldsymbol{K_5}, \boldsymbol{K_6}, \boldsymbol{K_7}\}$. We know the standard unit vectors $\boldsymbol{e_1}$ to $\boldsymbol{e_3}$ are mapped by $\boldsymbol{A}$ into the range vectors $\boldsymbol{V_1}$ to $\boldsymbol{V_3}$. Hence, we know we can write any $\boldsymbol{x}$ in the input space as

$$\boldsymbol{x} = \sum_{i=1}^{3} < \boldsymbol{x}, \boldsymbol{e_i} > \boldsymbol{e_i} + \sum_{j=1}^{7} < \boldsymbol{x}, \boldsymbol{K_j} > \boldsymbol{K_j}$$

The second sum is clearly in the subspace $\boldsymbol{K}$ and we can use the first sum to define $\boldsymbol{x_p} = \sum_{i=1}^{3} < \boldsymbol{x}, \boldsymbol{e_i} > \boldsymbol{e_i}$. We can define an equivalence relationship between vectors $\boldsymbol{x}$ and $\boldsymbol{y}$ by saying $\boldsymbol{x}$ is related to $\boldsymbol{y}$, $\boldsymbol{x} \sim \boldsymbol{y}$ if and only if $\boldsymbol{x} - \boldsymbol{y}$ lies in $\boldsymbol{K}$. The collection of all vectors equivalent to $\boldsymbol{x}$ under the relationship $\sim$ is called the equivalence class of $\boldsymbol{x}$ which is denoted by $[\boldsymbol{x}]$; i.e.

$$[\boldsymbol{x}] = \boldsymbol{x_p} + \boldsymbol{K}$$

This is the first step. Next, the projection of $\boldsymbol{e_1}$ to $\boldsymbol{e_3}$ onto the kernel, $\boldsymbol{W_1}$ to $\boldsymbol{W_3}$, can easily be calculated using

$$\boldsymbol{W_i} = \sum_{j=1}^{7} < \boldsymbol{e_i}, \boldsymbol{K_j} > \boldsymbol{K_j}.$$

The vectors $\boldsymbol{W_1}$ to $\boldsymbol{W_3}$ are simply projected to $\boldsymbol{K}$ and this is not an orthogonal projection. In this case, these vectors are linearly independent. The vector $\boldsymbol{x_p}$ is a linear combination of the vectors $\boldsymbol{e_1}$ through $\boldsymbol{e_3}$ and we note each $\boldsymbol{e_i}$ can be written as

$$\boldsymbol{e_i} = (\boldsymbol{e_i} - \boldsymbol{W_i}) + \boldsymbol{W_i}.$$

Now look at the projection of $\boldsymbol{W_1}$, $\boldsymbol{W_2}$ and $\boldsymbol{W_3}$ onto the subspace spanned by $\boldsymbol{K_4}$ through $\boldsymbol{K_7}$, $\boldsymbol{Z_1}$, $\boldsymbol{Z_2}$ and $\boldsymbol{Z_3}$, using

$$\boldsymbol{Z_i} = \sum_{j=4}^{7} < \boldsymbol{W_i}, \boldsymbol{K_j} > \boldsymbol{K_j}.$$

The vectors $\boldsymbol{Z_1}$ through $\boldsymbol{Z_3}$ live in the subspace spanned by $\boldsymbol{K_4}$ through $\boldsymbol{K_7}$. The vectors $\boldsymbol{W_i} - \boldsymbol{Z_i}$ live outside that subspace but are still in the kernel. So now we have a collection of vectors satisfying

- $\{\boldsymbol{e_1} - \boldsymbol{W_1}, \boldsymbol{e_2} - \boldsymbol{W_2}, \boldsymbol{e_3} - \boldsymbol{W_3}\}$ is a basis for the part of $\boldsymbol{x_p}$ not in K. Call this subspace $\boldsymbol{E} - \boldsymbol{W}$. This is 3 dimensional.

- $\{\boldsymbol{W_1} - \boldsymbol{Z_1}, \boldsymbol{W_2} - \boldsymbol{Z_2}, \boldsymbol{W_3} - \boldsymbol{Z_3}\}$ is a basis for the part of the projection of $\boldsymbol{x_p}$ found in K but outside the subspace spanned by $\{\boldsymbol{Z_1}, \boldsymbol{Z_2}, \boldsymbol{Z_3}\}$. Call this subspace $\boldsymbol{W} - \boldsymbol{Z}$ and call the subspace spanned by the $\boldsymbol{Z_i}$, $\boldsymbol{Z}$. By construction $\boldsymbol{Z}$ is a subspace of the span of $\{\boldsymbol{K_4}, \boldsymbol{K_5}, \boldsymbol{K_6}, \boldsymbol{K_7}\}$.

- $\{\boldsymbol{Z_1}, \boldsymbol{Z_2}, \boldsymbol{Z_3}\}$, which is a subspace of the span of $\{\boldsymbol{K_4}, \boldsymbol{K_5}, \boldsymbol{K_6}, \boldsymbol{K_7}\}$. Since each the vectors that span $\boldsymbol{Z}$ are independent, this part is 3 dimensional also.

- Thus, 6 dimensions of the 7 dimensional space $\boldsymbol{K}$ have been used up by this projection process. One of the vectors $\boldsymbol{K_4}$ to $\boldsymbol{K_7}$ spans the one dimensional subspace of $\boldsymbol{K}$ left. We will call this vector $\boldsymbol{\Omega}$ and the subspace generated by $\boldsymbol{\Omega}$, $\boldsymbol{K_\Omega}$.

From the discussions above, it is clear we have found a decomposition of the input space of the form

$$\Re^{10} = (\boldsymbol{E} - \boldsymbol{W}) \oplus (\boldsymbol{W} - \boldsymbol{Z}) \oplus \boldsymbol{Z} \oplus \boldsymbol{K}_{\Omega}.$$

We can easily check a basis for the input space $\Re^{10}$ is given by

$$\boldsymbol{I_B} = \{\{\boldsymbol{e_1} - \boldsymbol{W_1}, \boldsymbol{e_2} - \boldsymbol{W_2}, \boldsymbol{e_3} - \boldsymbol{W_3}\}, \{\boldsymbol{W_1} - \boldsymbol{Z_1}, \boldsymbol{W_2} - \boldsymbol{Z_2}, \boldsymbol{W_3} - \boldsymbol{Z_3}\}, \{\boldsymbol{Z_1}, \boldsymbol{Z_2}, \boldsymbol{Z_3}\}, \boldsymbol{K_7}\}$$

So given $\boldsymbol{x}$ in $\Re^{10}$ as an input, let $\boldsymbol{x} = \boldsymbol{x_p} + \boldsymbol{x_k}$ where $\boldsymbol{x_p}$ is in the span of $\boldsymbol{e_1}$ to $\boldsymbol{e_3}$ and $\boldsymbol{x_k}$ is in $\boldsymbol{K}$. Then the equivalence class $[\boldsymbol{x}] = \boldsymbol{x_p} + \boldsymbol{K}$ is the first decomposition of the input signal $\boldsymbol{x}$. Thus at this point, we can say the input space can be written as the direct sum $\Re^{10} = \Re^3 \oplus \boldsymbol{K}$. We know $\boldsymbol{e_1} \Longrightarrow \boldsymbol{V_1}, \boldsymbol{e_2} \Longrightarrow \boldsymbol{V_2}$ and $\boldsymbol{e_3} \Longrightarrow \boldsymbol{V_3}$, under $\boldsymbol{A}$. Further, we have seen we can project $\boldsymbol{e_1}$ to $\boldsymbol{e_3}$ onto $\boldsymbol{K}$ giving vectors $\boldsymbol{W_1}$ to $\boldsymbol{W_3}$. So we can say $\boldsymbol{x_p} = \boldsymbol{y_p} + \boldsymbol{W}$, where $\boldsymbol{W}$ is the span of $\boldsymbol{W_1}$ to $\boldsymbol{W_3}$ and $\boldsymbol{y_p}$ is in $\boldsymbol{K}/\boldsymbol{W}$. We further decompose the signal using the vectors $\boldsymbol{Z_1}$, $\boldsymbol{Z_2}$ and $\boldsymbol{Z_3}$ which are the projections of the $\boldsymbol{W_i}$ onto the subspace spanned by $\boldsymbol{K_4}$ through $\boldsymbol{K_7}$. The subspace determined by the vectors $\boldsymbol{Z_i}$ is called $\boldsymbol{Z}$ for convenience. Then, the differences $\boldsymbol{W_i} - \boldsymbol{Z_i}$ lie outside of the span of $\boldsymbol{K_4}$ through $\boldsymbol{K_7}$, which we will denote by $\boldsymbol{L}$. Then we have $\boldsymbol{y_p} = \boldsymbol{z_p} + \boldsymbol{Z}$, where z_p is in the part of $\boldsymbol{Z}$ not in $\boldsymbol{L}$. We see $\boldsymbol{L}$ is three dimensional. Putting it all together, we have

$$\boldsymbol{x} = \boldsymbol{x_p} + \boldsymbol{K} = \boldsymbol{y_p} + \boldsymbol{W} + \boldsymbol{K}/\boldsymbol{W} = \boldsymbol{z_p} + \boldsymbol{Z} + (\boldsymbol{W} - \boldsymbol{Z}) + \boldsymbol{K}/(\boldsymbol{W} - \boldsymbol{Z})/\boldsymbol{Z}$$

That is

$$\Re^{10} = \Re^4 \oplus \Re^3 \oplus \Re^3 \oplus \Re^1 = <\boldsymbol{e_i} - \boldsymbol{W_i}> \oplus <\boldsymbol{W_i} - \boldsymbol{Z_i}> \oplus <\boldsymbol{Z_i}> \oplus <\boldsymbol{\Omega}>$$

Hence, a signal $\boldsymbol{x}$ in $\Re^{10}$ has a representation $(\mathscr{C}_1, \mathscr{C}_2, \mathscr{C}_3, \mathscr{C}_4)$. Then

$$\begin{aligned}
\mathscr{C}_1 &\in span\{\boldsymbol{e_1} - \boldsymbol{W_1}, \boldsymbol{e_2} - \boldsymbol{W_2}, \boldsymbol{e_3} - \boldsymbol{W_3}\}, \ \mathscr{C}_2 \in span\{\boldsymbol{W_1} - \boldsymbol{Z_1}, \boldsymbol{W_2} - \boldsymbol{Z_2}, \boldsymbol{W_3} - \boldsymbol{Z_3}\} \\
\mathscr{C}_3 &\in span\{\boldsymbol{Z_1}, \boldsymbol{Z_2}, \boldsymbol{Z_3}\}, \ \mathscr{C}_4 \in span\{\Omega\}
\end{aligned}$$

Next, truncate and discretize the signals on each axis of $\Re^{10}$ into $[-N, N]$ for a choice of positive integer N. Then

$$\mathscr{C}_1 \in [-N, N]^3, \ \mathscr{C}_2 \in [-N, N]^3, \ \mathscr{C}_3 \in [-N, N]^3, \ \mathscr{C}_4 \in [-N, N].$$

For example, say $N = 10$ and we have $s_1 = (-3, 5, 2)$, $s_2 = (1, 5, -8)$, $s_3 = (0, 0, 3)$ and $s_4 = -4$. Now convert to a positive integer by taking the integer j in each slot and converting to $j + 10$. This maps the integers from $-N$ to N into the integers in $[0, 2N]$. This gives

$$\begin{aligned}
\mathscr{C}_1 &= (-3, 5, 2) \longrightarrow (7, 15, 12), \quad \mathscr{C}_2 = (1, 5, -8) \longrightarrow (11, 15, 12) \\
\mathscr{C}_3 &= (0, 0, 3) \longrightarrow (10, 10, 13), \quad \mathscr{C}_4 = -4 \longrightarrow 6
\end{aligned}$$

This leads to the **bar code** which consists of 4 codes which can be determined a variety of ways. Here, we take the integer entry, say -13, which was the raw value, convert to $[0, 20]$ and then use the value 7 to set a 1 in the column associated with that value. This gives us a 20×10 matrix which we call a bar code, $\mathscr{B}_{\boldsymbol{x}}$, which consists of 4 blocks of messages, M_1 to M_4. Hence, the signal $\boldsymbol{x}$ is mapped into a bar code $\mathscr{B}_{\boldsymbol{x}}$. This is an example of a signal decomposition determined by this particular matrix $\boldsymbol{A}$. A nice way to interpret this decomposition strategy, is to go back to the original matrix $\boldsymbol{A}$. The input space is $\Re^{10}$ and imagine that we have 10 independent signals $\boldsymbol{T_1}$ to $\boldsymbol{T_{10}}$ which we can identify with the generic basis vectors $\boldsymbol{e_1}$ to $\boldsymbol{e_{10}}$. The only combinations of these 10 signals that are valid for the purpose of generating a signal response are the linearly independent vectors $\boldsymbol{V_1}$, $\boldsymbol{V_2}$, and $\boldsymbol{V_3}$. The 7 dimensional kernel we see represents the fact that most signals do not occur in useful combinations for the purpose of triggering a response. The useful signals $\boldsymbol{x}$ then have a decomposition as we have discussed. From that we can build a model of the probability of signal response. Let p_x denote the number of columns that is different from 10 (the original 0 response in the raw discretization) in the barcode $\mathscr{B}$. Then we define the **probability** of signal response, $\mathscr{P}_{\boldsymbol{x}}$ to be $\mathscr{P}_{\boldsymbol{x}} = 10p_x$. This is a simple model, but it says that there is a guaranteed response if all 10 columns do not contain a 10 component, which indicates the barcode is fully generated. For each column that does not contain a 10, we reduce the probability of response by 10%.

We can take this a step further. The mapping into $[-10, 10]$ is replaced by a mapping into Z_k, the set of integers mod k. For example, we could take the original signal decomposition and do a different discretization. Letting $\lfloor t \rfloor$ denote the floor of the real number t, we first do this:

$$
\begin{aligned}
\mathscr{C}_{\mathbf{1}} = (s_1, s_2, s_3) &\longrightarrow (\lfloor s_1 \rfloor, \lfloor s_2 \rfloor, \lfloor s_3 \rfloor), \quad \mathscr{C}_{\mathbf{2}} = (s_4, s_5, s_6) \longrightarrow (\lfloor s_4 \rfloor, \lfloor s_5 \rfloor, \lfloor s_6 \rfloor) \\
\mathscr{C}_{\mathbf{3}} = (s_7, s_8, s_9) &\longrightarrow (\lfloor s_7 \rfloor, \lfloor s_8 \rfloor, \lfloor s_9 \rfloor), \quad \mathscr{C}_{\mathbf{4}} = \quad (s_{10}) \longrightarrow \lfloor s_{10} \rfloor
\end{aligned}
$$

Now we have replaced all the raw signal components by integer values. Next, assume

$$
\begin{aligned}
(\lfloor s_1 \rfloor, \lfloor s_2 \rfloor, \lfloor s_3 \rfloor) &\in (Z_7, Z_7, Z_7), \quad (\lfloor s_4 \rfloor, \lfloor s_5 \rfloor, \lfloor s_6 \rfloor) \in (Z_9, Z_9, Z_9), \\
(\lfloor s_7 \rfloor, \lfloor s_8 \rfloor, \lfloor s_9 \rfloor) &\in (Z_4, Z_4, Z_4), \quad \lfloor s_{10} \rfloor \in Z_5
\end{aligned}
$$

Then, after the processing $\lfloor m \rfloor \longrightarrow m\%k$, we have

$$
\begin{aligned}
(\lfloor s_1 \rfloor, \lfloor s_2 \rfloor, \lfloor s_3 \rfloor) &\longrightarrow (\lfloor s_1 \rfloor\%7, \lfloor s_2 \rfloor\%7, \lfloor s_2 \rfloor\%7) = (\lfloor -3 \rfloor\%7, \lfloor 5 \rfloor\%7, \lfloor 2 \rfloor\%7) = (4, 5, 2) \\
(\lfloor s_4 \rfloor, \lfloor s_5 \rfloor, \lfloor s_6 \rfloor) &\longrightarrow (\lfloor s_4 \rfloor\%9, \lfloor s_5 \rfloor\%9, \lfloor s_6 \rfloor\%9) = (\lfloor 1 \rfloor\%9, \lfloor 5 \rfloor\%9, \lfloor -8 \rfloor\%9) = (1, 5, 1) \\
(\lfloor s_7 \rfloor, \lfloor s_8 \rfloor, \lfloor s_9 \rfloor) &\longrightarrow (\lfloor s_7 \rfloor\%4, \lfloor s_8 \rfloor\%4, \lfloor s_9 \rfloor\%4) = (\lfloor 0 \rfloor\%4, \lfloor 0 \rfloor\%4, \lfloor 3 \rfloor\%4) = (0, 0, 3) \\
\lfloor s_{10} \rfloor &\longrightarrow \lfloor s_{10} \rfloor\%5 = \lfloor -4 \rfloor\%5 = (1).
\end{aligned}
$$

In this case, the resulting bar code is not a matrix as the columns are of different sizes. It helps to see this visually. In Figure 13.1, we denote the *active* part of the code in each column with a closed circle. There are four blocks here, so we use four colors. In a Z_7 column, there are 6 ways to fill the column with just one 1 and a seventh choice which is all zeros. For our example, there are $7^3 \times 9^3 \times 4^3 \times 5$ or $343 \times 729 \times 64 \times 5 \approx 80$ million valid signals. Some of these signals have columns that are all zeros. There are $6^3 \times 8^3 \times 3^3 \times 4$ or $216 \times 512 \times 27 \times 4 \approx 12$ million signals having no zero columns which is about 15%. There are also signals which have at least one zero column. The calculations for how many there are of these are quite complicated in our example as our columns are not equal sizes. But just to give you the feel of it, let's look at how many ways there are to have one column of zeros. There are $\binom{10}{1} = 10$ ways to have one column be all zeros. The probability of activating a target here is 90% in this case where there are 10 columns.

- Column 1, 2 or 3 is zero giving $6^2 \times 8^3 \times 4^3 \times 4$ or $36 \times 512 \times 64 \times 4 \approx 5$ million signals.
- Column 4, 5 or 6 is zero giving $6^3 \times 8^2 \times 4^3 \times 4$ or $216 \times 64 \times 64 \times 4 \approx 3.5$ million signals.
- Column 7, 8 or 9 is zero giving $6^3 \times 8^3 \times 4^2 \times 4$ or $216 \times 512 \times 16 \times 4 \approx 7.0$ million signals.
- Column 10 is zero giving $6^3 \times 8^3 \times 4^3$ or $216 \times 512 \times 64 \approx 7.0$ million signals.

Thus, there are about 22.5 million signals with one column of zeros or about 28%. The remaining possibilities of signals having more than one zero column fill out the remainder of the signals. We see that about 85% of the signals correspond to signals with at least one column of zeros. A signal which is not valid might correspond to actions that should not be taken. If we assume that all signals with nonzero components potentially correspond to an action, albeit one we may not wish to take, we note we have a correspondence mapping of the form $\boldsymbol{x} \in \Re^{10} \longrightarrow \mathscr{T} \cup \mathscr{T}^c$ where $\mathscr{T}$ is the set of signal targets with all nonzero columns and $\mathscr{T}^c$ are the rest of the signals. Signals in $\mathscr{T}^c$ should not be used to create target responses. Hence, a signal processing error related to such an improper signal being used to create a target response would be a type of security violation. A simple decomposition component calculation error could take an invalid signal and inappropriately add a 1 to enough columns to cause the signal to create a target activation signal. This would mean a part of the signal space that should not be used to create responses is actually being used to do that. A partial sketch

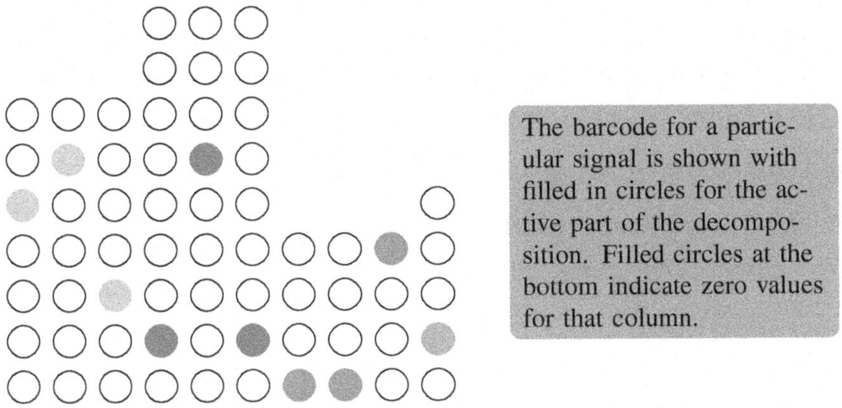

Figure 13.1: A visual barcode.

of the modeling process is as follows. An R^n sensory input signal collection can be discretized into a cubical set. In a neurological/ biological situation the choice of discretization would be guided by the underlying science. This gives rise to the computational sequence where we denote the signal by $\boldsymbol{S}$ and the cubical sets by the family $\boldsymbol{C}$. We then have $\boldsymbol{S} \Longrightarrow \boldsymbol{C} \Longrightarrow \sum_{i=1}^{N} \oplus \boldsymbol{G_i}$ where $\boldsymbol{G_i}$ is a group generated by an element $< y_i >$ which can be a variety of objects. Given the decomposition, algorithms in (Kaczynski et al. (66) 2004) will decompose the signal chain into $x \in \sum_{i=1}^{N} \oplus \boldsymbol{G_i}$. This technique is used to experimentally explore what the group decomposition could be for a given set of $\Re^n$ data as discussed in (G. Carlsson (35) 2009).

Homework

Exercise 13.0.1 *We start with three specific cytokine signals,* `V1`, `V2` *and* `V3`, *and assume a given cytokine is identified with a parameter vector in $\Re^{10}$ which comes from a flow cytometry measurement. All of the cytokine signals are linear combinations of these three signals above. In MATLAB, we set up the ten signals as follows.*

Listing 13.1: **Cytokine Family**

```
V1 = [1;2; -1;4;1.3; -2.5;7.8;10.1; -2.6;11.4];
V2 = [ -1;4.5;1.3;2.7;11.3; -22.5;70.8; -12.1; -23.4;1.4];
V3 = [9;1.15;21.3;11.7;1.95; -2.5;9.8;24.1; -1.14;10.6];
V4 = 4.5*V1;
V5 = 2*V2 + 3*V3;
V6 = 11*V4+2*V5;
V7 = 2*V1 + 3*V2 - 5*V5;
V8 = -2*V4;
V9 = 6*V7;
V10 = 4*V4 - 2*V5 + 8*V8;
A = [V1,V2,V3,V4,V5,V6,V7,V8,V9,V10];
```

Use standard linear algebra codes to do the analysis of these signals as described in the text.

Exercise 13.0.2 *Assume a signal x in $\Re^{10}$ has a representation $\{x_1, x_2, x_3, x_4\}$ in (V_1, V_2, V_3, V_4) where V_1 is three dimensional, V_2 is 3 dimensional, V_3 is 2 dimensional and V_4 is 2 dimensional. Truncate and discretize the signals on each axis of $\Re^{10}$ into $[-10, 10]$. Then*

$$x_1 \in [-10, 10]^3, \; x_2 \in [-10, 10]^3, \; x_3 \in [-10, 10]^2, \; x_4 \in [-10, 10]^2$$

Let $x_1 = (-4, 6.5, 21)$, $x_2 = (11, 5, -18)$, $x_3 = (1, 2.5)$ and $x_4 = (-4, 7)$. Convert each x_i to a positive integer in $[0, 20]$ following the procedure in the text giving c_1, c_2, c_3 and c_4. This leads to the **barcode** *which is a 20×10 matrix whose value is 0 in a column unless the row value of the column is the value of c_i. These columns are the messages M_1 to M_4. Find the barcode for this example.*

Exercise 13.0.3 *Redo the exercise above using $N = 25$.*

Exercise 13.0.4 *Redo the exercise above using $N = 100$.*

Exercise 13.0.5 *We use the example signal from before. First, map each component using the floor function.*

$$\begin{aligned} c_1 = (\lfloor s_1 \rfloor, \lfloor s_2 \rfloor, \lfloor s_3 \rfloor), \quad c_2 = (\lfloor s_4 \rfloor, \lfloor s_5 \rfloor, \lfloor s_6 \rfloor) \\ c_3 = (\lfloor s_7 \rfloor, \lfloor s_8 \rfloor), \quad c_4 = (\lfloor s_9 \rfloor, \lfloor s_{10} \rfloor) \end{aligned}$$

We have then replaced all the raw signal components by integers. Next, assume these signals live in some Z_p scheme:

$$\begin{aligned} (\lfloor s_1 \rfloor, \lfloor s_2 \rfloor, \lfloor s_3 \rfloor) &\in (Z_7, Z_7, Z_7), \quad (\lfloor s_4 \rfloor, \lfloor s_5 \rfloor, \lfloor s_6 \rfloor) \in (Z_9, Z_9, Z_9), \\ (\lfloor s_7 \rfloor, \lfloor s_8 \rfloor) &\in (Z_4, Z_4), \quad (\lfloor s_9 \rfloor, \lfloor s_{10} \rfloor) \in (Z_5, Z_5) \end{aligned}$$

The resulting barcode is not a matrix as the columns are of different sizes. For the signal we used before, find this barcode.

Exercise 13.0.6 *Do the exercise above except the signal is now $x_1 = (-4, 6.5, 21)$, $x_2 = (11, 5, -18)$, $x_3 = (1, 2.5)$ and $x_4 = (-4, 7)$ and the decomposition is*

$$\begin{aligned} (\lfloor s_1 \rfloor, \lfloor s_2 \rfloor, \lfloor s_3 \rfloor) &\in (Z_5, Z_5, Z_5), \quad (\lfloor s_4 \rfloor, \lfloor s_5 \rfloor, \lfloor s_6 \rfloor) \in (Z_3, Z_3, Z_3), \\ (\lfloor s_7 \rfloor, \lfloor s_8 \rfloor, \lfloor s_9 \rfloor) &\in (Z_6, Z_6), \quad \lfloor s_{10} \rfloor \in (Z_3, Z_3) \end{aligned}$$

This example shows one way to find a subgraph type structure within the data which is quite limited in scope as it uses a simple $Ax = b$ type model and nullspace decompositions. But the $\Re^n$ point cloud data collected might be better modeled by a mathematical structure which is locally $\Re^n$. By this, we mean that open sets in our data are homeomorphic to an open set in some $\Re^m$. The dimension m need not match n, of course. This kind of structure is called a manifold and we need to discuss this very carefully. It is by no means obvious how to make this modeling choice. The above remarks are similar in spirit to principle component analysis and we are simply saying that maybe this is not the best choice to make. Let's begin.

13.1 Manifolds: Definitions and Properties

Based on what we have said, a manifold is a topological space that is locally Euclidean. Of course, we have to define this carefully and precisely and work out all the subsequent properties. Here is the formal definition whiich you will notice is quite abstract. However, do not be put off by that as we will gain much explanatory power from this level of abstraction. Note to understand this properly it is important you have a good understanding of how to choose open sets in a topological space to form an appropriate topology. Also, for each manifold we have to choose a useful topological space.

Definition 13.1.1 Manifolds

A manifold is a 3-tuple of elements $(M, \mathcal{T}, \{(\phi_\gamma, U_\gamma)\}_{\gamma\in\Gamma})$, where M is a Hausdorff, second countable topological space with topology $\mathcal{T}$, and $\{U_\gamma\}_{\gamma\in\Gamma}$ is a collection of neighborhoods, indexed by some set, Γ, with a corresponding collection of mappings, $\{\phi_\gamma\}_{\gamma\in\Gamma}$, satisfying the following properties.

1. *The collection $\{U_\gamma\}$ is an open covering of M.*
2. *Each map, ϕ_γ, is a homeomorphism, or a continuous bijection with a continuous inverse, from U_γ to an open subset of some Euclidean space.*

Since the manifold is second countable, we can assume that the collection of neighborhoods, $\{U_\gamma\}_{\gamma\in\Gamma}$, and the collection of mappings, $\{\phi_\gamma\}_{\gamma\in\Gamma}$, are countable. So, the collection of pairs of neighborhoods and maps can be denoted as $\{(\phi_i, U_i)\}_{i\geq 1}$. Usually, however, we will just refer to a single neighborhood U and its corresponding mapping ϕ_U. For us, we assume that the Euclidean spaces containing the open sets $\phi_U(U)$ are of finite dimension and are endowed with the usual Euclidean coordinate system. This is, strictly speaking, not necessary, but we do not wish to go into that level of generality here.

The sets $\{U_i\}$ are called coordinate neighborhoods, and the mappings $\{\phi_i\}$ are called coordinate mappings. Each pair (ϕ, U) is called a coordinate chart, or a coordinate system, and the collection $\{(\phi_i, U_i)\}_{i\geq 1}$ is called a coordinate covering, or atlas. The reason for this terminology is that these charts can be used to establish local coordinate systems around any point $p \in M$. If $p \in M$, then there is some chart (ϕ, U) such that $p \in U$ and $\phi(U)$ is an open subset of $\Re^m$ for some m. We can define the coordinates of $q \in M$ to be the coordinates of $\phi(q) \in \phi(U)$; i.e. the manifold is locally Euclidean.

The **dimension** of a manifold, M, at a point $p \in M$, is defined to be the dimension of the Euclidean space containing $\phi(U)$, where (ϕ, U) is a chart such that $p \in U$. The point p may lie in several different coordinate neighborhoods, so in order to show that dimension is a well-defined concept, we must show that each one is homeomorphic to a subset of the same Euclidean space. Now, assume there are two coordinate charts, (ϕ, U) and (ψ, V), such that $p \in U \cap V$, and suppose $\phi(U) \subset \Re^n$ and $\psi(V) \subset \Re^m$. Then $U \cap V$ is a neighborhood of p on which we have two coordinate mappings, ϕ and ψ. Moreover, $\psi(U \cap V)$ and $\phi(U \cap V)$ are open subsets of $\Re^m$ and $\Re^n$, respectively. The mapping $\phi \circ \psi^{-1} : \psi(U \cap V) \to \phi(U \cap V)$ is a homeomorphism, as it is a composition of homeomorphisms. That is, $\phi \circ \psi^{-1}$ maps an open subset of $\Re^m$ homeomorphically to an open subset of $\Re^n$. Two open subsets of Euclidean spaces cannot be homeomorphic unless the Euclidean spaces are of the same dimension as you can prove in the exercises. Hence, we must have $m = n$, and the dimension of a manifold is well-defined. In other words, the dimension of a manifold does not depend on the particular coordinate system chosen around a point. Coordinate independence is a very important aspect of global differential geometry. Local coordinate systems are useful, but it is better to derive results that are independent of any particular coordinate system

The dimension of a manifold is thus only defined locally and a manifold need not be of constant dimension. For example, the open strip $M = \{(x, y) \in \Re^2 \ : \ 0 < x < 1\}$ is a manifold with only one coordinate chart, namely the inclusion map from M to $\Re^2$. It is also of constant dimension, as M is everywhere locally homeomorphic to $\Re^2$. Likewise, the set $M = \{(x, y) \in \Re^2 \ : \ 0 < x < 1\} \bigcup \{(x, y) \in \Re^2 \ : \ x = 2\}$ is also a manifold, this time with only two coordinate charts. However, M is not of constant dimension. The point $(.5, .5)$ has a neighborhood that is homeomorphic to $\Re^2$, but the point $(2, 1)$ has a neighborhood that is homeomorphic to $\Re$.

In geometry and physics, most manifolds are of constant dimension. It is not clear whether manifolds in autoimmune or cognitive modeling share that property. However, let's assume for the moment, the manifolds of interest to us are of constant dimension. Consequently, we will use the following convention when referring to manifolds. We will denote a manifold by capital letters M, N, L, etc., and this is meant to imply that these manifolds are of constant dimension m, n, l, etc., respectively. Thus, we will refer to an m-dimensional manifold, M, and n-dimensional manifold, N, etc.

Homework

Exercise 13.1.1 *Consider the surface $z = x^2 + y^2$ in $\Re^3$. If you pick any circle $B(p, r)$ in the plane $\Re^2$, where $p = (x_0, y_0)$, the part of the surface lying above this circle is an open set in the manifold determined by the surface. Conversely, any open set V you draw on this surface projects down to an open set in the plane U. There are many C^∞ homeomorphisms $\phi : V \to U$. Find two such homeomorphisms for some V and U.*

Exercise 13.1.2 *Consider the surface $x^2 + y^2 + z^2 = 10$ in $\Re^3$. If you pick any circle $B(p, r)$ in the plane $\Re^2$, where $p = (x_0, y_0)$, the part of the surface lying above this circle is an open set in the manifold determined by the surface. Also, the part of the surface lying below this circle is an open set in the manifold determined by the surface. Conversely, any open set V you draw on this surface projects to an open set in one of the coordinate planes: $x - y$, $x - z$ or $y - z$. There are many C^∞ homeomorphisms $\phi : V \to U$. Find two such homeomorphisms for some V and U.*

Exercise 13.1.3 *Consider the surface $x^2 + y^2 = z^2$ in $\Re^3$. If you pick any circle $B(p, r)$ in the plane $\Re^2$, where $p = (x_0, y_0)$, the part of the surface lying above this circle is an open set in the manifold determined by the surface. Also, the part of the surface lying below this circle is an open set in the manifold determined by the surface. Conversely, any open set V you draw on this surface projects to an open set in the plane U. Note an open set in this manifold could consist of an open set on the top sheet and an open set on the bottom sheet. So this manifold can't be connected. There are many C^∞ homeomorphisms $\phi : V \to U$. Find two such homeomorphisms for some V and U.*

A manifold with a boundary is a space with both interior and boundary points. For example,

1. A rectangle in $\Re^2$, which includes its interior points, is a 2-manifold with a one dimensional boundary.

2. A cube in $\Re^3$, which includes the interior of each face, is a 2- manifold with a one dimensional boundary.

Every interior point of a manifold with a boundary has a neighborhood which is homeomorphic to the open ball centered at 0 in $\Re^n$, $\sum_{i=1}^n x_i^2 < 1$. Every boundary point homeomorphic to what is called a half-ball centered at 0 in $\Re^n$, $\sum_{i=1}^n x_i^2 < 1$ with $x_1 \geq 0$ and each boundary point corresponds to a point in the ball with $x_1 = 0$.

Exercise 13.1.4 *Consider the surface formed by the intersection of $z = x^2 + y^2$ and $z = 10$. Show this is a manifold with a boundary and explicitly find the homeomorphisms that correspond to the boundary points.*

Exercise 13.1.5 *Consider the set formed by the interior and boundary of a rectangle in $\Re^2$. Show this is a manifold with a boundary and explicitly find the homeomorphisms that correspond to the boundary points. What happens at the corners?*

Exercise 13.1.6 *Consider the surface formed by the interior of the faces of a cube in $\Re^3$ and its boundary. Show this is a manifold with a boundary and explicitly find the homeomorphisms that correspond to the boundary points. What happens at the corners?*

Manifolds have certain topological properties. The first is that a manifold is locally compact (we defined this property for topological spaces earlier).

Theorem 13.1.1 Manifolds are Locally Compact

A manifold, M, is locally compact.

Proof 13.1.1

Recall a Hausdorff space, X, is locally compact if and only if for any point, $x \in X$, and any neighborhood, U, of x, there is a neighborhood, V, of x, so that $\bar{V}$ is compact and $\bar{V} \subset U$. So, let p be any point in M. By the definition of a manifold, there is a coordinate chart (ϕ, U) such that $p \in U$. Let V be any neighborhood of p. Then $U \cap V \neq \emptyset$, $U \cap V$ is a neighborhood of p, and $V \cap U \subset U$. Thus, $\phi(U \cap V)$ is an open subset of $\phi(U)$ containing $\phi(p) = x$. Hence, there is some $\epsilon > 0$ such that $\overline{B(x,\epsilon)} \subset \phi(U \cap V)$. We claim that $p \in \phi^{-1}(B(x,\epsilon/2)) \subset \overline{\phi^{-1}(B(x,\epsilon/2))} \subset U \cap V \subset V$.

We can see $p \in \phi^{-1}(B(x,\epsilon/2))$ is obvious, since $\phi(p) = x \in B(x,\epsilon/2)$. Also, $\phi^{-1}(B(x,\epsilon/2)) \subset \overline{\phi^{-1}(B(x,\epsilon/2))}$ from the definition of closure.

We will show that $\overline{\phi^{-1}(B(x,\epsilon/2))} \subset U \cap V$ by showing that $\phi^{-1}(\overline{B(x,\epsilon/2)}) = \overline{\phi^{-1}(B(x,\epsilon/2))}$. Now, if $f : X \to Y$ is a continuous map between topological spaces, then $f(\bar{A}) \subset \overline{f(A)}$ for any $A \subset X$. So, since ϕ^{-1} is a continuous map between topological spaces, we know $\phi^{-1}(\overline{B(x,\epsilon/2)}) \subset \overline{\phi^{-1}(B(x,\epsilon/2))}$.

Likewise, we can show $\phi(\overline{\phi^{-1}(B(x,\epsilon/2))}) \subset \overline{B(x,\epsilon/2)}$. We note $\overline{\phi^{-1}(B(x,\epsilon/2))}$ is a subset of U, since ϕ is only defined on U. This follows because $\phi^{-1}(B(x,\epsilon/2)) \subset \phi^{-1}(\overline{B(x,\epsilon)}) \subset U \cap V \subset U$. Thus, $\overline{\phi^{-1}(B(x,\epsilon/2))} \subset U$. It follows that

$$\begin{aligned} \phi(\overline{\phi^{-1}(B(x,\epsilon/2))}) &\subset \overline{\phi(\phi^{-1}(B(x,\epsilon/2)))} = \overline{B(x,\epsilon/2)} \\ &\Rightarrow \overline{\phi^{-1}(B(x,\epsilon/2))} \subset \phi^{-1}(\overline{B(x,\epsilon/2)}) \end{aligned}$$

Therefore,

$$\phi^{-1}(\overline{B(x,\epsilon/2)}) = \overline{\phi^{-1}(B(x,\epsilon/2))}$$

and we have

$$p \in \phi^{-1}(B(x,\epsilon/2)) \subset \overline{\phi^{-1}(B(x,\epsilon/2))} \subset U \cap V \subset V$$

So, M is locally compact, since $\overline{\phi^{-1}(B(x,\epsilon/2))}$ is compact. ■

We can prove further properties a manifold satisfies.

Theorem 13.1.2 Manifolds are Normal, Separable, Metrizable and Linderlöf spaces

A manifold, M, is normal, separable, metrizable, and is a Linderlöf space, i.e., every open cover has a countable subcover.

Proof 13.1.2

*Recall, a topological space X is regular is if given any $x \in X$ and any closed set C which does not contain x, there are open sets U and V so that $x \in U$ and $C \subset V$ with $U \cap V = \emptyset$. If it is also a Hausdorff space (T_2), we say it is T_3-**separable** or just T_3. Hence, a T_3 space is a regular Hausdorff space. Our manifold is therefore a regular Hausdorff space.*

A topological space X is said to be **normal**, *if given any two disjoint closed sets C and D in X, there are open sets U and V so that $C \subset U$, $D \subset V$ with $U \cap V = \emptyset$. If the space is also Hausdorff (T_2) it is called T_4 separable or a T_4 space.*

Every regular space with a countable base is normal (this is an exercise for you). M is regular because, as we showed in the proof of local compactness, for any $p \in M$ and any neighborhood U, of p, there is a neighborhood, V, of p such that $\bar{V}$ is compact and $p \in V \subset \bar{V} \subset U$. Thus, M is normal.

Now, by hypothesis, there is a countable basis $\{B_n\}_{n \geq 1}$ for M. For each n, let p_n be a point in B_n. Define $D = \{p_n : p_n \in U_n, n \geq 1\}$. Given any nonempty open set $\Omega \subset M$, let q be in Ω. Then there is some basis set, B_n, such that $q \in B_n \subset \Omega$. It follows that $p_n \in \Omega \Rightarrow \Omega \cap D \neq \emptyset$. Hence, D intersects every open set in M, implying that D is a countable dense subset of M. Hence, M is separable.

Since M is second countable and normal, Urysohn's metrization theorem implies that M is metrizable.

Finally, let S be an open covering of M. For each $p \in M$, there is an open set, $\Omega_p \in S$ such that $p \in \Omega_p$. So, there is a basis set, B_{n_p}, such that $p \in B_{n_p} \subset \Omega_p$. Let $I = \{n_p : p \in M\}$. Then I is countable. For each $k \in I$, choose a set $\Omega_k \in S$ such that $B_k \subset \Omega_k$. The collection $\{\Omega_k : k \in I\}$ covers M and is countable. Thus, S has a countable subcover, implying that M is a Linderlöf space.
■

Finally, we prove that a manifold can be covered by a countable collection of compact sets.

Theorem 13.1.3 Manifolds can be Covered by a Countable Collection of Compact Subsets

A manifold, M, can be covered by a countable collection of compact subsets.

Proof 13.1.3
Since M is locally compact, given $p \in M$ and any neighborhood, U, of p, there is a neighborhood, V, of p such that $\bar{V}$ is compact and $\bar{V} \subset U$. So, if $p \in M$, there is a basis set, B_k, such that $p \in B_k$, and B_k is a neighborhood of p. Applying the local compactness, there is a neighborhood, V, of p such that $\bar{V}$ is compact and $p \in V \subset \bar{V} \subset B_k$. There is a basis set, B_j, such that $p \in B_j \subset V$. It follows that $\bar{B}_j \subset \bar{V}$. Moreover, since $\bar{V}$ is compact, $\bar{B}_j$ is compact, as it is a closed subspace of a compact space. Thus, given $p \in M$, we have found a basis element, B_j, such that $p \in \bar{B}_j$ and $\bar{B}_j$ is compact. Let B_c be the set of the closures of all such basis elements. Then B_c is a countable compact covering of M. ■

Thus far, we have only worked with manifolds as topological spaces with certain special properties. In order to use calculus to study the geometrical properties of manifolds, we need to add one more condition to our definition of a manifold.

Homework

Exercise 13.1.7 *We say $f : X \rightarrow Y$ is an embedding if f is a homeomorphism from X onto $f(X)$. Let $x = Y = \Re^n$. Prove if U is an open subset of $\Re^n$, then $f(U)$ is an open subset of $\Re^n$ also.*

Exercise 13.1.8 *Prove $\Re^n$ is homeomorphic to $\Re^m$ if and only if $n = m$. Note the only case to prove is when $n < m$. If $\phi : \Re^n \rightarrow \Re^m$, $g^{-1}(U)$ is open if U is open in their respective spaces. Fix $a_{n+1}, \ldots, a_m$ in $\Re^m$ and define $c : \Re^n \rightarrow \Re^m$ by $c(x_1, \ldots, x_n) = (x_1, \ldots, x_n, a_{n+1}, \ldots, a_m)$. Then prove c is an embedding, $c \circ \phi^{-1}$ is an embedding, and by the previous problem, $c \circ \phi^{-1}(U)$ is open in $\Re^m$. Show this is not possible as $c \circ \phi^{-1}(U)$ has no interior points.*

Exercise 13.1.9 *Let X be a normal topological space which is regular and second countable. Let F and G be disjoint closed sets in X.*

- *Show for all $x \in G$, there are open sets U_x and V_x with $x \in V_x$ and $F \subset U_x$ and $U_x \cap V_x = \emptyset$.*
- *$V = \{V_x : x \in G\}$ is an open cover of G and $V \cup G^C$ is an open cover of X. Since X is second countable, we can assume this is a countable open cover. So there are a countable number of points $x_i \in G$ and sets V_{x_i} which we can relabel as V_i.*
- *Thus, V is a countable open cover of G.*
- *$V_x \subset (U_x)^C$, so $\overline{V_x} \subset \overline{(U_x)^C} = U_x^C$. We also have $U_x^C \subset F^C$ so the closures of each V_x do not intersect F. Thus $\overline{V_i} \cap F = \emptyset$ for all i.*
- *Use a similar argument to show there is a countable cover $U = \{U_i\}$ of F with $\overline{U_i} \cap G = \emptyset$.*
- *At this point, the sets $\cup_i U_i$ and $\cup_i V_i$ need not be disjoint, so we need to modify the sets to construct disjoint unions. Do this:*
- *$V_1^* = V_1 \setminus \overline{U_1}$ and $V_n^* = V_n \setminus (\overline{U_1} \cup \ldots \cup \overline{U_n})$ for $n \geq 2$. All these sets are open and the family $\{V_n^*\}$ is still an open cover of G.*
- *$U_1^* = U_1 \setminus \overline{V_1}$ and $U_n^* = U_n \setminus (\overline{V_1} \cup \ldots \cup \overline{V_n})$ for $n \geq 2$. All these sets are open and the family $\{U_n^*\}$ is still an open cover of F.*
- *$\cup_i U_i^*$ and $\cup_i V_i^*$ are disjoint with $F \subset \cup_i U_i^*$ and $G \subset \cup_i V_i^*$.*

This shows X is normal.

Now, let's add more smoothness to the homeomorphisms.

Definition 13.1.2 Differentiable Manifolds

A differentiable, or smooth, manifold is a manifold, M, whose coordinate covering $\{(\phi_i, U_i)\}_{i \geq 1}$ satisfies the following additional properties.

(i): For any charts (ϕ, U) and (ψ, V) such that $U \cap V \neq \emptyset$, the map $\phi \circ \psi^{-1} : \psi(U \cap V) \to \phi(U \cap V)$ is C^∞, meaning that its component functions have continuous partial derivatives of all orders. Such pairs of charts are said to be C^∞-compatible.

(ii): The coordinate covering is maximal in the sense that any chart (ψ, V) that is C^∞ compatible with all charts in $\{(\phi_i, U_i)\}_{i \geq 1}$ is included in the coordinate covering.

A coordinate covering satisfying these properties is said to be a differentiable, or smooth, structure on M. In our discussions, we will use the term smooth as a synonym for C^∞. Such smoothness is, of course, not required. Differentiable manifolds can be defined with C^r smoothness meaning the composition of compatible coordinate mappings is C^r for a given $r \geq 1$. We have even seen C^0 manifolds in some discussions of physics-related models , but the use of manifolds in analysis, differential geometry and modeling usually requires at least several degrees of differentiability. Since C^∞ functions are dense in C^r spaces, it is convenient to assume C^∞ manifolds for our work here. One question is whether such differentiable structures even exist on an arbitrary manifold and what conditions are necessary and sufficient for their existence. Another question is whether or not there were manifolds for which distinct differentiable structures existed. That is, there was a question of whether two coordinate coverings existed on the same manifold, one consisting of charts that are not all C^∞ compatible with those in the other. It is indeed possible for such a manifold to exist, and this was shown in (Milnor (88) 1956) in the construction of the exotic sphere. Also, condition (ii) in the definition of a smooth manifold seems very technical. For the most part, it is only of importance in purely topological discussions, such as in the construction of the exotic sphere. We are usually only concerned with whether a smooth structure exists, and, even then, we will often assume that it does. So, we can stop worrying about condition (ii) by applying the following theorem.

Theorem 13.1.4 Unique C^∞ Structures on a Manifold

Let M be a manifold with a coordinate covering $\{(\phi_U, U)\}_{i\geq 1}$ such that any two charts in the covering are C^∞-compatible. Then there is a unique C^∞ structure on M containing these coordinate charts.

Proof 13.1.4

We define our differentiable structure, Λ, to be the collection of all charts, (ψ, V), that are C^∞ compatible with all charts in the given coordinate covering. Then Λ is nonempty by hypothesis and is a coordinate covering of M. Suppose (ρ, W) and (ψ, V) are two charts in Λ such that $W \cap V \neq \emptyset$. If at least one of them is in the original coordinate covering, then they are C^∞-compatible by hypothesis. So, we assume that they are not. The maps $\rho \circ \psi^{-1}$ and $\psi \circ \rho^{-1}$ are well-defined homeomorphisms between open sets in $\Re^m$, so we need only show that these maps are smooth. Let $x = \rho(p)$ be an arbitrary point of $\rho(W \cap V)$. Then there is some chart (ϕ, U) in the original coordinate covering such that $p \in U$. So, $N = U \cap V \cap W$ is a neighborhood of p and $\psi(N)$ is a neighborhood of x. On $\rho(N)$, we have $\psi \circ \rho^{-1} = \psi \circ \phi^{-1} \circ \phi \circ \rho^{-1}$. But $\psi \circ \phi^{-1}$ and $\phi \circ \rho^{-1}$ are C^∞ by hypothesis. Hence, their composition is also C^∞ on $\rho(N)$. Since p was arbitrary, it follows that for every point, q, in $\rho(W \cap V)$, there is a neighborhood of q on which $\psi \circ \rho^{-1}$ is C^∞, implying that this map is C^∞. A similar argument shows that $\rho \circ \psi^{-1}$ is C^∞ on $\psi(W \cap V)$. Hence, condition (1) of the definition of smooth manifold is satisfied. As for property (ii), it is satisfied by our definition of Λ. ■

Thus, any coordinate covering in which the charts are C^∞-compatible can be extended to a maximal covering. Hence, in verifying that a set is a smooth manifold, we need only show that it is a manifold and has a C^∞ coordinate covering. Let's look at some examples.

Example 13.1.1 *Any Euclidean space, $\Re^m$, is a smooth manifold. The single chart $(\mathbf{id}, \Re^{\mathbf{m}})$, where $\mathbf{id}$ is the identity map $\mathbf{id} : \Re^{\mathbf{m}} \rightarrow \Re^{\mathbf{m}}$, is a smooth structure on $\Re^m$*

Example 13.1.2 *The Graph of $f : \Re^2 \rightarrow \Re$:*
In multivariable calculus, the graph of a smooth function $f : \Re^2 \rightarrow \Re$, viewed as a subset of $\Re^3$, is known to induce a surface. As manifolds are generalizations of surfaces, it stands to reason that surfaces should certainly be manifolds. Suppose Ω is an open set in $\Re^2$, and $f : \Omega \rightarrow \Re$ is a smooth function. Let $M = \{(x, y, f(x, y)) \in \Re^3 \; : \; (x, y) \in \Omega\}$. We will show that M is a manifold. First, since M is a subspace of a Euclidean space, M is Hausdorff and second countable. Define a map $\phi : M \rightarrow \Omega$ by $\phi(x, y, f(x, y)) = (x, y)$. This map is clearly surjective, and it is injective, since $\phi(x_1, y_1, f(x_1, y_1)) = \phi(x_2, y_2, f(x_2, y_2)) \Rightarrow x_1 = x_2$ and $y_1 = y_2$. This clearly implies that $f(x_1, y_1) = f(x_2, y_2)$. Moreover, ϕ is smooth as its component functions are smooth. The inverse mapping, $\phi^{-1} : \Omega \rightarrow M$, is given by $\phi^{-1}(x, y) = (x, y, f(x, y))$. Since f is smooth by hypothesis, so is ϕ^{-1}. Thus, M, with the single chart (ϕ, M), is a smooth manifold.

Example 13.1.3 *The One-dimensional Sphere:*
Let $S^1 = \{(x, y) \in \Re^2 \; : \; x^2 + y^2 = 1\}$. This is, of course, the unit circle in $\Re^2$. It is a subspace of $\Re^2$, so it is Hausdorff and second countable. It is a curve, or a 1-dimensional surface, but it cannot be represented as the graph of a single function. Define the following mappings.

1. *Let $U_1 = \{(x, y) \in S^1 \; : \; y > 0\}$. Define $\phi_1 : U_1 \rightarrow (-1, 1)$ by $\phi_1(x, y) = x$.*
2. *Let $U_2 = \{(x, y) \in S^1 \; : \; y < 0\}$. Define $\phi_2 : U_2 \rightarrow (-1, 1)$ by $\phi_2(x, y) = x$.*

3. *Let* $U_3 = \{(x,y) \in S^1 \ : \ x > 0\}$. *Define* $\phi_3 : U_3 \to (-1,1)$ *by* $\phi_3(x,y) = y$.

4. *Let* $U_4 = \{(x,y) \in S^1 \ : \ x < 0\}$. *Define* $\phi_4 : U_4 \to (-1,1)$ *by* $\phi_4(x,y) = y$.

Noting that U_1 *can be represented as the set of all points in* S^1 *of the form* $(x, \sqrt{1-x^2})$, *we see that* ${\phi_1}^{-1}(x) = (x, \sqrt{1-x^2})$. *The inverses of the other maps are defined similarly. Note that we are simply projecting quadrants of the circle onto an open interval. Given their respective domains, the maps* ϕ_i, $1 \leq i \leq 4$ *are easily seen to be homeomorphisms. The charts* (ϕ_i, U_i) *are also all* C^∞*-compatible. To see this, consider the case* $\phi_1 \circ {\phi_3}^{-1} : \phi_3(U_1 \cap U_3) \to \phi_1(U_1 \cap U_3)$. $U_1 \cap U_3$ *is the set of all points,* (x,y), *on the circle such that* $x > 0$ *and* $y > 0$. *We can represent such a point as* $(x, \sqrt{1-x^2})$ *or* $(\sqrt{1-y^2}, y)$. *So, if* y *is a point in* $\phi_3(U_1 \cap U_3)$, *then* $0 < y < 1$, *and* ${\phi_3}^{-1}(y) = (\sqrt{1-y^2}, y)$. *Thus,* $\phi_1 \circ {\phi_3}^{-1}(y) = \phi_1(\sqrt{1-y^2}, y) = \sqrt{1-y^2}$. *Since* $0 < y < 1$, *this map is* C^∞. *The other possible compositions are computed in a similar way, and are also* C^∞. *So,* S^1, *with the coordinate covering consisting of the charts* (ϕ_i, U_i), *is a smooth manifold.*

In a similar way, we can show that the set $S^{n-1} = \{(x_1, x_2, ..., x_n) \in \Re^n \ : \ {x_1}^2 + \cdots + {x_n}^2 = 1\}$, the $(n-1)$-dimensional unit sphere, is a smooth manifold. However, it is possible to define, on any sphere, a smooth structure consisting of just two coordinate charts. This is called the stereographic projection, and is used frequently in complex analysis to define Riemann surfaces.

Example 13.1.4 *Open submanifolds of a manifold,* M*:*
Let M *be a smooth manifold, and let* Ω *be an open subset of* M. *Then* Ω *is both Hausdorff and second countable, as it is a subspace of a manifold. Suppose the coordinate covering of* M *is given by* $\{(\phi_i, U_i)\}_{i \geq 1}$. *Define a coordinate covering on* Ω *by* $\{(\phi_i|_{U_i \cap \Omega}, U_i \cap \Omega)\}$. *The restriction of a homeomorphism to an open subset of its domain is still a homeomorphism. Moreover, if* $[U_i \cap \Omega] \cap [U_j \cap \Omega] = U_i \cap U_j \cap \Omega \neq \emptyset$, *then* $\phi_i \circ \phi_j^{-1} : \phi_j(U_j \cap U_i \cap \Omega) \to \phi_i(U_j \cap U_i \cap \Omega)$ *is a* C^∞ *map of the open set* $\phi_j(U_i \cap U_j \Omega) \subset \phi_j(U_j)$ *onto the open set* $\phi_i(U_i \cap U_j \cap \Omega) \subset \phi_i(U_i)$, *as it is just a restriction of the* C^∞ *map* $\phi_i \circ \phi_j^{-1}$ *to an open subset of* $\phi_j(U_i \cap U_j)$.

Hence, Ω *is a smooth manifold, and we call* Ω *a submanifold of* M. *Consequently, it follows that every coordinate neighborhood,* U, *is a smooth submanifold of* M.

Homework

Exercise 13.1.10 *Work through the ellipse example:* $E^1 = \{(x,y) \in \Re^2 \ : \ x^2 + 2y^2 = 1\}$

Exercise 13.1.11 *Work through the ellipse example:* $E^2 = \{(x,y,z) \in \Re^3 \ : \ x^2 + 2y^2 + 3z^2 = 1\}$

Exercise 13.1.12 *Work through the sphere example:* $S^3 = \{(x_1, x_2, x_3, x_4) \in \Re^4 \ : \ x_1^2 + x_2^2 + x_3^2 + x_4^2 = 1\}$.

Exercise 13.1.13 *Work through the ellipse example:* $E^3 = \{(x_1, x_2, x_3, x_4) \in \Re^4 \ : \ 2x_1^2 + 4x_2^2 + 5x_3^2 + 3x_4^2 = 1\}$.

Exercise 13.1.14 *Work through the ellipse example:* $E^3 = \{(x_1, x_2, x_3, x_4) \in \Re^4 \ : \ 5x_1^2 + 7x_2^2 + 3x_3^2 + x_4^2 = 1\}$.

13.1.1 Implicit Function Manifolds

Manifolds defined by the Implicit Function Theorem are important in many applications. A version of the Implicit Function Theorem that is suited for our purposes can be stated as follows. You can look back at the discussions in (Peterson (101) 2020) for a lot more detail. We encourage you to do that as learning how to do mathematical reasoning at this level is not a spectator sport; active participation is required!

Theorem The Implicit Function Theorem

Let $\boldsymbol{D} \subset \Re^n \times \Re^m$ be an open set with $(a, b) \in \boldsymbol{D}$. Suppose $f : D \rightarrow \Re^m$ is C^∞ on $\boldsymbol{D}$ and $f(a, b) = 0$. Suppose also that the linear map $v \longmapsto Df(a, b)(0, v)$, for $v \in \Re^m$, is a bijection of $\Re^m$ onto $\Re^m$. Then:

(i): There exists a neighborhood, U, of $a \in \Re^n$ and a C^1 function, $\psi : U \rightarrow \Re^m$ such that $\psi(a) = b$ and $f(x, \psi(x)) = 0 \,\forall\, x \in U$.

(ii): There exists a neighborhood, W, of (a, b) in $\Re^n \times \Re^m$ such that the pair $(x, y) \in W$ satisfies $f(x, y) = 0$ if and only if $x \in U$ and $y = \psi(x)$.

We will also need the following definition in our construction. Recall, we used regular values in our work on Brouwer degree, but we restate it here for convenience.

Definition 13.1.3 Regular Values of a Function

Let $f : \Re^n \rightarrow \Re^m$ be differentiable, and let y be in the range of f. Then y is a regular value of f if for all $x \in f^{-1}(y)$, $Df(x) : \Re^n \rightarrow \Re^m$ has rank m (i.e. is surjective).

We will state our construction as a theorem.

Theorem 13.1.5 Inverse Images as Manifolds

Let $f : \Re^n \rightarrow \Re^m$ be a smooth function, and suppose $0 \in Range(f)$ is a regular value. Then $f^{-1}(0)$ is an $n - m$ dimensional smooth manifold.

Proof 13.1.5
First, $f^{-1}(0)$ is a subspace of $\Re^n$, so it is Hausdorff and second countable. Let p be in $f^{-1}(0)$. Then $Df(p) : \Re^n \rightarrow \Re^m$ has rank m. (Note that this requires $n \geq m$.) For simplicity of exposition, we assume that the last m columns of $Df(p)$ are linearly independent. Any other case is done in exactly the same manner except for notational differences.

So, we think of f as a function from $\Re^{n-m} \times \Re^m$ to $\Re^m$, and we denote p by $p = (x_p, y_p) \in f^{-1}(0) \subset \Re^{n-m} \times \Re^m$. Then the linear map $v \longmapsto Df(x_p, y_p)(0, v)$, for $v \in \Re^m$, is a bijection of $\Re^m$ onto $\Re^m$. So, there is a neighborhood, U, of $x_p \in \Re^{n-m}$ and a C^1 function, $\psi : U \rightarrow \Re^m$, such that $\psi(x_p) = y_p$ and $f(x, \psi(x)) = 0$ for all $x \in U$. There is also a neighborhood, W, of $p = (x_p, y_p) \in \Re^{n-m} \times \Re^m$ such that the pair $(x, y) \in W$ satisfies $f(x, y) = 0$ if and only if $x \in U$ and $y = \psi(x)$.

Now, let $N_p = f^{-1}(0) \cap W$. Then N_p is a neighborhood of $p = (x_p, y_p)$ in the subspace topology on $f^{-1}(0)$. Define a map $\phi_p : U \rightarrow N_p$ by $\phi_p(x) = (x, \psi(x))$. To see that ϕ_p is onto, let (x, y) be in $N_p \subset \Re^{n-m} \times \Re^m$. Then $f(x, y) = 0$ and $(x, y) \in W$. The Implicit Function Theorem implies that $x \in U$ and $y = \psi(x)$. So, $\phi_p(x) = (x, \psi(x)) = (x, y)$. To see that ϕ_p is injective, suppose $\phi_p(x_1) = \phi_p(x_2)$. Then $({x_1}^1, {x_1}^2, \ldots, {x_1}^{n-m}, \psi(x_1)) = ({x_2}^1, {x_2}^2, \ldots, {x_2}^{n-m}, \psi(x_2)) \Rightarrow x_1 = x_2$. So, ϕ_p is a continuous bijection of an open set in $\Re^{n-m}$ onto a neighborhood of $p = (x_p, y_p)$. Note that the inverse of ϕ_p, ${\phi_p}^{-1} : N_p \rightarrow U$, is just the projection map of $\Re^n$ onto the first $n - m$ coordinates. That is, if $\pi : \Re^{n-m} \times \Re^m \rightarrow \Re^{n-m}$ is defined by $\pi(x) = (x^1, \ldots, x^{n-m})$, then $\pi = {\phi_p}^{-1}$. This follows from the fact that $(x, y) \in N_p \Rightarrow (\phi_p \circ \pi)(x, y) = \phi_p(x) = (x, \psi(x)) = (x, y)$ and $x \in U \Rightarrow (\pi \circ \phi_p)(x) = \pi(x, y) = x$. Thus, since the projection map is continuous, ϕ_p is a homeomorphism.

Now, the point $p \in f^{-1}(0)$ was arbitrary, so for any $p \in f^{-1}(0)$ we can find an open set, V, in $\Re^{n-m}$, a neighborhood, U, of p, $U \subset f^{-1}(0)$, and a homeomorphism $\phi_p : U \to V$. Thus, $f^{-1}(0)$ is a manifold. We need only verify that the coordinate covering forms a smooth structure. But, by our construction, each map ϕ_p and its inverse, ${\phi_p}^{-1}$, are smooth. Hence, any possible composition will be smooth as well. Thus, $f^{-1}(0)$ is a smooth manifold. ■

Homework

Exercise 13.1.15 *Discuss the implicit manifold defined by $2x^2 + 4xy + 5x^4y^4 = 2$ in $\Re^2$.*

Exercise 13.1.16 *Discuss the implicit manifold defined by $2x_1^2 + 4x_1x_2 + 5x_3^4x_4^4 = 20$ in $\Re^4$.*

Exercise 13.1.17 *Discuss the implicit manifold defined by $2x_1^2 \sin(x_2^2 + x_3^2) = 1$ in $\Re^3$.*

13.1.2 Projective Space

Our final example is quite lengthy and more abstract in nature than the previous ones. We always had a hard time understanding this one!

Definition 13.1.4 Projective n Space

Let $X = \Re^{n+1} - \{0\}$. Define an equivalence relation, $\sim$, on X as follows. For $x, y \in X$, $x \sim y$ if and only if there exists a real number, $\lambda \neq 0$, such that $x = \lambda y$. Denote the equivalence class containing x by $[x]$. Let $\mathbb{P}^n = \{[x] : x \in X\}$. Then $\mathbb{P}^n$ is the quotient space $\mathbb{X}/\sim$, and we assume it is endowed with the usual quotient space topology. We denote the quotient map by $p : X \to \mathbb{P}^n$, where $p(x) = [x]$ for $x \in X = \Re^{n+1} - \{0\}$. We will show that $\mathbb{P}^n$ is a smooth n-dimensional manifold. Note that, geometrically speaking, we can identify $\mathbb{P}^n$ with the set of lines in R^{n+1} passing through the origin.

To show that $\mathbb{P}^n$ is a manifold, we must show that $\mathbb{P}^n$ is Hausdorff and second countable. These results are not trivial, as topological properties are not necessarily preserved under the formation of quotient spaces. So, to verify these properties, we will use the following definition and lemmas, common to general topology.

Definition 13.1.5 Open Equivalence Relations

An equivalence relation $\sim$ on a space X is said to be open if, for any open subset, $A \subset X$, the set $[A]$ is open in X, where $[A] = \bigcup_{x \in A}[x]$.

We can then prove a characterization of open equivalence relations.

Lemma 13.1.6 Equivalence Relation is Open if and only if the Quotient map is an Open Mapping. Also, if the Equivalence Relation is Open, the Quotient Space is Second Countable

An equivalence relation $\sim$ on X is open if and only if the quotient map, p, is an open mapping. When $\sim$ is open and X is second countable, then $X/\sim$ is second countable also.

Proof 13.1.6
First, suppose $\sim$ is open. To show that p is an open mapping, we must show that it maps open sets in X to open sets in $X/\sim$. By definition of the quotient topology, a set $\Omega \subset X/\sim$ is open if $p^{-1}(\Omega)$ is open. Let A be an open subset of X. Then the set $[A]$ is also open in X. Note, however, that $p^{-1}(p(A)) = [A]$. Thus, it follows that $p^{-1}(p(A))$ is open, which, in turn, implies that $p(A)$ is open

in $X/\sim$. Since A was arbitrary, this shows that p is an open map.

Conversely, suppose p is an open map. Let A be any open subset of X. Then $p(A)$ is open in $X/\sim$. Since p is continuous, this implies that $p^{-1}(p(A))$ is open in X. Since $p^{-1}(p(A)) = [A]$, it follows that $[A]$ is open. So, $\sim$ is an open equivalence relation.

Finally, suppose $\sim$ is open, and suppose that X is second countable. Let $\{U_i\}_{i=1}^{\infty}$ be a countable basis of open sets for X. Then the collection $\{p(U_i)\}_{i=1}^{\infty}$ is a countable collection of open sets in $X/\sim$. Let W be an open subset of $X/\sim$. Then $p^{-1}(W)$ is an open subset of X, implying that $p^{-1}(W)$ can be expressed as a union of basis sets U_i. Suppose

$$p^{-1}(W) = \bigcup_{j=1}^{\infty} U_{i_j}.$$

Then $W = p(p^{-1}(W)) = p(\cup_{j\geq 1}U_{i_j}) = \cup_{j\geq 1}p(U_{i_j})$. Hence, W can be expressed as the union of sets in the collection $\{p(U_i)\}$. Since W was arbitrary, this shows that this collection forms a basis for $X/\sim$. Thus, the quotient space is second countable. ■

Now let's look at quotient spaces.

Lemma 13.1.7 Conditions on the Quotient Space Being Hausdorff

Let $\sim$ be an open equivalence relation on X. Then $R = \{(x,y) \in X \times X \ : \ x \sim y\}$ is a closed subset of $X \times X$ (with the standard product topology) if and only if $X/\sim$ is Hausdorff.

Proof 13.1.7

First, suppose $X/\sim$ is Hausdorff. We will show that R is closed by showing that its complement is open. Suppose $(x,y) \notin R$. Then $x \nsim y$, which implies that $p(x) \neq p(y)$. Hence, there are disjoint open subsets, U and V, of $X/\sim$ such that $p(x) \in U$ and $p(y) \in V$. Then $p^{-1}(U)$ and $p^{-1}(V)$ are open sets in X containing x and y, respectively. So, the set $p^{-1}(U) \times p^{-1}(V)$ is an open set in $X \times X$ that contains (x,y). Suppose this set intersects R. Then it must contain some element (x',y') such that $x' \sim y'$. This implies that $p(x') = p(y')$. But (x',y') is also in $p^{-1}(U)\times p^{-1}(V)$, implying that $p(x') \in U$ and $p(y') \in V$. It follows that U and V have a nonempty intersection, contradicting the fact that they are disjoint. Thus, the set $p^{-1}(U) \times p^{-1}(V)$ does not intersect R. Therefore, for any $(x,y) \in R^c$, there is an open set containing (x,y) that does not intersect R. So, R^c is open, implying that R is closed.

Conversely, suppose R is closed. Since any point in $X/\sim$ is the image of some element under the map p, we can consider two elements of $X/\sim$ as $p(x)$ and $p(y)$. Suppose $p(x)$ and $p(y)$ are two distinct points of $X/\sim$. Then we cannot have $x \sim y$. Thus, the element (x,y) must be in R^c, which is open. Hence, there are open sets U and V in X such that $(x,y) \in U \times V$ and the set $U \times V$ does not intersect R. Consider the sets $p(U)$ and $p(V)$ in $X/\sim$. If there was an element, say $[z]$, in both of these sets, then we would have $x \sim z$ and $y \sim z$, implying that $x \sim y$. Hence, this contradiction shows that $p(U)$ and $p(V)$ are disjoint. The previous lemma shows that they are open, and we have $p(x) \in p(U)$ and $p(y) \in p(V)$. Since these two elements were arbitrary, this shows that $X/\sim$ is Hausdorff. ■

A very important and very abstract manifold is called projective n space. It is worth your time to understand this. It is not a lot more difficult than understanding the construction of the real numbers from the rationals as we do in (Peterson (100) 2020).

Theorem 13.1.8 Projective n Space is a Manifold

Projective n space is a manifold.

Proof 13.1.8

Note that, for $[x] = [(x^1, x^2, ..., x^{n+1})] \in \mathbb{P}^n$, if $x^i \neq 0$ for some $i = 1, 2, ..., n+1$, then $[(x^1, ..., x^i, ..., x^{n+1})] = [(x^1/x^i, ..., x^{i-1}/x^i, 1, x^{i+1}/x^i, ..., x^{n+1}/x^i)]$. Also, for $[x] \in \mathbb{P}^n$, if any element $(x^1, ..., x^{n+1}) \in [x]$ satisfies $x^i = 0$, then every element of $[x]$ has a 0 i^{th} component. So, we can speak of an equivalence class $[x] \in \mathbb{P}^n$ as having a zero or nonzero i^{th} component, $[x]_i$.

Now, since $\mathbb{P}^n$ is a quotient space, a set $U \subset \mathbb{P}^n$ is open if and only if $p^{-1}(U)$ is open in X. Since $p^{-1}(U) = \bigcup_{[x]\in U}[x]$, it follows that a set $U \subset \mathbb{P}^n$ is open if and only if the union of all the equivalence classes in U is open in X. Define sets U_i, $i = 1, ..., n+1$, by $U_i = \{[x] \in \mathbb{P}^n \ : \ [x]_i \neq 0\}$. If $[x] \in \mathbb{P}^n$, then at least one of the components must be nonzero. So, the sets U_i cover $\mathbb{P}^n$. They are also open. Consider $p^{-1}(U_i) = \bigcup_{[x]\in U_i}[x]$. This is just the union of all the equivalence classes that have a nonzero i^{th} component. Let $x = (x^1, ..., x^i, ..., x^{n+1})$ be in $p^{-1}(U_i)$. Then $x^i \neq 0$. Choose $\epsilon > 0$ such that $0 < \epsilon < \frac{|x^i|}{2}$. The set $\{y \in \Re^{n+1} \ : \ \|x - y\| < \epsilon\} = B(x, \epsilon)$ is an open ball centered at x. If $y \in B(x, \epsilon)$, then $|x_i - y_i| \leq \|x - y\| \Rightarrow |x_i - y_i| < \epsilon \Rightarrow -\frac{|x_i|}{2} < y_i - x_i < \frac{|x_i|}{2}$. It follows that we must have $y_i \neq 0$. Thus $B(x, \epsilon) \subset p^{-1}(U_i)$, implying that $p^{-1}(U_i)$ is open in X. Hence, each set U_i is open. We will use this collection of sets as a coordinate covering.

For each $i = 1, ..., n+1$, define the map $\phi_i : U_i \rightarrow \Re^n$ by

$$\phi_i([x]) \quad = \quad \phi_i\Big(\big[(x^1, ..., x^i, ..., x^{n+1})\big]\Big) = \left(\frac{x^1}{x^i}, \frac{x^2}{x^i}, ..., \frac{x^{i-1}}{x^i}, \frac{x^{i+1}}{x^i}, ..., \frac{x^{n+1}}{x^i}\right)$$

Note that ϕ_i is well-defined, since, if $x = (x^1, ..., x^{n+1})$ and $y = (y^1, ..., y^{n+1})$ are in $[x]$, then $y = \lambda x$ for some $\lambda \neq 0$. Thus,

$$\begin{aligned} \left(\frac{y^1}{y^i}, ..., \frac{y^{i-1}}{y^i}, \frac{y^{i+1}}{y^i}, ..., \frac{y^{n+1}}{y^i}\right) &= \left(\frac{\lambda x^1}{\lambda x^i}, ..., \frac{\lambda x^{i-1}}{\lambda x^i}, \frac{\lambda x^{i+1}}{\lambda x^i}, ..., \frac{\lambda x^{n+1}}{\lambda x^i}\right) \\ &= \left(\frac{x^1}{x^i}, ..., \frac{x^{i-1}}{x^i}, \frac{x^{i+1}}{x^i}, ..., \frac{x^{n+1}}{x^i}\right). \end{aligned}$$

Now, if $\phi_i([x]) = \phi_i([y])$ for $[x], [y] \in U_i$, then for any elements $x \in [x]$ and $y \in [y]$, we have

$$\begin{aligned} & \left(\frac{x^1}{x^i}, ..., \frac{x^{i-1}}{x^i}, \frac{x^{i+1}}{x^i}, ..., \frac{x^{n+1}}{x^i}\right) = \left(\frac{y^1}{y^i}, ..., \frac{y^{i-1}}{y^i}, \frac{y^{i+1}}{y^i}, ..., \frac{y^{n+1}}{y^i}\right) \\ \Rightarrow \quad & (x^1, ..., x^{i-1}, x^{i+1}, ..., x^{n+1}) = \frac{x^i}{y^i}(y^1, ..., y^{i-1}, y^{i+1}, ..., y^{n+1}) \\ \Rightarrow \quad & (x^1, ..., x^{i-1}, x^i, x^{i+1}, ..., x^{n+1}) = \frac{x^i}{y^i}(y^1, ..., y^{i-1}, y^i, y^{i+1}, ..., y^{n+1}) \end{aligned}$$

from which it follows that $x = \frac{x^i}{y^i}y$. So $[x] = [y]$, and each ϕ_i is injective. If $y \in \Re^n$, consider the element in X given by $\tilde{y} = (y^1, ..., y^{i-1}, 1, y^i, ..., y^n)$. That is, we put a 1 in the ith position and shift the remaining components up one index value. If $i = n+1$, we simply attach a 1 on the end of y. Then $[\tilde{y}] \in U_i$ and $\phi_i([\tilde{y}]) = (y^1, ..., y^{i-1}, y^i, ..., y^n) = y$. So, each ϕ_i is also surjective.

To show that each ϕ_i is continuous, it suffices to show that ${\phi_i}^{-1}(B(y,\epsilon))$ is open in $\mathbb{P}^n$ for arbitrary ϵ and $y \in \Re^n$. Hence, we want to show that $p^{-1}({\phi_i}^{-1}(B(y,\epsilon)))$ is open in X. But $p^{-1}({\phi_i}^{-1}(B(y,\epsilon))) = (\phi_i \circ p)^{-1}(B(y,\epsilon))$, so, if the map $\phi_i \circ p : p^{-1}(U_i) \to \Re^n$ is continuous, this will prove the result. Now, $p^{-1}(U_i)$ consists of all those points in X such that $x^i \neq 0$. If x is such a point, then $p(x) = [x]$ and

$$\phi_i(p(x)) = \left(\frac{x^1}{x^i}, \ldots, \frac{x^{i-1}}{x^i}, \frac{x^{i+1}}{x^i}, \ldots, \frac{x^{n+1}}{x^i} \right).$$

Hence, the map $\phi_i \circ p$ is continuous on $p^{-1}(U_i)$, as each of the component functions is continuous. Since ϕ_i is a bijection, it has a well-defined inverse. Explicitly, the inverse mapping ${\phi_i}^{-1} : \Re^n \to U_i$ satisfies

$${\phi_i}^{-1}(y_1, \ldots, y_n) = [y^1, \ldots, y^{i-1}, 1, y^i, \ldots, y^n].$$

To see that ${\phi_i}^{-1}$ is continuous, define a map $f : \Re^n \to \Re^{n+1}$ by $f(y^1, \ldots, y^n) = (y^1, \ldots, y^{i-1}, 1, y^i, \ldots, y^n)$. This map is continuous and its range lies in $p^{-1}(U_i)$. Moreover, ${\phi_i}^{-1} = p \circ f$. Thus, since the quotient map, p, is continuous, ${\phi_i}^{-1}$ is continuous. So, each ϕ_i is a homeomorphism.

To show that $\mathbb{P}^n$ is a manifold, we still must show that $\mathbb{P}^n$ is Hausdorff and second countable. Recall that, in our case, $X = \Re^{n+1} - \{0\}$. For $t \in \Re$, $t \neq 0$, consider $\alpha_t : X \to X$ defined by $\alpha_t(x) = tx$. This is just a scaling map, so it is a homeomorphism with ${\alpha_t}^{-1} = \alpha_{\frac{1}{t}}$. If $U \subset X$ is open, then $\alpha_t(U)$ is open in X also. Moreover, if $y \in [U]$, then $y \in [x]$ for some $x \in U$, implying that $y = tx$ for some $t \neq 0$. So, $y \in \alpha_t(U)$, and we have

$$[U] \subset \bigcup_{\substack{t \in \Re \\ t \neq 0}} \alpha_t(U).$$

Conversely, if $y \in \bigcup_{t\in\Re, t\neq 0} \alpha_t(U)$, then $y \in \alpha_t(U)$ for some t. So, $y = tx$ for some $x \in U$, implying that $y \sim x \Rightarrow y \in [U]$. Thus, we have

$$[U] = \bigcup_{\substack{t \in \Re \\ t \neq 0}} \alpha_t(U)$$

Finally, since each $\alpha_t(U)$ is open, $[U]$ is open. So, by Lemma 13.1.6, $X/\sim = \mathbb{P}^n$ is second countable. We now apply Lemma 13.1.7 to prove the Hausdorff condition. Define a function $f : X \times X \to \Re$ by

$$f(x,y) = f(x^1, \ldots, x^{n+1}, y^1, \ldots, y^{n+1}) = \sum_{i,j=1}^{n} (x^i y^j - x^j y^i)^2.$$

Then f is continuous on $X \times X$. We can also show that $f(x,y) = 0 \Leftrightarrow y = tx$ for some $t \neq 0$. If $y = tx$, then it is clear that f vanishes. Conversely, suppose $f(x,y) = 0$. Then, viewing x and y as vectors in $\Re^{n+1}$, this implies

$$\begin{aligned}
&\sum_{i=1}^{n}\sum_{j=1}^{n}(x^i)^2(y^j)^2 - 2\sum_{i=1}^{n}\sum_{j=1}^{n} x^i y^i x^j y^j + \sum_{i=1}^{n}\sum_{j=1}^{n}(x^j)^2(y^j)^2 = 0 \\
&\Rightarrow \Big(\sum_{i=1}^{n}(x^i)^2\Big)\Big(\sum_{j=1}^{n}(y^j)^2\Big) - 2\Big(\sum_{i=1}^{n} x^i y^i\Big)\Big(\sum_{j=1}^{n} x^j y^j\Big) + \Big(\sum_{i=1}^{n}(x^i)^2\Big)\Big(\sum_{j=1}^{n}(y^j)^2\Big) = 0 \\
&\Rightarrow \|x\|^2\|y\|^2 = \langle x, y\rangle^2
\end{aligned}$$

$$\Rightarrow \|x\|\|y\| = \langle x, y \rangle$$

from which it follows that x and y must be linearly dependent. That is, $y = tx$ for some $t \neq 0$. Thus, $f^{-1}(0)$ consists of all those pairs, (x, y), in $X \times X$ such that $y = tx$ for some $t \neq 0$. Hence, $f^{-1}(0) = R = \{(x, y) : x \sim y\}$ is closed, and Lemma 13.1.7 implies that $\Re^n$ is Hausdorff. So, $\mathbb{P}^n$ is a manifold. All that remains is to show that the maps ϕ_i are C^∞-compatible.

Suppose $U_i \cap U_j \neq \emptyset$ and $x \in \phi_i(U_i \cap U_j) \subset \Re^n$. Then $x = \phi_i([y])$ for some $[y] \in U_i \cap U_j$. This $[y]$ must, then, have nonzero i and j components. We can assume without loss of generality that $i < j$. The map ϕ_i will remove the ith component of $[y]$ and shift the jth component to the $(j-1)^{st}$ position. Hence, x has a nonzero $(j-1)^{st}$ component, and it follows that

$$\phi_i{}^{-1}(x) \quad = \quad [x^1, ..., x^{i-1}, 1, x^i, ..., x^{j-1}, ..., x^n]$$

where the component x^{j-1} is nonzero and in the jth position of the equivalence class. We then have

$$\begin{aligned} \phi_j(\phi_i{}^{-1}(x)) &= \phi_j\left([x^1, ..., x^{i-1}, 1, x^i, ..., x^{j-1}, ..., x^n]\right) \\ &= \left(\frac{x^1}{x^{j-1}}, ..., \frac{x^{i-1}}{x^{j-1}}, \frac{1}{x^{j-1}}, \frac{x^i}{x^{j-1}}, ..., \frac{x^{j-2}}{x^{j-1}}, \frac{x^j}{x^{j-1}}, ..., \frac{x^n}{x^{j-1}}\right) \end{aligned}$$

This map is defined on $\phi(U_i \cap U_j)$, which consists only of points whose $(j-1)^{st}$ component is nonzero. It is, therefore, C^∞. The indices i and j were arbitrary, so this shows that the coordinate charts (ϕ_i, U_i) are all C^∞-compatible. So, $\mathbb{P}^n$ is a smooth n-dimensional manifold. ■

This last example shows that manifolds are much more than mere generalizations of surfaces. Projective spaces, and quotient manifolds in general, are quite complex. Even the two-dimensional projective space cannot be easily viewed as a subset of $\Re^3$ like classical surfaces can, so we generally have to sacrifice visual intuition.

There are many references to this material. The elementary work by (M. DoCarmo (76) 1976) on the geometry of curves and surfaces is a nice start on classical differential geometry. However, we were first exposed to this material in (B. O'Neill (10) 1966) which used differential form language throughout. We have already discussed integration in $\Re^n$ and differentiability in $\Re^n$ in (Peterson (101) 2020) but another interesting choice is the classic book, (M. Spivak (81) 1965), although it is hard to read as it is indeed terse. The longer treatment of (M. Spivak (82) 1979) and (M. Spivak (83) 1979) contains much historical material also and is a good read once you are ready. Once that material is understood, there are many works on the more advanced subjects within the geometry of manifolds. Few present all aspects of the subject in the detail you might need if this is an area of intense interest to you. All of them, however, have particular strengths. So you have to get used to looking at many sources to get a good understanding.

Homework

Exercise 13.1.18 *Let $X = \Re^{1+1} - \{0\}$. Define an equivalence relation, $\sim$, on X as follows. For $x, y \in X$, $x \sim y$ if and only if there exists a real number, $\lambda \neq 0$, such that $x = \lambda y$. Denote the equivalence class containing x by $[x]$. Let $\mathbb{P}^1 = \{[x] : x \in X\}$. Then $\mathbb{P}^1$ is the quotient space $\mathbb{X}/\sim$, and we assume it is endowed with the usual quotient space topology. We denote the quotient map by $p : X \rightarrow \mathbb{P}^1$, where $p(x) = [x]$ for $x \in X = \Re^{1+1} - \{0\}$. Go through the details from the proof that shows $\mathbb{P}^1$ is a smooth 1-dimensional manifold. Note that, geometrically speaking, we can identify $\mathbb{P}^1$ with the set of lines in R^{1+1} passing through the origin. Can you visualize this case?*

Exercise 13.1.19 *Let* $X = \Re^{2+1} - \{0\}$. *Define an equivalence relation,* $\sim$, *on* X *as follows. For* $x, y \in X$, $x \sim y$ *if and only if there exists a real number,* $\lambda \neq 0$, *such that* $x = \lambda y$. *Denote the equivalence class containing* x *by* $[x]$. *Let* $\mathbb{P}^2 = \{[x] \,:\, x \in X\}$. *Then* $\mathbb{P}^1$ *is the quotient space* $\mathbb{X}/\sim$, *and we assume it is endowed with the usual quotient space topology. We denote the quotient map by* $p : X \to \mathbb{P}^2$, *where* $p(x) = [x]$ *for* $x \in X = \Re^{2+1} - \{0\}$. *Go through the details from the proof that shows* $\mathbb{P}^2$ *is a smooth* 2*-dimensional manifold. Note that, geometrically speaking, we can identify* $\mathbb{P}^2$ *with the set of lines in* R^{2+1} *passing through the origin. Can you visualize this case?*

Exercise 13.1.20 *Let* $X = \Re^{3+1} - \{0\}$. *Define an equivalence relation,* $\sim$, *on* X *as follows. For* $x, y \in X$, $x \sim y$ *if and only if there exists a real number,* $\lambda \neq 0$, *such that* $x = \lambda y$. *Denote the equivalence class containing* x *by* $[x]$. *Let* $\mathbb{P}^3 = \{[x] \,:\, x \in X\}$. *Then* $\mathbb{P}^1$ *is the quotient space* $\mathbb{X}/\sim$, *and we assume it is endowed with the usual quotient space topology. We denote the quotient map by* $p : X \to \mathbb{P}^3$, *where* $p(x) = [x]$ *for* $x \in X = \Re^{3+1} - \{0\}$. *Go through the details from the proof that shows* $\mathbb{P}^3$ *is a smooth* 3*-dimensional manifold. Note that, geometrically speaking, we can identify* $\mathbb{P}^3$ *with the set of lines in* R^{3+1} *passing through the origin. But we can't visualize this case!*

Chapter 14

Smooth Functions on Manifolds

Next, we want to discuss coordinate changes around p and for that, we need to introduce the ideas of smooth functions on manifolds. Since M is a topological space, we have a well-defined notion of what it means for a function to be continuous, however, differentiation is another story. We need a means of differentiating functions on the manifold and it is the locally Euclidean structure of a manifold that makes this possible.

Definition 14.0.1 Smooth Real-Valued Functions on Manifolds

Let f be a real-valued function defined on an open subset, Ω_f, of a smooth manifold, M, possibly all of M. We say that f is a C^∞, or **smooth**, *function on Ω_f if, for each $p \in \Omega_f$, there is a coordinate chart (ϕ, U) such that $p \in U$ and the function $f \circ \phi^{-1}$ is C^∞ on $\phi(U \cap \Omega_f)$. We call $f \circ {\phi_U}^{-1}$ the coordinate representation of f with respect to the chart (ϕ_U, U).*

It turns out that even though the definition here uses local coordinate systems, the smoothness of a function will be coordinate independent. If $p \in \Omega_f$, and (ϕ, U) and (ψ, V) are charts such that $p \in U \cap V$, then $\phi \circ \psi^{-1} : \psi(U \cap V \cap \Omega_f) \to \phi(U \cap V \cap \Omega_f)$ is a smooth map. Hence, $f \circ \psi^{-1} = f \circ \phi^{-1} \circ \phi \circ \psi^{-1}$ is C^∞. So, if $f \circ \phi^{-1}$ is smooth for any coordinate system around p, it is smooth for all such coordinate systems. This definition of a smooth real-valued function leads us to the definition of a smooth mapping between manifolds.

Definition 14.0.2 Smooth Mappings between Manifolds

Let M and N be smooth manifolds, and let $F : M \to N$ be a mapping of M into N. We say that F is a smooth mapping if for every $p \in M$ there exist coordinate charts (ϕ, U) and (ψ, V) such that $p \in U$, $F(p) \in V$, $F(U) \subset V$, and the mapping $\psi \circ F \circ \phi^{-1} : \phi(U) \to \psi(V)$ is C^∞. We call $\psi \circ F \circ \phi^{-1}$ the coordinate representation of F.

This definition of smoothness is also coordinate independent. If (ϕ', U') and (ψ', V') are two other coordinate systems around p and $F(p)$, respectively, then

$$\psi' \circ F \circ {\phi'}^{-1} \quad = \quad \psi' \circ \psi^{-1} \circ \psi \circ F \circ \phi^{-1} \circ \phi \circ {\phi'}^{-1}$$

on $\phi'(U \cap U')$. Since the maps $\psi' \circ \psi^{-1}$, $\psi \circ F \circ \phi^{-1}$, and $\phi \circ {\phi'}^{-1}$ are C^∞, so is $\psi' \circ F \circ {\phi'}^{-1}$.

There is a consequence of these definitions that is very useful. Each coordinate mapping $\phi_U : U \to \phi_U(U)$ is a smooth function. Clearly, the function $\phi_U \circ {\phi_U}^{-1}$ is smooth, as it is just the identity mapping of $\phi_U(U)$. If p is any point in U and (ψ_V, V) is any other coordinate chart around p, then

$$\phi_U \circ \psi_V{}^{-1} : \psi_V(V \cap U) \to \phi_U(V \cap U)$$

is C^∞ because of the smooth structure on M.

When we discuss tangent vectors and derivatives of maps, we need to find a way to represent the derivatives of these smooth functions. Let's start with this simple point of view for now. Let I be an interval in $\Re$, and let $\gamma : I \to M$ be a smooth function. This is a curve in the manifold M. If $p \in \gamma(I)$ and (ϕ, U) induces a coordinate system around p, then the coordinate representation of γ on $I \cap \phi^{-1}(U)$ is given by

$$\phi \circ \gamma(t) \quad = \quad (x^1(t), x^2(t), ..., x^m(t))$$

where each x^i is a real-valued coordinate function on $I \cap \phi^{-1}$. That is, if π^i is the ith coordinate projection on $\Re^m$, then each x^i is defined by $x^i = \pi^i \circ \phi \circ \gamma$. This coordinate representation is just a usual curve in $\Re^m$, and it has derivative $((x^1)'(t), ..., (x^m)'(t))$. This will be the coordinate representation of the tangent vector to the curve γ.

Finally, consider a point, $p \in M$, and suppose there are two charts, (ϕ, U) and (ψ, V), such that $p \in U \cap V$. Then we have two different means of representing points near p by local coordinates. Each coordinate mapping defines m coordinate functions from M to $\Re$. That is, we can represent the map ϕ on U by $\phi(q) = (x^1(q), ..., x^m(q))$, where each x^i maps U into $\Re$. We refer to x^i as the ith coordinate function of ϕ. There are also functions y^j such that $\psi(q) = (y^1(q), ..., y^m(q))$ for $q \in V$.

Now, a point $q \in U \cap V$ will have coordinates $\phi(q) \in \phi(U \cap V)$ and $\psi(q) \in \psi(U \cap V)$. Suppose we want to change from one coordinate system to another, say from ψ-coordinates to ϕ-coordinates. We need only apply to $\psi(q)$ the map $\phi \circ \psi^{-1}$, which will give us $\phi(p)$. We can change the other way too by applying to $\phi(q)$ the map $\psi \circ \phi^{-1}$. In terms of the coordinate functions, these mappings indicate that we can think of the functions x^i as each being functions of the y^j's, and vice versa. In short, we have, via the mapping $\phi \circ \psi^{-1}$, that

$$(y^1, ..., y^m) \quad \longmapsto \quad (x^1(y^1, ..., y^m), ..., x^m(y^1, ..., y^m)),$$

and, via the map $\psi \circ \phi^{-1}$, we have

$$(x^1, ..., x^m) \quad \longmapsto \quad (y^1(x^1, ..., x^m), ..., y^m(x^1, ..., x^m)).$$

Moreover, since the manifold is smooth, the coordinate transformations are C^∞. Thus, we can smoothly change from one coordinate system to another around any point $p \in M$. Except for the more abstract setting, this is no different than changing between, say, rectangular and polar coordinates in two-dimensional Euclidean space.

Homework

Exercise 14.0.1 *For the manifold $\Re^2$, look at the coordinate systems Cartesian, (x, y), and polar, (r, θ), in the context of coordinate transformations as we discuss in this section.*

Exercise 14.0.2 *For the manifold $\Re^3$, look at the coordinate systems cartesian, (x, y, z), and cylindrical, (r, θ, z) in the context of coordinate transformations as we discuss in this section.*

Exercise 14.0.3 *For the manifold $\Re^3$, look at the coordinate systems cartesian, (x, y, z), and spherical, (ρ, θ, ϕ) in the context of coordinate transformations as we discuss in this section.*

14.1 The Tangent Space

The tangent space is the first additional structure we add to smooth manifolds. In multivariable calculus, the notion of a tangent plane to a surface is a well-defined concept, and it can be identified with a 2-dimensional subspace of $\Re^3$, thus giving us the idea of the tangent pane or space. However, in that context, the tangent space at a particular point was defined by means of tangent vectors to curves in the surface passing through that point. Curves on a surface through a point are mappings from an interval of $\Re$ to the surface. We would like to find a way to avoid thinking of it that way and instead have the entire idea rooted in the manifold itself. This will take some time to achieve, so be patient. The first idea is really about mappings from one manifold (the one corresponding to the interval in $\Re$) to another (our actual manifold). To motivate the new interpretation we seek, consider a classical surface defined by a function of multiple variables and the plane tangent to the surface at a given point, p. A tangent vector in this particular tangent space can be identified as a vector in the tangent plane anchored at the point, p. Now, consider a smooth real-valued function, f, defined on the surface in some neighborhood of p. Finding the gradient of this function on the surface at p simply means that we are finding a directional derivative of f in the direction of one of these tangent vectors. Vectors in the tangent plane that lie along the same direction but have different magnitudes will produce different directional derivatives. So, each vector in the tangent plane defines a unique directional derivative operator on the set of smooth functions defined in a neighborhood of p. This will turn out to be a useful way of redefining a tangent vector. Let's go through the details.

Let p be a point in the smooth manifold, M. Let $C^\infty(p)$ be the set of all real-valued functions that are defined and smooth on some neighborhood of p. Thus,

$$C^\infty(p) \quad = \quad \{f : N_f \to \Re \mid N_f \text{ neighborhood of } p,\ f \text{ is } C^\infty \text{ on } N_f\}$$

Note, we can always assume that a neighborhood, N_f, associated with a function, f, lies within a coordinate neighborhood, U. Define a relation $\sim$ on $C^\infty(p)$ as follows. We say $f \sim g$ if and only if there exists some neighborhood, V of p such that $f = g$ on V. It is straightforward to see that this is an equivalence relation on $C^\infty(p)$. We denote the equivalence class containing f by $\tilde{f}$. Denote the quotient space $C^\infty(p)/\sim$ by $\mathscr{F}_p$.

Lemma 14.1.1 $\mathscr{F}_p$ is a Commutative Algebra with a Multiplicative Identity

> *$\mathscr{F}_p$ is a commutative algebra with a multiplicative identity.*

Proof 14.1.1
We often use $[f]$ to denote an equivalence class, but this time we will use the notation $\tilde{f}$ which is often used in these sorts of discussions. You should be comfortable with both types of notation, of course. In the discussion that follows, you can see why $[\cdot]$ is sometimes preferred: $[f+g]$ is more pleasing to the eye that $\widetilde{f+g}$!

For $\tilde{f}, \tilde{g} \in \mathscr{F}_p$, define $\tilde{f} + \tilde{g}$ to be the equivalence class of the function $f + g$. That is, define $\tilde{f} + \tilde{g} = \widetilde{f+g}$. We need to show that this is independent of the choice of $f \in \tilde{f}$ and $g \in \tilde{g}$. Suppose f_1 and f_2 are in $\tilde{f}$. Then there is some neighborhood, N_f, of p such that $f_1 = f_2$ on N_f. Let N_g be a neighborhood of p on which $g \in \tilde{g}$ is defined. Now set $W = N_f \cap N_g$. Then, for $q \in N_f \cap N_g$, we have $f_1(q) = f_2(q) \Rightarrow f_1(q)+g(q) = f_2(q)+g(q)$. So, $f_1+g \sim f_2+g \Rightarrow \widetilde{f_1+g} = \widetilde{f_2+g}$. Hence, the addition operation does not depend on the choice of $f \in \tilde{f}$. A similar argument shows that it is also independent of the choice of $g \in \tilde{g}$. Thus, we have a well-defined notion of vector addition in $\mathscr{F}_p$.

Now let α be a real scalar. Define $\alpha\tilde{f} = \widetilde{\alpha f}$. To see that this is well-defined, suppose f_1 and f_2 are in $\tilde{f}$. Let N_f be as in the previous paragraph. Then $\alpha f_1 = \alpha f_2$ on N_f, so $\widetilde{\alpha f_1} = \widetilde{\alpha f_2}$, and scalar multiplication is well-defined. The vector space properties are easily established from here. For example, the zero element, $\tilde{0}$, is the set of all functions in $C^\infty(p)$ that vanish in some neighborhood of p.

Finally, define $\tilde{f}\tilde{g} = \widetilde{fg}$. Let f_1, f_2, N_f, and N_g be defined as before. Then $f_1 g = f_2 g$ on $N_f \cap N_g$, so multiplication is well-defined. Associativity and commutativity of multiplication follow easily, and distributivity of multiplication over addition is also clear. The multiplicative identity is just the set of all functions that are equal to the constant, 1, on some neighborhood of p. ■

The reason for defining the equivalence classes above is one of convenience. We will generally only be concerned with derivatives of smooth functions at p, and functions that are equal on some neighborhood of p will have the same derivatives. This brings us to our definition of the tangent space at p. Note how reminiscent this is of the identification of Cauchy sequences of rationals with what we call real numbers.

Definition 14.1.1 Tangent Vectors

> **A tangent vector** *at p is a map, $X_p : \mathscr{F}_p \to \Re$ such that, for any $\tilde{f}$ and $\tilde{g} \in \mathscr{F}_p$ and α and $\beta \in \Re$,*
>
> *(i): $X_p(\alpha\tilde{f} + \beta\tilde{g}) = \alpha X_p(\tilde{f}) + \beta X_p(\tilde{g})$; this is the* **linearity** *property.*
>
> *(ii): $X_p(\tilde{f}\tilde{g}) = \tilde{f}(p)X_p(\tilde{g}) + X_p(\tilde{f})\tilde{g}(p)$; this is the* **Leibnitz property.**

Note, these are the defining characteristics of the derivative operator. In general, a map on an algebra of functions satisfying properties i and ii is called a **derivation**. So, a tangent vector is just a derivation on $\mathscr{F}_p$. The set of all tangent vectors at p is called the tangent space at p, and is denoted T_pM. Since a tangent vector is a local object, we will often denote an element of T_pM by X_p, to indicate that X_p is a tangent vector at p.

Clearly, T_pM is nonempty, as the zero derivation that maps all functions to 0 will be in this set. However, it is not clear that there are nontrivial derivations. Next, note T_pM is made into a linear space in the obvious way. For X_p and $X_{p'}$ in T_pM, any $\tilde{f} \in \mathscr{F}_p$, and any $\alpha \in \Re$, we define

$$(X_p + X_p')(\tilde{f}) = X_p(\tilde{f}) + X_p'(\tilde{f}), \; (\alpha X_p)(\tilde{f}) = \alpha X_p(\tilde{f})$$

The existence and linear structure of T_pM is coordinate independent. We have not used any local coordinate system around p to construct anything so far. This shows that the tangent space is a well-defined geometrical concept. However, to construct a basis and represent tangent vectors in explicit form, a coordinate system is necessary. Thus, representation of tangent vectors is not unique, as it is coordinate dependent. This will be evident in our construction of a basis for T_pM. This is not so unusual. If V is an n dimensional vector space, the actual representation of a vector $v \in V$ depends on our choice of basis for V.

Example 14.1.1 *Let's go back and think about tangent vectors classically just for fun. Consider the surface $z = x^2 + y^2$ in $\Re^3$. Pick the point $p = (x_0, y_0, z_0)$ on the surface. Let $f(x, y, z) = x^2 + y^2 - z$. Let C be a curve on this surface that passes through p. Then C determines a locus of points $(x, y, x^2 + y^2)$ on the surface. Choose any parameterization of the locus of points which gives $\Gamma : [0, 1] \to \Re$ by*

$$\Gamma(t) = (x(t), y(t), z(t)) = (x(t), y(t), x^2(t) + y^2(t)), \; 0 \leq t \leq 1$$

The velocity vector $T(t)$ to this curve is given by

$$T(t) = \Gamma'(t) = \begin{bmatrix} x'(t) \\ y'(t) \\ z'(t) \end{bmatrix} = \begin{bmatrix} x'(t) \\ y'(t) \\ 2x(t)x'(t) + 2y(t)y'(t) \end{bmatrix}$$

We know

$$\nabla(f) = \begin{bmatrix} 2x \\ 2y \\ -1 \end{bmatrix}$$

From standard multivariable calculus, we know that $< \nabla(f)(p), T(0) >= 0$; i.e.

$$\left\langle \begin{bmatrix} 2x_0 \\ 2y_0 \\ -1 \end{bmatrix}, \begin{bmatrix} x'(0) \\ y'(0) \\ 2x(0)x'(0) + 2y(0)y'(0) \end{bmatrix} \right. = 2x_0x'(0) + 2y_0y'(0) - 2x_0x'(0) - 2y_0y'(0) = 0$$

Hence, $\nabla(f)(p)$ is the normal vector to a plane and this plane is the span of the independent vectors $\begin{bmatrix} 1 \\ 0 \\ \frac{\partial f}{\partial x}(p) \end{bmatrix}$ and $\begin{bmatrix} 0 \\ 1 \\ \frac{\partial f}{\partial y}(p) \end{bmatrix}$. This is the tangent plane to the surface at p. It is clear any vector in the tangent plane $v = \alpha \begin{bmatrix} 1 \\ 0 \\ \frac{\partial f}{\partial x}(p) \end{bmatrix} + \beta \begin{bmatrix} 0 \\ 1 \\ \frac{\partial f}{\partial y}(p) \end{bmatrix}$ can be associated with a curve Γ whose tangent vector $T(0) = v = \begin{bmatrix} x'(0) \\ y'(0) \\ z'(0) \end{bmatrix}$. So a surface $z = f(x, y)$ like this determines a mapping $\phi(p)$ from $\Re^2$ to $\Re$ by $\phi(p)(v) =< \nabla(f)(p), v >$. At this point, we don't know how to set this up in the new tangent vector formulation because we are not dealing with real functions g acting on a neighborhood of p on the surface. Instead we are dealing with the mapping γ. This is the setting of a mapping between manifolds we discussed earlier.

Homework

Exercise 14.1.1 *Repeat the discussion of the example above for $2x^2 + 3y^2 = z$.*

Exercise 14.1.2 *Repeat the discussion example above for $2x^2 + 3y^2 + 3z^2 = 100$.*

Exercise 14.1.3 *If we define $E_i(p)(\tilde{f})$ by $E_i(p)(\tilde{f}) = \frac{\partial f}{\partial x_i}(p)$, show $E_i(p)$ is a derivation on $\mathscr{F}_p$. Show this corresponds to a particular tangent vector.*

14.1.1 Basis Vectors for the Tangent Space

Let (ϕ, U) be a coordinate system around p. In Euclidean spaces, we can identify the standard basis vectors with the partial derivative operators that give the directional derivatives in the direction of the coordinate axes. So, it is natural for us to ask whether we can construct a basis for our tangent space using the same idea. For $i = 1, 2, ..., m$ define mappings $E_i : \mathscr{F}_p \to \Re$ by

$$E_i(\tilde{f}) = \frac{\partial}{\partial x^i}(f \circ \phi^{-1})\Big|_{\phi(p)}.$$

Note this is a well-defined mapping, since any f_1 and f_2 in $\tilde{f}$ are equal on some neighborhood of p. Thus, $f_1 \circ \phi^{-1}$ and $f_2 \circ \phi^{-1}$ are equal on some neighborhood of $\phi(p)$. Hence, E_i just takes $\tilde{f}$ to the ith partial derivative of the coordinate representation of f at $\phi(p)$.

Lemma 14.1.2 Each E_i is a Tangent Vector

Each E_i is a tangent vector. That is, each E_i is a derivation on $\mathscr{F}_p$.

Proof 14.1.2

Let $\tilde{f}$ and $\tilde{g}$ be in $\mathscr{F}_p$. Then

$$E_i(\tilde{f} + \tilde{g}) = E_i(\widetilde{f+g}) = \frac{\partial}{\partial x^i}((f+g)\circ\phi^{-1})\Big|_{\phi(p)}.$$

But $(f+g)\circ\phi^{-1} = f\circ\phi^{-1} + g\circ\phi^{-1}$, so

$$\begin{aligned} E_i(\tilde{f} + \tilde{g}) &= \frac{\partial}{\partial x^i}(f\circ\phi^{-1} + g\circ\phi^{-1})\Big|_{\phi(p)} = \frac{\partial}{\partial x^i}(f\circ\phi^{-1})\Big|_{\phi(p)} + \frac{\partial}{\partial x^i}(g\circ\phi^{-1})\Big|_{\phi(p)} \\ &= E_i(\tilde{f}) + E_i(\tilde{g}) \end{aligned}$$

Now, if $\alpha \in \Re$, then we have

$$\begin{aligned} E_i(\alpha\tilde{f}) &= E_i(\widetilde{\alpha f}) = \frac{\partial}{\partial x^i}((\alpha f)\circ\phi^{-1})\Big|_{\phi(p)} = \frac{\partial}{\partial x^i}(\alpha(f\circ\phi^{-1}))\Big|_{\phi(p)} \\ &= \alpha\frac{\partial}{\partial x^i}(f\circ\phi^{-1})\Big|_{\phi(p)} = \alpha E_i(\tilde{f}). \end{aligned}$$

Finally, we have

$$\begin{aligned} E_i(\tilde{f}\tilde{g}) &= E_i(\widetilde{fg}) = \frac{\partial}{\partial x^i}((fg)\circ\phi^{-1})\Big|_{\phi(p)} = \frac{\partial}{\partial x^i}((f\circ\phi^{-1})(g\circ\phi^{-1}))\Big|_{\phi(p)} \\ &= (f\circ\phi^{-1})(\phi(p))\frac{\partial}{\partial x^i}(g\circ\phi^{-1})\Big|_{\phi(p)} + (g\circ\phi^{-1})(\phi(p))\frac{\partial}{\partial x^i}(f\circ\phi^{-1})\Big|_{\phi(p)} \\ &= f(p)E_i(\tilde{g}) + g(p)E_i(\tilde{f}). \end{aligned}$$

Thus, each E_i is a tangent vector. ■

Note, we have been using the partial derivative notation with respect to the variable x^i. There is a specific meaning behind this. These are the coordinate functions representing ϕ. As in Chapter 13, if π^i is the ith coordinate projection from $\Re^m$ to $\Re$, then, for $q \in U$, $x^i(q) = \pi^i(\phi(q))$. Note these coordinate functions are smooth real-valued functions on U. Our tangent vectors, E_i, then, are just the partial derivatives of $f \circ \phi^{-1}$ with respect to these coordinate functions. We will use these coordinate functions extensively in our construction of the cotangent space in a bit. Now, we prove the tangent vectors $\{E_i\}$ are linearly independent. Let 0_p denote the zero element of T_pM. That is, $0_p(\tilde{f}) = 0$ for all $\tilde{f} \in \mathscr{F}_p$. Suppose there are scalars $c_1, \ldots, c_m$ such that

$$\sum_{i=1}^{m} c_i E_i = 0_p$$

Consider the element $\sum_{i=1}^m c_i E_i$ acting on $\tilde{x}^k$, the equivalence class containing the kth coordinate function. We have

$$\begin{aligned}\sum_{i=1}^m c_i E_i(\tilde{x}^k) &= \sum_{i=1}^m c_i \frac{\partial}{\partial x^i}\left(x^k \circ \phi^{-1}\right)\Big|_{\phi(p)} = \sum_{i=1}^m c_i \frac{\partial}{\partial x^i}\left(\pi^k \circ \phi \circ \phi^{-1}\right)\Big|_{\phi(p)} \\ &= \sum_{i=1}^m c_i \frac{\partial}{\partial x^i}\pi^k\Big|_{\phi(p)} = \sum_{i=1}^m c_i \delta_i^k = c_k = 0\end{aligned}$$

This shows that $c_k = 0$. Applying this sum to each i, in turn, shows that $c_1 = c_2 = \cdots = c_m = 0$. Hence, these tangent vectors are linearly independent. To prove that they span T_pM, we need a preliminary result from multivariable calculus. This is a consequence of the multivariable mean value theorem.

Lemma 14.1.3 Expansion of a Real-Valued Function

Let f be a real-valued function that is differentiable in some neighborhood, U, of $a \in \Re^m$. Then there are functions $g_k : U \to \Re$, for $k = 1, 2, ..., m$, such that

$$f(x) = f(a) + \sum_{k=1}^m (x^k - a^k) g_k(x) \text{ and } g_k(a) = \frac{\partial f}{\partial x^k}\Big|_a.$$

Proof 14.1.3

For each $x \in [0,1]$, define $h_x : [0,1] \to \Re$ by $h_x(t) = f(a + t(x-a))$. Then we have

$$\frac{dh_x(t)}{dt} = \sum_{k=1}^m (x^k - a^k) D_k f(a + t(x-a)).$$

By the fundamental theorem of calculus, we know that

$$\int_0^1 h_x'(t)dt = h_x(1) - h_x(0) = f(x) - f(a)$$

so it follows that

$$f(x) - f(a) = \sum_{k=1}^m (x^k - a^k) \int_0^1 D_k f(a + t(x-a))dt \;\forall\, x \in U.$$

Letting $g_k(x)$ be the integral in the above expression, we have $g_k(a) = D_k f(a)$ and $f(x) = f(a) + \sum_{k=1}^m (x^k - a^k) g_k(x)$. ■

To use Lemma 14.1.3, we must translate it to the manifold. Suppose f is a C^∞ function defined on some neighborhood, N_f, of p. This implies that $f \circ \phi^{-1}$ is a C^∞ function on $\phi(N_f) \subset \phi(U)$which is a neighborhood of $\phi(p)$. Since $\phi(N_f)$ is an open set in $\Re^m$, we can apply Lemma 14.1.3 to conclude there are functions $g_k : N_f \to \Re$, so that, for all $x \in \phi(N_f)$,

$$(f \circ \phi^{-1})(x) = (f \circ \phi^{-1})(\phi(p)) + \sum_{k=1}^m (x^k - [\phi(p)]^k) g_k(x)$$

where $[\phi(p)]^k$ is the kth coordinate of $\phi(p) \in \Re^m$ and

$$g_k(\phi(p)) = \frac{\partial}{\partial x^k}(f \circ \phi^{-1})\Big|_{\phi(p)}.$$

Since ϕ is a homeomorphism, for each $x \in \phi(N_f)$, there is a unique point $q \in N_f$ such that $\phi(q) = x$. Hence, we have $(f \circ \phi^{-1})(x) = (f \circ \phi^{-1})(\phi(q)) = f(q)$. Thus, for all $q \in N_f$,

$$f(q) = f(p) + \sum_{k=1}^{m} ((\pi^k \circ \phi)(q) - (\pi^k \circ \phi)(p)) g_k(\phi(q)).$$

We emphasize, again, that q is the variable in this representation, while p is a fixed point of the manifold. This is a useful representation of the function f on N_f.

Lemma 14.1.4 Elements $\{E_i : 1 \leq i \leq m\}$ Span T_pM

The elements $\{E_i : 1 \leq i \leq m\}$ span T_pM.

Proof 14.1.4

Let X_p be an arbitrary tangent vector, and let $\tilde{f}$ be in $\mathscr{F}_p$. Then, for any $f \in \tilde{f}$,

$$\begin{aligned}
X_p(f) &= X_p\left(f(p) + \sum_{k=1}^{m}\left[(\pi^k \circ \phi) - (\pi^k \circ \phi)(p)\right](g_k \circ \phi)\right) \\
&= X_p(f(p)) + \sum_{k=1}^{m} X_p\left[\left((\pi^k \circ \phi) - (\pi^k \circ \phi)(p)\right)(g_k \circ \phi)\right] \\
&= X_p(f(p)) + \sum_{k=1}^{m} X_p\left[(\pi^k \circ \phi) - (\pi^k \circ \phi)(p)\right] g_k(\phi(p)) \\
&\quad + \sum_{k=1}^{m}\left((\pi^k \circ \phi)(p) - (\pi^k \circ \phi)(p)\right) X_p(g_k \circ \phi) \\
&= X_p(f(p)) + \sum_{k=1}^{m} X_p((\pi^k \circ \phi) - (\pi^k \circ \phi)(p)) g_k(\phi(p)).
\end{aligned}$$

Since X_p is a derivation, it must map constants to 0. Moreover, by our previous result, we have

$$g_k(\phi(p)) = \frac{\partial}{\partial x^k}(f \circ \phi^{-1})\Big|_{\phi(p)} = E_k(f).$$

Hence, we have

$$X_p(f) = \sum_{k=1}^{m} X_p(\pi^k \circ \phi) E_k(f).$$

Now, if f_1 and f_2 are in $\tilde{f}$, then $f_1 = f_2$ on some neighborhood, N, of p. Thus $f_1 \circ \phi^{-1} = f_2 \circ \phi^{-1}$ on $\phi(U \cap N)$. It follows the expressions we derive for f_1 and f_2 using Lemma 14.1.3 are the same on $\phi(U \cap N)$. Hence, $X_p(f_1) = X_p(f_2)$. Since f_1 and f_2 were arbitrary elements of $\tilde{f}$, it makes *sense to say that $X_p(\tilde{f}) = X_p(f)$ for any $f \in \tilde{f}$. So, we have*

$$X_p(\tilde{f}) = \sum_{k=1}^{m} X_p(\pi^k \circ \phi)E_k(\tilde{f}),\ \forall\, \tilde{f} \in F_p \Longrightarrow X_p = \sum_{k=1}^{m} X_p(\tilde{x}^k)E_k$$

This shows that the set $\{E_i : 1 \leq i \leq m\}$ spans T_pM. Moreover, it also shows that the components of a tangent vector, X, with respect to the basis $\{E_i\}$ are the values obtained by X acting on the coordinate functions representing ϕ. ■

Homework

Exercise 14.1.4 *Find the expansion discussed in Lemma 14.1.3 for $f(x,y) = x^2 + y^2$ at $(1,1)$.*

Exercise 14.1.5 *Find the expansion discussed in Lemma 14.1.3 for $f(x,y,z) = 2x^2 + 3y^2 + 4z^2$ at $(1,1,2)$.*

14.1.2 Change of Basis Results

Our results imply the dimension of the tangent space, T_pM, is the same as that of the manifold at p. Remember though, the basis constructed depends upon the coordinate system (ϕ, U). If we had used another chart, (ψ, V), such that $p \in V$, then we would still have obtained a basis for an m-dimensional vector space, but that basis would, in general, be different. While the tangent space, T_pM, exists and has linear structure independent of any coordinate system, in order to obtain a basis and, consequently, an explicit representation for each tangent vector, we must choose a coordinate system around p. This is just like having to choose a basis for an m-dimensional space in order to get a representation! Hence, a single tangent vector, X_p, will have as many representations as there are coordinate neighborhoods of p. We call the basis $\{E_i : 1 \leq i \leq m\}$ obtained at p from a particular chart, the standard coordinate frame, or standard coordinate basis, induced by the coordinate system. Although, our goal is to obtain coordinate independent results, the coordinate dependent representation of T_pM is not a setback, for it is only a representation. When referring to a tangent vector at a point p, its explicit representation with respect to a particular basis will usually not matter. When such a representation is used, though, there is a convenient formula for switching between different basis representations. Before presenting this change of basis theorem, however, it is time to introduce two convenient notational conventions to make our expressions much more concise and elegant in the results that follow.

First, since we have shown the action of a tangent vector on $\tilde{f}$ is independent of the function f, we choose from $\tilde{f}$, we will omit the tilde in our notation whenever convenient. For example, the operator equation we derived for the representation of a tangent vector with respect to a particular coordinate system can be more succinctly written as $X_p = \sum_{k=1}^{m} X_p(x^k)E_k$, where it is implied that x^k actually represents the equivalence class $\tilde{x}^k$. This alternate notation will not always be used. In particular, when we discuss the cotangent space later on, we will use the tilde again, since referring to the entire equivalence class will be necessary. However, it should be clear from the context which particular notation is appropriate. In general, all of the mappings and operators we define will be independent of any particular element chosen from an equivalence class. So, there will be no difference in the results regardless of which notation we use. This is really just a matter of convenient notation, which is very useful in differential geometry. We use the same convenience of expression when we describe irrational numbers, of course. In general, we will try to simplify expressions, both visually and mathematically, if we can.

Second, the following transformation theorem, along with its analogue in the case of the dual of the tangent space or cotangent space which we will discuss later, requires we take partial derivatives of the change of variable maps. Recall that if (ϕ, U) and (ψ, V) are two coordinate systems such that $p \in U \cap V$, then we can change from ϕ-coordinates to ψ-coordinates via the map $\psi \circ \phi^{-1}$, as well as from ψ-coordinates to ϕ-coordinates via the inverse map, $\phi \circ \psi^{-1}$. Each of these is a map from $\Re^m$ into $\Re^m$. As before, we will denote the coordinate functions of ϕ by x^i, meaning that, for $p \in U$, we have $x^i(p) = \pi^i(\phi(p))$, where π^i is the ith coordinate projection in $\Re^m$. We will also denote the coordinate functions of ψ by y^i. Hence, in forming the composition $\phi \circ \psi^{-1}$, we see that each coordinate map, x^i, becomes a function of the coordinates $y^1, \ldots, y^m$, since $\phi \circ \psi^{-1}$ maps points described by ψ-coordinates to points described by ϕ-coordinates. In other words, we can represent this map in terms of its input and output coordinates by the expression

$$\phi \circ \psi^{-1}(y^1, \ldots, y^m) = (x^1(y^1, \ldots, y^m), \ldots, x^m(y^1, \ldots, y^m))$$

Similarly, in forming the composition $\psi \circ \phi^{-1}$, we see that each coordinate map, y^i, becomes a function of the coordinates $x^1, \ldots, x^m$, since $\psi \circ \phi^{-1}$ maps points described by ϕ-coordinates to points described by ψ-coordinates. That is, we obtain a representation of the form

$$\psi \circ \phi^{-1}(x^1, \ldots, x^m) = (y^1(x^1, \ldots, x^m), \ldots, y^m(x^1, \ldots, x^m))$$

Now, suppose we wish to compute the ith partial derivative of the jth component of the map $\phi \circ \psi^{-1}$ at the point $\psi(p)$. The correct, coordinate-free notation, of course, would be to write this partial derivative as

$$D_i\left(\phi \circ \psi^{-1}\right)^j\Big|_{\psi(p)}$$

where D_i represents the general partial derivative operator with respect to the ith variable of the function in question, and the superscript j indicates the jth component function of the map $\phi \circ \psi^{-1}$. Note that this is equivalent to the classically motivated expression

$$\frac{\partial\left(\pi^j \circ \phi \circ \psi^{-1}\right)}{\partial y^i}\Bigg|_{\psi(p)},$$

since composing π^j with $\phi \circ \psi^{-1}$ just gives us the jth component function of this map. The classical partial derivative operator, $\partial / \partial y^i$, simply indicates that the derivative is to be taken with respect to the ith variable. Since the composition operation is associative, consider composing π^j with ϕ in this expression, and leaving ψ^{-1} as is. We know that $\pi^j \circ \phi$ is just x^j. So, the previous expression is further equivalent to

$$\frac{\partial\left(x^j \circ \psi^{-1}\right)}{\partial y^i}\Bigg|_{\psi(p)}.$$

So, in a very real sense, we are simply computing the ith partial derivative, with respect to the ψ-coordinate system, of the jth ϕ-coordinate. These particular partial derivatives are often simply written in the abbreviated form $\frac{\partial x^j}{\partial y^i}\Big|_{\psi(p)}$ since the fact that we are composing x^j with ψ^{-1} is implied by the fact that we are taking the partial derivative with respect to one of the ψ-coordinates. The point of all of this is to alleviate any confusion that might arise later on. This abbreviated form is just a notational device. It is very elegant and convenient, but it should always be taken to represent the formal derivative $D_i(\phi \circ \psi^{-1})^j|_{\psi(p)}$. In the exact same fashion, we will use the abbreviated expression $\frac{\partial y^j}{\partial x^i}\Big|_{\phi(p)}$ to represent the formal partial derivative $D_i(\psi \circ \phi^{-1})^j|_{\phi(p)}$, which is the ith partial derivative of the jth component of the map $\psi \circ \phi^{-1}$ at $\phi(p)$. With these conventions in mind, we can finally present the following result.

Theorem 14.1.5 Comparing Expansions in Two Coordinate systems

Suppose $p \in U \cap V$, where (ϕ, U) and (ψ, V) are two coordinate systems around p. Let $\{x^i\}$ denote the coordinate functions making up the coordinate representation of ϕ, and let $\{y^i\}$ denote the coordinates of ψ. Let $\{E_i^x\}$ and $\{E_i^y\}$ denote the coordinate frames induced by (ϕ, U) and (ψ, V), respectively, on T_pM. Then

$$E_i^x = \sum_{k=1}^m \frac{\partial y^k}{\partial x^i}\Big|_{\phi(p)} E_k^y \quad \text{and} \quad E_j^y = \sum_{\ell=1}^m \frac{\partial x^\ell}{\partial y^j}\Big|_{\psi(p)} E_l^x.$$

If $X_p = \sum \alpha^i E_i^x = \sum \beta^j E_j^y$ is a tangent vector at p, then

$$\alpha^i = \sum_{j=1}^m \beta^j \frac{\partial x^i}{\partial y^j}\Big|_{\psi(p)} \quad \text{and} \quad \beta^j = \sum_{i=1}^m \alpha^i \frac{\partial y^j}{\partial x^i}\Big|_{\phi(p)}.$$

Proof 14.1.5

The proof is just an application of change of coordinates at p. By Lemma 14.1.4, we know $E_i^x = \sum_k E_i^x(y^k) E_k^y$. Thus, using the abbreviated notation,

$$E_i^x(y^k) = \frac{\partial}{\partial x^i}\left(y^k \circ \phi^{-1}\right)\Big|_{\phi(p)} = \frac{\partial\left(\pi^k \circ \psi \circ \phi^{-1}\right)}{\partial x^i}\Big|_{\phi(p)} = D_i\left(\psi \circ \phi^{-1}\right)^k\Big|_{\phi(p)} = \frac{\partial y^k}{\partial x^i}\Big|_{\phi(p)}$$

It follows that

$$E_i^x = \sum_{k=1}^m \frac{\partial y^k}{\partial x^i}\Big|_{\phi(p)} E_k^y$$

and the second formula follows similarly. Once those formulas are established, given X_p, we know that its ith component, α_i, with respect to the basis $\{E_i^x\}$, is given by $X_p(x^i)$. So, we have

$$\alpha^i = X_p(x^i) = \sum_{j=1}^m \beta^j E_j^y(x^i) = \sum_{j=1}^m \beta^j \frac{\partial x^i}{\partial y^j}\Big|_{\psi(p)}$$

and this verifies the third formula. The last one follows in the same manner. ■

Now, recall from elementary linear algebra that, if $\mathscr{B}_1 = \{e_1, \ldots, e_m\}$ and $\mathscr{B}_2 = \{f_1, \ldots, f_m\}$ are two bases for an m-dimensional vector space, V, then we can easily construct a change of basis matrix, $P_{1,2}$. Given any vector, $v \in V$, if $[v]_1$ represents the vector of the components of v with respect to $\mathscr{B}_1$, then $P_{1,2}(v)$ will give us $[v]_2$, the vector of the components of v with respect to $\mathscr{B}_2$. The ith column of $P_{1,2}$ is given by $[e_i]_2$, the vector of the components of e_i with respect to $\mathscr{B}_2$. The inverse of this matrix, denoted $P_{2,1}$, maps $[v]_2$ to $[v]_1$. Now, consider applying this idea to the results given in Lemma 14.1.4. If X_p is an arbitrary tangent vector in T_pM, with representations $X_p = \sum \alpha^i E_i^x = \sum \beta^j E_j^y$, then we can transform between the two component vectors $[X_p]_\phi =$

$[\alpha^1 \cdots \alpha^m]^T$ and $[X_p]_\psi = [\beta^1 \cdots \beta^m]^T$. The matrix that maps $[X_p]_\phi$ to $[X_p]_\psi$ is then given by

$$\begin{bmatrix} \frac{\partial y^1}{\partial x^1} & \cdots & \frac{\partial y^1}{\partial x^m} \\ \vdots & \ddots & \vdots \\ \frac{\partial y^m}{\partial x^1} & \cdots & \frac{\partial y^m}{\partial x^m} \end{bmatrix}$$

where the partial derivatives are evaluated at $\phi(p)$. Thus, the change of basis matrix that maps tangent vectors in span$\{E_i^x\}$ to tangent vectors in span$\{E_j^y\}$ is just the Jacobian of the change of coordinate mapping $\psi \circ \phi^{-1} : \phi(U \cap V) \to \psi(U \cap V)$ evaluated at $\phi(p)$. The change of basis matrix that maps tangent vectors in span$\{E_j^y\}$ to tangent vectors in span$\{E_i^x\}$ is the inverse of that matrix, which is nothing more than the Jacobian of the change of coordinate mapping $\phi \circ \psi^{-1} : \psi(U \cap V) \to \phi(U \cap V)$ evaluated at $\psi(p)$. These Jacobians are often given in shorthand notation by

$$D(\psi \circ \phi^{-1})(p) \qquad \text{and} \qquad D(\phi \circ \psi^{-1})(p)$$

Homework

Exercise 14.1.6 *Prove the relation $\sim$ on $C^\infty(p)$ defined by $f \sim g \Leftrightarrow$ there exists some neighborhood, V, of p such that $f = g$ on V is an equivalence relation.*

Exercise 14.1.7 *In Theorem 14.1.5, prove*

$$E_j^y = \sum_{\ell=1}^{m} \frac{\partial x^\ell}{\partial y^j}\bigg|_{\psi(p)} E_l^x$$

Exercise 14.1.8 *In Theorem 14.1.5, prove*

$$\beta^j = \sum_{i=1}^{m} \alpha^i \frac{\partial y^j}{\partial x^i}\bigg|_{\phi(p)}$$

14.2 The Cotangent Space

We now know the tangent space T_pM has the same dimension as the manifold, which we assume to be finite. Hence, after choosing a basis, the tangent space is isomorphic to $\Re^m$. Consequently, it has a well-defined dual space of the same dimension. This dual space is called the cotangent space, and it is the set of all linear functionals on T_pM. We denote the cotangent space by T_p^*M. Many authors choose not to show the details of the construction of the cotangent space, but we think there is value in going through this explicitly. Remember, the point of all of this is to carefully explain this so that you can be confident in using ideas similar to this in your own work. So, additional discussion helps you along that path. We will also show in detail the pairing between the two, showing that they are dual spaces of each other. Given what we have constructed so far, we can show that the cotangent space at a point p is actually a quotient space of $\mathscr{F}_p$.

Let Γ_p be the set of all smooth curves in M through p. By a suitable translation, we can assume that each curve maps $0 \in \Re$ to p and is defined on an interval of the form $(-\epsilon, \epsilon)$. So, precisely, we have

$$\Gamma_p = \{\gamma : (-\epsilon, \epsilon) \to M \mid \gamma(0) = p,\ \gamma \textit{ is } C^\infty\}.$$

Note that ϵ depends on γ in this definition. Now, define a map $T : \mathscr{F}_p \times \Gamma_p \to \Re$ by

$$T(\tilde{f},\gamma) = \frac{d}{dt}(f \circ \gamma)\Big|_{t=0}.$$

This map is well-defined, for if f_1 and f_2 are in $\tilde{f}$, then $f_1 = f_2$ on some neighborhood, N, of p, implying that $f_1 = f_2$ on $N \cap \gamma(-\epsilon,\epsilon)$. T is also linear in the first variable, since if $\tilde{f},\tilde{g} \in \mathscr{F}_p$ and $\alpha, \beta \in \Re$, then

$$\begin{aligned} T(\alpha\tilde{f} + \beta\tilde{g},\gamma) &= T(\widetilde{\alpha f + \beta g},\gamma) = \frac{d}{dt}((\alpha f + \beta g) \circ \gamma)\Big|_{t=0} \\ &= \frac{d}{dt}(\alpha f \circ \gamma + \beta g \circ \gamma)\Big|_{t=0} \\ &= \alpha\frac{d}{dt}(f \circ \gamma)\Big|_{t=0} + \beta\frac{d}{dt}(g \circ \gamma)\Big|_{t=0} = \alpha T(\tilde{f},\gamma) + \beta T(\tilde{g},\gamma) \end{aligned}$$

Next, define a set $\mathscr{H}_p$ by

$$\mathscr{H}_p = \{\tilde{f} \in \mathscr{F}_p : T(\tilde{f},\gamma) = 0 \ \forall\, \gamma \in \Gamma_p\}.$$

Since T is linear in the first variable, it follows that $\mathscr{H}_p$ is a linear subspace of $\mathscr{F}_p$. In fact, we can completely characterize those elements of $\mathscr{F}_p$ that are in $\mathscr{H}_p$.

Lemma 14.2.1 Characterization of $\mathscr{H}_p$

Let $\tilde{f}$ be in $\mathscr{F}_p$. Then $\tilde{f} \in \mathscr{H}_p$ if and only if for any coordinate chart, (ϕ, U), with $p \in U$, we have

$$\frac{\partial}{\partial x^i}(f \circ \phi^{-1})\Big|_{\phi(p)} = 0$$

for all $i = 1, 2, ..., m$. In short, $\tilde{f} \in \mathscr{H}_p$ if and only if all coordinate partial derivatives of f vanish at p.

Proof 14.2.1

Let (ϕ, U) be any coordinate system around p, and let γ be any curve in Γ_p. We can write the coordinate representation of γ as

$$(\phi \circ \gamma)(t) = (x^1(t), ..., x^m(t)),\ t \in (\epsilon, \epsilon).$$

Then for any $f \in \tilde{f}$, we have

$$\begin{aligned} T(\tilde{f},\gamma) &= \frac{d}{dt}(f \circ \gamma)\Big|_{t=0} = \frac{d}{dt}(f \circ \phi^{-1} \circ \phi \circ \gamma)\Big|_{t=0} \\ &= \frac{d}{dt}(f \circ \phi^{-1})(x^1(t), ..., x^m(t))\Big|_{t=0} \\ &= \sum_{i=1}^{m} \frac{\partial}{\partial x^i}(f \circ \phi^{-1})\Big|_{\phi(p)} \frac{dx^i}{dt}\Big|_{t=0} \end{aligned} \tag{14.1}$$

So, if all the partial derivatives of $f \circ \phi^{-1}$ vanish at p, we must have $T(\tilde{f},\gamma) = 0$. Conversely, suppose $\tilde{f} \in \mathscr{H}_p$, so that $T(\tilde{f},\gamma) = 0$ for all $\gamma \in \Gamma_p$. For each $i = 1, 2, ..., m$, define a particular *curve in Γ_p as follows. Let $\lambda_i : \Re \rightarrow \Re^m$ be given by*

$$\lambda_i(t) = ((\phi(p))^1, ..., (\phi(p))^{i-1}, (\phi(p))^i + t, (\phi(p))^{i+1}, ..., (\phi(p))^m)$$

where $(\phi(p))^k$ is the kth component of $\phi(p) \in \Re^m$. Then define $\gamma_i : \Re \rightarrow M$ by $\gamma_i = \phi^{-1} \circ \lambda_i$. Each curve, γ_i, is smooth and satisfies $\gamma_i(0) = p$. Then the coordinate representation, $\phi \circ \gamma_i = (x^1(t), ..., x^m(t))$, satisfies

$$\frac{dx^k}{dt}\bigg|_{t=0} = \begin{cases} 1 & k = i \\ 0 & k \neq i \end{cases}.$$

Thus,

$$T(\tilde{f}, \gamma_i) = 0 \Rightarrow \frac{\partial}{\partial x^i}(f \circ \phi^{-1})\bigg|_{\phi(p)} = 0.$$

Applying this to each $i = 1, 2, ..., m$ shows that $\frac{\partial}{\partial x^i}(f \circ \phi^{-1}) = 0$ at $\phi(p)$ for all i. Moreover, the coordinate system (ϕ, U) was arbitrary. This proves the result. ■

We are now ready to define the cotangent space. Recall that, for a vector space, V, and a subspace $W \subset V$, the quotient space V/W is defined as follows. We define an equivalence relation on V by $x \approx y$ if and only if $x - y \in W$. Then V/W is the set of all equivalence classes under this relation. It is a linear space with the operations $\tilde{x} + \tilde{y} = \widetilde{x + y}$ and $\alpha\tilde{x} = \widetilde{\alpha x}$. The zero element of V/W is the entire subspace W. Using this concept, we have the following.

Definition 14.2.1 Cotangent Space

The quotient space $\mathscr{F}_p/\mathscr{H}_p$ is called the **cotangent space** *and is denoted by T_p^*M. The $\mathscr{H}_p$ equivalence class of $\tilde{f}$ is denoted $d\tilde{f}_p$, and is called a* **cotangent vector** *at p.*

We see immediately T_p^*M is a linear space, as it is the quotient space defined by a subspace of $\mathscr{F}_p$. Moreover, by the general structure of a quotient space, we know that for any $\tilde{f}, \tilde{g} \in \mathscr{F}_p$ and $\alpha \in \Re$, we have $d\tilde{f}_p + d\tilde{g}_p = d(\widetilde{f + g})_p$ and $\alpha d\tilde{f}_p = d(\widetilde{\alpha f})_p$.

As a remark on notation, we will usually denote a cotangent vector simply by $d\tilde{f}$, if it is clear from the context to which specific point we are referring. Since all of our discussions here are focused on the single point $p \in M$, we will use this abbreviation frequently.

The definition of the cotangent space may not be as immediately transparent as that of the tangent space. However, it is easy to see two elements $\tilde{f}, \tilde{g} \in \mathscr{F}_p$ are in the same $\mathscr{H}_p$ equivalence class if and only if each of the coordinate partial derivatives of f, in terms of any coordinate system around p, equals the corresponding coordinate partial derivative of g. This characterization does not, of course, depend on which functions $f \in \tilde{f}$ and $g \in \tilde{g}$. We choose to compute such partial derivatives. Hence, a cotangent vector, $d\tilde{f}$, can be thought of as a set of smooth functions defined in a neighborhood of p that all have identical coordinate partial derivatives at p.

Next, we show the dimension of T_p^*M is m. To do this, we will proceed as in the case of the tangent space, choosing a local coordinate system and constructing a basis. As we said before, this basis will be coordinate dependent, but that will only affect the specific representation of cotangent vectors. Our construction thus far has not made use of any coordinate system, so the existence and linear structure of T_p^*M is well-defined and independent of any coordinate system around p.

Let's find a natural choice for a basis. For example, let $M = \Re^2$, and $p = (0,0)$. Suppose two functions, f and g, defined and smooth on respective neighborhoods of p, have the same coordinate partial derivatives at the origin. Then we would equate these functions as a cotangent vector. In essence, their gradients, ∇f and ∇g, are equal at p. Any other function whose gradient equaled these would also be identified with f and g. Hence, we can think of the cotangent vector $d\tilde{f}$ as simply defining a vector in $\Re^2$ emanating from the origin, namely the vector defined by the gradient of any function identified with f. Conversely, given a vector in $\Re^2$, is straightforward to construct a smooth function having this vector as its gradient. Hence, we can actually identify the cotangent space with $\Re^2$. This space is spanned by the standard basis vectors $e_1 = (1,0)$ and $e_2 = (0,1)$. So, to think of these vectors as a basis for the cotangent space at p, we need to consider what functions have these vectors as gradients. The simple choice, of course, is the set of coordinate functions $f_1(x,y) = x$ and $f_2(x,y) = y$. That is, the coordinate functions are the natural choice for a basis for the cotangent space. Let's use this insight to find a basis for the cotangent space in general.

Homework

Exercise 14.2.1 *Assume the gradient of f at $p = (1,2)$ is $\begin{bmatrix} 2 \\ -3 \end{bmatrix}$. Find all other functions with this gradient. Explain why these functions give the cotangent space at p.*

Exercise 14.2.2 *Assume the gradient of f at $p = (-1,2)$ is $\begin{bmatrix} -2 \\ 1 \end{bmatrix}$. Find all other functions with this gradient. Explain why these functions give the cotangent space at p.*

Exercise 14.2.3 *Assume the gradient of f at $p = (-1,2,3)$ is $\begin{bmatrix} -2 \\ 1 \\ 5 \end{bmatrix}$. Find all other functions with this gradient. Explain why these functions give the cotangent space at p.*

Exercise 14.2.4 *Assume the gradient of f at $p = (-1,2,-4,3)$ is $\begin{bmatrix} -2 \\ 1 \\ 5 \\ -4 \end{bmatrix}$. Find all other functions with this gradient. Explain why these functions give the cotangent space at p.*

Let (ϕ, U) be a coordinate system around p. Then we have local coordinates on U defined by $\phi(q) = (x^1(q), ..., x^m(q))$ for $q \in U$. The functions $x^k : U \to \Re$ are the coordinate functions on U. As before, we will also use the fact that each x^i can be obtained by composing the ith component projection on $\Re^m$, π^i, with ϕ. That is, $x^i = \pi^i \circ \phi$. Each coordinate function x^i is also smooth, so it is an element of $C^\infty(p)$. We show that $\{d\tilde{x}^i : 1 \leq i \leq m\}$ is a basis for $T_p^* M$ induced by the coordinate system (ϕ, U).

We need the following technical lemma to aid us in the construction.

Lemma 14.2.2 Derivatives of Compositions

Suppose $f_1, ..., f_k$ are in $C^\infty(p)$, and let $F(y^1, ..., y^k)$ be a C^∞ real-valued function defined in a neighborhood of $(f_1(p), ..., f_k(p))$. Then $f = F(f_1, ..., f_k)$ is in $C^\infty(p)$ and

$$d\tilde{f} = \sum_{i=1}^{k} \frac{\partial F}{\partial f_i}(f_1(p), ..., f_k(p)) d\tilde{f}_i.$$

Proof 14.2.2
If U_j is the neighborhood of p on which f_j is smooth, then all the functions f_j are smooth on $N = \cap_{j=1}^k U_j$. It follows that $f = F(f_1, ..., f_k)$ is smooth on N. So, $f \in C^\infty(p)$, and we can consider $\tilde{f} \in F_p$ and $d\tilde{f} \in T_p^* M$. Let γ be any curve in Γ_p. Then, using the linearity of the map T in its first variable, we have

$$\begin{aligned} T(\tilde{f}, \gamma) &= \frac{d}{dt}(f \circ \gamma)\Big|_{t=0} = \frac{d}{dt}\Big(F(f_1 \circ \gamma, ..., f_k \circ \gamma)\Big)\Big|_{t=0} \\ &= \sum_{i=1}^k \frac{\partial F}{\partial f_i}(f_1(p), ..., f_k(p)) \frac{d}{dt}(f_i \circ \gamma)\Big|_{t=0} = \sum_{i=1}^k \frac{\partial F}{\partial f_i}(f_1(p), ..., f_k(p)) T(\tilde{f}_i, \gamma) \\ &= T\left(\sum_{i=1}^k \frac{\partial F}{\partial f_i}(f_1(p), ..., f_k(p)) \tilde{f}_i, \gamma\right) \end{aligned}$$

This implies that

$$T\left(\tilde{f} - \sum_{i=1}^k \frac{\partial F}{\partial f_i}(f_1(p), ..., f_k(p)) \tilde{f}_i, \gamma\right) = 0.$$

Since γ was arbitrary, this implies that

$$\tilde{f} - \sum_{i=1}^k \frac{\partial F}{\partial f_i}(f_1(p), ..., f_k(p)) \tilde{f}_i \in \mathscr{H}_p$$

Applying the linear structure of the cotangent space, it follows that $d\tilde{f} = \sum_{i=1}^k \frac{\partial F}{\partial f_i}(f_1(p), ..., f_k(p)) d\tilde{f}_i$. ∎

Now, to show that the cotangent vectors $d\tilde{x}^i$ form a basis for $T_p^* M$, we start by showing they form a set of m distinct elements. We want to know that if $i \neq j$, then $d\tilde{x}^i \neq d\tilde{x}^j$. This result is obvious in the Euclidean case, but it may not be in an arbitrary coordinate system. So, suppose i and j are indices such that $1 \leq i < j \leq m$. If $\tilde{x}^i = \tilde{x}^j$, then there is a neighborhood, N, of p on which $x^i = x^j$. Thus, every point $(y^1, ..., y^m) \in \phi(N)$ has the same i^{th} and j^{th} coordinates, implying that there is an open ball, $B(\phi(p), \epsilon)$, such that every point in this ball, including $\phi(p)$, has the same i^{th} and j^{th} coordinates. Consider, however, the point $y = \phi(p) + \frac{\epsilon}{2} e_i$, where e_i is the ith standard basis vector in $\Re^m$. Then this point is in the ball $B(\phi(p), \epsilon)$, but its ith and j^{th} coordinates are not the same, as we have shifted one by a positive amount while leaving the other fixed. Hence we cannot have $\tilde{x}^i = \tilde{x}^j$ when $i \neq j$. This shows that the elements $\{\tilde{x}^i \in F_p \ : \ 1 \leq i \leq m\}$ are all distinct.

Next, with indices i and j as before, suppose $d\tilde{x}^i = d\tilde{x}^j$. Then for any $k = 1, ..., m$, we have

$$\frac{\partial}{\partial x^k}(x^i \circ \phi^{-1})\Big|_{\phi(p)} = \frac{\partial}{\partial x^k}(x^j \circ \phi^{-1})\Big|_{\phi(p)}$$

But $x^i = \pi^i \circ \phi$, so $x^i \circ \phi^{-1} = \pi^i$, and also for x^j. Thus, letting $k = i$, the previous equation implies that

$$\frac{\partial}{\partial x^i}\pi^i\Big|_{\phi(p)} = \frac{\partial}{\partial x^i}\pi^j\Big|_{\phi(p)}$$

from which it follows that $1 = 0$, as $i \neq j$. The contradiction shows that, if $i \neq j$, then $d\tilde{x}^i \neq d\tilde{x}^j$. Hence, the set $\{d\tilde{x}^i \ : \ 1 \leq i \leq m\}$ does, indeed, contain m distinct elements. Thus, we can prove the following result.

Theorem 14.2.3 Basis for T_p^*M with respect to the Coordinate System (ϕ, U)

The set $\{d\tilde{x}^i \, : \, 1 \leq i \leq m\}$ forms a basis for T_p^*M with respect to the coordinate system (ϕ, U).

Proof 14.2.3

*(**Spanning**): We let $k = m$ and let the functions $f_1, ..., f_m$ be the coordinate functions $x^1, ..., x^m$. Then, for any element $d\tilde{f} \in T_p^*M$, we let F be $f \circ \phi^{-1}$. Then the function $g = F(x^1, ..., x^m)$ satisfies $g = f \circ \phi^{-1} \circ \phi = f$. Hence, $d\tilde{g} = d\tilde{f}$, and Lemma 14.2.2 implies that*

$$d\tilde{f} = \sum_{i=1}^{m} \frac{\partial F}{\partial x^i}(x^1(p), ..., x^m(p)) d\tilde{x}^i = \sum_{i=1}^{m} \frac{\partial (f \circ \phi^{-1})}{\partial x^i}\bigg|_{\phi(p)} d\tilde{x}^i.$$

So, the set $\{d\tilde{x}^i \, : \, 1 \leq i \leq m\}$ spans T_p^*M.

*(**Independence**): Now, let $d\tilde{0}$ denote the zero element of T_p^*M, and suppose there exist scalars $c_1, c_2, ..., c_m$ such that*

$$\sum_{i=1}^{m} c_i d\tilde{x}^i = d\Big(\sum_{i=1}^{m} c_i \tilde{x}^i\Big) = d\tilde{0}.$$

Then, using the fact that T_p^*M is a quotient space, it follows that $\sum_{i=1}^{m} c_i \tilde{x}^i \in \mathscr{H}_p$. So, for any $\gamma \in \Gamma_p$, we have

$$T\Big(\sum_{i=1}^{m} c_i \tilde{x}^i, \gamma\Big) = \sum_{i=1}^{m} c_i T(\tilde{x}^i, \gamma) = \sum_{i=1}^{m} c_i \frac{d}{dt}(x^i \circ \gamma)\Big|_{t=0} = 0$$

Now, for each $i = 1, ..., m$, define curves γ_i as in the proof of Lemma 14.2.1. Then

$$\begin{aligned} T\Big(\sum_{i=1}^{m} c_i \tilde{x}^i, \gamma_j\Big) &= \sum_{i=1}^{m} c_i \frac{d}{dt}(x^i \circ \phi^{-1} \circ \phi \circ \gamma_j)\Big|_{t=0} = \sum_{i=1}^{m} c_i \frac{d}{dt}(x^i \circ \phi^{-1} \circ \lambda_j)\Big|_{t=0} \\ &= \sum_{i=1}^{m} c_i \frac{d}{dt} \pi^i\big((\phi(p))^1, ..., (\phi(p))^{j-1}, (\phi(p))^j + t, (\phi(p))^{j+1}, ..., (\phi(p))^m\big)\Big|_{t=0} \\ &= \sum_{i=1}^{m} c_i \delta_j^i = c_j \end{aligned}$$

Thus, since T vanishes for any curve γ, it follows that $c_j = 0$. Letting $j = 1, ..., m$, in turn, shows that $c_1 = c_2 = \cdots = c_m = 0$. Hence, the set $\{d\tilde{x}^i \, : \, 1 \leq i \leq m\}$ is linearly independent and forms a basis for T_p^*M with respect to the coordinate system (ϕ, U). ■

Note that the components of the cotangent vector $d\tilde{f}$ with respect to this basis are the partial derivatives

$$\frac{\partial (f \circ \phi^{-1})}{\partial x^i}\bigg|_{\phi(p)} \qquad i = 1, 2, \ldots, m.$$

These are just the partial derivatives of the coordinate representation of f with respect to the coordinate system (ϕ, U). This gives us another way to think of a cotangent vector. A cotangent vector

$d\tilde{f}$ is essentially just a coordinate gradient of f at p. It is even more interesting to note that the ith component of the cotangent vector, $d\tilde{f}$, is just the value obtained by the ith tangent basis vector, E_i, acting on $\tilde{f}$. That is, it follows directly from the definition of E_i that

$$\frac{\partial(f\circ\phi^{-1})}{\partial x^i}\bigg|_{\phi(p)} = E_i(\tilde{f}).$$

Hence, we can represent a cotangent vector with respect to the standard cotangent basis by

$$d\tilde{f} = \sum_{i=1}^{m} E_i(\tilde{f})d\tilde{x}^i.$$

Of course, $E_i(\tilde{f})$ is independent of the choice of $f\in d\tilde{f}$, so we usually write this expansion as

$$d\tilde{f} = \sum_{i=1}^{m} E_i(f)d\tilde{x}^i.$$

It should be pointed out, as an interesting consequence of Lemma 14.2.2, that we can define a Leibnitz rule for cotangent vectors just as we have for tangent vectors. In the Lemma, just let F be defined by $F(y^1,y^2)=y^1y^2$. Then, for any two functions f_1 and f_2 are in $C^\infty(p)$, we have $f=F(f_1,f_2)=f_1f_2$. It follows that

$$\begin{aligned} d(\tilde{f}_1\tilde{f}_2) &= d(\widetilde{f_1f_2}) = d\tilde{f} = \frac{\partial F}{\partial f_1}(f_1(p),f_2(p))d\tilde{f}_1 + \frac{\partial F}{\partial f_2}(f_1(p),f_2(p))d\tilde{f}_2 \\ &= \tilde{f}_2(p)d\tilde{f}_1 + \tilde{f}_1(p)d\tilde{f}_2. \end{aligned}$$

That is, for $\tilde{f},\tilde{g}\in\mathscr{F}_p$, we have $d(\tilde{f}\tilde{g})=\tilde{f}(p)d\tilde{g}+\tilde{g}(p)d\tilde{f}$.

We can also derive a change of basis formula for the representation of a cotangent vector as we did with the tangent vectors. Let (ϕ,U) and (ψ,V) be two coordinate systems around p. We denote ϕ-coordinates by x^i and ψ-coordinates by y^i. Thus, the standard bases for T_p^*M with respect to these coordinate systems are, respectively, $\{d\tilde{x}^i\}$ and $\{d\tilde{y}^i\}$. Similarly, the standard bases for T_pM with respect to these coordinate systems will be denoted, respectively, by $\{E_i^x\}$ and $\{E_i^y\}$. Recalling how we can express the components of a cotangent vector in terms of the action of the tangent vectors, it follows that $d\tilde{x}^i=\sum_{j=1}^m E_j^y(x^i)d\tilde{y}^j$. So, if the representation of an arbitrary cotangent vector, $d\tilde{f}$, with respect to the coordinate system (ϕ,U) is $d\tilde{f}=\sum_i E_i^x(f)d\tilde{x}^i$, it follows that

$$d\tilde{f} = \sum_{i=1}^m E_i^x(f)\sum_{j=1}^m E_j^y(x^i)d\tilde{y}^j = \sum_{j=1}^m\sum_{i=1}^m E_j^y(x^i)E_i^x(f)d\tilde{y}^j = \sum_{j=1}^m\left(\sum_{i=1}^m E_j^y(x^i)E_i^x(f)\right)d\tilde{y}^j$$

But $E_j^y(x^i)$ is just the jth partial derivative of $x^i\circ\psi^{-1}$ at $\psi(p)$, for which we have an abbreviated notation. Using this notation, the previous equality reduces to

$$d\tilde{f} = \sum_{j=1}^m\left(\sum_{i=1}^m \frac{\partial x^i}{\partial y^j}\bigg|_{\psi(p)} E_i^x(f)\right)d\tilde{y}^j$$

But we also know that $d\tilde{f}=\sum_j E_j^y(f)d\tilde{y}^j$, so we must have

$$E_j^y(f) = \sum_{i=1}^m \frac{\partial x^i}{\partial y^j}\bigg|_{\psi(p)} E_i^x(f)$$

This is the transformation law for the components of the cotangent vector. In terms of matrices, with all components considered, this equation takes the form

$$\begin{bmatrix} E_1^y(f) \\ \vdots \\ E_m^y(f) \end{bmatrix} = \begin{bmatrix} \frac{\partial x^1}{\partial y^1} & \cdots & \frac{\partial x^m}{\partial y^1} \\ \vdots & \ddots & \vdots \\ \frac{\partial x^1}{\partial y^m} & \cdots & \frac{\partial x^m}{\partial y^m} \end{bmatrix} \begin{bmatrix} E_1^x(f) \\ \vdots \\ E_m^x(f) \end{bmatrix}$$

where the partial derivatives in the matrix are evaluated at $\psi(p)$. Note that this matrix is the transpose of the Jacobian of the map $\phi \circ \psi^{-1}$ at $\psi(p)$. Thus, the change of basis transformation that maps the ϕ-components of a cotangent vector to its ψ-components is given in matrix form by the transpose of $D(\phi \circ \psi^{-1})$.

Homework

Exercise 14.2.5 *Let* $M = \Re^2$, $(u_1, v_1) = \phi(x, y) = (x^2 + y^2, x^2 - y^2)$ *and* $(u_2, v_2) = \psi(x, y) = (2x^2 + 3y^2, 3x^2 - 4y^2)$.

- *Show* ϕ *and* ψ *are homeomorphisms for any point p other than* $(0, 0)$.
- *Show*

$$\begin{aligned} \phi^{-1}(u_1, v_1) &= (\sqrt{(u_1 + v_1)/2}, \sqrt{(u_1 - v_1)/2}) \\ \psi^{-1}(u_2, v_2) &= (\sqrt{(5u_2 + 3v_2)/22}, \sqrt{(2u_2 - v_1)/11}) \end{aligned}$$

  except at $(0, 0)$.

- *Show* $(u_1, v_1) = \phi\psi^{-1}(u_2, v_2) = ((11u_2 + v_2)/22, (u_2 + 5v_2)/22)$. *This is* $\begin{bmatrix} \frac{\partial u_1}{\partial u_2} & \frac{\partial u_1}{\partial v_2} \\ \frac{\partial v_1}{\partial u_2} & \frac{\partial v_1}{\partial v_2} \end{bmatrix}$.

  Note, we have $u_1 = \pi^1 \circ \phi$ *and* $v_1 = \pi^2 \circ \phi$ *and* $d\tilde{u_1} = \frac{\partial}{\partial u_1} \pi^1 |_{\phi(p)}$ *and so forth.*

Exercise 14.2.6 *Let* $M = \Re^2$, $(u_1, v_1) = \phi(x, y) = (x^2 + 3y^2, x^2 + 4y^2)$ *and* $(u_2, v_2) = \psi(x, y) = (2x^2 + y^2, 7x^2 - y^2)$.

- *Show* ϕ *and* ψ *are homeomorphisms for any point p other than* $(0, 0)$.
- *Find* ϕ^{-1} *and* ψ^{-1}.
- *Find* $\begin{bmatrix} \frac{\partial u_1}{\partial u_2} & \frac{\partial u_1}{\partial v_2} \\ \frac{\partial v_1}{\partial u_2} & \frac{\partial v_1}{\partial v_2} \end{bmatrix}$. *Note, we have* $u_1 = \pi^1 \circ \phi$ *and* $v_1 = \pi^2 \circ \phi$ *and* $d\tilde{u_1} = \frac{\partial}{\partial u_1} \pi^1 |_{\phi(p)}$ *and so forth.*

14.3 The Duality between the Tangent and Cotangent Space

We have independently constructed both the tangent space and the cotangent space. Based on our constructions and the bases we derived for each space, it seems intuitive that there is some connection between the two. In fact, pairing between the two is quite natural. A tangent vector is an operator, acting on smooth functions defined at p. A cotangent vector is an equivalence class of functions that all have identical coordinate derivatives. The natural pairing, then, should be the action of a tangent vector on the functions that make up a cotangent vector. More precisely, we define a map $\omega : T_pM \times T_p^*M \to \Re$ by

$$\omega(X, d\tilde{f}) = X(f).$$

Theorem 14.3.1 Canonical Mapping between the Tangent and Cotangent space

The mapping ω is well-defined, meaning that it does not depend on the particular $f \in d\tilde{f}$ we choose to compute it. Moreover, it is bilinear, and, if (ϕ, U) is any coordinate system around p, inducing the bases $\{E_i\}$ and $\{d\tilde{x}^i\}$ for T_pM and T_p^*M, respectively, then we have $\omega(E_i, d\tilde{x}^j) = E_i(x^j) = \delta_i^j$.

Proof 14.3.1
First, the mapping is clearly linear in the second variable, as any $X \in T_pM$ is a linear operator. Also, using the linear structure of the tangent space, we have $\omega(X_1 + X_2, d\tilde{f}) = (X_1 + X_2)(f) = X_1(f) + X_1(f)$. Scalar multiplication holds similarly, showing that ω is bilinear.

To show that the mapping is well-defined, we use local coordinate systems. Let (ϕ, U) be a coordinate system around p. Suppose X is a tangent vector, $d\tilde{f}$ a cotangent vector, and let f_1 and f_2 be any two functions in $d\tilde{f}$. Then $X = \sum_i \alpha_i E_i$. So,

$$\begin{aligned} X(f_1) &= \Big(\sum_{i=1}^m \alpha_i E_i\Big)(f_1) = \sum_{i=1}^m \alpha_i E_i(f_1) = \sum_{i=1}^m \alpha_i \frac{\partial}{\partial x^i}(f_1 \circ \phi^{-1})\Big|_{\phi(p)} \\ &= \sum_{i=1}^m \alpha_i \frac{\partial}{\partial x^i}(f_2 \circ \phi^{-1})\Big|_{\phi(p)} = \sum_{i=1}^m \alpha_i E_i(f_2) = \Big(\sum_{i=1}^m \alpha_i E_i\Big)(f_2) = X(f_2) \end{aligned}$$

Finally, since the choice of function $f \in d\tilde{f}$ does not matter, we have

$$E_i(d\tilde{x}^j) = \frac{\partial}{\partial x^i}(x^j \circ \phi^{-1})\Big|_{\phi(p)} = \frac{\partial}{\partial x^i}(\pi^j \circ \phi \circ \phi^{-1})\Big|_{\phi(p)} = \frac{\partial \pi^j}{\partial x^i}\Big|_{\phi(p)} = \delta_i^j$$

■

Thus, since the dual of a given vector space is unique up to isomorphism, we can conclude that the cotangent space is the dual of the tangent space. Moreover, given a coordinate system, (ϕ, U), around p, this shows that $\{d\tilde{x}^i : 1 \leq i \leq m\}$ is the dual basis to $\{E_i : 1 \leq i \leq m\}$. In fact, since the tangent and cotangent spaces are of finite dimension, when a coordinate system is chosen, the mapping, ω, is just a bilinear form on $\Re^m$. Given a tangent vector X_p and a cotangent vector $d\tilde{f}_p$, we call the real number $\omega(X_p, d\tilde{f})$ the directional derivative of f at p in the direction of X_p with respect to the chosen coordinate system. The actual value of this derivative is coordinate dependent, but its bilinearity is not. Mappings of this sort, whose specific values are coordinate dependent but whose multilinear structure is not, are used frequently in mathematics and physics. They are called tensors and we study such mappings in detail later.
We use the mapping ω to define the action of cotangent vectors on tangent vectors. Recall T_p^*M is the dual space of T_pM, so it must consist of linear functionals on T_pM. This mapping gives us the means for defining how a cotangent vector operates on a tangent vector. For $X_p \in T_pM$ and $d\tilde{f} \in T_p^*M$, we define $d\tilde{f}(X_p)$ to be $\omega(X_p, d\tilde{f}) = X_p(f)$. This is a well-defined linear functional on T_pM.

As a final note on the map, ω, there is a common and intuitive notational convenience employed in most differential geometry texts. Once bases are chosen for T_pM and T_p^*M, both spaces are isomorphic to $\Re^m$, inducing an identification between tangent and cotangent vectors. Hence, the map ω behaves just like the usual inner product on a Euclidean space. Consequently, given $X_p \in T_pM$ and $d\tilde{f} \in T_p^*M$, the map, $\omega : T_pM \times T_p^*M \to \Re$, is typically denoted by $\langle X_p, d\tilde{f}\rangle$.

Thus, for example, the directional derivative of f at p in the direction of X_p is given by the real number $\langle X_p, d\tilde{f}\rangle$. Henceforth, we will use this notation whenever the duality map is required.
To conclude this section on the duality of the tangent and cotangent spaces, we point out an interesting consequence of the formula $d(\tilde{f}\tilde{g}) = \tilde{f}(p)d\tilde{g} + \tilde{g}(p)d\tilde{f}$. The directional derivative also satisfies a Leibnitz rule. If X_p is a tangent vector, then the action of X_p on the product $\tilde{f}\tilde{g}$ is given by

$$\langle X_p, d(\tilde{f}\tilde{g})\rangle = \langle X_p, \tilde{f}(p)d\tilde{g} + \tilde{g}(p)d\tilde{f}\rangle = \tilde{f}(p)\langle X_p, d\tilde{g}\rangle + \tilde{g}(p)\langle X_p, d\tilde{f}\rangle.$$

Since $d(\tilde{f}\tilde{g}) = d(\widetilde{fg})$, we can rewrite $\langle X_p, d(\tilde{f}\tilde{g})\rangle$ using the definition of the directional derivative as $X_p(fg)$. Hence, we have the differential of a map $X_p(fg) = f(p)X_p(g) + g(p)X_p(f)$.

Homework

Exercise 14.3.1 *Let* $M = \Re^2$ *and* $p = \begin{bmatrix}1\\2\end{bmatrix}$. *We know* $\omega : T_pM \times T_p^*M \rightarrow \Re$ *by* $\omega(X, d\tilde{f}) = X(f)$ *for any* $X \in T_pM$ *and* $f \in T_p^*M$.

- *Explain why we know if* $\Re^2$ *is interpreted as column vectors, then* $T_pM = p + \Re^2$, *a copy of* $\Re^2$ *rooted at p.*
- *Any element in* T_p^*M, *is a linear functional on* T_pM. *Explain* $d\tilde{f}(X) =< \nabla(f)(p), X - p >$ *and so then* $d\tilde{f}$ *is a linear functional on* T_pM.
- *Explain why* $E_i(f) = \frac{\partial f}{\partial x_i}|_p$.
- *Explain why* $\omega(X, d\tilde{f}) =< \nabla(f)(p), X - p >$.

Exercise 14.3.2 *Let* $M = \Re^3$ *and* $p = \begin{bmatrix}1\\2\\-1\end{bmatrix}$ *and redo the problem above.*

Exercise 14.3.3 *Let* $M = \Re^4$ *and* $p = \begin{bmatrix}1\\2\\-1\\5\end{bmatrix}$ *and redo the problem above.*

14.4 The Differential of a Map

An important application of the tangent and cotangent spaces is the notion of the differential of a map. The differential of a map is the natural generalization of the classical derivative of a function $f : \Re^n \rightarrow \Re^m$. The derivative of such a function is usually defined to be a linear map $Df : \Re^n \rightarrow \Re^m$. We want to translate this concept to functions mapping manifolds to manifolds. To do so, it is necessary to look at the classical definition in a new way.

We know the Euclidean spaces $\Re^n$ and $\Re^m$ are manifolds. It is easy to see that the tangent space of a Euclidean space at any point, p, is simply another copy of that Euclidean space. That is, for any $p \in \Re^n$, we have $T_p\Re^n = \Re^n$. This is a consequence of the unique structure of the Euclidean spaces. We can think of any vector in a Euclidean space as being anchored at the origin, and, in this way, we identify all of the tangent spaces of $\Re^n$, at any point, with $\Re^n$ itself. So, a more precise way of defining the derivative of a map $f : \Re^n \rightarrow \Re^m$ at $p \in \Re^n$ would be as a linear map from $T_p\Re^n$ to $T_{f(p)}\Re^m$. This distinction is not made in classical analysis for the reason we just gave. These tangent spaces are nothing more than copies of $\Re^n$ and $\Re^m$ respectively. This is not the case for manifolds with arbitrary coordinate systems. Hence, we must define the derivative of a map in terms of the

tangent spaces.
We begin by first defining the cotangent map.

Definition 14.4.1 Cotangent Map

Let M and N be smooth manifolds, and let $F : M \to N$ be a smooth map. For $p \in M$ and $q = F(p) \in N$, the **cotangent map** *induced by F at q is the map $F^* : T_q^* N \to T_p^* M$ defined by $F^*(d\tilde{f}_q) = d(\widetilde{f \circ F})_p$.*

Note that the direction of this mapping is opposite the direction of F. The cotangent map is linear, for if $d\tilde{f}_q, d\tilde{g}_q \in T_q^* N$ and $\alpha, \beta \in \Re$, then

$$\begin{aligned} F^*(\alpha d\tilde{f}_q + \beta d\tilde{g}_q) &= F^*(d(\widetilde{\alpha f + \beta g})_q) = d((\widetilde{\alpha f + \beta g}) \circ F)_p = d(\widetilde{\alpha f \circ F})_p + d(\widetilde{\beta g \circ F})_p \\ &= \alpha d(\widetilde{f \circ F})_p + \beta d(\widetilde{g \circ F})_p = \alpha F^*(d\tilde{f}_q) + \beta F^*(d\tilde{g}_q) \end{aligned}$$

We now use this map to define the differential of F.

Definition 14.4.2 Differential or Tangent Map

Let M, N, and F be as before. The **differential**, *or* **tangent map**, *induced by F at p is the map $F_* : T_p M \to T_q N$ defined by $F_*(X_p)(f) = X_p(F^*(d\tilde{f}_q))$ for $f \in C^\infty(q)$.*

This definition does make sense. Since f is a smooth function in a neighborhood of q, there is a cotangent vector $d\tilde{f}_q$. The map F^* will map this cotangent vector to a cotangent vector at $p \in M$. A cotangent vector is nothing more than an equivalence class of functions all having the same coordinate partial derivatives. Hence, X_p can act on any function in the cotangent vector $F^*(d\tilde{f}_q)$ and produce the same value. We will usually denote the tangent map by DF instead of F_* and refer to it as the differential. The reason for the * notation is to distinguish between the tangent and cotangent maps, although we do not need to do this often. So, to emphasize that the differential is a generalization of the classical derivative, we will use the familiar nomenclature and notation.

As expected, the differential is linear also. If $X_p, {X_p}' \in T_p M$ and $\alpha, \beta \in \Re$, then, for any $f \in C^\infty(q)$, we have

$$\begin{aligned} D(\alpha X_p + \beta {X_p}')(f) &= \big(\alpha X_p + \beta {X_p}'\big)\big(F^*(d\tilde{f}_q)\big) = \alpha X_p\big(F^*(d\tilde{f}_q)\big) + \beta {X_p}'\big(F^*(d\tilde{f}_q)\big) \\ &= \alpha D(X_p)(f) + \beta D({X_p}')(f) \end{aligned}$$

So, just as in the classical case, the differential of a map, $F : M \to N$, establishes a homomorphism between the tangent spaces at p and at $q = F(p)$.

Theorem 14.4.1 Chain Rule

Let $F : M \to N$ be a smooth map between smooth manifolds. For $p \in M$, let $q = F(p)$. Then, if F is a diffeomorphism of some neighborhood of p onto some neighborhood of q, then the differential $DF : T_p M \to T_q N$ is an isomorphism. Also, if $G : N \to L$ is another smooth map between manifolds, then $D(G \circ F) = DG \circ DF$ (the chain rule).

Proof 14.4.1
We prove the chain rule first, and then the first result will follow from that.

$G \circ F$ is a map from M to L. Let p be in M and let $q = F(p)$ and $r = G(F(p))$. Suppose X_p is a tangent vector at p and $f \in C^\infty(r)$. Then, by definition we have

$$D(G \circ F)(X_p)(f) \quad = \quad X_p((G \circ F)^*(d\tilde{f}_r))$$

Now, $(G \circ F)^*(d\tilde{f}_r) = d(\widetilde{f \circ G \circ F})_p$, so

$$D(G \circ F)(X_p)(f) \quad = \quad X_p\Big(d(\widetilde{f \circ G \circ F})_p\Big).$$

On the other hand, we have, for any function $h \in C^\infty(q)$,

$$DF(X_p)(h) \quad = \quad X_p(F^*(d\tilde{h}_q)) = X_p(d(\widetilde{h \circ F})_p)$$

from which it follows that

$$\begin{aligned} DG(DF(X_p))(f) &= DF(X_p)(G^*(d\tilde{f}_r)) = DF(X_p)(d(\widetilde{f \circ G})_q) = DF(X_p)(f \circ G) \\ &= X_p\Big(F^*(d(\widetilde{f \circ G})_q)\Big) = X_p(d(\widetilde{f \circ G \circ F})_p) \end{aligned}$$

Thus, $(DG \circ DF)(X_p) = D(G \circ F)(X_p)$, and this proves the chain rule.

Now, suppose $F : M \to N$ is a diffeomorphism of a neighborhood, V, of p onto a neighborhood, W, of q. Then $F^{-1} : N \to M$ is also a well-defined smooth map between manifolds. Let Y_q be a tangent vector in T_qN. Then $DF^{-1}(Y_q)$ is a tangent vector, X_p, in T_pM, and the chain rule implies that $DF(X_p) = DF(DF^{-1}(Y_p)) = D(F \circ F^{-1})(Y_q) = Y_q$. Hence, DF is surjective. If $DF(X_{p_1}) = DF(X_{p_2})$, then $DF^{-1}(DF(X_{p_1})) = DF^{-1}(DF(X_{p_2}))$, implying that

$$D(F^{-1} \circ F)(X_{p_1}) = D(F^{-1} \circ F)(X_{p_2}) \Rightarrow X_{p_1} \quad = \quad X_{p_2}.$$

So DF is injective. Hence, it is an isomorphism. Note that this also requires the dimensions of M at p and N at q to be equal. ■

In terms of local coordinates, we can establish a familiar expression for the differential of a map.

Theorem 14.4.2 Local Coordinate Expressions for the Differential Map

Let $F : M \to N$ be a smooth map between manifolds, and let p be a point in M with $q = F(p)$. Suppose (ϕ, U) and (ψ, V) are coordinate systems around p and q respectively, with respective coordinate functions denoted by $\{x^i\}$ and $\{y^j\}$. Let $\{E_i^x\}$ and $\{E_j^y\}$ be the bases induced by each coordinate system for T_pM and T_qN, respectively. Then, for each $i = 1, \ldots, m$,

$$DF(E_i^x) \quad = \quad \sum_{j=1}^{n} \left(\frac{\partial F^j}{\partial x^i}\right)\Bigg|_{\phi(p)} E_j^y,$$

where F^j denotes the jth coordinate function of the coordinate representation of F, $\psi \circ F \circ \phi^{-1}$. Moreover, if $X_p = \sum_i \alpha_i E_i^x$ and $DF(X_p) = \sum_j \beta_j E_j^y$, then

$$\beta_j \quad = \quad \sum_{i=1}^{m} \alpha_i \left(\frac{\partial F^j}{\partial x^i}\right)\Bigg|_{\phi(p)}.$$

Proof 14.4.2
The proof of the first formula is a computation. The components of a tangent vector with respect to a given coordinate system are given by the values obtained by that tangent vector acting on the coordinate functions. That is, the jth component of $DF(E_i^x)$ is given by $DF(E_i^x)(y^j)$. Thus, we have

$$
\begin{aligned}
DF(E_i^x)(y^j) &= E_i^x\left(F^*(d\tilde{y}_q^j)\right) = E_i^x\left(d(\widetilde{y^j \circ F})_p\right) \\
&= E_i^x(y^j \circ F) = \frac{\partial}{\partial x^i}(y^j \circ F \circ \phi^{-1}).
\end{aligned}
$$

But $y^j \circ F \circ \phi^{-1}$ is just the jth coordinate function of the coordinate representation, $\psi \circ F \circ \phi^{-1}$, of F. Denoting this component function by F^j, we get

$$
DF(E_i^x)(y^j) = \frac{\partial F^j}{\partial x^i}
$$

and this is the jth component of $DF(E_i^x)$. Hence,

$$
DF(E_i^x) = \sum_{j=1}^{n} \left(\frac{\partial F^j}{\partial x^i}\right)\Bigg|_{\phi(p)} E_j^y.
$$

Now, suppose $X_p = \sum_i \alpha_i E_i^x$ and $DF(X_p) = \sum_j \beta_j E_j^y$. Then, we just use the linearity of the differential to obtain

$$
\begin{aligned}
DF(X_p) &= DF\Big(\sum_{i=1}^{m} \alpha_i E_i^x\Big) = \sum_{i=1}^{m} \alpha_i DF(E_i^x) = \sum_{i=1}^{m} \alpha_i \sum_{j=1}^{n} \Big(\frac{\partial F^j}{\partial x^i}\Big)\Big|_{\phi(p)} E_j^y \\
&= \sum_{j=1}^{n} \Big(\sum_{i=1}^{m} \alpha_i \Big(\frac{\partial F^j}{\partial x^i}\Big)\Big|_{\phi(p)}\Big) E_j^y
\end{aligned}
$$

So, if β_j is the jth component of $DF(X_p)$, then this shows that

$$
\beta_j = \sum_{i=1}^{m} \alpha_i \Big(\frac{\partial F^j}{\partial x^i}\Big)\Big|_{\phi(p)}.
$$

■

Notice that this theorem states that the local coordinate representation of DF is just the Jacobian of the coordinate representation $\psi \circ F \circ \phi^{-1}$. Moreover, for the basis vector E_i^x, the components of the image $DF(E_i^x)$ are just the entries of the ith column of this Jacobian matrix.

Homework

Let S be the unit sphere in $\Re^3$. So S is just the unit vectors in $\Re^3$. Assume S is given coordinates by stereographic projection (if you need to, look this up and go through the details).

Exercise 14.4.1 *Show the mapping given by*

$$
\begin{aligned}
\phi_0(x,y) &= \left(\frac{2x}{1+x^2+y^2}, \frac{2y}{1+x^2+y^2}, \frac{1-x^2-y^2}{1+x^2+y^2}\right) \\
\phi_1(x,y) &= \left(\frac{2x}{1+x^2+y^2}, \frac{2y}{1+x^2+y^2}, \frac{x^2+y^2-1}{1+x^2+y^2}\right)
\end{aligned}
$$

cover a neighborhood U_0 of the north pole N and U_1 of the south pole S, respectively.

Exercise 14.4.2 *Let X, Y, Z be the coordinates of a point on S. Prove ϕ_0 and ϕ_1 have inverses*

$$\phi_0^{-1}(X,Y,Z) = \left(\frac{X}{Z+1}, \frac{Y}{Z+1}\right), \quad \phi_1^{-1}(X,Y,Z) = \left(\frac{-X}{Z-1}, \frac{-Y}{Z-1}\right)$$

Exercise 14.4.3 *Prove the coordinate transformation function here is*

$$\phi_{01}(x,y) = \phi_0^{-1} \circ \phi_1(x,y) = \left(\frac{x}{x^2+y^2}, \frac{y}{x^2+y^2}\right)$$

which is inversion in the circle.

Exercise 14.4.4 *Let v be a vector field on S so that at each P in S, $v(P)$ is a tangent vector. Show if P is a point in U_0, then v can be represented by the pushforward of a vector field $\boldsymbol{v_0}$ on $\Re^2$:*

$$v(P) = J_{\phi_0}(\phi_0^{-1}(P))\, \boldsymbol{v_0}(\phi_0^{-1}(P))$$

where J_{ϕ_0} is the Jacobian matrix of ϕ_0.

On the overlap $U_0 \cap U_1$, we can also represent v by

$$v(P) = J_{\phi_1}(\phi_1^{-1}(P))\, \boldsymbol{v_1}(\phi_1^{-1}(P))$$

by the pushforward of a vector field $\boldsymbol{v_1}$ on $\Re^2$:

Exercise 14.4.5 *Apply the chain rule to $\phi_1 = \phi_0 \circ \phi_{01}$ and show*

$$J_{\phi_1}(\phi_1^{-1}(P)) = J_{\phi_0}(\phi_0^{-1}(P)) \circ J_{\phi_{01}}(\phi_1^{-1}(P))$$

Apply this to $\boldsymbol{v_1}(\phi^{-1}(P))$ to find

$$\boldsymbol{v_0}(\phi_0^{-1}(P)) = J_{\phi_{01}}(\phi_1^{-1}(P))\boldsymbol{v_1}(\phi_1^{-1}(P))$$

Exercise 14.4.6 *Now we use all the above work. Let $P(t)$ be a curve in S. Let's say a vector field is parallel if the coordinate vectors of the vector field are constant along the curve. You can see the problem: in which coordinate systems should these components be constant? Assume $\boldsymbol{v_1}(\phi_1^{-1}(P(t)))$ is constant. Show*

$$\frac{d}{dt}\left(\boldsymbol{v_0}(\phi_0^{-1}(P(t)))\right) = \left(\frac{d}{dt}J_{\phi_{01}}(\phi_1^{-1}(P(t)))\right)\boldsymbol{v_1}(\phi_1^{-1}(P(t)))$$

which tells us $\boldsymbol{v_1}$ and $\boldsymbol{v_0}$ cannot be simultaneously constant along the curve.

Exercise 14.4.7 *The problem we illustrate here is that the usual directional derivative of vector calculus does not behave well under changes in the coordinate system when applied to the components of vector fields. Transporting a vector field along a curve so the components do not change, thus requires a different approach. We will not do anything with this here but we will suggest you do some reading as an exercise: look up covariant derivatives and also the Levi - Civita connection which each provide a solution.*

14.4.1 Tangent Vectors of Curves in a Manifold

As an application of the differential concept, let's look at the tangent vectors of curves in the manifold M. This is a notion that is needed to construct a metric on M in terms of the arc lengths of curves. Let $\gamma : (a, b) \to M$ be a smooth map between $(a, b) \subset \Re$ and M. Looking at (a, b) as a manifold, we see that its tangent space is 1-dimensional. So, once we choose a basis, the tangent space $T_{t_0}(a, b)$ at any point $t_0 \in (a, b)$ is just the set of real numbers, thought of as a vector space. This tangent space is spanned by the tangent vector E_{t_0} defined by

$$E_{t_0}(f) \quad = \quad \frac{df}{dt}\Big|_{t_0},$$

for any smooth function, f, defined on some neighborhood of t_0. Hence, E_{t_0} is just the ordinary derivative operator on $\Re$ evaluated at t_0. Denote E_{t_0} by the more intuitive symbol d/dt, where the point t_0 at which the derivative is evaluated is assumed to be known. Suppose $\gamma(t_0) = p$. The differential of γ at t_0, which we will denote $D_{t_0}\gamma$, is a linear map from $T_{t_0}(a, b)$ to T_pM. For any function $f \in C^\infty(p)$, we have

$$D_{t_0}\gamma\Big(\frac{d}{dt}\Big)f \quad = \quad \frac{d}{dt}(f \circ \gamma)\Big|_{t_0}.$$

This is simply the directional derivative of f in the direction of the curve γ. More formally, this is the image under $D_{t_0}\gamma$ of d/dt, and we will call this tangent vector in T_pM the tangent vector to the curve γ at $p = \gamma(t_0)$. Since the curve, γ, is smooth, we can let t vary over (a, b), and we obtain a tangent vector to γ at every point $p \in \gamma(a, b)$. The field of tangent vectors we obtain is called the tangent vector field of γ.

Now, suppose (ϕ, U) is a coordinate system around $p = \gamma(t_0)$. The local coordinate representation of γ on $\gamma(a, b) \cap U$ is given by $\phi \circ \gamma(t) = (x^1 \circ \gamma(t), \ldots, x^m \circ \gamma(t))$. Let $\{E_i\}$ denote the basis induced by this coordinate system at T_pM. The ith component of $D_{t_0}\gamma$ with respect to this basis is the action of this tangent vector on x^i. Thus, this component is given by

$$D_{t_0}\gamma\Big(\frac{d}{dt}\Big)x^i \quad = \quad \frac{d}{dt}(x^i \circ \gamma)\Big|_{t_0},$$

which is just the ordinary derivative of the ith component function of the coordinate representation of γ. If, for brevity of notation, we denote this ordinary derivative, evaluated at t_0, by

$$\frac{d}{dt}(x^i \circ \gamma)\Big|_{t_0} \quad = \quad \dot{x}^i(t_0),$$

then we have

$$D_{t_0}\gamma\Big(\frac{d}{dt}\Big) \quad = \quad \sum_{i=1}^{m} \dot{x}^i(t_0)E_i.$$

This is the local coordinate representation of the tangent vector to the curve, γ at $p = \gamma(t_0)$.

Example 14.4.1 *Let's think about tangent vectors in the way we just described. Consider the surface $z = x^2 + y^2 = f(x, y)$ in $\Re^3$. Pick the point $p = (x_0, y_0, z_0)$ on the surface. Let $f(x, y, z) = x^2 + y^2 - z$. Let C be a curve on this surface that passes through p. Then C determines a locus of points $(x, y, x^2 + y^2)$ on the surface. Choose any parameterization of the locus of points which gives*

$\Gamma : [0,1] \rightarrow \Re$ *by*

$$\Gamma(t) = (x(t), y(t), z(t)) = (x(t), y(t), x^2(t) + y^2(t)),\ 0 \le t \le 1$$

The velocity vector $T(t)$ *to this curve is given by*

$$T(t) = \Gamma'(t) = \begin{bmatrix} x'(t) \\ y'(t) \\ z'(t) \end{bmatrix} = \begin{bmatrix} x'(t) \\ y'(t) \\ 2x(t)x'(t) + 2y(t)y'(t) \end{bmatrix}$$

Again, we know $\nabla(f) = \begin{bmatrix} 2x \\ 2y \\ -1 \end{bmatrix}$ *and* $\nabla(f)(p)$ *is the normal vector to the tangent plane, which is the span of the independent vectors* $\begin{bmatrix} 1 \\ 0 \\ \frac{\partial f}{\partial x}(p) \end{bmatrix}$ *and* $\begin{bmatrix} 0 \\ 1 \\ \frac{\partial f}{\partial y}(p) \end{bmatrix}$. *Then we have* $D_0\gamma\left(\frac{d}{dt}\right) = \sum_{i=1}^{2} \dot{x}^i(0) E_i$ *This is the local coordinate representation of the tangent vector to the curve,* Γ *at* $p = \Gamma(0)$.

Homework

Exercise 14.4.8 *Repeat the discussion of the example above for* $2x^2 + 3y^2 = z$ *so that you have some experience at interpreting the tangent space in this way.*

Exercise 14.4.9 *Repeat the discussion example above for* $2x^2 + 3y^2 + 3z^2 = 100$ *so that you have some experience at interpreting the tangent space in this way.*

Exercise 14.4.10 *Repeat the discussion example above for* $3x + 2y^2 + 3z^2 = 100$ *so that you have some experience at interpreting the tangent space in this way.*

For simplicity of notation, we typically denote cotangent vectors simply by *df*, or, if we need to reference the point at which the cotangent space exists, df_p and we stop explicitly denoting these objects as equivalence classes $d\tilde{f}$. We have shown all definitions and results are independent of the choice of $f \in \tilde{f}$ or $d\tilde{f}$. Indeed, any two functions f_1 and f_2 are in $\tilde{f}$ will be equal near *p*, so they will certainly be in the same $\mathscr{H}_p$-equivalence class $d\tilde{f}$. Moreover, any computations we perform with cotangent (or tangent) vectors are independent of the choice of function from the equivalence class $d\tilde{f}$. Thus, referencing a cotangent vector as *df* will simply refer to the cotangent vector induced by the equivalence class $\tilde{f}$.

14.5 The Tangent Space Using Curves

We have used E_{p_i} to denote the tangent vector at p giving the partial derivative of a smooth function at p with respect to its ith coordinate. However, there is a more intuitively pleasing notation that is used more often even though this notation is not coordinate independent. The interpretation of a tangent vector as an operator on smooth functions is the most concise and useful when doing analysis on a manifold, but the geometric intuition of the operators is lacking. There is another means of constructing the tangent space in terms of the tangent vectors to curves passing through points on the manifold which is entirely equivalent and it is easier to visualize.

When constructing a basis for the tangent space, T_pM, at a point $p \in M$, we denoted the basis vectors induced by a particular coordinate system, (ϕ, U), by E_i, or E_{p_i}, where these operators are defined on smooth functions in a neighborhood of p by

$$E_i(f) = \frac{\partial(f \circ \phi^{-1})}{\partial x^i}\Big|_{\phi(p)}.$$

In this relation, x^i denotes the ith coordinate function of the map ϕ. Thus, E_i is just the partial derivative operator at p with respect to the ith coordinate. Consequently, it is common in differential geometry to denote this basis vector simply by the partial derivative operator symbol

$$\frac{\partial}{\partial x^i}.$$

We chose not to use this notation during our construction of the basis, since it explicitly refers to a coordinate system by using the coordinate function x^i. However, once the nature of the tangent vectors is understood, this representation is very useful. It tells us a tangent vector is a partial derivative operator, whereas the operator E_i does not. Using this notation, the basis for the tangent space T_pM induced by the coordinate system (ϕ, U), where the coordinate functions of ϕ are denoted by x^i, can be written as

$$\left\{ \frac{\partial}{\partial x^1}, \ldots, \frac{\partial}{\partial x^m} \right\}.$$

Recall that, using our standard notation, the basis for T_p^*M is denoted $\{dx^i\}$. We have shown that this is the dual basis to the standard tangent basis. Thus, in our new notation, we have the relations

$$dx^i\left(\frac{\partial}{\partial x^j}\right) = \delta^i_j.$$

With this new notation, we can give another interpretation of a tangent vector that is distinct from, but equivalent to, our original one. Recall that in our construction of the cotangent space, we used the set Γ_p, the set of all smooth curves, $\gamma : I \to M$ such that $\gamma(0) = p$. We will define an equivalence relation, $\sim$, on Γ_p as follows. Let $\gamma_1 \sim \gamma_2$ if and only if there is a coordinate system, (ϕ, U) around p such that $(\phi \circ \gamma_1)'(0) = (\phi \circ \gamma_2)'(0)$. This is easily seen to be an equivalence relation, and it simply states that two curves are equivalent if their coordinate tangent vectors at p are the same. We will denote the equivalence class containing γ by $\tilde{\gamma}$.

One consequence of this definition is that if two curves are equivalent with respect to one chart (ϕ, U), then they are also equivalent with respect to any other chart containing p. To see this, suppose $\gamma_1 \sim \gamma_2$. Then there is a chart (ϕ, U) around p such that $(\phi \circ \gamma_1)'(0) = (\phi \circ \gamma_2)'(0)$. Let (ψ, V) be another coordinate system around p. Then

$$\begin{aligned}(\psi \circ \gamma_1)'(0) &= (\psi \circ \phi^{-1} \circ \phi \circ \gamma_1)'(0) = D_{\phi(p)}(\psi \circ \phi^{-1})(\phi \circ \gamma_1)'(0) \\ &= D_{\phi(p)}(\psi \circ \phi^{-1})(\phi \circ \gamma_2)'(0) = (\psi \circ \phi^{-1} \circ \phi \circ \gamma_2)'(0) = (\psi \circ \gamma_2)'(0)\end{aligned}$$

Thus, we need only verify that two curves are equivalent with respect to a single chart at p, and it will follow that they have the same coordinate derivative with respect to any coordinate system.

Consider the set $\Gamma_p/\sim$. We will continue to let T_pM denote the tangent space at $p \in M$ as we have already constructed it. So, elements of T_pM are still considered as operators on $\mathscr{F}_p$, which are smooth functions defined on a neighborhood of p. Our goal is to show that $\Gamma_p/\sim\, \cong T_pM$ by defining an identification between equivalence classes of curves and derivations on $\mathscr{F}_p$.

We begin by uniquely associating to each $\tilde{\gamma} \in \Gamma_p$, a derivation on $\mathscr{F}_p$, or, simply, a tangent vector. For $\tilde{\gamma} \in \Gamma_p/\sim$, define an element $T_\gamma \in T_pM$ such that for $\tilde{f} \in \mathscr{F}_p$,

$$T_\gamma(\tilde{f}) \quad = \quad \frac{d}{dt}(f \circ \gamma)\Big|_{t=0}.$$

This definition is independent of the choice of curve $\gamma \in \tilde{\gamma}$ and of the choice of function $f \in \tilde{f}$, since any two functions in $\tilde{f}$ will be equal on a neighborhood of p and any two curves in $\tilde{\gamma}$ will have the same coordinate derivatives at p with respect to any chart. Thus, it is not ambiguous to denote the operator by T_γ, without the tilde. Of course, we still must show that our definition makes sense. That is, we must know that T_γ actually is a tangent vector. Each T_γ is a linear operator on $\mathscr{F}_p$, since for $\alpha, \beta \in \Re$ and $\tilde{f}, \tilde{g} \in \mathscr{F}_p$, we have

$$\begin{aligned} T_\gamma(\alpha\tilde{f} + \beta\tilde{g}) &= T_\gamma(\widetilde{\alpha f + \beta g}) = \frac{d}{dt}((\alpha f + \beta g) \circ \gamma)\Big|_{t=0} \\ &= \frac{d}{dt}(\alpha(f \circ \gamma) + \beta(g \circ \gamma))\Big|_{t=0} = \alpha\frac{d}{dt}(f \circ \gamma)|_{t=0} + \beta\frac{d}{dt}(g \circ \gamma)|_{t=0} \\ &= \alpha T_\gamma(\tilde{f}) + \beta T_\gamma(\tilde{g}) \end{aligned}$$

We also see T_γ satisfies the Leibnitz property, since for $\tilde{f}, \tilde{g} \in \mathscr{F}_p$, we have

$$\begin{aligned} T_\gamma(\tilde{f}\tilde{g}) &= T_\gamma(\widetilde{fg}) = \frac{d}{dt}((fg) \circ \gamma)\Big|_{t=0} \\ &= \frac{d}{dt}((f \circ \gamma)(g \circ \gamma))\Big|_{t=0} = f(p)\frac{d}{dt}(g \circ \gamma)\Big|_{t=0} + g(p)\frac{d}{dt}(f \circ \gamma)\Big|_{t=0} \\ &= f(p)T_\gamma(\tilde{g}) + g(p)T_\gamma(\tilde{f}) \end{aligned}$$

Thus, each T_γ is a tangent vector. Moreover, this association is unique. If $\tilde{\gamma_1}$ induces T_{γ_1} and $\tilde{\gamma_2}$ induces T_{γ_2} with $T_{\gamma_1} = T_{\gamma_2}$, then for any $\tilde{f} \in \mathscr{F}_p$ we have

$$\frac{d}{dt}(f \circ \gamma_1)\Big|_{t=0} \quad = \quad \frac{d}{dt}(f \circ \gamma_2)\Big|_{t=0}.$$

Let (ϕ, U) be a coordinate system around p. Since the above equality must hold for any smooth function f, we can let $f = x^i$, the ith coordinate function of the mapping ϕ. Recall that this mapping is given by $\pi^i \circ \phi$, where π^i is the ith coordinate projection on $\Re^m$. We then obtain

$$\begin{aligned} \frac{d}{dt}\left(x^i \circ \phi^{-1} \circ \phi \circ \gamma_1 - x^i \circ \phi^{-1} \circ \phi \circ \gamma_2\right)\Big|_{t=0} &= \frac{d}{dt}\left(\pi^i \circ \phi \circ \gamma_1 - \pi^i \circ \phi \circ \gamma_2\right)\Big|_{t=0} \\ &= \frac{d}{dt}\left(\pi^i \circ (\phi \circ \gamma_1 - \phi \circ \gamma_2)\right)\Big|_{t=0} \\ &= \dot{x}_1^i(0) - \dot{x}_2^i(0) \end{aligned}$$

where $\dot{x}_k^i(0)$ denotes the derivative if the ith coordinate function of $\phi \circ \gamma_k$ at $t = 0$. It follows that $\dot{x}_1^i(0) = \dot{x}_2^i(0)$ for each $i = 1, \ldots, m$. That is, we have $(\phi \circ \gamma_1)'(0) = (\phi \circ \gamma_2)'(0)$, implying that $\tilde{\gamma_1} = \tilde{\gamma_2}$.

Now, let T_p^cM be the set defined by

$$T_p^cM \quad = \quad \{T_\gamma \,:\, \tilde{\gamma} \in \Gamma_P/\sim\}.$$

Then, by what we have shown, we know that $T_p^cM \subset T_pM$, and that the association between elements of T_p^cM and elements of T_pM is one to one. To conclude, we will show that, in fact,

$T_p^c M = T_p M$, implying that each tangent vector can be identified with a unique equivalence class of curves. To do so, we will need to use local coordinate systems. Thus, we also need to show the identification does not depend on the particular coordinate system chosen.

Let X be any tangent vector in $T_p M$, and let (ϕ, U) be any coordinate system around p. We must find a curve, $\gamma \in \Gamma_p$ such that $T_\gamma = X$. Our coordinate system induces the standard basis for $T_p M$, which, in our new notation, is denoted

$$\left\{\frac{\partial}{\partial x^1}, \ldots, \frac{\partial}{\partial x^m}\right\}.$$

Then X has a representation with respect to this basis given by

$$X = \sum_{i=1}^{m} \alpha^i \frac{\partial}{\partial x^i},$$

for some set of scalars $\alpha^1, \ldots, \alpha^m$. Define a curve $\lambda : \Re \to \Re^m$ by

$$\lambda(t) = \Big(\alpha^1 t + (\phi(p))^1, \ldots, \alpha^m t + (\phi(p))^m\Big).$$

Then define a curve, γ, in M by $\gamma(t) = (\phi^{-1} \circ \lambda)(t)$. This is a smooth curve in M and it satisfies $\gamma(0) = \phi^{-1}(\lambda(0)) = \phi^{-1}(\phi(p)) = p$. So, γ defines an equivalence class $\tilde{\gamma} \in \Gamma_p / \sim$, and we have $(\phi \circ \gamma)'(0) = (\alpha^1, \ldots, \alpha^m)$. The operator T_γ induced by this equivalence class is given by

$$\begin{aligned} T_\gamma(f) &= \frac{d}{dt}(f \circ \gamma)\Big|_{t=0} = \frac{d}{dt}(f \circ \phi^{-1} \circ \phi \circ \gamma)\Big|_{t=0} \\ &= D_{\phi(p)}(f \circ \phi^{-1})(\phi \circ \gamma)'(0) = \sum_{i=1}^{m} \alpha^i \frac{\partial (f \circ \phi^{-1})}{\partial x^i}\Big|_{\phi(p)} \end{aligned}$$

Recalling the definition of our tangent basis vectors and our new notation, this last equality is simply

$$T_\gamma(f) = \sum_{i=1}^{m} \alpha^i \frac{\partial}{\partial x^i}(f),$$

or, in operator form,

$$T_\gamma = \sum_{i=1}^{m} \alpha^i \frac{\partial}{\partial x^i} = X.$$

Thus, every tangent vector $X \in T_p M$ can be associated with an equivalence class of curves in $\Gamma_p / \sim$. However, we used only one particular coordinate system in this proof. In order to show that the identification is natural, we must show that if X is associated to $\tilde{\gamma}_1$ under the coordinate system (ϕ, U) and to $\tilde{\gamma}_2$ under another coordinate system, say (ψ, V), then $\tilde{\gamma}_1 = \tilde{\gamma}_2$.

Let the coordinate functions of ϕ be denoted by x^i and let the coordinate functions of ψ be denoted by y^i. Further, let the bases for $T_p M$ induced by these coordinate systems be denoted, respectively, by

$$\left\{\frac{\partial}{\partial x^1}, \ldots, \frac{\partial}{\partial x^m}\right\} \quad \text{and} \quad \left\{\frac{\partial}{\partial y^1}, \ldots, \frac{\partial}{\partial y^m}\right\}.$$

Suppose that with respect to these two bases, X has representations

$$X = \sum_{i=1}^{m} \alpha^i \frac{\partial}{\partial x^i} = \sum_{j=1}^{m} \beta^j \frac{\partial}{\partial y^j}.$$

We know that the association between tangent vectors and equivalence classes of curves under a particular coordinate system is one to one. Thus, with respect to the coordinate system (ϕ, U), we associate X with the equivalence class $\tilde{\gamma_1}$, where $\gamma_1 = \phi^{-1} \circ \lambda_1$ and

$$\lambda_1(t) = \left(\alpha^1 t + (\phi(p))^1, \ldots, \alpha^m t + (\phi(p))^m\right).$$

With respect to the coordinate system (ψ, V), we associate X with the equivalence class $\tilde{\gamma_2}$, where $\gamma_2 = \psi^{-1} \circ \lambda_2$ and

$$\lambda_2(t) = \left(\beta^1 t + (\psi(p))^1, \ldots, \beta^m t + (\psi(p))^m\right).$$

Since equivalence of two curves with respect to one chart implies equivalence with respect to all charts, we need only show that $(\phi \circ \gamma_1)'(0) = (\phi \circ \gamma_2)'(0)$. Now, we know that $(\phi \circ \gamma_1)'(0) = (\alpha^1, \ldots, \alpha^m)$. So, consider $(\phi \circ \gamma_2)'(0)$. This can be rewritten as

$$\begin{aligned} (\phi \circ \gamma_2)'(0) &= (\phi \circ \psi^{-1} \circ \lambda_2)'(0) \\ &= D_{\psi(p)}(\phi \circ \psi^{-1})(\beta^1, \ldots, \beta^m), \end{aligned}$$

where $D_{\psi(p)}(\phi \circ \psi^{-1})$ is the Jacobian of the change of coordinates map, $\phi \circ \psi^{-1}$, evaluated at $\psi(p)$. We have already discussed the structure of this Jacobian and changes of coordinates. The $(ij)^{th}$ entry of this matrix is just the coordinate partial derivative $\partial x^i / \partial y^j$. Thus, we see that

$$(\phi \circ \gamma_2)'(0) = \left(\sum_{i=1}^{m} \frac{\partial x^1}{\partial y^i} \beta^i, \ldots, \sum_{i=1}^{m} \frac{\partial x^m}{\partial y^i} \beta^i\right).$$

But, by Theorem 14.1.5, where we showed how components of a tangent vector transform under a change of coordinates, the sum

$$\sum_{i=1}^{m} \frac{\partial x^j}{\partial y^i} \beta^i$$

is equal to the component, α^j, of X. Thus, we see that $(\phi \circ \gamma_2)'(0) = (\alpha^1, \ldots, \alpha^m) = (\phi \circ \gamma_1)'(0)$. This implies that $\tilde{\gamma_1} = \tilde{\gamma_2}$, and our association of equivalence classes of curves with tangent vectors is not coordinate dependent.

So, we can uniquely associate to each tangent vector, X, an equivalence class of curves, $\tilde{\gamma}$, such that the operator T_γ satisfies $X(\tilde{f}) = T_\gamma(\tilde{f})$ for all $\tilde{f} \in \mathscr{F}_p$. Moreover, we see by our construction that the association has an intuitively geometric basis. The equivalence class of curves, $\tilde{\gamma}$, that we associate to $X \in T_pM$ are simply the curves whose tangent vectors denote the direction in which X gives us the directional derivative of a function f at p. Conversely, the tangent vector X gives us the directional derivative of a function f in the direction of the tangent vectors of the curves in $\tilde{\gamma}$ at p.

Homework

Exercise 14.5.1 *Recall Γ_p is the set of all smooth curves, $\gamma : I \to M$ such that $\gamma(0) = p$. Define the* *relation,* $\sim$*, on* Γ_p *by* $\gamma_1 \sim \gamma_2$ *if and only if there is a coordinate system,* (ϕ, U) *around* p *such that* $(\phi \circ \gamma_1)'(0) = (\phi \circ \gamma_2)'(0)$*. Prove this is an equivalence relation.*

Chapter 15

The Global Structure of Manifolds

Everything we have done so far has been a local construction which was restricted to a single point, p, in the manifold, M. The first global construction is that of tangent vector fields and what is called the tangent bundle.

15.1 Vector Fields and the Tangent Bundle

Let's begin with the definition of a tangent vector field.

Definition 15.1.1 Tangent Vector Field

A tangent vector field, X, on a smooth manifold, M, is an assignment of a tangent vector, $X_p \in T_pM$, to each $p \in M$; i.e. at each p, $X(p)$ is an element of T_pM. We usually denote vector fields by letters X, Y, etc., and denote the vector at particular point, p, by X_p or $X(p)$.

We will usually simply refer to a tangent vector field as a vector field. Note that a vector field, X, defines a field of operators that act on smooth functions defined on the manifold. Let $C^\infty(M)$ denote the collection of smooth functions defined on M. For any $f \in C^\infty(M)$, define a real-valued function on M by $(Xf)(p) = X_p(f)$. That is, the value of Xf at p is the value of the tangent vector, X_p, acting on f. We now need to discuss what we mean by the smoothness of a vector field.

Definition 15.1.2 Smooth Vector Fields I

Let X be a vector field on M. We say that X is a smooth vector field if for any $f \in C^\infty(M)$, we have $Xf \in C^\infty(M)$.

It then follows that a smooth vector field defines an operator from $C^\infty(M)$ to itself. That is, we can think of X as mapping $f \in C^\infty(M)$ to $Xf \in C^\infty(M)$. Moreover, by the properties of tangent vectors, this operator satisfies $X(\alpha f + \beta g) = \alpha Xf + \beta Xg$ and $X(fg) = f \cdot Xg + g \cdot Xf$, where $f, g \in C^\infty(M)$ and $\alpha, \beta \in \Re$.

Observe that, restricted to a particular coordinate neighborhood, U, with corresponding coordinate mapping, ϕ, we can express a vector field, X, in local coordinates. For each $p \in U$, this coordinate system induces the standard coordinate frame, $\{E_{p_i}\}$. The vector field on U whose tangent vector at each $p \in U$ is E_{p_i} is smooth, since E_{p_i} is just a partial derivative operator acting on smooth functions. Hence, we have a standard coordinate basis field on U, denoted by $\{E_i\}_{i=1}^m$, which is a set of vector fields on U such that at each $p \in U$, the set $\{E_{p_i}\}$ is the standard coordinate frame at for T_pM. Moreover, each vector field in this standard basis field is smooth. Hence, at each $p \in U$, there are scalars $\alpha_1, \ldots, \alpha_m$ so that $X(p)$ has representation

$$X(p) = \sum_{i=1}^{m} \alpha_i E_{p_i}.$$

Letting p vary over all of U, we obtain m real-valued functions, $\alpha_i : U \to \Re$, $i = 1, \ldots, m$, such that $\alpha_i(p)$ is the ith coordinate of $X(p)$ with respect to the basis $\{E_{p_i}\}$. Hence, we can represent the vector field, X, on U by $X = \sum_{i=1}^{m} \alpha_i E_i$, where we interpret each α_i as a real-valued function on U and each E_i as the vector field on U such that $E_i(p) = E_{p_i}$. Using this local representation, we can give another characterization of smoothness based on local coordinates.

Lemma 15.1.1 Characterization of a Smooth Vector Field in Terms of Coordinate Charts

A vector field, X, is smooth if and only if for every coordinate chart, (ϕ, U), on M, the coordinate representation of X restricted to U is given by $\sum_i \alpha_i E_i$ where each α_i is a smooth function on U.

Proof 15.1.1

Suppose, first, that for every coordinate chart, (ϕ, U), on M, we have $X|_U = \sum_i \alpha_i E_i$ where each α_i is a smooth function on U. Let f be in $C^\infty(M)$. Then $f|_U$ is a smooth function on U. Hence,

$$Xf|_U(p) = \sum_{i=1}^{m} \alpha_i(p) E_{p_i}(f) = \sum_{i=1}^{m} \alpha_i(p) \frac{\partial}{\partial x^i}(f \circ \phi^{-1})\Big|_{\phi(p)}$$

But, since f is smooth, the function

$$p \mapsto \frac{\partial}{\partial x^i}(f \circ \phi^{-1})\Big|_{\phi(p)}$$

is smooth on U. Hence, $Xf|_U$ is a finite sum of smooth functions on U, implying that $Xf|_U$ is a smooth function on U.

The coordinate chart, (ϕ, U), was arbitrary, though, so this implies that Xf is a smooth function on a collection of open sets covering M, namely the collection of coordinate neighborhoods. Hence, Xf must be smooth on M. Since $f \in C^\infty(M)$ was arbitrary, this shows that X is a smooth vector field.

Conversely, suppose X is smooth according to the definition, and let (ϕ, U) be any coordinate chart. Then X restricted to U is a smooth vector field on U. We can express $X|_U$ as

$$X|_U = \sum_{i=1}^{m} \alpha_i E_i$$

where each α_i is a real-valued function on U and $E_i(p) = E_{p_i}$. The coordinate functions $x^j : U \to \Re$ are smooth functions on U, so, by hypothesis, $X|_U(x^j) : U \to \Re$ is a smooth function for each $j = 1, \ldots, m$. But, for any $p \in U$, we have

$$X|_U(x^j)(p) = \sum_{i=1}^{m} \alpha_i(p) E_{p_i}(x^j) = \sum_{i=1}^{m} \alpha_i(p) \delta_j^i = \alpha_j(p).$$

Thus, $X|_U(x^j) = \alpha_j$ for each $j = 1, \ldots, m$, implying that each α_j is a smooth function on U. ■

Characterizing the smoothness of vector fields in terms of local coordinates is useful and the characterization involving smooth functions, while not coordinate dependent, still relies on an external set of objects to verify smoothness. Moreover, we need to show a vector field, X, induces a smooth function, Xf, for every $f \in C^\infty(M)$, which may be difficult to do. After we construct the tangent bundle, we can give restatements of the previous definitions, which are better.

Homework

Exercise 15.1.1 *Using the properties of tangent vectors, prove the operator X solves $X(\alpha f + \beta g) = \alpha Xf + \beta Xg$ and $X(fg) = f \cdot Xg + g \cdot Xf$, where $f, g \in C^\infty(M)$ and $\alpha, \beta \in \Re$.*

Exercise 15.1.2 *Using the properties of tangent vectors, prove the operator X solves $X(\sum_{i=1}^n c_i f_i) = \sum_{i=1}^n c_i X f_i$ where $f_i \in C^\infty(M)$ and $c_i \in \Re$.*

15.2 The Tangent Bundle

The tangent bundle is just the union of all the tangent spaces over a manifold, M. Formally, we have the following.

Definition 15.2.1 Tangent Bundle

The tangent bundle of M, denoted TM, is defined by

$$TM = \bigcup_{p \in M} T_p M = \{X : X \in T_p M \; for \; some \; p \in M\}.$$

We will define a topology on TM, making it a second countable Hausdorff space. We can also define a coordinate covering for TM, making it a $2m$-dimensional manifold if m is the dimension of the manifold M. This dimension makes sense intuitively. We usually denote an element of the tangent bundle by (p, X), where $X \in T_p M$. This will save us from continually having to refer to the tangent space in which a tangent vector lies. Thus, given this representation, we need $2m$ coordinates to distinguish an element of TM, m for the point, p, and m for the basis coordinates of $X \in T_p M$.

Let $\tau : TM \to M$ be the surjective projection mapping such that $\tau(p, X) = p$. So, if U is any subset of M, $\tau^{-1}(U) = \cup_{p \in U} T_p M$. That is, $\tau^{-1}(U)$ is the union of all the tangent spaces $T_p M$ for $p \in U$. For each chart, (ϕ, U) in the coordinate covering of M, we can think of the mapping ϕ as a smooth mapping between the manifolds U and $\phi(U) \subset \Re^m$. By the way we have defined our smooth manifolds, this map is actually a diffeomorphism. Hence, the differential $D_p\phi : T_p M \to \Re^m$, where $D_p\phi$ indicates that this is the differential of ϕ at p, is an isomorphism. Likewise, the mapping $\phi^{-1} : \phi(U) \to U$ is a diffeomorphism, and, for $x \in \phi(U)$, the differential $D_x\phi^{-1} : \Re^m \to T_{\phi^{-1}(x)} M$ is an isomorphism. We use these facts in the construction of both the topology and the coordinate covering on TM.

Define a mapping $T\phi : \tau^{-1}(U) \to \phi(U) \times \Re^m$ by

$$T\phi(p, X) = \big(\phi(p), D_p\phi(X)\big).$$

If $T\phi(p, X_p) = T\phi(q, X_q)$, then $(\phi(p), D_p\phi(X_p)) = (\phi(q), D_q\phi(X_q)) \Rightarrow \phi(p) = \phi(q)$ and $D_p\phi(X_p) = D_q\phi(X_q)$. But ϕ is injective, so this implies $p = q$. This, in turn, implies that

$D_p\phi = D_q\phi$, so $(p, X_p) = (q, X_q)$, and $T\phi$ is injective. If $(a, b) \in \phi(U) \times \Re^m$, then

$$T\phi(\phi^{-1}(a), D_a\phi^{-1}(b)) = (\phi(\phi^{-1}(a)), D_{\phi^{-1}(a)}\phi(D_a\phi^{-1}(b))) = (a, D_a(\phi \circ \phi^{-1})(b)) = (a, b)$$

Hence, $T\phi$ is surjective, and we see that this map is a bijection of $\tau^{-1}(U)$ onto $\phi(U) \times \Re^m$.

Now, let

$$\mathbb{T} = \{W \subset TM | \text{ for each chart } (\phi, U) \text{ on } M,\ T\phi(W \cap \tau^{-1}(U)) \text{ is open in } \Re^m \times \Re^m\}$$

Theorem 15.2.1 $\mathbb{T}$ is a Topology on TM

$\mathbb{T}$ is a topology on TM.

Proof 15.2.1
The empty set, $\emptyset$, is in $\mathbb{T}$, because, for any chart (ϕ, U), we have $\emptyset \cap \tau^{-1}(U) = \emptyset$ and $T\phi(\emptyset) = \emptyset$.

TM is in $\mathbb{T}$ because $TM \cap \tau^{-1}(U) = \tau^{-1}(U) \Rightarrow T\phi(\tau^{-1}(U)) = \phi(U) \times \Re^m$, which is open in $\Re^m \times \Re^m$.

Now, let $\{W_\alpha\}_{\alpha \in A}$ be an arbitrary collection of sets in $\mathbb{T}$ indexed by some set A. Then, for any chart, (ϕ, U), we have

$$T\phi\Big((\bigcup_\alpha W_\alpha) \cap \tau^{-1}(U)\Big) = T\phi\Big(\bigcup_\alpha [W_\alpha \cap \tau^{-1}(U)]\Big) = \bigcup_\alpha T\phi\Big(W_\alpha \cap \tau^{-1}(U)\Big) \in \mathbb{T}$$

Finally, if $\{W_i\}_{i=1}^n$ is a finite collection of sets in $\mathbb{T}$, then

$$T\phi\Big(\bigcap_{i=1}^n W_i \cap \tau^{-1}(U)\Big) = \bigcap_{i=1}^n T\phi\Big(W_i \cap \tau^{-1}(U)\Big) \in \mathbb{T}.$$

So, $\mathbb{T}$ is a topology on TM. ■

Defining the coordinate covering on TM is fairly straightforward. The following theorem does most of the technical work.

Theorem 15.2.2 Coordinate Covering on TM

If (ϕ, U) is a coordinate chart on M, then $\tau^{-1}(U)$ is open in TM. Also, the mapping

$$T\phi : \tau^{-1}(U) \to \phi(U) \times \Re^m$$

is a homeomorphism, and, for any other chart (ψ, V) on M, the map

$$T\phi \circ (T\psi)^{-1} : T\psi(\tau^{-1}(U \cap V)) \to T\phi(\tau^{-1}(U \cap V))$$

is C^∞.

Proof 15.2.2
To show that $\tau^{-1}(U)$ is open, we need to show that for any chart, (ψ, V), the set $T\psi(\tau^{-1}(U) \cap \tau^{-1}(V))$ is open. But

$$T\psi(\tau^{-1}(U) \cap \tau^{-1}(V)) = T\psi(\tau^{-1}(U \cap V)) = \psi(U \cap V) \times \Re^m,$$

which is open in $\Re^m \times \Re^m$. So, $\tau^{-1}(U)$ is open for any chart (ϕ, U).

Next, we show that for any two charts, (ϕ, U) and (ψ, V), the map $T\phi \circ (T\psi)^{-1}$ is C^∞. Note that the inverse of the map, $T\psi$, is the map $(T\psi)^{-1} : \psi(V) \times \Re^m \to \tau^{-1}(V)$ defined explicitly by

$$(T\psi)^{-1}(a^1 \ldots, a^m, b^1, \ldots, b^m) = (\psi^{-1}(a), D_a\psi^{-1}(b)),$$

for $(a, b) = (a^1, \ldots, a^m, b^1, \ldots, b^m) \in \psi(V) \times \Re^m$. So, the map $T\phi \circ (T\psi)^{-1}$ will map $T\psi(\tau^{-1}(U \cap V))$ to $T\phi(\tau^{-1}(U \cap V))$. Denote the coordinate functions of ϕ and ψ by x^i and y^j, respectively. Then, for $(a, b) \in T\psi(\tau^{-1}(U \cap V))$, we have

$$\begin{aligned} T\phi \circ (T\psi)^{-1}(a, b) &= T\phi\Big(\psi^{-1}(a), D_a\psi^{-1}(b)\Big) = \Big((\phi \circ \psi^{-1})(a), D_{\psi^{-1}(a)}\phi\big(D_a\psi^{-1}(b)\big)\Big) \\ &= \Big((\phi \circ \psi^{-1})(a), D_a(\phi \circ \psi^{-1})(b)\Big) \\ &= \left((\phi \circ \psi^{-1})(a), \sum_{i=1}^m \frac{\partial x^1}{\partial y^i} b^i, \sum_{i=1}^m \frac{\partial x^2}{\partial y^i} b^i, \ldots, \sum_{i=1}^m \frac{\partial x^m}{\partial y^i} b^i, \right) \end{aligned}$$

where the last equality follows from the fact that $D_a(\phi \circ \psi^{-1})(b)$ is just the differential of the change of coordinates map, $\phi \circ \psi^{-1}$. It is implied that these partial derivatives are to be evaluated at $a \in \psi(U \cap V)$. Now, the map $\phi \circ \psi^{-1}$ is C^∞ because M is a smooth manifold. By this same fact, the partial derivatives $\frac{\partial x^i}{\partial y^j}$ are smooth functions. Hence, each component function of this map is smooth, and we can conclude that $T\phi \circ (T\psi)^{-1}$ is smooth. It follows by almost the same argument that $T\psi \circ (T\phi)^{-1}$ is C^∞ also. Hence, since these charts were arbitrary, it follows that all possible compositions of the form $T\phi \circ (T\psi)^{-1}$ are smooth. In fact, since each map, $T\phi$, is bijective, this shows that the maps $T\phi \circ (T\psi)^{-1}$ are diffeomorphisms.

Finally, we will show that $T\phi : \tau^{-1}(U) \to \phi(U) \times \Re^m$ is a homeomorphism for any chart (ϕ, U). Suppose W is open in $\tau^{-1}(U)$. Since $\tau^{-1}(U)$ is open in TM, it follows that W is also open in TM, for there is some open set, $\Omega \subset TM$, such that $W = \Omega \cap \tau^{-1}(U)$. So, for any chart, (ψ, V), the set $T\psi(W \cap \tau^{-1}(V))$ is open in $\Re^m \times \Re^m$. Hence, $T\phi(W \cap \tau^{-1}(U)) = T\phi(W)$ is open in $\phi(U) \times \Re^m$. So, $T\phi$ maps open sets to open sets.

Conversely, suppose $Z \subset \phi(U) \times \Re^m$ is open. We must show that $T\psi(\tau^{-1}(V) \cap (T\phi)^{-1}(Z))$ is open for any chart (ψ, V). We first note that

$$T\phi\Big(\tau^{-1}(V) \cap (T\phi)^{-1}(Z)\Big) = Z \cap \Big(T\phi(\tau^{-1}(V))\Big) = Z \cap \Big(T\phi(\tau^{-1}(V) \cap \tau^{-1}(U))\Big),$$

where this last line follows from the fact that $T\phi$ is only defined on $\tau^{-1}(U)$. So, the set $T\phi(\tau^{-1}(V))$ is empty unless $\tau^{-1}(V)$ intersects $\tau^{-1}(U)$, and if it does intersect $\tau^{-1}(U)$, then we are only concerned with the intersection. Hence, we have

$$T\phi\Big(\tau^{-1}(V) \cap (T\phi)^{-1}(Z)\Big) = Z \cap \Big(T\phi(\tau^{-1}(U \cap V))\Big) = Z \cap \Big(\phi(U \cap V) \times \Re^m\Big)$$

which implies that

$$\begin{aligned} &\tau^{-1}(V) \cap (T\phi)^{-1}(Z) = (T\phi)^{-1}\Big(Z \cap (\phi(U \cap V) \times \Re^m)\Big) \\ \Rightarrow\ & T\psi\Big(\tau^{-1}(V) \cap (T\phi)^{-1}(Z)\Big) = T\psi \circ (T\phi)^{-1}\Big(Z \cap (\phi(U \cap V) \times \Re^m)\Big) \end{aligned}$$

Since $T\psi \circ (T\phi)^{-1}$ is a diffeomorphism, the right side of this last equality is an open set. Hence, the set $T\psi(\tau^{-1}(V) \cap (T\phi)^{-1}(Z))$ is open, and $T\phi$ is a homeomorphism. ■

We can now prove the projection map τ is continuous.

Theorem 15.2.3 $\tau : TM \to M$ is Continuous

> *$\tau : TM \to M$ is continuous.*

Proof 15.2.3
Let $\Omega \subset M$ be open, and let (ϕ, U) be any chart on M. We have

$$T\phi(\tau^{-1}(\Omega) \cap \tau^{-1}(U)) = T\phi(\tau^{-1}(\Omega \cap U))$$

If $\Omega \cap U = \emptyset$, then $T\phi(\tau^{-1}(\Omega \cap U)) = \emptyset$, which is open. If $\Omega \cap U \neq \emptyset$, then $\Omega \cap U$ is an open subset of U, so $T\phi(\tau^{-1}(\Omega \cap U)) = \phi(\Omega \cap U) \times \Re^m$, which is open. ■

We can now prove that TM is a smooth manifold. The most difficult part about proving that TM is a manifold is proving that it has the correct topological structure.

Theorem 15.2.4 Tangent Bundle is a Smooth Manifold

> *The tangent bundle, TM, is a smooth manifold of dimension $2m$ with coordinate covering $A = \{(T\phi, \tau^{-1}(U))\}$ where (ϕ, U) is a chart on M.*

Proof 15.2.4
For any $(p, X) \in TM$, we must have $p \in U$ for some chart (ϕ, U), so $(p, X) \in \tau^{-1}(U)$. Moreover, we have shown that each $T\phi : \tau^{-1}(U) \to \phi(U) \times \Re^m$ is a homeomorphism. So, A is a coordinate covering of TM. We have also shown that, for any two charts, (ϕ, U) and (ψ, V), the composition $T\phi \circ (T\psi)^{-1}$ is C^∞. Hence, TM will be a smooth manifold if we can show that it is both Hausdorff and second countable. We will prove the Hausdorff condition first.

Let (p, X_p) and (q, X_q) be any two distinct points of TM. If $p = q$, then we must have $X_p \neq X_q$, which implies that $D_p\phi(X_p) \neq D_p\phi(X_q)$. Since $\Re^m$ is Hausdorff, there are open balls, B_1 and B_2, around $D_p\phi(X_p)$ and $D_p\phi(X_q)$, respectively, such that $B_1 \cap B_2 = \emptyset$. Then $(T\phi)^{-1}(\phi(U) \times B_1)$ and $(T\phi)^{-1}(\phi(U) \times B_2)$ are disjoint open sets containing (p, X_p) and (p, X_q) respectively. If $p \neq q$, there are two mutually exclusive possibilities.

(i): Suppose there exists a single coordinate chart, (ϕ, U), such that $p, q \in U$. Since M is Hausdorff, we can find open sets Ω_p and Ω_q such that $p \in \Omega_p \subset U$, $q \in \Omega_q \subset U$, and $\Omega_p \cap \Omega_q = \emptyset$. Then $\phi(\Omega_p)$ and $\phi(\Omega_q)$ are disjoint open subsets of $\phi(U)$. It follows that $\phi(\Omega_p) \times \Re^m$ and $\phi(\Omega_q) \times \Re^m$ are disjoint open sets containing $(\phi(p), D_p\phi(X_p))$ and $(\phi(q), D_q\phi(X_q))$ respectively. Since $T\phi$ is a homeomorphism, it follows that

$$(T\phi)^{-1}(\phi(\Omega_p) \times \Re^m) \text{ and } (T\phi)^{-1}(\phi(\Omega_q) \times \Re^m)$$

are disjoint open sets in TM such that $(p, X_p) \in (T\phi)^{-1}(\phi(\Omega_p) \times \Re^m)$ and $(q, X_q) \in (T\phi)^{-1}(\phi(\Omega_q) \times \Re^m)$.

(ii): Now, suppose there is no single coordinate neighborhood containing p and q. Let (ϕ, U) and (ψ, V) be charts on M such that $p \in U$ and $q \in V$. There are open sets $\Omega_p \subset U$ and $\Omega_q \subset V$ such that $p \in \Omega_p$, $q \in \Omega_q$, and $\Omega_p \cap \Omega_q = \emptyset$. Since τ is continuous, the sets $\tau^{-1}(\Omega_p)$ and $\tau^{-1}(\Omega_q)$ are disjoint open sets in TM containing (p, X_p) and (q, X_q), respectively.

Hence, TM is Hausdorff.

Finally, we will construct a countable basis for TM. Define a collection, S, of subsets of TM as follows. The Euclidean space $\Re^m$ is second countable, so there is a countable collection of open balls, $\{B_i\}_{i=1}^{\infty}$, that form a basis for $\Re^m$. Moreover, for any chart, (ϕ, U), in the coordinate covering of M, the set $\phi(U)$ is an open subspace of $\Re^m$, so there is a countable collection of open balls, $\{B_{U_i}\}_{i=1}^{\infty}$, that form a basis for $\phi(U)$. Hence, all the sets of the form $B_{U_i} \times B_j$, $i, j \geq 1$, form a basis for the subspace $\phi(U) \times \Re^m \subset \Re^m \times \Re^m$. We define S as

$$S = \{(T\phi)^{-1}(B_{U_i} \times B_j) : (\phi, U) \text{ is a chart on } M,\ i, j \geq 1\}.$$

That is, S is the set of all images of the basis elements of $\phi(U) \times \Re^m$ under $(T\phi)^{-1}$ for all charts, (ϕ, U), on M. The basis, $\{B_{U_i} \times B_j\}_{i,j\geq 1}$, for $\phi(U) \times \Re^m$ is countable. Moreover, the coordinate covering of M is countable. Thus, S is a countable collection of open subsets of TM. If $(p, X) \in TM$, then we must have $p \in U$ for some chart (ϕ, U). Hence, $(p, X) \in \tau^{-1}(U) = (T\phi)^{-1}(\phi(U) \times \Re^m)$. There are sets $B_{U_i} \subset \phi(U)$ and $B_j \subset \Re^m$ such that $(\phi(p), D_p\phi(X)) \in B_{U_i} \times B_j$, implying that $(p, X) \in (T\phi)^{-1}(B_{U_i} \times B_j)$. So, S covers TM. Now, let W be any open set in TM, and let (p, X) be any point in W. Let (ϕ, U) be a chart such that $p \in U$. Then $(p, X) \in W \cap \tau^{-1}(U)$, so this set is nonempty and open. Hence, the set $T\phi(W \cap \tau^{-1}(U))$ is open in $\phi(U) \times \Re^m$ and contains $T\phi(p, X) = (\phi(p), D_p\phi(X))$. Thus, there are sets $B_{U_i} \subset \phi(U)$ and $B_j \subset \Re^m$ such that

$$B_{U_i} \times B_j \subset T\phi(W \cap \tau^{-1}(U)) \text{ and } T\phi(p, X) = (\phi(p), D_p\phi(X)) \in B_{U_i} \times B_j$$

It follows that $(p, X) \in (T\phi)^{-1}(B_{U_j} \times B_j) \subset W \cap \tau^{-1}(U) \subset W$. Since $(p, X) \in W$ was arbitrary, this implies that we can express any open set, $W \subset TM$, as a union of elements from S. Hence, S must be a basis, and we can conclude that TM is second countable. ■

Thus, the tangent bundle of a smooth manifold is, itself, a smooth manifold. It is tempting to simply think that, once a basis for each tangent space is chosen and each tangent space is, consequently, isomorphic to $\Re^m$, that the tangent bundle of a manifold can simply be described as $M \times \Re^m$. This is not true. Manifolds whose tangent bundles are of this structure are said to have a trivial bundle structure. But many manifolds have a tangent bundle that cannot be globally described by a single Cartesian product like this. Instead, such products must be pasted together over the manifold to create the tangent bundle. In purely geometric and topological terms, the tangent bundle is a typical example of a more general class of structures called vector bundles. A vector bundle of rank k over a manifold, M, is a $4-tuple$, (E, M, V, τ), where E and M are smooth manifolds, $\tau : E \to M$ is a smooth surjective map, and V is a k-dimensional vector space, with the following properties.

(i): For each $p \in M$, $\tau^{-1}(p) = V$. V is called the typical fiber of the vector bundle.

(ii): If (ϕ, U) is a coordinate chart on M, then there is a corresponding map, $\tilde{\phi}$, such that $\tilde{\phi} : \tau^{-1}(U) \to U \times V$ is a homeomorphism. For any single point, $p \in U$, $\tilde{\phi} : \tau^{-1}(p) \to p \times V$ is a vector space isomorphism.

(iii): If (ϕ, U) and (ψ, W) are two charts on M such that $U \cap W \neq \emptyset$, then for any $p \in U \cap W$, the map $G_{UW_p} = \tilde{\psi} \circ \tilde{\phi}^{-1} : V \to V$ is an automorphism and depends smoothly on p.

So, loosely speaking, a vector bundle is just a collection of vector spaces parameterized by points of some base space (a manifold in our case), such that there is some notion of transferring smoothly from one fiber to another. In the case of the tangent bundle, the typical fiber at p is the tangent space, T_pM, which is m-dimensional. So, we can identify the typical fiber with $\Re^m$. The mappings $\tilde{\phi}$ are just the mappings $T\phi$, and these mappings induce the smooth structure on TM. Restricted to just a single point, $p \in U$, where (ϕ, U) is a chart on M, then $T\phi(p, T_pM) = \phi(p) \times \Re^m$. Since $T\phi$ is a homeomorphism, and the differential $D_p\phi$ is linear, this induces an isomorphism between T_pM and

$\phi(p) \times \Re^m$, as we have shown.

The tangent bundle is the most widely used example of a vector bundle. We can also construct the **cotangent bundle** using similar arguments. We will leave that discussion to other textual sources for now.

Homework

Exercise 15.2.1 *Let M be a surface in $\Re^3$. Let TM be the collection of all tangent vectors to M. Hence, M has dimension 2 and for each (x_1, x_2) in M, each tangent plane $T_{(x_1,x_2)}(M)$ has dimension 2 so TM will have dimension 4. We want a collection of open sets for TM, $\mathcal{P}$, that will make TM into a manifold. We will find for each chart $(\phi, U) \in M$, a chart $(\tilde{\phi}, \tilde{U})$ in TM. Let W be the open set in $\Re^4$ consisting of the points (p_1, p_2, p_3, p_4) for which $(p_1, p_2) = \phi(x_1, x_2) \in \phi(U)$. Define $\psi : \phi(U) \rightarrow TM$ by $\psi(p_1, p_2, p_3, p_4) = p_3 x_{u_1}(p_1, p_2) + p_4 x_{u_2}(p_1, p_2)$ where the local coordinates of the tangent plane $T_{(x_1,x_2)}$ are (x_{u_1}, x_{u_2}). Note $\psi(p_1, p_2, p_3, p_4) \in TM$. Prove ψ is $1-1$ and so (ψ^{-1}, W) is a chart in TM. Go through the arguments that show TM is a manifold for practice.*

Exercise 15.2.2 *Let M be the surface given by $z = 2x^2 + 3y^2$ in $\Re^3$. Go through all the arguments from Exercise 15.2.1. Also, draw the pictures that show the mappings we use as we can do this in our $\Re^3$ setting.*

Exercise 15.2.3 *Let M be the surface given by $z = x^2 + 5y^2$ in $\Re^3$. Go through all the arguments from Exercise 15.2.1. Also, draw the pictures that show the mappings we use as we can do this in our $\Re^3$ setting.*

Exercise 15.2.4 *Let M be the surface given by $x^2 + 5y^2 + 4z^2$ in $\Re^3$. Go through all the arguments from Exercise 15.2.1. Also, draw the pictures that show the mappings we use as we can do this in our $\Re^3$ setting.*

15.3 Vector Fields Revisited

We can now revisit the notion of a smooth vector field on a manifold and reformulate the ideas in a more concise fashion.

Definition 15.3.1 Smooth Vector Fields II

A smooth vector field, X, on a smooth manifold, M, is a C^∞ mapping $X : M \rightarrow TM$.

We can prove this definition is equivalent to the previous one.

Theorem 15.3.1 Smooth Vector Fields II is equivalent to Smooth Vector Fields I

Let X be a smooth vector field in the sense of Definition 15.1.2 and let X_p be the tangent vector at p defined by this vector field. Then the mapping from M to TM defined by $X(p) = X_p$ is a smooth mapping. Conversely, if $X : M \rightarrow TM$ is a smooth map, then, for any smooth function $f \in C^\infty(M)$, we have $Xf \in C^\infty(M)$.

Proof 15.3.1
Let $X : M \rightarrow TM$ be a smooth map. We will actually show that the alternate characterization of smoothness given in Lemma 15.1.1 holds. Since this is equivalent to our original definition, this will prove the first implication.

Let (ϕ, U) be any chart on M, and let $\{E_i\}_{i=1}^m$ be the standard coordinate basis field induced by this chart on U. For a particular point, $p \in U$, we will denote the basis of T_pM by $\{E_{p_i}\}$, so that the vector field E_i satisfies $E_i(p) = E_{p_i}$. Now, we can observe the restriction of X to U, which must be smooth. For $p \in U$, we have $X(p) \in T_pM = span\{E_{p_i}\}$, so there are scalars $\alpha_i, \ldots, \alpha_m$ such that $X(p) = \sum_i \alpha_i E_{p_i}$. Letting p vary over U, we obtain functions $\alpha_i : U \to \Re$ such that the restriction of X to U can be represented as $X|_U = \sum_{i=1}^m \alpha_i E_i$. We need to show that each α_i is a C^∞ function on U. By the definition of a smooth map between manifolds, the smoothness of $X : M \to TM$ implies that the coordinate representation $T\phi \circ X \circ \phi^{-1} : \phi(U) \to T\phi(\tau^{-1}(U))$ is a smooth mapping. We will denote the ϕ-coordinate functions by x^i. For a point $x = (x^1, \ldots, x^m) \in \phi(U)$ with $p = \phi^{-1}(x)$, we have

$$T\phi \circ X \circ \phi^{-1}(x) = T\phi(X(\phi^{-1}(x))) = T\phi\left(\phi^{-1}(x), \sum_{i=1}^m \alpha_i(\phi^{-1}(x)) E_{\phi^{-1}(x)_i}\right)$$

where we have simply represented the point $X(\phi^{-1}(x)) \in TM$ in our usual form. Thus, we have

$$T\phi \circ X \circ \phi^{-1}(x) = \left(\phi(\phi^{-1}(x)), D_p\phi\Big(\sum_{i=1}^m \alpha_i(p) E_{p_i}\Big)\right) = \left(x, \sum_{i=1}^m \alpha_i(p) D_p\phi(E_{p_i})\right)$$

Now, the differential of ϕ at p maps E_{p_i} to the partial derivative operator $\frac{\partial}{\partial x^i}$, which can be identified with the ith standard basis vector e_i in $\Re^m$. Hence,

$$\begin{aligned} T\phi \circ X \circ \phi^{-1}(x) &= \Big(x, \sum_{i=1}^m \alpha_i(p) e_i\Big) = (x, \alpha_1(p), \alpha_2(p), \ldots, \alpha_m(p)) \\ &= (x, \alpha_1(\phi^{-1}(x)), \ldots, \alpha_m(\phi^{-1}(x))) \end{aligned}$$

Hence, we can express this map as $T\phi \circ X \circ \phi^{-1} = (\mathbf{id}, \alpha_1 \circ \phi^{-1}, \ldots, \alpha_m \circ \phi^{-1})$, where $\mathbf{id}$ represents the identity function. Since this is a C^∞ function, each of the real-valued functions $\alpha_i \circ \phi^{-1}$ is C^∞. So, by definition, each of the functions α_i is smooth on U. This proves that the new characterization implies the old one.

Now, suppose X is a smooth vector field in the sense of Definition 15.1.2 and Lemma 15.1.1. We have the induced mapping $X(p) = X_p$. Let p be any point on M, and let (ϕ, U) be a coordinate chart on M such that $p \in U$. Then $(T\phi, \tau^{-1}(U))$ is a coordinate system around $X(p)$. The local representation of X on U is

$$X|_U = \sum_{i=1}^m \alpha_i E_i,$$

where each α_i is a smooth function on U. Going through the same process we just completed, we show that, for $x \in \phi(U)$,

$$T\phi \circ X \circ \phi^{-1}(x) = (x, \alpha_1(\phi^{-1}(x)), \ldots, \alpha_m(\phi^{-1}(x))).$$

Since each α_i is smooth by hypothesis, this is a C^∞ map. Finally, let $(T\psi, \tau^{-1}(V))$ be any coordinate chart containing $(p, X(p))$, and consider the coordinate representation $T\psi \circ X \circ \phi^{-1}$. We must have $p \in U \cap V$, so $T\psi \circ X \circ \phi^{-1}$ maps $\phi(U \cap V)$ to $T\psi(\tau^{-1}(U \cap V))$. Moreover, we also see that

$$T\psi \circ X \circ \phi^{-1} = T\psi \circ (T\phi)^{-1} \circ T\phi \circ X \circ \phi^{-1},$$

which is C^∞ on $U \cap V$. But the point $p \in M$ was arbitrary, as were the charts (ϕ, U) and (ψ, V).

So, we can conclude that X is a C^∞ mapping from M to TM. This shows that the old definition implies the new one. ■

Hence, we lose nothing with our new definitions. In fact, we gain a more elegant and concise means of working with vector fields on a manifold. Moreover, this shows that the tangent bundle is the natural setting in which to develop the calculus of vector fields.

Homework

Exercise 15.3.1 *Our traditional vector fields from advanced calculus defined by $F(x_1, x_2, x_3) = \big[F_1(x_1, x_2, x_3), F_2(x_1, x_2, x_3), F_3(x_1, x_2, x_3)\big]$ assigns to each point in $\Re^3$ a three dimensional vector. Note how we are using $\Re^3$ in two different ways here: first, it is the collection of three-dimensional points and second, it is the collection of three dimensional vectors. We know $\Re^3$ is a manifold and we can identify $T\Re^3$ with $\Re^3$ also. Both are three dimensional manifolds. Hence, these traditional vector fields are all examples of smooth vector fields on a manifold.*

Exercise 15.3.2 *Let M be the surface given by $z = 2x^2 + 3y^2$ in $\Re^3$. We know TM is a manifold. A vector field on M must assign a tangent vector to each point in the manifold. Characterize the tangent space at a point as usual and find examples of such vector fields.*

Exercise 15.3.3 *Let M be the surface given by $z = x^2 + 5y^2$ in $\Re^3$. We know TM is a manifold. A vector field on M must assign a tangent vector to each point in the manifold. Characterize the tangent space at a point as usual and find examples of such vector fields.*

Exercise 15.3.4 *Let M be the surface given by $x^2 + 5y^2 + 4z^2$ in $\Re^3$. We know TM is a manifold. A vector field on M must assign a tangent vector to each point in the manifold. Characterize the tangent space at a point as usual and find examples of such vector fields.*

Exercise 15.3.5 *Let X and Y be vector fields on the manifold M of dimension n. Then, show we can write $X = \sum_{i=1}^n X_i \frac{\partial}{\partial x_i}$ and $Y = \sum_{i=1}^n Y_i \frac{\partial}{\partial x_i}$. The commutator or Lie bracket of X and Y, $[X, Y]$ is defined by*

$$([X,Y]f)(p) \quad = \quad (X(Yf))(p) - (Y(Xf))(p) = \sum_{i=1}^{n}\sum_{k=1}^{n}\left(X_k \frac{\partial Y_i}{\partial x_k} - Y_k \frac{\partial X_i}{\partial x_k}\right)\frac{\partial f}{\partial x_i}(p)$$

Prove

- $[X, Y] = -[Y, X]$.
- $[[X, Y], Z] + [[Y, Z], X] + [[Z, X], Y] = 0$, *where Z is another vector field on M.*

Exercise 15.3.6 *If X is the velocity vector field for particle A in $\Re^3$ and Y is the velocity vector field for particle B in $\Re^3$, find $[X, Y]$.*

15.4 Tensor Analysis on Manifolds

Next, we will discuss tensors on manifolds. Like vector fields, tensors are also most naturally described within vector bundles over manifolds, but we will approach it in a simpler fashion. We will begin by discussing tensors on an arbitrary finite dimensional vector space, V, pointing out results that will be useful for our purposes. Later on, we will transfer these results to the case where the vector space in question is the tangent space of a manifold.

Definition 15.4.1 Tensors on Vector Spaces

Let V be a vector space of dimension n, and let V^* denote its dual space. An (r, s) – type tensor, σ, over V is a multilinear mapping

$$\sigma : \underbrace{V^* \times \cdots \times V^*}_{r\ copies} \times \underbrace{V \times \cdots \times V}_{s\ copies} \rightarrow \Re.$$

As a note on this definition, tensors can, in general, map to any field $\mathbb{F}$. We will restrict ourselves to real-valued tensors, though. Let σ_1, σ_2, and σ be (r, s)-type tensors over V, and let α be in $\Re$. If $v_1^*, \ldots, v_r^*$ are dual vectors in V^*, and if $v_1, \ldots, v_s$ are vectors in V, then we can define pointwise addition and scalar multiplication of tensors by the relations

$$\begin{aligned}(\sigma_1 + \sigma_2)(v_1^*, \ldots, v_r^*, v_1, \ldots, v_s) &= \sigma_1(v_1^*, \ldots, v_r^*, v_1, \ldots, v_s) + \sigma_2(v_1^*, \ldots, v_r^*, v_1, \ldots, v_s) \\ (\alpha\sigma)(v_1^*, \ldots, v_r^*, v_1, \ldots, v_s) &= \alpha\sigma(v_1^*, \ldots, v_r^*, v_1, \ldots, v_s).\end{aligned}$$

In this way, the set of (r, s)-type tensors over V becomes a vector space. We denote this vector space by

$$\underbrace{V \otimes \cdots \otimes V}_{r\ copies} \otimes \underbrace{V^* \otimes \cdots \otimes V^*}_{s\ copies},$$

and we call this space the *tensor product* of the vector spaces $V, \ldots, V, V^*, \ldots, V^*$. For brevity of notation, we will usually denote this vector space with the shorthand notation V_s^r.

The number r is called the *contravariant order* of σ, and s is called the *covariant order*. An $(r, 0)$-type tensor is called a *contravariant tensor* of order r, and a $(0, s)$-type tensor is called a *covariant tensor* of order s.

As a final remark on notation, observe that the domain of σ is defined to be

$$V^* \times \cdots \times V^* \times V \times \cdots \times V$$

while the vector space containing these tensors is denoted by

$$V \otimes \cdots \otimes V \otimes V^* \otimes \cdots \otimes V^*$$

The purpose for reversing the order of the spaces V and V^* in the vector space notation is to emphasize **the vectors that are performing the action**, as opposed to **those that are being acted upon**. For example, the first input variable of σ is a dual vector from V^*. Since σ is a multilinear map, the notation $V \otimes \cdots \otimes V \otimes V^* \otimes \cdots \otimes V^*$ is used to signify that the functional acting on this particular variable is a vector from V. Likewise, the $(r+s)^{th}$ input variable of σ is a vector from V. Hence, the vector space notation emphasizes the fact that the functional acting on this particular variable is a dual vector from V^*.

We can also define the product of two tensors. If σ_1 and σ_2 are (r, s) and (k, l)-type tensors, respectively, then we define their **tensor product**, $\sigma_1 \otimes \sigma_2$, to be the $(r + k, s + l)$-type tensor given by

$$\begin{aligned}&(\sigma_1 \otimes \sigma_2)(v_1^*, \ldots, v_r^*, v_{r+1}^*, \ldots, v_{r+k}^*, v_1, \ldots, v_s, v_{s+1}, \ldots, v_{s+l}) \\ &= \quad \sigma_1(v_1^*, \ldots, v_r^*, v_1, \ldots, v_s)\sigma_2(v_{r+1}^*, \ldots, v_{r+k}^*, v_{s+1}, \ldots, v_{s+l}).\end{aligned}$$

Moreover, the tensor product is associative. The proof of the associativity property is not difficult. It is just a tedious computation, so we will only illustrate associativity in the case $\sigma_1, \sigma_2, \sigma_3 \in V_1^1$. Let v^{*1}, v^{*2}, v^{*3} be dual vectors and let v_1, v_2, v_3 be vectors. Then

$$\begin{aligned}
&\sigma_1 \otimes (\sigma_2 \otimes \sigma_3)(v^{*1}, v^{*2}, v^{*3}, v_1, v_2, v_3)\\
&= \sigma_1(v^{*1}, v_1)(\sigma_2 \otimes \sigma_3)(v^{*2}, v^{*3}, v_2, v_3) = \sigma_1(v^{*1}, v_1)\sigma_2(v^{*2}, v_2)\sigma_3(v^{*3}, v_3)\\
&= (\sigma_1(v^{*1}, v_1)\sigma_2(v^{*2}, v_2))\sigma_3(v^{*3}, v_3) = (\sigma_1 \otimes \sigma_2)(v^{*1}, v^{*2}, v_1, v_2)\sigma_3(v^{*3}, v_3)\\
&= (\sigma_1 \otimes \sigma_2) \otimes \sigma_3(v^{*1}, v^{*2}, v^{*3}, v_1, v_2, v_3).
\end{aligned}$$

Hence, we can give meaning to expressions like $\sigma_1 \otimes \sigma_2 \otimes \sigma_3$, or any tensor product of a finite number of tensors.

Note the set of $(0,1)$-type tensors is just the dual space of V. That is, $V_1^0 = V^*$. Likewise, since V is finite dimensional and can be identified with its double dual, V^{**}, we see the set of $(1,0)$-type tensors is just V itself, since each vector $v \in V$ can be naturally identified with a linear functional on V^*. Thus, $V_0^1 = V$.

Before going on, we will give some examples of common tensors.

Example 15.4.1 *An inner product, $\langle , \rangle$, on a real linear space, V, is a $(0,2)$-type tensor, or a covariant tensor of order 2, over V, as it is real-valued and linear in each variable. This tensor has the additional properties of symmetry, $\langle v_1, v_2 \rangle = \langle v_2, v_1 \rangle$ for $v_1, v_2 \in V$, and positive definiteness, $\langle v, v \rangle \geq 0$ for $v \in V$ and equality holds if and only if $v = 0$.*

Example 15.4.2 *The determinant function, det : $(\Re^n)^n \to \Re$, thought of as acting on the cartesian product $(\Re^n)^n$, is a $(0,n)$-type tensor, or a covariant tensor of order n, over $\Re^n$, since the determinant is linear in each row of a matrix. If we identify $\Re^n$ with row vectors and the dual space $\Re^{n*}$ with column vectors, we can also think of the determinant function as an $(n,0)$-type tensor, since it is linear in each column as well.*

Example 15.4.3 *For a vector space, V, let $T : V \to V$ be a linear transformation. Define a $(1,1)$-type tensor, σ, as follows. For $v \in V, w^* \in V^*$, let $\sigma(w^*, v) = w^*(T(v))$. This a $(1,1)$ tensor over V. Note that this is nothing more than a bilinear form on V, determined by the linear transformation, T. If $V = \Re^n$, then, assuming we have chosen a basis, we can represent T by a matrix, A, and we can identify V with V^* (this identification is only possible after a basis has been chosen). Hence, we simply think of w^* as its corresponding element $w \in V$, so we have $\sigma(w^*, v) = w^T A v$.*

The following theorem tells us that we can use a basis of V and its dual basis to construct a basis for V_s^r.

Theorem 15.4.1 Canonical Basis for Tensors on Vector Spaces

Let V be an m-dimensional vector space. Let $\{e_1, \ldots, e_m\}$ be a basis for V, and let $\{e^{*1}, \ldots, e^{*m}\}$ be its corresponding dual basis, so that $e^{*i}(e_j) = \delta^i_j$. Then the set

$$e_{i_1} \otimes \cdots e_{i_r} \otimes e^{*j_1} \otimes \cdots \otimes e^{*j_s}, \qquad 1 \leq i_1, \ldots, i_r, j_1, \ldots, j_s \leq m$$

forms a basis for V_s^r. In other words, the m^{r+s} possible tensor products, consisting of all possible permutations (with repetitions allowed) of size r from the set $\{e_1, \ldots, e_m\}$ and all possible permutations (with repetitions allowed) of size s from the set $\{e^{*1}, \ldots, e^{*m}\}$, form a basis for the set V_s^r.

Proof 15.4.1
Suppose there are scalars, $c^{i_1\cdots i_r}_{j_1\cdots j_s}$, such that

$$\sum_{\substack{i_1,\ldots,i_r,\\ j_1,\ldots,j_s=1}}^{m} c^{i_1\ldots i_r}_{j_1\ldots j_s} e_{i_1}\otimes\cdots e_{i_r}\otimes e^{*j_1}\otimes\cdots\otimes e^{*j_s} = 0_T,$$

where 0_T indicates the zero element of the space V^r_s. Then, for fixed indices

$$1\le k_1,\ldots,k_s,l_1,\ldots,l_r\le m$$

we can evaluate this tensor at $(e^{*l_1},\ldots,e^{*l_r},e_{k_1},\ldots,e_{k_s})$ to obtain

$$\sum_{\substack{i_1,\ldots,i_r,\\ j_1,\ldots,j_s=1}}^{m} c^{i_1\ldots i_r}_{j_1\ldots j_s} e_{i_1}\otimes\cdots e_{i_r}\otimes e^{*j_1}\otimes\cdots\otimes e^{*j_s}(e^{*l_1},\ldots,e^{*l_r},e_{k_1},\ldots,e_{k_s}) = 0.$$

But this implies

$$\begin{aligned} 0 &= \sum_{\substack{i_1,\ldots,i_r,\\ j_1,\ldots,j_s=1}}^{m} c^{i_1\ldots i_r}_{j_1\ldots j_s} e_{i_1}(e^{*l_1})\cdots e_{i_r}(e^{*l_r})e^{*j_1}(e_{k_1})\cdots e^{*j_s}(e_{k_s}) \\ &= \sum_{\substack{i_1,\ldots,i_r,\\ j_1,\ldots,j_s=1}}^{m} c^{i_1\ldots i_r}_{j_1\ldots j_s}\delta^{l_1}_{i_1}\cdots\delta^{l_r}_{i_r}\delta^{k_1}_{j_1}\cdots\delta^{k_s}_{j_s} = c^{l_1\ldots l_r}_{k_1\ldots k_s} \end{aligned}$$

Thus, evaluating this tensor at each of the elements, $e_{i_1}\otimes\cdots e_{i_r}\otimes e^{*j_1}\otimes\cdots\otimes e^{*j_s}$, in turn, shows that each of the scalars must be zero. So, these tensors are linearly independent.

Now, let σ be an arbitrary (r,s)-type tensor. Let $v^{*1},\ldots,v^{*r}$ be r dual vectors, and let $v_1,\ldots,v_s$ be s vectors. Each dual vector, v^{*i} can be expressed in terms of the dual basis as $v^{*i}=\sum_{k=1}^m c^i_k e^{*k}$ for some scalars c^i_k. Likewise, each vector v_i can be expressed in the form $v_i=\sum_{l=1}^m c^l_i e_l$ for scalars c^l_i. The multilinearity of σ then implies

$$\begin{aligned} &\sigma(v^{*1},\ldots,v^{*r},v_1,\ldots,v_s) \\ &= \sigma\left(\sum_{k_1=1}^m c^1_{k_1}e^{*k_1},\ldots,\sum_{k_r=1}^m c^r_{k_r}e^{*k_r},\sum_{l_1=1}^m c^{l_1}_1 e_{l_1},\ldots,\sum_{l_s=1}^m c^{l_s}_s e_{l_s}\right) \\ &= \sum_{\substack{k_1,\ldots,k_r\\ l_1,\ldots,l_s=1}}^{m} c^1_{k_1}\cdots c^r_{k_r}c^{l_1}_1\cdots c^{l_s}_s\sigma(e^{*k_1},\ldots,e^{*k_r},e_{l_1},\ldots,e_{l_s}) \end{aligned}$$

But, for $i=1,\ldots,r$, the scalar $c^i_{k_n}$, for $1\le k_n\le m$, is given by

$$e_{k_n}(v^{*i}) = e_{k_n}\left(\sum_{k_i=1}^m c^i_{k_i}e^{*k_i}\right) = \sum_{k_i=1}^m c^i_{k_i}e_{k_n}(e^{*k_i}) = \sum_{k_i=1}^m c^i_{k_i}\delta^{k_i}_{k_n} = c^i_{k_n}.$$

Similarly, for $i=1,\ldots,s$, the scalar $c^{l_n}_i$, for $1\le l_n\le m$, is given by $e^{*l_n}(v_i)$. Thus, we have

$$\sigma(v^{*1},\ldots,v^{*r},v_1,\ldots,v_s)$$

$$= \sum_{\substack{k_1,\ldots,k_r \\ l_1,\ldots,l_s=1}}^{m} \sigma(e^{*k_1},\ldots,e^{*k_r},e_{l_1},\ldots,e_{l_s})e_{k_1}(v^{*1})\cdots e_{k_r}(v^{*r})e^{*l_1}(v_1)\cdots e^{*l_s}(v_s)$$

$$= \sum_{\substack{k_1,\ldots,k_r \\ l_1,\ldots,l_s=1}}^{m} \sigma(e^{*k_1},...,e^{*k_r},e_{l_1},...,e_{l_s})e_{k_1}\otimes\cdots\otimes e_{k_r}\otimes e^{*l_1}\otimes\cdots\otimes e^{*l_s}(v^{*1},...,v^{*r},v_1,...,v_s)$$

Hence, these elements span V^r_s. Moreover, this shows that the dimension of V^r_s is m^{r+s}, and the components of a tensor, σ, with respect to this basis are the values obtained by σ acting on the basis elements. ■

How do the components of a tensor change with respect to a change in basis?

Theorem 15.4.2 Component Transformation for Tensors with a Change in Basis

Let σ be an (r,s)-type tensor over an m-dimensional vector space, V. Let $\{e_i\}_{i=1}^m$ and $\{f_i\}_{i=1}^m$ be two bases for V, and let $\{e^{*i}\}$ and $\{f^{*i}\}$ be their corresponding dual bases of V^*. Suppose σ has representations

$$\sigma = \sum_{\substack{i_1,\ldots,i_r \\ j_1,\ldots,j_s=1}}^{m} c^{i_1\cdots i_r}_{j_1\cdots j_s} e_{i_1}\otimes\cdots\otimes e_{i_r}\otimes e^{*j_1}\otimes\cdots\otimes e^{*j_s}$$

$$\sigma = \sum_{\substack{k_1,\ldots,k_r \\ l_1,\ldots,l_s=1}}^{m} d^{k_1\cdots k_r}_{l_1\cdots l_s} f_{k_1}\otimes\cdots\otimes f_{k_r}\otimes f^{*l_1}\otimes\cdots\otimes f^{*l_s}$$

with respect the two bases induced on V^r_s. Further, suppose that we have, for each $i = 1,\ldots,m$, $e_i = \sum_{k=1}^m \alpha_i^k f_k$ and $e^{*i} = \sum_{k=1}^m \beta_k^i f^{*k}$ for scalars $\{\alpha_i^k\}$ and $\{\beta_k^i\}$. Then the components of σ transform according to the formula

$$c^{i_1\cdots i_r}_{j_1\cdots j_s} = \sum_{\substack{k_1,\ldots,k_r \\ l_1,\ldots,l_s=1}}^{m} \beta^{i_1}_{k_1}\cdots\beta^{i_r}_{k_r}\alpha^{l_1}_{j_1}\cdots\alpha^{l_s}_{j_s} d^{k_1\cdots k_r}_{l_1\cdots l_s}.$$

Proof 15.4.2

The component $c^{i_1\cdots i_r}_{j_1\cdots j_s}$ is given by $\sigma(e^{*i_1},\ldots,e^{*i_r},e_{j_1},\ldots,e_{j_s})$, and the component $d^{k_1\cdots k_r}_{l_1\cdots l_s}$ is given by $\sigma(f^{*k_1},\ldots,f^{*k_r},f_{l_1},\ldots,f_{l_s})$. Thus, we see that

$$\begin{aligned} c^{i_1\cdots i_r}_{j_1\cdots j_s} &= \sigma\left(\sum_{k_1=1}^m \beta^{i_1}_{k_1} f^{*k_1},...,\sum_{k_r=1}^m \beta^{i_r}_{k_r} f^{*k_r},\sum_{l_1=1}^m \alpha^{l_1}_{j_1} f_{l_1},...,\sum_{l_s=1}^m \alpha^{l_s}_{j_s} f_{l_s}\right) \\ &= \sum_{\substack{k_1,\ldots,k_r \\ l_1,\ldots,l_s=1}}^{m} \beta^{i_1}_{k_1}\cdots\beta^{i_r}_{k_r}\alpha^{l_1}_{j_1}\cdots\alpha^{l_s}_{j_s}\sigma(f^{*k_1},...,f^{*k_r},f_{l_1},...,f_{l_s}) \end{aligned}$$

which implies that

$$c^{i_r\cdots i_r}_{j_1\cdots j_s} = \sum_{\substack{k_1,\ldots,k_r \\ l_1,\ldots,l_s=1}}^{m} \beta^{i_1}_{k_1}\cdots\beta^{i_r}_{k_r}\alpha^{l_1}_{j_1}\cdots\alpha^{l_s}_{j_s} d^{k_1\cdots k_r}_{l_1\cdots l_s}.$$

■

Now consider what these results look like when the underlying vector space is the tangent space, T_pM, at $p \in M$. Let (ϕ, U) be a coordinate system around p, and let $\{E_i\}$ and $\{dx^i\}$ denote the bases of T_pM and T_p^*M, respectively. Then, these bases induce a basis for $(T_pM)^r_s$ given by the set

$$\{E_{i_1} \otimes \cdots \otimes E_{i_r} \otimes dx^{j_1} \otimes \cdots \otimes dx^{j_s} \;:\; 1 \leq i_1, \ldots, i_r, j_1, \ldots, j_s \leq m\}.$$

Thus, an (r, s)-type tensor over $(T_pM)^r_s$ has representation

$$\sigma \;=\; \sum_{\substack{i_1,\ldots,i_r\\ j_1,\ldots,j_s=1}}^{m} c^{i_1\cdots i_r}_{j_1\cdots j_s} E_{i_1} \otimes \cdots \otimes E_{i_r} \otimes dx^{j_1} \otimes \cdots \otimes dx^{j_s},$$

where $c^{i_1\cdots i_r}_{j_1\cdots j_s} = \sigma(dx^{i_1}, \ldots, dx^{i_r}, dx^{j_1}, \ldots, dx^{j_s})$.

Even more interesting is the change of basis formula in this case. Let (ϕ, U) and (ψ, V) be two coordinate systems such that $p \in U \cap V$, and let us denote the coordinate functions of ϕ and ψ by $\{x^i\}$ and $\{y^i\}$, respectively. Let $\{E^x_i\}$ and $\{dx^i\}$ be the bases for T_pM and T^*_pM induced by ϕ, with $\{E^y_i\}$ and $\{dy^i\}$ the corresponding bases induced by ψ. Further, suppose, for each $i = 1, \ldots, m$, $E^x_i = \sum_{k=1}^m \alpha^k_i E^y_k$ and $dx^i = \sum_{k=1}^m \beta^i_k dy^k$ for scalars $\{\alpha^k_i\}$ and $\{\beta^i_k\}$. Then, if

$$\sigma \;=\; \sum_{\substack{i_1,\ldots,i_r\\ j_1,\ldots,j_s=1}}^{m} a^{i_1\cdots i_r}_{j_1\cdots j_s} E^x_{i_1} \otimes \cdots \otimes E^x_{i_r} \otimes dx^{j_1} \otimes \cdots \otimes dx^{j_s}$$

$$\sigma \;=\; \sum_{\substack{k_1,\ldots,k_r\\ l_1,\ldots,l_s=1}}^{m} b^{k_1\cdots k_r}_{l_1\cdots l_s} E^y_{k_1} \otimes \cdots \otimes E^y_{k_r} \otimes dx^{l_1} \otimes \cdots \otimes dx^{l_s}$$

are the representations of σ with respect to the coordinate systems (ϕ, U) and (ψ, V), respectively, we have

$$a^{i_r\cdots i_r}_{j_1\cdots j_s} \;=\; \sum_{\substack{k_1,\ldots,k_r\\ l_1,\ldots,l_s=1}}^{m} \beta^{i_1}_{k_1} \cdots \beta^{i_r}_{k_r} \alpha^{l_1}_{j_1} \cdots \alpha^{l_s}_{j_s} b^{k_1\cdots k_r}_{l_1\cdots l_s}.$$

But, for $1 \leq n \leq r$ and any $1 \leq i_n, k_n \leq m$, the scalar $\beta^{i_n}_{k_n}$ is given by $E^y_{k_n}(dx^{i_n})$. Likewise, for $1 \leq n \leq s$ and any $1 \leq j_n, l_n \leq m$, the scalar $\alpha^{l_n}_{j_n}$ is given by $dy^{l_n}(E^x_{j_n})$. Thus, the previous equality becomes

$$\begin{aligned} a^{i_r\cdots i_r}_{j_1\cdots j_s} &= \sum_{\substack{k_1,\ldots,k_r\\ l_1,\ldots,l_s=1}}^{m} E^y_{k_1}(dx^{i_1}) \cdots E^y_{k_r}(dx^{i_r}) dy^{l_1}(E^x_{j_1}) \cdots dy^{l_s}(E^x_{j_s}) b^{k_1\cdots k_r}_{l_1\cdots l_s} \\ &= \sum_{\substack{k_1,\ldots,k_r\\ l_1,\ldots,l_s=1}}^{m} \left.\frac{\partial x^{i_1}}{\partial y^{k_1}}\right|_{\psi(p)} \cdots \left.\frac{\partial x^{i_r}}{\partial y^{k_r}}\right|_{\psi(p)} \left.\frac{\partial y^{l_1}}{\partial x^{j_1}}\right|_{\phi(p)} \cdots \left.\frac{\partial y^{l_s}}{\partial x^{j_s}}\right|_{\phi(p)} b^{k_1\cdots k_r}_{l_1\cdots l_s}, \end{aligned}$$

where we have used the fact that the action of a cotangent vector, dy^{l_n}, on a tangent vector, $E^x_{j_n}$, is given by the pairing $\langle E^x_{j_n}, dy^{l_n}\rangle = E^x_{j_n}(dy^{l_n})$. In classical tensor analysis, this relation was used to define tensors. The definition was later refined, however, when it was discovered that tensors whose transformation formula was defined by the partial derivatives of the coordinate functions were simply special cases of a larger class of multilinear mappings.

Homework

For a vector space V of dimension n, we say a tensor ω in V_0^k is called alternating if for any $i \neq j$, we have

$$\omega(v_1, v_2, \ldots, \boldsymbol{v_i}, \ldots, \boldsymbol{v_j}, \ldots, v_k) = -\omega(v_1, v_2, \ldots, \boldsymbol{v_j}, \ldots, \boldsymbol{v_i}, \ldots, v_k)$$

Let the set of all alternating tensors in V_0^k be called $\Lambda^k(V)$. Recall the sign of a permutation σ, $sign\, \sigma$, is $+1$ if the permutation is even and -1 if the permutation is odd. The set of all permutations on k symbols is denoted S_k as usual. Define for $\omega \in V_0^k$

$$Alt(\omega)(v_1, \ldots, v_k) = \frac{1}{k!} \sum_{\sigma \in S_k} sign\, \sigma\, \omega(v_{\sigma(1)}, \ldots, v_{\sigma(k)})$$

Exercise 15.4.1 *Prove $\omega \in V_0^k$ implies $Alt(\omega) \in \Lambda^k(V)$.*

Exercise 15.4.2 *Prove $\eta \in \Lambda^k(V)$ implies $Alt(\eta) = \eta$.*

Exercise 15.4.3 *Prove $\omega \in V_0^k$ implies $Alt(Alt(\omega)) = Alt(\omega)$.*

Exercise 15.4.4 *Define the wedge product $\wedge$ on $\Lambda^k(V) \times \Lambda^\ell(V) \to \Lambda^{k+\ell}(V)$ by*

$$\omega \wedge \eta = \frac{(k+\ell)!}{k!\ell!} Alt(\omega \otimes \eta)$$

If $\{\boldsymbol{V_1}, \ldots, \boldsymbol{V_n}\}$ is a basis for V and $\{\boldsymbol{\Phi_1}, \ldots, \boldsymbol{\Phi_n}\}$ is the dual basis for V^* prove $\{\boldsymbol{\Phi_{i_1}} \wedge \ldots \wedge \boldsymbol{\Phi_{i_k}}\}$, where $1 \leq i_1 < i_2 < \ldots < i_k \leq n$ is a basis for $\Lambda^k(V)$. Further, prove $\Lambda^k(V)$ has dimension $\binom{n}{k}$.

Exercise 15.4.5 *From the previous exercise, the dimension of $\Lambda^n(V)$ is $\binom{n}{n} = 1$. Prove the determinant on $n \times n$ matrices is an alternating tensor in $\Lambda^n(V)$ where V is n-dimensional. So all alternating n-tensors on V are multiples of one nonzero one. Prove if $\{\boldsymbol{V_1}, \ldots, \boldsymbol{V_n}\}$ is a basis for V and $\omega \in \Lambda^n(V)$, then if $y_i = \sum_{i=1}^n y_{ij} \boldsymbol{V_j}$,*

$$\omega(y_1, \ldots, y_n) = det(y_{ij})\, \omega(\boldsymbol{V_1}, \ldots, \boldsymbol{V_n})$$

Hence, a nonzero $\omega \in \Lambda^n(V)$, splits the basis of V into two disjoint groups: those with $\omega(\boldsymbol{V_1}, \ldots, \boldsymbol{V_n}) > 0$ and those with $\omega(\boldsymbol{V_1}, \ldots, \boldsymbol{V_n}) < 0$. The determinant function is the unique ω with $\omega(\boldsymbol{V_1}, \ldots, \boldsymbol{V_n}) = 1$, of course.

15.5 Tensor Fields

We can extend the idea of a tensor over T_pM to a field of tensors over M. At each point, $p \in M$, this tensor field will induce an (r, s)-type tensor over T_pM, a real-valued mapping on r cotangent vectors and s tangent vectors at p. As p varies over M, we have a field of tensors acting on r fields of cotangent vectors and s fields of tangent vectors.

We already have a well-defined notion of a field of tangent vectors, or a vector field. To complete our definition of a tensor field, we need the definition of a cotangent vector field. This is actually quite a bit more abstract and you should think about these ideas carefully. All of this involves what is best described as a notational nightmare, of course. Each of our symbols is really a convention for composition of mappings so it is easy to get confused. Just keep at it!

Definition 15.5.1 Smooth Cotangent Vector or Covector Field

A smooth cotangent vector field, or covector field, ω, is an assignment of a cotangent vector, df_p, to each point $p \in M$, such that, for any smooth vector field, X, on M, the function $\omega(X) : M \to \Re$ defined by $\omega(X)(p) = df_p(X_p) = X_p(f)$, is C^∞ on M.

For notational purposes, we will denote a covector field by ω, and its cotangent vector at a particular point, p, by $\omega_p = df_p$. Note that, on a given coordinate system, (ϕ, U), a covector field, ω, can be represented in terms of the coordinate frame induced for the cotangent spaces at points $p \in U$. This coordinate frame is denoted $\{dx^i\}_{i=1}^m$, where x^i is the ith coordinate function of the mapping ϕ. That is, for any $i = 1, \ldots, m$, dx^i is the covector field on U such that, for any $p \in U$, we denote $dx^i(p)$ by dx^i_p. Hence, there are real-valued functions, β_i, on U such that $\omega|_U = \sum_i \beta_i dx^i$.

As with the tangent vectors, it is more elegant to construct the cotangent bundle, T^*M, and define a covector field as a mapping $\omega : M \to T^*M$. We will not construct the cotangent bundle here,as we are not interested in that level of generality. It is straightforward to show, analogous to the proof of Lemma 15.1.1, that a covector field, ω, is smooth if and only if its component functions on every coordinate neighborhood are smooth. We will leave that proof to you as an exercise. The following characterization of smoothness is like the one given in Definition 15.1.1 for vector fields.

Definition 15.5.2 Smooth Tensor Fields

A smooth, (r, s)-type tensor field, σ, on a smooth manifold, M, is an assignment of an (r, s)-type tensor, σ_p, to each point $p \in M$, such that for any smooth covector fields $\omega_1, \ldots, \omega_r$ and any smooth vector fields $X_1, \ldots, X_s$ the function $\sigma(\omega_1, \ldots, \omega_r, X_1, \ldots, X_s) : M \to \Re$ defined by

$$\sigma(\omega_1, \ldots, \omega_r, X_1, \ldots, X_s)(p) = \sigma_p(\omega_{1_p}, \ldots, \omega_{r_p}, X_{1_p}, \ldots, X_{s_p})$$

is C^∞. We call this collection of tensor fields M^r_s.

Now, let (ϕ, U) be a coordinate system on M. Let $\{E_i\}$ and $\{dx^i\}$ denote the coordinate frames for T_pM and T^*_pM, respectively, that are induced by this coordinate chart. Thus, notationally speaking, for any $i = 1, \ldots, m$ and any $q \in U$, the vector field, E_i, and covector field, dx^i, satisfy $E_i(q) = E_{q_i}$ and $dx^i(q) = dx^i_q$. Suppose that with respect to these coordinate frames, the covector fields ω_i have representation

$$\omega_i = \sum_{k=1}^m \beta^i_k dx^k \tag{15.1}$$

and the vector fields X_i have representation

$$X_j = \sum_{l=1}^m \alpha^l_j E_l. \tag{15.2}$$

Then, for all $1 \leq i \leq r$, $1 \leq j \leq s$, and $1 \leq k \leq m$, the functions β^i_k and α^k_j are smooth functions on U. Moreover, on the coordinate neighborhood, U, the expression $\sigma(\omega_1, \ldots, \omega_r, X_1, \ldots, X_s)$, for an (r, s)-type tensor field, σ, takes the form

$$\sigma\left(\sum_{k_1=1}^m \beta^1_{k_1} dx^{k_1}, \ldots, \sum_{k_r=1}^m \beta^r_{k_r} dx^{k_r}, \sum_{l_1=1}^m \alpha^{l_1}_1 E_{l_1}, \ldots, \sum_{l_s=1}^m \alpha^{l_s}_s E_{l_s}\right),$$

which yields, using the multilinearity of σ,

$$\begin{aligned}
&\sigma(\omega_1, \ldots, \omega_r, X_1, \ldots, X_s) \\
&= \sum_{\substack{k_1,\ldots,k_r \\ l_1,\ldots,l_s=1}}^{m} \beta^1_{k_1} \cdots \beta^r_{k_r} \alpha^{l_1}_1 \cdots \alpha^{l_s}_s \sigma(dx^{k_1}, \ldots, dx^{k_r}, E_{l_1}, \ldots, E_{l_s}) \\
&= \sum_{\substack{k_1,\ldots,k_r \\ l_1,\ldots,l_s=1}}^{m} E_{k_1}(\omega_1) \cdots E_{k_r}(\omega_r) dx^{l_1}(X_1) \cdots dx^{l_s}(X_s) \sigma(dx^{k_1}, \ldots, dx^{k_r}, E_{l_1}, \ldots, E_{l_s}) \\
&= \sum_{\substack{k_1,\ldots,k_r \\ l_1,\ldots,l_s=1}}^{m} \sigma(dx^{k_1}, ..., dx^{k_r}, E_{l_1}, ..., E_{l_s}) E_{k_1} \otimes \cdots \otimes E_{k_r} \otimes \\
&dx^{l_1} \otimes \cdots \otimes dx^{l_s}(\omega_1, ..., \omega_r, X_1, ..., X_s)
\end{aligned}$$

This shows that the (r, s)-type tensor fields

$$E_{k_1} \otimes \cdots \otimes E_{k_r} \otimes dx^{l_1} \otimes \cdots \otimes dx^{l_s}$$

span the set of (r, s)-type tensor fields on the coordinate neighborhood U. Thus, for any (r, s)-type tensor field, σ, on U,

$$\sigma = \sum_{\substack{k_1,\ldots,k_r \\ l_1,\ldots,l_s=1}}^{m} \sigma(dx^{k_1}, ..., dx^{k_r}, E_{l_1}, ..., E_{l_s}) E_{k_1} \otimes \cdots \otimes E_{k_r} \otimes dx^{l_1} \otimes \cdots \otimes dx^{l_s}.$$

Now, if σ is smooth according to Definition 15.5.2, then $\sigma|_U$ is smooth on this coordinate neighborhood. Hence, the local representation given above is smooth. If we evaluate σ at the covector fields $dx^{i_1}, \ldots, dx^{i_r}$ and the vector fields $E_{j_1}, \ldots, E_{j_s}$, all of which are smooth, we simply obtain the component of σ given by

$$\sigma(dx^{i_1}, \ldots, dx^{i_r}, E_{j_1}, \ldots, E_{j_s}),$$

and this must be a C^∞ real-valued function on U, in the sense that the mapping

$$p \mapsto \sigma_p(dx^{i_1}_p, \ldots, dx^{i_r}_p, E_{p_{j_1}}, \ldots, E_{p_{j_s}})$$

for $p \in U$, is smooth. Applying this to each component in turn shows that if σ is smooth according to the definition, then each of the component functions on an arbitrary coordinate neighborhood is also smooth. Conversely, suppose the local coordinate representations of σ on any coordinate neighborhood are smooth. Let $\omega_1, \ldots, \omega_r$ and $X_1, \ldots, X_s$ be smooth covector and vector fields, respectively, on M. Then, we obtain smooth covector and vector fields by restricting these to U. Suppose the restrictions of these covector and vector fields have local coordinate representations given by Equation 15.1 and Equation 15.2. It follows that

$$\begin{aligned}
&\sigma|_U(\omega_1|_U, \ldots, \omega_r|_U, X_1|_U, \ldots, X_s|_U) \\
&= \sigma|_U\left(\sum_{k_1=1}^{m} \beta^1_{k_1} dx^{k_1}, ..., \sum_{k_r=1}^{m} \beta^r_{k_r} dx^{k_r}, \sum_{l_1=1}^{m} \alpha^{l_1}_1 E_{l_1}, \ldots, \sum_{l_s=1}^{m} \alpha^{l_1}_s E_{l_s}\right) \\
&= \sum_{\substack{k_1,\ldots,k_r \\ l_1,\ldots,l_s=1}}^{m} \beta^1_{k_1} \cdots \beta^r_{k_r} \alpha^{l_1}_1 \cdots \alpha^{l_s}_s \sigma|_U(dx^{k_1}, ..., dx^{k_r}, E_{l_1}, ..., E_{l_s}).
\end{aligned}$$

But the functions $\sigma|_U(dx^{k_1}, \ldots, dx^{k_r}, E_{l_1}, \ldots, E_{l_s})$, for any indices $1 \leq k_1,...,k_r, l_1, ..., l_s \leq m$, are smooth on U by hypothesis, as are the functions $\beta^i_{k_i}$ and $\alpha^{l_j}_j$ for any $1 \leq i \leq r$ and $1 \leq j \leq s$. Hence, the function $\sigma|_U$ is smooth on U. Since the chart (ϕ, U) was arbitrary, it follows that σ is smooth on a collection of open sets covering M. Thus, σ is smooth on M, and we have shown the following.

Theorem 15.5.1 Smooth Tensor Fields and Local Coordinate Chart Representations

Let σ be an (r, s)-type tensor field on a smooth manifold, M. Then σ is smooth if and only if for any coordinate chart, (ϕ, U), on M, the local coordinate representation of σ on U is given by

$$\sigma|_U = \sum_{\substack{k_1,...,k_r \\ l_1,...,l_s}}^{m} \sigma|_U(dx^{k_1}, ..., dx^{k_r}, E_{l_1}, ..., E_{l_s}) E_{k_1} \otimes \cdots \otimes E_{k_r} \otimes dx^{l_1} \otimes \cdots \otimes dx^{l_s}$$

where the component functions $\sigma|_U(dx^{k_1}, \ldots, dx^{k_r}, E_{l_1}, \ldots, E_{l_s})$ are smooth functions on U.

Proof 15.5.1
This was proven in the discussions above. ■

Homework

For a manifold M of dimension n, we say a tensor ω in the $(k, 0)$ tensor field on M, M^k_0 is called alternating if at each point $p \in M$, the $(k, 0)$ tensor $\omega(p)$ satisfies for any $i \neq j$, we have

$$\omega(p)(v_1, v_2, \ldots, \boldsymbol{v_i},\ ldots, \boldsymbol{v_j}, \ldots, v_k) = -\omega(p)(v_1, v_2, \ldots, \boldsymbol{v_j},\ ldots, \boldsymbol{v_i}, \ldots, v_k)$$

where the coordinates v_i come from the tangent space T_p. Let the set of all alternating tensors in M^k_0 be called $\Lambda^k(M)$. Define for $\omega(p) \in M^k_0$

$$Alt(\omega)(v_1, \ldots, v_k) = \frac{1}{k!} \sum_{\sigma \in S_k} sign\, \sigma\, \omega(p)(v_{\sigma(1)}, \ldots, v_{\sigma(k)})$$

Exercise 15.5.1 *Prove $\omega \in M^k_0$ implies $Alt(\omega) \in \Lambda^k(M)$.*

Exercise 15.5.2 *Prove $\eta \in \Lambda^k(M)$ implies $Alt(\eta) = \eta$.*

Exercise 15.5.3 *Prove $\omega \in M^k_0$ implies $Alt(Alt(\omega)) = Alt(\omega)$.*

Exercise 15.5.4 *Define the wedge product $\wedge$ on $\Lambda^k(M) \times \Lambda^\ell(M) \to \Lambda^{k+\ell}(M)$ by*

$$\omega \wedge \eta = \frac{(k+\ell)!}{k!\ell!} Alt(\omega \otimes \eta)$$

If $\{\boldsymbol{V_1}, \ldots, \boldsymbol{V_n}\}$ is a basis for T_pM and $\{\boldsymbol{\Phi_1}, \ldots, \boldsymbol{\Phi_n}\}$ is the dual basis for T^*_p prove $\{\boldsymbol{\Phi_{i_1}} \wedge \ldots \wedge \boldsymbol{\Phi_{i_k}}\}$, where $1 \leq i_1 < i_2 < \ldots < i_k \leq n$ is a basis for $\Lambda^k(M)$. Further, prove $\Lambda^k(M)$ has dimension $\binom{n}{k}$.

Exercise 15.5.5 *Following the ideas in the proof of Lemma 15.1.1,show that a covector field, ω, is smooth if and only if its component functions on every coordinate neighborhood are smooth.*

15.6 Metric Tensors

In order to study geometrical properties of a space or a phenomena within a space, we usually need some means of measurement. Such a construction is not intrinsic to the space itself. There are usually many means of taking measurements within any particular space. It is our job in modeling to try to find a measurement protocol that is useful for our needs. Hence, we simply want to know that such a means does exist and that there is some kind of a metric structure on the space. The construction of this metric structure requires a specific type of tensor, namely a $(0, 2)$-type tensor, or a second order covariant tensor. At a point, p, on a manifold, M, such a tensor will be a real-valued, bilinear operator on T_pM.
The motivation for this is not difficult to see. For example, consider the problem of measuring the length of a curve in three dimensional space. If the curve is given by $\gamma(t) = (x^1(t), x^2(t), x^3(t))$, for $t \in (a, b)$, then the arc length, L, is defined to be

$$L = \int_a^b \sqrt{\left(\frac{dx^1}{dt}\right)^2 + \left(\frac{dx^2}{dt}\right)^2 + \left(\frac{dx^3}{dt}\right)^2}\, dt = \int_a^b \sqrt{\langle \dot{\gamma}(t), \dot{\gamma}(t)\rangle} dt.$$

Thus, our classical definition of arc length requires the notion of an inner product on $\Re^3$, which, as we have seen, is just a $(0, 2)$-type tensor on $\Re^3$. Other classical geometrical quantities, such as surface area, curvature, etc., can also be formulated using inner products. Hence, it makes sense to base our construction of a means of measurement on these particular types of tensors. Of course, an inner product, in addition to being bilinear, is symmetric and positive definite. So, we will require a bit more of our second order covariant tensors than just their bilinearity.

Definition 15.6.1 Symmetric Second Order Covariant Tensors

A second order covariant tensor, ρ, over a vector space, V, is said to be symmetric if, for every $v_1, v_2 \in V$, we have $\rho(v_1, v_2) = \rho(v_2, v_1)$. If σ is a second order covariant tensor field over a smooth manifold, M, we say σ is symmetric if, for every point, $p \in M$, the tensor σ_p is symmetric over T_pM.

We can now define the metric tensor.

Definition 15.6.2 Metric Tensor

A metric tensor, G, on a smooth manifold, M, is a smooth, symmetric, second order covariant tensor field over M. At a point, $p \in M$, we will denote the $(0, 2)$-type tensor induced by this field by G_p.

Note that we have not required G to be positive definite. This restriction will come later. Also, we should technically refer to G as a metric tensor field, but this is to be implied, as we want G to be defined on each tangent space of M. It is also common to abuse these ideas even more by referring to this tensor field as a metric. We do not resort to this, as we show in a bit, the metric tensor is used to define an actual metric on the manifold. Throughout the remainder of this chapter, we will derive consequences of this definition, assuming that such a tensor field exists. This is not at all obvious, but we will do this later.

Intuitively, a metric tensor just defines a symmetric bilinear form on each tangent space of M, such that the transition from one point to another is smooth. Consider the local coordinate representation of G on a coordinate chart (ϕ, U). As usual, let $\{E_i\}$ and $\{dx^i\}$ denote the coordinate frames for the tangent and cotangent spaces, respectively, on U. Then, from our general results, we can express the metric tensor $G|_U$ in the form

$$G = \sum_{i,j=1}^{m} g_{ij} dx^i \otimes dx^j,$$

where each g_{ij} is a smooth real-valued function on U. For any point, $p \in U$, this actually defines a bilinear form in the classical sense. If X_1 and X_2 are smooth vector fields on M, then they have local coordinate representations on U given by

$$X_1 = \sum_{k=1}^{m} \alpha_1^k E_k, \quad X_2 = \sum_{l=2}^{m} \alpha_2^l E_l.$$

For any point, $p \in U$, we have

$$\begin{aligned}
G_p(X_{1_p}, X_{2_p}) &= \sum_{i,j=1}^{m} g_{ij}(p) dx_p^i \otimes dx_p^j \left(\sum_{k=1}^{m} \alpha_1^k(p) E_{p_k}, \sum_{l=1}^{m} \alpha_2^l(p) E_{p_l} \right) \\
&= \sum_{i,j=1}^{m} g_{ij}(p) dx_p^i \left(\sum_{k=1}^{m} \alpha_1^k(p) E_{p_k} \right) dx_p^j \left(\sum_{l=1}^{m} \alpha_2^l(p) E_{p_l} \right) \\
&= \sum_{i,j=1}^{m} g_{ij}(p) \left(\sum_{k=1}^{m} \alpha_1^k(p) dx_p^i(E_{p_k}) \right) \left(\sum_{l=1}^{m} \alpha_2^l(p) dx_p^j(E_{p_l}) \right) \\
&= \sum_{i,j=1}^{m} g_{ij}(p) \sum_{k,l=1}^{m} \alpha_1^k(p) \alpha_2^l(p) \delta_k^i \delta_l^j = \sum_{i,j=1}^{m} g_{ij}(p) \alpha_1^i(p) \alpha_2^j(p)
\end{aligned}$$

Let $[G_p]$ be the $m \times m$ matrix whose entries are defined by $[G_p]_{ij} = g_{ij}(p)$, and let $v(p)$ and $w(p)$ be the $m \times 1$ vectors whose components are given, respectively, by $(v(p))_j = \alpha_2^j$ and $(w(p))_i = \alpha_1^i$. Then we have

$$G_p(X_{1_p}, X_{2_p}) = v(p)^T [G_p] w(p).$$

Letting p vary over U, we have the local coordinate representation of G on U, given in this matrix form by

$$G|_U(X_1|_U, X_2|_U) = \sum_{i,j=1}^{m} g_{ij} \alpha_1^i \alpha_2^j = v^T [G] w$$

where X_1 and X_2 are arbitrary smooth vector fields on M with local representations given by relations (6) and (7), $[G]$ is the matrix whose entries are the smooth functions $g_{ij} : U \to \Re$, and v and w are the vector-valued functions on U whose images at a point $p \in U$ are the components of X_{1_p} and X_{2_p} respectively. We now wish to construct a particular type of metric tensor that is positive definite at each point in the following sense.

Definition 15.6.3 Positive Definite Second Order Covariant Tensors

A second order covariant tensor, ρ, over a vector space, V, is said to be positive definite over V if $\rho(v, v) \geq 0$ for all $v \in V$, and equality holds if and only if $v = 0$. A metric tensor, G, on a smooth manifold is said to be positive definite if, for each point, $p \in M$, the tensor G_p is positive definite over T_pM.

This brings us to the definition that lies at the foundation of differential geometry.

Definition 15.6.4 Riemannian Manifolds

Let M be a smooth manifold. A metric tensor, G, on M is a Riemannian Metric Tensor if it is positive definite. If this is the case, we call M a Riemannian manifold.

Recalling that the tangent space of $\Re^m$ is just $\Re^m$, itself, we see right away that there is a Riemannian metric tensor on the manifold $M = \Re^m$. There is only one coordinate chart, namely the identity map $\mathbf{id} : \Re^{\mathbf{m}} \rightarrow \Re^{\mathbf{m}}$, so we simply define G by $G_p(x, y) = \langle x, y \rangle$, where p is any point in $\Re^m$ and x and y are any two vectors in $\Re^m$. Thus, on manifolds that are globally Euclidean, the typical Riemannian metric tensor is simply the usual inner product on the space.

We now prove every smooth manifold has a metric tensor.

Theorem 15.6.1 Every Smooth Manifold Admits a Riemannian Metric Tensor

Every smooth manifold, M, admits a Riemannian metric tensor.

Proof 15.6.1
We refer you to any of a wide variety of books that prove this result such as (M. Spivak (82) 1979), (M. Spivak (83) 1979) as well as (Wilkins (120) 2005) which is particularly nice as it tries to be as elementary as possible. ■

Part of the proof above requires transferring, locally at least, the structure of the Euclidean metric tensor, $\langle , \rangle$, to M. That is, we will use the coordinate charts, (ϕ, U), to transfer the Euclidean metric tensor on $\phi(U)$ to $U \subset M$. To facilitate this, we will need the following results.

Definition 15.6.5 Immersions

Let $F : M \rightarrow N$ be a smooth map between smooth manifolds. We say that F is an immersion if for each $p \in M$, the rank of F at p equals m, the dimension of M.

The rank of a mapping $F : M \rightarrow N$ at $p \in M$ is defined as one would expect from classical analysis. It is the rank of the Jacobian matrix of the coordinate representation $\psi \circ F \circ \phi^{-1}$, for coordinate systems (ψ, V) and (ϕ, U) around $F(p)$ and p, respectively.

Theorem 15.6.2 Pulling Back Riemannian Tensors

Let $F : M \rightarrow N$ be an immersion, and suppose G is a Riemannian metric tensor on N. Define a tensor field, F^*G, on M by $(F^*G)_p(X_{1_p}, X_{2_p}) = G_{F(p)}(D_pF(X_{1_p}), D_pF(X_{2_p}))$, for each $p \in M$ and any smooth vector fields, X_1 and X_2 on M. Then F^*G is a Riemannian metric tensor on M.

Proof 15.6.2
We first show that F^*G is smooth. We will do this by using the local coordinate characterization. Let p be any point in M, and let $q = F(p)$. Let (ϕ, U) and (ψ, V) be coordinate systems around p and q, respectively. Denote the coordinate functions of ϕ and ψ by x^i and y^i respectively. On V, we can represent G locally in the form

$$G = \sum_{i,j=1}^{m} g_{ij} dy^i \otimes dy^j .$$

Let X_1 and X_2 be smooth vector fields on M, and consider their representations on U, given by relations (6) and (7) above. Then

$$
\begin{aligned}
(F^*G)_p(X_{1_p}, X_{2_p}) &= G_q(D_pF(X_{1_p}), D_pF(X_{2_p})) \\
&= G_q\left(D_pF\left(\sum_{k=1}^{m} \alpha_1^k(p)E_{p_k}^x\right), D_pF\left(\sum_{l=1}^{m} \alpha_2^l(p)E_{p_l}^x\right)\right) \\
&= G_q\left(\sum_{k=1}^{m} \alpha_1^k(p)D_pF(E_{p_k}^x), \sum_{l=1}^{m} \alpha_2^l(p)D_pF(E_{p_l}^x)\right) \\
&= \sum_{k,l=1}^{m} \alpha_1^k(p)\alpha_2^l(p)G_q\big(D_pF(E_{p_k}^x), D_pF(E_{p_l}^x)\big) \\
&= \sum_{k,l=1}^{m} \alpha_1^k(p)\alpha_2^l(p)G_q\left(\sum_{i=1}^{n} \frac{\partial y^i}{\partial x^k}\bigg|_{\phi(p)} E_{q_i}^y, \sum_{j=1}^{n} \frac{\partial y^j}{\partial x^l}\bigg|_{\phi(p)} E_{q_j}^y\right) \\
&= \sum_{k,l=1}^{m} \alpha_1^k(p)\alpha_2^l(p) \sum_{i,j=1}^{n} \frac{\partial y^i}{\partial x^k}\bigg|_{\phi(p)} \frac{\partial y^j}{\partial x^l}\bigg|_{\phi(p)} G_q(E_{q_i}^y, E_{q_j}^y) \\
&= \sum_{k,l=1}^{m} \sum_{i,j=1}^{n} \frac{\partial y^i}{\partial x^k}\bigg|_{\phi(p)} \frac{\partial y^j}{\partial x^l}\bigg|_{\phi(p)} G_q(E_{q_i}^y, E_{q_j}^y) dx_p^k \otimes dx_p^l(X_{1_p}, X_{2_p})
\end{aligned}
$$

But this actually holds for any point in U, so we have

$$
(F^*G)|_U(X_1|_U, X_2|_U) = \sum_{k,l=1}^{m} \sum_{i,j=1}^{n} \frac{\partial y^i}{\partial x^k} \frac{\partial y^j}{\partial x^l} G(E_i^y, E_j^y) dx^k \otimes dx^l(X_1|_U, X_2|_U)
$$

where the partial derivatives are to be evaluated at the point $\phi(p)$ in question. This shows that the component functions of F^*G on U are given by

$$
\sum_{i,j=1}^{n} \frac{\partial y^i}{\partial x^k} \frac{\partial y^j}{\partial x^l} G(E_i^y, E_j^y).
$$

Since F is a smooth mapping, the mappings $p \mapsto \frac{\partial y^i}{\partial x^k}|_{\phi(p)}$, for all $i = 1, \ldots, n$ and $j = 1, \ldots, m$ are smooth. Moreover, since G is smooth by hypothesis, its components, $G(E_i^y, E_j^y)$, are smooth on V, implying that the mappings $p \mapsto G_q(E_{q_i}^y, E_{q_j}^y)$, for $p \in U$ and $q = F(p)$, are smooth. Hence, the components of F^*G are smooth on U. The chart, (ϕ, U), was arbitrary, so, by Theorem 5.3, we can conclude that F^*G is smooth on M.

To see that F^*G is symmetric, let p be any point in M, and consider $(F^*G)_p$. If X_p and X'_p are any two tangent vectors at p, then the symmetry of G implies

$$
\begin{aligned}
(F^*G)_p(X_p, X'_p) &= G_{F(p)}\big(D_pF(X_p), D_pF(X'_p)\big) = G_{F(p)}\big(D_pF(X'_p), D_pF(X_p)\big) \\
&= (F^*G)_p(X'_p, X_p).
\end{aligned}
$$

Finally, to see that F^*G is positive definite, let p be any point in M, and let X_p be any tangent vector at p. Then

$$
(F^*G)_p(X_p, X_p) = G_{F(p)}\big(D_pF(X_p), D_pF(X_p)\big).
$$

But G is positive definite, so $(F^*G)_p(X_p, X_p) \geq 0$ for all $X_p \in T_pM$. Moreover, if $X_p = 0$, then

$D_pF(X_p) = 0 \Rightarrow G_{F(p)}(D_pF(X_p), D_pF(X_p)) = 0 \Rightarrow (F^*G)(X_p, X_p) = 0$. Conversely, suppose $(F^*G)_p(X_p, X_p) = 0$. Then, since G is positive definite, we have $G_{F(p)}(D_pF(X_p), D_pF(X_p)) = 0 \Rightarrow D_pF(X_p) = 0$. But $F : M \to N$ is an immersion, so the rank of F at p is m. This implies that the dimension of $D_pF(T_pM)$ is m. Hence, D_pF is injective, and it follows that $D_pF(X_p) = 0 \Rightarrow X_p = 0$. ■

Like vector fields, tensor fields over a smooth manifold are more naturally described in the context of vector bundles. We know that the set of (r, s)-type tensors over a finite dimensional vector space is, itself, a vector space of finite dimension. So, as we used the tangent space at each point $p \in M$ as the typical fiber of the tangent bundle, we can take the set of all (r, s)-type tensors over T_pM, for each $p \in M$, to be the typical fiber of another vector bundle, called the (r, s)-type tensor bundle. This vector bundle is usually denoted T^r_s. Using a proof that is not much different from the tangent bundle case, one can show that T^r_s is a smooth manifold of dimension m^{r+s}. A smooth (r, s)-type tensor field, σ, is then defined to be a smooth map $\sigma : M \to T^s_r$. It is simply a map that defines an (r, s)-type tensor at each point $p \in M$ such that the transition between points p and q in M is smooth. We have, of course, not followed this method of construction, as the necessary excursion into differential topology would take us too far off course. All of the material in the last three chapters can be generalized considerably and, if you are interested, you might want to check out (S. Chern and W. Chen and K. Lam (107) 1999) and (N. Steenrod (91) 1999). Their approach is much more theoretical and abstract than the one we have used in our discussions. But as your mathematical sophistication grows, these are good sources for further study.

Theorem 15.6.2 is used in most applications to induce metric tensors on manifolds. If $F : M \to N$ is a smooth mapping between smooth manifolds, and if F is an immersion, then we say that $F(M)$ is an immersed submanifold of N. Note that an immersion must be locally injective since the rank of F at any point $p \in M$ is m. This implies the differential DF at each point p is an injective map between T_pM and $T_{F(p)}N$. This requires that m be less than or equal to n. So, the idea of immersing a manifold, M, into another manifold, N, is similar to the notion of embedding one linear space into another. However, even if F is an immersion, it need not be globally injective. A smooth map $F : M \to N$ such that F is an immersion and F is a homeomorphism is called an embedding and we say that $F(M)$ is an embedded submanifold of N.

Most manifolds that occur in applications are embedded or immersed submanifolds of another manifold which usually has a simpler structure. Given a smooth manifold, M, we may be able to embed or immerse M into a Euclidean space, $\Re^n$. Using Theorem 15.6.2, this allows us to induce a Riemannian metric tensor on M by transferring the Euclidean metric structure of $\Re^n$ to M.

Example 15.6.1 Parameterized Surfaces in $\Re^3$:

If M and N are smooth manifolds, we say that the map $F : \Omega \subset M \to N$ is a parameterization of $\Omega \subset M$ in N if Ω is an open submanifold of M, F is a diffeomorphism, and $F(\Omega)$ is an open submanifold of N. The most common examples of parameterizations are mappings from two-dimensional manifolds into $\Re^3$. Let M be a smooth two-dimensional manifold, and let (ϕ, U) be a coordinate chart on M. Denote the coordinate functions of ϕ by u^1 and u^2, and suppose $r : U \to \Re^3$ is a parameterization. The Euclidean metric tensor on $\Re^3$ is just the usual inner product, which, in tensor notation is just $G = \sum_{i=1}^3 dx^i \otimes dx^i$. If X and Y are two smooth vector fields on M, and we consider their restrictions to U, then, applying Theorem 15.6.2, we have a Riemannian metric tensor, r^*G, on U defined for $p \in U$ by

$$(r^*G)_p(X_p, Y_p) \quad = \quad G_{r(p)}(D_pr(X_p), D_pr(Y_p)). \tag{15.3}$$

Now, if we think of r as a function of its local coordinates on U, we can represent r by $r(u^1, u^2) = (x^1(u^1, u^2), x^2(u^1, u^2), x^3(u^1, u^2))$. This is nothing more than the local coordinate representation *of r at any point $p \in U$. Thus, the differential of r at $p \in U$ takes the form*

$$Dr(u^1, u^2) = \begin{bmatrix} D_1x^1 & D_2x^1 \\ D_1x^2 & D_2x^2 \\ D_1x^3 & D_2x^3 \end{bmatrix},$$

where we have used the shorthand notation D_ix^j for the partial derivative $\frac{\partial x^j}{\partial u^i}$. Since, for any $p \in U$, the basis vectors for the tangent space, T_pM, are just the partial derivative operators with respect to the coordinate functions u^1 and u^2, let us denote the basis field induced on U by the more intuitive form

$$\left\{\frac{\partial}{\partial u^1}, \frac{\partial}{\partial u^2}\right\}.$$

Likewise, we can represent the basis of $T_{r(p)}\Re^3$ by

$$\left\{\frac{\partial}{\partial x^1}, \frac{\partial}{\partial x^2}, \frac{\partial}{\partial x^3}\right\}.$$

Then we can represent $X|_U$ and $Y|_U$ by

$$\begin{aligned} X &= \alpha^1\frac{\partial}{\partial u^1} + \alpha^2\frac{\partial}{\partial u^2} \\ Y &= \beta^1\frac{\partial}{\partial u^1} + \beta^2\frac{\partial}{\partial u^2}. \end{aligned}$$

Then, at any point $p \in U$, Equation 15.3 takes the form

$$\begin{aligned} (r^*G)(X_p, Y_p) &= G_{r(p)}\big(D_pr(X_p), D_pr(Y_p)\big) \\ &= \sum_{k=1}^{3} dx^k \otimes dx^k\big(D_pr(X_p), D_pr(Y_p)\big) \\ &= \sum_{k=1}^{3} dx^k\left(D_pr\left(\sum_{i=1}^{2}\alpha^i\frac{\partial}{\partial u^i}\right)\right)dx^k\left(D_pr\left(\sum_{j=1}^{2}\beta^j\frac{\partial}{\partial u^j}\right)\right) \\ &= \sum_{k=1}^{3}\sum_{i=1}^{2}\alpha^i dx^k\left(D_pr\left(\frac{\partial}{\partial u^i}\right)\right)\sum_{j=1}^{2}\beta^j dx^k\left(D_pr\left(\frac{\partial}{\partial u^j}\right)\right) \\ &= \sum_{k=1}^{3}\sum_{i,j=1}^{2}\alpha^i\beta^j dx^k\left(\sum_{s=1}^{3}\frac{\partial x^s}{\partial u^i}\frac{\partial}{\partial x^s}\right)dx^k\left(\sum_{t=1}^{3}\frac{\partial x^t}{\partial u^j}\frac{\partial}{\partial x^t}\right) \\ &= \sum_{i,j=1}^{2}\sum_{k=1}^{3}\alpha^i\beta^j\frac{\partial x^k}{\partial u^i}\frac{\partial x^k}{\partial u^j}. \end{aligned}$$

But, α^i and β^j are just $du^i(X_p)$ and $du^j(Y_p)$, respectively, so we have

$$(r^*G)(X_p, Y_p) = \sum_{i,j=1}^{2}\sum_{k=1}^{3}\frac{\partial x^k}{\partial u^i}\frac{\partial x^k}{\partial u^j}du^i \otimes du^j(X_p, Y_p),$$

or, in operator form,

$$r^*G = \sum_{i,j=1}^{2} \left(\sum_{k=1}^{3} \frac{\partial x^k}{\partial u^i} \frac{\partial x^k}{\partial u^j} \right) du^i \otimes du^j .$$

Note, however, that, for fixed i *and* j*, the coefficient of* $du^i \otimes du^j$*, given by*

$$\sum_{k=1}^{3} \frac{\partial x^k}{\partial u^i} \frac{\partial x^k}{\partial u^j},$$

is just the usual dot product of the ith *and* j^{th} *columns of* $D_p r$*. This differential is just a matrix of real numbers, namely the values of the partial derivatives of the coordinate functions* x^i *with respect to the coordinates* u^j *at* $\phi(p)$*. Hence, it makes sense to use the term dot product, if we think of the columns of the matrix as vectors in* $\Re^3$*. If we denote the differential,* $D_p r$*, by the abbreviated form* $D_p r = [r_1 \; r_2]$*, where* r_i *denotes the column vector whose entries are* $D_i x^j$*, for* $j = 1, 2, 3$*, then we can rewrite as*

$$r^*G = \sum_{i,j=1}^{2} \langle r_i, r_j \rangle du^i \otimes du^j,$$

which is an expression of what is called the **first fundamental form** *of a surface. In classical differential geometry, the first fundamental form of a surface is just the inner product that is defined on the tangent plane at each point. This is nothing more than the metric tensor field over* M*. In other words, the Riemannian metric tensor on a two dimensional manifold (i.e. a surface) embedded into* $\Re^3$ *is the same thing as the first fundamental form of the surface. However, we have not discussed classical differential geometry at all! As you can see, there is always more to read!*

Example 15.6.2 *The Metric Tensor is Spherical Coordinates:*

The previous example allows us to construct Riemannian metric tensors on all of the classical surfaces in differential geometry. Consider S^2*, the two dimensional sphere, as a surface embedded in* $\Re^3$*, and consider the family of charts on* S^2 *given by the following.*

1. *Let* $U_1 = \{(x, y, z) \in S^2 \; : \; x > 0\}$*. Define* $\psi_1 : U_1 \rightarrow (-\frac{\pi}{2}, \frac{\pi}{2}) \times (0, \pi)$ *by*

   $$\psi_1(x, y, z) = (\tan^{-1}(\frac{y}{x}), \cos^{-1} z) \text{ with } \psi_1^{-1}(\theta, \phi) = (\sin\phi\cos\theta, \sin\phi\sin\theta, \cos\phi)$$

2. *Let* $U_2 = \{(x, y, z) \in S^2 \; : \; x < 0\}$*. Define* $\psi_2 : U_2 \rightarrow (\frac{\pi}{2}, \frac{3\pi}{2}) \times (0, \pi)$ *by*

   $$\psi_2(x, y, z) = (\pi + \tan^{-1}(\frac{y}{x}), \cos^{-1} z) \text{ with } \psi_2^{-1}(\theta, \phi) = (\sin\phi\cos\theta, \sin\phi\sin\theta, \cos\phi)$$

3. *Let* $U_3 = \{(x, y, z) \in S^2 \; : \; y > 0\}$*. Define* $\psi_3 : U_3 \rightarrow (0, \pi) \times (0, \pi)$ *by*

   $$\psi_3(x, y, z) = (\cot^{-1}(\frac{x}{y}), \cos^{-1} z) \text{ with } \psi_3^{-1}(\theta, \phi) = (\sin\phi\cos\theta, \sin\phi\sin\theta, \cos\phi)$$

4. *Let* $U_4 = \{(x, y, z) \in S^2 \; : \; y < 0\}$*. Define* $\psi_4 : U_4 \rightarrow (\pi, 2\pi) \times (0, \pi)$ *by*

   $$\psi_4(x, y, z) = (\pi + \cot^{-1}(\frac{x}{y}), \cos^{-1} z) \text{ with } \psi_4^{-1}(\theta, \phi) = (\sin\phi\cos\theta, \sin\phi\sin\theta, \cos\phi)$$

These charts do not cover S^2, as the north and south poles are not in any of the sets U_i. The poles, however, are degenerate points in the standard spherical coordinate system, since they are not uniquely defined in terms of the parameters θ and ϕ. Nevertheless, these charts are C^∞ compatible with all other admissible charts one can define on S^2. Moreover, these charts uniquely associate to every point in $S^2 - \{(0,0,1),(0,0,-1)\}$ a point in $[0,2\pi) \times (0,\pi)$. Consequently, we can consider the mapping $s : (0,2\pi) \times (0,\pi) \to \Re^3$ defined by

$$s(\theta,\phi) = (\sin\phi\cos\theta, \sin\phi\sin\theta, \cos\phi).$$

This is the coordinate representation of a parameterization of an open submanifold of S^2 into $\Re^3$. The differential of this map at (θ,ϕ) is the linear map given by

$$Ds(\theta,\phi) = \begin{bmatrix} -\sin\phi\sin\theta & \cos\phi\cos\theta \\ \sin\phi\cos\theta & \cos\phi\sin\theta \\ 0 & -\sin\phi \end{bmatrix},$$

and this map has rank 2 for all (θ,ϕ) in $(0,2\pi)\times(0,\pi)$. Hence, s is an immersion. In fact, since we have restricted the domain, it is an embedding. Using the result derived in the previous example, we see that the metric tensor on S^2 in the spherical coordinate system, which we will denote by G, is given by

$$\begin{aligned} G &= (\sin^2\phi\sin^2\theta + \sin^2\phi\cos^2\theta)d\theta\otimes d\theta \\ &\quad + 2(-\sin\phi\sin\theta\cos\phi\cos\theta + \sin\phi\sin\theta\cos\phi\cos\theta)d\theta\otimes d\phi \\ &\quad + (\cos^2\phi\cos^2\theta + \cos^2\phi\sin^2\theta + \sin^2\phi)d\phi\otimes d\phi \\ &= \sin^2\phi d\theta\otimes d\theta + d\phi\otimes d\phi. \end{aligned}$$

This is the first fundamental form given for the sphere in classical geometry. You can look up this result.

Example 15.6.3 The infinite cylinder in cylindrical coordinates:

Consider the set $C^2 = \{(x,y,z) \in \Re^3 : x^2 + y^2 = 1\}$. This is the infinite cylinder of radius 1 centered around the z-axis in $\Re^3$. As a manifold, one can easily define C^2 to be the product manifold $S^1 \times \Re$. The standard family of charts, though, in cylindrical coordinates is given by the following.

1. *Let $U_1 = \{(x,y,z) \in C^2 : x > 0\}$ and let $\phi_1 : U_1 \to (-\frac{\pi}{2}, \frac{\pi}{2}) \times \Re$ be defined by $\phi_1(x,y,z) = (\tan^{-1}(\frac{y}{x}), z)$.*
2. *Let $U_2 = \{(x,y,z) \in C^2 : x < 0\}$ and let $\phi_2 : U_2 \to (\frac{\pi}{2}, \frac{3\pi}{2}) \times \Re$ be defined by $\phi_2(x,y,z) = (\pi + \tan^{-1}(\frac{y}{x}), z)$.*
3. *Let $U_3 = \{(x,y,z) \in C^2 : y > 0\}$ and let $\phi_3 : U_3 \to (0,\pi) \times \Re$ be defined by $\phi_3(x,y,z) = (\cot^{-1}(\frac{x}{y}), z)$.*
4. *Let $U_4 = \{(x,y,z) \in C^2 : y < 0\}$ and let $\phi_4 : U_4 \to (\pi, 2\pi) \times \Re$ be defined by $\phi_4(x,y,z) = (\pi + \cot^{-1}(\frac{x}{y}), z)$.*

The map $c : (0,2\pi) \times \Re \to \Re^3$ defined by $c(\theta,z) = (\cos\theta, \sin\theta, z)$ is a parameterization of an open submanifold of C^2, and, by symmetry, we could use this same map to cover any portion of C^2. The differential of this map at a point (θ, z) is given by

$$Dc(\theta,z) = \begin{bmatrix} -\sin\theta & 0 \\ \cos\theta & 0 \\ 0 & 1 \end{bmatrix}.$$

The rank of this mapping is 2 for all $(\theta, z) \in (0, 2\pi) \times \Re$. Hence, this is an immersion. Using our derivation in the example for parameterized surfaces in $\Re^3$, we see that the standard Riemannian metric tensor on C^2, denoted by G, is given by

$$\begin{aligned} G &= (\sin^2\theta + \cos^2\theta) d\theta \otimes d\theta + dz \otimes dz \\ &= d\theta \otimes d\theta + dz \otimes dz. \end{aligned}$$

This is the first fundamental form of the cylinder in classical geometry. Again, something to read about!

Homework

Exercise 15.6.1 *Consider the set $C = \{(x, y, z) \in \Re^3 \ : \ x^2 + y^2 = z\}$. Find the standard family of charts and determine the metric tensor.*

Exercise 15.6.2 *Consider the set $C = \{(x, y, z) \in \Re^3 \ : \ 2x^2 + 3y^2 + 4z^2 = 1\}$. Find the standard family of charts and determine the metric tensor.*

Exercise 15.6.3 *Consider the set $C = \{(x, y, z) \in \Re^3 \ : \ 2x^2 + 3y^4 + 4z^6 = 1\}$. Find the standard family of charts and determine the metric tensor.*

15.7 The Riemannian Metric

Given the Riemannian metric tensor, we can construct an actual metric on a smooth manifold. We know that manifolds are always metrizable from our earlier results. However, any metric space will always have infinitely many possible metrics defined on it. Thus, the question naturally arises as to whether there is a metric that is more reasonable than others with respect to the study of geometry and physics. As we have said several times already, no means of measurement can be preferred over another. Nevertheless, falling back on our mathematical and scientific intuition, there is a metric that seems to arise naturally in some contexts.

Consider measuring the distance between two points on the 2-dimensional sphere, S^2, where we picture this surface as an embedded subset of $\Re^3$. We would not measure the distance between these two points by drawing a straight (in the Euclidean sense) line between them. This line does not even lie on the space we are considering. Instead, we would measure the length of curves on the sphere connecting these two points. Consequently, it seems natural to define the distance between these points to be the length of the curve of shortest arc length connecting them. Note that we used the word *define* here for a significant reason. The notion of distance is not something that is provided for us as was thought in Euclid's geometry. We must define what we mean by distance in terms of the metric we choose. In Euclidean space, our intuition is that the curve of shortest length is the straight line connecting the two points. There are words that are hard to define here, such as straight line, but we can prove this result in $\Re^m$. This seems intuitively obvious, but if we ignore our Euclidean intuition, take $\Re^m$ to be a manifold, and define a metric on it using the arc length of curves, the result is not at all trivial. Here is a proof of the argument. This discussion is from (Wilkins (120) 2005).

Theorem 15.7.1 Metric value in $\Re^m$ is the straight line distance

Let p and q be two points in $\Re^m$ with the metric, d, defined as before. Then the value of $d(p, q)$ is attained by the length of straight line connecting p and q. That is $d(p, q) = \|p - q\|$.

Proof 15.7.1

We can assume without loss of generality that the initial point is the origin, since arc length is invariant under translations. Let $\gamma : [0,1] \to \Re^m$ be a curve in $\Re^m$ such that $\gamma(0) = 0$ and $\gamma(1) = p$. Note that by a suitable reparameterization, we can assume that all curves connecting 0 and p are defined on $[0,1]$. We will assume that γ is piecewise smooth and regular. Moreover, we will make the assumption that $\|\gamma(t)\| > 0$ for all $t > 0$. That is, we do not want γ to circle back to its initial point before reaching the final point, p. If a curve does do so, that case can be reduced to this one.

Let $\lambda : [0,1] \to \Re$ be a real-valued function such that $\lambda(t) = \|\gamma(t)\|$. Then, if $v(t)$ denotes a unit vector function on $[0,1]$, we can write $\gamma(t) = \lambda(t)v(t)$. Since γ is piecewise smooth, so are the functions λ and v. We want to compute $\langle\gamma'(t), \gamma'(t)\rangle = \|\gamma'(t)\|^2$. First, note that

$$\gamma'(t) = \lambda'(t)v(t) + \lambda(t)v'(t).$$

Since, γ is only piecewise smooth, this derivative will not be defined for all t, but it will be defined almost everywhere. In fact, by our conditions on the curves connecting two points, the set of discontinuities of $\gamma'(t)$ will be finite, and γ' will be piecewise continuous. Hence, it will still make sense to integrate $\|\gamma'(t)\|$ using the basic Riemann integral. Thus, for almost every $t \in [0,1]$, we have

$$\begin{aligned}
\langle\gamma'(t), \gamma'(t)\rangle &= \langle\lambda'(t)v(t) + \lambda(t)v'(t), \lambda'(t)v(t) + \lambda(t)v'(t)\rangle \\
&= \langle\lambda'(t)v(t), \lambda'(t)v(t)\rangle + 2\langle\lambda'(t)v(t), \lambda(t)v'(t)\rangle + \langle\lambda(t)v'(t), \lambda(t)v'(t)\rangle \\
&= (\lambda'(t))^2\|v(t)\|^2 + 2\lambda'(t)\lambda(t)\langle v(t), v'(t)\rangle + (\lambda(t))^2\|v'(t)\|^2 \\
&= (\lambda'(t))^2 + (\lambda(t))^2\|v'(t)\|^2
\end{aligned}$$

Now, the length of the curve, γ is given by

$$L_\gamma = \int_0^1 \|\gamma'(t)\|dt = \int_0^1 \sqrt{\langle\gamma'(t), \gamma'(t)\rangle}dt.$$

Hence, we see that

$$\begin{aligned}
L_\gamma &= \int_0^1 \sqrt{(\lambda'(t))^2 + (\lambda(t))^2\|v'(t)\|^2}dt \ge \int_0^1 \sqrt{(\lambda'(t))^2}dt \ge \int_0^1 |\lambda'(t)|dt \\
&\ge \left|\int_0^1 \lambda'(t)dt\right| \ge |\lambda(1) - \lambda(0)| \ge \|\gamma(1)\| \quad \textit{but } \lambda(0) = \|\gamma(0)\| = 0 \\
&\ge \|\gamma(1) - \gamma(0)\| \ge \|p - 0\|
\end{aligned}$$

So, this shows that $L_\gamma \ge \|\gamma(1) - \gamma(0)\| = \|p - 0\|$, and, since the curve γ was arbitrary, it follows that $\|p - 0\| \le d(0,p)$. But the straight line connecting 0 and p is an admissible curve, and its arc length is just $\|p - 0\|$, so we also have $d(0,p) \le \|p - 0\|$. Hence, we have $d(0,p) = \|p - 0\|$. ■

The method we use above to define the distance between two points in $\Re^m$ can be used to define a metric on M in terms of the lengths of curves between two points. Furthermore, we can show that the topology induced by this metric agrees with the original manifold topology, lending more credibility to our choice of metric as being natural in some sense. There are two restrictions on this construction we must point out.

First, we have not yet required our manifolds to be connected. We must do so to define this metric. If M is not connected, there will be points on M that cannot be connected by any curve. This is necessary for our definition. Second, in our example of the sphere, we mentioned the curve of shortest arc length connecting two points on the surface. Such a curve need not exist, but the arc lengths will always be bounded below by 0. Thus, our definition will be in terms of the infimum of the arc lengths of all curves connecting the two points. There need not be any curve that actually attains this infimum. In fact, there is a famous result in differential geometry called the Hopf-Rinow Theorem that gives necessary and sufficient conditions for a manifold to have the property that there is always a curve between any two points that attains the minimum arc length of curves between them. We will not prove this result here, but we will mention it again in our conclusion.

To begin our construction of the metric, let M be a Riemannian manifold with Riemannian metric tensor, G. A curve connecting two points, p and q, is a smooth map $\gamma : [a, b] \to M$ such that $\gamma(a) = p$ and $\gamma(p) = b$. The domain of γ could, of course, be an open or half open interval as well. We will simply use closed intervals throughout for consistency of notation. Recalling our previous discussion of the tangent vectors of a curve, we know that the tangent vector of γ at $p = \gamma(t_0)$, for some $t_0 \in [a, b]$, is given by $D_{t_0}\gamma(d/dt)$. (If $t_0 = a$ or b, we will assume that γ can be smoothly extended so that the tangent vectors at the endpoints may be well-defined.) We will abuse our notation slightly and denote this tangent vector by the more intuitive form

$$D_{t_0}\gamma\Big(\frac{d}{dt}\Big) = \frac{d\gamma}{dt},$$

and the particular point to which we are referring should be clear from the context. We define the length of the curve, γ, to be the real number

$$L = \int_a^b \left(G\Big(\frac{d\gamma}{dt}, \frac{d\gamma}{dt}\Big)\right)^{1/2} dt.$$

First, note that this is nothing more than an integral of a real-valued function on $[a, b]$. For each $t \in [a, b]$, $d\gamma/dt$ is just a tangent vector in $T_{\gamma(t)}M$. The metric tensor, G, then maps $T_pM \times T_pM$ to $\Re$. Moreover, since G and γ are both smooth, this is a smooth function, so its integral is defined just as in the classical sense of the Riemann integral. Taking the square root is always valid, as G is positive definite. Hence, this integral does make sense. Note, this is a generalization of the classical definition of arc length involving the inner product of the tangent vectors with themselves.

We can then show the value of this integral is independent of the parameterization of γ, which tells us the length of the curve is a true geometric property that depends only on the set of points making up the curve and not the particular parameterization. The independence of parameterization allows us to define the *arclength function* for the curve γ. For $t \in [a, b]$, we see that the arc length of γ between $\gamma(a)$ and $\gamma(t)$ is given by

$$L(t) = \int_a^t \left(G\Big(\frac{d\gamma}{du}, \frac{d\gamma}{du}\Big)\right)^{1/2} du.$$

This is a continuous strictly increasing function of t and so it defines a new parameter, s, called the arc length parameter. If we let $s(t) = L(t)$, then we can change between the parameter t and the arc length parameter s. By the definition of $s(t)$, we see that

$$\Big(\frac{ds}{dt}\Big)^2 = G\Big(\frac{d\gamma}{dt}, \frac{d\gamma}{dt}\Big).$$

This is just one expression for the *element of arclength* defined by the metric tensor G.

Next, consider what the arc length integral looks like in a local coordinate system. That is, suppose that $\gamma([a, b])$ lies within a single coordinate system (ϕ, U). Let $\{E_i\}$ denote the basis field induced by this coordinate system on the tangent spaces over U, and denote the coordinate functions of ϕ by x^i. Then, for any $p = \gamma(t) \in U \cap \gamma([a,b])$, we can represent the tangent vector $d\gamma/dt$ at p as we discussed in Chapter 14 by the expression

$$\frac{d\gamma}{dt} = \sum_{i=1}^{m} \dot{x}^i(t) E_{p_i},$$

where $\dot{x}^i(t)$ simply represents the ordinary derivative of the coordinate function $x^i \circ \gamma$ at t. Hence, for any point $p = \gamma(t) \in U \cap \gamma([a,b])$, we have

$$\begin{aligned} G_p\Big(\frac{d\gamma}{dt}, \frac{d\gamma}{dt}\Big) &= G_p\Bigg(\sum_{i=1}^{m} \dot{x}^i(t) E_{p_i}, \sum_{j=1}^{m} \dot{x}^j(t) E_{p_j}\Bigg) = \sum_{i,j=1}^{m} \dot{x}^i(t)\dot{x}^j(t) G_p\Big(E_{p_i}, E_{p_j}\Big) \\ &= \sum_{i,j=1}^{m} g_{ij}(p)\dot{x}^i(t)\dot{x}^j(t). \end{aligned}$$

Thus, the arc length integral, in this coordinate system, can be expressed as

$$s(t) = \int_a^t \left(\sum_{i,j=1}^{m} g_{ij}(\gamma(t)) \frac{d(x^i \circ \gamma)}{dt} \frac{d(x^j \circ \gamma)}{dt} \right)^{1/2} dt.$$

Now, the metric tensor components, g_{ij}, are real-valued functions on U and, thus, on $U \cap \gamma([a,b])$. However, because the coordinate map, ϕ, is a homeomorphism, we can uniquely associate each point $\gamma(t)$ with its coordinates $\phi \circ \gamma(t) = (x^1(t), \ldots, x^m(t))$. In this way, we can think of g_{ij} as being a function of the local coordinates of the curve. That is, we will implicitly identify g_{ij} with its coordinate representation $g_{ij} \circ \phi^{-1}$. Consequently, we can rewrite the previous equation in a form that will be useful to us, namely

$$s(t) = \int_a^t \left(\sum_{i,j=1}^{m} g_{ij}(x(t)) \frac{d(x^i \circ \gamma)}{dt} \frac{d(x^j \circ \gamma)}{dt} \right)^{1/2} dt,$$

where $g_{ij}(x(t))$ denotes g_{ij} as a function of all the coordinate functions $x^i(t)$.

From here on, we assume that M is connected. Let p and q be two fixed points in M. Let $\Gamma_{p,q}$ be the set of all piecewise smooth curves connecting p and q. Thus, $\Gamma_{p,q}$ is the set of curves, $\gamma : [a,b] \to M$ such that $\gamma(a) = p$, $\gamma(b) = q$, and there is a partition of $[a,b]$, $a = t_0 < t_1 < \cdots < t_{n-1} < t_n = b$, such that $\gamma(t_i) = \gamma(t_{i+1})$ for $1 \leq i \leq n-2$ and the curve $\gamma : [t_{i-1}, t_i] \to M$ is smooth for $1 \leq i \leq n$. For a curve, $\gamma \in \Gamma_{p,q}$, let L_γ denote the arc length of the curve as we have defined it. Define a function $d : M \times M \to \Re$ by

$$d(p,q) = \inf\{L_\gamma \ : \ \gamma \in \Gamma_{p,q}\}.$$

First, note that the function d is well-defined. For any curve, there will be, at most, a finite number of points where the integrand of the arc length integral is not differentiable. It will always be continuous, though, so the integral is well-defined. Moreover, since G is positive definite, the integrand in the arc length function is always nonnegative, so the integral is bounded below by zero. Hence the infimum always exists and is nonnegative. The symmetry and the triangle inequality are easily verified and with more work we can show it is positive definite. A special case in this argument requires our understanding that metric value is the straight line distance in $\Re^n$. This sketch of the proof can be fleshed out leading to the theorem we state next.

Theorem 15.7.2 Metric on a Connected Riemannian Manifold

Let M be a connected Riemannian manifold with metric tensor G. Then the function $d : M \times M \to \Re$ defined by

$$d(p, q) = \inf\{L_\gamma : \gamma \in \Gamma_{p,q}\}$$

defines a metric on M.

Proof 15.7.2

Again, we refer you to any of a wide variety of books that prove this result such as (M. Spivak (82) 1979), (M. Spivak (83) 1979) and (Wilkins (120) 2005). ■

To conclude, we can show this metric is natural in the sense that its induced topology is equivalent to the original manifold topology. Thus, in the case of Riemannian manifolds, there is a distinct connection between the original manifold structure and the smooth metric structure we add to it.

Theorem 15.7.3 Metric topology is Equivalent to the Original Manifold Topology

The topology induced on a Riemannian manifold, M, by the metric d is equivalent to the original topology on M.

Proof 15.7.3

We refer you to the standard sources for this proof. ■

Part VI

Emerging Topologies

Chapter 16

Asynchronous Computation

When we build a model, we have to find a way to assign abstractions to the interactions of the data we measure. Our ability to gain insight from our models is very dependent on the choice of abstraction we use. In the previous chapters, we have told a number of stories about how our perception of what we mean by smoothness, functional relationships between objects and the space the objects come from, determines the tools we bring to the modeling process. In this final part of this text, we are going to discuss the modeling of parts of science that are truly difficult. Our upcoming discussions will be in the realm of cognitive science and immune systems and their interaction. What mathematical tools would be useful to us here? We are going to try to give some partial answers to that question in what follows. First, you should realize we are used to making abstractions already.

- The construction of the reals by completing the rationals. Understanding what we really mean by $\sqrt{3}$ is quite complicated.
- Completing the integrable functions using the d_1 metric to create $\mathbb{L}_1([a,b])$.
- Completing the integrable functions using the d_2 metric to create $\mathbb{L}_2([a,b])$.
- Using Lebesgue measure to construct $\mathcal{L}_2(X, \mathcal{S}, \mu)$.

In addition, to understand reality, several useful abstractions have been made which are very useful and still not completely understood.

Quantum Mechanics: One model of the world we experience is the one we call Quantum Mechanics. To understand better what common sense variables such as position and velocity mean given the outcomes of experiments with very, very small mass particles and also light speed interactions with photons, required a radical rethinking of these concepts. We simply want to mention the postulates of Quantum Mechanics and note that a better understanding of how physical particles interact with one another means we have to reinterpret what a particle is and what its attributes are. We will follow (Weinberg (119) 2015).

To understand the various types of particles in the world, more abstractions have been developed. You should note these increasingly more abstract ways of looking at the world were required to explain data. So expanding our world view to include such incredibly abstract notions is something we must do. And it does introduce bewildering new points of view!

Here is Schrödinger's Equation: Here V is the potential energy of the particle (a function of x, y, z and t), $\boldsymbol{i} = \sqrt{-1}$, m is the mass of the particle and h is Planck's constant. The solution to this equation is the wave function $\Psi(x, y, z, t)$.

$$\frac{\boldsymbol{i}h}{2\pi}\frac{\partial \boldsymbol{Psi}}{\partial t} = -\frac{h^2}{8\pi^2 m}\nabla^2\boldsymbol{\Psi} + \boldsymbol{V}\boldsymbol{\Psi}$$

Note, we can think of this as an equation defined on the space $\Re^3 \times [0, \infty)$ where time is considered to be non-negative. Of course, we need to think of this equation defined on space-time manifold that is either Riemannian or Lorentzian or something else. We then know this requires a connection and so forth.

The postulates of Quantum Mechanics are:

Q1: The state of a physical quantity can be represented as an object in a Hilbert Space. Note, since a prototypical Hilbert space is $\mathcal{L}_2(X, \mathcal{S}, \mu)$, such objects are going to be equivalence classes and so once we choose a representation for a physical quantity, we know it is really a member of an equivalence class and hence other representations are available.

Q2: The observable physical quantities such as position, momentum, energy and so forth as represented as Hermitian operators on the Hilbert Space. It is assumed that the eigenvectors associated with the Hermitian operator for an observable form a complete orthonormal set. This is hard to prove in general, so it usually an assumption. We have a nice theory of self-adjoint operators on Hilbert spaces and since the eigenvalues are real and form a countable sequence with at most one limit point, in many cases it is possible to associate the real-valued eigenvalue with the observed value of position, etc. There are many caveats, of course, but this postulate leads inevitably to the Heisenberg Uncertainty Principle, which says we cannot know both position and velocity both to perfect accuracy.

You should note that all of this discussion is set in the context of square summable Lebesgue integration. Hence, we never really know position and velocity; instead, we know them as equivalence classes. Note, to obtain better insight into data such as the fact that some observables occurred in discrete multiples of a constant factor, the second postulate showed why this was possible. It was simply a consequence of the self-adjoint operator representation of the observable.

Homework

Exercise 16.0.1 *Do some reading about the quantum measurement problem. This will help you understand that interpreting what our abstractions mean is very difficult.*

Exercise 16.0.2 *Do some reading about Lorentzian space-time. The proper setting to add relativistic effects is to use an inner product that is not positive definite. That means the abstract framework setup in the discussion above is not right as we will not have a Hilbert space structure.*

The Standard Model: Our previous discussions of manifolds set the appropriate foundation for a better understanding of physical reality. We did not talk about the notion of differentiation of vector fields defined on manifolds. This is not a trivial notion, particularly if we work with manifolds that are not embedded submanifolds of Euclidean spaces. If this is the case, the only space in which we can work with a vector field is the tangent bundle of a manifold. Outside of the simple Euclidean cases, there is no apriori reason to think that the derivative of a vector field should lie in the tangent bundle. We would like it to do so, however, so that we might study it. Hence, there must be a means of differentiating a vector field on a manifold so that its derivative is another vector field over the manifold. The means for doing this is what is called a **connection**. A connection provides a means of differentiating vector fields on manifolds that generalizes the notion of differentiating vector fields in Euclidean spaces. Like metric tensors, connections are not intrinsically determined by the structure of the manifold. So, their existence must be verified, and they are not unique. However, given any particular metric tensor on a smooth manifold, M, there is always a connection that is naturally associated with that metric. If the metric tensor is Riemannian, then the connection associated with it is unique, meaning that there is a natural choice of connection on any Riemannian manifold, called a Riemannian connection.

The notion of connection can be extended to develop a means for differentiating tensor fields of arbitrary type. This, in turn, leads to the discussion of three of the most important topics in advanced geometry: parallelism, curvature, and geodesics. These topics are all related to one another in very interesting ways. Parallelism is an old idea in geometry, dating back to a controversy over whether or not Euclid's fifth postulate, the so-called parallel postulate, was necessary in constructing in a geometrical system. The concept of parallelism is more or less intuitive in Euclidean geometry. On spaces that bend in curve in unusual ways, however, it is not at all intuitive as to how the notion of parallelism should be defined. It was Tullio Levi-Cevita who used the existence and unique character of the Riemannian connection to clearly define a notion of parallelism on a Riemannian manifold.

This definition of parallelism allows one to define a precise means of curvature. Curvature was a difficult topic to discuss in classical differential geometry. The curvature of plane curves is relatively easy to understand, but once one moves into higher dimensions and begins discussing the curvature of surfaces as opposed to curves, intuition is not readily available. The classical curvature of a surface was defined locally in terms of the curvature of curves passing through a particular point. Later on, Gauss developed a means of defining curvature that proved its intrinsic nature. That is, curvature is a geometric property of a manifold. Defining the curvature of a general smooth manifold is even more difficult, and is only possible after the notion of parallelism has been well-defined. In modern geometry, curvature is defined in terms of a particular tensor field on a manifold, called the curvature tensor.

Once the concept of curvature has been developed, one can discuss the existence of geodesics on manifolds, which, intuitively speaking, are curves of minimal arc length and curvature. Geometrically, the geodesics of a manifold are the curves in the manifold that curve as little as possible. Moreover, for any two points that lie on a geodesic, the distance between them, in terms of the metric we discussed in Section 15.7 is the arc length of the segment of the geodesic connecting the two points. It is not quite correct, however, to define geodesics as curves of minimal arc length, for there may not be a curve that actually attains the distance between the two points. The Hopf-Rinow theorem gives necessary and sufficient conditions for a Riemannian manifold to have the property that the distance between two points is actually attained by a curve connecting the two points. These manifolds are actually complete as topological spaces, meaning that Cauchy sequences converge with respect to the metric induced by the Riemannian metric tensor. On such a manifold, it follows that the shortest distance (i.e. arc length of a curve) between two points is manifested by a geodesic connecting the two points. Thus, the geodesics of a manifold are the lines of the space, the curves that provide the shortest path from one point to another. In Euclidean spaces, the geodesics are straight lines in the classical sense. On S^2, the sphere, the geodesics are the great circles, or any curve that divides the sphere into two symmetric hemispheres. In terms of physical models, geodesics are the inertial lines of motion in a particular space, or the paths along which particles move with the least energy. The existence of geodesics can be used to establish useful coordinate systems in a neighborhood of any point on a complete Riemannian manifold. A normal coordinate system is one in which the natural coordinate directions extending from a point (i.e. the coordinate axes) are in the direction of geodesic curves passing through that point.

Let's look at a short introduction to the standard model in physics as described in (Hermann (48) 1980). There are many aspects of this theoretical background that are still not understood as the way physics at the large scale modeled by general relativity couples to quantum mechanics and the standard model still have difficulties in interpretation. We want to model immunology, the gut biome and consciousness in the last part of this text, and it is not clear at all what abstract principles we should use to build those models. So this brief description of the standard model should help you understand how the quest to build a valid model of what we measure leads to ideas that are just as far from simple things like integers and rational numbers as a correct model of the real numbers requires. So let's begin. We are freely adapting from Hermann's discussion here as it is short and relatively easy to digest. It also lays out the basics as known in 1980 and that is enough for us here, even though more has been done since.

The old way of looking at an atom is something that has a nucleus that is made up of elementary components called protons and neutrons and the nucleus has electrons in orbit around the nucleus. Electromagnetic radiation is emitted and absorbed in interactions and these interactions are thought of as mediated by the exchange of a new particle called the photon. However, this is a picture that needs to be changed to understand data that was being obtained in experiments. In the new quantum picture, the particles (neutrons, protons, electrons and photons) are now described by fields. We have been learning how to understand fields properly in our discussions of manifolds, so we are poised to use this language now. We have also introduced the idea of fiber bundles and we can think of space-time as some sort of 4 dimensional manifold whether Riemannian or Lorentzian in nature. These fields, then, are cross-sections of fiber bundles over space-time to be more exact with our description.

The fibers have components along with groups of actions that operate on the component. In turn, other groups also act on the bundles. The fields, since they are cross-sections of bundles, satisfy certain nonlinear partial differential equations. Hence, the need to understand differentiation in this abstract setting! There is a special partial differential equation which models interactions which have no field associated with them. These are called the field-free equations and they are linear. Field-free equations, in general, are those that are modeled by linear partial differential equations. The classic Maxwell's equations are an example. The important variables are:

- $\boldsymbol{E}$ is the electric field.
- $\boldsymbol{H}$ is the magnetic field strength.
- $\boldsymbol{D}$ is the electric displacement field.
- $\boldsymbol{B}$ is the magnetic flux density.
- ρ is the free electric charge density.
- $\boldsymbol{J}$ is the free current density.

It is not important to understand all these terms and it is easy enough to do some background reading. These variables are related to each other using the divergence operator $\nabla\cdot$ and the curl operator $\nabla\times$. The field-free equations are then

$$\begin{aligned}
\nabla \cdot \boldsymbol{D} &= \rho, && \textbf{Gauss's Law} \\
\nabla \cdot \boldsymbol{B} &= 0, && \textbf{Gauss's Law for Magnetism} \\
\nabla \times \boldsymbol{E} &= -\frac{\partial \boldsymbol{B}}{\partial t}, && \textbf{Faraday's Law of Induction} \\
\nabla \times \boldsymbol{H} &= \boldsymbol{J} + \frac{\partial \boldsymbol{D}}{\partial t}, && \textbf{Ampére's Law}
\end{aligned}$$

There are also integral formulations which we do not show. All of these ideas are most easily done in $\Re^3$ for space and positive or zero t for time: i.e. in the space $\Re^3 \times [0, \infty)$ but we know this should really be put in a manifold setting which requires a connection and so forth. So just read these equations as place-keepers for more abstract concepts.

The interactions between the fields are then represented by nonlinear perturbations of the linear field-free equation such as the ones above. Another free-field equation involved here is the generalization of Schrödinger's equation to the relativistic case, which is called Dirac's equation. The original Dirac equation for the electron of rest mass m with coordinates $\boldsymbol{x} \in \Re^3$ and time t is

$$\left(\beta\, m\, c^2 + \sum_{in=1}^{3} \boldsymbol{\alpha_n} p_n\right) \boldsymbol{\Psi} \quad = \quad \frac{\boldsymbol{i}h}{2\pi} \frac{\partial \boldsymbol{\Psi}}{\partial t}$$

where $\boldsymbol{p}$ is the momentum and $\boldsymbol{\beta}$ and $\boldsymbol{\alpha_n}$ are 4×4 matrices. Dirac was trying to explain the behavior of the relativistically moving electron and so to allow the atom to be treated in a manner consistent with relativity. Dirac's equation has very important implications for the structure of matter and introduced new mathematical classes of objects that are now essential elements of fundamental physics. The matrices $\boldsymbol{\alpha_n}$ and β are Hermitian and satisfy

$$\alpha_n^2 \quad = \quad \beta^2 = \boldsymbol{I}$$

where $\boldsymbol{I}$ is the identity matrix on $\Re^4$ and for distinct i and j

$$\boldsymbol{\alpha_i \alpha_j + \alpha_j \alpha_i} \quad = \quad \boldsymbol{0}$$

and also

$$\boldsymbol{\alpha_i \beta + \beta \alpha_i} \quad = \quad \boldsymbol{0}$$

Dirac's equation can then be expressed as four coupled linear first order partial differential equations. The algebraic ideas here have profound consequences on our understanding of reality.

The free equations governing protons, neutrons and electrons are then basically the same: the Dirac equation. Now the proton and neutron also have some internal symmetry structure called isotopic spin which has been verified by experiments which are typically modeled by a group called $SU(2)$, which has special structure, which makes it into what is called a Lie group. Again, for our purposes, the details are not important here. The interaction between Maxwell and Dirac fields uses what is called a minimal coupling. In more mathematical detail, the Dirac field is a cross-section of a vector bundle over space-time and the Maxwell field is a linear connection for this vector bundle. The minimal coupling is a form of differentiation which is called covariant differentiation with respect to the connection.

Experiments have also shown more symmetries. This is a group associated with these bundles which is $SO(2, \Re)$. This group is called the commutative (i.e. abelian) compact one-dimensional group. It has an associated Lie algebra which turns out to be $\Re$. A connection for the bundle is determined locally by a one dimensional differential form which has values in a Lie algebra on the base of the bundle. This base turns out to be a four dimensional Euclidean space so it has four components.

So, the connection (i.e. the electromagnetic field) is determined by four real-valued functions of four real-valued variables which are called the electromagnetic potentials. Maxwell's equations are the partial differential equations these potentials satisfy. These equations do not change (i.e. are invariant) under the action of any autoisomorphism of $SO(2, \Re)$.

So, in a quantum setting, the Dirac field and the Maxwell field give rise to certain general features:

1. The free particles corresponding to the Maxwell field have zero mass and intrinsic spin 1: these are the photons.
2. There is no way to naturally define a unique state of lowest energy. Hence, this is imposed by adding some other condition which scientists can argue about.
3. The corresponding system of Maxwell - Dirac equations needs to satisfy a technical condition called renormalizability, which allows us to ensure the perturbation expansions we use to solve problems converge in some sense. Although this is still not understood really, this technical condition works very well and allows us to determine important parameters to great accuracy that matches what we measure experimentally.

This is just the start of this abstraction process. In 1954, Yang and Mills, (Yang and Mills (123) 1954), generalized the fiber bundle setting again to a non-abelian or non-commutative setting. These new approaches led to the development of the full Standard model, which incorporates all of the particles we have discussed and added new ones which are used to build the proton, neutron and so forth. It is not our intent here to go into any of this detail. The point for you is that by choosing the right generalization of the data we measure leads us to models that give better insight. This is the point of view we are taking in the next part where we talk about building models of immune systems and cognition. You should know we are mostly interested in the modeling process and there is a lot of mathematical machinery we can bring to bear on our problem, but the very hard question to answer, is what machinery should we use?

Homework

Exercise 16.0.3 *Do some reading about connections on manifolds which prepares you to understand how to take the derivative of a map between manifolds. Our discussions in this book do not explore these ideas, but you have enough background to do that yourself.*

Exercise 16.0.4 *Look at the paper (Yang and Mills (123) 1954) yourself so you can see what a gap there is between the style of writing we use in a mathematics oriented text and a physics paper. You should learn to be able to handle that because being able to do so opens up a whole world of interesting papers and their ideas.*

We want to model complex systems and to do that we will have to make our own choices of abstractions. The first thing we want to do is to look at complex interactions using asynchronous computation. We start with two simple biological models.

16.1 Gene Viability

Consider a model of natural selection called **viability selection**, from (R. McElreath and R. Boyd (103) 2007). We are interested in understanding the long-term effects of genes in a population. Obviously, it is very hard to even frame questions about this. One of the benefits of our use of mathematics is that it allows us to build a very simplified model which nevertheless helps us understand general principles. These are biological versions of the famous Einstein *gedanken* experiments: i.e. thought experiments which help develop intuition and clarity.

Assume we have a population of N individuals at a given time. It also seems reasonable to think of our time unit as **generations**. So we would say the population at generation t is given by $N(t)$. Hence, the number of individuals in our population is not necessarily fixed but can change from one generation to the next. To understand this change, we need to know how our population reproduces so as to create the next generation. We want a very simple model here, so we will also assume each individual in our population is **haploid**, which means new individuals are produced without sex or any sort of genetic recombination. Usually, an adult has a certain number of chromosomes; call this number $2P$. The gametes are the cells with half of the genetic material and therefore have only P chromosomes. Then in sexual reproduction there is a complicated process by which the sperm and the egg interact to create a new cell called a *zygote*, which has $2P$ chromosomes. The gametes are considered **haploid** as they each have half of the chromosomes of the adult. Sexual reproduction allows a mixing of the chromosomes from two adults to form a new individual having $2P$ chromosomes. The cell formed by the union of the sperm and egg is called a **zygote** and it is diploid as it has $2P$ chromosomes. So we are making a very big simplifying assumption. Essentially, we are saying each adult has Q chromosomes and the reproduction process does not mix genetic information from another adult and hence the zygote formed by what is evidently some form of asexual reproduction also has Q chromosomes. Note, calling the cell formed by this reproductive process a zygote is a bit odd as usually that word is reserved for the cell formed by a sexual reproduction. So we have a really simplistic population dynamic here and the term *zygote* here is used loosely!

We also assume in each generation individuals go through their life cycle exactly the same: all individuals are born at the same time and all individuals reproduce at the same time; i.e. we have a **discrete** dynamic. Note a zygote does not have to live long enough to survive to an adult. Since we want to develop a very simple model we assume there are only two genotypes, type **A** and type **B**. We also assume **A** is more likely to survive to an adult. We need to start defining quantities of interest: i.e variables now.

- N_A is the number of individuals of phenotype A and N_B is the number of individuals of phenotype B in a given generation t. We start at generation 0, so $t = 0$ initially. Then the first generation is $t = 1$ and so forth. We are interested in what happens in the population as t increases. So we have the population equation

$$N(t) = N_A(t) + N_B(t)$$

Also, note generations here are integers, not numbers like $1/2$ or 2.4.

- It is also convenient to keep track of the fraction of individuals in the population that are genotype **A** or **B**. This fraction is also called the **frequency** of type **A** and **B** respectively. We use new variables for this which are functions of t also.

$$P_A(t) = \frac{N_A(t)}{N_A(t) + N_B(t)}, \quad P_B(t) = \frac{N_B(t)}{N_A(t) + N_B(t)}$$

- Finally, as we said, individuals of each phenotype do not necessarily survive to adulthood. We will assume each phenotype survives to adulthood with a certain probability, V_A and V_B, respectively.

16.1.0.1 The Next Generation

Now consider how the population moves from one generation to the next.

- The number of zygotes from individuals of genotype **A** at generation t is assumed to be $z\, \boldsymbol{N_A}(t)$ where z is the number of zygotes each individual of type **A** produces. Note that z plays the role of the **fertility** of individuals of type **A**. We will assume that individuals of type **B** also create z zygotes. Hence, their fertility is also z. We could also have assumed these fertilities are different and labeled them as $\boldsymbol{z_A}$ and $\boldsymbol{z_B}$, respectively, but we aren't doing that here. So the number of zygotes of **B** type individuals is $z\, \boldsymbol{N_B}(t)$.

- The frequency of **A** zygotes at generation t is then

$$\boldsymbol{P_{AZ}}(t) = \frac{z\, \boldsymbol{N_A}(t)}{z\, \boldsymbol{N_A}(t) + z\, \boldsymbol{N_B}(t)}$$

  where we add an additional subscript to indicate we are looking at zygote frequencies. Note the z's cancel to show us that the frequency of **A** zygotes in the population does not depend on the value of z at all. We have

$$\boldsymbol{P_{AZ}}(t) = \frac{\boldsymbol{N_A}(t)}{\boldsymbol{N_A}(t) + \boldsymbol{N_B}(t)} = \boldsymbol{P_A}(t).$$

- But not all zygotes survive to adulthood. If we multiply numbers of zygotes by their probability of survival, $\boldsymbol{V_A}$ or $\boldsymbol{V_B}$, the number of **A** zygotes that survive to adulthood is $\boldsymbol{V_A}\, \boldsymbol{N_A}(t)$ and the number of **B** zygotes that survive to adulthood is $\boldsymbol{V_B}\, \boldsymbol{N_B}(t)$. We see the frequency of **A** zygotes that survive to adulthood to give the generation $t+1$ must be

$$\boldsymbol{P_{AZS}}(t+1) = \frac{\boldsymbol{V_A}\, \boldsymbol{N_A}(t)}{\boldsymbol{V_A}\, \boldsymbol{N_A}(t) + \boldsymbol{V_B}\, \boldsymbol{N_B}(t)}$$

  where we have added yet another subscript $\boldsymbol{S}$ to indicate survival. Note we add the generation label $t+1$ to $\boldsymbol{P_{AZS}}$ because this number is the frequency of adults that start generation $t+1$. Also, note this fraction is exactly how we define our usual frequency of **A** at generation $t+1$. Hence, we can say

$$\boldsymbol{P_A}(t+1) = \frac{\boldsymbol{V_A}\, \boldsymbol{N_A}(t)}{\boldsymbol{V_A}\, \boldsymbol{N_A}(t) + \boldsymbol{V_B}\, \boldsymbol{N_B}(t)}.$$

Now for the final step. From the way we define stuff, notice that

$$\boldsymbol{P_A}(t) = \frac{\boldsymbol{N_A}(t)}{\boldsymbol{N_A}(t) + \boldsymbol{N_B}(t)} \Longrightarrow \boldsymbol{N}(t)\, \boldsymbol{P_A}(t) = \boldsymbol{N_A}(t)$$

Then consider the frequency for **B**. Note

$$1 - \boldsymbol{P_A}(t) = 1 - \frac{\boldsymbol{N_A}(t)}{\boldsymbol{N_A}(t) + \boldsymbol{N_B}(t)} = 1 - \frac{\boldsymbol{N_A}(t)}{\boldsymbol{N}(t)} = \frac{\boldsymbol{N}(t) - \boldsymbol{N_A}(t)}{\boldsymbol{N}(t)}.$$

But $\boldsymbol{N}(t) - \boldsymbol{N_A}(t) = \boldsymbol{N_B}(t)$ and so

$$1 - \boldsymbol{P_A}(t) = \frac{\boldsymbol{N_B}(t)}{\boldsymbol{N}(t)}$$

which leads to the identity we wanted: $\boldsymbol{N_B}(t) = (1 - \boldsymbol{P_A}(t))\, \boldsymbol{N}(t)$. This analysis works just fine at generation $t+1$ too, but at that generation, we have

$$\boldsymbol{N_A}(t+1) = \boldsymbol{N}(t)\, \boldsymbol{V_A}\, \boldsymbol{P_A}(t), \quad \boldsymbol{N_B}(t+1) = \boldsymbol{N}(t)\, \boldsymbol{V_B}\, \boldsymbol{P_B}(t) = \boldsymbol{N}(t)\, \boldsymbol{V_B}\, (1 - \boldsymbol{P_A}(t))$$

So we can say

$$\boldsymbol{P_A}(t+1) \;=\; \frac{\boldsymbol{N_A}(t+1)}{\boldsymbol{N_A}(t+1) \,+\, \boldsymbol{N_B}(t+1)} = \frac{\boldsymbol{P_A}(t)\,\boldsymbol{N}(t)\,\boldsymbol{V_A}}{\boldsymbol{P_A}(t)\,\boldsymbol{N}(t)\,\boldsymbol{V_A} \,+\, (1 \,-\, \boldsymbol{P_A}(t))\boldsymbol{N}(t)\,\boldsymbol{V_B}}$$

Since $\boldsymbol{N}(t)$ is common in both the numerator and denominator, we can cancel them to get

$$\boldsymbol{P_A}(t+1) \;=\; \frac{\boldsymbol{V_A}\,\boldsymbol{P_A}(t)}{\boldsymbol{P_A}(t)\,\boldsymbol{V_A} \,+\, (1 \,-\, \boldsymbol{P_A}(t))\,\boldsymbol{V_B}}.$$

This tells us how the frequency of the **A** genotype changes each generation.

16.1.0.2 A Difference Equation

We can also derive a formula for the change in frequency at each generation by doing a subtraction. We consider

$$\begin{aligned}
\boldsymbol{P_A}(t+1) \,-\, \boldsymbol{P_A}(t) &= \frac{\boldsymbol{V_A}\,\boldsymbol{P_A}(t)}{\boldsymbol{P_A}(t)\,\boldsymbol{V_A} \,+\, (1 \,-\, \boldsymbol{P_A}(t))\,\boldsymbol{V_B}} \,-\, \boldsymbol{P_A}(t)\\
&= \frac{\boldsymbol{V_A}\,\boldsymbol{P_A}(t) - \boldsymbol{P_A}(t)\boldsymbol{P_A}(t)\,\boldsymbol{V_A} \,+\, (1 \,-\, \boldsymbol{P_A}(t))\,\boldsymbol{V_B}}{\boldsymbol{P_A}(t)\,\boldsymbol{V_A} \,+\, (1 \,-\, \boldsymbol{P_A}(t))\,\boldsymbol{V_B}}\\
&= \frac{\boldsymbol{V_A}\,\boldsymbol{P_A}(t) \,-\, \boldsymbol{P_A}(t)\left(\boldsymbol{P_A}(t)\,\boldsymbol{V_A} \,+\, (1 \,-\, \boldsymbol{P_A}(t))\,\boldsymbol{V_B}\right)}{\boldsymbol{P_A}(t)\,\boldsymbol{V_A} \,+\, (1 \,-\, \boldsymbol{P_A}(t))\,\boldsymbol{V_B}}\\
&= \frac{\boldsymbol{V_A}\,\boldsymbol{P_A}(t) \,-\, (\boldsymbol{P_A}(t))^2\,\boldsymbol{V_A} \,-\, \left(\boldsymbol{P_A}(t)\,(1 \,-\, \boldsymbol{P_A}(t))\,\boldsymbol{V_B}\right)}{\boldsymbol{P_A}(t)\,\boldsymbol{V_A} \,+\, (1 \,-\, \boldsymbol{P_A}(t))\,\boldsymbol{V_B}}\\
&= \frac{\left(\boldsymbol{P_A}(t)\,(1 \,-\, \boldsymbol{P_A}(t))\right)\left(\boldsymbol{V_A} \,-\, \boldsymbol{V_B}\right)}{\boldsymbol{P_A}(t)\,\boldsymbol{V_A} \,+\, (1 \,-\, \boldsymbol{P_A}(t))\,\boldsymbol{V_B}}.
\end{aligned}$$

This is the **recursion** equation for P_A. An important thing to notice is that the idea of **generation** is a fluid concept as **generation** means very different things in different species. So when we use t to represent a generation, the time interval to get to generation $t+1$ can be years, months, days, hours and even less. Our severely odd thought experiment creature of this model can be replaced by other creatures with sexual reproduction and all sorts of other more accurate assumptions. But the basic questions will still be the same. We can use this model to find how the frequencies change in each generation, and we ask the big question: what is the limiting behavior? Also, note if we allowed each member of the population to be an individual node that accepts inputs and computes output results locally, the P_A recursion relation is an inferred consequence of these local interactions under our assumptions.

Homework

Exercise 16.1.1 *The difference equation implies a differential equation as long as we can treat generation size as a continuous variable that can go to zero. Derive these dynamics and comment on all the assumptions you make.*

Exercise 16.1.2 *Redo the analysis above, but this time, assume the fertilities are different and label them $\boldsymbol{z_A}$ and $\boldsymbol{z_B}$, respectively.*

Exercise 16.1.3 *Derive the dynamics for the case where the fertilities are different and comment on all the assumptions you make.*

Exercise 16.1.4 *Write down a graph model of N nodes. Each node can have state $\boldsymbol{A}$ or $\boldsymbol{B}$. For a small N, draw the graph and discuss how you would define the edge $E_{i \rightarrow j}$ for the interaction of node i with node j. Do not think of a global time clock here and instead think of all interactions as being asynchronous.*

16.2 SIR Disease Models

We can also build a simple model of an infectious disease called the SIR model. Assume the total population we are studying is fixed at N individuals just like in the viability model. This population is then divided into three separate pieces. We have individuals:

- That are susceptible to becoming infected are called **Susceptible** and are labeled by the variable S. Hence, $S(t)$ is the number that are capable of becoming infected at time t.
- That can infect others. They are called **Infectious** and the number that are infectious at time t is given by $I(t)$.
- That have been removed from the general population. These are called **Removed** and their number at time t is labeled by $R(t)$.

We make a number of key assumptions about how these population pools interact.

- Individuals stop being *infectious* at a positive rate γ, which is proportional to the number of individuals that are in the *infectious* pool. If an individual stops being infectious, this means this individual has been *removed* from the population. This could mean they have died, the infection has progressed to the point where they can no longer pass the infection on to others or they have been put into quarantine in a hospital so that further interaction with the general population is not possible. In all of these cases, these individuals are not infectious or can't cause infections and so they have been *removed* from the part of the population N which can be infected or is susceptible. Mathematically, this means we assume

$$I'_{loss} \quad = \quad -\gamma\, I.$$

- Susceptible individuals are those capable of catching an infection. We model the interaction of infectious and susceptible individuals in the same way we handled the interaction of food fish and predator fish in the Predator - Prey model. We assume this interaction is proportional to the product of their population sizes: i.e. SI. We assume the rate of change of **Infectious** is proportional to this interaction with positive proportionality constant r. Hence, mathematically, we assume

$$I'_{gain} \quad = \quad r\, S\, I.$$

We can then figure out the net rates of change of the three populations. The infectious population gains at the rate $r\, S\, I$ and loses at the rate $\gamma\, I$. Hence, the net gain is $I'_{gain} + I'_{loss}$ or

$$I' \quad = \quad r\, S\, I \; - \; \gamma\, I.$$

The net change of susceptibles is that of simple decay. Susceptibles are lost at the rate $-r\, S\, I$. Thus, we have

$$S' \quad = \quad -\, r\, S\, I.$$

Finally, the removed population increases at the same rate the infectious population decreases. We have

$$R' = \gamma I.$$

We also know that $R(t) + S(t) + I(t) = N$ for all time t because our population is constant. So only two of the three variables here are independent. We will focus on the variables I and S from now on. Our complete model is then

$$\begin{aligned} I' &= r\,S\,I - \gamma\,I & (16.1)\\ S' &= -r\,S\,I & (16.2)\\ I(0) &= I_0 & (16.3)\\ S(0) &= S_0 & (16.4) \end{aligned}$$

where we can compute $R(t)$ as $N - I(t) - S(t)$.

When we set I' and S' to zero, we obtain the usual nullcline equations.

$$I' = 0 = I\,(r\,S - \gamma), \quad S' = 0 = -r\,S\,I$$

We see $I' = 0$ when $I = 0$ or when $S = \gamma/r$. The nullcline information for $I' = 0$ and $S' = 0$ can be combined into one picture which we show in Figure 16.1. We can show the only biologically

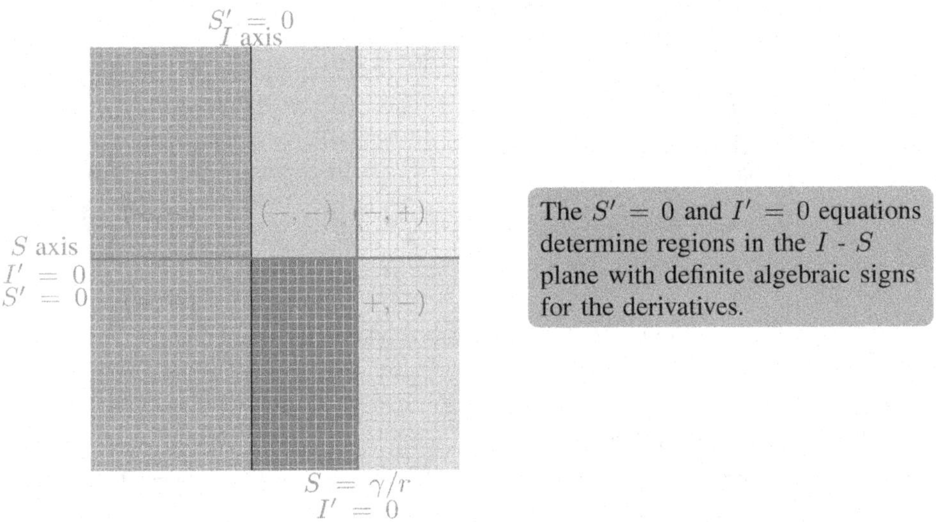

Figure 16.1: Finding the (I', S') algebraic sign regions for the disease model.

reasonable trajectories are the ones starting in Quadrant 1 with $I_0 > 0$. We thus know biologically reasonable solutions occur with initial conditions starting in Quadrant 1 and we know that solutions satisfy $S' < 0$ always with both S and I positive until we hit the S axis. Let the time where we hit the S axis be given by t^*. Then, we can manipulate the disease model as follows. For any $t < t^*$, we can divide to obtain

$$\frac{I'(t)}{S'(t)} = \frac{r\,S(t)\,I(t) - \gamma\,I(t)}{-r\,S(t)\,I(t)} = -1 + \frac{\gamma}{r}\,\frac{1}{S(t)}.$$

Thus,

$$\frac{dI}{dS} = -1 + \frac{\gamma}{r}\frac{1}{S} \implies I(t) = I_0 - \Big(S(t) - S_0\Big) + \frac{\gamma}{r}\ln\left(\frac{S(t)}{S_0}\right)$$
$$\implies I(t) = I_0 + S_0 - S(t) + \frac{\gamma}{r}\ln\left(\frac{S(t)}{S_0}\right)$$

Dropping the dependence on time t for convenience of notation, we see in Equation 16.5, the functional dependence of I on S.

$$I = I_0 + S_0 - S\,\frac{\gamma}{r}\ln\left(\frac{S}{S_0}\right). \tag{16.5}$$

It is clear that this curve has a maximum at the critical value γ/r. This value is very important in infectious disease modeling and we call it the infectious to susceptible rate ρ. We can use ρ to introduce the idea of an **epidemic**. We show these solutions in Figure 16.2.

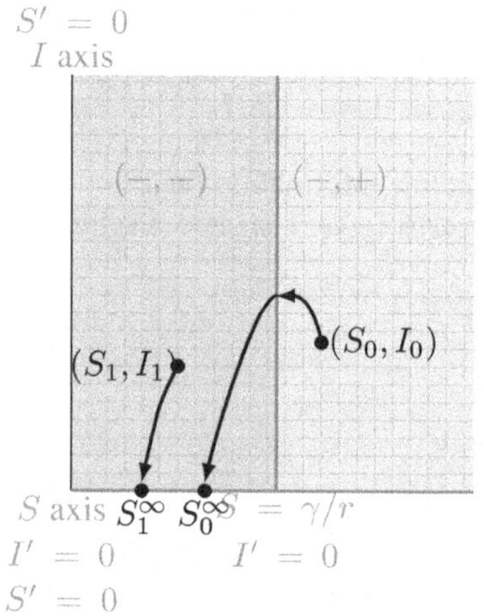

A plausible trajectory starting at the point $S_0 > \gamma/r$ and $I_0 > 0$. Another trajectory starting at $S_1 < \gamma/r$ and $I_0 > 0$ is also shown. In addition, the intersections with the S axis are labeled S_0^∞ and S_1^∞, respectively.

Figure 16.2: The disease model in quadrant one.

Definition 16.2.1 Disease Epidemic

For the disease model

$$I' = r\,S\,I - \gamma\,I, \quad S' = -r\,S\,I, \quad I(0) = I_0, \quad S(0) = S_0$$

the dependence of I on S is given by $I = I_0 + S_0 - S\,\rho\,\ln(S/S_0)$. For this model, we say the infection becomes an **epidemic** *if the initial value of susceptibles, S_0 exceeds the critical infectious to susceptible ratio $\rho = \frac{\gamma}{r}$.*

Also, note if we allowed each member of the population to be an individual node that accepts inputs and computes output results locally, the assumed dynamics can be interpreted as asynchronous interactions and the critical ratio $\rho = \frac{\gamma}{r}$ is then an inferred consequence of these local interactions under our assumptions. All of the analysis above is based on the assumption that we can do nullcline analysis which, of course, assumes the variables here interact in a differentiable manner. Rethinking this model in terms of interactions between nodes in a graph let us think about such dynamics in a very different way.

Homework

Exercise 16.2.1 *Rewrite the differential equation as a difference equation. What do you think iteration time means now? Derive these dynamics and comment on all the assumptions you make.*

Exercise 16.2.2 *Write down a graph model of N nodes. Each node can have three states: infected, susceptible and removed. For a small N, draw the graph and discuss how you would define the edge $E_{i \to j}$ for the interaction of node i with node j. Do not think of a global time clock here and instead think of all interactions as being asynchronous.*

16.3 Associations in Complex Graph Models

As the examples above show, we are therefore interested in ways to model how associations are made in a functioning complex system. The system could be an immune model or a brain model which allows us to ask questions about the nature of consciousness and possibly to shed insight into what such a high-level concept might mean. Let's focus on brain models for now. Brains and immune systems interact, and although we acknowledge this, for the moment we will focus on examples of how a functioning neural network known as a brain can be altered via a specialized signal to function very differently. Whether the brain model is from a small animal such as a ctenophore, a spider or a more complex organism such as a squid or human, we feel there is a lot of commonality in how a functioning brain takes raw sensory input and transforms it into outputs which help the organism thrive. We can outline an approach to how associations are formed and indicate graph-based neural models that can take advantage of this formalism. A general model, which is not specialized to biological processing, would thus be a directed graph architecture consisting of computational nodes $\mathcal{V}$ and edge functions $\mathcal{E}$ which mediate the transfer of information between two nodes or vertices. Hence, if $\boldsymbol{V_i}$ and $\boldsymbol{V_j}$ are two computational nodes, then $\boldsymbol{E_{i \to j}}$ would be the corresponding edge function that handles information transfer from node $\boldsymbol{V_i}$ and node $\boldsymbol{V_j}$. The symbol $\mathcal{G}(\mathcal{V}, \mathcal{E})$ denotes this graph.

A typical graph architecture is that of a chained feedforward architecture or **CFFN**. This is not a general graph model, as in this network model, all information is propagated forward and feedback connections are not allowed. Consider the function $H : \Re^{n_I} \to \Re^{n_O}$, that has a very special nonlinear structure consisting of a chain of computational elements, generally referred to as *neurons* since a very simple model of neural processing models the action potential spike as a sigmoid function which transitions rapidly from a binary 0 (no spike) to 1 (spike). This sigmoid is called a transfer function, and since it cannot exceed 1 because 1 is a horizontal asymptote, it is called a *saturating* transfer function also. This model is known as a lumped sum model of post-synaptic potential and it is simplistic.

Each *neuron* processes a summed collection of weighted inputs via a saturating transfer function with bounded output range (i.e. $[0, 1]$). The neurons whose outputs connect to a given target or **postsynaptic** neuron are called **presynaptic** neurons. Each presynaptic neuron has an output Y which is modified by the synaptic weight $W_{pre,post}$ connecting the presynaptic neuron to the postsynaptic neuron. This gives a contribution $W_{pre,post}\, Y$ to the input of the postsynaptic neuron.

The chained model then consists of a string of N neurons or processing elements or nodes, labeled from 0 to $N-1$. Some of these nodes can accept external input and some have their outputs compared

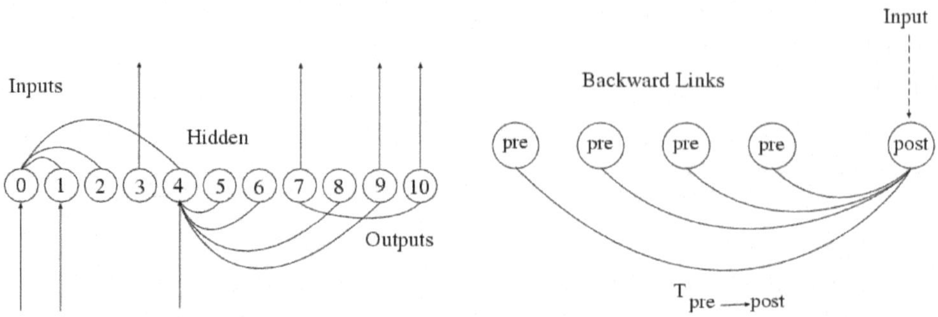

(a) Chained feedforward.

(b) Postsynaptic output calculation.

Figure 16.3: Feedforward networks.

to external targets. We let

$$\begin{aligned}\mathcal{U} &= \text{indices } i \text{ where neuron } i \text{ is an input} = \{u_0, \ldots, u_{n_I - 1}\} \quad (16.6)\\ \mathcal{V} &= \text{indices } i \text{ where neuron } i \text{ is an output} = \{v_0, \ldots, v_{n_O - 1}\} \quad (16.7)\end{aligned}$$

For now, let our nodes be called neurons. We will let n_I and n_O denote the size of $\mathcal{U}$ and $\mathcal{V}$ respectively. The remaining neurons in the chain which have no external role are sometimes called hidden neurons with dimension n_H. Note in a chain, it is also possible for an input neuron to be an output neuron; hence $\mathcal{U}$ and $\mathcal{V}$ need not be disjoint sets. The chain is thus divided by function into three possibly overlapping types of processing elements: n_I input neurons, n_O output neurons and n_H internal or hidden neurons (the ones not connected to input or sending their outputs to external monitors). In Figure 16.3(a), we see a prototypical chain of eleven neurons. For clarity only a few synaptic links from pre to post neurons are shown. We see three input neurons (neurons 0, 1 and 4) and four output neurons (neurons 3, 7, 9 and 10). Note input neuron 0 feeds its output forward to input neuron 1 in addition to feeding forward to other postsynaptic neurons. The set of postsynaptic neurons for neuron 0 can be denoted by the symbol $\mathcal{F}(0)$ which here is the set $\mathcal{F}(0) = \{1, 2, 4\}$. Similarly, we see $\mathcal{F}(4) = \{5, 6, 8, 9\}$. We will let the set of postsynaptic neurons for neuron i be denoted by $\mathcal{F}(i)$, the set of **forward links** for neuron i. Note also that each neuron can be viewed as a postsynaptic neuron with a set of presynaptic neurons feeding into it: thus, each neuron i has associated with it a set of backward links which will be denoted by $\mathcal{B}(i)$. In our example, $\mathcal{B}(0) = \{\}$ and $\mathcal{B}(4) = \{0\}$, where in general, the backward link sets will be much richer in connections than these simple examples indicate. The weight of the synaptic link connecting the presynaptic neuron i to the postsynaptic neuron j is denoted by $W_{i \to j}$. For a feedforward architecture, we will have $j > i$, however, this is not true in more general chain architectures. The input of a typical postsynaptic neuron therefore requires summing over the backward link set of the postsynaptic neuron in the following way:

$$y^{post} = x + \sum_{pre \in \mathcal{B}(post)} W_{pre \to post} Y^{pre}$$

where the term x is the external input term, which is only used if the post neuron is an input neuron. This is illustrated in Figure 16.3(b). We will use the following notation to describe the various elements of the chained architecture. Each node accepts an input and its output is modulated by what is called an offset and gain parameter. These are used to shape the output curve of the node's calculation. This kind of detail is very specific to this model and is just an example of the kind of more general processing we want the asynchronous graphs to be capable of.

x^i	External input to the ith input neuron	y^i	Summed input to the ith neuron
o^i	Offset of the ith neuron	g^i	Gain of the ith neuron
σ^i	Transfer function of the ith neuron	Y^i	Output of the ith neuron
$T_{i\to j}$	Synaptic efficacy of link	$\mathcal{F}(i)$	Forward link set for neuron i
$\mathcal{B}(i)$	Backward link set for neuron i		

The chain FFN then processes an arbitrary input vector $\boldsymbol{x} \in R^{n_I}$ via an iterative process as shown below.

$$
\begin{aligned}
&\text{for } i = 0,\ i < N \\
y^i &= \begin{cases} x^i + \sum_{j\in\mathcal{B}(i)} E_{j\to i} Y^j, & i \in \mathcal{U} \\ \sum_{j\in\mathcal{B}(i)} E_{j\to i} Y^j, & i \notin \mathcal{U} \end{cases} \\
Y^i &= \sigma^i(y^i, o^i, g^i)
\end{aligned}
$$

The output of the CFFN is therefore a vector in $\Re^{n_O}$ defined by $H(\boldsymbol{x}) = \{Y^i \mid i \in \mathcal{V}\}$. We see $H : \Re^{n_I} \to \Re^{n_O}$ is a highly nonlinear function that is built out of chains of nonlinearities. The parameters that control the value of $H(\boldsymbol{x})$ are the link values, the offsets and the gains for each neuron. Note computations are performed on each node's input queue and that all these computations are done using a global time clock as we sweep through the graph following the iterative scheme given.

Homework

Exercise 16.3.1 *Choose a CFFN network of* 4 *nodes and write down the nonlinear function* H *for this network. Also, write down explicitly all the forward and backward sets for each node.*

Exercise 16.3.2 *Choose a CFFN network of* 6 *nodes and write down the nonlinear function* H *for this network. Also, write down explicitly all the forward and backward sets for each node.*

Exercise 16.3.3 *Look at the CFFN code in (Peterson (98) 2019) and think about the various ways you can implement these ideas in MATLAB and C++.*

16.4 Comments on Feedback Graph Models

Now let's consider general CFFN models with feedback, although for convenience, we will not allow self-feedback, which is when the output of a node connects back to itself. This is not really a problem, as we always add an additional node which accepts the output and sends it back without change. We developed an implementation of this kind of CFFN in MATLAB to build models using graphs in a hierarchical fashion as part of our work on building complex models and first discussed it in the context of bioinformation processing in (Peterson (96) 2016). We also discussed it carefully in the context of learning how to code efficiently in (Peterson (98) 2019). Here, we will just mention these codes in passing and let you do background reading as you see fit. In general, a dynamic graph model will consist of many nodes interacting asynchronously. Usually, simulations use a global time clock which forces the interactions to be serial and at multiple time scales. We are going to develop a true asynchronous architecture which we can use in simulations using a good asynchronous computer language such as Erlang ((Armstrong (7) 2013) and (Hébert (46) 2013)) or Haskell ((Lipovača (74) 2011) and (Kurt (69) 2018)). Progress in building large-scale models that involve many cooperating nodes will certainly involve making suitable abstractions in the way they handle information processing.

Two fundamental circuits in the cortex of a brain are the FFP and OCOS. We will not discuss these here and instead refer you to background that can be found in (Peterson (96) 2016). The FFP and

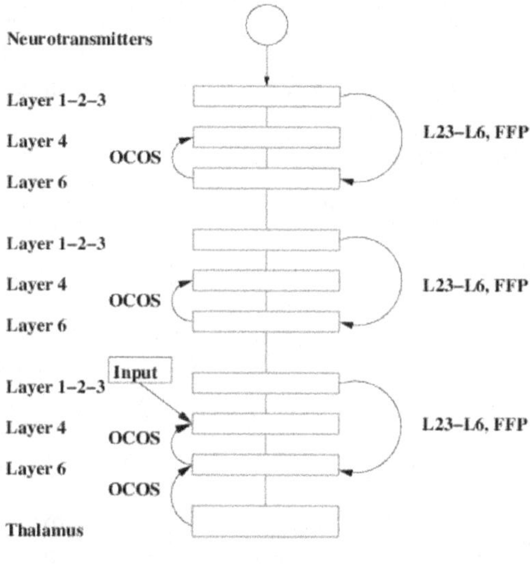

Figure 16.4: The OCOS/ FFP cortical model.

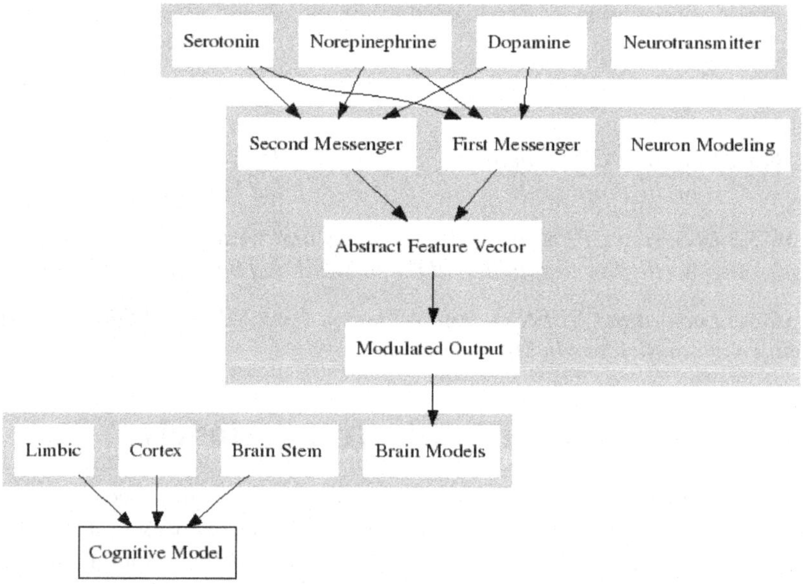

Figure 16.5: Basic cognitive software architecture design.

OCOS cortical circuits can be combined into a multi-column model (105) as seen in Figure 16.4. This model can be implemented using neuron processing as shown in Figure 16.5. In Figure 16.4, each cortical stack consists of 3 cortical columns. If each cortical layer consists of 9 neurons, this allows for 3 OCOS/ FFP pathways inside each column. This is roughly 27 neurons per column. In Figure 16.6, a more complete architecture is shown. Here, for convenience, each cortical module consists of 4 cortical stacks, all of which are interconnected. Hence, the 4 column structure illustrated in Figure 16.6 can be implemented with approximately 108 neurons each. In a typical brain model, we use five cortical modules (Frontal, Parietal, Occipital, Temporal and Limbic) which would require a total of 540 artificial neurons for their implementation. One convenient assumption is that all cortex modules start with a common architecture called isocortex where the number of stacks is a constant. Of course, we can allow more OCOS/FFP circuits per layer. Letting $N_{OCOS/FFP}$ denote

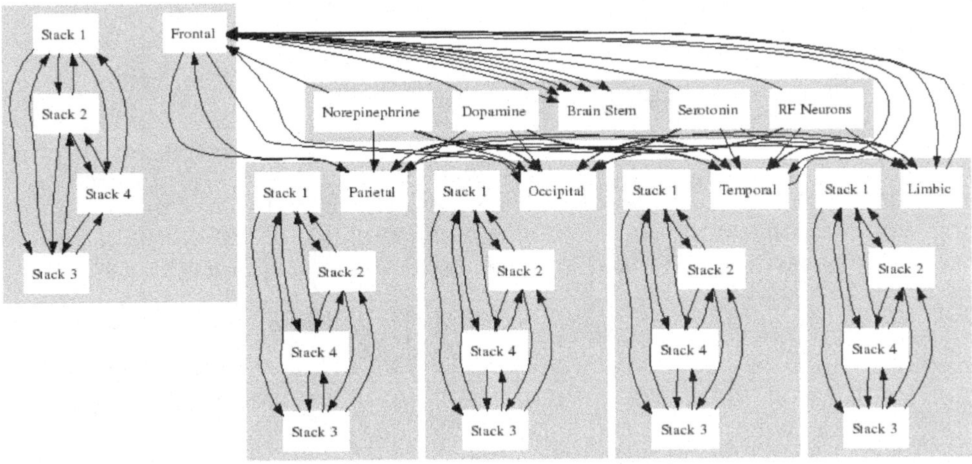

Figure 16.6: Cognitive model components.

the number of OCOS/FFP circuits per layer, we need $3N_{OCOS/FFP}$ neurons per stack. Thus, if we use N_S cortical stacks for each cortical module, we will use $3N_{OCOS/FFP}N_S$ neurons for each of the five cortical modules (Frontal, Parietal, Occipital, Temporal and Limbic). This leads to a total of $15N_{OCOS/FFP}N_S$ for the simulation of the cortical model.

The evaluation algorithms must be interpreted carefully in the case of feedback connections. We now show the nodal processing is done by a function f_i at each node which depends on a parameter vector p. The parameter vector earlier consisted on the gain and offset values for each node; however, in general there could be other parameters of interest.

$$
\begin{aligned}
&\text{for } i = 0,\ i < N \\
&y^i(t+1) \;=\; \begin{cases} x^i(t) + \sum_{j\in\mathcal{B}(i)} E_{j\to i} f_j(t), & i \in \mathcal{U} \\ \sum_{j\in\mathcal{B}(i)} E_{j\to i}(t) f_j(t), & i \notin \mathcal{U} \end{cases} \\
&f_i(t+1) \;=\; \sigma^i(y^i(t+1), p)
\end{aligned}
$$

There is also what is called a Hebbian update, which means synaptic values are changed if the correlation between input and output values at a node are significant. This is modeled using tunable parameters ϵ and ζ which, of course, are difficult to choose.

$$
\begin{aligned}
&\text{for } i = 0,\ i < N \\
&\quad \text{for } j \in \mathcal{B}(i) \\
&\qquad y^p(t) \;=\; f_i(t)\boldsymbol{E}_{j\to i}(t) \\
&\qquad\qquad \text{if } \; y^p(t) > \epsilon, \boldsymbol{E}_{j\to i}(t) = \zeta \boldsymbol{E}_{j\to i}(t)
\end{aligned}
$$

If the backward set $\boldsymbol{\mathcal{B}}(i)$ contains feedback, then how should one do the evaluation? An example will make this clear. Take a standard **Folded Feedback Pathway** cortical circuit model from (R. Raizada and S. Grossberg (105) 2003) as shown in Figure 16.7. This is the usual biological circuit model and the nodes are labeled backwards for our purposes – as we have mentioned before. We relabel them in with the nodes starting at $N1$ in the redone figure on the left. There are 11 nodes here and 11 edges with $\boldsymbol{E}_{11\to 10}$, $\boldsymbol{E}_{10\to 9}$ and $\boldsymbol{E}_{10\to 9}$ being explicit feedback. Let's assume external input comes into $\boldsymbol{N}_8$ from the thalamus and into $\boldsymbol{N}_{11}$ from the cortical column above it. The input for the nodal

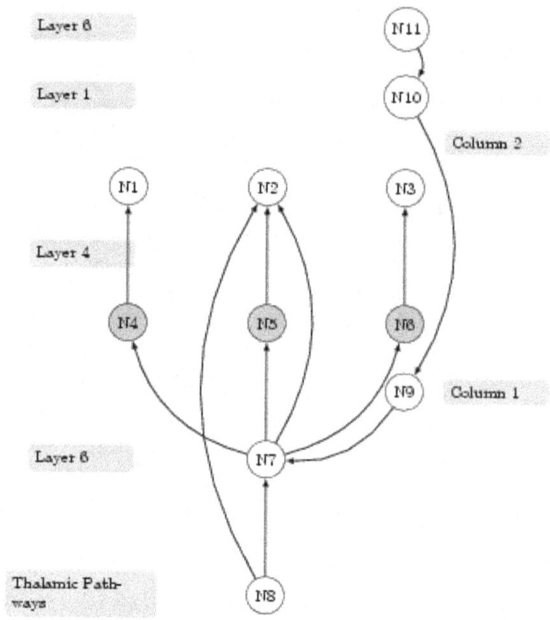

Figure 16.7: The folded feedback pathway graph.

calculation for $\boldsymbol{N_7}$ in the graph in Figure 16.8 is then

$$y^7_{new} = \boldsymbol{E_{8\to 7}} Y^8_{old} + \boldsymbol{E_{9\to 7}} Y^9_{old}$$

which in terms of a global clock would be implemented as

$$y^7(t+1) = \boldsymbol{E_{8\to 7}} Y^8(t) + \boldsymbol{E_{9\to 7}} Y^9(t)$$

where we can initialize the nodal outputs a variety of ways: for example, by calculating all the outputs without feedback initially, and hence, setting $Y^9(0) = Y^{10}(0) = 0$. When we relabel this graph as shown in Figure 16.8, it is now clear this is a feed forward chain of computational nodes and the only feedback is the relabeled edge $\boldsymbol{E_{11\to 2}}$. The external input comes into $\boldsymbol{N_1}$ from the thalamus and into $\boldsymbol{N_9}$ from the cortical column above it. The input for the nodal calculation for $\boldsymbol{N_2}$ is in terms of time ticks on our clock

$$y^2_{new} = \boldsymbol{E_{1\to 2}} Y^1_{old} + \boldsymbol{E_{11\to 2}} Y^{11}_{old} \Longrightarrow y^2(t+1) = \boldsymbol{E_{1\to 2}} Y_1(t) + \boldsymbol{E_{11\to 2}} Y_{11}(t)$$

However we choose to implement the graph, there will be feedback terms and so we must interpret the evaluation and update equations appropriately. Then, once a graph structure is chosen, we apply the usual update equations.

Homework

Exercise 16.4.1 *Let* $\boldsymbol{E_{1\to 2}} = 0.05$ *and* $\boldsymbol{E_{11\to 2}} = 0.9$ *and initialize* $Y^1 = 1.0$ *and* $Y^{11} = -1$. *Compute* 5 *iterations of* $y^2(t+1) = \boldsymbol{E_{1\to 2}} Y_1(t) + \boldsymbol{E_{11\to 2}} Y_{11}(t)$.

Exercise 16.4.2 *Let* $\boldsymbol{E_{1\to 2}} = -0.75$ *and* $\boldsymbol{E_{11\to 2}} = 0.3$ *and initialize* $Y^1 = 0.4$ *and* $Y^{11} = 0.3$. *Use MATLAB to compute* 25 *iterations of* $y^2(t+1) = \boldsymbol{E_{1\to 2}} Y_1(t) + \boldsymbol{E_{11\to 2}} Y_{11}(t)$.

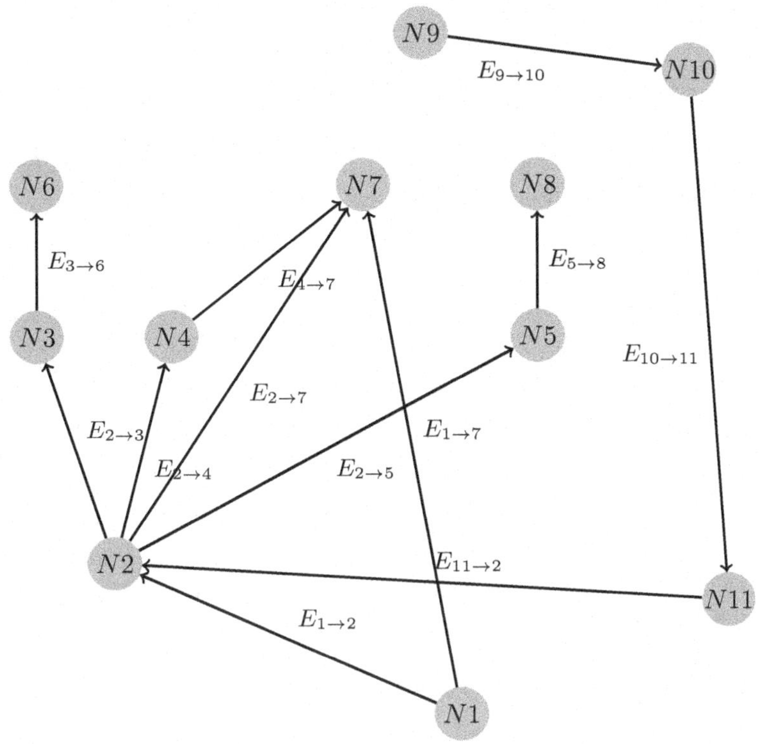

Figure 16.8: The relabeled folded feedback pathway graph.

16.4.1 Information Flow

For the given graph architecture, there is an incidence matrix. If N is the number of nodes and V is the number of edges of the graph, the incidence matrix is an $N \times V$ matrix K where $K_{ij} = 0$ except as follows: given node n and edge v, we assign the value $+1$ if E_v has *from* value n and assign the value -1 is E_v has *to* value n.

Here, each edge is modeled by the vector $\begin{bmatrix} \text{from} \\ \text{to} \end{bmatrix}$ where *to is the node the edge goes to and* from is the node the edge starts from. So a -1 value at K_{nv} means the edge v goes into the node n. The *gradient* of $\boldsymbol{f}$ is then defined by $\nabla\, \boldsymbol{f} = K^T\, f$ and the Laplacian, by $\nabla^2\, \boldsymbol{f}$. A reasonable way to model information flow through $\boldsymbol{\mathcal{G}}$ is given by the graph-based *partial differential equation* $\nabla^2 \boldsymbol{f} - \boldsymbol{\alpha} \frac{\partial \boldsymbol{f}}{\partial t} - \boldsymbol{\beta}\, \boldsymbol{f} = -\boldsymbol{\mathcal{I}}$, where $\boldsymbol{\mathcal{I}}$ is the external input. This is similar to a standard cable equation. We interpret the $\frac{\partial \boldsymbol{f}}{\partial t}$ using a standard forward difference $\boldsymbol{\Delta f}$ which is defined at each iteration time t by $\boldsymbol{\Delta f}(0) = 0$ and otherwise $\boldsymbol{\Delta f}(t) = \boldsymbol{f}(t) - \boldsymbol{f}(t-1)$. This gives, using a finite difference for the $\frac{\partial f}{\partial t}$ term, Equation 16.8, where we define the finite difference $\boldsymbol{\Delta f_n}(t)$ as $f_n(t+1) \; - \; f_n(t)$.

$$\nabla^2\, \boldsymbol{f} - \alpha\, \boldsymbol{\Delta f} - \beta\, \boldsymbol{f} \;\; = \;\; -\boldsymbol{\mathcal{I}} \tag{16.8}$$

The update equation is then

$$\boldsymbol{K}\boldsymbol{K}^T\, \boldsymbol{f} - \alpha\, \boldsymbol{\Delta f} - \beta\, \boldsymbol{f} \;\; = \;\; -\boldsymbol{\mathcal{I}}$$

For iteration t, we then have the update equation

$$\boldsymbol{K}\boldsymbol{K}^T\, \boldsymbol{f}(t+1) - \alpha\, \boldsymbol{\Delta f}(t) - \beta\, \boldsymbol{f}(t+1) \;\; = \;\; -\boldsymbol{\mathcal{I}}(t+1),$$

which gives

$$\left(\boldsymbol{K}\boldsymbol{K}^T - (\alpha+\beta)\boldsymbol{Id}\right)\, \boldsymbol{f}(t+1) + \beta\, \boldsymbol{f}(t) \quad = \quad -\boldsymbol{\mathcal{I}}(t+1),$$

where $\boldsymbol{Id}$ is the appropriate identity matrix. Let $\boldsymbol{\mathcal{H}}_{\alpha,\beta}$ denote the operator $\boldsymbol{K}\boldsymbol{K}^T - (\alpha+\beta)\boldsymbol{Id}$. We thus have the iterative equation

$$\boldsymbol{\mathcal{H}}_{\alpha,\beta}\, \boldsymbol{f}(t+1) \quad = \quad -\beta\, \boldsymbol{f}(t) - \boldsymbol{\mathcal{I}}(t+1)$$

or since $\boldsymbol{\mathcal{H}}_{\alpha,\beta}$ is invertible,

$$\boldsymbol{f}(t+1) \quad = \quad -\boldsymbol{\mathcal{H}}_{\alpha,\beta}^{-1}\, \beta\, \boldsymbol{f}(t) - \boldsymbol{\mathcal{H}}_{\alpha,\beta}^{-1}\, \boldsymbol{\mathcal{I}}(t+1).$$

For convenience, let $\boldsymbol{\Lambda}(t+1) = -\boldsymbol{\mathcal{H}}_{\alpha,\beta}^{-1}\, \boldsymbol{\mathcal{I}}(t+1)$. Then, we have

$$\boldsymbol{f}(t+1) \quad = \quad \boldsymbol{\Lambda}(t+1).$$

Now let's switch to a more typical nodal processing formulation. Recall, each node $\boldsymbol{N}_i$ has an input $\boldsymbol{y_i}(t)$ which is processed by the node using the function $\boldsymbol{\sigma_i}$ as $\boldsymbol{Y}_i(t) = \boldsymbol{\sigma_i}(\boldsymbol{y_i}(t))$. The node processing could also depend on parameters, but here we will only assume an offset and a gain is used in each computation. Thus, the node values $\boldsymbol{f}$ are equivalent to the node outputs $\boldsymbol{Y}$. This allows us to write the update as $\boldsymbol{Y}_i(t+1) = \boldsymbol{\sigma_i}(\boldsymbol{y_i}(t)) \ = \ \boldsymbol{\Lambda}_i(t+1)$ or $\boldsymbol{y_i}(t) = \boldsymbol{\sigma_i}^{-1}\,(\boldsymbol{\Lambda}_i(t+1))$.

16.4.2 A Constructive Example

First, we build an OCOS graph directly using the MATLAB codes we mentioned previously. This is just to illustrate the process, so we do not give many details of this implementation. This OCOS circuit includes a thalamus connection which we later remove, as the thalamus is eventually implemented separately. We define the node and edge vectors using integers for each node. We will label the nodes in ascending order now so the resulting graph is feedforward.

Listing 16.1: **Define the OCOS nodes and edges data**

```
% define the OCOS nodes
V = [1;2;3;4;5;6;7;8];
% define the OCOS edges
E = {[1;2],[2;3],[2;4],[2;5],[3;6],[4;7],[5;8],[2;7],[1;7]};
% construct the OCOS edge object
```

We then construct edge and vertices objects.

Listing 16.2: **Construct the OCOS edges and vertices object**

```
e = edges(E);
% construct the OCOS vertices object
v = vertices(V);
```

Next, we construct the graph object and check to see if our overloaded `subsref.m` code is working correctly.

Listing 16.3: **Construct the OCOS graph**

```
% construct the OCOS graph
OCOS=graphs(v,e);
% verify the OCOS edge list
OCOS.v
ans =
   1   2   3   4   5   6   7   8
% verify the OCOS edges list
OCOS.e
ans =
     1     2     2     2     3     4     5     2     1
     2     3     4     5     6     7     8     7     7
```

We then explicitly add nodes and edges to build an FFP graph. Note for a small graph this is not very expensive, but the way nodes and edges are added is actually very inefficient as we have to rebuild the node and edge lists from scratch with each add.

Listing 16.4: **Create FFP by adding nodes and edges to OCOS**

```
% add the FFP nodes 9, 10 and 11
FFP = addnode(OCOS,9);
FFP = addnode(FFP,10);
FFP = addnode(FFP,11);
% add the FFP edges
FFP = addedge(FFP,11,2);
FFP = addedge(FFP,10,11);
FFP = addedge(FFP,9,10);
```

Next, let's find incidence matrices. The incidence matrix for the FFP graph is then

Listing 16.5: **Get Incidence Matrix**

```
% find the FFP incidence matrix
K2 = incidence(FFP);
```

This gives

$$K2 = \begin{bmatrix} 1 & 0 & 0 & 0 & 0 & 0 & 0 & 0 & 1 & 0 & 0 & 0 \\ -1 & 1 & 1 & 1 & 0 & 0 & 0 & 1 & 0 & -1 & 0 & 0 \\ 0 & -1 & 0 & 0 & 1 & 0 & 0 & 0 & 0 & 0 & 0 & 0 \\ 0 & 0 & -1 & 0 & 0 & 1 & 0 & 0 & 0 & 0 & 0 & 0 \\ 0 & 0 & 0 & -1 & 0 & 0 & 1 & 0 & 0 & 0 & 0 & 0 \\ 0 & 0 & 0 & 0 & -1 & 0 & 0 & 0 & 0 & 0 & 0 & 0 \\ 0 & 0 & 0 & 0 & 0 & -1 & 0 & -1 & -1 & 0 & 0 & 0 \\ 0 & 0 & 0 & 0 & 0 & 0 & -1 & 0 & 0 & 0 & 0 & 0 \\ 0 & 0 & 0 & 0 & 0 & 0 & 0 & 0 & 0 & 0 & 0 & 1 \\ 0 & 0 & 0 & 0 & 0 & 0 & 0 & 0 & 0 & 0 & 1 & -1 \\ 0 & 0 & 0 & 0 & 0 & 0 & 0 & 0 & 0 & 1 & -1 & 0 \end{bmatrix}$$

We can then find the Laplacian of the FFP graph which is $K2 \times K2^T$.

Listing 16.6: **Get Laplacian**

```
% find the FFP Laplacian
L2 = laplacian(FFP);
```

Thus, we have

$$
L2 = \begin{bmatrix}
2 & -1 & 0 & 0 & 0 & 0 & -1 & 0 & 0 & 0 & 0 \\
-1 & 6 & -1 & -1 & -1 & 0 & -1 & 0 & 0 & 0 & -1 \\
0 & -1 & 2 & 0 & 0 & -1 & 0 & 0 & 0 & 0 & 0 \\
0 & -1 & 0 & 2 & 0 & 0 & -1 & 0 & 0 & 0 & 0 \\
0 & -1 & 0 & 0 & 2 & 0 & 0 & -1 & 0 & 0 & 0 \\
0 & 0 & -1 & 0 & 0 & 1 & 0 & 0 & 0 & 0 & 0 \\
-1 & -1 & 0 & -1 & 0 & 0 & 3 & 0 & 0 & 0 & 0 \\
0 & 0 & 0 & 0 & -1 & 0 & 0 & 1 & 0 & 0 & 0 \\
0 & 0 & 0 & 0 & 0 & 0 & 0 & 0 & 1 & -1 & 0 \\
0 & 0 & 0 & 0 & 0 & 0 & 0 & 0 & -1 & 2 & -1 \\
0 & -1 & 0 & 0 & 0 & 0 & 0 & 0 & 0 & -1 & 2
\end{bmatrix}
$$

Once we have these matrices, we can use standard tools to find their corresponding eigenvalues and eigenvectors, which we might want to do now and then. Note the graph has a structure, captured succinctly by the incidence matrix, which is separate from the processing of nodal information determined by the edge and node processing functions. Hence, the dynamics of information flow suggested by the graph cable equation could be a multiply connected set of solutions. However, all of this discussion is still serial based and involves a global time clock.

Homework

Exercise 16.4.3 *Using the MATLAB codes in (Peterson (98) 2019), construct your own graph built from two subgraphs.*

Exercise 16.4.4 *Design a CFFN and compute its cable graph equation.*

Exercise 16.4.5 *Design a CFFN and compute the eigenvalues and eigenvectors of its incidence matrix.*

16.5 Sudden Complex Model Changes Due to an External Signal

In anesthesia, a percentage of patients continue to experience the trauma of the surgery despite being anesthetized. Such patients are a type of zombie and there is a need for brain models which can detect this state using measurements that are available in the operating theater. The altered state of consciousness is obtained by the careful administration of a variety of drugs and in many respects is similar to the altered state of behavior induced by a predatory wasp injection of a potent neural cocktail into their cockroach or spider prey. These external events reprogram the host into a new behavioral pattern. Since all of the usual neural modules are present, we can posit that these external inputs alter the usual connections between the functioning neural modules allowing the full brain outputs to change.

We can discuss several examples of a change in consciousness state which can be quantitatively measured. First, let's look at the hard problem of consciousness. Many discussions of consciousness focus on conceptual and theoretical analysis only. But we want to approach this from the point of known data on such consciousness change. An interesting problem focuses on detecting what are called **iZombies**. An **iZombie** is an **inverse zombie** which is a creature that appears to be unconscious when in fact it is conscious (G. Mashour and E. LaRock (37) 2008).

16.5.1 Zombie Creation in Anesthesia

Characteristics of an **iZombie** are unresponsiveness to verbal commands, absence of spontaneous or evoked vocalization or speech, absence of spontaneous or evoked movement and unresponsiveness to noxious stimulus. A subset of patients experiencing awareness during general anesthesia which is called *anesthesia awareness* are **iZombies**. Such patients have explicit recall of intraoperative events. Studies have shown awareness with explicit recall in about 0.13% of operations involving anesthesia. There are about 30 million such anesthesia events per year in just the US, so this means as many as 39 thousand patients have not had adequate suppression of awareness. The condition of anesthesia awareness is a problem of consciousness. This can also occur in patients with neurological injury leading to vegetative states or locked-in syndrome. So the question is, how can such undetected consciousness occur and how can we find out if it is occurring. There are many references here. The basics are covered in (S. Laureys (109) 2005), (A. Owen and N. Schiff and S. Laureys (5) 2009), The four goals of general anesthesia are hypnosis (suppressed consciousness), amnesia, analgesia and immobility. To check if there has been adequate dosing, there are three tools: anesthetic stages, minimum alveolar concentration (MAC) and techniques based on encephalography (EEG). These techniques are focused on determining anesthesia-induced unconsciousness. After discovery of ether anesthesia, various stages of anesthesia were defined based on respiratory rate, muscle tone and size and function of the pupils. MAC was developed in the 1960's to measure anesthetic depth. MAC is the minimum alveolar concentration needed to prevent movement in response to a noxious stimulus in 50% of the population. But it turns out, this prevention of movement is mostly a reflection of spinal cord function and not the brain. **Hence, this does not tell us anything about the state of consciousness**. EEG techniques were developed to assess anesthetic depth and perhaps detect consciousness. But there is not a unique EEG signature common to all agents. Apparatus to measure EEGs is big, labor intensive and needs a dedicated observer in the operating room. Many processed EEG modules that relay on the Fourier Transform have been developed and are called **awareness monitors**. Raw EEG data is collected, Fourier transformed, and then parameters that are thought to best represent consciousness are analyzed. The output is a scalar given as 100 wide wake to 0 clinically dead.

This processing still does not address other features of interest such as amnesia, analgesia and immobility. Also, these modules are insensitive to anesthetics like NO, ketamine and xenon; the effect these agents have on the NMDA receptor may be pharmacologically equivalent. This processing can also be distorted by β blockers and others. Hence, like all tools, there are caveats. So we have no completely reliable way to ensure the absence of consciousness in a patient undergoing anesthesia and there is a class of individuals who appear unconsciousness but are actually consciousness. Real-time monitoring in the operating room is just not possible. **iZombies** are not only possible or probable, they are known to exist.

16.5.2 Zombie Creation from Parasites

There are many examples of how an iZombie is created due to some sort of signaling induced by a parasite into its targeted host. All of these examples give us clues as to how states of consciousness can be disrupted by an external signal and also give us a structure on which to build a potential answer to the question of what is consciousness and how do we define its states. An introduction to the topic of parasitic manipulation in general by a variety of organisms is given in (K. Weinersmith and Z. Faulkes (65) 2014). This is one of several papers from a special issue of the journal **Integrative and Comparative Biology** devoted to this topic. For example, the details of a particular predatory wasp's zombie creation are given in (F. Libersat and R. Gal (29) 2014). Infection results in changes to the phenotype of the host through three basic routes. First, infection induces pathology that may or may not benefit the parasite. Second, the host induces adaptive changes in the phenotype reducing the cost of infection and even eliminating the parasite and third, parasites can adaptively manipulate the phenotype in ways that increase the probability of transmission. For example, the protozoan parasite *Toxoplasma gondii* infects cats as the hosts in which their sexual reproduction can occur, but many other species can serve as intermediary hosts. In rodents, infection is associated with decreased memory and learning, increased activity, increased time spent in the open. There is also increased *neophobia*, which is the tendency of an animal to avoid or retreat from an unfamiliar object or situation. Finally, infected rats no longer avoid the smell of a predator's urine (i.e., a cat) and can even be attracted to the scent leading to a higher probability of predation by the cat, the preferred host, because the parasite reproduces in the cat. The alterations to the rat's behavior involves long lasting changes to the limbic system. *We can't reproduce this behavioral modification in the lab though*, which means a simulation environment might be helpful here. Parasites, like anesthesia drugs, manipulate the host's behavior and hence give us new sets of data from which we can learn more about the links between the brain, behavior and the immune system.

Hence, we clearly need brain models that allow us to study how the introduction of an environmental signal, such as the ones coming from the parasitic infection, an external drug, or a sting from a predatory wasp can lead to a fundamental alteration of brain state. Neurobiologists often see animal behavior as the output of the nervous system, so parasites probably manipulate the host by affecting the nervous system. The more precise the manipulation, the more likely it is to involve the nervous system. Nervous system output is determined by electrical and chemical signals between computational nodes called neurons. Electrophysiological techniques can control neurons and hence behavior and there are some documented cases of parasites affecting the host by altering electrical signals, but many more cases of parasites controlling host behavior by exploiting chemical signals. A model to gain insight into this kind of behavior modification should therefore be capable of using both chemical and electrical signals in the simulation. Thus the ability to model second messenger systems is crucial here.

16.5.2.1 The Jewel Wasp's Manipulation of a Cockroach

In (F. Libersat and R. Gal (29) 2014), Libersat and Gal study the Jewel wasp's manipulation of their cockroach host. It is probably the best example of how a nervous system can be hijacked by another animal using a cocktail of neuroactive chemicals, i.e. second messengers. Adult jewel wasps manipulate the host in ways to benefit their offspring. They lay eggs within the host and alter the host behavior to create an environment where the egg is nurtured properly. The neuroactive chemicals are delivered by a precise and sudden injection of venom. This amounts to a massive introduction of a mix of second messengers that spread from the injection site to alter the behavior that results from the interconnection of many neural modules. The cocktail therefore changes many computational nodes in the modules effectively simultaneously. The venom cocktail acts on the dopaminergic system in the subesophageal ganglion and probably the brain via the supraesophageal ganglion.

The injected cockroach first engages in grooming behavior for about 30 minutes. Dopamine in the wasp's venom probably causes this. The wasp leaves during this stage. A long-lasting lethargic state is induced during which the cockroach has a dramatically reduced drive to self-initiate movement. The venom does not appear to affect the motor centers directly but instead affects the motivation to do movement. The venom manipulates neuronal centers within the cerebral ganglia that are involved in the initiation and maintenance of walking and also blocks opiate receptors which reduces the cockroach's responsiveness to stimuli. After grooming is complete, the wasp returns and for the next few days, the cockroach becomes an **iZombie** that is docile and completely available for the wasp's needs. The wasp cuts off the cockroach's antenna and drinks the liquid exuding from the stumps, and during this, the cockroach does not flee or fight and lets the wasp grab it by one of the antenna stumps and lead it off to the wasp's nest. In the nest, the cockroach remains immobile while the egg is implanted. The cockroach is not paralyzed but remains immobile while the wasp seals up the entrance to the nest.

16.5.2.2 Wasp Manipulation of Spider Hosts

- In (K. Takasuka and T. Yasui and T. Ishigami and K. Nakata and R. Matsumoto and K. Ikeda and K. Maeto (64) 2015), host manipulation by a spider is studied. The ichneumonid spider ectoparasitoid *Reclinervellus nielseni* turns its host spider *Cyclosa argenteoalba* into an iZombie which modifies its web structure into a more persistent cocoon web so that the wasp larvae can grow to maturity after the spider's death. The iZombie spider has significant modifications to its web structure and wasp-derived components might be responsible for the manipulation. The modified webs also reflect UV light which might prevent damage by flying web-destroyers such as birds or large insects. Furthermore, iZombies create webs which have more silk threads than normal indicating a web that is harder to destroy, which increases the probability the wasp larvae survives.

- The parasitoid wasp genus *Zatypota* transforms the social spider *Anelosimus eximius*, into an iZombie that abandons its colony to do the wasp's bidding. The spider *Anelosimus eximius* forms large colonies and since the spiders rarely leave their nest, they coordinate on the capture of prey and they share parental duties. Spiders infected with the parasite wander several feet away from the colony so they can spin a web of densely spun silk and bits of leaves from their environment. When examined, these webs were found to contain a parasitoid wasp species of the Zatypota genus. After an adult female wasp lays an egg on the abdomen of a spider, the larva hatches and attaches itself to its host. It then feeds on the spider and slowly takes over its body. The iZombie spider then leaves the colony, spins a cocoon for the wasp larva and waits to be killed and consumed. After feasting on the spider, the larva emerges fully formed nine to eleven days later. As said in (Fernadez et al. (32) 2018), "The wasp completely hijacks the spider's behavior and brain and makes it do something it would never do, like leave its nest and spinning a completely different structure. That's very dangerous for these tiny spiders." It's not known how the iZombie is created, but may be caused by an injection of hormones that make the spider think it's in a different life-stage or cause it to disperse from the colony.

16.5.2.3 The Crypt Keeper

To deposit its eggs, the parasitic oak gall wasp pierces a leaf or stem with its ovipositor. The plant then forms tumor-like growths called galls which serve as nurseries (called crypts) for the wasps. Within each crypt, a wasp egg develops until it has matured enough to chew a hole in the skin of the gall and emerge an adult. A species of the genus Euderus Haliday, *Euderus set*, parasitizes the crypt gall wasp, *Bassettia pallida*. The wasp is very small, the size of a pin, and once it finds the gall created by other wasps, it punctures the gall and injects its eggs either next to or inside a young wasp. As these eggs develop in the crypt, the crypt keep baby feeds off the young of the gall wasp. This wasp was first described in (Egan et al. (28) 2017). Its behavior was more fully discussed in (Ward et al. (118) 2019). The parasitic wasp changes the behavior of the gall wasp so that the gall wasp chews a much smaller exit hole in the gall than usual and seals the resulting hole with its head before dying. The skin of the gall is quite tough but the skull of the gall wasp is not as difficult to get through. The parasitic wasp therefore benefits from this behavior as it is easier to escape the gall. This parasitic wasp is also known to *attack and manipulate the behavior of at least six additional gall wasp species and that these hosts are taxonomically diverse* (Ward et al. (118) 2019). However, the prey all produce galls vulnerable to attack. *The specialization required to behaviorally manipulate hosts may be less important in determining the range of hosts in this parasitoid system than other dimensions of the host-parasitoid interaction, like the host's physical defenses* (Ward et al. (118) 2019).

16.5.2.4 Ant Fungus Host Manipulation

The fungus *Ophiocordyceps* is called the zombie ant fungi because it alters the behavior of ants and other insects. Ants infected exhibit abnormal behavior such as randomly walking around and falling down. The parasite grows inside the body of the ant and brain and alters muscle movements and central nervous system function. The infected ant seeks out a cool, damp place and bites down on the underside of a leaf. The underside of the leaf is an ideal environment for the reproduction of the fungus. The parasite also causes the infected ant's jaw muscles to lock, so at this point the ant is permanently attached to the bottom of the leaf. *The fungal infection kills the ant and the fungus grows through the ant's head. The growing fungal stroma has reproducing structures that produce spores. Once the fungal spores are released, they spread and are picked up by other ants* (Hughes et al. (51) 2011). This type of infection is prevented from overtaking the entire ant colony by another fungus which is called a *hyperparasitic* fungus. This fungus attacks the parasite and prevents infected ants from spreading spores which limits the infected population. Since fewer spores grow to maturity, fewer ants can be infected. This is also discussed in (C. de Bekker and M. Merrow and D. Hughes (16) 2014).

16.5.3 iZombie Creation from an Epileptic Episode

Impaired consciousness is a hallmark of epileptic seizures and seizures are characterized by a variety of altered conscious states, and hence a person who experiences an epileptic seizure can therefore be characterized as an iZombie as we have discussed. However, instead of a chemical signal such as is used by parasites to accomplish the creation of the iZombie state, this state comes about due to an electrical signal in portions of the brain that disrupts the normal assembly of computational modules. In (A. and Monaco (1) 2009), it is noted that

> Complete loss of consciousness occurs when epileptic activity involves both cortical and subcortical structures, as in tonic-clonic seizures and absence seizures. Medial temporal lobe discharges can selectively impair experience in complex partial seizures (with affected responsiveness) and certain simple partial seizures (with unaffected responsiveness). ... The spread of epileptic discharges from the medial temporal lobe to the same subcortical structures can ultimately cause impairment in the level of consciousness in the late ictal and immediate postictal phase of complex partial seizures.

Hence, epileptic states are another rich source of data to help us understand consciousness and the states of consciousness.

Homework

Exercise 16.5.1 *Do a literature search to find recent papers about assessment of consciousness state in anesthesia patients in surgery.*

Exercise 16.5.2 *Do a literature search to find additional examples of zombie creation via the injection of toxin into prey by a predator.*

Exercise 16.5.3 *Do a literature search of the sudden creation of cognitive abilities after a lightning strike or other traumatic brain injury. There are known examples of a person suddenly becoming able to write music, paint and so forth. Think about how such a reorganization could occur.*

16.5.4 Changes in Cognitive Processing Due to External Drug Injection

Recently, there have been reports of patients whose brain functioning has been radically altered by the ingestion of massive dosages of LSD or controlled dosages of psilocybin and ketamine. The LSD massive dosages, see (Haden and Woods (44) 2020), can be described as follows:

1. A girl with bipolar II disorder and a history of paranoia, severe depression and disruptive behavior leading to a turbulent life, accidentally received a dose of LSD 10 times normal. A normal recreational dose was 100 micrograms and she received 1000. Her initial response was almost as if she had a seizure but after being placed in the hospital, she told her father she was not free of her bipolar symptoms. She remained free of her previous symptoms for 13 years until she experienced post-partum depression. It is interesting to note, the large changes in personality after the birth of her baby followed a massive change in hormone patterns; i.e. another type of sudden signaling event.

2. A woman who snorted 550 times the usual recreational dose of LSD. Her initial response was also close to a catatonic state for about 12 hours. When she returned to a normal state, she found she no longer needed to take morphine for crippling foot pain. The foot pain did come back later, but microdosing with LSD helped manage the pain and eventually she did not need morphine at all.

In addition, there have been controlled experiments with psilocybin, another hallucinogen. There appears to be evidence that psilocybin alters functional connectivity in the brain (Barrett et al. (11) 2020). Two useful surveys of the effects can be found in (Carbonaro et al. (18) 2020) and (Davis et al. (23) 2020). We can summarize some of the case studies as follows:

1. A woman was a heavy smoker for 46 years and was unable to quit. After her participation in the controlled psilocybin trial, she quit smoking and hasn't gone back to the habit.

2. A man was a heavy drinker with up to 20 cocktails a night. After the psilocybin trial in 2016, he stopped drinking and has no desire for a drink even now.

3. A woman in stage four terminal cancer was completely crippled by anxiety and the psilocybin trial enabled her to find peace, which is no small thing to achieve.

In these studies, the people involved experienced profound world change. This is also similar to zombie creation as it is perhaps a global reorganization of how the computational units of their brain interact. In (Barrett et al. (11) 2020), there is discussion about how this might be done, so we encourage you to read that for background.

Finally, ketamine has been used recently to treat intractable depression. Although typically used an anesthetic, it has been found useful in rapidly reducing suicide impulses leading to life-threatening thoughts and acts and for treating depression that is combined with anxiety. Many other treatments take weeks to months to be useful. These treatments include antidepressant medicines, transcranial magnetic stimulation and electroconvulsive therapy that are used to attempt to treat major depression that is not responding to other therapeutic options. Ketamine comes in two forms, which are mirror-image molecules: the R and the S form. These two forms interact differently in the brain and it is unknown which form is the best to use. Also, although it is very effective, it is not known how it works. But it does exert a powerful antidepressant effect through a new mechanism. A likely target is the NMDA receptor in the brain and when ketamine binds to this receptor, it seems to increase glutamate in the fluid and structures surrounding neurons. Glutamate then binds to the AMPA receptor, which probably causes the release of other molecules that aid in neuronal communication along possibly new pathways. It might also have an anti-inflammatory effect. A nice study is the one of (Tiger et al. (116) 2020).

Homework

Exercise 16.5.4 *Do a literature search of the use of ketamine for the treatment of depression untreatable by other methods. Read about the neurobiological pathways that might be involved and* *think about how it is possible for such a radical reorganization of the functioning of the brain can be modeled.*

Exercise 16.5.5 *Do a literature search about the treatment of schizophrenics. Read about the neurobiological pathways that might be involved and read about how the actual brain structure of the patient might be different from normal. Also read about how some designer drugs, applied as injections, seem to give rise to symptoms that are similar to schizophrenia. Do you think this means an injectable trigger might be useful here as a therapy? How could we model this sort of thing?*

16.5.5 What Does This Mean?

This sort of behavioral modification of the prey in these parasitic attacks is very reminiscent of the brain state alteration we seek to obtain, in a reversible way, in anesthesia. So all give us additional examples of **iZombie** creation. Some parasites rely more on altering the connections between the host's immune system and the nervous system. Hence, studying these sorts of parasites could potentially help us learn how alterations in these connections can lead to autoimmune disease, as that is another place where such immune and nervous system interaction are very important. Also note Parasitic manipulation of the nervous system has risks as there must be a tradeoff between control and damage. The neurons must function well enough to do essential tasks and altering them has the risk of going too far and causing host death. In general, each parasitic attack is different, so each one we study gives us new mechanisms of manipulation. This in turn is data for building both better brain models and models of consciousness because if we understand how these inputs can alter the outputs of a mature brain system, we are closer to understanding how a brain model should be built.

We note that in building models of autoimmune disease, it is also important to recognize that ideas arising from the creation of **iZombie** states, either by anesthesia drugs or the venom cocktails of a parasite, shed important light on how a mature nervous system responds to outside signals and also how outside signals determine cognitive and hence consciousness state. We are therefore interested in ways to model how associations are made in a functioning neural system as we feel this leads to the creation of consciousness and consciousness states. Whether the brain model is from a small animal such as a ctenophore, a spider, or a more complex organism such as a squid or human, we feel there is a lot of commonality in how a functioning brain takes raw sensory input and transforms it into outputs which help the organism thrive.

Homework

Exercise 16.5.6 *Find papers that discuss how inflammatory signals can cross the blood-brain barrier to get into the brain and disrupt the normal functioning of the brain neurobiological networks. You can find papers that indicate that such signals originate in the gut biome interactions with ingested substances or foreign proteins from exposure to insect bites or infection. Hence, cognition appears to be influenced by interacting complex systems that control immune reactions.*

Exercise 16.5.7 *Find papers about the LRRK2 regulatory gene and how it can function as a multi-input and multi-output controller that determines immune response and cognitive performance.*

16.5.6 Lesion Studies

It is possible to modify only part of a graph architecture to achieve an input-output recognition. We discuss this algorithm in (Peterson (96) 2016) so we will discuss this idea briefly here. Once we have a given graph structure for our model, we will probably want to update it by adding module interconnections, additional modules and so forth. We find this is our normal pathway to building a better model. Our understanding of the neural computations we need to approximate will change with new data and we then need to add additional complexity to our model. However, just to illustrate what can happen, let's just look at what happens when we add a new subgraph to an existing graph. We assume we have a model $\mathcal{G}_1(\mathcal{V}_1, \mathcal{E}_1)$ with incidence matrix $\boldsymbol{K_1}$. We then add to that model the subgraph $\mathcal{G}_2(\mathcal{V}_2, \mathcal{E}_2)$ with incidence matrix $\boldsymbol{K_2}$. When we combine these neural modules, we must decide on how to connect the neurons of $\mathcal{G}_1$ to the neurons of $\mathcal{G}_2$. The combined graph has $N_1 + N_2$ nodes and $E_1 + E_2$ edges plus the additional edges from the connections between the modules. The combined graph has an incidence matrix $\boldsymbol{K}$ and the connections between $\mathcal{G}_1$ to $\mathcal{G}_2$ will give rise to submatrices in $\boldsymbol{K}$ of the form

$$\boldsymbol{K} \quad = \quad \left[\begin{array}{c|c|c} \boldsymbol{K}_1 & \boldsymbol{O}_1 & \boldsymbol{C} \\ (N_1 \times E_1) & (N_1 \times E_2) & (N_1 \times N) \\ \hline \boldsymbol{O}_2 & \boldsymbol{K}_2 & \boldsymbol{D} \\ (N_2 \times E_1) & (N_2 \times E_2) & (N_2 \times N) \end{array}\right]$$

where N is the number of intermodule edges we have added. The new incidence matrix is thus $(N_1 + N_2) \times (E_1 + E_2 + N)$. The matrix $\boldsymbol{K}^T$ is then

$$\boldsymbol{K}^T \quad = \quad \left[\begin{array}{c|c} \boldsymbol{K}_1^T & \boldsymbol{O}_2^T \\ (E_1 \times N_1) & (E_1 \times N_2) \\ \hline \boldsymbol{O}_1^T & \boldsymbol{K}_2 \\ (E_2 \times N_1) & (E_2 \times N_1) \\ \hline \boldsymbol{C}^T & \boldsymbol{D}^T \\ (N \times N_1) & (N \times N_2) \end{array}\right]$$

and the new Laplacian is

$$\boldsymbol{K}\boldsymbol{K}^T \quad = \quad \left[\begin{array}{c|c|c} \boldsymbol{K}_1 & \boldsymbol{O}_1 & \boldsymbol{C} \\ (N_1 \times E_1) & (N_1 \times E_2) & (N_1 \times N) \\ \hline \boldsymbol{O}_2 & \boldsymbol{K}_2 & \boldsymbol{D} \\ (N_2 \times E_1) & (N_2 \times E_2) & (N_2 \times N) \end{array}\right] \left[\begin{array}{c|c} \boldsymbol{K}_1^T & \boldsymbol{O}_2^T \\ (E_1 \times N_1) & (E_1 \times N_2) \\ \hline \boldsymbol{O}_1^T & \boldsymbol{K}_2 \\ (E_2 \times N_1) & (E_2 \times N_2) \\ \hline \boldsymbol{C}^T & \boldsymbol{D}^T \\ (N \times N_1) & (N \times N_2) \end{array}\right]$$

After multiplying, we have (multiplies involving our various zero matrices $\boldsymbol{O}_1$ and $\boldsymbol{O}_2$ vanish)

$$\boldsymbol{K}\boldsymbol{K}^T \quad = \quad \left[\begin{array}{c|c} \boldsymbol{K}_1\boldsymbol{K}_1^T + \boldsymbol{C}\boldsymbol{C}^T & \boldsymbol{C}\boldsymbol{D}^T \\ (N_1 \times N_1) & (N_1 \times N_2) \\ \hline \boldsymbol{D}\boldsymbol{C}^T & \boldsymbol{K}_1\boldsymbol{K}_1^T + \boldsymbol{C}\boldsymbol{C}^T \\ (N_2 \times N_1) & (N_2 \times N_2) \end{array}\right]$$

This can be rewritten as

$$\boldsymbol{K}\boldsymbol{K}^T \quad = \quad \begin{bmatrix} \boldsymbol{K}_1\boldsymbol{K}_1^T & \boldsymbol{O} \\ \boldsymbol{O} & \boldsymbol{K}_2\boldsymbol{K}_2^T \end{bmatrix} + \begin{bmatrix} \boldsymbol{C}\boldsymbol{C}^T & \boldsymbol{O} \\ \boldsymbol{O} & \boldsymbol{D}\boldsymbol{D}^T \end{bmatrix} + \begin{bmatrix} \boldsymbol{O} & \boldsymbol{C}\boldsymbol{D}^T \\ \boldsymbol{D}\boldsymbol{C}^T & \boldsymbol{O} \end{bmatrix}$$

for zero submatrices of appropriate sizes in each matrix all labeled $\boldsymbol{O}$. Recall the Laplacian updating algorithm is given by

$$\boldsymbol{K}\boldsymbol{K}^T \boldsymbol{f} - \alpha \frac{\partial \boldsymbol{f}}{\partial t} - \beta \boldsymbol{f} \quad = \quad -\boldsymbol{\mathcal{I}},$$

which can be written in finite difference form using $\boldsymbol{f}(t)$ for the value of the node values at time t as

$$\boldsymbol{K}\boldsymbol{K}^T \boldsymbol{f}(t) - \alpha \boldsymbol{\Delta f}(t) - \beta \boldsymbol{f}(t) \quad = \quad -\boldsymbol{\mathcal{I}}(t),$$

where $\boldsymbol{I}(t)$ is the external input at time t and $\boldsymbol{\Delta f}(0) = 0$ and otherwise $\boldsymbol{\Delta f}(t) = \boldsymbol{f}(t) - \boldsymbol{f}(t-1)$. Now divide the node vectors $\boldsymbol{f}$ and $\boldsymbol{\mathcal{I}}$ into the components $\boldsymbol{f_I}$, $\boldsymbol{f_{II}}$, $\boldsymbol{\mathcal{I}_I}$ and $\boldsymbol{\mathcal{I}_{II}}$ to denote the nodes for subgraph $\boldsymbol{\mathcal{G}_1}$ and $\boldsymbol{\mathcal{G}_2}$, respectively. Then, we see we can write the Laplacian for the full graph in terms of the components used to assemble the graph from its modules and module interconnections.

$$\begin{aligned} \boldsymbol{K}\boldsymbol{K}^T \begin{bmatrix} \boldsymbol{f_I} \\ \boldsymbol{f_{II}} \end{bmatrix} &= \begin{bmatrix} \boldsymbol{K}_1\boldsymbol{K}_1^T & \boldsymbol{O} \\ \boldsymbol{O} & \boldsymbol{K}_2\boldsymbol{K}_2^T \end{bmatrix} \begin{bmatrix} \boldsymbol{f_I} \\ \boldsymbol{f_{II}} \end{bmatrix} + \begin{bmatrix} \boldsymbol{C}\boldsymbol{C}^T & \boldsymbol{O} \\ \boldsymbol{O} & \boldsymbol{D}\boldsymbol{D}^T \end{bmatrix} \begin{bmatrix} \boldsymbol{f_I} \\ \boldsymbol{f_{II}} \end{bmatrix} \\ &+ \begin{bmatrix} \boldsymbol{O} & \boldsymbol{C}\boldsymbol{D}^T \\ \boldsymbol{D}\boldsymbol{C}^T & \boldsymbol{O} \end{bmatrix} \begin{bmatrix} \boldsymbol{f_I} \\ \boldsymbol{f_{II}} \end{bmatrix} \end{aligned}$$

This can be rewritten as

$$\boldsymbol{K}\boldsymbol{K}^T \begin{bmatrix} \boldsymbol{f_I} \\ \boldsymbol{f_{II}} \end{bmatrix} = \begin{bmatrix} \boldsymbol{K}_1\boldsymbol{K}_1^T + \boldsymbol{C}\boldsymbol{C}^T \\ \boldsymbol{K}_2\boldsymbol{K}_2^T + \boldsymbol{D}\boldsymbol{D}^T \end{bmatrix} \begin{bmatrix} \boldsymbol{f_I} \\ \boldsymbol{f_{II}} \end{bmatrix} + \begin{bmatrix} \boldsymbol{C}\boldsymbol{D}^T \\ \boldsymbol{D}\boldsymbol{C}^T \end{bmatrix} \begin{bmatrix} \boldsymbol{f_{II}} \\ \boldsymbol{f_I} \end{bmatrix}$$

Now substitute this into the update equation to find

$$\begin{bmatrix} \boldsymbol{K}_1\boldsymbol{K}_1^T + \boldsymbol{C}\boldsymbol{C}^T \\ \boldsymbol{K}_2\boldsymbol{K}_2^T + \boldsymbol{D}\boldsymbol{D}^T \end{bmatrix} \begin{bmatrix} \boldsymbol{f_I} \\ \boldsymbol{f_{II}} \end{bmatrix} + \begin{bmatrix} \boldsymbol{C}\boldsymbol{D}^T \\ \boldsymbol{D}\boldsymbol{C}^T \end{bmatrix} \begin{bmatrix} \boldsymbol{f_{II}} \\ \boldsymbol{f_I} \end{bmatrix} - \alpha \begin{bmatrix} \boldsymbol{\Delta f_I} \\ \boldsymbol{\Delta f_{II}} \end{bmatrix} - \beta \begin{bmatrix} \boldsymbol{f_I} \\ \boldsymbol{f_{II}} \end{bmatrix} = - \begin{bmatrix} \boldsymbol{\mathcal{I}_I} \\ \boldsymbol{\mathcal{I}_{II}} \end{bmatrix}$$

This can be then further rewritten as follows:

$$\begin{aligned} \boldsymbol{K}_1\boldsymbol{K}_1^T\, \boldsymbol{f_I} - \alpha\, \boldsymbol{\Delta f_I} - \beta \boldsymbol{f_I} + \boldsymbol{C}\boldsymbol{C}^T\, \boldsymbol{f_I} + \boldsymbol{C}\boldsymbol{D}^T\, \boldsymbol{f_{II}} &= -\boldsymbol{\mathcal{I}_I} \\ \boldsymbol{K}_2\boldsymbol{K}_2^T\, \boldsymbol{f_{II}} - \alpha\, \boldsymbol{\Delta f_{II}} - \beta \boldsymbol{f_{II}} + \boldsymbol{D}\boldsymbol{D}^T\, \boldsymbol{f_{II}} + \boldsymbol{D}\boldsymbol{C}^T\, \boldsymbol{f_I} &= -\boldsymbol{\mathcal{I}_{II}} \end{aligned}$$

Hence, the terms $\left(\boldsymbol{C}\boldsymbol{C}^T + \boldsymbol{C}\boldsymbol{D}^T\right)\, \boldsymbol{f_I}$, $\left(\boldsymbol{D}\boldsymbol{D}^T + \boldsymbol{D}\boldsymbol{C}^T\right)\, \boldsymbol{f_{II}}$, $\boldsymbol{C}\boldsymbol{D}^T \boldsymbol{f_{II}}$ and $\boldsymbol{D}\boldsymbol{C}^T\, \boldsymbol{f_I}$ represent the mixing of signals between modules. We see that if we have an existing model, we can use these update equations to add to an existing graph, a new graph. Thus, we can take a trained brain module and add a new cortical submodule and so forth and to some extent retain the training effort we have already undertaken. Now add time indices and expand the difference terms to get

$$\begin{aligned} \boldsymbol{K}_1\boldsymbol{K}_1^T\, \boldsymbol{f_I^t} - \alpha\, (\boldsymbol{f_I^t} - \boldsymbol{f_I^{t-1}}) - \beta \boldsymbol{f_I^t} + \boldsymbol{C}\boldsymbol{C}^T\, \boldsymbol{f_I^t} + \boldsymbol{C}\boldsymbol{D}^T\, \boldsymbol{f_{II}^t} &= -\boldsymbol{\mathcal{I}_I^t} \\ \boldsymbol{K}_2\boldsymbol{K}_2^T\, \boldsymbol{f_{II}^t} - \alpha\, (\boldsymbol{f_{II}^t} - \boldsymbol{f_{II}^{t-1}}) - \beta \boldsymbol{f_{II}^t} + \boldsymbol{D}\boldsymbol{D}^T\, \boldsymbol{f_{II}^t} + \boldsymbol{D}\boldsymbol{C}^T\, \boldsymbol{f_I^t} &= -\boldsymbol{\mathcal{I}_{II}^t} \end{aligned}$$

If we let $\boldsymbol{\Lambda f} = \boldsymbol{K}\boldsymbol{K}^T \boldsymbol{f} - \alpha \boldsymbol{f} - \beta \boldsymbol{f}$, for the graph with incidence matrix $\boldsymbol{K}$, we can rewrite the update as

$$\begin{aligned} \boldsymbol{\Lambda_I}\, \boldsymbol{f_I^t} + \boldsymbol{C}\boldsymbol{C}^T\, \boldsymbol{f_I^t} + \boldsymbol{C}\boldsymbol{D}^T\, \boldsymbol{f_{II}^t} &= \alpha \boldsymbol{f_I^{t-1}} - \boldsymbol{\mathcal{I}_I^t} \\ \boldsymbol{\Lambda_{II}}\, \boldsymbol{f_{II}^t} + \boldsymbol{D}\boldsymbol{D}^T\, \boldsymbol{f_{II}^t} + \boldsymbol{D}\boldsymbol{C}^T\, \boldsymbol{f_I^t} &= \alpha\, \boldsymbol{f_{II}^{t-1}} - \boldsymbol{\mathcal{I}_{II}^t} \end{aligned}$$

Thus, if we can do partial updates, we can use this to shed some light on some of high-level questions.

Homework

Exercise 16.5.8 *Assume we have a model $\boldsymbol{\mathcal{G}_1}(\boldsymbol{\mathcal{V}_1}, \boldsymbol{\mathcal{E}_1})$ with incidence matrix $\boldsymbol{K_1}$ of size 3×5 and we add to that model the subgraph $\boldsymbol{\mathcal{G}_2}(\boldsymbol{\mathcal{V}_2}, \boldsymbol{\mathcal{E}_2})$ with incidence matrix $\boldsymbol{K_2}$ of size 5×6. Assume you have been given the cross connections which connect the neurons of $\boldsymbol{\mathcal{G}_1}$ to the neurons of $\boldsymbol{\mathcal{G}_2}$. Work out the update equations for the combined graph. Make up specific graphs of these sizes so you can see all the details.*

Exercise 16.5.9 *Assume we have a model $\boldsymbol{\mathcal{G}_1}(\boldsymbol{\mathcal{V}_1}, \boldsymbol{\mathcal{E}_1})$ with incidence matrix $\boldsymbol{K_1}$ of size 3×5 and* *we add to that model the subgraph $\mathcal{G}_2(\mathcal{V}_2, \mathcal{E}_2)$ with incidence matrix $\boldsymbol{K}_2$ of size 5×6. Assume you have been given the cross connections which connect the neurons of $\mathcal{G}_1$ to the neurons of $\mathcal{G}_2$. Work out the update equations for the combined graph.*

Exercise 16.5.10 *Assume we have a model $\mathcal{G}_1(\mathcal{V}_1, \mathcal{E}_1)$ with incidence matrix $\boldsymbol{K}_1$ of size 9×6 and we add to that model the subgraph $\mathcal{G}_2(\mathcal{V}_2, \mathcal{E}_2)$ with incidence matrix $\boldsymbol{K}_2$ of size 8×5. Assume you have been given the cross connections which connect the neurons of $\mathcal{G}_1$ to the neurons of $\mathcal{G}_2$. Work out the update equations for the combined graph.*

We can use these ideas to ask how we might model schizophrenia with a left and right brain and a connecting corpus callosum. Two modules would be the left and right brain subgraph and the third, the corpus callosum subgraph. Let's look at this as a two-graph system where the first graph contains the left and the right brain and the second graph contains the corpus callosum. Assume we have trained on $\boldsymbol{T}$ samples using partial update training equations

$$\begin{aligned}(\boldsymbol{\Lambda}_I + \boldsymbol{\mathcal{F}}_I^T)\, \boldsymbol{f}_I^t + \left(\boldsymbol{\mathcal{I}}_I^t + \boldsymbol{g}_I^t - \alpha\, \boldsymbol{f}_I^{t-1}\right) &= 0\\ (\boldsymbol{\Lambda}_{II} + \boldsymbol{\mathcal{F}}_{II}^T)\, \boldsymbol{f}_{II}^t + \left(\boldsymbol{\mathcal{I}}_{II}^t + \boldsymbol{g}_{II}^t - \alpha\, \boldsymbol{f}_{II}^{t-1}\right) &= 0\end{aligned}$$

At convergence, we obtain stable values for $\boldsymbol{f}_I^t$ and $\boldsymbol{f}_{II}^t$, labeled as $\boldsymbol{f}_I^\infty$ and $\boldsymbol{f}_{II}^\infty$. Then we must have

$$\begin{aligned}(\boldsymbol{\Lambda}_I + \boldsymbol{\mathcal{F}}_I^T)\, \boldsymbol{f}_I^\infty + (\boldsymbol{\mathcal{I}}_I^t + \boldsymbol{g}_I^\infty - \alpha\, \boldsymbol{f}_I^\infty) &= 0\\ (\boldsymbol{\Lambda}_{II} + \boldsymbol{\mathcal{F}}_{II}^T)\, \boldsymbol{f}_{II}^\infty + (\boldsymbol{\mathcal{I}}_{II}^t + \boldsymbol{g}_{II}^\infty - \alpha\, \boldsymbol{f}_{II}^\infty) &= 0\end{aligned}$$

A change in the connectivity from the corpus callosum to the left and right brains and vice versa implies a change $\boldsymbol{C} \rightarrow \boldsymbol{C}'$ and $\boldsymbol{D} \rightarrow \boldsymbol{D}'$. This changes terms which are built from the matrices $\boldsymbol{C}$ and $\boldsymbol{D}$, such as $\boldsymbol{\mathcal{F}}_I^T = \boldsymbol{D}\boldsymbol{D}^T\boldsymbol{E}_{II}$, which changes to $\boldsymbol{D}'(\boldsymbol{D}')^T\boldsymbol{E}_{II}$.

After the connectivity changes are made, the graph can evaluate all $\boldsymbol{T}$ inputs to generate a collection of new outputs $\boldsymbol{S}_I' = \{(\boldsymbol{f}_I^1)', \ldots, (\boldsymbol{f}_I^T)'\}$ and $\boldsymbol{S}_{II}' = \{(\boldsymbol{f}_{II}^1)', \ldots, (\boldsymbol{f}_{II}^T)'\}$. The original outputs due to the trained graph are $\boldsymbol{S}_I$ and $\boldsymbol{S}_{II}$. Hence, any measure of the entries of these matrices gives us a way to compare the performance of the graph model of the brain with a *correct* set of corpus callosum connectivities to the graph model of a brain with dysfunctional connectivity between the corpus callosum and the two halves of the brain.

A simple ratio gives us a useful measure (here, the norm symbols indicate any choice of matrix measurement such as a simple Frobenius norm). Choosing a set of tolerances ϵ_I and ϵ_{II}, we obtain a tool for deciding if there is dysfunction: for example, $\frac{||\boldsymbol{S}_I'||}{||\boldsymbol{S}_I||} > \epsilon_I$ and $\frac{||\boldsymbol{S}_{II}'||}{||\boldsymbol{S}_{II}||} > \epsilon_{II}$.

In general, if we let $\boldsymbol{\Phi}$ denote our tool for measuring the brain model's effectiveness as a function of the measured outputs for a sample set, we would check to see if $\frac{\boldsymbol{\Phi}((\boldsymbol{S}_I'))}{\boldsymbol{\Phi}(\boldsymbol{S}_I)} > \epsilon_I$ and $\frac{\boldsymbol{\Phi}(\boldsymbol{S}_{II}')}{\boldsymbol{\Phi}(\boldsymbol{S}_{II})} > \epsilon_{II}$. The choice of $\boldsymbol{\Phi}$ is, of course, critical and it is not easy to determine a good choice. Another issue is that we need a way to determine what the *normal* output of a brain model should be so we can make a reasonable comparison. However, despite these obstacles, we see this procedure of partial update training allows us to identify theoretical consequences of changes in corpus callosum communication pathways (i.e. the cross connections to the left and right brain) and perhaps gives some illumination to this difficult problem. And, of course, we are still trapped into an analysis based on serial processing and global time clocks.

Let's look at a little more detail. Information passing is modulated by neurotransmitter levels in a brain. Let's focus on three neurotransmitters here: dopamine, serotonin and norepinephrine. Each has an associated triple $(r_u, r_d, r_r) \equiv T$ which we label with a superscript for each neurotransmitter.

Hence, T^D, T^S and T^N are the respective triples for our three neurotransmitters dopamine, serotonin and norepinephrine. Consider a simple graph model $\mathcal{G}(\mathcal{V}, \mathcal{E})$ as shown in Figure 16.9.

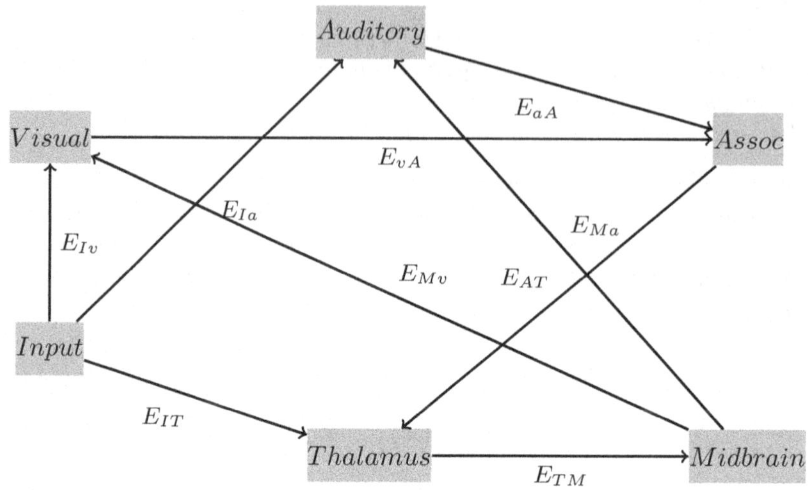

Figure 16.9: A simple cognitive processing map focusing on a few salient modules. The edge connections are of the form $E_{\alpha,\beta}$ where α and β can be a selection from v (Visual), a (Auditory), A (Associative), T (Thalamus), M (Midbrain) and I (Input).

We can design auditory and visual training data in four emotional labelings; **neutral**, **sad**, **happy** and **angry**. We discuss carefully how we design this sort of data in (Peterson (96) 2016) and we refer you there for details. We assume that each emotional state corresponds to neurotransmitter triples (T_N^D, T_N^S, T_N^N) (**neutral**), (T_S^D, T_S^S, T_S^N) (**sad**), (T_H^D, T_H^S, T_H^N) (**happy**) and (T_A^D, T_A^S, T_A^N) (**angry**), In addition, we assume that these triple states can be different in the visual, auditory and associative cortex and hence, we add an additional subscript label to denote that. We let $(T_{\alpha,\beta}^D, T_{\alpha,\beta}^S, T_{\alpha,\beta}^N)$ be the triple for cortex β ($\beta = 0$ is visual cortex, $\beta = 1$ is auditory cortex and $\beta = 2$ is associative cortex) and emotional state α ($\alpha = 0$ is neutral, $\alpha = 1$ is sad, $\alpha = 2$ is happy and $\alpha = 3$ is angry). Hence, we are using our emotionally labeled data as a way of constructing a simplistic normal brain model. Of course, it is much more complicated than this. Emotion, in general, is something that is hard to quantify. It is also difficult to find good measures that determine what emotional state a person is in. A good review of that problem is in (I. Mauss and M. Robinson (52) 2009). Cortical activity that is asymmetric probably plays a role here, see (E. Harmon-Jones and P. Gable and C. Peterson (25) 2010), and what is nice about our modeling choices is that we have the tools to model such asymmetry using the graph models. However, our first model is much simpler.

The graph encodes the information about emotional states in both the nodal processing and the edge processing functions. Let's start with auditory **neutral** data and consider the graph shown in Figure 16.10. The neurotransmitter triples can exist in four states: **neutral**, **sad**, **happy** and **angry** which are indicated by the toggle α in the triples $T_{\alpha,\beta}^{\gamma}$ where γ denotes the neurotransmitter choice ($\gamma = 0$ is dopamine, $\gamma = 1$ is serotonin and $\gamma = 2$ is norepinephrine).

For auditory neutral data, there is a choice of triple $T_{0,1}^{\gamma}$ and $T_{0,2}^{\gamma}$ for each neurotransmitter which corresponds to how the brain labels this data as emotionally neutral. Hence, we are identifying a collection of six triples or eighteen numbers with neutral auditory data. This is a vector in $\Re^{18}$ which we will call $\boldsymbol{V}_{0,1,2}$. We need to do this for the other emotional states, which gives us three additional vectors $\boldsymbol{V}_{1,1,2}$, $\boldsymbol{V}_{2,1,2}$ and $\boldsymbol{V}_{3,1,2}$. Thus, emotional processing for emotionally labeled auditory data requires us to set four vectors in $\Re^{18}$. We need to do the same thing for the processing of emotionally

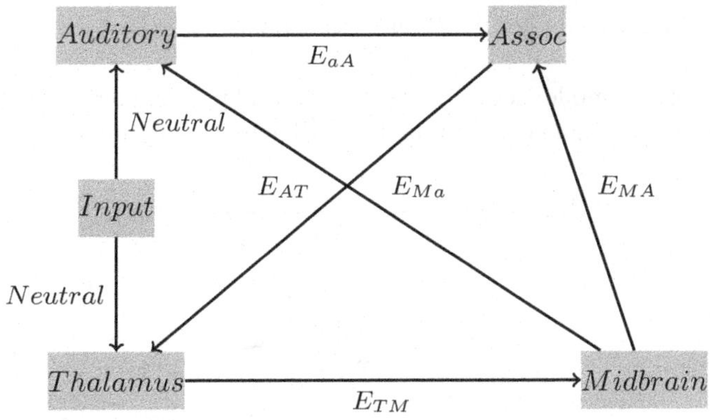

Figure 16.10: Neural processing for neutral auditory data.

labeled visual data which will give us the four vectors $\boldsymbol{V}_{0,0,2}$, $\boldsymbol{V}_{1,0,2}$, $\boldsymbol{V}_{2,0,2}$ and $\boldsymbol{V}_{3,0,2}$.

To process the data for both types of sensory cortex thus requires eight vectors in $\Re^{18}$. Choose 8 orthonormal vectors in $\Re^{18}$ to correspond to these states. To train the full graph $\mathcal{G}(\mathcal{V}, \mathcal{E})$ to understand neutral auditory data, it is a question of what internal outputs of the graph should be fixed or clamped and which should be allowed to alter. For neutral auditory data, we clamp or fix the neurotransmitter triples in the auditory cortex and associative cortex using the vector $\boldsymbol{V}_{0,1,2}$ and we force the output of the associative cortex to be the same as the incoming auditory data. The midbrain module makes the neurotransmitter connections to each cortex module, but how these connections are used is determined by the triples and the usual nodal processing. In a sense, we are identifying the neutral auditory emotional state with the vector $\boldsymbol{V}_{0,1,2}$. We assume the thalamus module can shape processing in the midbrain (we show a simple model of thalamus processing later). Hence, we will let $\boldsymbol{V}_{0,1,2}$ be the desired output for the thalamus module for neutral auditory data. The midbrain module accepts the input $\boldsymbol{V}_{0,1,2}$ and uses it to set the triple states in the auditory and associative cortex. In effect, this is an abstraction of what are called second messenger systems that affect the neural processing in these cortical modules. We can do this for each emotional state. For example, Figure 16.11 shows the requisite processing for the cortical submodules.

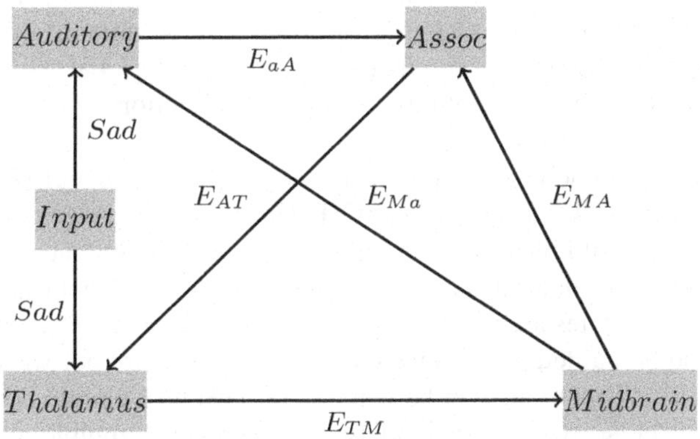

Figure 16.11: Neural processing for sad auditory data.

Processing the sad auditory data clamps the triples in auditory and associative cortex using $\boldsymbol{V}_{1,1,2}$ and clamps the associative cortex output to the incoming sad auditory data. We then train the graph to have a Thalamus output of $\boldsymbol{V}_{1,1,2}$. We do the same thing for the other emotional states for auditory and for all the emotional states for the visual cortex data.

Now consider the graph model accepting new music and painting data. We see the requisite graph in Figure 16.12. The music data and painting data will generate an output vector $\boldsymbol{W}$ from the thalamus

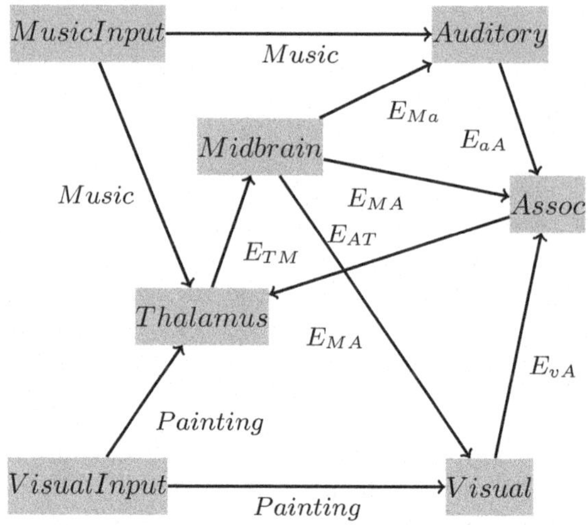

Figure 16.12: Neural processing for new data.

module in addition to generating associative cortex outputs corresponding to a music and painting fragment on the basis of the encoded edge and nodal processing that has been set by the training. Hence, we can classify the new combined auditory and visual data as corresponding to an emotional labeling of the form shown in Equation 16.9.

$$\boldsymbol{W} \quad = \quad \sum_{i=1}^{4} < \boldsymbol{W}, \boldsymbol{V}_{i,1,2} > \boldsymbol{V}_{i,1,2} + \sum_{i=1}^{4} < \boldsymbol{W}, \boldsymbol{V}_{i,0,2} > \boldsymbol{V}_{i,0,2} \tag{16.9}$$

This is the projection of $\boldsymbol{W}$ to the subspace of $\Re^{18}$ spanned by our neurotransmitter triples. We can then perform standard cluster analysis to determine which emotional labeling is closest. The training process we have gone through generates what we will call the **normal** brain. This is, of course, an abstraction of many things and it is a serial analysis based on a global time clock, but we think it has enough value that it can give us insight into cognitive dysfunction.

Let's summarize this approach at this point. For a given brain model $\mathcal{G}(\mathcal{V}, \mathcal{E})$ which consists of auditory sensory cortex, **SCI**, visual cortex, **SCII**, Thalamus, **Th**, Associative Cortex, **AC**, and Midbrain, **MB**, we can now build its normal emotional state. We assign neurotransmitter triples $T^{\gamma}_{\alpha,\beta}$ for the four emotional states of neutral, sad, happy and angry for each neurotransmitter. These eighteen values for eight possible states are chosen as orthonormal vectors in $\Re^{18}$, but our choice of values for these triples can also be guided by the biopsychological literature but we will not discuss that here. The auditory and visual input data can be mapped into exclusive regions of skin galvanic response and an fMRI parameter, so part of our modeling is to create a suitable abstraction of this mapping from the inputs into the two dimensional skin galvanic response and fMRI space, which is $\Re^2$. To construct the normal model, we do the following:

- For neutral music data, we train the graph to imprint the emotional states as follows: for neutral music inputs to **SCI**, the clamped **AC** output is the same neutral music input and the clamped **Th** to **MB** outputs are the neutral triples $T^{\gamma}_{0,1}$ and $T^{\gamma}_{0,2}$ encoded as $\boldsymbol{V}_{0,1,2}$. When training is finished, the neutral music samples are recognized as neutral music with the neutral neurotransmitter triples engaged.

- For neutral painting data, use the combined evaluation/Hebbian update loop to imprint the emotional states by clamping **AC** output to the neutral paint input to **SCII** and clamping the **Th** to **MB** outputs to the neutral triples $T^{\gamma}_{0,1}$ and $T^{\gamma}_{0,2}$ encoded as $\boldsymbol{V}_{0,0,2}$. When training is finished, the neutral painting samples are recognized as neutral paintings with the neutral neurotransmitter triples engaged.
- Do the same training loop for the sad, happy and angry music and painting data samples using the sad, happy and angry neurotransmitter triples. The **AC** output is clamped to the appropriate data input values and the **Th** to **MB** outputs are the required neurotransmitter triples.

At this point, the model $\mathcal{G}(\mathcal{V}, \mathcal{E})$ assigns known emotionally labeled data in two sensory modalities to their correct outputs in associative cortex. A new music and painting input would then be processed by the model and the **Th** to **MB** connections would assign the inputs to a set of neurotransmitter triple states $\boldsymbol{W}$ and generate a set of **AC** outputs for both the music and painting inputs. These outputs would be interpreted as a new musical sequence and a new painting. We can think of this trained model $\mathcal{G}(V, E)$ as defined the *normal* state. Also, as shown in Equation 16.9, we can assign an emotional labeling to the new data.

We can model dysfunction at this point by allowing changes in midbrain processing.

- Alter T^D, T^S and T^N in the auditory, visual and associative cortex to create a new vector $\boldsymbol{W}$ and therefore a new model $\mathcal{G}^{new}(\mathcal{V}, \mathcal{E})$. The nodes and edges do not change but the way computations are processed does change. The change in model can be labeled $\delta\mathcal{G}$. Each triple change $(\delta \boldsymbol{T^D}, \delta \boldsymbol{T^S}, \delta \boldsymbol{T^N})$ gives a potential mapping to cognitive dysfunction.
- Using the normal model $\mathcal{G}(\mathcal{V}, \mathcal{E})$, map all neutral music and painting data to sad data. Hence, we clamp **AC** outputs to sad states but do not clamp the **Th** to **MB** mapping. This will generate new neurotransmitter triple states in **MB** which we can label $\boldsymbol{W}^{sad}$. The exhibited changes $(\delta \boldsymbol{T^D}, \delta \boldsymbol{T^S}, \delta \boldsymbol{T^N})$ give us a quantitative model of a cognitive dysfunction having some of the characteristics of depression. Note all 18 parameters from all 3 neurotransmitters are potentially involved. Note also standard drugs such as *Abilify* only make adjustments to 2 or 3 of these parameters which suggests our current pharmacological treatments are too narrow in scope. If we know the normal state, this model would suggest to restore normalcy, we apply drugs which alter neurotransmitter activity by $(-\delta \boldsymbol{T^D}, -\delta \boldsymbol{T^S}, -\delta \boldsymbol{T^N})$. The idea of applying an external stimulus to alter brain functionality is discussed carefully in Section 16.5. In that section, we give examples from biology and surgery that support the premise that such triggers are possible and occur naturally.
- Lesions in a brain can be simulated by the removal of nodes and edges from the graph model $\mathcal{G}(\mathcal{V}, \mathcal{E})$. This allows us to study when the normal emotional responses alter and lead to dysfunction.
- If we construct a model $\mathcal{G}_{\mathcal{L}}(\mathcal{V}_{\mathcal{L}}, \mathcal{E}_{\mathcal{L}})$ for the left half of the brain, a model $\mathcal{G}_{\mathcal{R}}(\mathcal{V}_{\mathcal{R}}, \mathcal{E}_{\mathcal{R}})$ and a model of the corpus callosum, $\mathcal{G}_{\mathcal{C}}(\mathcal{V}_{\mathcal{C}}, \mathcal{E}_{\mathcal{C}})$, we can combine these modules using the ideas to create a brain model $\mathcal{G}(\mathcal{V}, \mathcal{E})$, which can be used to model problems with right and left half miscomunnication problems such as schizophrenia which could be very useful. The interconnections between the right and left half of the brain through the corpus callosum modules and neurotransmitter connections can be altered and the responses of the altered model to a normal brain model can be compared for possible insight.

From our discussions so far, we see we can make progress in a quest to obtain insight from graph architectures that are modeled serially with a global time clock. We can model a shuffling of the computational modules that comprise cognition to some degree. However, these models make severe assumptions: serial computation and global time clocks. Can we do better?

Homework

The high-level questions we are thinking about here are of great interest to us, yet it is not clear at all how to develop models that allow us to probe how such complex systems operate. Here are some questions designed to make you think carefully about such things.

Exercise 16.5.11 *In a complex system, there are low-level building blocks and groupings of such components into larger modules. The interactions of the low-level components are usually not amenable to being described by variables that depend smoothly on other variables. However, much modeling is done via differential equations. Is it possible to reimagine a general graph model as a differential equation? What would* $\Delta\mathcal{G}(\mathcal{V},\mathcal{E})$ *mean here? To implement as a differential equation, wouldn't you have to have a global simulation counter which implies a global time clock?*

Exercise 16.5.12 *If the proper framework to use is that of a manifold, think about what that implies in the modeling. We are assuming the topological structure of the graph is a manifold. Do you think it is reasonable that this manifold will possess a Riemannian metric?*

Exercise 16.5.13 *If the proper framework to use is that of a topological vector space, the topology could be constructed from functionals. It is straightforward to define addition of graphs, but what should scalar multiplication mean? We would need to define scalar multiplication to think about linear functionals. Can you think of ways to define a linear functional on* $\mathcal{G}$*?*

Exercise 16.5.14 *Can you think of a way to define the construction of a consciousness state as the result of a* $I-T$ *computation where* T *is a compact perturbation of the identity? To do this, we would need to frame the space of graphs as a normed space. We do know how to determine if a topological space is normable. Note the Leray - Schauder degree is constant on components. Is it possible to use this to show that the proper injection of a signal can move the solution to a different component?*

Exercise 16.5.15 *Can you think of a way to define the construction of a consciousness state as the result of a* $L+G$ *computation where* L *is a linear Fredholm map of index* 0 *and* G *is* L *- compact? To do this, we would need to frame the space of graphs as a normed space. We do know how to determine if a topological space is normable. Note coincidence degree is also constant on components. Is it possible to use this to show that the proper injection of a signal can move the solution to a different component?*

16.6 Message Passing Architectures

Based on the discussions above, we can propose a general model of consciousness and immune modulation based on asynchronous models of message passing in a graph with signals. Data from iZombie creation from anesthesia records and creation from parasites as well as the reassembly of what we would call normal consciousness from an epileptic episode and modern experiments with psilocybin, LSD and ketamine in behavioral modification in intractable cases, suggest computational modules whose interaction is determined by the topology of the graph in its usual signal ecosystem are a useful model with explanatory capability. From this known data, it is clear external events can reprogram a host into new behavioral patterns and states of consciousness. Since all of the usual neural modules are present, we can posit that these external inputs alter the usual connections and mappings between the functioning neural modules allowing the full brain outputs to change. The real question is, how can we build a model that shows this kind of behavior? Eventually, we will use a simplified model of cortex-thalamus information processing for the consciousness models. However, this model can interact with an associated immune model to create complex scenarios. So all of this structure is best organized in conjunction with a message passing/signal enabled asynchronous graph architecture.

Our approach is based on ideas from theoretical immunology such as in (A. Rot and U. von Andrian (6) 2004) that involve the bar codes associated with cytokine messages and from homology, degree theory and differential geometry. We will discuss the detail of this approach later. For now, we will provide highlights only. A signal from a member of a set of specialized messenger molecules called cytokines is associated with a code we call a bar code, which represents an abstraction of the protein patch the cytokine constructs on the surface of its target cell. These signals are then used to model send and receive signal processing functions which determine the topology and hence dynamics of signal chains processed by the graph.

Determining the correct way to model the data we can measure for this process is difficult. For example, what is the appropriate set of objects that forms our space X? Is X a topological space, a topological vector space or a manifold? What topology is useful? One technique is to take point cloud data determined by measurements of cytokines via methods such as flow cytometry to find the right decomposition of the signals as a direct sum of subgraphs or submodules. This approach is discussed in (G. Carlsson (35) 2009), (S. Lafon and A. Lee (108) 2006), and (G. Singh and F. Memoli and T. Ishkhanov and G. Sapiro and G. Carlsson and D. Ringach (39) 2008), (H. Edelsbrunner and D. Letscher and A. Zomorordian (41) 2002) and (X. Li and Z. Xie and D. Yi (121) 2012) but we do not follow that approach here. For the moment, assume the asynchronous graph structure, $\mathcal{G}(\mathcal{V}, \mathcal{E})$, is known.

We now look carefully at a general abstract model of asynchronous computation in a graph architecture with message passing architecture. We explore carefully an architecture which assumes there is a cost of both sending and receiving messages. This leads to some nodes becoming permanently inaccessible in the sense that these nodes cannot send messages nor receive them. We then add to this model signals exchanged between nodes with content using edge and node processing functions. These computational graphs then give insight into the structure of immune and cognitive models for autoimmune disease and consciousness. An understanding of the kinds of topologies that emerge during their evolution, which is shaped by their dynamics, is essential. When these networks are first initialized, we should be wary of paying too much attention to their positional information and instead should focus on the kind of manifold structure that can arise due to the signals that drive the network. Insights from the asynchronous graph model can then be used to explain aspects of an architectural blueprint for biological information processing. These insights, rephrased for immunosynapse networks, can also be used to study central nervous system (CNS) damage models due to viral, bacterial and other toxins and pathogens. We have discussed these lesion models already for the serial computation case, but we are developing a new suite of tools here. It can also be used to study other complex systems such as economic and neuro-economic systems (C. Camerer and J. Cohen and E. Fehr and P. Glimcher and D. Laibson (15) 2016) as well. It might also be useful in building a blueprint for information processing which can be applied to real-time assessments of consciousness to determine proper levels of anesthesia.

So in what follows, we explore neural and immune system computation from the point of view of asynchronous computation and local time clocks which can be implemented within the powerful framework of Erlang processes (Armstrong (7) 2013) and the zeroMQ implementations of threads (P. Hintjens (94) 2013), although we do not discuss those implementation details here. It is important to look carefully at this level of detail so that the algorithms we develop are reasonably faithful to a variety of information processing challenges.

16.6.1 Computational Graph Models for Information Processing

We begin our analysis using traditional serial processing. Here, we do use the notion of a global time clock and examine edge and node interactions using fairly traditional simulation loops in terms of simulation counter t which increments in unit amounts.

Let's look carefully now at graphs where both the nodes and the edges process information. In the biological context, this processing is quite complex. A general model, without specializing to biological processing, would thus be a directed graph architecture consisting of computational nodes $\mathcal{V}$ and edge functions $\mathcal{E}$ which mediate the transfer of information between two nodes or vertices. Hence, if $\boldsymbol{V_i}$ and $\boldsymbol{V_j}$ are two computational nodes, then $\boldsymbol{E_{i \rightarrow j}}$ would be the corresponding edge function that handles information transfer from node $\boldsymbol{V_i}$ and node $\boldsymbol{V_j}$. The symbol $\mathcal{G}(\mathcal{V}, \mathcal{E})$ denotes this graph. For our purposes, we may assume the maximum number of nodes is finite, albeit quite large, labeled as $\boldsymbol{N}^{\max}$. In the cognitive models, $\boldsymbol{N}^{\max} \approx 10^{10}$ and the number of edges is perhaps 100000 times larger, i.e. $\approx 10^{14}$ with similar sizes for immune models. For fixed $\boldsymbol{N}^{\max}$, the graph whose nodes are all connected has $\boldsymbol{N}^{\max}(\boldsymbol{N}^{\max} - 1)/2$ edges which we denote by $\mathbb{H}(\boldsymbol{N}^{\max})$ which is identified with $\mathbb{H}$ if the size choice of $\boldsymbol{N}^{\max}$ is clear from context. Each of our graphs, $\mathcal{G}(\mathcal{V}, \mathcal{E})$ then corresponds to a graph of $\boldsymbol{N_V}$ nodes and $\boldsymbol{N_E}$ edges. We can destroy the edge between two nodes in $\mathbb{H}$ by simply setting the edge function to be zero and we can remove a node from $\mathbb{H}$ by setting the edge functions connecting to that node to be zero and the edge functions coming out of the node to be zero as well. Hence, each graph $\mathcal{G}(\mathcal{V}, \mathcal{E})$ can be obtained from $\mathbb{H}$ by applying an edge deletion map $\mathscr{E}$ to $\mathbb{H}$. It is convenient to split this mapping into two pieces: the first deletes nodes and the second deletes edges between active nodes. Given a node $\boldsymbol{V}_i$, the set of nodes that connect to it is the *backward set* of the node $\mathscr{B}(\boldsymbol{V_i})$ and the set of nodes it connects to is the *forward set* of the node, $\mathscr{F}(\boldsymbol{V_i})$. Then if $\boldsymbol{V_i}$ is to be deleted, we set the edges $\{\boldsymbol{E_{j \rightarrow i}} : \boldsymbol{V_j} \in \mathscr{F}(\boldsymbol{V_i}) \text{ or } \boldsymbol{V_i} \in \mathscr{B}(\boldsymbol{V_j})\}$ to zero. We let $\mathscr{E}_v$ denote the mapping that deletes the nodes and $\mathscr{E}_e$, the mapping that deletes edges from active nodes. Thus, given a graph $\mathcal{G}(\mathcal{V}, \mathcal{E})$ there are two mappings, $\mathscr{E}_v(\mathcal{G}(\mathcal{V}, \mathcal{E}))$ and $\mathscr{E}_e(\mathcal{G}(\mathcal{V}, \mathcal{E}))$ that create the graph $\mathcal{G}(\mathcal{V}, \mathcal{E})$. Hence,

$$\mathcal{G}(\mathcal{V}, \mathcal{E}) = \mathscr{E}_v(\mathcal{G}(\mathcal{V}, \mathcal{E}))\Big(\mathscr{E}_e(\mathcal{G}(\mathcal{V}, \mathcal{E}))(\mathbb{H})\Big).$$

This is cumbersome notation, so we will simply write $\mathscr{E}_{ve}(\mathcal{G}) = \mathscr{E}_v(\mathcal{G})\, \mathscr{E}_e(\mathcal{G})$ and say

$$\mathcal{G}(\mathcal{V}, \mathcal{E}) = \mathscr{E}_{ve}(\mathcal{G})(\mathbb{H}).$$

Our universe of potential graphs is thus chosen from the base graph $\mathbb{H}$ as described above. At this point, the edge functions are quite simple: an edge function is 1 if the nodes are connected and 0 if not. Hence, no matter what the edge and node functions are, the physical architecture that comprises the edges and nodes can be considered fixed within the universe of $\mathbb{H}$.

The computations performed by nodes are usually quite complex and the edge functions $\boldsymbol{E_{j \rightarrow i}}$ between the nodes $\boldsymbol{V_j}$ and $\boldsymbol{V_i}$ are equally complicated in general. Here is a sample of such a calculation in a cognitive model. At a given neuron $\boldsymbol{V_i}$, which is the node in our graph, the following simplified events occur:

- The axons from the neurons in $\mathscr{B}(\boldsymbol{V_i})$ connect to the input side of neuron $\boldsymbol{V_i}$ as follows. The output signal from neuron $\boldsymbol{V_j}$ is a time dependent voltage spike which is sent undiminished to the synaptic bulb where the voltage signal, if it is in the proper format, initiates a Ca^{+2} current spike inside the synaptic vesicle which initiates the movement of packets of neurotransmitter to move to the cellular wall of the vesicle where they pass through the cell wall and enter the synaptic cleft. This neurotransmitter release is quantal. The neurotransmitter released then either initiates a voltage spike of the input side of neuron $\boldsymbol{V_i}$ or binds to a specialized gate on the cell wall of the input side of neuron $\boldsymbol{V_i}$, which in turns generates a molecular signal which alters the physical structure of neuron $\boldsymbol{V_i}$ thereby changing how it generates outputs. Note, the output signal is a voltage signal, which is transduced into a current signal at the synaptic bulb and then further transduced into a quantal chemical packet signal which then either induces a voltage signal or a molecular signal which goes to the nucleus of neuron $\boldsymbol{V_i}$ to initiate protein creation to alter the physical structure of neuron $\boldsymbol{V_i}$. This is a simplified view of this computation. An interaction leading to a voltage signal on the input side of the neuron is called a **First Messenger System** response while the molecular signal response is in the **Second Messenger System** category. The background on the chemical physics behind these signals can be found in many books and papers.

- There are many ways to abstract this process. The easiest is to set the edge functions to be scalars whose size is dependent on the strength of the neurotransmitter signal from the synaptic bulb. The node function is then the voltage signal response, which is often modeled as a simple sigmoid response. But other choices are possible. Each edge function accepts the output of a neuron, processes it and passes its computational result to the neuron it is connecting to. There are many further ideas on how to abstract these computations in this context in the open literature.

Hence, if we let Φ_e denote a function space of edge functions and Φ_v a function space of node functions, the graph model $\mathcal{G}(\mathcal{V},\mathcal{E})$ which is constructed from $\mathbb{H}$ must also have its corresponding edge and node functions applied. Hence, the computational graph we use requires a choice of edge function from Φ_e and a node function from Φ_v to be complete.

Consider the set of objects defined as follows:

$$\mathscr{G} = \{\mathscr{H}(\mathcal{V},\mathcal{E}) \mid \boldsymbol{E}_{j\to i} \in \Phi_e, \forall \boldsymbol{V}_i, \boldsymbol{V}_j \in \Phi_v\}$$

The set $\mathscr{G}$ contains the computational graphs we need in our modeling. Consider a sequence of such graphs $\{\mathscr{G}_{\alpha\in J}\}$ for index set J.

- This set could comprise a sequence of time points or iterations of the model building process. For example, a first model $\mathscr{H}_0$ could be built and then additional edges and nodes could be added to create a second model $\mathscr{H}_1$ and so on. For convenience of exposition, we are not adding the nodes and edge arguments here to keep the notation cleaner. After P iterations from the base, the model $\mathscr{H}_P$ has been formed and we are now ready to use this model in simulations. Given the graph $\mathscr{H}_P$, we can look at some optimization questions related to the training of this graph to match data. For a typical least squares error setting, assuming we start with the graph $\mathscr{H}_P$, this is the problem of minimizing

$$\sum_{i\in\Omega}(\boldsymbol{O}_i - \boldsymbol{T}_i)^2 \quad \text{subject to } \boldsymbol{E}_{j\to i} \in \Phi_e,\ \boldsymbol{V}_j,\ \boldsymbol{V}_i \in \Phi_v$$

where $\boldsymbol{I}_i\mathscr{H}_P$ is the output $\boldsymbol{O}_i$ obtained when $\boldsymbol{I}_i$ is applied to the model. The set Ω is the index set for the set of all possible inputs which are to be paired with outputs. If the edge functions are scalars and the nodal functions are simple function computational engines of the form $\sigma(x,g,o) = 0.5(1+\tanh((x-o)/g))$ this optimization is difficult and computationally expensive. However, in this setting, it is an example of optimization over $\Re^n$ where n is the number of tunable parameters. The setting we use here is an optimization over function spaces and has many challenges. The least squares setting is not necessarily the most appropriate and a non-differentiable measures of error could be used. The nodal and edge computations could involve discontinuous calculation and hence, more general optimization tools set in general metric space should be used (F. Schlief and P. Tino (31) 2004).

There is clearly an interest in understanding what we mean by letting $P \rightarrow \infty$. The sequence $\{\mathscr{H}_{p=0}^{\infty}\} \subset \mathscr{G}$ and at every step in the creation process, a viable graph model is built and can be used in simulation. To discuss the convergence process, the set of objects $\mathscr{G}$ must be given a topological structure, which could be metric or norm based but need not be. The limiting object, $\mathscr{H}^{\infty}$, will then belong to the completion of $\mathscr{G}$, $\mathscr{G}^{\infty}$. In the context of cognitive modeling, this can be used to create a **normal** brain model, $\mathscr{H}_{Brain}^{N,\infty}$, where here the superscript N indicates *normal*. If the context is building a model of autoimmune disease, the goal is to build a model of normal immune response, $\mathscr{H}_{Immune}^{N,\infty}$. In both cases, the simulations of how to deal with *damage* then begin by looking for deviations from the baseline *normal* models.

- Once normal models have been constructed, the next type of sequence is based on simulation time. For a finite sequence of time points $\{t_i\}_{i=1}^{Q}$, external signals, $\{\boldsymbol{S}_i\}$, are applied to the normal model creating a new version of the model. The external signal has several effects: one, it can alter the edge and node function processing but leave the physical graph alone; second, it can remove some of the edge processing between two nodes and third, it can remove all the edge processing for a particular node effectively removing the node. Clearly, node removal is a common occurrence in all of the three modeling scenarios we have described. Hence, we can write the signal $\boldsymbol{S}_i$ as a composition: $\boldsymbol{S}_i = \mathscr{E}_v \circ \mathscr{E}_e \circ \mathscr{E}_f$, where we let $\mathscr{E}_f$ denote the edge and node function alteration mapping.

From now on, we will let $\mathscr{E}_{evf} = \mathscr{E}_v \circ \mathscr{E}_e \circ \mathscr{E}_f$. Hence, each signal applied to the graph model has the potential to alter everything about it, even if the edge and nodal functions are simple scalars and sigmoids respectively. The mapping $\mathscr{E}_f$ in a cognitive model is due to second messenger effects: the neurotransmitter packets in the synaptic cleft bind to a gate in the input cellular membrane and initiate a signal that results in protein construction or destruction. The proteins created could alter the molecular machinery on the output side of the neuron which generates the output pulse, which in turn is part of the input to other neurons. In an immune model, the mapping $\mathscr{E}_f$ changes the way T-Cells interact with MHC complexes on a cell, potentially leading to an incorrect signal to destroy the cell and cause self-damage.

At any time point t_j in the simulation, we therefore have a graph model, $\mathcal{H}_j^t$, which has $\boldsymbol{V}_j$ node functions and $\boldsymbol{E}_j$ edge functions. Note $\mathscr{H}_p$ denotes the pth iteration in the model building process, which is an approximation to the limiting object $\mathscr{H}^{\infty}$ and $\mathcal{H}_j^t$ is the result of applying signals $\{\boldsymbol{S}_1, \ldots, \boldsymbol{S}_j\}$ to the approximate model $\mathscr{H}_p$. This result gives insight into what the signals $\{\boldsymbol{S}_1, \ldots, \boldsymbol{S}_j\}$ applied to the converged model $\mathscr{H}^{\infty}$ would do. It is useful to look at this training sequence in greater detail. Signal $\boldsymbol{S}_1$ is applied to the model $\mathscr{H}_p$ to create the new model $\mathcal{H}_1^t$, which is the result of applying a $\mathscr{E}_{evf}^1$ to $\mathscr{H}_p$. The mapping $\mathscr{E}_{evf}^1$ alters edges and nodes in the physical structure of the graph as well as altering edge and nodal function via the mapping $\mathscr{E}_f^1$. Thus, the first step is $\boldsymbol{S}_1 \rightarrow \mathscr{H}_p = \mathscr{E}_{evf}^1 \mathscr{H}_p = \mathcal{H}_1^t$. In general, after Q time steps, we have

$$\boldsymbol{S}_Q \rightarrow \mathcal{H}_Q^t \quad = \quad \mathscr{E}_{evf}^Q \mathcal{H}_{Q-1}^t = \mathscr{E}_{evf}^Q \cdots \mathscr{E}_{evf}^1 \mathscr{H}_p = \mathcal{H}_Q^t$$

Since all the structural components of the graph model and the edge and node processing functions are potentially being altered, this process is dependent on the order in which the signals are applied. Thus, $\mathcal{H}_Q^t = \mathscr{E}_{evf}^Q \cdots \mathscr{E}_{evf}^1 \mathscr{H}_p$ is dependent on the signal sequence order. Note all graphs that are built are in $\mathbb{H}$ and hence can be added or subtracted.

The differences $\Delta_{\textbf{Signal}} \mathcal{H}_i^t = \mathcal{H}_{i+1}^t - \mathcal{H}_i^t$ and $\Delta_{\textbf{Structural}} \mathscr{H}_{i+1} = \mathscr{H}_{i+1} - \mathscr{H}_i$ due to a particular signal chain $\{\boldsymbol{S}_{i+1}, \ldots, \boldsymbol{S}_1\}$ or steps in the construction process of the base model are, respectively, the **signal based** and **structural based** changes in the graph structure between iterations. In general, there is much *similarity* between the steps from i to $i+1$ in the sense that $\Delta_{\textbf{Signal}} \mathcal{H}_i^t$ and

$\Delta_{\textbf{Structural}}\mathscr{H}_{i+1}$ possess structure that is qualitatively similar at each step. Following (G. Edgar (36) 1990), for each of our structural graphs, $\mathscr{H}_i$, it is possible to endow it with a metric space topology based on paths through the graph of various lengths. When we apply edge and nodal functions to $\mathscr{H}_i$ we are constructing a topological space based on applying signals. Hence, we wish to exploit this facet of the structure when we make decisions about the kind of topological space our graphs lie inside of ((G. Edgar (36) 1990), (M. Barlow (75) 1998), (H. Tuominen (42) 2014) and (J. Heinonen (54) 2007)).

Let $\mathbb{S}$ be the space of all possible signals. Define *addition* and *multiplication* of signals as follows: $\boldsymbol{S}_\alpha + \boldsymbol{S}_\beta$ is the signal obtained when the two signals are added. This means $\boldsymbol{S}_\alpha + \boldsymbol{S}_\beta = \mathscr{E}^{\alpha}_{evf} + \mathscr{E}^{\beta}_{evf}$. The operation $\boldsymbol{S}_\alpha \boldsymbol{S}_\beta$ is the result of the two signals applied iteratively which means $\boldsymbol{S}_\alpha \boldsymbol{S}_\beta = \mathscr{E}^{\alpha}_{evf}\mathscr{E}^{\beta}_{evf}$. Since all of the graphs are in $\mathbb{H}$, it is easy to add the graphs that result from the application of signals. Hence, all of these terms are defined. Clearly, the addition operation is abelian as order does not matter when applying the edge, node and processing function updates in this manner. However, the multiplication operation is non-abelian because the operation is order dependent. Also it is easy to see both addition and multiplication are associative. The distributive law is also satisfied as

$$(\boldsymbol{S}_\alpha + \boldsymbol{S}_\beta)\boldsymbol{S}_\gamma \quad = \quad (\mathscr{E}^{\alpha}_{evf} + \mathscr{E}^{\beta}_{evf})\mathscr{E}^{\gamma}_{evf} = \mathscr{E}^{\alpha}_{evf}\mathscr{E}^{\gamma}_{evf} + \mathscr{E}^{\beta}_{evf}\mathscr{E}^{\gamma}_{evf}$$

which is the same as $\boldsymbol{S}_\alpha \boldsymbol{S}_\gamma + \boldsymbol{S}_\beta \boldsymbol{S}_\gamma$. The set of objects we wish to understand is then

$$\mathbb{T} \quad = \quad \{\boldsymbol{S}\mathscr{H}(\mathcal{V}, \mathcal{E}) \mid \boldsymbol{E}_{j\to i} \in \Phi_e,\ \boldsymbol{V}_i \in \Phi_v, \boldsymbol{S} \in \mathbb{S}\}$$

where $\mathbb{S}$ is a non-abelian division ring. Our search for the proper topological space setting for the computational graph models therefore must reflect this reality. These details do not hamper our abilities to construct approximate models; however, our goal is also to understand the objects $\mathscr{H}^{N,\infty}_{Brain}$ and $\mathscr{H}^{N,\infty}_{Immune}$.

In general, once a suitable approximate to our complex system has been obtained, $\mathscr{H}_P$, we can use a signal family of external signals, $\boldsymbol{S}_i$, in two ways. The first is to take a model that is functional as a version of *normalcy* and apply signals to it to measure the effects of the external triggers. For example, a cocktail of injected neurotoxins or an anesthetic agent could be used to alter the edge and node functions to change consciousness state or simulate cognitive dysfunction. An external trigger could also be used to simulate terrain or resource damage in a complicated geographical environment. We have discussed examples of this already. However, in order to fully build a *normal* model, it is often necessary to supply training signals, $\boldsymbol{I}_i$ which must be coupled to desired targets or outputs, $\boldsymbol{T}_i$. This is the optimization problem which tries to minimize a measure of error over the summed outputs of the graph model subject to the tunable parameters of the model which we have mentioned earlier.

There has been some work in understanding how gravity and quantum field theory paradigms for understanding the universe might have emerged from discrete geometries ((Konopka et al. (67) 2008)). These approaches start with the fully connected graph of $\boldsymbol{N}^{max}$ nodes $\mathbb{H}$. The nodes are various types of elementary physical entities and connections between them are the edges. The edge between nodes $\boldsymbol{V_i}$ and $\boldsymbol{V_j}$ is in a state characterized by $\{(0,0),(1,-1),(1,0),(1,1)\}$ with a severed edge given the value $\{(0,0)\}$. When the universe is initialized, no edges are severed and as the universe evolves, edges between nodes are removed by setting their edge value to $\{(0,0)\}$. The important questions are what kind of topological space does this graph move towards and what dynamics between nodes and what topological structure, differentiable or otherwise, emerge? The use of discrete geometric ideas here is intended to try to better understand how the laws that appear to work at the quantum level change into the laws we see in four dimensional space-time as evidenced by general relativity. Hence, it is an attempt to understand how to bring quantum theory and general relativity together. The issues in building interesting network models here are similar. In this work, each computational graph $\mathcal{G}(\mathcal{V}, \mathcal{E})$ is also formed from $\mathbb{H}$. Each $\mathcal{G}(\mathcal{V}, \mathcal{E})$ then corresponds to a graph of N_V nodes and N_E edges. As the computational graph model is built, for a given set of nodes, many edges are added in a quasi-fractal manner and other edges are deleted. The topology that emerges as the number of connected nodes and edges increase is similar to what happens in the fully connected physics model (occurring at very high temperature) as the temperature decreases and connections break. In the physics case, we are obtaining topological structure as we move from the fully connected topology to versions having disconnected nodes. In our case, we can start either with no connections and build a graph hierarchically as we have done with our MATLAB tools or we can start with a fully connected graph and remove nodes and edges. In either case, we are interested in what topological structure emerges as it determines convergence for sequences and the connectivity of the solutions to the system over a given domain. As topological structure emerges, corresponding laws of information transfer emerge also which give us the appropriate equations we can use to optimize how the models handle input to output transformations.

Homework

C:

Exercise 16.6.1 *How would you design a general graph in* **C**? *In our tutorial programming lectures we discuss just the rudiments of* **C** *programming and in our new studies we have been looking at asynchronous programming using* **zeroMQ**. *The discussions in this section are all about nodes with backward and forward edge sets whose state values would be organized using a global simulation counter. Think about how you would code this and organize it into an effective library for use.*

Exercise 16.6.2 *How would you implement delete edge and node functions for the graph? Since all graphs are subgraphs of a fully connected graph of N nodes, it is possible to compare graphs with one another. How would you write such comparison functions? Could differences of such graphs indicate a kind of information flow?*

Exercise 16.6.3 *Think about the incidence matrix of these graphs and the graph Laplacian and its possible use as an information processing tool similar to a discrete diffusion equation. How would you implement these sorts of functions?*

Exercise 16.6.4 *Add report functions to these codes. Can you think of a good way to collect statistics on the graph's performance?*

Exercise 16.6.5 *The graphs above are structural and do not have interesting edge and node processing. Now consider how to add that functionality so that we can carefully discuss signal based graphs with useful edge and node processing functions. How do we write code to do this?*

C++:

Exercise 16.6.6 *How would you design a general graph in* **C++**? *In our tutorial programming lectures we discuss some possible implementation ideas in* **C++** *carefully and there is also a large body of layout information for this type of code in* **MATLAB** *which is interesting but not pointer based and so not so useful for our discussions here. The discussions in this section are, again, all about nodes with backward and forward edge sets whose state values would be organized using a global simulation counter. Think about how you would code this into useful classes and organize it into an effective library for use.*

Exercise 16.6.7 *How would you implement delete edge and node functions for the graph? Since all graphs are subgraphs of a fully connected graph of N nodes, it is possible to compare graphs with* *one another. How would you write such comparison functions? Could differences of such graphs indicate a kind of information flow?*

Exercise 16.6.8 *Think about the incidence matrix of these graphs and the graph Laplacian and its possible use as an information processing tool similar to a discrete diffusion equation. How would you implement these sorts of functions?*

Exercise 16.6.9 *Add report functions to these codes. Can you think of a good way to collect statistics on the graph's performance?*

Exercise 16.6.10 *The graphs above are structural and do not have interesting edge and node processing. Now consider how to add that functionality so that we can carefully discuss signal based graphs with useful edge and node processing functions. How do we write code to do this?*

16.6.2 Asynchronous Graph Models

The discussion above about computational graphs is tightly tied to the notion of an external time clock and all simulation related issues are phrased in terms of signals applied at a time t_i, which are used to generate responses at time t_{i+1} where i is the index of the global time clock. We want to redo all of this in the context of asynchronous computation where each node simply communicates with others by messages which are then processed and responded to. Hence, there is no global time clock and the computational graph framework above is not the correct model we seek. To describe a general model of graph computation which is asynchronous requires using ideas from the theory of asynchronous message passing as well as the implementation of these ideas into software tools. A cogent explanation of message passing ideas is found in (A. Downey (2) 2008). The implementation into a robust set of C tools is described in (P. Hintjens (94) 2013). The Erlang language is superb at implementing these ideas into large-scale systems and a good introduction to its tool set is in (Armstrong (7) 2013).

Consider a collection of N nodes (here N is assumed to be very large, say 10^{10} or more) but for the moment we won't specify the size of N. This population of nodes is distributed uniformly between nodes of two types which can both send and receive messages. We posit a number of things about these nodes. For convenience of exposition, we are going to set our discussion in the context of cosmology even though we have no particular expertise in that area. However, it has been useful to think in terms of mass and energy as we were developing these thoughts. Once we are finished with our *toy* cosmological model and have explored the consequences for the topology of the graph of nodes and edges we are working with, we will return to the realms of autoimmune and cognition models and show how the toy cosmological model gives us insight into their particular structure. The graph has these properties:

- A node n has a **value** $\boxed{m_n || E_n}$ which is interpreted as the node value is split between a mass and energy form. From physics, we know we can convert the mass m_n into energy and we let this conversion function be ϕ and we assume it is linear: i.e. $\phi(au + bv) = a\phi(u) + b\phi(v)$. There is a corresponding energy to mass conversion function called ψ which is also assumed to be a linear map. Clearly $\phi(\psi) = \psi(\phi) = I$ where I is the identity map. The *state* of the node can either be in the mass form or the energy form. Hence, it can be $\boxed{\textbf{On } m_n || \textbf{ Off } E_n}$ or $\boxed{\textbf{Off } m_n || \textbf{ On } E_n}$ which we indicate more succinctly as $\boxed{+||-}$ and $\boxed{-||+}$. To be even more brief, we will simply let $(+,-)$ and $(-,+)$ denote these states.
- Nodes can send messages and this is shown for the two types $(+,-)$ and $(-,+)$ in Figure 16.13 and Figure 16.14. Nodes can also receive messages and this is shown in Figure 16.15 and Figure 16.16. Note in these figures, we explicitly show the mass and energy changes due to

send and receive messages for each possible state transition. We also assume the total energy of the graph is assumed to be a constant $\mathscr{E}$, which is defined as

$$\mathscr{E} = \mathscr{E}_A + \mathscr{E}_I + \mathscr{E}_B$$

where $\mathscr{E}_A$ is the energy of all the nodes which can send and receive messages (accessible nodes), $\mathscr{E}_I$ is the energy of all the nodes which cannot send and receive messages (inaccessible nodes), and $\mathscr{E}_B$ is the background energy. Since the total energy of the graph is assumed to be a constant $\mathscr{E}$, we have

$$\begin{aligned} \mathscr{E} &= \sum_{i \in \mathscr{V}} (\phi(m_n) + E_n) + \sum_{i \in \mathscr{U}} (\phi(m^c) + E^c) \\ &+ \sum_{i \in \mathscr{V}} \left(c_{\text{send}}^{(+,-)} + c_{\text{send}}^{(-,+)} + c_{\text{receive}}^{(+,-)} + c_{\text{receive}}^{(-,+)} \right) \end{aligned}$$

where $\mathscr{U}$ is the set of node indices corresponding to nodes which cannot send and receive messages and are therefore inaccessible, and $\mathscr{V}$ is the set of nodes which can send and receive and are therefore accessible nodes in the graph (see (F. Markopoulou (30) 2007) for a background discussion of such conserved quantities). When we first encounter the graph all nodes are accessible and so $\mathscr{U} = \emptyset$. The mass and energy values m^c and E^c are graph constants which we interpret as the minimum mass and energy a node can have; i.e. a node cannot drop its mass and energy value below this threshold. If all the nodes were inaccessible then we would have

$$\mathscr{E} = \sum_{i=1}^{N} (\phi(m^c) + E^c) = N\phi(m^c) + NE^c \Longrightarrow \phi(m^c) + E^c = \frac{\mathscr{E}}{N}.$$

We define

$$\begin{aligned} \mathscr{E}_A &= \sum_{i \in \mathscr{V}} (\phi(m_n) + E_n), \text{ accessible energy}, \quad \mathscr{D}_M = \sum_{i \in \mathscr{U}} \phi(m^c), \text{ dark matter} \\ \mathscr{D}_E &= \sum_{i \in \mathscr{U}} E^c, \text{ dark energy} \end{aligned}$$

The send and receive message costs are written in terms of the accessible energy $\mathscr{E}_A$.

Figure 16.13: Sending messages from $(+,-)$.

Figure 16.14: Sending messages from $(-,+)$.

Note, before messages have been exchanged, $|\mathscr{V}| = N$.

- We assume each node can send and receive messages from all other nodes and we explicitly do not allow self-communication. Hence, this is a total connected graph of N nodes and $N(N-1)$

Figure 16.15: Receiving messages into $(+, -)$.

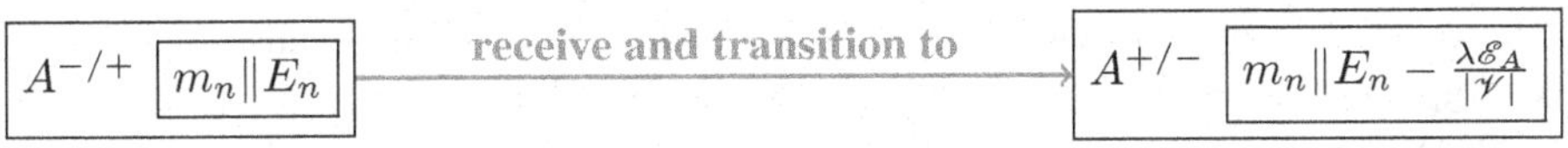

Figure 16.16: Receiving messages into $(-, +)$.

edges. In our model, it is possible for a node to be unable to send and receive messages which would make it an **inaccessible** node, but we assume the node value of such a node is still important to the behavior of the graph. Note we simply have a graph here. There is no notion of time as a way to assess the change of the graph as time progresses, and there is no notion of positional information where each node is assigned a position coordinate in some $\Re^p$ where p is the dimension of a cartesian manifold in which this graph is embedded. We have to be scrupulous in rooting this sort of bias out of our thinking.

- Messages are processed **asynchronously** as they come in. A node processes a send or receive as needed which alters its state and the value of its mass and energy.

16.6.3 Breaking the Initial Symmetry

Although we have no notion of time at this point, it is still convenient to say that we start with the graph in its initial state. Since all nodes freely communicate to all other nodes, the graph is in a state of symmetry and the loss of a node to inaccessibility is a form of symmetry breaking. We want to understand how a node becomes inaccessible and to do that we analyze the state of a node after sending and receiving messages.

Sending messages We have assumed we cannot send a message if $m_n - \psi\left(\alpha \frac{\mathscr{E}_A}{|\mathscr{V}|}\right) < m^c$ and we cannot send a message if $E_n - \beta \frac{\mathscr{E}_A}{|\mathscr{V}|} < E^c$.

- In the state $(+, -)$, if sending a message drops m_n below m^c, the node cannot send messages and so the rest of the graph cannot receive this node's state information. In this case, although the node cannot send, perhaps it can receive. From our transition diagrams, we see that a receive message would switch the state to $(-, +)$ and drop the mass to $m_n - \psi\left(\mu \frac{\mathscr{E}_A}{|\mathscr{V}|}\right)$. We already know $m_n - \psi\left(\alpha \frac{\mathscr{E}_A}{|\mathscr{V}|}\right) < m^c$. So we have

$$m_n - \psi\left(\mu \frac{\mathscr{E}_A}{|\mathscr{V}|}\right) \quad < \quad m^c + \psi\left(\alpha \frac{\mathscr{E}_A}{|\mathscr{V}|}\right) - \psi\left(\mu \frac{\mathscr{E}_A}{|\mathscr{V}|}\right) = m^c + (\alpha - \mu)\psi\left(\frac{\mathscr{E}_A}{|\mathscr{V}|}\right)$$

  This is less than m^c if $\alpha < \mu$. Thus, the receive is also blocked in this case.

- In the state $(-, +)$, if sending a message drops E_n below E^c, the node also cannot send a message and hence the rest of the graph is unaware of the node's state. Is it possible for the node to receive? From our transition diagrams we see a receive message would switch the state to $(+, -)$ and drop the energy to $E_n - \lambda \frac{\mathscr{E}_A}{|\mathscr{V}|}$. We already know $E_n - \beta \frac{\mathscr{E}_A}{|\mathscr{V}|} < E^c$.

So we have

$$E_n - \lambda \frac{\mathscr{E}_A}{|\mathscr{V}|} \quad < \quad E^c + \beta \frac{\mathscr{E}_A}{|\mathscr{V}|} - \lambda \frac{\mathscr{E}_A}{|\mathscr{V}|} = E^c + (\beta - \lambda) \frac{\mathscr{E}_A}{|\mathscr{V}|}$$

This is less than E^c if $\beta < \lambda$. Thus, the receive is also blocked in this case.

Hence, if both $\alpha < \mu$ and $\beta < \lambda$ once a node cannot send because of minimum mass and energy constraints, the node becomes inaccessible with energy value $\phi(m^c) + E^c$ and it is in either state.

Receiving Messages We have assumed we cannot receive a message if $m_n - \psi\left(\mu \frac{\mathscr{E}_A}{|\mathscr{V}|}\right) < m^c$ and we cannot receive a message if $E_n - \lambda \frac{\mathscr{E}_A}{|\mathscr{V}|} < E^c$.

- In the state $(+, -)$, if receiving a message drops m_n below m^c, the node cannot receive messages and so the rest of the graph cannot send this node state information. In this case, although the node cannot receive, perhaps it can send. From our transition diagrams, we see that a send message would switch the state to $(-, +)$ and drop the mass to $m_n - \psi\left(\alpha \frac{\mathscr{E}_A}{|\mathscr{V}|}\right)$. We already know $m_n - \psi\left(\mu \frac{\mathscr{E}_A}{|\mathscr{V}|}\right) < m^c$. So we have

$$m_n - \psi\left(\alpha \frac{\mathscr{E}_A}{|\mathscr{V}|}\right) \quad < \quad m^c + \psi\left(\mu \frac{\mathscr{E}_A}{|\mathscr{V}|}\right) - \psi\left(\alpha \frac{\mathscr{E}_A}{|\mathscr{V}|}\right) = m^c + (\mu - \alpha)\psi\left(\frac{\mathscr{E}_A}{|\mathscr{V}|}\right)$$

  This is less than m^c if $\mu < \alpha$. Thus, the send is also blocked in this case.

- In the state $(-, +)$, if receiving a message drops E_n below E^c, the node also cannot receive a message and hence the rest of the graph cannot send information to this node. Is it possible for the node to send? From our transition diagrams we see a send message would switch the state to $(+, -)$ and drop the energy to $E_n - \beta \frac{\mathscr{E}_A}{|\mathscr{V}|}$. We already know $E_n - \lambda \frac{\mathscr{E}_A}{|\mathscr{V}|} < E^c$. So we have

$$E_n - \beta \frac{\mathscr{E}_A}{|\mathscr{V}|} \quad < \quad E^c + \lambda \frac{\mathscr{E}_A}{|\mathscr{V}|} - \beta \frac{\mathscr{E}_A}{|\mathscr{V}|} = E^c + (\lambda - \beta) \frac{\mathscr{E}_A}{|\mathscr{V}|}$$

  This is less than E^c if $\lambda < \beta$. Thus, the send is also blocked in this case.

Hence, if both $\alpha > \mu$ and $\beta > \lambda$ once a node cannot receive because of minimum mass and energy constraints, the node becomes inaccessible with energy value $\phi(m^c) + E^c$ and it is in either state.

From the discussions above, we can conclude that if $\alpha < \mu$ and $\beta < \lambda$, once a node tries to send when its mass or energy are too low, the node becomes inaccessible and we add that index to $\mathscr{U}$, On the other hand, if $\alpha > \mu$ and $\beta > \lambda$, the same thing happens when a node tries to send when its mass or energy are too low. So we only have to assume one of these cases, as eventually a node can become inaccessible through either incompatible route. We therefore will assume $\alpha < \mu$ and $\beta < \lambda$ for the rest of this discussion.

Homework

Exercise 16.6.11 *Take a graph of size* 10×10 *and write a MATLAB simulation that uses the message passing scheme discussed. Note this will be a serial, global time clock simulation but it will give you insight.*

Exercise 16.6.12 *What would happen if the distribution of mass and energy was not uniform in this thought experiment?*

Exercise 16.6.13 *Write simulation code functions that allow you to remove all edges to a node so as to make it inaccessible. Experiment with how these resulting graphs look. Also compute the incidence matrix after each experiment and do comparisons.*

16.6.4 Message Sequences

Now let's look at how a node is triggered into becoming inaccessible. Let's consider a sequence of P transactions for a node n where each transaction is labeled T_i and hence the transaction string is $\{T_1, T_2, \ldots, T_P\}$. Each transaction is either a receive R or a send S request.

- Each receive implies a $\psi\left(\alpha \frac{\mathscr{E}_A}{|\mathscr{V}|}\right)$ loss of energy if in state $(+,-)$. Let I_1 be the set of indices from $\{1,\ldots,P\}$ where T_i is a receive with state $(+,-)$.
- Each receive implies a $\beta \frac{\mathscr{E}_A}{|\mathscr{V}|}$ loss of energy if in state $(-,+)$. Let I_2 be the set of indices from $\{1,\ldots,P\}$ where T_i is a receive with state $(-,+)$.
- Each send implies a $\psi\left(\mu \frac{\mathscr{E}_A}{|\mathscr{V}|}\right)$ loss of energy if in state $(+,-)$. Let I_3 be the set of indices from $\{1,\ldots,P\}$ where T_i is a send with state $(+,-)$.
- Each send implies a $\lambda \frac{\mathscr{E}_A}{|\mathscr{V}|}$ loss of energy if in state $(-,+)$. Let I_4 be the set of indices from $\{1,\ldots,P\}$ where T_i is a send with state $(-,+)$.

Then the total loss is

$$\begin{aligned}
\mathscr{L} &= \sum_{i\in I_1} \alpha\psi\left(\frac{\mathscr{E}_A}{|\mathscr{V}|}\right) + \sum_{i\in I_2} \beta\frac{\mathscr{E}_A}{|\mathscr{V}|} + \sum_{i\in I_3} \mu\psi\left(\frac{\mathscr{E}_A}{|\mathscr{V}|}\right) + \sum_{i\in I_4} \lambda\frac{\mathscr{E}_A}{|\mathscr{V}|} \\
&= \text{Mass Loss } + \text{ Energy Loss} \\
&= \left(\sum_{i\in I_1} \alpha\psi\left(\frac{\mathscr{E}_A}{|\mathscr{V}|}\right) + \sum_{i\in I_3} \mu\psi\left(\frac{\mathscr{E}_A}{|\mathscr{V}|}\right)\right) + \left(\sum_{i\in I_2} \beta\frac{\mathscr{E}_A}{|\mathscr{V}|} + \sum_{i\in I_4} \lambda\frac{\mathscr{E}_A}{|\mathscr{V}|}\right)
\end{aligned}$$

Thus

$$\mathscr{L} = (|I_1|\alpha + |I_3|\mu)\psi\left(\frac{\mathscr{E}_A}{|\mathscr{V}|}\right) + (|I_2|\beta + |I_4|\lambda)\frac{\mathscr{E}_A}{|\mathscr{V}|}$$

Once a node is in $(+,-)$, it cannot send if $m_n - (|I_1|\alpha + |I_3|\mu)\psi\left(\frac{\mathscr{E}_A}{|\mathscr{V}|}\right) < m^c$ and it cannot receive if $E_n - (|I_2|\beta + |I_4|\lambda)\frac{\mathscr{E}_A}{|\mathscr{V}|} < E^c$. Thus, a good estimate for the occurrence of the first isolation event when a node becomes inaccessible is

$$\phi\left(m_n - (|I_1|\alpha + |I_3|\mu)\psi\left(\frac{\mathscr{E}_A}{|\mathscr{V}|}\right)\right) + E_n - (|I_2|\beta + |I_4|\lambda)\frac{\mathscr{E}_A}{|\mathscr{V}|} \quad < \quad \phi(m^c) + E^c$$

where remember, since we have not had an isolation event yet, $\frac{\mathscr{E}_A}{|\mathscr{V}|} = \frac{\mathscr{E}}{N}$. We can then rewrite as

$$\phi(m_n) + E_n - (|I_1|\alpha + |I_3|\mu)\,\frac{\mathscr{E}_A}{|\mathscr{V}|} - (|I_2|\beta + |I_4|\lambda)\,\frac{\mathscr{E}_A}{|\mathscr{V}|} < \phi(m^c) + E^c$$

Thus, canceling

$$(|I_1|\alpha + |I_3|\mu) + (|I_2|\beta + |I_4|\lambda)\frac{\mathscr{E}_A}{|\mathscr{V}|} \quad > \quad (\phi(m_n) + E_n) - (\phi(m^c) + E^c)$$

Since $\alpha < \mu$ and $\beta < \lambda$, we see

$$(|I_1| + |I_2| + |I_3| + |I_4|)\max(\{\mu, \lambda\}) \quad > \quad |\mathscr{V}|\, \frac{(\phi(m_n) + E_n) - (\phi(m^c) + E^c)}{\mathscr{E}_A}$$

But $|P| = |I_1| + |I_2| + |I_3| + |I_4|$, so the defining equation for the creation of an inaccessible node is

$$|P| \max\{\mu, \lambda\} \quad \approx \quad |\mathscr{V}|\, \frac{(\phi(m_n) + E_n) - (\phi(m^c) + E^c)}{\mathscr{E}_A}$$

The distribution of the values of the state m_n and E_n of a node at the start of communication will then determine how the inaccessible nodes are created. We assume the states follow a simple discrete distribution. The set of nodes $\mathscr{N} = \{1, \ldots, N\}$ is written as a countable union of disjoint sets of indices $\mathscr{N} = \cup_{n=1}^{\infty} \mathscr{W}_n$ where $|\mathscr{W}_n| = \frac{N}{2^n+1}$. Hence, the cardinality of the sets is $\{\mathscr{W} = \frac{N}{3}, \mathscr{W} = \frac{N}{5}, \mathscr{W} = \frac{N}{9}, \ldots\}$. The value of $\phi(m_n) + E^n$ in each set $\mathscr{W}_n$ is $(1 + \frac{1}{2^n})(\phi(m^c) + E^c)$. Then, we have the total energy before message passing begins is

$$\begin{aligned}
\sum_{n=1}^{\infty}(\phi(m_n) + E^n) &= \sum_{n=1}^{\infty}\sum_{j \in \mathscr{W}_n}(\phi(m_j) + E^j) = \sum_{n=1}^{\infty} \frac{N}{2^n + 1}\left(1 + \frac{1}{2^n}\right)(\phi(m^c) + E^c) \\
&= \sum_{n=1}^{\infty} \frac{N}{2^n}(\phi(m^c) + E^c) = N(\phi(m^c) + E^c) = N\frac{\mathscr{E}}{N} = \mathscr{E}
\end{aligned}$$

Thus, for $j \in \cup_{n=m_0+1}^{\infty} \mathscr{W}_n$, the energy before message passing is

$$\begin{aligned}
\sum_{n=m_0+1}^{\infty}\sum_{j \in \mathscr{W}_n}\left(1 + \frac{1}{2^n}\right)(\phi(m_j) + E^j) &= \sum_{n=m_0+1}^{\infty} \frac{N}{2^n + 1}\left(1 + \frac{1}{2^n}\right)(\phi(m^c) + E^c) \\
&= \sum_{n=m_0+1}^{\infty} \frac{N}{2^n}(\phi(m^c) + E^c) = \frac{1}{2^{m_0}} N(\phi(m^c) + E^c) \\
&= \frac{1}{2^{m_0}}\mathscr{E}
\end{aligned}$$

and this corresponds to

$$\sum_{n=m_0+1}^{\infty} |\mathscr{W}_n| = \sum_{n=m_0+1}^{\infty} \frac{N}{2^n + 1} \quad \approx \quad \frac{N}{2^{m_0}}$$

nodes. For each of these nodes, the defining equation for inaccessibility is

$$|P| \max\{\mu, \lambda\} \quad \approx \quad (\phi(m^c) + E^c)\frac{(1 + \frac{1}{2^n}) - 1}{\frac{\mathscr{E}_A}{|\mathscr{V}|}} = \frac{\frac{\mathscr{E}}{N}}{\frac{\mathscr{E}_A}{|\mathscr{V}|}}\,\frac{1}{2^n} \leq \frac{\frac{\mathscr{E}}{N}}{\frac{\mathscr{E}_A}{|\mathscr{V}|}}\,\frac{1}{2^{m_0}}$$

as $n > m_0$. Before message passing, $\mathscr{E}_A = \mathscr{E}$ and $|\mathscr{V}| = N$, so this becomes

$$|P| \max\{\mu, \lambda\} \quad \approx \quad \frac{1}{2^{m_0}}$$

Thus, $|P| \approx 1$ if $\frac{1}{2^{m_0}} \approx \mu \vee \lambda$. Thus, in our model of a graph with a minimum mass and energy state, about $\frac{N}{2^{m_0}} = N\mu \vee \lambda$ nodes become isolated after one message being passed. We then have that $|\mathscr{V}|$ decreases to $(1 - \frac{1}{2^{m_0}})N = (1 - \mu \vee \lambda)N$ and $|\mathscr{U}|$ increases to $N\frac{1}{2^{m_0}} = N\mu \vee \lambda$. Hence, if $\mu \vee \lambda = 0.7$, $|\mathscr{U}| = 0.7N$ or 70% of the nodes become isolated after one message passing event. The fully connected graph of nodes is now profoundly altered in just one message passing event and 70% of the nodes are inaccessible although still connected by edges. Further the nodes that are inaccessible influence the next round of message passing.

Homework

Exercise 16.6.14 *Suppose we start with the number of nodes being $N_0 = 2^{300} \approx 10^{90}$. The number of particles in the universe is estimated to be about $N = 2^{133} \approx 10^{40}$. If we lose 70% of the nodes due to inaccessibility, each message passing event, how many nodes are accessible after 5 events?*

Exercise 16.6.15 *Suppose we start with the number of nodes being $N_0 = 2^{300} \approx 10^{90}$. The number of particles in the universe is estimated to be about $N = 2^{133} \approx 10^{40}$. If we lose 70% of the nodes due to inaccessibility, each message passing event, how many messaging passing events have to occur for the number of accessible nodes to drop to $N \approx 10^40$?*

Exercise 16.6.16 *Suppose we start with the number of nodes being $N_0 = 2^{300} \approx 10^{90}$. The number of particles in the universe is estimated to be about $N = 2^{133} \approx 10^{40}$. If we lose 60% of the nodes due to inaccessibility, each message passing event, how many nodes are accessible after 5 events?*

Exercise 16.6.17 *Suppose we start with the number of nodes being $N_0 = 2^{300} \approx 10^{90}$. The number of particles in the universe is estimated to be about $N = 2^{133} \approx 10^{40}$. If we lose 40% of the nodes due to inaccessibility, each message passing event, how many messaging passing events have to occur for the number of accessible nodes to drop to $N \approx 10^40$?*

Exercise 16.6.18 *Let $J = \begin{bmatrix} 0 & -1 \\ 1 & 0 \end{bmatrix}$. Then $J^2 = -I$. Let $V = span\{I, J\}$ be the two dimensional subspace of 2×2 real matrices. Define $\phi : V \to \mathbb{C}$, where $\mathbb{C}$ is the complex numbers, by $\phi(aI+bJ) = a+bi$. Prove ϕ is a bijection which is an isometry for a suitable choice of norm on V. Hence $V \cong \mathbb{C}$. Can you design a graph $\mathcal{G}(\mathcal{V}, \mathcal{E})$ which implements this version of the complex numbers?*

16.6.5 Breaking the Symmetry Again: Version One

For convenience, let the accessible nodes initially be $\mathscr{V}_0$ with $|\mathscr{V}_0| = N_0 = N$ and after the first symmetry breaking, the accessible nodes are $\mathscr{V}_1$ with $|\mathscr{V}_1| = (1-\mu\vee\lambda)|\mathscr{V}_0|$. Energy balance requires that after the first symmetry breaking, there is also a dark background component $\mathscr{D}_B$ which we won't concern ourselves with here. Also, the energy and mass amounts m_n and E^n are what we have after the initial round of message passing and hence they have been altered from their previous values. We will still use the labels m_n and E^n for convenience and remind you to make the mental adjustment. The inaccessible nodes are initially $|\mathscr{U}_0|$ with $|\mathscr{U}_0| = 0$ and after the first symmetry breaking, the accessible nodes are $\mathscr{U}_1$ with $|\mathscr{U}_1| = |\mathscr{U}_0| + (1 - \mu \vee \lambda)|\mathscr{U}_0|$. Before symmetry breaking, let's label accessible energy as $\mathscr{E}_A^0$ where $\mathscr{E}_A^0 = \sum_{j\in\mathscr{V}_0}(\phi(m_n) + E_n)$ and after the first symmetry breaking, the accessible energy $\mathscr{E}_A^1 = \sum_{j\in\mathscr{V}_1}(\phi(m_n) + E_n)$.

In our discussion of the message passing sequence, we divided the transaction string into disjoint subsets of indices and, using our new notation, and defined

- Each receive implies a $\psi\left(\alpha\frac{\mathscr{E}_A^0}{|\mathscr{V}_0|}\right)$ loss of energy if in state $(+,-)$. Let I_1 be the set of indices from $\{1,\ldots,P\}$ where T_i is a receive with state $(+,-)$.

- Each receive implies a $\beta \frac{\mathscr{E}_A^0}{|\mathscr{V}_0|}$ loss of energy if in state $(-,+)$. Let I_2 be the set of indices from $\{1,\ldots,P\}$ where T_i is a receive with state $(-,+)$.

- Each send implies a $\psi\left(\mu \frac{\mathscr{E}_A^0}{|\mathscr{V}_0|}\right)$ loss of energy if in state $(+,-)$. Let I_3 be the set of indices from $\{1,\ldots,P\}$ where T_i is a send with state $(+,-)$.

- Each send implies a $\lambda \frac{\mathscr{E}_A^0}{|\mathscr{V}_0|}$ loss of energy if in state $(-,+)$. Let I_4 be the set of indices from $\{1,\ldots,P\}$ where T_i is a send with state $(-,+)$.

After the first symmetry breaking, we assume the costs are now proportional to $\mathscr{E}_A^1/N$ which means the costs have decreased. We will continue to use $\mathscr{V}_0 = N$ in our formulae. Note there are other things we could have done and we will pursue this in later sections. The new transaction costs are thus:

- Each receive implies a $\psi\left(\alpha \frac{\mathscr{E}_A^1}{|\mathscr{V}_0|}\right)$ loss of energy if in state $(+,-)$. Let I_1 be the set of indices from $\{1,\ldots,P\}$ where T_i is a receive with state $(+,-)$.

- Each receive implies a $\beta \frac{\mathscr{E}_A^1}{|\mathscr{V}_0|}$ loss of energy if in state $(-,+)$. Let I_2 be the set of indices from $\{1,\ldots,P\}$ where T_i is a receive with state $(-,+)$.

- Each send implies a $\psi\left(\mu \frac{\mathscr{E}_A^1}{|\mathscr{V}_0|}\right)$ loss of energy if in state $(+,-)$. Let I_3 be the set of indices from $\{1,\ldots,P\}$ where T_i is a send with state $(+,-)$.

- Each send implies a $\lambda \frac{\mathscr{E}_A^1}{|\mathscr{V}_0|}$ loss of energy if in state $(-,+)$. Let I_4 be the set of indices from $\{1,\ldots,P\}$ where T_i is a send with state $(-,+)$.

Then the total loss for the message string after symmetric breaking is then

$$\begin{aligned}
\mathscr{L}^1 &= \sum_{i\in I_1} \alpha\psi\left(\frac{\mathscr{E}_A^1}{|\mathscr{V}_0|}\right) + \sum_{i\in I_2} \beta \frac{\mathscr{E}_A^1}{|\mathscr{V}_0|} + \sum_{i\in I_3} \mu\psi\left(\frac{\mathscr{E}_A^1}{|\mathscr{V}_0|}\right) + \sum_{i\in I_4} \lambda \frac{\mathscr{E}_A^1}{|\mathscr{V}_0|} \\
&= \left(\sum_{i\in I_1} \alpha\psi\left(\frac{\mathscr{E}_A^1}{|\mathscr{V}_0|}\right) + \sum_{i\in I_3} \mu\psi\left(\frac{\mathscr{E}_A^1}{|\mathscr{V}_0|}\right)\right) + \left(\sum_{i\in I_2} \beta \frac{\mathscr{E}_A^1}{|\mathscr{V}_0|} + \sum_{i\in I_4} \lambda \frac{\mathscr{E}_A^1}{|\mathscr{V}_0|}\right)
\end{aligned}$$

Thus

$$\mathscr{L}^1 = (|I_1|\alpha + |I_3|\mu)\psi\left(\frac{\mathscr{E}_A^1}{|\mathscr{V}_0|}\right) + (|I_2|\beta + |I_4|\lambda)\frac{\mathscr{E}_A^1}{|\mathscr{V}_0|}$$

Once a node is in $(+,-)$, it cannot send if $m_n - (|I_1|\alpha + |I_3|\mu)\psi\left(\frac{\mathscr{E}_A^1}{|\mathscr{V}_0|}\right) < m^c$ and it cannot receive if $E_n - (|I_2|\beta + |I_4|\lambda)\frac{\mathscr{E}_A^1}{|\mathscr{V}_0|} < E^c$. Thus, a good estimate for the occurrence of an isolation event after the first symmetry breaking and a node becomes inaccessible is

$$\phi\left(m_n - (|I_1|\alpha + |I_3|\mu)\psi\left(\frac{\mathscr{E}_A^1}{|\mathscr{V}_0|}\right)\right) + E_n - (|I_2|\beta + |I_4|\lambda)\frac{\mathscr{E}_A^1}{|\mathscr{V}_0|} < \phi(m^c) + E^c$$

We can then rewrite as

$$\phi(m_n) + E_n - (|I_1|\alpha + |I_3|\mu)\,\frac{\mathscr{E}_A^1}{|\mathscr{V}_0|} - (|I_2|\beta + |I_4|\lambda)\,\frac{\mathscr{E}_A^1}{|\mathscr{V}_0|} < \phi(m^c) + E^c$$

Thus, canceling

$$(|I_1|\alpha + |I_3|\mu) + (|I_2|\beta + |I_4|\lambda)\frac{\mathscr{E}_A^1}{|\mathscr{V}_0|} \quad > \quad (\phi(m_n) + E_n) - (\phi(m^c) + E^c)$$

Since $\alpha < \mu$ and $\beta < \lambda$, we see

$$(|I_1| + |I_2| + |I_3| + |I_4|)\max(\{\mu, \lambda\}) \quad > \quad |\mathscr{V}_0|\,\frac{(\phi(m_n) + E_n) - (\phi(m^c) + E^c)}{\mathscr{E}_A^1}$$

But $|P| = |I_1| + |I_2| + |I_3| + |I_4|$, so the defining equation for the creation of an inaccessible node is

$$|P|\max\{\mu, \lambda\} \quad \approx \quad |\mathscr{V}_1|\,\frac{(\phi(m_n) + E_n) - (\phi(m^c) + E^c)}{\mathscr{E}_A^1}$$

The distribution of the values of the state m_n and E_n of a node after the first symmetry breaking again follows a simple discrete distribution. The set of accessible nodes $\mathscr{N}_1 = N(1-\mu\vee\lambda) = \{1, \ldots, N_1\}$ is written as a countable union of disjoint sets of indices $\mathscr{N}_1 = \cup_{n=1}^{\infty}\mathscr{W}_n$ where $|\mathscr{W}_n| = \frac{N_1}{2^n+1}$. The value of $\phi(m_n) + E^n$ in each set $\mathscr{W}_n$ is $(1 + \frac{1}{2^n})(\phi(m^c) + E^c)$. The total energy before message passing begins in this phase is then

$$\begin{aligned}
\mathscr{E}_A^1 &= \sum_{n=1}^{\infty}\sum_{j\in\mathscr{W}_n}(\phi(m_j) + E^j) = \sum_{n=1}^{\infty}\frac{N_1}{2^n+1}\left(1 + \frac{1}{2^n}\right)(\phi(m^c) + E^c) \\
&= \sum_{n=1}^{\infty}\frac{N_1}{2^n}(\phi(m^c) + E^c) = N_1(\phi(m^c) + E^c) = N_1\frac{\mathscr{E}}{N} = \mathscr{E}(1 - \mu\vee\lambda)
\end{aligned}$$

Thus, for $j \in \cup_{n=m_1+1}^{\infty}\mathscr{V}_n$, the energy before message passing is

$$\begin{aligned}
&\sum_{n=m_1+1}^{\infty}\sum_{j\in\mathscr{V}_n}\left(1 + \frac{1}{2^n}\right)(\phi(m_j) + E^j) = \sum_{n=m_1+1}^{\infty}\frac{N_1}{2^n+1}\left(1 + \frac{1}{2^n}\right)(\phi(m^c) + E^c) \\
&= \sum_{n=m+1}^{\infty}\frac{N_1}{2^n}(\phi(m^c) + E^c) = \frac{1}{2^m}N_1(\phi(m^c) + E^c) = \frac{1}{2^m}\mathscr{E}(1 - \mu\vee\lambda)
\end{aligned}$$

and this corresponds to

$$\sum_{n=m_1+1}^{\infty}|\mathscr{W}_n| = \sum_{n=m_1+1}^{\infty}\frac{N_1}{2^n+1} \quad\approx\quad \frac{N_1}{2^{m_1}} = N(1 - \mu\vee\lambda)\frac{1}{2^{m_1}}$$

nodes. For each of these nodes, the defining equation for inaccessibility is

$$\begin{aligned}
|P|\max\{\mu, \lambda\} &\approx (\phi(m^c) + E^c)\frac{(1 + \frac{1}{2^{m_1}}) - 1}{\frac{\mathscr{E}_A^1}{|\mathscr{V}_0|}} = \frac{\frac{\mathscr{E}}{N}}{\frac{\mathscr{E}_A^1}{|\mathscr{V}_0|}}\frac{1}{2^{m_1}} \\
&= \frac{\mathscr{E}}{\mathscr{E}_A^1}\frac{1}{2^{m_1}} = \frac{1}{(1 - \mu\vee\lambda)}\frac{1}{2^{m_1}} \Longrightarrow |P| \approx \frac{1}{(1 - \mu\vee\lambda)(\mu\vee\lambda)}\frac{1}{2^{m_1}}
\end{aligned}$$

Thus, $|P| \approx 1$ if $\frac{1}{2^{m_1}} = (1 - \mu\vee\lambda)(\mu\vee\lambda)$. The number of nodes which can be isolated after one more message is

$$\mathscr{U}_2 = N(1 - \mu\vee\lambda)\frac{1}{2^{m_1}} = N(1 - \mu\vee\lambda)\,(1 - \mu\vee\lambda)(\mu\vee\lambda) = N(1 - \mu\vee\lambda)^2(\mu\vee\lambda)$$

nodes. The number of accessible nodes after the second message passing cycle is then

$$\begin{aligned}|\mathscr{V}_2| &= N_1 - |\mathscr{U}_2| = N(1-\mu\vee\lambda) - N(1-\mu\vee\lambda)^2(\mu\vee\lambda)\\ &= N(1-\mu\vee\lambda)\,(1-(1-\mu\vee\lambda)(\mu\vee\lambda))\end{aligned}$$

Thus, the total number of isolated nodes after two message passing cycles is

$$|\mathscr{U}_1| + |\mathscr{U}_2| = N\left((\mu\vee\lambda) + (1-\mu\vee\lambda)^2(\mu\vee\lambda)\right) = N(\mu\vee\lambda)(1+(1-(\mu\vee\lambda)^2))$$

Note, we can rewrite our arguments in a recursive form. Here we add the next step in the symmetry breaking even though we have not gone through the details. We start with $|\mathscr{U}_0| = 0$, $|\mathscr{V}_0| = N$ and $\mathscr{E}_A^0 = \mathscr{E}|\mathscr{V}_0|$.

- Symmetry breaking one: $|P_0| \approx 1$ if $\frac{1}{2^{m_0}} = \frac{\mathscr{E}_A^0}{\mathscr{E}}(\mu\vee\lambda) = \frac{|\mathscr{V}_0|}{N}\mu\vee\lambda$:

$$\begin{aligned}|\mathscr{U}_1| &= \frac{(|\mathscr{V}_0|)^2}{N}(\mu\vee\lambda) = N(\mu\vee\lambda)\\ \mathscr{V}_1 &= N - |\mathscr{U}_1| = \frac{N^2-(|\mathscr{V}_0|)^2(\mu\vee\lambda)}{N} = N(1-\mu\vee\lambda)\\ \mathscr{E}_A^1 &= \mathscr{E}\frac{|\mathscr{V}_1|}{N} = \mathscr{E}(1-\mu\vee\lambda)\end{aligned}$$

- Symmetry breaking two: $|P_1| \approx 1$ if $\frac{1}{2^{m_1}} = \frac{\mathscr{E}_A^1}{\mathscr{E}}(\mu\vee\lambda) = (1-\mu\vee\lambda)(\mu\vee\lambda) = \frac{|\mathscr{V}_1|}{N}\,(\mu\vee\lambda)$:

$$\begin{aligned}|\mathscr{U}_2| &= \frac{(|\mathscr{V}_1|)^2}{N}(\mu\vee\lambda) = N(1-\mu\vee\lambda)^2(\mu\vee\lambda)\\ |\mathscr{V}_2| &= N - |\mathscr{U}_2| = \frac{N^2-|\mathscr{V}_1|^2(\mu\vee\lambda)}{N} = N(1-(1-\mu\vee\lambda)^2)(\mu\vee\lambda)\\ \mathscr{E}_A^2 &= \mathscr{E}\frac{|\mathscr{V}_2|}{N} = \mathscr{E}\left(1-(1-\mu\vee\lambda)^2(\mu\vee\lambda)\right)\end{aligned}$$

- Symmetry breaking three: $|P_2| \approx 1$ if $\frac{1}{2^{m_2}} = \frac{\mathscr{E}_A^2}{\mathscr{E}}(\mu\vee\lambda) = \frac{|\mathscr{V}_2|}{N}\,(\mu\vee\lambda)$:

$$\begin{aligned}\mathscr{U}_3| &= \frac{(|\mathscr{V}_2|)^2}{N}(\mu\vee\lambda)\\ |\mathscr{V}_2| &= N - |\mathscr{U}_3| = \frac{N^2-(|\mathscr{V}_2|)^2(\mu\vee\lambda)}{N}\\ \mathscr{E}_A^2 &= \mathscr{E}\frac{|\mathscr{V}_3|}{N}\end{aligned}$$

We can compute as many as allowed. The conditions to check are thus

$$\begin{aligned}|P_0| \approx 1 \text{ if } \frac{1}{2^{m_0}} &= \frac{\mathscr{E}_A^0}{\mathscr{E}}(\mu\vee\lambda) = \frac{|\mathscr{V}|_0}{N}\mu\vee\lambda\\ |P_1| \approx 1 \text{ if } \frac{1}{2^{m_1}} &= \frac{\mathscr{E}_A^1}{\mathscr{E}}(\mu\vee\lambda) = (1-\mu\vee\lambda)(\mu\vee\lambda)\\ &= \frac{|\mathscr{V}_1|}{N}\,(\mu\vee\lambda)\\ |P_2| \approx 1 \text{ if } \frac{1}{2^{m_2}} &= \frac{\mathscr{E}_A^2}{\mathscr{E}}(\mu\vee\lambda) = \frac{|\mathscr{V}_2|}{N}\,(\mu\vee\lambda)\end{aligned}$$

Homework

Exercise 16.6.19 *Suppose we start with the number of nodes being* $N_0 = 2^{300} \approx 10^{90}$. *The number of particles in the universe is estimated to be about* $N = 2^{133} \approx 10^{40}$. *If we lose* 70% *of the nodes due to inaccessibility, each message passing event, how many nodes are accessible after* 5 *events?*

Exercise 16.6.20 *Suppose we start with the number of nodes being* $N_0 = 2^{300} \approx 10^{90}$. *The number of particles in the universe is estimated to be about* $N = 2^{133} \approx 10^{40}$. *If we lose* 70% *of the nodes due to inaccessibility, each message passing event, how many messaging passing events have to occur for the number of accessible nodes to drop to* $N \approx 10^40$*?*

Exercise 16.6.21 *Suppose we start with the number of nodes being* $N_0 = 2^{300} \approx 10^{90}$. *The number of particles in the universe is estimated to be about* $N = 2^{133} \approx 10^{40}$. *If we lose* 60% *of the nodes due to inaccessibility, each message passing event, how many nodes are accessible after* 5 *events?*

Exercise 16.6.22 *Suppose we start with the number of nodes being* $N_0 = 2^{300} \approx 10^{90}$. *The number of particles in the universe is estimated to be about* $N = 2^{133} \approx 10^{40}$. *If we lose* 40% *of the nodes due to inaccessibility, each message passing event, how many messaging passing events have to occur for the number of accessible nodes to drop to* $N \approx 10^40$*?*

16.6.6 Breaking the Symmetry Again: Version Two

The accessible nodes initially are still $\mathscr{V}_0$ with $|\mathscr{V}_0| = N_0 = N$ and after the first symmetry breaking, the accessible nodes are $\mathscr{V}_1$ with $|\mathscr{V}_1| = M_0 = (1 - \mu \vee \lambda)|\mathscr{V}_0|$. Energy balance requires that after the first symmetry breaking, there is also a dark background component $\mathscr{D}_B$, which again, we won't concern ourselves with here. As before, the energy and mass amounts m_n and E^n are what we have after the initial round of message passing and hence they have been altered from their previous values. We will still use the labels m_n and E^n for convenience and remind you to make the mental adjustment. After the first symmetry breaking, we now assume the costs are proportional to $\frac{\mathscr{E}_A^1}{M_0}$, which is an increase over what we had before. Note

$$\frac{\mathscr{E}_A^1}{M_0} - \frac{\mathscr{E}_A^1}{N} = \frac{\mathscr{E}_A^1}{N}\left(\frac{N}{M_0} - 1\right) = \mathscr{E}_A^1 \frac{N - M_0}{NM_0}$$

The new model is that $\mathscr{E}N$ is roughly the energy needed for the sending and receiving of messages. The amount left over goes to increase the mass or energy of the state of the node. The new send message costs and mass and energy increases are shown in Figure 16.17 and Figure 16.18.

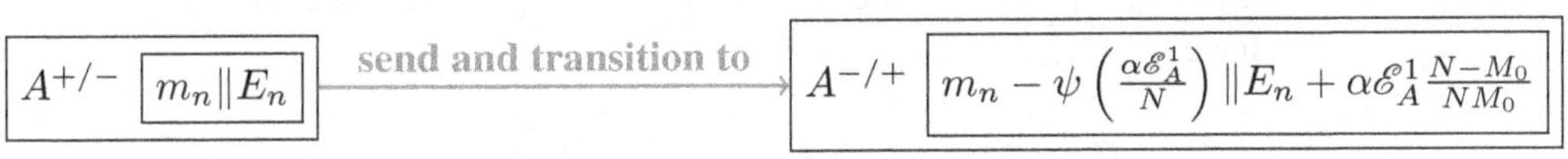

Figure 16.17: Sending messages from $(+,-)$ with energy increase.

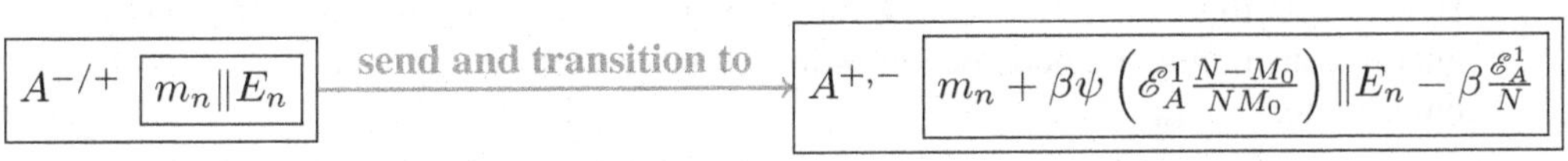

Figure 16.18: Sending messages from $(-,+)$ with mass increase.

The new receive message costs and mass and energy increases are shown in Figure 16.19 and Figure 16.20.

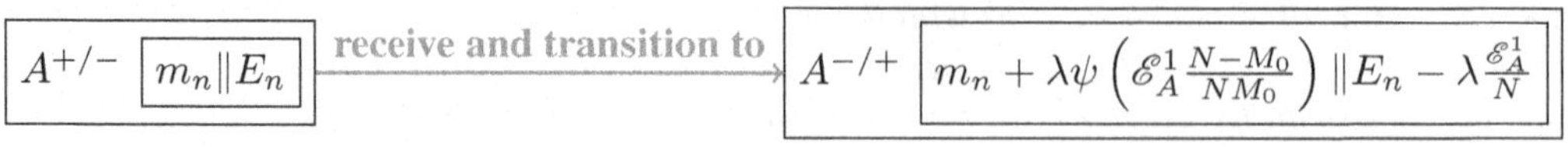

Figure 16.19: Receiving messages into $(+,-)$ with mass increase.

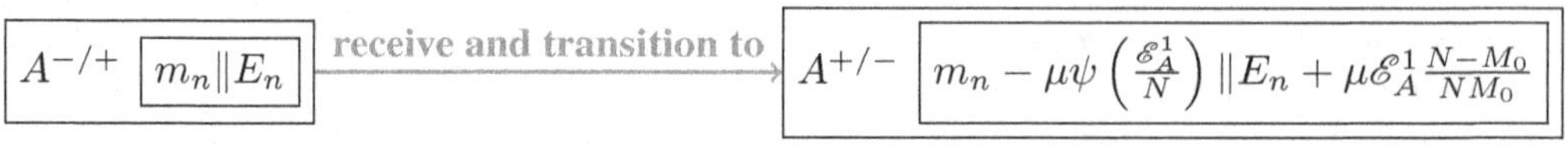

Figure 16.20: Receiving messages into $(-,+)$ with energy increase.

Let's figure out how a node becomes inaccessible under these new rules.

Sending messages We have assumed we cannot send a message if $m_n - \alpha\psi\left(\frac{\mathscr{E}^1_A}{N}\right) < m^c$ and we cannot send a message if $E_n - \beta\frac{\mathscr{E}^1_A}{N} < E^c$.

- In the state $(+,-)$, if sending a message drops m_n below m^c, the node cannot send messages and so the rest of the graph cannot receive this node's state information. Can this node receive? We see a receive message would switch the state to $(-,+)$ and drop the mass to $m_n - \mu\psi\left(\frac{\mathscr{E}^1_A}{N}\right)$. We already know $m_n - \alpha\psi\left(\frac{\mathscr{E}^1_A}{N}\right) < m^c$. So we have

$$m_n - \mu\psi\left(\frac{\mathscr{E}^1_A}{N}\right) \quad < \quad m^c + \alpha\psi\left(\frac{\mathscr{E}^1_A}{N}\right) - \psi\left(\mu\frac{\mathscr{E}_A}{|\mathscr{V}|}\right) = m^c + (\alpha-\mu)\psi\left(\frac{\mathscr{E}^1_A}{N}\right)$$

  This is less than m^c if $\alpha < \mu$. Thus, the receive is also blocked in this case. Hence, this node becomes inaccessible.

- In the state $(-,+)$, if sending a message drops E_n below E^c, the node also cannot send a message and hence the rest of the graph is unaware of the node's state. Is it possible for the node to receive? From our transition diagrams we see a receive message would switch the state to $(+,-)$ and drop the energy to $E_n - \lambda\frac{\mathscr{E}^1_A}{|N}$. We already know $E_n - \beta\frac{\mathscr{E}^1_A}{N} < E^c$. So we have

$$E_n - \lambda\frac{\mathscr{E}^1_A}{N} \quad < \quad E^c + \beta\frac{\mathscr{E}^1_A}{N} - \lambda\frac{\mathscr{E}^1_A}{N} = E^c + (\beta-\lambda)\frac{\mathscr{E}^1_A}{N}$$

  This is less than E^c if $\beta < \lambda$. Thus, the receive is also blocked in this case. This node also becomes inaccessible.

Hence, if both $\alpha < \mu$ and $\beta < \lambda$ once a node cannot send because of minimum mass and energy constraints, the node becomes inaccessible with energy value $\phi(m^c) + E^c$ and it is in either state.

Receiving Messages We have assumed we cannot receive a message if $m_n - \mu\psi\left(\mu\frac{\mathscr{E}^1_A}{N}\right) < m^c$ and we cannot receive a message if $E_n - \lambda\frac{\mathscr{E}^1_A}{N} < E^c$.

- In the state $(+,-)$, if receiving a message drops m_n below m^c, the node cannot receive messages and so the rest of the graph cannot send this node state information. In this case, although the node cannot receive, perhaps it can send. From our transition diagrams we see that a send message would switch the state to $(-,+)$ and drop the mass to $m_n - \alpha\psi\left(\frac{\mathscr{E}_A^1}{N}\right)$. We already know $m_n - \mu\psi\left(\frac{\mathscr{E}_A^1}{N}\right) < m^c$. So we have

$$m_n - \alpha\psi\left(\frac{\mathscr{E}_A^1}{N}\right) \quad < \quad m^c + \mu\psi\left(\frac{\mathscr{E}_A^1}{N}\right) - \alpha\psi\left(\frac{\mathscr{E}_A^1}{N}\right) = m^c + (\mu - \alpha)\psi\left(\frac{\mathscr{E}_A^1}{N}\right)$$

  This is less than m^c if $\mu < \alpha$. Thus, the send is also blocked in this case and the node is inaccessible.

- In the state $(-,+)$, if receiving a message drops E_n below E^c, the node also cannot receive a message and hence the rest of the graph cannot send information to this node. Is it possible for the node to send? From our transition diagrams we see a send message would switch the state to $(+,-)$ and drop the energy to $E_n - \beta\frac{\mathscr{E}_A^1}{N}$. We already know $E_n - \lambda\frac{\mathscr{E}_A^1}{N} < E^c$. So we have

$$E_n - \beta\frac{\mathscr{E}_A^1}{N} \quad < \quad E^c + \lambda\frac{\mathscr{E}_A^1}{N} - \beta\frac{\mathscr{E}_A^1}{N} = E^c + (\lambda - \beta)\frac{\mathscr{E}_A}{N}$$

  This is less than E^c if $\lambda < \beta$. Thus, the send is also blocked in this case and the node is inaccessible.

Hence, if both $\alpha > \mu$ and $\beta > \lambda$ once a node cannot receive because of minimum mass and energy constraints, the node becomes inaccessible with energy value $\phi(m^c) + E^c$ and it is in either state.

From the discussions above, we can conclude that if $\alpha < \mu$ and $\beta < \lambda$, once a node tries to send when its mass or energy are too low, the node becomes inaccessible and we add that index to $\mathscr{U}_1$, On the other hand, if $\alpha > \mu$ and $\beta > \lambda$, the same thing happens when a node tries to receive when its mass or energy are too low. So we only have to assume one of these cases as eventually a node can become inaccessible through either incompatible route. We therefore will assume $\alpha < \mu$ and $\beta < \lambda$ for the rest of this discussion.

We can now look at message strings again. Given a sequence of P transactions for a node n where each transaction is labeled T_i and hence the transaction string is $\{T_1, T_2, \ldots, T_P\}$. Each transaction is either a receive R or a send S request.

- Each receive implies a $\alpha\psi\left(\frac{\mathscr{E}_A^1}{N}\right)$ loss of energy if in state $(+,-)$. Let I_1 be the set of indices from $\{1, \ldots, P\}$ where T_i is a receive with state $(+,-)$.

- Each receive implies a $\beta\frac{\mathscr{E}_A^1}{N}$ loss of energy if in state $(-,+)$. Let I_2 be the set of indices from $\{1, \ldots, P\}$ where T_i is a receive with state $(-,+)$.

- Each send implies a $\mu\psi\left(\frac{\mathscr{E}_A^1}{N}\right)$ loss of energy if in state $(+,-)$. Let I_3 be the set of indices from $\{1, \ldots, P\}$ where T_i is a send with state $(+,-)$.

- Each send implies a $\lambda\frac{\mathscr{E}_A^1}{N}$ loss of energy if in state $(-,+)$. Let I_4 be the set of indices from $\{1, \ldots, P\}$ where T_i is a send with state $(-,+)$.

The analysis then follows exactly as we have done before. The recursive calculations are just like we laid out earlier. $|\mathscr{U}_0| = 0$, $|\mathscr{V}_0| = N$ and $\mathscr{E}_A^0 = \mathscr{E}|\mathscr{V}_0|$.

- Symmetry breaking one: $|P_0| \approx 1$ if $\frac{1}{2^{m_0}} = \frac{\mathscr{E}_A^0}{\mathscr{E}}(\mu \vee \lambda) = \frac{|\mathscr{V}_0|}{N}\mu \vee \lambda$:

$$\begin{aligned} |\mathscr{U}_1| &= \frac{(|\mathscr{V}_0|)^2}{N}(\mu \vee \lambda) = N(\mu \vee \lambda) \\ \mathscr{V}_1 &= N - |\mathscr{U}_1| = \frac{N^2 - (|\mathscr{V}_0|)^2(\mu \vee \lambda)}{N} = N(1 - \mu \vee \lambda) \\ \mathscr{E}_A^1 &= \mathscr{E}\frac{|\mathscr{V}_1|}{N} = \mathscr{E}(1 - \mu \vee \lambda) \end{aligned}$$

Note there are no paths between nodes of lengths larger than one as we are only looking at the first message. However, the nodes that are accessible can now accept another message implying they are processing paths of length two.

- Symmetry breaking two: $|P_1| \approx 1$ if $\frac{1}{2^{m_1}} = \frac{\mathscr{E}_A^1}{\mathscr{E}}(\mu \vee \lambda) = (1 - \mu \vee \lambda)(\mu \vee \lambda) = \frac{|\mathscr{V}_1|}{N}(\mu \vee \lambda)$:

$$\begin{aligned} |\mathscr{U}_2| &= \frac{(|\mathscr{V}_1|)^2}{N}(\mu \vee \lambda) = N(1 - \mu \vee \lambda)^2(\mu \vee \lambda) \\ |\mathscr{V}_2| &= N - |\mathscr{U}_2| = \frac{N^2 - |\mathscr{V}_1|^2(\mu \vee \lambda)}{N} = N\left(1 - (1 - \mu \vee \lambda)^2(\mu \vee \lambda)\right) \\ \mathscr{E}_A^2 &= \mathscr{E}\frac{|\mathscr{V}_2|}{N} = \mathscr{E}\left(1 - (1 - \mu \vee \lambda)^2(\mu \vee \lambda)\right) \end{aligned}$$

We have now removed two groups of nodes as inaccessible. The accessible nodes now have processed paths of length two and are ready to process a path of length three. Note there are no paths between nodes of lengths larger than two here as we are only looking at the first two messages. However, the nodes that are accessible can now accept another message implying they are processing paths of length three. This is important as for the first time, we can have loops where a path starts and ends on a given node. Also, note, since mass and energy now have the ability to increase, the edge connections between nodes in paths of length three could be considered as either $+$ or $-$ edge weights which allows for interesting dynamics. We still do not have any idea about positional coordinates attached to a node or any notion of time.

- Symmetry breaking three: $|P_2| \approx 1$ if $\frac{1}{2^{m_2}} = \frac{\mathscr{E}_A^2}{\mathscr{E}}(\mu \vee \lambda) = \frac{|\mathscr{V}_2|}{N}(\mu \vee \lambda)$:

$$\begin{aligned} \mathscr{U}_3| &= \frac{(|\mathscr{V}_2|)^2}{N}(\mu \vee \lambda) \\ |\mathscr{V}_2| &= N - |\mathscr{U}_3| = \frac{N^2 - (|\mathscr{V}_2|)^2(\mu \vee \lambda)}{N} \\ \mathscr{E}_A^2 &= \mathscr{E}\frac{|\mathscr{V}_3|}{N} \end{aligned}$$

We are now processing paths of length four which allows for further complexity.

Since mass and energy at the nodes can increase now, we will introduce an additional constraint. We now assume there is a maximum mass M^{max} and maximum energy E^{max} possible at each node. The arguments we have presented work exactly the same and we can have symmetry breaking and nodes become inaccessible due to crossing the maximum mass or energy barrier. Hence the inaccessible nodes in this version of the model after the first symmetry breaking can include contributions from the maximum mass and maximum energy states. You can think of all these newest inaccessible states as a form of black hole if you wish.

Homework

Exercise 16.6.23 *Suppose we start with the number of nodes being* $N_0 = 2^{300} \approx 10^{90}$. *The number of particles in the universe is estimated to be about* $N = 2^{133} \approx 10^{40}$. *If we lose* 70% *of the nodes due to inaccessibility, each message passing event, how many nodes are accessible after* 5 *events?*

Exercise 16.6.24 *Suppose we start with the number of nodes being* $N_0 = 2^{300} \approx 10^{90}$. *The number of particles in the universe is estimated to be about* $N = 2^{133} \approx 10^{40}$. *If we lose* 70% *of the nodes due to inaccessibility, each message passing event, how many message passing events have to occur for the number of accessible nodes to drop to* $N \approx 10^40$*?*

Exercise 16.6.25 *Suppose we start with the number of nodes being* $N_0 = 2^{300} \approx 10^{90}$. *The number of particles in the universe is estimated to be about* $N = 2^{133} \approx 10^{40}$. *If we lose* 60% *of the nodes due to inaccessibility, each message passing event, how many nodes are accessible after* 5 *events?*

Exercise 16.6.26 *Suppose we start with the number of nodes being* $N_0 = 2^{300} \approx 10^{90}$. *The number of particles in the universe is estimated to be about* $N = 2^{133} \approx 10^{40}$. *If we lose* 40% *of the nodes due to inaccessibility, each message passing event, how many message passing events have to occur for the number of accessible nodes to drop to* $N \approx 10^40$*?*

16.6.7 Topological Considerations

After a finite number Q of message passings, the graph has $|\mathscr{U}_Q|$ nodes. The graph now has nodes which are fully connected to the graph and whose values play a role in the function of the graph but which cannot accept or send messages. However, since their state is still part of graph computation, the new graph is not the same as a graph which has fewer nodes because nodes and their incoming and outgoing edges have been excised. The paths can be loops too and the edge connections can be $+$ or $-$. Let's look at these details now for a loop of length N. Since the only change in state value for a node occurs when the node does a send/receive or a receive/send, the number of intermediate nodes between the start and final node, which is the start node again, do not matter. There are only two cases to consider.

N **Loop** $(+,-)$ **to** $(+,-)$

- Node i in $(+,-)$ state sends a message. Node i's state is now $(-,+)$ with value

$$m_n - \alpha\psi\left(\frac{\mathscr{E}_A^1}{N}\right) \|E_n + \alpha\mathscr{E}_A^1 \frac{N - M_0}{NM_0}$$

- After some intermediate chains in the loop, a message is sent back to Node i. Node i moves back to state $(+,-)$ with value

$$m_n - \alpha\psi\left(\frac{\mathscr{E}_A^1}{N}\right) - \mu\psi\left(\frac{\mathscr{E}_A^1}{N}\right) \|E_n + \alpha\mathscr{E}_A^1 \frac{N - M_0}{NM_0} + \mu\mathscr{E}_A^1 \frac{N - M_0}{NM_0}$$

N **Loop** $(-,+)$ **through** $(-,+)$

- Node i in $(-,+)$ state sends a message. Node i's state is now $(+,-)$ with value

$$m_n + \beta\psi\left(\mathscr{E}_A^1 \frac{N - M_0}{NM_0}\right) \|E_n - \beta\frac{\mathscr{E}_A^1}{N}$$

- After some intermediate chains in the loop, a message is sent back to Node i. Node i moves back to state $(-,+)$ with value

$$m_n + \beta\psi\left(\mathscr{E}_A^1 \frac{N-M_0}{NM_0}\right) + \lambda\psi\left(\mathscr{E}_A^1 \frac{N-M_0}{NM_0}\right) || E_n - \beta\frac{\mathscr{E}_A^1}{N} - \lambda\frac{\mathscr{E}_A^1}{N}$$

We conclude

- For an N loop that starts on a $(+,-)$ node and ends on the same node, the value of the state decreases m_n $m_n - (\alpha+\mu)\psi\left(\frac{\mathscr{E}_A^1}{N}\right)$ and increases E_n to $E_n + (\alpha+\mu)\mathscr{E}_A^1 \frac{N-M_0}{NM_0}$.
- For an N loop that starts on a $(,+)$ node and ends on the same node, the value of the state increases m_n to $m_n + (\beta+\lambda)\psi\left(\mathscr{E}_A^1 \frac{N-M_0}{NM_0}\right)$ and decreases E_n to $E_n - (\beta+\lambda)\frac{\mathscr{E}_A^1}{N}$.

These simple results follow because in our model, nodes only pay a cost for sending and receiving messages. Our model does not allow a message to contain a signal, which also adds to the change in the value of the state. In either the increasing mass or increasing energy cases, the other part of the state value is decreasing. Hence, this process will terminate using the minimum mass or minimum energy condition. However, as we mentioned, we have introduced a maximum mass M^{max} and maximum energy E^{max} condition also. So one or the other of these conditions will terminate the N loop process. Of course, there are other N length paths that do not self-terminate we have not discussed. At this point, we have introduced a useful model of a graph of computation nodes which involves only the structural components of the graph. Edges between nodes carry no information yet and signals are not yet involved. However, from the message passing assumptions we have made, we already see topological changes in the graph being introduced due to symmetry breaking. Now we add signals to the mix.

We can graph a simple example that shows the inaccessible nodes that arise from the symmetry breaking. The code is simple

Listing 16.7: **A Simple Graph**

```
function graphtwo(tol,M,sz,a,b)
%
X = randperm(M);
Y = randperm(M);
U = linspace(a,b,M);
V = linspace(a,b,M);
Z = 0.2*rand(M,M);
color={'r','k'};
clf
hold on
for i = 1:M
  for j = 1:M
    c = color{1};
    if Z(X(i),Y(j)) < tol
     c = color{2};
    end
    scatter(U(X(i)),V(Y(j)),sz,c,'filled');
  end
end
hold off
end
```

and we generate Figure 16.21 using the line `graphtwo(0.01,200,4,0,12)`. Note, this graph-

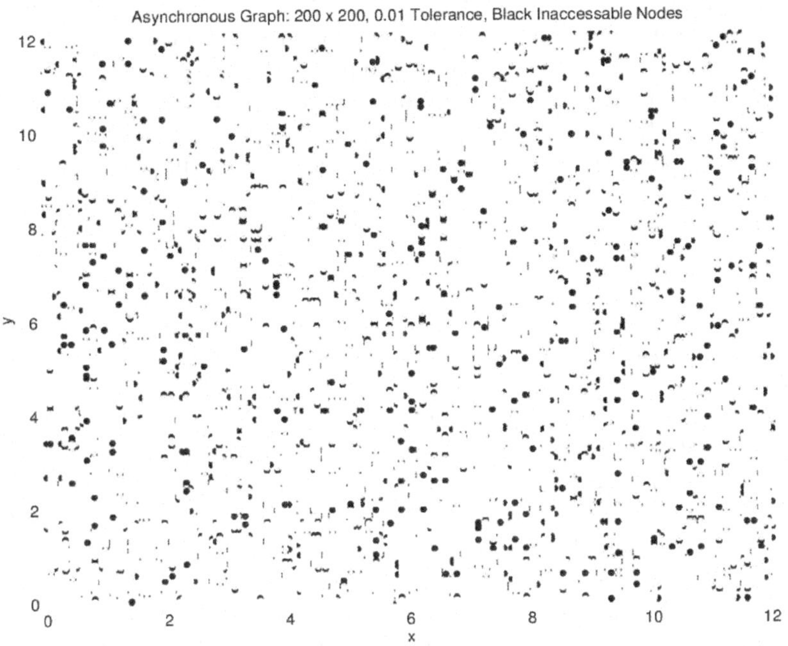

Figure 16.21: An asynchronous graph example.

ing experiment is, of course, completely wrong as it assumes positional information for the nodes so that you can see where the inaccessible nodes land up in the rectangle the nodes live within. Thus,

- Since the graph has no positional information, so in this code we are imposing that structure. The positions associated with the nodes could be in $\Re^k$ for any k and we wouldn't be able to display them at all. Imagine a sparsely connected graph which we code using edges and nodes in C++. The software architecture needed to process information flow through the graph does not require us to state where each node should be placed in the plane or $\Re^3$. The display aspects of the graph are generally things we impose to help us visualize. For example, the graphing program `dot` makes it difficult to specify coordinates for node placement and instead uses an algorithm which moves nodes and edges around so that there are fewer edge crossings.

- Our graph shows nicely the random distribution of the inaccessible nodes but that is an artificial construction. Since positional information is not available we have no way of knowing how *close* one node is to an inaccessible node. Hence, we also cannot visualize closed paths in the graph although there will certainly be such cycles.

Homework

Exercise 16.6.27 *Redo the discussion here using* `Z = p*rand(M,M);` *for* $p = 0.1$ *and* $M = 100$.

Exercise 16.6.28 *Redo the discussion here using* `Z = p*rand(M,M);` *for* $p = 0.05$ *and* $M = 50$.

Exercise 16.6.29 *Redo the discussion here using* `Z = p*rand(M,M);` *for* $p = 0.135$ *and* $M = 200$.

16.6.8 The Signal Network

We will assume the setup for the graph following the version two model of symmetry breaking. We use this message passing formalism as a base and assume that once the message sending and receiving costs are paid, the nodes are free to also send and receive content in the messages without incurring further costs. If you want, you can add extra costs for the content into the m_n and E_n values for the state and even split the costs into pieces comprising fractions of the cost for content versus communication, but we will not do that here as it adds only unnecessary complexity. Note by adding signals to the model, we are breaking with our toy cosmological model, but that was only intended as a *gedanken* experiment to help focus our thoughts. In the cosmological model, there is no mechanism at this point in its development to understand how to add signals. This would have to come from additional symmetry breaking and the resulting changes in the topology the graph has, but that is not our interest here.

We now want to let signals carrying messages to come to nodes in such a way that some value associated with the nodes, possibly independent of (m_n, E_n), to change with both receiving and sending signals. The accessible nodes initially are still $\mathscr{V}_0$ with $|\mathscr{V}_0| = N_0 = N$ and after the first symmetry breaking, the accessible nodes are $\mathscr{V}_1$ with $|\mathscr{V}_1| = M_0 = (1 - \mu \vee \lambda)|\mathscr{V}_0|$. Energy balance requires that after the first symmetry breaking, there is also a dark background component $\mathscr{D}_B$ which again we won't concern ourselves with here. As before, the energy and mass amounts m_n and E^n are what we have after the initial round of message passing, and hence they have been altered from their previous values. We will still use the labels m_n and E^n for convenience and remind you to make the mental adjustment. The new send message costs and mass and energy increases and content are shown in Figure 16.22 and Figure 16.23. We have added a new state value S_n to each node and each send adds the content c_n and each receive adds the content c_n also. After sending content c_n, we assume the state alters to $\sigma_n^s(S_n, c_n)$ where σ_n^s is a processing function which in general will be determined by the context; i.e., cognitive, immune models or something else. There is similar processing for received content. For content c_n, the new value is $\sigma_n^r(S_n, c_n)$.

$$A^{+/-}\;\; m_n \| E_n \| S_n \xrightarrow{\text{send } c_n} A^{-/+}\;\; m_n - \psi\left(\frac{\alpha \mathscr{E}_A^1}{N}\right) \| E_n + \alpha \mathscr{E}_A^1 \frac{N - M_0}{N M_0} \| \sigma_n^s(S_n, c_n)$$

Figure 16.22: Sending messages from $(+,-)$ with energy increase and content.

$$A^{-/+}\;\; m_n \| E_n \xrightarrow{\text{send } c_n} A^{+,-}\;\; m_n + \beta\psi\left(\mathscr{E}_A^1 \frac{N - M_0}{N M_0}\right) \| E_n - \beta \frac{\mathscr{E}_A^1}{N} \| \sigma_n^s(S_n, c_n)$$

Figure 16.23: Sending messages from $(-,+)$ with mass increase and content.

The new receive message costs and mass and energy increases are shown in Figure 16.24 and Figure 16.25.

$$A^{+/-}\;\; m_n \| E_n \| S_n \xrightarrow{\text{receive } c_n} A^{-/+}\;\; m_n + \lambda\psi\left(\mathscr{E}_A^1 \frac{N - M_0}{N M_0}\right) \| E_n - \lambda \frac{\mathscr{E}_A^1}{N} \| \sigma_n^r(S_n, c_n)$$

Figure 16.24: Receiving messages into $(+,-)$ with mass increase and content.

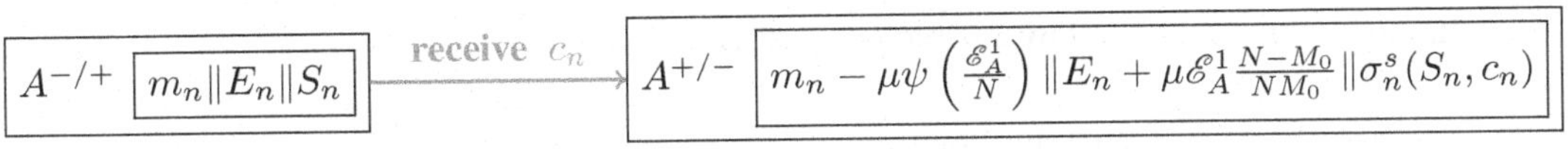

Figure 16.25: Receiving messages into $(-, +)$ with energy increase and content.

All of our derivations involving creating inaccessible nodes are still valid, however the addition of content and content processing functions creates much more variability. After the first symmetry breaking, message strings of length one have been processed and used to sequester inaccessible nodes. We can represent the new graph structure in a convenient matrix form as

$$\begin{bmatrix} [\mathcal{G}(\mathcal{V}_{\mathscr{V}_1}, \mathcal{E}_{\mathscr{V}_1 \to \mathscr{V}_1})] & [\mathcal{G}(\mathcal{V}_{\mathscr{V}_1}, \mathcal{E}_{\mathscr{U}_1 \to \mathscr{V}_1})] \\ [\mathcal{G}(\mathcal{V}_{\mathscr{U}_1}, \mathcal{E}_{\mathscr{V}_1 \to \mathscr{U}_1})] & [\mathcal{G}(\mathcal{V}_{\mathscr{U}_1}, \mathcal{E}_{\mathscr{U}_1 \to \mathscr{U}_1})] \end{bmatrix}$$

and after the second symmetry breaking, we have now processed loops of length two giving

$$\begin{bmatrix} \begin{bmatrix} \mathcal{G}(\mathcal{V}_{\mathscr{V}_2}, \mathcal{E}_{\mathscr{V}_2 \to \mathscr{V}_2}) & \mathcal{G}(\mathcal{V}_{\mathscr{V}_2}, \mathcal{E}_{\mathscr{U}_2 \to \mathscr{V}_2}) \\ \mathcal{G}(\mathcal{V}_{\mathscr{U}_2}, \mathcal{E}_{\mathscr{V}_2 \to \mathscr{U}_2}) & \mathcal{G}(\mathcal{V}_{\mathscr{U}_2}, \mathcal{E}_{\mathscr{U}_2 \to \mathscr{U}_2}) \end{bmatrix} & [\mathcal{G}(\mathcal{V}_{\mathscr{V}_1}, \mathcal{E}_{\mathscr{U}_1 \to \mathscr{V}_1})] \\ [\mathcal{G}(\mathcal{V}_{\mathscr{U}_1}, \mathcal{E}_{\mathscr{V}_1 \to \mathscr{U}_1})] & [\mathcal{G}(\mathcal{V}_{\mathscr{U}_1}, \mathcal{E}_{\mathscr{U}_1 \to \mathscr{U}_1})] \end{bmatrix}$$

As more messages are processed asynchronously, additional nodes may be labeled as inaccessible and now that paths of length three are processed, there can be loops that start and end at the same node. Note the partitioning of the graphs into disjoint subsets looks like this

$$\begin{bmatrix} \textbf{ACCESSIBLE} & \textbf{INACCESSIBLE} \\ [\mathcal{G}(\mathcal{V}_{\mathscr{V}_1}, \mathcal{E}_{\mathscr{V}_1 \to \mathscr{V}_1})] & [\mathcal{G}(\mathcal{V}_{\mathscr{V}_1}, \mathcal{E}_{\mathscr{U}_1 \to \mathscr{V}_1})] \\ \textbf{INACCESSIBLE} & \textbf{INACCESSIBLE} \\ [\mathcal{G}(\mathcal{V}_{\mathscr{U}_1}, \mathcal{E}_{\mathscr{V}_1 \to \mathscr{U}_1})] & [\mathcal{G}(\mathcal{V}_{\mathscr{U}_1}, \mathcal{E}_{\mathscr{U}_1 \to \mathscr{U}_1})] \end{bmatrix}$$

and after the second symmetry breaking, we have now processed loops of length two and the graph structure is now

$$\begin{bmatrix} \begin{bmatrix} \textbf{ACCESSIBLE} & \textbf{INACCESSIBLE} \\ \mathcal{G}(\mathcal{V}_{\mathscr{V}_2}, \mathcal{E}_{\mathscr{V}_2 \to \mathscr{V}_2}) & \mathcal{G}(\mathcal{V}_{\mathscr{V}_2}, \mathcal{E}_{\mathscr{U}_2 \to \mathscr{V}_2}) \\ \textbf{INACCESSIBLE} & \textbf{INACCESSIBLE} \\ \mathcal{G}(\mathcal{V}_{\mathscr{U}_2}, \mathcal{E}_{\mathscr{V}_2 \to \mathscr{U}_2}) & \mathcal{G}(\mathcal{V}_{\mathscr{U}_2}, \mathcal{E}_{\mathscr{U}_2 \to \mathscr{U}_2}) \end{bmatrix} & \begin{matrix} \textbf{INACCESSIBLE} \\ [\mathcal{G}(\mathcal{V}_{\mathscr{V}_1}, \mathcal{E}_{\mathscr{U}_1 \to \mathscr{V}_1})] \end{matrix} \\ \begin{matrix} \textbf{INACCESSIBLE} \\ [\mathcal{G}(\mathcal{V}_{\mathscr{U}_1}, \mathcal{E}_{\mathscr{V}_1 \to \mathscr{U}_1})] \end{matrix} & \begin{matrix} \textbf{INACCESSIBLE} \\ [\mathcal{G}(\mathcal{V}_{\mathscr{U}_1}, \mathcal{E}_{\mathscr{U}_1 \to \mathscr{U}_1})] \end{matrix} \end{bmatrix}$$

However, the inaccessible nodes still play a role in the message processing of the graph.

After a finite number of symmetry breakings, we are left with a graph in the upper left-hand corner which processes paths of length at least three which includes the possibility of loops, where a path starts and stops on the same node. Computations around closed paths allow for the possibility that classes of paths become equivalent. The exact nature of these equivalence classes of paths and the topology they determine on the subgraph $\mathcal{G}(\mathcal{V}_{\mathscr{V}_2}, \mathcal{E}_{\mathscr{V}_2 \to \mathscr{V}_2})$ will be determined by the receive and send processing functions. For convenience of exposition, let's denote the collection of signal processing capability by $\mathcal{C}$. Hence, the topology of the graph induced by the graph structure and processing functions can be represented by a triple $(\mathcal{C}, \mathcal{G}(\mathcal{V}, \mathcal{E}))$.

We will assume this topology has the form

$$\begin{aligned} &\mathcal{G}\left(\mathcal{V}_{\mathscr{V}_2}, \mathcal{E}_{\mathscr{V}_2 \to \mathscr{V}_2}\right) \\ &= \mathcal{G}_\infty\left(\mathcal{V}_{\mathscr{V}_{21}}, \mathcal{E}_{\mathscr{V}_{21} \to \mathscr{V}_{21}}\right) \oplus \mathcal{G}_\in\left(\mathcal{V}_{\mathscr{V}_{22}}, \mathcal{E}_{\mathscr{V}_{22} \to \mathscr{V}_{22}}\right) \oplus \ldots \oplus \mathcal{G}_{\text{II}}\left(\mathcal{V}_{\mathscr{V}_{2q}}, \mathcal{E}_{\mathscr{V}_{2q} \to \mathscr{V}_{2q}}\right) \end{aligned}$$

To make the notation simpler, we will drop some of the details and write

$$\mathcal{G}\left(\mathcal{V}_{\mathscr{V}_2}, \mathcal{E}_{\mathscr{V}_2 \to \mathscr{V}_2}\right) = \mathcal{G}_\infty\left(\mathcal{V}_{21}, \mathcal{E}_{21}\right) \oplus \mathcal{G}_\in\left(\mathcal{V}_{22}, \mathcal{E}_{22}\right) \oplus \ldots \oplus \mathcal{G}_{\text{II}}\left(\mathcal{V}_{2q}, \mathcal{E}_{2q}\right)$$

which is shown in diagram form as

$$\begin{bmatrix} [\mathcal{G}_\infty(\mathcal{V}_{21}, \mathcal{E}_{21})] & [\mathbb{O}] & [\mathbb{O}] & [\mathbb{O}] \\ [\mathbb{O}] & [\mathcal{G}_\in(\mathcal{V}_{22}, \mathcal{E}_{22})] & [\mathbb{O}] & [\mathbb{O}] \\ & & \ddots & \\ [\mathbb{O}] & [\mathbb{O}] & [\mathbb{O}] & [\mathcal{G}_{\text{II}}(\mathcal{V}_{2q}, \mathcal{E}_{2q})] \end{bmatrix}$$

In other words the nodes in the index set $\mathscr{V}$ have a decomposition $\mathscr{V}_2 = \boldsymbol{V}_{21} \oplus \ldots \oplus \boldsymbol{V}_{2q}$ with corresponding edges only from $\boldsymbol{V}_{2i}$ to $\boldsymbol{V}_{2i}$ for all $1 \leq i \leq q$ allowing for now cross talk between summands. Finally, since all of this notation can easily be understood in context, we will simply write

$$\mathcal{G}\left(\mathcal{V}_{\mathscr{V}_2}, \mathcal{E}_{\mathscr{V}_2 \to \mathscr{V}_2}\right) = \mathcal{G}_\infty \oplus \mathcal{G}_\in \oplus \ldots \oplus \mathcal{G}_{\text{II}}$$

and all of the direct sum decompositions are implied by the notation.

An example would help at this point. Let's assume the graph decomposition is $(\mathscr{C}_1, \mathscr{C}_2, \mathscr{C}_3, \mathscr{C}_4)$ where each subgraph $\mathcal{G}_\rangle$ is identified with a vector space $\mathscr{C}_i$. For our example, assume $\mathscr{C}_1 \equiv \Re^3$, $\mathscr{C}_2 \equiv \Re^3$, $\mathscr{C}_3 \equiv \Re^3$ and $\mathscr{C}_4 \equiv \Re$. The state of the graph after a signal chain is applied is

$$s_1 = \begin{bmatrix} 7 \\ 15 \\ 12 \end{bmatrix} \in \mathscr{C}_1, \quad s_2 = \begin{bmatrix} 11 \\ 15 \\ 12 \end{bmatrix} \in \mathscr{C}_2, \quad s_3 = \begin{bmatrix} 10 \\ 10 \\ 13 \end{bmatrix} \in \mathscr{C}_3, \quad s_4 = \begin{bmatrix} 6 \end{bmatrix} \in \mathscr{C}_4$$

This leads to a matrix representation we can call a bar code corresponding to the response of the graph to the signal chain

$$\boldsymbol{B} = \begin{bmatrix} \begin{bmatrix} 7 \\ 15 \\ 12 \end{bmatrix} & \begin{bmatrix} 11 \\ 15 \\ 12 \end{bmatrix} & \begin{bmatrix} 7 \\ 15 \\ 12 \end{bmatrix} & \begin{bmatrix} 6 \end{bmatrix} \end{bmatrix}$$

We can take this a step further. The mappings could be replaced by a mapping into Z_k rather than $\Re^k$. In Figure 16.26, we show such a bar code for the graph response to a signal chain when the graph has the decomposition

$$\mathcal{G}\left(\mathcal{V}_{\mathscr{V}_2}, \mathcal{E}_{\mathscr{V}_2 \to \mathscr{V}_2}\right) = Z_7 \oplus Z_7 \oplus Z_7 \oplus Z_9 \oplus Z_9 \oplus Z_4 \oplus Z_4 \oplus Z_4 \oplus Z_5$$

Now only one entry in each column of the bar code will be nonzero and will consist of a 1 with all the other entries 0. The signal family the graph is exposed to will determine this topology. Hence, the structure of $\boldsymbol{C}$ is critical to understanding the functioning of the asynchronous graph. In cognitive models, we can see this structure in a simplified version of cortical processing. Figure 16.27 shows such a simplified portion of a cortical-thalamus model (S. Murray Sherman and R. Guillery

(110) 2006). This version of the thalamus circuitry consists of one neuron in the thalamic reticular nucleus (**TRN**), a higher order relay (**HO**), a first order relay (**FO**), an interneuron (**IN**) and inhibitory neurons $\boldsymbol{i_1}$ and $\boldsymbol{i_2}$. It is clear there is a dynamical loop formed from the cortical can and the thalamus circuitry, the **CT** construct. We assume each cortical column consists of 6 cortical cans comprised of standard building block circuits: **OCOS** On Center - Off Surround, **FFP** Folded Feedback Pathway, and **2 − 3** Layer Two - Three circuit elements. Input from primary sensors enters through the first order relay where it may be processed into burst mode or essentially passed with minimal change to the **OCOS** neurons $N1$ and $N3$. Cortical can processed information is sent back to the thalamus through the **TRN** into the higher order relay **HO** and then **HO**'s output is fed back as input to **OCOS** neurons $N1$ and $N7$. This is a full feedback computational loop. The outputs of this loop are sent out for action, which we indicate here with an edge to the motor cortex. Note that neuron $T2$ in the 2/3 circuit of the can gets input from $N7$ of the **OCOS** and sends its output via the cerebellum to the motor cortex as well. This is an additional feedback loop as neuron $F2$ of the folded feedback pathway of the can feeds back to $N7$. We can associate each **CT** construct with a computation based on the equivalence classes that the signals entering the **CT** determine. These equivalence classes give us a graph decomposition which, for example, might be modeled by direct sums of simple finite groups such as $\boldsymbol{Z_{p_i}}$ for a primes p_i. The computation of the **CT** circuit, then corresponds to choosing an integer value from $\boldsymbol{Z_p}$ or in general an entry from some finite group $\boldsymbol{G}$. For example, if the underlying group was $\boldsymbol{Z_3}$, the choices are from the set $\{0, 1, 2\}$. For a stack of three cans, each of the three **CT** circuits, $\{\boldsymbol{CT_1}, \boldsymbol{CT_2}, \boldsymbol{CT_3}\}$ would generate $\{i_1, i_2, i_3\}$ from the underlying $\boldsymbol{Z_{p_i}}$ finite group which would be identified with the cortical column computation. We can think of this computation as generating a matrix output $\boldsymbol{M_o}$ of this form

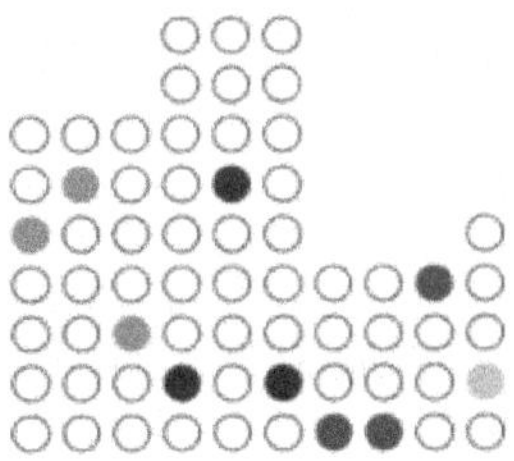

Figure 16.26: The barcode for a particular signal.

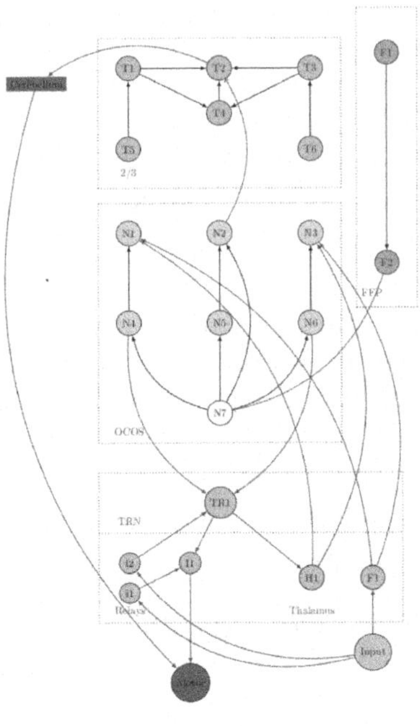

Figure 16.27: Typical cortical can/thalamus circuitry.

$$\boldsymbol{M_o} = \begin{bmatrix} \begin{bmatrix} 0 \\ \cdots \\ 1 \\ \cdots \\ 0 \end{bmatrix} & \begin{bmatrix} 0 \\ \cdots \\ 1 \\ \cdots \\ 0 \end{bmatrix} & \begin{bmatrix} 0 \\ \cdots \\ 1 \\ \cdots \\ 0 \end{bmatrix} \end{bmatrix}$$

where instead of thinking of the output as i_i, we instead place a **1** in the ith row of the column in $\boldsymbol{M}$ associated with this can. This is a **barcode** representation of the computation. There would be a matrix barcode $\boldsymbol{M_{ij}}$ for each cortical column (i, j) in the cortex model and hence there is a matrix of barcodes which represents the computational output of the entire cortical model. This interpretation of the decomposition of the graph cortical signals uses ideas from algebraic topology (W. Fulton (117) 1995) and Betti Node decompositions of high dimensional data as discussed in (R. Ghrist (102) 2008) and (Kaczynski et al. (66) 2004). At an equilibrium state, environmental signals form the standard cytokine signaling environment which induces this topology for signal parsing. Note the graph we see here is the graph $\mathcal{G}\left(\mathcal{V}_{\mathcal{V}_2}, \mathcal{E}_{\mathcal{V}_2 \rightarrow \mathcal{V}_2}\right)$ we have been discussing that corresponds to the accessible nodes after symmetry breaking. The dynamics we are studying are thus the ones that arise from signal chains that include closed loop computations.

Homework

Exercise 16.6.30 *Suppose we start with the number of nodes being $N_0 = 2^{300} \approx 10^{90}$. The number of particles in the universe is estimated to be about $N = 2^{133} \approx 10^{40}$. If we lose 70% of the nodes due to inaccessibility, each message passing event, how many nodes are accessible after 5 events?*

Exercise 16.6.31 *Suppose we start with the number of nodes being $N_0 = 2^{300} \approx 10^{90}$. The number of particles in the universe is estimated to be about $N = 2^{133} \approx 10^{40}$. If we lose 70% of the nodes due to inaccessibility, each message passing event, how many message passing events have to occur for the number of accessible nodes to drop to $N \approx 10^40$?*

Exercise 16.6.32 *Suppose we start with the number of nodes being $N_0 = 2^{300} \approx 10^{90}$. The number of particles in the universe is estimated to be about $N = 2^{133} \approx 10^{40}$. If we lose 60% of the nodes due to inaccessibility, each message passing event, how many nodes are accessible after 5 events?*

Exercise 16.6.33 *Suppose we start with the number of nodes being $N_0 = 2^{300} \approx 10^{90}$. The number of particles in the universe is estimated to be about $N = 2^{133} \approx 10^{40}$. If we lose 40% of the nodes due to inaccessibility each message passing event, how many message passing events have to occur for the number of accessible nodes to drop to $N \approx 10^40$?*

Now that we have a reasonable framework of an asynchronous message passing framework with signal capability loosely based on a toy cosmological model, it is fun to notice some consequences. The smallest change in the graph occurs with the cost of a send or receive message which is on the order of $\frac{\mathscr{E}}{N}$. Since energy is related to frequency via $E = h\nu$, this suggests that a measure of minimal time would be $t_{min} = \frac{hN}{\mathscr{E}}$. Since we also know $f = \frac{v}{d}$ where f is frequency, v is velocity and d is distance, we can infer other characteristics of our graph. We also know t_{min} satisfies the relationship $\frac{1}{t_{min}} = \frac{v_{max}}{d_{min}}$. Here d_{min} represents the *minimum distance* we can associate with our model and v_{max} is the maximum velocity. Of course, these are just abstractions and there is no such metric and time information associated with the graph in actuality.

After the symmetry breakings have settled down, the cost of send and receive increases to $\frac{\mathscr{E}}{Q}$ where Q is the size of the accessible node list after the symmetry breakings. The graph has paths of size $Q - 1$ in length so there are closed loop possibilities and with signals, the algebraic sign of the edge cost between nodes can be either plus or minus. Since $h \approx 2 \times 10^{-25}$ joule - sec, if $\mathscr{E} \approx 10^{40}$ joules and $N = 10^{40}$, we find $t_{min} = 2 \times 10^{-25}$ seconds. Let's assume $m_c = 10^{-20}$ kg and assume relativistic mass energy conversions so that $\sqrt{\frac{\text{Energy}}{\text{Mass}}} = \text{Velocity}$. For this calculation we will use the new energy after the symmetry breakings of $\frac{\mathscr{E}}{Q}$, instead of the energy initially $\frac{\mathscr{E}}{N}$. To determine a maximum velocity much like the speed of light constraint, $v_{\text{max}} = \sqrt{\frac{\mathscr{E}}{Q\, m_c}}$ If $Q \approx .3N$, then

$$
\begin{aligned}
v_{max} &= \sqrt{\frac{\mathscr{E}}{0.3N\, 10^{-20}}} \approx 3.3 \times 10^{10}\, \frac{\text{m}}{\text{s}}, \quad t_{min} \approx 2 \times 10^{-25}\text{s} \\
d_{min} &\approx v_{max} t_{min} = 3.3 \times 10^{10} \times \left(2 \times 10^{-25}\right) \text{m} = 6.6 \times 10^{-15}\text{m}
\end{aligned}
$$

Note that an estimate of the diameter of the *universe*, the graph is embedded in, is

$$Nd_{min} \approx 10^{40} \times 6.6 \times 10^{-15}\text{m} = 6.6 \times 10^{25}\text{m}$$

after the symmetry breakings have been completed and signaling has begun with paths at least 3 in length. From our discussions, we note we do not really have a unit of distance and a unit of time in this abstract model, although it is interesting to use standard physics relationships to try to define such units as we have done here.

Exercise 16.6.34 *Redo the discussion above using $Q \approx .3N$ and $N = 10^{90}$.*

Exercise 16.6.35 *Redo the discussion above using $Q \approx .2N$ and $N = 10^{40}$.*

Exercise 16.6.36 *Redo the discussion above using $Q \approx .2N$ and $N = 10^{90}$.*

16.6.9 Asynchronous Computational Graph Models

Now let's consider our graph model $\mathcal{G}(\mathcal{V}, \mathcal{E})$ from an asynchronous point of view. Now assume the maximum number of nodes $\boldsymbol{N}^{\mathbf{max}}$ is still finite, but perhaps even larger than before; i.e., $\boldsymbol{N}^{\mathbf{max}} \approx 10^{10} - 10^{30}$ and the number of edges is perhaps 100 times larger, i.e. $\approx 10^{12} - 10^{32}$. Immune, biome, cognitive and neurodecision models all fit well within this size structure. For fixed $\boldsymbol{N}^{\mathbf{max}}$, the graph whose nodes are all connected has $\boldsymbol{N}^{\mathbf{max}}(\boldsymbol{N}^{\mathbf{max}} - 1)/2$ edges and is denoted by $\mathbb{H}(\boldsymbol{N}^{\mathbf{max}})$, which is identified with $\mathbb{H}$ if the size choice of $\boldsymbol{N}^{\mathbf{max}}$ is clear from context. Each of our graphs, $\mathcal{G}(\mathcal{V}, \mathcal{E})$, then corresponds to a graph of $\boldsymbol{N_V}$ nodes and $\boldsymbol{N_E}$ edges. With a global time clock point of view, our reasoning starts with the fully connected graph and removed edges and nodes to create the working network for our model. We now follow the asynchronous message passing graph model in which we start with the fully connected graph, with some constant graph energy $\mathscr{E}$ choice and some choice of state $(+, -)$ and $(-, +)$ determined by a state (u_n, v_n) at node n where message passing and message receiving have a cost. There is a minimum (u^c, v^c) cost pair, which cannot be gone below, following the discussions we have had for the asynchronous message passing graph architectures. Hence, this model will generate inaccessible nodes still connected to the graph which help shape information passing and processing. The inaccessible nodes determine the initial topological structure also, which is underneath the new topologies that emerge as signal processing begins to become dominant, once message sequences achieve a minimal length of three. The longer message sequences allow for complex loops and the addition of signaling allows for the usual edge and node destruction and creation functions we have previously mentioned.

We can still destroy the edge between two nodes in $\mathbb{H}$ by simply setting the edge function to be zero and we can remove a node from $\mathbb{H}$ by setting the edge functions connecting to that node to be zero and the edge functions coming out of the node to be zero as well. We can also do this by actually deleting the edge and the node from the graph. If we wish all of our subgraphs to be subsets of the fully connected graph, we do not want to take this second approach. Also, note this reset of edge values to zero for a node is quite similar to the process of creating an inaccessible node. The model can then continue to use state information from that node for graph-level computations too. Each graph $\mathcal{G}(\mathcal{V}, \mathcal{E})$ can be obtained from $\mathbb{H}$ by applying an edge deletion map $\mathscr{E}$ to $\mathbb{H}$, which is split into the part that deletes nodes and the part that deletes edges between active nodes. The computations performed by nodes are quite complex and the edge functions $\boldsymbol{E}_{j \to i}$ between the nodes $\boldsymbol{V_j}$ and $\boldsymbol{V_i}$ can be equally complicated. Hence, in the context of the asynchronous message passing networks we are discussing, we will be choosing edge and node processing functions from the previously defined classes of functions Φ_e and Φ_v for our particular application domain.

The signal states (u_n, v_n) of the nodes of the graph are then, as before, used to create a set of objects

$$\mathscr{G} \quad = \quad \{\mathscr{H}(\mathbb{V}, \mathbb{E}) \mid \boldsymbol{E}_{j \to i} \in \Phi_e, \; \forall \, \boldsymbol{V}_i, \boldsymbol{V}_j \; \in \; \Phi_v\}$$

The set $\mathscr{G}$ contains the computational graphs we need in our modeling. This is where this discussion deviates from the one we had before using global time clocks. Each node now possesses its own unique local time clock and other nodes are not aware of it at all. This is the correct way to think about the way information is processed in all of our application domains, whether it is the immune system, a cognitive system, an economic system, a model of the bacterial biome of the gut or something else. There are really no message queues to process in a serial manner using backward sets and message sending as a serial process either. It is really accepting messages one at a time and sending messages one at a time. So simulation models based on the idea of a global simulation time clock are just not correct although they are very convenient. We create report functions which output the state of the graph according to a global time clock that is used for report purposes, but we must understand this global notion of time is completely artificial.

From the point of view of the global time clock for reporting, there is thus still a sequence of such graphs $\{\mathscr{G}_{\alpha \in J}\}$ for index set J. We can ask the usual questions. After reporting for P iterations of the global time clock from the base, the model $\mathscr{H}_P$ has been formed. There is clearly still an interest in understanding what we mean by letting $P \to \infty$. The sequence $\{\mathscr{H}_{p=0}^{\infty}\} \subset \mathscr{G}$ and at every step in the creation process, a viable graph model is built and can be used in simulation. To discuss the convergence process, the set of objects $\mathscr{G}$ must still be given a topological structure and in our remarks here we have tried to show a few of the processes that shape the topology of the graph starting with the formation of inaccessible nodes even before there is useful signal processing functionality. The limiting object, $\mathscr{H}^{\infty}$ then belong to the completion of $\mathscr{G}$, $\mathscr{G}^{\infty}$. In the context of our application domains, we are still interested in

- A **normal** brain model, $\mathscr{H}_{Brain}^{N,\infty}$ where here the superscript N indicates *normal.*

- If the context of a model of autoimmune disease, the goal is to build a model of normal immune response, $\mathscr{H}_{Immune}^{N,\infty}$.

- In the context of a model of the gut biome, we want a model of the **normal** gut biome, $\mathscr{H}_{Gut}^{N,\infty}$, which can then interact with $\mathscr{H}_{Brain}^{N,\infty}$ and $\mathscr{H}_{Immune}^{N,\infty}$. so we can study health when there are perturbations to this normal limiting graph structure.

- In the context of economic decision making using cognitive models for the decision process coupled to graph models of consumers and other actors, the goal is to build a model of normal decision response, $\mathscr{H}_{Decision}^{N,\infty}$, which can then interact with $\mathscr{H}_{Brain}^{N,\infty}$ so that we can study economic decision making when there are perturbations to this normal limiting graph structure.

In all cases, the report sequences help us learn how to deal with *damage* by looking for deviations from the baseline *normal* models.

Once normal models have been constructed, the next type of sequence is another round of collected report sequences based on the global reporting clock. For a finite sequence of time points $\{t_i\}_{i=1}^{Q}$ determined by the global reporting clock or another global clock, external signals, $\{\boldsymbol{S}_i\}$, are applied to the normal model creating a new version of the model. The external signal has the same effects: one, it can alter the edge and node function processing but leave the physical graph alone; second, it can remove some of the edge processing between two nodes, and third, it can remove all the edge processing for a particular node effectively creating an inaccessible node. Clearly, node removal is a common occurrence in all of the modeling scenarios we have described. Hence, as before, we can still write the signal $\boldsymbol{S}_i$ as a composition: $\boldsymbol{S}_i = \mathscr{E}_v \circ \mathscr{E}_e \circ \mathscr{E}_f$, where we let $\mathscr{E}_f$ denote the edge and node function alteration mapping. Hence, signals are of the functional form $\mathscr{E}_{evf} = \mathscr{E}_v \circ \mathscr{E}_e \circ \mathscr{E}_f$ and each signal applied to the graph model therefore has the potential to alter everything about it.

At any reporting time point t_j in the simulation, we therefore have a graph model, $\mathcal{H}_j^t$ which has $\boldsymbol{V}_j$ node functions and $\boldsymbol{E}_j$ edge functions. Note, $\mathscr{H}_p$ denotes the pth iteration in the model building process which is an approximation to the limiting object $\mathscr{H}^\infty$ and $\mathcal{H}_j^t$ is the result of applying signals $\{\boldsymbol{S}_1, \ldots, \boldsymbol{S}_j\}$ to the approximate model $\mathscr{H}_p$. This result gives insight into what the signals $\{\boldsymbol{S}_1, \ldots, \boldsymbol{S}_j\}$ applied to the converged model $\mathscr{H}^\infty$ would do. As we discussed before in the context of a global time clock, in this asynchronous time version, after Q signals are applied, a report query would find

$$\boldsymbol{S}_Q \rightarrow \mathcal{H}_Q^t \quad = \quad \mathscr{E}_{evf}^Q \mathcal{H}_{Q-1}^t = \mathscr{E}_{evf}^Q \cdots \mathscr{E}_{evf}^1 \mathscr{H}_p = \mathcal{H}_Q^t$$

Since all the structural components of the graph model and the edge and node processing functions are potentially being altered, this process is still dependent of the order in which the signal are applied. Thus, $\mathcal{H}_Q^t = \mathscr{E}_{evf}^Q \cdots \mathscr{E}_{evf}^1 \mathscr{H}_p$ will be dependent of the signal sequence order. However, all graphs that are built are in $\mathbb{H}$ and hence can be added or subtracted which gives us a vector space structure.

Just as in the global time clock case, we can define the differences $\Delta_{\textbf{Signal}} \mathcal{H}_i^t = \mathcal{H}_{i+1}^t - \mathcal{H}_i^t$ and $\Delta_{\textbf{Structural}} \mathscr{H}_{i+1} = \mathscr{H}_{i+1} - \mathscr{H}_i$ due to a particular signal chain $\{\boldsymbol{S}_{i+1}, \ldots, \boldsymbol{S}_1\}$ or steps in the construction process of the base model as the **signal based** and **structural based** changes in the graph structure between global report time iterations. We can then use these differences to study possible discrete analogs of differentiable structures in this space for the kinds of topologies and metrics that are arising.

In a cognitive model, we know that initially there is a massive die off of neurons and connections as part of the maturation of the system. This is similar to the first and second symmetry breaking we discuss here. Also, in an immune model, there is a large-scale pruning of self-cognate T-Cells as part of the development of the immune system, which can also be associated with the symmetry breaking we see in the message passing asynchronous models. Of course, we still need to work out the details of how we model a general system of cytokine signals and how they interact with what we choose to use as computational nodes in the models. The addition of signals to the simple toy cosmological model suggests that the topology, which determines how dynamics of the computational graph responds to signals, emerges based on the choice of signals, their content and the signal receiving and sending processing functions. What we have done here, then, is to lay down a proper theoretical foundation for further investigations.

Chapter 17

Signal Models and Autoimmune Disease

We now try to develop a general model of cytokine/chemokine signaling networks for use in our models of autoimmune disease and neurodegeneration. It is based on the generic asynchronous message passing model we have already discussed in Chapter 16 with added signaling components. This generic model is specialized here for immune system interactions. Valid cytokine signal combinations form a specific pattern, a barcode, of proteins and other molecules on the surface of a target cell. The barcodes generate a grammar of signal choices and an immune system response and these signals in conjunction with a message passing/signal aware asynchronous graph determines the dynamics of the response for a T-Cell interacting with a pMHC structure on the surface of a cell. This immunosynaptic response is used as the signaling content and edge and node processing in the generic asynchronous model with message passing and signals. This model explicitly models T-Cell and neuron pruning as part of determining inaccessible nodes in the computational graph model. This immunosynaptic response is modeled by an affinity/avidity threshold equation whose dependence on the relevant barcodes is discussed. The cytokine signaling cascade initiated by antigen recognition is also influenced by mutations of the gene LRRK2, which plays a crucial part in influencing consequences downstream for the primary event triggered by the antigen presentation. These mutations can give rise to cytokine storms which can lead to altered T-Cell-pMHC responses. These altered responses may allow for an autoimmune event due to a flow of cytokine messengers crossing the blood brain-barrier to cause inflammatory damage in microglia containing LRRK2 mutations, which leads to Parkinson's Disease and to the flows of cytokine messengers caused by epithelial West Nile Virus infections and gut biome interactions. So let's get started and try to apply the asynchronous message passing graph architectures to the specific problem of trying to understand autoimmune events.

For our models here, we will assume the general asynchronous message passing computational graph architecture with signal content and edge and node processing discussed in Chapter 16. However, we will use this model specialized for a state of (u_n, v_n) instead of the (m_n, E_n) now. Here, we replace the (m_n, E_n) arguments with similar ones involving the new states (u_n, v_n). The next step is to choose state pairs (u_n, v_n) and signals S_n with content c_n and associated edge and node processing functions $\sigma_n^r(S_n, c_n)$ and $\sigma_n^s(S_n, c_n)$ relevant for our autoimmune modeling. We also need to develop a model of LRRK2 mutation, cytokine signaling and gut biome interaction. In this immunological setting, we are interested in how T-Cells interact with MHC cradles carrying peptide fragments of both self and antigen protein. The formation of the structure which determines the fate of a cell being probed by a T-Cell is called the immunosynapse and this will be what we use as nodes in the model. For our purposes, for the states in these nodes, we will assume there is a minimum and maximum affinity of the immunosynapse. Hence, $u_c = (1 + g_0)\mathscr{A}_{min}$, where $\mathscr{A}_{min}$ is the minimum affinity of an immunosynapse and g_0 is a small positive parameter. This plays the role of m_c in the message passing graph model. We also assume there is a maximum affinity, $v_c = g_1 \mathscr{A}_{max}$, where $\mathscr{A}_{max}$ is the maximum affinity of an immunosynapse and g_1 is a small positive parameter. This plays the role of E_c in the message passing graph model. We let

$$f(\mathscr{A}) = \frac{\mathscr{A}_{max} + \mathscr{A}_{min}}{2} + \frac{\mathscr{A}_{max} - \mathscr{A}_{min}}{2} tanh(\mathscr{A} - \mathscr{O})$$

which maps a given affinity $\mathscr{A} \geq \mathscr{O}$ into the interval $(\mathscr{A}_{min}, \mathscr{A}_{max})$. We let $u_n = f(\mathscr{A}_n)$, where $\mathscr{A}_n$ is the affinity of immunosynapse and $v_n = \mathscr{A}_{max} - f(\mathscr{A})$. We note $u_n + v_n = \mathscr{A}_{max}$. The symmetry breaking here creates inaccessible immunosynapses when the affinity is below $(1+g_0)\mathscr{A}_{min}$ or above $(1 - g_1)\mathscr{A}_{max}$. Thus, T-Cells generating affinities that are too low are removed as well as T-Cells that generate affinities that are too tightly tuned to a given MHC cradle protein. This reflects what the immune system does. In addition to this choice of (u_n, v_n) we also must choose signal processing functions, which are very important once the inaccessible nodes are removed. In the usual message passing asynchronous derivations, for our (u_n, v_n) choice, all of the derivations involving creating inaccessible nodes are still valid. After the first symmetry breaking, message strings of length one have been processed and used to sequester inaccessible nodes. The new graph structure in matrix form is

$$\begin{bmatrix} [\mathcal{G}(\mathcal{V}_{\mathscr{V}_1}, \mathcal{E}_{\mathscr{V}_1 \to \mathscr{V}_1})] & [\mathcal{G}(\mathcal{V}_{\mathscr{V}_1}, \mathcal{E}_{\mathscr{U}_1 \to \mathscr{V}_1})] \\ [\mathcal{G}(\mathcal{V}_{\mathscr{U}_1}, \mathcal{E}_{\mathscr{V}_1 \to \mathscr{U}_1})] & [\mathcal{G}(\mathcal{V}_{\mathscr{U}_1}, \mathcal{E}_{\mathscr{U}_1 \to \mathscr{U}_1})] \end{bmatrix}$$

and after the second symmetry breaking, we have now processed loops of length two and the graph structure is now

$$\begin{bmatrix} \begin{bmatrix} \mathcal{G}(\mathcal{V}_{\mathscr{V}_2}, \mathcal{E}_{\mathscr{V}_2 \to \mathscr{V}_2}) & \mathcal{G}(\mathcal{V}_{\mathscr{V}_2}, \mathcal{E}_{\mathscr{U}_2 \to \mathscr{V}_2}) \\ \mathcal{G}(\mathcal{V}_{\mathscr{U}_2}, \mathcal{E}_{\mathscr{V}_2 \to \mathscr{U}_2}) & \mathcal{G}(\mathcal{V}_{\mathscr{U}_2}, \mathcal{E}_{\mathscr{U}_2 \to \mathscr{U}_2}) \end{bmatrix} & [\mathcal{G}(\mathcal{V}_{\mathscr{V}_1}, \mathcal{E}_{\mathscr{U}_1 \to \mathscr{V}_1})] \\ [\mathcal{G}(\mathcal{V}_{\mathscr{U}_1}, \mathcal{E}_{\mathscr{V}_1 \to \mathscr{U}_1})] & [\mathcal{G}(\mathcal{V}_{\mathscr{U}_1}, \mathcal{E}_{\mathscr{U}_1 \to \mathscr{U}_1})] \end{bmatrix}$$

At this point, as more messages are processed asynchronously, there may or may not be additional sequestering of nodes as inaccessible, but what really changes, is now that paths of length three are processed, there can be loops that start and end at the same node. Also, note the partitioning of the graphs into disjoint subsets of edges and nodes does not show that the graphs are structured as this

$$\begin{bmatrix} \textbf{ACCESSIBLE} & \textbf{INACCESSIBLE} \\ [\mathcal{G}(\mathcal{V}_{\mathscr{V}_1}, \mathcal{E}_{\mathscr{V}_1 \to \mathscr{V}_1})] & [\mathcal{G}(\mathcal{V}_{\mathscr{V}_1}, \mathcal{E}_{\mathscr{U}_1 \to \mathscr{V}_1})] \\ \textbf{INACCESSIBLE} & \textbf{INACCESSIBLE} \\ [\mathcal{G}(\mathcal{V}_{\mathscr{U}_1}, \mathcal{E}_{\mathscr{V}_1 \to \mathscr{U}_1})] & [\mathcal{G}(\mathcal{V}_{\mathscr{U}_1}, \mathcal{E}_{\mathscr{U}_1 \to \mathscr{U}_1})] \end{bmatrix}$$

and after the second symmetry breaking, we have now processed loops of length two and the graph structure is now

$$\begin{bmatrix} \begin{bmatrix} \textbf{ACCESSIBLE} & \textbf{INACCESSIBLE} \\ \mathcal{G}(\mathcal{V}_{\mathscr{V}_2}, \mathcal{E}_{\mathscr{V}_2 \to \mathscr{V}_2}) & \mathcal{G}(\mathcal{V}_{\mathscr{V}_2}, \mathcal{E}_{\mathscr{U}_2 \to \mathscr{V}_2}) \\ \textbf{INACCESSIBLE} & \textbf{INACCESSIBLE} \\ \mathcal{G}(\mathcal{V}_{\mathscr{U}_2}, \mathcal{E}_{\mathscr{V}_2 \to \mathscr{U}_2}) & \mathcal{G}(\mathcal{V}_{\mathscr{U}_2}, \mathcal{E}_{\mathscr{U}_2 \to \mathscr{U}_2}) \end{bmatrix} & \begin{matrix} \textbf{INACCESSIBLE} \\ [\mathcal{G}(\mathcal{V}_{\mathscr{V}_1}, \mathcal{E}_{\mathscr{U}_1 \to \mathscr{V}_1})] \end{matrix} \\ \begin{matrix} \textbf{INACCESSIBLE} \\ [\mathcal{G}(\mathcal{V}_{\mathscr{U}_1}, \mathcal{E}_{\mathscr{V}_1 \to \mathscr{U}_1})] \end{matrix} & \begin{matrix} \textbf{INACCESSIBLE} \\ [\mathcal{G}(\mathcal{V}_{\mathscr{U}_1}, \mathcal{E}_{\mathscr{U}_1 \to \mathscr{U}_1})] \end{matrix} \end{bmatrix}$$

However, the inaccessible nodes still play a role in the message processing of the graph.

After a finite number of symmetry breakings, we are left with a graph in the upper left-hand corner which is processing paths of length at least three which includes the possibility of loops where a path starts and stops on the same node. Computations around closed paths allow for the possibility that classes of paths become equivalent. The exact nature of these equivalence classes of paths and the topology they determine on the subgraph $\mathcal{G}\left(\mathcal{V}_{\mathscr{V}_2}, \mathcal{E}_{\mathscr{V}_2 \rightarrow \mathscr{V}_2}\right)$ will be determined by the receive and send processing functions. We denote the collection of signal processing capability by $\mathcal{C}$. Hence, the topology of the graph induced by the graph structure and processing functions can be represented by a triple $(\mathcal{C}, \mathcal{G}(\mathcal{V}, \mathcal{E}))$. We will assume this topology has the form

$$\mathcal{G}\left(\mathcal{V}_{\mathscr{V}_2}, \mathcal{E}_{\mathscr{V}_2 \rightarrow \mathscr{V}_2}\right) = \mathcal{G}_1\left(\mathcal{V}_{\mathscr{V}_{21}}, \mathcal{E}_{\mathscr{V}_{21} \rightarrow \mathscr{V}_{21}}\right) \oplus \mathcal{G}_2\left(\mathcal{V}_{\mathscr{V}_{22}}, \mathcal{E}_{\mathscr{V}_{22} \rightarrow \mathscr{V}_{22}}\right) \oplus \ldots \oplus \mathcal{G}_q\left(\mathcal{V}_{\mathscr{V}_{2q}}, \mathcal{E}_{\mathscr{V}_{2q} \rightarrow \mathscr{V}_{2q}}\right)$$

To make the notation simpler, we will drop some of the details and write

$$\mathcal{G}\left(\mathcal{V}_{\mathscr{V}_2}, \mathcal{E}_{\mathscr{V}_2 \rightarrow \mathscr{V}_2}\right) = \mathcal{G}_1\left(\mathcal{V}_{21}, \mathcal{E}_{21}\right) \oplus \mathcal{G}_2\left(\mathcal{V}_{22}, \mathcal{E}_{22}\right) \oplus \ldots \oplus \mathcal{G}_q\left(\mathcal{V}_{2q}, \mathcal{E}_{2q}\right)$$

which is shown in diagram form as

$$\begin{bmatrix} [\mathcal{G}_1(\mathcal{V}_{21}, \mathcal{E}_{21})] & [\mathbb{O}] & [\mathbb{O}] & [\mathbb{O}] \\ [\mathbb{O}] & [\mathcal{G}_2(\mathcal{V}_{22}, \mathcal{E}_{22})] & [\mathbb{O}] & [\mathbb{O}] \\ & & \ddots & \\ [\mathbb{O}] & [\mathbb{O}] & [\mathbb{O}] & [\mathcal{G}_q(\mathcal{V}_{2q}, \mathcal{E}_{2q})] \end{bmatrix}$$

In other words, the nodes in the index set $\mathscr{V}$ have a decomposition $\mathscr{V}_2 = \boldsymbol{V}_{21} \oplus \ldots \oplus \boldsymbol{V}_{2q}$ with corresponding edges only from $\boldsymbol{V}_{2i}$ to $\boldsymbol{V}_{2i}$ for all $1 \leq i \leq q$ allowing for no cross talk between summands. Finally, since all of this notation can easily be understood in context, we will simply write

$$\mathcal{G}\left(\mathcal{V}_{\mathscr{V}_2}, \mathcal{E}_{\mathscr{V}_2 \rightarrow \mathscr{V}_2}\right) = \mathcal{G}_1 \oplus \mathcal{G}_2 \oplus \ldots \oplus \mathcal{G}_q$$

and all of the direct sum decompositions are implied by the notation. Our next step is to try to understand the details of this decomposition for our immune system model.

Homework

Exercise 17.0.1 *Given*

$$\mathcal{G}\left(\mathcal{V}_{\mathscr{V}_2}, \mathcal{E}_{\mathscr{V}_2 \rightarrow \mathscr{V}_2}\right) = \mathcal{G}_1 \oplus \mathcal{G}_2$$

choose particular $\mathcal{G}_i$ graphs and work out all the details for this representation.

Exercise 17.0.2 *Given*

$$\mathcal{G}\left(\mathcal{V}_{\mathscr{V}_2}, \mathcal{E}_{\mathscr{V}_2 \rightarrow \mathscr{V}_2}\right) = \mathcal{G}_1 \oplus \mathcal{G}_2 \oplus \mathcal{G}_3$$

choose particular $\mathcal{G}_i$ graphs and work out all the details for this representation.

Exercise 17.0.3 *Given*

$$\mathcal{G}\left(\mathcal{V}_{\mathscr{V}_2}, \mathcal{E}_{\mathscr{V}_2 \rightarrow \mathscr{V}_2}\right) = \mathcal{G}_1 \oplus \mathcal{G}_2 \oplus \mathcal{G}_3 \oplus \mathcal{G}_4$$

choose particular $\mathcal{G}_i$ graphs and work out all the details for this representation.

17.1 Antigen Pathway Models

To help understand the models we are building here, we need an abstract version of the way antigen is processed by the immune system. We begin with the structure of the Immunoglobulin (**Ig**) molecule which is made from four chains and consists of heavy (**H**) and light **L** chains. The $\boldsymbol{Ig}$ complexes are built from two light chains and two heavy chains. The heavy chains are constructed from three sequence pieces and the light chains are made from two, and so historically, the heavy chains were indeed *heavier* and so were tagged with that label. In Figure 17.1(a), we show the typical layout of a immunoglobulin molecule. The chains are bound together by disulfide bonds and there is a hinge in the heavy chain that allows for movement. The antigen is captured between the light and heavy chains. The antigen epitope is positioned between the light and heavy chains as shown in Figure 17.1(b). There are three regions whose molecular composition has high variability in the $\boldsymbol{Ig}$ molecule: L_1, L_2 and L_3 on the light chain and H_1, H_2 and H_3 on the heavy chain and these occur on both sides of the hinge. The antigen epitope is captured between the light and heavy chain and the makeup of the L_i and H_i pieces is complementary to the makeup of the antigen epitope and so there is a weak bond between them. We have genes which generate the high variability regions of the light and heavy chains using a combinatorial approach we will discuss in a bit. The heavy chains

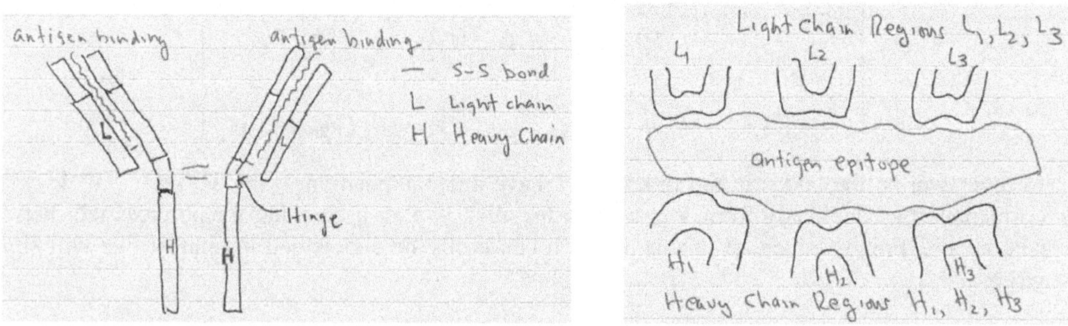

(a) A typical immunoglobulin molecule consisting of two light chains and two heavy chains.

(b) A typical immunoglobulin antigen epitope capture.

Figure 17.1: Immunoglobulin structure.

are made using a sequence $\boldsymbol{C_H}$ which has multiple components.

$$\boldsymbol{C_H} = \{\boldsymbol{\mu}, \boldsymbol{\delta}, \boldsymbol{\gamma_3}, \boldsymbol{\gamma_1}, \boldsymbol{\alpha_1}, \boldsymbol{\gamma_2}, \boldsymbol{\gamma_4}, \boldsymbol{\epsilon}, \boldsymbol{\alpha_2}\}$$

The light chains are built using two choices: $\boldsymbol{L_\kappa}$ and $\boldsymbol{L_\lambda}$. The $\boldsymbol{Ig}$ complexes are then built using two heavy and two light chains in the following combinations.

- $\boldsymbol{IgM}$: $\boldsymbol{H_\mu}$ and $\boldsymbol{L_\kappa}$ in the form $\kappa 2 \mu 2$, meaning the two light chains are $\boldsymbol{L_\kappa}$ and the two heavy chains are $\boldsymbol{H_\mu}$, or $\boldsymbol{H_\mu}$ and $\boldsymbol{L_\lambda}$ in the form $\lambda\ 2\ \mu\ 2$, meaning the two light chains are $\boldsymbol{L_\lambda}$ and the two heavy chains are $\boldsymbol{H_\mu}$.

- $\boldsymbol{IgD}$: $\boldsymbol{H_\delta}$ and $\boldsymbol{L_\kappa}$ in the form $\kappa\, 2\, \delta\, 2$, meaning the two light chains are $\boldsymbol{L_\kappa}$ and the two heavy chains are $\boldsymbol{H_\delta}$, or $\boldsymbol{H_\delta}$ and $\boldsymbol{L_\lambda}$ in the form $\lambda\, 2\, \delta\, 2$, meaning the two light chains are $\boldsymbol{L_\lambda}$ and the two heavy chains are $\boldsymbol{H_\delta}$.

- $\boldsymbol{IgG}$: $\boldsymbol{H_\gamma}$ and $\boldsymbol{L_\kappa}$ in the form $\kappa\, 2\, \gamma\, 2$, meaning the two light chains are $\boldsymbol{L_\kappa}$ and the two heavy chains are $\boldsymbol{H_\gamma}$, or $\boldsymbol{H_\gamma}$ and $\boldsymbol{L_\lambda}$ in the form $\lambda\ 2\ \gamma\ 2$, meaning the two light chains are $\boldsymbol{L_\lambda}$ and the two heavy chains are $\boldsymbol{H_\gamma}$.

- $\boldsymbol{IgA}$: $\boldsymbol{H_\alpha}$ and $\boldsymbol{L_\kappa}$ in the form $\kappa\,2\,\alpha\,2$, meaning the two light chains are $\boldsymbol{L_\kappa}$ and the two heavy chains are $\boldsymbol{H_\alpha}$, or $\boldsymbol{H_\alpha}$ and $\boldsymbol{L_\lambda}$ in the form $\lambda\,2\,\alpha\,2$, meaning the two light chains are $\boldsymbol{L_\lambda}$ and the two heavy chains are $\boldsymbol{H_\alpha}$.

- $\boldsymbol{IgE}$: $\boldsymbol{H_\epsilon}$ and $\boldsymbol{L_\kappa}$ in the form $\kappa\,2\,\epsilon\,2$, meaning the two light chains are $\boldsymbol{L_\kappa}$ and the two heavy chains are $\boldsymbol{H_\epsilon}$, or $\boldsymbol{H_\mu}$ and $\boldsymbol{L_\lambda}$ in the form $\lambda\,2\,\epsilon\,2$, meaning the two light chains are $\boldsymbol{L_\lambda}$ and the two heavy chains are $\boldsymbol{H_\epsilon}$.

Here the γ comes in four varieties and α comes in two. The construction for the light chain shows how the variability in the L_1, L_2 and L_3 portions of the chain are possible. There are two genes which code for the light chains called κ and λ. One gene has two segments which code for the **L** chains which are made up of a variable region, $\boldsymbol{V_L}$, of about 108 residues and another segment of the gene codes for a constant region $\boldsymbol{C_L}$. There are also different genes that code for a transitional protein sequence called the joining sequence $\boldsymbol{J}$. There is molecular machinery called the $\boldsymbol{V(D)J}$ polymerase that cuts the $\boldsymbol{L}$ chain coding segment and the joining segment $\boldsymbol{J}$ and puts the pieces together. The assembled $\boldsymbol{L}$ and $\boldsymbol{J}$ pieces are then combined with the constant sequence $\boldsymbol{C}$ to create a complete $\boldsymbol{L}$ chain.

The sequence to assemble a $\boldsymbol{V_\kappa}$ chain from a κ chain uses genes from Chromosome 2.

$$\begin{array}{rcl}
5'\,\boldsymbol{V_1} \equiv \boldsymbol{V_2} \equiv \cdots \equiv \boldsymbol{V_{40}} & \equiv & \boldsymbol{J_1} \equiv \boldsymbol{J_2} \equiv \cdots \equiv \boldsymbol{J_5} \equiv \boldsymbol{C_\kappa}\,3' \quad \textbf{GERMLINE} \\
 & \Downarrow & \text{polymerase } \boldsymbol{V(D)J}, \text{ select } \boldsymbol{V_j J_k} \\
5'\,\boldsymbol{V_1} \equiv \cdots \equiv \boldsymbol{V_{j-1}} & \equiv & \boldsymbol{V_j} \equiv \boldsymbol{J_k} \equiv \cdots \equiv \boldsymbol{J_5} \equiv \boldsymbol{C_\kappa}\,3' \quad \textbf{CHANGE B-CELL} \\
 & \Downarrow & \text{first or primary DNA transcript is made: still has introns} \\
 & \Downarrow & \text{splicing occurs: introns removed } \boldsymbol{V_j}, \boldsymbol{J_k} \text{ and } \boldsymbol{C_\kappa} \text{ combined} \\
 & \Downarrow & \text{mature messenger RNA } \boldsymbol{V_j J_k C_\kappa} \\
 & \Downarrow & \text{translation to protein} \\
 & \Downarrow & NH_2 - \mathbb{V}_j \mathbb{J}_k \mathbb{C}_\kappa - COOH \quad \textbf{A PARTICULAR SELECTION}
\end{array}$$

where $\boldsymbol{C_\kappa}$ is the constant sequence for the $\boldsymbol{V_\kappa}$ gene. We would use $\boldsymbol{C_\lambda}$ for the constant sequence for the $\boldsymbol{V_\lambda}$ gene. If we leave out this detail, we see we have the sequence

$$\boldsymbol{V_1} \cdots \boldsymbol{V_{40}}\, \boldsymbol{J_1} \cdots \boldsymbol{J_5}\, \boldsymbol{C_\kappa} \quad \rightarrow \quad \boldsymbol{V_j J_k C_\kappa} \rightarrow \mathbb{V}_j \mathbb{J}_k \mathbb{C}_\kappa$$

leaving off the amine and carboxyl groups attached to the $5'$ and $3'$ terminals of the chain. Hence, we can be briefer and say the transcription of the target protein uses the sequence

$$\boldsymbol{V_1} \cdots \boldsymbol{V_{40}}\, \boldsymbol{J_1} \cdots \boldsymbol{J_5}\, \boldsymbol{C_\kappa} \quad \rightarrow \quad \mathbb{V}_j \mathbb{J}_k \mathbb{C}_\kappa$$

We usually simply identify the RNA sequence $\boldsymbol{V_j J_k C_\kappa}$ with the transcribed protein $\mathbb{V}_j \mathbb{J}_k \mathbb{C}_\kappa$. Hence, this combinatorial process has the mnemonic $\boldsymbol{V_1} \cdots \boldsymbol{V_{40}}\, \boldsymbol{J_1} \cdots \boldsymbol{J_5}\, \boldsymbol{C_\kappa} \rightarrow \boldsymbol{V_j J_k C_\kappa}$. Thus there are $40^5 \approx 10^8$ possibilities. The particular $\boldsymbol{V_j J_k}$ determine the antigen specificity.

If the κ chain above is not constructed successfully, the λ version is made. Success here is related to how the κ chain binds to proteins. The gene segment responsible for constructing the $\boldsymbol{V_\lambda}$ chain from the λ gene is similar to what we just described for the $\boldsymbol{V_\kappa}$ chain. The κ gene is located in a different chromosome from the κ gene. In this construction, there are four different constant segments $\boldsymbol{C_\lambda^1}$ through $\boldsymbol{C_\lambda^4}$ with $\boldsymbol{C_\lambda^i}$ associated with $\boldsymbol{J_i}$. We indicate this in the sequence below using the symbol $\boldsymbol{C_\lambda^i}$ for convenience. After the editing steps the correct constant sequence will be added. There appears to be little functional difference between the $\boldsymbol{C_\lambda^i}$ selections, so this choice does not add to the combinatorial complexity here. We have

$$\begin{aligned} 5'\, V_1 \equiv V_2 \equiv \cdots \equiv V_{30} &\equiv J_1 \equiv J_2 \equiv \cdots \equiv J_4 \equiv C_\lambda^i\, 3' \textbf{ GERMLINE} \\ &\Downarrow \text{ polymerase } V(D)J, \text{ select } V_j J_k \\ 5'\, V_1 \equiv \cdots \equiv V_{j-1} &\equiv V_j \equiv J_k \equiv \cdots \equiv J_5 \equiv C_\kappa\, 3' \textbf{ CHANGE B-CELL} \end{aligned}$$

and then there is transcription to protein

$\Downarrow$ first or primary DNA transcript is made: still has introns
$\Downarrow$ splicing occurs: introns removed and V_j, J_k and C_λ^k brought together
$\Downarrow$ mature messenger RNA $V_j J_k C_\lambda^k$
$\Downarrow$ translation to protein
$\Downarrow$ $NH_2 - \mathbb{V}_j \mathbb{J}_k \mathbb{C}_\lambda^k - COOH$ **A PARTICULAR SELECTION**

If we leave out this detail, we see we have the sequence

$$V_1 \cdots V_{30}\, J_1 \cdots J_4\, C_\lambda^k \quad \rightarrow \quad V_j J_k C_\kappa \rightarrow \mathbb{V}_j \mathbb{J}_k \mathbb{C}_\lambda^k$$

As before, we simply identify the RNA sequence $V_j J_k C_\lambda^k$ with the transcribed protein $\mathbb{V}_j \mathbb{J}_k \mathbb{C}_\lambda^k$. Hence, this combinatorial process has the mnemonic $V_1 \cdots V_{30}\, J_1 \cdots J_4\, C_\lambda^k \rightarrow V_j J_k C_\lambda^k$. Thus there are $30^4 \approx 10^6$ possibilities. The B-Cell precursor now has a selected V_κ and V_λ chain. Again, the particular $V_j J_k$ determine the antigen specificity. We will call these the light chains: $L_\kappa = V_\kappa$ and $L_\lambda = V_\lambda$. So no matter which of the light chains is made, the light chain used has the $V_j J_k$ determining the antigen specificity derived from the genes on Chromosome 2. The choices V_j and J_k here should be labeled with a superscript L to indicate these are sequences from Chromosome 2 which are different from the heavy chain choices which are on Chromosome 14. But for clarity of exposition we did not do this.

Next, consider how the Ig heavy chains are created. The genes for this construction are on Chromosome 14 and this process will also give us $V_j J_k$ determining the antigen specificity derived from the genes on Chromosome 14. These choices will determine a *different* antigen specificity than what we obtain in the light chain. There are three gene segments that code for the variable portion of the heavy chain: $V_H = \{V_1, \ldots, V_{40}\}$, $D_H = \{D_1, \ldots, D_{25}\}$ and $J_H = \{J_1, \ldots, J_6\}$. In addition, there is a C region,

$$C_H = \{\mu, \delta, \gamma - 3, \gamma_1, \alpha_1, \gamma_2, \gamma_4, \epsilon, \alpha_2\}$$

The C region lists nine possibilities. In humans there are nine different C_H genes and each of the nine choices given codes for a different type of Ig. The choices are IgM IgD, IgG, IgA and IgE. The μ encodes the IgM and the δ encodes the IgD. Each of these Ig choices has a different biological function and so this combinatorial process generates heavy chains with a variety of immunoglobulin possibilities. We have

$$\begin{aligned} 5'\, V_1 \equiv \cdots \equiv V_{40} \equiv D_1 \equiv \cdots \equiv D_{20} &\equiv J_1 \equiv \cdots \equiv J_6 \equiv C_H\, 3' \textbf{ GERMLINE} \\ &\Downarrow V(D)J, \text{ select } V_j D_p J_k \textbf{ CHANGE B-CELL} \\ 5' : V_1 \equiv \cdots \equiv V_{j-1} &\equiv V_j \equiv D_p \equiv J_k \equiv \cdots \equiv J_6 \equiv C_H\, 3' \end{aligned}$$

The choices V_j, D_p and J_k here should be labeled with a superscript H to indicate these are sequences for Chromosome 40 which are different from the light chain choices which are on Chromosome 2. But for clarity of exposition we again did not do this. At this stage of the assembly, only the

μ and δ part of C_H are used. Then we have

$$\begin{array}{rcl} & \Downarrow & \text{first or primary DNA transcript is made: still has introns} \\ V_j \equiv D_p \equiv J_k & \equiv & \cdots \equiv J_6 \equiv \mu\delta \\ & \Downarrow & \text{splicing occurs: introns removed and } V_j, D_pJ_k, C_H \text{ assembled} \\ & \Downarrow & \text{mature messenger RNA} \end{array}$$

Then the resulting mRNA must be split and transcribed.

$$\begin{array}{rcl} V_jD_pJ_k\mu\delta \equiv \text{poly A} & & \textbf{IMMATURE B-CELL: ONLY } IgM \\ & \Downarrow & \text{splitting} \\ V_jD_pJ_k\mu \equiv \text{poly A} & \text{and} & V_jD_pJ_k\delta \equiv \text{poly A} \\ & \Downarrow & \text{translation to } \mu, \delta \text{ heavy chains: } \textbf{MATURE B-CELL} \\ NH_2 - \mathbb{V}_j\mathbb{D}_p\mathbb{J}_k\mu - COOH & \text{and} & NH_2 - \mathbb{V}_j\mathbb{D}_p\mathbb{J}_k\delta - COOH \quad \textbf{TWO SEQUENCES} \end{array}$$

The selection $V_jD_pJ_k$ fixes the antigen specificity of the heavy chain. The mature B-Cell synthesizes both of these chains, which implies it can express two immunoglobulins: IgM and IgD which have identical antigenic specificity as they are built from the same $V_jD_pJ_k$ choice. Hence, we have a H_μ and a H_δ heavy chain. For convenience, we will again identify the proteins with their sequences before transcription.

These heavy chains combine inside the B-Cell with light chains to form the IgM and IgD immunoglobulins. In this process, there $50^6 \approx 1.6 \times 10^{10}$ possible combinations. So in-between the light and heavy chains of an Ig molecule there are about 10^8 possible antigen specificities from the light chain side and about 10^{10} choices for the heavy chain side. The process of building the Ig out of light and heavy chains thus allows for the construction of an enormous number of possible binding choices for an antigen epitope.

A vast number of potential B-Cells are therefore created via combinatorial strategies to express antibodies which come in different varieties each of which responds to the universe of potential antigens differently. The created B-Cells are in an environment that exposes them to cells carrying peptide fragments of both self and non-self (antigens) proteins. The B-Cells that bind with high affinity to a self-protein fragment are removed in what is called negative selection. This leaves mature or primed B-Cells that bind with a reasonable affinity to antigens and should not bind to peptide chains from self-proteins. A similar process is used to create a large pool of mature T-Cells which also undergo negative selection to remove T-Cells that respond highly to self-protein. A major set of antigens called Thymus-dependent (TD) antigens requires cooperation between mature B- and T-Cells that both respond to the same antigen. These are called cognate B and T-Cells.

An antigen that gets into the body and enters tissue arrives at a draining lymph node. Naive B-Cells are B-Cells that have never been exposed to an antigen. They are the ones circulating through the node which are exposed to the antigen and some will mature into a B-Cell that responds to this antigen. The node is compartmentalized into a B-Cell part and a T-Cell part. Once a B-Cell is primed for the antigen, it essentially captures the antigen via an immunoglobulin (Ig) protein expressed on its surface. The B-Cell, Ig and antigen complex then moves towards the boundary of the region of the node that separates the B and T-Cells. The T-Cells interact with dendritic cells that bear antigen and some T-Cells will interact with the same antigen we have been discussing for the B-Cells. At the boundary in the node between the B and T-Cells, the B-Cell bearing the antigen acts like an antigen-presenting cell (such as a dendritic cell) to the primed T-Cell. Essentially, the B-Cell pulls the antigen into itself, processes it and presents it on its surface via a MHC complex. The interaction between the B-Cell and the T-Cell creates an activated T-Cell and the used B-Cell turns into a plasma cell which synthesizes immunoglobulin. The cells then move back into the B-Cell area of the node and form a germinal center. In the germinal center, the antigen-activated B-Cell is used to rapidly create a large population of B-Cells whose affinities are close to the affinity of the initial activated B-Cell. Following this rapid expansion, B-Cells enter another area of the germinal center where they interact with T-Cells specific for this antigen and dendritic cells carrying this antigen. The B-Cells that interact with the highest affinity for the T-Cells and dendritic cells carrying the antigen are selected for a rapid expansion. **B-Cells with weak affinities should die, but they may not**.

The T-Cells activated via B-Cell interactions in the draining lymph node are called $CD4^+$ T-Cells and are also known as **helper T-Cells** and the B-Cells they respond to present antigen in pMHC II complexes. The activated $CD4^+$ T-Cells release a large amount of signaling molecules called cytokines, which in general enhance the production of the MHC I molecules on the surfaces of many cells. Another class of T-Cells called the $CD8^+$ T-Cells interact with the pMHC I complexes to initiate lysis of an infected cell if the affinity is strong enough. Of course, an autoimmune event lyses a healthy cell even though the affinity is not very strong, so it is hard to explain the event from the mechanisms presented here. The $CD8^+$ T-Cells are created in the usual combinatorial way and also undergo negative selection. Hence, we see the B-Cell and $CD4^+$ interaction enhances the ability of the $CD8^+$ T-Cells to find and destroy infected cells.

Homework

You should note how many abstractions and approximations we are making in the discussions above. In the questions below, we will let you explore these things in more detail.

Exercise 17.1.1 *To put all of this in perspective, go and read the abstract model of (A. Rot and U. von Andrian (6) 2004). It is very rare to find biologists and life scientists, in general, who are willing to write speculative models of this sort. To try to understand the many interactions of the components of the immune system using cytokine messengers, Rot makes an attempt at stripping away detail and concentrating on higher-level ideas. Do you think Rot was successful? Can you see how Rot's attempts help us find a framework for mathematical abstraction?*

Exercise 17.1.2 *Note the combinatorial complexity that this short and simplified discussion reveals. The number of possible light and heavy chains is too large to handle computationally. Think about this carefully. Many of our abstractions forced upon us as brute force calculations that are based on tracing our way through all possible computational paths are simply impossible. So abstraction is needed. Work out the numbers of light and heavy chains that are possible for various choices of the parameters we use to model the antigen specificity.*

Exercise 17.1.3 *Do some additional reading on B- and T-Cells. You'll quickly see our discussions are very simplified. Do you think such simplified ideas are useful? If you don't, what other courses of modeling choices could you try?*

Exercise 17.1.4 *Molecules do not move in a cell according to a diffusion law only, as there are many protein strands that help guide their movement. Read about diffusion in the cell and you will see better how that particular partial differential equation point of view is itself an abstraction.*

17.2 Two Allele LRRK2 Mutation Models

There are known connections between the LRRK2 regulatory gene and a variety of immune and cognitive diseases. The background can be found in a sequence of papers which are useful to absorb. (P. Lewis and C. Manzoni (95) 2012) provides an overview of LRRK2 and human disease, (J. Thëvenet and R. Gobert and R. van Huijsduijnen and C. Wiessner and Y. Sagot (62) 2011) discusses how regulation of LRRK2 expression is linked to monocyte maturation, (Z. Liu and J. Lee and S. Krummery and W. Lu and H. Cai and M. Lenardo (124) 2011) discusses how LRRK2 regulates the transcription factor modulating inflammatory bowel disease, and (Hakimi et al. (45) 2011) show LRRK2 links to Parkinson's disease. (A. Gardet and Y. Benita and C. Li and B. Sands and I. Ballester and C. Stevens and J. Korzenik and J. Rioux and M. Daly and R. Xavier and D. Podolsky (3) 2010) goes over how LRRK2 is involved in IFN-γ response to pathogens. How mutant LRRK2 mediates immune response in neurodegeneration as discussed in (Kozina et al. (68) 2018) is an important paper for our purposes also. The survey paper (D. Berwick and K. Harvey (20) 2011) explains how LRRK2 interaction may be a key to understanding neurodegeneration. In addition, (I. Smets and B. Fiddes and J. Garcia-Perez and D. He and K. Mallants and W. Liao and J. Dooley and G. Wang and S. Humblet-Baron and B. Dubois and A. Compston and J. Jones and A. Coles and A. Liston and M. Ban and A. Goris and S. Sawcer (53) 2018) discusses multiple sclerosis risk variants in general. (M. Kubo and R. Nagashima and E. Ohta and T. Maekawa and Y. Isobe and M. Kurihara and K. Eshima and K. Iwabuchi and T. Sasaoka and S. Azuma and H. Melrose and M. Farrer and F. Obata (78) 2016) then sketches out how leucine-rich repeats in kinase 2 affect t1 antigen response and (J. Milosevic and S. Schwarz and V. Ogunlade and A. Meyer and A. Storch and J. Schwarz (56) 2009) observe that LRRK2 plays a role in neural cell cycle progression.

To understand LRRK2 gene signaling better, we will build a model which is a variant of a general model of how tumor repression genes lose alleles in a model of colon cancer based on colon crypts as discussed in (M. Novak (79) 2006). We are built of individual cells that have their own reproductive machinery. These cells can contain mutant copies of important genes and these mutations can influence the functioning not only of that cell, but of others in the body due to complicated dynamic interactions. Now, a given gene that occupies a certain position on a chromosome (this position is called the *locus* of the gene) can have a number of alternate forms. These alternate forms are called *alleles*. The number of alleles a gene has for an individual is called that individual's *genotype* for that gene. Note, the number of alleles a gene has is therefore the number of viable DNA codings for that gene. For example, mutations in mismatch repair genes lead to 50-100 fold increases in point mutation rates. These usually occur in repetitive stretches of short sequences of DNA. Such regions are called *micro satellite regions* of the genome. These regions are used as genetic markers to track inheritance in families. They are short sequences of nucleotides (i.e. **ATCG**) which are repeated over and over. Changes can occur such as increasing or decreasing the number of repeats. This type of instability is thus called a *micro satellite* or **MIN** instability. Another instability is *chromosomal instability* or **CIN**. This means an increase or decrease in the rate of gaining or losing whole chromosomes or large fractions of chromosomes during cell division. If the first allele is activated by a point mutation while the second allele is inactivated by a loss of one parent's contribution to part of the cell's genome, the second allele is inactivated because it is lost in the copying process. This is an example of CIN and in this case is called *loss of heterozygosity* or **LOH**.

A general model of enhanced cytokine signaling production based on LRRK2 mutation is as follows. The altered cytokine response starts with the activation of an allele called **A**, in a small compartment of cells. Initially, all cells have a correct version of the **LRRK2** gene. We will denote this by $\boldsymbol{A}^{-/-}$ where the superscript "$-/-$" indicates there are no LRRK2 mutations. One of the mutant alleles becomes activated at mutation rate u_1 to generate a cell type denoted by $\boldsymbol{A}^{+/-}$. The superscript $+/-$ tells us one allele is activated. The second allele becomes activated at rate $\hat{u}_2$ to become the cell type $\boldsymbol{A}^{+/+}$. In addition, $\boldsymbol{A}^{-/-}$ cells can also receive mutations that trigger **CIN**. This happens at the rate u_c resulting in the cell type $\boldsymbol{A}^{-/-\,CIN}$. This kind of a cell can activate the first allele of the LRRK2 gene with normal mutation rate u_1 to produce a cell with one activated allele (i.e. a $+/-$) which started from a CIN state. We denote these cells as $\boldsymbol{A}^{+/-\,CIN}$. We can also get a cell of type $\boldsymbol{A}^{+/-\,CIN}$ when a cell of type $\boldsymbol{A}^{+/-}$ receives a mutation which triggers **CIN**. We will assume this happens at the same rate u_c as before. The $\boldsymbol{A}^{+/-\,CIN}$ cell then rapidly undergoes **LOH** at rate $\hat{u}_3$ to produce cells having the second allele of LRRK2 which is of type $\boldsymbol{A}^{+/+\,CIN}$. Finally, $\boldsymbol{A}^{+/+}$

cells can experience **CIN** at rate u_c to generate $\boldsymbol{A}^{+/+\,CIN}$ cells. We show this information in Figure 17.2. Let N be the population size within which the LRRK2 mutations occur. We will assume a

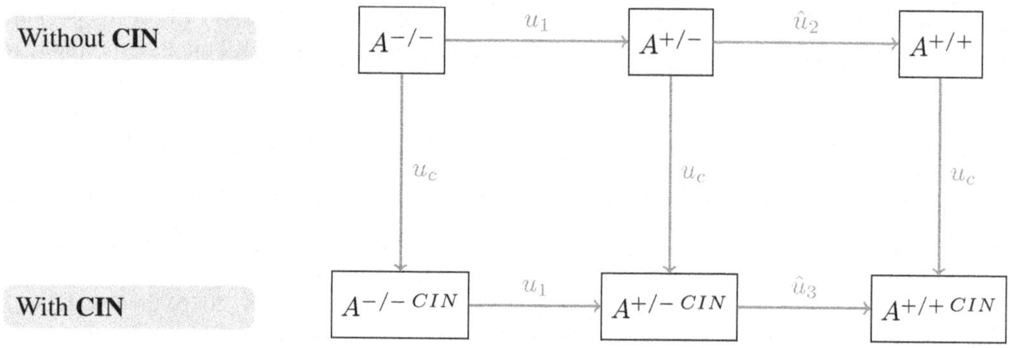

Figure 17.2: The pathways for the **LRRK2** allele gains.

typical value of N is 10^3 to 10^4. The first allele is activated by a point mutation. The rate at which this occurs is modeled by the rate u_1 as shown in Figure 17.2. We make the following assumptions:

- The mutations governed by the rates u_1 and u_c are **neutral**. This means that these rates do not depend on the size of the population N.
- The events governed by $\hat{u}_2$ and $\hat{u}_3$ give what is called **selective advantage**. This means that the size of the population size does matter.

Using these assumptions, we will model $\hat{u}_2$ and $\hat{u}_3$ as $\hat{u}_2 = N\,u_2$ and $\hat{u}_3 = N\,u_3$, where u_2 and u_3 are neutral rates. We can thus redraw our figure as Figure 17.3. The mathematical model is then set

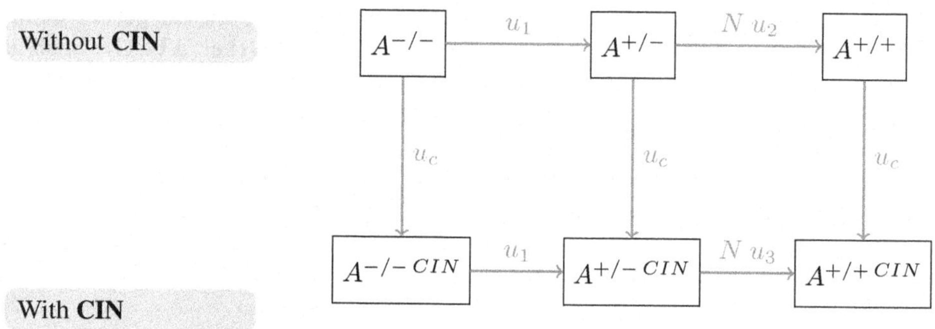

Figure 17.3: The pathways for the **LRRK2** allele gains rewritten using selective advantage.

up as follows. Let:

$X_0(t)$ is the probability a cell in in cell type $A^{-/-}$ at time t.

$X_1(t)$ is the probability a cell in in cell type $A^{+/-}$ at time t.

$X_2(t)$ is the probability a cell in in cell type $A^{+/+}$ at time t.

$Y_0(t)$ is the probability a cell in in cell type $A^{-/-\,CIN}$ at time t.

$Y_1(t)$ is the probability a cell in in cell type $A^{+/-\,CIN}$ at time t.

$Y_2(t)$ is the probability a cell in in cell type $A^{+/+\,CIN}$ at time t.

Looking at Figure 17.3, we can generate rate equations. Rewrite Figure 17.3 using our variables as Figure 17.4. To generate the equations we need, note each box has arrows coming into it and

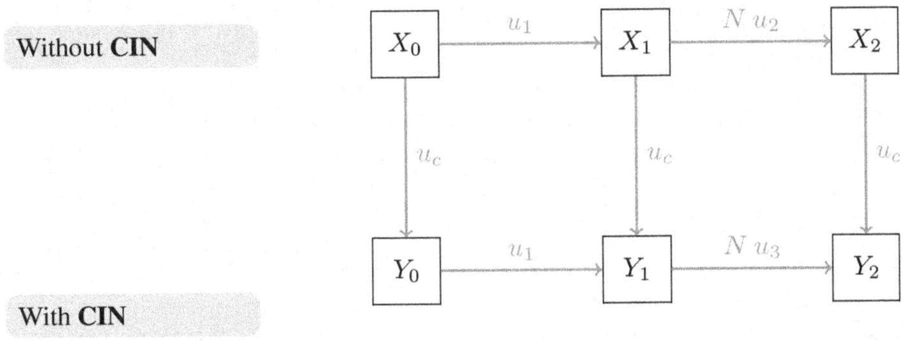

Figure 17.4: The pathways for the **LRRK2** allele gains rewritten using mathematical variables.

arrows coming out of it. The **arrows in** are **growth** terms for the net change of the variable in the box and the **arrows out** are the **decay or loss** terms. We model **growth** as **exponential growth** and **loss** as **exponential decay**. So X_0 only has arrows going out, which tells us it only has **loss** terms. So we would say $(X_0')_{loss} = -u_1 X_0 - u_c X_0$ which implies $X_0' = -(u_1 + u_c)X_0$. Further, X_1 has arrows going in and out, which tells us it has **growth** and **loss** terms. So we would say $(X_1')_{loss} = -Nu_2X_1 - u_cX_1$ and $(X_1')_{growth} = u_1X_0$, which implies $X_1' = u_1X_0 - (Nu_2 + u_c)X1$. We can continue in this way to find all the model equations. We can then see the rate equations are

$$X_0' = -(u_1 + u_c)\,X_0 \tag{17.1}$$
$$X_1' = u_1\,X_0 - (u_c + N\,u_2)\,X_1 \tag{17.2}$$
$$X_2' = N\,u_2\,X_1 - u_c\,X_2 \tag{17.3}$$
$$Y_0' = u_c\,X_0 - u_1\,Y_0 \tag{17.4}$$
$$Y_1' = u_c\,X_1 + u_1\,Y_0 - N\,u_3\,Y_1 \tag{17.5}$$
$$Y_2' = N\,u_3\,Y_1 + u_c\,X_2 \tag{17.6}$$

Initially, at time 0, all the cells are in the state X_0, so we have

$$X_0(0) = 1,\; X_1(0) = 0,\; X_2(0) = 0 \tag{17.7}$$
$$Y_0(0) = 0,\; Y_1(0) = 0,\; Y_2(0) = 0. \tag{17.8}$$

Now, under what circumstances is the CIN pathway to LRRK2 mutations the dominant one? In order to answer this, we need to analyze the trajectories of this model. Note, if we were interested in the asymptotic behavior of this model as t goes to infinity, then it is clear everything ends up with value 0. However, our interest is over the typical lifetime of a human being and thus, we never reach the asymptotic state. Thus, our analysis is always concerned with the values of our six variables before the end of a human life span. Since our interest in these variables is over some fraction of the typical lifetime of a human being, we need to pick a maximum typical lifetime. We assume the average human life span is 100 years. We also assume that cells divide once per day and so a good choice of time unit is *days*. The effect of LRRK2 mutation will probably show up around 30-50 years old and if we let human life time be $T_H = 3.65 \times 10^4$ days, our final time will be $T = f \cdot 3.65 \times 10^4$ days where f is about $0.3 - 0.5$. For our calculations, we will use $f = 0.5$ so that $T = 1.83 \times 10^4$ days. We will assume LRRK2 mutation occurs in small clusters of cells which means we assume N

is from 10^2 to 10^3 cells. For estimation purposes, we often think of N as the upper value, $N = 10^3$. We also assume $u_1 \approx 10^{-7}$ and $u_1 = u_2$. We will assume the rate $N\,u_3$ is quite rapid and so it is close to 1. We will set u_3 by $N\,u_3 \approx 1 - r$ for small positive values of r. Hence, once a cell reaches the Y_1 state, it will rapidly transition to the end state Y_2 if r is sufficiently small. We are not yet sure how to set the magnitude of u_c, but certainly it is at least u_1. For convenience, we will assume $u_c = R\,u_1$ where R is a number at least 1. For example, if $u_c = 10^{-5}$, this would mean $R = 100$.

Homework

Exercise 17.2.1 *Redo the model we discuss here, allowing for cross connections between nodes in the top and bottom levels that are* diagonal. *You will have to make decisions about new weighting parameters for these new links.*

Exercise 17.2.2 *Redo the model we discuss here, allowing for three LRRK2 alleles. You will have to make decisions about new weighting parameters for these new links.*

Exercise 17.2.3 *Redo the model we discuss here, allowing for two LRRK2 alleles and an additional gene called B which also occurs in two alleles. Now we have two regulatory genes that control outputs. You will have to make decisions about new weighting parameters for these new links.*

The development of this model is quite similar to the discussion in (J. Peterson (58) 2016) and (J. Peterson (59) 2016) which follows the lead of (M. Novak (79) 2006) but adds all the complicated error analysis. This was done in the context of calculus for biologists and cognitive scientists training, so these references have a lot of detail worked out. Using these results to see if the CIN pathway dominates, we can look at the ratio of the Y_2 output to the X_2 output. The ratio of Y_2 to X_2 tells us how likely the gain of both alleles is due to CIN or without CIN. We have, for $R < 23.5$, that

$$\frac{Y_2(T)}{X_2(T)} = \frac{u_1 u_c T^2 + E(T)}{(1/2) N u_1 u_2 T^2 + F(T)}$$

where $E(T)$ and $F(T)$ are the errors associated with our approximations for X_2 and Y_2. We assume $u_1 = u_2$ and so we can rewrite this as

$$\frac{Y_2(T)}{X_2(T)} = \frac{2\frac{R}{N} + \frac{2E(T)}{Nu^2T^2}}{1 + \frac{2F(T)}{Nu^2T^2}}$$

For $N = 1000$ and $u_1 = u_2 = 10^{-7}$ and 50 years, we find $Nu^2T^2 = 0.0033$ and so $1/(Nu^2T^2) \approx 330$ and hence we have

$$\frac{Y_2(T)}{X_2(T)} = \frac{2\frac{R}{N} + 330\,E(T)}{1 + 330\,F(T)}$$

Now $E(T) \approx 0.000027$ and $F(T) \approx 2.4 \times 10^{-5}$ and so

$$\frac{Y_2(T)}{X_2(T)} \approx \frac{(2R/N) + 0.01}{1.0008} \Longrightarrow \frac{Y_2(T)}{X_2(T)} \approx \frac{2R}{N} + 0.01$$

Thus, we see the error we make in using $(2R/N)$ as an estimate for $Y_2(T)/X_2(T)$ is fairly small. Also, note we can get CIN dominance if $R \approx 23.5$ and $N < 50$ as $2R > N$. But, it is much more reasonable to think R is much smaller, perhaps less than 10. Then, even for N as low as 25, the CIN pathway is still not dominant. Hence, we can be reasonably confident that the critical ratio $(2R)/N$

is the same as the ratio $Y_2(T)/X_2(T)$ as the error over 50 years is small. Hence, we can say

$$\frac{Y_2(T)}{X_2(T)} \approx \frac{2\,R}{N}.$$

Hence, the pathway to Y_2 is the most important if $2\,R > N$. This implies the CIN pathway is dominant if

$$R > \frac{N}{2}. \tag{17.9}$$

For the fixed value of $u_1 = 10^{-7}$, we calculate in Table 17.1 possible u_c values for various choices of N. We have CIN dominance if $R > \frac{N}{2}$. For LRRK2 models we will use $R \approx 10$ and cell

N	R Lower Bnd	CIN Dominance $R = 23$	CIN Dominance $R = 10$	CIN Dominance $R = 5$
5	2.5	Yes	Yes	Yes
10	5	Yes	Yes	No
15	7.5	Yes	Yes	No
20	10	Yes	No	No
25	12.5	Yes	No	No
50	25	No	No	No
100	50	No	No	No
500	250	No	No	No
1000	500	No	No	No

Table 17.1: The CIN decay rates, with $u_1 = u_2 = 10^{-7}$ and $u_c = R\,u_1$.

populations of about 100. Hence, in our models we will never have the CIN pathway dominant.

Thus, for our purposes, we can assume that the LRRK2 mutations are due to point mutation rate events and not due to CIN issues. The LRRK2 mutations can occur in location dependent pockets though, and depending on their position could have different effects. For the rest of our discussion here, we will assume there are sub-populations of LRRK2 mutations that affect the cytokine signaling results due to an external event such as antigen recognition. Note, the analysis we just presented could be extended to three alleles or more and the result will be the same.

LRRK2 mutations can alter signaling pathways if one or both alleles have been gained. We have $X_1(t) \approx u_1 t$ and $X_2(t) \approx N\,u_1\,u_2\,\frac{t^2}{2}$. Let's assume we track the effects of these mutations over a fraction of the human lifetime given by βT for $\beta \in (0,1)$. Then from now on, we have the output $X_1(\beta T) \approx 3.65\beta \times 10^{-3}$ and $X_2(\beta T) \approx N\,u_1\,u_2\,\beta^2 \frac{T^2}{2}$. If we assume seed populations for the mutations are on the order of $N \approx 100$, we have

$$X_1(\beta T) \approx 3.65\beta \times 10^{-3}, \quad X_2(\beta T) \approx N \times 0.067\beta^2$$

Thus after a time period of fifty years, $\beta = 0.5$ and we have $X_1(\beta T) \approx 1.83 \times 10^{-3}$ and for a population $N = 100$ cells $X_2(\beta T) \approx 6.7 \times 0.25 = 1.68$. The population of A^{+-} and A^{++} cells created is then on the order of $|A^{+-}| \approx X_1(\beta T)N = 0.18$ and $|A^{++}| \approx X_2(\beta T)N \approx 168$ cells in the collection have mutated both alleles. So this gives 168 mutated cells over 50 years. We can use the value of $X_1(t) + X_2(t)$ to drive the cytokine storm multiplier we use to model the affinity of the T-Cell-pMHC interactions as discussed in Section 17.4. We have α is the affinity whose strength is parameterized by the scalar $\mathscr{A}$ and if we model to affinity for cells having the LRRK2 mutations by $(X_1(t) + X_2(t))\mathscr{A}$, we have a way of quantitatively modeling the cytokine storms. Over a time period of fifty years, this amounts to 168 T-Cells with altered affinity processing, which can lead to autoimmune events, neurodegeneration and WNV brain inflammation.

We can also look at the solution space to a LRRK2 model using linear ordinary differential equation systems. We discussed this in (Peterson (101) 2020) and it is useful to go through it again. The eigenvalues of this system are the roots of the polynomial $p(\lambda) = det(\lambda \boldsymbol{I} - \boldsymbol{A})$ where $\boldsymbol{A}$ is the 6×6 coefficient matrix above. We find

$$p(\lambda) = det \begin{bmatrix} \lambda + (u_1 + u_c) & 0 & 0 & 0 & 0 & 0 \\ -u_1 & \lambda + (u_c + Nu_2) & 0 & 0 & 0 & 0 \\ 0 & -Nu_2 & \lambda + u_c & 0 & 0 & 0 \\ -u_c & 0 & 0 & \lambda + u_1 & 0 & 0 \\ 0 & -u_c & 0 & -u_1 & \lambda + Nu_3 & 0 \\ 0 & 0 & -u_c & 0 & -Nu_3 & \lambda \end{bmatrix}$$

This is easily expanded using the properties of determinants to give

$$p(\lambda) = (\lambda + (u_1 + u_c))\,(\lambda + (u_c + Nu_2))\,(\lambda + u_c)(\lambda + u_1)\,(\lambda + Nu_3)\,(\lambda)$$

Hence, the eigenvalues are

$$\begin{bmatrix} \lambda_1 \\ \lambda_2 \\ \lambda_3 \\ \lambda_4 \\ \lambda_5 \\ \lambda_6 \end{bmatrix} = \begin{bmatrix} -(u_1 + u_c) \\ -(u_c + Nu_2) \\ -u_c \\ -u_1 \\ -Nu_3 \\ 0 \end{bmatrix}$$

We find the eigenvectors are

$$\boldsymbol{E_1} = \begin{bmatrix} Nu_2 - u_1 \\ u_1 \\ -Nu_2 \\ -(Nu_2 - u_1) \\ -u_1 \frac{Nu_2 - u_1 - u_c}{Nu_3 - u_1 - u_c} \\ \frac{Nu_3(Nu_2 - u_1) - Nu_2 u_c}{Nu_3 - u_1 - u_c} \end{bmatrix}, \quad \boldsymbol{E_2} = \begin{bmatrix} 0 \\ 1 \\ -1 \\ 0 \\ \frac{-u_c}{Nu_2 + u_c - Nu_3} \\ \frac{u_c}{Nu_2 + u_c - Nu_3} \end{bmatrix}, \quad \boldsymbol{E_3} = \begin{bmatrix} 0 \\ 0 \\ 1 \\ 0 \\ 0 \\ 0 \end{bmatrix},$$

$$\boldsymbol{E_4} = \begin{bmatrix} 0 \\ 0 \\ 0 \\ 1 \\ \frac{u_1}{Nu_3 - u_1} \\ \frac{Nu_3}{Nu_3 - u_1} \end{bmatrix}, \quad \boldsymbol{E_5} = \begin{bmatrix} 0 \\ 0 \\ 0 \\ 0 \\ 1 \\ -1 \end{bmatrix}, \quad \boldsymbol{E_6} = \begin{bmatrix} 0 \\ 0 \\ 0 \\ 0 \\ 0 \\ 1 \end{bmatrix}$$

The general solution is thus any linear combination of the form

$$c_1 \boldsymbol{E_1} e^{-(u_1+u_c)t} + c_2 \boldsymbol{E_2} e^{-(u_c+Nu_2)t} + c_3 \boldsymbol{E_3} e^{-u_c t} + c_4 \boldsymbol{E_4} e^{-u_1 t} + c_5 \boldsymbol{E_5} e^{-Nu_3 t} + c_6 \boldsymbol{E_6} \boldsymbol{1}$$

where $\boldsymbol{1}$ denotes the constant function $e^{0t} = 1$. If we let these six solutions be denoted by $\boldsymbol{y_i}(t) = \boldsymbol{E_i} e^{\lambda_i t}$, we see the general solution is a member of the span of $\{\boldsymbol{y_1}, \ldots, \boldsymbol{y_6}\}$ and since these functions are linearly independent in the space $\boldsymbol{X} = C^1([0,T]) \times C^1([0,T]) \times C^1([0,T]) \times C^1([0,T]) \times C^1([0,T]) \times C^1([0,T])$ for any appropriate T, we know this solution space is a six dimensional subspace of $\boldsymbol{X}$ with the basis $\{\boldsymbol{y_1}, \ldots, \boldsymbol{y_6}\}$. If $\boldsymbol{F}$ and $\boldsymbol{G}$ are two elements in $\boldsymbol{X}$, an inner product on $\boldsymbol{X}$ is given by $\int_0^T < \boldsymbol{F(t)}, \boldsymbol{G(t)} > dt$. We can construct an orthonormal basis for the solution space by applying GSO to the eigenvectors $\boldsymbol{E_i}$ to create the orthonormal basis $\boldsymbol{G}$. Then the new functions $\boldsymbol{w_i} = \boldsymbol{G_i}e^{\lambda_i t}$ are solutions to the ODE system which are mutually orthogonal. So it is straightforward to find an orthonormal basis here.

Homework

Exercise 17.2.4 *Redo the model we discuss here, allowing for cross connections between nodes in the top and bottom levels that are* diagonal. *Redo the analysis we gave above to get some insight into which pathway is dominant.*

Exercise 17.2.5 *Redo the model we discuss here, allowing for three LRRK2 alleles. Redo the analysis we gave above to get some insight into which pathway is dominant.*

Exercise 17.2.6 *Redo the model we discuss here, allowing for two LRRK2 alleles and an additional gene called B which also occurs in two alleles. Now we have two regulatory genes that control outputs. Redo the analysis we gave above to get some insight into which pathway is dominant.*

Exercise 17.2.7 *Redo the model we discuss here, allowing for cross connections between nodes in the top and bottom levels that are* diagonal *using eigenvector and eigenvalue analysis.*

Exercise 17.2.8 *Redo the model we discuss here, allowing for three LRRK2 alleles using eigenvector and eigenvalue analysis.*

Exercise 17.2.9 *Redo the model we discuss here, allowing for two LRRK2 alleles and an additional gene called B which also occurs in two alleles, using eigenvector and eigenvalue analysis.*

17.3 Signaling Models

We now discuss how to model cytokine signals that influence an immune response. We will focus on the kinds of signals we could get from **chemokines** and **cytokines**. A good introduction to the chemokine family are the papers by Vela et al. (M. Vela and M. Aris and M. Llorente and J. Garcia-Sanz and L. Kremer (84) 2015) and Rossi et al. (D. Rossi and A. Zlotnik (22) 2000). The chemokine family is a subset of the larger collection of signaling molecules called **cytokines** and a good introduction to the larger family can be found in Deverman et al. (B. Deverman and P. Patterson (9) 2009) and Capuron et al. (L. Capuron and A. Miller (70) 2004) (these papers discuss cytokines and the central nervous system (CNS) and psychopathology) and Peters (M. Peters (80) 1996) (the role of cytokines in immune response).

The chemokines are small molecules which have very specific *cysteine* motifs in their amino acid sequences (D. Rossi and A. Zlotnik (22) 2000). They have been classified into a number of families **CXC**, **CC**, **C** and **CX3C** based on the motif displayed by the first two cysteines. Structurally, disulphide bonds are formed between the first and third and between the second and fourth cysteines. The chemokines bind with a corresponding family of receptors which are G-protein coupled to seven transmembrane types. From (D. Rossi and A. Zlotnik (22) 2000), we present a basic table, Table 17.2 of chemokine ligands and receptors. Our purpose is to try to understand, in an abstract way, signal chains of cytokine messengers, which may be appropriate in developing an understanding of immunopathology. Now, an attempt at developing a general abstract model of chemokine signals as a type of language is found in (A. Rot and U. von Andrian (6) 2004). Let F_1 through F_P be P cytokine families and let C_1 to C_P denote cytokine signals generated by these P distinct families. We will assume there is a common unit with which we can measure cytokine signal strength which we will denote by ϵ_0. A cytokine from family F_i would occur in a signal with strength $N_i\ \epsilon_0$ and a

Chemokine Receptors	Ligands
CXCR1	IL-8, GCP-2
CXCR2	IL-8, GCP-2, Gro α Gro β Gro γ, ENA-78, PBP
CXCR3	MIG, IP-10, 1-TAC
CXCR4	SDF-1/PBSF
CXCR5	BLC/ BCA-1
CCR1	MIP-1α, MIP-1β, RANTES, HCC-1, 2, 3 and 4
CCR2	MCP-1, 2, 3, and 4
CCR3	eotaxin-1, eotaxin-2, MCP-3
CCR4	TARC, MDC, MIP-1β, RANTES
CCR5	MIP-1α, MIP-1β, RANTES
CCR6	MIP-3α/ LARC
CCR7	MIP-3/β/ ELC, 6 Ckine/ LC
CCR8	I-309
CCR9	TECK
XCR1	Lymphotactin
CX3CR1	Fractalkine/ neurotactin

Table 17.2: Chemokine receptors and their human ligands

typical signal would be comprised of a linear combination of these signals of the from

$$N_1\,\epsilon_0\,C_1 \;+\; N_2\,\epsilon_0\,C_2 \;+\; \cdots \;+\; N_P\,\epsilon_0\,C_P$$

where we interpret the symbol $+$ in a formal way. This mathematical string represents a cytokine signal which is made of individual cytokine signals occurring at the strengths $N_i\,\epsilon_0$. Since the common strength factor ϵ_0 is not really needed to see how this signal is composed, we will now drop it and refer to the general cytokine signal as the formal sum $\sum_{i=1}^{P} N_i C_i$.

We need to think closely about the chemokine signaling that is done in the TcR-pMHC scenario. We will use the nice review by (A. Rot and U. von Andrian (6) 2004) to sharpen our thoughts. From (A. Rot and U. von Andrian (6) 2004)

> Chemokines are building blocks of the most versatile, coherently functioning system of intercellular communication signals. Chemokine messages are decoded through specific cell-surface receptors. Upon binding, these receptors unleash cascades of intracellular secondary mediators that turn on cell-specific intrinsic functional programs...The sum of individual “cell reflexes” results in complex system responses during both steady-state conditions and innate and adaptive immune reactions.

Let’s assume the maturing T-Cell, which is being activated via the serial signaling cascade we have been discussing, has associated with it a set of important chemokine signals. Following (A. Rot and U. von Andrian (6) 2004), we make the following assumptions. If a cell has no foreign antigen, think of it as covered with a web of chemokines and their cell-surface G protein coupled receptors or GPCRs. When a foreign antigen is introduced into this surface landscape, it changes this landscape by adding a web of chemokines and associated GPCRs which help attract the appropriate TcR to that location. For simplicity, let’s assume there are three chemokines and associated receptors. Many GPCRs bind with multiple chemokine signals and each of these binding strengths is given by the equations above. Let’s normalize these external efficacies so that each binding has range $(-1, 1)$, We haven’t yet thought of the efficacy being negative, but it certainly can be, as it can be interpreted as actively discouraging a binding. Let’s assume for our simple 3 chemokine example, the table or matrix of possible bindings at a given time t is given by

$$\boldsymbol{ChS}(t) = \begin{bmatrix} & \boldsymbol{RS_1} & \boldsymbol{RS_2} & \boldsymbol{RS_3} & \\ \boldsymbol{CS_2} & \alpha^S_{11}(t) & 0 & \alpha^S_{13}(t) & \\ \boldsymbol{CS_3} & 0 & \alpha^S_{22}(t) & 0 & 0 \\ \boldsymbol{CS_4} & \alpha^S_{31}(t) & \alpha^S_{32}(t) & 0 & 0 \end{bmatrix}$$

where $\boldsymbol{CS}$ refers to the matrix of chemokine-GPCR interactions near the foreign pMHC site on the surface of the cell. The individual chemokines are labeled $\boldsymbol{CS_i}$ and the receptors are $\boldsymbol{RS_j}$. The values α^s_{ij} are then the efficacies of the bindings on the surface. Next, let's assume the maturing IS releases chemokine signals that interact with receptors at another cellular surface such as the thymus. For convenience, we assume there are four chemokine signals. Let's assume for our simple 4 chemokine example, the table or matrix of possible bindings at a given time t is given by

$$\boldsymbol{ChT}(t) = \begin{bmatrix} & \boldsymbol{RT_1} & \boldsymbol{RT_2} & \boldsymbol{RT_3} & \boldsymbol{RT_4} \\ \boldsymbol{CT_1} & \alpha^T_{11}(t) & 0 & 0 & \alpha^T_{14}(t) \\ \boldsymbol{CT_2} & \alpha^T_{21}(t) & \alpha^T_{22}(t) & 0 & 0 \\ \boldsymbol{CT_3} & 0 & \alpha^T_{32}(t) & 0 & 0 \\ \boldsymbol{CT_4} & 0 & 0 & \alpha^T_{43}(t) & 0 \end{bmatrix}$$

where $\boldsymbol{RT_i}$ is the ith GPCR receptor, $\boldsymbol{CT_j}$ is the jth chemokine signal and $\alpha^T_{jk}(t)$ is the efficacy of the binding at time t at the target cell surface for these signals. Then the serial cascade of the TcR and pMHC binding generates a sequence of chemokine and GPCR associations which is rather uniquely associated to this binding. On the surface, the chemokine-receptor matrix is initialized with $\boldsymbol{ChS}(0)$. We assume the serial self-pMHC bindings that strengthen the TcR-pMHC interaction that leads to the mature immunosynapse occur at the times Δt, $2\Delta t$ out to $N\Delta t$. Further, for simplicity, let's assume the binding strengths for the entries in the surface chemokine-receptor matrix stay constant for each Δt. Then the building of the mature IS has associated with it a sequence of chemokine-receptor arrays on the surface of the cell containing the pMHC given by

$$\{\boldsymbol{ChS}(0), \boldsymbol{ChS}(\Delta t), \boldsymbol{ChS}(2\Delta t), \ldots, \boldsymbol{ChS}(N\Delta t)\}$$

and the corresponding sequence of chemokine-receptor arrays on the target surface are given by

$$\{\boldsymbol{ChT}(0), \boldsymbol{ChT}(\Delta t), \boldsymbol{ChT}(2\Delta t), \ldots, \boldsymbol{ChT}(N\Delta t)\}$$

For example, the target sequence would look like

$$\begin{bmatrix} \alpha^T_{11}(0) & 0 & 0 & \alpha^T_{14}(0) \\ \alpha^T_{21}(0) & \alpha^T_{22}(0) & 0 & 0 \\ 0 & \alpha^T_{32}(0) & 0 & 0 \\ 0 & 0 & \alpha^T_{43}(0) & 0 \end{bmatrix}, \begin{bmatrix} \alpha^T_{11}(\Delta t) & 0 & 0 & \alpha^T_{14}(\Delta t) \\ \alpha^T_{21}(\Delta t) & \alpha^T_{22}(\Delta t) & 0 & 0 \\ 0 & \alpha^T_{32}(\Delta t) & 0 & 0 \\ 0 & 0 & \alpha^T_{43}(\Delta t) & 0 \end{bmatrix},$$

$$\begin{bmatrix} \alpha^T_{11}(2\Delta t) & 0 & 0 & \alpha^T_{14}(2\Delta t) \\ \alpha^T_{21}(2\Delta t) & \alpha^T_{22}(2\Delta t) & 0 & 0 \\ 0 & \alpha^T_{32}(2\Delta t) & 0 & 0 \\ 0 & 0 & \alpha^T_{43}(2\Delta t) & 0 \end{bmatrix},$$

$$\vdots$$

$$\begin{bmatrix} \alpha^T_{11}(N\Delta t) & 0 & 0 & \alpha^T_{14}(N\Delta t) \\ \alpha^T_{21}(N\Delta t) & \alpha^T_{22}(N\Delta t) & 0 & 0 \\ 0 & \alpha^T_{32}(3\Delta t) & 0 & 0 \\ 0 & 0 & \alpha^T_{43}(N\Delta t) & 0 \end{bmatrix}$$

where N is the number of discrete time steps we need to achieve full immunosynapse maturity. This can be more efficiently encoded, but we wanted you to see the full flow of the action. So, to trigger a self interaction inappropriately, we need to not only achieve a binding that exceeds the usual threshold value, but also generates the right sequence of chemokine signals.

To summarize,

- We assume the presence of the foreign peptide in the pMHC initiates a surface chemokine array sequence,

$$\{\boldsymbol{ChS}(0), \boldsymbol{ChS}(\Delta t), \boldsymbol{ChS}(2\Delta t), \ldots, \boldsymbol{ChS}(N\Delta t)\}$$

- We assume there is an associated target surface chemokine-receptor sequence,

$$\{\boldsymbol{ChT}(0), \boldsymbol{ChT}(\Delta t), \boldsymbol{ChT}(2\Delta t), \ldots, \boldsymbol{ChT}(N\Delta t)\}$$

Homework

Exercise 17.3.1 *Assume for a 3 chemokine example, the matrix of possible bindings at a given time t is given by*

$$\boldsymbol{ChS}(t) = \begin{bmatrix} & \boldsymbol{RS_1} & \boldsymbol{RS_2} & \boldsymbol{RS_3} & \\ \boldsymbol{CS_2} & \alpha^S_{11}(t) & 0 & \alpha^S_{13}(t) & \\ \boldsymbol{CS_3} & 0 & \alpha^S_{22}(t) & 0 & 0 \\ \boldsymbol{CS_4} & \alpha^S_{31}(t) & \alpha^S_{32}(t) & 0 & 0 \end{bmatrix}$$

where $\boldsymbol{CS}$ refers to the matrix of chemokine-GPCR interactions near the foreign pMHC site on the surface of the cell. Assume the maturing IS releases chemokine signals and there are four chemokine signals and the matrix of possible bindings at a given time t is given by

$$\boldsymbol{ChT}(t) = \begin{bmatrix} & \boldsymbol{RT_1} & \boldsymbol{RT_2} & \boldsymbol{RT_3} & \boldsymbol{RT_4} \\ \boldsymbol{CT_1} & \alpha^T_{11}(t) & 0 & 0 & \alpha^T_{14}(t) \\ \boldsymbol{CT_2} & \alpha^T_{21}(t) & \alpha^T_{22}(t) & 0 & 0 \\ \boldsymbol{CT_3} & 0 & \alpha^T_{32}(t) & 0 & 0 \\ \boldsymbol{CT_4} & 0 & 0 & \alpha^T_{43}(t) & 0 \end{bmatrix}$$

Assign constant efficacy values

$$\begin{aligned} \alpha^S_{11} &= 0.4,\ \alpha^S_{13} = 0.2,\ \alpha^S_{22} = 0.1,\ \alpha^S_{31} = 0.35,\ \alpha^S_{32} = 0.23 \\ \alpha^T_{11} &= 0.1,\ \alpha^T_{14} = 0.15,\ \alpha^T_{21} = 0.3\ \alpha^T_{22} = 0.23,\ \alpha^T_{32} = 0.41,\ \alpha^T_{43} = 0.07 \end{aligned}$$

and set $\Delta t = 0.1$. Initialize all $\boldsymbol{ChS}$ and $\boldsymbol{ChT}$ values to be 0.001. Use MATLAB to compute the $\boldsymbol{ChS}$ and $\boldsymbol{ChT}$ sequence values for 5 time steps.

Exercise 17.3.2 *Use the setup of the previous problem but add time dependency. So, set efficacy value functions*

$$\begin{aligned} \alpha^S_{11} &= 0.4e^{-.005t},\ \alpha^S_{13} = 0.2e^{-.004t},\ \alpha^S_{22} = 0.1e^{-.003t} \\ \alpha^S_{31} &= 0.35e^{-.001t},\ \alpha^S_{32} = 0.23e^{-.002t} \\ \alpha^T_{11} &= 0.1e^{-.006t},\ \alpha^T_{14} = 0.15e^{-.007t},\ \alpha^T_{21} = 0.3e^{-.003t} \\ \alpha^T_{22} &= 0.23e^{-.004t},\ \alpha^T_{32} = 0.41e^{-.006t},\ \alpha^T_{43} = 0.07e^{-.001t} \end{aligned}$$

Now use MATLAB to compute the $\boldsymbol{ChS}$ and $\boldsymbol{ChT}$ sequence values for 5 time steps.

17.4 The Avidity Calculation

Let's repeat for completeness the avidity calculation derived in (J. Peterson (60) 2018). We find the *external efficacy* e_E is given by

$$e_E = (\xi - 1)(1 + K_{act})[\boldsymbol{sc}] + K_P[\boldsymbol{P}]\xi[\boldsymbol{sc}] \begin{bmatrix} \alpha\gamma\delta - \xi^{-1} \\ \alpha\gamma\delta - \alpha\xi^{-1} \\ \alpha\gamma\delta - \gamma \\ 0 \end{bmatrix} \cdot \begin{bmatrix} 1 \\ K_{act} \\ K_{Sc}[\boldsymbol{sc}] \\ 0 \end{bmatrix}$$

where $[\boldsymbol{sc}]$ denotes the concentration of the scaffolding system $\boldsymbol{Sc}$ prior to the pMHC binding to the receptor cloud. We refer you to (J. Peterson (60) 2018) for detailed explanations of the many terms in this expression.

To create the mature immunosynapse, the initial value e_E^1 must exceed the threshold for activation. That means after the foreign peptide catalyst event, we must have

$$\begin{aligned} e_E^1 &= (\xi_1 - 1)(1 + K_{act}^1)[(\boldsymbol{sc})^{\boldsymbol{1}}] + K_P^1[\boldsymbol{P}^{\boldsymbol{1}}]\xi_1[(\boldsymbol{sc})^{\boldsymbol{1}}] \begin{bmatrix} \alpha^1\gamma^1\delta^1 - \xi_1^{-1} \\ \alpha^1\gamma^1\delta^1 - \alpha^1\xi_1^{-1} \\ \alpha^1\gamma^1\delta^1 - \gamma^1 \\ 0 \end{bmatrix} \cdot \begin{bmatrix} 1 \\ K_{act}^1 \\ K_{Sc}^1[(\boldsymbol{sc})^1] \\ 0 \end{bmatrix} \\ &> (\xi_{1*} - 1)(1 + K_{act}^{1*})[(\boldsymbol{sc})^{\boldsymbol{1*}}] \\ &+ K_P^{1*}[\boldsymbol{P}^{\boldsymbol{1*}}]\xi_{1*}[(\boldsymbol{sc})^{\boldsymbol{1*}}] \begin{bmatrix} \alpha^{1*}\gamma^{1*}\delta^{1*} - \xi_{1*}^{-1} \\ \alpha^{1*}\gamma^{1*}\delta^{1*} - \alpha^{1*}\xi_1^{-1} \\ \alpha^{1*}\gamma^{1*}\delta^{1*} - \gamma^{1*} \\ 0 \end{bmatrix} \cdot \begin{bmatrix} 1 \\ K_{act}^{1*} \\ K_{Sc}^{1*}[(\boldsymbol{sc})^{\boldsymbol{1*}}] \\ 0 \end{bmatrix} \end{aligned} \tag{17.10}$$

We need a simpler equation for our signal receive and send processing functions. Simplify by assuming the only parameters that matter are δ and ξ. Also, redefine all the units here so that all the other parameters have value 1. Then, we can write Equation 17.10 as

$$e_E^1 = (\xi_1 - 1)(2) + \xi_1 \begin{bmatrix} \delta^1 - \xi_1^{-1} \\ \delta^1 - \xi_1^{-1} \\ \delta^1 - 1 \\ 0 \end{bmatrix} \cdot \begin{bmatrix} 1 \\ 1 \\ 1 \\ 0 \end{bmatrix} > e_E^* = (\xi_* - 1)(2) + \xi_* \begin{bmatrix} \delta^* - \xi_*^{-1} \\ \delta^* - \xi_*^{-1} \\ \delta^* - 1 \\ 0 \end{bmatrix} \cdot \begin{bmatrix} 1 \\ 1 \\ 1 \\ 0 \end{bmatrix}$$

or

$$\xi_1 - 3 + 3\xi_1\delta^1 > \xi_* - 3 + 3\xi_*\delta^*$$

Thus, to get over the threshold, we need

$$\xi_1(1 + 3\delta^1) > \xi_*(1 + 3\delta^*). \tag{17.11}$$

Clearly, this occurs if $\xi^1 > \xi^*$ and $\delta^1 > \delta^*$ which is the easiest possibility. Now ξ^1 is the multiplier for the concentration of the scaffolding proteins we have prior to the binding of the ligand, i.e. the pMHC. The term δ^1 is related to the coupling of the ligand pMHC and the scaffolding. If $\delta^1 > 1$, this means the coupling of the pMHC enhances the scaffolding protein complexes we lump together as $\boldsymbol{Sc}$. So these two terms are related. Since we assume $\xi^1 > \xi^*$, we know this means the concentration of the scaffolding complex is upregulated due to the pMHC binding over the threshold value for the self-pMHC. Further, since $\delta^1 > \delta^*$, we know the binding of this self-pMHC enhances the coupling between the scaffolding a bit over the threshold amount for this self-pMHC. How could this happen? Here is a possible scenario. There is an initial viral infection which creates an inflammatory response which upregulates the scaffolding protein complexes around a self-pMHC. This pushes both ξ^1 and δ^1 up and allows the serial method of T-Cell activation to create a mature IS. Although simplified enormously, this thought experiment allows us to focus on what we think is really important here. What is the mechanism or process that allows the self-pMHC and TcR interaction to exceed the putative threshold value? Once this happens, the formation of the IS is essentially assured and self-damage ensues. This clearly does not happen except in people who are susceptible. There is a nice numerical example in (J. Peterson (60) 2018) which you can look at for specific avidity results.

Homework

Exercise 17.4.1 *Consider the threshold equation*

$$\xi_1(1+3\delta^1) \quad > \quad \xi_*(1+3\delta^*)$$

and assume the base state is $\xi^* = \delta^* = 0$. *In this case, the threshold equation is*

$$\xi_1(1+3\delta^1) \quad > \quad 0$$

Plot this in the $\delta^1 - \xi^1$ *plane and you can see* δ^1 *and* ξ^1 *threshold dependence.*

Exercise 17.4.2 *Set all the important parameters to* 2 *and rederive the threshold equation.*

17.5 A Simple Cytokine Signaling Model

For example, consider the result of a cytokine signal $\boldsymbol{c}$ applied to the graph. From our discussions, we know the particular pattern of proteins created on the target cell can be organized into a matrix of different sized columns. If this patch on the surface of the cell corresponds to the decomposition $\boldsymbol{c} \in= Z_7 \oplus Z_7 \oplus Z_7 \oplus Z_9 \oplus Z_9 \oplus Z_4 \oplus Z_4 \oplus Z_4 \oplus Z_5$, only one entry in each column will be a 1 with all other entries zero. In Figure 17.5, we denote the *active* part of the code in each column with a closed circle. For our example, there are $7^3 \times 9^3 \times 4^3 \times 5$ or $343 \times 729 \times 64 \times 5 \approx 80$ million possible signals. Some of these signals have columns that are all zeros. There are $6^3 \times 8^3 \times 3^3 \times 4$ or $216 \times 512 \times 27 \times 4 \approx 12$ million signals having no zero columns which is about 15%. There are also signals which have at least one zero column. The calculations for how many there are of these are quite complicated in our example as our columns are not equal sizes. But just to give you the feel of it, let's look at how many ways there are to have one column of zeros. There are $\binom{10}{1} = 10$ ways to have one column be all zeros. The probability of activating a target here is 90% in this case where there are 10 columns.

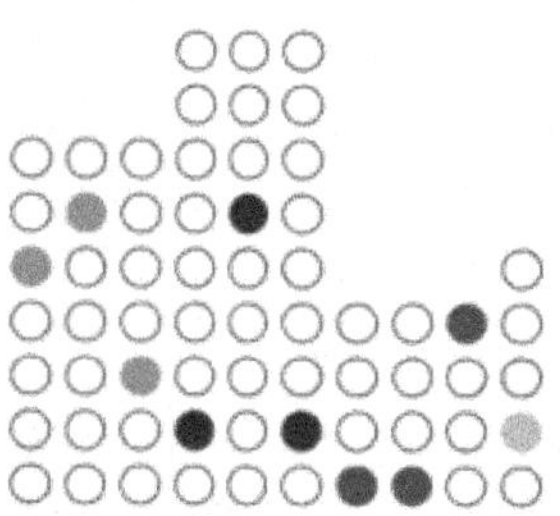

Figure 17.5: The cytokine signal barcode.

- Column 1, 2 or 3 is zero giving $6^2 \times 8^3 \times 4^3 \times 4$ or $36 \times 512 \times 64 \times 4 \approx 5$ million signals.
- Column 4, 5 or 6 is zero giving $6^3 \times 8^2 \times 4^3 \times 4$ or $216 \times 64 \times 64 \times 4 \approx 3.5$ million signals.
- Column 7, 8 or 9 is zero giving $6^3 \times 8^3 \times 4^2 \times 4$ or $216 \times 512 \times 16 \times 4 \approx 7.0$ million signals.
- Column 10 is zero giving $6^3 \times 8^3 \times 4^3$ or $216 \times 512 \times 64 \approx 7.0$ million signals.

Thus, there are about 22.5 million signals with one column of zeros or about 28%. The remaining possibilities of signals having more than one zero column fill out the remainder of the signals. We see that about 85% of the signals correspond to signals with at least one column of zeros. These barcodes would not correspond to valid cytokine-target cell interactions as the right proteins are not created and attached to the cell's surface. Hence, a non-valid signal might correspond to actions that should not be taken. If we assume that all signals with nonzero components potentially correspond to an action, albeit one we may not wish to take, we note we have a correspondence mapping of the form $\boldsymbol{x} \in \Re^{10} \longrightarrow \mathscr{T} \cup \mathscr{T}^c$ where $\mathscr{T}$ is the set of signal targets with all nonzero columns and $\mathscr{T}^c$ are the rest of the signals. Signals in $\mathscr{T}^c$ should not be used to create target responses. Hence, a signal processing error that related to such an improper signal being used to create a target response would be a type of security violation in the sense that an improper follow up cytokine response is generated for the network of cytokine signals. A simple decomposition component calculation error could take an invalid signal and inappropriately add a 1 to enough columns to cause the signal to create a target activation signal. This would mean a part of the signal space that should not be used to create responses is actually being used to do that. In essence, since the barcode can be thought of as indicating the binding sites of proteins created by cytokine messages, we see for a proper response, a patch on the surface of the cell must contain proteins bound in *specific locations*. Wrong messages due to errors in cytokine signaling, mistakes due to the interactions of antigens that alter the cytokine signaling process, and so forth, can give rise to an incorrect surface patch. This would be represented by a loss of a component in the barcode giving a column with a value of zero. In the immune system response, this means a mistake caused by, say, viral interference with cytokine signaling machinery could have profound consequences as incorrect cytokine signals are created downstream of the error. This could lead to incorrect T-Cell maturation and the potential for self-damage due to an autoimmune reaction.

The signal processing can be extended to processing chains. If we concatenate p signal decomposition engines together, we form a chain $\boldsymbol{x} \longrightarrow \mathscr{B}_1 \longrightarrow \mathscr{B}_2 \longrightarrow \cdots \longrightarrow \mathscr{B}_p$. Let's look at something similar to what is shown in Figure 17.8(b) with three concatenated decompositions. The signal family which generates the barcode family $\mathscr{B}_1$ generates target outputs for each viable signal; that is, a signal whose barcode has no zero columns in it. We will stick to our example for now. This first family of barcodes is represented by the direct sum

$$\boldsymbol{x} \in (Z^7 \oplus Z^7 \oplus Z^7) \oplus (Z^9 \oplus Z^9 \oplus Z^9) \oplus (Z^4 \oplus Z^4 \oplus Z^4) \oplus (Z^5).$$

This family of barcodes can be used to generate another family of signals in a variety of ways and this will generate another family of barcodes. For convenience of discussion, let's assume this intermediate step gives signals $\boldsymbol{y}$ which satisfy

$$\boldsymbol{y} \in (Z^6 \oplus Z^6) \oplus (Z^8 \oplus Z^8 \oplus Z^8 \oplus Z^8) \oplus (Z^4 \oplus Z^4 \oplus Z^4).$$

This second family of barcodes, $\mathscr{B}_2$, then generates a third family of signals, $\boldsymbol{z}$, satisfying

$$\boldsymbol{z} \in (Z^3 \oplus Z^3 \oplus Z^3 \oplus Z^3 \oplus Z^3) \oplus (Z^2 \oplus Z^2 \oplus Z^2 \oplus Z^2).$$

We illustrate this process in Figure 17.6, where we show the first barcode from $\mathscr{B}_1$ is

$$((4,5,2),(1,5,1),(0,0,3),(1)),$$

the second barcode from $\mathscr{B}_2$ is

$$((2,3),(5,1,2,7),(1,1,2))$$

and the third barcode from $\mathscr{B}_3$ is

$$((2,0,2,1,1),(1,1,0,1)).$$

The numbers here are for convenience: the $(4, 5, 2)$ means the first entry in $4 \in Z^7$ which shows up as a colored circle in position 5 in the column as the column starts from 0 and goes to 6. Similarly, the next entry 5 gives a closed circle in position 6. Note, using a grammatical construction

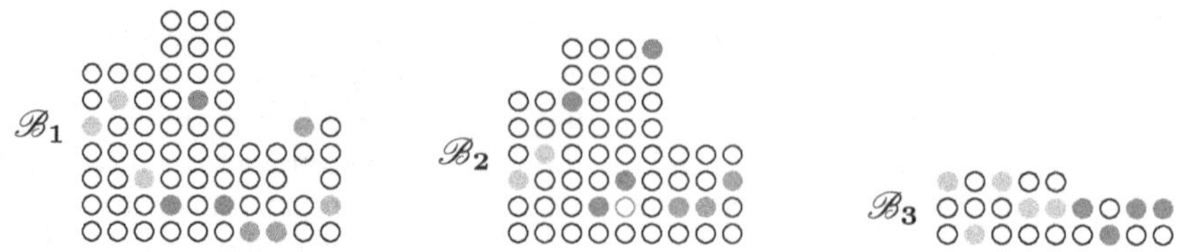

Three barcodes corresponding to a sentence in a signaling grammar.

Figure 17.6: A barcode sentence.

is a way to ensure that a signal $\boldsymbol{x}$ is parsed without error. Our intended target is the outcome from the tertiary signal $\boldsymbol{z}$ and errors in $\boldsymbol{x}$ and or $\boldsymbol{y}$ have a lower probability of altering the correct final output wanted for $\boldsymbol{z}$. It is clear we can construct graphs of immunosynapses connected by cytokine signals mediated by barcodes. In the message passing/signal enabled asynchronous model, after a finite number of symmetry breakings, we are left with a graph corresponding to the accessible nodes which are processing paths of length at least three, which includes the possibility of loops where a path starts and stops on the same node. Computations around closed paths allow for the possibility that classes of paths become equivalent. The exact nature of these equivalence classes of paths and the topology they determine on the subgraph $\mathcal{G}\left(\mathcal{V}_{\mathcal{V}_2}, \mathcal{E}_{\mathcal{V}_2 \to \mathcal{V}_2}\right)$ will be determined by the receive and send processing functions. The barcodes used to model the cytokine signals are then part of the signaling added to the graph model, and the details of these interactions are part of the send and receive signal processing functions. We illustrate a very simple snapshot of such an immunosynapse network in Figure 17.7(a). Consider a cytokine signal of the form $S = \sum_{i=1}^{N_1} n_i F_i$. Let $\mathscr{F}$ denote the family of all possible cytokines. This is a finite collection of signaling molecules. The signal S is shown in Figure 17.7(b). This valid signal creates a pattern of proteins on the surface of the target cell, which is indicated by a simple matrix with five columns. Each column has one closed dot in it which corresponds to the created protein. The target cell will generate a new signal only if the five proteins shown are created. For simplicity, we think of each column as having only one possible *dot*. The columns in Figure 17.7(b) need not be of the same height, of course. If each column has p possible positions at which the protein can be built, then a valid protein for each column can be represented as the integer m in the collection of numbers called Z_p which is defined by $Z_p = \{\text{ integers } i \,:\, 0 \leq i \leq p-1\}$. Usually p is a prime number. For example, $Z_5 = \{\text{ integers } i : 0 \leq i \leq 4\}$. These collections of numbers are called fields. The barcode in Figure 17.7(b) can then be represented by $B_S = m_1 z^{p_1} \oplus m_2 z^{p_2} \oplus m_3 z^{p_3} \oplus m_4 z^{p_4} \oplus m_5 z^{p_5} \oplus$. As long as we remember the underlying Z_{p_i} field, we can also denote S as the five tuple $(m_1, m_2, m_3, m_4, m_5)$. Hence, in general a barcode has the form $(m_1, m_2, m_3, m_4, m_5) \in \sum_{i=1}^{N} \oplus Z^{p_i}$. Of interest to us would be a simple T-Cell-pMHC system as shown in Figure 17.8(a). The T-Cell receives a valid signal S which generates a barcode B_S. The cytokine represented by this barcode is a cytokine $G \in \mathscr{F}$. The cytokine G generates its own barcode B_G on the surface of the cell containing the pMHC containing the protein snippet the T-Cell might recognize. A successful activation and the generation of a mature immunosynapse generates a cytokine signal H back to the T-Cell which creates the barcode B_H on the immunosynapse complex that is being formed. In Figure 17.8(a), we do not show the immunosynapse being created and instead keep the diagram as a simple T-Cell and pMHC in-

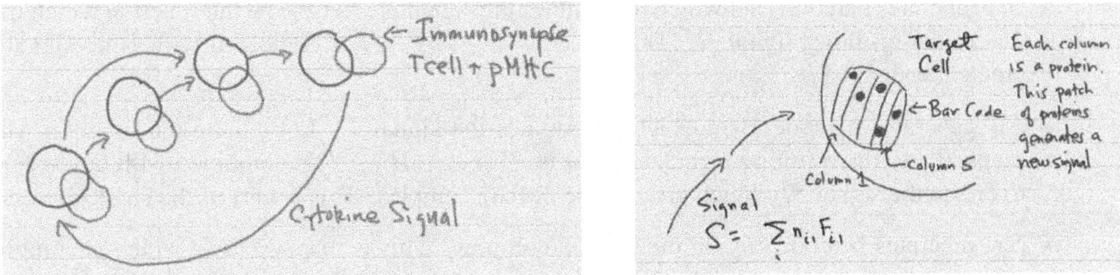

(a) A subset of an immunosynapse network.

(b) A cytokine signal and its barcode.

Figure 17.7: Signals and barcodes.

teraction. Note the interchange of cytokine signals. The symbol e represents the affinity threshold calculation as discussed earlier. As an example of a longer chain of signals, consider what we show in Figure 17.8(b). In Figure 17.8(b), the infectious agent generates a signal S, which creates a valid

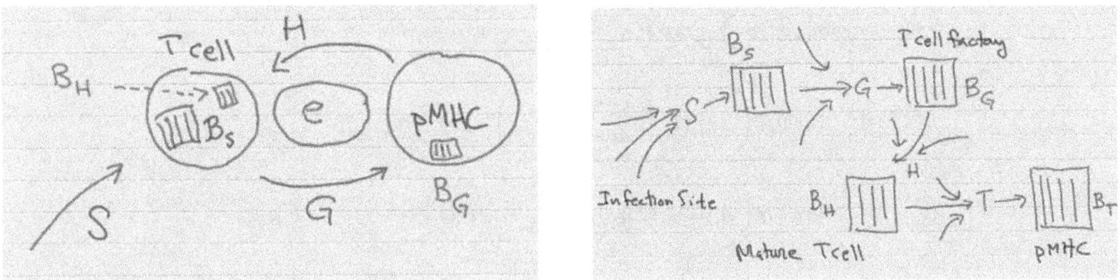

(a) A sample T-Cell and pMHC system.

(b) A sample infection, T-Cell factory, mature T-Cell and pMHC cell signal chain.

Figure 17.8: T-Cell interactions.

barcode B_S on a cell, which begins the infection. This in turn generates a signal G, which creates a valid barcode B_G on the appropriate cell in the T-Cell factory machinery. This generates another signal H and barcode B_H on a mature T-Cell. We can think of this signal as causing the changes in a precursor T-Cell that turn it into a mature T-Cell primed for antigen recognition. If the e affinity threshold is crossed, then there is another signal T from the T-Cell to the pMHC complex generating the barcode B_T. Then, just as in Figure 17.8(a), there is an interchange of cytokine signals leading to the formation of a mature immunosynapse. We can also cast this in the form of equations. The signal $S = \sum_{i=1}^{N_S} n_{i1} F_{i1}$ and the resulting barcode B_S generate the cytokine $G \in \mathscr{F}$. The signal G and others give a second signal of the form $S_G = n_{21} G + \sum_{i=2}^{n_G} n_{2i} F_{2i}$. This generates a new barcode B_G and a resulting cytokine H. Then the cytokine H and other signals form the generated signal $S_H = n_{31} H + \sum_{i=2}^{n_H} n_{3i} F_{3i}$, which forms the signal T leaving the T-Cell. This combines with other signals to give the signal $S_T = n_{41} T + \sum_{i=2}^{n_T} n_{4i} F_{4i}$, which creates the barcode B_T on the cell containing the pMHC.

Let's set up some notation:

- S generates barcode B_S, which determines the signal G. Let ϕ_S be the function which maps B_S to the resulting signal G. That is, $G = \phi_S(B_S)$. G is combined with other signals to create the signal S_G.

- S_G generates barcode B_G, which determines the signal H. Let ϕ_G be the function which maps B_G to the resulting signal H. That is, $H = \phi_G(B_G)$. H is combined with other signals to create the signal S_H.
- S_H generates barcode B_H, which determines the signal T. Let ϕ_H be the function which maps B_H to the resulting signal T. That is, $T = \phi_H(B_H)$. T is combined with other signals to create the signal S_Q which goes to the pMHC complex. This is part of the e calculation.
- S_Q generates barcode B_Q on the pMHC complex. This is mapped to Q which is combined with other signals to create the signal S_P back to the T-Cell. Let ϕ_Q be the function which maps B_Q to the resulting signal Q. That is, $Q = \phi_Q(B_Q)$.
- S_P generates barcode B_P, which determines the signal U. This is combined with other signals to create the signal S_U back to the pMHC. Let ϕ_P be the function which maps B_P to the resulting signal U. That is, $U = \phi_P(B_P)$. This generates the barcode B_U on the pMHC complex.

We show the T-Cell-pMHC interactions in detail in Figure 17.9(a). Note, if we desired to, we could do expansions of this form

$$
\begin{aligned}
S_Q &= n_{41}T + \sum_{i=2}^{n_T} n_{4i}F_{4i} = n_{41}\phi_H(B_H) + \sum_{i=2}^{n_T} n_{4i}F_{4i} \\
&= n_{41}\phi_H\left(n_{31}H + \sum_{i=2}^{n_H} n_{3i}F_{3i}\right) + \sum_{i=2}^{n_T} n_{4i}F_{4i} \\
&= n_{41}\phi_H\left(n_{31}\phi_G(B_G) + \sum_{i=2}^{n_H} n_{3i}F_{3i}\right) + \sum_{i=2}^{n_T} n_{4i}F_{4i} \\
&= n_{41}\phi_H\left(n_{31}\left\{n_{21}G + \sum_{i=2}^{n_G} n_{2i}F_{2i}\right\} + \sum_{i=2}^{n_H} n_{3i}F_{3i}\right) + \sum_{i=2}^{n_T} n_{4i}F_{4i}
\end{aligned}
$$

We can clearly see the recursive nature of these computations in the chain above. These computations are best organized using a graph of computational nodes connected by edges, which we will discuss shortly. The interaction between the T-Cell and the cell containing the pMHC complex is mediated by the threshold calculation e. It is clear now that if ϵ is the value of the threshold function e, we have $\epsilon = e(S_Q, S_P, S_U)$. This can also be written in terms of barcodes, which will be more convenient. We have $\epsilon = e(B_H, B_Q, B_P)$ as B_H determines S_Q, B_Q determines S_P, and B_P determines S_U. Recall the threshold computation is given by

$$
e_E = e(B_H, B_Q, B_P) = (\xi - 1)(1 + K_{act})[\boldsymbol{sc}] + K_P[\boldsymbol{P}]\xi[\boldsymbol{sc}] \begin{bmatrix} \alpha\gamma\delta - \xi^{-1} \\ \alpha\gamma\delta - \alpha\xi^{-1} \\ \alpha\gamma\delta - \gamma \\ 0 \end{bmatrix} \cdot \begin{bmatrix} 1 \\ K_{act} \\ K_{Sc}[\boldsymbol{sc}] \\ 0 \end{bmatrix}
$$

where $[\boldsymbol{sc}]$ denotes the concentration of the scaffolding system $\boldsymbol{Sc}$ prior to the pMHC binding to the receptor cloud. The constants that occur in e_E have connections to the barcodes B_H, B_Q and B_U as follows:

- We assume the concentration of scaffolding proteins after the ligand P (pMHC) is introduced is $\xi[Sc]$, where $[Sc]$ is their concentration before P. If $\xi > 1$, we see $e_E > 0$. Both of the feedback barcodes B_P and B_U can affect this. Hence, we can replace ξ by $\xi(B_P, B_U)$.
- α scales the conversion $T_i \rightarrow T_a$ due to P being added. This is dependent on B_H; so $\alpha \rightarrow \alpha(B_H)$.

- γ determines how the conversion $T_i \rightarrow T_i Sc$ is altered by the presence of P. This is dependent on B_P and B_U. Thus, $\gamma \rightarrow \gamma(B_p, B_U)$.
- δ determines how effective Sc coupling to P is and is hence dependent on both the initial activation barcode B_H and the feedback barcode B_U. So $\delta \rightarrow \delta(B_H, B_U)$.
- K_P is the $T_i \rightarrow PT_i$ rate; i.e. the rate at which inactivated T_i binds to P. This could be important in autoimmune disease. This is dependent on B_P and B_U. Hence, $K_P \rightarrow K_P(B_P, B_U)$.
- K_{act} is the rate of $T_i \rightarrow T_a$, which is determined by B_H. So $K_{act} \rightarrow K_{act}(B_H)$.
- K_{Sc} is the rate at which scaffolding proteins are added to components. This is initially set by B_H. So $KSc \rightarrow K_{Sc}(B_H)$.

We can then rewrite e_E using these dependencies.

$$
\begin{aligned}
e(B_H, B_Q, B_P) &= (\xi - 1)(1 + K_{act}(B_H))[\boldsymbol{sc}] \\
&+ K_P(B_P, B_U)[\boldsymbol{P}]\xi[\boldsymbol{sc}] \begin{bmatrix} \alpha(B_H)\gamma(B_p, B_U)\delta(B_H, B_U) - \xi(B_P, B_U)^{-1} \\ \alpha(B_H)\gamma(B_p, B_U)\delta(B_H, B_U) - \alpha\xi(B_P, B_U)^{-1} \\ \alpha(B_H)\gamma(B_p, B_U)\delta(B_H, B_U) - \gamma(B_p, B_U) \\ 0 \end{bmatrix} \cdot \begin{bmatrix} 1 \\ K_{act}(B_H) \\ K_{Sc}(B_H)[\boldsymbol{sc}] \\ 0 \end{bmatrix}
\end{aligned}
$$

The comments above show clearly how nonlinear the dependence of e_E is on the barcodes B_H, B_P and B_U. From a less detailed perspective, we can come to this conclusion by noting:

- The cytokine signal to activate the T-Cell is controlled by B_H. Since $T_i \rightarrow T_a$ with rate constant K_{act}, we see the structure of the barcode B_H determines K_{act}.
- T-Cell and pMHC binding induces B_P on the T-Cell due to the feedback from the pMHC to the T-Cell. The conversion $T_a \rightarrow T_a Sc$ mediated by βK_{Sc} is thus related to the structure of the protein patch determined by B_P.
- T-Cell and pMHC binding also induces B_U on the pMHC complex due to the feedback form the T-Cell to the pMHC. The pathway $PT_a \rightarrow PT_a Sc$ with rate constant $\delta\gamma\beta K_{Sc}$ is therefore due to the barcode B_U.

It follows that the cytokine signaling model gives another quantitative mechanism for computing the self-affinity computation. We can express ϵ in terms of detailed parameters related to T-Cell-pMHC interactions or cytokine signals in terms of signaling barcodes. We show the T-Cell-pMHC interac-

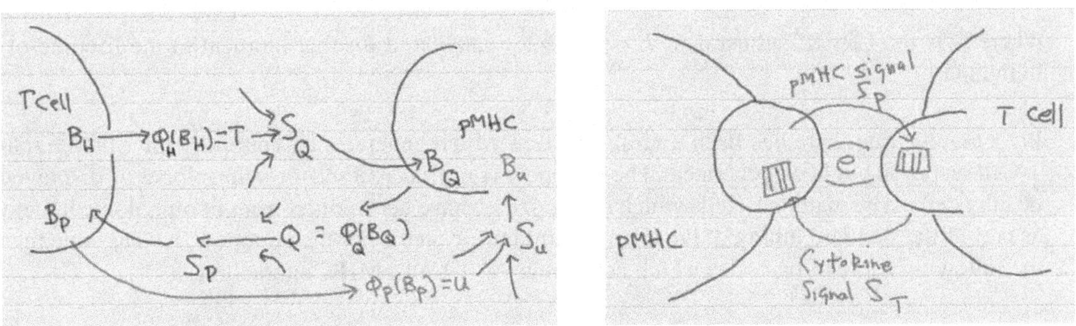

(a) A sample T-Cell-pMHC interaction.

(b) A sample T-Cell and pMHC interaction using signals.

Figure 17.9: T-Cell interactions with signals.

tion again in Figure 17.9(b). The T-Cell is attempting to recognize the antigen in the pMHC complex and in this process, cytokine messages are sent back and forth between the T-Cell and the pMHC complex. This allows the threshold affinity for the interaction to be computed, $e(S_Q, S_P, S_U)$. The symbol e in Figure 17.9(b) hides the S_Q signal which is transmitted from the T-Cell to the barcode B_Q, which then generates the signal S_P back to the T-Cell from the pMHC complex.

Homework

Exercise 17.5.1 *Assume a patch on the surface of the cell corresponds to the decomposition* $\boldsymbol{c} \in= Z_5 \oplus Z_5 \oplus Z_4 \oplus Z_8 \oplus Z_7$ *where only one entry in each column will be a* 1 *with all others zero. Draw various barcodes for this decomposition.*

Exercise 17.5.2 *Assume a patch on the surface of the cell corresponds to the decomposition* $\boldsymbol{c} \in= Z_6 \oplus Z_6 \oplus Z_4 \oplus Z_4 \oplus Z_7 oplus Z_7$ *where only one entry in each column will be a* 1 *with all others zero. Draw various barcodes for this decomposition.*

Exercise 17.5.3 *Draw a sample interaction with two T-Cells and pMHC complexes using barcodes.*

Exercise 17.5.4 *Draw a sample interaction with two T-Cells and pMHC complexes using signals.*

17.6 Sample Self-Damage Scenarios

Let's consider a T-Cell-pMHC interaction as shown in Figure 17.9(a). This is excised from a larger immunosynaptic network so that we can focus clearly on affinity computations. The affinity calculation is given by

$$\begin{aligned} e(B_H, B_Q, B_P) &= (\xi - 1)(1 + K_{act}(B_H))[\boldsymbol{sc}] \\ &+K_P(B_P, B_U)[\boldsymbol{P}]\xi[\boldsymbol{sc}] \begin{bmatrix} \alpha(B_H)\gamma(B_p, B_U)\delta(B_H, B_U) - \xi(B_P, B_U)^{-1} \\ \alpha(B_H)\gamma(B_p, B_U)\delta(B_H, B_U) - \alpha\xi(B_P, B_U)^{-1} \\ \alpha(B_H)\gamma(B_p, B_U)\delta(B_H, B_U) - \gamma(B_p, B_U) \\ 0 \end{bmatrix} \cdot \begin{bmatrix} 1 \\ K_{act}(B_H) \\ K_{Sc}(B_H)[\boldsymbol{sc}] \\ 0 \end{bmatrix} \end{aligned}$$

This is clearly our required signal receive processing function at each immunosynapse. We can also add to this the LRRK2 adjustment given by $x_1(t) + x_2(t)$. The signal processing functions thus have the form

$$\sigma_n^r(L, B) = e(B_H, B_Q, B_P), \quad \sigma_n^s(L, B) = L\, e(B_H, B_Q, B_P)$$

where L is the LRRK2 adjustment $L = x_1 + x_2$ calculated for that moment in the lifetime of the individual.

Let's assume that there has been a viral infection which generates an epitope v for which a mature T-Cell response has been generated. The epitope v is *similar* to a self-protein snippet v^* displayed by healthy cells. The mature T-Cells which recognize v^* have been edited from immunological memory as part of the standard tuning of the adaptive immune system. For the epitope v, let's assume there is the following triple of barcodes which determine the T-Cell-pMHC interaction.

$$\begin{aligned} B_H &= (1, 2, 3, 4) \in Z^4 \oplus Z^4 \oplus Z^5 \oplus Z^5 \\ B_P &= (3, 5, 2) \in Z^6 \oplus Z^8 \oplus Z^3 \\ B_U &= (1, 2, 1) \in Z^3 \oplus Z^3 \oplus Z^2 \end{aligned}$$

For this example, we will assume the barcode triple corresponding to v^* differs only in the first component. We have

$$\begin{aligned} B_H^* &= (s_1, s_2, s_3, s_4) \in Z^4 \oplus Z^4 \oplus Z^5 \oplus Z^5 \\ B_P^* = B_P &= (3,5,2) \in Z^6 \oplus Z^8 \oplus Z^3 \\ B_U^* = B_U &= (1,2,1) \in Z^3 \oplus Z^3 \oplus Z^2 \end{aligned}$$

The affinity calculation e_v here is large enough to initiate lysis. Since B_P and B_U do not change in the barcodes for v^*, differences are encoded by B_H^*. There are $2^4 \cdot 2^4 \cdot 2^5 \cdot 2^5$ or 262,144 possible ways to fill a barcode in $Z^4 \oplus Z^4 \oplus Z^5 \oplus Z^5$. We have assumed a valid barcode which will generate the next cytokine signal (i.e. the target cell's response) must have only one 1 in each of its columns. Here, there are $4 \cdot 4 \cdot 5 \cdot 5$ or 400 ways to create a valid B_H barcode which has a 1 in only one component in each column. Thus, only 0.00153 or .15% of the possibilities are valid protein patches.

Now, since affinity computations follow a probability distribution, it is not necessary to the perfect barcode B_H to be present to make a match to v. Barcodes which are *close* to B_H will generate a weaker, but still potentially useful match. For our discussion here, we will assume the first two columns of B_H must be correct, but any valid code for the remaining two columns will create a weaker match. If e_v is the affinity for the perfect match, assume that a perfect match in the first two columns only will diminish the affinity by e^{-2r} for some r. On the other hand if three columns match, leaving only one column not matched, the affinity is reduced by e^{-r}. The number of matches including partials is thus $5 \cdot 5$ or 25. Hence, 25 partial B_H matches will generate an affinity value.

- The viral infection could be a recurring one which generates a pool of printed T-Cells for epitope v which are always there. In the absence of a flareup in the viral infection, T-Cells primed for v can still bind to v^*. The resulting affinity is $e^{-r}e_v$ or $e^{-2r}e_v$. This is less than the value e_v required for a lysis event for the cell displaying v in a pMHC complex. However, the barcodes B_P and B_U initiate feedback pathways that *strengthen* the scaffolding proteins Sc which affect the calculation of the affinity. If there is a sufficient population of primed T-Cells for v, say M in number, the affinity rises to $Me^{-r}e_v$ or $Me^{-2r}e_v$. Even if the barcode match is only a two-column match, if M is large enough, the aggregate affinity could exceed e_v and initiate lysis.

- Note the barcodes B_P and B_U can be interpreted as introducing alterations in the ligand-G protein binding rates which for the immunosynapse setting means altering the rates at which the T-Cell-pMHC complex recruits additional structural proteins which stabilize the growing immunosynapse interface. This can increase the size of the pMHC rafts that are known to occur on the surface of the cell displaying v in a pMHC cradle or alter the geometry by decreasing the angle ϕ between the ligand and receptor space. Note the affinity calculation is maximized when $\cos(\phi) = 0$ or $\phi = 0$ which means the ligand and receptor space are collinear in the sense that their normal vectors are parallel. This angle is very dependent on the scaffolding G proteins used to create the mature immunosynapse.

- The situation becomes more complicated therefore if we allow partial matches in both B_P and B_U, as the stabilizing influence of the original B_P and B_U is diminished. However, the multiplier effect still holds.

A chain of immunosynapse interactions within an immunosynaptic graph model would give us even more insight.

Homework

Redo the above discussions for the following choices:

Exercise 17.6.1

$$\begin{aligned} B_H &= (1,2,3,4) \in Z^3 \oplus Z^3 \oplus Z^6 \\ B_P &= (3,5,2) \in Z^6 \oplus Z^6 \oplus Z^5 \\ B_U &= (3,1,1) \in Z^5 \oplus Z^2 \oplus Z^2 \end{aligned}$$

Exercise 17.6.2

$$\begin{aligned} B_H &= (3,2,4,1) \in Z^4 \oplus Z^4 \oplus Z^5 \oplus Z^5 \\ B_P &= (6,2) \in Z^6 \oplus Z^4 \\ B_U &= (4,2,1) \in Z^5 \oplus Z^7 \oplus Z^8 \end{aligned}$$

Exercise 17.6.3

$$\begin{aligned} B_H &= (1,1,3,3,9) \in Z^4 \oplus Z^4 \oplus Z^5 \oplus Z^5 \oplus Z_{11} \\ B_P &= (3,5,4) \in Z^6 \oplus Z^6 \oplus Z^7 \\ B_U &= (3,2) \in Z^4 \oplus Z^4 \end{aligned}$$

17.7 The Asynchronous Graph Neurodegeneration Model

We note another source of interaction for neurodegeneration comes from an infection such as a flavivirus or a gut biome interaction. Background papers on West Vile Virus (WNV) infection interactions include (Arnold et al. (8) 2004), which discusses how antigen processing is regulated in WNV infected fibroblasts, (S. Bao and N. J. C. King and R. Dos (106) 1992), which presents a model of MHC antigen on myoblasts, (A. M. Kesson and Y. Cheng and N. J. C. King (4) 2002), which outlines a theory of how immune recognition molecules are regulated by WNV, and (N. J. C. King and D. R. Getts and M. T. Getts and S. Rana and B. Shrestha and A. M. Kesson (90) 2007), which discusses the immunopathology of flavivirus infections. In addition, (C. Wacher and M. Muller and M. J. Hofer and D. R. Getts and R. Zabaras and S. S. Ousman and F. Terenzi and G. C. Sen and N. J. C. King and I. L. Campbell (17) 2007) organizes what we know about interferon genes and their regulation after various viral infections. Also, there are interactions between antigen in the gut, the gut immune system, the blood brain barrier and the brain. A good overview of gut biome interactions can be found in (H. Wekerle (43) 2017). With all this said, our models can now include all this information as follows:

- Our discussion of cytokine models, an affinity computational engine, and barcode calculational chains have informed our choice of state pairs (u_n, v_n) and signals S_n with content c_n and processing functions $\sigma_n^r(S_n, c_n)$ and $\sigma_n^s(S_n, c_n)$ for the immune component. We have also added an additional signaling component for the LRRK2 allele loss and its influence on the immune system via microglial interaction, although we have not discussed in detail the effects of the LRRK2 mutations on possible inflammatory cascades.

Future questions to address include:

- How do we add an additional asynchronous message passing and signal enhanced graph model for the gut biome using specialized state pairs (g_n, h_n) and signals T_n with content d_n and processing functions $\rho_n^r(T_n, d_n)$ and $\rho_n^s(T_n, d_n)$?

- How do we add an additional asynchronous message passing and signal enhanced graph model for the gut immune system using specialized state pairs (x_n, y_n) and signals Z_n with content e_n processing functions $\zeta_n^r(Z_n, e_n)$ and $\zeta_n^s(Z_n, e_n)$?

A more complete model would then consist of all the components above interacting. The autoimmune responses we seek could then emerge from the fact that the topology of these graphs splits incoming signals into direct sums of subgraphs, which essentially form equivalence classes of signal possibilities. The right antigen signals mediated by LRRK2 and bacterial generated cytokines along with LRRK2 mutations might then allow for normally low affinity responses to be upgraded to a lysis event of healthy tissue.

The work presented here is the theoretical underpinning for a new stage of simulations. We have been trying to get you to not think of these interactions as serial in nature and instead focus on their asynchronous nature. Of course, the thread that binds this all together is this: what mathematical abstractions should we use to get a better understanding of these complex systems?

Chapter 18

Bar Code Computations in Consciousness Models

Let's look at some simple models of the kind of information processing that occurs in a cognitive system. We would like appropriate mathematical formalisms to help explain zombie creation, of course, and we want a mechanism that helps explain how an external trigger can alter the functioning of a cognitive system in such a way that it appears to be a reordering of computational submodules. Hence, let's discuss a cognitive architecture blueprint for information processing which can be applied to the construction of cognitive models and real-time assessments of consciousness to determine proper levels of anesthesia. Note, these same ideas, rephrased for immunosynapse networks can also be used to study central nervous system (CNS) damage models due to viral, bacterial and other toxins and pathogens. We laid the foundation for this in Chapter 17. We also have hopes that these ideas can also be used to study economic systems (see (C. Camerer and J. Cohen and E. Fehr and P. Glimcher and D. Laibson (15) 2016)).

18.1 General Graph Models for Information Processing

Let's consider some node update strategies based on Cortical/Can (CT) architectures. We will be thinking of this in terms of subgraph decompositions as we discussed with barcodes in Chapter 17. We will call a subgraph decomposition model a **Betti** node for convenience. We also call the type of calculations we do using such decompositions, Betti calculations. For us, a typical cortical column consists of six stacked **FFP**, **OCOS** and **2-3** circuits we call cans with each having connections to a thalamic module **Th**. Hence, this is a stack of six **CT** computational modules which can all be modeled by a Betti node **BN**. The column then has the structure comprising six cortical cans

$$\begin{bmatrix} \boldsymbol{CT_6} \\ \boldsymbol{CT_5} \\ \boldsymbol{CT_4} \\ \boldsymbol{CT_3} \\ \boldsymbol{CT_2} \\ \boldsymbol{CT_1} \end{bmatrix} \equiv \begin{bmatrix} \boldsymbol{BN_6} \\ \boldsymbol{BN_5} \\ \boldsymbol{BN_4} \\ \boldsymbol{BN_3} \\ \boldsymbol{BN_2} \\ \boldsymbol{BN_1} \end{bmatrix} = \begin{bmatrix} \sum_{i=1}^{n_6} \oplus \boldsymbol{Z_{p_{6i}}} \\ \sum_{i=1}^{n_5} \oplus \boldsymbol{Z_{p_{5i}}} \\ \sum_{i=1}^{n_4} \oplus \boldsymbol{Z_{p_{4i}}} \\ \sum_{i=1}^{n_3} \oplus \boldsymbol{Z_{p_{3i}}} \\ \sum_{i=1}^{n_2} \oplus \boldsymbol{Z_{p_{2i}}} \\ \sum_{i=1}^{n_1} \oplus \boldsymbol{Z_{p_{1i}}} \end{bmatrix} \equiv \begin{bmatrix} \oplus \boldsymbol{Z}(\boldsymbol{P_6}, n_6) \\ \oplus \boldsymbol{Z}(\boldsymbol{P_5}, n_5) \\ \oplus \boldsymbol{Z}(\boldsymbol{P_4}, n_4) \\ \oplus \boldsymbol{Z}(\boldsymbol{P_3}, n_3) \\ \oplus \boldsymbol{Z}(\boldsymbol{P_2}, n_2) \\ \oplus \boldsymbol{Z}(\boldsymbol{P_1}, n_1) \end{bmatrix}$$

where $\boldsymbol{P_i}$ is the multi-index $\{p_{i1}, \ldots, p_{i,n_i}\}$ and $\oplus \boldsymbol{Z}(\boldsymbol{P_i}, n_i)$ is a shorthand for the full direct sum. We can further reduce the notational complexity by letting $\boldsymbol{q_i} \equiv (\boldsymbol{P_i}, n_i)$. Hence, a cortical can with thalamic connections can be represented by the compact notation $\oplus \boldsymbol{Z}(\boldsymbol{q_i})$ and the cortical column by $\sum_{i=1}^{6} \oplus \boldsymbol{Z}(\boldsymbol{q_i})$. Each input $\boldsymbol{I}$ into a column is also represented as a direct sum $\boldsymbol{I} = \sum_{i=1}^{6} \oplus \boldsymbol{Z}(\boldsymbol{u_i})$ where $\boldsymbol{u_i} \equiv (\boldsymbol{U_i}, m_i)$ with $\boldsymbol{U_i}$ the multi-index associated with the direct sum decomposition of the input and m_i the number of terms in the input direct sum.

Let's consider a simple example. Let $\boldsymbol{I} = \boldsymbol{Z_3} \oplus \boldsymbol{Z_5} \oplus \boldsymbol{Z_4}$. This BN can be implemented in MATLAB by writing a function which takes a given element of $\boldsymbol{Z_{p_i}}$ and computes its corresponding value. Consider the code of `bn`.

Listing 18.1: **The Betti Node Structure**

```
function [Q,z] = bn(p)
% p is an integer vector used for the sum Z mod p(i)
[m,n] = size(p);
z = cell(1,n);
for j = 1:n
  x = linspace(0,p(j)-1,p(j));
  z{j}(1,:) = x;
end
L = 0;
for i = 1:n
  k = n-i;
  L = L + power(2,k)*(p(i)-1);
end
Q = log2(L/power(2,n-1));
end
```

The code takes the defined direct sum $\oplus \boldsymbol{Z_{p_i}}$ and determines the output of the BN using a simple equation. We want the output due to the maximum values of each $\boldsymbol{Z_{p_i}}$ to be scaled so as to add up to 1. Thus,

$$\frac{(p_1 - 1)}{2^\ell} + \frac{(p_2 - 1)}{2^\ell + 1} + \ldots + \frac{(p_n - 1)}{2^\ell + n - 1} = 1$$

This implies

$$L = 2^{n-1}(p_1 - 1) + 2^{n-2}(p_2 - 1) + \ldots + 2^1(p_{n-1} - 1) + (p_n - 1) = 2^{\ell+n-1}$$

Thus, we want $\ell = \ln_2(L/2^{n-1})$. These are the equations implemented in `bn`. Note we return ℓ as `Q` and we return the possible choices for each $\boldsymbol{Z_{p_i}}$ in the cell `z`. For our example, we would have

Listing 18.2: **A 3,5,4 BN**

```
>> [Qp,zp] = bn(p);
>> zp
zp =
{
  [1,1] =
     0   1   2
  [1,2] =
     0   1   2   3   4
  [1,3] =
     0   1   2   3
}
```

We then want to compute all the possible values for a given BN. We do this in `bnsigma`. There are many index possibilities. For $\oplus \boldsymbol{Z_{p_i}}$ there are $N = p_1 \cdot p_2 \cdots p_n$ such choices which is a large number potentially. We typically assume our direct sum decompositions are limited to a small number of summands and therefore have a reasonable N. Of course, the direct sum model is determined by the biology and so in principle we cannot be sure of this. The code uses the functions `vectimesvec` and `celltimesvec` to compute N. The function `vectimesvec` creates a cell which holds all the possible index choices coming from two vectors. So if the vectors were size 5 and 7, this would create a cell structure of 35 items. If the BN has three summands or more, the remaining calculations use `celltimesvec` to take the last computed cell and find all the possible index values for that cell and the next vector. The code is organized as follows:

- In the `z` returned from `bn`, for the first summand possibilities, z_i, compute and store $w_1 = z_1/(2^Q)$. Then for the second set of possibilities, compute $w_i = z_i/(2^{Q+i-1})$. This is stored in the cell `w`.
- Compute all possible index choices using functions `vectimesvec` and `celltimesvec` and store in the cell `Address`.
- Compute all possible w choices using functions `vectimesvec` and `celltimesvec` and store in the cell `A`.
- Using `A`, set up a structure `BN` with fields `Address`, `components` and `value` where `value` is the sum of the vector of values in a given w index.
- Then sort these values and return the sorted structure as `BSN`.

Listing 18.3: **The Betti Node Values**

```
function [N,max,w,A,v,Address,BN,BSN] = bnsigma(p,z,Q)
% p is an integer vector used for the sum Z mod p(i)
[m,n] = size(p);
value = @(u) sum(u);
w = cell(1,n);
for i=1:n
  Base = Q+i-1;
  w{i} = z{i}/power(2,Base);
end
s=[];
N = p(1);
s = [s,w{1}(p(1))];
for i=2:n
   N *= p(i);
   s = [s,w{i}(p(i))];
end
max = value(s);
% create addresses
x = linspace(0,p(1)-1,p(1));
y = linspace(0,p(2)-1,p(2));
% create first address cell
Address = vectimesvec(x,y);
for i = 3:n
  x = linspace(0,p(i)-1,p(i));
  baddress = celltimesvec(Address,x);
  Address = baddress;
end
% create BN values
V = zeros(1,N);
x = w{1};
y = w{2};
% create first value cell
A = vectimesvec(x,y);
for i = 3:n
  x = w{i};
  B = celltimesvec(A,x);
  A = B;
end
% create struct to hold addresses and values
BN = struct();
for i = 1:N
  BN(i).address = Address{i};
  BN(i).components = A{i};
  BN(i).value = value(A{i});
  V(i) = value(A{i});
end
[v,I] = sort(V);
for i = 1:N
  BSN(i) = BN(I(i));
end
end
```

Applying this to our example, we have

Listing 18.4: **345 BN Values**

```
[Qp,zp] = bn(p);
[Np,maxp,wp,Ap,vp,Addressp,BNp,BSNp] = bnsigma(p,zp,Qp);
% number of possibilities
>> Np
```

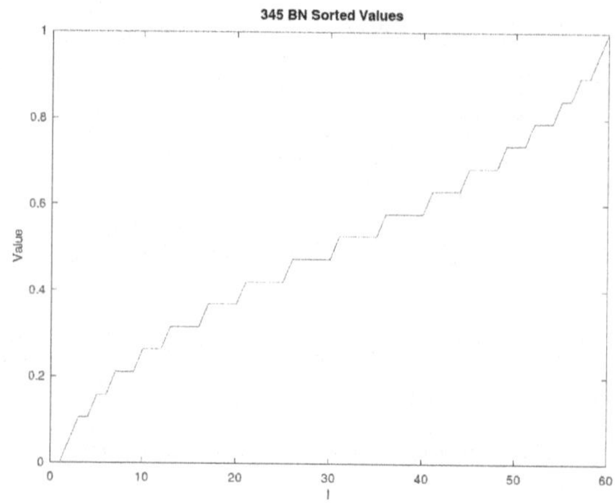

Figure 18.1: The 354 BN.

```
Np =  60
% w possibilities
>> wp
wp =
{
  [1,1] =
     0.00000    0.21053    0.42105
  [1,2] =
     0.00000    0.10526    0.21053    0.31579    0.42105
  [1,3] =
     0.00000    0.05263    0.10526    0.15789
>> BNp
BNp =
  1x60 struct array containing the fields:
    address
    components
    value
% for example
>> BNp(45)
ans =
  scalar structure containing the fields:
    address =
       2   1   0
    components =
       0.42105    0.10526    0.00000
    value =  0.52632
}
```

We return the sorted values in $\mathbf{vp}$, which we can plot as shown in Figure 18.1. Assume this is the input to can one which has representation $\boldsymbol{Z_5} \oplus \boldsymbol{Z_5} \oplus \boldsymbol{Z_5} \oplus \boldsymbol{Z_7}$. Convert the input into the integer vector $[\alpha, \beta, \gamma] \in \boldsymbol{Z_3} \oplus \boldsymbol{Z_5} \oplus \boldsymbol{Z_4}$. Then convert the integer vector into a vector as follows:

$$[\alpha, \quad \beta, \quad \gamma] \quad \rightarrow \quad \alpha(3) + \beta(5) + \gamma(4) = M$$

For example

$$\begin{aligned} [1, \quad 3, \quad 1] \quad &\rightarrow \quad 1(3) + 3(5) + 1(4) = 22 \\ [0, \quad 3, \quad 1] \quad &\rightarrow \quad 0(3) + 3(5) + 1(4) = 19 \end{aligned}$$

Note $0 \leq M \leq 6 + 20 + 12 = 38$. We can invert this process to convert any integer input M into an integer vector as well that is in can one: $M = [u, v, w, z] \in \boldsymbol{Z_5} \oplus \boldsymbol{Z_5} \oplus \boldsymbol{Z_5} \oplus \boldsymbol{Z_7}$ using integer division of say m by n written as $m|n$ is a whole number plus a remainder; i.e. if $m/n = s+t$ where t is less than n. We will write this as $m/n = s + rt$ for clarity. For example,

$$38 \quad = \quad [u, v, w, z] \in \boldsymbol{Z_5} \oplus \boldsymbol{Z_5} \oplus \boldsymbol{Z_5} \oplus \boldsymbol{Z_7} = (u + v + w)(5) + z(7)$$

We assume we want $u + v + w = 7$ so that the higher levels of the can are populated to reflect the fact that the high levels of the input are populated. The remainder is then put into the last slot; i.e. $z = 3$. The choices are many: i.e. 18 are possible.

$$\begin{bmatrix}4\\3\\0\\3\end{bmatrix}, \begin{bmatrix}4\\1\\2\\3\end{bmatrix}, \begin{bmatrix}4\\2\\1\\3\end{bmatrix}, \begin{bmatrix}4\\0\\3\\3\end{bmatrix}, \begin{bmatrix}3\\4\\0\\3\end{bmatrix}, \begin{bmatrix}3\\3\\1\\3\end{bmatrix}, \begin{bmatrix}3\\2\\2\\3\end{bmatrix}, \begin{bmatrix}3\\1\\1\\3\end{bmatrix}, \begin{bmatrix}3\\0\\4\\3\end{bmatrix}$$

$$\begin{bmatrix}2\\4\\1\\3\end{bmatrix}, \begin{bmatrix}2\\3\\2\\3\end{bmatrix}, \begin{bmatrix}2\\2\\3\\3\end{bmatrix}, \begin{bmatrix}2\\1\\4\\3\end{bmatrix}, \begin{bmatrix}1\\4\\2\\3\end{bmatrix}, \begin{bmatrix}1\\3\\3\\3\end{bmatrix}, \begin{bmatrix}1\\2\\4\\3\end{bmatrix}, \begin{bmatrix}0\\4\\3\\3\end{bmatrix}, \begin{bmatrix}0\\3\\4\\3\end{bmatrix}$$

We assume the choices that populate the higher levels of can one are the only ones that lead to useful outputs. Hence, the top is 4 or the level from the top is 4. This leads to the 8 possibilities:

$$\begin{bmatrix}4\\3\\0\\3\end{bmatrix}, \begin{bmatrix}4\\1\\2\\3\end{bmatrix}, \begin{bmatrix}4\\2\\1\\3\end{bmatrix}, \begin{bmatrix}4\\0\\3\\3\end{bmatrix}, \begin{bmatrix}3\\4\\0\\3\end{bmatrix}, \begin{bmatrix}2\\4\\1\\3\end{bmatrix}, \begin{bmatrix}1\\4\\2\\3\end{bmatrix}, \begin{bmatrix}0\\4\\3\\3\end{bmatrix}$$

The 8 can one outputs then are used as inputs to can two and so on. This leads to a combinatorial explosion. In general, if the input generates M_i outputs from the ith can, there are $M_1 M_2 M_3 M_4 M_5 M_6$ output possibilities from the cortical column.

Therefore, if the outputs of the graph were to be used to match particular targets (as they would in a supervised learning task) or if they were to be chosen to enhance correlations between inputs and outputs (as they would in an unsupervised learning task using Hebbian learning approaches), the large number of possibilities we have derived in the discussion above indicates that there are many choices to find pseudo-optimal solutions. For example, if we had two inputs $\boldsymbol{I_1} = \begin{bmatrix}2, & 4, & 3\end{bmatrix}$ from $\boldsymbol{Z_3} \oplus \boldsymbol{Z_5} \oplus \boldsymbol{Z_4}$ and $\boldsymbol{I_2} = \begin{bmatrix}1, & 5, & 2\end{bmatrix}$ from $\boldsymbol{Z_6} \oplus \boldsymbol{Z_7} \oplus \boldsymbol{Z_3}$, we have

- $\boldsymbol{I_1}$ gives a collection of outputs from the column.
- $\boldsymbol{I_2}$ is an input into each of the outputs above.

We could reverse this process and find the column outputs for input $\boldsymbol{I_1}$ given input $\boldsymbol{I_2}$.

Homework

Exercise 18.1.1 *Work out* $2-5-7$ *Betti node values by hand and by MATLAB.*

Exercise 18.1.2 *Assume the input comes from a* $2-5-7$ *Betti node. Follow the discussion in this section to convert the input into an integer which is then converted into an element of* $\boldsymbol{Z_4} \oplus \boldsymbol{Z_4} \oplus \boldsymbol{Z_5}$*. Do this in MATLAB also.*

Exercise 18.1.3 *Assume the input comes from a* $2-5-7$ *Betti node and goes into can one, which has a representation, which is a* $5-4-3$ *Betti node.*

- *Convert all possible inputs* $\begin{bmatrix}\alpha, & \beta, & \gamma\end{bmatrix}$ *into integers.*
- *Convert the integers above into a set of possible elements of the* $5-4-3$ *Betti node. Assume the choices that populate the higher levels are the only ones that lead to useful outputs. Find all these choices.*

Exercise 18.1.4 *Assume the input comes from a* $4-3-3$ *Betti node and goes into can one, which has a representation, which is a* $3-4-6$ *Betti node.*

- *Convert all possible inputs* $\begin{bmatrix}\alpha, & \beta, & \gamma\end{bmatrix}$ *into integers.*
- *Convert the integers above into a set of possible elements of the* $3-4-6$ *Betti node. Assume the choices that populate the higher levels are the only ones that lead to useful outputs. Find all these choices.*

18.2 The Asynchronous Immune Graph Model

For our models here, we will assume the general asynchronous message passing computational graph architecture with signal content and edge and node processing discussed in Chapter 16. However, we will use this model specialized for a state of (u_n, v_n) instead of the (m_n, E_n). The next step is to choose state pairs (u_n, v_n) and signals S_n with content c_n and associated edge and node processing functions $\sigma_n^r(S_n, c_n)$ and $\sigma_n^s(S_n, c_n)$ relevant for our modeling. Eventually, a model of LRRK2 mutation, cytokine signaling and gut biome interaction within the brain is also needed and we have developed a general abstract understanding of how to do this in Chapter 17. In a cognitive model, we are interested in how cytokine signals interact with a functioning simple brain model to shape the cortical-thalamic loop. Of course, there are other effects of cytokine signal cascades. For example, two important ones are as follows:

- LRRK2 mutations can cause problems that might lead to neural processing defects such as loss of dopamine and a resulting diagnosis of Parkinson's disease.
- Inflammatory molecules originating in gut biome and gut immune interactions with antigen can cross the blood-brain barrier causing an enhanced inflammatory state within the brain which could lead to autoimmune events.

To understand how all these complex interactions unfold, there is a need for asynchronous graph models as are discussed in Chapter 17 which use the theoretical structure of asynchronous message passing models for all of these interacting subsystems. That is a much larger model than the one we want to discuss here. Our purposes here are narrower as we want to focus on cortical-thalamic loop processing using cytokine signal models. This approach will give us a way to map cortical activity cleanly into a definition of cognitive state. In the immune model of Chapter 17, T-Cells interact with MHC cradles carrying peptide fragments of both self and antigen protein. The formation of the structure which determines the fate of a cell being probed by a T-Cell is called the immunosynapse, and this will be what we use as nodes in the model. For our purposes, for the states in these nodes, we will assume the affinity of the immunosynapse is modeled by $u_c = (1 + g_0)\mathscr{A}_{min}$, where $\mathscr{A}_{min}$ is the minimum affinity of an immunosynapse and g_0 is a small positive parameter. This plays the role of m_c in the message passing graph model. We also assume there is a maximum affinity, $v_c = g_1\mathscr{A}_{max}$, where $\mathscr{A}_{max}$ is the maximum affinity of an immunosynapse and g_1 is a small positive parameter. This plays the role of E_c in the message passing graph model in the immune model. We let $f(\mathscr{A}) = \mathscr{A}_{min} + (\mathscr{A}_{max} - \mathscr{A}_{min} tanh(\mathscr{A} - \mathscr{O}))$, which mapped a given affinity $\mathscr{A} \geq \mathscr{O}$ into the interval $[\mathscr{A}_{min}, \mathscr{A}_{max}]$. We let $u_n = f(\mathscr{A}_n)$, where $\mathscr{A}_n$ is the affinity of immunosynapse and $v_n = \mathscr{A}_{max} - f(\mathscr{A})$. We note $u_n + v_n = \mathscr{A}_{max}$. The symmetry breaking here creates inaccessible immunosynapses when the affinity is below $(1 + g_0)\mathscr{A}_{min}$ or above $(1 - g_1)\mathscr{A}_{max}$. Thus, T-Cells generating affinities that are too low are removed as well as T-Cells that generate affinities that are too tightly tuned to a given MHC cradle protein. This reflects what the immune system does. In addition to this choice of (u_n, v_n) we choose signal processing functions.

We need to choose a different pair (u_n, v_n) for the computational nodes which are now neurons, as the needs for the cognitive model are not the same as the immune model. However, the precise choice of (u_n, v_n) in the cognitive context is not important here. Note, in development, there is a massive die-off of synaptic connections, which in our theory would correspond to the formation of inaccessible nodes after symmetry breaking.

Following our usual asynchronous graph discussion, after a finite number of symmetry breakings, we are left with a graph in the upper left-hand corner which is processing paths of length at least three which includes the possibility of loops where a path starts and stops on the same node. Computations around closed paths allow for the possibility that classes of paths become equivalent. The exact nature of these equivalence classes of paths and the topology they determine on the subgraph $\mathcal{G}(\mathcal{V}_{\mathscr{V}_2}, \mathcal{E}_{\mathscr{V}_2 \rightarrow \mathscr{V}_2})$ will be determined by the receive and send processing functions. We denote the collection of signal processing capability by $\boldsymbol{C}$. Hence, the topology of the graph induced by the graph structure and processing functions can be represented by a triple $(\mathcal{C}, \mathcal{G}(\mathcal{V}, \mathcal{E}))$. We will assume this topology has the form

$$\begin{aligned}&\mathcal{G}(\mathcal{V}_{\mathscr{V}_2}, \mathcal{E}_{\mathscr{V}_2 \rightarrow \mathscr{V}_2})\\&= \boldsymbol{G_1}(\boldsymbol{V}_{\mathscr{V}_{21}}, \boldsymbol{E}_{\mathscr{V}_{21} \rightarrow \mathscr{V}_{21}}) \oplus \boldsymbol{G_2}(\boldsymbol{V}_{\mathscr{V}_{22}}, \boldsymbol{E}_{\mathscr{V}_{22} \rightarrow \mathscr{V}_{22}}) \oplus \ldots \oplus \boldsymbol{G_q}(\boldsymbol{V}_{\mathscr{V}_{2q}}, \boldsymbol{E}_{\mathscr{V}_{2q} \rightarrow \mathscr{V}_{2q}})\end{aligned}$$

To make the notation simpler, we will drop some of the details and write

$$\mathcal{G}(\mathcal{V}_{\mathscr{V}_2}, \mathcal{E}_{\mathscr{V}_2 \rightarrow \mathscr{V}_2}) = \boldsymbol{G_1}(\boldsymbol{V}_{21}, \boldsymbol{E}_{21}) \oplus \boldsymbol{G_2}(\boldsymbol{V}_{22}, \boldsymbol{E}_{22}) \oplus \ldots \oplus \boldsymbol{G_q}(\boldsymbol{V}_{2q}, \boldsymbol{E}_{2q})$$

which is shown in diagram form as

$$\begin{bmatrix} [\boldsymbol{G_1}(\boldsymbol{V}_{21}, \boldsymbol{E}_{21})] & [\mathbb{O}] & [\mathbb{O}] & [\mathbb{O}] \\ [\mathbb{O}] & [\boldsymbol{G_2}(\boldsymbol{V}_{22}, \boldsymbol{E}_{22})] & [\mathbb{O}] & [\mathbb{O}] \\ & & \ddots & \\ [\mathbb{O}] & [\mathbb{O}] & [\mathbb{O}] & [\boldsymbol{G_q}(\boldsymbol{V}_{2q}, \boldsymbol{E}_{2q})] \end{bmatrix}$$

In other words, the nodes in the index set $\mathscr{V}$ have a decomposition $\mathscr{V}_2 = \boldsymbol{V}_{21} \oplus \ldots \oplus \boldsymbol{V}_{2q}$ with corresponding edges only from $\boldsymbol{V}_{2i}$ to $\boldsymbol{V}_{2i}$ for all $1 \leq i \leq q$, allowing for no cross talk between summands. Finally, since all of this notation can easily be understood in context, we will simply write

$$\mathcal{G}(\mathcal{V}_{\mathscr{V}_2}, \mathcal{E}_{\mathscr{V}_2 \rightarrow \mathscr{V}_2}) = \boldsymbol{G_1} \oplus \boldsymbol{G_2} \oplus \ldots \oplus \boldsymbol{G_q}$$

and all of the direct sum decompositions are implied by the notation. Our next step is to try to understand the details of this decomposition for our cognitive system model.

Homework

Exercise 18.2.1 *Assume*

$$\mathcal{G}(\mathcal{V}_{\mathscr{V}_2}, \mathcal{E}_{\mathscr{V}_2 \rightarrow \mathscr{V}_2}) = \boldsymbol{G_1} \oplus \boldsymbol{G_2} \oplus \ldots \oplus \boldsymbol{G_q}$$

for $q = 3$. Assume $\boldsymbol{G_1}$ is 3×4, $\boldsymbol{G_2}$ is 4×5 and $\boldsymbol{G_3}$ is 2×3. Draw this graph showing the inaccessible nodes for various choices of the graphs $\boldsymbol{G_i}$.

Exercise 18.2.2 *Assume*

$$\mathcal{G}(\mathcal{V}_{\mathscr{V}_2}, \mathcal{E}_{\mathscr{V}_2 \rightarrow \mathscr{V}_2}) = \boldsymbol{G_1} \oplus \boldsymbol{G_2} \oplus \ldots \oplus \boldsymbol{G_q}$$

for $q = 4$. Assume $\boldsymbol{G_1}$ is 2×3, $\boldsymbol{G_2}$ is 5×4, $\boldsymbol{G_3}$ is 6×4 and $\boldsymbol{G_3}$ is 4×3. Draw this graph showing the inaccessible nodes for various choices of the graphs $\boldsymbol{G_i}$.

18.3 The Cortex-Thalamus Computational Loop

We now focus on a subset of cortical processing called the Cortex - Thalamus Loop (CT). We assume this gives rise to a graph model with decomposition $\mathcal{G}\left(\mathcal{V}_{\mathcal{V}_2}, \mathcal{E}_{\mathcal{V}_2 \to \mathcal{V}_2}\right) = \boldsymbol{G_1} \oplus \boldsymbol{G_2} \oplus \ldots \oplus \boldsymbol{G_q}$. To better understand this signal decomposition idea, let's look at a simple model of cortical processing using the neural circuitry building blocks called the On Center-Off Surround (**OCOS**), the Folded Feedback Pathway (**FFP**) and the Level Two-Three Circuit (**2/3**) as discussed in (R. Raizada and S. Grossberg (105) 2003). We will model the cortex as a sheet of columns each made of N cortical cans. A typical cortical can is shown in Figure 18.3(a). Neurons $N1$ and $N3$ in the **OCOS** receive input from the first order and higher order relays of the thalamus but these inputs are not shown. Neurons $N4$ and $N6$ of the **OCOS** send outputs back to the higher order relays of the thalamus. Hence, there is a computational feedback loop here. These outputs are not shown either. There are also connections to the motor cortex so that decisions based on the associations computed in this loop can activate movement and other things. The motor cortex outputs are from neuron $T2$ of the **2/3** circuit, which goes to Purkinje neurons in the cerebellum, whose output then goes to the motor cortex. Also, there are interneurons in the thalamus which connect to the motor cortex as well. None of these connections are shown here. Typically, a cortical column contains six cortical cans. We show a stacked architecture of three cans in Figure 18.3(b). We have added inter-can connections now showing the linkage between the **FFP** of a can to the **OCOS** neuron $N7$ of the can above it. These are in green. We also show feedback pathways from **OCOS** neuron $N7$ of a lower can to **2/3** neuron $T2$ in the can above which are in blue. This embedded figure is already almost too large to see well so you will have to use your imagination to extend this picture to five cans stacked on top of one another. However, the general feedback strategy of computation is still followed. Also, we do not show inhibitory neurons in these figures and instead focus on the wiring diagrams. In Figure 18.2, we show a very simplified portion of a thalamus model. We are using the ideas on how the thalamus contributes to information processing found in (S. Sherman (111) 2004) and (S. Murray Sherman and R. Guillery (110) 2006). This version of the thalamus circuitry consists of one neuron in the thalamic reticular nucleus (**TRN**), a higher order relay (**HO**), a first order relay (**FO**), an interneuron (**IN**) and inhibitory neurons $\boldsymbol{i_1}$ and $\boldsymbol{i_2}$. If you look closely, you can see a dynamical loop formed from the cortical can and the thalamus circuitry, the **CT** construct. Input from primary sensors enters through the first order relay where it may be processed into burst mode or essentially passed with minimal change to the **OCOS** neurons $N1$ and $N3$. Cortical can processed information is sent back to the thalamus through the **TRN** into the higher order relay **HO** and then **HO**'s output is fed back as input to **OCOS** neurons $N1$ and $N7$. This is a full feedback computational loop. The outputs of this loop are sent out for action which we indicate here with an edge to the motor cortex. Note neuron $T2$ in the 2/3 circuit of the can gets input from $N7$ of the **OCOS** and sends its output via the cerebellum to the motor cortex as well. This is an additional feedback loop as neuron $F2$ of the folded feedback pathway of the can feeds back to $N7$. We will associate each **CT** construct with a computation based on equivalence classes of the signals entering the **CT**. These signals can be decomposed into a set of equivalence classes which can be modeled by a simple finite group such as $\boldsymbol{Z_p}$ for a prime p. The computation of the **CT** circuit, then corresponds to choosing an integer value from $\boldsymbol{Z_p}$ or in general an entry from some finite group $\boldsymbol{G}$. For example, if the underlying group was $\boldsymbol{Z_3}$, the choices are from the set $\{0, 1, 2\}$. For a stack of three cans, each of the three **CT** circuits, $\{\boldsymbol{CT_1}, \boldsymbol{CT_2}, \boldsymbol{CT_3}\}$ would generate an integer result $\{i_1, i_2, i_3\}$ from the underlying $\boldsymbol{Z_{p_i}}$ finite group which would be identified with the cortical column computation. We can think of this computation as generating a matrix output $\boldsymbol{M}$ where instead of thinking of the output as i_i we instead place a $\boldsymbol{1}$ in the ith row of the column in $\boldsymbol{M}$ associated with this can. This matrix representation is the usual barcode representation of the computation. There would be a matrix barcode $\boldsymbol{M_{ij}}$ for each cortical column (i, j) in the cortex model and hence there is a matrix of barcodes which represents the computational output of the entire cortical model. The thalamus has a complicated wiring diagram and a simple version of it is shown in Figure 18.4. In this figure, we have shown the topographic map we find associated

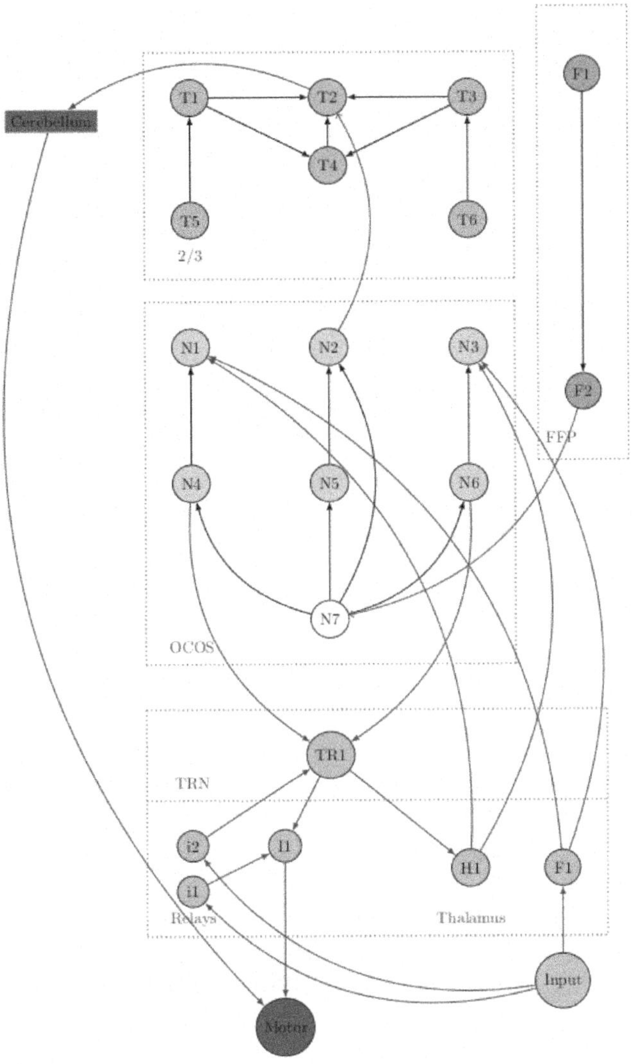

Figure 18.2: Cortical can/thalamus circuitry.

with the thalamus-cortex connections. Our simple model shows six higher order relays, $H1L$, $H1R$, $H2L$, $H42R$, $H3L$ and $H3R$. The pair $H1L$ and $H1R$ connect their outputs to layer 5 of can 1 in column 3 of the cortex. The pair $H2L$ and $H2R$ connect their outputs to layer 5 of can 1 in column 2 and column 4 of the cortex. Finally, the pair $H3L$ and $H3R$ connect their outputs to layer 5 of can 1 in column 1 and column 5 of the cortex. Each pair receives inputs from first order relays routed through a TRN neuron. There are three here: $U1$ accepts inputs from layer 4 of can 1 of column 3 and sends them to the pair $H1L$ and $H1R$ as well as to interneuron $I1$. $U2$ accepts inputs from layer 4 of can 1 of column 2 and can 1 of column 4 and sends them to the pair $H2L$ and $H2R$ as well as to interneuron $I2$. Finally, $U3$ accepts inputs from layer 4 of can 1 of column 1 and can 1 of column 5 and sends them to the pair $H3L$ and $H3R$ as well as to interneuron $I3$. This is quite simplified, of course, but shows some of the basic feedback circuits. Hence, to build a simplified model of cortex and thalamus processing, we would use the circuitry of Figure 18.4 with the understanding that the details of Figure 18.2 would also be used. We would then construct sensory cortex models for each

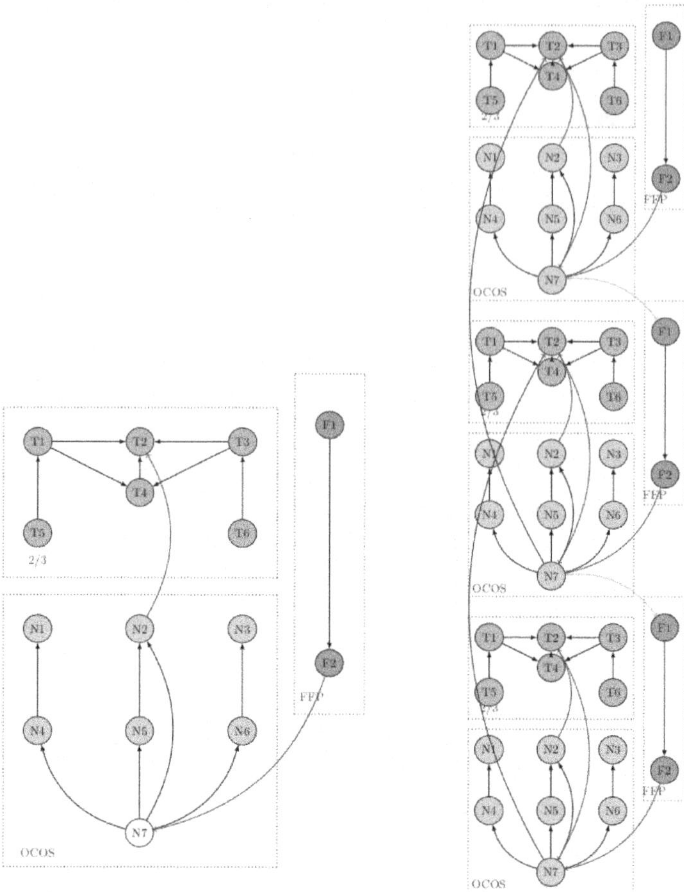

(a) A Typical cortical can. (b) A typical 3 can column.

Figure 18.3: Cortical architectures.

sense we wish to include and wire them to the thalamus. The thalamus structure we have shown has a building block of six pairs of higher order relays with three corresponding TRN neurons and some additional connections. A larger model would have the structure of Figure 18.5. We are showing direct sensory cortex to motor cortex connections, although in a simple model we could drop them and the associative cortex and look at a larger model of cortex and thalamus interactions as shown in Figure 18.6. We will be using the simplified model shown in Figure 18.6 for the rest of this paper. This model is informed by the details of the circuitry of Figure 18.4 and Figure 18.2. These ideas have been illustrated with cortical columns with at most three cans. The general cortical column of N cans as described by simple circuitry ((R. Raizada and S. Grossberg (105) 2003), (S. Murray Sherman and R. Guillery (110) 2006)) then corresponds to a $\sum_{i=1}^{N} \oplus \boldsymbol{G_i}$ for some choice of subgraph. The choice of finite group can be found using computational homology algorithms but we will not go into those details here. We also know that there is a debate at various levels about the existence of minicolumns in the cortex but we think there is compelling evidence for it as seen in (D. Buxhoeveden and M. Casanova (21) 2002), although a dissenting view is presented in (J. Horton and D. Adams (55) 2005). Hence, we are positing that the cortical column model is reasonable for our discussions. Each can therefore corresponds to a $\boldsymbol{G_i}$ computation and the structure of $\boldsymbol{G_i}$ comes from what is called a Betti decomposition in homology. For this reason, we will call the computations performed

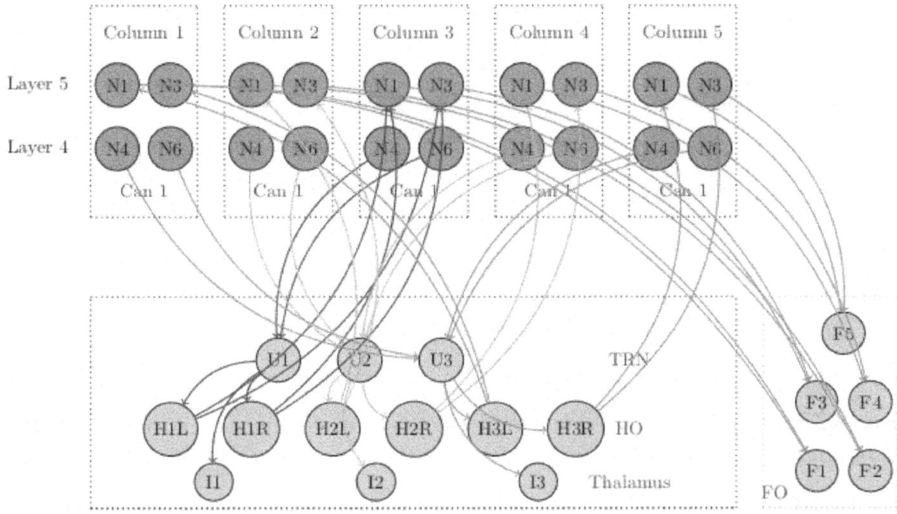

Figure 18.4: Wiring the cortex to the thalamus.

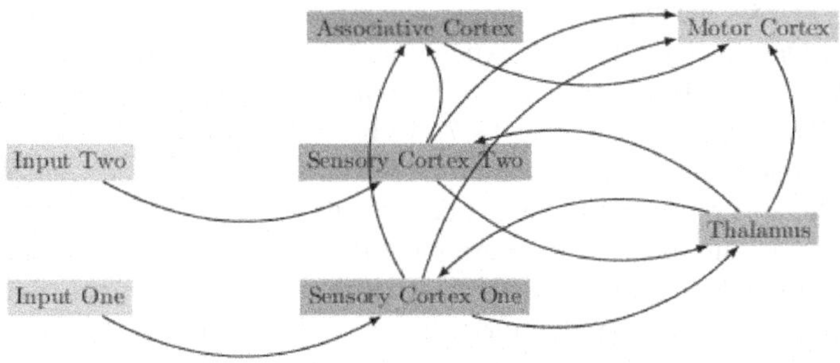

Figure 18.5: A Fuller model.

on the signal chains by the process above, **Betti** calculations and the engine which does this sort of nodal processing will be called a **Betti** node. Hence, we can replace an entire cortical column of nodal processing by either a **Betti** node or a sequence of Betti nodes, one for each can in the column. If a lateral input goes into a cortical can at level i, we would treat this as an input from a $\boldsymbol{G_i}$ value in one can to the $\boldsymbol{G_i}$ level of another can. If our finite groups are simply generated, when we choose to consider signal decompositions of the form $\sum_{i=1}^{N} \oplus \boldsymbol{G_i}$, we also must choose generators $\{y_i\}$ which should be informed by the neurobiology of the computations. Most of what we have discussed here is informed and specialized by a study of human neurobiology, but there are similar laminar information processing structures in the wasp and cockroach and so with modifications, these ideas can be applied there as well.

Homework

Exercise 18.3.1 *Draw Figure 18.5 when there are only two columns. Do this in complete detail.*

Exercise 18.3.2 *Draw Figure 18.5 when there are only two columns but each column has two cans. Use the standard connections for can* 1 *to can* 2 *in each case. Do this in complete detail.*

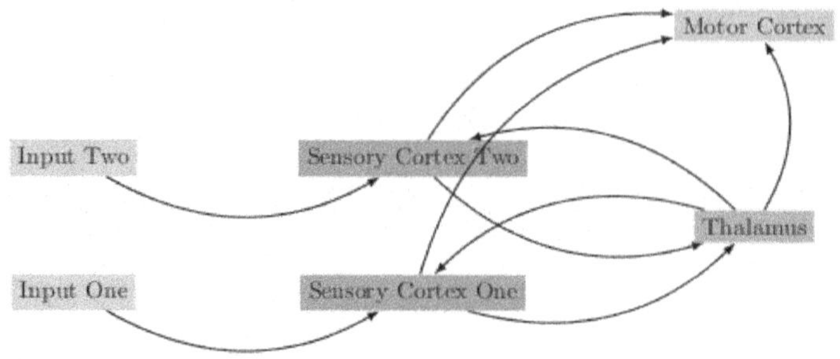

Figure 18.6: A Fuller model of only cortex and thalamus connections.

Exercise 18.3.3 *Draw Figure 18.5 when there are only two columns but each column has three cans. Use the standard connections for can i to can j in each case. Do this in complete detail.*

Exercise 18.3.4 *Do a connection count for a model with 6 cans per column based on Figure 18.5 and a cortical sheet of $M \times N$ columns.*

To give you a rough idea what these decomposition models might mean for a model of consciousness, let's assume the cortical-thalamus computational loop is represented by a simple column of six modules. In Figure 18.4 we showed a simple version of the cortical-thalamic loop processing. We only showed Can One for simplicity of exposition but there are also additional cans involved from each column. Hence Can Two through Can 6 would have these connections as well. To make the diagram more understandable, in Figure 18.7, we have dropped all the wiring linkages but retained some of the neuron structure. We show a block for both Can One and Can Six. If we assume Can One plus the Thalamus processing block are equivalent to a subgraph $\boldsymbol{\mathcal{G}_1}$ of simple structure and the additional cans plus thalamus processing are modeled by the subgraphs $\boldsymbol{\mathcal{G}_2}$ through $\boldsymbol{\mathcal{G}_6}$, we can see how a typical subgraph decomposition, which is determined by the signal environment the cognitive model graph is exposed to, can be identified. It clearly simplifies much of the low-level processing. At the bottom of the column is the thalamus input and the stacked cans of the columns all receive input from the can below as well as inputs from neighboring column stacks. We are abstracting and removing most of the complicated structure we have seen in our previous discussions, of course, but this thought experiment should give us some insight. The outputs of the stack model will all be modeled as bar codes determined by entries from $Z_6 \oplus Z_6 \oplus Z_6 \oplus Z_6 \oplus Z_6 \oplus Z_6$ for convenience. Remember, this is an abstract representation of a patch of protein bindings generated by a cytokine signal sequence. So given a barcode that is an input to the stack model, we will generate six more barcodes from the stack, one for each can. We can compute different measures of cortical activity for each stack. We have not really modeled cortical activity using the cortical-thalamic loop of course, but assuming this decomposition we can assign a random barcode output to each of the six can barcode computations. This gives a $\boldsymbol{\mathcal{G}_1} \oplus \ldots \oplus \boldsymbol{\mathcal{G}_6}$ output, which is a stack of six barcode matrices whose columns encode the calculation at each can level. For a barcode corresponding to a permutation of $\{1, \ldots, M\}$, this generates a matrix representation of size $M \times M$ for each can in the CT loop which consists of five separate cortical stacks tied together. If the cortical stack has N cans in it, the full matrix representation of the N generated barcodes is C of size $M \times M\,N$. One measure of cortical activity is then

$$V = \ln\left(\sum_{i=1}^{N} \sum_{j=(i-1)M+1}^{iM} C_{ij} 2^{i+j}\right)$$

We can then compute the cortical column values for an entire cortical sheet. Our conjecture is that

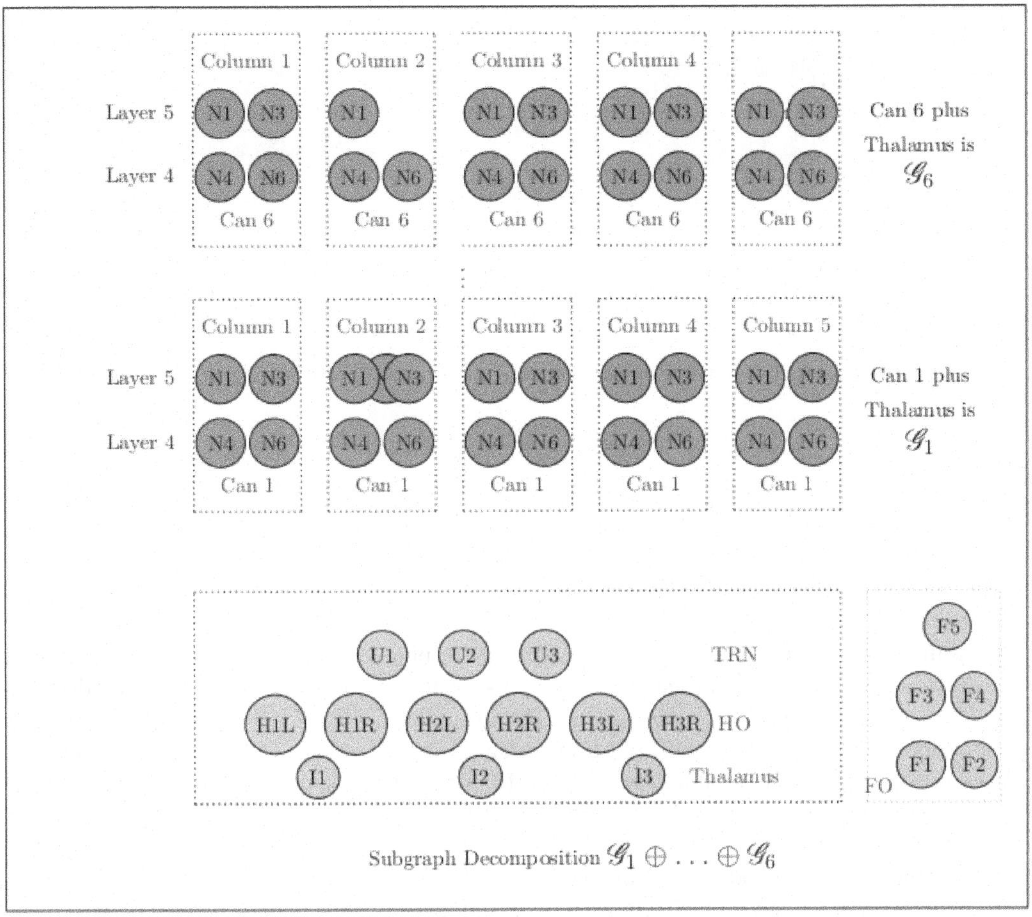

Figure 18.7: A possible cortical-thalamus loop subgraph decomposition.

if a cocktail of neurotoxins is injected into the brain it will attenuate the topmost level of the cortical column barcode outputs, leading to a change in how the computational modules of the brain are assembled into working behavioral patterns. This would explain the creation of iZombie states in both anesthesia and in parasitic wasp toxin injections into prey. We can calculate the results of a toxin injection and visualize it for a toxin centered at $(35, 30)$ of radius 20 injected into an 80×80 cortical sheet. Note the attenuation this causes is shown in Figure 18.8(a). It is harder to see the attenuation zones in the surface plot, but if you plot this yourself you can spin the surface and clearly see them. It is also possible to see what might happen if the cortex was subject to multiple attenuation events of smaller scope due to disease or multiple toxin injection events. For example, we can inject a toxin cocktail at 4 different centers for a fixed radius of 5. This plays the role of our random injections. For this example, the damaged cortical output is shown in Figure 18.8(b). You can clearly see the four damage areas, but as you can imagine, it would be hard to see them if the radius of damage was smaller. We have run this experiment with many injections with small radii of influence (say $1 \leq r \leq 5$) and what you see is a speckling of damage in the top of the cortical activity surface, which is hard to see in the surface plot even though it is there. It is interesting to note that any scenario that causes cortical damage to the higher levels of the barcode outputs with a small radius

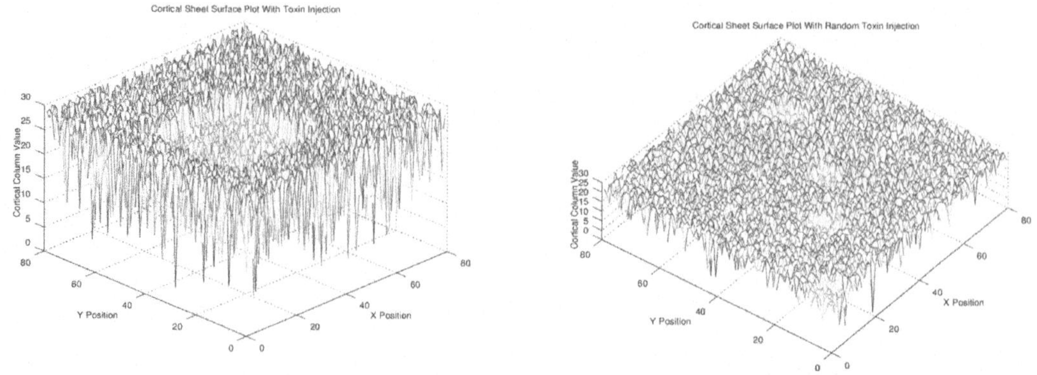

(a) 80×80 cortical sheet with toxin injection.

(b) 80×80 cortical sheet with random toxin injection

Figure 18.8: Damaged cortical areas.

of activity will be difficult to see but will probably have a profound impact on how computational modules in the brain are assembled.

We could also have made a more complicated subgraph decomposition by using individual decompositions for each cortical column consisting of six cans and then embed that into the decomposition $\mathcal{G}_1 \oplus \ldots \oplus \mathcal{G}_6$ we have been discussing. For example, in Figure 18.9, we show a standard cortical column with just three cans and no thalamus connections and we have indicated how each can would give rise to a subgraph decomposition. In this picture, we do not show any of the connections to the thalamus which are really there, of course. The cortical surface values would be computed in a similar manner and we would obtain similar results for a toxin injection. Building the model which replaces the CT loops with this sort of processing is complicated but straightforward. Hence, we are confident we are on the right track with this model. This approach does provide insight into why a toxin injection or an anesthetic drug injection generates a change in consciousness. **Hence, we posit that reasonable measure of consciousness state is related to the presence of nonzero components in the can six subgraph representations of cortical activity. The collection of such component activity over the entire cortical sheet encodes how the computational modules of the brain are assembled into higher level constructs**. It is also clear from our discussions above that we are predicting that small radii disturbances for whatever reason should comprise the level of consciousness in an animal host which is an interesting thought.

Homework

Exercise 18.3.5 *If the cortical stack has N cans in it, the full matrix representation of the N generated barcodes is C of size $M \times M\ N$. We have posited a measure of cortical activity is $V = \ln\left(\sum_{i=1}^{N} \sum_{j=(i-1)M+1}^{iM} C_{ij} 2^{i+j}\right)$. If we alter C_{ij} by one unit (this is complicated, of course, by the Z_p nature of these values), compute ΔV. This is a type of partial derivative, $\frac{\partial V}{\partial C_{ij}}$.*

Exercise 18.3.6 *These experiments are done with the MATLAB code we have discussed. First, calculate the results of a toxin injection and visualize it for a toxin centered at $(15, 40)$ of radius 10 injected into a 100×80 cortical sheet and graph the result, then rerun the experiment with two injections of radii 5 centered at $(25, 35)$ and $(75, 15)$ and graph the results.*

Exercise 18.3.7 *These experiments are also done with the MATLAB code we have discussed. First,* *calculate the results of a toxin injection and visualize it for a toxin centered at* $(35, 20)$ *of radius* 30 *injected into an* 80×80 *cortical sheet and graph the result, then rerun the experiment with two injections of radii* 10 *centered at* $(35, 15)$ *and* $(60, 35)$ *and graph the results.*

18.4 Cortical Representation and Cognitive Models

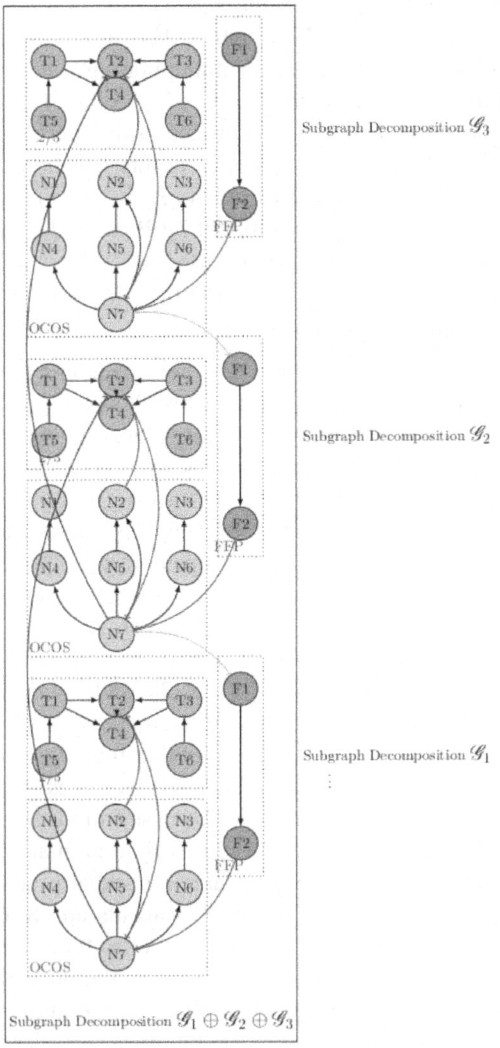

Figure 18.9: 3 can Betti decomposition.

Let's summarize our discussions so far. First, note whether the brain model is from a small animal such as a ctenophore, a spider, or a more complex organism such as a squid or a human, there is a lot of commonality in how a functioning brain takes raw sensory input and transforms it into outputs which help the organism thrive and perhaps develop consciousness. Forming associations is probably key to this process and hence models of consciousness are closely tied to models of sensor fusion. We have been taking the position that each can in a cortical column of N cans corresponds to $\sum_{i=1}^{N} \oplus \boldsymbol{G_i}$, where $\boldsymbol{G_i}$ can be a subgraph or a group. The mathematical structure of $\boldsymbol{G}$, as developed by our asynchronous graph approach, develops from signals, edge processing and nodal processing functions and could be modeled as a manifold or some other type of topological space. The way different submodules of $\boldsymbol{G}$ assemble into this structure might then be explored by tools that can detect separate components in these topologies. Each can, therefore, corresponds to a $\boldsymbol{G_i}$ computation; in essence, the outer node calculations correspond to finding the modded-out subgroup and the inner node calculations give the particular equivalence class. We have already discussed algorithms to parse an input signal chain into the form $(x_1, x_2, ..., x_N)$ where each $x_i \in \boldsymbol{G_i}$ and shown how to take a $\boldsymbol{G_i}$ input to $\boldsymbol{G_{i+1}}$ and how to deal with the outputs from each $\boldsymbol{G_i}$ as well. In Figure 18.9 we show the subgroup/subgraph structure clearly. You can compare this to the details we showed in Figure 18.4 for a portion of a cortical sheet too.

As we have mentioned, we model brain networks, human or otherwise, using a message passing/signal enabled asynchronous graph model. We note brain models connect computational nodes to other computational nodes using edges between the nodes. Although computation is asynchronous, a global time clock of reports allows us to compare snapshots of the asynchronous graph between report time t and $t+1$. We can therefore think of a useful brain model as a sequence of such graphs of nodes and edges. For simulation purposes, this means there is a finite sequence of times, $\{t_1, t_2, \ldots, t_n\}$ and associated graphs $\mathcal{G}_i(\mathcal{V}_i, \mathcal{E}_i)$, where the subscript i denotes the time point t_i. Between these times, the graph has fixed connectivity and we can use a variety of tools to train the graph to meet input to output objectives. At each time point, we can therefore build our cognitive model using many different approaches (e.g., (S. Sherman (111) 2004) and (K. Friston (63) 2005)). In the graph context, the update equations for a given node are then given as an input/output pair. For the node $\boldsymbol{N_i}$, let $\boldsymbol{y_i}$ and $\boldsymbol{Y_i}$ denote the input and output from the node, respectively. Then, we use the serial update equations, $\boldsymbol{y_i}(t+1) = \boldsymbol{I_i} + \sum_{j \in \boldsymbol{\mathcal{B}}(i)} \boldsymbol{E_{j \rightarrow i}}(t)\, \boldsymbol{Y_j}(t)$ where $\boldsymbol{Y_i}(t+1) = \boldsymbol{\sigma_i}(t)(\boldsymbol{y_i}(t))$ and $\boldsymbol{I_i}$ is a possible external input, which are perfectly adequate for manipulating a static snapshot of the graph even though the computation is asynchronous. Finally, $\boldsymbol{\mathcal{B}}(i)$ is the list of nodes which connect to the input side of node $\boldsymbol{N_i}$ and $\boldsymbol{\sigma_i}(t)$ is the function which processes the inputs to the node into outputs. Clearly $\boldsymbol{\sigma_i}(t)$ is related to our send and receive signal processing functions described in the asynchronous graph architecture with signals. We emphasize we replace the cortical modules or cortical clusters such as shown in Figure 18.7 in such a graph model by **Betti** nodes, allowing a cortical object to be replaced by $\sum_{i=1}^{N} \oplus \boldsymbol{G_i}$ computations. If the cortical sheet consists of an $m \times n$ array of cortical columns ϕ_{ij} each of which is modeled by the subgraph ϕ_{ij}^G, we assume the correspondence $\phi_{ij}^G \Longrightarrow \sum_{k=1}^{n_{ij}} \oplus \boldsymbol{G_{ij}^{\phi}}$ with lateral inputs between the cortical cans corresponding to links between Betti nodes $\sum_{k=1}^{n_{ij}} \oplus \boldsymbol{G_{ij}^{\phi}}$ at the same level. The last three dots below the last marked subgraph indicate that this column could be extended to more subgraphs. This allows us to construct a new way of doing information processing in a brain model. A new training algorithm would amount to using an iterative scheme to alter the $\sum_{k=1}^{n_{ij}} \oplus \boldsymbol{G_{ij}^{\phi}}$ choices in various ways including changing the generators to achieve goals. This leads to a number of conjectures. Cortical architectures are attempts by evolutionary processes to evolve a computational homology approach to association. In order to perform sensor fusion, also called association, animals have evolved various types of neural circuitry. For example, in birds there is the *nidopallium caudolaterale* (Herrold et al. (49) 2011) and the structure of the auditory cortex (Y. Wang and A. Brzozowsha-Prechtl and H. Karten (122) 2010) and the honeybee is sometimes used as a model of cognition (R. Menzel (104) 2012). The newly discovered genome of Ctenophores (L. Moroz and K. Kocot and M. Citarella and S. Dosung and T. Norekian and I. Povolotskaya and A. Grigorenko and C. Dailey and E. Berezikov and K. Buskely and A. Ptitsyn and D. Reshetov et al. (71) 2014) has made us understand that there was probably a second neural system evolution. In fact, a study of different organisms leads us to design principles for the neural systems that have arisen to solve various problems as discussed in (J. Sanes and S. Zipursky (61) 2010) and (S. Sprecher and H. Reichert (112) 2003). However all animals need to do sensor fusion and this is what we wish to develop a model for. For human cortical processing, we note the neural architectures for cortical processing as suggested by (R. Raizada and S. Grossberg (105) 2003) and others, use an on center-off surround type processing element to focus attention on the relevant aspect of the incoming signal. This can be thought of as follows: the outer part of the circuit determines the subgroup placement $\boldsymbol{G_i}$ with the focusing part of the circuit finding the particular equivalence class, giving the representation $[j] + \boldsymbol{G_i}$ where $[j]$ is the equivalence class the input belongs to at that level. The Betti node decomposition captures the idea that a signal gives rise to progressively more *complicated* representations. Hence, it is possible that modeling association with a Betti node approach could have value in understanding how associations are formed via computation. Indeed, this addresses the question of what the function of the cortical column is, as we posit the minicolumn structure which arose in many species as a solution to sensor fusion because the Betti node graph model is an efficient solution to the problem of finding associations. Note this gives a function to the minicolumn which answers the objections of (J. Horton and D. Adams (55) 2005). Hence, a signal $x \in \boldsymbol{S}$, the collection of all signals, going into a Betti node $\sum_{i=1}^{n} \oplus \boldsymbol{G_i}$ corresponds to the decomposition $(x_1, \ldots, x_n)$ with $x_i \in \boldsymbol{G_i}$. The signal decomposes into the following tree:

$$\begin{array}{ll} x_1 + \boldsymbol{G_1} & [x_1] \text{ coset of } \boldsymbol{S}/\boldsymbol{G_1}, \text{ level 1 representation} \\ x_2 + \boldsymbol{G_2} & [x_2] \text{ coset pf } \boldsymbol{G_1}/\boldsymbol{G_2}, \text{ level 2 representation, a chain of level 1 representations} \\ \vdots & \vdots \\ x_n + \boldsymbol{G_n} & [x_n] \text{ coset of } \boldsymbol{G_{n-1}}/\boldsymbol{G_n}, \text{ level n representation, a chain of level n-1 representations} \end{array}$$

Here, we are interpreting the computations in terms of group decompositions and equivalence classes. We could just as easily redraw this equation in terms of subgraphs also. We note the levels i in the above tree correlate with higher levels of associations. Also, this helps explain how understanding of *meaning* has discontinuous jumps as a signal input might suddenly cross the threshold to move to a new representation $(x_1, \ldots, x_n)$ with x_i changing to $x_i' \in \boldsymbol{G_i}$ due the new information. These changes are integer valued and so are essentially quantal, as the integer values we use are simply integer scales of information processing units we pick.

Higher organisms have occasionally developed cortical architectures that enable facile language processing. Assume for the moment there is a fundamental basic cortical architecture which can be modeled as

$$\sum \oplus \boldsymbol{G_i} \quad = \quad \boldsymbol{G_1} \oplus \ldots \oplus \boldsymbol{G_{i-1}} \oplus \boldsymbol{G_i}, \ldots, \oplus \boldsymbol{G_n}.$$

Then the cortical architecture that enables facile language processing (some birds, porpoises, whales etc.) could amount to the insertion of a language processing group $\boldsymbol{\mathcal{L}}$ into this given the decomposition

$$\sum \oplus \boldsymbol{G_i} \quad = \quad \boldsymbol{G_1} \oplus \ldots \oplus \boldsymbol{G_{i-1}} \oplus \boldsymbol{\mathcal{L}} \oplus \boldsymbol{G_i} \ldots \oplus \boldsymbol{G_n}.$$

The models presented here give us, at any instant of time, a graph architecture $\mathcal{G}(\mathcal{V}, \mathcal{E})$. The graph can be used to add as much detail as we wish and so from a certain perspective, if we construct a sequence of graphs $\mathcal{G}^k(\mathcal{V}_k, \mathcal{E}_k)$, we have a quantitative way of asking what happens as we move outward in this sequence. Do we approach a stable mathematical structure that represents a true brain model, $\mathcal{G}^\infty$? Each graph $\mathcal{G}^k(\mathcal{V}_k, \mathcal{E}_k)$ defines a topological space $\boldsymbol{\mathcal{B}}$, which in turn can be used to determine conditions under which the graph sequence is a Cauchy Sequence in $\boldsymbol{\mathcal{B}}$. If we can do this, we can begin to ask interesting questions about how the brain sequence converged states $\boldsymbol{\mathcal{B}}$ relate in different species. We note the graph models are quite complicated mathematically, as they are a mixture of edge processing functions and nodes that can be very complex.

Homework

Exercise 18.4.1 *Redraw Figure 18.9 for two columns.*

Exercise 18.4.2 *Redraw Figure 18.9 for six columns.*

Exercise 18.4.3 *Given a graph $\mathcal{G}(\mathcal{V}, \mathcal{E})$, can you think of a way to define the scalar multiplication operator of graphs? Can you think of a way to define a linear function on the set of all graphs? Can you think of a way to define a seminorm on the set of all graphs?*

18.5 What Does This Mean?

Let's start with a model of consciousness. The question is, what is a state of consciousness? There are clues to conscious states and their inactivation in how an **iZombie** state is induced into prey by the predatory wasps as well as the induction of such states in anesthesia. A model to gain insight into this kind of behavior modification must be capable of using abstractions of how both chemical and electrical signals are modified due to second messenger systems. To model anesthesia iZombies and the behavioral changes induced in a host by a parasitic wasp, the asynchronous graph models give us a set of tools that enable us to understand these events more quantitatively than before. Assume both anesthesia and the administration of neurotoxins in a wasp's sting correspond to a spatial constrained innervation of a subset $\boldsymbol{M_I}$ of the original cortical sheet $\boldsymbol{M}_0$. For convenience assume this subset is a submatrix of $\boldsymbol{M}_0$. This induces a coarse tiling of the sheet into subsets. There is a matrix barcode $\boldsymbol{M_{ij}}$ for each cortical column (i, j) in the cortex model and hence there is a matrix of barcodes which represents the computational output of the entire cortical model. The computations performed here are thus equivalent to assigning a signal to the components $\boldsymbol{A} = \{a_1, a_2, \ldots, a_6\}$ which correspond to the column's processing. The individual components a_i can be scalars or vectors depending on how we approach the modeling. If they are scalars, we posit normal operation of the organism's brain corresponds to the output signals of the cortical sheet having a critical percentage of the **higher** levels of each decomposition $\boldsymbol{A}_5^{\alpha\beta}$ and $\boldsymbol{A}_6^{\alpha\beta}$ having **nonzero** entries for each of the (α, β) sites from $\boldsymbol{M_I}$. The neurotoxin signal causes a drop in this critical percentage flooding the higher levels with zero entries. This corresponds to a high-level *rewiring* of the way cognitive modules interact to create behavior and could offer a reasonable first approximation to explaining Zombie creation. In general, the neurotoxin injection causes a cytokine family change $\boldsymbol{\Delta}\mathcal{C}_I$, which in turn creates a new topology $(\mathcal{C}_0 + \boldsymbol{\Delta}\mathcal{C}_I, \mathcal{G}(\mathcal{V}, \mathcal{E}))$. This also allows us to think of positive reversals for this action via injections leading to additional cytokine changes $\boldsymbol{\Delta}\mathcal{C}_R$. The new topology would then be $(\mathcal{C}_0 + \boldsymbol{\Delta}\mathcal{C}_I + \boldsymbol{\Delta}\mathcal{C}_R, \mathcal{G}(\mathcal{V}, \mathcal{E}))$. We can then model dysfunction at this point by allowing changes in midbrain processing as discussed in Section 16.5.6.

Our discussion of **iZombie** creation leads us to a blueprint for the construction of a useful cognitive model which is rooted in some ideas about consciousness. We can create a small scale human model with minimal neural submodules using emotionally labeled auditory and visual data. Although this model is based on normal human emotional responses to specific visual and auditory data and hence, we are more confident it represents a simple *normal* model of cognition, we can do more. We can use the discussions about the creation of *iZombie* states from the normal state to alter this model as needed to make its use as a normal brain model more probable. Hence, the created model is now checked to see if it can achieve the kind of state we would see in the anesthesia test events for consciousness. This kind of test can easily be abstracted into a general purpose computational suite. If the simple brain model we have discussed does not pass the test, then we create a second model which consists of a left brain, a right brain and a connecting corpus collosum module. We then test again. If the test is still not passed, we add asymmetric processing and test again. If we are still not successful, we add additional intermodule links (i.e. edges in the graph model) until the test is passed. We can move from one built model to another using the Laplacian-based training. For example, if we construct a model $\boldsymbol{\mathcal{G}_L(\mathcal{V}_L, \mathcal{E}_L)}$ for the left half of the brain, a model $\boldsymbol{\mathcal{G}_R(\mathcal{V}_R, \mathcal{E}_R)}$ and a model of the corpus callosum, $\boldsymbol{\mathcal{G}_C(\mathcal{V}_C, \mathcal{E}_C)}$, we can combine these modules to create a brain model $\boldsymbol{\mathcal{G}(\mathcal{V}, \mathcal{E})}$, which can be used to model problems with right and left half miscomunnication problems such as schizophrenia which could be very useful. The interconnections between the right and left half of the brain through the corpus callosum modules and neurotransmitter connections can be altered and the responses of the altered model to a normal brain model can be compared for possible insight. Of course, starting with a normal brain model is essential for these investigations.

The normal brain model thus achieved can then be altered by adding an external agent to mimic the neurotoxin inputs of the predatory wasp (massive and global in nature) and the more focused anesthesia inputs which alter brain state. We adjust the working normal model using training to make sure these inputs move from the chaotic state to an iZombie state. At this point, we can be reasonably confident we have a functional normal brain, model as not only does it achieve a normal working state under normal operating conditions, but we can also switch it into a lower consciousness state using external triggers similar to those we see in biology and surgery.

The models presented here, give us at any instant of time, a graph architecture $\mathcal{G}(\mathcal{V}, \mathcal{E})$ for the purpose of sensor fusion to build higher levels of meaning. The graph can be used to add as much detail as we wish and so from a certain perspective, if we construct a sequence of graphs $\mathcal{G}^k(\mathcal{V}_k, \mathcal{E}_k)$, we have a quantitative way of asking what happens as we move outward in this sequence. Do we approach a stable mathematical structure $\mathcal{G}^\infty$? Each graph $\mathcal{G}^k(\mathcal{V}_k, \mathcal{E}_k)$ defines a topological space $\mathcal{B}$, which in turn can be used to determine conditions under which the graph sequence is a Cauchy Sequence in $\mathcal{B}$. We note the graph models are quite complicated mathematically, as they are a mixture of edge processing functions and nodes that can be simple sigmoidal units to full Betti computational units. A full Betti node cortical architecture would consist of a full sheet of Betti nodes with corresponding connections.

Homework

Exercise 18.5.1 *Sketch out a simplified model of three subgraphs:* $\mathcal{G}_1(\mathcal{V}_1, \mathcal{E}_1)$ *for the gut biome,* $\mathcal{G}_2(\mathcal{V}_2, \mathcal{E}_2)$ *for the gut immune system and* $\mathcal{G}_3(\mathcal{V}_3, \mathcal{E}_3)$ *for the brain cognitive system. What kind of intermodular connections should be used? How does the LRRK2 pathway fit into this model?*

Exercise 18.5.2 *Sketch out a simplified model of three subgraphs:* $\mathcal{G}_1(\mathcal{V}_1, \mathcal{E}_1)$ *for the gut biome,* $\mathcal{G}_2(\mathcal{V}_2, \mathcal{E}_2)$ *for the gut immune system and* $\mathcal{G}_3(\mathcal{V}_3, \mathcal{E}_3)$ *for the brain cognitive system. This time, interpret them as systems of nonlinear differential equations. What kind of intermodular connections are used now? How does the LRRK2 pathway fit into this model?*

Exercise 18.5.3 *We have not discussed the gut biome at all, but it is known the gut biome could consist of* $10,000$ *interacting species. These species exist in several zones: no oxygen, low oxygen and high oxygen. Assume the number of species is* N *and* N *can be decomposed into* $N = N_1 + N_2 + N_3$ *where* N_i *denotes the number of species living in zone* i*. Assume each zone connects to the brain cognitive system via the LRRK2 pathways. Sketch out a simplified two graph model of this with the subgraph model for the gut biome setup as a direct sum of three subgraphs: one for each zone.*

Part VII

Summing It All Up

Chapter 19

Summing It All Up

We have now come to the end of this series of lecture notes. We have not covered all of the things we wanted to, but we view that as a plus: there is more to look forward to! However, we have written a series of five primers on analysis that we think would be a solid foundation for a practicing mathematician's and modelist's toolkit. This material can be learned from both a classroom course and self-study. Learning this material gives the fundamental core knowledge that makes it straightforward to pass any sort of preliminary or qualifying exam for advanced graduate study for those of you who need to do that. Also, we are mostly interested in developing abstractions of complicated science (biology, immunology, physics and so forth) from first principles and to do that we use a lot of mathematics and computation. So coding skills are equally important but probably the most essential thing is the ability to think clearly and to always understand the consequences of assumptions. Most of the colleagues we work with outside of mathematics do not know much of the abstract framework which we apply when we think about problems, so we spend a lot of time thinking of ways to help people outside of mathematics learn these sorts of abstractions. Also, we never learn enough from the courses we take and we always need to learn new things.

The texts we have written, although we have used personal notes in teaching a variety of analysis courses over the last 30 years or so, are carefully designed for self-study too. Much of the material is standard but there are also things we discuss off the beaten path that are fun and interesting. After you get through these five books you can learn even more analysis, algebra and topology! Here is a short recap of what you have learned in this set of texts.

Basic Analysis I: Functions of a Real Variable We study basic analysis in this text. This is the analysis a mathematics major or a student from another discipline who needs this background should learn in their early years. Learning how to think in this way changes you for life. You will always know how to look carefully at assumptions and parse their consequences. That is a set of tools you can use for your own profit in your future. In this text, you learn a lot about how functions which map numbers to numbers behave in the context of the usual ideas from calculus such as limits, continuity, differentiation and integration. However, these ideas are firmly rooted in the special properties of the numbers themselves and we take great pains in this set of notes to place these ideas into the wider world of mappings on sets of objects. This is done anecdotally, as we cannot study those things in careful detail yet, but we want you exposed to more general things. The text also covers ideas from sequences of functions, series of functions and Fourier Series. We teach a two-semester senior level analysis course out of this book. The book evolved from handwritten notes and I have been using this take on the material for about 30 years, so it has been tested in the classroom a lot. This volume is 590 pages, has 680 exercises and 1207 indexed entries.

Basic Analysis II: A Modern Calculus in Many Variables This book introduces you to more ideas from multidimensional calculus. In addition to classical approaches, we will also discuss things like 1-forms, a little algebraic topology and a reasonable coverage of the multidimensional versions of the single variable Fundamental Theorem of Calculus ideas. We also continue your training in the *abstract* way of looking at the world. We feel that is a most important skill to have when your life's work will involve quantitative modeling to gain insight into the real world. This is material we don't teach in a class now in most universities, so I think of this as a transitional set of notes that leads to graduate-level analysis ideas. Our target audience is almost anyone ready to learn more about vectors and multidimensional calculus. There is good coverage of the implicit and inverse function theorem and the use of vectors in applied work as well. We have tried to put a lot of interesting things in here. Most of this material has been taught as pieces of one or another course I have offered over the years, so it has also been tested. This book has 525 pages, 634 exercises and 609 indexed entries.

Basic Analysis III: Mappings on Infinite Dimensional Spaces This book introduces you to more about analysis by looking at spaces of various types. Spaces are collections of objects with certain properties and here we will focus on three types: metric, normed and inner product spaces. Our emphasis is on the new things we find when the underlying vector spaces are not finite dimensional. Hence, this text continues your training in the *abstract* way of looking at the world. As always, we feel that is a most important skill to have when your life's work will involve quantitative modeling to gain insight into the real world. In addition to that, we discuss some of the basic ideas from linear functional analysis: the Hahn - Banach Theorem, the open and closed graph theorems and the Baire Category Theorem. We also work through a lot of material involving dual and double dual spaces. When we teach this course, most students have not yet been exposed to measure theory and so our most important Hilbert spaces of functions are not as accessible as desired. To help with that, we build the real field from scratch and then complete the space of summable and square summable functions. We can't use ideas from Lebesgue integration at this point but we try hard to handle these metric completions in a way that makes them useful. There is a lot of nice thinking and analysis here. We have taught this material from these notes a fair bit, so there has been a lot of classroom testing. This book has 464 pages, 648 exercises and 802 indexed entries.

Basic Analysis IV: Measure Theory and Integration This book introduces you to concepts from measure theory and also, continues your training in the *abstract* way of looking at the world. We feel that is a most important skill to have when your life's work will involve quantitative modeling to gain insight into the real world. We cover the construction of measures from outer measure and the construction of outer measures from covering families and premeasures carefully so you can have a good command of these ideas. The book concludes with sections on typical exams for the material covered in the first and third books also, so you can check how your assimilation of all these ideas is going. Finally, we show you some examples of typical qualifying examinations in analysis for those of you going that route! We have also taught this material from these notes a fair bit, so there has been a lot of classroom testing. This book has 496 pages, 595 exercises and 790 indexed entries.

Basic Analysis V: Functional Analysis and Topology This is the text you have just finished which has added a thorough treatment of topology and applications of topological ideas such as degree theory and a good discussion of manifolds. This book has 582 pages, 615 exercises and 1165 indexed entries.

In this book, in addition to spending time on the mechanics of building a model to gives us insight into some aspect of biology or consciousness, we have also been allowing ourselves to talk about what we feel in the *art* of modeling. This *art* is mostly about finding the right level of abstraction to bring to bear which will engender the insight we seek. But how do we teach our students and readers how to do this? We are afraid much of the way that we learn how to abstract things is disappearing. The real question is how do we teach interdisciplinary things? We had some thoughts about this in (J. Peterson (57) 2014), which we think are important and we encourage all of you to read them when you get a chance.

An expert in science has the ability to give us a wonderful roadmap through complicated material. For example, in mathematics, each book we try to read is chock full of complicated things and when we first start out it is hard to assess what parts are the important ones. Learning how to outline material as we discussed earlier does help too, but so does an illustration. If you look at a photograph or a printout of a high-resolution biological event, it is hard to see all the structures that the expert knows are there. It is really helpful to have the expert extract from the real data, a clean black and white drawing with labeled parts. Of course this is an abstraction but it is useful! In mathematics, we do the same thing. For mappings in infinite dimensional spaces, we routinely draw a picture with a two-axis coordinate system: one axis is for the kernel or nullspace of the mapping and the other might be the range. And yes, the picture is *nonsensical* but it gives us *insight*! Illustrations coming from a fervent brew of scientists and illustrators working together to create such insight have been essentially lost due to budget cuts. We want to tell you a story of what has been lost. It is similar to what has been lost to not learning abstraction via integration by substitution or learning abstraction via the use of outlines and the search for the nugget in the rough in a document.

The story comes from the writing of the texts (Bullock and Horridge (13) 1965) and (Bullock and Horridge (14) 1965). The story comes from Horridge's introductory remarks in (Schmdt-Rhaesa et al. (113) 2016) titled '*How to write and invertebrate anatomy book*'. Bullock brought Horridge on board to help with the writing and soon after they hired 15 artists to do the illustrations for the book. Imagine that! Also they hired two people to help with the huge bibliography and when they needed it, they could find people at the University of California library who could translate scientific articles from Polish, Russian and so forth to English. Double imagine that! Such resources are hard to find now. The new publisher for the two books was going to be W. H. Freeman and over dinner, Freeman himself made sure the funding was available so that the books would be both scientifically accurate and works of art. That was important at the time. The printing was done on acid-free paper using cast metal plates. Our copy of this great set of books was bought used (both volumes for $40 or so as cast offs from the library of Ohio State) a few years ago. Not valued now despite the fact that it is filled with wonderful examples of black and white drawings where the master helps us by drawing out the abstractions we seek. Their ideas are brought to life by the talented artists they had the money to hire!

As Bullock says in his introduction to (Bullock and Horridge (13) 1965)

> Illustrations carry a special importance in a work of this kind and no pains have spared to provide them, both in number and in quality. Since this book is a statement of the status of the literature, we have whenever possible selected from the figures of original authors. Most of the drawings have been redrawn to improve the rendering. In special cases new figures have been created. Defrayment of the cost of preparing figures has been another major contribution of the funding agencies.

As we have discussed, the ability to hire qualified artists to make sure the important abstractions of the individual scientists are made clear was very important. Now switch to the introductory remarks in (Nieuwenhuys et al. (92) 1998). When their three-volume treatise was started, more resources were available. The authors note some of the most important supplementary elements that enabled the book to be made including

> The presence of an illustration department (founded by Prof. H. J. Lammers) with no less than ten competent illustrators led by Christian van Huijzen and later by Joep de Bekker. Almost all of the half-tome drawings, illustrating the gross anatomy of the brains of 17 species, and the India-ink drawings showing the microscopical structure at representative levels of the brains and spinal

> cords of 18 species, which form the central core of the present work, were fortunately prepared before stringent budget cuts essentially brought the illustration department to an end. The fact that one of the artists, Mr. J. P. M. Maas, long after his retirement and until the conclusion of the project, continued preparing high quality illustrations for the work, deserves a special mention.

The business of transferring the expert observations and expertise of the scientists rested firmly, in part, on the excellent service these artists provided. Such a collaboration is now gone and it is a tragedy of the first magnitude. We need this sort of art and science interaction even more today. Just more food for thought as it brings to mind the question of how do we generate such artistic insights now? What if we do not have artistic ability ourselves and cannot afford to hire the necessary artists to help? Also, notice that an artist had to donate his time and talent for free to the project to get it finished. The scientists were paid but he was not. Sad statement there and it shows the hurdles we face now.

One thing you should notice is that much of the mathematics we develop uses a building block point of view. In the development of extensions to the Riemann integral we learn how to build measures from core collections of sets. If this core collection is not *rich* enough, we still build a measure but it is less useful. The same thing happens in topology. We can build a topology on a set of objects by choosing a core collection of subsets we consider to be open. If this collection of open sets is not *rich* enough, we again can build a topology but it is less helpful. Remember, the goal of the topology is to find a good collection of subsets called open which make the functions we are interested in studying have reasonable smoothness. We also find we can build core collections of sets we can call open using the inverse images of critical sets of functions such as seminorms. We can also pick as our open sets the homeomorphic images of open sets in $\Re^n$, for some n, and develop an large theory of manifolds.

If all this is given to us, we can learn how to turn the mathematical crank and study the interactions between objects that are interesting to us when we try to model the real world. But the question that really interests us is how do we choose the mathematical structure for our model? Each choice leads to mathematical results about how our interactions behave that may or may not give us insight into the data we are trying to understand. We find this journey fascinating and we give you a taste of it in the last part on immunology and consciousness models.

Now that you are finished with this five-volume sequence on increasingly sophisticated ideas in mathematics, we hope you are poised to begin your own investigations into the models you wish to build to gain understanding. So enjoy your journey and get started!

Part VIII

References

References

[1] Cavanna A. and F. Monaco. Brain Mechanisms of Altered Conscious States During Epileptic Seizures. *Nature Reviews Neurology*, 5:267–276, 2009.

[2] A. Downey. *The Little Book of Semaphores: The Ins and Outs of Concurrency Control and Common Mistakes: Understanding Semaphores and Learning How to Apply Them.* SoHo Books, Lexington, KY, 2008.

[3] A. Gardet and Y. Benita and C. Li and B. Sands and I. Ballester and C. Stevens and J. Korzenik and J. Rioux and M. Daly and R. Xavier and D. Podolsky. LRRK2 Is Involved in the IFN-γ Response and Host Response to Pathogens. *The J. of Immunology*, 185:5577–5585, 2010.

[4] A. M. Kesson and Y. Cheng and N. J. C. King. Regulation of Immune Recognition Molecules by Flavivirus and West Nile Virus. *Viral Immunol*, 15:273–283, 2002.

[5] A. Owen and N. Schiff and S. Laureys. *A New Era of Coma and Consciousness Science*, pages 399–411. Elseveir, 2009. Editors S. Laureys et al.

[6] A. Rot and U. von Andrian. Chemokines in Innate and Adaptive Host Defense: Basic Chemokinese Grammar for Immune Cells. In *Annual Review Immunology*, volume 22, pages 891–928. Annual Reviews, 2004.

[7] J. Armstrong. *Programming Erlang Second Edition: Software for a Concurrent World.* The Pragmatic Bookshelf, Dallas, TX, 2013.

[8] S. Arnold, S. Osvath, R. Hall, N. J. C. King, and L. Sedger. Regulation of antigen processing and presentation molecules in West Nile virus-infected human skin fibroblasts. *Virology*, 324: 286–296, 2004.

[9] B. Deverman and P. Patterson. Cytokines and CNS Development. *Neuron*, 64:61–78, 2009.

[10] B. O'Neill. *Elementary Differential Geometry*. Academic Press, 1966.

[11] F. Barrett, S. Krimmel, R. Griffiths, D. Seminowicz, and B. Mathur. Psilocybin acutely alters the functional connectivity of the claustrum with brain networks that support perception, memory, and attention. *NeuroImage*, 218:1–10, 2020.

[12] G. Birkoff and O. Kellog. Invariant Points in Function Spaces. *Trans. American Math. Soc.*, pages 96–115, 1922.

[13] T. Bullock and G. Horridge. *Structure and Function in the Nervous System of Invertebrates: Volume 1*. W. H. Freeman and Company, 1965.

[14] T. Bullock and G. Horridge. *Structure and Function in the Nervous System of Invertebrates: Volume 2*. W. H. Freeman and Company, 1965.

[15] C. Camerer and J. Cohen and E. Fehr and P. Glimcher and D. Laibson. Neuroeconomics. In J. Kagel and A. Roth, editor, *Handbook of Experimental Economics*, volume 2, pages 153–216. Princeton University Press, 2016.

[16] C. de Bekker and M. Merrow and D. Hughes. From Behavior to Mechanism: An Integrative Approach to the Manipulation by a Parasitic Fungus (Ophiocordyceps unilateralis s. l.) of Its Host Ants (Camponotus spp. *Integrative and Comparative Biology*, 54(2):166–176, 2014.

[17] C. Wacher and M. Muller and M. J. Hofer and D. R. Getts and R. Zabaras and S. S. Ousman and F. Terenzi and G. C. Sen and N. J. C. King and I. L. Campbell. Coordinated Regulation and Widespread Cellular Expression of Interferon Stimulated Genes (ISG) ISG-49 and ISG-54 and ISG-56 in the Central Nervous system after Infection with Distinct Viruses. *J. Virology*, 81:860–871, 2007.

[18] T. Carbonaro, M. Johnson, and R. Griffiths. Subjective features of the psilocybin experience that may account for its self-administration by humans: a double-blind comparison of psilocybin and dextromethorphan. *Psychopharmacology*, 237:1–12, 2020.

[19] S. Crawford. Topology and Distributions. Master's thesis, Clemson University, Clemson, SC, 2003.

[20] D. Berwick and K. Harvey. LRRK2 Signalling Pathways: the Key to Unlocking Neurodegeneration? *Trends in Cell Biology*, 21:257–265, 2011.

[21] D. Buxhoeveden and M. Casanova. The Minicolumn Hypothesis in Neuroscience. *Brain*, 125: 935–951, 2002.

[22] D. Rossi and A. Zlotnik. The Biology of Chemokines and Their Receptors. In *Annual Review Immunology*, volume 18, pages 217–242. Annual Reviews, 2000.

[23] A. Davis, J. Clifotn, E. Weaver, E. Hurwitz, M. Johnson, and R. Griffiths. Survey of entity encounter experiences occasioned by inhaled N,N-dimethyltryptamine: Phenomenology, interpretation, and enduring effects. *Journal of Psychopharmacology*, 34:1–9, 2020.

[24] K. Deimling. *Nonlinear Functional Analysis*. Dover Publications, Inc., 2010.

[25] E. Harmon-Jones and P. Gable and C. Peterson. The Role of Asymmetric Frontal Cortical Activity in Emotion-related Phenomena: a Review and Update. *Biological Psychology*, 84: 451–462, 2010.

[26] E. Kreyszig. *Introductory Functional Analysis with Applications*. John Wiley and Sons NY, 1989.

[27] John W. Eaton, David Bateman, Søren Hauberg, and Rik Wehbring. *GNU Octave: version 5.2.0 manual: a high-level interactive language for numerical computations*, 2020. URL `https://www.gnu.org/software/octave/doc/v5.2.0/`.

[28] S. Egan, K. Weinersmith, S. Liu, R. Ridenbaugh, Y. Zhang, and A. Forbes. Description of a New Species of Euderus Haliday from the Southeastern United States (Hymenoptera, Chalcidoidea, Eulophidae): the Crypt-keeper Wasp. *ZooKeys*, 2017. URL `https://doi.org/10.3897/zookeys.645.11117`.

[29] F. Libersat and R. Gal. Wasp Voodoo Rituals, Venom-Cocktails, and the Zombification of Cockroach Hosts. *Integrative and Comparative Biology*, 54(2):129–142, 2014.

[30] F. Markopoulou. Conserved Quantities in Background Independent Theories. *arXiv.org: General Relativity-Quantum Cosmology*, arXiv.org:0703027v1 [hep-th]:1–11, 2007.

[31] F. Schlief and P. Tino. Indefinite Proximity Learning: A Review. *Neural Computation*, 27: 2039–2096, 2004.

[32] F. Fernadez, S. Straus, R. Sharpe, and Avil'es. Behavioural Modification of a Social Spider by a Parasitoid Wasp. *Ecological Entomology*, 2018. URL `https://doi.org/10.1111/een.12698`.

[33] Free Software Foundation. *GNU General Public License Version 3*, 2020. URL `http://www.gnu.org/licenses/gpl.html`.

[34] S. Fučik, J. Nečas, J. Souček, and V. Souček. *Spectral Analysis of Nonlinear Operators: Lecture Notes in Mathematics 346*. Springer, 1973.

[35] G. Carlsson. Topology and Data. *Bulletin (New Series) of the American Mathematical Society*, 46(2):255–308, 2009.

[36] G. Edgar. *Measure, Topology and Fractal Geometry*. Springer-Verlag, 1990.

[37] G. Mashour and E. LaRock. Inverse zombies, Anesthesia awareness, and the Hard problem of Unconsciousness. *Consciousness and Cognition*, 17:1163–1168, 2008.

[38] G. Simmons. *Introduction to Topology and Modern Analysis*. McGraw-Hill Book Company, 1963.

[39] G. Singh and F. Memoli and T. Ishkhanov and G. Sapiro and G. Carlsson and D. Ringach. Topological analysis of Population Activity in Visual Cortex. *Journal of Vision*, 8:1–18, 2008.

[40] R. Gaines and J. Mawhin. *Coincidence Degree and Nonlinear Differential Equations: Lecture Notes in Mathematics, 568*. Springer-Verlag, 1977.

[41] H. Edelsbrunner and D. Letscher and A. Zomorordian. Topological persistence and Simplification. *Discrete Computational Geometry*, 28:511–533, 2002.

[42] H. Tuominen. Analysis in Metric Spaces. *Unpublished Lecture Notes*, pages 1–87, 2014. URL `https://helituominen.files.wordpress.com/2014/01/luennot23.pdf`.

[43] H. Wekerle. Brain Autoimmunity and Intestinal Microbiota: 100 Trillion Game Changers. *Trends in Immunology*, 38(7):483–497, 2017.

[44] M. Haden and B. Woods. LSD Overdoses: Three Case Reports. *Journal of Studies on Alcohol and Drugs*, 81(1):115–118, 2020.

[45] M. Hakimi, T. Selvanantham, E. Swinton, R. Padmore, Y. Tong, G. Kabbach, K. Venderiva, S. Giradin, and D. Bulman et al. Parkinson's Disease-linked LRRK2 is Expressed in Circulating and Tissue Immune Cells and Upregulated Following Recognition of Microbial Structures. *J. Neural Transm.*, 118:795–808, 2011.

[46] F. Hébert. *Learn You Some Erlang for Great Good*. No Starch Press, San Francisco, CA, 2013.

[47] E. Heinz. An Elementary Analytic Theory of the Degree of Mapping in n-Dimensional Space. *J. Math. and Mechanics*, 8:231–247, 1959.

[48] R. Hermann. *Cartanian Geometry, Nonlinear Waves, and Control Theory, Part B*. Math. Sci. Press, 1980.

[49] C. Herrold, N. Palomero-Gallagher, B. Hellman, S. Kröner, C. Theiss, O. Güntürkün, and K. Zilles. The receptor architecture of the pigeons nidopallium caudolaterale: and avian analogue to the mammalian prefrontal cortex. *Brain Structure Function*, 216:239–254, 2011.

[50] E. Hewitt and K. Stromberg. *Real and Abstract Analysis*. Spring-Verlag, 1965.

[51] D. Hughes, S. Andersen, N. Hywel-Jones, W. Himaman, J. Billen, and Boomsma. J. Behavioral Mechanisms and Morphological Symptoms of Zombie Ants Dying from Fungal Infection. *BMC Ecology*, 2011. URL `http://dx.doi.org/10.1186/1472-6785-11-13`.

[52] I. Mauss and M. Robinson. Measures of Emotion: a Review. *Cognition and Emotion*, 23: 209–237, 2009.

[53] I. Smets and B. Fiddes and J. Garcia-Perez and D. He and K. Mallants and W. Liao and J. Dooley and G. Wang and S. Humblet-Baron and B. Dubois and A. Compston and J. Jones and A. Coles and A. Liston and M. Ban and A. Goris and S. Sawcer. Multiple Sclerosis Risk Variants Alter Expression of Co-stimulatory Genes in B cells. *Brain*, 141:786–796, 2018.

[54] J. Heinonen. Nonsmooth Calculus. *The American Mathematical Society Bulletin (New Series)*, 44(2):163–201, 2007.

[55] J. Horton and D. Adams. The Cortical Column: a Structure without a Function. *Phil. Trans. Royal Society B*, 360:837–862, 2005.

[56] J. Milosevic and S. Schwarz and V. Ogunlade and A. Meyer and A. Storch and J. Schwarz. Emerging Role of LRRK2 in Human Neural Progenitor Cell Cycle Progression, Survival and Differentiation. *Molecular Neurodegeneration*, 4:1–9, 2009.

[57] J. Peterson. Some Thoughts on BioMathematics Education. *Biology International*, 54:130–179, 2014.

[58] J. Peterson. *Calculus for Cognitive Scientists: Derivatives, Integration and Modeling*. Springer Series on Cognitive Science and Technology, Springer Science+Business Media Singapore Pte Ltd. 152 Beach Road, #22-06/08 Gateway East Singapore 189721, Singapore, 2016. URL `http://dx.doi.org/10.1007/978-981-287-874-8`.

[59] J. Peterson. *Calculus for Cognitive Scientists: Higher Order Models and Their Analysis*. Springer Series on Cognitive Science and Technology, Springer Science+Business Media Singapore Pte Ltd. 152 Beach Road, #22-06/08 Gateway East Singapore 189721, Singapore, 2016. URL `http://dx.doi.org/10.1007/978-981-287-877-9`.

[60] J. Peterson. Affinity and Avidity Models in Autoimmune Disease. *AIMS Allergy and Immunology*, 2(1):45–81, 2018. URL `http://dx.doi.org/10.3934/Allergy.2018.1.45`.

[61] J. Sanes and S. Zipursky. Design Principles of Insect and Vertebrate Visual Systems. *Neuron*, 66(1):15–36, 2010.

[62] J. Thëvenet and R. Gobert and R. van Huijsduijnen and C. Wiessner and Y. Sagot. Regulation of LRRK2 Expression Points to a Functional Role in Human Monocyte Maturation. *PLoS ONE*, 6:1–14, 2011.

[63] K. Friston. A Theory of Cortical Responses. *Phil. Trans. R. Soc. B.*, 360:815–836, 2005.

[64] K. Takasuka and T. Yasui and T. Ishigami and K. Nakata and R. Matsumoto and K. Ikeda and K. Maeto. Host Manipulation by an Ichneumonid Spider Extoparasitoid That Takes Advantage of Preprogrammed Web-building Behavior for its Cocoon Protection. *The Journal of Experimental Biology*, 218:2326–2332: doi:10.1242/jeb.122739, 2015.

[65] K. Weinersmith and Z. Faulkes. Parasitic Manipulation of Host's Phenotype, or How to Make a Zombie-An Introduction to the Symposium. *Integrative and Comparative Biology*, 54(2): 93–100, 2014.

[66] T. Kaczynski, K. Mischaikow, and M. Mrozek. *Computational Homology*. Springer, 2004.

[67] T. Konopka, F. Markopoulou, and S. Severini. Quantum Graphity: a Model of Emergent Locality. *arXiv.org: High Energy Physics-Theory*, arXiv.org:0801.0861v2 [hep-th]:1–14, 2008.

[68] E. Kozina, S. Sadasivan, Y. Jiao, Y. Dou, Z. Ma, H. Tan, K. Kodali, T. Shaw, J. Peng, and R. Smeyne. Mutant LRRK2 Mediates Peripheral and Central Immune Responses Leading to Neurodegeneration in Vivo. *BRAIN*, 6:1–17, 2018.

[69] W. Kurt. *Get Programming with Haskell*. Manning, 2018.

[70] L. Capuron and A. Miller. Cytokines and Psychopathology: Lessons from Interferon-α. *Biol. Psychiatry*, 56:819–824, 2004.

[71] L. Moroz and K. Kocot and M. Citarella and S. Dosung and T. Norekian and I. Povolotskaya and A. Grigorenko and C. Dailey and E. Berezikov and K. Buskely and A. Ptitsyn and D. Reshetov et al. The Ctenophore Genome and the Evolutionary Origins of Neural Systems. *Nature*, 510:109–120, 2014.

[72] J. Leray. Topologie et Équations Fonctionnelles. *Ann. Ecole. Norm. Sup. 3*, 51:45–78, 1934.

[73] J. Leray. Les Problėms Non Linéares. *Enseign. Math.*, pages 139–151, 1936.

[74] M. Lipovača. *Learn You a Haskell for Great Good*. No Starch Press, 2011.

[75] M. Barlow. Diffusions on fractals. In P. Bernard, editor, *Lectures on Probability Theory and Statistics: Ecole d'Eté de Probabilités de Saint-Flour XXV—1995*, pages 1–121. Springer Berlin Heidelberg, Berlin, Heidelberg, 1998. ISBN 978-3-540-69228-7. doi: 10.1007/BFb0092537. URL `http://dx.doi.org/10.1007/BFb0092537`.

[76] M. DoCarmo. *Differential Geometry of Curves and Surfaces*. Springer Verlag, 1976.

[77] M. Gemignani. *Elementary Topology: Second Edition*. Dover NY, 1972.

[78] M. Kubo and R. Nagashima and E. Ohta and T. Maekawa and Y. Isobe and M. Kurihara and K. Eshima and K. Iwabuchi and T. Sasaoka and S. Azuma and H. Melrose and M. Farrer and F. Obata. Leucine-rich Repeat Kinase 2 is a Regulator of B Cell Function, Affecting Homeostasis, BCR Signalling, IgA Production, and T1 Antigen Response. *J. of Neuroimmunology*, 292:1–8, 2016.

[79] M. Novak. *Evolutionary Dynamics: Exploring the Equations of Life*. Belknap Press, 2006.

[80] M. Peters. Actions of Cytokines on the Immune Response and Viral Interactions: An Overview. *Hepatology*, 23:909–916, 1996.

[81] M. Spivak. *Calculus on Manifolds*. Perseus Press, 1965.

[82] M. Spivak. *A Comprehensive Introduction to Differential Geometry: Volume One*. Publish or Perish, Inc., 1979.

[83] M. Spivak. *A Comprehensive Introduction to Differential Geometry: Volume Two*. Publish or Perish, Inc., 1979.

[84] M. Vela and M. Aris and M. Llorente and J. Garcia-Sanz and L. Kremer. Chemokine Receptor-specific Antibodies in Cancer Immunotherapy: Achievements and Challenges. *Frontiers in Immunology*, 6(12), 2015. URL `http://dx.doi.org/10.3389/fimmu.2015.00012`.

[85] MATLAB. *Version Various (R2010a)-(R2019b)*, 2018-2020. URL `https://www.mathworks.com/products/matlab.html`.

[86] J. Mawhin. *Topological Degree Methods in Nonlinear Boundary Value Problems*. CBMS Regional Conference Series in Mathematics, Vol. 40, 1979.

[87] J. Mawhin. Leray-Schauder Degree: A Half Century of Extensions and Applications. *Topological Methods in Nonlinear Analysis*, 14:195–228, 1999.

[88] J. Milnor. On Manifolds Homeomorphic to the 7-Sphere. *Annals of Mathematics: Second Series*, 64(2):399–405, 1956.

[89] N. Haaser and J. Sullivan. *Real Analysis*. Dover NY, 1991.

[90] N. J. C. King and D. R. Getts and M. T. Getts and S. Rana and B. Shrestha and A. M. Kesson. Immunopathology of Flavivirus Infections. *Immunol. Cell. Biol.*, 85:33–42, 2007.

[91] N. Steenrod. *The Topology of Fiber Bundles*. Princeton University Press, 1999.

[92] R. Nieuwenhuys, H. Donkelaar, and C. Nicholson. *The Central Nervous System of Vertebrates: Volume 1*. Springer Verlag, 1998.

[93] L. Nirenberg. *Topics in Nonlinear Functional Analysis*. American Mathematical Society, 2001.

[94] P. Hintjens. *ZeroMQ*. O'Reilly Media Inc., Sebastopol, CA, 2013.

[95] P. Lewis and C. Manzoni. LRRK2 and Human Disease: A Complicated Question or a Question of Complexes? *Sci. Signal*, 5:1–4, 2012.

[96] J. Peterson. *BioInformation Processing: A Primer on Computational Cognitive Science*. Springer Series on Cognitive Science and Technology, Springer Science+Business Media Singapore Pte Ltd. 152 Beach Road, #22-06/08 Gateway East Singapore 189721, Singapore, 2016. doi: 10.1007/978-981-287-871-7.

[97] J. Peterson. Modeling Associations: Sensor Fusion and Signaling Bar Codes. In S. Saha, A. Mandal, A. MN, S. Sangam, and S. Ram, editors, *Handbook of Research on Applied Cybernetics and Systems Science*, pages 1–35. IGI Global, 2016.

[98] J. Peterson. *Complex Models on Graph Based Topological Spaces IIB: The Implementation of Neural Codes in MatLab, C and C++*. Gneural Gnome Press, Lulu.com, Clemson, SC USA, 2019. URL `http://www.lulu.com/shop/james-peterson`.

[99] J. Peterson. *Basic Analysis I: Functions of a Real Variable*. CRC Press, Boca Raton, Florida 33487, 2020.

[100] J. Peterson. *Basic Analysis III: Mappings on Infinite Dimensional Spaces*. CRC Press, Boca Raton, Florida 33487, 2020.

[101] J. Peterson. *Basic Analysis II: A Modern Calculus in Many Variables*. CRC Press, Boca Raton, Florida 33487, 2020.

[102] R. Ghrist. Barcodes: The Persistent Topology of Data. *Bulletin (New Series) of the American Mathematical Society*, 45(1):61–75, 2008.

[103] R. McElreath and R. Boyd. *Mathematical Models of Social Evolution: A Guide for the Perplexed*. University of Chicago Press, 2007.

[104] R. Menzel. The honeybee as a Model for Understanding the Basis of Cognition. *Nature Reviews: Neuroscience*, 13:758–768, 2012.

[105] R. Raizada and S. Grossberg. Towards a Theory of the Laminar Architecture of Cerebral Cortex: Computational Clues from the Visual System. *Cerebral Cortex*, pages 100–113, 2003.

[106] S. Bao and N. J. C. King and R. Dos. Flavivirus Induces MHC Antigen on Human Myoblasts: a Model of Autoimmune Myositis? *Muscle Nerve*, 15:1271–1277, 1992.

[107] S. Chern and W. Chen and K. Lam. *Lectures on Differential Geometry*. World Scientific, 1999.

[108] S. Lafon and A. Lee. Diffusion Maps and Coarse-Graining: A Unified Framework for Dimensionality Reduction, Graph Partitioning, and Data Set Parameterization. *IEEE Trans. on Pattern Analysis and Machine Intelligence*, 28(9):1393–1403, 2006.

[109] S. Laureys. The Neural Correlate of (Un)awareness: Lessons from the Vegetative State. *Trends in Cognitive Science*, 9(12):556–559, 2005.

[110] S. Murray Sherman and R. Guillery. *Exploring the Thalamus and Its Role in Cortical Function*. The MIT Press, 2006.

[111] S. Sherman. Interneurons and Triadic Circuitry of the Thalamus. *Trends in Neuroscience*, 27 (11):670–675, 2004.

[112] S. Sprecher and H. Reichert. The Urbilaterian Brain: Developmental Insights into the Evolutionary Origin of the Brain in Insects and Vertebrates. *Arthopod Structure and Development*, 32:141–156, 2003.

[113] A. Schmdt-Rhaesa, S. Harzsch, and G. Purschke. *Structure and Evolution of Invertebrate Nervous Systems*. Oxford University Press, 2016.

[114] J. Schwartz. *Nonlinear Functional Analysis*. Gordon and Breach, 1968.

[115] T. Seki. An Elementary Proof of the Brouwer's Fixed Point Theorem. *Tohoku Math. J.*, 9: 105–109, 1957.

[116] M. Tiger, E. Veldman, C. Ekman, C. Halldin, P. Svenningsson, and J. Jundberg. A randomized placebo-controlled PET study of ketamine's effect on serotonin1B receptor binding in patients with SSRI-resistant depression. *Translational Psychiatry*, 10(159):1–8, 2020.

[117] W. Fulton. *Algebraic Topology: A First Course*. Graduate Texts in Mathematics 153, Springer NY, 1995.

[118] A. Ward, O. Khodor, S. Egan, K. Weinersmith, and A. Forbes. A Keeper of Many Crypts: a Behaviour-manipulating Parasite Attacks a Taxonomically Diverse Array of Oak Gall Wasp Species. *Biol. Lett.*, 2019.

[119] S. Weinberg. *Lectures on Quantum Mechanics, Second Edition*. Cambridge University Press, 2015.

[120] L. Wilkins. Topological Structure of Manifolds. Master's thesis, Clemson University, Clemson, SC, 2005.

[121] X. Li and Z. Xie and D. Yi. A Fast Algorithm for Constructing Topological Structure in Large Data. *Homology, Homotopy and Applications*, 14:221–238, 2012.

[122] Y. Wang and A. Brzozowsha-Prechtl and H. Karten. Laminary and Columnar Auditory Cortex in Avian Brain. *Proceedings of the National Academy of Sciences*, 107(28):12676–12681, 2010.

[123] C. Yang and R. Mills. Conservation of Isotopic Spin and Isotopic Gauge Invariance. *Physical Review*, 96(1):191–195, 1954.

[124] Z. Liu and J. Lee and S. Krummery and W. Lu and H. Cai and M. Lenardo. The Kinase LRRK2 is a Regulator of the Transcription Factor NFAT That Modulates the Severity of Inflammatory Bowel Disease. *Nature Immunology*, 12:1063–1070, 2011.

Part IX

Detailed Index

Index